Graduate Texts in Physics

Series Editors

Kurt H. Becker, NYU Polytechnic School of Engineering, Brooklyn, USA

Jean-Marc Di Meglio, Matière et Systèmes Complexes, Bâtiment Condorcet, Université Paris Diderot, Paris, France

Sadri Hassani, Department of Physics, Illinois State University, Normal, USA

Morten Hjorth-Jensen, Department of Physics, Blindern, University of Oslo, Oslo, Norway

Bill Munro, Okinawa Institute of Science and Technology Graduate University, Onna-son, Japan

Richard Needs, Cavendish Laboratory, University of Cambridge, Cambridge, UK

William T. Rhodes, Department of Computer and Electrical Engineering and Computer Science, Florida Atlantic University, Boca Raton, USA

Susan Scott, Australian National University, Acton, Australia

H. Eugene Stanley, Department of Physics, Center for Polymer Studies, Boston University, Boston, MA, USA

Martin Stutzmann, Walter Schottky Institute, Technical University of Munich, Garching, Germany

Andreas Wipf, Institute of Theoretical Physics, Friedrich-Schiller-University Jena, Jena, Germany

Graduate Texts in Physics publishes core learning/teaching material for graduate- and advanced-level undergraduate courses on topics of current and emerging fields within physics, both pure and applied. These textbooks serve students at the MS- or PhD-level and their instructors as comprehensive sources of principles, definitions, derivations, experiments and applications (as relevant) for their mastery and teaching, respectively. International in scope and relevance, the textbooks correspond to course syllabi sufficiently to serve as required reading. Their didactic style, comprehensiveness and coverage of fundamental material also make them suitable as introductions or references for scientists entering, or requiring timely knowledge of, a research field.

Gerald Eigen

Detectors in High-Energy Physics Experiments

Underlying Physics, Working Principles, and Realizations

 Springer

Gerald Eigen
Department of Physics
University of Göttingen/Caltech
Göttingen, Germany

ISSN 1868-4513 ISSN 1868-4521 (electronic)
Graduate Texts in Physics
ISBN 978-3-031-67335-1 ISBN 978-3-031-67336-8 (eBook)
https://doi.org/10.1007/978-3-031-67336-8

This work was supported by CERN.

© The Editor(s) (if applicable) and The Author(s) 2025. This book is an open access publication.

Open Access This book is licensed under the terms of the Creative Commons Attribution 4.0 International License (http://creativecommons.org/licenses/by/4.0/), which permits use, sharing, adaptation, distribution and reproduction in any medium or format, as long as you give appropriate credit to the original author(s) and the source, provide a link to the Creative Commons license and indicate if changes were made.
The images or other third party material in this book are included in the book's Creative Commons license, unless indicated otherwise in a credit line to the material. If material is not included in the book's Creative Commons license and your intended use is not permitted by statutory regulation or exceeds the permitted use, you will need to obtain permission directly from the copyright holder.
The use of general descriptive names, registered names, trademarks, service marks, etc. in this publication does not imply, even in the absence of a specific statement, that such names are exempt from the relevant protective laws and regulations and therefore free for general use.
The publisher, the authors and the editors are safe to assume that the advice and information in this book are believed to be true and accurate at the date of publication. Neither the publisher nor the authors or the editors give a warranty, expressed or implied, with respect to the material contained herein or for any errors or omissions that may have been made. The publisher remains neutral with regard to jurisdictional claims in published maps and institutional affiliations.

This Springer imprint is published by the registered company Springer Nature Switzerland AG
The registered company address is: Gewerbestrasse 11, 6330 Cham, Switzerland

If disposing of this product, please recycle the paper.

Dedicated to my wife Dr. Andrea Christine Eigen

Preface

The book evolved from lectures I have been giving at the University of Bergen from 1995 until 2020 in the course "Experimental techniques in particle and nuclear physics". This is an introductory course on detector physics for master students and early Ph.D. students in high-energy and heavy-ion physics. It also served useful for students in medical physics. The lectures started with material from textbooks by Kleinknecht and Leo as well as review articles from the Particle Data Group. They evolved by adding material from original publications, review articles, and the web. I have updated the slides annually by adding new technology developments.

The goal of the course was to teach the physics and working principles of various detectors in high-energy physics and show examples of their realization. The focus lied on sub-detectors in colliding-beam and accelerator experiments. In addition, a few non-accelerator sub-detectors are discussed. Each chapter has a problem set of typically 10 questions. Exceptions are Chaps. 10 and 11, which have five problems each. The book is organized into 11 chapters and two appendices. In order to minimize mistakes, I have asked expert colleagues and students in the field to proofread chapters.

Göttingen, Germany   Gerald Eigen

Acknowledgments I would like to acknowledge the following colleagues and students for their useful comments, J. Brau, Y. Buch, T. Crane, M. Demarteau, D. E. Groom, D. G. Hitlin, K. Jakobs, M. Moll, T. Mori, F. C. Porter, B. N. Ratcliff, M. Roney, A. Seiden, J. Va'vra, and Th. Wengler. I would like to thank my sons Julian and Christoph for helping with technical issues and reading part of the chapters.

Contents

1 Introduction 1
   1.1 Historic Overview 2
   1.2 Some General Remarks 10
   1.3 Accelerators 13
       1.3.1 Hadron Machines 17
       1.3.2 Electron Machines 19
       1.3.3 Luminosity 24
   1.4 Detector Characteristics 25
       1.4.1 Detector Efficiency 26
       1.4.2 Mean Value and Resolution 27
       1.4.3 Readout Chain 29
       1.4.4 Kinematic Observables 30
       1.4.5 Decay Quantities 33
       1.4.6 A Comment on Units 34
   1.5 Effects of Radiation 34
   Exercises 35
   References 37

2 Interactions of Charged Particles and Photons with Matter and Fields 43
   2.1 Passage of Heavy Charged Particles Through Matter 43
       2.1.1 Energy Loss by Excitation and Ionization 50
       2.1.2 Instantaneous Energy Loss Distribution 61
       2.1.3 Restricted Energy Loss 67
       2.1.4 Energy Loss in Thick Absorbers 67
       2.1.5 Particle Range 68
       2.1.6 Properties of the Ionization Process 71
       2.1.7 Production of Knock-on Electrons or δ-Electrons 71
   2.2 Energy Loss of Electrons and Positrons 73
       2.2.1 Energy Loss by Collision 73
       2.2.2 Energy Loss by Bremsstrahlung 74
       2.2.3 Radiation Length 80
       2.2.4 Critical Energy 81

		2.2.5	Range of Electrons	84
		2.2.6	Muon Energy Loss at High Energies	86
		2.2.7	Summarizing Remarks	87
	2.3		Multiple Scattering	87
	2.4		The Cherenkov Effect	92
	2.5		Transition Radiation	93
	2.6		Detection of Photons	97
		2.6.1	Photoelectric Absorption	98
		2.6.2	Compton Effect	102
		2.6.3	Pair Creation	104
		2.6.4	Electron-Photon Showers	106
	2.7		Electrons and Ions in Gases	109
		2.7.1	Production of Electron-Ion Pairs	109
		2.7.2	Mobility of Ions	110
		2.7.3	Diffusion of Ions and Electrons in a Field-Free Gas	112
		2.7.4	Drift of Electrons in Electric Fields	116
		2.7.5	Drift of Electrons and Ions in Electric and Magnetic Fields	121
		2.7.6	Diffusion of Electrons in Electric and Magnetic Fields	124
	2.8		Scintillation Processes	127
		2.8.1	Energy Levels in Solids and Liquid Noble Gases	127
		2.8.2	Scintillation Process in Liquid Noble Gases	129
		2.8.3	Scintillation Mechanism in Inorganic Crystals	131
		2.8.4	Scintillation Process in Organic Scintillators	131
		2.8.5	Birks' Law	133
	Exercises			135
	References			137
3	**Measurements of Energy Loss of Charged Particles**			141
	3.1		Energy Loss Measurements in Gases	142
		3.1.1	Pulse Mode Ionization Chamber	142
		3.1.2	Proportional Counters	145
		3.1.3	Geiger-Müller Counters	153
	3.2		Ionization in Liquids	154
		3.2.1	Ionization Process	155
		3.2.2	Charge Distribution	157
		3.2.3	Recombination and Electron Trapping	158
		3.2.4	Scintillation	162
		3.2.5	Drift Velocity	166
		3.2.6	Diffusion	169
		3.2.7	Energy Resolution	170

	3.3	Ionization in Solid-State Detectors	173
		3.3.1 Basic Properties of Semi-Conductors	173
		3.3.2 The *pn* Junction	178
		3.3.3 Depletion Layer	182
	Exercises		184
	References		185
4	**Gaseous Tracking Detectors**	189
	4.1	Multi-Wire Proportional Chambers	190
		4.1.1 Principles	190
		4.1.2 Design Considerations	194
		4.1.3 Electric-Field Distortions	196
		4.1.4 Performance	199
		4.1.5 Position Measurement	202
		4.1.6 Gas Chamber Aging	205
	4.2	Planar Drift Chambers	207
		4.2.1 Drift Chamber Properties	210
		4.2.2 Drift Chamber Performance	212
	4.3	Cylindrical Drift Chambers	213
		4.3.1 Conceptual Chamber Design	214
		4.3.2 Chamber Properties	215
		4.3.3 The TASSO Drift Chamber	215
		4.3.4 The B*A*B*AR* Drift Chamber	218
		4.3.5 Drift Chamber Alignment	221
	4.4	Jet Drift Chambers	223
		4.4.1 Small Jet Drift Chambers	228
	4.5	Time Projection Chambers	228
		4.5.1 Conceptual TPC Design	228
		4.5.2 The ALEPH and ALICE Time Projection Chambers	232
		4.5.3 Calibration and Monitoring	233
		4.5.4 Performance	233
	4.6	Micro Strip Gas Chambers	236
	4.7	Gas Electron Multiplier	238
	4.8	Micro Mesh Gaseous Structure	244
	4.9	Resistive Plate Chambers	252
	4.10	Limited Streamer Tubes	257
	4.11	Drift Tubes	263
	4.12	Straw Tube Chambers	267
	4.13	Cathode Strip Chambers	269
	4.14	Thin Gap Chambers	270
	4.15	Bubble Chambers	274
	4.16	Other Detectors	276
	4.17	Position Resolution of Tracking Detectors	276

	Exercises		278
	References		279
5	**Momentum Measurements**		**287**
5.1	Deflections in Magnetic Fields		287
5.2	Particle Motion in a Magnetic Field		289
	5.2.1	The Particle Trajectory	290
	5.2.2	Impact Parameter Determination	292
	5.2.3	Position Resolution and Impact Parameter Resolution	293
	5.2.4	Track Finding and Reconstruction	293
5.3	Magnet Shapes for Fixed-Target Experiments		296
	5.3.1	Properties of Dipole Magnets	298
5.4	Momentum Measurements in Fixed-Target Experiments		298
5.5	Magnet Shapes for Storage Ring Experiments		300
	5.5.1	Solenoidal Magnet	300
	5.5.2	Toroidal Magnet	302
	5.5.3	Dipole Magnet with Two Compensator Magnets	303
	5.5.4	Split Field Dipole Magnet	304
5.6	Momentum Measurements in Solenoids		304
	Exercises		307
	References		309
6	**Detectors for Vertex Measurements**		**311**
6.1	Silicon Microstrip Detectors		312
	6.1.1	Fabrication of Mono-crystalline Silicon	312
	6.1.2	Layout of a Silicon Microstrip Detector	314
	6.1.3	Noise, Current Flow and Collection Time	315
	6.1.4	Fabrication of a Silicon Microstrip Detector	319
	6.1.5	The First Microstrip Detector	320
	6.1.6	Double-Sided Microstrip Detectors	322
	6.1.7	Position Resolution and Impact Parameter Resolution	324
6.2	Recently Used Microstrip Detectors		326
	6.2.1	The B$_A$B$_{AR}$ Silicon Vertex Tracker	327
	6.2.2	The ATLAS Semi-conductor Tracker	331
	6.2.3	Examples of Microstrip Detectors	336
	6.2.4	The Power of Microstrip Detectors	336
6.3	Charge-Coupled Devices		336
	6.3.1	The SLD VXD3 Vertex Detector	341
6.4	Pixel Detectors		343
	6.4.1	Pixel Detectors in the WA97 and DELPHI Experiments	345
	6.4.2	Pixel Detectors in the ATLAS Experiment	347
	6.4.3	Pixel Detectors in the CMS Experiment	354

6.5		Radiation Hardness of Silicon Pixel Sensors	356
6.6		Vertex Detectors for the LHC Phase II Upgrade	370
6.7		New Structures	371
	6.7.1	The Silicon Drift Chamber	372
	6.7.2	Monolithic Active Pixel Sensors	373
	6.7.3	DEpleted P-Channel Field Effect Transistors	377
	6.7.4	Low-Gain Avalanche Detectors	384
	6.7.5	3D Silicon Detectors	389
	6.7.6	Silicon-on-Insulator Sensors	395
6.8		Diamond Detectors	397
Exercises			402
References			403

7 Detectors for Energy Measurements ... 409

- 7.1 Characteristics of the Electron-Photon Shower ... 409
  - 7.1.1 Longitudinal Shower Profile ... 410
  - 7.1.2 Transverse Shower Profile ... 412
  - 7.1.3 Energy Reconstruction ... 414
  - 7.1.4 Cluster Shapes ... 415
  - 7.1.5 Electromagnetic Energy Resolution ... 416
  - 7.1.6 Noise ... 419
  - 7.1.7 Timing ... 421
  - 7.1.8 Radiation Hardness ... 422
- 7.2 Homogeneous Shower Counters ... 423
  - 7.2.1 Inorganic Scintillating Crystals ... 423
  - 7.2.2 Examples of Crystal Calorimeters ... 425
  - 7.2.3 Beam Radiation Monitor ... 432
  - 7.2.4 Lead Glass Calorimeters ... 434
  - 7.2.5 Liquid Noble Gases ... 437
- 7.3 Sampling Shower Detectors ... 442
  - 7.3.1 Properties of Sampling Calorimeters ... 443
  - 7.3.2 Energy Resolution of Electromagnetic Sampling Calorimeters ... 446
  - 7.3.3 Position Resolution of Electromagnetic Sampling Calorimeters ... 449
  - 7.3.4 Examples of Electromagnetic Sampling Calorimeters ... 449
- 7.4 Hadron Shower Measurements ... 459
  - 7.4.1 Characteristics of Hadron Showers ... 461
  - 7.4.2 Intrinsic Energy Resolution ... 473
  - 7.4.3 Spatial Resolution ... 495
  - 7.4.4 Examples of Hadron Calorimeters ... 495
  - 7.4.5 Properties of Different Hadron Sampling Calorimeters ... 500

	7.5	New Calorimeter Concepts	501
		7.5.1 Calorimeters Based on Particle Flow	501
		7.5.2 Dual Readout Calorimetry	513
	7.6	Concluding Remarks	515
	Exercises		515
	References		517
8	**Photodetectors, Plastic Scintillators and Time Measurements**		**525**
	8.1	The Photomultiplier	525
		8.1.1 Photomultiplier Layout	526
		8.1.2 Dynode Configurations	530
		8.1.3 Photomultiplier Properties	532
		8.1.4 Fine-Mesh Photomultiplier Tube	537
		8.1.5 Photomultipliers in Subdetector Systems	539
		8.1.6 Multi-anode Photomultiplier Tube	540
		8.1.7 Vacuum Phototriodes	541
	8.2	Micro-channel Plates	542
		8.2.1 Detection Properties	543
		8.2.2 Timing Measurements	545
		8.2.3 Examples of MCP-PMTs	546
		8.2.4 Lifetime Issues and Performance in Magnetic Fields	546
		8.2.5 Applications of MCP-PMTs	548
	8.3	Silicon-Based Photodetectors	549
		8.3.1 PIN Diodes	549
		8.3.2 Avalanche PhotoDiodes	555
		8.3.3 Silicon PhotoMultipliers	562
		8.3.4 Digital Silicon Photomultipliers	573
		8.3.5 Visible Light Photon Counters	573
		8.3.6 Hybrid PhotoDiodes	574
	8.4	Gaseous Photon Detectors	578
	8.5	Comparison of Photodetectors	578
	8.6	Timing Detectors at LHC Experiments	579
	8.7	Plastic Scintillators	582
		8.7.1 Wavelength Shifters	584
		8.7.2 Usage of Plastic Scintillators	589
		8.7.3 Collection of Scintillation Light and Light Guides	590
		8.7.4 Winston Cone	591
		8.7.5 Wavelength-Shifting Fibers	593
	8.8	The LHCb Scintillating-Fiber Tracker	594
	Exercises		596
	References		598

9 Particle Identification Techniques and Devices ... 603
- 9.1 Energy Loss Measurements ... 604
- 9.2 Time-of-Flight Measurements ... 610
- 9.3 Cherenkov Counters ... 617
  - 9.3.1 Threshold Cherenkov Counters ... 618
  - 9.3.2 Differential and DISC Cherenkov Counters ... 622
  - 9.3.3 Ring-Imaging Cherenkov Counters ... 624
  - 9.3.4 Other RICH Detectors ... 631
  - 9.3.5 The BABAR DIRC ... 632
  - 9.3.6 Next Generation DIRC Counters ... 636
- 9.4 Transition Radiation Measurements ... 638
  - 9.4.1 The ATLAS Transition Radiation Tracker ... 641
- 9.5 Electron Identification via the E/p Method ... 645
- 9.6 Muon Identification ... 647
- 9.7 Neutron Identification ... 649
- Exercises ... 650
- References ... 652

10 Trigger and Data Acquisition ... 657
- 10.1 Overview ... 657
- 10.2 Trigger Layout ... 660
- 10.3 Trigger Signatures ... 662
  - 10.3.1 The ATLAS Trigger Systems in Run 1 and Run 2 ... 662
  - 10.3.2 The ATLAS Level 1 Calorimeter Trigger System ... 664
  - 10.3.3 The ATLAS Level 1 Muon Trigger ... 667
  - 10.3.4 The ATLAS Level-1 Topological Trigger ... 668
  - 10.3.5 The ATLAS High-Level Trigger ... 670
  - 10.3.6 Trigger Performance in the ATLAS Experiment ... 670
- 10.4 The CMS Trigger System ... 672
- 10.5 The LHCb Trigger System ... 676
- 10.6 New Developments ... 678
- Exercises ... 681
- References ... 681

11 Examples of Full Detector Systems ... 683
- 11.1 Colliding-Beam Detectors ... 683
  - 11.1.1 The BABAR Detector ... 683
  - 11.1.2 The Belle II Detector ... 685
  - 11.1.3 The ALEPH Detector ... 687
  - 11.1.4 The DELPHI Detector ... 689
  - 11.1.5 The SLD Detector ... 691
  - 11.1.6 The Collider Detector at Fermilab ... 693
  - 11.1.7 The ATLAS Detector ... 696
  - 11.1.8 The CMS Detector ... 700
  - 11.1.9 The LHCb Detector ... 706
  - 11.1.10 The ALICE Detector ... 712

	11.2	Non-collider Detectors	714
		11.2.1 The Mu2e Detector	714
		11.2.2 The Super-Kamiokande Detector	717
		11.2.3 The Ice Cube Detector	718
		11.2.4 The LUX-ZEPLIN Detector	720
		11.2.5 The Cherenkov Telescope Array	722
	11.3	Applications in other Fields	723
		11.3.1 Geophysics Applications	723
		11.3.2 Positron Emission Tomography	724
	11.4	Conclusion	726
	Exercises		726
	References		728

Appendix A: Types and Sources of Radiation 731

Appendix B: Advanced Statistical Techniques 735

About the Author

Gerald Eigen is a retired professor in high-energy physics from the University of Bergen. He studied physics at the University of Freiburg graduating with a thesis on the polarization correlation measurement of the two photon system produced in electron positron annihilation. In his Ph.D. thesis conducted in the CUSB experiment he discovered four of the six χ_b / χ_b' states in the $b\bar{b}$ system. After his Ph.D. from the Ludwig Maximilians University Munich, he went to Caltech for ten years as a Lynen fellow, senior research fellow and Heisenberg professor before becoming full professor in Bergen, Norway in 1995. At Caltech he worked on the experiments Mark III, SLD, CLEO II and *BABAR*. In Bergen he joined the DELPHI experiment at CERN. He further became visiting faculty at Caltech continuing with physics analyses in the *BABAR* experiment. After the completion of the DELPHI experiment he joined the ATLAS experiment performing physics analyses and new readout studies of the Tile Calorimeter. He designed an electromagnetic calorimeter backward endcap for the SuperB experiment and started to build a prototype before the experiment was cancelled. He took sabbaticals at DESY and the University of Göttingen. At DESY he joined the analog hadron calorimeter activities becoming a member of the CALICE collaboration. This lead to a participation in the European networks EUDET, AIDA, AIDA2020 and AIDAinnova. He is now a visitor at the University of Göttingen and Caltech continuing with work on the CKM matrix, calorimeter readout and Belle II physics.

The author's main detector expertise is calorimetry and readout of various scintillators and silica aerogel with different photodetectors including photomultiplier tubes, PIN diodes, avalanche photodiodes, silicon photomultipliers and silicon photomultiplier arrays.

The author was involved in the construction of half of the electromagnetic liquid argon sampling calorimeter modules for SLD and the commission of the entire calorimeter. He worked on CsI readout for *BABAR* and performed R&D with LYSO crystals. In the CALICE collaboration he worked on analog hadron calorimeter prototypes for a linear collider detector. During his career he also helped with building strip chambers for the CUSB experiment and repairing the Mark III drift chamber. Besides detector work the author participated in various physics analyses. Thus, his physics expertise ranges from entangled photons over bottomonium spectroscopy, glueball searches, CKM matrix work, rare semileptonic decays and dark photons to Higgs physics.

The author worked five years on the LHCC, reviewing the LHCb experiment of which he served the last two years as chief reviewer. He was the Norwegian representative on the restricted ECFA committee for six years and served as Norwegian representative on the recent European Strategy Update. He is still an active reviewer for Physical Review Letters, Physical Review D, Nuclear Instrumentation and Methods A and JINST. Though the book is open access, the usage of figures and tables from refereed journals requires permission.

Acronyms

2D	Two-Dimensional
3D	Three-Dimensional
3HF	3-Hydroxyflavone
α	Fine-structure constant, which is (1/137) at low energies; electromagnetic coupling constant
α radiation	Emission of He nuclei
α_s	Strong coupling constant
AC	Alternating Current
ACC	Belle II Aerogel threshold Cherenkov Counter
ACO	Electron positron Collider at Orsay laboratory, France
ACORDE	ALICE Cosmic Ray Detector
ADC	Analog-to-Digital Converter
ADONE	Electron Positron Collider at Frascati, Italy
AFP	ATLAS Forward Proton detector
AGS	Alternating Gradient Synchrotron
AHCAL	Analog Hadron Calorimeter
ALD	Atomic Layer Deposition
ALEPH	Apparatus for LEP PHysics, one detector at the LEP collider at CERN
ALFA	Absolute Luminosity for ATLAS
ALICE	A Large Ion Collider Experiment, experiment at the LHC at CERN
AMANDA	Antarctic Muon and Neutrino Detector Array in Antartica
AMS	Alpha Magnetic Spectrometer on the International Space Station
AMY	Experiment at the TRISTAN collider at KEK
APD	Avalanche Photodiode
AR	Accumulation Ring
ARGUS	Experiment at the DORIS II storage ring at DESY
ASIC	Application-Specific Integrated Circuit

ATLAS	A Toroidal LHC Apparatus, one of the LHC experiments at CERN
AUGER	Cosmic Ray Observatory in Argentina
β	Relative velocity
β radiation	Emission of electrons
BABAR	Experiment at the PEP II B factory at SLAC, Palo Alto, CA, USA
BaF	Bariumfluoride
BBQ	Benzimidazo-benzi-sochinolin-7-one
BCM	Beam Conditions Monitor
Belle	Experiment at the KEKB factory at KEK, Tsukuba, Japan
Belle2	Experiment at the SuperKEKB factory at KEK, Tsukuba
BEPC	Beijing Electron Positron Collider in Beijing, China
BERT	Bertini cascade
BES III	Experiment at the BEPC collider in Beijing, China
Bevatron	Synchrotron at the Lawrence Berkeley Laboratory
BGO	Bismuth Germanate
BH	CMS Backward Hadron Calorimeter
BIC	Binary Cascade
bis-MSB	1,4-Bis(2-methylstyryl)benzol
BNL	Brookhaven National Laboratory in Brookhaven, NY, USA
Bs	Boron interstitial
BTL	Barrel Timing Layer
c	Speed of light
CALET	CALorimetric Electron Telescope on the International Space Station
CCD	Charge-Coupled Device
CCE	Charge Collection Efficiency
CDC	Central Drift Chamber
CDF	Collider Detector at Fermilab, one experiment at the Tevatron
CDHS	CERN, Dortmund Heidelberg, Saclay neutrino experiment at CERN
CE-E	CMS Electromagnetic Endcap calorimeter
CE-H	CMS Hadronic Endcap calorimeter
CELLO	Experiment at the PETRA Collider at DESY
CEPC	Circular Electron Positron Collider
CERN	European Organization for Nuclear Research, Geneva, Switzerland
CESR	Cornell Electron Storage Ring
CGS	Electrostatic units, grams, centimeters, seconds
CHIPS	Chiral Invariant Phase Space
CiS	Electronic company in Germany

CLEO	Experiment at the CESR Collider at Cornell University
CLEO II	Upgraded B physics experiment at the CESR Collider at Cornell University
CLEO III	Charm physics experiment at the CESR Collider at Cornell University
CLIC	Compact Linear Collider
CM	Center-of-Mass
CMOS	Complementary Metal-Oxide Semiconductor
CMS	Compact Muon Solenoid, one of the LHC experiments at CERN
CNM	Centro Nacional de Microelectrónica
CO	Carbon-Oxygen complex
COG	Center-of-Gravity
COMET	Coherent Muon to Electron Transition at JPARC Tokai, Japan
COMPASS	The Common Muon and Proton Apparatus for Structure and Spectroscopy experiment at CERN
Cosmotron	Synchrotron at BNL
COT	Central Outer Tracker in CDF
CP	Symmetry of charge conjugation and parity transformation
CRID	Cherenkov Ring-Imaging Detector
Crystal Ball	Experiment at the SPEAR collider at SLAC and DORIS II collider at DESY
Crystal Barrel	Experiment at the LEAR Storage ring
CSC	Cathode Strip Chamber
CsI	Cesium Iodide
CTA	Cherenkov Telescope Array
CTP	Central Trigger Processors
CPU	Central Processing Unit
Cuore	Cryogenic Underground Observatory for Rare Events at the Gran Sasso, Italy
CUSB	B physics experiment at the CESR collider at Cornell University
CVD	Chemical Vapor Deposition
Cz	Czochralski
ΔE_{mp}	Most probable energy loss
D0	One experiment at the Tevatron Collider at Fermilab
d_0	Transverse impact parameter
DAMIC	Dark Matter in CCDs at Modane, France
DAPhNe	Electron Positron collider at Frascati, Italy
DARWIN	Dark matter WIMP search experiment with liquid xenon
DASP	Experiment at the DORIS collider at DESY
DAQ	Data Acquisition
Daya Bay	Neutrino experiment in China

DBM	Diamond Beam Monitor
DC	Direct current
DC	Drift chamber
DCI	Electron positron collider at Orsay laboratory, France
DD	Thermal double donor
D_E	Energy dose
DELCO	Experiment at the PEP Collider at SLAC
DELPHI	DEtector with Lepton, Photon and Hadron Identification, experiment at the LEP collider at CERN
DEPFET	DEpleted P-channel Field Effect Transistor
DHCAL	Digital Hadron CALorimeter
D_I	Ion dose
DIRC	Detector of Internally Reflected Cherenkov light
D_k	Diffusion coefficient of particle k, electron (e) and ion (ion)
DME	Dimethyl ether
DESY	German Electron Synchrotron Laboratory DESY in Hamburg
DM2	Experiment at DCI
DONUT	Experiment at the Tevatron at Fermilab
DORIS	Electron positron Collider at DESY, Hamburg
DORIS II	Upgraded electron positron collider to CM energy of 10 GeV at DESY
DPO	Diphenyl phosphine oxide
D_q	Equivalent dose
DQM	Data Quality Monitoring
DT	Drift Tube
Dune	Deep Underground Neutrino Experiment, a next generation long-baseline neutrino experiment in the Sanford mine, South Dakota, USA
ϵ	Dielectric constant
ϵ_c	Collection efficiency
ϵ_{det}	Detection efficiency
ϵ_{geo}	Ratio of sensitive area to total area
ϵ_{ph}	Photon detection efficiency
ϵ_{QM}	Quantum efficiency
η	Pseudorapidity
E_c	Critical energy for electrons
$E_{\mu c}$	Critical energy for muons
E_{cut}	Cut-off energy
E_T	Transverse energy
$E_{T,miss}$	Missing transverse energy
E989	Muon g-2 experiment at Fermilab
E_B	Binding energy
EC	Endcap

ECAL	Electromagnetic CALorimeter
EF	Event Filter
E_k	Kinetic energy of particle k, for electron (e), for ion (i)
EM	ElectroMagnetic
EMC	ElectroMagnetic Calorimeter
EMCAL	ElectroMagnetic CALorimeter in ALICE
EMV	Variations of the standard EM package
$\overline{ENC}$	Mean Equivalent Noise Charge
EPI	EPItaxial layer, deposition of a higher-purity layer on a substrate of the same material
EW	Electroweak
ϕ	Azimuth angle
Φ	Fluence
FAIR	Facility for Antiproton and Ion Research in Darmstadt, Germany
FBK	Bruno Kessler Institute
FCAL	Forward CALorimeter
FCC-ee	Future Circular Collider for electron positron collisions
FCC-hh	Future Circular Collider for hadron hadron collisions
FE	Frontend
Fermilab	Fermi National Laboratory in Batavia, Il., USA
Fermilab-LAT	Fermi Large-Area Telescope
FH	CMS Forward Hadron Calorimeter
FPGA	Field-Programmable Gate Array
FTFP	Fritof model
FWHM	Full Width at Half Maximum
FZ	Float Zone
γ	Lorentz factor
γ radiation	Energetic photons
Γ_i	Partial width for decay into channel i
Γ_{tot}	Total decay width
G-APD	Geiger Avalanche PhotoDetector
GEM	Gas Electron Multiplier
GPD	Gaseius Photon Detector
GPU	Graphical Processing Unit
$\hbar\omega$	Photon energy
H1	One experiment at the HERA collider
HCAL	Hadron CALorimeter
HEC	Hadron Endcap Calorimeter
HELIOS	Two fixed-target experiments at the SPS at CERN
HGCAL	CSM High-Granularity Calorimeter
HEP	High-Energy Parametrization model
HERA	Electron proton collider at DESY
HESS	High-Energy Stereoscopic System in Namibia
HL-LHC	High-Luminosity LHC

HLT	High-Level Trigger
HMPID	High-Momentum Particle Identification Detector
HP	High-Precision neutron model
HPD	Hybrid PhotoDiode
HPK	Hamamatsu Photonics K. K.
HyperKamiokande	Succesor experiment of SuperKamiokande
IBL	Insertable B Layer, Inner layer of the ATLAS pixel detector
IceCube	Neutrino experiment at the Amundsen–Scott South Pole Station, Antarctica
ID	Inner Detector
IFR	Instrumented Flux Return
ILC	International Linear Collider
INFN	Instituto Nazionale Fisica Nucleare, Italy
IP	Interaction Point
IPT	Inner Positioning Tube
IR	Interaction Region
ISL	Third silicon strip detector in CDF
Isochrone	Contours of equal drift time
ISR	Intersecting Storage Ring at CERN
IST	Inner Support Tube
ITS	Inner Tracker in the ALICE experiment
JADE	One experiments at the PETRA Collider and TRISTAN collider
JES	Jet Energy Scale
JFET	Junction Field Effect Transistor
JUNO	Jiangmen Underground Neutrino Observatory, medium baseline reactor neutrino experiment in China
κ	Average energy loss of a charged particle to the maximum energy transfer
K2K	KEK to Kamioka Neutrino beam
KamLand	Long-baseline neutrino experiment at Kamioka mine
KamLand-Zen	Neutrinoless double beta experiment in the KamLand detector
KATRIN	Karlsruhe Tritium Neutrino Experiment
KEK	High-Energy Accelerator Research Organization in Tsukuba, Japan
KEKB	Asymmetric electron positron collider at KEK in Tsukuba, Japan
KLOE	Experiment at the DAPhNE collider at Fracati, Italy
KOTO	Rare neutral Kaon decay experiment at J-Park, Japan
KTeV	Fixed-target experiment at Fermilab
λ	Photon wavelength
λ_a	Absorption length
λ_d	Decay constant

λ_{dip}	Dip angle
λ_I	Interaction length
λ_k	Mean free path of particle k, electron (e) and ion (ion)
L1, 2, 3	Level 1, Level 2, Level 3
L1Calo	Level 1 calorimeter trigger system
L1Muon	Level 1 muon trigger system
L1TOPO	L1 topological trigger modules
L3	Experiment at the LEP collider at CERN
Labview	Engineering software developed by National Instruments
LAr	Liquid argon
LASS	Large Area Solenoid Spectrometer at SLAC
LBNL	Lawrence Berkeley National Laboratory in Berkeley, CA, USA
LEAR	Low Energy Antiproton Ring at CERN
LED	Light-emitting diode
LGAD	Low-Gain Avalanche Diode
LIGO	The Laser Interferometer Gravitational-wave Observatory at Hanford, WA and Livingston, LA, USA
LKr	Liquid Krypton
LEP	Large Electron Positron collider at CERN
LEP	Low-Energy Parametrization model
LHC	Large Hadron Collider at CERN
LHCb	One of the LHC experiments at CERN
LHEP	Low-Energy and High-energy Parametrization model
ℓ_I	Path in units of interaction length
LSO	Lutetium Oxyorthosilicate
LYSO	Lutetium-Yttrium Oxyorthosilicate
LSST	Legacy Survey of Space and Time at the Vera C. Rubin Observatory in Chile
LST	Limited Streamer Tube
LXe	Liquid xenon
LUCID	LUminosity measurement using Cherenkov Integrating Detector
LUX	Dark-matter experiment in the Sanford mine, South Dakota
LUX-LZ	Next-generation dark-matter experiment in the Sanford mine
μ	Muon or mean value of a Gaussian/Poisson distribution
μ_k	Mobility of particle k, for electron (e), for ion (ion)
m_k	Mass of particle k, for electron (e), kaon (K), pion (π), proton (p), ion (ion)
m_{miss}	Missing mass
m_T	Transverse mass
MAC	Experiment at the PEP Collider at SLAC

MAGIC	Major Atmospheric Gamma Imaging Cherenkov Telescopes in La Palma, Canary Island, Spain
MaPMT	Multi-anode PMT
MAPS	Monolithic Active Pixel Sensor
Mark II	Experiment at the PEP and SLC Colliders at SLAC
Mark III	Experiment at the Spear Collider at SLAC
Mark J	One experiments at the PETRA Collider
MARS	Modeling of the radiation environment with Mokhov's code
MC	Monte Carlo
MCC	Module Controler Chip
MCP	Microchannel Plate
MDT	Monitored Drift Tubes
MEG/MEGII	Muon to Electron Gamma experiment at PSI
Micromegas	Micro-mesh gaseous structure
MiniBoone	Cherenkov detector experiment at Fermilab designed to observe neutrino oscillations
MIP	Minimum-Ionizing Particle
MOSFET	Metal-Oxide-Semiconductor Field-Effect Transistor
MPGD	Micro-Pattern Gas Detector
MPPC	Multiple Pixel Photon Counter
MR	Main ring
mrad	milliradian
MRPC	multi-gap Resistive Plate Chamber
MS	Multiple Scattering
MSGC	Micro-strip Gas Chamber
m_T	Transverse mass
Mu2e	Muon-to-electron conversion experiment at Fermilab
MWPC	Multi-Wire Proportional Chamber
n	Index of refraction
N_e	Electron density
NA32	Charm production experiment at CERN, also known as ACCMOR
Na48	Fixed-target experiment at CERN
Na62	A fixed-target rare kaon decay experiment at CERN
NaI	Sodium Iodide
NCE	Nuclear Counter Effect
nEXO	Neutroless double beta experiment in the SNO laboratory
NIEL	Non-ionizing Energy Loss
NMOS	N-channel Metal-Oxide Semiconductors
NTC	Negative Temperature Coefficient
NTP	Normal Temperature and Pressure, 20 °C and 1 atmosphere
ω_B	Cyclotron frequency, 17.6 MHz
ω_p	Plasma frequency

OD	Outer Detector
ODF	Online Data Flow
OPAL	Omni-Purpose Apparatus for LEP, experiment at the LEP collider at CERN
ORCA	Oscillation Research with Cosmics in the Abyss, Toulon, France
p_T	Transverse momentum
$p_{T,miss}$	Missing transverse momentum
Ps	Phosphorus interstitial
PAI	PhotoAbsorption Ionization model
PAIR	Extended PhotoAbsorption Ionization model
PAMELA	Payload for Antimatter Matter Exploration and Light-nuclei Astrophysics mounted on a satellite
PANDA	Experiment at the FAIR accelerator facility in Darmstadt
PBD	2-phenyl-5(4-biphenyl-1,3,4-oxadizole
PbWO4	Lead Tungstate
PCB	Printed Circuit Board
PDF	Probability Density Function
pe	Photoelectron
PEB	Peripheral Electronics Board
PEP	Positron Electron Collider at SLAC
PEP II	Asymmetric energy $e^+ e^-$ collider at 10 GeV CM at SLAC
PET	Positron Electron Tomography
PETRA	Positron Electron Collider at DESY
PFA	Particle Flow Algorithms
PFlow	Particle Flow
P_g	Probability to trigger a Geiger avalanche
PHOS	PHOton Spectrometer in ALICE
PID	Particle IDentification
PINGU	Precision IceCube Next Generation Upgrade
PLUTO	Experiment at the DORIS and PETRA colliders
PMM	Pre-Processor Module
PMMA	Polymethyl Methacryate
PMOS	P-channel Metal-Oxide Semiconductor
PMT	PhotoMultiplier Tube
POPOP	1,4-bis(5-phenyloxazol-2-yl) benzene
PPD	Pixelated Photon Detector
PPs	Precision Proton Spectrometer in the CMS eperiment
PPO	2,5-Diphenyloxazole
PRECO	PRE-COmpound Model
PS	Proton Synchrotron at CERN
PSI	Paul Scherror Institute in Villingen, Switzerland
QE	Quantum efficiency
QCD	Quantum Chromodynamics

QED	Quantum Electrodynamics
QGSP	Quark-gluon String Model
ρ	Density
r_e	Electron radius
RF	Radio frequency
RHIC	Relativistic Heavy Ion Collider at Brookhaven, USA
RICH	Ring-Imaging Cherenkov Counter
RMS	Root-Mean Square
ROB	ReadOut Buffer
ROD	ReadOut Drivers
RoI	Regions of Interest
RPC	Resistive Plate Chamber
σ_γ	Photoabsorption ionization cross section
σ_ϕ	Angular resolution in the azimuth angle
σ_θ	Angular resolution in the polar angle
σ_B	Born cross section
σ_C	Compton cross section
σ_{d_0}	Transverse impact parameter resolution
σ_E	Energy resolution
σ_p	Momentum resolution
σ_{pair}	Pair creation cross section
σ_{ph}	Photoabsorption cross section
σ_R	Rutherford cross section
σ_{Th}	Thomsom cross section
$\sigma_{r\phi}$	Position resolution in $r\phi$
σ_z	Position resolution in z
σ_{z_0}	Longitudinal impact parameter resolution
SCT	SemiConductor Tracker
SDD	Silicon Drift Detector in ALICE
SENSE	Sub-Electron-Noise Skipper CCD Experimental Instrument at SNOLAB
SiPM	Silicon Photomultiplier
SLAC	Stanford Linear Accelerator Center
SLC	Linear electron positron collider at SLAC
SLD	Detector at SLC
SM	Standard Model
S/N	Signal-to-Noise
SNO	Sudbury Neutrino Observatory
SOI	Silicon on Insulator
SPACAL	Spaghetti Calorimeter
SPD	Silicon Pixel Detector
sr	Steradian
SSC	Superconducting Supercollider
SPS	Super Proton Sychrotron at CERN
SP$\bar{\text{P}}$S	Proton antiproton collider at CERN

SSD	Silicon Strip Detector in ALICE
STAR	Relativistic heavy-ion collision experiment at RHIC, Brookhaven
STIC	Small-angle TIle Calorimeter in DELPHI
Superkamiokande	Neutrino experiment at the Kamioka mine, Japan
SuperKEKB	High-luminosity asymmetric electron positron collider at KEK in Tsukuba, Japan
STP	Standard Temperature and Pressure, 0 °C and 1 atmosphere
SUSY	SUperSYmmetry
SVT	Silicon Vertex Tracker
θ_c	Cherenkov angle
T2K	Tokai to Kamiokande experiment
TASSO	One experiments at the PETRA Collider
TDD	Thermal Double Donor
TEA	Triethyl amine
TEC	Time Epansion Chamber
TEC	Tracker EndCap in CMS
Tevatron	Proton Antiproton Collider at Fermilab
TGC	Thin-Gap Chamber
TIB	Tracker Inner Barrel in CMS
TID	Tracker Inner Disk in CMS
TileCal	ATLAS Hadronic Tile Calorimeter
TMAE	Tetrakis diMethyl Amino Ethylene
TMP	TetraMethylPenthane
TMT	Thermal Management Tile
TOB	Tracker Outer Barrel in CMS
TOF	Time-of-Flight
TOP	Time-of-Propagation Counter
TOPAZ	Experiment at the TRISTAN collider
TORCH	Timing Of internally Reflected Cherenkov light, PID detector at LHCb
ToT	Time over Threshold
TPB	TetraPhenyl Butadiene
TPC-2gamma	Experiment at the PEP Collider at SLAC
TPC	Time Projection Chamber
TPG	Trigger Primitive Generation
TR	Transition Radiation
t_r	Path in units of radiation lengths
TRD	Transistion Radiation Detector in ALICE
TRISTAN	Transposable Ring Intersecting Storage Accelerator in Nippon, electron positron collider at KEK
TRIUMF	Canadian National Laboratory for Particle and Nuclear Physics, Vancouver, Canada
TRK	CMS Outer Tracking Detector

TRT	Transition Radiation Tracker
UA1	Experiment at the SPPS at CERN
UA2	Experiment at the SPPS at CERN
u_k	Thermal velocity of particle k, for electron (e), for ion (ion)
UT	Upstream Tracker
UV	Ultra-violet
v_d^k	General drift velocity, for particle k, electron (e), hole (h), ion (ion)
VELO	LHCb Vertex Locator
VENUS	Experiment at the TRISTAN collider
VEPP 1,2,3,4	Electron positron colliders at the Budker Institute in Novosibirsk
VERITAS	Very Energetic Radiation Imaging Telescope Array System at the Fred Lawrence Whipple Observatory in Arizona, USA
VIRGO	The European Gravitational Observatory in Italy
VLPC	Visible Light Photon Counter
VME	VERSA Module European, system with parallel readout architecture
VO	Vacency-Oxygen complex
VSAT	Very Small Angle Tagger in DELPHI
VUV	Vacuum Ultra-Violet
VV	Amphoteric Di-vacency
WA78	Heavy-ion experiment at the SPS at CERN
WIMP	Weakly Interacting Massive Particle
WLS	WaveLength-Shifter
ξ	Screening variable
X_0	Radiation length
XENON1T	Dark-matter experiment in the Gran Sasso tunnel
XENONnT	Next-generation dark-matter experiment in the Gran Sasso tunnel
y_r	Rapidity
Y7, Y8, Y11	Wavelenght-shifting fibers from Kuraray
YS1, YS2, YS4, YS6	New wavelength-shifting fibers from Kuraray
z_0	Longitudinal impact parameter
ZDC	Zero-Degree Calorimeter
ZEUS	One experiment at the HERA collider at DESY

Introduction

Research in nuclear, particle and astro-particle physics has come a long way. Today's experiments use highly sophisticated instruments utilizing the latest advances in technology or having developed new technologies themselves. In this book we will discuss the underlying physics of measurement techniques, detector technologies and sub-detectors. We assume that the reader has basic knowledge in particle physics, statistics and electronics. Excellent introductions to high-energy physics are given in the textbooks [1,2]. We provide a short summary on statistics and multivariate analyses in Appendix B. For more information we recommend the reader to read the review articles in the Particle Data Group web page [3]. An excellent textbook on electronics is that by Horowitz and Hill [4].

The book is organized into 11 chapters and two appendices. Chapter 1 starts with a brief historic overview of particle physics and the development of accelerators followed by basic kinematic properties, general detector characteristics and effects of radiation. In Chap. 2 interactions of charged particles and photons with matter and fields are discussed with some focus on electron and ion properties. Chapter 3 presents ionization measurements in gases, liquids and solids. Chapter 4 gives a discussion of various gaseous detectors that provide position measurements and Chap. 5 covers propagation in magnetic fields, different magnet types and momentum measurements. Chapter 6 presents different solid-state detectors that provide vertex measurements. Chapter 7 focuses on energy measurements of electromagnetic and hadron showers discussing various calorimeter technologies. Chapter 8 deals with time measurements presenting various photodetectors, organic scintillators and scintillating fibers. In Chap. 9 different methods used for particle identification are discussed and various sub-detectors are shown. Chapter 10 talks about triggers and data acquisition. Finally, Chap. 11 presents several complete particle physics detectors. Appendix A summarizes types and sources of radiation and Appendix B presents advanced statistical techniques.

1.1 Historic Overview

Modern physics essentially started in the late nineteenth century with a few fundamental experiments. In 1887, Hertz observed the photoelectric effect [5]. He had set up a coil with a spark gap that was illuminated with ultraviolet (UV) light. He observed sparks confirming that UV light produced a discharge. We know today that the photon is fully absorbed and most of its energy is transferred to an electron that produces a discharge. In 1905, Einstein interpreted the photoelectric effect as emission of electrons caused by the absorption of light quanta [6]. In 1887, Michelson and Morley measured the fine structure of the spectral lines of hydrogen [7]. They observed two closely spaced lines, which are separated by spin-orbit interactions, an interaction of the electron spin with the angular momentum of the atom. The measurements were performed with an interferometer that had measured the speed of light to a high degree of precision. Much later, in 1916, Sommerfeld introduced the fine-structure constant α to explain the observed fine structure of the energy levels of the hydrogen atom [8]. In 1896, Zeeman measured the splitting of spectral lines by a strong magnetic field [9]. He noticed a broadening of the sodium yellow D-lines in a flame placed in a strong magnetic field. In 1913, Stark noticed a splitting of spectral lines in strong electric fields [10]. Another important milestone happened in 1901, when Planck predicted the spectrum of black-body radiation correctly [11]. His prediction was based on the postulate that the photon energy $E = \hbar\omega$ is quantized, where $\hbar$ is Planck's constant. In 1914, Franck and Hertz clearly demonstrated the quantum nature of atoms [12]. By colliding energetic electrons with mercury atoms, they noticed that the electrons lost 4.9 eV of their kinetic energy after the collision. The amount of energy loss was independent of the initial kinetic energy and so incompatible with classical physics.

In 1895, Röntgen discovered X-rays when experimenting with various types of vacuum tubes [13]. He noticed some shimmering on a barium platinocyanide screen when an electrical discharge passed through a vacuum tube. He speculated that this effect was caused by a new type of radiation. In 1896, Becquerel noticed in studies with uranium that three different types of ionization radiation emerged: positively charged, negatively charged and neutral radiation [14]. He had discovered α, β and γ radiation, respectively. In addition, we have radiation from neutrons, which we discuss later. Madame Curie studied the radiation from uranium compounds, such as pitchblende and torbernite, which were much more active than uranium itself. In 1898, she discovered the chemical elements thorium and together with her husband, polonium and radium [15]. Madame Curie is seen as the founder of nuclear physics, which developed into a large, rich field with many discoveries. For example, Hahn and Strassmann discovered nuclear fission in 1938 [16]. Bethe developed the carbon-nitrogen-oxygen cycle [17] in 1938 that produces 7% of the energy in the sun. Today, nuclear physics studies are still ongoing but we focus in the following more on the high-energy physics aspects.

In 1897, Thomson discovered the electron [18]. Experimenting with cathode rays that were developed by Lenard [19], he showed that the rays behaved like negatively charged particles. He concluded that the particles must be smaller than atoms and

1.1 Historic Overview

have a rather large charge-to-mass ratio. In 1909, Millikan measured the charge of the electron in the famous oil drop experiment [20]. In 1912, Hess discovered cosmic rays in balloon experiments [21] and Wilson had perfected the cloud chamber [22] that was used in several experiments from the 1920's to the 1950's. For example, Blanckett discovered the pair creation process, $\gamma \to e^+e^-$, in a cloud chamber in 1933 [23]. In the years 1908–1913, Geiger and Marsden performed the "gold foil" experiments in which α particles scattered off a gold foil [24]. The results disagreed with the plum pudding model by Kelvin and Thomson [25], since some of the α particles were scattered backwards. Rutherford concluded that the particles had encountered an electrostatic force. To explain the result, Rutherford assumed that the positive charge of the atom was concentrated in a tiny nucleus at its center [26]. In 1917, Rutherford performed the first artificially induced nuclear reaction by bombarding nitrogen nuclei with α particles [27]. He observed the emission of a particle he first called a hydrogen atom, but in 1920 he named it more accurately the proton. Davisson and Germer showed in an experiment 1924 to 1927 that electrons also behave like waves [28]. Scattering electrons off a crystal they observed a diffractive pattern that is typical for waves. This confirmed the particle-wave duality proposed by de Broglie in 1924 [29]. Based on Rutherford's measurements Bohr proposed a simple model of the hydrogen atom in 1913 that accounted for the new quantum physical interpretation [30]. Many of the previously mentioned measurements could not be explained in terms of classical physics and required a new theory leading to the development of Quantum Mechanics in the mid 1920s by Bohr, Schrödinger, Heisenberg, Born and others [31–34].

In 1925, Klein and Gordon formulated the first relativistic wave equation for spinless bosons [35,36]. Three years later, Dirac formulated a relativistic wave equation for fermions and predicted the existence of anti-matter [37]. These two equations were the first quantum mechanical results that fully accounted for special relativity, which had been developed by Einstein in 1905. In 1930, Pauli postulated an invisible new neutral particle that later was called the neutrino [38]. He did not accept Bohr's explanation that in nuclear β decay energy-momentum conservation was violated. Cowan and Reines detected the anti-neutrino in the Savannah River reactor experiment in 1953 [39]. Anti-neutrinos from a reactor interacted with protons in a target, producing a neutron and a positron, which annihilated with an electron into two 511 keV photons. In 1932, Chadwick discovered the neutron [40] conducting an experiment in which beryllium hit a paraffin wax target. He noticed that neutral particles with a mass similar to that of a proton were emitted. Now, the constituents of atoms were complete, proton, neutron and electron. In 1932, Anderson discovered the positron by studying cosmic rays in a cloud chamber [41]. He observed a particle with the same mass as the electron but with the opposite charge. This established the first evidence of anti-matter that was predicted by Dirac in 1931. It took another 23 years until Segrè and Chamberlain discovered the antiproton at the Bevatron particle accelerator in Berkeley, California [42]. The discovery of the antineutron by Cork occurred at the Bevatron in 1957 [43].

In 1934, Fermi established an effective theory of weak interactions [44] in order to explain the nuclear β decay in terms of a four-fermion interaction involving a contact force with no range. Cherenkov discovered an electromagnetic radiation now known as Cherenkov radiation [45] by observing blue light emitted from a liquid adjacent to a radioactive source emitting relativistic electrons. The radiation was produced by charged particles that move through matter at velocities greater than the phase velocity of light.

In 1935, Yukawa published his theory of mesons, which explained the interaction between protons and neutrons as well as predicted the pion [46]. In 1936, Anderson discovered a particle in studies of cosmic rays [47]. This particle was first mistaken as the pion; however, it turned out to be the muon, a new lepton with a mass 207 times heavier than the electron mass. In response to the discovery of the muon Rabi asked "who ordered that"? In 1947, Powell, Occhialini and Lattes utilized Powel's emulsion technique that revealed charged particle tracks to study high-energy cosmic rays [48]. Here, they observed the charged pion. However, it was not until 1950, that Bjorklund and collaborators observed the neutral pion in $\pi^0 \to \gamma\gamma$ at the Berkeley cyclotron [49]. In 1950, Rochester and Butler published two cloud chamber photographs of cosmic ray-induced events, one showing what appeared to be a neutral particle decaying into two charged pions and one, which appeared to be a charged particle decaying into a charged pion and something neutral [50]. It turned out that they had discovered kaon decays, the neutral kaon decay $K_S^0 \to \pi^+\pi^-$ and the charged kaon decay $K^+ \to \pi^+\pi^0$. The decays proceeded rather slowly, with typical lifetimes of the order of 10^{-10} s for K^0 and 10^{-8} s for the charged kaon. However, production in pion-proton reactions proceeds much faster, at a time scale of 10^{-23} s. This time discrepancy was solved by Gell-Mann, Pais and Nishijima who proposed a new quantum number called "strangeness" [51–53], which is conserved in strong interactions but is violated in weak interactions. Thus, the production of kaons is fast but the decay of kaons is slow.

By 1950, a quantum theory of electromagnetism, the Quantum Electrodynamics (QED), was developed by Feynman, Schwinger and Tomonaga [54–56] who showed that charged particles interact via photon exchange. This explained the experiments by Millikan [57] in 1916 and Compton [58] in 1923 who discovered the inelastic scattering of photons in atoms. Quantum Electrodynamics is a relativistic quantum field theory for the electromagnetic interaction of photons with charged particles. The novelty is that QED is a so-called gauge theory based on the Abelian group $U(1)$ that is associated with the photon field. This was an important step in the development of the Standard Model (SM).

In 1954, Yang and Mills extended the concept of the QED gauge theory to non-Abelian groups to develop similar theories for weak interactions as well as strong interactions [59]. In 1956, Wu conducted the ^{60}Co experiment in which she showed that parity (left-right symmetry) is violated in weak interactions [60] as had been predicted by Lee and Yang [61]. Furthermore, Goldhaber showed in 1958 that neutrinos are left-handed [62]. In parallel, Gell-Mann and Feynman as well as Sudershan and Marshak discovered the chiral structures of weak interactions and developed the V-A theory [63,64], which has been tested successfully in many experiments.

From 1947–1960, a large number of new mesons and baryons with and without the strangeness quantum number was observed [3]. This lead Gell-Mann in 1961 to sort these mesons and baryons into a scheme called the Eightfold-Way [65]. One baryon with three strange quarks, the Ω^-, was missing. So, Gell-Mann predicted it. In 1964, the Ω^- particle was discovered in a bubble chamber experiment at the alternating gradient synchrotron (AGS) in the Brookhaven National Laboratory [66] confirming the Eightfold-Way. The same year, Gell-Mann and Zweig developed the quark model [67,68]. In 1965, Struminsky, Bogolyubov and Nambu postulated a new quantum number, later named color by Gell-Mann and Fritsch, to make the wave function of the Δ^{++} baryon antisymmetric [69,70]. Color is a strong charge that is mediated by gluons, the exchange particles of strong interactions. The corresponding charge of weak interactions is the weak isospin that is mediated by weak bosons. Evidence for gluons was found by the PETRA experiments PLUTO [71], Mark J [72] and TASSO [73] in 1979, and later by JADE [74]. The signature consisted of observing three-jet events that come from two-quark jets with gluon radiation. This topology had been suggested by Gaillard and Ross in 1976 [75]. In 1968-69, Friedman, Kendall and Taylor found evidence in deep inelastic scattering experiments at SLAC that protons had internal structure [76]. They confirmed the idea of the parton model Feynman had proposed in 1969 [77]. Among many other things Feynman also contributed to the development of quantum chromodynamics, the theory of strong interactions. In 1973–1974, Politzer, Gross and Wilczek discovered asymptotic freedom of quarks confined in the strong interaction potential [78,79]. At this time, quantum chromodynamics (QCD) acquired its modern form. In 1968, experiments at SLAC confirmed that hadrons were composed of fractionally charged quarks.

In 1961, Glashow combined the electromagnetic and weak interactions [80]. A year later Goldstone, Weinberg and Salam incorporated symmetry breaking into Glashow's electroweak theory and showed that in spontaneously broken symmetries zero-mass particles must exist [81]. After Englert and Higgs developed the Higgs mechanism [82,83], Weinberg and Salam incorporated it into the electroweak theory [84]. This theory is the Standard Model of weak and electromagnetic interactions. Strong interactions were added later. Further discoveries completed the picture of electroweak interactions. In 1962, Steinberger, Lederman and Schwartz discovered two types of neutrinos at the AGS in Brookhaven [85]. So, there is an electron-neutrino associated with the electron and a muon-neutrino associated with the muon. Other measurements indicated that this lepton flavor quantum number was conserved until the discovery of neutrino mixing. Though parity violation in weak interactions had been observed, it was believed that the combination of charge conjugation and parity transformation (CP) was conserved until Cronin and Fitch observed its violation in an experiment at the Brookhaven AGS in 1964 [86] that studied decays of the long-lived neutral kaon K_L^0. If CP is conserved, the K_L^0 can decay only to three pions while the short-lived K_S^0 can only decay to two pion. However, Cronin and Fitch also observed two-pion decays of the K_L^0, establishing CP violation in weak interactions. In 1964, Cabbibo postulating weak universality introduced a mixing angle (the Cabibbo angle) between the down quark and strange quark [87]. This means that the charged weak boson couples with 100% to a weak eigenstate that is a mixture of d and s quarks. In 1970, Glashow, Illiopolous and Maiani introduced a mechanism that explained the absence of so-called flavor-changing neutral currents in tree level decays [88], which are decays like $b \to s$ or $s \to d$. They expressed up-type and down-type quarks in terms of left-handed doublets while right-handed quarks were arranged into singlets. Since only the up, down and strange quarks were known at that time and two orthogonal mixtures of d and s quarks can be formed, one up-type partner was missing. Thus, they predicted the existence of a fourth quark, the charm quark. It took another four years until Richter at SLAC [89] and Ting at Brookhaven [90] simultaneously discovered the charm quark by observing the J/ψ resonance (a charm anti-charm bound state) in e^+e^- collisions and p-Be interactions, respectively. A year later, Perl discovered the tau lepton at SLAC [91], which is the charged lepton of the third family. This implied another doublet of quarks and another neutrino. In 1973, Kobayashi and Maskawa extended the concept of family mixing to three families since they noticed that CP violation could not be explained in a four-quark model [92]. They introduced weak eigenstates and mass eigenstates that were connected through a 3×3 mixing matrix, which has four independent parameters, three mixing angles and one phase that introduces CP violation into the SM.

Another important step in the development of the electroweak theory was the discovery of neutral currents at CERN by Haidt and Pulliain in the Gargamelle bubble chamber in 1973 [93]. This indicated the existence of a new heavy neutral gauge boson, the Z^0 boson, which was discovered along with a heavy charged gauge boson, the $W^{\pm}$ bosons, at CERN in the SP$\bar{\text{P}}$S collider in 1983. The experiment was proposed by Rubbia [94]. One of the technical challenges of keeping the transverse momenta of the hadrons very small was solved by van der Meer by introducing the stochastic cooling for anti-protons [95]. The $W^{\pm}$ and Z^0 bosons were found exactly at the masses that were predicted by the electroweak theory, which was a great success for the electroweak theory. The term "Standard Model" was first introduced by Pais and Treiman in 1975, with reference to the electroweak theory with four quarks [96]. In 1976, four quarks, three charged leptons, two neutrinos, the photon and three weak bosons were known. The third charged lepton already indicated the existence of a third family. In 1977, Lederman discovered the b quark in the E288 experiment at Fermilab [97]. This implied the existence of another up-type quark and another neutrino. In 1995, both the CDF and D0 experiments at Fermilab discovered the top quark with a mass of 175 GeV/c^2 [98,99]. Previously, lower-energy searches at DESY, SLAC and KEK had found nothing. The tau neutrino was finally found in the DONUT experiment at Fermilab in 2000 [100].

The discovery of the b quark led to the production of B mesons in the CLEO experiment at the Cornell Electron Storage Ring (CESR) and the experiments Argus [101] and Crystal Ball [102] at the e^+e^- collider DORIS at DESY. The B mesons are produced in pairs in the decay of the $\Upsilon(4S)$ resonance, either as B^+B^- or $B^0\bar{B}^0$. The Argus experiment discovered $B^0\bar{B}^0$ mixing in 1987 [103]. Here, a neutral B^0 meson oscillates into its antiparticle and back. The discovery of $B_s^0\bar{B}_s^0$ oscillations by the CDF experiment at the Tevatron followed in 2006 [104]. The CLEO experiment was the first to observe flavor-changing neutral currents produced in loop processes in 1993 [105]. The observed decay was $B^0 \to K^{*0}\gamma$, which is a so-called "penguin" decay in which a b quark couples to an s quark via a loop diagram. The long B life times and the large $B^0\bar{B}^0$ mixing provided excellent conditions for the observation of CP violation in the B system. This task consisted of measuring the decay rate asymmetry as a function of the time difference between the B^0 and $\bar{B}^0$ decays since the time-integrated CP asymmetry vanishes. However, since B mesons at the $\Upsilon(4S)$ are produced nearly at rest, the center-of-mass (CM) system had to be boosted to observe CP violation. Thus, SLAC and KEK built the asymmetric e^+e^- storage rings PEPII and KEKB, respectively. In 2001, the BABAR experiment at PEP II and the Belle experiment at KEK B observed CP violation in the neutral B system [106, 107] by studying the decay $B^0 \to J/\psi K_S^0$. In 1999, the KTeV experiment at Fermilab and NA48 at CERN finally established direct CP violation in the neutral kaon system [108, 109]. They observed the decays $K_L^0 \to \pi^+\pi^-$ and $K_L^0 \to \pi^0\pi^0$ and determined the parameter ϵ'/ϵ to be non-zero.

In 2005, the four LEP experiments showed that the Z^0 line shape agrees with three light neutrinos [110]. This confirms that only three families with light neutrinos exist. For quite some time neutrinos were considered to be massless fermions until Superkamiokande and the Sudbury Neutrino Observatories (SNO) showed in 1998-99 that they mix and in turn have a small mass [111, 112]. Since then several experiments have measured the mixing angles and neutrino mass differences. At the moment the focus is to measure CP violation in the neutrino sector. To explain the light neutrino masses the "see-saw" mechanism was proposed by Gell-Mann, Mohapatra and others [113, 114]. On July 4th 2012, the discovery of the final building block of the SM, the Higgs boson, was announced by the ATLAS and CMS experiments at CERN [115, 116]. The Higgs boson has a mass of 125 GeV$/c^2$.

To conclude this overview we mention a few other discoveries. In 1964, Penzias and Wilson measured the cosmic microwave background and determined a temperature of the universe of about 3 K [117]. In 1993, Smoot and Mather discovered the anisotropy of the cosmic microwave background radiation [118]. In 1998, Perlmutter, Schmidt and Riess found evidence for an accelerated expansion of our universe by studying supernova explosions in a neighbor galaxy [119, 120]. In 2016, the two LIGO sites in the US built by Barish, Drever, Thorne and Weiss, announced the first observation of gravitational waves by seeing the merger of two black holes [121]. Since then LIGO and VIRGO in Italy observed several other mergers of two black holes as well as neutron star black hole mergers. Gravitational effects are not included in the SM. They play a role at the Planck mass scale (10^{19} GeV). Studies of gravitational waves may help with understanding dark matter.

The SM is complete now and provides excellent predictions of all measurements we have done so far. Figure 1.1 shows the building blocks of the SM and Fig. 1.2 shows the forces. The building blocks consist of six quarks (up, down, strange charm, bottom and top), three charged leptons (electron, muon and tau-lepton) plus three neutrinos ν_e, ν_μ and ν_τ, which are arranged into left-handed doublets and right-handed singlets in three families. The quarks come in three colors and carry fractional charge. There are twelve spin-1 gauge bosons (8 gluons, 1 photon, $W^\pm$ and Z^0), which are the exchange particles of the forces (strong and electromagnetic and weak

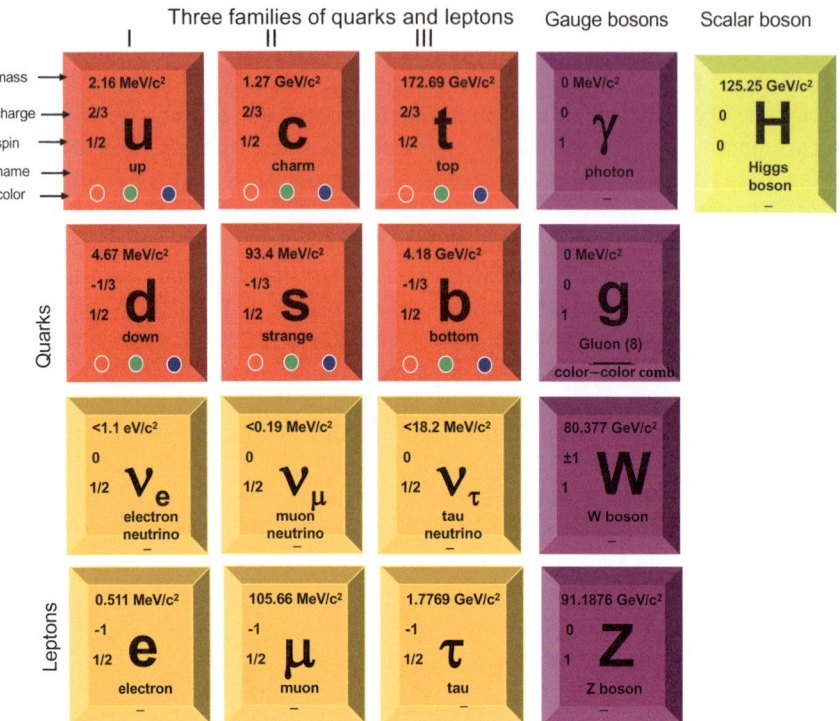

Fig. 1.1 Building blocks of the SM. The quarks come in three colors (red, green and blue) and all particles have antiparticles. The quarks and leptons come in three families. Besides the name, mass, charge and spin, we show the color quantum number for quarks (red, green, blue) and gluons (color-anticolor)

interactions) and one scalar particle, the Higgs boson. The gravitational force, with its spin-2 exchange particle, the graviton, is not part of the SM.

The SM cannot give answers to many interesting questions and does not predict fundamental parameters such as masses, couplings, mixing angles and *CP* violation phases. We, therefore, believe that an extended theory exists that embeds the SM and provides answers to all outstanding questions as well as predictions for the fundamental parameters. In the new theory we expect new particles to appear. There may be supersymmetric particles [122], dark matter particles [123], extra dimensions [124], leptoquarks [125], or excited $W^{\pm}$ and Z^0 bosons [126]. Despite active searches at the LHC and other experiments, so far no new phenomena have been observed.

Figure 1.3 shows the evolution of our universe, which was created in a big bang. The universe subsequently expanded slowly up to 10^{-37} s. During this time it had cooled by four orders of magnitude. Then inflation started and the universe expanded rapidly. This process stopped around 10^{-32} s. The universe inflated by a factor of 10^{26}. After that it continued to expand but at a much slower rate. Here, all known fundamental particles and antiparticles were produced. This period was the age of

1.1 Historic Overview

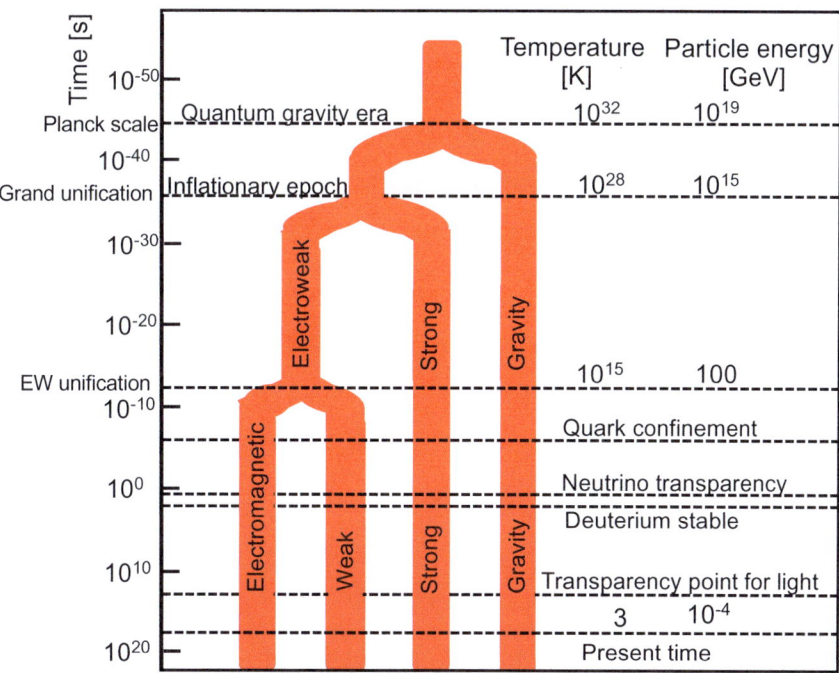

Fig. 1.2 Forces in the SM. At 100 GeV, the electroweak unification scale, the weak and electromagnetic forces are unified to the electroweak force. At 10^{15} GeV, the grand unification scale, the strong and electroweak forces are unified. At the Planck scale (10^{19} GeV), all forces merge

the leptons. At a time of 10^{-10} s corresponding to a temperature of 10^{15} K and an energy of 100 GeV, there may have been a possible freeze out of dark matter particles. The mix of fundamental particles remained to a time scale of 10^{-5} s. Here, the temperature had cooled sufficiently to produce mesons and baryons defining the age of the nucleons. Around 10^2 s, where the energy had decreased to about 100 MeV, first ions were formed specifying the age of nuclear synthesis that ended around 1,000 y. From 1,000 y to 3,000 y, was the age of ions and from 3,000 y to 300,000 y was the age of atoms. Around a time of 3×10^5 y the cosmic microwave background became visible. Around 10^9 y gaseous clusters, stars and galaxies began to form. Around 12×10^9 y planetary systems like our solar system formed and black holes were produced. The LHC may reach a scale of a few thousand GeV. However, high-energy cosmic rays may probe energies up to $10^{11} - 10^{12}$ GeV.

With higher and higher energies of accelerators we can look backwards in time and access eras of the earlier universe. The knowledge we have gathered was a result of excellent interplay between theory and experiment, the development of higher-energy and higher-intensity colliders, as well as improvements in detector performance and the development of new experimental techniques. We discuss accelerators and their development in Sect. 1.3. First, we classify the known particles according to their lifetimes.

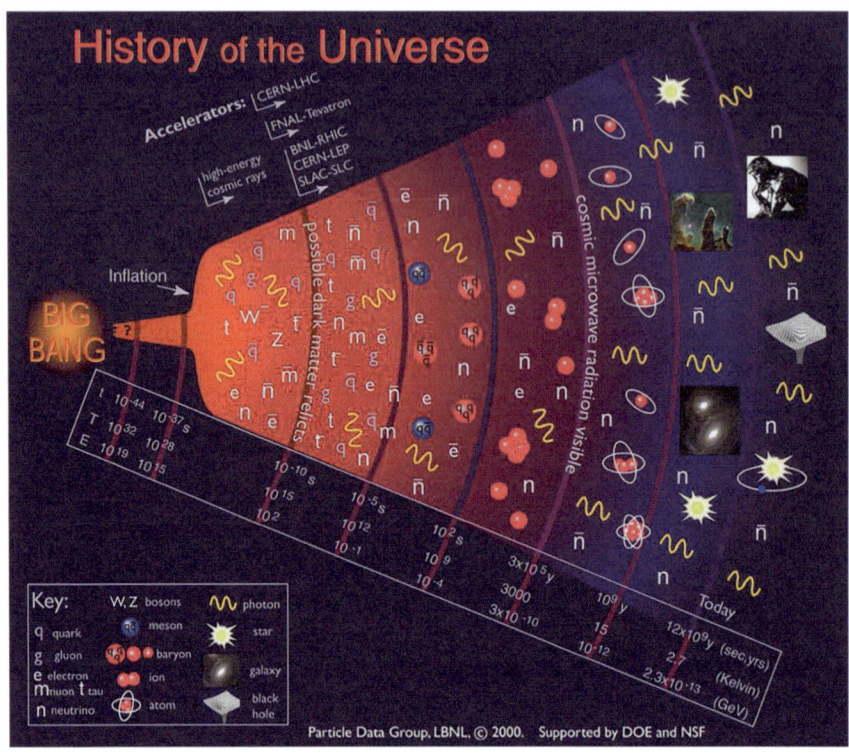

Fig. 1.3 Evolution of the universe as a function of time, temperature and energy showing also the energy range that present accelerators can cover. Reprinted with kind permission from [127], © 2000, the Particle Data Group LBNL. All rights reserved

1.2 Some General Remarks

Today's experiments can be sorted into two classes, (i) multipurpose experiments, which cover many different physics topics with optimal sub-detectors, and (ii) dedicated experiments that focus on individual physics questions. Multipurpose experiments include the LHC experiments ATLAS [128] and CMS [129], Tevatron experiments CDF [130] and D0[131], LEP experiments ALEPH [132], DELPHI [133], L3 [134] and OPAL [135], SLC experiments Mark II [136] and SLD [137], HERA experiments ZEUS [138] and H1 [139] as well as the B factory experiments CLEO II [140], BABAR [141] and Belle/Belle II [142,143]. In addition, there is the BES III experiment in Beijing [144] and the KLOE experiment at Frascati [145]. Furthermore, there were several experiments at the PETRA e^+e^- collider (PLUTO [146], MARK J [147], TASSO [148], JADE [149] and CELLO [150]) and the TRISTAN e^+e^- storage ring (AMY [151], TOPAZ [152] and VENUS [153]).

At the LHC, the dedicated experiments are the bottom and charm experiment LHCb [154] and the heavy ion experiment ALICE [155]. Other dedicated experiments, for example, include neutrino physics experiments (e.g.: Superkamiokande [156], T2K [157], Daya Bay [158]), neutrinoless double beta decay experiments (e.g.: Cuore [159], KamLand-ZEN [160], nEXO [161]), experiments searching for weakly interacting massive particles, WIMPs, (e.g.: Xenon 1T [162], LUX-LZ [163], Darwin [164]), and experiments looking for charged lepton flavor violation (e.g.: Mu2e at Fermilab [165], MEG II at the Paul Scherrer Institute [166]). Note that some experiments designed for specific tasks expanded analyses to other topics. For example, *BABAR* expanded analyses to dark photon, axion-like particles and dark higgs searches, while LHCb expanded the physics program to studies of heavy ions and searches for dark matter. In Chap. 11, we will show some of these experiments in more detail. In addition, there are dedicated kaon experiments. At CERN, NA62 measures the rare decay $K_L^+ \to \pi^+ \nu \bar{\nu}$ and at JPARK, KOTO measures the corresponding neutral kaon decay $K_L^0 \to \pi^0 \nu \bar{\nu}$.

Why do we do these experiments? Well, we want to measure certain properties of particular particles to a high degree of precision, so we can check if their proprties agree with predictions. The other task is to search for new particles or new phenomena that are predicted by theorists. Both tasks may lead to new physics effects that require new terms in the Lagrangian. Since we do not know where and how large new physics effects are, it may require several iterations of detectors with higher and higher sensitivity, which may involve new technologies or new detection approaches.

By now we have a huge zoo of particles that we can organize into five categories.

1. Six leptons: electron (e), muon (μ), electron neutrino (ν_e) and muon neutrino (ν_μ), tau lepton (τ) and tau neutrino (ν_τ), which are fundamental spin one-half point-like particles.
2. Mesons: pion (π), kaon (K), D-meson (D), B-meson (B), etc., which are composite particles made of a quark and an anti-quark with integer spin.
3. Baryons: proton (p), neutron (n), Lambda (Λ), etc., which are composite particles made from three quarks with spins 1/2 or 3/2.
4. 12 exchange particles: photon (γ), gluons (g), W-boson ($W^\pm$), Z-boson (Z), which are gauge bosons with spin 1.
5. The Higgs boson, which is a scalar particle (*i.e.* spin zero).

With respect to particle detection it is useful to classify them according to their decay length (speed of light c times decay time τ) since most of these particles decay. There are only a few stable and long-lived particles.

1. Stable particles ($c\tau = \infty$): electron, photon, proton and neutrinos.[1]
2. Long-lived ($c\tau \geq 1$ m): $\pi^\pm$, $K^\pm$, K_L^0, muon and neutron.
3. Medium-lived ($c\tau \geq 2.5$ cm): K_S^0, Λ.

---

[1] The proton may decay and the neutrinos may oscillate into a different flavor neutrino, but for our purpose the time scales are large so we can consider them infinite.

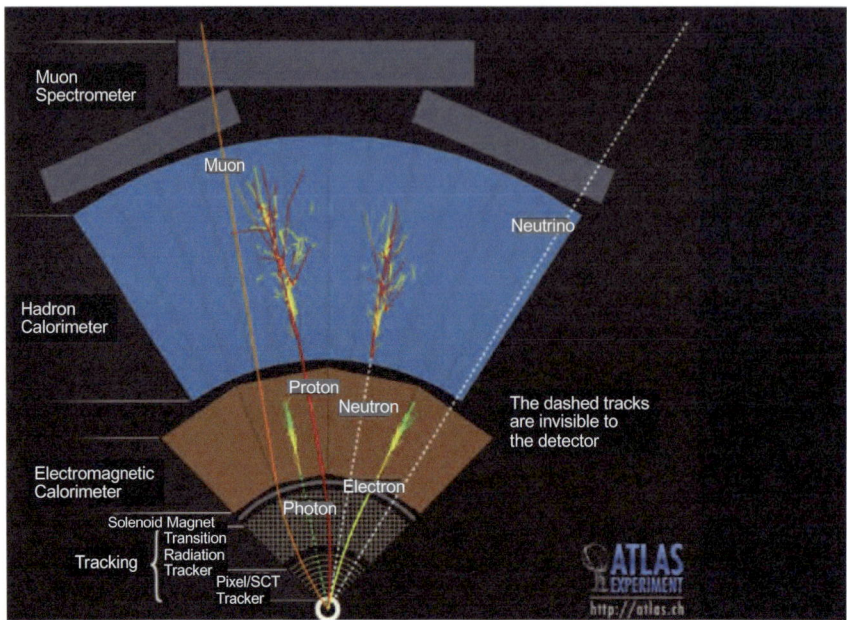

Fig. 1.4 Imprints left by different particle species in the sub-detectors of the ATLAS experiment. Charged (neutral) particles are shown by solid (dashed) trajectories. Reprinted with kind permission from [167], © 2006, the ATLAS Collaboration. All rights reserved

4. Short-lived ($c\tau \geq 88 - 390$ μm): τ, D, B.
5. Very-short-lived: (prompt decay, $c\tau$ is not observable): π^0, J/ψ, Υ and many others.

Only particles from classes 1 and 2 leave direct imprints in detectors. Other particles have to be reconstructed via their decay products, which consist of particles in classes 1 and 2. So, in detectors we see directly the charged particles (pions, kaons, protons/antiprotons, muons and electrons/positrons) and the neutral particles (photons and neutrons/antineutrons). Figure 1.4 illustrates this in a schematic slice of the ATLAS detector. Outside the beam pipe, there are the three tracking detectors inside the superconducting coil, Pixel, SemiConductor Tracker and Transition Radiation Tracker. Outside the coil are the electromagnetic calorimeter, the hadron calorimeter and the muon system. A muon penetrates the entire detector leaving imprints in all sub-detectors. A photon passes the tracking detectors without any imprints and is stopped in the electromagnetic calorimeter where it typically leaves all its energy. A proton leaves imprints in the tracking detectors and electromagnetic calorimeter and is stopped in the hadron calorimeter where it looses all its energy if it is sufficiently deep. Most neutrons or K_L^0s leave no imprints in the tracking detectors and electromagnetic calorimeter. They are typically stopped in the hadron calorimeter where they loose all their energy except for some that may already have interacted in the electromagnetic calorimeter. An electron leaves imprints in the tracking detector before it is stopped in the electromagnetic calorimeter where it deposits all its energy. A neutrino passes the entire detector without leaving any imprints.

The imprints are caused by electromagnetic interactions of the particle with the medium. Charged particles interact via ionization of the medium or by emission of electromagnetic radiation. High-energy photons produce electromagnetic showers. Low-energy photons interact via photo-absorption, Compton scattering or e^+e^- pair creation. Typical processes of neutral hadrons are scattering and nuclear interactions. Thus, a typical multipurpose detector is laid out to measure track positions, particle vertices, particle momenta, particle energies, timing and particle types. The measurement of these observables is performed by dedicated sub-detectors that are arranged in onion-like shells around the interaction region.

1.3 Accelerators

First, radioactive sources were used to produce particles scattering off targets. We summarize these in appendix A. Today, radioactive sources and cosmic rays still provide useful tools since they are used, for example, to perform calibrations of detectors, monitor the detector performance, trace tracks and perform system checks. Since the energies of the source particles were rather limited, particle accelerators became a crucial instrument in producing the high energies needed to provide the manifold of discoveries in high-energy physics. Over the past 70 years, hadron and e^+e^- machines have been developed in parallel, achieving higher-and-higher energies. Here, some basics are presented. For further details on accelerator physics the book "The Physics of Particle Accelerators" by Wille is recommended [168].

The first accelerator was the cathode ray tube depicted schematically in Fig. 1.5, which was used in 1897 by Thomson [18]. Electrons are evaporated from the cathode by heat and are accelerated in the electric field. They pass through the anode and are deflected by a magnetic field onto a screen at the end of the tube. In 1932, Cockcroft and Walton developed an electrostatic accelerator that produced energies up to 800 keV [169]. Figure 1.6 shows a schematic layout (left) and a photograph

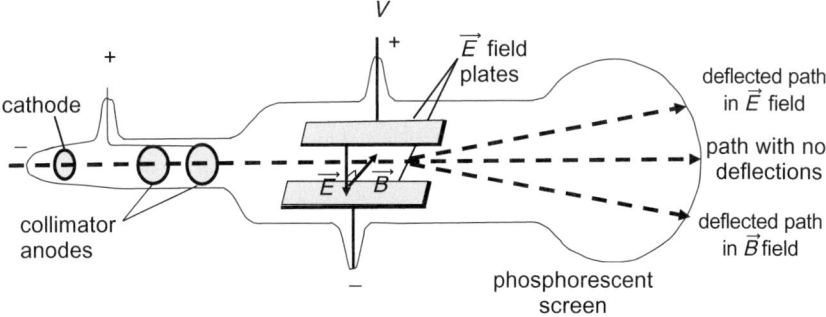

Fig. 1.5 Schematic view of Thomson's cathode ray tube experiment. The accelerated electrons can be steered with electric and magnetic fields

Fig. 1.6 Left: Schematic layout of a Cockcroft-Walton accelerator. Center: Photograph of a Cockcroft-Walton accelerator. Both reprinted with kind permission from [176], © 2019, J. L. Conradi. All rights reserved. Right: Schematic view of a van de Graaff accelerator. Positive charge is carried by a belt on to a conductor. The accumulated charge is used to accelerate a beam

(right) of a Cockcroft-Walton accelerator. A high electrostatic field produces a positive high DC voltage. The acceleration of ions is accomplished through intermediate potentials created by a voltage divider shown by the metal donuts in the photograph. The energy limit was raised with the development of van de Graaff accelerators in which the ladder of rectified voltages were replaced by an insulating belt that transports charge to a conductor. Figure 1.6 (right) illustrates this principle, which produces energies of about 10 MeV. Note that this is higher than energies produced in radioactive decays. After constructing the first machine in 1929, van de Graaff built an accelerator that reached 7 MeV in 1933 [170]. The van de Graaff accelerator became the most used device until Lawrence developed the cyclotron in 1930 that reached higher energies [171].

Figure 1.7 (left) shows a schematic layout of a cyclotron, which consists of two half disks called Dees that are connected to a radio frequency (RF) field. The entire arrangement is placed into a magnetic field whose configuration is depicted in Fig. 1.7 (right). Protons enter in the center of the accelerator, where they are accelerated from one Dee to the other. The RF changes accordingly so the protons always see an accelerating field. With the increased proton momentum the radius of the orbit increases in the constant magnetic field. Once the protons reach the outer radius of the Dee they are steered to a target. The final proton energy and momentum depend on the size of the Dee's and the magnetic-field strength. In his first machine, Lawrence accelerated protons to an energy of 1 MeV. Cyclotrons became the most powerful particle accelerators until the 1950s when they were superseded by the synchrotron [172, 173]. Though the largest cyclotron was built at TRIUMF producing 500 MeV protons, however, the weak-focusing synchrotron at the University of California, Berkeley (Bevatron) could accelerate protons up to 730 MeV.

In order to build synchrotrons that produce a pencil-thin beam of particles with high energies, engineers had to overcome major technical hurdles. A major step occurred in 1944, when Veksler and McMillan developed the "weak focusing" [174]. Particles, which have orbit radii slightly larger or smaller than the optimum radius of the center of the beam tube can be managed by designing fringe magnetic fields,

1.3 Accelerators

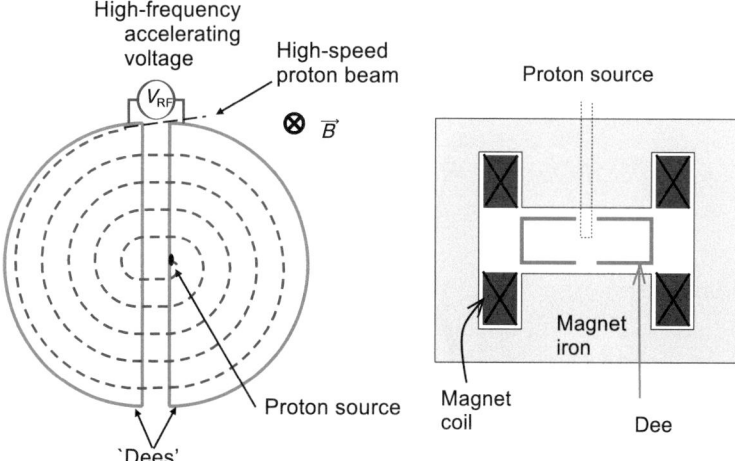

Fig. 1.7 Left: Schematic view of a cyclotron showing the two Dees and the RF power. The dashed line indicates the particle trajectory. The magnetic field is perpendicular to the Dees. Right: Configuration of the magnets in a cyclotron. Both reprinted with kind permission from [179], © 2014, A. Variola. All rights reserved

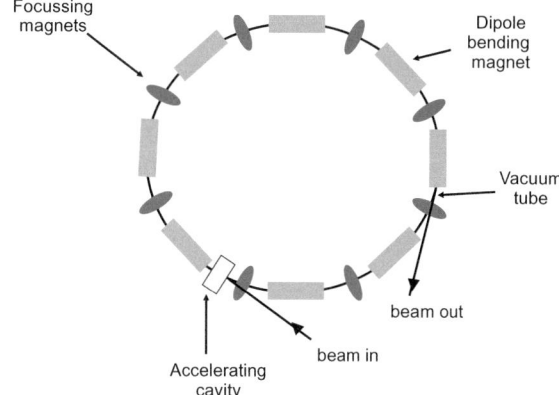

Fig. 1.8 Schematic layout of a synchrotron showing the vacuum tube (solid ring), bending magnets (dark grey boxes), focusing magnets (ellipses) and the accelerating RF cavity (grey boxes)

which provide larger or smaller bending so that these particles arrive in phase with the accelerating mechanism. The first electron synchrotron was built at General Electric in 1947 [175]. A second major step was the development of the "strong focusing", which was developed by Christofilos [177] in 1950 and Courant, Livingston and Snyder in 1952 [178], who built the first proton synchrotron at Brookhaven. A combination of dipole and quadrupole magnets alternately focus and defocus the beam in both the horizontal and vertical beam directions. The combination of magnets can be arranged to achieve a net collimation or "focusing" of the beam. Strong focusing is important in high-energy accelerators to keep the beam size small to save on magnet costs. For proton synchrotrons, the high-energy limit is determined by the strength of the bending magnets, which keeps the particle on a fixed orbit. Figure 1.8 shows a

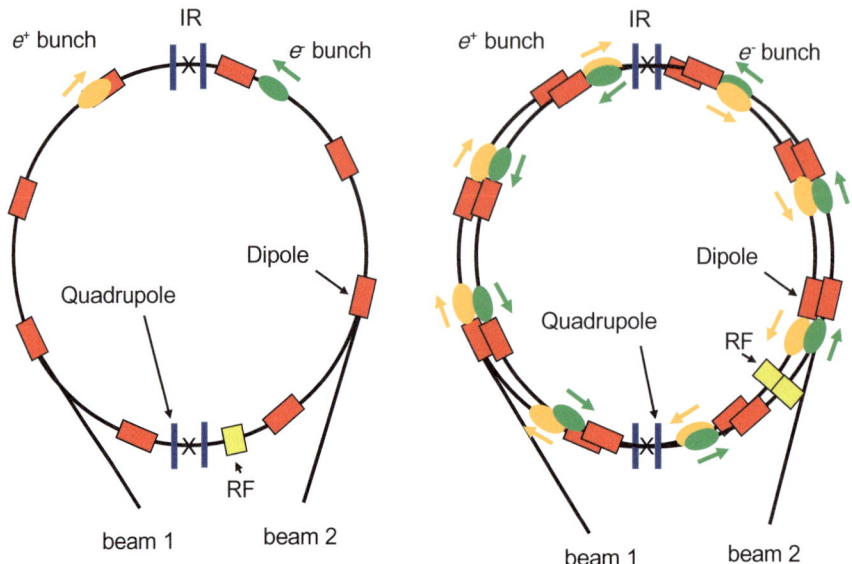

Fig. 1.9 Left: Schematic view of a one-ring storage ring. Right: Schematic view of a two-ring storage ring. The crosses show the interaction regions. The arrows indicate the directions of the bunches

schematic layout of a synchrotron. An injecting line inserts the beam into the vacuum chamber. Bending magnets keep the charged particles on a fixed orbit. The RF cavity accelerates the particles. The extraction line sends the beam to the experiments.

To increase the center-of-mass (CM) energy, storage rings were developed in which two beams circulate in opposite directions and are brought into collision at the interaction regions. The beams are steered in either one-ring or two-ring colliders as depicted in Fig. 1.9 (left, right), respectively. The particles are collected in bunches that circulate in the ring. Typically, several bunches are placed in a bunch train and several bunch trains are placed into the storage ring. Most colliders are one-ring storage rings since for symmetric beam energies and opposite-charge particles the beam optics can manage both beams in a single ring. Two-ring machines are necessary for asymmetric B factories, the e-p collider Hera and the LHC. Figure 1.10 shows a comparison of hadron and electron colliders at the energy frontier including proposed projects. The LHC with 7 TeV proton beam energies is presently the highest-energy collider.

The step from a single beam hitting a stationary target to two colliding beams brought a substantial increase in CM energies. In a single-beam experiment, where the beam from a synchrotron or linear accelerator hits a proton in a stationary target, the CM energy s is given by

$$\sqrt{s} = \sqrt{2m_p c^2 E}, \qquad (1.1)$$

where E is the beam energy and $m_p c^2$ is the proton mass. Let us look at an example. For an 800 GeV proton hitting a stationary proton the CM energy is $\sqrt{s} = 40$ GeV.

1.3 Accelerators

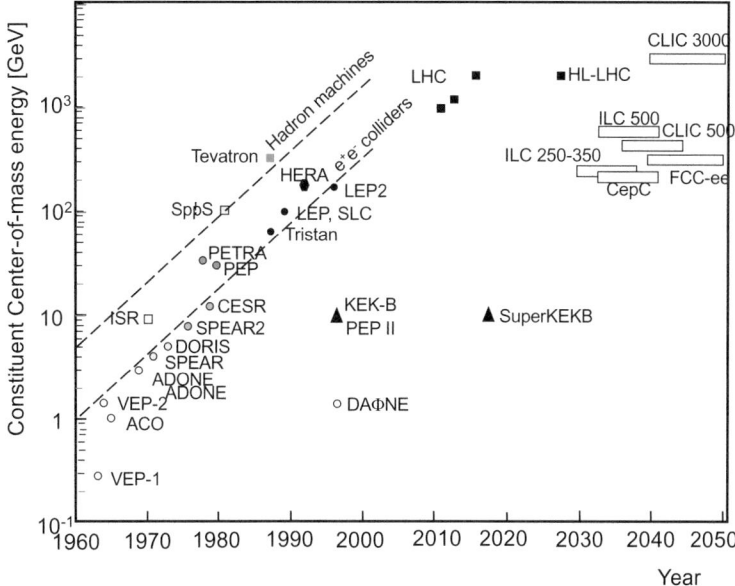

Fig. 1.10 Constituent center-of-mass energy as a function of accelerator start time for different hadron and electron accelerators. Note that the ILC, CLIC, CEPC and FCC-ee colliders are planned machines for which the operation period is estimated. Reprinted and adapted with kind permission from [185], © 2012, AIP Publishing. All rights reserved

Now, let us consider two colliding beams in a storage ring or linear collider with symmetric beam energies (e^+e^-, $p\bar{p}$, pp) the CM energy is

$$\sqrt{s} = 2E. \tag{1.2}$$

So, for a 800 GeV proton hitting an 800 GeV proton, the CM energy increases to $\sqrt{s} = 1600$ GeV. This is 40 times larger than the CM energy in a fixed target experiment. If the beam energies are asymmetric (E_1 and E_2) we get

$$\sqrt{s} = \sqrt{(E_1 + E_2)^2 - (E_1 - E_2)^2}. \tag{1.3}$$

As an example we look at PEP II. For $E_1 = 9.0$ GeV and $E_2 = 3.1$ GeV, we obtain $\sqrt{s} = 10.564$ GeV, which is the $\Upsilon(4S)$ mass.

1.3.1 Hadron Machines

The CERN laboratory was established in 1954. The first accelerator, the proton synchrotron (PS), started operation in 1959 [180] with a proton energy of up to 25 GeV. Internal targets produced secondary beams, which were used by several fixed-target experiments. In 1971, the Intersection Storage Ring (ISR) began with the collisions of two proton beams yielding a maximum CM energy of 64 GeV [181]. The ISR ran until 1984. In 1976 the super proton synchrotron (SPS) started operation [182] with a designed beam energy of 300 GeV. However, the SPS reached energies of 400 GeV. Now, it serves as injector for the LHC and for producing secondary beams for fixed-target experiments and test beams.

From 1981 to 1991, the SPS was converted into a proton-antiproton collider (SP$\bar{\text{P}}$S) with a CM energy around 100 GeV. The two experiments UA1 [183] and UA2 [184] discovered the weak bosons $W^{\pm}$ and Z^0 [94]. To explore the properties of the $W^{\pm}$ and Z^0 bosons, CERN built the first Large Electron Positron (LEP) collider [186], discussed in the next section. The LEP tunnel with a circumference of 26.66 km was reused to house the Large Hadron Collider (LHC [187]), which produced first collisions in 2010 at a CM energy of 7 TeV. Presently the CM energy has been increased to 13 TeV, which is close to the design value of 14 TeV. As already mentioned, four major experiments operate at the LHC: ATLAS [128], CMS [129], LHCb [154] and ALICE, [155]. Figure 1.11 shows the arrangement of present experiments at CERN. In 2012, ATLAS [115] and CMS [116] discovered the Higgs boson and started measurements of its properties. The LHC and the detectors have been upgraded for Run 3, which is ongoing. During this run the beam energies may be increased to 7 TeV. The high-luminosity upgrade of the LHC machine and the ATLAS and CMS experiments will take place in the next long shut down planned for 2025. After that the LHC will deliver a luminosity of 3000 fb^{-1}. The laboratory is looking into the feasibility of building a 100 TeV hadron collider (FCC-hh), which needs a new approximately 100 km long tunnel.

Fig. 1.11 Present operation of accelerators at CERN. Reprinted with kind permission from [188], © 2010, CERN. All rights reserved

1.3 Accelerators

Brookhaven National Laboratory (BNL) started in 1947 with the construction of the first nuclear reactor, the Brookhaven Graphite Research Reactor, which began operation in 1950. The first particle physics accelerator was the Cosmotron that started in 1952 and ran until 1966 [172]. It was superseded in 1960 by the alternating gradient synchrotron (AGS) [190]. Three important discoveries were made at the AGS, namely the discovery of the muon neutrino, the charm quark and *CP* violation in the kaon system. In 2000, the Relativistic Heavy Ion Collider (RHIC) started operation [191] that produced many heavy ion collisions. In early 2020, BNL was selected to build an electron-ion collider. The goal is to upgrade the existing Relativistic Heavy Ion Collider, which will collide beams of light to heavy ions including polarized protons, with a polarized electron beam that will be housed in the same tunnel.

Fermilab was founded in 1969 and produced its first particle beam in 1972. A linear accelerator (Linac) produced 200 MeV protons that were accelerated in the booster ring up to 8 GeV before entering the main ring, which accelerated them up to 400 GeV. The beam was used for a rich fixed-target physics program. For example, in 1977 the *b* quark was discovered. To remain competitive the proton-antiproton collider Tevatron was built [192]. By 1983, the beam energy reached 512 GeV, which was increased to 800 GeV by 1985. Two experiments, the Colliding Detector at Fermilab (CDF) [130] and D0 [131] harvested the physics at the Tevatron. In 1993, the main injector was designed, which started operation in 1999. It allowed to boost the beam energies up to 980 GeV producing a CM energy of 1.96 TeV, the highest energy reached in a storage ring before the LHC. The upgraded CDF [193] and D0 [194] detectors continued to harvest rich physics. Figure 1.12 shows the Fermilab site at the time of the Tevatron operation. Now, Fermilab is developing high-power beams from the main injector to the Deep Underground Neutrino Experiment (DUNE) target at 120 GeV [195]. While the near detector will be on the Fermilab site, the far detector will be located in the Sanford Underground Research Facility in South Dakota. Table 1.1 summarizes past, present and future hadron machines, hadron colliders and the electron-hadron collider.

1.3.2 Electron Machines

In 1989, CERN started the operation of the Large Electron Positron (LEP) collider [186], the only e^+e^- machine ever built at CERN. In the first phase the machine ran at beam energies of about 45 GeV to produce large samples of Z^0 bosons. As mentioned before, four experiments (ALEPH [132], DELPHI [133], L3 [134] and OPAL [135]) measured properties of the weak bosons. In the second phase, LEP was upgraded to produce W^+W^-- and Z^0Z^0- pairs and eventually reach a CM energy of 209 GeV to search for the Higgs boson. In the year 2000, LEP was shut down to install the LHC in the LEP tunnel [187].

The first electron synchrotron (Deutsches Elektronen SYnchrotron) started operation at the DESY site in 1964. It produced beam energies up to 7 GeV [196]. In 1974, the e^+e^- storage ring DORIS started operation with 3 GeV beam energies

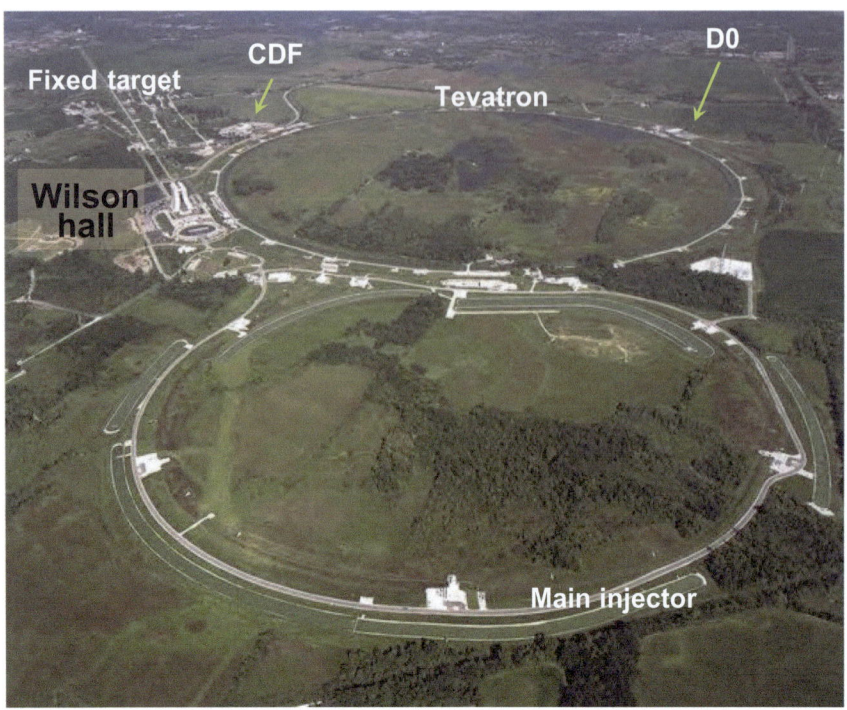

Fig. 1.12 Areal view of the Fermilab site at the time of the Tevatron operation. Extracted from [189], © 2007, Wikipedia free Licence. All rights reserved

Table 1.1 Examples of proton synchrotrons, hadron colliders and e-p machines [3]. †Energy reached and design in parentheses

Machine	Type	Laboratory	$\int dl$ [km]	E_{beam} [GeV]	$\mathcal{L}$ [cm^{-2}s^{-1}]	Date
PS	Synchrotron	CERN	0.628	26	–	1959–now
Cosmotron	Synchrotron	Brookhaven	0.0701	3.3	–	1952–1966
Bevatron	Synchrotron	Berkeley	0.122	6.2	–	1954–1993
AGS	Synchrotron	Brookhaven	0.807	24	–	1960–now
SPS	Synchrotron	CERN	6.9	450	–	1976–now
Tevatron	Synchrotron	Fermilab	6.28	1000	–	1985–2011
ISR	pp	CERN	0.943	26	4×10^{30}	1971–1978
SPPS	$p\bar{p}$	CERN	6.9	450	6×10^{30}	1981–190
Tevatron	$p\bar{p}$	Fermilab	6.28	980	4.31×10^{32}	1987–2011
LHC	pp	CERN	26.659	6500 (7000)†	2.1×10^{34}	2009–2030s
FCC$_{\text{hh}}$	pp	CERN	97.75	50	5–30×10^{34}	?
HERA	ep	DESY	6.336	30×920	7.5×10^{31}	1992–2007

1.3 Accelerators

supplying data to the PLUTO [146] and DASP [197] experiments. The goal was to study the newly discovered J/ψ, τ and charmed particles. In 1981, the machine was upgraded to DORIS II producing B mesons. It served beams to two experiments, Crystal Ball [102] and Argus [101]. In 1978, the PETRA e^+e^- storage ring went into operation producing a CM energy of 38 GeV [198]. It was built to find the top quark. The PLUTO experiment moved to PETRA. From 1980–1982 PLUTO shared the interaction region with CELLO, which continued until 1986. Thus, five experiments were taking data at PETRA (Mark J [147], JADE [149], PLUTO [146], TASSO [148] and later CELLO [150]). The most important result was the discovery of the gluon in three-jet events. Since Cello started later it missed out on the discovery. In 1990, the first electron-proton collider, HERA, started operation [199]. The multipurpose experiments ZEUS [138] and H1 [139] operated until 2007 when HERA was closed down for high-energy physics.

After the discovery of the c quark, Cornell University started to build the Cornell Electron Storage Ring (CESR) to collide electrons and positron with beam energies up to 8 GeV [200]. With the discovery of the b quark, CESR was in a splendid situation to produce B mesons at the $\Upsilon(4S)$ resonance. In 1979, first collisions were observed. In the beginning two detectors operated, CLEO [201] on the south interaction region and CUSB [202] on the north interaction region. The latter experiment finished data taking in the late 1980s. The CESR machine ran at the $\Upsilon(4S)$ until 1999. During this time, the CLEO detector was upgraded four times and the collaboration harvested many results on B and D mesons, (see e.g. CLEO II [203]). Then, it switched to operate at the $\psi(3770)$ for a couple of years using the latest upgraded version of the CLEO detector (CLEO III) [204] improving D physics measurements.

At the Stanford Linear Accelerator Center (SLAC) the largest (two mile long) linear accelerator was built in 1962. Operation started in 1967. Highly inelastic electron proton scattering data revealed that protons are composite objects [205]. In 1972, the Stanford Positron Electron Asymmetry Ring (SPEAR) started operation [206] at CM energies of around 4 GeV. The SLAC/LBL magnetic detector was the first experiment to take data at SPEAR [207]. It discovered the J/ψ particle in 1974 and the τ lepton in 1975. The next generation experiments were Crystal Ball [102] and Mark II [136]. When the PEP collider opened in 1980 [208] with CM energies of up to 29 GeV, the Mark II detector moved over to PEP while the Chrystal Ball detector moved to DORIS II at DESY. At the PEP collider, the other experiments were DELCO, [209] MAC [210] and TPC-2γ [211]. At SPEAR the Mark III experiment started data taking in 1983 [212]. SLAC built the first linear collider called SLAC Linear Collider (SLC), which started operation in 1989 with the Mark II experiment. In 1992, Mark II was succeeded by the SLD detector [137], which ran until 1998. In 1999, the PEP ring was converted in an asymmetric B factory (PEP II) operating at the $\Upsilon(4S)$ resonance colliding 3.1 GeV positrons on 9.0 GeV electrons [213]. The *BABAR* detector was the only experiment at PEP II [141], which was closed down in 2008. Figure 1.13 shows the configuration of SLAC at the time of the PEP II operation.

In Japan the National Laboratory for High Energy Physics (KEK) was established in 1971. The KEK laboratory started with a proton synchrotron. In 1984, the

Fig. 1.13 Areal view of the SLAC site at 2020. Reprinted under CC-BY-4.0 Licence from [214], © 2022, F. Gross

Transposable Ring Intersecting Storage Accelerator in Nippon (TRISTAN) Accumulation Ring (AR) started running [215] and in 1986, the TRISTAN Main Ring (MR) was completed accelerating both electron and positron beams to 25.5 GeV. It was upgraded to 30 GeV two years later. Four experiments (AMY [151], JADE [149], TOPAZ [152] and VENUS [153]) were operating until 1995. After that KEK built a B factory with asymmetric beam energies. In 1999, the KEKB storage ring started operation colliding 3.5 GeV positrons with 8.0 GeV electrons [216]. Belle was the only experiment [142], which ran until 2009 when the upgrade of the KEKB ring began [216]. In 1999, KEK also delivered beam to the K2K Long-baseline Neutrino Oscillation experiment [217]. In 2017, the Super KEKB asymmetric e^+e^- collider was completed. The Belle II experiment started operation reporting first collisions in 2018 [143]. The KEKB machine is expected to deliver a luminosity of 50 ab^{-1} in the next decade.

The Frascati laboratory near Rome was founded in 1954 operating an electron synchrotron with 1.1 GeV beam energy. In 1969, the ADONE e^+e^- storage ring started operation with 1.5 GeV beam energies [218], which ran until 1995 serving various experiments. In 2002, the DAΦNE machine [219] housed in the same hall as ADONE started operation with beam energies that produced the ϕ resonance with

1.3 Accelerators

Table 1.2 Examples of e^+e^- Storage Rings [3]. The luminosities for FCC$_{ee}$ (CEPC) refer to the three (two) beam energies. †The machine was laid out to operate with beam energies of 2.5 – 4.0 GeV (e^+) × 7.0 – 12 GeV ($e-$)

Machine	Laboratory	$\int dl$ [km]	E_{beam} [GeV]	$\mathcal{L}$ [cm^{-2}s^{-1}]	Date
ADONE	Frascati	0.105	1.5	1.7×10^{29}	1969–1993
DAΦNE	Frascati	0.098	0.510	4.53×10^{32}	1999-now
BEBC	Beijing	0.2404	2.5	5×10^{30} @ 1.55 GeV	1989–2005
BEBC II	Beijing	0.23753	1.89	1×10^{33} @ 1.89 GeV	2008-now
SPEAR	SLAC	0.234	2.4-4.0	$1. \times 10^{31}$ @ 3.0 GeV	1972–1988
DORIS	DESY	0.288	5.6	3.3×10^{31} @ 5.3 GeV	1974–1992
CESR	Cornell	0.768	6	1.28×10^{33} @ 5.3 GeV	1979–2002
CESR C	Cornell	0.768	6	0.076×10^{33}	2002–2008
KEKB	KEK	3.016	3.51×8.0	2.1×10^{34}	1999–2010
SuperKEKB	KEK	3.016	4.0×7.0	8.0×10^{35}	2018-now
PEP II	SLAC	2.2	$3.1 \times 9.0^{\dagger}$	1.21×10^{34}	1999–2008
PETRA	DESY	2.304	23.4	2.4×10^{31} @ 17.5 GeV	1978–1986
PEP	SLAC	2.2	15	6×10^{31}	1980–1990
TRISTAN	KEK	3.018	32	3.5×10^{31}	1986–1995
SLC	SLAC	1.45/1.47	50	2.5×10^{30}	1989–1993
LEP	CERN	26.659	55	2.4×10^{31}	1989–1996
LEP200	CERN	26.659	104.6	1×10^{32}	1996–2000
VEPP 4	Novosibirsk	0.366	6.0	5×10^{31}	1994-now
ILC	Japan	20.5-31	125-500	1×10^{34}	?
CLIC	CERN	11-50	few 1000	1×10^{34}	?
FCC$_{ee}$	CERN	97.75	46, 120, 183	$(230, 8.5, 1.6) \times 10^{34}$	?
CEPC	China	53.6	46, 120	$(32, 3) \times 10^{34}$	?

incredibly intense beams. The goal was to study *CP* violation in the neutral kaon system with the KLOE detector [220]. The DAΦNE accelerator is still operating today.

The Linear Accelerator Laboratory in Orsay started with a linear accelerator that produced electrons with a beam energy of 1.1 GeV. In 1967, the ACO storage ring started operation [221]. It was replaced by the DCI (Dispositif de Collisions dans l'Igloo) storage ring in 1978, which produced beam energies up to 1.8 GeV [222]. The DM2 experiment ran at the J/ψ resonance until 1990 [223] as a competitor to the Mark III experiment.

In Novosibirsk, the Budker Institute for Nuclear Physics Novosibirsk was founded in 1959. The first machine was VEPP 1, an e^+e^- storage ring with beam energies of 0.16 GeV [224]. It was followed by VEPP 2 in 1966 with 0.5 GeV beam energies. The VEPP 3 storage ring was operated with beam energies of 0.4-2.0 GeV. In 2000, the VEPP 4 storage ring started operation with beam energies of 1.0 to 1.9 GeV [224]. Since 2016 beam energies up to 5.2 GeV are possible due to a new injector.

The Beijing Electron Positron Collider (BEPC) [225] started operation at CM energies of the charmonium states in 1988 with the BES experiment [226]. In 2004–2005 the accelerator was upgraded to BEPC II [227]. Using two separate rings with beam energies up to 1.89 GeV the luminosity was increased by a factor of 100 with respect to that of BEPC. In 2008, the detector was upgraded to BES III [228], which continues to do charm and τ physics.

Concerning the future, planning is ongoing to build a linear e^+e^- collider, the International Linear Collider (ILC), in Japan. At the beginning, the beam energies will be 125 GeV. The goal is to build a Higgs factory that provides precision measurements of Higgs boson properties. The machine should be upgradable to beam energies of 250 GeV and 500 GeV at a later stage. At CERN, a design for a Compact Linear Collider (CLIC) exists that provides beam energies of 190 GeV in the first stage, but is upgradable to energies of 750 GeV and 1.5 TeV. In parallel, a proposal exists for a circular e^+e^- collider hosted in the FCC-hh tunnel. In China, designs of a similar circular e^+e^- collider (CEPC) are under discussion. Table 1.2 shows a comparison of different past, present and future e^+e^- colliders. Besides past and present projects also future projects are displayed.

1.3.3 Luminosity

In colliding beam storage rings, the particle production rate is given by

$$\frac{dN}{dt} = \sigma_p \mathcal{L}, \tag{1.4}$$

where σ_p is the interaction cross section and $\mathcal{L}$ is the luminosity describing how effectively an accelerator performs. The luminosity depends only on machine parameters and has units of

$$\mathcal{L}\left[\frac{1}{\text{cm}^2\text{s}}\right] = \mathcal{L}\left[\frac{10^{33}}{\text{nb} \cdot \text{s}}\right] = \frac{dN/dt}{\sigma_p}. \tag{1.5}$$

The time integral over $\mathcal{L}$ is called integrated luminosity in units of inverse barns (b^{-1}),

$$\mathcal{L}_{\text{tot}} = \int \mathcal{L} dt. \tag{1.6}$$

PEP II achieved a peak luminosity of $\mathcal{L}_{\text{peak}} > 1 \times 10^{34}$ cm^{-2}s^{-1}, yielding an integrated luminosity of $\mathcal{L} = 0.73$ fb$^{-1}/day$.

To calculate the luminosity we project the positrons in the positron bunch onto the x-y plane as shown in Fig. 1.14. So the electrons will interact with positrons inside the ellipse. This reduces the problem to two dimensions. The calculation yields [168]

$$\mathcal{L} = \frac{n_b}{4\pi} \frac{f_{\text{rev}} N_1 N_2}{\sigma_x^* \sigma_y^*}, \tag{1.7}$$

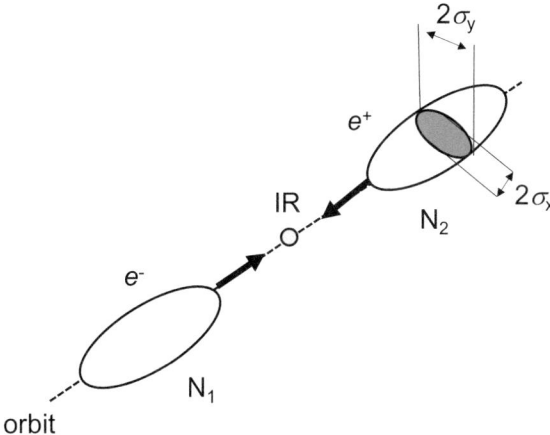

Fig. 1.14 An electron bunch approaching from the left and a positron bunch from the right. When projecting the positrons in the positron bunch onto the x-y plane, we obtain the black ellipse, which has dimensions of $2\sigma_x$ and $2\sigma_y$. So the electrons effectively will interact with positrons inside the ellipse

where N_1, N_2 are respectively the number of particles in a bunch of beam one and beam two, n_b is the number of bunches, f_{rev} is the revolution frequency, and σ_x^* and σ_y^* are the beam cross sections in the two directions at the interaction point. It is more convenient to use average beam currents,

$$I_i = N_i q_e n_b f_{\text{rev}}, \qquad (1.8)$$

where q_e is the elementary charge. Inserting (1.8) into (1.7) yields

$$\mathcal{L} = \frac{1}{4\pi q_e^2 f_{\text{rev}} n_b} \frac{I_1 I_2}{\sigma_x^* \sigma_y^*}. \qquad (1.9)$$

To achieve high luminosity, we need high beam currents and small beam cross sections. In e^+e^- colliders, the probability of collisions in each bunch crossing is rather small and only a few particles in each bunch will interact. This is different, for example, at the hadron collider LHC, where many interactions occur in one bunch crossing. At the present operation of the LHC, the average number of interactions per bunch crossing is around 90. In order to achieve small beam cross sections one places focussing magnets called MiniBeta insertions in the detector near the interaction region.

1.4 Detector Characteristics

The fiducial volume or geometric acceptance of a detector is determined by the solid angle,

$$\Delta \Omega = \int_{\text{detector}} (-\text{d}\cos\theta)\text{d}\phi, \qquad (1.10)$$

where the integral runs over the detector area specified in polar and azimuth angle coordinates seen from the interaction region. The polar angle denotes the deflection in the x-y plane from the z axis while the azimuth angle denotes the position in the x-y plane with respect to the x axis. In fixed-target experiments the detector covers a small part of the solid angle while in colliding beam experiments it covers nearly the entire solid angle around the interaction region. The unit of the solid angle is steradian (sr).

1.4.1 Detector Efficiency

An important quantity is the detector efficiency, which is simply

$$\epsilon_{tot} = \frac{N_{rec}}{N_{tot}} = \frac{\text{number of recorded events}}{\text{number of produced events}}, \quad (1.11)$$

where the produced events are given by the cross section times luminosity. The recorded events have to satisfy the detector fiducial volume, trigger requirements, event reconstruction, particle identification and several selection criteria used in the analysis. It is customary to factor out the geometric acceptance,

$$\epsilon_{tot} = \frac{\Delta\Omega}{4\pi} \cdot \epsilon_{int}, \quad (1.12)$$

where

$$\epsilon_{int} = \frac{N_{rec}}{N_{fid}} = \frac{\text{number of recorded events}}{\text{number of events in the fiducial detector volume}}. \quad (1.13)$$

We can determine an efficiency for each sub-detector, which may depend on kinematic observables such as transverse momentum. We try to measure the intrinsic efficiencies in data. For analysis, we often calculate the intrinsic efficiency using simulations, where the simulated sample is treated in the same way as the data sample. Throughout the book, we will encounter various sub-detector efficiencies. In a sub-detector the detection efficiency may depend on several detection processes. Let us consider, for example, a silicon photomultiplier discussed in Sect. 8.3.3, which is a pixelated photon detector operating in the Geiger mode. First, the photon has to hit the active pixel area. Second, it has to produce a photoelectron, which finally has to trigger a Geiger-Müller avalanche. Each of these processes has an efficiency. The total efficiency for the photon to be detected is the product of the three individual efficiencies.

Another important contribution to the intrinsic efficiency results from the fact that the detector is unable to process an event because it is still busy with the previous event. This effect typically occurs at high counting rates and is called dead time. We distinguish between non-paralyzable and paralyzable detectors. In non-paralyzable detectors, each recorded signal is followed by a time interval τ_d during which no new event is accepted. Figure 1.15 (left) shows the deadtime for a non-paralyzable detector. So, the third and fourth event will not be recorded. If R_t is the true rate and R'_t is the accepted rate, $R'_t \tau_d$ is the fraction of time in which detector is dead and

1.4 Detector Characteristics

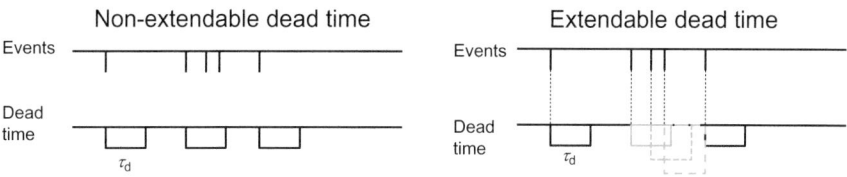

Fig. 1.15 Left: Sketch of the dead time in a non-paralyzable detector. Right: Sketch of the dead time in a paralyzable detector

$R_t R'_t \tau_d = R_t - R'_t$ is the rate of true events lost by this dead time. Solving for the true rate yields

$$R_t = \frac{R'_t}{1 - R'_t \tau_d}. \qquad (1.14)$$

In a paralyzable detector a new blocking cycle with length τ_d is started if a true event occurs during the dead time of the preceding event as shown in Fig. 1.15 (right). Here the dead time has variable length. The rate of recorded events R'_t is same as the rate at which time intervals larger than τ_d occur in real events. Using Poisson statistics the probability of obtaining an interval larger than τ_d is $e^{-R_t \tau_d}$ and the rate at which this occurs is $R'_t = R_t e^{-R_t \tau_d}$. This equation cannot be solved analytically. For low rates, we expand the exponential function,

$$R'_t = R_t(1 - R_t \tau_d), \qquad (1.15)$$

and in turn

$$R_t = R'_t(1 + R'_t \tau_d). \qquad (1.16)$$

Figure 1.16 (left) shows the observed count rate versus the true count rate in a paralyzable detector. Note that for most observed rates, an ambiguity for the true rate exists. Intrinsic detector efficiencies are measured if possible or are determined from simulations.

1.4.2 Mean Value and Resolution

In an experiment we typically measure an observable, such as charge, energy, momentum, position, time, etc. We denote the mean value of the measured observable by $\langle \mathcal{O} \rangle$, which results by detecting the input observable $\mathcal{O}_{in}$ with an apparatus. The latter is typically distributed according to a Delta function $\delta(\mathcal{O}_{in} - \langle \mathcal{O}_{in} \rangle)$ while the measured quantity is smeared out according to a statistical distribution $D(\mathcal{O})$, which usually is a Poisson or Gaussian distribution. In most cases, a linear relation holds between the input and output, $\langle \mathcal{O} \rangle = c \cdot \mathcal{O}_{in}$ with constant c. However, there are some detectors that produce non linear outputs. For example, a silicon photomultiplier (SiPM) has a finite number of pixels that can be fired by impinging photons. If two photons impinge on the same pixel at most one photoelectron is produced. Thus, if the number of incoming photons becomes large enough not every photon is

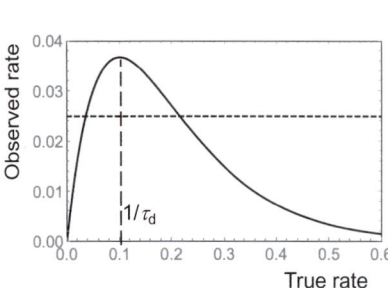 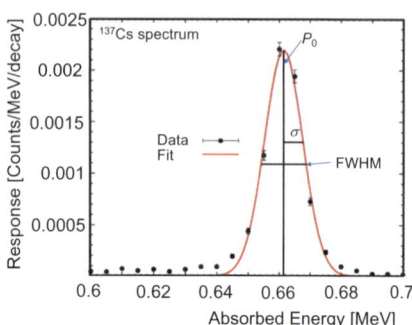

Fig. 1.16 Left: Numerical solution for (1.16) to extract the true count rate for a paralyzable detector. Note that there are typically two solutions. Right: Pulse height distribution of ^{137}Cs photons in a CdZnTe crystal. Beside the mean value $P_0 = 0.667$ MeV, the full-width at half maximum (FWHM) and the standard deviation (σ) are shown. The relative resolution is $\sigma/P_0 = 0.9\%$. The solid line is a fit to a Gaussian function. Reprinted under CC-BY 4.0 Licence from [229], © 2019, Penerbit UTM Press

detected and the dependence of observed photons with respect to incoming photons is no longer linear. Here, we need to include a non-linearity term $dc/d\mathcal{O}_{in}$ if we want to describe the response adequately.

For observable $\mathcal{O}$ we define the mean value $\langle \mathcal{O} \rangle$ and its variance,

$$\langle \mathcal{O} \rangle = \int \mathcal{O} D(\mathcal{O}) d\mathcal{O}, \tag{1.17}$$

and

$$\sigma_{\mathcal{O}}^2 = \int (\mathcal{O} - \langle \mathcal{O} \rangle)^2 D(\mathcal{O}) d\mathcal{O}, \tag{1.18}$$

respectively. Taking the square root of the variance yields the standard deviation or resolution ($\sigma_{\mathcal{O}}$). In a measured distribution, we frequently extract the value of full-width-at-half-maximum (FWHM), $\Delta\mathcal{O}$, which is related to the standard deviation by

$$\sigma_{\mathcal{O}} = \Delta\mathcal{O}/2.35. \tag{1.19}$$

Figure 1.16 (right) shows a measured pulse height distribution. The resolution is typically obtained from a fit to a Gaussian function. However, we can determine it also without a fit by measuring the FWHM. This method is particularly useful if the fit function is not known or if a fitting tool is not available. The quantity $\sigma_{\mathcal{O}}/\mathcal{O}$ is called the relative resolution. If the only source of fluctuation is of statistical nature such as the number N of charge carriers and if their formation is uncorrelated, the response function $D(\mathcal{O})$ is a Poisson distribution, which for large $N > 20$ approaches a Gaussian function. In principle, this should be the lower limit of the resolution. For some detectors the contrary is found, where the resolution is up to a factor of four smaller. The assumption of Poisson statistics is wrong since a correlation

1.4 Detector Characteristics

exists between the individual charge production processes. This is called the Fano effect [230],

$$F = \left(\frac{\text{observed resolution}}{\text{resolution expected from Poisson statistics}}\right)^2. \qquad (1.20)$$

The Fano factor F ranges from ~ 0.06 for semiconductors to 0.17 for noble gases and to 1 for scintillators.

The mean value and the variance are the first and second moments of the probability distribution. Higher moments (k) are given by

$$\langle \mathcal{O}^k \rangle = \int \mathcal{O}^k D(\mathcal{O}) \mathrm{d}\mathcal{O}. \qquad (1.21)$$

For example, the third moment is the skewness of the distribution. In appendix B we present different distributions that are frequently used for describing measurements. In addition, we discuss different tools that are used in data analysis.

1.4.3 Readout Chain

In many particle detectors, we determine an observable from measurements of the charge Q a particle liberates during its passage. With the help of electric and magnetic fields the charge is guided towards an electrode where it is recorded. The collection time t_c varies from tens of picoseconds in semiconductors and photodetectors to a few milliseconds in ionization chambers. The instrument will produce a current flow I and the charge is obtained by integrating I over the time interval t_0 to t_c,

$$Q = \int_{t_0}^{t_c} I \mathrm{d}t. \qquad (1.22)$$

The time interval $\Delta t = t_c - t_0$ could reflect the time difference between the stopping time of a sub-detector and the starting time of a trigger signal. Another time interval selection is the length of the gate for collecting the charge in a sub-detector. At the LHC often the time above threshold is used, which is the time for which the charge of the signal is larger than a given threshold. The simplest way consists of measuring an average DC output by the detector, which is typically used if timing is unimportant. In case timing information is necessary, the detector output records each individual particle (pulse mode). Here, the output current is transformed into a voltage signal, where the time structure is determined by the input impedance with time constant $\tau_{im} = R_i C_i$ as depicted in Fig. 1.17 (left).

If $\tau_{im} \ll t_c$, the signal follows the detector output closely, while for $\tau_{im} \gg t_c$ the voltage on C_i rises until Q is completely collected at $t = t_c$ reaching a maximum voltage $V_{max} = Q/C_i$. The rise time of the voltage pulse is determined by the charge collection time of the detector and the decay time is determined by the time constant

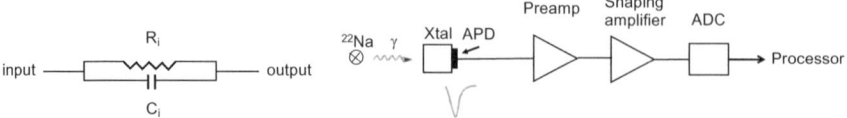

Fig. 1.17 Left: Sketch of a circuit for time constant $\tau_i = R_i C_i$. Right: Readout chain for detecting photons from ^{22}Na source with a crystal coupled to a photodetector such as an avalanche photodiode. The photodetector signal passes through a preamplifier, the main amplifier into the ADC, where it is digitized

τ_{im} of the circuit. Note that for some detectors, such as semiconductors, C_i cannot be kept constant. In this case we use a charge-sensitive preamplifier, which removes the dependence of C_i on the voltage. This is typically an inverting amplifier with a feed-back loop through the capacitance C_f. The amplification A is large compared to $(C_f + C_i)/C_f$. Thus, the output voltage is independent of C_i,

$$V_{\text{out}} = -A \frac{Q}{C_f(A+1) + C_i} \sim -\frac{Q}{C_f}. \tag{1.23}$$

For each traversing particle the associated electronics of the detector delivers a voltage pulse that is proportional to the collected charge. The analog voltage signal is usually converted into a digital signal via an analog-to-digital converter (ADC). The resulting spectrum yields a pulse height distribution that is stored and that can be converted into the desired observable, for example energy, after applying appropriate conversion factors. Figure 1.17 (right) shows an example of a readout chain used for measuring the energy of a radioactive source. We will talk more about data collection in Chap. 10 since in today's experiments the readout electronics is rather sophisticated.

1.4.4 Kinematic Observables

In an interaction, several particles are produced, which are characterized by their mass m, energy E and momentum $\vec{p}$. Energy and momentum are related by

$$E^2 = |\vec{p}|^2 c^2 + m^2 c^4, \tag{1.24}$$

where $c = 2.99792458 \times 10^8$ m/s is the speed of light. Since particles typically are relativistic, it is useful to define the relative velocity

$$\vec{\beta} = \frac{\vec{v}}{c}, \tag{1.25}$$

where $\vec{v}$ is the velocity of the particle that can be determined from the particle energy and momentum

$$|\vec{\beta}| = \beta = |\vec{p}|/E. \tag{1.26}$$

1.4 Detector Characteristics

The relation of the energy to the particle mass m is

$$E = \gamma m c^2, \tag{1.27}$$

where γ is called the Lorentz factor and is given by

$$\gamma = \frac{1}{\sqrt{1-\beta^2}}. \tag{1.28}$$

Thus, we have

$$|\vec{p}| = \beta \gamma m c. \tag{1.29}$$

The kinetic energy of the particle is given by

$$E_k = \sqrt{(\vec{p}c)^2 + m^2 c^4} - mc^2 = mc^2(\sqrt{\beta^2 \gamma^2 + 1} - 1) = mc^2(\gamma - 1). \tag{1.30}$$

Typically, we use four-momenta,

$$p = (E/c, \vec{p}), \tag{1.31}$$

to describe particles. Note that p^2,

$$p^2 = E^2/c^2 - |\vec{p}|^2 = m^2 c^2. \tag{1.32}$$

is a Lorentz invariant while E and $\vec{p}$ are not. Therefore, the four-momentum length squared is equal to the mass squared in all reference frames.

In both single beam and colliding beam experiments the beam direction is parallel to the z axis. Thus, the beam momentum essentially has no transverse components. In fixed-target experiments the sub-detectors typically consist of rectangular stacks centered around the beam pipe in the x-y direction placed downstream from the target and positions are specified in Cartesian coordinates. The deflection from the z axis is described by the polar angle θ. In colliding beam experiments solenoidal magnetic fields that have cylindrical geometry are typically used. Thus, we use a cylindrical coordinate system defined by the radial coordinate r, the azimuth angle ϕ and the longitudinal position z. Note that the scattering angle θ is defined with respect to the z-axis through $\tan \theta = r/z$. The dip angle λ_{dip} is defined with respect to the orthogonal direction to z,

$$\lambda_{\text{dip}} = \pi/2 - \theta. \tag{1.33}$$

In symmetric $e^+ e^-$ colliders the initial conditions are $E = \sqrt{s}$ and $\vec{p} = (0, 0, 0)$. The interaction products satisfy energy-momentum conservation, which is an important analysis tool. In pp and $p\bar{p}$ colliders, only one parton, either a quark or a gluon in each hadron interacts. Each parton carries a fraction of the momentum of the parent proton or antiproton. For interacting quarks, the remaining quarks have to hadronize with quarks formed in the interaction. So, the initial conditions are not well specified except that the initial transverse momentum $|\vec{p}_{T,in}|$ is zero,

$$|\vec{p}_{T,in}| = p_{T,in} = \sqrt{p_{x,in}^2 + p_{y,in}^2} = 0. \tag{1.34}$$

Since the transverse momentum is conserved, the particles produced in an interaction also satisfy,

$$p_T = \sum_{i=1}^{n} \sqrt{p_{x,i}^2 + p_{y,i}^2} = 0. \tag{1.35}$$

Since some neutral particles may escape, such as neutrinos or new particles, it is useful to define the missing transverse momentum $p_{T,miss}$ and missing transverse energy $E_{T,miss}$,

$$p_{T,miss} = -\sum_{i=1}^{n} p_{T,i}, \tag{1.36}$$

$$E_{T,miss} = E_{tot} - \sum_{i=1}^{n} E_{T,i}, \tag{1.37}$$

where $p_{T,i}$ and $E_{T,i}$ are the observed transverse momenta and transverse energies of each charged and neutral particle in the event, respectively. So, $p_{T,miss}$ and $E_{T,miss}$ are non-zero if particles in the interaction remain undetected such as neutrinos. Since many particles are boosted in the forward and backward directions, we use the pseudo-rapidity instead of θ,

$$\eta = -\ln \tan(\theta/2). \tag{1.38}$$

Another observable is the rapidity,

$$y_r = \frac{1}{2} \ln \left(\frac{E + p_z c}{E - p_z c} \right), \tag{1.39}$$

where E and p_z are the energy and longitudinal momentum component of the particle. In the limit of zero-mass particles, rapidity and pseudorapidity are identical. Another important observable is the distance between two objects,

$$\Delta R = \sqrt{\Delta \eta^2 + \Delta \phi^2}, \tag{1.40}$$

where $\Delta \eta$ and $\Delta \phi$ are the differences in pseudo-rapidity and azimuth angle of the two objects, respectively. In the massless limit, ΔR is a Lorentz invariant such as $\Delta \eta$ and $\Delta \phi$. We can now define the transverse mass as

$$m_T^2 c^4 = E^2 - p_z^2 c^2 = m^2 c^4 + p_T^2 c^2, \tag{1.41}$$

1.4 Detector Characteristics

which is used in many analyses at hadron colliders. Note that we have the following relation,

$$E = m_T c^2 \cosh y_r. \tag{1.42}$$

The maximum value of y_r at fixed energy E occurs at $p_T = 0$, yielding $\cosh y_r^{max} = \gamma$. At the Tevatron for $\sqrt{s} = 2$ TeV, we obtain $y_r^{max} = 7.7$ while at the LHC at $\sqrt{s} = 14$ TeV, we obtain $y_r^{max} = 9.6$.

1.4.5 Decay Quantities

The activity of a radioactive source gives the number of decays per second,

$$A = dN/dt. \tag{1.43}$$

It is related to the decay constant λ_d via

$$A = dN/dt = -\lambda_d N, \tag{1.44}$$

yielding the exponential decay law for decays of radioactive sources and particles

$$N = N_0 \exp(-\lambda_d t), \tag{1.45}$$

where N_0 is the initial number of decaying particles. The mean lifetime τ_d of a radioactive isotope or a particle is the time interval after which N_0 has decreased to N_0/e, where Euler's number is $e = 2.71828$. The life time is related to the decay width Γ_{tot},

$$\tau_d = \hbar/\Gamma_{tot} = 1/\lambda_d. \tag{1.46}$$

Radioactive sources are characterized often by the half life ($t_{1/2}$), which is the time interval after which half of the nuclei have decayed. The half life is related to the lifetime by

$$t_{1/2} = \tau_d \cdot \ln 2 = 0.693 \cdot \tau_d. \tag{1.47}$$

The functional form of a resonance is given by the Breit-Wigner line shape. The non-relativistic form is given by

$$\frac{dN}{dE} \propto \frac{\Gamma_i \Gamma_f}{(E - mc^2)^2 - \Gamma_{tot}^2/4}, \tag{1.48}$$

where m is resonance mass, Γ_i is the partial width for the production channel, Γ_f is the partial width for the decay channel and the total width Γ_{tot} is the sum of all partial widths Γ_f,

$$\Gamma_{tot} = \sum_f \Gamma_f. \tag{1.49}$$

The relativistic form is given by

$$\frac{dN}{dE} = 12\pi(\hbar c)^2 \frac{\Gamma_i \Gamma_f}{(E^2 - m^2 c^4)^2 - \Gamma_{\text{tot}}^2 m^2 c^2}. \quad (1.50)$$

In hadron machines, some detector parts are exposed to higher levels of radiation. The characteristic quantity is the fluence, which is the ratio of the sum of the particle track lengths and the volume,

$$\Phi = d\ell/dV. \quad (1.51)$$

It can be expressed also in terms of the number of particles dN incident upon a small sphere with a cross section area da,

$$\Phi = dN/da. \quad (1.52)$$

1.4.6 A Comment on Units

We measure energy in units of electron volts (eV). An electron accelerated in a potential of 1 V gains an energy of 1 eV, which corresponds to 1.6×10^{-19} J. It is rather useful to introduce new units for higher energies: 1 keV = 10^3 eV, 1 MeV = 10^6 eV, 1 GeV = 10^9 eV, 1 TeV = 10^{12} eV and 1 PeV = 10^{15} eV. Mass is related to the rest energy energy by $E_0 = mc^2$. Thus, the unit is eV/c^2 and 1 eV/c^2 = 1.78×10^{-38} kg. For example, the mass of an electron is $m_e = 9.1 \times 10^{-31}$ kg or 511 keV/c^2. Note, that it is custom to set $\hbar = c = 1$ and quote masses and momenta in units of eV. We measure the activity of a radioactive source in units of Bequerel (Bq), where 1 Bq corresponds to one decay per second. The old unit was Curie (Ci), which corresponded to 3.7×10^{10} Bq. The momentum of 1 eV/c = 0.535×10^{-27} kgs/m. The strength of an interaction is measured by the cross section. The unit is 1 barn = 10^{-28} m^2 = 10^{-24} cm^2. Since many cross sections are much smaller than 1 barn [b], we introduce 1 mb = 10^{-27} cm^2, 1 μb = 10^{-30} cm^2, 1 nb = 10^{-33} cm^2, 1 pb = 10^{-36} cm^2, 1 fb = 10^{-39} cm^2 and 1 ab = 10^{-42} cm^2. Two standards are used to specify gas densities. The first is at standard temperature and standard pressure (STP), which correspond to a temperature of 273.15 K and a pressure of one atmosphere. The second choice is normal pressure and normal temperature (NTP), which corresponds to a pressure of one atmosphere and room temperature (293.15 K).

1.5 Effects of Radiation

Three quantities measure the effects of radiation on matter, the energy dose, the ion dose and the equivalent dose:

1. The energy dose D_E is the energy dW absorbed in the material with volume dV and density ρ,

$$D_E = \frac{dW}{\rho dV}. \tag{1.53}$$

The old unit is 1 rad = 10^{-2} J/kg. The new unit is gray, 1 Gy = 1 J/kg = 10^2 rad.

2. The ion dose D_I is the charge dQ liberated by radiation in the volume dV and density ρ_A,

$$D_I = \frac{dQ}{\rho_A dV}. \tag{1.54}$$

The unit is roentgen, 1 $R = 2.58 \times 10^{-4}$ C/kg in air. An ion dose of 1 R/q_e in air corresponds to 1.61×10^{15} ions/kg and an energy dose of $1R/q_e \cdot w_i$, where w_i is mean effective energy needed to liberate an electron-ion pair in air ($w_i = 33.7$ eV) yielding $D_E = 0.87$ rad.

3. The equivalent dose D_q is a measure for the effect of radiation on the human body,

$$D_q = q_m \cdot D_E, \tag{1.55}$$

where q_m is a quality factor for biological effects from different types of radiation on the human tissue. The old unit is Rem (roentgen equivalent for man). The new unit is sievert, 1 Sv = 100 Rem. The quality factor is $q_m = 1$ for photons and electrons of all energies, $q_m = 10$ for α particles, protons and deuterons, $q_m = 20$ for heavy nuclear fragments and $q_m = 5 - 20$ for neutrons (depending on energy). For $E < 10$ keV or $E > 20$ MeV, $q_m = 5$, for energies of $E = 10\text{-}100$ keV and $E = 2 - 20$ MeV, $q_m = 10$ and for energies $E = 100$ keV-2 MeV, $q_m = 20$.

The natural background is < 4 mSv. At CERN radiation workers are allowed to accumulate < 15 mSv per year. In the US the limit is < 50 mSv. Note that the lethal dose is 2.5-4.5 Gy received over a short period. The mortality here is 50% in 30 days.

Exercises

1.1 (a) Show that the center-of-mass energy in a fixed target experiment with $p_1 = (E, \vec{p})$, $p_2 = (m_t, \vec{0})$ is $\sqrt{s} = \sqrt{2Em_t c^2}$, while that for a colliding beam experiment with asymmetric energies E_1 and E_2 is $\sqrt{s} = \sqrt{((E_1 + E_2)^2 - (E_2 - E_1)^2)}$. (b) What is the center-of-mass energy for head-on collisions of a 3.1 GeV e^+ beam and a 9.0 GeV e^- beam? (c) Assume two heavy particles with mass $m = 5.28$ GeV/c^2 are produced, which have a lifetime of $\tau_d = 1.54$ ps. How large is their decay length, which is defined by $\ell_d = \beta\gamma c\tau_d$, where $\beta\gamma$ is the boost and c is the velocity of light in vacuum?

1.2 Consider the decay of particle A at rest into particles B and C ($A \to B + C$); (a) Find the energy of the outgoing particles, in terms of various masses; (b) Find the magnitudes of the outgoing momenta.

1.3 A ^{22}Na source has an activity of 5.6×10^6 Bq. If you handle the source carefully, how big is the minimum equivalent dose on your hands, if all the radiation were absorbed? Assume the volume and density to be $10 \times 5 \times 1$ cm^3 and ~ 1 g/cm^3, respectively.

1.4 A ^{64}Cu radioactive source has a half life of 12.8 hours and produces two 511 keV photons 19% of the time. You want to do an angular correlation experiment, (detect both photons simultaneously). The efficiency for your setup is $\epsilon = 7 \times 10^{-5}$. You have beam time for three days and want to measure 24 points with an average statistics of 2,500 events per point. What activity should the source have to complete your task?

1.5 An e^+e^- collider runs with a luminosity of $\mathcal{L} = 5 \times 10^{35}$ cm^{-2}sec^{-1} at the $\Upsilon(4S)$ resonance, which produces pairs of B mesons with a cross section of $\sigma = 1$ nb. How long do you have to run the machine to collect 1×10^{10} pairs of B mesons?

1.6 In order to reach a luminosity of $\mathcal{L} = 1 \times 10^{34}$ cm^{-2}sec^{-1}, how big do the currents have to be? Typically, machines are operated with flat beams ($\sigma_x \gg \sigma_y$), where $\sigma_x = 136$ μm and $\sigma_y = 5.2$ μm. The circumference of the ring is 2.2 km and the particles fly with the speed of light.

1.7 What is the geometric acceptance of a circular surface with radius r placed at a distance R? Assume that an accelerator produces 100 GeV/c pions, which are dominantly flying in the z direction (longitudinal direction). If your detector is located 10 m downstream and covers a 1 m radius, what range of transverse momenta can be recorded in your experiment?

1.8 Determine the rms of a box distribution with width Δz. In Si strip detectors one measures the position of particle tracks. Assume the detector is an array of parallel strips; each strip is 40 μm wide and is separated from its neighbor by 10 μm. If only one strip records a signal what position resolution do you measure?

1.9 The classical method for measuring the polarization of a particle, such as a proton or neutron, consists of scattering it off a suitable analyzing target. We measure the asymmetry in the scattered particle distribution, by counting, for example, the number of particles scattered to the left of beam at a certain angle and those scattered to the right at the same corresponding angle. If we call particles scattered to the right by (R) and those scattered to the left by (L), the asymmetry is given by $\epsilon = \frac{R-L}{R+L}$. Calculate the error on ϵ as a function of the counts R and L.

1.10 Two counters are non-paralyzable with dead times of 30 μs and 100 μs, respectively. At which true event rate will the dead time losses in counter B be twice as big as those for counter A?

References

1. D.J. Griffiths, *Introduction to Elementary Particles* (J. Wiley & Sons, New York, 1987)
2. J.W. Rohlf, *Modern Physics from α to Z* (J. Wiley & Sons, New York, 1994)
3. The Particle Data Group (S. Navas et al.), Phys. Rev D. **110**, 030001 (2024)
4. P. Horowitz, W. Hill, *The Art of Electronics*, 3rd edn. (Cambridge University Press, 2015)
5. G. Hertz, Annalen der Physik. **267**-8, 983 (1887)
6. A. Einstein, Annalen der Physik **17**, 132 (1905)
7. A.A. Michelson and E. W. Morley American Journal of Science. **34**: 427 (1887); Philosophical Magazine **24**, 463 (1887)
8. A. Sommerfeld, Annalen der Physik. **51**, 1 (1916)
9. P. Zeeman, Nature **55**, 347 (1897)
10. J. Stark, Annalen der Physik **43**, 965 (1914)
11. M.K.E.L. Planck, Annalen der Physik. **4**-3, 553 (1901)
12. J. Franck and G. Hertz, Verhandlungen der Deutschen Physikalischen Gesellschaft **16**, 457 (1914); ibid **16**, 512 (1914)
13. W. Röntgen, Aus den Sitzungsberichten der Würzburger Physik.-medic. Gesellschaft Würzburg, **137** (1895)
14. H. Becquerel, Compt. Rend. Hebd. Seances Acad. Sci. **122**-8, 420 (1896)
15. M.S. Curie, Comptes Rendus Acad. Sciences **126**, 1101 (1898); P. Curie and M. Curie, Comptes Rendus Acad. Sciences **127**, 175 (1898)
16. O. Hahn, F. Strassmann, Naturwissenschaften **27**-1, 11 (1939)
17. H.A. Bethe, Phys. Rev. **55**-1, 541 (1939); ibid **55**-5, 434 (1939)
18. J.J. Thomson, Phil. Mag. Ser. **5**(44), 293 (1897)
19. P. Lenard, Annalen der Physik. **287**-2, 225 (1894)
20. R.A. Millikan, Phil. Mag. Ser. **6**(19), 209 (1910)
21. V. Hess, Z. Phys. **13**, 1084 (1912)
22. C.T.R. Wilson, Proc. Roy. Soc. Lond. A **87**-595, 277 (1912)
23. P.M.S. Blanckett, G.P.S. Occhialini, Proc. R. Soc. Lond. A **139**-839, 699 (1933)
24. H. Geiger, E. Marsden, Proceedings of the Royal Society of London A. **82**-557, 495 (1909)
25. J.J. Thomson, Phil. Mag. Ser. **6**(7–39), 237 (1904)
26. E. Rutherford, Phil. Mag. Ser. **6**(21–125), 669 (1911)
27. E. Rutherford, Phil. Mag. Ser. **6**(37), 581 (1919)
28. C.J. Davisson, L.H. Germer, Nature **119**-2998, 558 (1927); Proc. of the Nat. Acad. of Scien. of the USA. **14**-4, 317 (1928)
29. L.V.P.R. de Broglie, Annals Phys. **2**, 22 (1925)
30. N. Bohr, Phil. Mag. Ser. **6**-26, 1 (1913); ibid 476 (1913)
31. N. Bohr, Nature **121**-3050, 580 (1928)
32. E. Schrödinger, Annalen Phys. **384**-6, 489 (1926); Annalen Phys. **79** Ser. IV,, 489 (1926)
33. W. Heisenberg, Z. Phys. **33**, 879 (1925)
34. M. Born, Z. Phys. **26**(1), 379 (1924)
35. O. Klein, Z. Phys **37**, 895 (1926)
36. W. Gordon, Z. Phys **40**, 117 (1926)
37. P. Dirac, Proc. Roy. Soc. Lond. **A117**, 610 (1928)
38. W. Pauli, Phys. Today **31**-N9, 27 (1978)
39. F. Reines, C.L. Cowan, Phys. Rev. **92**, 830 (1953)
40. J. Chadwick, Nature **129**, 312 (1932)
41. C.D. Anderson, Science **76**, 238 (1932)
42. O. Chamberlain, E. Segrè, C. Wiegand, Phys. Rev. **100**, 947 (1955)
43. B. Cork et al., Phys. Rev. **104**, 1193 (1957)
44. E. Fermi, Z. Phys. **88**, 161 (1934)
45. P.A. Cherenkov, Dokl. Akad. Nauk SSSR **2**-8, 451 (1934)
46. H. Yukawa, Proc. Phys. Math. Soc. Jap. **17**, 48 (1935)
47. C.D. Anderson, S.H. Neddermeyer, Phys. Rev. **50**, 263 (1936)

48. C.M.G. Lattes, G.P.S. Occhialini, C.F. Powell, Nature **160**, 453 (1947)
49. R. Bjorklund et al., Phys. Rev. **77**-2, 213 (1950)
50. G.D. Rochester, C.C. Butler, Nature **160**, 855 (1947)
51. A. Pais, Phys. Rev. **86**, 663 (1952)
52. M. Gell-Mann, Phys. Rev. **92**, 883 (1953)
53. T. Nakano, K. Nishijima, Prog. Theor. Phys. **10**, 581 (1953)
54. R. Feynman, Phys. Rev. **74**, 1430 (1948); Phys. Rev. **80**, 440 (1950)
55. J. Schwinger, Phys. Rev. **74**, 1439 (1948)
56. S. Tomonaga, Progress of Theoretical Physics **1**-2, 27 (1946)
57. R.A. Millikan, Phys. Rev. **7**, 355 (1916)
58. A. Compton, Phys. Rev. **21**, 483 (1923)
59. C.N. Yang, R.L. Mills, Phys. Rev. **96**, 191 (1954)
60. C.S. Wu et al., Phys. Rev. **105**, 1413 (1957)
61. T.D. Lee, Phys. Rev. **104**, 254 (1956)
62. M. Goldhaber, L. Grodzins, A.W. Sunyar, Phys. Rev. **109**, 1015 (1958)
63. R.P. Feynman, M. Gell-Mann, Phys. Rev. **109**, 193 (1958)
64. E.C.G. Sudershan, R.E. Marshak, Phys. Rev. **109**, 1860 (1958)
65. M. Gell-Mann, report CTSL-**20** (1961)
66. V.E. Barnes et al., Phys. Rev. Lett. **12**, 204 (1964)
67. M. Gell-Mann, Phys. Lett. **8**, 214 (1964)
68. G. Zweig, CERN Preprint TH **401**. (1964)
69. N. Bogolubov, B. Struminsky and A. Tavkhelidze, JINR Preprint D **1968**. Dubna, (1965)
70. M.Y. Han, Y. Nambu, Phys. Rev. **139**-4B, B1006 (1965)
71. The PLUTO Collaboration (C. Berger et al.), Phys. Lett. B **86**, 418 (1979)
72. The MARKJ Collaboration (D. P. Barber et al.), Phys. Rev. Lett. **43**, 830 (1979)
73. The TASSO Collaboration (G. Wolf et al.), eConf C **790823**, 34 (1979); The Tasso Collaboration (R. Brandelik et al.), Phys. Lett. B **97**, 453 (1980)
74. The JADE Collaboration (W. Bartels et al.), Phys. Lett. B **91**, 142 (1980)
75. J.R. Ellis, M.K. Gaillard, G.G. Ross, Nucl. Phys. B **111**, 253 (1976); Nucl. Phys. B **130**, 516 (erratum) (1977)
76. E. Bloom et al., Phys. Rev. Lett. **23**, 930 (1969); M. Breidenbach et al., Phys. Rev. Lett. **23**, 935 (1969)
77. R.P. Feynman, Conf. Proc. C **700414**, 773 (1970)
78. H.D. Politzer, Phys. Rev. D **9**, 2174 (1974)
79. D.J. Gross, F. Wilczek, Phys. Rev. D **8**, 3633 (1973)
80. S.L. Glashow, Nucl. Phys. **22**, 579 (1961); Annals Phys. **15**, 437 (1961)
81. J. Goldstone, A. Salam, S. Weinberg, Phys. Rev. **127**, 965 (1962)
82. F. Englert, R. Brout, Phys. Rev. Lett. **13**-9, 321 (1964)
83. P.W. Higgs, Phys. Rev. Lett. **13**-16, 508 (1964)
84. S. Weinberg, Phys. Rev. Lett. **19**-21, 1264 (1967)
85. G. Danby et al., Phys. Rev. Lett. **9**, 36 (1962)
86. J.H. Christenson et al., Phys. Rev. Lett. **13**, 138 (1964)
87. N. Cabibbo, Phys. Rev. Lett. **12**, 62 (1964)
88. S.L. Glashow, J. Iliopoulos, L. Maiani, Phys. Rev. D **2**, 1285 (1970)
89. J.E. Augustin et al., Phys. Rev. Lett. **33**, 1406 (1974)
90. J.J. Aubert et al., Phys. Rev. Lett. **33**, 1404 (1974)
91. M.L. Perl et al., Phys. Rev. Lett. **35**, 1489 (1975)
92. M. Kobayashi, T. Maskawa, Prog. Theor. Phys. **49**, 652 (1973)
93. F.J. Hasert et al., Phys. Lett. B **46**, 138 (1973)
94. The UA1 Collaboration (G. Arnison et al), Phys. Lett. B **122**, 103 (1983); ibid Phys. Lett. B **126**, 398 (1983); UA2 Collaboration (M. Banner et al.), Phys. Lett. B **122**, 476 (1983); ibid Phys. Lett. B **129**, 130 (1983)
95. G. Carron et al., IEEE Trans. Nucl. Sci. **30**, 2587 (1983)
96. A. Pais, S.B. Treiman, Phys. Rev. D **9**, 1459 (1974)

References

97. S.W. Herb et al., Phys. Rev. Lett. **39**, 252 (1977)
98. The CDF Collaboration (F. Abe et al.), Phys. Rev. Lett. **74**, 2626 (1995)
99. The D0 Collaboration (S. Abachi et al.), Phys. Rev. Lett. **74**, 2632 (1995)
100. The DONUT Collaboration (K. Kodama et al.), Phys. Lett. B **504**, 218 (2001)
101. The Argus Collaboration (H. Albrecht et al.), Nucl. Instrum. Meth. A **275**, 1 (1989)
102. E. Bloom, C. Peck, Ann. Rev. Nucl. Part. Sci. **33**, 149 (1983); W. Bartel, 1975 PEP Summer Study, PEP-0192 (1875)
103. The ARGUS Collaboration (H. Albrecht et al.), Phys. Lett. B **192**, 245 (1987)
104. The CDF Collaboration (A. Abulencia et al.), Phys. Rev. Lett. **97**, 242003 (2006)
105. The CLEO Collaboration (R. Ammar et al.), Phys. Rev. Lett. **71**, 674 (1993)
106. The *BABAR* Collaboration (B. Aubert et al.), Phys. Rev. Lett. **86**, 2515 (2001)
107. The Belle Collaboration (A. Abashian *et at.*), Phys. Rev. Lett. **86**, 2509 (2001)
108. The KTeV Collaboration (A. Alavi-Harati *et al.*), Phys. Rev. Lett. **83**, 22 (1999)
109. The NA48 Collaboration (V. Fanti et al.), Phys. Lett. B **465**, 335 (1999)
110. The ALEPH, DELPHI, L3 and OPAL Collaborations (S. Schael et al.), Phys. Rept. **427**, 257 (2006)
111. The Superkamiokande Collaboration (O. Yasuda et al.), Phys. Rev. D **58**, 091301 (1998)
112. Q.R. Ahmad et al., Phys. Rev. Lett. **89**, 011302 (2002)
113. M. Gell-Mann, P. Ramond, R. Slansky, in Supergravity, eds. P. van Niewen-huizen and D.Z. Freedman (North Holland 1979); T. Yanagida, in Proceedings of Workshop on Unified Theory and Baryon number in the Universe, eds. O. Sawada and A. Sugamoto (KEK 1979)
114. R.N. Mohapatra, G. Senjanovic, Phys. Rev. Lett. **44**, 912 (1980); Phys. Rev. D **23**, 165 (1981)
115. The ATLAS Collaboration (G. Aad et al.), Phys. Lett. B **716**, 1 (2012); Phys. Lett. B **710**, 383 (2012); Phys. Rev. Lett. **108**, 111803 (2012); Science **338**, 1576 (2012)
116. The CMS Collaboration (S. Chatrchyan et al.), Phys. Lett. B **710**, 403 (2012); Phys. Rev. Lett. **108**, 111804 (2012)
117. A. Penzias, R.W. Wilson, Astrophys. J. **142**, 419 (1965)
118. G.F. Smoot et al., Phys. Rev. Lett. **39**, 898 (1977); G. F. Smoot, Class. Quant. Grav. **10**, S3 (1993)
119. S. Perlmutter, Astrophys. J. **517**-2, 565 (1999)
120. A.G. Riess et al., Astron. J. **116**-3, 1009 (1998)
121. The LIGO Scientific and Virgo Collaborations (B. P. Abbott et al.), Phys. Rev. D **93**, 12 (2016)
122. H. Baer, X. Tata, Weak Scale Supersymmetry, Cambridge University Press (2006)
123. M. Bauer, T. Plehn, Lect. Notes Phys. 959 (2019); O. Buchmüller, C. Doglioni and L. T. Wang, Nature Phys. **13**-3, 217 (2017); F. Iocco et al., Nature Phys. **11**, 245 (2015); D. W. Sciama, Cambridge Lect. Notes Phys. **3**, 1 (1993)
124. M. Shifman, Int. J. Mod. Phys. A **25**, 199 (2010); G. C. Cho et al., Phys. Rev. D **91**-11, 115015 (2015)
125. C. Pati and A. Salam, Phys. Rev. D **10**, 275 (1974); H. Georgi and S. L. Glashow, Phys. Rev. Lett. **32**, 438 (1974)
126. P. Langacker, Rev. Mod. Phys. **81**, 1199 (2009); J. L. Hewett and T. G. Rizzo, Int. J. Mod. Phys. A **2**, 1189 (1987)
127. https://www.pinterest.com/pin/374713631467395809/
128. The ATLAS Collaboration (G. Aad et al.), HEP **09**, 056 (2010)
129. The CMS Collaboration (S. Chatrchyan et al.), JINST **5**, T03019 (2010)
130. The CDF Collaboration (D. Ayres et al.), Technical Report, FERMILAB-DESIGN-**1981**-02 (1981)
131. The D0 Collaboration (B. Pifer et al.), FERMILAB-PUB-**83**-111-E; PRINT-84-0306 (FERMILAB); FERMILAB-DESIGN-**1983**-04 (1983)
132. The ALEPH Collaboration, Technical Report, CERN-LEPC-**83**-2 ; LEPC-P-1 (1983); The ALEPH Collaboration (D. Decamp et al.), Nucl. Instrum. Meth. A **294**, 121 (1990)
133. The DELPHI Collaboration, Technical Report, CERN-LEPC-**83**-3; DELPHI-83-66; LEPC-P-**2** (1983); The DELPHI Collaboration (P. A. Aarnio *et at.*), Nucl. Instrum. Meth. A **303**, 233 (1991)

134. The L3 Collaboration, Technical Report, CERN-LEPC-**83**-5; LEPC-P-4. (1983)
135. The OPAL Collaboration, Technical Report, CERN-LEPC-**83**-4; LEPC-P-3 (1983); The OPAL Collaboration (K. Ahmet et al.), Nucl. Instrum. Meth. A **305**, 275 (1991)
136. S. Abrams et al., Phys. Rev. Lett. **43**, 477 (1979)
137. SLD Design Report, SLAC-**0273** (1984); M. Breidenbach, IEEE Trans. Nucl. Sci. **33**, 46 (1986)
138. The ZEUS Collaboration (G. Wolf et al.), Technical Report, DESY-HERA-ZEUS-**1** (1986)
139. The H1 Collaboration, Technical Report, DESY PRC-**86**-02 (1986)
140. The CLEO II Collaboration (Y. Kubota et al.), Nucl. Instrum. Meth. A **320**, 66 (1992)
141. The *BABAR* Collaboration (B. Aubert et al.), Technical design reports, SLAC-R-**95**-457 (1995)
142. The Belle Collaboration (N. Toge et al.), KEK-Report-**95**-7 (1995);. J. Haba, Nucl. Instrum. Meth. A **368**, 74 (1995)
143. Belle II Technical Design Report, KEK-REPORT-**2010-1** (2010)
144. W.D. Li et al., Int. J. Mod. Phys. A **24**-S1, 9 (2009)
145. S. Dell'Agnello et al., Frascati Phys. Ser. **16**, 381 (1999)
146. The PLUTO Collaboration, (Ch. Berger et al.), DESY Report **79**/11 (1979)
147. The Mark J Collaboration, DESY Print-**79**-0301 (1979)
148. The TASSO Collaboration (R. Brandelik et al.), DESY Report **79**/11 (1979)
149. The JADE Collaboration (W. Bartl et al.), Phys. Lett. B **88**, 171 (1979)
150. H.J. Behrend et al., PETRA Proposal-**76**/13, 1 (1976)
151. H. Sagawa et al., Phys. Rev. Lett. **60**, 93 (1988)
152. I. Adachi et al., KEK-PREPRINT- **87**-59, 1 (1987)
153. F. Takasaki, KEK-PREPRINT- **84**-16, 1 (1984)
154. The LHCb Collaboration (A. Alves et al.), JINST **3**, S08005 (2008)
155. The ALICE Collaboration (K. Aamodt et al.), JINST **3**, S08002 (2008)
156. The Superkamiokande Collaboration (S. Fukuda et al.), Nucl. Instrum. Meth. A **501**, 418 (2003)
157. The T2K Collaboration, https://t2k-experiment.org/t2k/collaboration/
158. The Daya Bay Collaboration (F. P. An et al.), Nucl. Instrum. Meth. A **811**, 133 (2016)
159. The Cuore Collaboration (C. Alduino et al.), JINST **11**- 07, P07009 (2016)
160. The KamLand Collaboration (K. Eguchi et al.), Phys. Rev. Lett. **90**, 021802 (2003)
161. The EXO Collaboration (M. Auger et al.), JINST **7**, P05010 (2012)
162. The Xenon1T Collaboration (E. Aprile et al.), Eur. Phys. J. C **77**-12, 881 (2017)
163. The LZ Collaboration (D. S. Akerib et al.), e-Print: 1509.02910 [physics.ins-det] (2015)
164. The DARWIN Collaboration (C. Macolino et al.), J. Phys. Conf. Ser. **1468**, 1 (2020)
165. The Mu2e Collaboration (R. J. Abrams et al.), TDR, arXiv:1211.7019 [phys.-ins-det.] (2012)
166. The MEG II Collaboration (A. M. Baldini et al.), Eur. Phys. J. C **78**, 5 (2018)
167. ATLAS event- how ATLAS detects particles, 2019 CERN. http://doi.org/10.17181/cds.1458883
168. K. Wille, The Physics of Particle Accelerators, Oxford University Press (2001)
169. J.D. Cockcroft, E.T.S. Walton, Proc. Roy. Soc. Lond. A **136**, 619 (1932)
170. R.J. van de Graaff, K.T. Compton, L.C. van Atta, Phys. Rev. **43**, 149 (1933)
171. E.O. Lawrence, M.S. Livingston, Phys. Rev. **38**, 834 (1931); Phys. Rev. **40**-1, 19 (1932)
172. L.W. Smith, HEACC **59**, 397 (1959)
173. E.J. Lofgren, Berkeley UCRL-**3369** (1956)
174. V.I. Veksler, Comptes Rendus de l'Académie des Sciences de l'URSS. **43**-8, 346 (1944)
175. F.R. Elder et al., Phys. Rev. **71**, 829 (1947)
176. J.L. Conradie, Electrostatic Accelerators, Electrostatic Accelerators, talk presented at the Joint ICTP-IAEA Workshop on Accelerator Technologies, Basic Instruments and Analytical Techniques, Trieste, 21-29 October 2019. https://indico.ictp.it/event/8728/session/2/contribution/2/material/slides/0.pdf
177. N. Christofilos, US patent 2736799 issued **1956**-02-28 (1956)
178. E.D. Courant, M.S. Livingston, H.S. Snyder, Phys. Rev. **88**, 1190 (1952)

179. A. Variola, Particle Accelerators Course, lecture presented at the EIPS School, IPLN Lyon 2014, https://indico.in2p3.fr/event/10396/sessions/419/attachments/1631/2065/Lyon_variola_accelerateur.pdf (2014)
180. E. Regenstreif, CERN Yellow Report 1 (1959). https://doi.org/10.5170/CERN-1959-029
181. K. Johnsen, Proc. Nat. Acad. Sci. **70**-2, 619 (1973)
182. R. Clifft and N. Doble, CERN/SPSC/**74**-12 (1974); J. Redfearn, Phys. Bull. **27**, 499 (1976)
183. The UA1 Collaboration (A. Astbury et al.), CERN-SPSC-**78**-06, 1 (1978)
184. The UA2 Collaboration, (M. Banner et al.), CERN/SPSC/**78**-08, 1 and CERN/SPSC/78-54, 1 (1978)
185. V. Shiltsev, AIP Conference Proceedings **1507**, 950 (2012)
186. J.R.J. Bennett et al., CERN Yellow Report **77**-14, 1 (1977)
187. L. Evans, P. Bryant, JINST **3**, S08001 (2007)
188. S. Myers, CERN Accelerators, talk at CSP meeting CERN (2010)
189. https://upload.wikimedia.org/wikipedia/commons/archive/3/3f/20070210022452 %21Fermilab.jpg
190. R.A. Beth, C. Lasky, Science **128**(3336), 919580 (1393)
191. N. Samios, BNL Report BNL- **51923**, 1 (1986)
192. R.R. Wilson, Fermilab. FERMILAB-TM-**0763** (1978)
193. The CDF Collaboration (L. Demortier et al.), FERMILAB-TM-**2084**,1 (1999)
194. The D0 Collaboration (V. M. Abazovet et al.), Nucl. Instrum. Meth. A **565**, 463 (2006)
195. The DUNE Collaboration (R. Acciarri et al.), arXiv:1512.06148 [phys.ins.det] (2016)
196. H. Wiedemann, Phys. Bl. **30**, 215 (1974)
197. The DASP-Collaboration (R. Brandelik et al.), Phys. Lett. B **73**, 109, (1978); DESY 77/81, 1 (1977)
198. PETRA Report, DESY, 1 (1976)
199. B.H. Wiik, Conf. Proc. C **8405141**, 23 (1984)
200. CESR Design Report, CLNS-**345**/76, 1 (1976)
201. E. Nordberg and A. Silverman, Laboratory of Nuclear Studies, CBX **79**-6 (1979)
202. P. Franzini, J. Lee-Franzini, Phys. Reports **81**, 240 (1982)
203. The CLEO Collaboration (Y. Kubota et al.), Cornell Preprint 91/1122 (1991); Nucl. Instrum. Meth. A **320**, 66 (1992)
204. The CLEO Collaboration (S. E. Kopp et al.), Nucl. Instrum. Meth. A **384**, 61 (1996)
205. M. Breidenbach et al., Phys. Rev. Lett. **23**, 930 (1969)
206. B. Richter, Kerntech. **12**, 531 (1970)
207. J.E. Augustin et al., Phys. Rev. Lett. **34**, 764 (1975)
208. J. Rees, eConf C **740805**, 7 (1974)
209. W. Bacino et al., Phys. Rev. Lett. **40**, 671 (1978)
210. The MAC Collaboration (W. T. Ford et al.), SLAC-PUB-**2894** (1982)
211. The PEP-4 TPC Collaboration, (H. Aihara et al.), IEEE Trans. NS-**30**, 63, 67, 76, 11 7, 1 53, 162 (1983); Nucl. Instrum. Meth. A **217**, 259 (1984)
212. R. Mozley, SLAC Beam Line **12**, 3 (1981)
213. A. Hutton et al., IEEE Conf. Proc. C **910506**, 84 (1991)
214. F. Gross et al., Eur. Phys. J. C **83**, 1125 (2023)
215. T. Nishikawa, Conf. Proc. C **760811**, 217 (1976)
216. KEKB B factory design report, KEK-REPORT-**95**-7 (1995)
217. P.F. Loverre, J. Phys. Conf. Ser. **39**, 323 (2006)
218. F. Amman, Kerntech. **12**, 528 (1970)
219. DAPHNE machine project, LNF-**92**-033 (1992)
220. KLOE detector technical proposal, LNF-**93**-002-IR, 1 (1993)
221. The ACO Storage Ring, CEA-R-**2929**, 1 (1966)
222. Z. Arzelier et al., eConf C **710920**, 150 (1971)
223. J.E. Augustin et al., Phys. Scripta **23**, 623 (1981)
224. A. Skrinsky, I.E.E.E. Conf. Proc. C **950501**, 14 (1996)
225. B.S. Feng, Z.Q. Yu, Conf. Proc. C **8405141**, 17 (1984)

226. The BES Collaboration (J. Z. Bai et al.), Nucl. Instrum. Meth. A **345**, 541 (1994)
227. C. Zhang et al., Chin. Phys. C **33**(S2), 60 (2009)
228. The BESIII Collaboration (M. Ablikim et al.), Nucl. Instrum. Meth. A **614**, 345 (2010)
229. M. Tatjudin et al., Mal. J. Fund. App. Sci. **15**-4, 580 (2019)
230. U. Fano, Phys. Rev. **72**, 26 (1947)

Open Access This chapter is licensed under the terms of the Creative Commons Attribution 4.0 International License (http://creativecommons.org/licenses/by/4.0/), which permits use, sharing, adaptation, distribution and reproduction in any medium or format, as long as you give appropriate credit to the original author(s) and the source, provide a link to the Creative Commons license and indicate if changes were made.

The images or other third party material in this chapter are included in the chapter's Creative Commons license, unless indicated otherwise in a credit line to the material. If material is not included in the chapter's Creative Commons license and your intended use is not permitted by statutory regulation or exceeds the permitted use, you will need to obtain permission directly from the copyright holder.

Interactions of Charged Particles and Photons with Matter and Fields

2

A charged particle passing through a medium typically interacts with nucleons and electrons of the atoms or molecules in the medium. The interaction probability depends on the interaction cross section, density of the medium and distance the particle travels inside the medium. The interaction cross section depends on particle properties like charge, spin, momentum and mass as well as specific properties of the medium. Most processes are based on the electromagnetic interaction and cause an energy loss of the particle.

The goal of this chapter is to introduce the basic processes and concepts that are necessary to understand the underlying physics of detectors in modern particle and nuclear physics experiments. Section 2.1 covers interactions of heavy charged particles with matter. Section 2.2 extends this to interactions of electrons and positrons introducing important observables like critical energy and radiation length. Section 2.3 is devoted to multiple scattering of charged particles. Section 2.4 covers the interactions of photons with matter and introduces the concept of the electron-photon shower. Section 2.5 presents properties of electrons and ions in gases introducing concepts such as drift velocity and diffusion. Section 2.6 discusses the scintillation process in inorganic crystals, liquid noble gases and organic scintillators concluding with Birks' law. The first part of this chapter is based on reviews of the Particle Data Group [1].

2.1 Passage of Heavy Charged Particles Through Matter

A charged particle traversing a medium may encounter the following interactions:

(i) **Elastic scattering**: a charged particle bounces off an electron or proton in an atom of the medium, leading to a change in the momentum direction of the

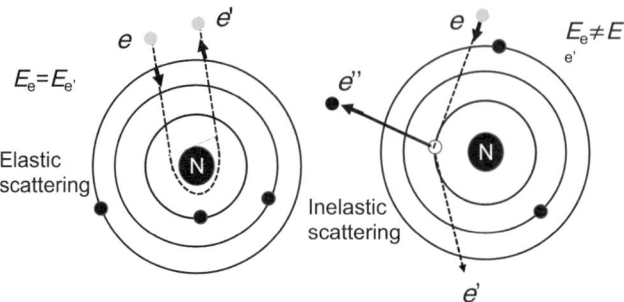

Fig. 2.1 Left: An electron that is elastically back-scattered from the nucleus. Right: An electron that is inelastically scattered (e') emitting a secondary electron (e'')

particle but conserves its energy. Figure 2.1 (left) shows a backscattered electron in an atom, which is the extreme case of elastic scattering with a change in angle of $180°$.

(ii) **Inelastic collisions**: the charged particle loses energy in the interaction that is transferred to an atom of the medium causing excitation or ionization. Figure 2.1 (right) shows the case of ionization where a secondary electron is produced.

(iii) **Cherenkov radiation**: a charged particle traversing a medium may have a velocity v smaller or larger than the speed of light c in the medium, as depicted in Fig. 2.2 (left, center). In the latter case the particle emits Cherenkov radiation in a cone with angle θ_c. The radiation is typically in the UV region.

(iv) **Transition radiation**: an energetic charged particle passing through detector layers with alternating high and low indices of refraction may radiate. This is different than Cherenkov radiation because transition radiation occurs for $v < c/n$. We may illustrate the effect by a simple model. The charge in the high-n medium forms a dipole with the image charge in the low-n medium, as illustrated in Fig. 2.2 (right). Since the dipole moves, it radiates photons that lie in the X-ray region. At present LHC experiments, this is relevant for energetic electrons and positrons.

(v) **Bremsstrahlung**: a sufficiently high-energy electron produces electromagnetic radiation as it decelerates in a medium (see Fig. 2.3 left). Mostly low-energy photons are radiated since the cross section depends on the radiated photon energy as $1/\hbar\omega$.

(vi) **Nuclear interactions**: The charged particle interacts with the nucleus, which may lead to a break up, ejection of nucleons or production of additional hadrons, typically pions. These processes are typically small and become relevant in hadron calorimeters where we discuss them further.

Thus, for charged particles heavier than electrons, the first two processes are the most dominant. Process iii is comparatively much smaller but still relevant. For electrons and positrons, processes iv and v are important as well, becoming dominant at high energies.

2.1 Passage of Heavy Charged Particles Through Matter

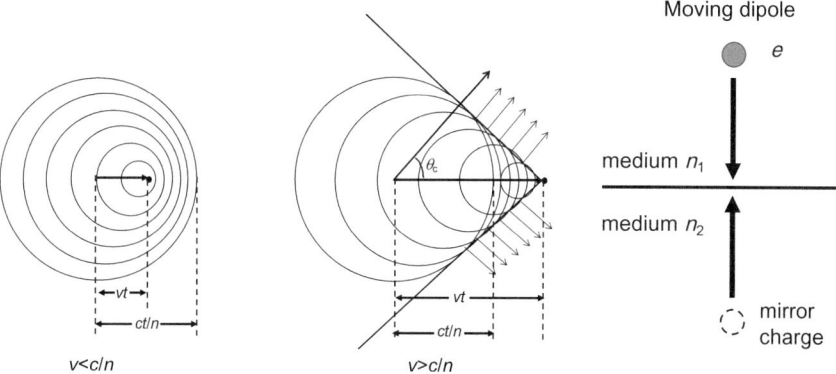

Fig. 2.2 Left, center: Spherical wavelets of fields of a charged particle traveling less than (left) and greater than (center) the speed of light in the medium. For $v > c/n$, an electromagnetic shock wave appears that moves in the direction of θ_c. Right: Explanation of the formation of transition radiation. An electron in medium n_1 produces a mirror charge in medium n_2. Since the electron and the mirror charge move towards each other, the system represents a moving dipole, which emits radiation

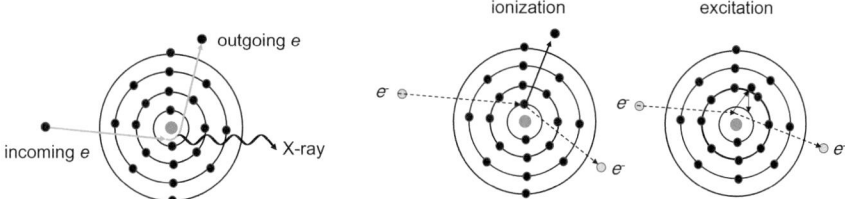

Fig. 2.3 Left: Production of bremsstrahlung of an electron in the electric field of the nucleus, which absorbs the recoil. Center: Ionization of an atom. Right: Excitation of an atom

In the interaction of a heavy charged particle with an electron in the atomic shell of an atom or molecule in the medium, a virtual photon is exchanged. The photon energy $\hbar\omega$ is equal to the energy loss W of the charged particle in a single collision [2]. If the photon energy is too small, the charged particle is scattered elastically. If the emitted virtual photon carries sufficient energy, the atom will be ionized or excited as displayed in Fig. 2.3 (center, right), respectively. The emitted electron has a kinetic energy that equals the photon energy minus the binding energy. The interaction properties of the medium are characterized by the dielectric constant, $\epsilon(\hbar\omega)$, which is a function of the photon energy.[1] Figure 2.4 shows the imaginary part Im(ϵ) (top) and real part Re(ϵ) − 1 (bottom) of the dielectric constant versus photon energy. We observe three different regions:

---

[1] In general, $\epsilon(\hbar\omega)$ is complex. For practical reasons we often leave off the dependence on the photon energy.

(i) For photon energies much smaller than excitation or ionization energies, $\hbar\omega \ll E_{ex} < E_{ion}$, the dielectric constant is real with $\text{Re}(\epsilon) > 1$. The value of E_{ion} depends on the medium ranging from 2–10 eV. This is the optical region, where the index of refraction is given by $n = \sqrt{\epsilon}$ for $\text{Im}(\epsilon) = 0$. If the particle velocity is $v > c/n$, Cherenkov radiation is produced as discussed in Sects. 2.4 and 9.3. The medium is transparent and absorption is small.

(ii) For photon energies in the range of a $E_{ion} < \hbar\omega <\sim 5$ keV, ϵ is complex with $\text{Im}(\epsilon) > 0$ and $\text{Re}(\epsilon)$ is typically below one. Both reflect the shell structure of the atom. The resonance structures are anti-correlated. Thus, this is called the resonance region. The photon energies are comparable with excitation energies. Here, only virtual photons are created that excite or ionize the medium -(see Sect. 2.1.1). The medium is opaque and absorption is high.

(iii) For photon energies with $\hbar\omega >\sim 5$ keV, absorption becomes small $\text{Im}(\epsilon) \ll 1$ but $\text{Re}(\epsilon) < 1$. The photon sees electrons as free particles and scattered electrons can have high energies. This is the X-ray region, where the threshold velocity for Cherenkov radiation is larger than c, so the charged particle cannot produce Cherenkov photons. However, transition radiation can be emitted if the medium has layers of alternating refractive indices, as discussed in Sects. 2.5 and 9.4.

In principle, we only need to know $\text{Im}(\epsilon(\hbar\omega))$ for momentum transfer q since $\text{Re}(\epsilon(\hbar\omega))$ can be calculated via the Kramers-Kronig relation [3],

$$\text{Re}(\epsilon(\hbar\omega)) = 1 + \frac{2}{\pi}\mathcal{P}\int_0^\infty \frac{w\,\text{Im}(\epsilon(\hbar\omega))}{w^2 - (\hbar\omega)^2}dw, \qquad (2.1)$$

where $\mathcal{P}$ denotes the Cauchy principle value, $\hbar\omega$ is the photon energy and w is an integration variable. Note that the integral has to be evaluated numerically. Since for certain situations the numerical integration gave inaccurate results, Bichsel came up with a modification [2,4].

The interaction of a fast charged spin-zero particle heavier than an electron with a free electron as a function of the energy loss W in a single collision is described by the differential Rutherford cross section [5–7],

$$\frac{d\sigma_R(W,\beta)}{dW} = \frac{k_R}{\beta^2}\frac{z^2}{W^2}\left(1 - \beta^2\frac{W}{W_{max}}\right), \qquad (2.2)$$

where $k_R = \frac{2\pi q_e^4}{m_e c^2} = 2\pi r_e^2 mc^2 = 2.5495 \times 10^{-19}$ eVcm^2 is a constant, $r_e = 2.817 \times 10^{-15}$ m is the electron radius, z is the charge of the heavy particle in units of the elementary charge $q_e = 1.6 \times 10^{-19}$ C, m_e is the electron mass and β is its relative velocity. For a spin-1/2 particle the Rutherford cross section is [5,8],

$$\frac{d\sigma_R(W,\beta)}{dW} = \frac{k_R}{\beta^2}\frac{z^2}{W^2}\left(1 - \beta^2\frac{W}{W_{max}} + \frac{W^2}{2E^2}\right), \qquad (2.3)$$

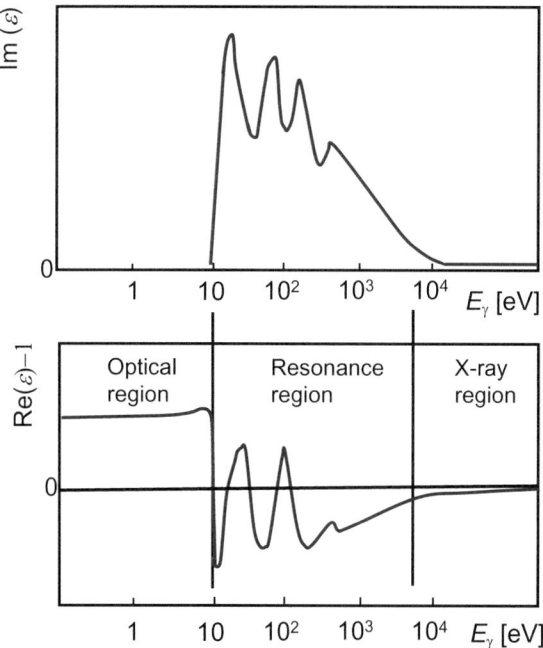

Fig. 2.4 Imaginary part of the dielectric constant, Im(ϵ) (top) and the real part Re(ϵ) − 1 (bottom) as a function of the photon energy for a generic medium (adapted from [9]). In the three energy regions (optical region, resonance region and X-ray region), Im(ϵ) and Re(ϵ) − 1 reveal rather distinctive shapes. In the optical region Im(ϵ) = 0 and Re(ϵ) > 1. In the resonance region Im(ϵ) > 0, while typically Re(ϵ) − 1 typically becomes negative. Both Im(ϵ) and Re(ϵ) reflect the structure of the atomic shell in an anti-correlated relation. In the X-ray region Im(ϵ) is rather small and Re(ϵ) is below one

where E is the total energy of the particle $E = E_k + mc^2$, E_k is the kinetic energy and m is the particle mass. For spin-1 particles the cross section is listed in (4) of [7].

The maximum energy loss transferred to the electron is given by

$$W_{max} = \frac{2m_e c^2 \beta^2 \gamma^2}{1 + \frac{2m_e}{m}\gamma + \frac{m_e^2}{m^2}}, \qquad (2.4)$$

where γ is the Lorentz factor. For heavy particles with $2\gamma m_e \ll m$, the maximum transferred energy is often simplified to $W'_{max} = 2m_e c^2 \beta^2 \gamma^2$. For a pion with $\beta\gamma = 50$ the approximation W'_{max} is a factor of 1.4 higher than W_{max} because $2\gamma m_e/m \simeq 0.5$, which does not satisfy the requirement.

Note that for most collisions the energy losses are small. For 90% of all collisions the energy loss is below 100 eV. As an example, Fig. 2.5 shows the probability for triggering a specific process as a function of energy transfer W for 100 keV electrons in water [10]. The maximum energy transfer here is 50 keV as indicated by the arrow. Energy transfers below $\sim$ 15 eV result from distant collisions and lead to excitations of the atoms. Energy transfers between $\sim$ 15 eV and $\sim$ 150 eV also come from

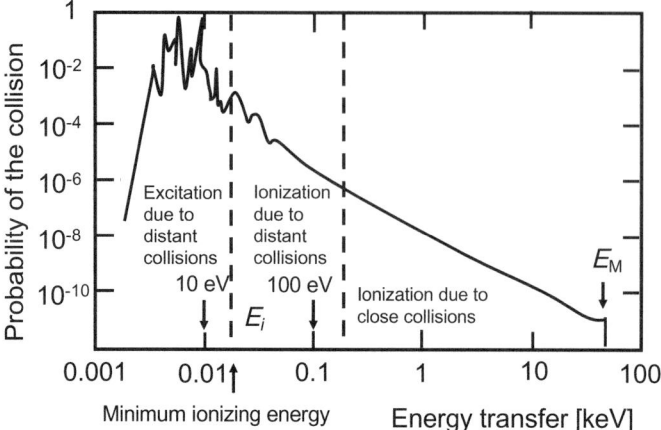

Fig. 2.5 Probability for an interaction as a function of energy transfer for 100 keV electrons in water. The maximum energy transfer (50 keV) is indicated by the arrow at E_M. Reprinted with kind permission from [27], © 1967, Elsevier. All rights reserved

distant collisions and lead to ionization of the atoms, while energy transfers above ~ 150 eV result fron close collisions and also lead to ionization of the medium. The energy loss of the charged particle is transferred to the electrons in the atomic shell. Basically, we can distinguish between two types of interactions, close-by collisions that involve large energy transfers leading to ionization of the material and distant collisions that involves smaller energy transfers yielding excitations and ionizations.

Since the electron is not free but is bound in an atom or molecule, we have to include a correction factor representing the binding effects $B(W)$ to obtain more accurate results. This is achieved in the Born approximation [11],

$$\frac{d\sigma_B(W,\beta)}{dW} = \frac{d\sigma_R(W,\beta)}{dW} B(W), \qquad (2.5)$$

where the correction factor $B(W)$ is parameterized in terms of the dielectric constant or the generalized oscillator strengths [4].

In 1915, Bohr performed a classical calculation of the energy loss of charged particles interacting with atoms [12]. In 1930, Bethe performed the first non-relativistic quantum-mechanical calculation using the first-order Born approximation [11]. He extended his calculation to include relativistic corrections in 1932 [13]. A year later, Bloch performed an independent relativistic quantum-mechanical calculation [14] and showed that the energy loss calculated in relativistic quantum theory agrees with Bethe's plane wave approximation for low-momentum transfers and with Bohr's classical impact parameter approach for high-momentum transfers. In 1944, Landau studied the energy loss of fast charged particles by ionization [15]. Fano published various extensions of Bethe's and Bloch's work including new terms for the shell and density corrections [16,17] (see Sect. 2.1.1). In 1952, Sternheimer started with calculations of density corrections for various materials [18,19]. Bloch [14] and

2.1 Passage of Heavy Charged Particles Through Matter

Barkas [20,21] calculated higher-order corrections. In 1980, Allison and Cobb [22–24] developed the photoabsorption ionization model building on the work of Landau-Lifshitz [25] and Fano [26]. The electric field of the particle is calculated at position $\vec{r} = \vec{\beta}ct$ at time t in which the electromagnetic potentials are defined in the Coulomb gauge [28] and the mean energy loss per unit time is described as the effect of the electric field doing work on the particle. Allison and Cobb computed the differential cross section $d\sigma/dW$ of a heavy charged particle interacting with the electrons of the medium, yielding

$$\frac{d\sigma}{dW} = \frac{z^2\alpha}{\beta^2\pi} \left\{ \frac{\sigma_\gamma(W)}{ZW} \left[\ln\left(\frac{1}{\sqrt{\left[1 - \beta^2\text{Re}(\epsilon)\right]^2 + \left[\beta^2\text{Im}(\epsilon)\right]^2}} \right) + \ln\left(\frac{2mc^2\beta^2}{W}\right) \right] \right. $$
$$\left. + \frac{1}{ZW^2} \int_0^W \sigma_\gamma(w)dw + \frac{1}{N\hbar c}\left(\beta^2 - \frac{\text{Re}(\epsilon)}{|\epsilon|^2}\right)\Theta \right\}, \tag{2.6}$$

where $\alpha = 1/137$ is the fine structure constant, Z is the atomic number of the medium, Θ is the phase of $1 - \text{Re}(\epsilon)\beta^2 + \text{Im}(\epsilon)\beta^2$, $\sigma_\gamma(W)$ is the cross section for the absorption of a photon with energy $W = \hbar\omega$ by the atoms of the medium and N is the atomic density, which is usually expressed in terms of Avogadro's number $N_A = 6.022 \times 10^{23}$/mol, the atomic weight A [g/mol] and the density of the medium ρ and is given by,

$$N = N_A \rho / A. \tag{2.7}$$

The other parameters are the same as those in (2.2). The first term includes as well the relativistic increase of the cross section and its saturation, while the second term represents the resonance region. These are identical to those calculated by Bethe [13]. The third term represents Rutherford scattering off quasi-free electrons and the last term is independent of the photoabsorption ionization cross section and describes the emission of Cherenkov radiation. The first and fourth terms in (2.6) arise from the magnetic vector potential in the Coulomb gauge for which the electric field is transverse to the direction of the momentum transfer $\hbar\vec{k}$. Therefore, these terms are also called the transverse cross section. The second and third terms in (2.6) arise from the electrostatic term in the Coulomb gauge for which the electric field is parallel to the momentum transfer. Thus, they are also called the longitudinal cross section.

The photoabsorption ionization cross section $\sigma_\gamma(W)$ depends on the photon energy $\hbar\omega$ and the imaginary part of the dielectric constant $\text{Im}(\epsilon)$ [24],

$$\sigma_\gamma(W) = \frac{ZW}{N\hbar c} \frac{\text{Im}(\epsilon)}{\sqrt{\text{Re}(\epsilon)}} \simeq \frac{ZW}{N\hbar c}\text{Im}(\epsilon). \tag{2.8}$$

The second relation results for low-density media, where $\text{Re}(\epsilon)$ is assumed to be one. The integral of $\sigma_\gamma(W)$ is obtained from the Thomas-Reiche-Kuhn sum rule [29],

$$\int_0^\infty \sigma_\gamma(w)dw = \frac{2\pi^2 q_e^2 Z}{m_e c}. \tag{2.9}$$

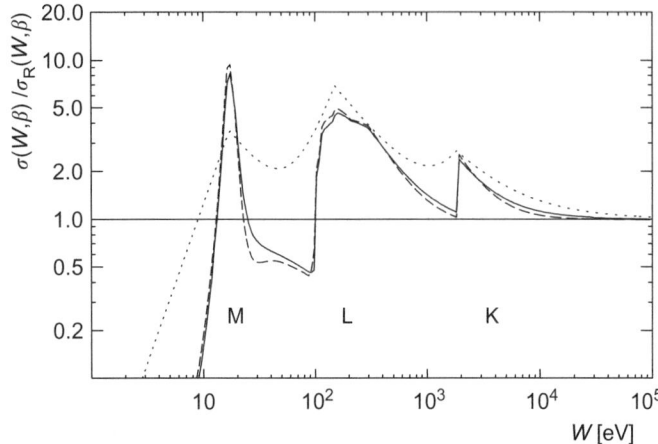

Fig. 2.6 The inelastic collision cross section $\sigma(W, \beta)$ for single collisions of heavy charged particles with $\beta\gamma = 4$ in silicon normalized to the Rutherford cross section [2] in the photoabsorption ionization model (dashed line) [24], the Bethe-Fano theory [26,30] (solid line) and a binary encounter algorithm (dotted line). The atomic shells are labeled by M, L and K. Reprinted with kind permission from [2], © 2006, Elsevier. All rights reserved

The photoabsorption ionization cross section decreases as $1/W^{2.5}$ [24]. When the energy is larger than the binding energy of electrons in the atomic shell that absorbs the photon, it asymptotically converges to the Rutherford cross section.

The accuracy of the cross section calculation depends on how well the details of the atomic structure are modeled in $B(W)$. Figure 2.6 shows three calculations of the inelastic collision cross section $\sigma(W, \beta)$: Bethe-Fano theory [26,30], photoabsorption ionization model by Allison and Cobb [24] and a binary encounter approximation [2]) with respect to the Rutherford cross section for single collisions of heavy charged particles in silicon as a function of the charged-particle energy loss. The normalization to the Rutherford cross section is performed to display the impact of the shell structures. The binary encounter approximation produces a rather different spectrum compared to the Bethe-Fano theory and photoabsorption ionization model. The latter two calculations agree reasonably well. Using measurements provides a better description of the details of the atomic structure. Other improvements result if we account for polarization of the medium and the density effect discussed in Sect. 2.1.1.

2.1.1 Energy Loss by Excitation and Ionization

We start with the discussion of the first and second terms in (2.6), which are relevant for the resonant region and describe the energy loss by excitation and ionization. The average number of collisions within a path length interval Δx in the medium is

2.1 Passage of Heavy Charged Particles Through Matter

$$\langle N_{coll}(\beta)\rangle = N_e \Delta x \int_{I_{ex}}^{W_{max}} \frac{d\sigma(W,\beta)}{dW} dW = N_e \Delta x \cdot \sigma(\beta), \tag{2.10}$$

where $d\sigma(W,\beta)/dW$ is the differential cross section given in (2.6), N_e is the electron density and I_{ex} is the mean excitation energy. Equation (2.10) can be interpreted as the zeroth moment of the energy loss distribution. It is also useful to define higher moments,

$$\mathcal{M}_k(\beta) = N_e \Delta x \int_{I}^{W_{max}} W^k \frac{d\sigma(W,\beta)}{dW} dW. \tag{2.11}$$

The first moment yields the average energy loss at distance Δx and $(\mathcal{M}_2 - \mathcal{M}_1)^2$ is the variance of the energy loss distribution. The number of collisions is described by a Poisson distribution with expectation value $\Delta x \cdot \mathcal{M}_0$,

$$P(n) = \frac{(\Delta x \cdot \mathcal{M}_0)^n}{n!} \exp(-\Delta x \cdot \mathcal{M}_0), \tag{2.12}$$

where $P(n)$ yields the probability for particles to have n collisions. Let us consider an example [2]. Figure 2.7 (left) shows the results of a Monte Carlo simulation of 10 particles with $\beta\gamma = 3.6$ traversing 1.8 mm of a gas mixture of argon-methane (90:10).[2] Since the average length between two interactions (mean free path (2.146), defined in Sect. 2.7.3), is 0.3 mm, we expect six interactions on average in the total path length. So, here $\mathcal{M}_0 = 200$ collisions/cm. The observed number of interaction varies between $n = 2$ and $n = 9$. We distinguish between interactions with energy loss smaller or larger than 33 eV. For two of the ten particles, the largest energy loss is below 33 eV. The largest energy loss of all ten particles is 774 eV, which is still small with respect to $W_{max} = 13$ MeV. Note that the probability for an energy loss > 50 keV is 0.002 per cm. Figure 2.7 (right) shows the observed probability distribution in comparison to a Poisson distribution with a mean value $\mu = 6$. The observed probability distribution agrees well with the Poisson distribution. So, we already see that the energy loss distribution in general is not symmetric.

Inserting the first two terms in (2.6) into (2.11) for the first moment and integrating over W from I_{ex} to W_{max} yields the average energy loss of the heavy charged particle $(m > m_e)$.[3] From this we get the average energy loss per unit length also called stopping power that is described by the Bethe-Bloch equation [24],

$$\left\langle -\frac{dE}{dx} \right\rangle = K z^2 \frac{Z}{A} \frac{1}{\beta^2} \left[\frac{1}{2} \ln\left(2 m_e c^2 \beta^2 \gamma^2 \frac{W_{max}}{I_{ex}^2}\right) - \beta^2 - \frac{\delta(\beta\gamma)}{2} - \frac{C(I,\beta\gamma)}{Z} \right], \tag{2.13}$$

---

[2] This is our standard notation for gas mixtures indicating that the argon concentration is 90% and the methane concentration is 10%.
[3] It is not valid for electrons (see Sect. 2.2). The electrons in the atomic shell are assumed to be stationary, which is valid for high velocities.

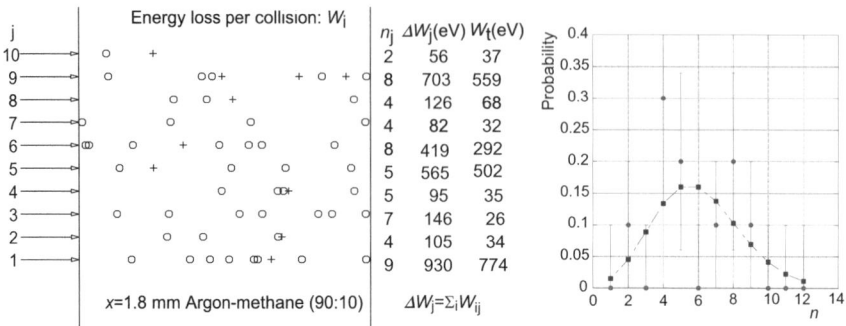

Fig. 2.7 Left: Monte Carlo simulation of the passage of 10 heavy charged particles with $\beta\gamma = 3.6$ through a 1.8 mm thick gas mixture consisting of argon-methane (90:10). The gas is operated at a pressure of 1 atmosphere and a temperature of 25°C. The arrows indicate the direction of motion. The position of an interaction is marked by two symbols, circles and "+" signs, which indicate if the energy loss is smaller or larger than 33 eV, respectively. The average length between two interactions is 0.3 mm. For interactions with energy loss greater 33 eV the average length between two interactions is 2.0 mm. At each interaction point, a random energy loss W_i is selected from the distribution function discussed in Sect. 2.1.2. For each particle, we show the number of interactions n_j, the total energy loss $\Delta W_j = \sum_i W_{ij}$ and the largest energy loss W_t. Reprinted with kind permission from [2], © 2006, Elsevier. All rights reserved. Right: Observed probability distribution (points with error bars) and a Poisson distribution (squares and solid curve) for a mean value of $\mu = 6$

where $K = 4\pi N_A r_e^2 m_e c^2 = 0.307$ MeVcm^2/g is a constant. The first two terms represent the excitation and energy loss of the heavy charged particle. The third term is the density correction and the fourth term is the shell correction that is relevant only for low Z materials. The latter two terms were not present in the original Bethe-Bloch equation. The unit of the stopping power is [MeV cm^2/g]. This definition is particularly useful for liquids and gases for which the density depends on the experimental setup. In the literature $W_{\max}$ is often replaced by $W'_{\max}$, implying larger energy losses as discussed in Sect. 2.1.

Parametrizing $\langle dE/dx \rangle$ in terms of $\beta\gamma$ instead of momentum yields the same average energy loss for all particles with the same $\beta\gamma$. So in turn, if we know the particle momentum, we can determine the particle mass from energy loss measurements in a certain momentum range (see Sect. 9.1). To obtain the average energy loss in units of [MeV/cm] we multiply (2.13) by the density

$$\langle -dE/d\hat{x} \rangle = \langle -dE/dx \rangle \rho. \tag{2.14}$$

This is also called the linear stopping power.

Figure 2.8 shows the average energy loss or mass stopping power for positively-charged muons in copper (solid line) over more than nine orders of magnitude in $\beta\gamma$. The Bethe-Bloch regime lies between the middle and right vertical bars ($0.04 < \beta\gamma < 800$). The dashed line represents $\langle -dE/dx \rangle$ without the density correction while the dot-dashed line shows $\langle -dE/dx \rangle$ with the $\delta(\beta\gamma)$ term included. For $\beta\gamma > 800$ radiative losses start to become dominant as shown by the dotted

2.1 Passage of Heavy Charged Particles Through Matter

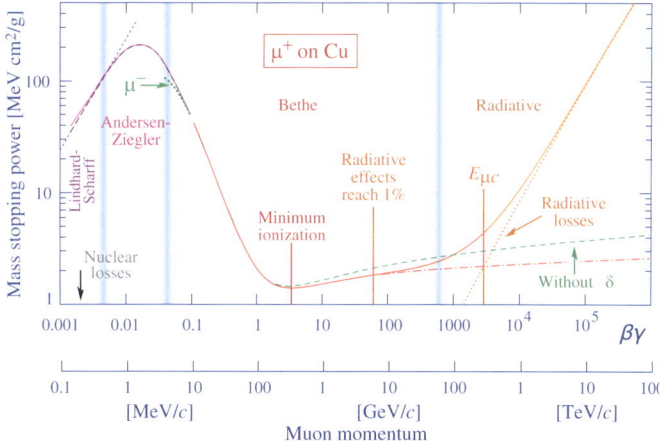

Fig. 2.8 Average energy loss per density and unit length also called stopping power for positively-charged muons in copper as a function of $\beta\gamma$ over nine orders of magnitude shown in a double logarithmic scale. The $\beta\gamma$ region is divided into four distinct regions indicated by the three vertical bars where different approximations hold. The Bethe-Bloch region lies between the middle and right vertical bars ($0.04 < \beta\gamma < 800$). The plot shows the total stopping power (solid curve), the relativistic rise without density correction (dashed curve), the relativistic rise with density correction (dashed-dotted curve) and the average energy loss due to radiation (dotted curve). The latter energy loss becomes dominant at large values of $\beta\gamma > 800$. At low values of $\beta\gamma$, the Lindhard-Scharff [31] approximation (dotted curve) and Andersen-Ziegler approximation (dashed curve) are depicted. For negatively-charged muons, a slight reduction in the stopping power is predicted in the low $\beta\gamma$ region. The symbol $E_{\mu c}$ denotes the muon critical energy (see (2.81)). Reprinted under CC-BY-NC-4.0 Licence with kind permission from [1], © 2024, the Particle Data Group LBNL

line. For values $\beta\gamma < 0.04$ the particle momentum is of the same order as electron momenta in the atoms. Here, the assumptions made for high velocities are no longer valid and higher-order Born corrections become important. The plot includes the Barkas correction [20] as well as the Bloch correction [14]. The former reduces the energy loss for negatively-charged particles with respect to that of positively-charged particles as indicated in Fig. 2.8. In addition, the details of the atomic structure, which are not accounted for in the Bethe-Bloch equation, become important. For $0.01 < \beta\gamma < 0.04$, there is no satisfactory theory. Instead the phenomenological parameterization by the Andersen and Ziegler [32,33] is used. For $\beta\gamma < 0.01$, the Lindhard-Scharff approximation [31] provides a successful description in which the energy loss is proportional to β.

Figure 2.9 shows the average energy loss $\langle -dE/dx \rangle$ of muons in different materials in units of [MeV cm^2/g] as a function of $\beta\gamma$. It is interesting to note that all curves are rather similar if the density is divided out. Broad minima occur around $\beta\gamma = 3 - 3.5$. Particles with $\beta\gamma$ close to the minimum are called minimum ionizing particles or MIPs. In many practical cases, relativistic particles like cosmic-ray muons are minimum ionizing particles. The average energy loss at the minimum

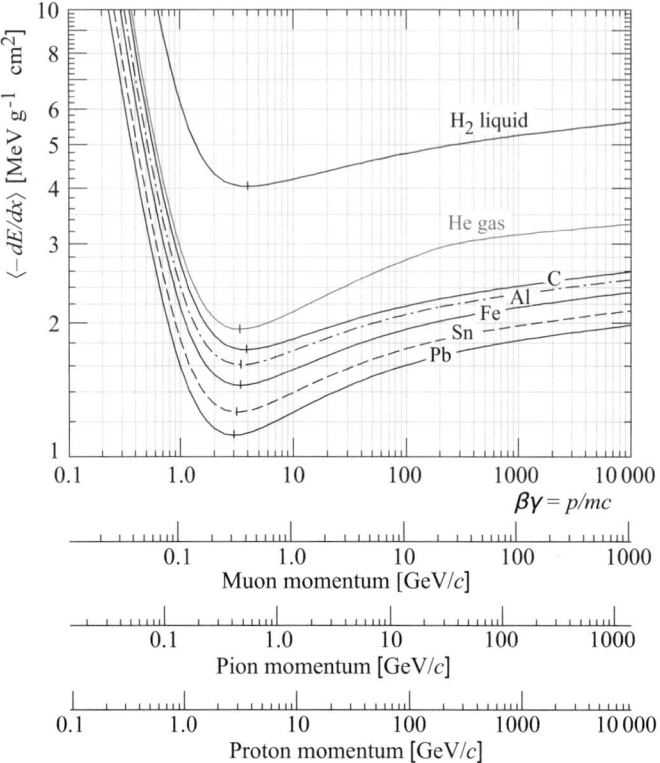

Fig. 2.9 Average energy loss per density and unit length in liquid hydrogen, gaseous helium, carbon, aluminum, iron, tin and lead plotted in a double logarithmic scale. Radiation losses are not shown. Reprinted under CC-BY-NC-4.0 Licence with kind permission from [1], © 2024, the Particle Data Group LBNL

$\langle dE/dx \rangle_{min}$ decreases logarithmically with Z. Figure 2.10 shows $\langle dE/dx \rangle_{min}$ for gases and solids as a function of the atomic number. The parameterization

$$\langle -dE/dx \rangle_{min} = 2.35 - 0.28 \ln(Z) \text{ [MeV cm}^2\text{/g]} \tag{2.15}$$

fits the calculated MIP values rather well, except for low-Z elements ($Z < 6$). This is amazing since the atomic structure is rather different in these atomic elements.

Figure 2.11 (left) shows calculations and measurements of the average energy loss normalized to the energy loss at the minimum versus $\beta\gamma$ for hadrons in an argon-methane gas mixture (95:5). By comparing the stopping-power curves in different materials, we arrive at the following conclusions.

- At low values of $\beta\gamma$, $\langle dE/dx \rangle$ decreases as $1/\beta^{1.4}$ to $1/\beta^{1.7}$. The $1/\beta^2$ term, which results from the non-relativistic Rutherford scattering, is softened by the logarithm in the first term.

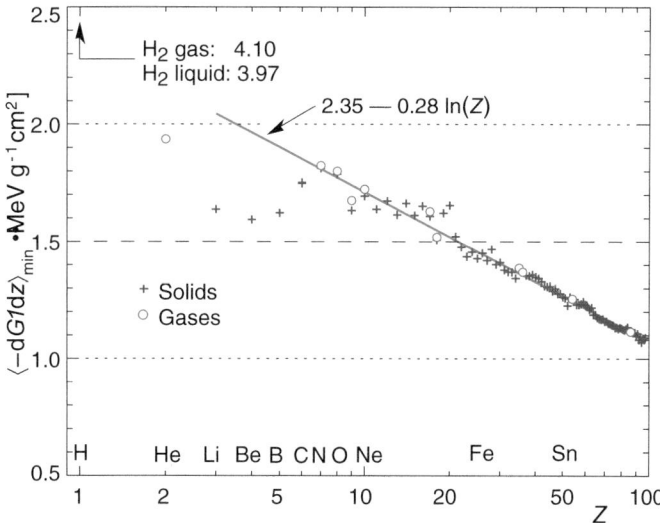

Fig. 2.10 Average mass stopping power at minimum ionization as a function of the atomic number for gases (circles) and solids (plus signs). The solid line represents a fit for $Z > 6$ with the function shown in (2.15). Reprinted under CC-BY-NC-4.0 Licence with kind permission from [1], © 2024, the Particle Data Group LBNL

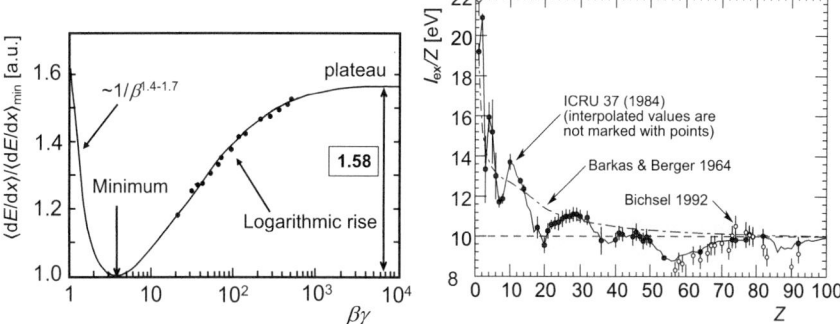

Fig. 2.11 Left: Average ionization loss divided by the minimum ionization loss of hadrons in an argon-methane gas mixture (95:5) at normal density as a function of $\beta\gamma$ for measurements (solid points) [34] and the photoabsorption ionization model [24] (solid line). Reprinted and adapted with kind permission from [34], © 1978, Elsevier. All rights reserved. Right: Average excitation energies divided by the atomic number as a function of Z [36]. Black points with error bars show measurements. The solid line represents an interpolation between data. The open circles show more recent calculations by Bichsel [37]. The black and grey points at $Z = 1$ are for gaseous and liquid H_2, respectively. The dash-dotted line is the approximation by Barkas and Berger [38]. Reprinted under CC-BY-NC-4.0 Licence with kind permission from [1], © 2024, the Particle Data Group LBNL

- A broad minimum lies around $\beta\gamma \simeq 3 - 3.5$, where lower values are found for higher Z materials.
- Above the minimum, we encounter the logarithmic rise towards the plateau, which is reduced due to the density effect.[4]
- For liquids and solids the plateau lies about 10–20% above the minimum while for gases it is increased to about 50–60%.

The data [34] are well described by the density effect computed in the photoabsorption ionization model [24]. The density correction given by the original Sternheimer parameterization was too small yielding a $\approx$ 10% too high stopping power [18,35]. Note that two independent mechanisms contribute to the relativistic rise. The explicit $\beta^2\gamma^2$ dependence via the relativistic flattening and extension of the particles' electric field produces two thirds of the rise. At high energies polarization of medium prevents ionization at greater distances. Thus, the increase in stopping power is reduced. This is accounted for in the density effect correction discussed below. The remaining one third comes from the $\beta^2\gamma^2$ dependence of $W_{\max}$. Hard collision events increase the tail of the energy loss distribution, which increases the mean energy loss without affecting the most probable energy loss.

The mean excitation energy I_{ex} increases with the atomic number Z. A simple parameterization is given by [38],

$$I_{\mathrm{ex}} = I_0 \cdot Z + I_c \quad (Z < 13) \qquad \text{with } I_0 = 12\,\mathrm{eV},\ I_c = 7.0\,\mathrm{eV}, \quad (2.16)$$
$$I_{\mathrm{ex}} = I_0 \cdot Z + I_1 \cdot Z^{-0.19} \ (Z \geq 13) \text{ with } I_0 = 9.8\,\mathrm{eV},\ I_1 = 58.8\,\mathrm{eV}.$$

Figure 2.11 (right) shows the individual mean excitation energies divided by the atomic number as a function of Z. The Barkas parameterization is shown by the dash-dotted curve. Since the structure of individual shells is not taken into account, this simple parameterization is not very accurate for some atoms. Note that I_{ex} can be obtained from the photoabsorption ionization cross section by [39],

$$\ln I_{\mathrm{ex}} = \int \sigma_\gamma(W) \ln W \,\mathrm{d}W. \quad (2.17)$$

Nowadays, the average excitation energies have been estimated for many materials using stopping-power measurements of protons, deuterons and α particles as well as oscillator strength distributions and dielectric response functions. The results are compiled in a report [40]. Table 2.1 lists mean excitation energies plus some other properties for some standard gases. Gas densities are given at normal conditions. To obtain densities of gases at other temperatures we can use the ideal gas equation,

$$PV_g = nRT = \frac{m}{m_m}RT, \quad (2.18)$$

---

[4] The density effect can be understood as a consequence of the transverse electromagnetic field of the moving charge reaching neighbor atoms. For high-density media, this occurs at lower particle velocities than for gases.

2.1 Passage of Heavy Charged Particles Through Matter

Table 2.1 Properties of gases at normal conditions: atomic number (Z), ratio (Z/A), density (ρ) at normal temperature and pressure ($T = 20°C$, $P=1$ atm), minimum energy for excitation (E_{ex}), minimum energy for ionization per electron (E_{ion}), ionization potential $I_0 = I_{ex}/Z$, energy loss (W_{ion}) per ion pair produced, energy loss at the minimum ($\langle dE/dx \rangle_{min}$), number of primary ions (n_p) per centimeter of path for minimum ionizing particles and total number of ion pairs (n_T). Reprinted under CC-BY-3.0 Licence with kind permission from [10], © 1977–2024, CERN. All rights reserved

Gas	Z	Z/A mol/g	$\rho \times 10^5$ [g/cm^3]	E_{ex} [eV]	E_{ion} [eV]	I_0 [eV]	W_{ion} [eV]	$\langle \frac{dE}{dx} \rangle_{min}$ [$\frac{MeV cm^2}{g}$]	$\langle \frac{dE}{dx} \rangle_{min}$ [$\frac{keV}{cm}$]	n_p [$\frac{1}{cm}$]	n_T [$\frac{1}{cm}$]
H$_2$	2	1	8.38	10.8	15.9	15.4	37	4.03	0.343	5.2	9.2
He	2	0.4997	16.6	19.8	24.5	24.6	41	1.937	0.32	5.9	7.8
N$_2$	14	0.49976	117	8.1	16.7	15.58	35	1.68	1.96	10	56
O$_2$	16	0.5000	133.	7.9	12.8	12.2	31	1.69	2.26	22	73
Ne	10	0.4955	83.9	16.6	21.5	21.6	36.3	1.68	1.41	12	39
Ar	18	0.4506	166.2	11.6	15.7	15.8	26	1.47	2.44	29.4	94
Kr	36	0.4296	349.0	10.0	13.9	14.0	24	1.32	4.60	22	192
Xe	54	0.4113	548	8.4	12.1	12.1	22	1.23	6.76	44	307
CO$_2$	22	0.49989	184	5.2	13.7	13.7	33	1.62	3.35	35.5	91
CH$_4$	10	0.62334	66.7	9.8	15.2	13.1	28	2.08	1.61	25	53
C$_2$H$_6$	18	0.59861	134	8.7	11.7	11.5	27.0	0.86	2.92	41	111
C$_4$H$_{10}$	34	0.59497	249	6.5	10.6	10.8	23	2.29	5.67	90	220
CF$_4$	42	0.47721	378	10	16	16.2	54	1.78	6.38	51	100
DME	26	0.5652	393	6.4	10	10.03	23.9	1.77	3.9	55	160

where P is the pressure, V_g is the volume, $n = m/m_m$ is the ratio of the mass of the gas divided by the mass of one mol, R is a constant ($8.314 J/(K \cdot mol)$) and T is the temperature. Since the density is defined as $\rho = m/V_g$, we obtain the scaling law,

$$\rho(T_2) = \rho(T_1) \frac{T_1}{T_2}, \tag{2.19}$$

where T_1 and T_2 are two temperatures.

When a charged particle traverses a medium, it produces electron-ion pairs in each collision. The sum of electron-ion pairs is called the number of primary electron-ion pairs (n_P). The emitted electrons may carry sufficient kinetic energy to produce additional electron-ion pairs. The sum of primary and additional electron-ion pairs is called the total number of electron-ion pairs (n_T). While the number of primary electron-ion pairs is difficult to determine, the total number of electron-ion pairs is simply given by

$$n_T = \frac{\Delta E}{W_{ion}}, \tag{2.20}$$

where $\Delta E = W$ is the energy loss of the charged particle per unit length and W_{ion} is the average energy needed to produce an electron-ion pair. The values given for

n_P in Table 2.1 are measurement averages for minimum ionizing particles per unit length at normal conditions. Uncertainties are of the order of 20% to 30%.

The shell correction becomes important at low energies, where it accounts for particle momenta that are of similar size as the electron momenta in the atomic cloud. Two independent approaches, the hydrogen wave function by Bichsel [41] and the local density approximation by Ziegler [32,33] based on the Lindhard-Winther formalism [31], yield similar results. An empirical formula for the shell correction was proposed by Barkas [38],

$$C(I_{\text{ex}}, \beta\gamma) = \left[\frac{0.422377}{(\beta\gamma)^2} + \frac{0.0304043}{(\beta\gamma)^4} - \frac{0.00038106}{(\beta\gamma)^6}\right] \times 10^{-6} I_{\text{ex}}^2 \, [\text{eV}^2] \quad (2.21)$$
$$+ \left[\frac{3.850190}{(\beta\gamma)^2} - \frac{0.1667989}{(\beta\gamma)^4} + \frac{0.00157955}{(\beta\gamma)^6}\right] \times 10^{-9} I_{\text{ex}}^3 \, [\text{eV}^3].$$

For example, for a 30 MeV/c proton traversing aluminum the shell correction is 0.6%. It increases to 9.9% for a proton with momentum of 0.3 MeV/c.

Originally, Swann [42] and Fermi [43] proposed the density correction. Their work was extended by Bohr [44], Sternheimer [19,45] and Crispin and Fowler [46]. With increasing particle energy the electric field of the heavy charged particle flattens and extends so that contributions from distant collisions need to be included. These increase as $\ln(\beta\gamma)$. Since the medium becomes polarized, the field extension is limited leading to a truncation of the relativistic rise. The density effect has been studied extensively [35,47,48]. The approach is to subtract a correction δ from distant-interaction contributions. At very high energies, we get [24],

$$\frac{\delta(\beta\gamma)}{2} \to \ln\left(\frac{\hbar\omega_p}{I_{\text{ex}}}\right) + \ln(\beta\gamma) - \frac{1}{2}, \quad (2.22)$$

where ω_p is the plasma frequency of the medium, which depends on the electron density N_e, electron radius $r_e = 2.818 \times 10^{-15}$ m, electron mass and the fine structure constant α,

$$\hbar\omega_p = \sqrt{4\pi N_e r_e^3 \frac{m_e c^2}{\alpha}} = \sqrt{4\pi N_A r_e^3 \rho \langle Z/A\rangle \frac{m_e c^2}{\alpha}} = 28.81 \, eV \sqrt{\rho\left[\frac{g}{\text{cm}^3}\right] \langle Z/A\rangle \left[\frac{\text{mol}}{g}\right]}. \quad (2.23)$$

In order to obtain the correct unit of eV, the density needs to be inserted in units of g/cm^3 and the atomic weight in units of g/mol. The other quantities are the same as those in (2.6). For materials like styrene or similar, $\hbar\omega_p \simeq 20$ eV while for air it is 0.7 eV.

Inserting (2.22) into (2.13) eliminates the explicit $\beta^2\gamma^2$ dependence in the first log term in the square bracket. The remaining relativistic rise in the stopping power comes from E_{max} defined in (2.4). The effect of the relativistic rise is indicated in Fig. 2.8. Since the plasma frequency is proportional to the square root of the density, the correction is much larger in liquids and solids than in gases. Note that at low energies, the density effect plays no significant role in nonconductors while

2.1 Passage of Heavy Charged Particles Through Matter

at rather high energies it is well represented by the parameterization given in (2.22). Sternheimer proposed the following parameterization [18],

$$\delta(x_\beta) = \begin{cases} 4.605 \cdot x_\beta - \bar{C} & \text{for } x_\beta > x_1, \\ 4.605 \cdot x_\beta - \bar{C} + a(x_1 - x)^k & \text{for } x_0 < x_\beta < x_1, \\ \delta_0 \cdot 10^{2(x_\beta - x_0)} & \text{for } x_\beta < x_0, \end{cases} \quad (2.24)$$

where $x_\beta = \log_{10}(\beta\gamma)$. The parameter $\bar{C}$ is obtained by equating (2.24) with the limit set in (2.22), yielding

$$\bar{C} = 2\ln(I_{\text{ex}}/\hbar\omega_p) + 1. \quad (2.25)$$

The parameters x_0, x_1, a and δ_0 are free and have to be determined for each material. Sternheimer determined the density parameters for nearly 300 different materials. The agreement with more detailed calculations or results obtained with other parameter sets is usually at the 0.5% level [49]. Examples are listed in Table 2.2. Sternheimer and Peirls have proposed an algorithm to determine the density-effect parameters for materials that have not been tabulated [35, 50].

The average stopping power given in (2.13) holds for atoms. However, typically we deal with compounds or materials that consist of different atoms or molecules. The stopping power for a mixture or compound that is made of i different atoms in which the ith type of atom occurs n_i times is given by

$$\left\langle \frac{dE}{dx} \right\rangle = \sum_i w_i \left\langle \frac{dE_i}{dx} \right\rangle. \quad (2.26)$$

The weight factor is defined by

$$w_i = \frac{n_i A_i}{\sum_k n_k A_k}, \quad (2.27)$$

where A_i is the atomic weight of element i in units of [g/mol] and n_i is the numerical factor of how frequently the element appears in the compound. For example, if we want to determine the stopping power of ammonia (NH_3) we have $n_H = 3$ and $n_N = 1$ yielding $w_H = 0.176$ and $w_N = 0.824$. Then, we compute the average stopping power for hydrogen and nitrogen gas and multiply them by the weights yielding $\langle -dE/dx \rangle = 2.227$ MeVcm^2/g. This agrees with the average ionization loss in ammonia ($\langle -dE/dx \rangle = 2.265$ MeVcm^2/g) listed in [1] within about 1.7%.

Note that it is experimentally difficult to measure the average energy loss exactly since the individual energy loss is drawn from an asymmetric distribution with a long tail towards large values. Large energy losses are rare but impose larger weights in the averaging procedure. Therefore, we typically use the most probable energy loss instead, which we introduce in the next Sect. 2.1.2.

Table 2.2 The ratio Z/A, density, mean effective excitation energy per electron and fit parameters of the Sternheimer density parametrization for different materials [47]. The values for carbon refer to graphite. †The density for gases is in [mg/cm³]

Material	Z/A	ρ [g/cm³]	I_{ex} [eV]	$-\bar{C}$	a	k	x_1	x_0	δ_0
Li	0.43221	0.534	40.0	3.1221	0.95136	2.4993	1.6397	0.1304	0.14
C	0.49954	2.0	78.0	2.9925	0.20240	3.0036	2.486	−0.0351	0.10
Mg	0.49373	1.74	156	4.5297	0.08163	3.6166	3.0668	0.1499	0.08
Al	0.48181	2.699	166	4.2395	0.08024	3.6345	3.0127	0.1708	0.12
Si	0.49848	2.33	173	4.435	0.14921	3.2546	2.8715	0.2014	0.14
P	0.48428	2.20	173	4.5214	0.2361	2.9158	2.7815	0.1696	0.14
Mn	0.45506	7.44	272	4.2702	0.14973	2.9796	3.1074	0.0447	0.14
Fe	0.46556	7.874	286	4.2911	0.1468	2.9632	3.1531	−0.0012	0.12
Cu	0.45636	8.96	322	4.419	0.14339	2.9044	3.2792	−0.0254	0.08
Ga	0.44464	5.90	334	4.9353	0.0944	3.1314	3.5434	0.2267	0.14
Ge	0.44083	5.323	350	5.1411	0.07188	3.3306	3.6096	0.3376	0.14
As	0.44046	5.73	347	5.0510	0.06633	3.4176	3.5702	0.1767	0.08
W	0.4025	19.30	727	5.4059	0.15509	2.8447	3.4960	0.2167	0.14
Au	0.40108	19.3	790	5.5747	0.09756	3.1101	3.6979	0.2021	0.14
Pb	0.39575	11.35	823	6.2018	0.09359	3.1608	3.8073	0.3776	0.14
U	0.38651	18.95	890	5.8694	0.19677	2.8171	3.3721	0.226	0.14
$H_2^\dagger$	0.99216	0.08375	19.2	9.5835	0.14092	5.7273	3.2718	1.8639	0.0
$N_2^\dagger$	0.49976	1.1653	82	10.5400	0.15349	3.2125	4.1323	1.7378	0.0
$O_2^\dagger$	0.50002	1.3315	95	10.7004	0.11778	3.2913	4.3213	1.7541	0.0
Ar†	0.45059	1.662	188	11.948	0.19714	2.9618	4.4855	1.7635	0.0
Kr†	0.42959	3.4783	352	12.5115	0.07446	3.4051	5.0748	1.7158	0.0
Xe†	0.41130	5.4854	482	12.7281	0.23314	2.7414	4.7371	1.5630	0.0
H_2O	0.55509	1	75	3.5017	0.09116	3.4773	2.8004	0.2400	0.0
Air†	0.49919	1.2048	85.7	10.5961	0.10914	3.3994	4.2759	1.7418	0.0
Methane†	0.62334	0.66715	41.7	9.5243	0.09253	3.6257	3.9716	1.6263	0.0
$CO_2^\dagger$	0.49989	1.842	85	10.1537	0.11768	3.3227	4.1825	1.6294	0.0
Lucite	0.53937	1.190	74.0	3.3297	0.11433	3.3836	2.6681	0.1824	0.0
Freon-12	0.47968	1.12	143	4.8251	0.07978	3.4626	3.2659	0.3035	0.0
Freon-13	0.47866	0.95	126.6	4.7483	0.07238	3.5551	3.2337	0.3659	0.0
Anthracene	0.5274	1.283	69.5	3.1514	0.14677	3.2831	2.5213	0.1146	0.0
CsI	0.41569	4.510	553.1	6.2807	0.25381	2.6657	3.3353	0.0395	0.0
NaI	0.42697	3.667	452	6.0572	0.12516	3.0398	3.592	0.1203	0.0
BGO	0.42065	7.13	534.1	5.7409	0.09569	3.0781	3.7816	0.0456	0.0
Pl. scint.	0.54141	1.032	64.7	3.1997	0.16101	3.2393	2.4855	0.1464	0.0
Concrete	0.50274	2.3	135.2	3.9464	0.07515	3.5467	3.0466	0.1301	0.0
Lead glass	0.42101	6.22	526.4	5.8476	0.09544	3.074	3.8146	0.0614	0.0

2.1.2 Instantaneous Energy Loss Distribution

The Bethe-Bloch equation predicts the average energy loss of a heavy charged particle. However, the energy loss distribution is not symmetric since it is possible to transfer a very large amount of energy to an atomic electron in a single collision. Typically, such large energy losses are caused by collisions at high-momentum transfers leading to δ-rays. Though the probability for large energy transfers is rather small, such processes carry a large weight in the average. Thus, the most probable energy loss is typically much smaller than the mean energy loss. This was first noticed by Landau [15]. For practical applications, it is also much easier to determine the most probable energy loss than the average energy loss.

Let us start to consider a thin absorber, where the number of collisions is small. Since all the energy may be lost in a single interaction, this case is hard to calculate. For example, for a 1 GeV/c pion about 5% of the energy may be lost in a single interaction, while for electrons up to half the energy may be lost. These events are rare but introduce a long tail on the high-energy side of energy-loss probability distribution.

Landau [15] performed the basic calculations of the instantaneous energy loss. The distinguishing parameter is

$$\kappa = \xi_E / W_{\max}, \tag{2.28}$$

where ξ_E is obtained from the Bethe-Bloch equation by just using the first multiplicative term ignoring the logarithm,

$$\langle \Delta E \rangle \simeq \xi_E = \frac{K}{2} \frac{Z}{A} \left(\frac{z}{\beta} \right)^2 x. \tag{2.29}$$

Note that the thickness x has units of g/cm^2. A thin absorber is one for which $\kappa < 0.01$. For absorbers of moderate thickness, we have $0.05 \leq \kappa \leq 0.1$. Absorbers with larger values of κ are thick. For $\kappa > 1$, the distribution approaches a Gaussian distribution. Landau's theory typically holds for thin absorbers. His calculations are based on the Rutherford cross section in which the atomic binding energy is approximated by I_{ex}. Landau made three assumptions:

(i) The maximum energy transfer permitted is $W_{\max} \to \infty$ when $\kappa \to 0$.
(ii) Individual energy transfers are sufficiently large such that the electron may be treated as free; small energy transfers from distant collisions are ignored.
(iii) The decrease in particle velocity is neglected assuming that the particle velocity is constant.

The probability density function (PDF) for the energy-loss is defined by

$$f(x, \Delta E) = \frac{\phi(\upsilon)}{\xi_E}, \tag{2.30}$$

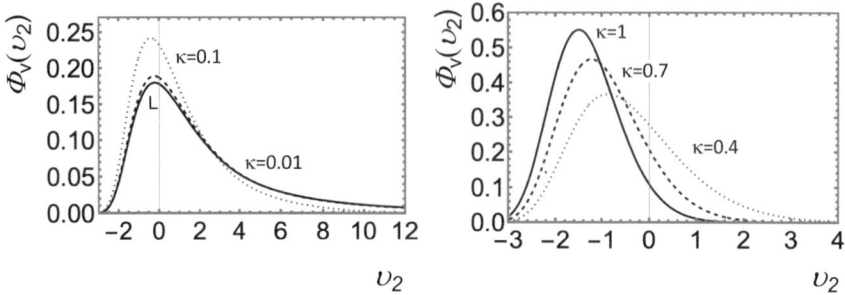

Fig. 2.12 Energy loss distribution of charged particles as a function of v in a thin absorber for small values of κ (left) and for large values of κ (right). The solid curve denoted by L is the Landau distribution ($\kappa \to 0$). The dashed curve above is for $\kappa = 0.01$

where

$$\phi(v) = \frac{1}{\pi} \int_0^\infty \exp(-u \ln u - uv) \sin(\pi u) du. \tag{2.31}$$

Note that $\phi(v)$ is a universal function that depends only on v, which must be evaluated numerically. The parameter v can be expressed by [51],

$$v = \frac{\Delta E}{\xi_E} - \ln \xi_E + \left[\ln \frac{(1-\beta^2)I_{ex}^2}{2mc^2\beta^2} + \beta^2 - 1 + 0.577 \right], \tag{2.32}$$

where ΔE is the actual energy loss and x is the material thickness.

Vavilov extended Landau's theory to intermediate values of κ by varying assumption 1 to account for the correct expression of the maximum allowable energy transfer [51,52]. The Vavilov universal functions depend on three parameters and are given by,

$$\phi_v(v_2, \kappa, \beta^2) = \frac{\kappa}{\pi} \exp[\kappa(1 + \beta^2 \cdot 0.577)] \int_0^\infty \exp(\kappa \cdot f_1(u)) \cos(u \cdot v_2 + \kappa f_2(u)) du, \tag{2.33}$$

where

$$f_1(u) = \beta^2 (\ln u - \text{Ci}(u)) - \cos u - u \cdot \text{Si}(u), \tag{2.34}$$

$$f_2(u) = u \cdot (\ln u - \text{Ci}(u)) + \sin u + \beta^2 \cdot \text{Si}(u), \tag{2.35}$$

where $\text{Ci}(u)$ and $\text{Si}(u)$ are the cosine and sine integral functions

$$\text{Ci}(u) = -\int_u^\infty \frac{\cos t \, dt}{t} \tag{2.36}$$

$$\text{Si}(u) = \int_0^u \frac{\sin t \, dt}{t}, \tag{2.37}$$

and $v_2 = \kappa \cdot v + \kappa \ln \kappa$. Figure 2.12 (left) shows the Landau universal function in comparison to the Vavilov universal functions for small values of κ. The Landau universal function is obtained for $\kappa \to 0$. As we see, for $\kappa = 0.01$ the Vavilov universal function already approaches the Landau universal function. The Vavilov PDF is given by

$$f(x, \Delta E) = \frac{\phi_v(v_2, \kappa, \beta^2)}{\xi_E}. \quad (2.38)$$

Figure 2.12 (right) shows Vavilov universal functions for larger values of κ. For $\kappa > 1$, the Vavilov PDF approaches a Gaussian function. In the Gaussian limit, the variance approaches $\sigma^2 = \frac{\langle \Delta E \rangle^2}{\kappa} \frac{1-\beta^2}{2}$ (see Sect. 2.1.4).

Figure 2.13 (left) shows the energy loss distribution for 10 GeV/c muons traversing 1.7 mm of silicon for the Landau-Vavilov theory and the Bethe-Fano-Bichsel theory. The Landau-Vavilov function uses a Rutherford cross section without atomic binding corrections but with a kinetic energy transfer limit of $W_{\max}$. This energy loss is equivalent to that of a heavy charged particle traversing 3.0 mm of scintillator. The PDFs are rather asymmetric and the two predictions lie close to each other. For thicker absorbers, the Landau-Vavilov theory provides accurate results. The most probable energy loss lies at $\Delta E_{mp} \simeq 0.525$ MeV while the mean energy loss is $\langle \Delta E \rangle \simeq 0.85$ MeV. So in this example, the most probable energy loss is 62% of the mean. Experimentally, it is quite easy to measure ΔE_{mp} even with a limited number of dE/dx measurements, while it is rather difficult to determine the average $\langle \Delta E \rangle$ even with a large data set due to the rare events with very high energy losses. Therefore, it is more practical to use ΔE_{mp} rather than $\langle \Delta E \rangle$. The full-width-half-maximum (FWHM) of the Landau-Vavilov distribution is about $4\langle \Delta E \rangle$ for detectors of moderate thickness.

Figure 2.13 (right) shows the measured energy loss distribution in an Ar-CO_2 gas mixture in comparison to the Landau prediction and a simple Gaussian distribution expected for the same average energy loss for fast heavy particles. It is obvious that the Gaussian model fails. Even increasing the counter thickness does not help since the number of energetic electrons will increase. Note that in these collisions energetic δ-rays may be produced, which are responsible for the large fluctuations in the energy loss distribution that produce the high-energy tail. Though the Landau distribution provides a much better description of the data than the simple Gaussian function, it is a bit too narrow. Note that the most probable energy loss and average energy loss are also different, though they lie closer together than in Fig. 2.13 (left). In practice, it is difficult to determine the average energy loss since a very large number of measurements is necessary whereas the most probable energy loss can be extracted from a few measurements.

A practical form of Landau-like functions that is useful for data fits is given by

$$\phi(v; \tilde{\mu}, w) = \frac{1}{\pi w} \int_0^\infty \exp(-t') \cos\left(t'\left(\frac{v - \tilde{\mu}}{w}\right) + \frac{2t'}{\pi} \ln\left(\frac{t'}{w}\right)\right) dt', \quad (2.39)$$

where $\tilde{\mu}$ is a location parameter and w is a scale parameter. For $\tilde{\mu} = 0$ and $w = \pi/2$, we recover the Landau distribution depicted in Fig. 2.14.

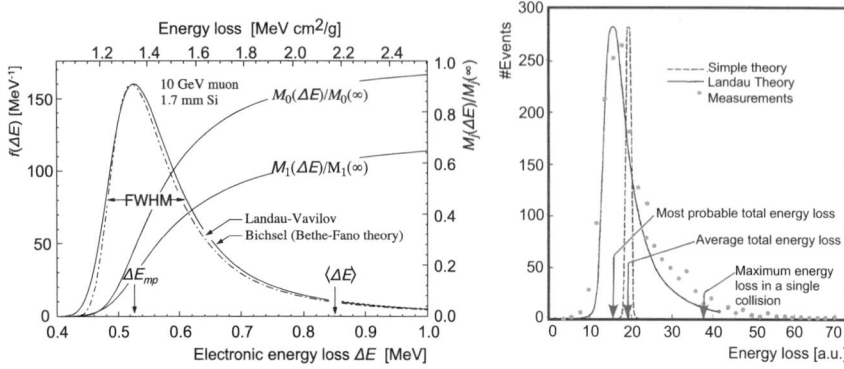

Fig. 2.13 Left: Energy loss distribution of a 10 GeV/c muon (left scale) traversing 1.7 mm of silicon calculated with the Landau-Vavilov theory (dash-dotted line) and with the Bethe-Fano-Bichsel theory (solid line) [1]. The functions $\mathcal{M}_0(\Delta E)$ and $\mathcal{M}_1(\Delta E)$ are, respectively, the cumulative zeroth (mean number of collisions) and first moments (mean energy loss) in crossing the silicon (right scale). The most probable energy loss and the average energy loss are indicated. Reprinted under CC-BY-NC-4.0 Licence with kind permission from [1], © 2024, the Particle Data Group LBNL. Right: The measured energy loss distribution of 31.5 MeV protons in a proportional counter filled with an Ar-CO_2 gas mixture in comparison to a simple Gaussian model of the thin counter and a Landau distribution predicted by Symon [53]. The width of the Gaussian distribution is determined by statistics of the number of ion pairs formed at 25 eV per electron-ion pair. The most probable energy loss and the average energy loss are marked. Reprinted with kind permission from [54], © 1952, AIP Publishing. All rights reserved

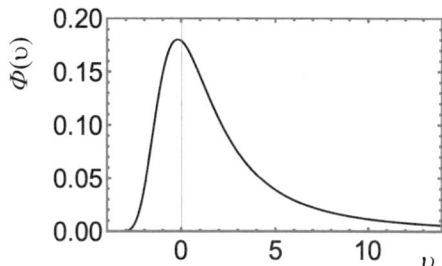

Fig. 2.14 Landau distribution parameterization shown in (2.39) for $\tilde{\mu} = 0$ and $w = \pi/2$

Corrections to the Landau distribution were performed by Blunk and Leisegang [55] by including approximate atomic shell corrections and by Allison and Cobb [24] who refined the procedure to include details of atomic structure. Similar corrections to the Vavilov [52] distributions were performed by Shulek [56]. Figure 2.15 shows the measured energy loss distributions of 3 GeV/c pions (left) and 3 GeV/c electrons (center) in a 1.5 cm thick layer of an argon-methane gas mixture (93:7) at normal conditions [57]. The dashed histogram shows the prediction of the photoabsorption ionization (PAI) model by Allison and Cobb while the solid histogram represents the prediction of Smirnov [57] who extended the PAI model. Both predictions agree well with the pion and electron energy loss measurements. Figure 2.15 (right) shows the energy loss distribution of 80 MeV protons in 15 cm path length in an argon-methane gas mixture (93:7) at NTP. The prediction

2.1 Passage of Heavy Charged Particles Through Matter

of Blunck and Leisegang represents the data well while the Vavilov prediction is off by $\sim 10\%$ [58]. The models for calculating the energy loss distributions of heavy charged particles are quite sophisticated and produce measurements rather well. This includes the modeling of the relativistic rise [47]. Detailed tables for about 300 media have been produced (e.g. see [1]). These calculations are implemented into the GEANT4 Monte Carlo program [59], which is used for simulations of detector responses.

Figure 2.16 (left) shows the energy loss distribution of 2 GeV/c positrons traversing a silicon detector of 32 μm thickness. In thin detectors we encounter different effects. The energy loss may be transferred to a single electron producing a δ-electron, which may leave the detector and thus does not deposit all its energy. The energy loss is of the order of binding energies or smaller. So, it becomes important to account for atomic details. In addition, collective excitations need to be accounted for, which are the major interaction. Bichsel included these effect in the Bethe-Fano theory, which describes the data reasonably well. The deviation between measurement and prediction for $\Delta E < 4$ keV is probably due to a loss of detector sensitivity near the edges. Bak who included the binding effects, produces the correct most probable energy loss [60]. However, the agreement with all the data is limited. The Landau-Vavilov theory produces a much too narrow probability density function. This problem was also noticed for thin gas cells in chambers.

Figure 2.16 (right) shows the energy loss distribution of 2.1 GeV/c protons traversing a path of 5 cm of argon gas. While the calculation by Talman represents the data well [61], which is a correction to the Landau distribution similar to that of Blunck and Leisegang, the Landau theory fails. The quantity $\Delta E_{mp}/x$ may be calculated adequately using (2.40), but the observed distribution is significantly wider than the Landau width of $4\langle\Delta E\rangle$. For detectors of moderate thickness like scintillators or cells of liquid argon (LAr), the highly skewed Landau-Vavilov distribution predicts the energy loss PDF reasonably well [15,52], while it does not for thin silicon and thin gas detectors. Bichsel made an empirical approach assuming that the ionization in silicon is proportional to the energy deposited by a charged particle [4].

For very thin layers of material, the number of collisions is small and the probability for a large energy loss is reduced. Thus, the average energy loss may be replaced by the most probable energy loss ΔE_{mp} [4], which is parameterized as[5]

$$\Delta E_{mp} = \xi_E \left[\ln\left(\frac{2m_e c^2 \beta^2 \gamma^2}{I_{ex}}\right) + \ln\frac{\xi_E}{I_{ex}} + j_{mp} - \beta^2 - \delta(\beta\gamma) \right], \qquad (2.40)$$

where $j_{mp} = 0.2$.[6] Note that while $\langle dE/dx \rangle$ is independent of thickness x, $\Delta E_{mp}/x$ scales as $f(x) = a \ln x + b$. In the original work by Landau and Vavilov, the density

---

[5] We have neglected the small shell correction here.
[6] Different authors give different values for j_{mp} [6,61]. However, the most probable energy loss is not sensitive to its value.

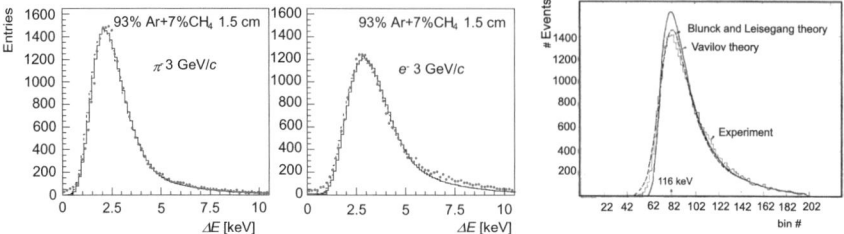

Fig. 2.15 Left: Energy loss distributions of 3 GeV pions. Center: Energy loss distributions of 3 GeV electrons. Both reprinted with kind permission from [57], © 2005, Elsevier. All rights reserved. Right: Energy loss distributions of 80 MeV protons. All particles pass through an argon-methane gas mixture (93:7) at NTP. The path lengths for pions and electrons are 1.5 cm while that for protons is 15 cm. Solid points show data for pions and electrons, while the corresponding dashed and solid histograms show predictions of the PAI model by Allison and Cobb and the extended PAI model (PAIR) by Smirnov, respectively. For protons, data are shown by the solid histogram while predictions by Vavilov and Blunck-Leisegang are represented by solid and dashed lines, respectively. Reprinted with kind permission from [58], © 1967, Elsevier. All rights reserved

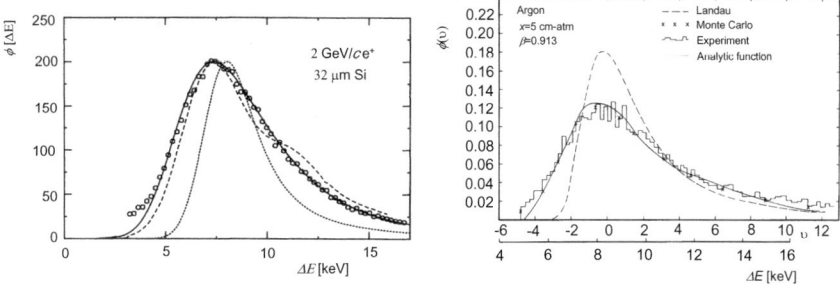

Fig. 2.16 Left: Energy loss distribution of 2 GeV/c positrons traversing a silicon detector of 32 μm thickness (open circles) in comparison to the Bethe-Fano-Bichsel theory (solid line) [4], the Vavilov theory convolved with electronic noise (dotted curve) and a theory function calculated by Bak (dashed line) [60]. Reprinted with kind permission from [4], © 1988, AIP Publishing. All rights reserved. Right: Observed energy loss distribution of 2.1 GeV/c protons traversing a path of 5 cm of argon gas (histogram [62]) in comparison to a simulation (x symbols) [63], Landau's theory (dashed line) [15] and an analytic calculation by Talman (solid line) [61]. The energy loss is expressed in terms of Landau's function versus both Landau's variable υ and the energy loss. Reprinted with kind permission from [61], © 1979, Elsevier. All rights reserved

effect correction was not included. It was later added by Bichsel [4]. The high-energy behavior of $\delta(\beta\gamma)$ is such that ΔE_{mp} reaches a Fermi plateau,

$$\Delta E_{mp} = \xi_E \left(\ln \frac{2m_e c^2 \xi_E}{(\hbar \omega_p)^2} + 0.2 \right), \quad (2.41)$$

where the energy loss of the particle no longer increases as the energy increases.

In summary, the average energy loss is given by the Bethe-Bloch equation ((2.13)). However, the most probable energy loss is more useful for many applications. For the instantaneous energy loss the Landau-Vavilov theory works well for detectors of moderate thickness like scintillators or liquid argon cells. Even better work the extensions by Blunck and Leisegang, Shulek, the photoabsorption ionization model by Allison and Cobb or Smirnov since they use cross sections that account for atomic structures. For thin silicon detectors or thin gas detector cells the Bethe-Fano-Bichsel theory represents the data well underlining that is important to account for details in the atomic structure and for δ-electrons. For thick absorbers, the energy loss distribution becomes Gaussian as discussed in Sect. 2.1.4.

2.1.3 Restricted Energy Loss

If we restrict the energy transfer of an ionizing particle to $E_k \leq W_c \leq W_{\max}$ where E_k is the kinetic energy, we get a restricted energy loss rate of

$$-\frac{dE}{dx}\bigg|_{E_k<W_c} = Kz^2\frac{Z}{A}\frac{1}{\beta^2}\left[\frac{1}{2}\ln\left(2m_ec^2\beta^2\gamma^2\frac{W_c}{I_{\text{ex}}^2}\right) - \frac{\beta^2}{2}\left(1+\frac{W_c}{W_{\max}}\right) - \frac{\delta(\beta\gamma)}{2}\right]. \tag{2.42}$$

For $W_c \to W_{\max}$, (2.42) approaches the Bethe Bloch equation.[7] Since W_c replaces $W_{\max}$ in the logarithmic term of the Bethe Bloch equation, the $\beta\gamma$ term producing the relativistic rise in the close-collision part is replaced by a constant. So, $dE/dx|_{E_k<W_c}$ approaches the constant Fermi plateau. Figure 2.17 illustrates this for two examples of W_c in comparison to Bethe-Bloch and the Landau-Vavilov most probable energy loss (2.40). Since W_c places a requirement on the total mean energy, the most probable energy loss is far more useful in situations where single-particle energy losses are observed.

2.1.4 Energy Loss in Thick Absorbers

In thick absorbers, the number of collisions is large but the total energy loss of all collisions is much smaller than the particle energy ($\Delta E \ll E$). The energy loss distribution becomes a Gaussian distribution. This follows from the Central Limit Theorem.

$$f(x, \Delta E) \propto \exp\left(-\frac{(\Delta E - \langle\Delta E\rangle)^2}{2\sigma^2}\right), \tag{2.43}$$

where x is the thickness of the absorber. For non-relativistic heavy particles we get

$$\sigma^2 = Km_ec^2\rho\frac{Z}{A}x = 0.1569\rho\frac{Z}{A}x\,[\text{MeV}^2] \equiv \sigma_0^2. \tag{2.44}$$

---

[7] The shell correction has been left out.

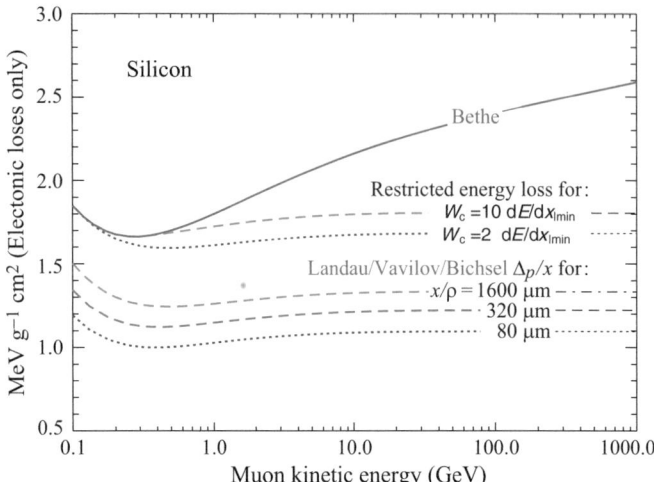

Fig. 2.17 Two examples of restricted energy loss for muons in comparison to the Bethe-Bloch equation and the Landau most probable energy loss per unit thickness in silicon given in (2.42) for three values of x/ρ. The change of $\Delta E_{mp}/x$ with thickness x illustrates the $a \ln x + b$ dependence. The minimum ionization is 1.664 MeVcm^2/g. Radiative losses are not considered. Reprinted under CC-BY-NC-4.0 Licence with kind permission from [1], © 2024, the Particle Data Group LBNL. All rights reserved

For relativistic particles we have to apply a correction factor of

$$\sigma_0 \rightarrow \sigma^2 = \sigma_0^2 \left(\frac{1 - \frac{1}{2}\beta^2}{1 - \beta^2} \right). \tag{2.45}$$

For very thick absorbers the assumption $\Delta E \sim W$ breaks down. The PDF is listed in [64, 65].

2.1.5 Particle Range

A heavy charged particle passing through matter will be slowed down. Once $\beta\gamma$ is below the minimum ionization loss, $\langle -dE/dx \rangle$ will increase rapidly with increasing path length (see Fig. 2.8) due to the $1/\beta^{1.4-1.7}$ dependence below the minimum. This process continues until the particle is slowed down to $\beta\gamma \simeq 0.01$. Here, the photoabsorption cross section becomes smaller because the energy spectrum of the virtual photons becomes softer and the probability for multiple scattering increases. The particle may loose the remaining energy in a few or in several collisions. The energy loss becomes a statistical process that affects the total path length. Thus, the $\langle dE/dx \rangle$ distribution versus penetration depth shows a characteristic shape, known as the Bragg curve, which is depicted in Fig. 2.18 (left). The peak of the Bragg curve corresponds to the peak at low $\beta\gamma$ in Fig. 2.8. Figure 2.18 (right) shows the transmission probability as a function of absorber depth. Due to the statistical

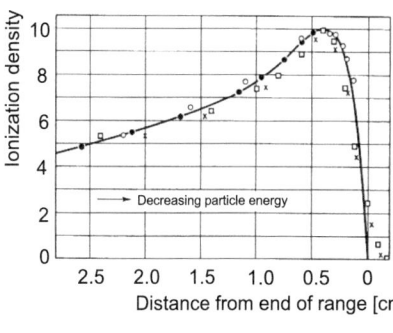 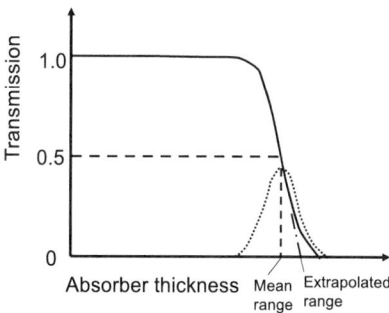

Fig. 2.18 Left: Expected specific ionization of an α particle as a function of penetration depth (solid line) and measurements from Holloway and Livingston (small black points) [66], Feather and Nimmo (open circles) [67], Stetter and Jentschke (squares) [68] and Schulze (crosses) [69]. Reprinted with kind permission from [66], © 1938, AIP Publishing. All rights reserved. Right: Expected transmission of particles as a function of absorber thickness. The depth where the transmission reaches 50% is called mean range. The extrapolated range is obtained by extrapolating the slope of the transmission curve at 50% transmission to zero transmission. Reprinted with kind permission from [70], © 1987, Springer. All rights reserved

fluctuations in the energy loss, the transmission does not drop abruptly but is smeared out. This effect is called straggling. The range where the transmission drops to 50% is called the mean range. The extrapolated range is obtained by extrapolating the slope of the transmission curve at the 50% transmission point to zero transmission (see Fig. 2.18 (right)). The total range is larger than the extrapolated range. At a given depth, the distribution of the penetrating particles is Gaussian (see Fig. 2.18 right) despite the fact that the individual energy loss is non-Gaussian, which again follows from the central limit theorem.

The mean range of a particle with kinetic energy E_k is appropriately written as

$$R(E_k) = R_0(E_{\min}) + \int_{E_{\min}}^{E_k} \left[\left\langle \frac{dE}{dx} \right\rangle\right]^{-1} dE, \qquad (2.46)$$

where $E_{\min}$ is the minimum kinetic energy for which the Bethe-Bloch equation is valid. The range $R_0(E_{min})$ is an empirically determined constant describing the region of straggling.

Figure 2.19 shows the range normalized to mass for different elements, ranging from H_2 liquid, He gas to solids up to lead. The curves, which are calculated numerically by integrating the Bethe-Bloch equation, yield accurate results within a few percent. The shapes can be characterized in the log-log scale by $R \propto E_k^b$ where $b \sim 2$, which follows from the first term in the Bethe-Bloch equation. Fits to the energy loss distribution yield values around $b = 1.4 - 1.7$.

Let us consider an example. A charged kaon with momentum $p = 1.0$ GeV/c traverses a lead shield. The kaon has a mass of $m_{K^+} = 0.494$ GeV/c^2 yielding $\beta\gamma = 2.02$. From Fig. 2.19 we read off a value of $R/m \simeq (810 \pm 20)$ g/cm^{2}GeV^{-1}

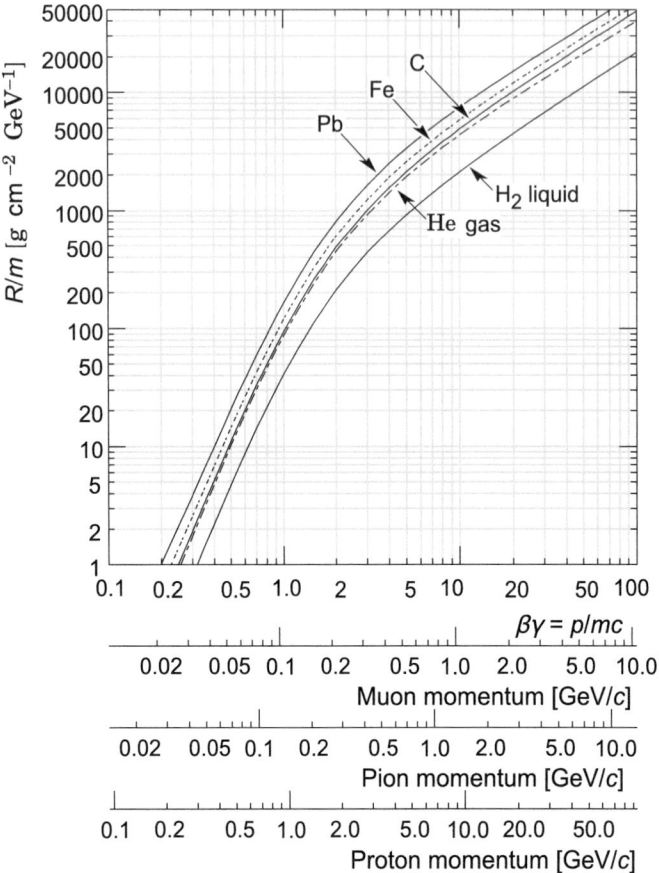

Fig. 2.19 Range per particle mass of heavy charged particles as a function of $\beta\gamma$ in liquid hydrogen, helium gas, carbon, iron and lead in units of [g cm^{-2} GeV^{-1}]. Reprinted under CC-BY-NC-4.0 Licence with kind permission from [1], © 2024, the Particle Data Group LBNL

yielding $R = (400 \pm 10)$ g/cm^2 or (35.1 ± 0.9) cm. It is straight forward to derive the following scaling law. Denoting the masses of two particles by m_2 and m_1 and their charges by z_2 and z_1 the range of particle 2 with kinetic energy E_{k_2} is related to the range of particle 1 with kinetic energy $E_{k_1} = E_{k_2} m_1/m_2$ by

$$R_2(E_{k_2}) = \frac{m_2}{m_1} \frac{z_1^2}{z_2^2} R_1\left(E_{k_2} \frac{m_1}{m_2}\right). \tag{2.47}$$

2.1.6 Properties of the Ionization Process

The following picture emerges about the energy loss process by ionization and excitation:

(i) In the primary process atoms are excited and ionized.
(ii) The energy distribution of ejected electrons is proportional to $1/W^{1.4-1.7}$.
(iii) Energetic electrons ($E_e > 100$ eV) are able to further ionize in a second collision.

The number of total ions is given by (2.20) and is typically two to seven times the number of primary ions. For example, in gases, $W_{ion} = 41$ eV for He, and $W_{ion} = 22$ eV for Xe; in liquid noble gases, $W_{ion} = 23.6$ eV for LAr and $W_{ion} = 16$ eV for LXe while in semiconductors, $W_{ion} = 3.5$ eV for Si and $W_{ion} = 2.85$ eV for Ge. Thus, for the same ΔE, more electron-ion pairs are produced in semiconductors leading in turn to smaller statistical fluctuations. Thus, in semi-conductors an excellent energy resolution is achieved, $\Delta E/E \simeq 10^{-3}$ to 10^{-4}. However, it has taken some technical developments to build large detector systems (see Chap. 5). Liquid noble gases are suited for large total absorbing shower counters and calorimeters due to their high densities with respect to that of gases. Here, the issues consist of building a system with a very high degree of purity for charge collection (see Sect. 3.2).

2.1.7 Production of Knock-on Electrons or δ-Electrons

The third term in (2.6) gives the probability for generating knock-on electrons also called δ-electrons. The distribution of secondary electrons N_δ with kinetic energy $E_e \simeq W$ much larger than the mean excitation energy I of the medium is

$$\frac{d^2 N_\delta}{dE_e dx} = \frac{1}{2} K z^2 \frac{Z}{A} \frac{1}{\beta^2} \frac{F(E_e)}{E_e^2}. \tag{2.48}$$

The function $F(E_e)$ is spin dependent but for small β, $F \sim 1$. For example, for spin zero particles

$$F(E_e) = 1 - \beta^2 \cdot \frac{E_e}{E_{max}}, \tag{2.49}$$

where the maximum kinetic energy of the electron can be approximated by $E_{max} = W_{max} - E_B$. Since the binding energy E_B is small with respect to the W_{max}, the binding energy can be neglected, so $E_{max} \simeq W_{max}$. Note that in this case (2.48) is equal to the Rutherford cross section. For spin-1/2 particles we get [6,7]

$$F(E_e) = 1 - \beta^2 \cdot \frac{E_e}{E_{max}} + \frac{E_e^2}{2 \cdot E^2}, \tag{2.50}$$

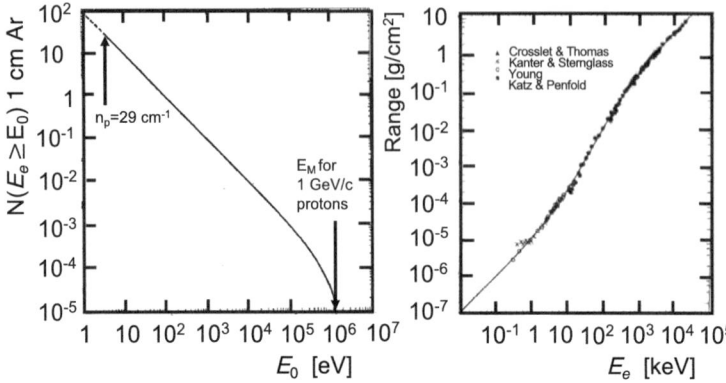

Fig. 2.20 Left: The calculated number of δ-electrons produced in 1 cm Ar gas at normal conditions with energies $E \geq E_0$ as a function of E_0 for a 1 GeV/c proton. The parameter n_p denotes the number of primary ionizing collisions per centimeter. Reprinted with kind permission under CC-BY-3.0 Licence from [10], © 1977–2024, CERN. All rights reserved. Right: The practical range versus energy for electrons in aluminum [72–76]. Each data point has a 10% uncertainty. Reprinted with kind permission from [72], © 1968, AIP Publishing. All rights reserved

where E is the energy of the heavy particle. For spin-1 particles $F(E_e)$ is listed in [6, 7].

If the kinetic energy is close to the mean excitation energy, (2.48) is not correct any longer [71]. For $2m_e\gamma \ll m$, the maximum kinetic energy can be approximated by $E_{\max} = 2m_e c^2 \beta^2 \gamma^2$. We can determine the number of δ-electrons by integrating (2.48) over dx and dW from W_0 to $W_{\max}$. Figure 2.20 (left) shows the calculated number of δ-electrons produced in 1 cm Ar gas at normal conditions with energies $E_e \geq E_0$ as a function of E_0 for a 1 GeV/c proton. Around 100 eV, we expect about 1 δ electron while around 1 eV we expect around 200 δ electrons produced by the 1 GeV/c proton. Note that δ electrons even with modest energies are rare. For a charged particle with $\beta \simeq 1$, one collision with $E_e > 1$ keV will occur on average along a path of 90 cm Ar gas. The δ-electrons are produced at an angle

$$\cos\theta = \frac{E_e}{E_{\max}} \frac{p_{\max}}{p_e}, \qquad (2.51)$$

where p_e and $p_{\max}$ are the typical momentum and maximum momentum of the struck electron, respectively.

Due to elastic and inelastic scattering, δ-electrons have a limited range. Figure 2.20 (right) shows measurements of the practical range in aluminum for electrons with energies between 3 keV and 3 MeV in comparison to a calculation by Weber [72, 77]. The data are parameterized by a semi-empirical range-energy-relation

$$R = \tilde{A} \cdot W \left[1 - \frac{\tilde{B}}{1 + \tilde{C} \cdot W} \right], \qquad (2.52)$$

A fit to the data in Fig. 2.20 (right) gave $\tilde{A} = 5.42 \times 10^{-4}$ g cm^{-2} keV^{-1}, $\tilde{B} = 0.9856$ and $\tilde{C} = 3.14 \times 10^{-3}$ keV^{-1}. Kobetich and Katz showed that by small adjustments, yielding $\tilde{A} = 5.37 \times 10^{-4}$ g cm^{-2} keV^{-1}, $\tilde{B} = 0.9815$ and $\tilde{C} = 3.1230 \times 10^{-3}$ keV^{-1}, the range was extended to energies between 0.3 keV and 20 MeV. Furthermore, it is used to approximate the practical range for all materials. A 100 keV δ-electron, will travel 8.15 cm in argon gas at normal conditions, while in silicon it will reach 58 μm.

2.2 Energy Loss of Electrons and Positrons

Electrons and positrons suffer energy losses by radiation in addition to those by ionization since they become highly relativistic rather fast. So the energy loss is

$$\left(\frac{dE}{dx}\right) = \left(\frac{dE}{dx}\right)_{\text{rad}} + \left(\frac{dE}{dx}\right)_{\text{coll}}. \tag{2.53}$$

Let us first look at the energy losses caused by ionization.

2.2.1 Energy Loss by Collision

The energy losses via collisions for electrons and positrons differ from each other and from that of heavy charged particles for three reasons:

(i) Due to the small electron and positron masses the incident particle may be deflected.
(ii) For e^-, we have collisions between identical particles. So, we have to account for indistinguishable particles.
(iii) Electrons and positrons are typically relativistic. Their interactions with electrons are described by Møller and Bhabha scattering, respectively.

The interaction of electrons with atomic electrons is described by the Møller cross section [7]. The maximum energy transfer in a single collision is half the kinetic energy, $E_{\text{max}} = m_e c^2 (\gamma - 1)/2$ due to identical particles [7]. The stopping power is calculated for the faster of the two electrons. The first moment of the Møller cross section divided by dx yields

$$\left\langle -\frac{dE}{dx} \right\rangle = \frac{K}{2} \frac{Z}{A} \frac{1}{\beta^2} \left[\ln \frac{m_e^2 c^4 \beta^2 \gamma^2 (\gamma-1)}{2 I_{\text{ex}}^2} + (1-\beta^2) - \frac{2\gamma - 1}{\gamma^2} \ln 2 + \frac{1}{8}\left(\frac{\gamma-1}{\gamma}\right)^2 - \delta(\beta\gamma) \right]. \tag{2.54}$$

The logarithmic term can be compared to that in the Bethe-Bloch equation by substituting $W_{\text{max}} = m_e c^2 (\gamma - 1)/2$. The two terms differ by $\ln 2$.

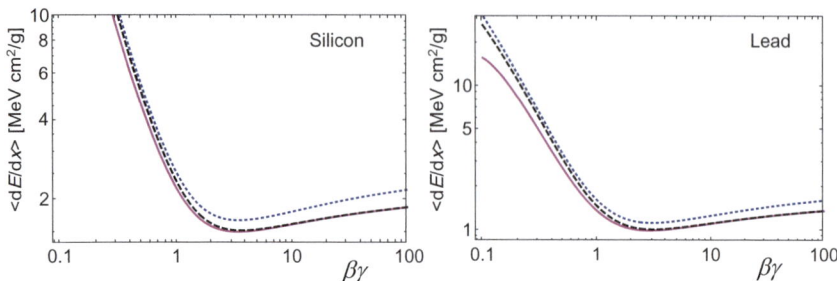

Fig. 2.21 Stopping power of electrons (magenta solid curve) and positrons (black dashed curve) as a function of $\beta\gamma$ in comparison to the average energy loss given by the Bethe-Bloch equation for muons (blue dotted curve) for silicon (left) and lead (right)

Electron positron scattering is described by the Bhabha cross section [7]. Since there are no identical particles, here $W_{\max} = m_e c^2 (\gamma - 1)$. The first moment of the Bhabha cross section yields

$$\left\langle -\frac{dE}{dx} \right\rangle = \frac{K}{2}\frac{Z}{A}\frac{1}{\beta^2}\left[\ln\frac{m_e^2 c^4 \beta^2 \gamma^2 (\gamma - 1)}{I_{ex}^2} + 2\ln 2 - \frac{\beta^2}{12}\left(23 + \frac{14}{\gamma+1} + \frac{10}{(\gamma+1)^2} + \frac{4}{(\gamma+1)^3}\right) - \delta(\beta\gamma)\right]. \quad (2.55)$$

Figure 2.21 shows the average stopping power of electrons and positrons as a function of $\beta\gamma$ in silicon (left) and lead (right) in comparison to that obtained for muons from the Bethe-Bloch equation. Below the minimum, the energy loss is smallest for electrons, followed by that of positrons and muons. For silicon the three curves are rather close. Above the minimum, the energy loss of electrons and positrons is similar while that for muons is higher. For values ($\beta\gamma > 800$), energy losses due to radiation become important. For practical reasons, electrons and positrons are essentially the only particles in which radiation contributes substantially to energy loss, except for very-high-energy muons at the LHC.

2.2.2 Energy Loss by Bremsstrahlung

The emission of bremsstrahlung by electrons depends on the strength of the $\vec{E}$-field the electrons feel. It consists of the electric field of the nucleus plus that of the surrounding electrons. The amount of screening from atomic electrons surrounding the nucleus plays an important role. The limiting cases are complete screening and no screening. For complete screening, the electric field of the atomic electrons neutralizes that of the protons. So, the traversing electron or positron feels no $\vec{E}$-field and, therefore, does not radiate. For no screening, the traversing electron or positron feels the full nuclear field and looses energy via radiation. Screening effects are parametrized by the screening variable,

$$\xi = \frac{100 m_e c^2 \hbar\omega}{E E_0 Z^{1/3}} = \frac{100 m_e c^2 \eta_r}{(1 - \eta_r) E_0 Z^{1/3}}, \quad (2.56)$$

2.2 Energy Loss of Electrons and Positrons

where E_0 is the initial $e^{\mp}$ energy, E is the final $e^{\mp}$ energy, $\hbar\omega = E_0 - E$ is the energy of the radiated photon and $\eta_r = \hbar\omega/E_0$. The screening variable is related to the radius of the Thomas-Fermi atom, yielding small values for complete screening ($\xi \simeq 0$ when $\eta_r = \hbar\omega/E_0 \simeq 0$) and large values for no screening ($\xi \gg 1$ when $\eta_r \simeq 1$). The bremsstrahlung cross section for relativistic $e^{\mp}$ ($E >$ few MeV) interacting with the electric field of the nucleus is given by [78][8]

$$\frac{d\sigma_{br}^{ep}}{d\eta_r} = Z^2 r_e^2 \frac{\alpha}{\eta_r} \left[\left(\frac{4}{3} - \frac{4}{3}\eta_r + \eta_r^2\right)\left(\Phi_1(\xi) - \frac{4}{3}\ln Z - 4f(Z)\right) + \frac{2}{3}(1-\eta_r)\left(\Phi_1(\xi) - \Phi_2(\xi)\right) \right], \quad (2.57)$$

where $f(Z)$ is a Coulomb correction to the one-photon exchange approximation and $\Phi_1(\xi)$, $\Phi_2(\xi)$ are screening functions introduced by Bethe-Heitler [79]. The approximation does not hold for low photon energies. For heavy elements ($Z > 5$) the screening functions $\Phi_1(\xi)$ and $\Phi_2(\xi)$ are usually calculated with the atomic Thomas-Fermi theory and their values are given numerically. A useful approximation that is precise to 0.5% is given by the following empirical formulas,

$$\Phi_1(\xi) = 4\ln(184.15) - 2\ln\left[1 + (0.55846 \cdot \xi)^2\right] - 4\left[1 - 0.6\exp(-0.9 \cdot \xi) - 0.4\exp(-1.5 \cdot \xi)\right],$$
$$\Phi_2(\xi) = \Phi_1(\xi) - \frac{2}{3}\frac{1}{(1 + 6.5 \cdot \xi + 6 \cdot \xi^2)}. \quad (2.58)$$

Besides radiation from the electric field of the nucleus, radiation is also produced in the electric field of the atomic electrons. The contribution to the bremsstrahlung cross section is [80],[9]

$$\frac{d\sigma_{br}^{ee}}{d\eta_r} = Z r_e^2 \alpha \frac{1}{\eta_r} \left[\left(\frac{4}{3} - \frac{4}{3}\eta_r + \eta_r^2\right)\left(\Psi_1(\xi') - \frac{8}{3}\ln Z\right) + \frac{2}{3}(1-\eta_r)\left(\Psi_1(\xi') - \Psi_2(\xi')\right) \right]. \quad (2.59)$$

The screening functions $\Psi_1(\xi')$ and $\Psi_2(\xi')$ have been introduced by Wheeler-Lamb [82],

$$\Psi_1(\xi') = 4\ln(1194) - 2\ln\left[1 + (3.621 \cdot \xi')^2\right] - 4\left[1 - 0.7\exp(-8 \cdot \xi') - 0.3\exp(-29.2 \cdot \xi')\right],$$
$$\Psi_2(\xi') = \Psi_1(\xi') - \frac{2}{3}\frac{1}{(1 + 40 \cdot \xi' + 400 \cdot \xi'^2)}, \quad (2.60)$$

where ξ' is defined as

$$\xi' = \frac{100 m_e c^2 \hbar\omega}{E E_0 Z^{2/3}} = \frac{100 m_e c^2 \eta_r}{(1-\eta_r) E_0 Z^{2/3}}. \quad (2.61)$$

---

[8] The first index e denotes $e^{\pm}$ while the second index p stands for the contributions due to the nucleus charge.
[9] The first index e denotes $e^{\pm}$ while the second index p stands for the contributions due to shell electrons.

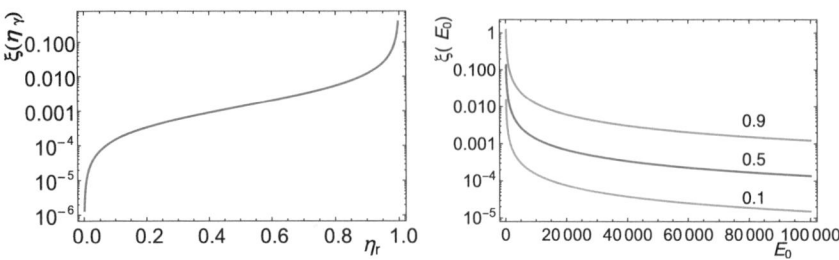

Fig. 2.22 Left: Screening variable as a function of relative electron energy η_r. Right: Screening variable as a function of the initial electron energy E_0 for three values of η_r

The entire bremsstrahlung cross section is the sum of (2.57) and (2.59), $d\sigma_{br}/d\eta_r = d\sigma_{br}^{ep}/d\eta_r + d\sigma_{br}^{ee}/d\eta_r$. Note that a good approximation results by replacing Z^2 with $Z(Z+1)$ in (2.57). Except for values of η_r near one, the approximation provides results that are accurate within 1% of the entire bremsstrahlung cross section. It is interesting to look at the asymptotic behavior of the screening functions for $\xi \to 0$ and $\xi \to \infty$,

$$\Phi_1(0) = \Phi_2(0) + 2/3 = 4\ln(184.15) = 20.863 \quad \text{for } \xi \to 0 \text{ (complete screening)},$$
$$\Psi_1(0) = \Psi_2(0) + 2/3 = 4\ln(1194) = 28.340 \quad \text{for } \xi' \to 0 \text{ (complete screening)},$$
$$\Phi_1(\infty) = \Phi_2(\infty) \to 4\big(\ln(184.15) - \ln(0.55846) - 1 - \ln\xi\big)$$
$$= 19.19 - 4\ln\xi \quad \text{for } \xi \to \infty \text{ (no screening)},$$
$$\Psi_1(\infty) = \Psi_2(\infty) \to 4\big(\ln(1194) - \ln(3.621) - 1 - \ln\xi'\big)$$
$$= 19.19 - 4\ln\xi' \quad \text{for } \xi' \to \infty \text{ (no screening)}. \quad (2.62)$$

Figure 2.22 (left) shows the screening variable ξ as a function of the relative photon energy for a 10 GeV electron in xenon. For $\eta_r \to 0$, the screening variable approaches zero (complete screening) while for $\eta_r \to 1$ the screening variable diverges (no screening). For low electron energies, ξ increases rapidly while for high electron energies ξ falls off to zero, as shown in Fig. 2.22 (right). Figure 2.23 shows the screening functions Φ_1 and Φ_2 versus the relative electron energy $1 - \eta_r$ (left) and versus the screening variable ξ (right). The approximations for complete screening and no screening are depicted also.

The Coulomb correction $f(Z)$ accounts for Coulomb interactions of the emitted electron in the electric field of the nucleus and is approximated by

$$f(Z) = a^2 \left(\frac{1}{1+a^2} + 0.20206 - 0.0369a^2 + 0.0083a^4 - 0.002a^6 \right), \quad (2.63)$$

where $a = \alpha Z$. Figure 2.24 (left) shows the Coulomb correction as a function of αZ. Note that the Coulomb correction is negligible in low-Z materials and yields small

2.2 Energy Loss of Electrons and Positrons

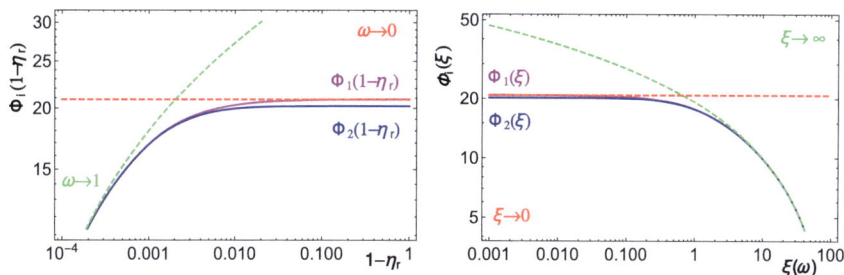

Fig. 2.23 Left: Screening functions Φ_1 and Φ_2 versus the relative electron energy $1 - \eta_r$. Right: The screening functions Φ_1 and Φ_2 as a function of the screening variable. Solid lines show the screening functions while dashed lines show the approximations for no screening and complete screening

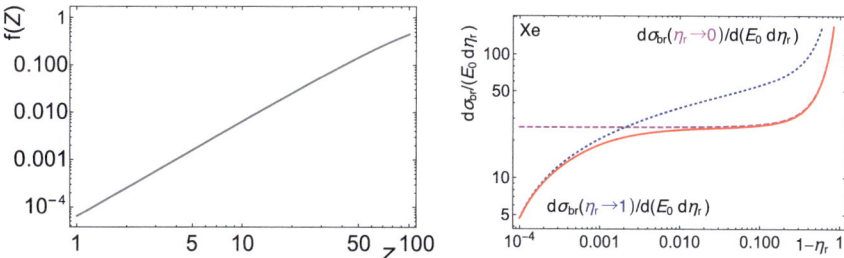

Fig. 2.24 Left: The Coulomb correction f(Z) as a function of Z. Right: The full differential bremsstrahlung cross section $d\sigma_{br}/d(\hbar\omega)$ for 10 GeV electrons and positrons in xenon (red solid line), the complete screening approximation ($\xi \to 0$, magenta dashed line) and that for no screening ($\xi \to \infty$, blue dotted line)

adjustments in heavy elements. The cross sections for complete screening and for no screening are

$$\frac{d\sigma_{br}}{d\eta_r} = 4r_e^2 \frac{\alpha}{\eta_r} \left\{ \left(1 - \frac{2}{3}\eta_r + \eta_r^2\right)\left[Z^2[\ln\frac{184.15}{Z^{1/3}} - f(Z)] + Z\ln\frac{1194}{Z^{2/3}}\right] + \frac{\eta_r}{9} Z(Z+1) \right\} \; (\xi \to 0)$$

$$\frac{d\sigma_{br}}{d\eta_r} = 4r_e^2 \frac{\alpha}{\eta_r} \left(1 - \frac{2}{3}\eta_r + \eta_r^2\right)\left(Z(Z+1)\left[\ln\frac{2E_0(1-\eta_r)}{m_e c^2 \eta_r} - \frac{1}{2}\right] - Z^2 f(Z)\right) \; (\xi \to \infty). \quad (2.64)$$

Figure 2.24 (right) shows the differential bremsstrahlung cross section for 10 GeV $e^\pm$ in Xe. In addition, the approximation for complete screening and no screening are depicted. For $\eta_r \to 0$, the approximation is better than 1%.

Let us look at the bremsstrahlung cross section at high photon energies except for the high-energy tip, which is represented well by the cross section for complete screening (2.64). Near $\eta_r = 1$ the screening may become incomplete and near $\eta_r = 0$ we have to deal with the infrared divergency, which is removed by the interference of bremsstrahlung amplitudes from nearby scattering centers called Landau-Pomeranchuk-Migdal (LPM) effect [83,84] and dielectric suppression [85,86]. Figure 2.25 shows the normalized bremsstrahlung cross section multiplied by the

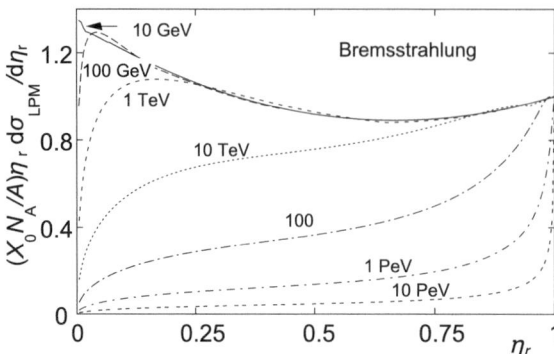

Fig. 2.25 The normalized bremsstrahlung cross section multiplied by the photon energy in lead versus the fractional photon energy $\eta_r = \hbar\omega/E_0$. The vertical axis has units of photons per radiation length. The solid curve is the approximation given in (2.65). Reprinted under CC-BY-NC-4.0 Licence with kind permission from [1], © 2024, the Particle Data Group LBNL

photon energy for lead versus the fractional photon energy $\eta_r = \hbar\omega/E_0$. In the intermediate photon energy range, the cross section may be approximated by

$$\frac{d\sigma_{br}}{d\eta_r} = \frac{A}{X_0 N_A \eta_r}\left(\frac{4}{3} - \frac{4}{3}\eta_r + \eta_r^2\right). \quad (2.65)$$

Here, the last term in (2.64 top) has been omitted, since for low photon energies it contributes only at the level of 1.7% and 2.5% for low Z and high Z, respectively. In addition, the square bracket is replaced by $1/X_0$, where X_0 is the radiation length, which will be introduced in Sect. 2.2.3 (see (2.73)). The number of photons radiated by an electron traveling at a distance $L_{path} \ll X_0$ at energies between $\hbar\omega_{min}$ and $\hbar\omega_{max}$ is

$$N_\gamma = \frac{L_{path}}{X_0}\left[\frac{4}{3}\ln\left(\frac{\omega_{max}}{\omega_{min}}\right) - \frac{4\hbar(\omega_{max} - \omega_{min})}{3E_0} + \frac{(\hbar\omega_{max})^2 - (\hbar\omega_{min})^2}{2E_0^2}\right]. \quad (2.66)$$

The energy loss due to radiation now results simply by integrating the differential cross section over the allowable photon energy range. Since $d\sigma_{br}/d(\hbar\omega) \propto (\hbar\omega)^{-1}$, the integral is nearly independent of $\hbar\omega$ in the region of complete screening ($\xi \to 0$) and thus is a function of the material parameters only. Thus, we can write

$$-\left(\frac{dE}{dx}\right)_{rad} = \frac{N_A}{A}\int_0^{\omega_0} \hbar\omega \frac{d\sigma_{br}}{d(\hbar\omega)}\left(E_0, \hbar\omega\right) d(\hbar\omega) = \frac{N_A}{A}\Phi_{rad} \cdot E_0. \quad (2.67)$$

The upper integration limit is the electron energy E_0. The units are again [MeV cm^2/g].

2.2 Energy Loss of Electrons and Positrons

Table 2.3 The radiation logarithms L_{rad} and L'_{rad} calculated by Tsai [80] for $\xi \to 0$. Reprinted with kind permission from [80], © 1974, AIP Publishing. All rights reserved

Element	Z	L_{rad}	L'_{rad}
H	1	5.31	6.144
He	2	4.79	5.621
Li	3	4.74	5.805
Be	4	4.71	5.924
B	5	4.679	6.012
C	6	4.619	5.891
N	7	4.567	5.788
Others	$Z \geq 5$	$\ln(184.15/Z^{1/3})$	$\ln(1194/Z^{2/3})$

For $E_0 \gg 137 m_e c^2 Z^{1/3}$, where complete-screening ($\xi \simeq 0$) holds, we can parametrize Φ_{rad} in terms of radiation logarithms for bremsstrahlung from the electric field of the nucleus, L_{rad}, and that from the electric field of the atomic electrons L'_{rad},

$$\Phi_{rad} = 4 r_e^2 \alpha \left\{ Z^2 \left[L_{rad} - f(Z) + \frac{1}{18} \right] + Z L'_{rad} \right\}. \quad (2.68)$$

For complete screening, L_{rad} and L'_{rad} are independent of the photon energy and are determined by properties of the material. Table 2.3 lists the values of L_{rad} and L'_{rad} for different materials. For $Z < 5$, they have been calculated numerically and for $Z \geq 5$, an approximation is listed.

Plugging in the values of L_{rad} and L'_{rad} for $Z \geq 5$ from Table 2.3 into (2.68) yields,

$$\Phi_{rad} = 4 r_e^2 \alpha \left[Z^2 \left(\ln\left(184.15\, Z^{-1/3}\right) - f(Z) + \frac{1}{18} \right) + Z \ln\left(1194\, Z^{-2/3}\right) \right]. \quad (2.69)$$

For $m_e c^2 \ll E_0 \ll 137 m_e c^2 Z^{1/3}$ we get the no-screening limit $\xi \gg 1$, where the integral yields

$$\Phi_{rad} = 4 r_e^2 \alpha \left[Z(Z+1)\left(\ln\left(\frac{2 E_0}{m_e c^2}\right) - \frac{1}{3} \right) - Z^2 f(Z) \right]. \quad (2.70)$$

At intermediate values of ξ, the integral needs to be calculated numerically.

Let us compare the energy loss by bremsstrahlung to that by ionization. The energy loss via ionization is proportional to $\ln E$ and Z, while the energy loss via radiation is proportional to E and $Z(Z+1)$. This explains the rapid rise in the energy loss due to radiation. Another difference is that all energy can be radiated via one or two photons (see Fig. 2.24 (right)) whereas the energy loss via ionization typically has small losses in multiple collisions.

2.2.3 Radiation Length

The radiation length X_0 is defined as the distance over which the electron energy is reduced on average by $1/e$ due to radiation loss only. Let us look at equation (2.67), which can be rewritten as a differential equation

$$-dE/E = N\Phi_{rad} dx, \qquad (2.71)$$

where $N = \rho N_A/A$. In the high-energy limit, where collisions can be ignored with respect to radiation, Φ_{rad} is independent of energy. Thus,

$$E = E_0 \exp\left(-\frac{x}{X_0}\right), \qquad (2.72)$$

where x is distance traveled and $X_0 = 1/(N\Phi_{rad})$. Note that typically the density is divided out, so x has dimensions of $[g/cm^2]$ as these are the units of X_0. Plugging the expression for Φ_{rad} into (2.68) and omitting the small constant term yields,

$$\frac{1}{X_0} = \left(4\alpha r_e^2 \frac{N_A}{A}\left\{Z^2[L_{rad} - f(Z)] + ZL'_{rad}\right\}\right), \qquad (2.73)$$

where A is the atomic weight in units of $[g/mol]$. For $A = 1$ g/mol, $4\alpha r_e^2 N_A = (716.408 \text{ g/cm}^2)^{-1}$. Thus, the radiation length has units of $[g/cm^2]$. Division by the density yields units of $[cm]$. Defining the dimensionless variable $t_r = x/X_0$, we can write (2.71) as

$$-\frac{dE}{dt_r} \simeq E_0. \qquad (2.74)$$

Thus, the radiation energy loss is approximately independent of the material type when expressed in terms of X_0. Tables 2.4, 2.5 and 2.6 list the radiation length and some other properties[10] of gases, atomic elements and compounds, respectively. For example, the radiation length is $X_0 = 63$ $[g/cm^2]$ for H_2, $X_0 = 24$ $[g/cm^2]$ for Al and $X_0 = 6.3$ $[g/cm^2]$ for Pb. With increasing Z, X_0 becomes smaller. For compounds the radiation length is computed by

$$\frac{1}{X_0} = \sum_i w_i \frac{1}{X_{0,i}}, \qquad (2.75)$$

where w_i is the weight defined in (2.27) and $X_{0,i}$ is the radiation length of the ith element.

---

[10] The nuclear (pion) interaction length, which is the mean distance a hadron (pion) travels before undergoing an inelastic nuclear (pion) interaction, is introduced in Sect. 7.4. The Moliére radius, which denotes the radius at which the transverse shower contains 90% of the electromagnetic shower energy, is discussed in Sect. 7.1.2.

2.2 Energy Loss of Electrons and Positrons

Table 2.4 The ratio Z/A, the plasma energy, nuclear interaction length λ_I, pion interaction length λ_π, radiation length X_0, critical energy E_c, Moliére radius R_M and density of gases and liquids. The density is given at normal conditions (NTP). †The Moliére radius will be introduced in Sect. 7.1.2. Reprinted under CC-BY-NC-4.0 Licence with kind permission from [1,4], © 2024, the Particle Data Group LBNL

Material	Z/A	$\hbar\omega_p$ [eV]	λ_I [g/cm^2]	λ_π [g/cm^2]	X_0 [g/cm^2]	E_c [MeV]	$R_M^\dagger$ [g/cm^2]	ρ [mg/cm^3]
H_2	0.99212	0.26	52.0	82.0	63.04	344.8	3.88	0.0838
D_2	0.49650	0.26	71.8	110.1	125.98	345.5	7.73	0.1677
He	0.499675	0.26	71.0	103.6	94.32	257.13	7.78	0.1663
N_2	0.49976	0.70	89.7	121.7	37.99	91.74	8.78	1.165
O_2	0.50002	0.74	90.2	121.9	34.24	81.45	8.91	1.332
F_2	0.4737	0.79	97.4	127.2	32.93	73.15	9.55	1.58
Ne	0.49554	0.59	99.0	128.7	28.93	67.02	9.15	0.8385
Cl	0.47951	1.09	115.7	144.9	19.28	40.05	10.21	2.98
Ar	0.45059	0.79	119.7	148.8	19.55	38.03	10.90	1.662
LAr	0.45059	22.85	119.7	148.8	19.55	32.84	12.62	1396
Kr	0.42960	1.12	149.4	177.4	11.37	18.61	12.96	3.486
LKr	0.42960	29.37	149.4	177.4	11.37	17.03	14.16	2418
Xe	0.41129	1.37	172.1	199.3	8.48	12.30	14.63	5.483
LXe	0.41129	31.76	172.1	199.3	8.48	11.66	15.43	2953
Air	0.49919	0.71	90.1	122.0	36.62	87.92	8.83	1.205
CH_4	0.62334	0.588	73.8	106.1	46.47	146.86	6.71	0.667
CO_2	0.49989	0.874	88.9	120.8	36.19	86.17	8.91	1.842
Freon 12	0.47968	21.12	105.8	135.8	23.65	43.24	11.60	1120
Freon 13	0.47966	19.45	101.3	131.4	27.12	50.36	11.42	950
CF_4	0.47221	1.22	95.7	125.8	33.99	75.12	9.60	3.78
C_2H_6	0.59861	0.79	75.9	108.3	45.66	135.97	7.12	1.263
C_3H_8	0.58962	0.96	76.7	109.1	45.37	131.58	7.31	1.868
C_4H_{10}	0.9497	1.11	77.1	109.5	45.23	131.26	7.31	2.489
C_5H_{12}	0.58212	17.4	77.4	109.7	45.14	106.25	9.01	626

2.2.4 Critical Energy

The energy for which the energy loss by ionization is identical to the energy loss by radiation is called critical energy $\tilde{E}_c$ [87]. Rossi introduced an alternative definition in which E_c is the energy where the energy loss by ionization is equal to the electron energy [6].[11] If the energy loss by bremsstrahlung is approximated

---

[11] We use Rossi's definition in this book.

Table 2.5 The ratio Z/A, plasma energy, nuclear interaction length, pion interaction length, radiation length, critical energy, Moliére radius and density of some atomic elements. Reprinted under CC-BY-NC-4.0 Licence with kind permission from [1,4], © 2024, the Particle Data Group LBNL

Material	Z/A	$\hbar\omega_p$ [eV]	λ_I [g/cm^2]	λ_π [g/cm^2]	X_0 [g/cm^2]	E_c [MeV]	R_M [g/cm^2]	ρ [g/cm^3]
Li	0.43221	13.84	71.3	103.0	82.77	149.04	11.78	0.534
Be	0.4384	26.10	77.8	109.7	65.19	113.70	12.16	1.848
C amorphous	0.49955	28.8	85.8	117.8	42.70	82.08	11.03	2.000
C graphite	0.49955	30.28	85.8	117.8	42.70	81.74	11.08	2.210
C diamond	0.49955	38.21	85.8	117.8	42.70	80.17	11.29	3.520
Mg	0.49372	26.71	104.1	133.6	25.03	46.55	11.4	1.740
Al	0.48181	32.86	107.2	136.6	24.01	42.70	11.93	2.699
Si	0.49848	31.05	108.4	137.7	21.82	40.19	11.51	2.329
P	0.48428	29.74	111.4	140.7	21.21	37.92	11.86	2.20
Ti	0.45960	41.63	126.2	155.0	16.16	26.01	13.18	4.540
Mn	0.45541	53.02	131.4	160.1	14.64	22.59	13.74	7.440
Fe	0.46550	55.17	132.1	160.7	13.84	21.68	13.53	7.874
Cu	0.45636	58.27	137.3	165.7	12.86	19.42	14.05	8.96
Zn	0.45886	52.13	138.5	166.9	12.43	18.93	13.92	7.133
Ga	0.44462	46.69	141.2	169.5	12.47	18.57	14.24	5.904
Ge	0.44053	44.13	143.0	171.2	12.25	18.16	14.31	5.323
As	0.44046	45.78	144.4	172.5	11.94	17.65	14.35	5.730
Ag	0.43572	61.64	161.7	120.3	8.97	12.36	15.39	10.50
Sn	0.42119	50.56	166.7	194.0	8.82	11.86	15.77	7.310
W	0.40252	80.32	191.9	218.4	6.76	7.97	18.00	19.30
Pt	0.39983	84.39	195.7	222.1	6.54	7.59	18.29	21.45
Au	0.40108	80.21	196.3	222.7	6.46	7.53	18.19	19.32
Pb	0.39575	61.07	199.6	225.9	6.37	7.43	18.18	11.35
U	0.38650	77.99	209.0	234.9	6.00	6.65	19.12	18.95

Table 2.6 The ratio Z/A, plasma energy, nuclear interaction length, pion interaction length, radiation length, critical energy, Moliére radius and density of compounds. Reprinted under CC-BY-NC-4.0 Licence with kind permission from [1,4], © 2024, the Particle Data Group LBNL

Material	Z/A	$\hbar\omega_p$ [eV]	λ_I [g/cm^2]	λ_π [g/cm^2]	X_0 [g/cm^2]	E_c [MeV]	R_M [g/cm^2]	ρ [g/L]
CsI	0.41569	39.46	171.5	198.7	8.39	11.17	15.92	4.51
NaI	0.42697	36.06	154.6	183.6	9.49	13.37	15.05	3.667
NaCl	0.47910	29.38	110.1	139.6	21.91	38.50	12.07	2.17
BGO	0.42065	49.9	159.1	190.5	7.97	10.5	16.1	7.130
BAF$_2$	0.42207	41.41	149.0	179.0	9.91	13.78	15.25	4.893
PbWO$_4$	0.41315	53.36	168.3	199.3	7.39	9.64	16.26	8.300
Lu$_2$SiO$_5$	0.42793	51.28	152.7	184.4	8.46	11.71	15.32	7.40
Fused quartz	0.49930	30.20	97.8	128.8	27.05	50.58	11.34	2.200
LiF	0.46262	31.82	88.7	119.6	39.26	68.82	12.09	2.635
Sapphire	0.49038	40.21	98.4	129.3	27.94	50.18	11.81	3.970
Dry Ice	0.49989	25.47	88.9	120.8	36.19	69.99	10.97	1.563
Lead glass	0.42101	46.63	158.0	189.8	7.87	10.41	16.04	6.220
Borosiliate glass	0.49707	30.34	96.5	127.7	28.17	52.54	11.37	2.230
Concrete	0.50274	30.99	97.5	128.7	26.57	49.90	11.29	2.300
Polycarbonate	0.52697	22.91	83.6	115.7	41.5	85.85	10.25	1.200
Polysterene	0.53768	21.75	81.7	113.9	43.79	93.11	9.97	1.060
Polyethylene	0.57034	20.53	78.5	110.8	44.77	101.79	9.33	0.89
Mylar	0.52037	24.60	84.9	117.0	39.95	80.95	10.46	1.400
Acrylic	0.53937	23.09	82.8	115.0	40.55	85.79	10.02	1.190
Teflon	0.47992	29.61	94.4	124.8	34.84	63.5	11.64	2.200
Anthracene	0.52740	23.70	82.8	114.9	43.49	89.98	10.25	1.283
Naphtalene	0.53053	22.46	82.4	114.6	43.58	91.13	10.14	1.145
Bakelite	0.52792	23.41	83.4	115.5	41.74	86.37	10.25	1.250
Silica aerogel	0.50093	9.12	97.3	128.4	27.25	56.07	10.31	0.200
Water	0.55509	21.47	83.3	115.6	36.08	78.33	9.77	1.00
Ethanol	0.56437	19.23	80.3	112.6	40.92	92.14	9.42	0.7893
Benzine	0.53769	19.81	81.7	113.9	43.79	93.82	9.90	0.8787
Cyclohexane	0.57034	19.21	78.5	110.8	44.77	102.33	9.28	0.7790
C$_6$H$_4$Cl$_2$	0.50339	23.35	96.1	127.8	27.07	52.20	11.00	1.305
C$_4$H$_8$Cl$_2$O	0.51744	22.89	95.0	126.9	26.49	52.55	10.69	1.220
C$_2$H$_4$Cl$_2$	0.50526	22.76	102.0	133.3	22.99	44.19	11.03	1.235
Diethyl ether	0.56663	18.33	79.6	111.9	42.30	96.15	9.33	0.7138
n-Hexane	0.59020	17.99	77.6	109.9	45.08	107.43	8.90	0.6603
Methanol	0.56176	19.21	81.1	113.5	39.43	88.24	9.48	0.7914
Parafin	0.57275	21.03	78.2	110.6	44.85	102.22	9.30	0.9300
n-propyl alcohol	0.56577	19.43	79.9	112.2	41.76	94.29	9.39	0.8035
Terphenyl	0.52148	23.12	83.4	115.5	43.32	88.74	10.35	1.234
Toluene	0.54265	19.76	81.2	113.4	43.94	95.06	9.80	0.8669
Trichloroethylene	0.48710	24.30	107.8	138.2	21.55	39.45	11.58	1.460
Xylene	0.54631	19.87	80.8	113.1	44.05	95.94	9.74	0.8700

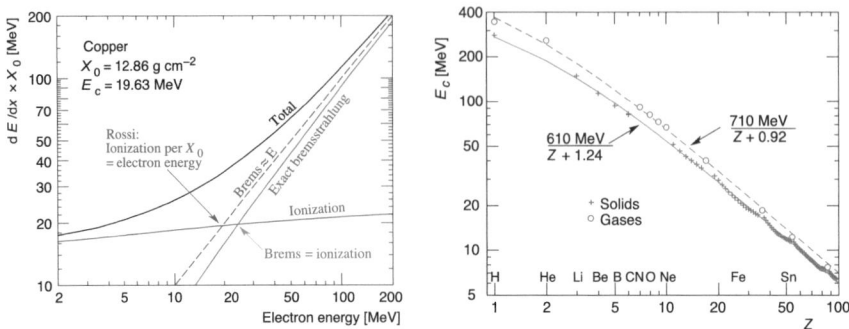

Fig. 2.26 Left: Two definitions of the critical energy. Unless stated otherwise, we use Rossi's definition. Right: The electron critical energy E_c determined with Rossi's definition as a function of the atomic number Z [6]. The points are calculations from (2.76), the solid curve shows a fit for solids and liquids while the dashed curve shows a fit for gases. The rms deviation is 2.2% for solids and 4% for gases. Both reprinted under CC-BY-NC-4.0 Licence with kind permission from [1], © 2024, the Particle Data Group LBNL

by $|dE/dx|_{\text{brems}} \sim E/X_0$ where X_0 is the radiation length, both definitions agree. Figure 2.26 (left) shows the two definitions of the critical energy. For the full $|dE/dx|_{\text{brems}}$ contribution, $\tilde{E}_c$ is somewhat larger than that obtained from Rossi's definition. Using Rossi's definition, the critical energy for solids and gases can be determined from these simple empirical formulas,

$$E_c = \frac{610}{Z + 1.24} \quad \text{for solids,}$$
$$E_c = \frac{710}{Z + 0.92} \quad \text{for gases.} \tag{2.76}$$

Figure 2.26 (right) shows the critical energy computed for solids and gases in comparison to the parameterizations given in (2.76). The critical energy drops with increasing Z. The rms deviation is 2.2% for solids and 4% for gases. For example, the critical energies for lead and air are $E_c = 7.3$ MeV and $E_c = 103$ MeV, respectively.

2.2.5 Range of Electrons

The range of electrons is rather different than that of heavy charged particles due to elastic and inelastic scattering. Depending on the energy and material it can vary between 20% and 400%. Fluctuations are larger due to higher possible energy transfers per collision and bremsstrahlung, which yields range straggling. Figure 2.27 shows the intensity versus the absorber thickness for different electron energies in aluminum, which are calculated on the assumption of continuous slowing-down. The fluctuations in the straggling of electrons are larger than those for other charged particles causing larger tails on the range distributions (Table 2.5).

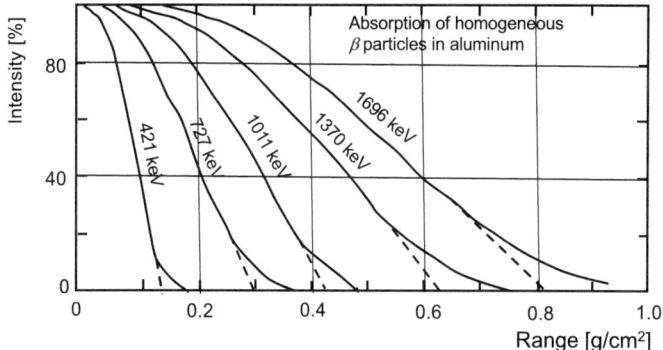

Fig. 2.27 Range curves for electrons of different energies in aluminum. Reprinted with kind permission from [90], © 1937, Canadian Science Publishing. All rights reserved

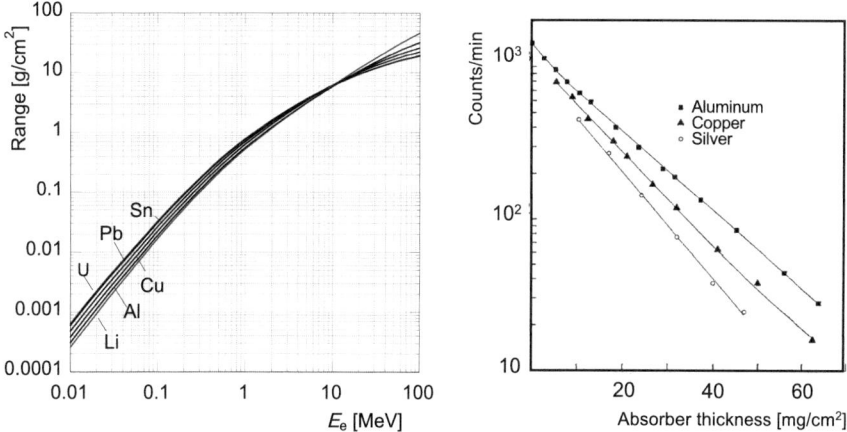

Fig. 2.28 Left: Range versus energy for electrons in several materials in the continuous slowing-down approximation. The curves are for uranium, lead, tin, copper, aluminum and lithium. Right: Absorption curves for electrons from β decay from ^{185}W in aluminum, copper and silver. Reprinted with kind permission from [89], © 1970, Elsevier. All rights reserved

Figure 2.28 (left) shows the electron range as function of energy for different materials [88]. The absorption spectrum is well approximated by

$$I_e = I_0 \exp(-\hat{\mu}^e x), \tag{2.77}$$

where $\hat{\mu}^e = \mu_a^e \rho$ is a linear absorption coefficient. Note that μ_a^e has units of cm^2/g while $\hat{\mu}^e$ has units of cm^{-1}. Figure 2.28 (right) shows absorption curves for β decays in different materials [89]. The data are well described by the exponential function ((2.77)).

2.2.6 Muon Energy Loss at High Energies

For all charged particles at sufficiently high energies, energy losses by radiation become more important than those by ionization. The critical energy for pions and muons occurs at several hundred GeV. For example, the critical energies for electrons in silicon and lead are 40.2 MeV and 7.43 MeV while those for muons are 575.3 GeV and 139.8 GeV, respectively. So, the critical energies differ by more than four orders of magnitude. Similar factors hold for gases. There is no simple scaling with mass. Bremsstrahlung becomes important for very energetic muons, which occur in cosmic rays or are produced at the LHC and future high-energy colliders. Lower energy muons loose energy via ionization. Concerning radiation losses, we distinguish among three processes. The dominant one is the production of e^+e^- pairs through $\mu A \to \mu A e^+ e^-$. The second largest process is bremsstrahlung while photonuclear interactions occur at smaller probability. Since cross sections are small, spectra are hard, energy fluctuations are large and associated electromagnetic and (in case of photonuclear interactions) hadronic showers are generated, the treatment of energy loss as a uniform and continuous process is, for many purposes, inadequate. The average energy loss for muons can be parameterized as

$$\left\langle -\frac{dE}{dx} \right\rangle = f_1(E) + f_2(E)E, \tag{2.78}$$

where $f_1(E)$ represents the ionization energy loss and $f_2(E)$ is the sum of e^+e^- pair production, bremsstrahlung and photonuclear contributions. For muons, the critical energy is defined as the energy at which radiative and ionization losses are equal, which is different from Rossi's definition of critical energy.

Approximating these slowly-varying functions by constants, the mean range $\bar{x}_\mu$ of a muon with initial energy E_0 is given by

$$\bar{x}_\mu \approx \frac{1}{f_2(E)} \ln\left(1 + \frac{E_0}{E_{\mu c}}\right), \tag{2.79}$$

where $E_{\mu c}$ is the muon critical energy, which can be determined from

$$E_{\mu c} = \frac{f_1(E_{\mu c})}{f_2(E_{\mu c})}. \tag{2.80}$$

The Z dependence of the critical energy is given by

$$E_{\mu c}(\text{gases}) = \frac{7980 \text{ GeV}}{(Z+2.03)^{0.879}}$$

$$E_{\mu c}(\text{solids}) = \frac{5700 \text{ GeV}}{(Z+1.47)^{0.838}}. \tag{2.81}$$

Figure 2.29 (left) shows the average energy loss of muons in hydrogen gas, iron and uranium. For more information, see [1]. Figure 2.29 (right) shows calculated muon critical energies for solids and gases with fit curves overlaid.

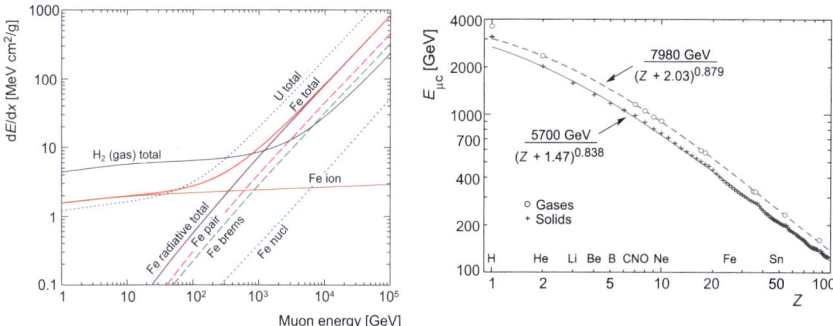

Fig. 2.29 Left: The average energy loss of a muon in hydrogen gas (black solid curve), iron (red solid curve) and uranium (blue dotted curve) as a function of the muon energy. For iron, the energy loss by ionization (red solid curve) and radiation (brown solid line) is shown as well as the contributions from pair production (magenta dashed line), bremsstrahlung (green dashed line) and photonuclear interactions (blue dotted line). Right: Muon critical energy as a function of the atomic number. Gases (open circles) have higher critical energies than liquids and solids (+ symbols). The solid and dashed curves are fits to solids/liquids and gases, respectively, excluding hydrogen. Note, that most solids lie within 2% of the fit curve, whereas alkali metals lie about 3–4% higher. Both reprinted under CC-BY-NC-4.0 Licence with kind permission from [1], © 2024, the Particle Data Group LBNL

2.2.7 Summarizing Remarks

The average energy loss of electrons and positrons below the critical energy is given by the modified Bethe-Bloch equations ((2.54) and (2.55)). However, more useful is the most probables energy loss since it can be measured with a few data points. Above the critical energy losses by bremsstrahlung become more important, which are determined by (2.67). We have defined two important quantities, the radiation length and the critical energy. For most experiments radiation of electrons and positrons are relevant. Higher-mass particles like muons have small radiation probabilities. At LHC energies this changes and muon and pion have higher probabilities of radiating photons. This becomes visible by the several orders of magnitude higher critical energy for muons.

2.3 Multiple Scattering

In addition to inelastic collisions with atomic electrons, charged particles undergo repeated elastic Coulomb scatterings off nuclei, but with smaller probability. If we neglect screening and spin effects, these collisions are described by the Rutherford cross section. The angular dependence for a single scattering in the non-relativistic approximation is given by

$$\frac{d\sigma}{d\Omega} = z^2 Z^2 r_e^2 \frac{m_e c}{p\beta} \frac{1}{4\sin^4(\theta/2)}, \qquad (2.82)$$

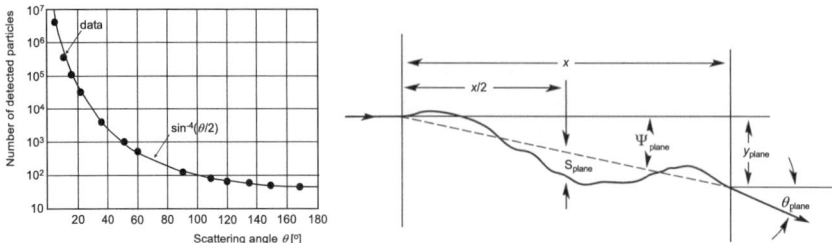

Fig. 2.30 Left: Single Coulomb scattering of α particles off a thin gold foil. Measurements (points) in comparison to the Rutherford cross section (solid line) [92]. Right: Schematic view of multiple Coulomb scattering showing the zig-zag path. Reprinted under CC-BY-NC-4.0 Licence with kind permission from [1], © 2024, the Particle Data Group LBNL

where θ is the scattering angle. The vast majority of particles experience small-angle deflections because of the $1/\sin^4(\theta/2)$ term. Figure 2.30 (left) shows single Coulomb scatterings of α particles off a thin gold foil. The data taken from the original Geiger-Marsden experiment [91] are in good agreement with the Rutherford cross section [92]. Large-angle scattering also occurs but with much smaller probability. With increasing thickness, the number of scatterings increase. Since the scattering angle is typically small, the particle trajectory is a zig-zag path, as illustrated in Fig. 2.30 (right).

If the number of small-angle scatters in the medium is sufficiently large ($N > 20$), the net scattering and the displacement distributions approach Gaussian distributions with a mean value of zero degrees, which follows from the central limit theorem. The rms is given in (2.83). However, for a few additional hard scatters, non-Gaussian tails appear. They are well represented by Molière's theory [93,94], which is discussed below. After a particle passed through the medium, its direction has changed. If the particle originally moves along the z-direction, it will have acquired components in the x- and y-direction. For small scattering angles the total deflection angle θ_{space} is approximately $\theta_{space} \sim \sqrt{(\theta_x^2 + \theta_y^2)}$, where $\theta_{x,y}$ are the deflection angles in the x-z and y-z planes and the solid angle is given by $d\Omega \sim d\theta_x d\theta_y$. Defining the rms scattering angle by $\theta_0 = \theta_{plane}^{rms} = \frac{1}{\sqrt{2}}\theta_{space}^{rms}$, the projected angular distribution is approximately Gaussian if a small probability for large-angle scattering ($\sim 2\%$) is neglected. Lynch and Dahl calculated the rms width to be [1,95]

$$\theta_0 = \frac{13.6 \text{ MeV}/c}{\beta c p} z \sqrt{t_r} \left\{1 + 0.088 \ln\left(t_r \frac{z^2}{\beta^2}\right)\right\} \text{ [radian]}, \qquad (2.83)$$

where $t_r = x/X_0$ is the thickness of the medium in units of radiation lengths. This takes into account the p and z dependence quite well at small Z, but for large Z and small x the β-dependence is not well represented. Improvements are discussed in [95].

Equation (2.83) describes the scattering in a single material. To determine the scattering of a particle traversing a compound, molecule or many layers of different elements we first have to determine x/X_0 for the medium, which is inserted then

2.3 Multiple Scattering

into (2.83). The approximate angular distributions for the deflection angle θ in a plane (projected) and in space, respectively, are given by

$$P(\theta)\mathrm{d}\Omega = \frac{1}{\sqrt{2\pi\theta_0^2}} \exp\left\{-\frac{\theta_{plane}^2}{2\theta_0^2}\right\} \mathrm{d}\Omega,$$

$$P(\theta)\mathrm{d}\Omega = \frac{1}{2\pi\theta_0^2} \exp\left\{-\frac{\theta_{space}^2}{2\theta_0^2}\right\} \mathrm{d}\Omega. \tag{2.84}$$

Here, $\theta_{space}^2 \simeq (\theta_{plane,x}^2 + \theta_{plane,y}^2)$ where the coordinates x and y are orthogonal to the direction of motion and $\mathrm{d}\Omega = \mathrm{d}\theta_{plane,x}\mathrm{d}\theta_{plane,y}$. Note that deflections into $\theta_{plane,x}$ and $\theta_{plane,y}$ are independent and have identical distributions.

General calculations of multiple scattering are rather complicated. Various calculations with different levels of sophistication exist in the literature as reviewed by Scott and Hemmer [96] and Farquahr [97]. The small-angle approximations by Molière and by Snyder and Scott are the most used calculations whose formulations are essentially equivalent. Comparison with measurements have shown that they are valid for all particles up to angles of $\theta \simeq 30°$ except for slow electrons ($\beta < 0.05$) and electrons in heavy elements.

We follow here the discussion by Bethe [94] and Scott [96] whose calculations are based on Molière's theory that depends just on a single parameter, the screening angle χ_a', defined by

$$\chi_a' = 1.08028\chi_\alpha = 1.08028\chi_0\sqrt{(1.13 + 3.76\alpha_B^2)}, \tag{2.85}$$

where α_B is the so-called Born parameter given by

$$\alpha_B = \frac{zZq_e^2}{\hbar\beta c} = \frac{zZ\alpha}{\beta} \tag{2.86}$$

and angle χ_0 defined as the ratio of the de-Broglie wavelength $\lambdabar_{\mathrm{dB}} = \hbar/p$ of the charged particle and the screening radius[12] $r_0 = 0.885 r_e Z^{-1/3}/\alpha^2$, yielding

$$\chi_0 = \frac{\lambdabar_{\mathrm{dB}}}{r_0} = 0.472 Z^{-\frac{1}{3}} \frac{m_e c}{p} \text{ [degrees]}. \tag{2.87}$$

For momentum $p \simeq m_e c$, which corresponds to an electron kinetic energy of 211 keV or $255(m_e/m)$ keV of a heavier particle with mass m, χ_0 is typically less than $2°$ for the heaviest elements [96]. The angular distribution depends only on the ratio of the characteristic angle to the screening angle. The characteristic angle denotes the

---

[12] This is taken as the Thomas-Fermi radius [29].

angle for which a scattering with angle χ_c or larger occurs with unit probability. For a homogeneous scatterer with no energy loss it is defined by

$$\chi_c^2 = \frac{4\pi q_e^4 z^2 Z(Z+1) N L_{\text{path}}}{p^2 \beta^2 c^2} \quad [\text{radians}^2] \tag{2.88}$$

for electrons, where N is the atomic density and L_{path} is the distance along the path. For heavy particles, we have to replace $Z(Z+1)$ by Z^2. Plugging $N = N_A \rho / A$ into (2.88) we can express χ_c in degrees by

$$\chi_c = \left[22.7 \sqrt{\frac{\rho L_{\text{path}} Z(Z+1)}{A}} \frac{z}{p\beta c} \quad [\text{degrees}] \right]. \tag{2.89}$$

Molière's theory, which is valid for $\chi_c/\chi_0 \geq 5$, expresses the angular distribution in terms of a power series in the parameter $1/B$, which is approximately given by

$$B = 2.583 \log_{10} \frac{\chi_c^2}{\chi_\alpha^2}. \tag{2.90}$$

For values $100 < \chi_c^2/\chi_\alpha^2 < 10^5$ (10^9) the accuracy is better than 0.5% (3%). In this range B increases from 6.32 to 24.4. Since Molière's theory is valid from $\chi_c/\chi_\alpha \geq 5$, the lowest value of B is 4.76. The expansion then yields

$$f(\theta)\theta d\theta = \vartheta d\vartheta \left[2\exp(-\vartheta^2) + \frac{f^{(1)}(\vartheta)}{B} + \frac{f^{(2)}(\vartheta)}{B^2} + \ldots \right], \tag{2.91}$$

where

$$\vartheta = \frac{\theta}{\chi_c \sqrt{B}}. \tag{2.92}$$

The first-order and second-order functions $f^{(1)}(\vartheta)$ and $f^{(2)}(\vartheta)$ are numerical integrals, which are tabulated in Table II in [94]. Here, we illustrate how well the Gaussian parameterization works with respect to Molière's theory. Figure 2.31 (left) shows the angular dependence of the first-order and second-order functions $f^{(1)}(\vartheta)$ and $f^{(2)}(\vartheta)$ in comparison to the main Gaussian contribution. Figure 2.31 (right) shows the Molière angular distribution in comparison to the Gaussian approximation. Note that the most probable scattering angle is $0°$.

Multiple scattering measurements have been performed by various groups. We select here examples from Bichsel [98], who measured multiple scattering of protons in the energy range 0.7 to 4.8 MeV in several metals (Al, Ni, Ag, Au). Figure 2.32 (left) shows the angular distribution $f(\theta^2)$ as a function of θ^2 for 2.18 MeV protons in Al foils with a thickness of 3.42 mg/cm^2. The data agree well with Molière's calculation whereas the Gaussian approximation fits rather poorly. The deviations are 10–20% at small angles ($\theta < 1.7°$) and become very large at larger angles.

2.3 Multiple Scattering

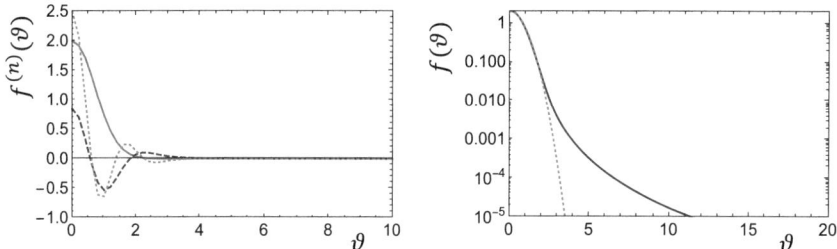

Fig. 2.31 Left: The first-order and second-order expansion functions $f^{(1)}(\vartheta)$ (dashed curve) and $f^{(2)}(\vartheta)$ (dotted curve) in the multiple scattering angular distribution in comparison with the Gaussian zeroth-order contribution (solid curve). Right: The complete multiple scattering angular distribution up to second order (solid curve) in comparison to the main Gaussian contribution (dashed curve)

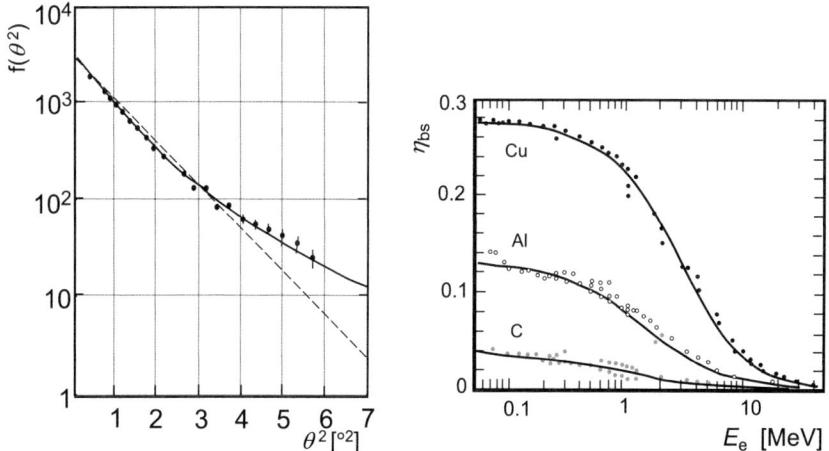

Fig. 2.32 Left: Measured angular distribution (points with error bars) as a function of θ^2 for 2.18 MeV/c protons scattering off aluminum foils with a thickness of 3.42 mg/cm^2 in comparison to the normalized Molière function (solid curve) and the Gaussian approximation (dashed curve). Reprinted with kind permission from [98], © 1958, AIP Publishing. All rights reserved. Right: The backscattering coefficient η_{bs} versus electron kinetic energy in carbon, aluminum, and copper. Reprinted with kind permission from [99], © 1971, Elsevier. All rights reserved

Note that electrons are particularly susceptible to large-angle scatterings because of their small mass. So the probability for back scattering is rather high. For low-energy electrons (< 1 MeV), the effect of backscattering is rather strong, particularly in high-Z materials. The ratio of backscattered electrons to incident electrons is called albedo or backscattering coefficient $\eta_{bs} = N_{bs}/N_{as}$, where N_{bs} is the number of backscattered electrons and N_{as} is the number of all electrons that enter the material. For any detector design, backscattering is an important consideration. Figure 2.32 (right) shows η_{bs} as a function of kinetic energy E_e for electrons in lighter materials (C, Al, Cu). For example, for non-collimated electrons on a NaI detector up to 80% of the electrons may be reflected back.

2.4 The Cherenkov Effect

We now turn our discussion to the fourth term in (2.6), which is independent of the photoabsorption cross section. Therefore, it is the only term that is present in the optical region where $\sigma_\gamma = 0$, thus representing Cherenkov radiation. Note that Cherenkov radiation occurs in radiators that have a thickness much larger than the wavelength of the radiation, $L_{\text{rad}} \gg \lambda$. Below the Cherenkov threshold, the phase Θ in (2.6) is zero jumping to π above threshold.

We denote the four-momenta of the heavy charged particle before and after photon emission by $p = (E/c, \vec{p})$ and $p' = (E'/c, \vec{p}')$, respectively. Energy-momentum conservation yields

$$p' = p - p_\gamma, \tag{2.93}$$

where $p_\gamma = (\hbar\omega/c, \hbar\vec{k})$ is the four-momentum carried away by the photon. While $\hbar\omega$ is the photon energy, $\hbar|\vec{k}| = \hbar 2\pi/\lambda$ is the magnitude of the photon momentum. Squaring (2.93) and inserting the energies and three-momenta explicitly, yields

$$\omega E = \vec{k} \cdot \vec{p} c^2. \tag{2.94}$$

For small photon energies ($\hbar\omega \ll \gamma mc^2$), this determines ω since $E = \gamma mc^2$ and $\vec{p} = \vec{\beta}\gamma mc$,

$$\omega = \vec{k} \cdot \vec{\beta} = |\vec{k}||\vec{\beta}|c\cos\theta_c, \tag{2.95}$$

where θ_c is the Cherenkov angle. The dispersion relation provides a link between the photon energy and the photon momentum in the medium,

$$\omega^2 = \frac{k^2 c^2}{\epsilon}. \tag{2.96}$$

Combining (2.95) and (2.96) yields the Cherenkov condition,

$$\cos\theta_c = \frac{1}{n(\lambda)\beta}. \tag{2.97}$$

Thus, the threshold velocity for Cherenkov radiation is $\beta_{\text{th}} = 1/n$. Note that $\cos\theta_c$ is also the ratio of the distance the wave travels, $ct/n(\lambda)$, and the distance the particle travels, βct. In a dispersive medium, the group velocity is

$$v_g = d\omega/dk = c/\left[n(\omega) + \omega dn/d\omega\right], \tag{2.98}$$

which becomes $v_g = c/n$ for a non dispersive medium.[13] For particle velocities where $\beta \simeq 1$, photons will be radiated along a cone with angle η_c, where

$$\cot \eta_c = \left[\frac{d}{d\omega} (\omega \tan \theta_c) \right]_{\omega_0} = \left[\tan \theta_c + \beta^2 \omega n(\omega) \frac{dn}{d\omega} \cot \theta_c \right]_{\omega_0}, \qquad (2.99)$$

where ω_0 is a central value of small frequency under consideration. Note that $\theta_c + \eta_c \neq 90°$ unless the medium is non dispersive. Thus, the larger the relative velocity, the larger the Cherenkov angle. The maximum angle is given at $\theta_c^{max} = \arccos(1/n(\lambda))$.

Multiplying the fourth term in (2.6) by the electron density, Z and path length L_{rad} yields the number of radiated Cherenkov photons dN_γ per frequency interval $d\omega$ for long radiators ($L_{rad} \gg \lambda$),

$$\frac{dN_\gamma}{\hbar d\omega} = \frac{z^2 \alpha}{\hbar c} \left(1 - \frac{1}{\beta^2 n^2(\hbar\omega)} \right) \cdot L_{rad}. \qquad (2.100)$$

Integration over the photon energy yields the total number of Cherenkov photons. In Sect. 9.3, we discuss different Cherenkov detectors.

2.5 Transition Radiation

Transition radiation occurs in short radiators where $L_{rad} \sim \lambda$. Here, the fourth term in (2.6) can be used to determine the intensity of transition radiation, which is typically radiated in the X-ray domain. The theory of transition radiation has been developed since 1946 [100–102]. Experimental evidence came many years later [103–105]. We focus here on the case where particles are highly relativistic,[14] i.e. $\gamma \gg 1$. The dielectric constant can be approximated in terms of the plasma frequency ω_p by

$$\epsilon = 1 - \omega_p^2/\omega^2 \simeq 1. \qquad (2.101)$$

The energy $\hbar\omega_p$ is typically at the order of tens of eV for solids and fractions of eV for gases. Specific values are listed in Tables 2.4, 2.5 and 2.6. Table 2.7 summarizes several properties of radiator materials. We see that LiH, lithium and beryllium are suitable solid radiators.

For a single interface and small emission angles, the differential transition radiation intensity radiated per frequency interval $d\omega$ and angle $d\theta$ is given by [100, 101],

$$\frac{d^2 S_0}{d(\hbar\omega) d\theta} = \frac{2\alpha\theta^3}{\pi} \left[\frac{1}{\frac{1}{\gamma^2} + \frac{\omega_1^2}{\omega^2} + \theta^2} - \frac{1}{\frac{1}{\gamma^2} + \frac{\omega_2^2}{\omega^2} + \theta^2} \right], \qquad (2.102)$$

---

[13] For convenience, we typically leave off the λ dependence from now on unless it is explicitly necessary.
[14] The particles are assumed to have a charge of $z = \pm 1$.

Table 2.7 Properties of radiator materials, density, plasma frequency ω_p, frequency ω_k where the dominant process becomes inelastic scattering with a uniform energy dependence, the natural threshold $\gamma_1 = 2.5 \cdot (\omega_p/[\text{eV}]) \cdot (\ell_1/[\mu\text{m}])$ for foil thickness ℓ_1, the signal to ionization ratio and the relative number of photons. †This is for radiation below the K-edge. Reprinted with kind permission from [107], © 1977, Elsevier. All rights reserved

Material	Density ρ	ω_p	ω_k	Natural Threshold	rel. $\frac{\text{Signal}}{\text{Ionization}}$	N_{photons}
	[g/cm^3]	[eV]	[keV]	γ_1	$= S/I = \left(\frac{\omega_{k,\text{Li}}}{\omega_k}\right)^{3.5}$	$\left(\frac{\rho}{\rho_{\text{Li}}}\right)^{1.25} \cdot S/I$
H$_2$	0.07	7.86	2.8	1,070	77	6.3
LiH	0.82	19.1	8.9	1,400	1.35	2.4
Be	1.85	26.9	13.6	1,520	0.30	1.45
Li	0.53	14.2	9.7	2,050	1.0	1.0
B	2.37	31.1	17	1,640	0.14	0.92
C$_{\text{diamond}}$	3.52	39.4	21	1,600	0.067	0.71
C$_{\text{graphite}}$	2.25	31.5	21	2,000	0.067	0.41
B$_4$C	2,52	32.3	19	1,760	0.095	0.67
Al$_2$O$_3$	3.97	41.0	45	3,300	0.0047	0.058
Al$_2$O$_3^\dagger$				110	0.05	0.62
CH$_2$	0.9	18.6	19.3	2,900	0.09	0.18

where ω_1 and ω_2 are the plasma frequencies in the two media. For vacuum, we have $\omega_2 = 0$. Typically, the gap between the foils in a transition radiation detector uses gas detectors for the detection of X-rays for which $\omega_2 \ll \omega_1$. So we use the approximation $\omega_2 = 0$ and define $\omega_p = \omega_1$. The general case is discussed in ref [106]. Integration over θ yields,

$$\frac{dS_0}{d(\hbar\omega)} = \frac{\alpha}{\pi}\left[\left(1 + 2\frac{\omega^2}{\omega_p^2\gamma^2}\right) \times \ln\left(1 + \frac{\gamma^2\omega_p^2}{\omega^2}\right) - 2\right]. \quad (2.103)$$

The differential transition radiation intensity increases logarithmically with γ and the plasma frequency but decreases logarithmically with the photon energy. Integration over all transition radiation frequencies yields the total radiation intensity,

$$S_0 = \frac{\alpha}{3}\hbar\omega_p\gamma. \quad (2.104)$$

In typical radiators the alternating dielectric media consist of thin foils and air gaps or noble-gas gaps for which $\omega_1 > \omega_2$ holds. The radiation shows a sharp peak in the forward direction at

$$\theta \simeq \frac{1}{\gamma}. \quad (2.105)$$

2.5 Transition Radiation

Generally, we cannot observe transition radiation from a single layer. In practice, we encounter at least two interfaces, the entry and exit faces of the thin foil. However, to observe a sizable signal, two interfaces are not sufficient. We need a large array of foils and gaps. Since the intensity of transition radiation from a single foil is very small ($S_0 = 10^{-2}\gamma$ eV), we need to stack a large array of foils. The double differential transition radiation intensity for a periodic radiator consisting of N foils with thickness ℓ_1 and spacing ℓ_2 is obtained by linear superposition of the single-interface,

$$\frac{d^2 S_N}{d(\hbar\omega)d\theta} = \frac{d^2 S_0}{d(\hbar\omega)d\theta} \times 4\sin^2\left(\frac{\ell_1}{\zeta_1(\omega)}\right) \frac{\sin^2[N(\ell_1/\zeta_1(\omega) + \ell_2/\zeta_2(\omega))]}{\sin^2(\ell_1/\zeta_1(\omega) + \ell_2/\zeta_2(\omega))}, \quad (2.106)$$

where $\zeta_1(\omega)$ and $\zeta_2(\omega)$ are the formation zones of the radiator and the gas gap, respectively,

$$\zeta_1(\omega) = \frac{4c}{\omega}\left(\frac{1}{\gamma^2} + \theta^2 + \frac{\omega_p^2}{\omega^2}\right)^{-1},$$

$$\zeta_2(\omega) = \frac{4c}{\omega}\left(\frac{1}{\gamma^2} + \theta^2\right)^{-1}. \quad (2.107)$$

The formation zone represents the distance along the charged particle trajectory in a given medium after which the separation between particle and generated photon is of the order of the photon wavelength. For $\theta = 1/\gamma$ and $\hbar\omega = \gamma\hbar\omega_p$ we get $\zeta_1(\gamma\omega_p) = \gamma c/(\sqrt{2}\omega_p)$, which is of the order of tens of microns for real detectors. For $N = 1$ we obtain the intensity of radiation in a single foil,

$$\frac{d^2 S_1}{d(\hbar\omega)d\theta} = \frac{d^2 S_0}{d(\hbar\omega)d\theta} \times 4\sin^2\left(\frac{\ell_1}{\zeta_1(\omega)}\right). \quad (2.108)$$

Note that the $\sin^2$ term comes from interference of waves produced at the entry and exit of the thin foil. If you call the two amplitudes by $\mathcal{A}$ and $-\mathcal{A}\exp[i\delta]$, respectively, the intensity is given by $4|\mathcal{A}|\sin^2[\delta/2]$, which has the same form as the second term in (2.108). The detailed calculations can be found in [106]. We present here the $dS_N/d(\hbar\omega)$ spectrum per interface as a function of the photon energy for a single interface as well as for a stack of 200 $\ell_1 = 25$ μm mylar foils separated by 1.5 mm air gaps depicted in Fig. 2.33. Without re-absorption, the radiation intensity of a periodic radiator shows oscillations around the transition radiation spectrum for a single interface. Including re-absorption the yield is heavily suppressed in the low-energy region. The curves are obtained by taking the absorption cross sections into account. Figure 2.33 indicates that most of the transition radiation is emitted near the last maximum expected in $dS_N/d(\hbar\omega)$. For this stack, the highest yield is observed in the 10–20 keV energy range.

In real detectors the formation zone is tens of microns. Figure 2.34 (left) shows the detected X-ray yield as a function of the electron energy for a radiator consisting of 188 $\ell_1 = 25$ μm thick mylar foils spaced in air at a distance of $\ell_2 = 1.5$ mm.

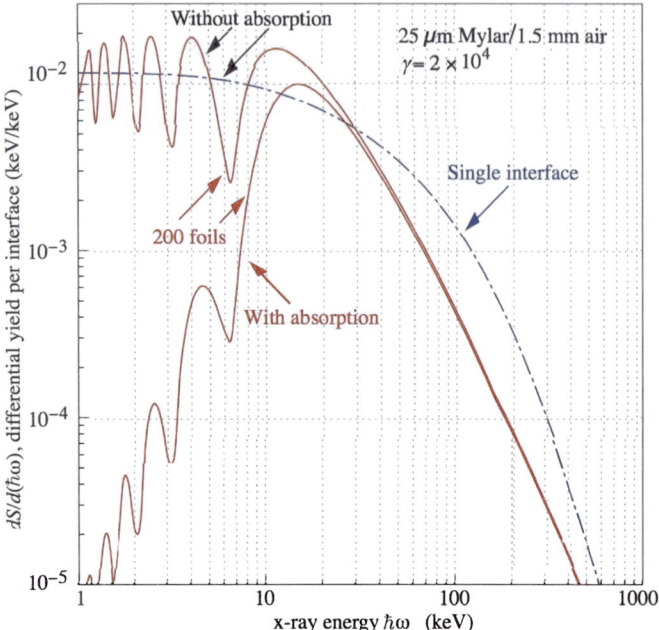

Fig. 2.33 Intensity distribution of transition radiation as a function of the X-ray energy for a single interface (dash-dotted line) specified by (2.103) and for a radiator consisting of 200 $\ell_1 = 25$ μm thick Mylar foils spaced by 1.5 mm air gaps (solid lines) for both without absorption and with absorption [106]. The particle energy is $\gamma = 2 \times 10^4$. Reprinted under CC-BY-NC-4.0 Licence with kind permission from [1], © 2024, the Particle Data Group LBNL

The predicted saturation is confirmed by the data measured with a krypton detector. Figure 2.34 (right) shows the detected X-ray yield in krypton/CO_2 chambers as a function of the foil thickness of polypropylene and mylar for 10 GeV electrons. The X-ray yield is similar for both polypropylene and mylar and the measurements are consistent with calculations.

About half of the energy is emitted in a range $0.1 \leq (\hbar\omega)/(\gamma\hbar\omega_p) \leq 1$. For a particle with $\gamma = 10^3$, the radiated photons are in the soft X-ray range of 2–40 keV. The γ dependence of the emitted energy thus comes from the hardening of the spectrum rather than from an increased quantum yield. The number of photons with energy $\hbar\omega > \hbar\omega_0$ is given by

$$N_\gamma(\hbar\omega > \hbar\omega_0) = \frac{\alpha z^2}{\pi}\left[\left(\ln\frac{\gamma\hbar\omega_p}{\hbar\omega_0} - 1\right)^2 + \frac{\pi^2}{12}\right]. \quad (2.109)$$

Corrections are of the order of $(\hbar\omega_0/\gamma\hbar\omega_p)^2$. The number of photons above a fixed energy $\hbar\omega_0 \ll \gamma\hbar\omega_p$ grows as $(\ln\gamma)^2$, however, the number above a fixed fraction of $\gamma\hbar\omega$ is constant. For example, for $\hbar\omega > \gamma\hbar\omega_p/10$, we get $N_\gamma = 2.519z^2\alpha/\pi = 0.59\%z^2$. In Sect. 9.4 we discuss transition radiation detectors and measurements.

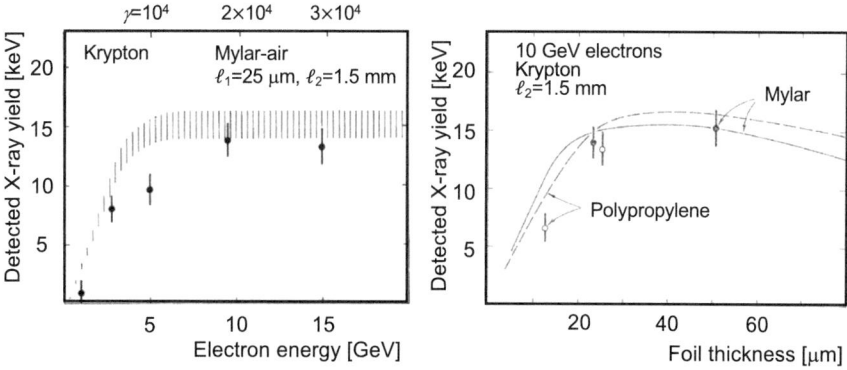

Fig. 2.34 Left: Detected X-ray yield (points with error bars) in a krypton chamber compared to a computation (grey shaded band) as a function of the electron energy for a radiator consisting of 188 $\ell_1 = 25$ μm thick mylar foils spaced in air at a distance of 1.5 mm. The computation has a $\pm 8\%$ uncertainty. Right: Detected X-ray yield in krypton/CO_2 chambers as a function of the foil thickness for 10 GeV electrons for polypropylene (open circles) and mylar foils (solid points) [106]. The foil spacing in air is 1.5 mm. The calculations for the two foils are shown by dashed and solid lines, respectively. The formation zones for 15 keV X-rays ($\gamma = 2 \times 10^4$) are 20 μm for mylar and 27 μm for polypropylene. Both reprinted with kind permission from [106], © 1974, AIP Publishing. All rights reserved

2.6 Detection of Photons

A photon traversing a medium can experience three different processes

(i) Photoelectric absorption.
(ii) Compton scattering and Rayleigh scattering.[15]
(iii) Pair creation.

Figure 2.35 shows illustrations of these processes, which all reduce the initial intensity exponentially with material depth x,[16]

$$I_{\text{ph}}(x)) = I_0 \exp\left(-\hat{\mu}^{\text{ph}} x\right), \tag{2.110}$$

where $\hat{\mu}^{\text{ph}}$ is the linear absorption coefficient that is related to the photon absorption cross section σ_{ph} by $\hat{\mu}^{\text{ph}} = \sigma_{\text{ph}} N_A \rho / A$. Note that frequently the parameter $\mu_a^{\text{ph}} = \hat{\mu}^{\text{ph}}/\rho$ is quoted. The linear absorption coefficient $\hat{\mu}^{\text{ph}}$ has units of $[1/\text{cm}]$ while μ_a^{ph} has units of $[\text{cm}^2/\text{g}]$. The linear absorption coefficient is related to the photon mass attenuation length λ_{ph} by $\lambda_{\text{ph}} = 1/\mu_a^{\text{ph}}$. So after passing $1 \cdot \lambda_{\text{ph}}$, the initial intensity

---

[15] This classical process describes the coherent scattering of photons by atoms where no energy is transferred to the medium.
[16] For photoelectric absorption and pair creation the photon disappears in the interaction, while for Compton scattering the photon survives with reduced energy and different direction.

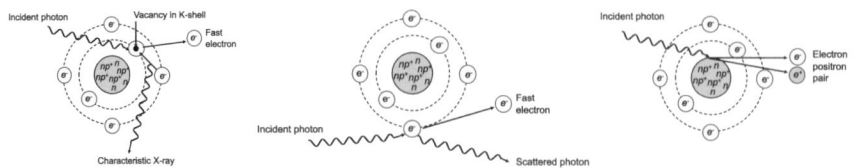

Fig. 2.35 Illustration of the photoelectric effect (left), Compton scattering (center) and pair creation (right)

is reduced by 63%. For compounds or chemical mixtures, the effective attenuation length is given by

$$\frac{1}{\lambda_{\text{eff}}} = \sum_i \frac{w_i}{\lambda_{\text{ph}}^i}, \qquad (2.111)$$

where w_i and λ_{ph}^i are weight and photon mass attenuation length for element i.

Figure 2.36 shows the photon total cross sections for carbon (top) and lead (bottom). At low photon energies, the photoelectric absorption is the dominant process although Compton scattering, Rayleigh scattering, and photonuclear absorption also contribute. With increasing E_γ the cross section σ_{ph} decreases rapidly. The photoelectric absorption cross section shows discontinuities (absorption edges) when the energy reaches thresholds for photoionization of various atomic levels. The highest electron binding energy is for the K-shell, which is the most inner shell containing two electrons. For lower photon energies, we observe a similar pattern for the higher-energy-level atomic shells (L, M, ...), called the L–absorption edge, M–absorption edge and so on. At intermediate energies particularly for low-Z materials, the Compton effect becomes sizable while at high energies the pair creation process dominates. Figure 2.37 shows the photon mass attenuation length as a function of the photon energy for hydrogen, carbon, silicon, iron, tin and lead. Here, the discontinuities in the cross section (K-edge, L-edge, M-edge, ...) are also clearly visible.

2.6.1 Photoelectric Absorption

In the 10 eV to a few 100 keV energy range, photons have sufficient energy to excite or ionize the material they traverse. The photon is absorbed transferring all its energy onto an electron in the atomic shell of the medium. The emitted electron has a kinetic energy $E_e = E_\gamma - E_B$, where E_B is the binding energy. Note that a free electron cannot absorb a photon because this would violate momentum conservation. For the bound electron, the nucleus absorbs the recoil momentum. After the emission of the photoelectron, the atom remains in an excited state. De-excitation to the ground state may proceed either via the Auger effect in which several electrons are involved in a radiationless internal rearrangement or via a transition of an electron from an outer shell with the emission of a photon. The fraction of de-excitations producing a photon is called the fluorescence yield. For the K-shell, the fluorescence yield increases with Z, as shown in Fig. 2.38 (left). For example in argon, about 15% of the

2.6 Detection of Photons

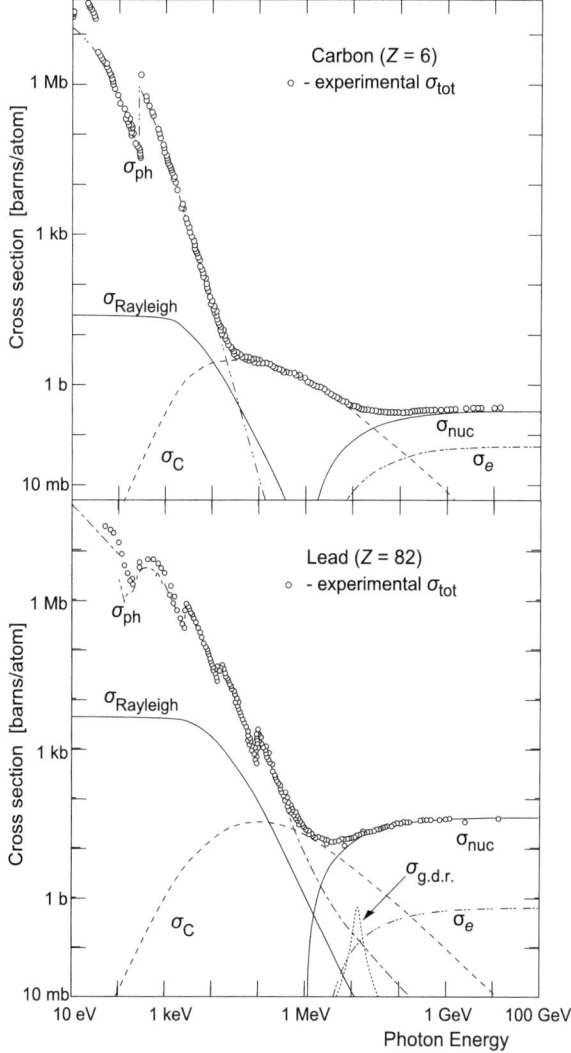

Fig. 2.36 Top: Photon total cross section as a function of photon energy in carbon in a double logarithmic scale. Bottom: Photon total cross section as a function of photon energy in lead in a double logarithmic scale [108]. The open circles show the measured cross sections. The individual contributions are: σ_{ph} = Atomic photoelectric effect in which the photon is absorbed and the electron is ejected (dash-dotted curves). $\sigma_{Rayleigh}$ = Coherent or Rayleigh scattering; the atom is neither excited nor ionized (solid curves). σ_C = Incoherent or Compton scattering; the photon looses energy that is transferred to an electron (dashed curves). σ_{nuc} = Pair production in the nuclear field (solid curves). σ_e = Pair production in the electron field (dash-double-dotted curves). $\sigma_{g.d.r}$ = Photonuclear interactions, most notably the Giant Dipole Resonance [109]; breakup of the nucleus. Reprinted under CC-BY-NC-4.0 Licence with kind permission from [1], © 2024, the Particle Data Group LBNL

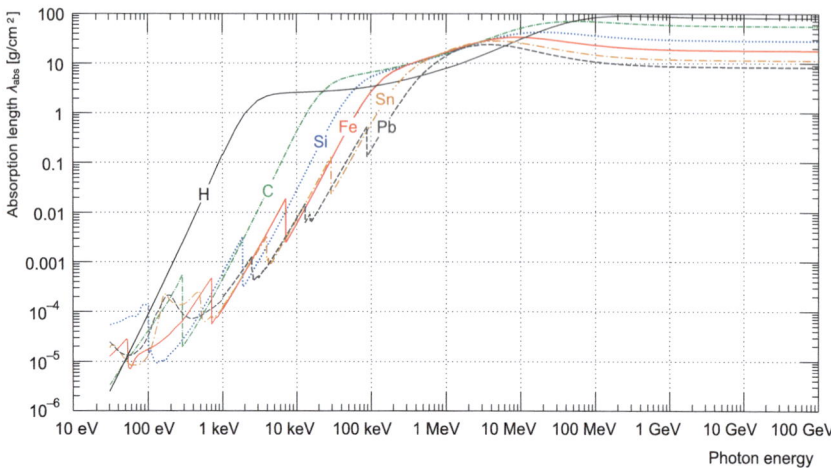

Fig. 2.37 The photon mass attenuation length λ_{abs} or mean free path $\lambda_{abs} = 1/(\mu_a^{ph}\rho)$ as a function of photon energy for hydrogen, carbon, silicon, iron, tin and lead. The data for photon energies 30 eV < E_γ < 1 keV come from [110] and for 1 keV < E_γ < 100 GeV from [111]. Reprinted under CC-BY-NC-4.0 Licence with kind permission from [1], © 2024, the Particle Data Group LBNL

photoelectric absorptions are followed by photon emission while for the other 85% the excitation energy is transferred to two electrons. The secondary photon emitted at an energy just below the K-edge has a very long mean free path and, therefore, escapes detection. The primary photoelectron is emitted in a preferential direction depending on its energy. Figure 2.38 (right) shows the probability for photoelectron emission in angular space with respect to the direction of the incoming photon. Up to a few tens of keV, the direction of the emitted photoelectron is approximately perpendicular to the direction of the incoming photon. Multiple scattering, however, quickly randomizes the direction of the ionizing photoelectron.

The theoretical calculation of the photoelectric effect in general is difficult because of the complexity of Dirac wave functions for atomic electrons. For energies above the K−shell, typically only electrons of the K−shell are involved. If $\eta_r = E_\gamma/m_e c^2 < 1$, we can perform a non-relativistic calculation using the Born approximation. If θ denotes the polar angle between the direction of the photon and that of the emitted electron and ϕ denotes the azimuth angle between the plane defined by the photon and electron momenta and the plane defined between the direction of the photon and the direction of the photon polarization vector, we obtain the differential cross section per atom [112],

$$\frac{d\sigma_{ph}}{d\Omega} = 4\sqrt{2}r_e^2 Z^5 \alpha^4 \frac{1}{\eta_r^{7/2}} \frac{\sin^2\theta \cos^2\phi}{(1-\beta\cos\theta)^4} \quad \text{for } \eta_K < \eta_r < 1, \quad (2.112)$$

where β is the relative velocity of the electron. So, most of the photoelectrons are emitted in the direction of the polarization of the primary photon (i.e. $\theta = \pi/2$, $\phi = 0$). In the photon direction, however, the cross section becomes zero unless $\beta = 1$.

2.6 Detection of Photons

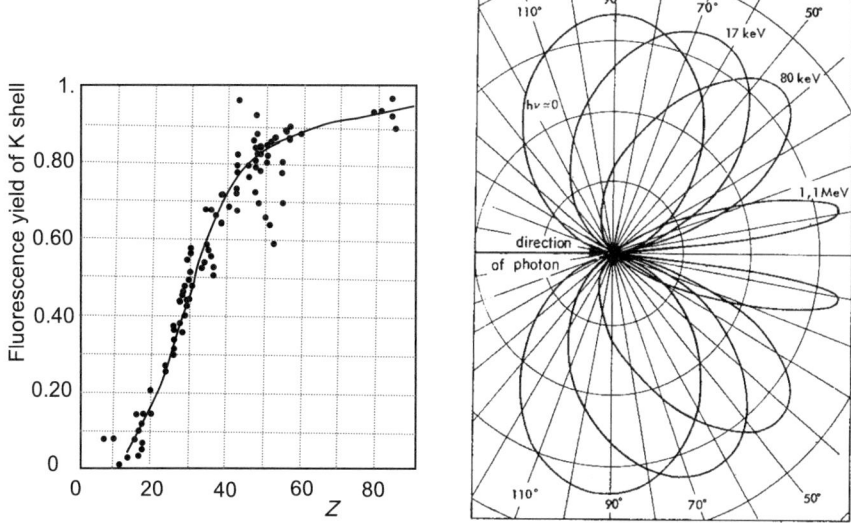

Fig. 2.38 Left: Fluorescence yield of the K-shell that produces a photon as a function of the atomic number. Reprinted with kind permission from [113], © 1953, AIP Publishing. All rights reserved. Right: Probability for an incoming photon to emit a photoelectron at a particular angle for different photon energies. Reprinted with kind permission from [114], © 1971, Bollati Boringhieri. All rights reserved

Multiplying by a factor of two to account for both electrons in the K-shell and integrating over the angles yields

$$\sigma_{\rm ph} = 4\sqrt{2}\sigma_{Th}Z^5\alpha^4 \frac{1}{\eta_r^{7/2}} \quad \text{for } \eta_K < \eta_r < 1, \tag{2.113}$$

where $E_K = (Z - 0.03)^2 m_e c^2 \alpha^2/2$ is the binding energy of K-shell electrons, $\eta_K = \sqrt{(E_K/(E_\gamma - E_K)}$ is the ratio of the binding (or ionization) energy of the K-shell to the electron kinetic energy and $\sigma_{\rm Th}$ denotes the classical Thomson scattering cross section per electron,

$$\sigma_{\rm Th} = \frac{8\pi}{3}r_e^2 = 0.665 \text{ [barn]}. \tag{2.114}$$

For example, for a 0.1 MeV photon in lead, the cross section is 1.19×10^4 barn while in carbon it is only 0.025 barn for the same photon energy. Multiplying the cross section by the atomic density N yields the absorption coefficient for energies near the K-shell,

$$\mu_K(\eta_r) = N\frac{32}{3}\sqrt{2\pi}r_e^2 Z^5 \alpha^4 \frac{1}{\eta_r^{7/2}}. \tag{2.115}$$

Formulas for the $L-$ and $M-$shells have been calculated also and can be found in [115].

For extremely high photon energies ($\eta_r \gg 1$), the cross section for photoelectric absorption is given by

$$\sigma_{\text{ph}} = 4\pi r_e^2 Z^5 \alpha^4 \frac{1}{\eta_r} \quad \text{for } \eta_r \gg 1. \tag{2.116}$$

Hulme and collaborators performed numerical calculations for a few heavy elements [116]. For very high photon energies the cross section is presented in [117]. It is important to note that the photoelectric absorption cross section increases as Z^5 and decreases as $\sim \eta_r^{-b}$ where $b \simeq 1$ at high energies and becomes much larger at lower energies, for example above the K shell $b \sim 3.5$.

2.6.2 Compton Effect

Photons also may scatter off an electron in the atomic shell. Initially, the electron is assumed to be at rest. Thus, its four-momentum is simply $p_e = (m_e c, \vec{0})$. The photon has a four-momentum of $p_\gamma = (E_\gamma/c, \vec{k})$. Denoting the four-momenta after the collision by p'_e and $p'_\gamma = (E'_\gamma/c, \vec{k}')$, energy-momentum conservation yields

$$p_e + p_\gamma = p'_e + p'_\gamma. \tag{2.117}$$

From this we can evaluate the energy of the scattered photon, $E'_\gamma = \hbar|\vec{k}'|c$, yielding

$$E'_\gamma = \frac{E_\gamma}{1 + \eta_r(1 - \cos\theta)}, \tag{2.118}$$

where θ is the angle of the scattered photon with respect to the direction of the initial photon. The kinetic energy of the electron is

$$E'_e = E_\gamma - E'_\gamma = E_\gamma \eta_r \frac{1 - \cos\theta}{1 + \eta_r(1 - \cos\theta)}. \tag{2.119}$$

There are two extreme cases:

i) Small angle scattering: $\theta \sim 0$; here $E_\gamma \simeq E'_\gamma$ and $E'_e \sim 0$.
ii) Back scattering at $\theta = \pi$. Here

$$E'_\gamma = \frac{E_\gamma}{1 + 2\eta_r} \to \frac{m_e c^2}{2} \quad \text{for } \eta_r \gg 1, \tag{2.120}$$

and the maximum electron energy, also called Compton edge, is

$$E'_e = E_\gamma \frac{2\eta_r}{1 + 2\eta_r} \to E_\gamma \quad \text{for } \eta_r \gg 1. \tag{2.121}$$

2.6 Detection of Photons

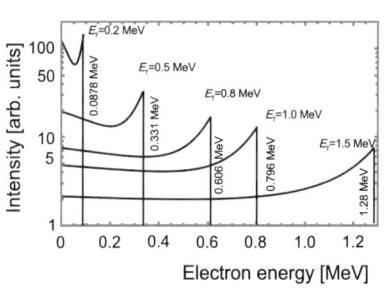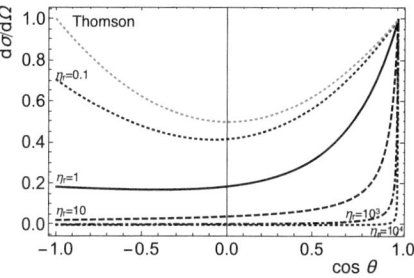

Fig. 2.39 Left: Compton cross section as a function of the electron kinetic energy for different initial photon energies. The energy of the Compton edge is shown for various photon energies by the vertical values. Right: Differential Compton cross section as a function of the scattering angle for different relative photon energies. For reference, the differential Thomson cross section is also given. With increasing values of η_r, the differential Compton cross section becomes highly peaked in the forward direction

The electron recoil spectrum in the detector is continuous between zero and the Compton edge. For large photon energies ($\eta_r \gg 1$) the gap between the Compton edge and the energy of the incident photon approaches the constant, $m_e c^2 / 2 = 0.256$ MeV. Figure 2.39 (left) shows the cross section as a function of the electron kinetic energy for different initial photon energies. At the maximum kinetic energy of the electrons, the Compton edge is clearly visible.

In the low-energy limit, Compton scattering approaches the classical limit of Thomson scattering, which represents scattering of photons by free electrons. Here, no energy is transferred to the medium. The angle of the emitted electron with respect to the direction of the initial photon is θ_e, which is determined by $\cot\theta_e = (1 + \eta_r)\tan(\theta/2)$. The Compton cross section as a function of the electron kinematic energy is given by

$$\frac{d\sigma_C}{d\eta_e} = \frac{\pi r_e^2 m_e c^2}{(\hbar\omega)^2}\left(2 + \frac{\eta_e^2 m_e^2 c^4}{(\hbar\omega)^2(1-\eta_e)^2} + \frac{\eta_e}{1-\eta_e}\left[\eta_e - \frac{2m_e c^2}{\hbar\omega}\right]\right), \quad (2.122)$$

where $\eta_e = E'_e/(\hbar\omega)$. Note, we have used $\hbar\omega/(m_e c^2)$ instead of η_r here to avoid confusion with η_e.

For relativistic photon energies, the Quantum Electrodynamics (QED) calculation leads to the Klein-Nishina cross section (see Fig. 2.36 (top, bottom) [118],

$$\sigma_C = 2\pi r_e^2\left[\left(\frac{1+\eta_r}{\eta_r^2}\right)\left\{\frac{2(1+\eta_r)}{1+2\eta_r} - \frac{1}{\eta_r}\ln(1+2\eta_r)\right\} + \frac{1}{2\eta_r}\ln(1+2\eta_r) - \frac{1+3\eta_r}{(1+2\eta_r)^2}\right]. \quad (2.123)$$

For Thomson scattering, the angular distribution of the scattered photon is forward-backward symmetric.

$$\frac{d\sigma_{Th}}{d\Omega} = \frac{1}{2}r_e^2(1+\cos^2\theta). \quad (2.124)$$

For relativistic Compton scattering, the angular distribution becomes asymmetric peaking at the forward direction. The Klein-Nishina cross section is

$$\frac{d\sigma_C}{d\Omega} = \frac{1}{4}r_e^2 \left(\frac{|\vec{k}'|}{|\vec{k}|}\right)^2 \left[\frac{|\vec{k}|}{|\vec{k}'|} + \frac{|\vec{k}'|}{|\vec{k}|} + 4(\vec{\epsilon}_\gamma \cdot \vec{\epsilon}_\gamma')^2 - 2\right], \quad (2.125)$$

where $\vec{\epsilon}_\gamma$ and $\vec{\epsilon}_\gamma'$ are the polarizations of the initial and scattered photons, respectively. For unpolarized photons, this leads to

$$\frac{d\sigma_C}{d\Omega} = \frac{1}{2}r_e^2 \left(\frac{1}{1 + \eta_r(1 - \cos\theta)}\right)^2 \left[1 + \cos^2\theta + \frac{\eta_r^2(1 - \cos\theta)^2}{1 + \eta_r(1 - \cos\theta)}\right]. \quad (2.126)$$

Figure 2.39 (right) shows the angular distribution of Compton scattering for several values of η_r. For $\eta_r \to 0$, we obtain the Thomson cross section while for increasing values of η_r, the angular distribution becomes more and more peaked in the forward direction, i.e. small-angle scattering dominates. In the limit $\eta_r \to \infty$, the scattering angle approaches zero. In summary, Compton scattering increases linearly with Z and becomes dominant in light materials for energies around 0.1–1 MeV (see Fig. 2.36). For example, the Compton cross section in carbon is 5.2 b at 0.1 MeV decreasing to 2.3 b at 1 MeV, which is much larger than the photoelectric absorption cross section. The corresponding cross sections in lead are 71.8 b and 30.8 b, which are smaller than the photoelectric absorption cross sections.

2.6.3 Pair Creation

For photon energies of $E_\gamma \geq 2m_e c^2 = 1.022$ MeV, the creation of an e^+e^- pair becomes possible. In order to conserve momentum, the conversion of the photon into an e^+e^- pair is possible only in the presence of a third object, typically a nucleus or, to a lesser extent, an electron. Note that pair creation is theoretically related to bremsstrahlung by simply interchanging particle four-momenta. As for bremsstrahlung, screening of the nuclear field by electrons in the atomic shells plays an important role. Here, the screening variables are defined by

$$\xi = \frac{100\eta_r}{\eta_+ \eta_- Z^{\frac{1}{3}}} \quad (2.127)$$

and

$$\xi' = \frac{100\eta_r}{\eta_+ \eta_- Z^{\frac{2}{3}}}, \quad (2.128)$$

where $\eta_+ = E_+/m_e c^2$ and $\eta_- = E_-/m_e c^2$ are the relative positron and electron energies and $\eta_r = E_\gamma/m_e c^2$ is the relative photon energy. If the recoil in the nucleus is negligible, we get $\eta_+ + \eta_- = \eta_r$. At extreme relativistic energies ($\eta_r \gg 1$) and

2.6 Detection of Photons

arbitrary screening, a Born approximation yields the differential pair creation cross section as a function of the relative positron energy η_+

$$\frac{d\sigma_{pair}}{d\eta_+} = \frac{\alpha r_e^2}{\eta_r^3} \left[Z^2\left((\eta_+^2 + \eta_-^2)\left(\Phi_1(\xi) - \tfrac{4}{3}\ln Z - 4f(Z)\right) + \tfrac{2}{3}\eta_+\eta_-\left(\Phi_2(\xi) - \tfrac{4}{3}\ln Z - 4f(Z)\right)\right) \right.$$
$$\left. + Z\left((\eta_+^2 + \eta_-^2)\left(\Psi_1(\xi') - \tfrac{8}{3}\ln Z\right) + \tfrac{2}{3}\eta_+\eta_-\left(\Psi_2(\xi') - \tfrac{8}{3}\ln Z\right)\right) \right]. \quad (2.129)$$

The functions $\Phi_1(\xi)$, $\Phi_2(\xi)$, $\Psi_1(\xi')$ and $\Psi_2(\xi')$ are the screening functions introduced in (2.58) and (2.60), respectively, and $f(Z)$ is the Coulomb correction. The first row was calculated by Bethe and Heitler [79] and the second row by Wheeler and Lamb [82]. For smaller energies the differential cross section is rather complicated. For further details we refer the reader to [119] or two review articles [80, 120].

Similarly as for bremsstrahlung, the cross section simplifies in the two limiting cases of no screening and complete screening. For no screening, where we have low energies ($2 \ll \eta_r \ll Z^{-1/3}/\alpha$), the cross section becomes [79],

$$\frac{d\sigma_{pair}}{d\eta_+} = 4\frac{\alpha r_e^2}{\eta_r^3}\left\{\left(\eta_+^2 + \eta_-^2 + \tfrac{2}{3}\eta_+\eta_-\right)\left[Z(Z+1)\left(\ln\frac{2\eta_+\eta_-}{\eta_r m_e c^2} - \tfrac{1}{2}\right) - Z^2 f(Z)\right]\right\}. \quad (2.130)$$

For complete screening ($\eta_r \gg Z^{-1/3}/\alpha$), we get [79],

$$\frac{d\sigma_{pair}}{d\eta_+} = 4\frac{\alpha r_e^2}{\eta_r^3}\left\{\left(\eta_+^2 + \eta_-^2 + \tfrac{2}{3}\eta_+\eta_-\right)\left[Z^2\left[\ln\frac{184.15}{Z^{1/3}} - f(Z)\right] + Z\ln\frac{1194}{Z^{2/3}}\right] - \frac{\eta_+\eta_-}{9}Z(Z+1)\right\}. \quad (2.131)$$

To obtain the total cross section for arbitrary screening, a numerical integration over η_+ is necessary. For no screening and complete screening, the integral can be solved analytically. For no screening, we get,

$$\sigma_{pair} = 4\alpha r_e^2\left[\tfrac{7}{9}\left(Z(Z+1)\ln(2\eta_r) - Z^2 f(Z)\right) - Z(Z+1)\frac{109}{54}\right]. \quad (2.132)$$

For complete screening, the total cross section becomes,

$$\sigma_{pair} = 4\alpha r_e^2\left[\tfrac{7}{9}\left(Z^2\left[L_{rad} - f(Z) + ZL'_{rad}\right]\right) - Z(Z+1)\frac{1}{54}\right], \quad (2.133)$$

where L_{rad} and L'_{rad} were defined in Chap. 2.2.2. Again, we can neglect the small last term in the square bracket.

For no screening, the cross section depends logarithmically on the photon energy while for compete screening, it just depends on properties of the material. For the latter, the mass absorption coefficient for pair creation $\mu_{pair} = \sigma_{pair} N_A/A$ approaches an asymptotic value μ^0_{pair} at high energies, which can be obtained from (2.73),

$$\mu^0_{pair} = 4\alpha r_e^2\frac{N_A}{A}\tfrac{7}{9}\left[Z^2[L_{rad} - f(Z)] + ZL'_{rad}\right] \simeq \frac{7}{9}\frac{1}{X_0}. \quad (2.134)$$

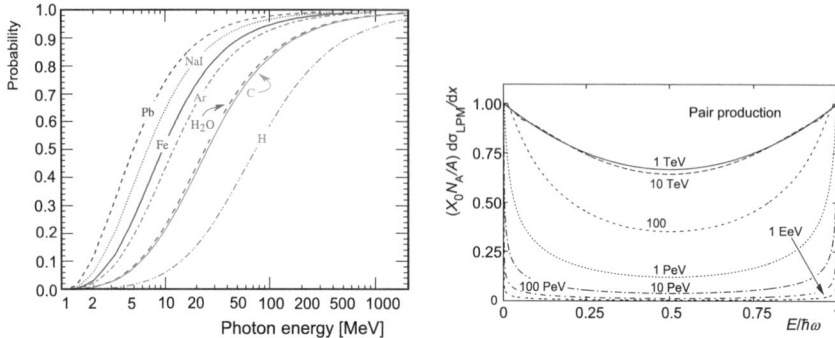

Fig. 2.40 Left: Probability P for photon conversion into an e^+e^- pair in different materials. Right: The normalized pair production cross section $d\sigma_{LPM}/dx$ versus fractional electron energy $E/(\hbar\omega)$. Both reprinted under CC-BY-NC-4.0 Licence with kind permission from [1], © 2024, the Particle Data Group LBNL

The factor 7/9 comes from the integration over the positron energy while the factor in the square brackets is the inverse of the radiation length. Thus, the probability that pair creation occurs at high energies within one radiation length is $P = 1 - \exp(-\frac{7}{9}) \sim$ 54%. Figure 2.40 (left) shows the probability for e^+e^- pair production as a function of the photon energy in different materials. At higher photon energies, pair creation becomes the dominant process. Figure 2.40 (right) shows the pair production cross section as a function of the relative electron energy for various initial photon energies. Note that asymmetric electron-positron energies are favored.

2.6.4 Electron-Photon Showers

The formation of an electron-photon shower is a statistical process. A high-energy photon converts within one radiation length into an e^+e^- pair with a probability of 54%. The created electrons and positrons loose energy via bremsstrahlung with a probability of 63% within $1X_0$. Positrons and electrons typically have different energies in the laboratory frame. The energy of bremsstrahlung photons is typically lower than half the energy of the radiating electron or positron. For photon energies above 1 MeV, the new bremsstrahlung photons will convert into e^+e^- pairs with high probability. Electrons and positrons with energies above the critical energy will continue to radiate. These processes produce an electromagnetic cascade or shower of electrons, positrons and photons. Figure 2.41 (left) illustrates the development of a photon shower.

Once the electrons and positrons reach the critical energy, the probability for radiation becomes small. Their energy loss by excitation and ionization becomes more probable. Positrons eventually will annihilate with an electron in the atomic shell into two photons. Since photons with energies below 1.022 MeV cannot create e^+e^- pairs any longer, the main processes become Compton scattering and photoelectric absorption. Since the Compton electrons have rather low energies, they loose energy

2.6 Detection of Photons

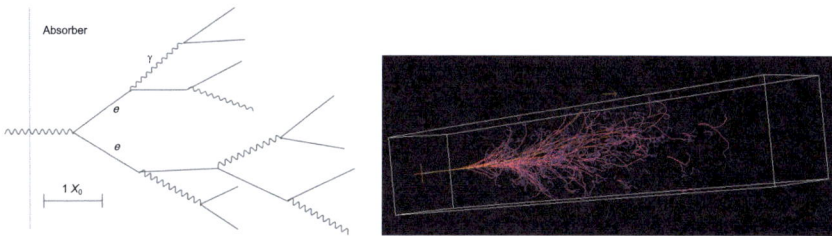

Fig. 2.41 Left: Schematic view of a photon shower as a function of radiation length. Solid lines represent electrons and positrons while squiggly lines represent photons. Reprinted with kind permission from [121], © 2009, R. M. Brown. All rights reserved. Right: Simulation of an 8.5 GeV photon shower in a CsI crystal [122]. Reprinted with kind permission from [122], © 1971, S. Menke. All rights reserved

by ionization and excitation. Thus, the cascade will eventually stop. The number of particles in the electromagnetic shower depends on the initial energy. A similar shower is obtained if we start with a high-energy electron or positron.

Let us consider a simple model of the shower development to illustrate some characteristic properties. We call the initial photon energy E_0 and assume that each produced particle has an interaction within one radiation length. So, after the first radiation length the photon has converted into an e^+e^- pair. For simplicity, we also assume that the electron and positron each carry half of the initial energy, $E_0/2$. After $2X_0$, both leptons will have radiated a photon. The evolving shower consists of two photons and an e^+e^- pair, where each particle carries 25% of the initial energy. After $3X_0$, the two photons have converted into e^+e^- pairs and the two leptons have radiated another photon. So, the shower now consists of three e^+e^- pairs and two photons, each of which carries an energy of $E_0/8$. This process will continue. After t_r radiation lengths, the shower contains 2^{t_r} particles, each carrying an energy of $E_0/2^{t_r}$. If the energy of electrons and positrons approaches the critical energy, $E_0/2^{t_r} \simeq E_c$, the energy loss via ionization will become the dominant process. Similarly, photons with energy below $2m_ec^2$ cannot produce e^+e^- pairs any longer and will be Compton scattered or absorbed. At this depth, particle production stops and the shower contains the maximum of particles. This point is called the shower maximum. In our simple model it is given by $t_{\max} = \ln(E_0/E_c)/\ln 2$ and the maximum number of particles is $n_{\max} = E_0/E_c$. After shower maximum, the cascade particles will continue to move through the medium eventually loosing all their energy. Note that this simple model just gives a qualitative view of the electromagnetic shower development. Detailed studies of electromagnetic showers have shown that we understand them fully and that we can reproduce them well in simulations. The "Electron-Gamma-Shower" (EGS) program was developed at SLAC [48]. The EGS4 version is integrated into the general detector simulation framework GEANT4 [59]. Figure 2.41 (right) shows the simulation of an 8.5 GeV photon shower in a CsI crystal with EGS4.

We can look at the longitudinal shower distribution obtained by summing the energy loss deposits of all produced particles at each selected t_r value. It increases rapidly with track length of $e^\pm$ up to the shower maximum and then slowly falls off. The track length of the charged particles is smaller than that of the total energy. From a study of many simulated longitudinal-shower distributions an empirical formula was developed that fits the shape well [1, 123],

$$\frac{dE}{dt_r} = E_0 b \frac{(bt_r)^{a-1} \exp(-bt_r)}{\Gamma(a)}, \qquad (2.135)$$

where $\Gamma(a)$ is the Gamma function while a and b are fit parameters that are determined for each measured or simulated longitudinal shower shape. The shower maximum occurs at

$$t_{\max} = \frac{a-1}{b} = 1.0 \times \left(\ln \frac{E}{E_c} \pm 0.5 \right), \qquad (2.136)$$

where the constant $+0.5$ is valid for photons and -0.5 for $e^\pm$. This parameterization has been tested well for energies between 1 GeV and 100 GeV. Note that in gas detectors the observed profile may be closer to the number of electrons, whereas in glass Cherenkov detectors or devices with a thick sensitive region the profile may look closer to that of the energy deposition or total track length.

Figure 2.42 (left) shows the EGS4 simulation of the fractional energy deposition per radiation length of a 30 GeV electron [48]. The fit function in (2.135) is overlaid. In addition, the number of electrons and photons with energy larger than 1.5 MeV crossing a plane is depicted. The number of particles crossing a plane is sensitive to the minimum selected energy, which is 1.5 MeV in our case (see Fig. 2.42 left). Note that the number of electrons falls off faster than the energy deposition because with increasing depth a larger fraction of the shower energy is carried by photons. What a calorimeter measures depends on the device but is typically something between the

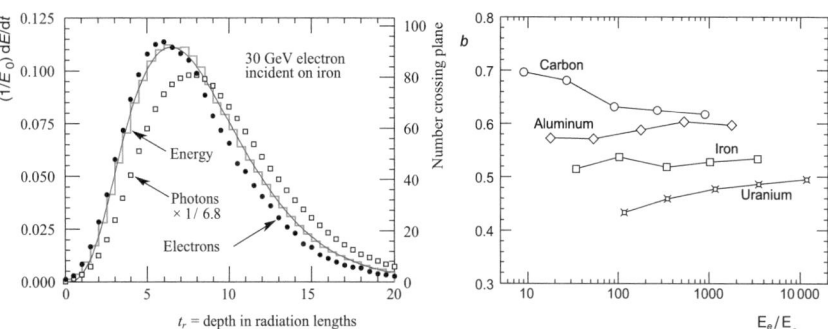

Fig. 2.42 Left: Simulation of a 30 GeV electron-induced shower in iron showing the fractional energy deposition per radiation length (histogram) with a fit curve (solid curve) overlaid that is based on (2.135) (left-hand scale). Solid points show the number of electrons with total energy greater than 1.5 MeV crossing planes at $X_0/2$ intervals (right-hand side scale) while squares show the number of photons with energy greater than 1.5 MeV crossing planes at X_0 intervals. The photon distribution has been scaled down to have the same area as that of the electrons. Right: Fitted values of parameter b in (2.135) obtained with EGS4 for different elements for incident electrons with $1 \leq E_e \leq 100$ GeV [1]. Values obtained for incident photons are essentially the same. Both reprinted under CC-BY-NC-4.0 Licence with kind permission from [1], © 2024, the Particle Data Group LBNL

charge track length and the total energy deposition length. Figure 2.42 (right) shows the fitted values of parameter b obtained from EGS4 simulations as a function of the electron energy divided by the critical energy for different elements. The values are slightly energy dependent and lie around 0.5.

2.7 Electrons and Ions in Gases

A positively charged ion cloud may be produced by a charged particle traversing a chamber. The electrons and ions are accelerated in the $\vec{E}$ field. The light electrons move rapidly towards the anode leaving the much heavier ions behind, which move slowly towards the cathode. On their path both electrons and ions typically collide with atoms in the medium. Fast electrons may further ionize the medium. With appropriate electric and magnetic fields the motion of the center-of-gravity (COG) of the cloud can be viewed as movement along the field lines with constant drift velocity. We will discuss this in more detail in Sects. 2.7.4 and 2.7.5.

2.7.1 Production of Electron-Ion Pairs

A charged particle traveling with $\beta\gamma$ through a medium produces electron-ion pairs. These ions are called primary ions. Their number depends on the momentum and type of the traversing particle. If the electric-field strength is sufficiently high, the electrons can produce new electron-ion pairs. The number of primary ions plus those produced in secondary processes is called total number of ions. Figure 2.43 shows the probability $P_e(E_e)$ of producing electrons with energy equal or greater than E_e in argon gas at normal conditions (NTP) [1] (left-hand scale) as well as the range of electrons (right-hand scale) as a function of energy. The probability approximately follows Rutherford's law,

$$P_e(E_e) \propto \frac{1}{E_e^2}. \tag{2.137}$$

Taking details of the electronic shell structure of the medium, e.g. (Ar) into account yields some variations depending on $\beta\gamma$. Note that the probability is high for producing low-energy (few eV) electrons. The range of energetic electrons increases exponentially with energy. Even without an electric field, energetic electrons can travel a few centimeters in a gas. The primary and total ionization yields produced by different particles at specific energies in argon gas at normal conditions (NTP) are shown in Table 2.8. We notice that charged particles produce higher ionization yields. For example, α particles may be detected in an ionization chamber without any amplification.

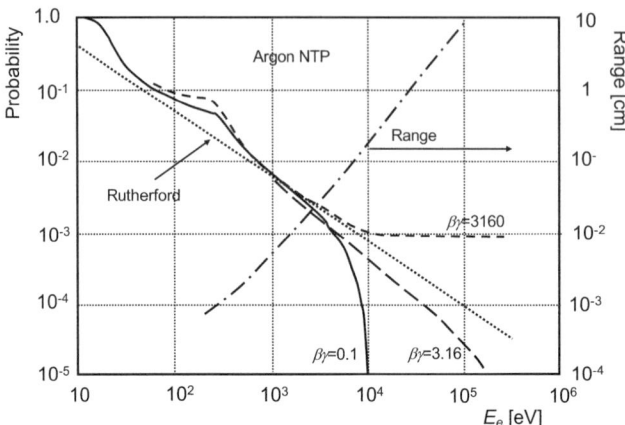

Fig. 2.43 Probability of single collisions of electrons with atomic electrons in an argon gas at NTP in which the released electrons have energy E or larger (left-hand scale). The dotted line shows Rutherford's law. More detailed estimates that account for the electronic structure of the medium are shown for $\beta\gamma = 0.1$ (solid line), $\beta\gamma = 3.16$ (long dashes) and $\beta\gamma = 3160$ (short dashes). The dash-dotted line shows the electron range (right-hand scale). Reprinted under CC-BY-NC-4.0 Licence with kind permission from [1], © 2024, the Particle Data Group LBNL

Table 2.8 Examples of ionization yields in argon at NTP for different types of radiation. Reprinted under CC-BY-NC-ND-4.0 Licence with kind permission from [124], © 2022, F. Sauli

Particle	Primary ions	Total ions
UV photon	1	1
1 keV X-ray	1	50
100 keV/c electron	1000 ion pairs/cm	3000 ion pairs/cm
1 GeV/c proton (MIP)	25 ion pairs/cm	100 ion pairs/cm
5 MeV/c α particle	$\sim 10^4$	$\sim 3 \times 10^4$

2.7.2 Mobility of Ions

According to measurements, the drift velocity $\vec{v}_d^e$ is proportional to $\vec{E}/P$ where $\vec{E}$ is the electric field and P is the gas pressure. For ions the drift velocity is given by

$$\vec{v}_d^{ion} = \mu_{ion} \vec{E} \frac{P_0}{P}, \tag{2.138}$$

where μ_{ion} is the ion mobility in units [cm^2/(Vs)] and $P_0 = 760$ Torr. Table 2.9 lists the mobility for common ions. The mobility is high for small atoms and molecules

2.7 Electrons and Ions in Gases

Table 2.9 Mobility of positive ions in different gases. Reprinted under CC-BY-3.0 Licence with kind permission from [10], © 1977–2024, CERN. All rights reserved

Ion	Gas	μ_{ion} [cm^2/(Vs)]
H_2^+	H_2	13.0
He^+	He	10.2
Ar^+	Ar	1.7
CH_4^+	Ar	1.87
$C_2H_6^+$	Ar	2.06
$C_3H_8^+$	Ar	2.08
CO_2^+	Ar	1.72
$IsoC_4H_{10}^+$	Ar	1.56
$(OCH_3)CH_2^+$	Ar	1.51
CH_4^+	CH_4	2.22
$C_2H_6^+$	C_2H_4	1.23
$C_3H_8^+$	C_3H_6	0.793
$IsoC_4H_{10}^+$	$IsoC_4H_{10}$	0.612
O_2^+	O_2	2.2
CO_2^+	CO_2	1.09
H_2O^+	H_2O	0.7
CF_4^+	CF_4	0.96
CF_4^+	CH_4	1.06
CF_4^+	C_2H_6	1.04
CF_4^+	C_3H_8	1.04
$(OCH_3)CH_2^+$	$(OCH_3)CH_2$ (Methylal)	0.26
$(OCH_3)CH_2^+$	$IsoC_4H_{10}$	0.55

and low for big atoms and molecules. For a mixture of different gases, the inverse of the ion mobility μ_{ion}, is given by

$$\frac{1}{\mu_{ion}} = \sum_{j=1}^{n} \frac{c_j}{\mu_{ion,j}}, \quad (2.139)$$

where c_j is the concentration of gas type j and $\mu_{ion,j}$ is the mobility of the ion in gas type j. If several kinds of ions are present, those with higher mean excitation energies are neutralized after $10^2 - 10^3$ collisions as electrons from atoms with lower mean excitation energies are removed.

2.7.3 Diffusion of Ions and Electrons in a Field-Free Gas

According to the equipartition law the average thermal energy of a gas molecule with three degrees of freedom is $E_k \simeq \frac{3}{2} k_B T$, where T is the temperature and $k_B = 8.617 \times 10^{-5}$ eV/K is the Boltzmann constant. At room temperature ($T = 298$K) the most probable thermal energy is $E_k \sim 0.0257$ eV. The kinetic energies E_k of the gas atoms, gas molecules and electrons as a function of temperature T are distributed according to a Maxwell-Boltzmann distribution,

$$F_E(E_k) = 2\sqrt{\frac{E_k}{\pi (k_B T)^3}} \exp\left(-\frac{E_k}{k_B T}\right). \tag{2.140}$$

As an example, Fig. 2.44 (left) shows Maxwell-Boltzmann distributions as a function of the electron kinetic energy E_e at different temperatures. At room temperature typical kinetic energies of the electrons are around $E_e = 0.03$ eV. Figure 2.44 (right) shows Maxwell-Boltzmann distributions at different temperatures as a function of the electron thermal velocity.

Let us look at a charge distribution that is localized at point $(0, 0, 0)$ at time t_0. By multiple scattering into the surrounding volume it will diffuse with time. According to the kinetic theory of gases, the charge density distribution at time t is Gaussian with its center at the origin. In one dimension we get

$$\frac{dN}{N} = \frac{1}{\sqrt{4\pi D_k t}} \exp\left(-\frac{x^2}{4 D_k t}\right) dx, \tag{2.141}$$

where D_k is the diffusion coefficient for ions ($k = ion$) and electrons ($k = e$). The *rms* spread is

$$\sigma_x^2 = 2 D_k t. \tag{2.142}$$

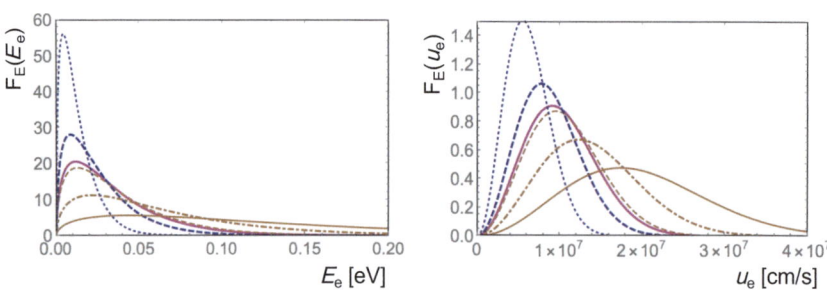

Fig. 2.44 Left: Maxwell-Boltzmann distributions as a function of the electron kinetic energy E_e for 100K (blue dotted curve), 200K (blue dashed curve), 273.15K (magenta solid curve), 298K (brown dashed curve), 500K (brown dash-dotted curve) and 1000K (brown solid curve). Right: Maxwell-Boltzmann distributions as a function of the electron thermal velocity for the same temperatures

2.7 Electrons and Ions in Gases

The diffusion coefficient can be expressed in terms of the mobility (μ_k) and the thermal energy,

$$D_k = \frac{\mu_k k_B T}{q_k}. \tag{2.143}$$

Using the kinetic theory of gases, we can express D_k in terms of the thermal velocity u_k and the mean free path λ_k, which is the average distance between two collisions,

$$D_k = \frac{1}{3} \int u_k \, \lambda_k(E_k) \, F_E(E_k) \, dE_k. \tag{2.144}$$

For energy-independent mean free paths we obtain

$$D_k = \frac{1}{3} u_k \lambda_k. \tag{2.145}$$

Thus, the diffusion coefficient decreases with increasing thermal velocity. Since $u_{rms} = \sqrt{(3k_B T/m_k)}$[17], D_k decreases as one over the square root of the mass of the moving particle m_k. For a classical gas, we have

$$\lambda_k(E_k) = \frac{1}{N\sigma(E_k)} = \frac{A}{N_A \rho \sigma(E_k)} \cdot \frac{P_0}{P} \frac{T}{T_0}, \tag{2.146}$$

where $\sigma(E_k)$ is the kinetic-energy-dependent collision cross section of the ion or electron with the atom. The latter term reflects the pressure and temperature dependence of the density. Table 2.10 summarizes the thermal velocity, diffusion coefficient, mobility, and mean free path for some typical gases. Sometimes it is useful to parameterize the Maxwell-Boltzmann distribution in terms of the thermal velocity u_k,

$$F_E(u_k) = \frac{4}{\sqrt{\pi}} \left(\frac{m_k}{2k_B T} \right)^{3/2} u_k^2 \exp\left(-\frac{m_k \cdot u_k^2}{2k_B T} \right). \tag{2.147}$$

At room temperature, electron drift velocities are $\sim 10^6 - 10^7$ cm/s while ion drift velocities are at least two orders of magnitude lower. Furthermore, the mean free paths for electrons are considerably larger, yielding

$$\lambda_e = 5.66 \lambda_{\text{ion}}. \tag{2.148}$$

In order to demonstrate this, let us consider a simple model. We represent the two interacting particles by spheres as sketched in Fig. 2.45 (left). Let a_1 and $\vec{v}_1$ be radius and velocity of particle 1, a_2 and $\vec{v}_2$ be those of particle 2 and N be the particle density. When both particles touch each other, the geometric cross section is $\sigma = \pi(a_1 + a_2)^2$. The relative velocity between the two particles is

---

[17] Note that we have set $c=1$ in the following discussion of the drift velocity. For explicit calculations the speed of light needs to be included.

Table 2.10 Thermal velocity u_{ion}, diffusion coefficient D_{ion}, mobility μ_{ion} and mean free path λ_{ion} of ions in their own gas at NTP [125,126]. Reprinted under CC-BY-3.0 Licence with kind permission from [10], © 1977–2024, CERN. All rights reserved

Gas	Mass	u_{ion} [cm/s]	D_{ion} [cm^2/s]	μ_{ion} [cm^2/(Vs)]	λ_{ion} [10^{-5} cm]
H_2	2.02	1.753×10^5	0.34	13.0	1.8
He	4.00	1.246×10^5	0.26	10.2	2.8
Ar	39.95	0.396×10^5	0.04	1.7	1.1
O_2	32	0.440×10^5	0.06	2.2	1.0
H_2O	18.2	0.584×10^5	0.02	0.7	0.86
CO_2	44	0.375×10^5	0.16	1.09	0.77
CH_4	16	0.623×10^5	0.217	2.26	0.87
CF_4	88	0.266×10^5	0.0647	1.06	
DME	46	0.367×10^5	0.06	0.56	0.42

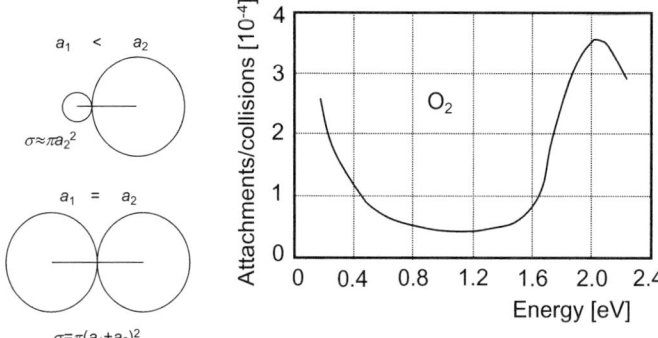

Fig. 2.45 Left: Illustration of the geometric cross section for two touching spheres with radii a_1 and a_2. Right: The ratio of attachment cross section per collision cross section for electron capture in O_2 as a function of electron energy. Reprinted and adapted with kind permission from [127], © 1933, AIP Publishing. All rights reserved

$$\left\langle \left| \vec{v}_1 - \vec{v}_2 \right|^2 \right\rangle = \left\langle \vec{v}_1^{\,2} - 2\vec{v}_1 \cdot \vec{v}_2 + \vec{v}_2^{\,2} \right\rangle = \left\langle \vec{v}_1^{\,2} + \vec{v}_2^{\,2} \right\rangle. \quad (2.149)$$

Since we have to average over all directions, the middle term vanishes. The mean free path for particle 1 is $\lambda_1 = |\vec{v}_1|\langle t \rangle$, where $\langle t \rangle$ is the mean time between two collisions.

Expressing λ_k in terms of the cross section, (2.146) yields

$$\langle t \rangle = \frac{1}{\pi (a_1 + a_2)^2 \, N} \frac{1}{\sqrt{[\langle \vec{v}_1^{\,2} \rangle + \langle \vec{v}_2^{\,2} \rangle]}}. \quad (2.150)$$

2.7 Electrons and Ions in Gases

Now, let us first consider the collision of two ions or of an ion and an atom. Here $a_1 \simeq a_2$ and $\langle v_1^2 \rangle = \langle v_2^2 \rangle$, yielding

$$\lambda_{\text{ion}} = \frac{1}{4\sqrt{2}N\pi a_2^2}. \tag{2.151}$$

For the collision of an electron with an ion, we have $a_1 \ll a_2$ and $\langle v_1^2 \rangle \gg \langle v_2^2 \rangle$ yielding

$$\lambda_e = \frac{1}{N\pi a_2^2}. \tag{2.152}$$

Thus, λ_e differs from λ_{ion} by the factor $4\sqrt{2} = 5.66$, confirming the result in (2.148). This factor is approximately observed for most gases.

Note that ions and electrons from an ionization process can be neutralized in a process called recombination before they are detected. The rate for i^+ to recombine with an electron is: $-dn^+/dt = \alpha_r n^+ n^-$ where n^+ and n^- are the charged particle densities and α_r is the recombination coefficient. For example, in gases like O_2 and CO_2 the recombination coefficients are rather large, $\mathcal{O}(10^{-6})$ for negative ions and $\mathcal{O}(10^{-7})$ for electrons.

Another process that removes electrons is electron capture. Gas molecules with several atoms can accumulate low-energy electrons. The probability for such an accumulation to occur during one collision is negligibly small for noble gases, and N_2, H_2 and CH_4. However, it is sizable for so-called electronegative gases, such as O_2, Cl_2^-, NH_3 and H_2O. The mean time for electron capture is

$$t_a = \frac{1}{p_a n_s} = \frac{1}{p_a} \frac{\lambda_e}{u_e^{rms}} \simeq \frac{1}{p_a} \frac{\lambda_e}{\sqrt{3k_B T/m_e}}, \tag{2.153}$$

where p_a is the accumulation probability, n_s is the number of collisions per time and u_e^{rms} is the mean thermal velocity of electrons. Table 2.11 shows the accumulation times t_a for the most electronegative gases. For strongly electronegative gases, the accumulation time can be as small as $t_a = 5$ ns. Figure 2.45 (right) shows the ratio of cross sections of electron attachments and collisions in O_2 as a function of electron energy. In the energy region 0.5–1.5 eV the ratio shows a broad minimum before rising again, which results from the Ramsauer effect discussed in Sect. 2.7.4. The electrons have a smaller collision cross section and also a smaller probability for attachments. For 2 eV electrons in O_2, the attachment parameter shows a peak. At higher energies the electrons excite or ionize the atom. Thus, the probability for attachments becomes smaller again.

If an electric field is applied, the electron kinetic energy increases. However, the accumulation probability is a function of the electron energy (see Fig. 2.45 (right)). If a counting gas contains a fraction of an electronegative gas, f_e, the number of collisions with the contaminant is $n_s' = f_e u_e / \lambda_e$ and the mean free path in the electronegative gas is $\lambda_a = v_d^e/(p_a \cdot n_s')$. For a 1% O_2 admixture in argon at $|\vec{E}| = $ 1kV/cm, we obtain $\lambda_a \simeq 5$ cm. For a 1% Cl contaminant for the same conditions,

Table 2.11 Probability p_a for electron attachment, number of collisions n_s per second and mean time for electron capture t_a at zero field and NTP for CO_2, O_2, H_2O and Cl gases. In addition, the mean free path λ_a in a 1% contaminant gas is listed for 1 eV electrons

Gas	p_a	n_s [s^{-1}]	t_a [ns]	λ_a [cm]
CO_2	6.2×10^{-9}	2.2×10^{11}	7.1×10^5	20,084
O_2	2.5×10^{-5}	2.1×10^{11}	190	5
H_2O	2.5×10^{-5}	2.8×10^{11}	140	5
Cl	4.8×10^{-4}	4.5×10^{11}	5	0.26

we obtain $\lambda_a \simeq 0.26$ cm. This effect cannot be neglected in large drift chambers. So here, a purification system needs to be installed.

2.7.4 Drift of Electrons in Electric Fields

Between two collisions in a gas, electrons can gain much more energy in an electric field than ions because their mean free path is longer. For energies close to 0.3 eV, the electron de Broglie wavelength is ~ 0.4 nm corresponding to approximately two diameters of bound electron orbits in noble gases ($r_{Ar} = 0.097$ nm). In this energy range, atoms become nearly transparent considering the cross section for electrons is minimal. This effect is called the Ramsauer effect. Figure 2.46 (left) shows the ionization probability in different gases as a function of the electron kinetic energy, which shows a peak at $E_e \simeq 100$ eV. Figure 2.46 (right) shows the cross section of electrons interacting with the electrons in argon. For energies near 0.3 eV the cross section shows a minimum.

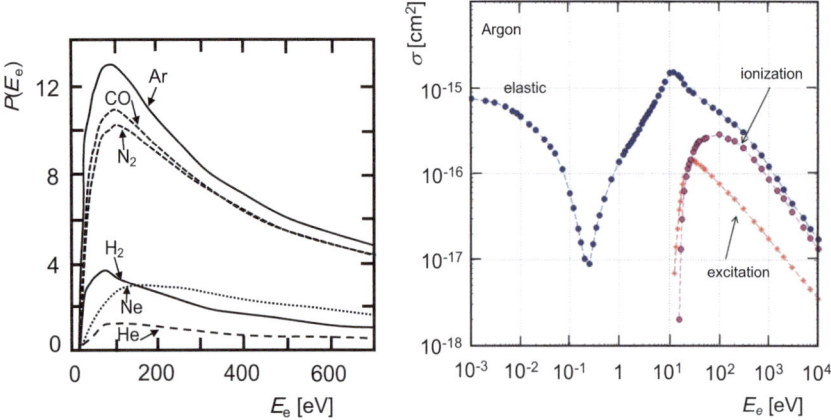

Fig. 2.46 Left: Ionization probability via the impact of electrons in different gases (He, Ne, H_2, N_2, CO and Ar) as a function of the electron kinetic energy. Right: Measured cross section of electrons with atomic electrons in argon as a function of energy. The Ramsauer minimum lies around 0.3 eV. Reprinted with kind permission from [128], © 2015, AGU. All rights reserved

2.7 Electrons and Ions in Gases

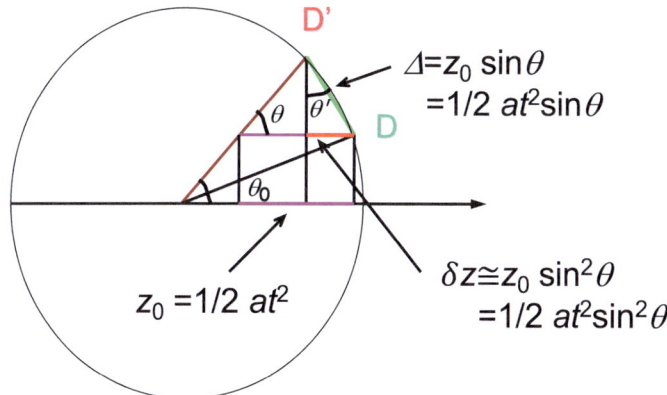

Fig. 2.47 Graphical derivation of the drift velocity. The $u \cdot t$ term is the diffusion contribution while the $(1/2)at^2$ term is the drift contribution. The angle θ is the angle between the $u_e \cdot t$ direction and the z axis

There is a simple method to extract the first term of the drift velocity ((2.159)), illustrated in Fig. 2.47. Consider an electron cloud with thermal velocity $u_e = \sqrt{2E_e/m_e}$ localized at $P = (0, 0, 0)$ at t_0. If no electric field is present, there is isotropic diffusion. The first collision occurs on average on a sphere with radius $r = \lambda_e$ at a penetration point $D' = (r \cos\theta, r \sin\theta)$, where θ is the polar angle with respect to the direction of the $\vec{E}$ field. Without loss of generality, we set the azimuth angle ϕ to zero. In the presence of a homogeneous electric field in the z-direction with $\vec{E} = (0, 0, E_z)$, the electron will follow a parabolic path due to acceleration $\vec{a} = q_e \vec{E}/m_e$ along the z-direction. The penetration point crosses the sphere now at position $D = (r \cos\theta_0, r \sin\theta_0)$, which is displaced on the surface of the sphere from D' by distance Δ, where θ_0 is the polar angle of D with respect to the z axis. We need to determine the length δz. From Fig. 2.47, we have two relations for Δ,

$$\Delta/\sin\theta = z_0/\cos(\theta' - \theta),$$
$$\Delta = \delta z/\sin\theta', \tag{2.154}$$

where θ' is the angle between Δ and the D' ray. We can rewrite it as $\theta + (\theta_0 - \theta)/2$. Thus,

$$\delta z = z_0 \left(\sin^2\theta + \frac{\sin\theta \cos\theta \sin[(\theta_0 - \theta)/2]}{\cos[(\theta_0 - \theta)/2]} \right), \tag{2.155}$$

where $z_0 = q_e |\vec{E}| t^2/(2m_e)$. Since θ can take any value between zero and π, we have to average (2.155) over θ. The second term also depends on $(\theta_0 - \theta)/2$, which we average in addition. Setting $\chi_0 = (\theta_0 - \theta)$, we get

$$\langle \sin^2 \theta \rangle = \frac{1}{2} \int_0^\pi -(1 - \cos\theta^2) \, \mathrm{d}\cos\theta = \frac{2}{3},$$

$$\langle \sin\theta \cos\theta \tan\chi_0/2 \rangle = \frac{1}{4} \int_0^\pi \int_0^\pi \sin\theta \cos\theta \tan\frac{\chi_0}{2} \, \mathrm{d}\cos\theta \, \mathrm{d}\cos\chi_0 = 0. \quad (2.156)$$

Note that the average displacement depends on the average of the time between two collisions δt^2, which is not constant. Calling the average collision time $\tau_e = \lambda_e/u_e$, the collision time is distributed as [129]

$$G[t] = \frac{1}{\tau_e} \exp\left(-\frac{t}{\tau_e}\right). \quad (2.157)$$

Assuming that the collision cross section σ_e and thus the mean free path λ_e are independent of the thermal velocity u_e, we get

$$\langle t^2 \rangle = 2\tau_e = \frac{2\lambda_e^2}{u_e^2}. \quad (2.158)$$

Inserting the averages of $\langle \delta z \rangle$, $\langle \sin^2\theta \rangle$, $\langle t^2 \rangle$ and $\langle t \rangle$ yields the electron drift velocity,[18]

$$v_\mathrm{d}^e = \frac{\langle \delta z \rangle}{\langle t \rangle} = \frac{2}{3} \frac{q_e |\vec{E}|}{m_e} \frac{\lambda_e}{u_e}, \quad (2.159)$$

which is the first term of the drift velocity (see (2.161)). In general, the collision cross section σ_e and the mean free path λ_e depend on u_e, yielding [129],

$$\langle t^2 \rangle = \frac{2\lambda_e^2}{u_e^2} + \frac{\lambda_e}{u_e} \frac{\mathrm{d}\lambda_e}{\mathrm{d}u_e}. \quad (2.160)$$

Accounting for this effect yields the generalized drift velocity,

$$\vec{v}_\mathrm{d}^e = \frac{q_e \vec{E}}{m_e} \left\{ \frac{2}{3} \left\langle \frac{\lambda_e(u_e)}{u_e} \right\rangle + \frac{1}{3} \left\langle \frac{\mathrm{d}\lambda_e(u_e)}{\mathrm{d}u_e} \right\rangle \right\}. \quad (2.161)$$

If we add a magnetic field in the x-direction (see Sect. 2.7.5), the drift velocity will get an additional component in the y-direction, orthogonal to the $\vec{E}$ and $\vec{B}$ fields.

$$\vec{v}_\mathrm{d}^{e\perp} = -\frac{q_e^2}{m_e^2} \vec{E} \times \vec{B} \left\{ \frac{1}{3} \left\langle \frac{\lambda_e^2(u_e)}{u_e^2} \right\rangle + \frac{2}{3} \left\langle \frac{\lambda_e}{u_e} \frac{\mathrm{d}\lambda_e(u_e)}{\mathrm{d}u_e} \right\rangle \right\}. \quad (2.162)$$

---

[18] We denote the generic drift velocity by v_d^k. If we refer to electrons, holes, positive ions and negative ions, we use v_d^e, v_d^h and v_d^{ion}, respectively.

2.7 Electrons and Ions in Gases

These equations apply to ions as well but the thermal velocities are reduced by $\sqrt{(m_e/m_{ion})}$. It is important to note that the drift velocity is constant with respect to time, depending linearly on the electric field and the mean free path and inversely on the mass of the drifting particle and the thermal velocity. In an electric field, electrons are accelerated until they loose the gained energy in a collision, which generally occurs after an average flight path of $\ell_e \sim \lambda_e$. This procedure of acceleration and energy loss continues. After about 10^{-11} s, an equilibrium is reached where the electrons reach a constant drift velocity described by (2.161). In a collision, the electrons loose the energy $\Delta(E_e) = f_\epsilon(E_e) \cdot E_e$, where $f_\epsilon(E_e)$ is the energy-dependent fraction of lost energy and E_e is the kinetic energy. The loss occurs within the time interval $\tau_e = \langle \lambda_e/u_e \rangle$. The energy gained in the electric field is $\langle q_e|\vec{E}|v_d^e\tau_e\rangle$, where $v_d^e = |\vec{v}_d^e|$ is the magnitude of the drift velocity. Thus,

$$q_e|\vec{E}|v_d^e = \langle f_\epsilon(E_e) \cdot E_e u_e/\lambda_e \rangle. \qquad (2.163)$$

For a narrow kinetic energy distribution, we use $\langle E_e \rangle = (1/2)m_e\langle u_e^2\rangle$ and get

$$(q_e/m_e)|\vec{E}|v_d^e = \frac{1}{2}\langle f_\epsilon(E_e)u_e^3/\lambda_e\rangle. \qquad (2.164)$$

If we express the thermal velocity in terms of the drift velocity given in (2.161), we obtain

$$v_d^e \simeq \sqrt{\frac{2}{3}\sqrt{\frac{\langle f_\epsilon(E_e)\rangle}{3}}\frac{q_e}{m_e}|\vec{E}|\langle\lambda_e\rangle}. \qquad (2.165)$$

So, the drift velocity increases with the square root of the electric field and mean free path as well as the fourth root of the energy loss fraction.

The energy loss fraction can be parameterized by $f_\epsilon(E_e) = f_0 E_e^m$ and the mean free path by $\lambda_e = \lambda_0 E_e^{-n}$, where f_0 and λ_0 are material-dependent constants. Furthermore, the electron energy is related to the electric field by $E_e \simeq E^{2/(m+2n+2)}$. Thus, the drift velocity has a dependence on the electric field as [129],

$$v_d^e \propto |\vec{E}|^{(m+1)/(m+2n+2)}. \qquad (2.166)$$

For low field strengths below the Ramsauer minimum, we get $n \sim -1$ in argon, yielding

$$v_d^e \propto |\vec{E}|^{\frac{m+1}{m-1}}. \qquad (2.167)$$

For $m > 1$, there is a rapid increase. For energies above the Ramsauer minimum, we have $n \sim +1$ yielding

$$v_d^e \propto |\vec{E}|^{\frac{m+1}{m+3}}. \qquad (2.168)$$

These dependences are qualitatively observed in noble gases. In molecular gases (CO_2, CH_4, etc.), inelastic collisions contribute significantly to the total cross

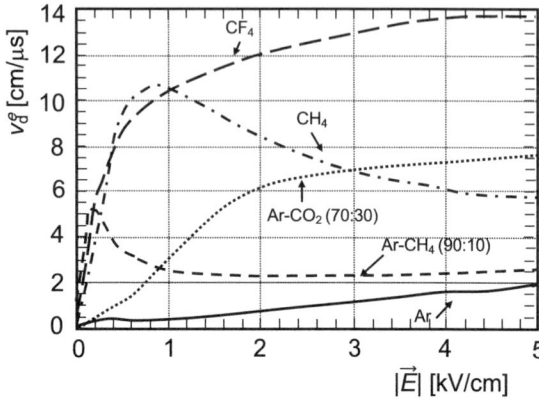

Fig. 2.48 Computed electron drift velocities as a function of electric field for argon, $Ar - CH_4$ (90:10), $Ar - CO_2$ (70:30), CH_4 and CF_4 at NTP. Reprinted with kind permission from [130], © 2012, Springer. All rights reserved

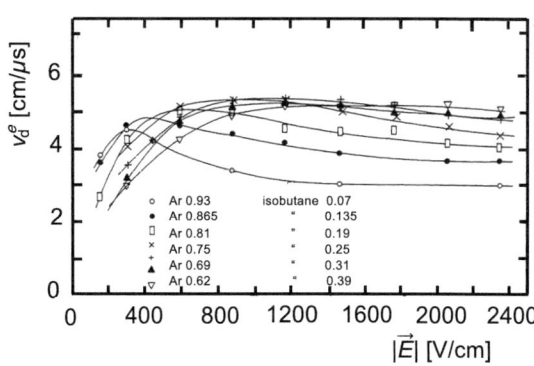

Fig. 2.49 Electron drift velocities as a function of the electric field in different Ar-isobutane gas mixtures versus electric-field strength. Reprinted with kind permission from [131], © 1974, Elsevier. All rights reserved

section. For example, in CO_2 molecular oscillations can be excited in the 0.1–1 eV range. The energy transfer $\Delta(E_e)$ from the electron to the molecule becomes large, but it decreases above maximum excitation energy as $f_\epsilon(E_e) \sim E_{max}/E_e$. For $E_e > E_{max}$, we get $m \sim -1$ in which case v_d^e becomes independent of $|\vec{E}|$. If $m < -1$, v_d^e decreases with increasing $|\vec{E}|$.

Figure 2.48 shows electron drift velocities versus electric-field strength for argon, $Ar - CH_4$ (90:10), $Ar - CO_2$ (70:30), CH_4 and CF_4 at NTP. Pure argon is a slow gas.[19] The drift velocity rises nearly linearly with the electric field and is ~ 1 cm/μs around 2.5 kV/cm and ~ 2 cm/μs around 5 kV/cm. On the other hand, CH_4 and CF_4 are fast gases. While CF_4 keeps rising with $|\vec{E}|$, CH_4 drops again for electric field strengths greater than 0.8 kV/cm. For $Ar - CH_4$ gas mixtures, $v_d^e = 5$ cm/μs at $|\vec{E}| = 0.2$ kV/cm before dropping again. Figure 2.49 shows the predicted electron drift velocities, calculated using (2.161) with measurements for different Ar-isobutane gas mixtures. Note that the observed drift velocities are well reproduced by the calculation. For higher concentrations of isobutane, the drift velocity is nearly independent of the electric field over a wide range. Figure 2.50 shows the drift veloc-

---

[19] A slow gas is one with a low drift velocity, a fast gas is one with a high drift velocity.

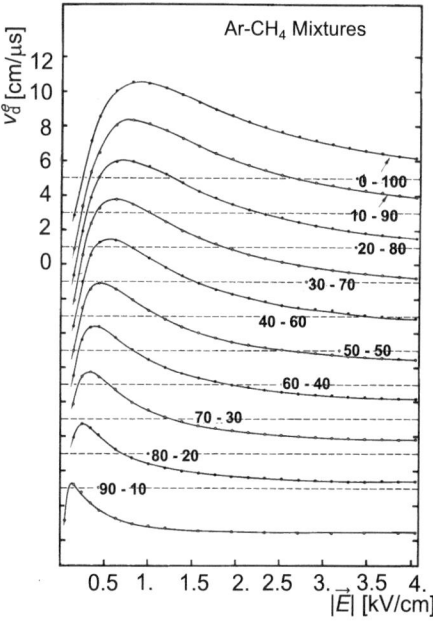

Fig. 2.50 Measured (points) and calculated (solid lines) electron drift velocities in Ar-methane gas mixtures at NTP versus electric-field strength. The labeled scale refers to pure methane gas. For each gas mixture with increasing argon contribution, the scale is shifted downwards by 2 cm/μs for each 10% argon concentration. For clarity, the dotted lines represent the 5 cm/μs point for each mixture. Reprinted with kind permission from [132], © 1979, Elsevier. All rights reserved

ity versus electric-field strength for ten Ar $-$ CH$_4$ gas mixtures, where the curves are stacked. The labeled scale refers to the pure methane gas. For each 10% increase in argon concentration, the scale is shifted downwards by 2 cm/μs. The dotted lines represent the 5 cm/μs points. For example for the (70:30) Ar-methane gas mixture, the drift velocity is ≥ 5 cm/μs for $|\vec{E}| < 1.25$ kV/cm. For all mixtures, the maximum drift velocity is > 5 cm/μs at some electric field.

2.7.5 Drift of Electrons and Ions in Electric and Magnetic Fields

In an electromagnetic field, a charged particle with charge q_k and mass m_k is subject to the Coulomb force $q_k \vec{E}$ and the Lorentz force $q_k \vec{v}_k \times \vec{B}$,

$$\vec{F} = m_k \dot{\vec{v}}_k = q_k (\vec{E} + \vec{v}_k \times \vec{B}), \qquad (2.169)$$

where $\vec{v}_k$ is the particle velocity, $\vec{E}$ is the electric field and $\vec{B}$ is the magnetic field. In a magnetic field, the particle energy remains unchanged and the particle follows a helical orbit (see Sect. 5.2.1) with angular velocity of

$$\vec{\omega}_B = -\frac{q_k}{m_k} \vec{B}, \qquad (2.170)$$

where $\omega_B = |\vec{\omega}_B|$ is the cyclotron frequency. For example, for an electron we find

$$\frac{\omega_B}{|\vec{B}|} = 17.6 \text{ MHz/G}. \qquad (2.171)$$

The velocity can be decomposed into a circular part with angular velocity $\vec{\omega}_B$ and a translation with drift velocity $\vec{v}_d^k$,

$$\vec{v}_k = \vec{v}_d^k + \vec{\omega}_B \times \vec{r}_B, \tag{2.172}$$

where $\vec{r}_B$ denotes the position of the particle in a plane perpendicular to $\vec{v}_d^k$. Expressing $\vec{v}_d^k$ in terms of components perpendicular and parallel to the magnetic field yields,

$$\vec{v}_d^k = \frac{\vec{E} \times \vec{B}}{|\vec{B}|^2} + \vec{v}_{k,\parallel}, \tag{2.173}$$

with

$$m_k \dot{\vec{v}}_{k,\parallel} = q_k \vec{E}_\parallel. \tag{2.174}$$

If the particle moves through a gas-filled volume, we need to add the time-dependent stochastic force $m_k \vec{A}(t)$, which describes collisions with gas molecules. Thus, the particle is exposed to the total force, called Langevin's force,

$$m_k \dot{\vec{v}}_k = q_k (\vec{E} + \vec{v}_k \times \vec{B}) + m_k \vec{A}(t). \tag{2.175}$$

It is useful to perform a time average. Since the solution is a translation with constant drift velocity, this implies that the average acceleration has to vanish. Thus, the stochastic acceleration averaged over time compensates for the translational acceleration,

$$\left\langle \vec{A}(t) \right\rangle = -\frac{\vec{v}_d^k}{\tau_k}, \tag{2.176}$$

where τ_k is the average time between two collisions yielding

$$\dot{\vec{v}}_d^k = \frac{q_k}{m_k} \vec{E} + \frac{q_k}{m_k} \vec{v}_d^k \times \vec{B} - \frac{\vec{v}_d^k}{\tau_k}. \tag{2.177}$$

In a constant $\vec{E}$ field where $d\vec{v}_d^k/dt = 0$, we obtain,

$$\frac{q_k}{m_k} \vec{E} = \frac{q_k}{m_k} \vec{B} \times \vec{v}_d^k + \frac{\vec{v}_d^k}{\tau_k}. \tag{2.178}$$

A solution to this equation is given by

$$\vec{v}_d^k = \frac{\mu_k}{1 + \omega_B^2 \tau_k^2} \left[\vec{E} + \frac{\vec{E} \times \vec{B}}{|\vec{B}|} \omega_B \tau_k + \frac{(\vec{E} \cdot \vec{B}) \cdot \vec{B}}{|\vec{B}|^2} \omega_B^2 \tau_k^2 \right], \tag{2.179}$$

where $\mu_k = q_k \tau_k / m_k c^2$ is the mobility of the particle. In the presence of $\vec{E}$ and $\vec{B}$ fields, the drift velocity has three components, one parallel to $\vec{E}$, one parallel to $\vec{B}$ and one perpendicular to the plane spanned by $\vec{E}$ and $\vec{B}$. So, let us focus now on

Fig. 2.51 Direction of the drift velocity projected into the x-y plane in the presence of a magnetic field in the z-direction and an arbitrary electric field

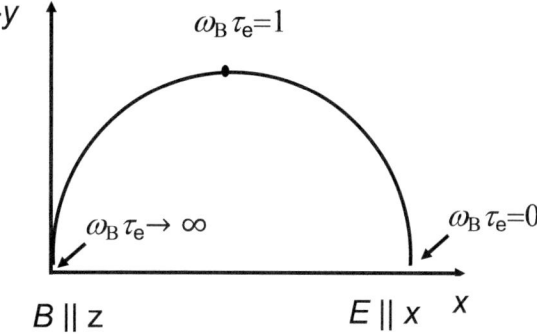

the electrons. If $\omega_B \tau_e = 0$, there is no $\vec{B}$ field and the drift velocity is parallel to the electric field. If $\omega_B \tau_e \gg 1$, the drift velocity is parallel to the magnetic field (z direction). For $\omega_B \tau_e = 1$, the drift velocity has equal transverse x and y components. For arbitrary finite values of $\omega_B \tau_e$, the penetration points of $\vec{v}_d^e$ lie on a semicircle in the transverse direction to the $\vec{B}$ field (x-y plane), where $\vec{B}$ points in the z-direction and $\vec{E}$ is in the x-direction as depicted in Fig. 2.51. If $\vec{E}$ and $\vec{B}$ are parallel, (2.177) simplifies to $\vec{v}_d^e = \mu_e \vec{E}$, in which $\vec{v}_d^e$ is parallel to $\vec{E}$.

Let us consider two examples. First, we choose an electric field in the x-z plane, $\vec{E} = (E_x, 0, E_z)$ with $E_x \ll E_z$, and a magnetic field in the z-direction, $\vec{B} = (0, 0, B_z)$. Here, $\vec{E}$ and $\vec{B}$ are nearly parallel.

$$v_x^e = \mu_e E_x \frac{1}{1 + \omega_B^2 \tau_e^2},$$
$$v_y^e = -\mu_e E_x \frac{\omega_B \tau_e}{1 + \omega_B^2 \tau_e^2},$$
$$v_z^e = \mu_e E_z. \tag{2.180}$$

The modulus of the drift velocity is nearly the same as that without any magnetic field. For a magnetic field of 20.4 kG in a CH$_4$ gas at a high electric field, we get $\omega_B \tau_e \simeq 4$. This, respectively, reduces the x and y components of the drift velocity by factors $17 E_z/E_x$ and $4.3 E_z/E_y$ with respect to the z component. In the second example, $\vec{E}$ and $\vec{B}$ are perpendicular to each other, where the electric field points in the x-direction and the B field in the z-direction. The drift velocity is given by

$$v_x^e = \mu_e E_x \frac{1}{1 + \omega_B^2 \tau_e^2},$$
$$v_y^e = -\mu_e E_x \frac{\omega_B \tau_e}{1 + \omega_B^2 \tau_e^2},$$
$$v_z^e = 0. \tag{2.181}$$

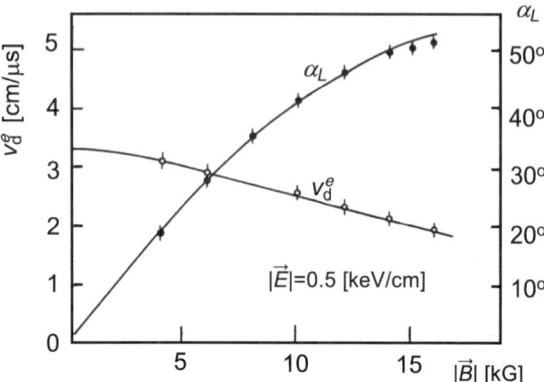

Fig. 2.52 Electron drift velocity and Lorentz angle as a function of magnetic field strength for orthogonal electric and magnetic fields. Points with error bars are data and solid curves are calculations using (2.179). Reprinted with kind permission from [133], © 1975, Elsevier. All rights reserved

In this case, the drift velocity strongly depends on the magnetic-field strength,

$$|\vec{v}_d^e| = \sqrt{v_x^2 + v_y^2} = \mu_e E_x \frac{1}{\sqrt{1 + \omega_B^2 \tau_e^2}}, \quad (2.182)$$

where no component is parallel to the magnetic field. For the above example, the drift velocity is a factor of 4.3 lower with respect to the drift velocity in the absence of a magnetic field. The angle between the drift velocity and the electric field is called Lorentz angle α_L and is given by

$$\tan \alpha_L = \omega_B \tau_e. \quad (2.183)$$

Figure 2.52 shows the electron drift velocity and Lorentz angle for crossed electric and magnetic fields. Note that the data for the drift velocity and Lorentz angle are described well by the predictions.

2.7.6 Diffusion of Electrons in Electric and Magnetic Fields

Once an electron-ion pair is produced, electrons and ions drift in opposite directions. If no magnetic field is turned on, electrons drift along the electric-field lines towards the anode. In addition to the drift, diffusion is present due to scattering of the electrons with atomic electrons. Both drift velocity and diffusion depend on the nature of the gas, especially on the inelastic cross section of the electron with the gas atom or molecule. In noble gases, the inelastic cross section is zero below the excitation threshold. In complex molecular gases, excitations into rotational and vibrational states increase the inelastic cross section at low kinetic energies. Since diffusion depends on the path length, it will worsen the transverse position resolution over long drift distances. Note that fast gases yield a smaller diffusion than slow gases and, therefore, better transverse position resolution. To produce a fast gas we need to add polyatomic gases such as CO_2, CH_4 or CF_4 to the noble gas since they have

2.7 Electrons and Ions in Gases

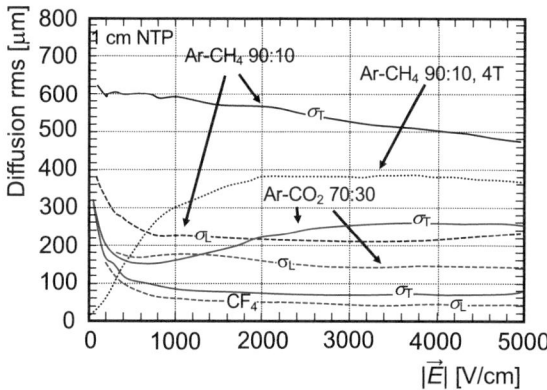

Fig. 2.53 Electron transverse diffusion (solid line, σ_T) and longitudinal diffusion (dashed line, σ_L) as a function of the electric-field strength for CF_4, $Ar - CH_4$ (90:10) and $Ar - CO_2$ (70:3) gas mixtures. The drift is 1 cm at NTP with no magnetic field present. The dotted line shows the electron transverse diffusion for $Ar - CH_4$ (90:10) at a magnetic field of $|\vec{B}| = 4\,T$ [134]. Reprinted under CC-BY-NC-4.0 Licence with kind permission from [1], © 2024, the Particle Data Group LBNL

large inelastic cross sections at moderate energies. This reduces the kinetic energy of the electrons into the regime of the Ramsauer minimum (~ 0.5 eV), which increases the mean free path and in turn leads to faster drift velocities. The diffusion can be split into longitudinal diffusion and transverse diffusion. We need to minimize the transverse diffusion. Figure 2.53 shows the electron longitudinal and transverse diffusion in different gases at NTP. These have been computed with the MAGBOLTZ program [134]. Note that CF_4 has the lowest transverse diffusion being nearly independent of the electric field. The $Ar - CH_4$ gas mixture has the largest transverse diffusion in the absence of a magnetic field. In a $4\,T$ magnetic field the transverse diffusion is significantly reduced, in particular at low electric-field strengths. The longitudinal diffusion is typically smaller than the transverse diffusion and has no effect on the transverse position resolution.

In a tracking detector, the goal is to minimize the diffusion transverse to the drift direction to obtain a sharp image of the primary ionization at the end of the path. The result depends upon whether the $\vec{B}$ field is parallel to the $\vec{E}$ field or not. For $\vec{B} = 0$, we had $\sigma_T = \sqrt{2 D_e t}$. Since $D_e = u_e \lambda_e / 3$ and $t = L_{\text{path}} / v_d^e$, we get

$$\sigma_T = \sqrt{\frac{2 L_{\text{path}}}{3 v_d^e}} \sqrt{u_e \lambda_e}. \tag{2.184}$$

Figure 2.54 (left) shows the standard deviation of the transverse diffusion broadening of a point-like electron cloud along an electric field in Ar-methane and Ar-CO_2 gas mixtures in the absence of a magnetic field. The transverse direction is minimized by choosing a gas with a very small mean-free path (e.g. CO_2). The diffusion parallel (L) and perpendicular (T) to the drift direction depends on the nature of the gas. Typically, faster gases yield a smaller diffusion than slower gases.

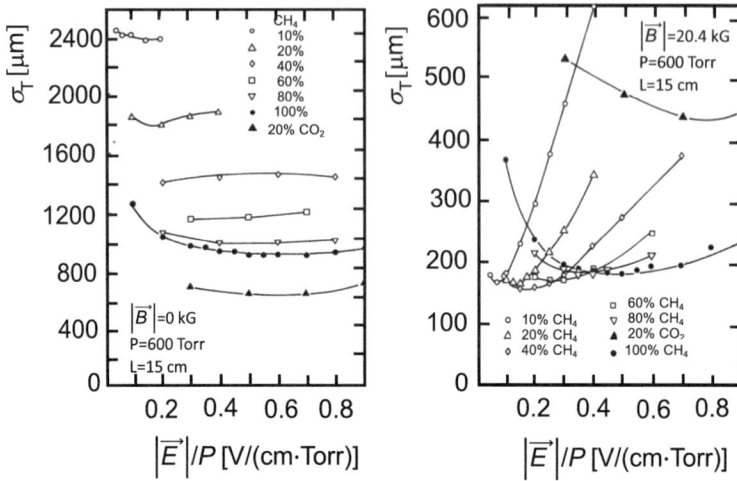

Fig. 2.54 Left: Standard deviation of the transverse diffusion as a function of the electric-field strength after 15 cm drift for electrons in argon-methane and argon-CO$_2$ gas mixtures with the magnetic field switched off. The methane contributions are 10%, 20%, 40%, 60%, 80% and 100% while the CO$_2$ contribution is 20%. Right: Same distributions in the presence of a ($|\vec{B}| = 20.4$ kG) magnetic field that is parallel to the electric field. Note the different scales in the two plots. Reprinted with kind permission from [135], © 1998, Springer. All rights reserved

If an additional magnetic field is switched on, the diffusion transverse to the $\vec{B}$ field is reduced by the factor

$$\left(1 + \omega_B^2 \tau_e^2\right)^{-1}. \tag{2.185}$$

If $\vec{B}$ is parallel to $\vec{E}$, we get for $\tau_e = \lambda_e/u_e$,

$$\sigma = \sqrt{\frac{2L_{\text{path}}}{3v_d^e}} \sqrt{\frac{u_e \lambda_e}{1 + \frac{\omega_B^2 \lambda_e^2}{u_e^2}}}. \tag{2.186}$$

Thus, for $\omega_B \tau_e \gg 1$, $\sigma(B)$ becomes minimal for the largest possible path length. The larger λ_e the more time is available for the magnetic field to keep the electron on helical orbits. Figure 2.54 (right) shows the same measurements as those in Fig. 2.54 (left) but in a magnetic field of magnitude $|\vec{B}| = 20.4$ kG parallel to the electric field. For argon-methane gas mixtures, reductions by one order of magnitude are observed.

Figure 2.55 shows the factor $\omega_B \tau_e$ for the measurements plotted in Fig. 2.54 (right). For $\omega_B \tau_e \gg 1$, $\sigma/\sigma(B) \sim \omega_B \tau_e$. The mean time between two collisions is $\sim 10^{-11}$ s. The maximum of $\omega_B \tau_e$ corresponds to the minimum collision cross section due to the Ramsauer effect. For ideal gases, the characteristic energy, where the Ramsauer minimum occurs, is

$$E_e = \frac{D_e q_e}{\mu_e}, \tag{2.187}$$

2.8 Scintillation Processes

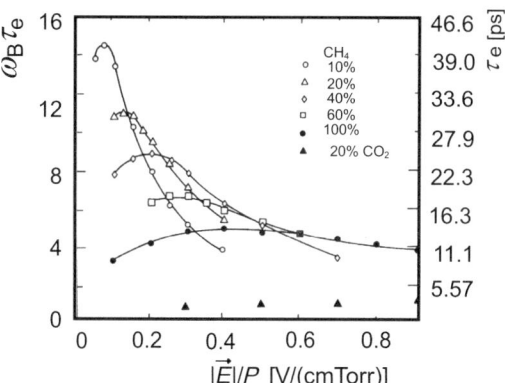

Fig. 2.55 The factor $\omega_B \tau_e$ reducing the standard deviation of the transverse diffusion as a function of the electric-field strength after 15 cm drift for electrons in argon-methane and argon-CO_2 gas mixtures for parallel electric and magnetic fields ($|\vec{B}| = 20.4$ kG), see Fig. 2.54 (right). Reprinted with kind permission from [135], © 1998, Springer. All rights reserved

which can be calculated from the value of the $\vec{E}$ field, $|\vec{E}_0|$, at this maximum of $\omega_B \tau_e$,

$$\frac{\sigma^2}{2L_{\text{path}}} = \frac{D_e}{v_d^e} = \frac{E_e \mu_e}{q_e v_d^e} = \frac{E_e}{q_e E_0}. \qquad (2.188)$$

For an Ar-methane gas mixture (90:10), we get $E_e = 0.15$ eV for $E_0 = 0.1$ V/(Torr cm) in agreement with the Ramsauer minimum in Ar.

2.8 Scintillation Processes

Besides measuring the charge liberated in an ionization process, we can also detect photons that are produced in scintillation processes. This property is observed in certain inorganic crystals, liquid noble gases and organic materials. Since the mechanism is different in each of these materials, we discuss them one after the other. We start with some basic properties of energy levels in solids.

2.8.1 Energy Levels in Solids and Liquid Noble Gases

In atoms, electrons are bound by the Coulomb field of the nucleus, where the allowed energy levels are discrete. In molecules, the electric field is already more complex since it is a superposition of the Coulomb fields of each atom. Thus, the valence electrons in a molecule are not associated with a single atom any longer, but with the entire molecule. Instead of discrete energy levels, they occupy orbitals. In solids, atoms are arranged in a lattice, which is a regular, three-dimensional array with a periodic electric-field configuration. The allowed energy levels are arranged into energy bands, filled bands below the Fermi energy and unfilled bands above the Fermi energy. The highest filled band is called the valence band and the lowest unfilled band is the conduction band. The two are separated by an energy gap.

At T=0 K, the electrons occupy all states in the valence band while the conduction band is empty. In this state, the current flow is zero. To conduct a current, electrons

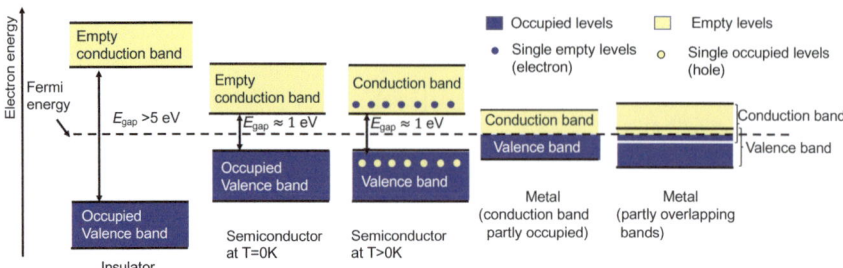

Fig. 2.56 The band structure in an insulator (left), a semiconductor at $T = 0$ K (center left), a semiconductor at room temperature (center), a metal in which the conduction and valence bands are adjacent (center right) and a metal with partly overlapping bands (right). Reprinted with kind permission from [137], © 2005, M. Krammer. All rights reserved

must be lifted from the valence band into the conduction band. The remaining hole in the valence band also can conduct a current. For metals, the valence band and conduction band may be adjacent or may partially overlap. In insulators, the two bands are separated by more than 5 eV while for semiconductors the gap is reduced to about 1 eV. So, at room temperature, semiconductors can conduct a current because electrons can be moved from the valence band to the conduction band due to thermal excitations. Figure 2.56 shows the configuration of the conduction and valence bands in an insulator, in a semi-conductor (for $T = 0$ K and $T > 0$ K) and in a metal. Since in an insulator the two bands are separated by a large energy gap, it is difficult to transfer electrons from the valence band into the conduction band at room temperature. Therefore, the conduction band remains empty. In a metal, the conduction band is partly occupied. In a semi-conductor at room temperature, electrons are lifted into the conduction band by thermal energy. For example, germanium detectors have to be operated at liquid nitrogen temperatures, since at room temperature the thermal noise is too large. This is a consequence of the small energy gap between conduction and valence bands. One important parameter is the Fermi level E_F, which denotes the top of the available energy states at low temperature.

We obtain the band structure of a crystal by solving the Schrödinger equation for an approximate one-electron problem with a periodic potential.

$$\left[-\frac{\hbar^2}{2m}\nabla^2 + V(r)\right]\psi_k(r) = E\psi_k(r). \quad (2.189)$$

The solutions $\psi_k(r)$ are the Bloch functions, which are factorized in terms of plane waves and a function $U_n(\vec{k}, r)$ that is periodic in r with the periodicity of the lattice and which depends on the band index n,

$$\psi_k(r) = \exp[ikr]U_n(\vec{k}, r). \quad (2.190)$$

Further details are given in [136].

2.8.2 Scintillation Process in Liquid Noble Gases

The scintillation of liquid noble gases is caused by excited dimers, which are formed both by direct excitation of atoms and electron-ion recombinations. Let us illustrate this for xenon. Excited dimers Xe_2^* are produced in two separate processes that produce scintillation light.

In the first process, a xenon atom is excited in a radiation process. The excited xenon atom interacts with two other xenon atoms in the liquid to form an excited xenon molecule called "excited dimer" Xe_2^*, which can be in a singlet Σ or triplet Σ state [138, 139]. The dimer decays to two xenon atoms plus an energetic photon in the VUV region.[20]

$$Xe^* + Xe + Xe \rightarrow Xe_2^* + Xe, \qquad (2.191)$$

$$Xe_2^* \rightarrow 2Xe + \hbar\omega. \qquad (2.192)$$

The photon wavelength is (175 ± 10) nm corresponding to an energy of 7 eV [140]. The singlet decays right away to the ground state with a lifetime of (2.2 ± 0.3) ns since this transition is allowed. The triplet state cannot decay directly to the singlet ground state. It has to mix with a singlet Π state. Then de-excitation occurs via spin-orbit interactions. Thus, the decay is delayed having a lifetime of (27 ± 1) ns. The second process is a recombination-induced excited-dimer formation, which occurs via the following steps,

$$Xe^+ + Xe \rightarrow Xe_2^+,$$
$$Xe_2^+ + e^- \rightarrow Xe^{**} + Xe,$$
$$Xe^{**} \rightarrow Xe^* + \text{heat}, \qquad (2.193)$$
$$Xe^* + Xe + Xe \rightarrow Xe_2^* + Xe,$$
$$Xe_2^* \rightarrow 2Xe + \hbar\omega. \qquad (2.194)$$

The xenon ions form ionized dimers, Xe_2^+, in one of the numerous collisions with the surrounding xenon atoms. The dimer ion picks up an electron and dissociates into one second excited xenon atom plus a non-excited xenon atom. The second excited xenon atom radiates heat before interacting with another xenon atom to become an excited dimer, which then decays in the same way as in the first process. Experimental evidence for the recombination-induced process came from measurements by Kubota et al. [141], which showed that the light output was reduced when an electrical field was applied. The reduction of scintillation light is explained by the following quenching process

$$Xe^* + Xe^* \rightarrow Xe + Xe^+ + e^- + \text{heat}. \qquad (2.195)$$

---

[20] The symbol VUV stands for vacuum UV light, which lies in the far UV range from 10–200 nm.

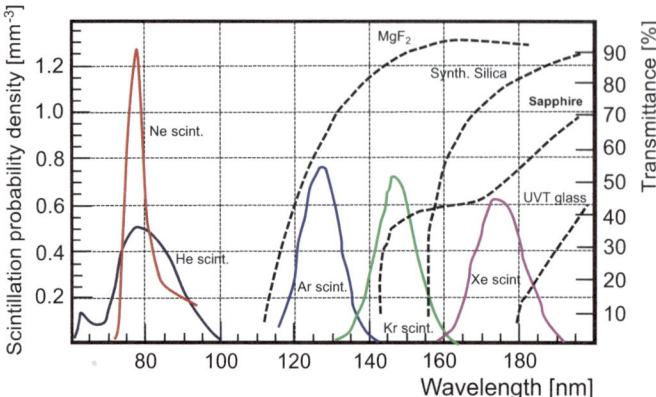

Fig. 2.57 Scintillation spectra for liquid helium, liquid neon, liquid argon, liquid krypton and liquid xenon (left-hand scale). The dotted lines show the percent transmittance of several common optical window materials (right-hand scale), MgF$_2$, synthetic silica, sapphire and UVT glass. Reprinted with kind permission from [146], © 2011, Elsevier. All rights reserved

Here, the electron caries away the excess energy, which is close to one excitation energy, before self trapping and recombination. A fraction of ionization electrons thermalize with the medium relatively far from the ionization track. Since this is at a different time scale than that of the prompt scintillation signal produced by direct excitation and prompt recombination, this light is not detected [142]. The scintillation light is also reduced if the ionization density is very high, such as for α radiation [143]. Liquid xenon is an efficient low-threshold scintillator and efficient e^- Xe$^+$ producer [144]. The valence and conduction bands are separated by a 9.28 eV. Due to the large gap, liquid xenon is a strong insulator. The average energy to produce an electron-ion pair is 14.5 eV. The electron mobility is $\mu_e = 100 \, \text{cm}^2/(\text{Vs})$ allowing for an efficient drift of electrons.

For liquid argon and liquid krypton, we encounter equivalent excited dimer processes. Figure 2.57 shows the scintillation spectra of liquid noble gases and the transmittance of several common optical window materials. In liquid argon, the emitted photons have a wavelength of 128 nm [145, 146]. In this wavelength region, the MgF$_2$ window is a perfect fit. While the singlet state decays with a lifetime of (5.0 ± 0.2) ns, the triplet state has a long lifetime of (860 ± 30) ns [147]. In liquid krypton, the emitted photon has a wavelength of 147 nm [146]. The optimal window material is still MgF$_2$ while sapphire has a lower transmittance. The lifetime of the singlet state is (2.1 ± 0.3) ns and that of the triplet state is (80 ± 3) ns [147]. In liquid xenon, the emitted photon has a wavelength of 175 nm. The MgF$_2$ and synthetic silica provide the best window materials with efficiencies over 90%.

2.8.3 Scintillation Mechanism in Inorganic Crystals

In scintillating inorganic crystals, the band structure is modified. In addition to the valence band, another band called exciton band is located just below the conduction band as depicted in Fig. 2.58. In addition, impurity traps are located in the gap near the exciton band. When a particle enters a crystal, two principle processes can occur. First, the particle can ionize the crystal by transferring electrons from the valence band to the conduction band thus creating a free electron and a free hole. Second, the particle can create an exciton by transferring an e^- to the exciton band. In this state, the electron-hole pair remains bound together. However, the pair can move freely through the crystal.

If the crystal contains impurities, electron levels in the forbidden energy gap can be locally created. A migrating hole or an exciton pair, which encounters an impurity level can ionize the impurity atom. If now a subsequent electron arrives, it can fall into the opening left by the hole and make a transition from excited state to ground state by emitting a photon. If the transition is radiationless, the impurity center becomes a trap and the energy is lost to other processes. The electron transferred into the conduction band can be de-excited by cascading down via the levels in the intermediate (impurity) gap emitting photons in the visible range. The excitation is typically done by electrons and positrons near the critical energy or photons below the pair creation energy. Since inorganic crystals are transparent, the light is transported from one side to the other and can be collected by a photodetector. The amount of dE/dx that is released by optical photons is given by Birks' law (see Chap. 2.8.5).

2.8.4 Scintillation Process in Organic Scintillators

In a plastic scintillator, the scintillation process is rather different with respect to that in an inorganic crystal. Plastic scintillators are organic compounds that are built

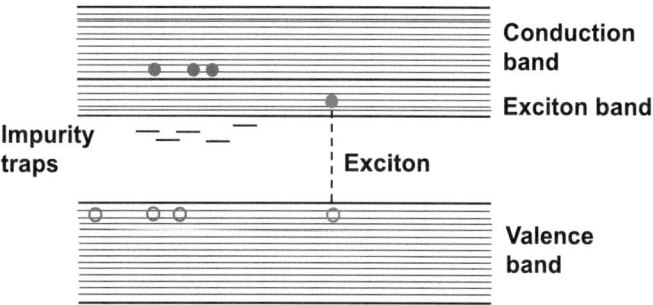

Fig. 2.58 Electronic band structure of inorganic scintillating crystals. In addition to the valence band and the conduction band, an exciton band is located just below the conduction band. In the gap, in addition, impurity traps are located near the exciton band. Besides the formation of free electrons and holes, loosely bound electron-hole pairs known as excitons are formed in the exciton-valence band. Excitons can migrate through the crystal and be captured by impurity centers

up from condensed or linked benzene-like structures [148]. Figure 2.59 shows the energy level diagram for the π-electron levels of an organic scintillator molecule. Organic scintillator compounds form molecular crystals in which the inter-molecular binding results from van-der-Waals forces that are weak. Thus, the crystal may be regarded as a system of oriented but quasi non-interacting molecules. Depending on the orientation of the electron spins, the molecule may be in a singlet (S) or in a triplet state (T). The ground state is the singlet state S_0 in which the electrons are fully paired. The first excited π-electron state is S^* and the second excited state is S^{**}, where the latter lies 3–4 eV above the ground state. The spacing between higher excited states becomes smaller. Higher singlet states are not relevant for the scintillation process. In addition, we have the triplet states in which the electrons are unpaired. The lowest state is T_0, which lies higher than the S_0 state. The excited triplet states are T^* and T^{**}, where T^{**} lies below T^*. A band of vibrational states is associated with each π-electron state, as depicted in Fig. 2.59. Note that photon absorption is constrained by the multiplicity selection rule for electric dipole transitions, which inhibits transitions between singlet and triplet states by a factor of 10^8 for the transition $S_0 \to T^{**}$. Thus, transitions can only occur to excited singlet states. In an ionization process, an electron in the singlet ground state, for example, is lifted into S^{**}, which decays immediately to the lowest vibrational state S^* by internal degradation (10^{-11} s). No photons are released in this process. This is true for all higher excited singlet states. From the excited state S^*, radiative or non-radiative transitions occur to a vibrational ground state within a few ns. Depending on the scintillator, the lifetime of the S^* state is 1–80 ns. For anthracene or stilbene, the probability for a radiative decay is larger than 80%.

It has been observed that excited triplet states T^{**} can be produced by double photon excitations and inter-system crossing $S^{**} \to T^{**}$. Like the S^{**} states, the T^{**} are found to decay rapidly to the T_0 state. Since the decay of T_0 to the ground state S_0 is strongly suppressed by the multiplicity selection rule, the T^{**} states have much longer lifetimes. The T_0 state decays instead by the triplet annihilation process $T_0 + T_0 \to S_0 + S^*$ plus phonons. Due to these different radiation processes, organic scintillators may have two decay times, where the fast component typically has a decay time of 2–3 ns.

The organic scintillator is dissolved in an organic solution. Often we add two fluors to the solvent to shift the scintillation light to higher wavelengths. Some solvents polymerize producing solid scintillators. Figure 2.60 shows a schematic diagram of the energy levels of the solvent, the first fluor and the second fluor. Energy released in an ionization process excites the solvent into a higher excited state, which then cascades down to the first excited states typically without emission of photons. The transition to the ground state releases a photon that is absorbed by the first fluor, which is placed into the second excited state. The excited first fluor then falls back down to the first excited state without emitting a photon. The transition to the ground state produces a photon that is either detected or excites the second fluor in which the molecule is transferred into the second exited state that falls back to the first excited state without photon emission. The succeeding transition to the ground states produces a photon at higher wavelengths that can be recorded by a photodetector.

2.8 Scintillation Processes

Fig. 2.59 Scintillation process in organic scintillators (see description in the text)

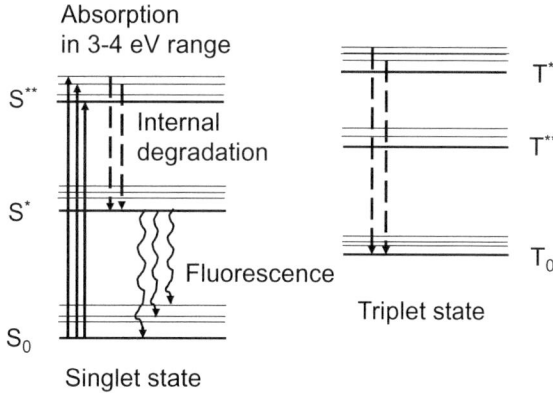

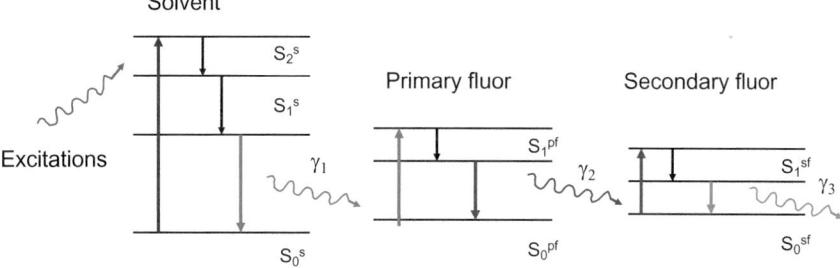

Fig. 2.60 Energy levels of a solvent, a primary fluor and a secondary fluor. The released photon of the solvent excites the first fluor. In the transition to the ground state, it releases a photon that excites the second fluor, which also emits a photon at higher wavelengths. Note that $\lambda_{\gamma_1} < \lambda_{\gamma_2} < \lambda_{\gamma_3}$

2.8.5 Birks' Law

In a plastic scintillator the energy loss of a particle is converted into scintillation light. For electrons there is a linear relation,

$$\frac{dL}{dx} = N_{sc} dE/dx, \tag{2.196}$$

where N_{sc} is the number of photons produced per MeV. For other particles a saturation sets in, which is parameterized by Birks law [149, 150],

$$\frac{dL_{sc}}{dx} = \frac{N_{sc} \, dE/dx}{1 + k_{Birks} \, dE/dx}, \tag{2.197}$$

where k_{Birks} is Birks' constant of the medium, which is of the order of 10^{-4}-10^{-2} g/cm^2/MeV. So Birks' law accounts for light quenching in the scintillator yielding a non-linear response. The total scintillation light is obtained from the integral of (2.197). Typically, k_{Birks} is treated as a free parameter. We fit the total light yield to

$$L_{sc}(E) = b + N_{sc} \int_0^E dE' \frac{1}{1 + k_{Birks} \, dE'/dx}, \tag{2.198}$$

where the constants N_{sc} and k_{Birks} as well as an integration constant b are extracted from the fit. As an example, we look at the light output for 3g/l PPO with and without 15 mg/l bis-MSB in linear alkyl benzene.

Figure 2.61 (left) shows the pulse height distribution versus the maximum Compton energy for two PPO samples (2 g/l and 3g/l), each either undoped or doped with 15 mg/L bis-MSB. The pulse height is slightly higher for the doped sample. For all displayed samples, the response is linear. Figure 2.61 (right) shows the light output of a 3g/l PPO sample doped with 15 mg/L bis-MSB for electrons and protons. For electrons, the response is linear. For protons, however, the response is non linear, but (2.197) gives a good representation with a constant $k_{Birks} = 0.0098 \pm 0.0003$ cm/MeV. Similar values are obtained for the 2g/l PPO sample with 15 mg/L bis-MSB and the samples without bis-MSB. Figure 2.62 (left) shows the light output dL_{sc}/dx as a function of energy loss for electrons, protons and α particles in an anthracene scintillator [149]. While the response for electrons is linear, that for protons and α particles shows quenching effects. Particularly for α particles, Birks' parameterization provides a good fit of the data. For protons, the Wright bi-molecular quenching model [152] provides a slightly better description than Birks' law. Figure 2.62 (right) shows the scintillation response of anthracene to different particles as a function of energy. At a given energy, the light output is rather different for the individual particles. The response for electrons is rather linear. For other particles, the response displays considerable quenching effects, which increase with the mass of the particle. In summary, Birks' parameterization agrees well with measurements accounting well for quenching effects in the scintillator. Birks' law also holds for inorganic scintillators. A generalization of Birks" law is given by [151],

$$L_{sc}(E) = b + N_{sc} \int_0^E dE' \frac{1}{1 + k_{Birks}\, dE'/dx + C_L\big(dE'/dx\big)^2}, \qquad (2.199)$$

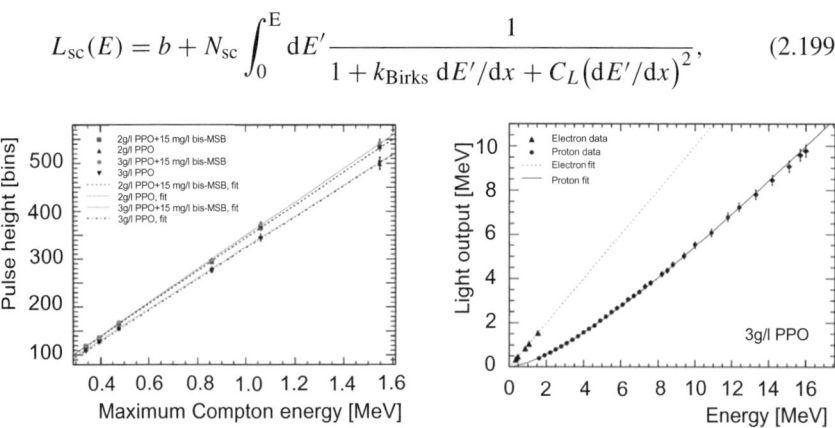

Fig. 2.61 Left: Pulse height for Compton-scattered electrons as a function of energy in samples of 2g/l PPO (upward triangles), 2g/l PPO plus 15mg/l bis-MSB (squares), 3g/l PPO (downward triangles) and 3g/l PPO plus 15mg/l bis-MSB (grey points). Right: Light output as a function of energy for electrons (triangles) and protons (solid points) in linear alkyl benzene with 3g/l PPO. The solid curve shows Birks' parameterization. Both reprinted under CC-BY-4.0 Licence with kind permission from [151], © 2013, B. von Krosigk

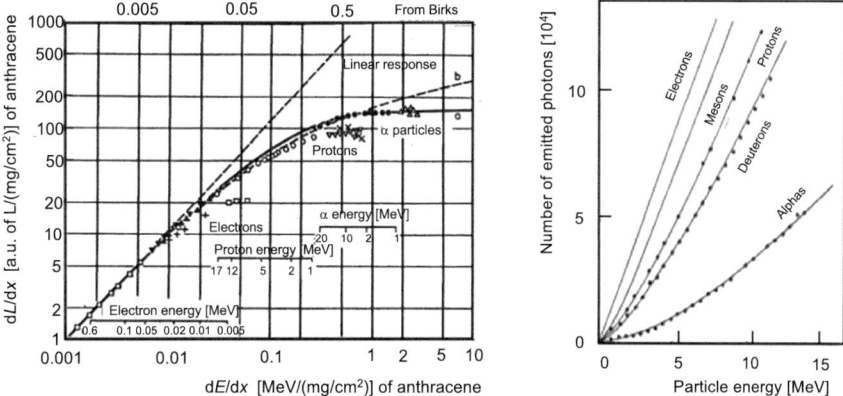

Fig. 2.62 Left: dL/dx as a function of dE/dx for electrons, protons and α particles in anthracene. The energy ranges of the different particles are indicated. Note that for all particles the energy loss is for energies below the minimum. The solid black curve shows Birks' parameterization [149]. The upper dashed line is for $dL/dx = dE/dx$, while the lower dashed curve shows the parameterization for bi-molecular quenching [152]. Reprinted under CC-BY-3.0 Licence with kind permission from [10], © 1977–2024, CERN. All rights reserved. Right: Scintillation response of anthracene to electrons, mesons, protons, deuterons and α-particles as a function of energy for data (solid points) and Birks' parameterization (solid curves). Reprinted with kind permission from [152], © 1953, AIP Publishing. All rights reserved

where a quadratic correction term has been included. Here, C_L is another constant that is extracted from the fit.

Exercises

2.1 A charged pion with $|\vec{p}| = 0.8$ GeV/c ($m_\pi = 139$ MeV/c^2) travels through a 25 cm long NaI crystal. What is the energy loss $\langle dE/dx \rangle$ per cm and in the total block due to ionization? Ignore the shell correction. Other parameters are $\rho = 4.51$ g/cm^3, $I_{\text{NaI}} = 452$ eV, $\delta(\beta\gamma) = 4.6 \log_{10}(\beta\gamma) - 6.1 + 0.1252 \cdot (3.59 - \log_{10}(\beta\gamma))^{3.04}$.

2.2 For a 10 GeV/c electron, sketch the bremsstrahlung differential cross section for complete and no screening in aluminum as a function of the relative electron energy $1 - \eta_r = E/E_0$. Explicitly write down the dependence of the differential cross section on $1 - \eta_r$ in the two limits. Include the contributions arising from interactions of the electric field of the atomic electrons in the approximation in which the atomic number is modified appropriately. What values will the screening variables take in the two limiting cases. Sketch the screening functions ϕ_1 and ϕ_2 in terms of $1 - \eta_r$ in the two limits. Sketch the dependence of the screening variable ξ in terms of $1 - \eta_r$. How big is the Coulomb correction?

2.3 You want to shield off hadrons with lead bricks. How far do 1 GeV/c pions, kaons ($m_K = 494$ MeV/c^2) and protons ($m_p = 938$ MeV/c^2) travel? Would iron or aluminum be a better shield?

2.4 Determine the radiation length X_0 for Xe gas, H$_2$O and W. Compare your calculations with the results listed in the Particle Data Group.

2.5 Two electrons with momenta of 40 MeV/c and 40 GeV/c penetrate Si strip detectors (300 μm thickness) in the ATLAS semiconductor tracker (SCT). Which mechanism yields the largest energy loss and how big is it? How much energy do the electrons loose in the ATLAS SCT (four layers). The Coulomb correction for silicon is 1.24%. The mean excitation energy is 173 eV. The density correction in Si is parameterized by $\delta_{Si}(\beta\gamma) = 4.6052 * \log_{10}[\beta\gamma] - 4.44 + 0.1492 * (2.87 - \log_{10}[\beta\gamma])^{3.25}$ valid for $0.2014 < \log_{10}[\beta\gamma] < 2.87$. For $\log_{10}[\beta\gamma] \geq 2.87$ only the first term (linear in $\log[\beta\gamma]$) contributes.

2.6 Using the small-angle approximation determine the *rms* scattering angle of a 1 GeV/c pion in a 5 cm thick block of iron. Sketch the *rms* angle as a function of the Fe thickness between 1 mm and 10 cm.

2.7 Consider Compton scattering of a photon with energy $E_\gamma = 0.511$ MeV in a plastic scintillator. Such photons are obtained for example from positron annihilation of a ^{22}Na source. Plot the energy of the electron as a function $\cos\theta$. What is the maximum energy transferred onto the electron? At which photon scattering angle θ is the cross section for Compton scattering minimal? Discuss the energy dependence of the Compton cross section for photon energies much larger than the electron mass.

2.8 a) Consider a cloud of electrons in a He gas. There is an electric field with $|\vec{E}| = 500$ V/cm and no magnetic field. What is the drift velocity for electrons? Assume that the mean free path for electrons is independent of the thermal velocity. The time between collisions is about 25 ps and the mean free path is $\lambda_e = 1.58 \times 10^{-4}$ cm. How fast do the He ions move in the same chamber? Compute this from the information you have for the electrons. b) If the electrons are collected after 2 cm, how big is the spread of the electrons in the transverse directions due to diffusion? If a 1.5 T magnetic field is switched on that is parallel to the electric field, what is the transverse spread due to diffusion?

2.9 You want to detect the photons (0.662 MeV) from a ^{137}Cs source with a cylindrical CsI crystal (dimensions: 2.5 cm diameter, 2.5 cm length) read out with a photon detector that has no gain (photodiode). Which is the most important interaction the photons experience? Which processes can occur? What is the strength of the dominant process with respect to the next largest one?

2.10 Consider a cloud of electrons, which has the following probability density function (pdf) for the thermal velocity,

$$F(u_e) = 4\pi \left(\frac{m_e}{2\pi k_B T}\right)^{3/2} u_e^2 \exp\left[-\frac{m_e u_e^2}{2k_B T}\right] du_e. \qquad (2.200)$$

Calculate the mean velocity of the electrons ($v_e = <u_e>$) and the *rms* velocity ($v_e = \sqrt{<u_e^2>}$). How do these compare to the most probable velocity? The individual steps in the calculation should be visible and the expressions should be expressed in terms of the appropriate integrals. Though the integrals are solvable, you can use results from tables or mathematics programs. Order the results from smallest to largest.

References

1. D.E. Groom, S.R. Klein, Passage of particles through matter, the particle data group (S. Navas et al.), Phys. Rev D **110**, 030001 (2024)
2. H. Bichsel, Nucl. Instrum. Meth. A **562**, 154 (2006)
3. E. Shiles et al., Phys. Rev. B **22**, 1612 (1980)
4. H. Bichsel, Rev. Mod. Phys. **60**, 663 (1988)
5. H.J. Bhabha, Proc. roy. Soc (London), A **164**, 257 (1938)
6. B. Rossi, *High-Energy Particles* (Prentice-Hall Inc, Englewood Cliffs, NJ, 1952)
7. E.A. Uehling, Ann. Rev. Nucl. Part. Sci. **4**, 315 (1954)
8. H.J. Massey, H.C. Corben, Proc. Cambridge Phil. Soc. **35**, 463 (1939)
9. https://slideplayer.com/slide/782134/
10. F. Sauli, CERN Yellow Report. CERN **77**-09, 1 (1977)
11. H.A. Bethe, Ann. Phys. **5**, 325 (1930)
12. N. Bohr. The London, Edinburgh, and Dublin Philosophical Magazine and Journal of Science, 25. Jg., Nr. **145**, S. 10 (1913)
13. H.A. Bethe, Z. Phys. **76**, 293 (1932)
14. F. Bloch, Annalen Phys. **408**, 285 (1933); Z. Phys. 81, 363 (1933)
15. L. Landau, J. Phys. (USSR) **8**, 241 (1944)
16. U. Fano, Phys. Rev. **72**, 26 (1947)
17. U. Fano, Phys. Rev. **102**, 385 (1956)
18. R.M. Sternheimer, Phys. Rev. **88**, 851 (1952)
19. R.M. Sternheimer, Phys. Rev. **117**, 485 (1960); R. M. Sternheimer, Phys. Rev. **145**, 247 (1966)
20. W.H. Barkas, W. Birnbaum, F.M. Smith, Phys. Rev. **101**, 778 (1956)
21. W.H. Barkas, N.J. Dyer, H.H. Heckmann, Phys. Rev. Lett. **11**, 26 (1963)
22. J.H. Cobb, W.W.M. Allison, J.N. Bunch, Nucl. Instrum. Meth. **133**, 315 (1976)
23. W.W.H. Allison et al., Nucl. Instrum. Meth. **133**, 325 (1976)
24. W.W.H. Allison, J.H. Cobb, Ann. Rev. Nucl. Part. Sci. **30**, 253 (1980)
25. D.L. Landau, E. Lifschitz (Soviet Union), Z. Phys. **8** 153 (1935)
26. U. Fano, Ann. Rev. Nucl. Part. Sci. **13**, 67 (1963)
27. R.L. Platzman, Energy spectrum of primary activations in the action of ionizing radiation, Radiation Research, G. Silini (ed.) Amsterdam, North Holland, p. 20 (1967)
28. M.E. Peskin, D.V. Schroeder, *An Introduction to Quantum Field Theory* (Addison-Wesley Publishing Company, Reading MA, 1995)
29. M. Inokuti, Rev. Mod. Phys. **43**, 297 (1971); M. Inokuti, Rev. Mod. Phys. **50**, 23 (1978)
30. H.A. Bethe, Annalen Phys. **3**(97), 325 (1930)
31. J. Lindhard and M. Scharff, Mat. Fys. Med. Dan. Vid. Selsk. **27**, No. 15 (1952); J. Lindhard, Mat. Fys. Med. Dan. Vid. Selsk. **28**, No. 8 (1954); J. Lindhard and A. Winther, Mat. Fys. Med. Dan. Vid. Selsk. **34**, No. 4 (1964)
32. H.H. Andersen and J. F. Ziegler, Pergamon, New York (1977)
33. J.F. Ziegler, Rev. Appl. Phys. **85**, 1249 (1999)
34. I. Lehraus et al., Nucl. Instrum. Meth. **153**, 347 (1978)
35. R.M. Sternheimer, R.F. Peierls, Phys. Rev. B **3**, 368 (1971)
36. Stopping Powers for Electrons and Positrons, ICRU Report No. **37** (1984)
37. H. Bichsel, Phys. Rev. A **46**, 5761 (1992)

38. W.H. Barkas, M.J. Berger, Tables of Energy Losses and Ranges of Heavy Charged Particles, NASA-SP-3013 (1964)
39. H. Bichsel, Charged-Particle-Matter interactions, in *Atomic, Molecular and Optical Physics Handbook (chapter 91)*. ed. by G. Drake (Springer, Berlin, 2005)
40. http://physics.nist.gov/PhysRefData/XrayMassCoef/tab1.html
41. H. Bichsel, Phys. Rev. A **28**, 1147 (1983); H. Bichsel, Phys. Rev. A **46**, 5761 (1992); includes citations to numerous prior calculations
42. W.F.G. Swann, J. Franklin, Inst. **226**, 598 (1938)
43. E. Fermi, Phys. Rev. **57**, 485 (1940)
44. N. Bohr, Kgl. Dansk, Vid. Sel. Mat.-Fys. Med. **18**, 1 (1948)
45. R.M. Sternheimer, S.M. Seltzer, M.J. Berger, Phys. Rev. B **26**, 6067 (1982)
46. A. Crispin, G.N. Fowler, Rev. Mod. Phys. **42**, 290 (1970)
47. R.M. Sternheimer, M.J. Berger, S.M. Seltzer, Atomic Data and Nuclear Data Table **30**, 261 (1984)
48. W.R. Nelson, H. Hirayama, D.W.O. Rogers, The EGS4 Code System, SLAC-265, Stanford Linear Accelerator Center (1985)
49. S.M. Seltzer, M.J. Berger, Int. J. Appl. Radiat. **35**, 665 (1984). This article corrects and extends results of S. M. Seltzer and M. J. Berger, Int. J. Appl. Radiat. **33**, 665 (1982)
50. D.E. Groom, N.V. Mokhov, S.I. Striganov, Atomic Data and Nuclear Data Tables **78**, 183 (2001)
51. H. Bichsel, R.P. Saxon, Phys. Rev. A **11**, 1286 (1975)
52. P.V. Vavilov, Sov. Phys, JETP **5**, 749 (1957)
53. K.R. Symon, thesis, Harvard University, (1948)
54. G.J. Igo et al., Phys. Rev. **89**, 879 (1952)
55. O. Blunck, S. Leisegang, Z. Physik **128**, 500 (1950)
56. P. Shulek et al., Yad. Fiz. **4**, 564 (1966)
57. I.B. Smirnov, Nucl. Instrum. Meth. A **554**, 474 (2005)
58. P.V. Ramana Murthy, G.D. Demeester, Nucl. Instrum. Meth. **56**, 93 (1967)
59. The GEANT4 Collaboration, J. Allison et al., Geant4 developments and applications, IEEE Trans. Nucl. Sci. **53**, 270 (2006)
60. J.F. Bak, Nucl. Phys. B **288**, 681 (1987)
61. R. Talman, Nucl. Instrum. Meth. **159**, 189 (1979)
62. E.A. Kopot et al., Phys. JETP **70**, 397 (1976)
63. V.C. Ermilova, L.P. Kotenko, G.J. Merzon, Nucl. Instrum. Meth. **145**, 555 (1977)
64. C. Tschalar, Nucl. Instrum. Meth. **64**, 237 (1968); Nucl. Instrum. Meth. **61**, 141 (1968)
65. Passage of Charged Particles through Matter, 3rd Ed., Mc Graw-Hill Book Co., New York 1972
66. M.G. Holloway, M.S. Livingston, Phys. Rev. **54**, 18 (1938)
67. N. Feather, R.R. Nimmo, Proc. Camb. Phil. Soc. **24**, 139 (1938)
68. G. Stetter, W. Jentschke, Phys. Zeits. **36**, 441 (1935)
69. R. Schulze, Z. Phys. **94**, 104 (1935)
70. W.R. Leo, *Techniques for Nuclear and Particle Physics Experiments* (Springer Verlag, Berlin, 1987)
71. N.F. Mott, H.S.W. Massey, *The Theory of Atomic Collisions* (Oxford Press, London, 1965)
72. E.J. Kobetich, R. Katz, Phys. Rev. **170**, 391 (1968)
73. L. Katz, A.S. Penfold, Rev. Mod. Phys. **24**, 28 (1952)
74. J.R. Young, Phys. Rev. **103**, 292 (1956)
75. H. Kanter, E.J. Sternglas, Phys. Rev. **126**, 620 (1962)
76. V.E. Cosslett, R.N. Thomas, Brit. J. Appl. Phys. **15**, 1283 (1964)
77. K.H. Weber, Nucl. Instrum. Meth. **25**, 261 (1964)
78. H.W. Koch, J.W. Motz, Rev. Mod. Phys. **31**-4, 920 (1959)
79. H. A. Bethe and W. Heitler, Proc. Roy. Soc. (London) Am, **83** (1934)
80. Y.S. Tsai, Rev. Mod. Phys. **46**, 815 (1974)
81. https://pdg.lbl.gov/2022/AtomicNuclearProperties/index.html (2022)

82. J.A. Wheeler, W.E. Lamb, Phys. Rev. **55**, 858 (1939); Phys. Rev. **101**, 1834 (1956)
83. L.D. Landau, I Pomeranchuk, Dokl. Akad. Nauk. Ser. Fiz. **92**, 535 (1953); **92**, 735 (1953); translated in A. B. Migdal, Phys. Rev. **103**, 1811 (1956)
84. S. Klein, Rev. Mod. Phys. **71**, 1501 (1999)
85. M.L. Ter-Mikaelian, *SSSR 94, 1033 (1954); High-Energy Electromagnetic Processes in Condensed Media* (John Wiley and Sons, New York, 1972)
86. P.L. Anthony et al., Phys. Rev. Lett. **76**, 3550 (1996)
87. M.J. Berger, S.M. Seltzer, Tables of Energy Losses and Ranges of Electrons and Positrons, National Aeronautics and Space Administration Report NASA-SP-3012 (Washington DC 1964)
88. L. Pages et al., Atomic Data **4**, 1 (1972)
89. T. Baltakmens, Nucl. Instrum. Meth. **82**, 264 (1970)
90. J. Marshall, A.G. Ward, Can. J. Research A **15**, 39 (1937)
91. H. Geiger, E. Marsden, Phil. Mag. **25**, 604 (1913)
92. E. Rutherford, Phil. Mag. Ser. **6**(21), 669 (1911)
93. G. Z. Molière, Z. Naturforsch. **2a**, 133 (1947); **3a**, 78 (1948); **10a**, 177 (1955)
94. H.A. Bethe, Phys. Rev. **89**, 1256 (1953)
95. G.R. Lynch, O.I. Dahl, Nucl. Instrum. Meth. B **58**, 6 (1991)
96. W.T. Scott, Rev. Mod. Phys. **35**, 231 (1963)
97. P.C. Hemmer, I.E. Farquahr, Phys. Rev. **168**, 294 (1968)
98. H. Bichsel, Phys. Rev. **112**, 182 (1958)
99. T. Tabata, R. Ito, S. Okabe, Nucl. Instrum. Methods **94**, 509 (1971)
100. V.L. Ginzburg, I.M. Frank, Zh. Eksp, Teor. Fiz. **16**, 15 (1946)
101. G.M. Garibian, Zh. Eksp, Teor. Fiz. **33**, 1403 (1957)
102. M.L. Ter-Mikaelian, Nucl. Phys. **24**, 43 (1961)
103. P. Goldsmith, J.V. Jelley, Philos. Mag. **4**, 836 (1959)
104. J. Oostens et al., Phys. Rev. Lett. **19**, 541 (1967)
105. L.C.L. Yuan et al., Phys. Rev. Lett. **23**, 496 (1969)
106. M.L. Cherry et al., Phys. Rev. D **10**, 3594 (1974); M. L. Cherry et al., Phys. Rev. D **17**, 2245 (1978)
107. J.H. Cobb et al., Nucl. Instrum. Meth. **140**, 413 (1977)
108. https://www.nist.gov/pml/xcom-photon-cross-sections-database
109. B.L. Berman, S.C. Fultz, Rev. Mod. Phys. **47**, 713 (1975)
110. http://www.cxro.LBNL.gov/optical_constants/pert_form.html
111. https://physics.nist.gov/PhysRefData/XrayMassCoef/tab3.html
112. W. Heitler, *"The Quantum Theory of Radiationâłž, Third ED, Dover Publications* (INC, New York, 1984)
113. C.D. Broyles, D.A. Thomas, S.K. Haynes, Phys. Rev. **89**, 715 (1953)
114. U. Amaldi, Fisica delle radiazion, (Bollati Boringhieri, 1971)
115. C. M. Davisson, "Interaction of Gamma Radiation with Matter" in Alpha, Beta and Gamma Ray Spectroscopy, ed. by K. Siegbahn (North Holland Publ. Co, Amsterdam 1968)
116. H.R. Hulme et al., Proc. Roy. Soc. **149**, 131 (1935)
117. R.H. Pratt, Phys. Rev. **117–4**, 1017 (1960)
118. O. Klein, Y. Nishina, Z. Phys. **52**, 853 (1928)
119. I. Øverbø, K.J. Mork, H.A. Olsen, Phys. Rev **175**, 1978 (1968)
120. J.W. Motz, H.A. Olsen, H.W. Koch, Rev. Mod. Phys. **41**, 581 (1969)
121. R.M. Brown, Graduate lectures 2009/10; https://slideplayer.com/slide/12089631/
122. S. Menke, https://www.mpp.mpg.de/~menke/elss/pic2.shtml
123. E. Longo, I. Sestili, Nucl. Instrum. Meth. **128**, 283 (1975)
124. F. Sauli, *Gaseous Radiation Detectors* (Cambridge University Press, Fundamentals and applications, 2022)
125. L.B. Loeb, *Basic processes of gaseous electronics* (University of California Press, Berkeley, 1961)

126. https://chem.libretexts.org/Bookshelves/Physical_and_Theoretical_Chemistry_Textbook_Maps/Supplemental_Modules_(Physical_and_Theoretical_Chemistry)/Kinetics/06%3A_Modeling_Reaction_Kinetics/6.01%3A_Collision_Theory/6.1.01%3A_Collisional_Cross_Section
127. N.E. Bradbury, Phys. Rev. **44**, 883 (1933)
128. E.N. Pusateri et al., J. Geophys. Res. Atmos. **120**, 7200 (2015)
129. V. Palladino and B. Sadoulet, Berkeley report LBNL-3013 (1974); V. Palladino and B. Sadoulet, Nucl. Instrum. Meth. **128**, 323 (1975)
130. M. Titov, *Gaseous Detectors, 26 p* (Springer, Handbook of Particle Detection and Imaging, 2012)
131. A. Breskin et al., Nucl. Instrum. Meth. **119**, 9 (1974)
132. B. Jean-Marie, V. Lepeltier, D. L'Hote, Nucl. Instrum. Meth. **159**, 213 (1979)
133. A. Breskin et al., Nucl. Instrum. Meth. **124**, 189 (1975)
134. https://magboltz.web.cern.ch/magboltz/
135. PEP–4 proposal, SLAC report PUB–5012 (1976); K. Kleinknecht, Detectors for Particle Radiation, Cambridge University Press, 2nd ed. (1998)
136. S.M. Sze, Physics of Semiconductor Devices, John Wiley and Sons (1981)
137. M. Krammer, Lecture Halbleiterdetektoren. Wien (2005)
138. S. Kubota, M. Hishida, J.Z. Raun, J. Phys. C: Solid State Phys. **11**, 2645 (1978)
139. S. Kubota et al., Nucl. Instrum. Meth. **196**, 101 (1982)
140. K. Fujii et al., Nucl. Instrum. Meth. A **795**, 293 (2015)
141. S. Kubota et al., Phys. Rev. B **17**-6, 2762 (1978)
142. T. Doke et al., Nucl. Instrum. Meth. **269**, 291 (1988)
143. A. Hitashi, Astropart. Phys. **24**-3, 247 (2005)
144. E. Aprile, T. Doke, Rev. Mod. Phys. **82**, 2053 (2010)
145. N. McFadden et al., arXiv:2006.09780 [physics.ins-det] (2020); D. E. Fields et al., arXiv:2009.10755 [physics.ins-det] (2020)
146. V.M. Gehman et al., Nucl. Instrum. Meth. A **654**, 116 (2011)
147. S. Kubota et al., Phys. Rev. B **20**-8, 3486 (1979)
148. F.D. Brooks, Nucl. Instrum. Meth. **163**, 331 (1979)
149. J.B. Birks, Proceedings of the Physical Society A **64**, 874 (1951); ibid A **64**, 511 (1951); Theory and Practice of Scintillation Counting, Pergamon Press (1964)
150. J.B. Birks, Scintillation Counters, Pergamon Press (1954)
151. B. von Krosigk et al., Eur. Phys. J. C **73**-4, 2390 (2013)
152. T. Wright, Phys. Rev. **91**, 1282 (1953)

Open Access This chapter is licensed under the terms of the Creative Commons Attribution 4.0 International License (http://creativecommons.org/licenses/by/4.0/), which permits use, sharing, adaptation, distribution and reproduction in any medium or format, as long as you give appropriate credit to the original author(s) and the source, provide a link to the Creative Commons license and indicate if changes were made.

The images or other third party material in this chapter are included in the chapter's Creative Commons license, unless indicated otherwise in a credit line to the material. If material is not included in the chapter's Creative Commons license and your intended use is not permitted by statutory regulation or exceeds the permitted use, you will need to obtain permission directly from the copyright holder.

Measurements of Energy Loss of Charged Particles

3

Charged particles passing through an appropriate path length of material will ionize the atoms. We saw, that the average energy loss of heavy charged particles is determined by the Bethe-Bloch equation (see (2.13)) and the most probable energy loss is given by (2.40). Thus, ionization measurements play a basic role for charged-particle detection. The simplest application discussed in this chapter is the usage of ionization measurements in detectors that record radiation. This is used to localize radioactive sources and measure their activity or to localize residual radiation and its intensity in accelerators. As we will see, the measurement of different particle properties depend on ionization measurements, such as identifying trajectories and vertices, measuring electromagnetic or hadronic energy and identifying particle types at low momenta. For example, from the measured particle energy loss we obtain the particle velocity as discussed in Chap. 2. Thus, tracking and vertex detectors provide particle identification in the low-momentum region in addition, if the particle momentum is known. This will be discussed in detail in Sect. 9.1.

We start our discussion with ionization in gases. We discuss pulse mode ionization chambers, followed by proportional counters and Geiger Müller counters. We present the concept of gas amplification, introducing the first Townsend coefficient. We also state the Raether condition that has to be satisfied to provide stable chamber operations. Next, we present ionization in liquids, in particular liquid noble gases. Finally, we extend the discussion to ionization in solid-state detectors. We introduce the *pn* junction and discuss its properties introducing the depletion layer.

3.1 Energy Loss Measurements in Gases

The energy loss of charged particles in gases typically leads to an ionization of the gas.

3.1.1 Pulse Mode Ionization Chamber

A pulse mode ionization chamber is basically a capacitor filled with a counting gas such as argon. A traversing charged particle ionizes the gas atoms along its path. To collect the electrons and ions, an electric field $\vec{E}$ is applied, which is sufficiently low such that primary electrons cannot gain sufficient kinetic energy to further ionize the gas in collisions with gas atoms. The primary electrons $N \cdot e^-$ and positive ions $N \cdot i^+$ are collected completely on the electrodes. As a first example, we consider a parallel-plate chamber sketched in Fig. 3.1 (left). The cathode is placed at high voltage $-V_0$ while the anode is kept at ground potential. A charged particle traversing the chamber produces electrons and ions along its path. In the electric field, electrons move towards the anode while the positive ions move towards the cathode.

The charge moves through a resistor R and can be measured as a voltage pulse. The energy of a parallel-plate capacitor is $\frac{1}{2} C_w V_0^2$, where C_w is the capacitance of the anode versus ground. Since the electrons are collected rapidly, the positive ions change the potential energy at z_0 to $\frac{1}{2} C_w V^2$. To move N charges from z_0 to z requires work in the electric field, $N \int_{z_0}^{z} q_k E_z dz$. Energy conservation yields

$$\frac{1}{2} C_w V^2 = \frac{1}{2} C_w V_0^2 - N \int_{z_0}^{z} q_k E_z dz, \tag{3.1}$$

where z_0 is the position of the electron-ion pair production, q_k is the charge (of the electron or ion) and d denotes the distance between anode and cathode. Since the electric field is uniform in parallel-plate chambers, we can write

$$\frac{1}{2} C_w [V_0 + V][V_0 - V] = -N q_k V_0 [z - z_0]/d. \tag{3.2}$$

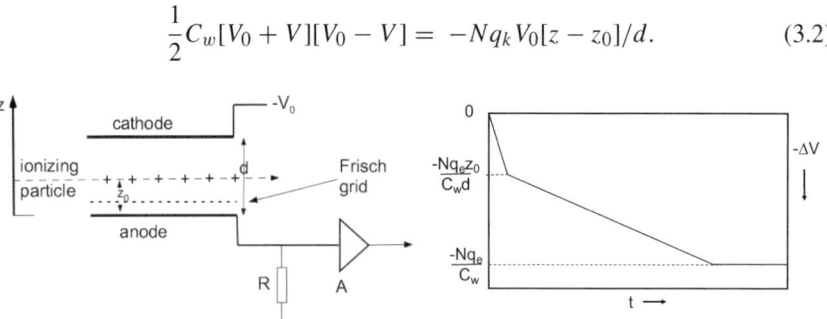

Fig. 3.1 Left: Schematic view of a parallel-plate ionization chamber. Right: Charge collection of electrons and ions produced in a parallel-plate ionization chamber as a function of time

3.1 Energy Loss Measurements in Gases

The work conducted in the electric field by the moving charges is small compared to the energy stored in the chamber. Thus, we introduce $\Delta V = V - V_0$ and approximate $V + V_0$ by $2V_0$ yielding

$$\Delta V = -\frac{Nq_k}{C_w d}[z - z_0]. \tag{3.3}$$

For constant drift velocities, v_d^{ion} for ions and v_d^e for electrons, we obtain changes in potential caused by ions and electrons,

$$\Delta V^+ = -\frac{Nq_{ion}}{C_w d} v_d^{ion} \Delta t^+, \tag{3.4}$$

$$\Delta V^- = -\frac{N(-q_e)}{C_w d}(-v_d^e)\Delta t^-, \tag{3.5}$$

where Δt^+ and q_{ion} are the collection time and charge of the ions while Δt^- and q_e are the corresponding observables of the electrons, respectively. Note that the polarity of the two contributions is the same. Since electrons have much faster drift velocities than ions, they are collected first. Figure 3.1 (right) schematically shows the charge collection versus time. After collection of the electrons the pulse height is $\Delta V = -Nq_e z_0/(C_w d)$. After collection of the much slower ions the pulse height is raised to $\Delta V = -Nq_e/C_w$. Let us consider a simple example of a counter filled with argon gas at NTP with a distance $d=5$ cm and a field strength of 500 V/cm. For the electrons, the collection time is $\Delta t^- \simeq 1$ μs while that of the ions is $\Delta t^+ \simeq 1$ ms.

The voltage pulse is independent of position z_0 at which the primary ionization occurred only, if $\tau = RC_w > \Delta t^+$. Such a long collection time is impractical for recording individual particles, especially if the counting rate is high. We can introduce an $R'C'_w$ coupling in front of the amplifier with $\Delta t^- < R'C'_w \ll \Delta t^+$. Here, the ion contribution is neglected and the voltage pulse is given by

$$\Delta V = -\frac{Nq_e}{C_w}\frac{z_0}{d}. \tag{3.6}$$

If $R'C'_w \ll \Delta t^-$ then only the first part of electron-induced pulse is recorded, which is nearly independent of z_0. In order to collect the entire electron signal independent of the production point z_0, we introduce a Frisch grid between the anode and cathode at an intermediate potential close to the anode. In this case, the pulse satisfies $\Delta t^- < R'C'_w \ll \Delta t^+$. By collimation, the incident particles are confined to a region between the cathode and the grid. The electrons drift to the grid, pass it and continue to drift towards the anode. If the pulse is collected only between the grid and the anode, we can get $\Delta t^- < R'C'_w \ll \Delta t^+$ and have a voltage pulse independent of z_0.

In the second example, we consider a cylindrical field as shown schematically in Fig. 3.2 (left, right). Here, the electric field is in the radial direction,

$$E(r) \equiv |\vec{E}(\vec{r})| = \frac{V_0}{r \ln(r_a/r_i)} = \frac{C_w V_0}{2\pi \epsilon_0}\frac{1}{r}, \tag{3.7}$$

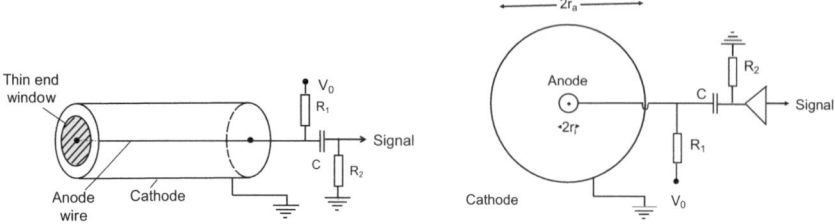

Fig. 3.2 Left: Three-dimensional schematic view of a cylindrical ionization chamber. Right: Cross section through a cylindrical ionization chamber. In some chambers, the cylinder is on negative high voltage and the anode wire is on ground potential

where $C_w = 2\pi\epsilon_0/\ln(r_a/r_i)$, $\epsilon_0 = 8.854 \times 10^{-12}$ F/m is the permittivity of free space, r_a is the radius of the cylinder and r_i is the radius of the wire.

The drift time for an electron formed at $r = r_0$ is

$$\Delta t^- = \int_{r_i}^{r_0} \frac{dr}{v_d^e}, \tag{3.8}$$

where $v_d^{e,\text{ion}} = |\vec{v}_d^{\,e,\text{ion}}|$ is the magnitude of the drift velocity of electrons and ion, respectively. For $E(r) = 0.1$ V/cm/Torr, $v_d^- = \mu_e E(r)$ such that the collection time is

$$\Delta t^- \simeq \int_{r_i}^{r_0} \frac{dr}{\mu_e E(r)} = \frac{\ln(r_a/r_i)}{\mu_e V_0} \int_{r_i}^{r_0} r\, dr = \frac{\ln(r_a/r_i)}{2\mu_e V_0}(r_0^2 - r_i^2). \tag{3.9}$$

Here, the voltage pulse coming from electrons is

$$\Delta V^- = -\frac{Nq_e}{C_w} \frac{\ln(r_0/r_i)}{\ln(r_a/r_i)}. \tag{3.10}$$

Note that the dependence on r_0 is just logarithmic. The contribution from the ions yields

$$\Delta V^+ = -\frac{Nq_e}{C_w} \frac{\ln(r_a/r_0)}{\ln(r_a/r_i)}. \tag{3.11}$$

To evaluate what the dominant contribution of the pulse is, we consider an example. We assume that $r_a \gg r_i$ and that the cylindrical chamber has homogeneous irradiation. If $r_a/r_i = 10^3$ and $r_0 = 1/2 r_a$, then $\Delta V^+/\Delta V^- = \ln 2/\ln 500 = 0.1$. Thus, the electron component dominates. For chambers with $RC_w > \Delta t^+$, it is impossible to measure individual pulses and only an average dc is measurable,

$$I = -\frac{Nq_e}{RC_w}. \tag{3.12}$$

This configuration is the current-mode ionization chamber.

3.1.2 Proportional Counters

Applying a high electric field of 10^4-10^5 V/cm to the ionization chamber depicted in Fig. 3.2 (left), electrons from the primary ionization may gain sufficient kinetic energy to cause further ionizations. In the high electric field near the anode wire, the primary electrons are accelerated. Between the radial points r_1 and r_2, they gain the kinetic energy,

$$\Delta E_e = q_e \int_{r_1}^{r_2} E(r) \mathrm{d}r = q_e V_0 \frac{\ln(r_2/r_1)}{\ln(r_a/r_i)}. \tag{3.13}$$

If ΔE_e is larger than the ionization potential of the gas, the electrons can ionize the gas in a collision. Since the electric field increases as $1/r$, the acceleration interval r_2-r_1 decreases such that secondary electrons are capable of producing tertiary electrons that themselves produce quartic electrons and so on. The process yields an avalanche of electrons and positive ions near the anode wire. Figure 3.3 schematically illustrates the production of primary electrons by a charged particle traversing the proportional counter and the subsequent avalanche formation near the anode wire.

Each avalanche produced from one primary electron induces a pulse on the anode wires whose main contribution comes from the ions that are slowly moving away from the anode wires. Figure 3.4 shows the time development of the avalanche formation in the vicinity of an anode wire. A primary electron moves towards the anode (a). On its way, the electron gains sufficient kinetic energy to perform secondary ionizations and multiplication begins (b). By secondary ionization, an electron cloud and an ion cloud is produced, which start to drift apart (c). The electron cloud drifts toward the anode wire and surrounds it producing more electron-ion pairs (d). The ion and electron charge distributions now both show this drop-like shape. For a short interval they show a large overlap. Since the ions move slowly away from the anode, there is a region around the anode wire that consists only of electrons. The electrons are collected at the anode and the electric field is enhanced by the drop-like positive ion cloud around the anode separated by a small gap from the wire (e). Now, the ions start to drift apart.

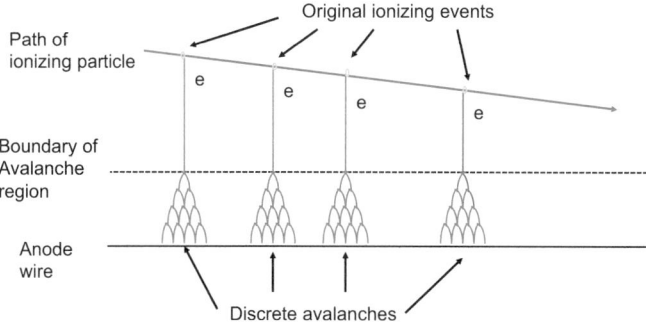

Fig. 3.3 Formation of discrete avalanches by a traversing charged particle

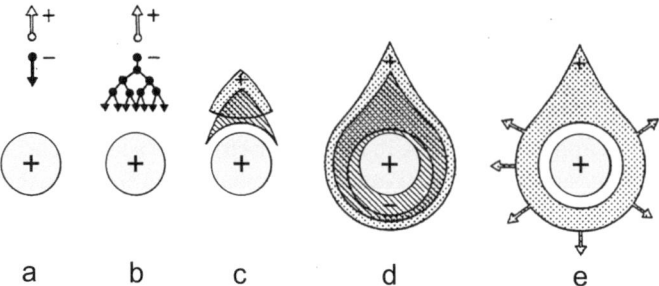

Fig. 3.4 Avalanche formation in the vicinity of an anode wire represented by the grey-shaded circle with a "+" sign: **a** a primary electron moves towards the anode wire; **b** the electron gains sufficient kinetic energy to perform secondary ionizations and multiplication begins; **c** electron and ion clouds drift apart; **d** the electron cloud drifts toward the anode wire and surrounds it producing more electron-ion pairs; **e** the electrons have been collected and the electric field is enhanced by the drop-like positive ion cloud around the anode wire separated by a small gap from the wire; the ions start to drift apart. Reprinted with kind permission from [1], © 1972, F. Sauli. All rights reserved

The amount of charge liberated is amplified by factor A_e with respect to the primary ionization charge density $-\mathbf{N} \cdot q_e$.[1] Now, the voltage pulse becomes

$$\Delta V = -A_e \frac{N q_e}{C_w}. \quad (3.14)$$

The gain is obtained from the first Townsend coefficient α_I, i.e. the number of electron-ion pairs formed by an electron along a 1 cm path, by

$$A_e = \exp\left(\int_{r_1}^{r_2} \alpha_I(r) \mathrm{d}r \right). \quad (3.15)$$

Note that the first Townsend coefficient is a function of the reduced electric-field strength $|\vec{E}|/P$, where P is the gas pressure. Several parametrization of α_I exist, which describe the data in some range of the electric field [2]. Korff proposed a simple approximation [3],

$$\frac{\alpha_I}{P} = c_1 \exp\left[-c_2 \frac{P}{|\vec{E}|} \right], \quad (3.16)$$

where c_1 and c_2 are parameters that depend on the gas and range of the electric field [4]. It has been shown that this parameterization is a satisfactory description of experimental data in a wide range of electric fields [5]. Table 3.1 lists c_1 and c_2 for noble and other gases and Tab. 3.2 lists these coefficients for argon gas mixtures at low reduced electric-field strengths [6].

---

[1] Please note that we denote particle densities by $\mathbf{N}$ and number of particles by N. For an ideal gas at standard conditions this is given by the Loschmidt number 2.69×10^{19} molecules/cm^3. For other temperatures and pressures it needs to be corrected by $P/760 \cdot 273/T$.

3.1 Energy Loss Measurements in Gases

Table 3.1 Parameters c_1 and c_2 for determining the first Townsend coefficient in different gases at reduced electric fields $|\vec{E}|/P = 30 - 500$ V/(cm · Torr), the reduced electric field $|\vec{E}|/P$ sufficient for Townsend breakdown of centimeter-sized gaps at atmospheric pressure [10,11] and parameter c_k for some gases

Gas	c_1 [Torr^{-1} · cm^{-1}]	c_2 [V/(cm · Torr)]	$|\vec{E}|/P$ [kV/(cm · Torr)]	c_k [10^{17} cm^2/V]
H_2	5	130	20	0.46
N_2	10	310	35	0.7
Air	15	365	32	
H_2O	13	290		
CO_2	20	466		
CH_4	8.08	212		1.24
He	3	34	10	0.11
Ne	4	100	1.4	0.14
Ar	12	180	2.7	1.81
Kr	17	240		
Xe	26	350		

On its way to the anode the electron is accelerated and undergoes multiple collisions, which may lead to further ionizations. The electric-field strength, where the electron gains sufficient energy to ionize the gas within λ_e is given by,

$$|\vec{E}_{\text{ion}}| = \frac{I_0}{q_e \lambda_e}, \tag{3.17}$$

where I_0 is the ionization potential. For high electric-field strengths we have

$$\alpha_I = \frac{1}{\lambda_e} \tag{3.18}$$

Since $\lambda_e = 1/(N\sigma(E_e))$, we may write for high electric fields,

$$\alpha_I = N\sigma(E_e). \tag{3.19}$$

Figure 3.5 (left) shows the ionization cross section for electrons in noble gases, which all have a broad maximum over a few hundred eV starting around 100 eV. Since typical ionization potentials lie around 10-15 eV, we get $|\vec{E}_{\text{ion}}| = 270 - 400$ kV/cm for normal gas pressure. In gaseous detectors, however, the gas pressure is low. For example at a pressure of 20 Torr, the electric field is reduced to $|\vec{E}_{\text{ion}}| = 7 - 10.5$ kV/cm.

To illustrate at what distance this condition is reached let us consider an example. Consider a single-wire cylindrical counter that has a 10 mm diameter cathode and

Table 3.2 Parameters c_1 and c_2 for determining the first Townsend coefficient in argon gas mixtures. In addition, the $(E/P_{\max})$ limit is listed for which the (3.16) holds. Extracted with kind permission from [6], © 1993, Elsevier. All rights reserved

Gas	Fraction	c_1 [$\frac{1}{\text{Torr·cm}}$]	c_2 [$\frac{V}{\text{Torr·cm}}$]	$(E/P)_{\max}$ [$\frac{V}{\text{Torr·cm}}$]
Ar-CH_4	(83.4:16.6)	2.7	81.7	19.7
Ar-CH_4	(44.4:55.6)	38.3	230.1	32.8
Ar-CH_4	(26:74)	111.4	298.5	36.8
Ar-CH_4	(17.8:82.2)	237.3	342.5	38.4
Ar-CH_4	(6.7:93.3)	244.7	347.8	38.9
Ar-CH_4	(0:100)	436.1	388.3	40.8
Ar-C_4H_{10}	(79.3:20.7)	70.7	193.2	25.5
Ar-C_4H_{10}	(47.3:52.7)	144.7	241.1	31.5
Ar-C_4H_{10}	(20.6:79.4)	200.1	314.2	3.5
Ar-C_4H_{10}	(9.3:90.7)	147.1	334.8	40.5
Ar-C_4H_{10}	(0:100)	176.3	360.0	41.8
Ar-C_2H_6	(68.3:31.7)	24.6	154,8	21.6
Ar-C_2H_6	(45.2:54.8)	40.0	237.1	29.7
Ar-C_2H_6	(17.9:72.1)	164.1	301.9	31.7
Ar-C_2H_6	(0:100)	242.6	340.8	34.4
Ar-DME	(91:9)	4.3	57.6	10.4
Ar-DME	(81.9:18.1)	7.9	107.5	13.5
Ar-DME	(0:100)	8.0	213.1	31.0
Ar-CO_2	(96.3:3.7)	5.04	90.82	16.2
Ar-CO_2	(77.2:22.8)	221.2	207.6	21.6
Ar-CO_2	(12.8:87.2)	158.3	291.8	32.9
Ar-CO_2	(0:100)	145.1	318.2	36.4

a 10 μm diameter anode wire. The counter is operated at a pressure of 20 Torr and a potential of 900 V. The gas amplification starts at a critical field of $|\vec{E}_{cr}|/P = 65$ V/(cm · Torr). This occurs at a radius of $r_c = 1.3$ mm. For a thicker anode wire of 50 μm diameter the amplification starts at 1.6 mm from the wire.

For certain electric-field strengths and gas pressures, the amplification A_e is independent of the amount of primary ionization. This means that the voltage pulse is proportional to the primary ionization. This domain of field strengths is called the proportional region. Here, the amplification is $A_e \sim 10^4\text{-}10^6$. The method of achieving high electric-field strengths consists of using very thin anode wires (10-100 μm). The amplification will start in close vicinity of the anode wire as indicated in Fig. 3.3.

3.1 Energy Loss Measurements in Gases

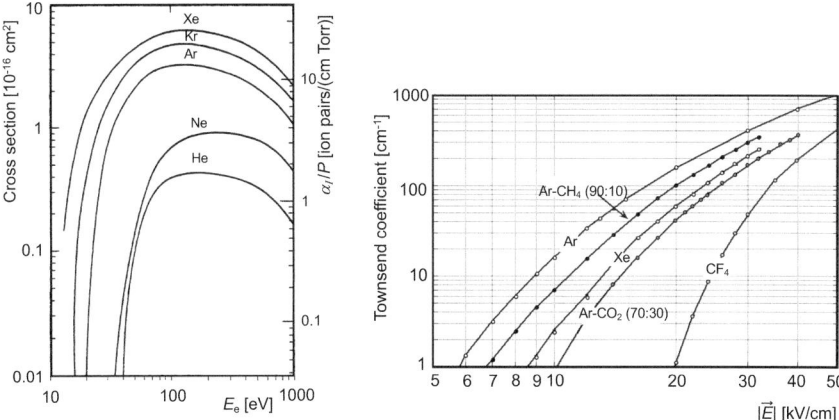

Fig. 3.5 Left: Ionization cross section as a function of the electron energy in noble gases. On the right-hand scale the corresponding α_I/P is shown. Right: Computed first Townsend coefficients α_I as a function of electric field in several gases at NTP. Both reprinted under CC-BY-NC-4.0 Licence with kind permission from [7], © 2024, the Particle Data Group LBNL

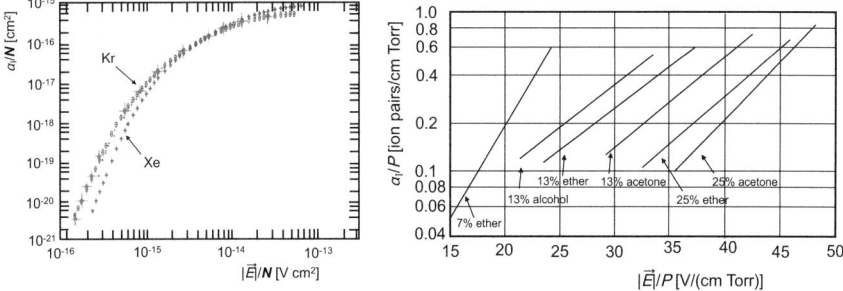

Fig. 3.6 Left: Measured first Townsend coefficients per ionization charge density for krypton and xenon as a function of the electric field per ionization charge density. Reprinted from [8]. Public domain. Right: Calculated first Townsend coefficients as a function of reduced electric field for Ar-ether, Ar-alcohol and Ar-acetone gas mixtures. Reprinted with kind permission from [9], © 1977, F. Sauli. All rights reserved

Figure 3.5 (right) shows computations of the first Townsend coefficients in several gases as a function of the electric-field strength. For fast gases, Townsend coefficients are considerably smaller than those for slow gases. We see that in argon we get about 20 secondary electrons per centimeter at an electric field of 10 kV/cm whereas in CF_4 we just get 1 secondary electron per centimeter at $|\vec{E}| = 20$ kV/cm. If we know the dependence of $\alpha_I(x)$ on the electric-field strength, we can compute the amplification for any electric field. Figure 3.6 (left) shows the normalized first Townsend coefficient as a function of the normalized electric field for krypton and xenon. Figure 3.6 (right) shows calculated first Townsend coefficients for several argon gas mixtures as a function of the reduced electric field. For these gases the first Townsend coefficients increase exponentially.

Furthermore, we can assume that α_I depends linearly on the electron kinetic energy ΔE_e,

$$\alpha_I = c_k N \Delta E_e / q_e, \tag{3.20}$$

where c_k is a gas-specific constant.

In a wire chamber, the radial electric field near the wire can be approximated by

$$E(r) = \frac{C_w V_0}{2\pi \epsilon_0} \frac{1}{r}, \tag{3.21}$$

where V_0 is the high voltage on the anode wire and C_w is the capacitance to ground. Accordingly, the potential is

$$V(r) = \frac{C_w V_0}{2\pi \epsilon_0} \ln\left(\frac{r}{r_i}\right). \tag{3.22}$$

The capacitance per unit length of a cylindrical counter is given by

$$C_w = \frac{2\pi \epsilon_0}{\ln(r_a/r_i)}. \tag{3.23}$$

We can determine the gas amplification based on the approximation in (3.20) [10, 12]. Since the mean free path is given by $1/\alpha_I$, the kinetic energy gained between two collisions is given by $|\vec{E}|/\alpha_I$. Using the approximation in (3.20) and the explicit form of the electric field yields

$$\Delta E_e(r) = q_e \sqrt{\frac{C_w V_0}{2\pi \epsilon_0 c_k N} \frac{1}{r}}, \tag{3.24}$$

from which we determine

$$\alpha_I(r) = \sqrt{\frac{C_w V_0 c_k N}{2\pi \epsilon_0} \frac{1}{r}}. \tag{3.25}$$

To determine the amplification we have to integrate (3.15) from the radius of the anode wire r_i to the position r_{th} where the avalanche production starts. We call V_{th} the potential at position r_{th}. The electric field here is given by

$$|\vec{E}_{\text{th}}| = \frac{C_w V_0}{2\pi \epsilon_0} \frac{1}{r_{\text{th}}} = \frac{C_w V_{\text{th}}}{2\pi \epsilon_0} \frac{1}{r_i}, \tag{3.26}$$

3.1 Energy Loss Measurements in Gases

yielding

$$\frac{V_0}{V_{th}} = \frac{r_{th}}{r_i}. \tag{3.27}$$

Replacing r_{th} by $r_i V_0/V_{th}$ yields

$$A_e(x) = \exp\left[2\sqrt{\frac{c_k N C_w V_0 r_i}{2\pi\epsilon_0}}\left(\sqrt{\frac{V_0}{V_{th}}}-1\right)\right] = \exp\left[\sqrt{2c_k N|\vec{E}_{th}|r_i}\sqrt{\frac{V_0}{V_{th}}}\left(\sqrt{\frac{V_0}{V_{th}}}-1\right)\right]. \tag{3.28}$$

Please note that we have neglected recombinations of electrons with positive ions and electron attachments. Furthermore, we have assumed that no ionization occurs via UV photons from excited atoms. So, for $V_{th} \ll V_0$, the amplification depends exponentially on V_0 or on the charge per unit length $Q_0 = C_w V_0$,

$$A_e(x) \propto \exp(C_w V_0). \tag{3.29}$$

The approximation works well for small amplification factors up to $A_e(x) \sim 10^3$. Figure 3.7 (left) shows the gas amplification as a function of applied voltage for two gas pressures in comparison to calculations based on (3.28). The measurements are in good agreement with the predictions. Note that this approximation is not valid in the region where ionization processes due to UV photons becomes sizable. These photons originate from de-excitations of atoms that are excited in collisions. They can produce electrons via the photo-electric effect in gas atoms or in the cathode.

Let us assume that in an avalanche formation $N_0 \cdot A_e(x)$ electrons are produced from N_0 primary electrons. Due to excitations, UV photons will be produced as

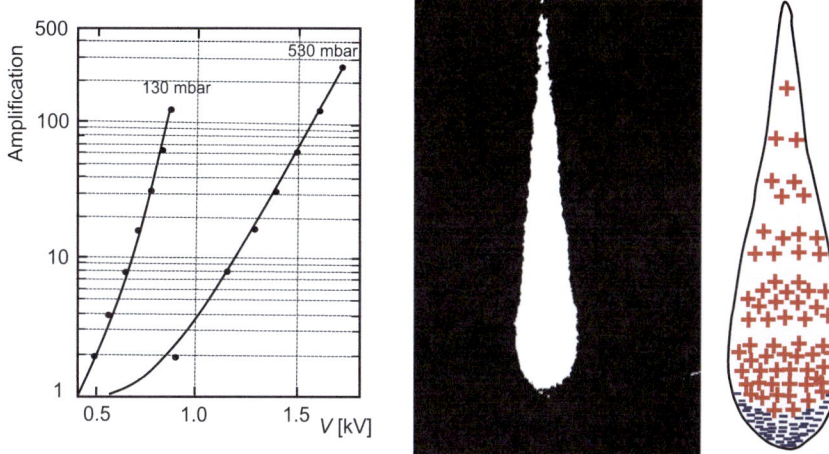

Fig. 3.7 Left: Gas amplification factor A_e as a function of applied voltage V in argon gas for two different gas pressures showing measurements (solid points taken from [13]) and predictions (solid curves from (3.28)). Center: Photograph of an avalanche in a cloud chamber revealing a drop-like shape, which is caused by the slowly moving ions [14]. Reprinted from [14]. Public Domain. Right: Charge composition of the drop-like charge distribution

well. Let us assume that they produce $N_0 \cdot A_e f_\gamma$ photoelectrons where the fraction f_γ is rather small ($A_e f_\gamma \ll 1$). By gas amplification, these photoelectrons give rise to $N_0 \cdot A_e^2 f_\gamma$ electrons and in addition $N_0 \cdot A_e^2 f_\gamma^2$ photoelectrons from UV photons, which in turn produce $N_0 \cdot A_e^3 f_\gamma^2$ electrons. This amplification continues. Summing up all terms yields

$$N_0 A_e(x) \sum_{n \geq 0} (A_e(x) f_\gamma)^n = \frac{N_0 A_e(x)}{1 - A_e \cdot f_\gamma} := N_0 A_\gamma(x) \qquad (3.30)$$

where $A_\gamma(x)$ is the gas amplification factor that includes all contributions from UV photons. The fraction f_γ is also called the second Townsend coefficient. For $A_e(x) \cdot f_\gamma \to 1$, (3.30) diverges and the signal no longer depends on the primary ionization. This regime is called Geiger-Müller region. Obviously, the gas amplification factor cannot be increased without a limit. At a certain value, avalanches triggered by the UV photon are spread over the entire gas volume and the space charge deformation of the electric field eventually results in a spark breakdown. The phenomenological limit for the gas amplification before breakdown is given by the Raether condition

$$\alpha_I \cdot x \sim 20, \qquad (3.31)$$

corresponding to an amplification factor of $A_e(x) \sim 5 \times 10^8$. Note that for an increased gap the Raether condition is already fulfilled at smaller values of α_I.

Now let us look at the pulse formation in a proportional counter. The charge carriers responsible for the pulse formation are those in avalanches near the anode. The radial position for them is a few mean free paths (j) away from the wire, i.e. $r_0 = r_i + j \cdot \lambda_e$. For $\lambda_e \ll r_i$ we get

$$R = \frac{\Delta V^+}{\Delta V^-} \sim \frac{\ln(r_a/r_i)}{\ln\left((r_i + j \cdot \lambda_e)/r_i\right)} \simeq \frac{\ln(r_a/r_i)}{j \lambda_e / r_i}, \qquad (3.32)$$

where ΔV^+ is again the voltage pulse from the ions and ΔV^- is that from the electrons. Let us consider an example. For $r_a = 20$ mm, $r_i = 0.1$ mm and $j \cdot \lambda_e = 0.02$ mm we obtain $R = 25$ in Argon gas at NTP. Thus, in proportional counters the main contribution to the signal comes from the positive ions drifting slowly away from the wire. The contribution of the electrons can be increased by reducing the gas pressure, which increases λ_e.

The electron rise time is

$$\Delta t^- = \frac{(r_0^2 - r_i^2)}{2\mu_e V_0} \ln \frac{r_a}{r_i}. \qquad (3.33)$$

Using $\mu_e \sim 10^2$-10^3 cm^2/(Vs), $V_0 = 100$ V and the dimensions listed above, we get $\Delta t^- \sim 10^{-8}$-10^{-9} s. The rise time of the positive ions is

$$\Delta t^+ = \frac{\ln r_a/r_i}{\mu_{\text{ion}} V_0} \int_{r_0}^{r_a} r \, dr \simeq \frac{(r_a^2 - r_0^2)}{2\mu_{\text{ion}} V_0} \ln \frac{r_a}{r_i}. \qquad (3.34)$$

3.1 Energy Loss Measurements in Gases

For $\mu_{ion} \sim 1\ cm^2/(Vs)$, we get $\Delta t^+ \sim 10$ ms. If a fast pulse is needed, as before, we introduce an RC coupling in front of the preamplifier. We choose $RC \ll \Delta t^+$. For $RC \sim 1$ ns, the fine structure of the anode pulse can be resolved yielding several short pulses.

The spatial distribution of the avalanche shows a drop-like distribution of Q^+ and Q^- near the anode caused by difference in drift velocities for electrons and ions. Figure 3.7 shows the spatial distribution of charges in an avalanche near the anode wire in a cloud chamber, photo (center) and charge composition (right). While the electrons move towards the anode wire the slowly moving ions appear stationary creating the drop-like shape.

3.1.3 Geiger-Müller Counters

Figure 3.8 shows the gas amplification as a function of the voltage applied to the proportional counter. Below 200 V no secondary ionization is possible. This region is that of an ionization chamber. Between 200 V and 600 V we have the proportional region. Between 600 V and 750 V we have the limited proportionality. Above 800 V we enter the Geiger-Müller region. Here, $A_e \cdot f_\gamma \sim 1$ so that $A_\gamma \to \infty$ yielding amplifications of 10^8. The avalanche spreads over the entire counter and leads to a complete discharge $Q = C_w V_0$ that is the same for each particle. The avalanche needs to be terminated. This is done by quenching. If the gas consists of a noble gas or two-atomic gases, the discharge process is stopped by a cloud of positive ions near the anode. The electric field is reduced to prevent the electrons to avalanche.

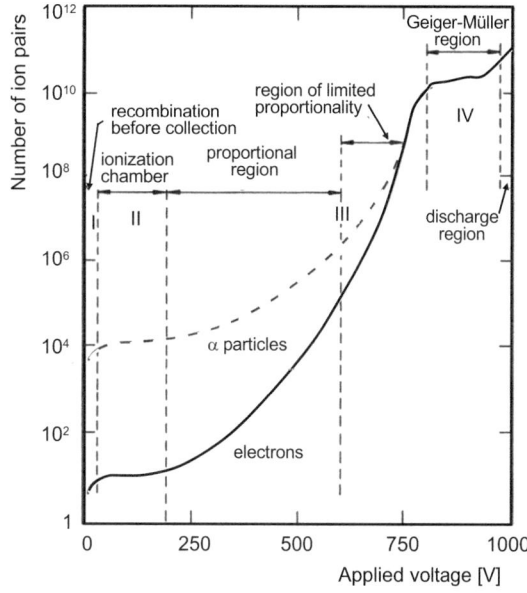

Fig. 3.8 Gas amplification factor as a function of applied voltage for electrons and α particles showing the different regions of chamber operation [15, 16]. Reprinted with kind permission from [15], © 1941, Elsevier. All rights reserved

But once the positive ions are moved to the cathode, new electrons can be produced, which can restart the avalanche formation.

The discharge can be quenched externally by choosing an external resistor that is sufficiently large to reduce the potential $V_0 - IR$ under the threshold required for Geiger-Müller operation. The propagation of UV photons can be prevented by adding a second gas with large molecules, such as C_2H_5OH, CH_4, C_2H_6, C_4H_{10} (isobutane) or $(OCH_3)_2CH_2$ (methylal). These molecules absorb the photons in 100 nm $\leq \lambda_e \leq$ 200 nm range. Using appropriate concentrations they reduce the range of UV photons to size $2r_i$. Along the wire a discharge can take place producing a tube of positive ions along the wire. However, the photons cannot reach the cathode and cannot produce new electrons. The electron production by positive ions in the cathode is also reduced, as positive ions of the counting gas transfer ionization to atoms of quenching gas since the latter has not enough energy for secondary ionizations. Thus, the avalanche stops by itself and the working resistor can be chosen much smaller for the Geiger-Müller regime yielding a dead time of $\tau = 10^{-6}$ s.

Geiger-Müller counters are yes-no counters that are highly efficient in recording individual particles. Since the avalanche needs to be stopped before the next particle can be recorded, the dead time is quite high. Therefore, these counters are of limited use for high counting rates. Proportional counters with very high gain are more appropriate.

3.2 Ionization in Liquids

The energy loss in liquids is significantly larger than that in gases since densities are three orders of magnitude higher. Our interest focuses on liquid noble gases such as liquid argon (LAr), liquid krypton (LKr) and liquid xenon (LXe). A unique feature of these materials is the production of both charge carriers and scintillation photons in response to radiation. The latter we discussed in Sect. 2.8.2. The signals from ionization and scintillation light are highly complementary and are anti-correlated. Simultaneous detection of both signals with high efficiency provides precise measurements of the particle's properties. The work on electrical conduction of dielectric liquids started with Jaffe [17] and Greinacher [18]. McLennan was the first to study cryogenic liquids [19]. Studies at Cavandish Laboratory [20] and Caltech [21] showed that electrons were the free carriers. This lead to the construction of the first liquid ionization detector by Marshall used in an experiment [22].

Many experiments have used liquid noble gases in homogeneous and sampling calorimeters, some of which we discuss in Chap. 7. They are also used in time projection chambers for WIMP searches [23,24]. Table 3.3 list properties of Ar, Kr and Xe in the gas and liquid forms. The ionization yield is the number of electron-ion pairs produced per unit absorbed energy [25–28]. The values for the liquid state are lower than those for the gaseous state. The high atomic number (54) and high density (3g/cm^3) of LXe make it very efficient to detect penetrating radiation. Figure 3.9 (left) shows the phase diagram for xenon. At atmospheric pressure, there is only a narrow temperature range (160-165 K) for producing the liquid phase. Figure 3.9 (right) shows high-resolution absorption spectra for solid Ar, Kr and Xe. The obser-

3.2 Ionization in Liquids

Table 3.3 Ionization properties of liquid noble gases listing the ionization potential I, the average energy needed to produce one electron-ion pair (W_{ion}), the Fano factor and $F \cdot W_{ion}$ [25,35,41]. For the liquid state we also report the density, index of refraction, gap energy, Moliére radius, radiation length, plasma energy, minimum ionization, critical energy, peak wavelength of emitted photons λ_{max}, attenuation length [42], the triple point and number of photons per MeV

Material (Z)	Ar (18)	Kr (36)	Xe (54)
Gas			
Ionization potential I [eV]	15.75	14.00	12.13
W_{ion}-values [eV]	26.4	24.2	22.0
Fano factor (F)	0.16	0.17	0.15
$F \cdot W_{ion}$ [eV]	4.22	4.11	3.3
Liquid			
Density [g/cm^3]	1.396	2.418	2.953
Refractive index	1.233	1.303	1.392
Gap energy [eV]	14.3	11.7	9.28
W_{ion}-values [eV]	23.6 ± 0.3	18.4 ± 0.3	15.6 ± 0.3
Fano factor (F)	0.116	0.07	0.059
$F \cdot W_{ion}$ [eV]	2.74	1.29	0.92
Moliére radius [cm]	9.043	5.857	5.224
Radiation length [cm]	14.00	4.703	2.872
Plasma energy [eV]	22.85	29.37	31.76
Minimum ionization [MeV/cm]	2.105	3.281	3.707
Critical energy (e^-) [MeV]	32.84	17.03	11.66
λ_{max} [nm]	128	147	174
Attenuation length [cm]	66 ± 3	82 ± 4	29 ± 2
Triple point [K]	83.8 ± 0.3	115.8 ± 0.3	161.4 ± 0.3
#photons /MeV	3.65×10^4	6.45×10^4	6.80×10^4

vation of exciton peaks in these spectra gives clear evidence for the band structure of the solid noble gases. Liquid noble gases also have a band structure since exciton levels were observed [29,30] and direct measurements of the band gap energy were performed [31–34]. Due to the large gap energies, liquid noble gases are excellent insulators.

3.2.1 Ionization Process

In liquid noble gases, the energy lost by a traversing charged particle (W) is used to produce a number of electron-ion pairs (N_i) at an average energy expenditure E_i,

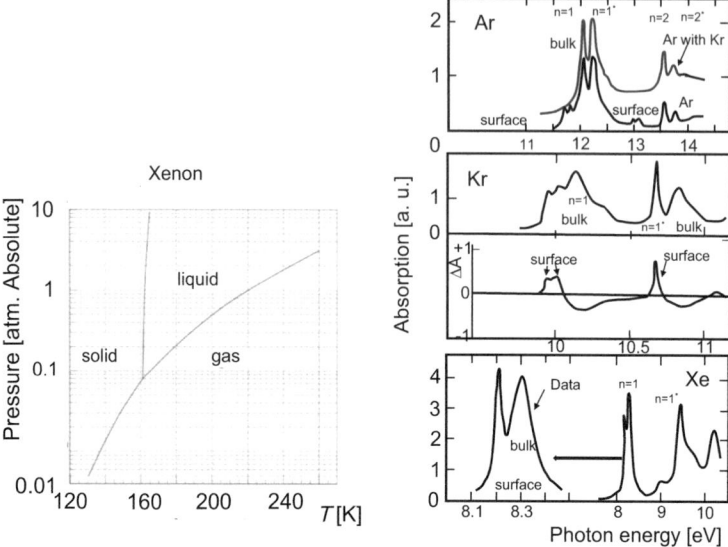

Fig. 3.9 Left: Phase diagram of xenon. Reprinted with kind permission from [35], © 2009, American Physical Society. All rights reserved. Right: High-resolution absorption spectra for solid Ar (top), Kr (center) and Xe (bottom) showing surface and bulk effects. The visible peaks result from excitons. Reprinted with kind permission from [36], © 1985, Springer. All rights reserved

excited atoms (N_{ex}) at an average energy expenditure E_{ex} and free electrons having an average kinetic energy E_e that is lower than the energy of the first excited level, called sub-excitation electrons. So

$$W = N_i E_i + N_{ex} E_{ex} + N_i E_e. \quad (3.35)$$

The average energy needed to produce one electron-ion pair is

$$W_{ion} = \langle W \rangle / \langle N_i \rangle = \langle E_i \rangle + (\langle N_{ex} \rangle / \langle N_i \rangle) \langle E_{ex} \rangle + \langle E_e \rangle. \quad (3.36)$$

Division by the gap energy yields

$$W_{ion}/E_g = \langle E_i \rangle / E_g + (\langle N_{ex} \rangle / \langle N_i \rangle) \langle E_{ex} \rangle / E_g + \langle E_e \rangle / E_g. \quad (3.37)$$

The ratio W_{ion}/E_g is calculated in the following way. The ratios $\langle E_{ex} \rangle / E_g$ and $\langle N_{ex} \rangle / \langle N_i \rangle$ are estimated in the optical approximation using the oscillator strength spectra of solid rare gases [28]. For $\langle E_i \rangle$, data are taken from reference [37] under the assumption that the width of the valence band is negligibly small. For an estimate of $\langle E_e \rangle$, the Alkhazov model was used [38, 39]. The measured and calculated ratios W_{ion}/E_g of all three liquid noble gases are listed in Table 3.4 together with averages of the energy of electron-ion pairs, energy of excited atoms, the ratio of number of excited atoms to number of electron-ion pairs and kinetic energy of electrons from the sub-excited atoms. Note that the Shockley model, which was constructed before the Alkhazov model, yields slightly lower kinetic energies [40].

3.2 Ionization in Liquids

Table 3.4 The measured and calculated ratios W_{ion}/E_g as well as the energy of electron ion pairs, energy of excited atoms, the ratio of the number of excited atoms to that of electron-ion pairs and the kinetic energy of electrons from the sub-excited atoms appearing in (3.37) for LAr, LKr and LXe [38,39]

Liquid Gas	Measured W_{ion}/E_g	Calculated W_{ion}/E_g	$\langle E_i \rangle$ [eV]	$\langle E_{ex} \rangle$ [eV]	$\langle N_{ex} \rangle / \langle N_i \rangle$	$\langle E_e \rangle$ [eV]
LAr	1.65 ± 0.02	1.71	15.4	12.7	0.21	7.7
LKr	1.57 ± 0.03	1.73	13.0	10.5	0.10	6.5
LXe	1.68 ± 0.03	1.69	10.5	8.4	0.06	5.25

3.2.2 Charge Distribution

Figure 3.10 (top left) shows a parallel-plate liquid ionization cell. A charged particle traversing the cell parallel to the plates produces N_e electron-ion pairs along its path at position z_0. The potential at z_0 with respect to the cathode is $V_b z_0/d$ where V_b is the applied high voltage across the plates. Thus, we have a constant electric field and in turn a constant electron drift velocity v_d^e. Let us just consider one electron-ion pair, which produces a constant current as depicted in Fig. 3.10 (top center),

$$I_s = q_e v_d^e / d = q_e / t_d. \tag{3.38}$$

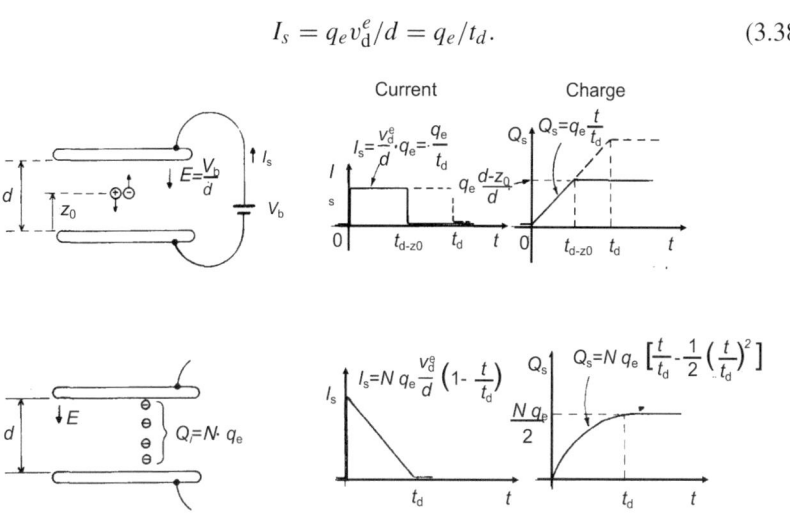

Fig. 3.10 Top left: A charged particle traversing a liquid ionization detector parallel to the plates. Top center: Current flow versus time for this configuration. Top right: Charge versus time for this configuration. Bottom left: A charged particle traversing a liquid ionization detector perpendicular to the plates. Bottom center: Current flow versus time for this configuration. Bottom right: Charge versus time for this configuration. All reprinted with kind permission from [43], © 1974, Elsevier. All rights reserved

The current persists until the electron reaches the anode, which takes time t_d. For $0 < t < t_d$, the charge distribution is given by,

$$Q_s(t) = I_s \cdot t = q_e t/t_d \quad \text{for } 0 < t < t_d = z_0/v_d^e,$$
$$Q_s(t) = \quad\quad q_e(d - z_0)/d \quad \text{for } t > t_d \quad (3.39)$$

as shown in Fig. 3.10 (top right). The observed charge of all produced electron-ion pairs is obtained by multiplying the value in (3.39) by N_e. Note that this is the same as for the parallel-plate ionization chamber.

Figure 3.10 (bottom left) shows the case for a charged particle traversing the cell perpendicular to the plates. Here, N_e electron-ion pairs are produced at the same x position uniformly along z. The current drops as a function of time as depicted in Fig. 3.10 (bottom center) since one electron after another reaches the anode. So, we get

$$I_s = N_e q_e \frac{v_d^e}{d} \left(1 - \frac{t}{t_d}\right), \quad (3.40)$$

where $t_d = d/v_d^e$. Integration over time yields the charge distribution as shown in Fig. 3.10 (bottom right),

$$Q_s(t) = N_e q_e \left[t/t_d - \frac{1}{2}(t/t_d)^2 \right] \quad \text{for } 0 < t < t_d,$$
$$Q_s(t) = \frac{1}{2} N_e q_e \quad\quad \text{for } t > t_d. \quad (3.41)$$

To measure the charge, we apply high voltage to a liquid xenon chamber in a similar way as for gaseous counters. Figure 3.11 shows the measured gain versus applied voltage in a cylindrical liquid xenon chamber. For low voltages, the gain is one. So, here the chamber acts like an ionization chamber. With increased high voltage, the gain increases proportional to the applied voltage. Here, the chamber acts like a proportional counter. It should be noted that the gain is much lower than that of gaseous proportional counters.

3.2.3 Recombination and Electron Trapping

A certain fraction of the electrons produced in an ionization process will recombine with the positive ions, even without impurities. Some of the recombined atoms produce excited dimers and in turn scintillation light. The fraction of recombined atoms is particularly high for slow α particles and fission products. Several predictions exist [17,44,45]. We will present here the results of the box model that is discussed in detail in reference [45]. The basis is the diffusion equation for electrons and positive ions, where small terms are neglected. The box model has the following

3.2 Ionization in Liquids

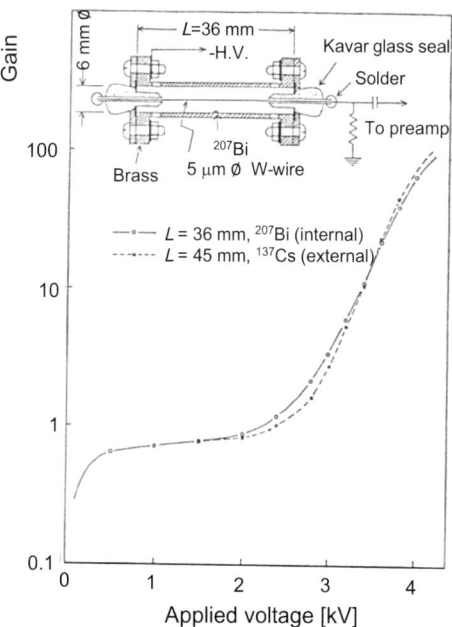

Fig. 3.11 Measured gain as a function of applied high voltage for a LXe cylindrical counter with a 5 μm diameter tungsten wire in the center. The counter is irradiated with γ rays from ^{207}Bi and from ^{137}Cs. A schematic of the cylindrical counter is shown in the inset. The counter is 6 mm in diameter and has a 36 mm length. Reprinted with kind permission from [46], © 1976, Elsevier. All rights reserved

boundary conditions. The electron-ion pairs are isolated where the initial distribution of electrons and ions uniformly populates a box of dimension a. The fraction of collected charge Q with respect to the produced charge Q_0 is given by

$$Q = Q_0 \frac{1}{\zeta_r} \log(1 + \zeta_r), \quad (3.42)$$

and

$$\zeta_r = \frac{N_0 \alpha_r}{4a^2 \mu_e |\vec{E}|}, \quad (3.43)$$

where N_0 is the number electrons in the box, α_r is the recombination coefficient, μ_e is the electron mobility and $|\vec{E}|$ is the electric-field strength. Note that $\zeta_r \to 0$ yields perfect charge collection and $\zeta_r \to \infty$ yields complete recombination. Figure 3.14 (left) shows the collected charge of α radiations in liquid argon as a function of the electric-field strength for two data sets. Fits to the box model are in good agreement with the data while the prediction of the Onsager model does not fit the data well. Figure 3.12 (right) shows the collected charge of a ^{113}Sn source as a function of the electric-field strength in pure LAr, LAr contaminated with 7 ppb O_2 and LXe. Again, the box model provides a good description of the data. Apart from the box model, also Birks' law has been used to fit the measured charge distributions [47]. However, the box model always provides a better fit. The recombination cross sections are $\sigma_{rec} = (7 \pm 2) \times 10^{-13}$ cm^2 for liquid argon, $\sigma_{rec} = (1.2 \pm 0.3) \times 10^{-14}$ cm^2 for liquid krypton and $(1 \pm 0.2) \times 10^{-15}$ cm^2 for liquid xenon [48]. Thus, recombination

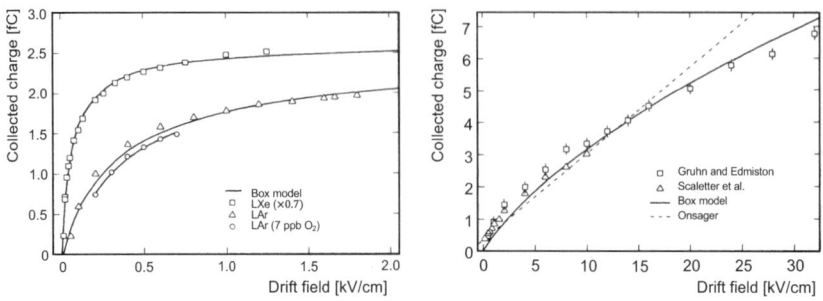

Fig. 3.12 Left: Collected charge as a function of the electric-field strength for a ^{113}Sn source in pure liquid argon (triangles), in liquid argon with 7 ppb O$_2$ (open circles) and in liquid xenon (squares). The curves are fits to the data using the box model with $\zeta_r|\vec{E}| = 0.84$ kV/cm for LAr and $\zeta_r|\vec{E}| = 0.15$ kV/cm for LXe. Both reprinted with kind permission from [45], © 1987, American Physical Society. All rights reserved. Right: Collected charge as a function of the electric-field strength for α radiation in liquid argon [45] for data by Gruen and Edmiston (squares) [49] and Scalettar (triangles) [50]. The solid curve is a fit to the data using the box model with $\zeta_r|\vec{E}| = 470$ kV/cm, while the dashed curve represents the Onsager model [44]

is largest in liquid argon. The recombination process yields excited dimers that produce scintillation light.

If impurities like oxygen or water are present, the electrons can be trapped and VUV light can be absorbed. The attenuation of the collected charge has a characteristic exponential shape,

$$Q = Q_0 \exp\left(-\frac{d_{\text{drift}}}{\lambda_{\text{imp}}}\right), \quad (3.44)$$

where d_{drift} is the drift distance and

$$\lambda_{\text{imp}} = \beta_{\text{LG}} \frac{|\vec{E}|}{\rho_{\text{imp}}}, \quad (3.45)$$

where $|\vec{E}|$ is the electric-field strength in units of [kV/cm], ρ_{imp} is the concentration of the impurity in units of [ppm] and $\beta_{\text{LG}} = 0.15$ for LAr [51]. We can express the VUV light attenuation as

$$I(x) = I_0 \exp\left(-\frac{x}{\lambda_{\text{att}}}\right), \quad (3.46)$$

where λ_{att} is the photon attenuation length, which has two contributions, the absorption length λ_{abs} accounting for true absorption and the scattering length λ_{scat} representing elastic scattering of photons without any loss. The latter component is dominated by Rayleigh scattering. So,

$$\frac{1}{\lambda_{\text{att}}} = \frac{1}{\lambda_{\text{abs}}} + \frac{1}{\lambda_{\text{scat}}}. \quad (3.47)$$

3.2 Ionization in Liquids

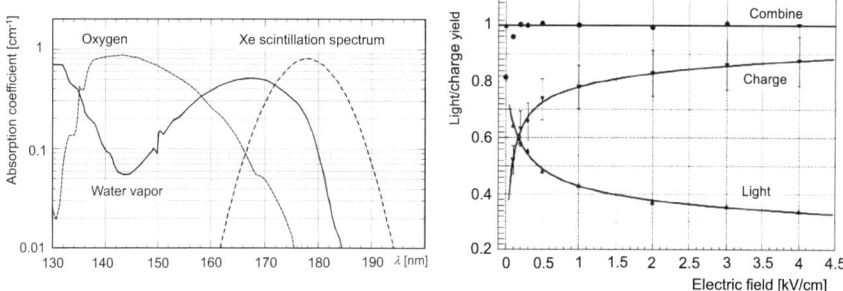

Fig. 3.13 Left: Absorption coefficients of VUV photons in 1 ppm water vapor and in 1 ppm oxygen superimposed onto the Xe emission spectrum. Reprinted with kind permission from [60], © 2005, the MEG Collaboration. All rights reserved. Right: Charge and scintillation light as a function of the applied electric field as well as their combination. The combined response is 100% if no electrons or VUV photons are absorbed. Reprinted with kind permission from [61], © 2007, Elsevier. All rights reserved

The measured attenuation lengths are summarized in Table 3.3. For example, in liquid xenon the measured attenuation length is $\lambda_{att} = 29 \pm 2$ cm [42] in excellent agreement with the prediction of 30 cm [52,53]. Figure 3.13 (left) shows the emission spectrum of LXe scintillation light and the absorption coefficients of far ultraviolet photons in 1 ppm water and in 1 ppm oxygen. Oxygen has a smaller impact on photon detection than water. The effect is worse in LKr and LAr since the emission spectra are shifted to lower wavelengths. Water results from the outgassing of the liquid containment vessel and other detector materials placed inside the liquid. To maintain stable observation of ionization signal electro-negative impurities (like oxygen) need to be kept below one part per billion (ppb).

The reaction of an electron with an impurity S leads to the formation of a negative ion, $e^- + S = S^-$. This decreases the electron concentration C_e,

$$\frac{dC_e}{dt} = -k_S C_e C_S, \quad (3.48)$$

where C_S is the concentration of impurity given in units of [mol/l] and k_S is the attachment rate constant in units of [l/(mol s)]. The time dependence of the electron concentration is then given by

$$C_e(t) = C_e(0) \exp(-k_S C_S t), \quad (3.49)$$

where the electron life time is given by

$$\tau_S = \frac{1}{k_S C_S}. \quad (3.50)$$

Note that in a sample of LXe typically both k_S and C_S are unknown. The exponential decay of the electron concentration is measurable as a function of time via the measurement of the direct current induced by the drift of the electrons. Attachment leads to a decrease of the current with time. This permits a determination of the electron attenuation length

$$\lambda_{\text{att}} = \mu_e |\vec{E}| \tau_S. \tag{3.51}$$

Note that typically two types of electro-negative impurities are found in LXe, those with an attachment cross section that decreases with increasing electric field and those with a cross section that increases with increasing $|\vec{E}|$ field [54]. For impurities such as O_2 and SF_6, the cross section decreases with increasing electric field strength while for N_2O the cross section increases. Note that a concentration of 1 ppb (oxygen equivalent) corresponds to an attenuation length of 1 meter. Several methods have been developed to remove impurities, They include adsorption and chemical reaction methods, filtration, separation, electrical discharges and irradiation with γ-rays [55–59].

3.2.4 Scintillation

The scintillation mechanism in liquid noble gases was discussed in Chap. 2, Sect. 2.8.2. The maximum scintillation yield is given by W/W_{ph}, where

$$W_{\text{ph}} = \frac{W_{ion}}{1 + N_{\text{ex}}/N_i}. \tag{3.52}$$

The number of photoelectrons is given by

$$N_{pe} = \epsilon_{QE} \epsilon_c W/W_{\text{ph}}, \tag{3.53}$$

where ϵ_{QE} is the quantum efficiency of the photomultiplier tube and ϵ_c is the light collection efficiency. The scintillation yield is largest at low electric fields. It is complementary to the charge recorded from ionization. Figure 3.13 (right), which schematically shows measurements of charge and scintillation light as a function of the applied electric field, illustrates this.

Figure 3.14 shows measurements of the relative luminescence intensity L_{rel} and collected charge Q of 1 MeV/c electrons as a function of the electric-field strength in LAr, LKr and LXe [48]. While the luminescence intensity decreases with increasing electric-field strength the collected charge increases correspondingly. From Fig. 3.14 we see that the relative luminescence intensity with respect to that at zero field is 64% for LAr at 6 kV/cm, 60% for LKr and 70% for LXe at 4 kV/cm. The luminescence intensity can be divided into two components, the luminescence intensity due to self-trapped exciton luminescence L_{ex} and luminescence due to recombination L_r. Under the condition of complete charge collection, L_r is zero. The observed ratio $L_r/(L_{\text{ex}} + L_r)$ is 64%, 67% and 71% for LAr, LKr and LXe, respectively. The corresponding range of 1 MeV/c electrons is 0.58 cm, 0.34 cm and 0.28 cm, respectively. At an electric-field strength of $|\vec{E}| = 4$ kV/cm, 92% of the produced charge in LAr, 94% in LKr and 96% in LXe is collected. Comparing the slopes of the luminescence

3.2 Ionization in Liquids

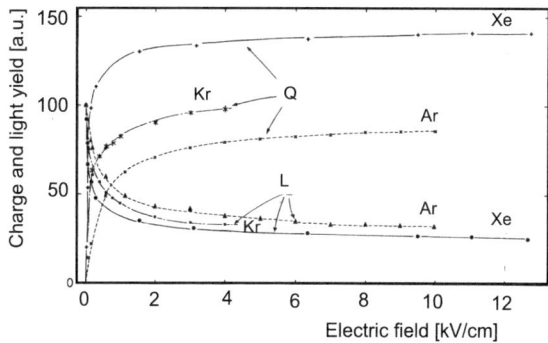

Fig. 3.14 Relative luminescence intensity L_{rel} and collected charge Q for 1 MeV/c electrons as a function of the electric-field strength in LAr (triangles and crosses), LKr (squares and stars) and LXe (solid points and plus signs). Reprinted with kind permission from [48], © 1979, American Physical Society. All rights reserved

intensity and collected charge for $|\vec{E}| < 3$ kV/cm, we notice that slopes for liquid xenon are steeper than those for liquid argon. This suggests that electrons are more easily separated from the positive ions in liquid xenon, which is mainly due to the smaller recombination cross section and larger electron mobility compared with those in liquid argon.

Figure 3.15 shows the anti-correlation between scintillation intensity S/S_0 and collected charge Q/Q_∞ in liquid argon measured with different ions, where S_0 is the scintillation light intensity at zero electric field and Q_∞ is the charge expected

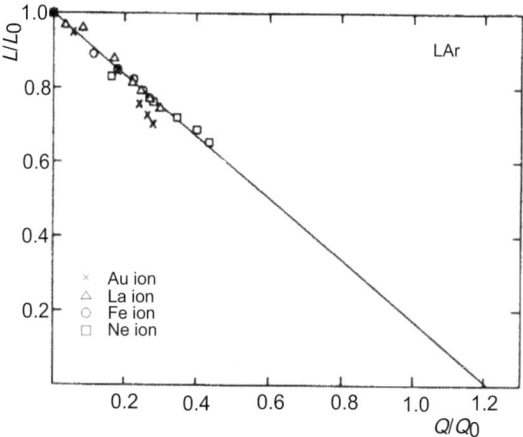

Fig. 3.15 Anti-correlation between scintillation intensity S normalized to the scintillation intensity at zero electric-field strength, S_0, and collected charge Q normalized to the charge expected for an infinite electric-field strength, Q_∞ in liquid argon for relativistic Ne ions (squares), Fe ions (points), LAr ions (triangles) and Au ions (crosses). The solid line is the theoretical estimate. Reprinted with kind permission from [62], © 1989, American Physical Society. All rights reserved

for infinite electric field. For relativistic Ne, Fe and LAr ions the scintillation yields are 100% while for Au ions due to quenching the scintillation yield is not 100%. The measurements agree well with theoretical estimates. Similar anti-correlations are observed for LKr and LXe.

Kubota *et al.* measured the decay times of LAr, LKr and LXe using conversion electrons from ^{207}Bi in zero $|\vec{E}|$ and in high $|\vec{E}|$ fields, where all of the observed decay characteristics have been attributed to the self-trapped exciton luminescence [63]. An applied electric field results in quenching of the luminescence from free electron recombination. The use of energetic electrons with low specific ionization density ensures complete collection of electron charges. In this way, the self-trapped exciton luminescence is free from the inter-carrier recombination luminescence. To detect the VUV photons, the inner surface of the glass window of the chamber was covered with POPOP scintillator.

Figures 3.16 (left, right) and 3.17 (left) respectively show typical decay curves for liquid argon, liquid krypton and liquid xenon with and without applied electric fields. The decay curves were normalized in area to intensities in the ratio of luminescence without applied electric field to that with applied electric field. The data are fitted with two lifetimes. Table 3.5 summarizes all results. Concerning the self-trapped

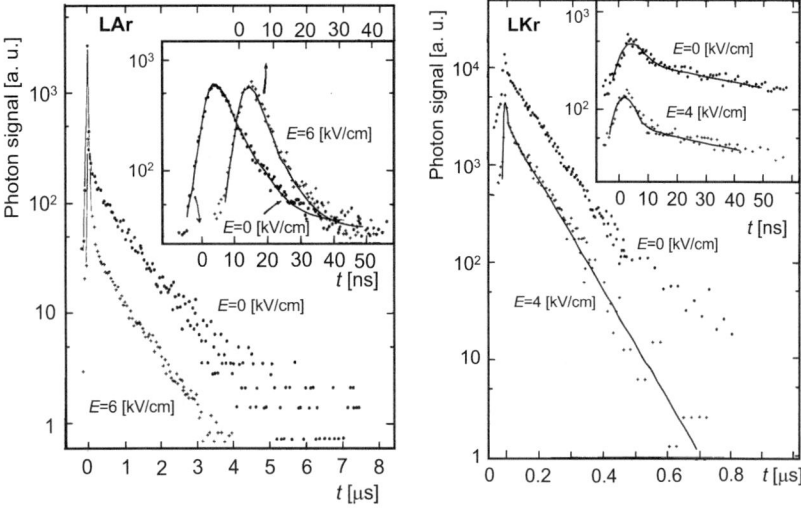

Fig. 3.16 Left: The luminescence intensity as a function of the decay time for liquid argon in a 6 kV/cm electric field (plus symbols) and no electric field (solid points) plotted in a semi-logarithmic scale. The inset shows a close-up view of the short range distribution. The solid lines represent fitted lifetime curves. The points are measurements. The lower scale is for zero electric field, while the upper shifted scale is for the 6 kV/cm electric field. Right: The luminescence intensity as a function of the decay time for liquid krypton with a 4 kV/cm electric field (plus symbols) and without any electric field (solid points) in a semi-logarithmic scale. The inset shows a close-up view of the short range distribution. The points are measurements and the solid curves represent lifetime fits. Both reprinted with kind permission from [63], © 1978, IOP Publishing. All rights reserved

3.2 Ionization in Liquids

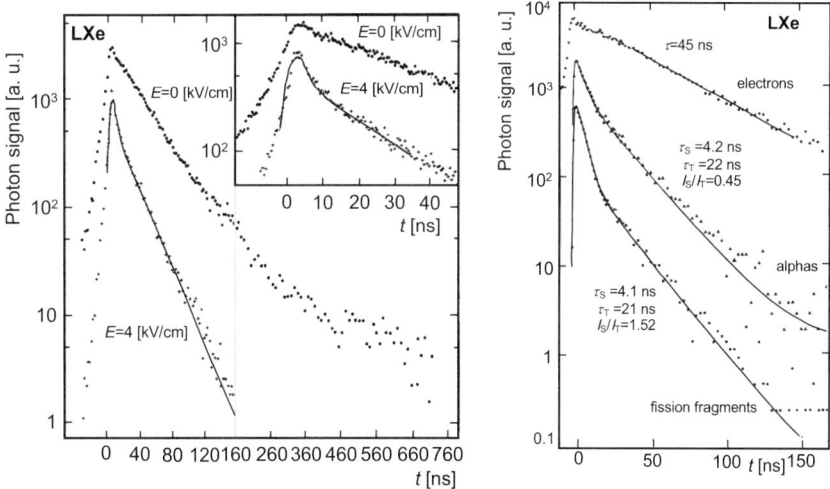

Fig. 3.17 Left: The luminescence intensity as a function of the decay time for liquid xenon in a 4 kV/cm electric field (plus symbols) and no electric field (solid points) plotted in a semi-logarithmic scale. The inset shows a close-up view of the short range distribution. The points are measurements and the solid curves represent lifetime fits. Right: Decay curves of scintillation light from liquid xenon excited by electrons, α-particles and fission fragments without an applied electric field. Both reprinted with kind permission from [63], © 1978, IOP Publishing. All rights reserved

luminescence, the decay times show a fast and a slow component. The fast component (2–5 ns) is nearly the same for all three liquid noble gases while the slow component decreases significantly from LAr to LXe. The intensity of the fast component $\tau_1 A_1$ is 20-50 times smaller than that of the slow component $\tau_2 A_2$. For LXe we see a drastic difference between the decays for recombination luminescence and that for self-trapped luminescence, suggesting that the recombination time is longer compared to the decay times for the excited molecular states. The observation of two decay times support the production of excited singlet and triplet states.

Table 3.5 Measured decay times and amplitudes of LAr, LKr and LXe with and without an electric field excited by 1 MeV/c electrons. Reprinted with kind permission from [63], © 1978, IOP Publishing. All rights reserved

Liquid gas	$\|\|\vec{E}\|\|$ [kV/cm]	τ_1 [ns]	τ_2 [ns]	A_1/A_2	$\tau_1 A_1/(\tau_2 A_2)$
LAr	6	5.0 ± 0.2	860 ± 30	7.8 ± 1	0.045
LAr	0	6.3 ± 0.2	$1,020 \pm 60$	13.5 ± 1.5	0.083
LKr	4	2.1 ± 0.3	80 ± 3	0.9 ± 0.2	0.02
LKr	0	2.0 ± 0.2	91 ± 2	0.4 ± 0.2	0.01
LXe	4	2.2 ± 0.3	27 ± 1	0.6 ± 0.2	0.05
LXe	0	-	34 ± 2		

The two lowest excited molecular states (singlet and triplet Σ states) responsible for the luminescence of liquid and solid argon, krypton and xenon excited by charged particles, have two main origins: direct excitation of exciton states by primary charged particles and secondary electrons and formation of singlet and triplet Σ states through a recombination process between free electrons and molecular ions. In the first process, excitation is followed by the self-trapping of free excitons producing two excited molecular states within 1 ps. In the second process primary particles and secondary electrons produce holes and electrons. The holes are immediately localized through the formation of rare-gas R_2^+ molecular ions within a picosecond. The secondary electrons lose kinetic energy promptly through excitation of excitons and production of electron-hole pairs as well as through the emission of phonons. These thermalized electrons finally recombine with localized R_2^+ ions to form the excited molecular states.

The predicted decay times for the slow component are 3 (LAr), 1.3 (LKr) and 2 (LXe) times bigger than the measurements [64]. One possible explanation of this discrepancy may be a mixing of the singlet and triplet excited states due to a phonon-activated depopulation leading to a decrease of the long lifetime. A measurement without an electric field by Hitachi et al. yielded a long lifetime of 1.6 ± 0.1 μs [65].

Figure 3.17 (right) shows the decay curves of scintillation light from liquid xenon excited by electrons, α-particles and fission fragments without an applied electric field. While for electrons only a slow component is measured, the decay distributions for α-particles and fission fragments show two components, a short one of 4.2/4.1 ns and a long one of 22/21 ns, respectively.

3.2.5 Drift Velocity

In low electric fields, the electron drift velocity is nearly proportional to the electric field strength,

$$v_d^e = \mu_e |\vec{E}|, \tag{3.54}$$

where μ_e is the electron mobility of the liquid noble gas. The electron mobility in LAr is $\mu_e(\text{LAr}) = 475$ cm^2/(Vs). In LKr it increases to $\mu_e(\text{LKr}) = 1800$ cm^2/(Vs) and in LXe it is $\mu_e(\text{LXe}) = 2200$ cm^2/(Vs). Note that the electron mobility in liquid xenon, which is about a factor of four larger than that of LAr, is similar to that in silicon. In liquid noble gases the positive carriers are holes. Their mobility is several orders of magnitude smaller than that of electrons. Figure 3.18 (left) shows the mobility of holes in liquid xenon as a function of temperature. The mobility is largest around 230 K yielding $\mu_h \simeq 4.6 \times 10^{-3}$ cm^2/(Vs).

Figure 3.18 (right) shows the drift velocity as a function of the reduced electric-field for gaseous and liquid argon and xenon. Note that the drift velocity in LAr for most electric-field strengths is about an order of magnitude larger than that in the gaseous phase. For LXe, at low electric fields the drift velocity is more than 2 orders of magnitude larger than that in the gaseous phase. This decreases to about one order of magnitude at high electric-field strengths. At low electric-field strengths, the drift

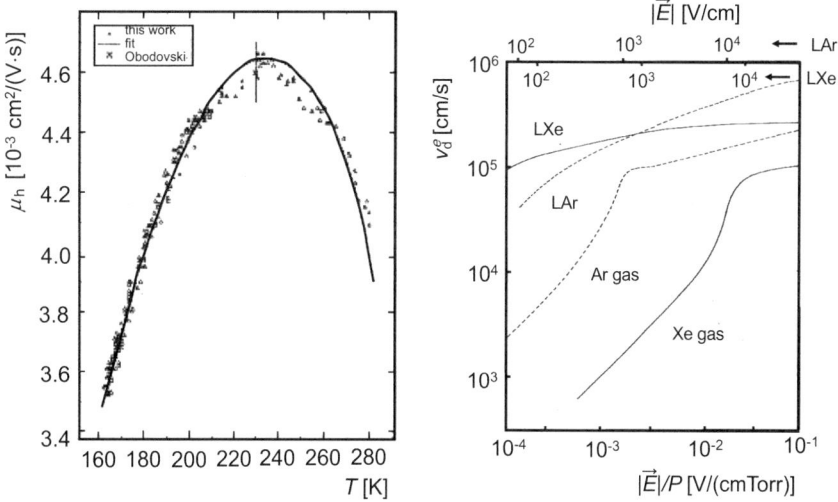

Fig. 3.18 Left: Mobility of positive holes in liquid xenon as a function of temperature. Reprinted with kind permission from [66], © 1994, IOP Publishing. All rights reserved. Right: Comparison of the electron drift velocity as a function of the reduced electric-field in liquid argon (dashed lines) and liquid xenon (solid lines) as well as gaseous argon and gaseous xenon [35,67–69]. The bottom (top) scale shows the electric-field strengths for liquid (gaseous) argon and liquid (gaseous) xenon. Reprinted with kind permission from [35], © 2009, American Physical Society. All rights reserved

velocity in LXe is larger than that in LAr. At high electric-field strengths this is the opposite, since the drift velocity becomes independent of $|\vec{E}|$ while that in LAr still rises.

Figure 3.19 shows measurements of the drift velocity in LAr as a function of electric-field strength for samples of different thickness by Miller [67]. In addition, predictions by Shockley and by Cohen and Lekner are shown as well. For low $|\vec{E}|$, Shockley's predictions fit the data well while for higher electric-field strengths up to 10 kV/cm the Cohen and Lekner theory fits well. The discrepancy for the Cohen and Lekner theory visible at very high $\vec{E}$ fields is due an insufficient accuracy in the cross section. At an electric-field strength of 10 kV/cm, the electron drift velocity is $v_d^e = 5 \times 10^5$ cm/s. At low electric-field strengths the drift velocity scales linearly with $|\vec{E}|$ while at intermediate electric-field strengths a transition from linear to $\sqrt{|\vec{E}|}$ dependence occurs. This happens when the drift velocity approaches the sound velocity in the medium. At high electric-field strengths, we observe a saturation of the drift velocity. This pattern is also true for LKr and LXe.

Figures 3.20 and 3.21 respectively show measurements of the drift velocity versus $|\vec{E}|$ in liquid krypton and liquid xenon for samples of different thickness by Miller [67]. For comparison, the drift velocities of the solid noble gases are depicted as well. Note that drift velocities in the solid-state are about a factor of two larger than those in the liquid form due to a larger mobility.

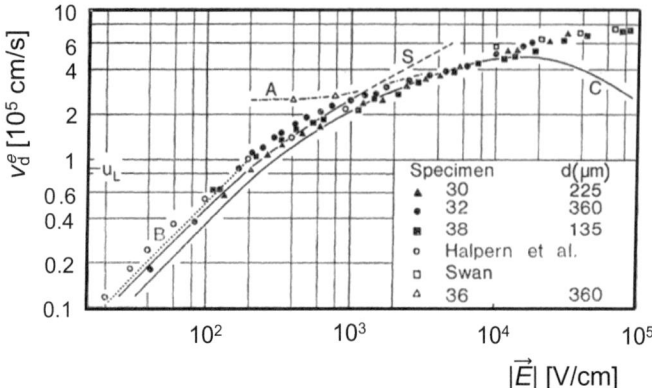

Fig. 3.19 Electron drift velocity as a function of electric-field strength in a double logarithmic scale for LAr at a temperature of 85 K for samples of different thickness. In addition, data from Halpern [70] and Swan [71] are shown. Curve S (dashed) represents fits to Shockley's theory, curve A (dash-dotted) shows results of a short electron lifetime ($\simeq 200$ ns), curve C shows fits to the theory of Cohen and Lekner [72] and curve B is corrected for multiple scattering and also indicates the linear electric-field dependence. The line for u_L shows the speed of sound. Reprinted with kind permission from [67], © 1968, American Physical Society. All rights reserved

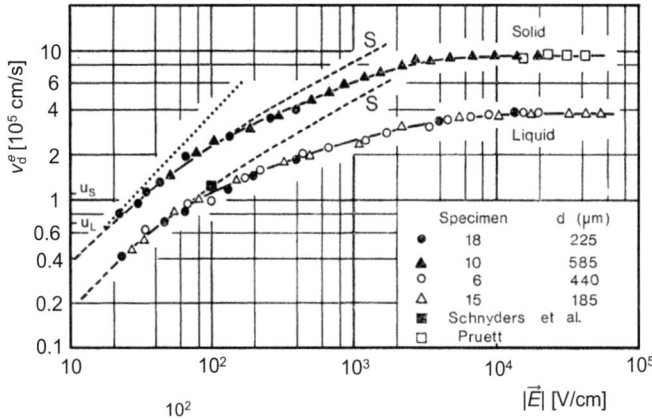

Fig. 3.20 Electron drift velocity as a function of electric-field strength in a double logarithmic scale for liquid/solid krypton at a temperature of 117/113 K for samples of different thickness. In addition, data from Schnyders *et al.* (black squares) [73] and Pruett (open squares) [67] are shown. The dotted line shows a linear field dependence whereas the dashed curves are fits to Shockley's theory. The lines u_S and u_L show the speed of sound in the solid and liquid phase, respectively. Reprinted with kind permission from [67], © 1968, American Physical Society. All rights reserved

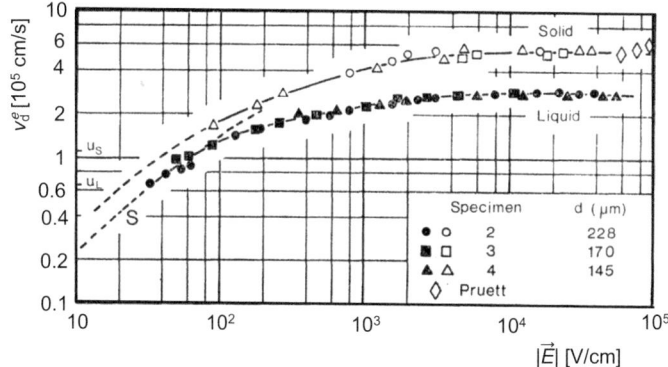

Fig. 3.21 Electron drift velocity as a function of electric-field strength in a double logarithmic scale for liquid and solid xenon at a temperature of 163/157 K for samples of different thickness. In addition, data from Pruett (diamonds) [67] are shown. The dashed curves show fits to Shockley's theory while the lines u_S and u_L show the speed of sound in the solid and liquid phase, respectively. Reprinted with kind permission from [67], © 1968, American Physical Society. All rights reserved

3.2.6 Diffusion

An electron cloud moving in the drift field in a liquid noble gas is broadened by diffusion. The electron diffusion coefficient has a longitudinal (D_L) and a transverse component (D_T). Just as in gases, D_L is smaller than D_T. In liquid xenon they differ by a factor of ten. Figure 3.22 shows the longitudinal and transverse diffusion coefficients as a function of electric field in liquid xenon. The transverse spread, which may affect the position resolution if it becomes too big, is given by

$$\sigma_{D_T} = \sqrt{(D_T t_{dr})}, \qquad (3.55)$$

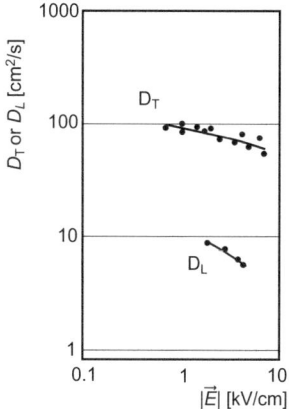

Fig. 3.22 Transverse (D_T) and longitudinal (D_L) diffusion coefficients for electrons in LXe as a function of the electric field. Reprinted with kind permission from [74], © 2024, E. Aprile

where $t_{\rm dr} = d_{\rm dr}/v_d^e$ is the drift time and $d_{\rm dr}$ is the drift distance. In a LXe detector the spread contributes a negligible amount to the position resolution, even over a long distance.

3.2.7 Energy Resolution

Liquid noble gases are used in electromagnetic and hadron calorimeters. The predicted energy resolution is much better than expected by Poisson statistics since the Fano factors are much less than one. Assuming that the energy loss W of an ionizing particle is absorbed completely in the material, we can express the Fano factor in the following way [38,75,76],[2]

$$\begin{aligned} F &= F_1 + F_2 + F_3 \\ &= \left(\langle N_{\rm ex}\rangle/\langle N_i\rangle\right)\left[1 + \left(\langle N_{\rm ex}\rangle/\langle N_i\rangle\right)\right]\left(\langle W_{\rm ex}\rangle^2/\langle W_{ion}\rangle^2\right) \\ &+ \left[\left(\langle E_e - W_i\rangle\right)^2/\langle W_{ion}\rangle^2\right] \\ &+ \left(\langle N_{\rm ex}\rangle/\langle N_i\rangle\right)\left[\left(\langle E_{\rm ex} - W_{\rm ex}\rangle\right)^2/\langle W_{ion}\rangle^2\right], \end{aligned} \quad (3.56)$$

where the factor F_1 accounts for the redistributions of the number of excited and ionized atoms, F_2 accounts for energy loss fluctuations in ionization, F_3 accounts for energy loss fluctuations in excitations, $W_{\rm ex} = \langle E_{\rm ex}\rangle$ and $W_i = \langle E_i\rangle + \langle E_e\rangle$. All other parameters are the same as those in (3.36) and are listed in Table 3.4. The Fano factors for LAr, LKr and LXe summarized in Table 3.6 are based on Alkhazov's model [38]. A few years earlier Shokley calculated the Fano factors using a different kinetic energy distribution, which gave lower values and in turn lower energy resolutions [40]. For example, for LXe the Fano factor is $F = 0.059$ in Alkhazov's model while it is $F = 0.041$ in Shokley's model. The corresponding total energy resolutions are 1.39 keV and 1.29 keV, respectively.

So, for liquid noble gases the error on the number of electron-ion pairs is given by

$$\sigma_N^2 = \langle \left(\langle N_i\rangle - N_i\right)^2\rangle = F \cdot \langle N_i\rangle. \quad (3.57)$$

Using the known Fano factor and W_{ion} value, we can write the ultimate energy resolution of a LXe detector as

$$\sigma_E \text{ [keV]} = \Delta E/2.35 = \sqrt{F \cdot W_{ion} \text{ [eV] } W \text{ [MeV]}}, \quad (3.58)$$

where the W_{ion}-value is inserted in units of [eV], the energy loss W is inserted in units of [MeV] and ΔE is the FWHM energy resolution. To obtain the total energy

---

[2] The three Fano factors are displayed in the three following lines.

3.2 Ionization in Liquids

Table 3.6 Fano factors F_1 accounting for the redistributions of the number of excited and ionized atoms, F_2 accounting for energy loss fluctuations in ionization, F_3 accounting for energy loss fluctuations in excitations, $F = F_1 + F_2 + F_3$, noise contribution σ_{noise}, energy resolution σ_E and total energy resolutions σ_{tot} including noise, all determined at $W = 1$ MeV [77]. These values are based on Alkhazov's model [38,78]

Liquid gas	F_1	F_2	F_3	F	σ_{noise} [keV]	σ_E [keV]	σ_{tot} [keV]
LAr	0.076	0.036	0.004	0.116	1.52	1.65	2.24
LKr	0.032	0.037	0.001	0.070	1.27	1.17	1.73
LXe	0.019	0.039	0.0006	0.059	1.0	0.96	1.39

resolution σ_{tot}, we have to add a noise term in quadrature that is given by

$$\sigma_{\text{noise}} \text{ [keV]} = W_{ion} \text{ [eV]} \, \overline{ENC} \times 10^{-3} \text{[keV]}, \qquad (3.59)$$

where $\overline{ENC}$ is the mean equivalent noise charge. For example the σ_{noise} value in table 3.6 for LXe is based on $\overline{ENC} = 65e$. For a typical chamber [43] gives explicit $\overline{ENC}$ calculations. The Fano-limit of the energy resolution of LXe is comparable to that measured with a germanium- or silicon detector. However, experimentally this has not been achieved yet. In fact, the measured energy resolution is even worse than the Poisson limit. Using 554 keV γ rays from ^{207}Bi, an energy resolution of 12.8 keV was measured in an electric field of 17 kV/cm [28,77] whereas the Poisson limit including noise is 3.1 keV. For summed signals of ionization and scintillation a similar energy resolution was measured at 1 kV/cm [81].

So far we have no reasonable explanation for the discrepancy between the expected and measured energy resolution of liquid noble gases. Egorov et al. proposed that large fluctuations in the number of δ electrons are the main source of the spread in the energy resolution [82]. Thomas et al. tried to explain the experimental energy resolution measured from the ionization signal in terms of a recombination model between electrons and ions produced along δ electron tracks [83]. Also Aprile et al. attempted to explain their observed energy resolution on the basis of the recombination model [84]. However, data from LXe doped with photo-ionizing molecules are in conflict with the model. Furthermore, the energy resolution of compressed Xe gas measured by Bolotnikov et al. also does not support the recombination model since the energy resolution improves at a lower gas density without any increase in the collected charge. Figure 3.23 (left), which shows the density dependence of the energy resolution (FWHM) measured with 662 keV γ-rays, illustrates this. For a density of 0.5g/cm^3 the energy is close to the Fano limit while for a density of 1.4g/cm^3 it increases by a factor of eight.

Figure 3.23 (right) shows the measured energy resolution (FWHM) of LAr ions as a function of the electric field for scintillation light, ionization charge and their combination [35]. The energy resolution for scintillation light is better than that for ionization charge. However, measuring the energy in both processes yields a

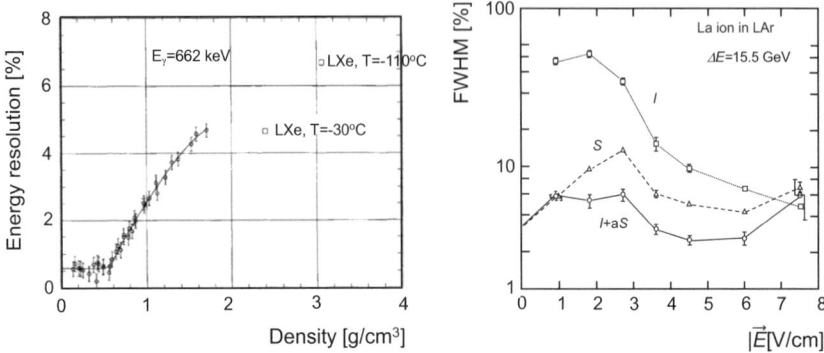

Fig. 3.23 Left: The measured energy resolution as a function of density for xenon gas measured with 662 keV γ rays (points). In addition, the measured values for LXe at 30 °C and −110 °C are shown for comparison (open squares). Reprinted with kind permission from [79], © 1997, Elsevier. All rights reserved. Right: Measured energy resolution (FWHM) of LAr ions in LAr as a function of the electric-field strength for scintillation light (S), ionization charge (I) and their sum (I+aS). The parameter a gives the conversion factor from the observed scintillation intensity at a given electric field to the absolute number of photons. Reprinted with kind permission from [80], © 1987, Elsevier. All rights reserved

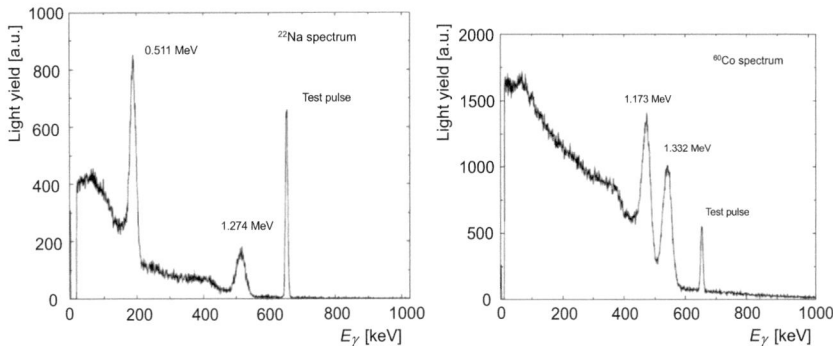

Fig. 3.24 Left: The energy spectrum of ^{22}Na γ-rays measured with a gridded liquid-xenon ionization chamber at 4 kV/cm drift field. Right: The energy spectrum of ^{60}Co γ-rays measured with a gridded liquid-xenon ionization chamber at 4 kV/cm drift field. Both reprinted with kind permission from [85], © 2002, Elsevier. All rights reserved

considerable improvement in the energy resolution (2% for electric-field strengths of 3-6 kV/cm) although only a small fraction of the scintillation light was collected.

The energy spectra of ^{22}Na and ^{60}Co were measured with a gridded LXe ionization chamber as respectively shown in Fig. 3.24 (left, right) [84,85]. The individual energy lines are well resolved. The energy resolution of the 511 keV line is about 3.4% and that of the ^{60}Co lines at 1.3 MeV is about 2.1% as expected from Fano theory.

3.3 Ionization in Solid-State Detectors

In Chap. 2, we introduced the band structures of solids and Fig. 2.56 showed the configuration of the conduction and valence bands in an insulator, in a semi-conductor (for $T = 0\,K$ and $T > 0\,K$) and in a metal. We mainly focus here on semiconductors, in particular on silicon.

3.3.1 Basic Properties of Semi-Conductors

Silicon has four valence electrons. In a silicon crystal, each atom is equidistant from each neighbor atom as illustrated in Fig. 3.25 (left). Each valence electron is coupled to a neighboring atom via a covalent bound producing a symmetric crystal structure. Such a semiconductor is called intrinsic. At $T = 0$ K all electrons are bound and cannot conduct a current. The valence band is filled while the conduction band is empty. The separation between the valence band and conduction band is 1.14 eV for silicon, 0.67 eV for germanium and 1.47 eV for gallium arsenide. At room temperature ($T = 300$ K), electrons can be shifted into the conduction band. This is illustrated in Fig. 3.25 (right). Some electrons have left the crystal structure and move through the lattice.

Extrinsic semi-conductors, however, in which controlled levels of impurities are added to the silicon have more interesting electronic properties. We can add group III elements, which have three valence electrons, like boron, gallium or indium. This material called p-type silicon is a hole carrier. At $T = 0$ K, the hole is confined to the impurity atom as displayed in Fig. 3.26 (left). At $T > 0$ K, such as room temperature ($T = 300$ K), the hole may be filled by an electron leaving behind a single negative ion as shown in Fig. 3.26 (right). The hole is mobile and may be moved through the crystals using an external electric field. Figure 3.27 (left) depicts the band structure of p-type silicon at $T = 0$ K. The acceptor impurity introduces an extra layer of acceptor levels at $\simeq 0.005$eV above the valence band. While the valence band is completely filled, the layer of acceptor levels and conduction band are empty. The situation changes for temperatures $T > 0$ K. Via thermic energy electrons from the valence band rapidly fill the layer of acceptor levels leaving holes in the valence band.

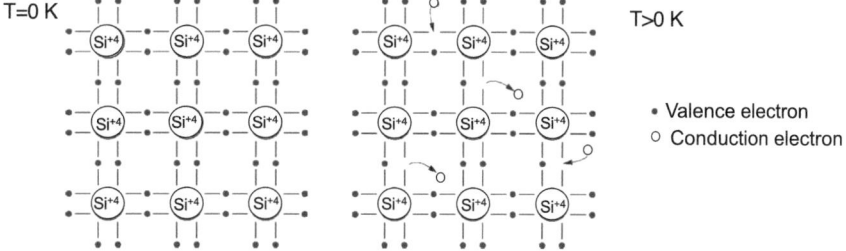

Fig. 3.25 Left: Crystal structure of pure silicon at $T = 0$ K. Right: Crystal structure of pure silicon at $T > 0$ K. Both reprinted with kind permission from [86], © 2005, M. Krammer. All rights reserved

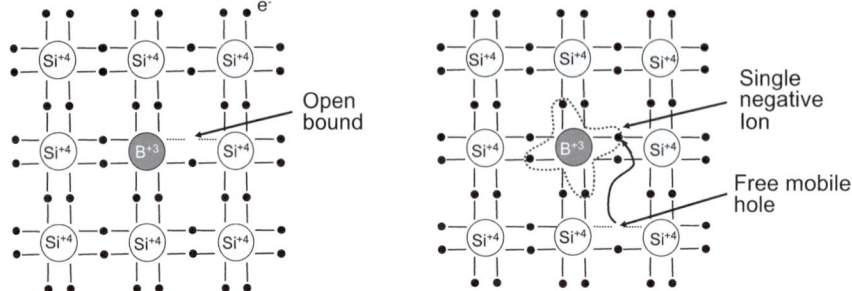

Fig. 3.26 Left: Crystal structure of p-type silicon at $T = 0$ K. Here, silicon is doped with boron atoms. At the boron atom we have an open bond. Right: Crystal structure of p-type silicon for $T > 0$ K. The open bonds may be filled by an electron from another silicon atom producing a single negative ion. The created hole is mobile. Both reprinted with kind permission from [86], © 2005, M. Krammer. All rights reserved

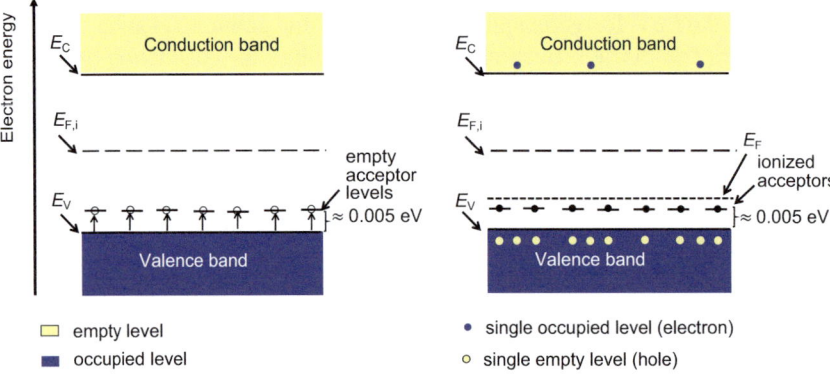

Fig. 3.27 Left: Band structure of p-type silicon for $T = 0$K. The energies E_V and E_C respectively denote the highest energy in the full valence band (dark shaded) and the lowest energy in the empty conduction band (light shaded), while $E_{F,i}$ represents the Fermi energy of the intrinsic semiconductor. The additional layer of acceptor levels is empty (open circles). Right: Band structure of p-type silicon for $T > 0$ K. Here, E_F denotes the Fermi energy of p-type silicon. The layer of acceptor levels is filled (solid points). Some electrons have been lifted to the conduction band (solid points) leaving holes in the valence band (open circles)

Some electrons even acquire enough energy to reach a level in the conduction band. This process is temperature dependent. Figure 3.27 (right) illustrates the electric properties of p-type silicon for $T > 0$ K. Due to charge conservation, the Fermi energy is lowered to an energy above the layer of acceptor levels.

The other possibility is to add group V elements, which have five valence electrons, like phosphorus, arsenic or antimony. In the silicon lattice we have one electron too much. This material called n-type silicon is an electron carrier. At $T = 0$ K, the additional electron is weakly bound as indicated in Fig. 3.28 (left). Figure 3.29 (left) shows the band structure. The impurity introduces a donor level of 0.005 eV below the conduction band. At $T = 0$ K, the valence band and the donor level are filled

3.3 Ionization in Solid-State Detectors

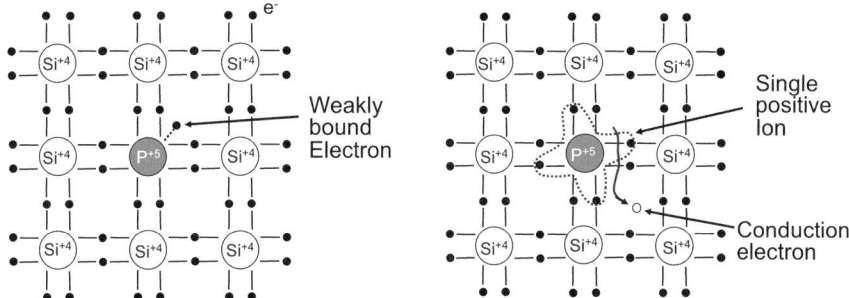

Fig. 3.28 Left: Crystal structure of n-type silicon at $T = 0$ K. Here, silicon is doped with phosphorus atoms. At the phosphorus atom we have an extra electron that does not fit into the crystal structure and thus is just weakly bound. Right: Crystal structure of n-type silicon for $T > 0$ K. The extra electron is lifted into the conduction band leaving behind a single positive ion. Both reprinted with kind permission from [86], © 2005, M. Krammer. All rights reserved

while the conduction band is empty. For $T > 0$ K, extra electrons are lifted into the conduction band as shown in Fig. 3.28 (right). Figure 3.29 (right) shows the resulting band structure. The electrons of the donor level and some in the valence band are lifted into the conduction band leaving behind an ionized donor atom and holes in the valence band. Due to charge conservation, the Fermi level E_F is raised to an energy just below the layer of donor levels. Note that n-type silicon plays the most important role for charged particle detection. Table 3.7 summarizes properties of the most commonly used semiconductors, Si, Ge, GaAs, CeTd and diamond. Figure 3.30 (left) shows the intrinsic carrier density as a function of $1/T$ for germanium, silicon and gallium arsenide. In silicon the concentration is $1.08 \times 10^{10}/\text{cm}^3$ at T=300 K doubling about every 11 K.

The probability that a state with energy E_e is occupied is given by the Fermi function,

$$f(E_e) = \frac{1}{1 + \exp\left(\frac{E_e - E_F}{k_B T}\right)}, \tag{3.60}$$

where E_F is the Fermi energy. For $E_e = E_F$, we obtain a probability of 0.5. For an intrinsic semiconductor, the Fermi energy is close to half the gap energy. Figure 3.30 (right) shows the probability of occupied energy levels as a function of energy. Note that at $T = 300$ K, $k_B T$ corresponds to an energy of 0.0259 eV. When we introduce impurities, the Fermi energy will adjust itself to preserve charge neutrality. At T=300 K, the number of carriers is much larger in extrinsic than in intrinsic semiconductors. The number of electrons (holes) in the conduction (valence) band for a given material depends on temperature. Calling the concentration of electrons n and that of holes p, we obtain due to charge conservation

$$\boldsymbol{n} \cdot \boldsymbol{p} = N_C N_V \exp\left(\frac{-E_g}{2k_B T}\right) = \text{constant}, \tag{3.61}$$

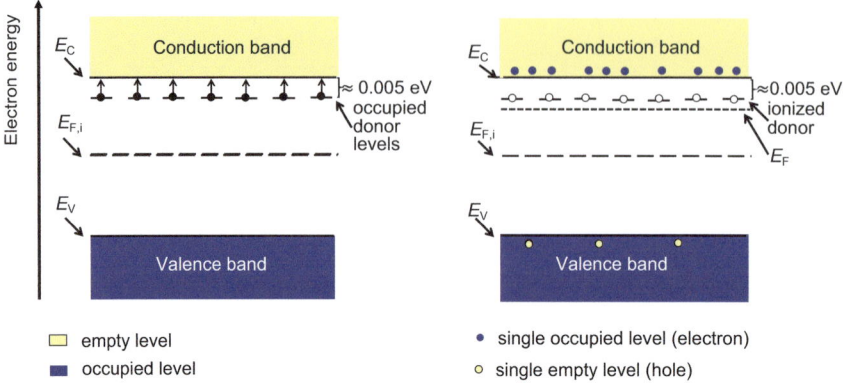

Fig. 3.29 Left: Band structure of n-type silicon for $T = 0$K. The energies E_V, E_C and $E_{F,i}$ are the same as those in Fig. 3.27 (left). The additional layer of donor levels just below the conduction band is filled (solid points). Right: Band structure of n-type silicon for $T > 0$ K. Here, E_F denotes the Fermi energy of n-type silicon just below the layer of donor levels, which is empty (solid points) leaving ionized donor atoms. Electrons in the conduction band and holes in the valence band are shown by solid points and open circles, respectively

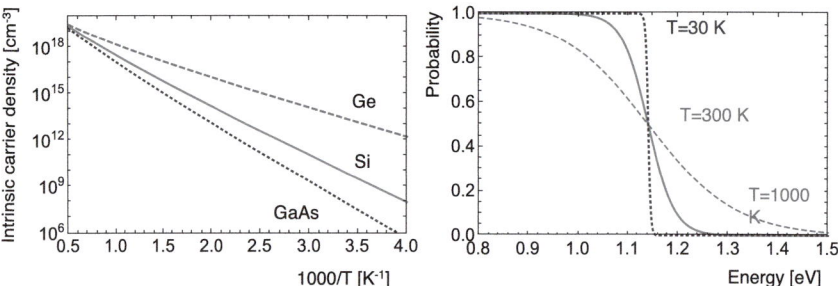

Fig. 3.30 Left: Intrinsic Carrier density as a function of $1000/T$ for germanium, silicon and gallium arsenide. Right: The probability of occupied energy levels as a function of energy for three different temperatures $T = 30$ K (dotted curve), $T = 300$ K (solid curve) and $T = 1000$ K (dashed curve) in silicon

where N_C is the effective density of states in the conduction band and, N_V is the effective density of states in the valence band and E_g is the energy gap at $T = 0$ K. The effective densities are given by

$$N_C = 2\left(\frac{m_e^* k_B T}{2\pi \hbar^2}\right)^{3/2} \cdot n_C^m,$$
$$N_V = 2\left(\frac{m_h^* k_B T}{2\pi \hbar^2}\right)^{3/2}, \qquad (3.62)$$

where $m_e^* = 0.603$ MeV [89] is the effective electron mass in silicon, $m_h^* = 0.414$ MeV [89] is the effective hole mass in silicon and n_C^m is the number of equivalent minima in the conduction band [90]. At $T = 300$ K the effective density of states

3.3 Ionization in Solid-State Detectors

Table 3.7 Properties of different semiconductor materials [7,87,88] including the atomic number Z, the atomic mass A, the lattice constant a, the density ρ, the radiation length X_0, thermal conductivity C_{th}, gap energy E_g, relative permittivity ϵ_{sc}, melting point T_{melt}, effective electron mass (m_e^*/m_e), effective hole mass (m_h/m_e), electron mobility μ_e at 300 K, hole mobility μ_h at 300 K, intrinsic charge carrier density ρ_{ch} at 300 K, intrinsic resistivity ρ_{in} at 300 K, mean energy to create an eh pair W_{eh} at 300 K and the average signal I_s

Material	Si	Ge	GaAs	CdTe	Diamond
Z	14	32	31+33	48+52	6
A [amu]	28.086	72.61	69.72+30.97	112.4+127.6	12.011
a [Å]	5.431	5.646	5.653	6.482	3.567
ρ [g/cm^3]	2.328	5.326	5.32	5.85	3.520
X_0 [g/cm^2]	9.37	2.302	2.296	1.52	12.13
C_{th} [$\frac{W}{\text{cmK}}$]	1.5	1.05	0.46	0.083	10-20
E_g [eV]	1.12	0.66	1.42	1.44	5.47
ϵ_{sc}	11.9	16.0	13.1	10.2	5.7
T_{melt} [°C]	1,414	938.2	1,238	1,040	3,527
(m_e^*/m_e)	1.18	0.55	0.067	0.111	0.2
(m_h/m_e)	0.81	0.4	0.53	0.117	0.25
μ_e [cm^2/Vs]	~1,450	3,900	≤8,500	1,050	1,900–2,300
μ_h [cm^2/Vs]	~450	1,900	≤400	90	1,500–2,300
ρ_{ch} [cm^{-3}]	1.1×10^{10}	2.4×10^{13}	2.1×10^6	10^7	$< 10^3$
ρ_{in} [Ω-cm]	3.1×10^5	47	10^8	$\approx 10^9$	$>10^{11}$
W_{eh} [eV]	3.65	2.96	4.35	4.43	13.1
I_s [e-h/μm]	110	260	130		36

in the valence band is $N_V = 1.05 \times 10^{19}$ cm^{-3} and that in the conduction band is $N_C = 2.82 \times 10^{19}$ cm^{-3}. It is important to note that the effective density of states increases with temperature as $T^{3/2}$. The concentrations in the conduction band and valence band can be calculated from Fermi-Dirac statistics yielding

$$n = N_C \exp\left(-\frac{E_C - E_F}{k_B T}\right),$$
$$p = N_V \exp\left(-\frac{E_F - E_V}{k_B T}\right). \qquad (3.63)$$

For an intrinsic semiconductor the Fermi energy is obtained from the charge balance

$$n = p, \qquad (3.64)$$

yielding

$$E_F = \frac{E_C + E_V}{2} + \frac{k_B T}{2} \log\left(\frac{N_V}{N_C}\right) = \frac{E_C + E_V}{2} + \frac{3k_B T}{4} \log\left(\frac{m_h^*}{m_e^*(n_C^m)^{2/3}}\right). \tag{3.65}$$

Thus, the Fermi energy lies close to the middle of the gap between valence band and conduction band.

For n-type silicon, we have to modify (3.64) to take the effect of the impurity atoms into account

$$n = p + N_D^+, \tag{3.66}$$

where N_D^+ is the density of ionized donor atoms, which is given by

$$N_D^+ = N_D\left[1 - \left(1 + \frac{1}{2}\exp\left(\frac{E_D - E_F}{k_B T}\right)\right)^{-1}\right]. \tag{3.67}$$

Note that we distinguish between the number N_D and density $\mathbf{N_D}$ of donor atoms and that E_D is the energy. The factor $1/2$ denotes that we have a ground-state degeneracy of two for the donor impurity level. Inserting this and the expressions for n and p into (3.66) yields

$$N_C \exp\left(-\frac{E_C - E_F}{k_B T}\right) = N_D\left[1 - \left(1 + \frac{1}{2}\exp\left(\frac{E_D - E_F}{k_B T}\right)\right)^{-1}\right] + N_V \exp\left(\frac{E_V - E_F}{k_B T}\right). \tag{3.68}$$

This needs to be solved numerically for given values of N_C, N_D, N_V, E_C, E_D, and E_V. Examples are shown in reference [87]. For typical concentrations of $N_D =$ few $\times\, 10^{12}/\text{cm}^3$ and room temperature, the Fermi level is close to the conduction band. We can write down an equivalent expression for p-type silicon. Here, we have to consider the acceptor ions. The degeneracy is four. At room temperature the Fermi level is close to the valence band.

3.3.2 The *pn* Junction

Let us look at what happens when we couple p-type silicon to n-type silicon, called a *pn* junction. Figure 3.31 (top left) depicts the electric structure in p-type and n-type silicon before coupling them. In p-type silicon we find acceptor ions and empty holes. In the n-type silicon we find donor ions and conduction electrons. Figure 3.31 (bottom left) shows the band structure for this configuration. In p-type silicon the holes populate the valence band. A few electrons may be lifted into the conduction band. The Fermi energy is slightly above the valence band. In the n-type silicon the extra electrons are in the conduction band. Due thermal interactions bound electrons may be lifted into the conduction band leaving behind holes in the valence band. The Fermi energy here lies slightly below the conductions band. So, the Fermi energies are different in the p-type and n-type silicon.

3.3 Ionization in Solid-State Detectors

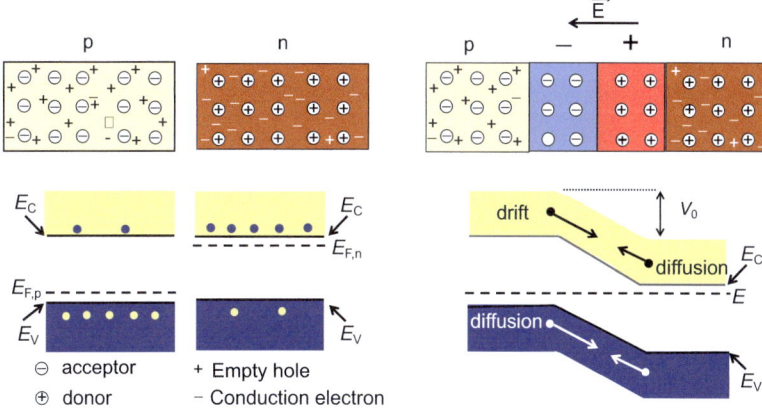

Fig. 3.31 Top left: Electric structure in separated *p*-type and *n*-type silicon showing acceptor ions (circles labeled with "−"), empty holes ("+"), donor ions (circles labeled with "+") and conduction electrons ("−"). Top right: Electric structure of a *pn* junction in the absence of applied bias voltage. Bottom left: Band structure of separate *p*-type and *n*-type silicon. Bottom right: Structure of the valence and conduction bands in a *pn* junction in the absence of applied bias voltage

Figure 3.31 (top right) depicts the electric structure in a *pn* junction in the absence of any applied bias voltage. We observe the following reactions. Caused by the difference in Fermi levels on the *n*-side and *p*-side, free electrons from the *n*-side diffuse into the *p*-side and fill the empty hole positions producing a region of stationary negative ions. Similarly, holes from the *p*-side diffuse into the *n*-side producing stationary positive ions. This process eventually stops when the negative charge density is sufficiently large to block electrons from diffusing further into the *p*-region. Now the Fermi levels are identical. The stationary ions produce an electric field pointing from the *n*-side to the *p*-side. The region of positive and negative ions is called depletion zone. The band structure is altered as shown in Fig. 3.31 (bottom right). The energy level on the *n*-side is lowered. The distribution of the ions has produced a contact potential $-V_0$ across the depletion zone. Both the valence and conduction bands in the undepleted region on the *n*-side are shifted by $-V_0$ with respect to those in the undepleted region on the *p*-side. Electrons in the conduction band can drift from the *p*-side to the *n*-side or diffuse by thermal energy in the opposite direction. Holes in the valence band can drift from the *n*-side to the *p*-side or diffuse by thermal energy in the opposite direction. Figures 3.32 (left and right) show the resulting charge density and electric-field configuration, respectively.

Let us look what happens when we apply a bias voltage V_b to the *pn* junction. First, we apply forward biasing for which the *p*-side is connected to the anode and the *n*-side to the cathode. Figure 3.33 (top left) shows the electric structure for this situation. As before electrons move from the *n*-side to the *p*-side while holes move into the opposite direction. However, the bias voltage reduces the contact potential producing a smaller depletion zone than the configuration without bias voltage. Figure 3.33 (bottom left) shows the structure of the energy bands for this case. Since the depletion layer is reduced, the energy shift of the two bands across

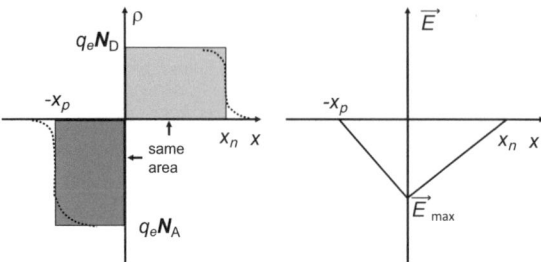

Fig. 3.32 Left: Charge density of a *pn* junction across the depletion layer. The dotted lines represent the actual density distribution while the solid line shows the assumed density distribution. Right: Electric-field configuration of a *pn* junction across the depletion layer. Both reprinted with kind permission from [91], © 2005, Dr. Bowden. All rights reserved

the depletion zone becomes smaller, $V_0 - V_b$, such that diffusion becomes larger increasing the leakage current. The Fermi energy shows a jump at the *p-n* contact position from a lower value to a higher value. The situation changes when we apply reverse bias voltage. Now the anode is connected to the *n*-side and the cathode to the *p*-side. Figure 3.33 (top right) shows the electric structure for this configuration. The electrons drift towards the *n*-side while holes drift towards the *p*-side creating an even larger depletion zone than that of a *pn* junction without bias voltage. Figure 3.33 (bottom right) shows the structure of the energy bands for this configuration. Since the depletion layer is increased, the shift of the energy bands becomes larger. Here, leakage currents are reduced because diffusion becomes less likely. The Fermi energy shows a jump at the *p-n* contact position from a low value to an even lower value. This configuration is the one we use for vertex tracking.

When ionization liberates a charge in the depletion zone, electrons and holes drift apart due to the strong internal field and produce a current. In order to detect a signal, the current has to be larger than the noise. At room temperature silicon is more advantageous than germanium since its resistivity is 200 kΩ-cm compared to 65 kΩ-cm while the energy gaps are 1.14 eV and 0.67 eV, respectively. The resistivity is given by $\rho_{np} = 1/\sigma_{np}$ with

$$\sigma_{np} = q_e(n\mu_e + p\mu_h), \tag{3.69}$$

where σ_{np} is the conductivity, μ_e is the electron mobility and μ_h is the hole mobility. Both μ_e and μ_h depend on the electric field. In *n*-type silicon the conductivity is determined by the electrons, thus $\sigma_{np} = \sigma_n \sim q_e n \mu_e$. In order to produce a detector for measuring minimum ionizing particles with high accuracy, we need a sufficiently thick depletion layer. The ideal current-voltage dependence of a *pn* junction is given by

$$I = I_0 \left[\exp\left(\frac{V_b + V_0}{k_B T}\right) - 1 \right], \tag{3.70}$$

where I_0 is the reverse saturation current. Figure 3.34 (left) shows the expected IV curve of a *pn* junction. In the forward direction, the current increases rapidly with

3.3 Ionization in Solid-State Detectors

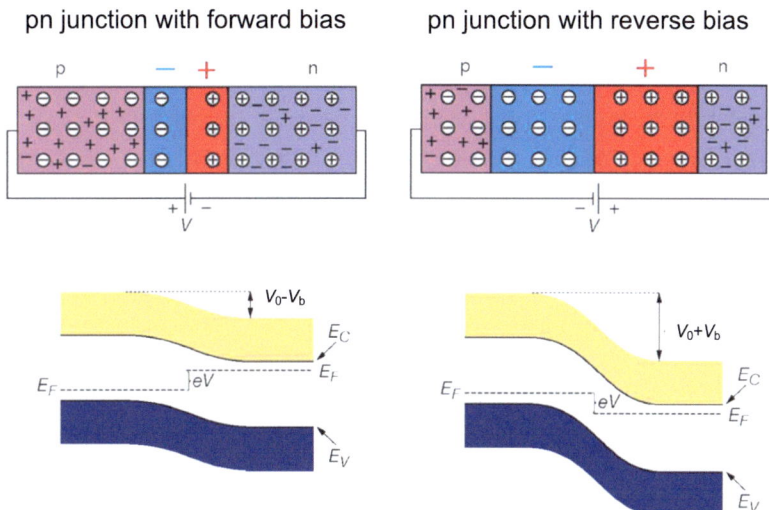

Fig. 3.33 Top left: Electric structure of a *pn* junction with forward bias voltage. Top right: Electric structure of a *pn* junction with reverse bias voltage. Bottom left: Structure of the valence and conduction bands in a *pn* junction with forward bias voltage. Bottom right: Structure of the valence and conduction bands in a *pn* junction with reverse bias voltage

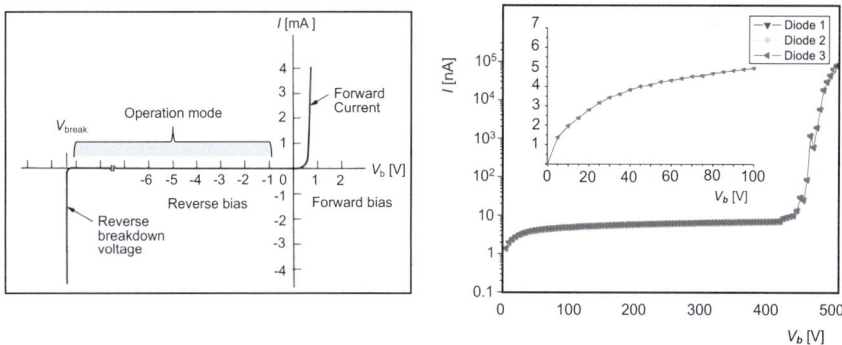

Fig. 3.34 Left: The ideal IV dependence of a *pn* junction. Reprinted with kind permission from [86], © 2005, M. Krammer. All rights reserved. Right: The leakage current versus the reverse bias voltage of a silicon diode [92]. Reprinted with kind permission from [92], © 2017, Springer. All rights reserved

the bias voltage. In the reverse direction, the current is low until the breakdown point, where it starts to increase rapidly.

Figure 3.34 (right) shows the leakage current versus the reverse bias voltage of a silicon diode. The current increases as $\sqrt{V_b}$ (shown in the inset).

3.3.3 Depletion Layer

To estimate the size of the depletion layer, we need to solve the one-dimensional Poisson equation,

$$-\frac{d^2V}{dx^2} = \frac{dE_x(x)}{dx} = \frac{\rho(x)}{\epsilon\epsilon_0}, \qquad (3.71)$$

where $\rho(x)$ is the charge distribution, ϵ is the dielectric constant of the material and ϵ_0 is the electric permittivity. If N_D and N_A are the densities of donor (n) and acceptor (p) impurities, then the asymmetric double layer of charge is simply given by

$$\rho(x) = \begin{cases} q_e N_D & \text{for } 0 < x \le x_n, \\ -q_e N_A & \text{for } -x_p \le x \le 0. \end{cases} \qquad (3.72)$$

Note that concentrations can be different but the total charge is conserved,

$$N_D \cdot x_n = N_A \cdot x_p. \qquad (3.73)$$

Double integration yields

$$V(x) = \begin{cases} -\frac{q_e N_D}{\epsilon_{sc}\epsilon_0}\left(\frac{x^2}{2} - x_n x\right) + C & \text{for } 0 < x \le x_n, \\ \frac{q_e N_A}{\epsilon_{sc}\epsilon_0}\left(\frac{x^2}{2} + x_p x\right) + C & \text{for } -x_p \le x \le 0, \end{cases} \qquad (3.74)$$

where x_n and x_p are determined from $dV/dx = 0$ at $x = x_n$ and $dV/dx = 0$ at $x = -x_p$, respectively. The integration constant C is the same since both solutions are equal at $x = 0$. So, C is determined from $V = 0$ at $x = -x_p$, yielding

$$C = \frac{q_e N_A}{2\epsilon_{sc}\epsilon_0} x_p^2. \qquad (3.75)$$

At $x = x_n$, we obtain the contact potential $V(x_n) = V_0$,

$$V_0 = \frac{q_e}{2\epsilon_{sc}\epsilon_0}\left(N_D \cdot x_n^2 + N_A \cdot x_p^2\right). \qquad (3.76)$$

Note that $|\vec{E}|$ is maximum at $x = 0$ taking a magnitude of

$$E_{\max} = \frac{q_e N_D x_n}{\epsilon_{sc}\epsilon_0} = \frac{q_e N_A x_p}{\epsilon_{sc}\epsilon_0}. \qquad (3.77)$$

Using (3.73) we get

$$x_n = \sqrt{\left[\frac{2\epsilon_{sc}\epsilon_0 V_0}{q_e N_D(1 + N_D/N_A)}\right]}, \qquad (3.78)$$

$$x_p = \sqrt{\left[\frac{2\epsilon_{sc}\epsilon_0 V_0}{q_e N_A(1 + N_A/N_D)}\right]}. \qquad (3.79)$$

3.3 Ionization in Solid-State Detectors

If $N_A \gg N_D$, then $x_p \ll x_n$ and depletion layer extends more into the lighter-doped side (n-type). This configuration is called p^+n structure. For the resistivities we find $\rho_p \ll \rho_n$.

The total depletion zone is given by

$$d = x_n + x_p = \sqrt{\left(\frac{2\epsilon_{sc}\epsilon_0 V_0}{q_e N_D N_A} \frac{N_A^2 + N_D^2}{N_A + N_D}\right)}. \tag{3.80}$$

For $N_A \gg N_D$, (3.80) reduces to

$$d = x_n = \sqrt{\left(\frac{2\epsilon_{sc}\epsilon_0 V_0}{q_e N_D}\right)}. \tag{3.81}$$

Since the resistivity ρ_n in the n region is approximately given by

$$\frac{1}{\rho_n} = \sigma_n \simeq q_e N_D \mu_e, \tag{3.82}$$

the total depletion zone can be written as

$$d = \sqrt{2\epsilon_{sc}\epsilon_0 \rho_n \mu_e V_0}. \tag{3.83}$$

If we apply a reverse bias voltage V_b to the pn junction, we have to replace V_0 with $V_0 + V_b$ in (3.76) to (3.83).

For silicon the depletion layer is

$$d_{Si} = \begin{cases} 0.53\sqrt{\rho_n(V_0 + V_b)} \; [\mu m] \text{ for n} - \text{type}, \\ 0.32\sqrt{\rho_p(V_0 + V_b)} \; [\mu m] \text{ for p} - \text{type}. \end{cases} \tag{3.84}$$

For example, for $V_0 = 1$ V and $V_b = 0$ V, we calculate $d_{Si} \simeq 75$ μm, while for $V_b = 300$ V we get $d_{Si} > 1$ mm. The depletion layer also has a capacitance that affects the noise. For planar geometry it is simply given by

$$C_{Si} = \epsilon_{sc}\epsilon_0 \frac{a_{Si}}{d_{Si}}, \tag{3.85}$$

where a_{Si} is the area of the silicon detector and d_{Si} is its thickness that is equal to the depletion zone. Plugging in the results for silicon, yields

$$C_{Si} = \begin{cases} 2.2\sqrt{\rho_n(V_0 + V_b)} \text{ for } n-\text{type}, \\ 3.7\sqrt{\rho_p(V_0 + V_b)} \text{ for } p-\text{type}. \end{cases} \tag{3.86}$$

Above a certain voltage, the capacitance becomes constant and the complete thickness of the silicon is depleted. This is found in most detectors, which act like solid-state ionization chambers.

Exercises

3.1 For a gas mixture Ar-CH$_4$ (90:10), the Townsend coefficient is $\alpha_I = 7$/cm at an electric field of 10 kV/cm. In the primary ionization 100 electrons are produced. How many electrons are produced after a path of 2 cm in the gas? For Xe gas the corresponding Townsend coefficient is 2.6/cm. How long does the Xe chamber have to be to observe the same number of electrons as that in the Ar $-$ CH$_4$ gas mixture?

3.2 Consider a cylindrical straw chamber with a radius of $r_a = 1$ cm filled with Ar gas at normal conditions. In the center a 50 μm thick wire is positioned. Electric fields of +100 V/cm and +10 kV/cm are applied to the wire. Signals recorded on the wire are read out. Assume that an e-ion pair is produced at a position $r_0 = 0.5 r_a$. If the electron would not loose any energy by collisions what would be the maximum kinetic energy for the two field configurations when it hits the wire. The observed pulse has an electron and an ion contribution. What are the relative contributions from electrons and ions for the two field configurations?

3.3 Let us consider a chamber with dimensions $x \times y = 2$ cm $\times$ 2 cm and a length $z = 5$ cm. What is the capacitance? The chamber is operated with a standard Ar-CH$_4$ (80:20) gas mixture. What voltage needs to be applied to achieve a collection time of the electrons within 0.5 μs? How many primary ions do we collect?

3.4 Let us consider the previous example in a cylindrical gas detector that is operated with a standard Ar-CH$_4$ (80:20) gas mixture at normal pressure and room temperature. The tube has a radius of 1 cm and a length of 5 cm. The anode wire has a radius of 10 μm. The anode is placed on a potential of 1 kV. What is the capacitance of the device? What is the electron collection time? What is the dominant component for the pulse collection? How does this change if we reduce the potential on the anode wire to 200 V?

3.5 Consider a cylindrical detector with a 1 cm radius filled with argon gas. The detector is operated with a potential of 100 V. A wire is placed at its center with a diameter of 20 μm. What is the kinetic energy the electrons gain and how big is the Townsend coefficient?

3.6 What is the Townsend coefficient for 100 eV electrons in argon gas at normal conditions? What potential is needed to produce this electron kinetic energy at 20 μm? For a 10 μm anode wire what is the gain?

3.7 Determine the Fano factor for liquid argon. Calculate the three factors $F1$, $F2$ and F_3 and the total factor. You can compare your results with those in Table 3.6. Determine the energy resolution for a heavy charged particle that looses an energy of 5 MeV.

3.8 Consider a cylindrical counter with an anode wire of 10 μm diameter in the center and an outer radius of 1 cm. The counter is operated with argon gas at a potential of 1 kV on the wire at normal conditions. Determine the distance from the wire for a kinetic energy gain of 2 eV, where the amplifications starts. Calculate the gain without UV photons. If 10% of the UV photons contribute to the amplification, how large is the gas amplification factor?

3.9 Consider a pn junction with $N_D = 5 \times 10^{15}$ /cm^3 and $N_A = 10^{18}$ /cm^3. The maximum electric field is 1 kV/cm. For $x_n = 490$ μm and $x_p = 10$ μm, how large is the contact potential? How big is the depletion layer if we apply a reverse bias voltage of 100 V? How large is the current at room temperature?

3.10 Consider a diamond detector operated at 300K and a potential of $V_0 = 200$ V. How thick is the depletion layer? For a thickness of $d_d = 100$ μm and an area of 1 mm $\times$ 1 mm, what is the capacitance? What is the thickness of the depletion layer and the capacitance for a GaAs sensor for the same dimensions operated under the same conditions?

References

1. G. Charpak, Decouverte **9**, 1 (1972)
2. T. Aoyama, Nucl. Instrum. Meth. A **234**, 125 (1985)
3. S.A. Korff, *Electron and Nuclear Counters* (Vav Nostrand, New York, 1955)
4. Yu.I. Davydov, IEEE Trans. Nucl. Sci. **53**, 2931 (2006)
5. A. Zastawny, Nucl. Instrum. Meth. A **385**, 239 (1997)
6. A. Sharma, F. Sauli, Nucl. Instrum. Meth. A **334**, 420 (1993)
7. S. Navas et al., The particle data group. Phys. Rev D. **110**, 030001 (2024)
8. L.K. Warne, R. E. Jorgenson, S. D. Nicolaysen, SAND2003-4078 Report (2003)
9. F. Sauli, CERN Yellow Report, CERN-77-09 (1977)
10. M.E. Rose, S.A. Korff, Phys. Rev. **59**, 850 (1941)
11. A. Fridman, A. Chirokov, A. Gutsol, J. Phys. D Appl. Phys. **38**, R1 (2005)
12. S.A. Korff, *Electrons and Nuclear Counters* (Van Nostrand, New York, 1955)
13. H. Staub, *Detection Methods in Nuclear Physics*. E. Segré (ed.) (Wiley, New York, 1953)
14. L.B. Loeb, *Basic Processes of Gaseous Electronics* (University of California Press, Berkeley, 1961)
15. G.C. Montgomery, D.D. Montgomery, Geiger-Müller. J. Franklin Inst. **231**, 447 (1941)
16. F. Sauli, *Gaseous Radiation Detectors* (Cambridge Monographs, 2014)
17. G. Jaffe, Le Radium 10, 136 (1913); Ann. Physik IV **42**, 303 (1913)
18. H. Greinacher, Phys. Z. **10**, 986 (1909)
19. J. C. Mc Lennan *et al.* Phil. Mag. **26**, 876 (1913)
20. G.W. Hutchinson, Nature **162**, 610 (1948)
21. N. Davidson, A.E. Larsh Jr., Phys. Rev. **77**, 706 (1949)
22. J. H. Marshall, Phys. Rev. **91**, 905 (1953); Rev. Sci. Instr. **25**, 232 (1954)
23. The XENON Collaboration (E. Aprile *et al.*, Astropart. Phys. **34**, 679 (2011)
24. The LUX Collaboration (D. S. Akerib *et al.*, Nucl. Instrum. Meth. A **704**, 111 (2012)
25. E. Aprile (et al.), Phys. Rev. A **48**, 1313 (1993)
26. T. Doke, Radiation Physics **1**, 24 (1977)
27. M. Miyajima, *et al.*, Phys. Rev. A **9**, 1438 (1974)
28. T. Takahashi *et al.*, Phys. Rev. A **12**, 1771 (1975)
29. D. Beaglehole, Phys. Rev. Lett. **15**, 55 (1965)
30. I.T. Steinberger, U. Asaf, Phys. Rev. B **8**, 914 (1973)
31. R. Reiniger et al., Phys. Rev. B **26**, 6294 (1982)
32. S. Bernstorff et al., Ann. Isr. Phys. Soc. **6**, 270 (1983)
33. R. Reiniger et al., Chem. Phys. **86**, 189 (1984)
34. W. B. Atwood *et al.* http://xxx.lanl.gov/abs/0902.1089 (2009)
35. E. Aprile, T. Doke, Rev. Mod. Phys. **82**, 205 (2010)

36. N. Schwenter, E.E. Koch, J. Jortner, *Electronic Excitations in Condensed Rare Gases* (Springer Tracts in Modern Physics (Springer-Verlag, New York, 1985)
37. V. Rössler, Phys. Status Solidi B **45**, 483 (1971)
38. G.D. Alkhazov, A.P. Komar, A.A. Vorobev, Nucl. Instrum. Meth. **48**, 1 (1967)
39. T. Doke et al., Nucl. Instrum. Meth. **134**, 353 (1976)
40. W. Shockley, Czech. J. Phys. B **ll**, 81 (1961)
41. E. Aprile et al., Nucl. Instrum. Meth. A **338**, 328 (1994)
42. N. Ishida et al., Nucl. Instrum. Meth. A **384**, 380 (1997)
43. W.J. Willis, V. Radeka, Nucl. Instrum. Meth. **120**, 221 (1974)
44. L. Onsager, Phys. Rev. **54**, 554 (1938)
45. J. Thomas, D.A. Imel, Phys. Rev. A **36**, 614 (1987)
46. M. Miyajima et al., Nucl. Instrum. Meth. **134**, 403 (1976)
47. R. Acciarri et al., JINST **8**, P08005 (2013)
48. S. Kubota et al., Phys. Rev. B **20**, 3486 (1979)
49. C.R. Gruen, M.D. Edmiston, Phys. Rev. Lett. **40**, 407 (1978)
50. R.T. Scalettar et al., Phys. Rev. A **25**, 2419 (1982)
51. W. Hofmann et al., Nucl. Instrum. Meth. **135**, 151 (1976)
52. A. Baldini et al., Nucl. Instrum. Meth. A **545**, 753 (2005)
53. A. Braem et al., Nucl. Instrum. Meth. A **320**, 228 (1992)
54. G. Bakale, U. Sowadand, W.F. Schmidt, J. Phys. Chem. **80**, 2556 (1976)
55. E. Aprile, R. Mukerjee, M. Suzuki, Nucl. Instr. Meth. A **300**, 343 (1991)
56. M. Chen et al., Nucl. Instrum. Meth. A **327**, 187 (1993)
57. M. Ichige ct al., Nucl. Instrum. Mcth. A **333**, 355 (1993)
58. M. Masuda et al., Nucl. Instrum. Meth. A **188**, 629 (1981)
59. J. Prunier, V.F. Pisarev, G.S.R. Revenko, Nucl. Instrum. Meth. **109**, 257 (1973)
60. K. Ozone, PhD thesis, U. Tokyo, http://meg.icepp.s.u-tokyo.ac.jp/docs/theses/ozoned.pdf (2005)
61. E. Aprile et al., Phys. Rev. B **76**, 014115 (2007)
62. K. Masuda et al., Phys. Rev. A **39**, 4732 (1989)
63. S. Kubota, M. Hishida, J. Ruan, J. Physics C **11**, 2645 (1978)
64. D.C. Lorents, D.J. Eckstrom, D.L. Huestis, *Stanford Research Inst* (MP, Rep. SRI, 1973), pp.73–2
65. A. Hitachi et al., Phys. Rev. B **27**, 5279 (1983)
66. O. Hilt, W.F. Schmidt, J. Phys. Condens. Matter **6**, L735 (1994)
67. L.S. Miller, S. Howe, W.E. Spear, Phys. Rev. **166**, 871 (1968)
68. J.L. Pack, R.E. Voshhall, A.V. Phelps, Phys. Rev. **127**, 2084 (1962)
69. Y. Yoshino, U. Spwada, W.F. Schmidt, Phys. Rev. A **14**, 438 (1976)
70. B. Halpern et al., Phys. Rev. **156**, 351 (1967)
71. D. W. Swan, Proc. Phys. Soc. (London) **83**, 659 (1964)
72. M.H. Cohen, J. Lekner, Phys. Rev. **158**, 305 (1967)
73. H. Schnyders, S.A. Rice, L. Meyers, Phys. Rev. Lett. **150**, 127 (1965)
74. E. Aprile (personal communication), (2024)
75. U. Fano, Phys. Rev. **72**, 26 (1947)
76. H. A. Bethe and J. Ashkin, (ed. E. Segré; Wiley, New York) Vol. 1, part 2 (1953)
77. T. Doke, Portugal Physics **12**, 9 (1981)
78. T. Doke et al., Nucl. Instrum. Meth. **134**, 353 (1976)
79. A. Bolotnikov, B. Ramsey, Nucl. Instrum. Meth. A **396**, 360 (1997)
80. H.J. Crawford et al., Nucl. Instrum. Meth. A **256**, 47 (1987)
81. E. Aprile et al., Phys. Rev. Lett. **95**, 081302 (2006)
82. V. V. Egorov, V. Ermilova, and B. Rodionov, preprint Lebedev Physical Institute **166** Moscow (1982)
83. J. Thomas, D.A. Imel, S. Biller, Phys. Rev. A **38**, 5793 (1988)
84. E. Aprile, R. Mukerjee, M. Suzuki, Nucl. Instrum. Meth. A **302**, 177 (1991)
85. E. Aprile et al., Nucl. Instr. Meth. A **480**, 636 (2002)

References

86. M. Krammer, *Halbleiterdetektoren* (Lectures, Wien, 2017)
87. S.M. Sze, *Physics of Semiconductor Devices*, 2nd edn. (John Wiley & Sons, New York, 1981)
88. U. K. Mishra and J. Singh, *Semiconductor Device Physics and Design* (Springer Verlag, 2008)
89. H.D. Barber, Solid-state Electron. **10**, 1039 (1967)
90. C.D. Thurmond, J. Electrochem. Soc. **122**, 1133 (1975)
91. www.pveducation.org (2019)
92. F. Hartmann, *Evolution of Silicon Sensor Technology in Particle Physics*, 2nd edn. (Springer Int. Pub., 2017)

Open Access This chapter is licensed under the terms of the Creative Commons Attribution 4.0 International License (http://creativecommons.org/licenses/by/4.0/), which permits use, sharing, adaptation, distribution and reproduction in any medium or format, as long as you give appropriate credit to the original author(s) and the source, provide a link to the Creative Commons license and indicate if changes were made.

The images or other third party material in this chapter are included in the chapter's Creative Commons license, unless indicated otherwise in a credit line to the material. If material is not included in the chapter's Creative Commons license and your intended use is not permitted by statutory regulation or exceeds the permitted use, you will need to obtain permission directly from the copyright holder.

Gaseous Tracking Detectors

4

This chapter describes various gaseous tracking detectors that are used to measure particle trajectories with high precision. First, we discuss planar multi-wire proportional chambers (Sect. 4.1) followed by planar drift chambers (Sect. 4.2). These chambers are still used in fixed-target experiments and in endcaps of colliding-beam detectors. For usage in the central region of colliding-beam experiments, cylindrical chambers are necessary. Thus, we continue our discussion with cylindrical drift chambers (Sect. 4.3), jet drift chambers (Sect. 4.4) and time projection chambers (Sect. 4.5). We then focus on some other developments such as the microstrip gas chambers (Sect. 4.6), the electron gas multiplication (Sect. 4.7), micromesh gaseous structures (Sect. 4.8), resistive plate chambers (Sect. 4.9), limited streamer tubes (Sect. 4.10), monitored drift tubes (Sect. 4.11) and thin gap chambers (Sect. 4.14). At the end we mention briefly the bubble chamber for historic reasons (Sect. 4.15), list some older technologies (Sect. 4.16) and present a comparison of position resolutions of tracking detectors (Sect. 4.17). We will not present here cloud, flash, spark and streamer chambers since they cannot be read out electronically and are no longer used. Cloud chambers, also called Wilson chambers [1], played an important role from the 1910s to 1950s. Note that the positron [2], muon [3] and kaon [4] were discovered using cloud chambers. Spark [5] and streamer [6] chambers were used from the 1930s to 1960s. While spark chambers consisted of a stack of metal plates housed in a gas volume streamer chambers just had two plates. Flash chamber were used in neutrino experiments [7,8].

4.1 Multi-Wire Proportional Chambers

In the previous chapter we saw that the ionization in proportional counters is localized rather precisely due to the slowly moving positive ions. Thus, by combining many proportional counters we can build large-area multi-wire proportional chambers (MWPC) to measure the position of traversing particles. This idea was proposed first by Charpak [9] for which he was awarded the Nobel prize in physics in 1992.

4.1.1 Principles

Figure 4.1 shows a schematic view (left) and a cross section (right) of a multi-wire proportional chamber. Many anode wires are arranged parallel to each other in a plane at fixed spacing called pitch. The anode wire plane is sandwiched between two cathode planes. A high electric field is applied between anode and cathode. While cathode planes typically are kept at ground potential, anode wires are connected to high voltage. Each anode wire acts as a separate proportional counter. Figure 4.2 (left) shows the electric-field lines and equipotential surfaces of an MWPC. The electric-field lines of two adjacent wires repel each other leading to a field-free region at the midpoint between the two wires. Away from the wire plane the electric field is constant in most parts of the chamber. If a wire is displaced, the electric field becomes distorted as depicted in Fig. 4.2 (right). We discuss the effects of displaced wires below.

We define a coordinate system in which the x-axis is perpendicular to the anode wires in the chamber plane, the y-axis is along the direction of the anode wires and the z-axis is perpendicular to the cathode plane as shown in Fig. 4.1 (left). In this coordinate system the beam direction is along the z axis. The analytic form of the electric field and electric potential is obtained by solving the Laplace equation in two

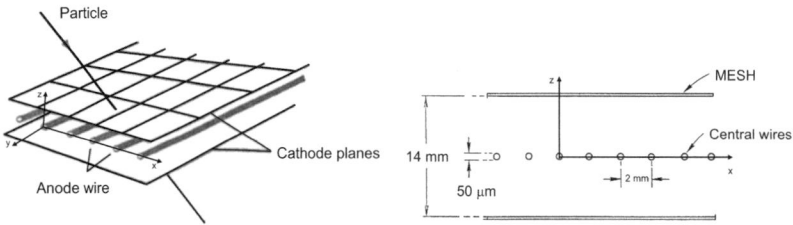

Fig. 4.1 Left: Schematic view of a multi-wire proportional chamber. Reprinted with kind permission from [10], © 1992, CERN. All rights reserved. Right: Cross section through a typical MWPC. Here, the anode wire pitch is 2 mm, the diameter of the gold-plated tungsten wire is 50 μm and the anode-cathode gap is 7 mm. Reprinted with kind permission from [11], © 1969, SLAC. All rights reserved

4.1 Multi-Wire Proportional Chambers

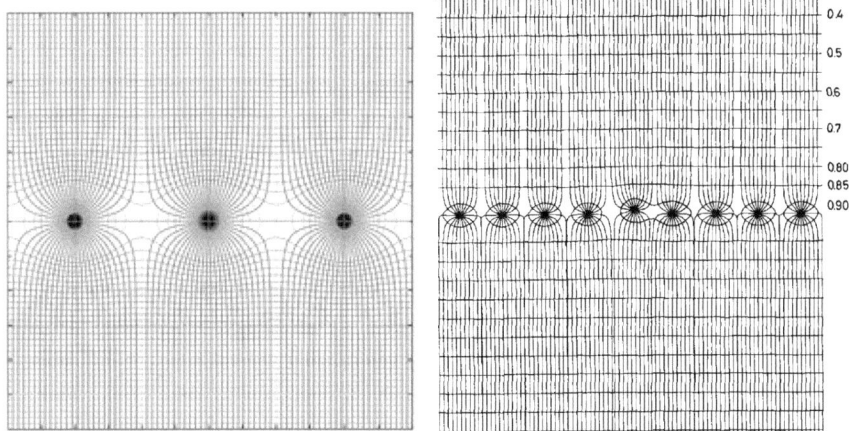

Fig. 4.2 Left: Schematic view of the electric-field lines (vertical lines) and equipotential surfaces (horizontal lines) in an MWPC in the vicinity of anode wires. Reprinted under CC-BY-NC-ND-4.0 Licence with kind permission from [12], © 2023, F. Sauli. Right: Schematic view of the electric-field lines and equipotential surfaces in an MWPC in which one wire is displaced. Reprinted with kind permission from [13], © 1970, Elsevier. All rights reserved

dimensions. The analytic solution is given in textbooks, e.g. Morse and Feshbach [14] or Buchholz [15]. A useful approximation is given by Erskine [16],

$$V(x,z) = \frac{C_w \cdot V_0}{4\pi \epsilon_0} \left\{ \frac{2\pi a_w}{s_w} - \ln\left[4\left(\sin^2\left(\frac{\pi x}{s_w}\right) + \sinh^2\left(\frac{\pi z}{s_w}\right) \right) \right] \right\}, \quad (4.1)$$

$$|\vec{E}(x,z)| = \frac{C_w \cdot V_0}{2\epsilon_0 s_w} \frac{\sqrt{1 + \tan^2\left(\frac{\pi x}{s_w}\right) \tanh^2\left(\frac{\pi z}{s_w}\right)}}{\sqrt{\tan^2\left(\frac{\pi x}{s_w}\right) + \tanh^2\left(\frac{\pi z}{s_w}\right)}}, \quad (4.2)$$

where s_w is the anode wire pitch, a_w is the gap between anode and cathode, d_w is the anode wire diameter, V_0 is the high voltage on the anode wire and C_w is the capacitance per unit length to ground determined by

$$C_w = \frac{2\pi \epsilon_0}{\frac{\pi a_w}{s_w} - \ln\left(\frac{\pi d_w}{s_w}\right)}. \quad (4.3)$$

Since $d_w/2 \ll s_w$ the capacitance per unit length in (4.3) is always smaller than that of a plane capacitor with the same surface ($2\epsilon_0 s_w/a_w$). Typical values for the anode-cathode gap are 4–8 mm and for the wire diameter are 20–100 μm. The anode wire pitch is around 2 mm. Figure 4.3 depicts the capacitance as a function of the wire spacing s_w. The capacitance increases with larger spacing up to ~ 1 cm before leveling off. For small wire spacings the dependence on the wire diameter is rather weak. Reducing the anode-cathode gap increases the capacitance. For example, for

Fig. 4.3 The capacitance per unit length as a function of the wire spacing s_w for $a_w = 8$ mm and $d_w = 0.02$ mm (solid curve), $a_w = 8$ mm and $d_w = 0.03$ mm (dotted curve), $a_w = 4$ mm and $d_w = 0.02$ mm (dash-dotted curve) and $a_w = 4$ mm and $d_w = 0.03$ mm (dashed curve)

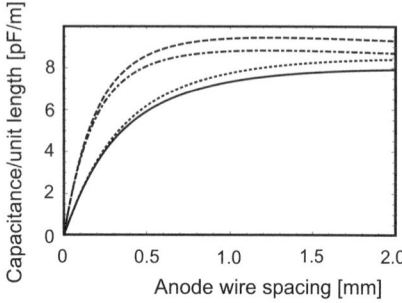

$a_w = 8$ mm, $d_w = 20\,\mu$m and $s_w = 2$ mm we get a capacitance of $C_w = 3.47$ pF/m. Reducing the gap to 4 mm increases the capacitance to 5.71 pF/m. More values are given in reference [17]. For $z = 0$ and $x = 0$ the electric field takes the forms

$$E_x = |\vec{E}(x, 0)| = \frac{C_w \cdot V_0}{2\epsilon_0 s_w} \cot \frac{\pi x}{s_w} \qquad (4.4)$$

$$E_z = |\vec{E}(0, z)| = \frac{C_w \cdot V_0}{2\epsilon_0 s_w} \coth \frac{\pi z}{s_w}, \qquad (4.5)$$

respectively. For a single anode wire, the electric field is $E(r) \propto 1/r$, where $r = \sqrt{x^2 + z^2}$. In fact, we recover the $1/r$ dependence for $z \ll s_w$

$$|\vec{E}(x, z)| = \frac{C_w V_0}{2\pi \epsilon_0} \frac{1}{r}. \qquad (4.6)$$

For large $z \geq s_w$, the electric field approaches a constant

$$|\vec{E}_0(x, z)| = \frac{C_w V_0}{2\epsilon_0 s_w}, \qquad (4.7)$$

because for large z values we get $\tanh(\pi z/s_w) = 1$.

Let us consider a chamber with dimensions $s_w = 2$ mm, $d_w = 0.02$ mm and $a_w = 8$ mm that is filled with an Ar-CO_2 (75:25) gas mixture. If we apply a voltage of $V_0 = 4000$ V, the electric field at the anode wire is $|\vec{E}| = 2.5 \times 10^5$ V/cm. Figure 4.4 shows the electric-field strength for this chamber in the x- and z-directions in a double logarithmic scale. Near the anode wire, the electric field decreases as $1/r$ like in a cylindrical proportional counter shown by the straight lines at small distances. In the x-direction, the electric field continues to drop becoming rather weak in the middle region between two anode wires. At the midpoint it is zero. In the z-direction, the electric field approaches a constant value of $|\vec{E}_0(x, z)|$ given in (4.7). This already occurs at a distance of $z \geq 0.5 s_w$.

A charged particle traversing the MWPC liberates electron-ion pairs along its path as sketched in Fig. 3.3. The electrons are accelerated along the electric-field lines towards the nearest anode wires while the positive ions move very slowly towards the cathode plane. As discussed in the previous chapter, each primary electron produces

4.1 Multi-Wire Proportional Chambers

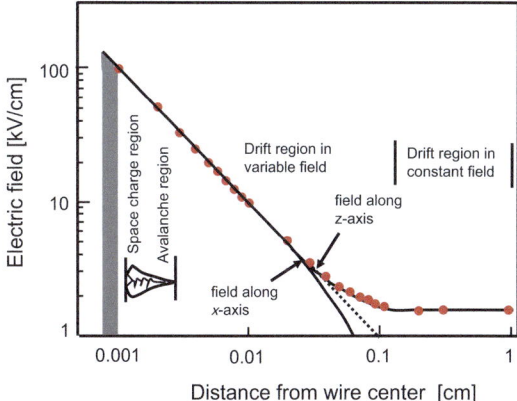

Fig. 4.4 Electric-field strengths E_x and E_z as a function of the anode wire distance perpendicular to the anode wire plane (z-direction) and that perpendicular to the anode wires in the anode wire plane (x-direction) for a chamber with $V_0 = 4000$ V, $s_w = 2$ mm, $d_w = 0.02$ mm and $a_w = 8$ mm. Solid points show measurements. The grey bar at 10 μm indicates the anode wire position. The space charge region, avalanche region, drift region in a variable electric field and the drift region in a constant field are marked. In the x-direction the electric-field strengths drops more rapidly than $1/r$ shown by the dashed line. At large z distances the electric field approaches the constant value of $|\vec{E}_0| = (C_w V_0)/(2 s_w \epsilon_0)$. Reprinted under CC-BY-3.0 Licence from [17], ©1977–2024, CERN

an avalanche near the closest anode wire if the gas amplification is sufficiently high. Since primary electrons are produced successively along the path of the traversing charged particle, the arrival time of each primary electron at each anode wire is slightly different. One primary electron after another reaches the high electric-field region and produces an individual avalanche and in turn a signal pulse. The observed signal on the anode wire consists of several individual pulses each produced by a separate avalanche. With good time resolution, these individual pulses can be resolved whose rise time is about 0.1 ns and whose decay time depends on the time constant $\tau_{ch} = R_{ch} C_{ch}$, where R_{ch} is the resistor and C_{ch} is the capacitance of the readout chain. Figure 4.5 (left) shows a picture of individual pulses taken from an oscilloscope and Fig. 4.5 (right) shows a simulation of individual pulses. If we sum over all individual pulses, a pulse with a long tail is observed as depicted in the inset of Fig. 4.5 (right). While the electron cloud formed in an avalanche develops into a tube around anode wire, the shape of the slowly moving ions still retains the memory of the origin of the avalanche by forming a drop-like shape. Each anode wire of an MWPC acts as an independent detector [9]. As Charpak noticed, positive signals induced in all electrodes by the positive ions are largely compensated by the negative signals produced by capacitive coupling.

The timing properties of proportional chambers are determined by the collection time of the electrons produced by ionizing tracks. Figure 4.6 (left) shows a typical signal recorded with an MWPC, which results from electrons collected in three regions (A, B, C) as shown in Fig. 4.6 (center) [19]. Electrons produced in region A are quickly collected because typical drift velocities in this high-field region are above 5 cm/μs. Electrons produced in the low field region B will produce a tail

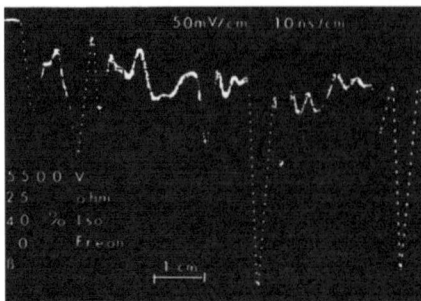 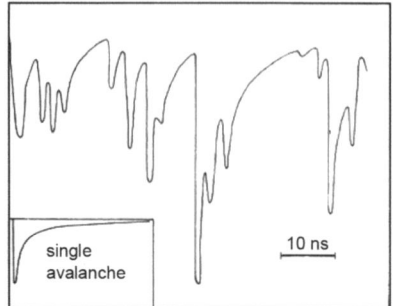

Fig. 4.5 Left: Voltage pulses from a proportional chamber recorded with an oscilloscope. Right: Simulation of avalanching secondary electrons expected in a multi-wire proportional chamber. Both reprinted with kind permission from [18], © 1975, CERN. All rights reserved

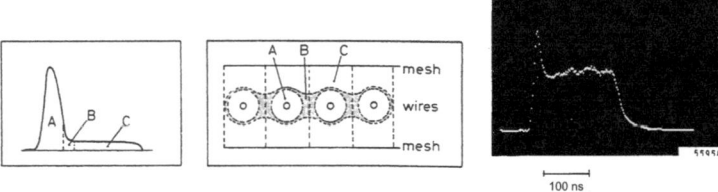

Fig. 4.6 Left: Typical signal pulse of an MWPC wire. Center: Three regions contribute to the signal pulse, the fast drift region (A), the intermediate region (B) and a smooth drift region (C). Both reprinted under CC-BY-3.0 Licence from [17], ©1977–2024, CERN. Right: Measured time distribution of a signal pulse for an inclined beam of minimum ionizing electrons. The observed signal on the anode is induced by the ions slowly moving away from the wire. Reprinted with kind permission from [20], © 1973, Elsevier. All rights reserved

while electrons produced in region C, on the other hand, smoothly drift to the anode where they are amplified and collected. The time resolution of a chamber is defined as the minimum gate width needed on the readout electronics for full efficiency. It is typically around 30 ns for 2 mm wire spacing. Figure 4.6 (right) shows the typical time distribution measured on a single wire for an inclined beam of minimum-ionizing electrons. The long uniform tail in this case corresponds to tracks crossing region C of the considered wire. Here, the gate was set to 160 ns. Optimizing the gate width reduced a large fraction of electrons from region C.

4.1.2 Design Considerations

The rule of thumb for the construction of MWPCs is to choose an anode wire diameter, which is about 1% of the anode wire pitch. Thus, for 2 mm pitch we choose a wire diameter of $d_w \simeq 20\ \mu$m. Above, we listed the capacitance per unit length for these dimensions and an 8 mm anode-cathode gap. Since the spatial resolution in an MWPC improves with smaller pitch (see Fig. 4.3), we would like to place anode wires as close

4.1 Multi-Wire Proportional Chambers

as possible next to each other. However, large MWPCs with an anode wire pitch of less than 2 mm become difficult to operate. For example, an MWPC with 1 mm anode wire pitch and anode wires 20 μm in diameter has $C_w = 1.99$ pF/m. To achieve a similar gain as for an MWPC with an anode wire pitch of 2 mm, we have to keep the charge per unit length $V_0 C_w$ constant. This would require doubling V_0, which would double the electric field in the drift region, bringing us closer to the Raether condition (3.31). Decreasing the anode wire diameter would help, but we would run quickly into mechanical instabilities and electrostatic limitations. Note that chambers with 30 cm^2 sensitive area, 1 mm anode wire spacing, 6–8 mm anode-cathode gaps and gold-plated W wires 20 μm in diameter were operated successfully using a gas mixture of argon-isobutane-freon (79.5:20.0:0.5) at $\sim$ 7 kV [21]. For large (> 1 m^2) MWPCs, typical parameters are [22]: anode wire spacing $s_w = 2$ mm, anode-cathode gap $a_w = 6$–8 mm and anode wire diameter $d_w = 20$ μm.

Standard anode wires are made of gold-plated tungsten or aluminum. The former material is less fragile but adds more inactive material into the chamber reducing the effective radiation length. This increases the probability for both photon conversions that produce new avalanches and elastic and inelastic scattering of the heavy charged particle. In most multipurpose detectors one tries to minimize the inactive material in order to optimize the momentum resolution. Cathode planes are made of stretched metal foil, metal strips or wires such as Cu-Be wires 50–100 μm in diameter. Many wires up to 100 may be ganged together to save readout channels. The chambers are held by rigid light frames typically made of honeycomb plates, G10 or fiber glass containing printed circuit boards. Aluminum frames serve as rigid support structures.

The counting gas in MWPCs typically consists of noble gases (Ar, Kr, Xe) with admixtures of CO_2, methane (CH_4), isobutane (iC_4H_{10}), ethylene (C_2H_4) or ethane (C_2H_6). In addition, often small portions of methylal ($C_3H_8O_2$) are added for quenching and freon (CF_3Br) or ethylbromide (C_2H_5Br) for reaching high amplifications $> 10^7$ without breakdown. The so-called magic gas is an argon-isobutane-freon mixture where the argon fraction is 70–80%, the isobutane fraction is 30–20% and the freon concentration is 0.15–0.4% [23].

The chamber is kept at a certain pressure. So, the volume has to be sealed and kept at constant temperature. The wires and pads are soldered to feedthroughs at the inner wall, which take out the signals and transfer them to preamplifiers. A high electric field is used to reach typical gas amplifications of $10^5 - 10^6$ in the proportional region. The detection efficiency depends on the high voltage and the gate width as shown in Fig. 4.7 (left). Above a certain field strength the detection efficiency reaches a constant value called plateau, which is typically several 100 V wide. Here, chambers are operated with efficiencies near 100%. The gas amplification also depends on the anode wire diameter as shown in Fig. 4.7 (right).

The starting value of the plateau depends on the readout such as the selected gate, a time window during which the signal is collected. Shorter gates yield a better signal-to-noise ratio at reduced efficiency. For wider gates, the plateau is reached at lower electric-field strengths than for shorter gates. The characteristics of the plateau also depends on the gas amplification and the threshold of the amplifier attached to each wire. Typical thresholds are 200–500 μV. Note that at very high electric-field

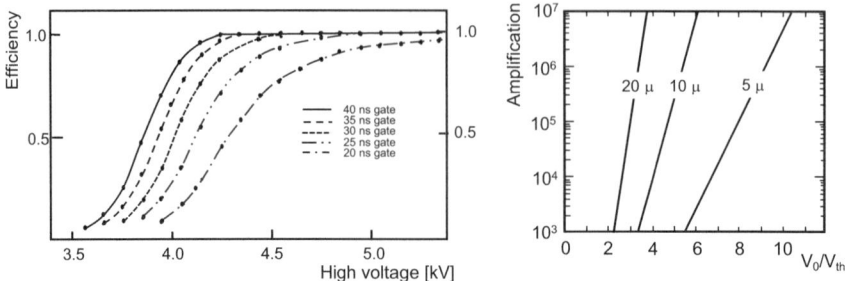

Fig. 4.7 Left: Detection efficiency as a function of applied high voltage for different gate widths of the signal readout electronics. Reprinted with kind permission from [24], © 1971, Elsevier. All rights reserved. Right: The amplification of a proportional chamber as a function of the ratio of operational voltage V_0 to threshold voltage V_{th} for different anode wire radii. Reprinted under CC-BY-3.0 License with kind permission from [17], © 1977–2024, CERN

strengths a chamber breakdown may occur via sparking. Sparking may burn carbon paths on the fiber glass chamber frame. If this occurs, the chamber is shorted and will not operate any longer. Discharges can also lead to anode wire breakage.

4.1.3 Electric-Field Distortions

Distortions of the electric-field due to displaced wires or varying wire diameter affect the performance. Figure 4.2 (right) shows the electric-field configuration and equipotential lines for a set of anode wires where the fifth wire is displaced upward modifying the anode-cathode gap. Displacements in the x direction change the pitch. Other effects include changes of the cathode position and variations of the strip/wire size. To estimate the gain changes caused by variations of the wire radius and wire positions, we differentiate (3.29), obtaining

$$\frac{\Delta A}{A} = \ln A \frac{\Delta Q}{Q}. \tag{4.8}$$

We can compute $\Delta Q/Q$ in terms of tolerances in the wire radius and variations in the anode-cathode gap. Recalling that $Q = C_w \cdot V_0$ with C_w given in (4.3), we obtain

$$\frac{\Delta Q}{Q}\bigg|_{d_w} = \frac{C_w}{2\pi\epsilon_0} \frac{\Delta d_w}{d_w}, \tag{4.9}$$

$$\frac{\Delta Q}{Q}\bigg|_{a_w} = \frac{C_w}{2\epsilon_0} \frac{a_w}{s_w} \frac{\Delta a_w}{a_w}, \tag{4.10}$$

$$\frac{\Delta Q}{Q}\bigg|_{s_w} = \frac{C_w}{2\pi\epsilon_0} \left(1 - \frac{\pi a_w}{s_w}\right) \frac{\Delta s_w}{s_w}. \tag{4.11}$$

4.1 Multi-Wire Proportional Chambers

Consider for example a typical chamber with parameters $a_w = 8$ mm, $d_w = 20$ μm, $s_w = 2$ mm and a gain of 10^6. The gain changes become

$$\frac{\Delta A}{A}\bigg|_{d_w} \approx 0.9 \frac{\Delta d_w}{d_w}, \quad (4.12)$$

$$\frac{\Delta A}{A}\bigg|_{a_w} \approx 11 \frac{\Delta a_w}{a_w}, \quad (4.13)$$

$$\frac{\Delta A}{A}\bigg|_{s_w} \approx 10 \frac{\Delta s_w}{s_w}. \quad (4.14)$$

Thus, a 1% change in the wire diameter leads to a 0.9% change in the gain while a 1% wire shift in the x and y directions yields a 11% and 10% gain change, respectively. Figure 4.8 (left) shows the effect of a displacement of wire 0 in the x-direction on $\Delta Q/Q$ for wire 0 and neighboring wires. The effects on wires ± 1 are largest. The relative charge change on wire 0 is smaller than that on wires ± 2. For a gain of 10^6 and a displacement of 0.1 mm, this introduces charge changes of about 0.8% on wires ± 1 and about 0.2% on wires ± 2, leading to gain changes of 9% and 2%, respectively. Figure 4.8 (right) shows the effect of the displacement of wire 0 in the y-direction on $\Delta Q/Q$ for wires 0 and 1. Here, the effect on wire 0 is largest. For a 0.25 mm displacement the change in $\Delta Q/Q$ is 1% leading to a gain change of 10%.

A problem specific to large chambers is the mechanical instability of anode wires due to electrostatic repulsion. If an anode wire is slightly displaced from the middle plane by $\delta_z > 0$ due to an inaccurately drilled hole, it will be pulled more into the positive than the negative z direction. The wire tension counteracts the electrostatic force. In large chambers at high potential, however, the wires will buckle even if the wires are positioned perfectly. If wire 0 is pulled up, wires ± 1 will be pulled down as illustrated in Fig. 4.9. The wires are no longer straight but take a parabolic shape, so that the displacement is largest at half the wire length. The charges induced on neighboring wires are opposite. The critical length of a chamber for wire stability was calculated in reference [25]. The magnitude of the force between two equal

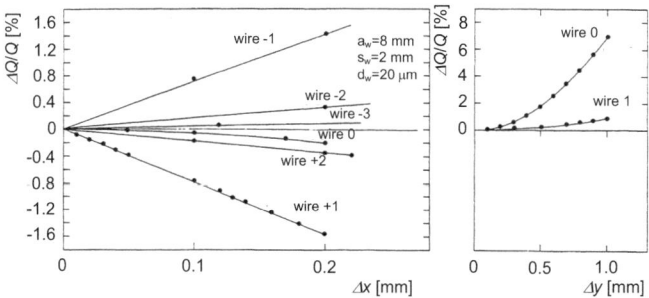

Fig. 4.8 Left: The $\Delta Q/Q$ variation in different wires caused by a displacement of wire 0 in the x-direction. Right: The $\Delta Q/Q$ variation in wires 0 and 1 caused by a displacement of wire 0 in the z-direction. Both reprinted under CC-BY-3.0 License with kind permission from [17], © 1977–2024, CERN

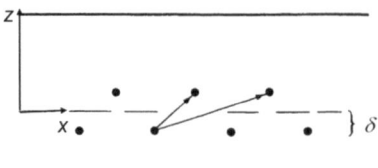

Fig. 4.9 Wire displacements due to electrostatic instabilities in a large MWPC (> 1 m) in the *x-z* plane. Reprinted under CC-BY-3.0 License with kind permission from [17], © 1977–2024, CERN

linear charges $C_w V_0$ at a distance r per unit length in which the electric field is given by (4.6) and is not modified due to the small displacement δ_z is

$$F(r) = \frac{(C_w V_0)^2}{2\pi \epsilon_0} \frac{1}{r}. \tag{4.15}$$

The total force on a given wire per unit length perpendicular to the wire plane is given by [25],

$$\sum F_\perp = 2 \frac{(C_w V_0)^2}{2\pi \epsilon_0} \left\{ \frac{1}{s_w} \frac{2\delta_z}{s_w} + \frac{1}{3 s_w} \frac{2\delta_z}{3 s_w} + \ldots \right\} = \frac{\pi (C_w V_0)^2}{4\epsilon_0} \frac{\delta_z}{s_w^2}. \tag{4.16}$$

If T_w is the mechanical tension of the wire, the restoring force perpendicular to the wire plane is

$$F_R = T_w \frac{d^2 \delta_z}{dy^2}, \tag{4.17}$$

where $\delta_z = \delta_z(y)$ is the displacement of the wire along its length with boundary conditions of $\delta_z(0) = \delta_z(\ell_w) = 0$ if ℓ_w is the total wire length. In equilibrium, the restoring force compensates $\sum F_\perp$, yielding

$$T_w \frac{d^2 \delta_z}{dy^2} = -\frac{\pi (C_w V_0)^2}{4\epsilon_0} \frac{\delta_z}{s_w^2}. \tag{4.18}$$

The solution of the second-order differential equation is

$$\delta_z(y) = \delta_0 \sin\left(\frac{C_w V_0}{2 s_w} \sqrt{\frac{\pi}{\epsilon_0 T_w}} y\right). \tag{4.19}$$

The boundary condition $\delta_z(\ell_w) = 0$ allows us to determine the required tension

$$\frac{C_w V_0}{2 s_w} \sqrt{\frac{\pi}{\epsilon_0 T_0}} \ell_w = \pi, \tag{4.20}$$

yielding

$$T_0 = \frac{1}{4\pi \epsilon_0} \left(\frac{C_w V_0 \ell_w}{s_w}\right)^2. \tag{4.21}$$

4.1 Multi-Wire Proportional Chambers

We achieve a stable condition if

$$T_m > T_w > T_0 = \frac{1}{4\pi\epsilon_0}\left(\frac{C_w V_0 \ell_w}{s_w}\right)^2, \quad (4.22)$$

where T_0 is the tension defined by the parameters of the chamber and T_m is the elastic limit. The other parameters are the same as those in (4.1). So, the critical stability length is given by

$$\ell_c = \frac{s_w}{C_w V_0}\sqrt{4\pi\epsilon_0 T_m}. \quad (4.23)$$

For example, for a typical MWPC with $d_w = 20$ μm, $s_w = 2$ mm and $a_w = 8$ mm the elastic limit of the wire is $T_m = 0.65$ N for $V_0 = 5$ kV yielding a critical length of about $\ell_c = 90$ cm. Since the elastic limit decreases with the diameter of the anode wire as d_w^2, the critical wire length is proportional to d_w. For larger chambers we need a mechanical support. One solution is to weave a thin nylon wire across the anode wires perpendicular to their direction. However, the electric field $\vec{E}$ of the anode wires is disturbed in the region near nylon wire causing an inefficient zone of 5 mm width around the nylon wire. In some very large chambers, solid support bars have been used.

Another effect of the electrostatic forces in an MWPC is the overall attraction of the cathode planes towards the anode planes reducing the gap in the center of the chamber. The calculation is tedious and thus omitted here. An approximation is given in [12]. For our typical chamber with parameters listed above, the maximum deflection is 220 μm for $V_0 = 4.5$ kV.

4.1.4 Performance

Multi-wire proportional chambers are robust, reliable, easy-to-build and rather inexpensive large-area detectors, which yield reasonable position resolutions. Thus, they have been used in many experiments. Figure 4.10 (left) shows a schematic view of a large self-supporting MWPC [26]. The frame consisted of fiber glass. This chamber used three layers with wires running horizontally in the first plane, vertically in the second plane and at an angle of 15° in the third plane. The counting gas was an argon-isobutane gas mixture (80:20) plus methylal as quencher. Using an amplifier threshold of 5 mV on a 3.3 kΩ resistor and a gate width of 60 ns, the plateau at 100% efficiency was larger than 700 V. The operating voltage was 4.7 kV. A spatial resolution of $\sigma_x = 0.6/\cos\lambda_d$ mm was measured where $\lambda_d = 90° - |\theta|$ is the dip angle. The gas gain depends on the ratio of operational voltage to threshold voltage and on the anode wire diameter. For thicker wires the same gain is achieved with a lower V_0/V_{th} ratio. An another large MWPC (200 cm $\times$ 50 cm) was built and tested at CERN by a different group [24]. Figure 4.10 (right) shows a photograph of an LHCb MWPC.

Table 4.1 summarizes parameters and properties of MWPCs used as tracking detectors in some high-energy physics experiments. In some experiments the

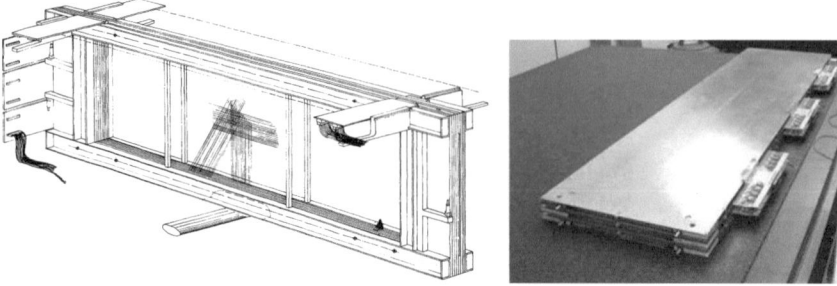

Fig. 4.10 Left: View of a large MWPC (39 cm × 154 cm). The chamber has three independent gaps with sense wires running horizontally, vertically and at 15° with respect to the vertical axis. Reprinted with kind permission from [22], © 1971, Elsevier. All rights reserved. Right: Photograph of an LHCb MWPC used for the muon system. This chamber is in the first muon layer (M1) in the outermost region (R4), which has dimensions of 20 cm × 96 cm. A total of 192 chambers covers the R4 region in layer M1. Reprinted with kind permission from [28], © 2008, IOP Publishing. All rights reserved

MWPCs are used for the trigger in the tracking or muon systems. So, here the spatial resolution is compromised for a reduced number of readout channels as seen in the SELEX experiments and the LHCb muon system. The best performance is achieved with Charpak's MWPC, which measured a spatial resolution of 45 μm in the direction along the anode wires. Next, we consider the Large-Area Solenoid Spectrometer (LASS) experiment [27] at SLAC.[1] The upgraded experiment used three planar and six cylindrical MWPCs. For planar chambers the wires were parallel to the upper frame edge while the cathode strips had a 45° stereo angle with respect to the anode wires. For the cylindrical chambers a small stereo angle was provided by helical cathode strips at ±10° for chambers 1–4 and ±15° for chambers 5 and 6. The LASS MWPCs achieved a reasonable spatial resolution. The second MWPC system was installed into the NA24 experiment [29] at CERN.[2] For tracking they used five layers of 1 m × 1 m and three layers of 2 m × 2 m MWPCs.

The SELEX experiment [30] at Fermilab[3] used three sets of MWPCs for tracking. Each set consisted of multiple chambers with multiple layers, having both axial (X,Y) and stereo (U,V) layers. The stereo angle was ±28.07°. The gold-plated tungsten sense wires were spaced at a 2 mm pitch, which was increased to 3 mm in some chambers. The HERMES experiment [31] at DESY[4] used three planes of MWPCs in the magnet region. The Cello experiment used cylindrical proportional chambers that were positioned in the tracking system [32]. The LHCb experiment used 1380

---

[1] LASS operated 1976–82 using 11 GeV/c charged kaon beams to study strange and non-strange lower-mass resonances.
[2] NA24 was a prompt photon production experiment operating with 300 GeV/c pion and proton beams 1980–84.
[3] SELEX studied properties of charmed baryons operating 1996–97.
[4] HERMES studied the proton spin at HERA running from 1995–2007.

4.1 Multi-Wire Proportional Chambers

Table 4.1 Properties of some MWPCs used in sub-detectors of different experiments listing the sensitive area in $x \times y$ for planar chambers and radius/length for cylindrical chambers, maximum unsupported sense wire length (ℓ_w), anode-cathode gap (a_w), sense wire diameter (d_w), sense wire spacing (s_w), cathode wire/strip spacing (c_w), high voltage, gas mixture and position resolution ($\sigma_{x(r\phi)}/\sigma_{y(rz)}$). The area denotes the sensitive area. The magic gas is a mixture of Ar-isobutane-methylal-freon in a ratio ($\sim$75:21:4:0.25). The upper part of the table shows planar chambers and the lower part cylindrical chambers. For planar chambers, the position resolution denotes σ_x and σ_y if the y position is measured. For cylindrical chambers, the position resolution refers to $\sigma_{r\phi}$ and σ_z. X and Y are vertical and horizontal planes while U and V are stereo layers with angles of $\pm 28.07°$. [†]The resolution is measured with X-rays; replacing argon with xenon improves the spatial resolution to 35 μm. [&]All chambers have two planes of cathode readout with strips arranged under $\pm 45°$ stereo angles. [#]In the central region three planes have anode wire spacing of 1 mm. [‡]The first pair is for the innermost region of station M3, the second pair is for the outermost region of M2; for all other regions σ_x and σ_y lie between these boundaries. [##]In muon stations M2 (M3) in the inner regions wires are combined to strips with sizes 6.3 mm (6.7 mm) and 12.5 mm (13.5 mm). [*]SELEX also has smaller MWPCs in magnets M2 and M3. [⋆]Listed are the smallest and largest pad sizes of the cathode planes; in the the different regions of the muon stations anode wires, cathode pads or both are read out. [§]This is for layers 1–4; for layers 5 and 6 the length is 87 cm

Experiment system	sens. area [cm × cm]	ℓ_w/a_w [cm]/[mm]	d_w [μm]	s_w/c_w [mm/mm]	HV [kV]	gas (fractions)	$\sigma_{x,r\phi}/\sigma_{y,z}$ [mm/mm]
Charpak [38,41] X, Y	small	–/8	10	1–2/0.5	3.9	Xe-HC(CH$_3$)$_3$-C$_3$H$_8$O$_2$ (55:38:7)	0.15–0.3/ 0.045[†]
LASS [27,44][&] gap PWC	155 × 155 3 chambers	31.5/5.1	20	2[#]/8.1	6.0	magic gas	≤ 0.3/ ≤ 0.3
NA24 [29] 3 × (X, Y, U, V)	200 × 200	66/6	25 75	3/–	2.97	Ar-CO$_2$-freon (80:20:0.15)	0.87 (wire)/
SELEX M1 [30] 3 × (X, Y, V, U)	200 × 200[*]	70/6	25	3/–	3.0	Ar-C$_4$H$_{10}$-freon (75:24.5:0.5)	0.6/–
LHCb [45] Muon M1–M5	30 × 25 to 129.6 × 27	30/2.5	30	1.5[##]/ 37.5 × 31.3[⋆] 54 × 270[⋆]	3.15	Ar-CO$_2$-CF$_4$ (40:55:5)	10/12[‡] to 60/100[‡]
HERMES MC1 –MC3 [31,46]	99.6 × 26.3 142.4 × 34.7	–/4	25	2/0.5	2.85	Ar-CO$_2$-CF$_4$ (65:30:5)	0.7/0.7
LASS [27,44] Cylindrical	6–49 × 100[§] 6 layers	50 & 43.5/ 4.8–5.8	20	2/5–10	6.0	magic gas	≤ 0.3/0.7–1.5 wire: 0.6
Cello [32] tracking	17–70 × 220 5 layers	220/4	20	2.09–2.86 /4.5	2.1	magic gas	0.179/ 0.44

MWPCs in five stations[5] of the muon system [28] with 276 chambers in each station, which is subdivided further into four regions. The cathode planes use pad or wire-pad readouts. Multi-wire proportional chambers have been used also as readout systems in other sub-detectors, for example, in Ring-Imaging Cherenkov detectors [33,34] and in time projection chambers [35–37].

4.1.5 Position Measurement

One plane of an MWPC in which the cathodes are not read out provides a position measurement only in the direction perpendicular to the wires (x). To measure x and y positions simultaneously, we need two chambers with anode wires planes placed perpendicular to each other. Another option is to make the cathode out of wires or metal strips that run orthogonal to the anode wires and that are also read out. Figure 4.11 (left) shows the layout of an MWPC with cathode readout [38,39]. The upper (lower) cathode has wires running parallel (perpendicular) to the anode wires. To reduce the number of readout channels, six cathode wires are ganged together while all anode wires are connected. Typically, anode wires are read out individually, in groups or not at all. In this chamber the x and y position measurements come from the cathode wires. Pulses are induced on the cathode wires by the positive ions. The magnitude varies with the distance between the avalanche position and the wire. On each readout line the size of the signal is sketched. Cathode wires produce positive pulses with signal heights decreasing with distance from the ionization point while the anode wires produce a negative pulse. Figure 4.11 (right) shows another layout of an MWPC with cathode readout. Here, the anode wires run under a stereo angle while the top and bottom cathode strips run in the x and y directions, respectively. In this chamber all wires and strips are read out. While only one anode wire records the signal, several x and y cathode strips see the signal allowing for a precise position measurement.

The x position is determined by the center-of-gravity (COG) method either using the charge recorded on individual anode wires or cathode strips running parallel to the wires [39],

$$x = \frac{\sum_i (Q_i - b) x_i}{\sum_i (Q_i - b)}, \tag{4.24}$$

where Q_i is the charge measured on wire or strip i at position x_i and b is a small bias level offset to correct for noise. Typically, the offset is determined from a fit to the data. Similarly, the y position is obtained by using the cathode strips running perpendicular to the anode wires,

$$y = \frac{\sum_j (\tilde{Q}_j - \tilde{b}) y_j}{\sum_j (\tilde{Q}_j - \tilde{b})}, \tag{4.25}$$

---

[5] For run 2 layer M1 was removed.

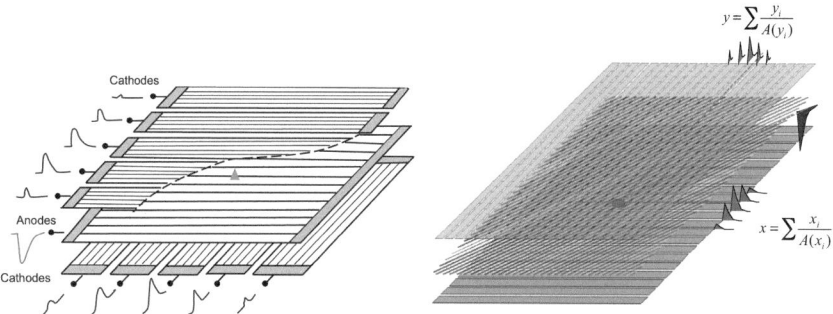

Fig. 4.11 Left: Schematic view of an MWPC with cathode readout [40,41]. The upper (lower) cathode strips run parallel (perpendicular) to the anode wires. To reduce the number of readout channels, all anode wires and rows of six cathode strips are combined. The triangle shows the position of the primary ionization. Reprinted with kind permission from [41], © 1978, Elsevier. All rights reserved. Right: Schematic view of an MWPC in which all anode wires running under a stereo angle and cathode strips running in x and y directions are read out. Note that the cathode strips see charge distributions on both strips in the x and y directions while only one anode wire records a signal. Reprinted with kind permission from [42], © 2014, F. Sauli. All rights reserved

where $\tilde{b}$ is another offset and y_i is the strip position in the y direction.

Another algorithm for determining the position is the head-tail algorithm, which is used for solid state detectors [43].

$$x = \frac{x_R + x_L}{2} + \frac{Q_R - Q_L}{2\bar{Q}} s_w, \qquad (4.26)$$

where $x_{R,L}$ are the positions of the right- and left-most pixel, $Q_{R,L}$ are their measured charges, $\bar{Q}$ is the average charge and s_w is the wire or strip pitch.

Charpak measured the position resolution using three chambers. The chambers were equipped with gold-plated tungsten wires 15 μm in diameter that were spaced at a 2.54 mm pitch [39,41]. The cathode planes used wires 100 μm in diameter separately spaced every 1.27 mm. The chambers were operated with a magic gas mixture at 3.9 kV. The two outer chambers defined the expected position. From the measured position in the central chamber residuals were extracted. The anode-cathode gap was 5 mm and the chamber had an active area of 100 mm × 100 mm. Charpak used 1.5 GeV/c test beam particles to study the chamber performance. Figure 4.12 (left) shows the residual distribution as a function of the y position. The distribution is uniform and has residuals of 100 μm at FWHM. This leads to a spatial resolution in y direction of 45 μm. Replacing argon with xenon leads to a further improvement in position resolution of 35 μm. Figure 4.12 (right) shows the charge distribution versus y position for the same MWPC operated with a xenon gas mixture. The MWPC recorded collimated soft X-rays that were moved twice by 200 μm.

Figure 4.13 (left) shows the correlation of the measured versus true center-of-gravity x position. There is clearly a clustering effect visible although charge sharing between adjacent wires introduces a certain amount of interpolation and a moderate

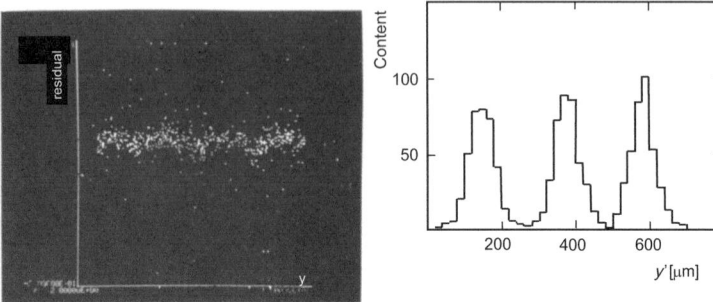

Fig. 4.12 Left: Measured residuals in the y direction along the anode wire for uniform irradiation of the chamber. The vertical dispersion has a FWHM of 100 μm. Reprinted with kind permission from [39], © 1979, Elsevier. All rights reserved. Right: Charge distribution of a proportional chamber with cathode readout irradiated with a 1.4 keV X-ray source at three positions separated by 200 μm. Reprinted with kind permission from [41], © 1978, Elsevier. All rights reserved

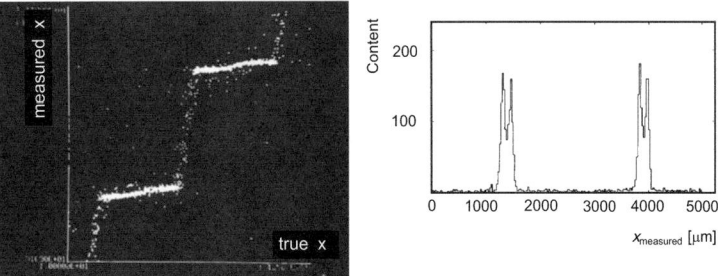

Fig. 4.13 Left: Correlation of the true (horizontal) and measured coordinate in the x direction for uniform irradiation perpendicular to the chamber. Note the quantization effect of the anode wires on the center-of-gravity. Right: Projection of the correlation plot onto the vertical axis showing clearly the left-right separation as two partially overlapping peaks. Both reprinted with kind permission from [39], © 1979, Elsevier. All rights reserved

slope of the clustering caused by the left-right effect. This becomes clearer if we look at the projection of the scatter plot on to the vertical axis as depicted in Fig. 4.13 (right). The two peaks correspond to tracks that cross the anode plane at the right-hand side and left-hand side of the wire, respectively. We can unfold the clustering effect by combining the left-right identification with the drift time measurement,

$$\Delta x = \pm v_d^e \Delta t, \qquad (4.27)$$

where Δt is the drift time measurement and $v_d^e = 5.2 \, \text{cm}/\mu\text{s}$ is the drift velocity. After correcting the measured position using (4.27) we obtain a linear relation between true and measured positions as shown in Fig. 4.14 (left), which in turn improves the spatial resolution to 120 μm as shown in Fig. 4.14 (right). If the spatial resolution is just taken from the COG distribution, we obtain $\sigma_x = 200 \, \mu\text{m}$ [39].

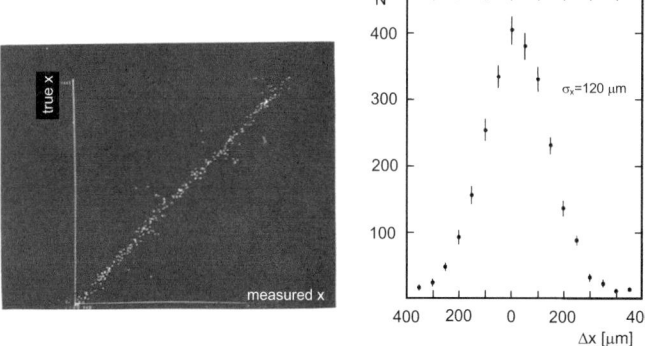

Fig. 4.14 Left: Correlation of the measured versus true x coordinate after corrections based on the drift time. Please note that the horizontal and vertical axes are interchanged with respect to the correlation plotted in Fig. 4.13 (left). Right: Distribution of measured minus true x position after combining the right-left identification given by the COG method with the drift-time measurement. Both reprinted with kind permission from [39], © 1979, Elsevier. All rights reserved

Without cathode readout, the position along the wire can be obtained via charge division requiring readout of both wire ends. For charge measurements Q_A and Q_B on the two ends of a wire, we obtain the position along the wire

$$y = \ell_w \frac{Q_A - b}{Q_A + Q_B - 2b}, \tag{4.28}$$

where ℓ_w is the wire length. The accuracy of the y position is $\sim 0.4\%$ of the wire length, which is much worse than reading out cathode strips directly.

4.1.6 Gas Chamber Aging

The operating gas of an MWPC typically contains organic components like methane or higher C_nH_{2n} molecules. Under the conditions of avalanche formation, the organic molecules dissociate and may form radicals. For example, electrons may trigger reactions like $e^- + CH_4 \to CH_2 + H_2 + e^-$ as depicted in Fig. 4.15 (left). The CH_2 radicals are ready to polymerize and form higher-complexity molecules containing chains of carbon and hydrogen, which can be liquid or solid. They may form whiskers on the cathode or polymers that get stuck on the anode wire. Figure 4.15 (right) schematically shows polymer chain formation by CH_2 radicals. Figure 4.16 (left, center) show whiskers growth on anode wires. Figure 4.16 (right) shows deposits on the cathode.

Figure 4.17 sketches the different aging effects on the anode and cathode. Deposits on the anode may be conductive (C) or insulating. This leads to an effective increase of the wire diameter, which in turn leads to a gain reduction (see Fig. 4.17 (left)). On the cathode the formation of a thin insulating layer causes an accumulation of i^+ that are produced in an avalanche (see Fig. 4.17 (center left)). The ion neutralization is slowed down depending on the resistivity. The high dipole field created by the ions may result in an emission of electrons from the cathode into the gas (see Fig. 4.17 (center)). This

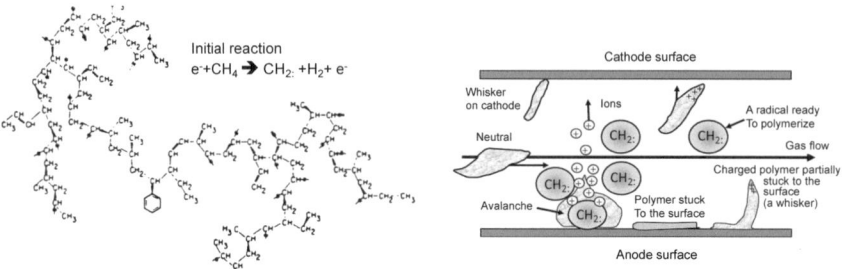

Fig. 4.15 Left: Initial reaction of $e^- + CH_4 \rightarrow CH_{2:} + H_2 + e^-$, where $CH_{2:}$ is a radical. Right: Polymer chain formation started by a CH_2 radical. Both reprinted with kind permission from [47], © 2001, J. Va'vra. All rights reserved

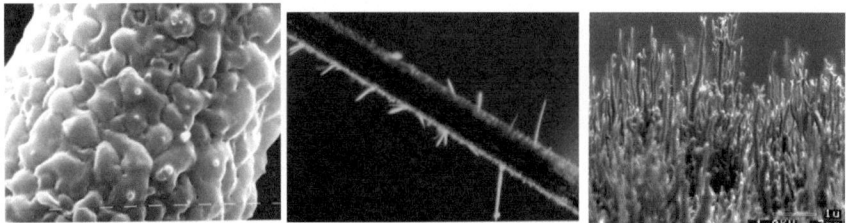

Fig. 4.16 Left: Formation of polymers on the anode wire. Reprinted with kind permission from [48], © 1986, LBNL. All rights reserved. Center: Growth of spikes on the anode wire. Reprinted with kind permission from [49], © 1986, LBNL. All rights reserved. Right: Silicon growth on the cathode. Reprinted with kind permission from [50], © 2003, Elsevier. All rights reserved

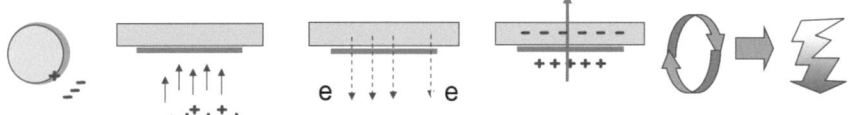

Fig. 4.17 Left: Deposits on the anode wire lead to an increase of the wire diameter, which in turn lead to a gain reduction. Center left: The formation of a thin insulating layer causes an accumulation of positive ions that are produced in an avalanche. Center: The high dipole field created by the ions may result in an emission of electrons from the cathode into the gas called Malter effect [51]. Center right: These electrons generate more avalanches near the anode and thus produce more ions that eventually reach the cathode accumulating on the deposit. Right: When the ion production rate exceeds the neutralization rate, the conditions become unstable and may lead to sparking. All reprinted with kind permission from [52], © 2009, F. Sauli. All rights reserved

is called the Malter effect [51]. These electrons generate more avalanches near the anode and thus produce more ions that eventually reach the cathode accumulating on the deposit (see Fig. 4.17 (center right)). When the ion production rate exceeds the neutralization rate, the conditions become unstable and may lead to sparking (see Fig. 4.17 (right)), which eventually destroys the chamber. So, the use of non-organic gases avoids the formation of Carbon-based polymers.

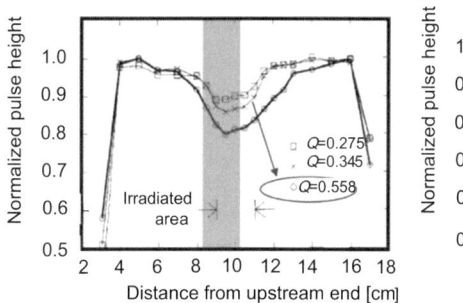 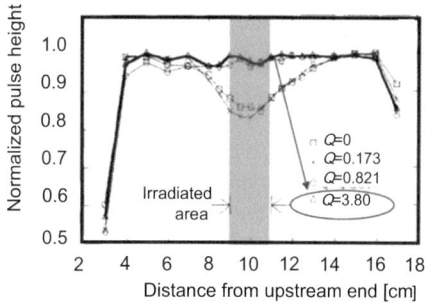

Fig. 4.18 Left: The measured pulse height in an Ar-C_2H_6 mixture after irradiation showing a loss in pulse height in the irradiated region already at low doses. Right: The measured pulse height in a CF_4-iC_4H_{10} gas mixture. Even for high irradiation levels the pulse height remains unaffected. Both reprinted with kind permission from [53], © 1991, Elsevier. All rights reserved

Adding carbon tetrafluoride (CF_4) into gas mixtures is attractive because of its fast drift velocity, low diffusion coefficient and low neutron cross section. Since CF_4 is very reactive, it prevents polymer formation and even removes existing deposits. The etching is particularly efficient for Si compounds. Note that CF_4 is used by the semiconductor industry to etch silicon and SiO_2 [54]. However, the relative balance between polymer formation and etching is difficult to control. Fluorine radicals released in avalanches can combine with water to form HF, which is a strong acid that attacks glass, epoxy and wires. Figure 4.18 (left) shows the measured pulse height in an Ar-C_2H_6 mixture after irradiation. Even for low dose, the pulse height in the irradiated area drops by 10%. For twice the dose it drops by 20%. This is different for a CF_4-iC_4H_{10} gas mixture as shown in Fig. 4.18 (right). Even for high irradiation the pulse height is unaffected. The chamber region that showed irradiation effects when operated without the CF_4 admixture, looks fine after adding CF_4. Further details of MWPCs are given in [42].

4.2 Planar Drift Chambers

In the development of MWPCs, it was realized that spatial information could be obtained also by measuring the drift time of electrons coming from an ionizing event. Note that Charpak used drift time information in improving the spatial resolution with respect to the center-of-gravity method. The start time t_0 is the time when the primary ionization occurs and it is correlated with the time when an interaction occurred. The stop time t_1 is time when the electron enters the high electric field generating an avalanche, which is correlated with the rising edge of the anode pulse. For the drift time $\Delta t = t_1 - t_0$ the drift distance is obtained by

$$z = \int_{t_0}^{t_1} v_d^e(t) dt, \qquad (4.29)$$

where $v_d^e(t)$ is the magnitude of the electron drift velocity. For constant drift velocity the integral simplifies to

$$z = v_d^e(t_1 - t_0) = v_d^e \Delta t. \qquad (4.30)$$

Constant drift velocity is obtained from a constant electric field. Note that this cannot be achieved in a conventional proportional chamber because of the zero field region between anode wires. We need to introduce a field wire at negative high voltage between anode wires. Here, we obtain almost a linear relation between drift time and drift path. Figure 4.19 shows the drift field and isochrones (contours of equal drift time) of a drift chamber (bottom) in comparison to that of an MWPC (top). Whereas the MWPC has no field lines in the middle between two anode wires, the drift chamber does. In a large region of a drift cell the electric-field lines are parallel producing a homogeneous field. This is achieved by inserting field wires around the anode wire. This principle was noticed by the Charpak [13] and Walenta-Heintze groups [55].

A complete linear relation requires an $\vec{E}$ field as constant as possible over the entire drift space. This is achieved with different configurations. Here, we present two examples. Figure 4.20 shows a grid of field wires that surround the anode wire. The potential is adjusted linearly from zero near the anode to -HV1 at both ends of the drift cell. The anode wire is kept at positive high voltage (few kV). As indicated by the equipotential lines this produces a constant electric field except in the vicinity of the anode wire. Figure 4.21 shows another configuration in which cathode planes are kept at a fixed potential but a shaping potential is created inside the cell by an appropriate number of field wires. The anode is kept at 3.7 kV while the cathode is at −2.0 kV. Figure 4.22 shows the equipotential lines in one quadrant of the latter drift cell. The electrons see a nearly constant field in the entire drift cell except near

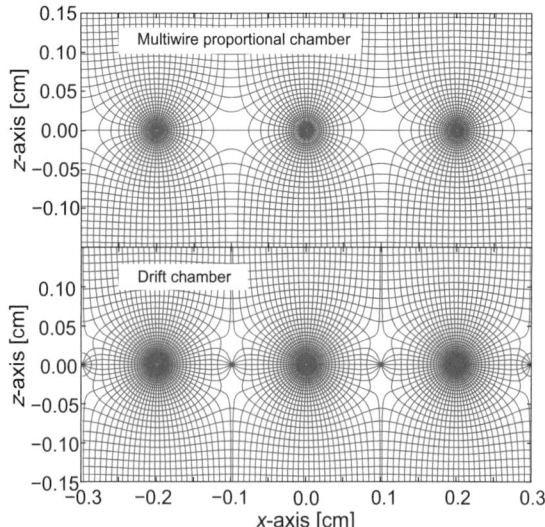

Fig. 4.19 Top: Electric-field lines and isochrones in an MWPC. Bottom: Electric-field lines and isochrones in a drift chamber. While in an MWPC there is a field-free region between two sense wires, this is not the case in a drift chamber due to the field wires. Reprinted under CC-BY-NC-4.0 Licence with kind permission from [56], © 2024, the Particle Data Group LBNL

4.2 Planar Drift Chambers

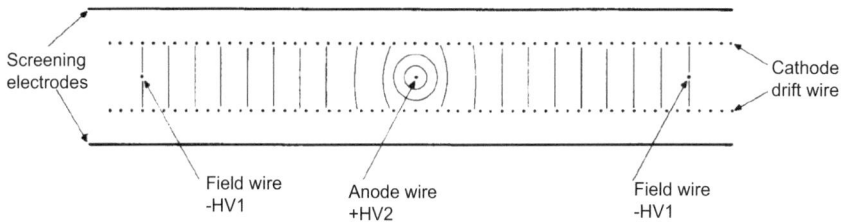

Fig. 4.20 Drift cell consisting of a grid of field wires that surround an anode wire. The potential is adjusted linearly from zero near the anode to -HV1 on the left and right ends of the drift cell. Reprinted with kind permission from [57], © 1975, Elsevier. All rights reserved

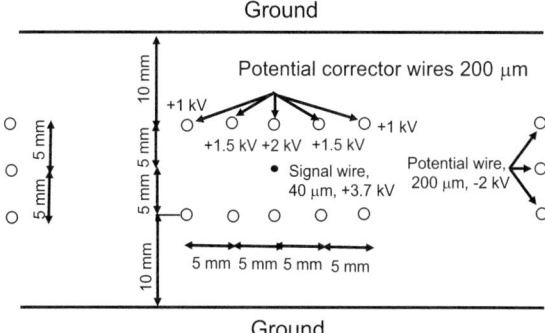

Fig. 4.21 Drift cell in which the cathode planes are kept at a fixed potential. The shaping potential inside the cell is built up by an appropriate number of field wires. Reprinted with kind permission from [58], © 1977, Elsevier. All rights reserved

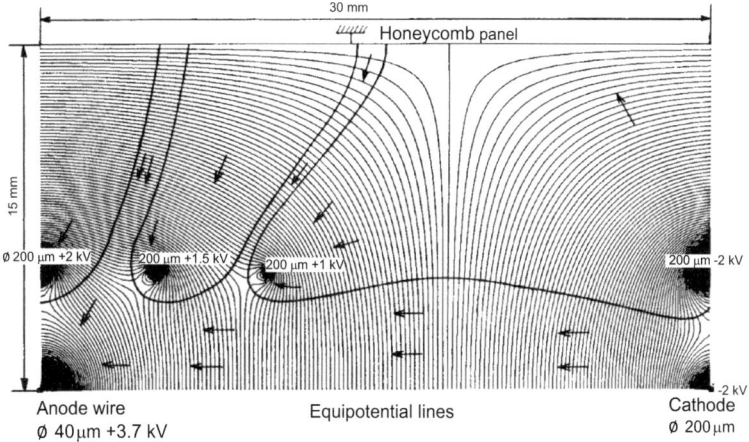

Fig. 4.22 Equipotential lines and electron drift directions (arrows) for the drift cell configuration in Fig. 4.21. Reprinted with kind permission from [58], © 1977, Elsevier. All rights reserved

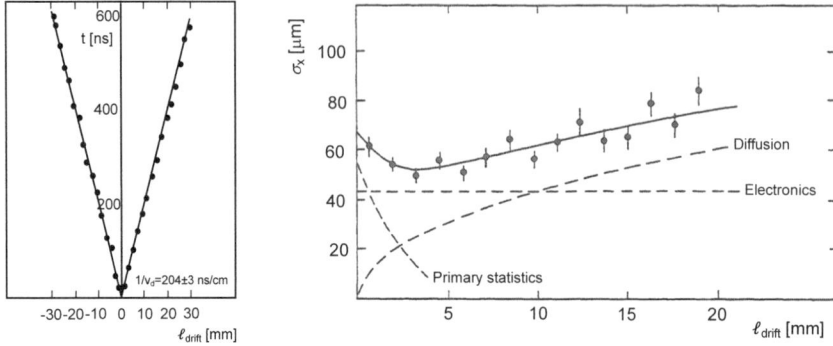

Fig. 4.23 Left: Time-to-distance relation for the drift cell configuration in Fig. 4.21. Reprinted with kind permission from [58], © 1977, Elsevier. All rights reserved. Right: Position resolution for the drift cell in Fig. 4.21 as a function of the drift distance for measurements (points) and a fit (solid line). The contributions from electron statistics, diffusion and electronic dispersion are shown also by dashed lines. Reprinted under CC-BY-3.0 License with kind permission from [17], © 1977–2024, CERN

the cell boundaries and close to the anode wire. The arrows indicate the electron drift direction. Figure 4.23 (left) shows measurements of the drift time versus drift distance. The measurements verify that the relation between drift time and drift distance is linear across the entire cell. While the field wires are just used to define the electric field inside the drift cell, the anode wires carrying the signal are read out.

Another important issue is the choice of gas for various reasons:

(i) The dependence of the drift velocity on the electric field: this is important for obtaining a linear relation between drift path and drift time. For some gas mixtures, v_d^e is nearly independent of $|\vec{E}|$ (see Fig. 2.48). For example, in a gas mixture of Ar-C_4H_{10} (80:20) the drift velocity is nearly independent of the electric field in the region $0.5 \leq |\vec{E}| \leq 2.0$ kV/cm. If we operate drift chambers in a configuration in which v_d^e does not depend on $|\vec{E}|$, field inhomogeneities and changes in temperature, pressure and operating voltage become less important.
(ii) Gas purity: we need to eliminate electro-negative gases.
(iii) Magnitude of the drift velocity: high drift velocities are useful for operating a drift chamber at high rates to minimize the dead time. So it is useful to use gas mixtures containing CF4. Slower drift velocities help to optimize the spatial resolution since timing errors are minimized. So here, gas mixtures containing DME, CO_2 or He-C_2H_6 are useful.

4.2.1 Drift Chamber Properties

Large planar drift chambers consist of many (100) such drift cells. Typically different layers are arranged in one chamber. The anode wires run in different directions, e.g. in x direction, y direction and under stereo angles. Thus, these chambers measure

4.2 Planar Drift Chambers

Table 4.2 Examples of planar drift chambers including the large planar drift chambers of the CDHS detector, three drift chamber planes of the NA32 experiment used in the electron spectrometer, the middle muon chambers (MM) of the L3 experiment and the planar drift chambers in the forward track detector (FTD) of the H1 experiment. The notation is $w \times \ell$ for width×length of an individual chamber, total area for the total coverage with chambers in the detector, number of sense wires per cell, number of cells per chamber, sense wire spacing s_w, diameter of a sense wire/grid wire d_w/d_c, maximum drift length ℓ_{drift}, type of z-measurement, gas type, gas type fractions, electric-field strength $|\vec{E}|$, magnetic-field strength $|\vec{B}|$, drift velocity v_d^e, efficiency, position resolution in x, double track resolution and resolution in the time-to-distance relation σ_t. [†]The drift chambers DC1 and DC2 have six layers (X, Y, U, V, X, Y) and (Y, X, U, V, Y, W) while DC3 has five layers (Y, XU, V, W), where the stereo angles are $U = 45°$, $V = -45°$ and $W = 15°$. [‡]This is the typical efficiency; individual channels were lower

Experiment	CDHS [58,59]	NA34[†] [60]	L3 [61,62]	H1 [63–66]
System	Tracking	e^- spectrometer	Muon MM	FTD (planar)
$w \times \ell$ [cm × cm]	400 × 400	DC1: 24 × 49.5	163 × 556	disk: $r_i = 14$
	hexagonal	DC2: 50.4 × 43	3 layers	$r_o = 78.5$
		DC3: 75.6 × 34		
Total area m^2	720	0.4	~ 730	5.6
# sense wires/cell	1	6 or 8	16 & 24	4
# cells/chamber	130	8–18	16	32
s_w [mm]	60	6	9	6.0
d_w/d_c [µm/µm]	40/200	20/100	30/75	40/125
ℓ_{drift} [mm]	30	15–21	50	28.1
z-coordinate	Stereo	Stereo	Spec. cham.	Spec. cham.
Gas type	Ar-iC$_4$H$_{10}$	CO$_2$-C$_2$H$_6$	Ar-C$_2$H$_6$	Ar-C$_3$H$_8$
Gas fractions	(70:30)	(90:10)	(62:38)	(90:10)
$\vec{E}$ field [kV/cm]	≥ 1.9	1.2	1.14	5.0
$\vec{B}$ field [T]	0	0	0.51	1.14
v_d^e [cm/µs]	4.9	0.89	4.9	0.42
Efficiency [%]	99.5	>99	>99	~96[‡]
σ_x [µm]	<700	60	110–210	150–170
$\sigma_{2-track}$ [mm]	10	0.6	–	<2
σ_t [ns]	12	2	–	–

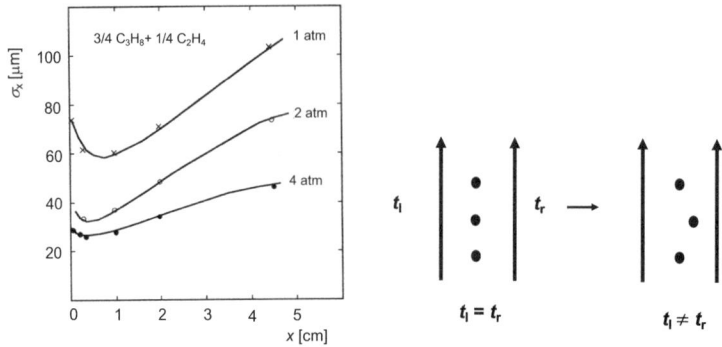

Fig. 4.24 Left: Position resolution as a function of the drift distance for different gas pressures. Reprinted with kind permission from [68], © 1978, Elsevier. All rights reserved. Right: Arrangement of sense wires to break the left-right ambiguity

x, y directions with similar spatial resolution. The construction of large planar drift chambers is done in a similar way as that of MWPCs. In a drift cell, we encounter a left-right ambiguity. A track passing the anode wire on the left-hand side cannot be distinguished from one passing on the right-hand side. One possibility to resolve this issues is to look at the field wires. The field wires on the side the track passed through will see an induced signal earlier than those on the other side [55]. So, by reading out the field wires we can resolve this ambiguity. Another possibility is to stagger the anode wires, which breaks the left-right symmetry (see Fig. 4.24 (right)). Table 4.2 summarizes properties of four planar drift chambers used in the CDHS,[6] NA32,[7] L3 and H1 experiments as examples.

4.2.2 Drift Chamber Performance

To estimate the efficiency of a chamber we can use a naive statistical argument. A charged particle traversing the chamber produces an average number n_c of collisions on its path. The probability for producing the actual number k is given by Poisson statistics [20]:

$$P_k^{n_c} = \frac{n_c^k}{k!} \exp(-n_c). \tag{4.31}$$

So, the inefficiency that not a single electron-ion pair is produced is given by $P_0^{n_c} = \exp(-n_c)$ and in turn the chamber efficiency

$$\epsilon_{ch} = 1 - \exp(-n_c). \tag{4.32}$$

---

[6] CDHS was a neutrino experiment at CERN taking data 1976–84.
[7] NA32 was a charm production experiment at CERN taking data 1982–86.

If we target an efficiency larger than 99%, this implies $\exp(-n_c) < 0.01$ yielding $n_c \geq 5$. So, for five collisions the efficiency is 99.3%. Since this is satisfied for most chambers, drift chambers typically operate with high efficiency. The first chamber built by Walenta [55] reached an efficiency of 98.6%. In many chambers operated in various experiments the efficiency is better than 99%. The aging issues we discussed with respect to MWPCs also hold for drift chambers. A first step is to add water vapor, which improved operations in several chambers.

In smaller chambers, the position resolution is determined by the knowledge of the time-to-distance relation, the diffusion of electrons during their migration towards the anode and a constant electronics dispersion. For large-area chambers ($\geq 100 \text{ m}^2$) mechanical tolerance in positioning the wires ($100 - 300 \ \mu\text{m}$) and sagging of the wires under their own weight come in addition. Thus, the spatial resolution of a small chamber is given by

$$\sigma_x^2 = \left(\frac{c_{ion}^2}{N_{ion}^2 \cdot x^2} \right) + \frac{2 D_e}{v_d^e} \cdot x + \sigma_{noise}^2, \tag{4.33}$$

where c_{ion} is a constant, N_{ion} gives the number of primary ions produced by a charged track in a cell, D_e is the electron diffusion coefficient defined in (2.144) and v_d^e is the magnitude of the electron drift velocity. The first term is determined by the number of produced primary ions, which dominates at small drift distances and drops inversely with the drift distance and N_{ion}. The second term accounts for the diffusion of electrons increasing with drift distance while the third term results from electronic noise. Note that in magnetic fields the diffusion coefficient is reduced by $(1 + \omega_B^2 \tau_e^2)^{-1}$.

Figure 4.23 (right) shows the position resolution for the drift cell in Fig. 4.21 as a function of the drift distance. The data were obtained by measuring the same track in a set of equal chambers, performing a fit to the data in each chamber, calculating the deviation between data and the fits and determining the standard deviation of the differences. The best position resolution is achieved at a drift distance of ~ 4 mm. The ultimate accuracy that can be obtained in a drift chamber depends both on the good knowledge of the time-to-distance relation and on the diffusion properties of electrons in gases. Figure 4.24 (left) shows the position resolution as a function of the drift distance for different gas pressures. At four atmospheres the position resolution improves significantly. The best average resolutions of 60 μm and 120 μm have been achieved in the Na34 [60] and H1 [67] experiments, respectively.

4.3 Cylindrical Drift Chambers

In colliding-beam experiments usually solenoidal magnetic fields around the beam pipe are used. In cylindrical coordinates the magnetic-field components are $B_r = B_\phi = 0$, $B_z \neq 0$. Since charged particles feel a force perpendicular to the $\vec{B}$ field, the charged-particle momentum is deflected in the $r\phi$ plane while their z component remains unaffected. Since the beam line is along the z direction, beam particles are left undisturbed. Detectors need to follow the cylindrical geometry. In planar drift chambers we use square cells. In cylindrical drift chambers, drift cells are arranged in cylindrical layers. The $\vec{E}$ field lies in the $r\phi$ plane, perpendicular to axial $\vec{B}$ field. It is generated by a suitable arrangement of potential wires, which run parallel to each other surrounding a single anode wire or a group of anode wires that are read out. A large fraction of layers (typically $\geq 50\%$) have wires running parallel to the magnetic field. They are called axial layers. The remaining layers have wires running skew under stereo angles, which are typically $\pm$ few degrees with $\vec{B}$ field axis and which are called stereo layers. For each signal wire we have to measure t_0, the beginning of the primary ionization and the time-to-distance relation. If signal wires are radially aligned, we encounter a left-right ambiguity. This problem is solved by staggering signal wires, thus breaking the radial symmetry as shown in Fig. 4.24 (right). To reduce the number of field wires, closure of the drift cell in radial direction can be omitted. Such a setup is called open-cell geometry. Here, the homogeneity of the $\vec{E}$ field and linear relation of the drift path to drift time deteriorate.

Axial wires only measure $r\phi$ positions. From the stereo layers we obtain the z information. The determination of the three-dimensional space point is a two-step process. First, we determine the $r\phi$ position from the axial wires and perform a track fit. In the second step, we use the measurements of the stereo layers. We move the z position until the $r\phi$ position matches that of the fit curve. Then the fit is repeated with the information of the stereo layers included. From the best fit the z-positions are obtained.

4.3.1 Conceptual Chamber Design

A cylindrical drift chamber conceptually consists of two circular endplates, an inner cylinder and an outer cylinder which are sealed and filled with gas. Typically, the endplates are made out of aluminum. Precision holes are drilled into the endplates into which feedthroughs are inserted. On the inside sense wires or field[8] wires are soldered. On the outside the sense wires are connected to the readout electronics and high voltage, while field wires are connected to high voltage. The inner cylinder typically consists of a light-weight material such as beryllium, fiber glass or carbon fiber, while the outer cylinder is made of carbon fiber or aluminum. The stringing of the chamber is done with a special winding machine. The endplates are fixed at the correct distance with the inner cylinder in place. The winding machine strings one wire at a time that is correctly tensioned and soldered to the feedthrough. The drift chamber consists of many cylindrical layers, some of which are axial while others are stereo layers. Sense wires typically consist of stainless steel, gold-plated tungsten, rhenium-plated tungsten, or aluminum. They have diameters of 20-30 μm while potential wires and guard wires are much thicker (40–170 μm), typically consisting of gold-plated tungsten, copper-beryllium or aluminum.

---

[8] Field wires are called also potential wires.

4.3.2 Chamber Properties

The cylindrical drift chamber has been the most widely used tracking detector in colliding-beam experiments. Large-area detectors can be built rather inexpensively, which provide good position resolution in the $r\phi$ plane, $\sigma_{r\phi}$. Using stereo layers, the resolution in the z direction is worsen by

$$\sigma_z = \sigma_{r\phi}/\sin(\alpha_{st}), \qquad (4.34)$$

where α_{st} is the stereo angle. Since for small stereo angles the sine term is of the order of 0.1, the rz resolution is typically a factor of ten worse than the $r\phi$ resolution.

The cell structure provides constant electric fields over nearly the entire drift cell. For example, the experiments MARK III [69,70], TASSO [71,72], ARGUS [73], SLD [74,75], CLEO [76], BABAR [77,78], Belle/Belle II [79,80], and CDF [81,82] used a central drift chamber as their main tracking detector. For these experiments, we list properties and performance parameters in Table 4.3. Note that all these experiments have been completed. Drift chambers are used also in several active experiments, such as BES III [83], COMET[9] [84] and the upgraded MEG II[10] detector [85]. However, we refer the interested reader to the cited publications.

Drift chamber efficiencies are very high and spatial resolutions of large chambers are typically 100–250 μm. Please note that the SLD and Belle II drift chambers achieve the best $r\phi$ resolutions of 55–110 μm and 50–120 μm, respectively. While the ARGUS experiment built the largest chamber, the BABAR experiment built the smallest. The CDF drift chamber had the largest number of measurements while the TASSO drift chamber had the smallest. The two-track separation is a few mm. Note that all drift chambers provide ionization measurements for each sense wire, which is used for particle identification. This information basically comes for free and will be discussed this in more detail in Sect. 9.1. Before we discuss alignment, track finding and track reconstruction procedures, we present some more details of the TASSO and BABAR drift chambers.

4.3.3 The TASSO Drift Chamber

One of the first cylindrical drift chambers, which was built for the Tasso experiment [71], had nine axial and six stereo layers using one-type of open cell structure shown in Fig. 4.25 (left). Three field wires defined the cell size in ϕ while cells were open in the radial direction. However, five separating cylinders made of aluminized Rohacell divided the chamber into six radial compartments. This enhanced

---

[9] COMET is a muon-to-electron conversion experiment at J-Park that has started data taking in 2022.
[10] The MEG II experiment at the Paul Scherrer Institute, which looks for the lepton-number-violating $\mu \to e\gamma$ decay, started data taking in 2021.

Table 4.3 Examples of cylindrical drift chambers showing the inner radius r_i, outer radius r_o, chamber length ℓ_{dc}, number of sense wires N_{sw}, number of sense wires per cell, cell structure, number of field wires N_{fw}, sense wire material, diameter of the sense wire d_w, field wire material, diameter of the field wire d_{fw}, maximum number of measurements N_{max}, stereo angle α_{st}, maximum drift distance ℓ_{drift}, method of z measurement, gas mixture, gas fraction, electric-field strength $|\vec{E}|$, magnetic-field strength $|\vec{B}|$, drift velocity v_d^e, spatial resolution range in $r\phi$, mean spatial resolution in $r\phi$, dE/dx resolution [86] and relative transverse momentum resolution σ_{p_T}/p_T^2. For the momentum resolution typically all tracking systems are used. *The first value refers to layer 2 and the second value to layers 3–8. †The first layer had 15 wires per cell (12 sense wires) whereas the remaining six layer had 5 wires (3 sense wires) per layer. ‡Both field wire types are gold-plated. §This holds for all but the innermost superlayer, where the electric field is 2.3 kV/cm. #The resolution depends on the drift distance and is best in the center of the cell and poor at the edges

Exp.	MARK III	TASSO	ARGUS	SLD	CLEO II	BABAR	Belle II	CDF II
Ref.	[69,70]	[71,72]	[73,87]	[74,75]	[76]	[77,78]	[79,80],	[81,82]
r_i [cm]	14.5	36.7	30	20	17.5	23.6	16	40
r_o [cm]	114.3	122.2	172	100	94.5	80.9	113	137
ℓ_{dc} [cm]	177.8/233.7*	352	200	200	215	276.4	232.5	310
N_{sw}	2,000	2,340	5,940	5,120	12,240	7,104	14,336	30,240
N_{sw}/cell	12/3†	1	1	8	1	1	1	12
Cell	Open	Closed	Closed	Open	Closed	Closed	Closed	Open
N_{fw}	2,000	7,020	24,588	32,640	36,240	45,000	42,240	39,840
Sense wire	Au-W	Au-W	Au-W	Au-W	Au-W	W-Re	Au-W	Au-W
d_w [μm]	20	30	30	25	20	20	30	40
Field wire	Cu-Be	Au-Mo	Cu-Be	Au-Al	Al/Cu-Be	Al‡	Al	Au-W
d_{fw} [μm]	175	120	75	150	110	120	126	40
N_{max}	30	15	36	80	51	40	56	96
$\pm\alpha_{st}$ [°]	+7.7 / −9.0	3.36–4.5	2.4–4.8	2.48	3.77–6.89	2.6–4.2	2.7–4.6	2.0
ℓ_{drift} [cm]	1.81–2.99	1.6	1	5	0.7	0.6–0.8	0.8–1.8	0.88
z_{meas}	stereo Q–div	Stereo	Stereo	stereo Q–div	Stereo	Stereo	Stereo	Stereo
Gas Mixture	Ar-CO_2-CH_4	Ar-CH_4	C_3H_8-C_3H_3OH	Ar-CO_2-iC_4H_{10}	Ar-C_2H_6	He-iC_4H_{10}	He-C_2H_6	Ar-CF_4-iC_3H_8
Fraction [%]	89:10:1	90:10	97:3	21:75:4	50:50	80:20	50:50	50:15:35
$|\vec{E}|$ [$\frac{kV}{cm}$]	0.8	1.54	1–2	0.9	1	1.96	1.26§	1.9–2.4
$|\vec{B}|$ [T]	0.4	0.5	0.76	0.6	1.5	1.5	1.5	1.41
v_d^e [$\frac{cm}{\mu s}$]	5.1	∼2.5	4–5	0.79	1	2.2	3.3	8.8
$\sigma_{r\phi}$ [μm]			100–400	55–110		100–250	50–120	100–200
$\langle\sigma_{r\phi}\rangle$ [μm]#	250/220*	220	190	92	290			140
$\frac{\sigma_{dE/dx}}{dE/dx}$ [%]	15.0	–	5.0	7.0	6.2	6.9	5.0	10.0
$\frac{\sigma_{p_T}}{p_T^2}$ [$\frac{c}{GeV}$]	0.015	0.02	0.009	0.005	0.0026	0.0013	0.00127	0.0015

4.3 Cylindrical Drift Chambers

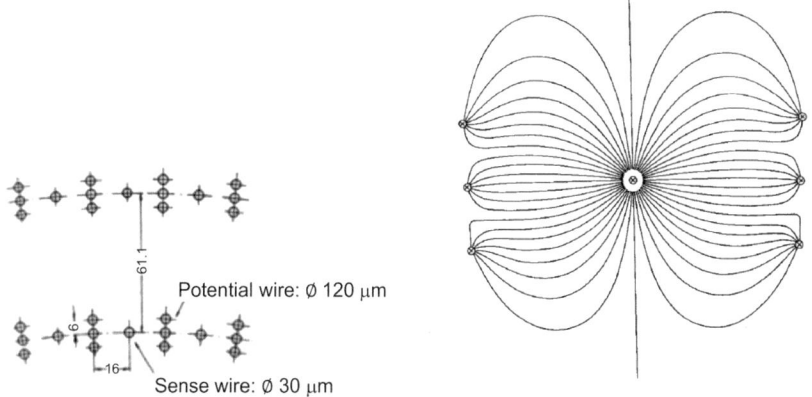

Fig. 4.25 Left: The open drift cell structure of the Tasso drift chamber. Right: The electric-field lines in the Tasso drift cell. Both reprinted with kind permission from [72], © 1978, Elsevier. All rights reserved

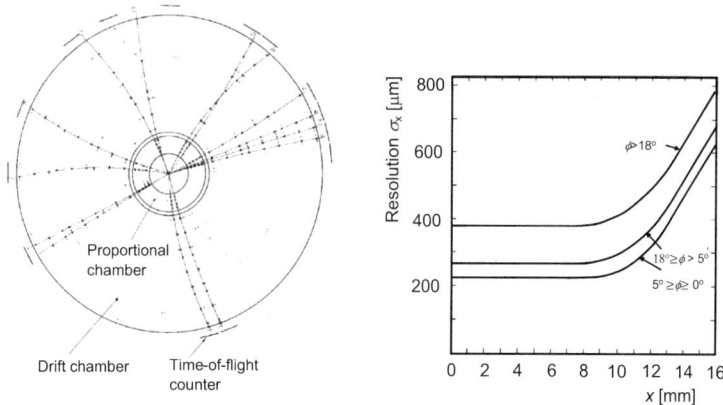

Fig. 4.26 Left: A hadronic event measured with the Tasso drift chamber. The points represent $r\phi$ position measurements and the solid curves show track fits. Right: The position resolution in the Tasso drift chamber as a function of drift length for different dip angles. Both reprinted with kind permission from [72], © 1980, Elsevier. All rights reserved

operational stability in case of broken wires as these would be confined to one compartment. The electric-field lines are displayed in Fig. 4.25 (right), which are quasi parallel over a wide range of the cell.

Figure 4.26 (left) depicts a hadronic event at 35.8 GeV in which only hits in the axial layers are displayed without background removal. Individual tracks from charged particles are visible. Events are rather clean. The solid lines show the fitted tracks. The transverse momentum is extracted from the curvature (see Chap. 6). Figure 4.26 (right) shows the position resolution as a function of drift distance for different entrance angles measured with Bhabhas and hadronic events. In the $r\phi$

plane a position resolution of $\sigma_{r\phi} \simeq 200$ μm is achieved for small entrance angles. For inclined tracks the position resolution becomes worse. For dip angles of $>18°$, the resolution is nearly twice as large.

4.3.4 The BABAR Drift Chamber

The BABAR drift chamber consists of 40 cylindrical layers arranged into 10 superlayers with four layers each, covering the radial region from 20 cm to 80 cm [77]. Figure 4.27 (left) shows a schematic layout of the ten superlayers. The innermost superlayer is an axial layer followed by a stereo layer with positive stereo angles (U-layer) and a stereo layer with negative stereo angles (V-layer). This pattern is repeated three times followed by an axial layer as outermost superlayer. Figure 4.27 (center) shows an enlarged view of the four innermost superlayers. Typically, a cell consists of one sense wire surrounded by six field wires where the sense wire is kept at 1.96 kV while the field wires are at ground potential. For inner and outer cells of each superlayer, the inner and outer field wires are replaced by two guard wires kept at 340 V. Figure 4.27 (right) shows the electric-field lines and isochrones of cells in layers 3 and 4 of an axial superlayer. Sense wires are tungsten-rhenium wires 20 μm in diameter and 2.75 m long while field wires are gold-plated aluminum wires 120 μm in diameter. The drift chamber is 2.8 m long and is operated with a He-iC$_4$H$_{10}$ (80:20) gas mixture plus 0.4% water vapor at a pressure of 4 mbar. The endplates, which carry an axial load of $\sim 32,000$ kN, are made of aluminum. The inner cylinder is a 1 mm thick beryllium structure, the outer shell consists of a 1.6 mm thick carbon-fiber structure. The inactive material in the BABAR drift chamber corresponds to $1.08\% \cdot X_0$. Some photographs of the BABAR drift chamber are depicted in Figure 4.28. The first photograph (top left) shows the stringing of the chamber. The wires are attached onto the feedthroughs on one endplate. The stringing machine drives them to the other endplate, where they are tensioned, cut and soldered to the feedthroughs. The arrangement of the feedthroughs is depicted on the bottom left. The photograph on the top right shows the completely strung drift chamber and that on the bottom right depicts the drift chamber after the outer shell had been put on.

Figure 4.29 (left) shows the drift distance versus drift time relation. Note that the dependence is not linear across the entire cell. For a drift distance less than 3 mm a linear relation gives a good approximation. In addition, there is a small asymmetry between the left-hand side and the right-hand side of the drift cell for drift times larger than 200 ns. However, since the time-to-distance relation is known, precise position measurements are possible. Figure 4.29 (right) shows the track efficiency as a function of transverse momentum (top right) and a function of the polar angle (bottom right). For an operating voltage of 1960 V, the efficiency is above 97% for $p_T > 400$ MeV/c and polar angles greater 0.5 radian.

Figure 4.30 (left) shows the position resolution inside a drift cell. The average resolution is $\sigma_{r\phi} = 125$ μm. The best resolution is found in the center of the chamber yielding $\sigma_{r\phi} = 100$ μm. Using the Kalman fitter method discussed in Sect. 5.2.4, we determine the track parameters, transverse impact parameter (d_0), longitudinal

4.3 Cylindrical Drift Chambers

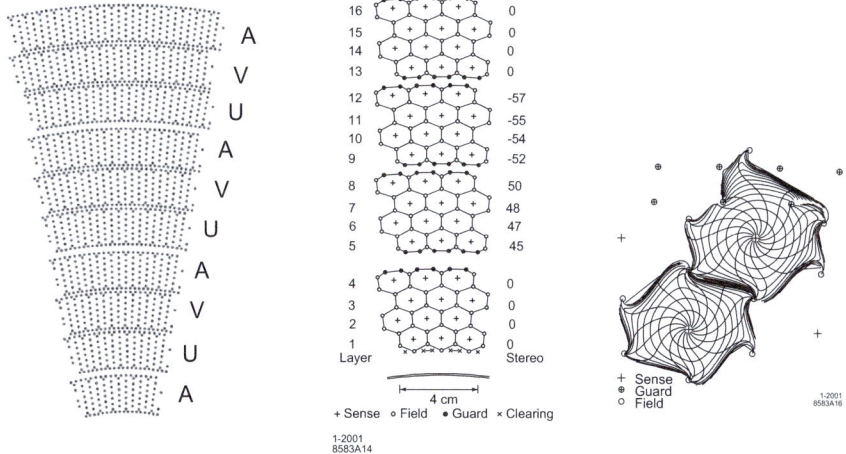

Fig. 4.27 Left: Radial layout of the $BABAR$ drift chamber, which consists of ten superlayers arranged in alternating axial A, stereo U and stereo V superlayers and so on. Center: Enlarged view of the first four superlayers. Right: Electric-field lines and isochrones of a drift cell in layers 3 and 4 of an axial superlayer. All cells are closed small drift cells that have six field wires. Inner and outer cells have additional guard wires. All reprinted with kind permission from [88], © 1998, Elsevier. All rights reserved

Fig. 4.28 Top left: Stringing of the $BABAR$ drift chamber. Top right: The $BABAR$ drift chamber after completion of stringing. Bottom left: A view of the feedthroughs on an endplate of the $BABAR$ drift chamber. Bottom right: The completed $BABAR$ drift chamber. All reprinted with kind permission from [89], © 2000, the $BABAR$ Collaboration. All rights reserved

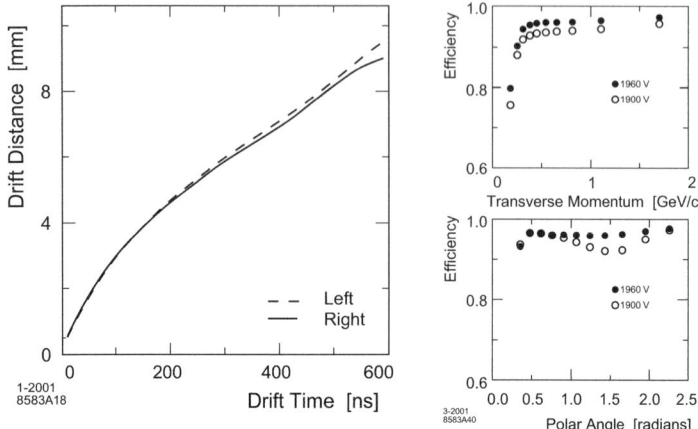

Fig. 4.29 Left: The time-to-distance relation for a BABAR drift cell. Top right: Efficiency versus transverse momentum. Bottom right: Efficiency versus polar angle. There is a slight asymmetry between the left-hand side (dashed) and the right-hand side (solid) of the BABAR drift cell. The solid (open) points show the efficiency for an operating voltage of 1960 V (1900 V). All reprinted with kind permission from [77], © 2002, Elsevier. All rights reserved

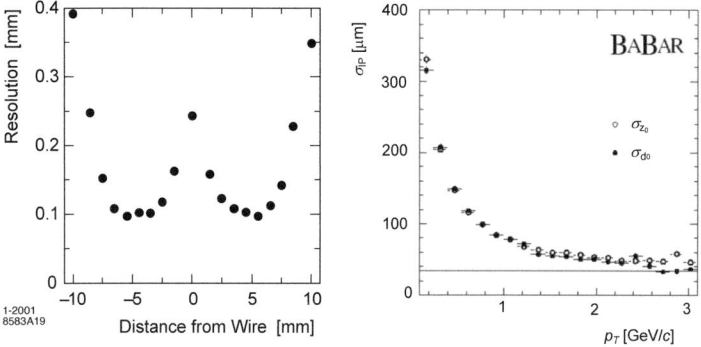

Fig. 4.30 Left: Position resolution as a function of drift distance in the BABAR drift chamber cell [77]. Right: Impact parameter resolution in σ_{d_0} (solid points) and σ_{z_0} (open circles) as a function of transverse momentum in the BABAR experiment. Both reprinted with kind permission from [77], © 2002, Elsevier. All rights reserved

impact parameter (z_0), azimuth angle (ϕ_0), tangent of the dip angle ($\tan \lambda_d$) and curvature (κ_t) from the measured fits. We will define the impact parameters in Sect. 5.2.2.

Figure 4.30 (right) shows the transverse and longitudinal impact parameter resolutions d_0 and z_0 for tracks in multi-hadron events as a function of the transverse momentum. Both have a rather similar dependence. For $p_T = 0.4$ GeV/c, the longitudinal and transverse impact parameter resolutions are both 200 μm approaching 30 μm at high momenta. For each hit on a track, we also measure the ionization. We perform pedestal subtraction, correct for gain variations and integrate the signal

4.3 Cylindrical Drift Chambers

over a period of approximately 1.8 μs. The ionization measurements are used for particle identification discussed in Chap. 9. The dE/dx resolution is 6.9%.

4.3.5 Drift Chamber Alignment

The construction of drift chambers is performed with great care keeping tight tolerance with positioning the wires on the endplates. Typically, the holes are drilled with an accuracy to better than 50 μm. Cosmic muon data are used to perform alignments. The procedure is discussed in detail in [80]. Here, we give a brief summary. The data are taken without the magnetic field and with the magnetic field turned on. Misaligned sense wire positions affect the spatial resolution. We can classify the misalignment of sense wire positions into three categories:

- Misalignment between layers of the drift chamber. This is described by shifts in the three coordinates, δx, δy and δz and a rotation around the z axis, $\delta \phi$.
- A chamber twist, i.e. a rotation of the forward endplate with respect to the backward endplate. The correction depends on the dip angle λ_d.
- Misalignment between the sense wires within a layer. Here, the corrections involve shifts in the x and y directions.

A drift chamber misalignment changes the time-to-distance relation

$$\Delta X(z) = X_{\text{true}}(z) - X_{\text{fit}}(z), \tag{4.35}$$

where $X_{\text{true}}(z)$ and $X_{\text{fit}}(z)$ are the true and fitted time-to-distance relations, respectively. Both depend on the drift distance z. From layer-by-layer misalignments we obtain

$$\Delta X(z) = -r_l \sin \delta\phi + \delta x(z) \sin \phi - \delta y(z) \cos \phi, \tag{4.36}$$

where r_l is the radius of the layer. The first term describes the effect of a rotation $\delta\phi$ while the second and third terms describe the effects of the δx and δy shifts, respectively. The δz shift affects the drift chamber performance for the stereo wires only and its effect is proportional to the sine of the stereo angle. Since the drift chamber has no external reference point, an appropriate internal reference point needs to be chosen for the first iteration. To avoid the effect of twist misalignments, we select vertical muons ($70° < \theta < 110°$). We fit the track with respect to the reference points and extrapolate it to the layers that we want to align. For these we calculate the residual $\Delta X(z)$. We perform this procedure step-by-step for each superlayer. The aligned layers are added to the reference. We use them for track fitting to align the next superlayer. We start with the inner layer moving out. Since alignments are different at the forward and backward planes, we perform the procedure at different z positions. Figure 4.31 (left) shows an example of the $\Delta X(z)$ dependence on z for the alignment of the Belle II drift chamber [80]. The residual shows a linear z dependence. We depict the ϕ dependence in Fig. 4.31 (right), which shows a sinusoidal ϕ dependence.

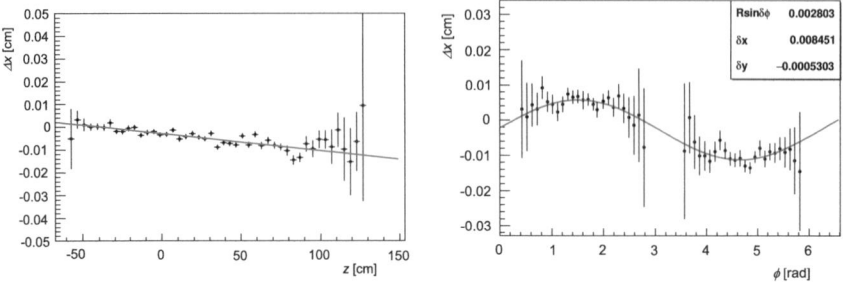

Fig. 4.31 Left: Example of the z dependence of the residual $\Delta X(z)$ in layer 38 determined in the alignment of the Belle II drift chamber. Right: Example of the ϕ dependence of the residual $\Delta X(z)$ in layer 38 determined in the alignment of the Belle II drift chamber. Both reprinted with kind permission from [80], © 2019, Elsevier. All rights reserved

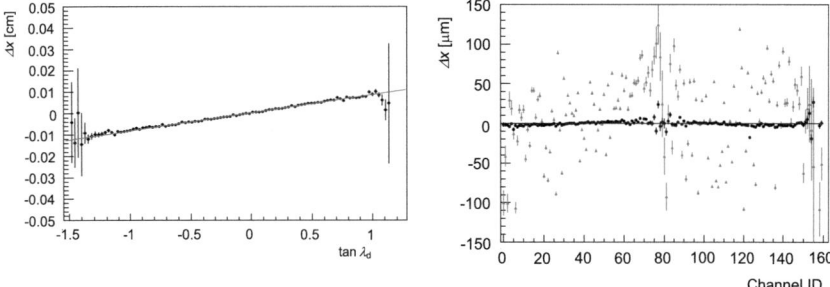

Fig. 4.32 Left: Example of the $\tan \lambda_d$ dependence of the residual $\Delta X(z)$ in layer 55 determined in the alignment of the Belle II drift chamber. Right: Residual distribution $\Delta X(z)$ as a function of the channel number in layer 12 showing data before alignment (triangles) and aligned data (points). Both reprinted with kind permission from [80], © 2019, Elsevier. All rights reserved

The twist misalignment is proportional to $\tan \lambda_d$ of the track,

$$\Delta X(z) = \tan \lambda_d \frac{r_l}{\ell_{dc}}(r_l - r_0) \sin \delta\phi + p_0, \tag{4.37}$$

where r_0 is the radius of the layer of reference, ℓ_{dc} is the length of the layer and p_0 is constant. We fit a cosmic muon track using the four innermost super-layers and extrapolate it to the outer layers. Figure 4.32 (left) shows the residual $\Delta X(z)$ as a function of $\tan \lambda_d$. The residual increases linearly from -100 μm at $\tan \lambda_d = -1.5$ to 100 μm at $\tan \lambda_d = 1.5$. The final twist angle is $+0.295 \pm 0.001$ mrad.

For the wire-to-wire alignment of sense wires in the ϕ-direction we use small-angle muons. Since we do not have sufficient data available for the incident angle around 90° we cannot perform alignments in the r direction. This can be done with beam data. We determine the residual at the endplate using a similar procedure as that for the layer-to-layer alignment and convert them into δx and δy shifts,

$$\delta x = -\Delta X(z) \sin \phi,$$
$$\delta y = -\Delta X(z) \cos \phi. \tag{4.38}$$

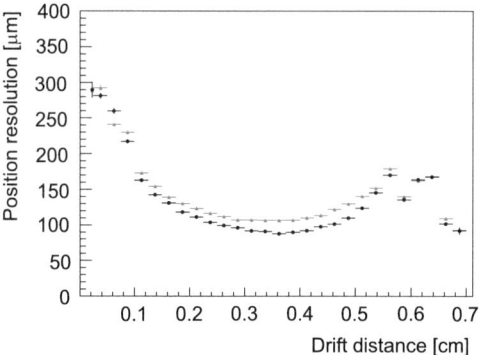

Fig. 4.33 Left: Position resolution as a function of the drift distance in layer 12 of the Belle II drift chamber for muon tracks with small incident angles ($|\alpha_{\text{tr}}| < 5°$) and polar angles of $120° < \theta < 150°$ for non-aligned wires (grey points) and aligned wires (black points). Reprinted with kind permission from [80], © 2019, Elsevier. All rights reserved

Figure 4.32 (right) shows the residual $\Delta X(z)$ as a function of the wire channel number. After aligning the wires, the residual distributions are close to zero. The wire-by-wire misalignment is smaller than 50 μm for most of the layers as expected from the hole position survey. Figure 4.33 shows the position resolution as a function of the drift distance. For drift distances of 1–6 mm, the position resolution after alignment is improved significantly. Between 0.3 mm and 0.4 mm drift distance the spatial resolution is 90 μm.

4.4 Jet Drift Chambers

In a jet drift chamber, the number of measured points along a radial track is considerably increased to about one measurement/cm. All signal wires run in the axial direction forming a middle plane of a sector-shaped cell in which the transverse $\vec{E}$ field points into the ϕ direction. The sense wires are closely spaced and are interleaved with potential wires. The drift paths of the electrons in the gas are much longer than those in cylindrical drift chambers. The first jet chamber was reconstructed for the JADE detector [90]. The Mark II [91], H1 [92] and the OPAL experiments also used jet chambers [93,94]. As examples we discuss the JADE and OPAL jet chambers below and summarize properties of all four chambers in Table 4.4. Jet chambers achieve very good spatial resolutions in $r\phi$ of $130 - -170$ μm and a two-track separation of a few mm. Due to the large number of measurements, they also measure ionization losses rather well.

Figure 4.34 (left) shows the layout of two sector-shaped segments of the JADE jet chamber [90], which operated at the PETRA e^+e^- storage ring to search for the top quark. The jet chamber had an inner radius of $r_i = 21$ cm, an outer radius of $r_o = 78$ cm and a length of $\ell_{\text{ch}} = 2.4$ m. The cylindrical volume was subdivided into 24 radial segments, each with a 15° opening angle. Each radial segment had 48 signal wires parallel to the magnetic field along the z direction arranged into

Table 4.4 Comparison of Jet Drift Chambers showing the inner radius r_i, outer radius r_o, chamber length ℓ_{ch}, number of sense wires N_{sw}, number of sense wires per cell, sense wire material, diameter of the sense wire d_w, sense wire spacing s_w, maximum number of measurements N_{max}, field wire material, diameter of the field wire, maximum drift distance ℓ_{drift}, method of z measurement, gas mixture, gas fraction, electric-field strength $|\vec{E}|$, magnetic-field strength $|\vec{B}|$, drift velocity v_d^e, Lorentz angle α_L, spatial resolution range in $r\phi$, average spatial resolution in $r\phi$, spatial resolution in z, two-track resolution Δ_{2-tr}, dE/dx resolution for minimum-ionizing particles, and transverse momentum resolution σ_{p_T}/p_T^2. [†]The drift cells are rotated by the expected Lorentz angle of 30°; the drift cells are arranged into an inner and an outer jet chamber. [‡]Tracks are constrained to originate from a single point. [#]This gas was used for standard e-p running; for other periods the gas mixture was Ar-CO_2-CH_4 (89.5:9.5:1). [§]A range indicates measured resolutions in a cell; a single value indicates the average resolution over the cell. [§§]They use a special chamber to get a more precise rz measurement than from charge division. [*]The latter value is for the end field wires

Experiment	JADE	Mark II	OPAL	H1[†]
Reference	[90,95]	[91,96]	[93,94,97]	[63,92]
r_i [cm]	2	19.2	25	20.4/45.2
r_o [cm]	80	151.9	185	53/84.4
ℓ_{ch} [cm]	240	230	320–400	220
N_{sw}	1,536	5,832	3,816	2,640
N_{sw}/cell	16	6	159	24/32
Sense wire	W-Re-Au	W-Au	W-Re-Au	W-Re-Au
d_w [μm]	20	30	25	20/20
s_w [mm]	10	8.33	10	5.1/5.1
N_{max}	48	72	159	24/32
Field wire	Cu-Be	Cu-Be	Cu-Be	Cu-Be
d_{fw} [μm]	100	178/304[*]	125	90/90
ℓ_{drift} [cm]	2.6–7.8	3.3	3–25	2.29–4.45/2.85–4.31
z_{meas}	Q div.	Stereo	Spec. cham.	Q-div.
Gas mixture	Ar-CH_4-iC_4H_{10}	Ar-CH_4-CO_2	Ar-CH_4-iC_4H_{10}[#]	Ar-C_2H_6
Fraction [%]	(88.7:8.5:2.8)	(89:1:10)	(88.2:9.8:2)	(50:50)
$|\vec{E}|$ field [$\frac{kV}{cm}$]	0.94	0.9	0.94	1.0
$|\vec{B}|$ field [T]	0.45	0.45	0.435	1.15
v_d^e [$\frac{cm}{\mu s}$]	5.04	5.2	5.29	5.0
α_L	18.5°	17.8°	20°	30°
$\sigma_{r\phi}$ [μm][§]	140–185	120–200	100–180	
$\langle\sigma_{r\phi}\rangle$ [μm][§]	160	170	135	170
σ_{rz} [mm]	26	0.12–0.20	1% $\cdot\ell_{wire}$[§§]	1% $\cdot\ell_{wire}$[§§]
Δ_{2-tr} [mm]	7	3.8	2	2
$\frac{\sigma_{dE/dx}}{dE/dx}$ [%]	8.1	6.9	3.1	10
$\frac{\sigma_{p_T}}{p_T^2}$ GeV/c^{-1}	0.022	0.0046 (0.0031)[‡]	0.0015	0.01

4.4 Jet Drift Chambers

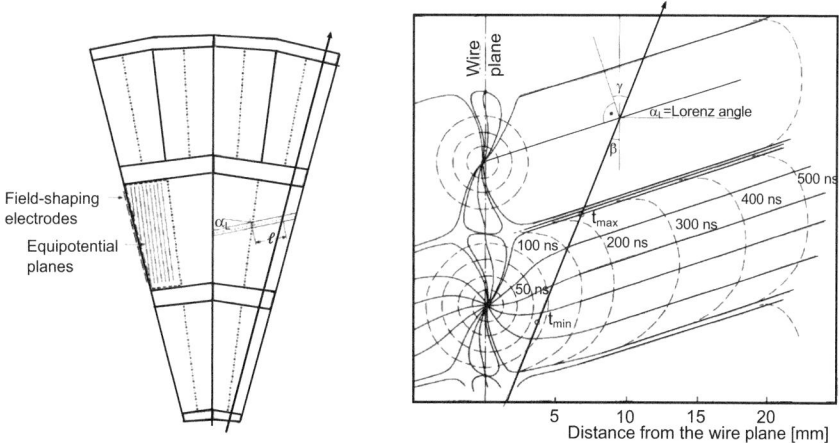

Fig. 4.34 Left: Cross section through two segments of the JADE jet chamber. The field-shaping electrodes are shown on the left-hand edge of the cell in the center partition. Both, equipotential lines and the Lorentz angle are indicated. Right: Drift trajectories (solid lines) and isochrones (dashed curves) of the JADE jet chamber. Both reprinted with kind permission from [90], © 1980, Elsevier. All rights reserved

three radial partitions. One partition had 16 sense wires in the center, which were staggered to break the left-right symmetry. So, a total of 48 measurements were performed over a 57 cm length. On the boundaries of each radial sector, field-shaping electrodes were positioned. Figure 4.34 (right) shows the isochrones and electron drift directions, which have complicated shapes near the signal wire. The jet chamber was operated with an argon-methane-isobutane gas mixture (88.7:8.5:2.8) at a pressure of 4 atmospheres. The electric-field strength was 0.94 kV/cm yielding a drift velocity of $v_d^e = 5.04$ cm/μs. Let us recall that the Lorentz angle is given by $\tan \alpha_L = \omega_B \tau_e \sim k(|\vec{E}|) \cdot v_d^e \cdot |\vec{B}|/|\vec{E}|$, where $k(|\vec{E}|)$ is a factor that depends on the gas mixture and the $\vec{E}$ field. The Jade experiments operated in a magnetic field of $|\vec{B}| = 0.45$ T, yielding a large Lorentz angle of $\alpha_L = 18.5°$ as shown in Fig. 4.34 (left).

Jet chambers yield good spatial resolutions. Figure 4.35 (left) shows the position resolution of the JADE jet chamber in the $r\phi$ plane as a function of drift length. Near the sense wire the position resolution was $\sigma_{r\phi} = 140 \,\mu$m increasing linearly with drift distance to $\sigma_{r\phi} \simeq 180 \,\mu$m at 6 cm, where the contribution from the electronic noise was 110 μm. The z position was obtained via charge division yielding a resolution of $\sigma_z = 16$ mm. Using 48 space points provided a momentum resolution of $\sigma_{p_T}/p_T^2 = 3.3\%$ (GeV/c)$^{-1}$. Figure 4.35 (right) shows an axial projection of an e^+e^- collision at a center-of-mass energy of 30 GeV in the JADE jet chamber. Since the hits are closely spaced track segments are clearly visible without a fit curve. All charged tracks are well resolved.

The OPAL jet chamber was the largest providing the most measurements and achieving a reasonable spatial resolution. Figure 4.36 (left) shows a schematic view of part of one sector. The jet chamber had an inner radius of $r_i = 0.25$ m, an outer

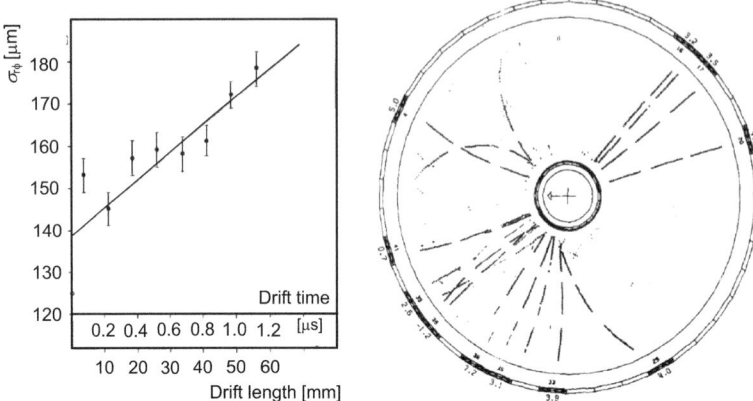

Fig. 4.35 Left: Mean position resolution of single-cell track fits in the $r\phi$-plane as function of the mean drift distance for the Jade jet chamber. Right: The axial projection of the collision process $e^+e^- \to hadrons$ in the Jade detector. The 16-hit measurements in each of the three partitions appear as three line segments. Both reprinted with kind permission from [90], © 1980, Elsevier. All rights reserved

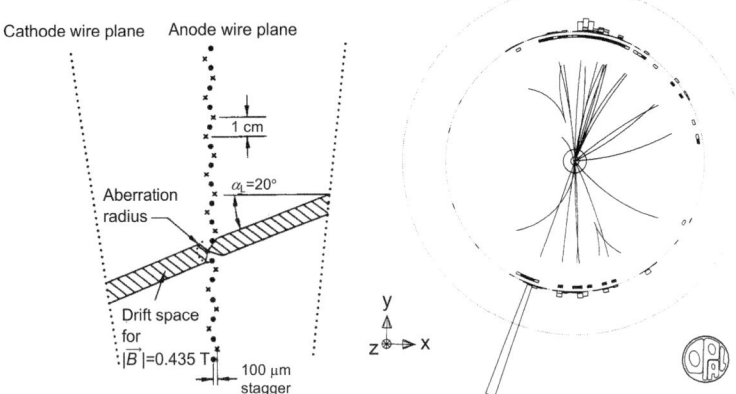

Fig. 4.36 Left: Schematic view of a part of one sector of the 24-sector OPAL jet chamber displaying the staggered sense wires (crosses) and potential wires (points) [94]. Furthermore, the drift space and Lorentz angles are shown. Right: A hadronic Z^0-decay. The lower jet contains a high-energy electron indicated by the straight track pointing to a high-energy shower in the electromagnetic calorimeter on the bottom. Note that the hits on the closely spaced wires already define the trajectories without a fit. The jet chamber only shows hits that belong to charged particles coming from the Z^0 decay. Both reprinted with kind permission from [94], © 1992, Elsevier. All rights reserved

4.4 Jet Drift Chambers

radius of $r_a = 1.85$ m and a length of $\ell_{ch} = 4$ m. It was segmented into 24 identical sectors each having 159 gold-plated tungsten-rhenium sense wires 25 μm in diameter equally spaced at 10 mm, which were kept at ground potential. The sense wires were interleaved with potential wires that were operated at a high voltage of −2.38 kV. At each end a 175 μm thick wire terminated the anode plane. The sense wires were staggered at 100 μm to break the left-right symmetry. Cathode wire planes formed boundaries between adjacent sectors. The cathodes consisted of copper-beryllium wires 125 μm in diameter spaced at 3.1 mm. The first wire in each plane had a diameter of 175 μm. The cathode planes were inclined by 7.5° with respect to the anode planes. Since the equipotential lines in the drift region are parallel to the anode plane, each wire of the cathode plane sat at a different potential. The voltages ranged from −25 kV at the outer radius to −5 kV at the inner radius. Figure 4.36 (right) shows an event display of a hadronic Z^0 decay. The charged tracks in each jet are clearly separated. Besides the hits of the charged tracks from the Z^0 decay, no extra hits are visible.

The chamber was operated with a gas mixture of Ar-CH$_4$-iC$_4$H$_{10}$ (88.2:9.8:2.0) in a magnetic field of 0.435 T. Electrons produced in an ionization process drifted with a drift velocity of 5.3 cm/μs in an electric field of 0.94 kV/cm under a Lorentz angle of $\alpha_L = 20°$. The maximum drift distance was 25 cm. Excluding bad channels, which were as low as 0.3%, a single-hit efficiency of >99.2% was measured by studying $Z^0 \to \mu^+\mu^-$ decays [94]. Figure 4.37 (left) shows the $r\phi$ position resolution as a function of drift distance. Close to the wire, a position resolution of $\sigma_{r\phi} = 100$ μm was achieved, while at 20 cm drift length the resolution increased to $\sigma_{r\phi} = 135$ μm. The z position was determined by charge division with a resolution of $\sigma_z = 4.5$ cm for di-muon events and 6 cm for hadron events. The transverse momentum resolution was $\sigma_{p_T}/p_T = 0.0015 \cdot p_T \oplus 0.02$. Figure 4.37 (right) shows the position resolution in the $r\phi$ plane as a function of the local track angle with respect to the anode plane. For larger angles the position resolution became significantly worse. The anode wires also recorded the ionization loss. The dE/dx resolution was 3.1% providing good K-π separation at low momenta (see Chap. 9).

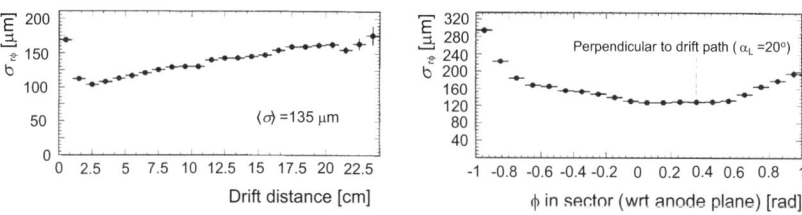

Fig. 4.37 Left: Position resolution as a function of the drift distance in the OPAL jet chamber. The average resolution is $\sigma = 135$ μm. Right: Position resolution as a function of the local track angle with respect to the anode wire plane. Both reprinted with kind permission from [94], © 1992, Elsevier. All rights reserved

4.4.1 Small Jet Drift Chambers

Before the development of solid state detectors small jet drift chambers were built to measure tracks more precisely close to the beam. For example, the Mark II experiment built a vertex jet drift chamber [98]. The chamber extended from 5.0 cm to 17.0 cm in the radial direction and 55 cm in the axial direction. It was divided into ten axial drift cells. To break the left-right symmetry individual cells were tilted with respect to the radial direction. There were 40 anode wires 20 μm in diameter spaced at 2.9 mm of which 38 were read out. They were interleaved with 40 potential wires 225 μm in diameter. The sense wire planes were sandwiched between two grid planes spaced at 1.8 mm and consisting of 150 μm diameter wires. The cell was closed with 59 cathode wires 225 μm in diameter spaced at 2.0 mm on each side of the sector. Extra electrodes at the innermost and outermost radii reduced edge effects. The chamber operated with a CO_2-C_2H_6 (92:8) gas mixture. Voltages on the cathode planes were graded between −3.5 kV and −11.7 kV. The sense wires were operated with +2.94 kV while the potential wires were grounded and the grid wires were held at −0.47 kV. The chamber operated in a 0.475 T magnetic field. The spatial resolution varied with the drift distance between 25 μm and 80 μm yielding an average of 50 μm. Properties and parameters of other small jet drift chambers are summarized in Table 11.3 in [87]. In small drift chambers that have been used as vertex chambers, the typical spatial resolutions were 40–55 μm [87]. In the mini-jet vertex chamber prototype a spatial resolution of 20–25 μm was achieved with a slow gas CO_2-isobutane (80:20) at a pressure of 2.5 bar [99]. These resolutions are rather large with respect to those reached in silicon detectors we discuss in Chap. 6.

4.5 Time Projection Chambers

The time projection chamber (TPC) combines the principles of a drift chamber with that of a proportional chamber. It was proposed first by Nygren for the TPC detector at PEP4 [100, 101]. The new idea consisted of making a direct three-dimensional position measurement.

4.5.1 Conceptual TPC Design

Figure 4.38 (left) shows a schematic view of the TPC built by Nygren [100]. A large cylindrical volume ($r = 1$ m, $\ell_{ch} = 2$ m) is filled with an Ar-CH_4 (80:20) gas mixture at a pressure of 8.5 bar. The cylinder is split into two half cylinders that are separated by the high voltage electrode. Field-defining electrodes are placed on the outer shell and the inner cylinder. The potential is defined by a voltage divider. Both endcaps of the barrel are equipped with one layer of MWPCs,[11] each subdivided

---

[11] In newer TPCs, the MWPCs are replaced with GEM detectors or Micromegas.

4.5 Time Projection Chambers

into six sectors. Each sector has 183 proportional signal wires for multiple ionization measurements along the $r\phi$ projection of the track. In addition, the cathode plane below the anode is segmented with pads that are read out. The size in the radial direction covers 15 anode wires. On each half cylinder, a high electric field is placed parallel to the beam axis pointing towards the center so electrons can drift towards the endcaps. A high magnetic field (1.5 T) lies parallel to the electric field so that the electrons are not exposed to any Lorentz force and $\vec{v}_d^e$ is parallel to $\vec{E}$. Since drift lengths are much longer than those in drift or jet chambers, the electric field has to be kept precisely uniform over the entire length requiring strict mechanical tolerances and precise manufacturing of field-defining electrodes on the outer shell and inner cylinder. The image of the electron cloud is broadened by diffusion during the drift process. The broadening is considerably reduced by the strong $|\vec{B}|$ field. The electrons move along a helical path around the magnetic-field lines. The transverse diffusion coefficient is reduced by $1/(1 + \omega_B^2 \tau_e^2)$.

Figure 4.38 (right) illustrates the layout of the endcaps. The anode plane is sandwiched between a cathode plane above and a pad-segmented cathode plane below that is also read out. In addition, a gating plane is placed above the upper cathode plane. When a trigger is accepted the gating plane is opened so the electrons can drift towards the anode. Once the electrons start to avalanche near the anode wires the gating plane is closed to prevent the positive ions to move back into the drift region. This is achieved by applying a positive high voltage to the gating plane. Signals recorded on the wires yield measurements of r while signals induced on the pads provide ϕ measurements. Note that the main contribution comes from the positive ions. The z position is obtained from the drift time the electrons need for traveling

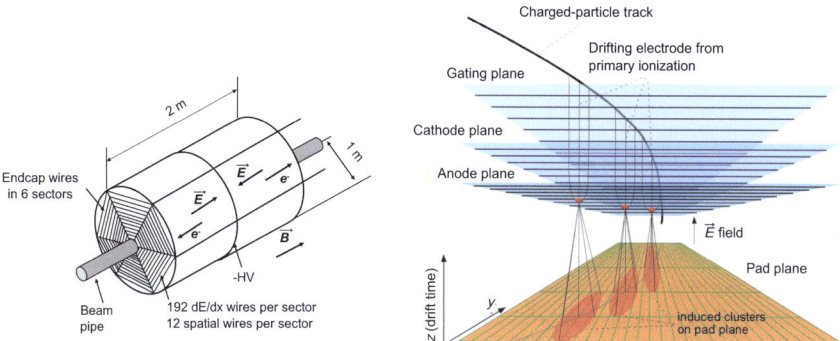

Fig. 4.38 Left: Schematic view of the time projection chamber built by Nygren. The cylindrical volume is segmented into two halves. The electric field in each half points to the center. The magnetic field is parallel or antiparallel to the electric field. Each endcap is segmented into six sectors equipped with one layer of multi-wire proportional chambers. Right: Schematic layout of the endcaps. The electrons produced by primary ionization of a traversing charged particle drift through a gating plane and a cathode plane towards the anode, where they are collected. The clusters on the anode wires induce charge clusters on a cathode pad plane located below the anode. Once the electrons have passed the gating plane it will be switched on to prevent the ions to drift back into the drift volume. Reprinted with kind permission from [102], © 2013, P. Christiansen. All rights reserved

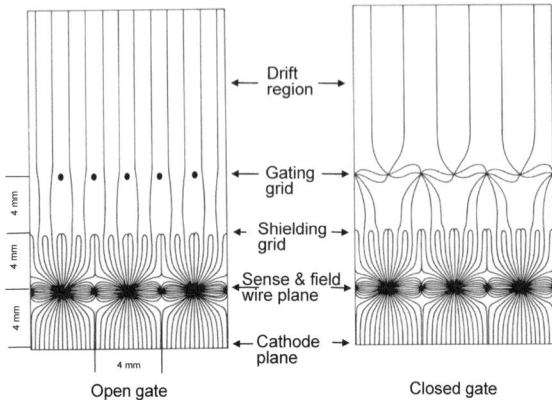

Fig. 4.39 The electric-field configuration for open (left) and closed gates (right). Reprinted with kind permission from [103], © 1985, Elsevier. All rights reserved

from the primary ionization point to the anode wires of the MWPCs. Figure 4.39 shows the electric-field configuration for open (left) and closed gates (right). In the open-gate configuration the electrons can move towards the anode. In the closed-gate configuration the ions are blocked from moving into the drift region.

At high event rates, field distortions will occur, which come from the space charge of positive ions formed in avalanches near the signal wires. The ion density is given by

$$\rho_{ion} = \frac{1}{\ell_w} \frac{f_{ion}}{\mu_{ion}} \frac{A_\gamma}{|\vec{E}|} \frac{N_e}{s_w}, \qquad (4.39)$$

where f_{ion} is the fraction of ions migrating back to the drift region, A_γ is the gas amplification at the anode, N_e is the number of electrons reaching the anode wire per unit time, ℓ_w is the wire length, μ_{ion} is ion mobility, $|\vec{E}|$ is the drift field strength and s_w is the wire pitch. The closed-gate configuration reduces the ion density considerably.

Like cylindrical drift chambers, TPCs have been used in various experiments, for example in TOPAZ [104–106], ALEPH [35, 107–109], DELPHI [36] and ALICE [110]. Properties of these TPCs and the PEP4 TPC are summarized in Table 4.5. Except for ALICE, all TPCs use an argon–methane gas mixture. The $r\phi$ position is measured by wires and pads in the endplate detectors while the z position is determined by the drift time. The drift velocities are all rather high. The PEP4 TPC, the smallest with a 1 m drift length, achieved the best spatial resolutions in $r\phi$ and rz. The ALICE TPC whose schematic layout is depicted in Fig. 4.40 is the largest with a drift length of 2.5 m. It has the poorest resolution in $r\phi$ and a rather poor resolution in rz. The ALEPH TPC with the second best rz spatial resolution achieved the best momentum resolution. We present some more details of the ALEPH and ALICE TPCs below. Before, we would like to mention two other types of TPCs. In searches for weakly-interacting massive particles the XENONnT [111] and LZ [112] experiments use a liquid xenon TPC, which is operated in dual phase. We discuss the principle in Sect. 7.2.5.2. The long-baseline neutrino experiment Dune will use a liquid argon TPC, both in the near and far detectors [113].

4.5 Time Projection Chambers

Table 4.5 Examples of time projection chambers used in high-energy physics experiments showing outer and inner radii (r_o, r_i), drift length ℓ_{drift}, number of sense wires N_{sense}, sense wire spacing s_w, number of pads N_{pad}, pad dimension, maximum number of pad measurements $N_{\text{max}}^{\text{pad}}$, maximum number of sense wire measurements $N_{\text{max}}^{\text{wire}}$, gas mixture, gas fractions, gas pressure, gas amplification, electric-field strength $|\vec{E}|$, magnetic-field strength $|\vec{B}|$, drift velocity v_{d}^e, factor $\omega_B \tau_e$, average spacial resolution in $r\phi$ from pads, spacial resolution in rz, two-track resolution in $r\phi$, relative transverse momentum resolution from the TPC alone, relative transverse momentum resolution from all trackers and dE/dx resolution. More examples are listed in [116]. [†]This included the DELPHI inner detector (ID) and outer detector (OD)

Experiment	PEP4	TOPAZ	ALEPH	DELPHI	ALICE
Reference	[100,107]	[104–107]	[35,107–109]	[36,107]	[110]
r_o [cm]	100	127	180	116	278
r_i [cm]	20	30	31	32	61
ℓ_{drift} [cm]	2 × 100	2 × 122	2 × 220	2 × 134	2 × 249.7
N_{sense}	2,196	2,816	6,336	1,152	No readout
s_w [mm]	4	4	4	4	2.5
N_{pad}	13,824	8,192	41,004	20,160	557,568
Pad $r \times r\phi$ [mm × mm]	7.0 × 7.5	12×9.1–10.7	30 × 6	8 × 7.	4 × 7.5 6 × 10(15)
$N_{\text{max}}^{\text{pad}}$	15	10	21	16	32 & 64
$N_{\text{max}}^{\text{wire}}$	183	175	338	192	–
Gas mixture	Ar-CH$_4$	Ar-CH$_4$	Ar-CH$_4$	Ar-CH$_4$	Ne-CO$_2$-N$_2$
Fractions	(80:20)	(90:10)	(90:10)	(80:20)	(85.7:9.5:4.8)
Pressure [bar]	8.5	3.5	1	1	1
Amplification	1,000		3,000–5,000	5,000	7,000–8,000
$|\vec{E}|$ [kV/cm]	0.55	0.353	0.125	0.15	0.4
$|\vec{B}|$ [T]	1.32	1.0	1.5	1.2	0.5
v_{d}^e [cm/μs]	5.0	5.3	5.24	6.7	2.65
$\omega_B \tau_e$	1.5	4.9	8.9	5.2	0.32
$\langle \sigma_{r\phi} \rangle$(pads) [$\mu$m]	150	185	173	180	1,100/800
σ_z [mm]	0.160	0.335	0.5	0.9	1.25/1.1
$\sigma_{2-\text{track}}$ [mm]	24	25	15	15	–
$\frac{\sigma_{p_T}^{\text{TPC}}}{p_T^2}$ [$\frac{\text{GeV}}{c}$]$^{-1}$	0.009	0.01	0.0012	–	0.002
$\frac{\sigma_{p_T}^{\text{all}}}{p_T^2}$ [$\frac{\text{GeV}}{c}$]$^{-1}$	0.0065	–	0.0008	0.0015	–
$\frac{\sigma_{dE/dx}}{dE/dx}$ [%]	2.7	4.6	4.5	5.7	5.0

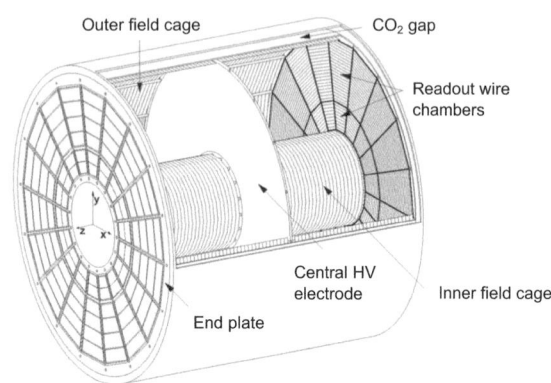

Fig. 4.40 Schematic view of the ALICE time projection chamber. The outer radius is 2.47 m and the inner radius is 0.85 m. The TPC has a 5.1 m length. Each endcap is segmented into 2 × 18 sectors. The electric-field strength is 400 V/cm. Reprinted with kind permission from [110], © 2010, Elsevier. All rights reserved

4.5.2 The ALEPH and ALICE Time Projection Chambers

The ALEPH experiment used a TPC as the main tracking detector, which consisted of a 2 × 2.2 m long segmented cylinder with circular endplates having an outer radius of 1.7 m and an inner radius of 0.33 m. The two halves were separated by a graphite-coated mylar membrane of 25 μm thickness. The endplates were segmented into 18 wire chamber sectors, six identical inner sectors plus 12 outer sectors coming in two shapes. The endplates contained 20,502 pads and 3,168 sense wires. The TPC was operated with an Ar-CH$_4$ (91:9) gas mixture at atmospheric pressure in a magnetic field of 1.5 T. The electric field was 125 V/cm, yielding a drift velocity of $v_d^e = 5.24$ cm/μs. The misalignment between $\vec{E}$ and $\vec{B}$ fields was found to be 0.03°. The factor affecting the transverse diffusions was $\omega_B \tau_e = 8.9 \pm 0.3$. Figure 4.41 (left) shows the ALEPH TPC from the inside. The endplate wire chamber sectors are well visible. The TPC performed very well. The spatial resolution in $r\phi$ was 173 μm while that in rz was 740 μm for tracks with polar angles 80° < θ < 100°. The energy loss was measured in each sense wire. With a 4 mm pitch this gave a total of 338 possible dE/dx measurements per track.

At the LHC, the ALICE experiment uses a TPC as central tracker, which is 5.1 m long (2 × 2.55 m) and has an inner (outer) radius of 0.845 m (2.466 m) (see Fig. 4.40). It operates with a Ne-CO$_2$-N$_2$ (85.7:9.5:4.8) gas mixture in a magnetic field of 0.5 T. The electric field is 400 V/cm yielding a drift velocity of $v_d^e = 2.65$ cm/μs, which is about a factor of two smaller than that in the ALPEH TPC. Each of the two endcaps is instrumented with 81 inner and 18 outer trapezoidal MWPCs containing a total of 557,568 pads that determine the $r\phi$ position. Figure 4.41 (right) shows a photograph of the field cage of the ALICE TPC. The MWPCs have four planes: the gating grid, cathode wire, anode wire and cathode pad planes that are separated by 3 mm each. While the gating grid has a wire spacing of 1.25 mm, the wires in the cathode and anode wire planes are separated by 2.5 mm. The sense wires are gold-plated tungsten wires 20 μm in diameter. All other wires are copper-beryllium wires 75 μ in diameter. For the inner chamber pad sizes are 4 mm × 7.5 mm while for the outer chambers two pad sizes are used, 6 mm × 10 mm and 6 mm × 15 mm for radii below and

4.5 Time Projection Chambers

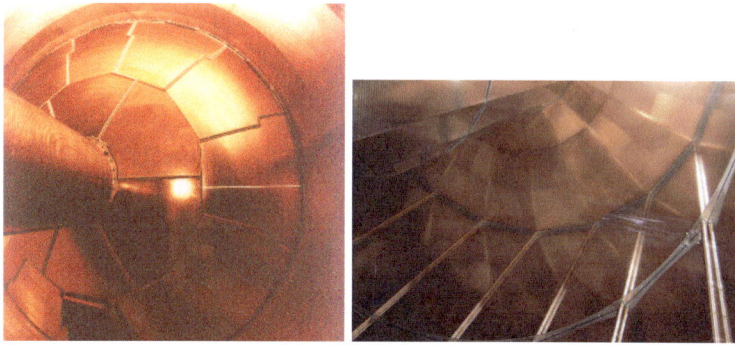

Fig. 4.41 Left: View of the inner side of the endcap in the ALEPH TPC showing the 18 sectors of the endplate [114]. Reprinted with kind permission from [114], © 1989, the ALEPH Collaboration. All rights reserved. Right: Field cage of the ALICE TPC. The trapezoidal sectors with the MWPCs are visible. Reprinted with kind permission from [110], © 2010, Elsevier. All rights reserved

above 1.99 m, respectively. Since $\omega_B \tau_e = 0.32$ the transverse diffusion is not much reduced by the magnetic field. The inactive material amounts to less than $3\% \cdot X_0$. For run 3 the MWPC readout will be replaced by four-GEM stacks [115].

4.5.3 Calibration and Monitoring

A system of UV lasers is used to calibrate the TPC, monitor its performance and measure distortions as well. For example in the ALEPH TPC, the laser calibration system allowed the generation of five laser beams on both sides of the TPC at three different azimuthal angles. This provided 30 straight ionization tracks that seemed to originate from the center of the TPC. It was used to measure and correct residual inhomogeneities of the electric and magnetic fields, which produce systematic displacements of reconstructed coordinates. It also served to measure the modulus of the drift velocity of electrons in the TPC gas.

In the ALICE TPC, each half has six rods, which are equipped with four micromirror bundles of which each contains seven small mirrors. The light of one laser in each half is split and guided into the six rods. The two lasers are synchronized to provide simultaneous laser pulses in the entire TPC yielding a total of 336 simultaneous narrow laser rays. The lasers are equipped with two frequency doublers producing UV light of 266 nm wavelength. The laser system runs every 30 min interspersed between physics events to measure the drift velocity and assess space charge effects.

4.5.4 Performance

The spatial reconstruction of ionizing tracks is performed by measuring two-dimensional projected images in the $r\phi$ plane on endcaps typically using the signal-induced charge on the cathode pads. The positions are determined using the center-of-gravity method. The z coordinate is obtained by measuring the arrival time of the primary electrons. So, with the $r\phi$ information from the MWPCs we get three-dimensional track segments. Thus, pattern recognition and track finding are considerably alleviated compared to chambers that have uncorrelated coordinate information as in drift chambers. In the PEP4 TPC, the spatial resolution in $r\phi$ with pad readout was $\sigma_{r\phi} = 100$ μm while the spatial resolution in z obtained from the drift distance was $\sigma_z = 300$ μm. The resolution of the COG cluster can be parameterized in the following way,

$$\sigma_{COG}^2 = \sigma_0^2 + p_L^2 \ell_{drift} + p_A \tan^2 \alpha_{tr}, \tag{4.40}$$

where σ_0 is the resolution for zero drift length ℓ_{drift} and zero crossing angle α_{tr}. The parameters p_L, p_A and σ_0 are typically obtained from a fit. Note that p_L increases with gas gain fluctuation as $\sqrt{A_{gas}}$ and the diffusion resolution σ_D and decreases with the pad length as $1/\sqrt{\ell_{pad}}$ while p_A increases with $\sqrt{\ell_{pad}}$ and the landau fluctuation of the ionization energy loss factor as $\sqrt{W_{landau}}$.

Figure 4.42 shows the spatial resolutions in $r\phi$ (left) and in rz (right) measured by the ALEPH TPC in $Z^0 \to \ell^+\ell^-$ events. The best spatial resolution in $r\phi$ was achieved for tracks with zero pad crossing angle yielding $\sigma_{r\phi} = 173$ μm. In the z direction the best spatial resolution of $\sigma_z = 0.74$ mm was obtained for a dip angle of zero degrees. Larger pad crossing angles and dip angles worsen the spatial resolution as Fig. 4.43 (left, right) illustrate.

Figure 4.44 shows the measured spatial resolutions in $r\phi$ (left) and in rz (right) as a function of the drift distance for the ALICE experiment. For 250 cm drift distance and small inclination angles, the spatial resolutions are $\sigma_{r\phi} = 800 \pm 80$ μm and $\sigma_z = 900 \pm 100$ μm. For a drift distance of 50 cm or smaller, the resolutions improve by a factor of ~ 2. For large inclination angles and large drift distances, the spatial resolutions increase to $\sigma_{r\phi} = 1.3$–2.0 mm and $\sigma_z = 1.5$–2.3 mm.

The separation of two close-by tracks is another important property. Figure 4.45 sketches the geometric layout of the cathode plane for separating two close-by tracks. The double track resolution is

$$\Delta z = v_d^e t_c + v_d^e t_d = v_d^e t_c + h/\tan\theta, \tag{4.41}$$

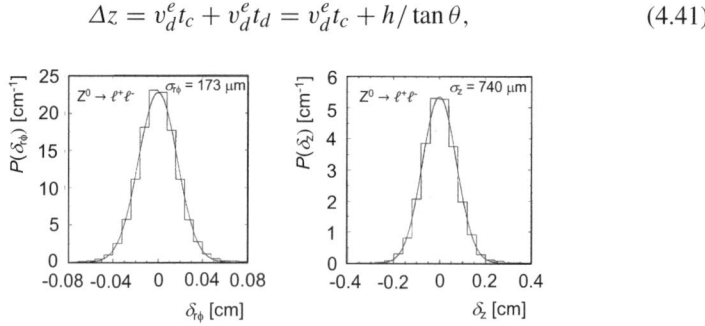

Fig. 4.42 The ALEPH position resolution in the $r\phi$ direction (left) and in the z direction (right) measured in $Z^0 \to \ell^+\ell^-$ decays. The $r\phi$ resolution is 173 μm while the z resolution is 740 μm. Reprinted with kind permission from [108], © 1991, Elsevier. All rights reserved

4.5 Time Projection Chambers

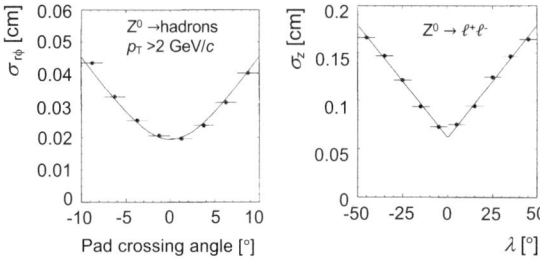

Fig. 4.43 Left: The $r\phi$ position resolution of the ALEPH TPC versus pad crossing angle measured in $Z^0 \rightarrow hadron$ events [108]. Right: The rz position resolution of the ALEPH TPC versus dip angle measured in $Z^0 \rightarrow \ell^+\ell^-$ decays. Both reprinted with kind permission from [108], © 1991, Elsevier. All rights reserved

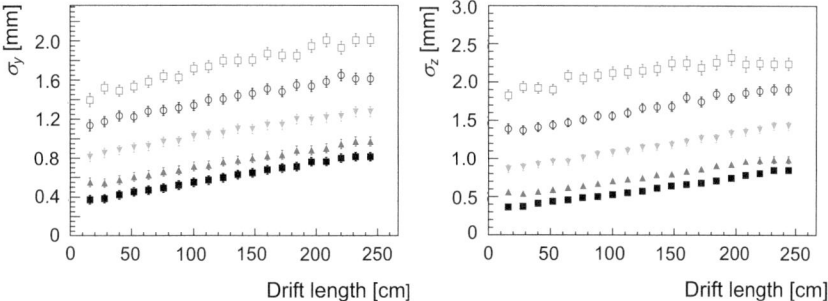

Fig. 4.44 The ALICE spatial resolution in the $r\phi$ (left) and z (right) directions as a function of the drift distance for different track inclination angles, $\tan(\alpha_{tr}) = 0$ (solid squares), $\tan(\alpha_{tr}) = 0.23$, (upward triangles), $\tan(\alpha_{tr}) = 0.46$ (downward triangles), $\tan(\alpha_{tr}) = 0.69$ (open circles) and $\tan(\alpha_{tr}) = 0.92$ (open squares). Reprinted with kind permission from [117], © 2010, Elsevier. All rights reserved

Fig. 4.45 Illustration of the separation of two-tracks

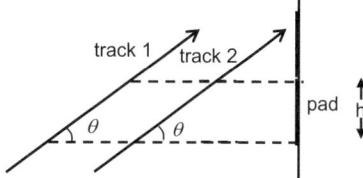

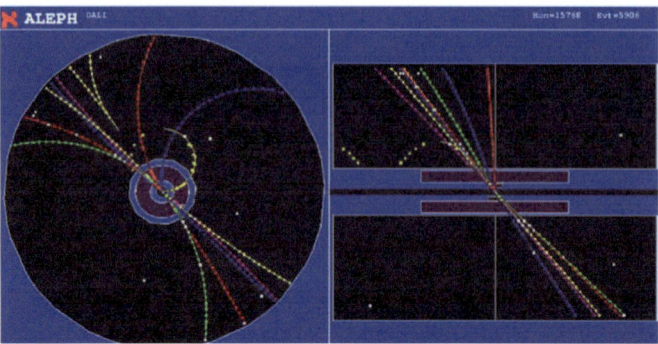

Fig. 4.46 One-event display of $Z \to hadrons$ in the ALPEH detector in the $r\phi$ and rz views, respectively. Reprinted with kind permission from [119], © 1996, the ALEPH Collaboration. All rights reserved

where v_d^e is the drift velocity, t_c is the intrinsic electronic pulse length, t_d is the drift time and h is the pad size. In the azimuth direction two tracks have to be separated by two to three pad widths in order to be resolved. We conclude this section with a one-event display from the ALEPH TPC in Fig. 4.46 showing the decay $Z^0 \to 2\ jets$. The individual charged tracks are well measured and clearly separated from each other.

4.6 Micro Strip Gas Chambers

Though the MWPC is one of the most widely used tracking detector it has two limitations. First, the spatial resolution in the direction perpendicular to the anode wires is limited by the wire spacing being ≥ 1 mm. Second, for rates $\geq 10^4$ particles/s/cm MWPCs develop an inefficiency that is caused by the slowly moving positive ions. To improve on these limitations, we have to reduce the wire spacing and the anode-cathode distance. However, wire spacings below 1 mm and anode-cathode gaps below 2 mm become difficult due electrostatic instabilities. Oed proposed a novel design by using photolithographic techniques that have accuracies of 0.1–0.2 μm [118]. This method allowed for the production of chambers with reduced anode wire pitch and anode-cathode gaps called MicroStrip Gas Chambers (MSGC). They consist of an array of parallel thin metallic strips, which are deposited by photography on an insulating support structure. Figure 4.47 (left) shows a schematic layout of an MSGC. The detector is made of two parts, a low electric-field drift region and a high electric-field amplification region as illustrated in Fig. 4.47 (right). Alternating anode strips and cathode strips are sandwiched between a conductive plane on the top kept at intermediate potential V_d and a cathode plane on the bottom at potential V_b that consists of strips orthogonal to the anode strips. The anode strips are kept at positive potential V_a while all cathode strips are kept at negative potential V_c.

Anode strips are typically 5–10 μm wide while cathodes strips are 60–100 μm wide and the anode pitch is 120–200 μm. The gap between the top plane and the

4.6 Micro Strip Gas Chambers

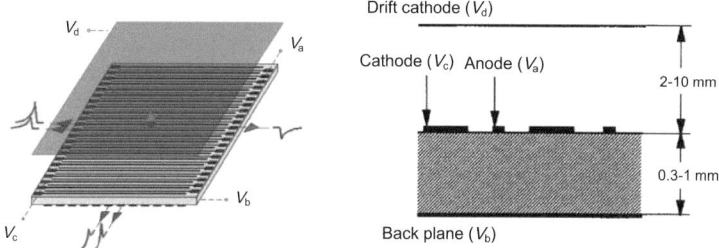

Fig. 4.47 Left: Schematic layout of a microstrip gas chamber [120]. Alternating anode strips and cathode strips are sandwiched between a drift cathode plane on top and cathode strips running perpendicular to the anode strips on the bottom. Reprinted with kind permission from [120], © 2004, IGI Publishing. All rights reserved. Right: A cross section through a microstrip gas chamber showing typical dimensions. Reprinted with kind permission from [121], © 1994, Elsevier. All rights reserved

plane containing the anode strips is the drift region and is 5–10 mm deep. The back cathode plane has wider strips (150 μm) and is separated from the anode plane by 500 μm thick glass, plastic or amorphous material, which has high resistivity (10^9-10^{15} Ω-cm).[12] This is necessary to prevent the build-up of ion clouds between strips at high rates. The chamber is operated with typical gas mixtures like Ar-CO_2, Ar-CH_4, Ar-DME, or Xe-DME. Primary electrons produced in an ionization process of a traversing charged particle produce an avalanche near the anode strips, which induces a negative signal on the anode strips and a positive signal on both neighboring cathode strips.

Several particle physics experiments have designed large arrays of MSGCs [31, 122]. The CMS experiment considered them as a baseline option. The HERMES experiment at DESY used them for tracking closest to the target [31]. The detector consisted of six planes, where four planes had strips running under angles of $\pm 30°$. The detector acceptance was 40 mrad $\leq |\theta_{vertical}| \leq$ 140 mrad and $|\theta_{horizontal}| \leq$ 170 mrad. For an anode strip width of 7 μm and a cathode strip width of 85 μm an anode pitch of 193 μm was chosen. The HERMES MSGC operated with a $\sim 94\%$ efficiency. The average resolution was 71 μm. Despite large development efforts, MSGCs appeared to develop a fast degradation and discharge issues when operated at gains needed to observe MIPs, which sometimes caused irreversible damage. Though the chambers worked well in the laboratory under very high fluxes of X rays, they experienced instabilities and discharge-related damage when operated in test beams [123]. Thus, other technologies were favored for tracking detectors.

---

[12] The term Ω-cm refers to the volume resistivity or bulk resistivity of the semi-conductive material.

4.7 Gas Electron Multiplier

In 1996, Sauli introduced a new concept of gas amplification, called the gas electron multiplier (GEM) [126]. Instead of wires he used kapton sheets with holes as shown in Fig. 4.48 (left). The holes have a typical size of 50–100 μm in diameter with a pitch of 100–200 μm. Figure 4.48 (right) shows a microscope picture of a GEM foil. A high-gradient field is applied across the hole such that electrons start to avalanche in the hole. Using 200 V, the field strength reaches 40 kV/cm. Typical chamber gas mixture are used such as Ar-CH$_4$ (70:30).

The production of a GEM is a simple process, depicted in Fig. 4.49. We start with a sheet of copper-clad kapton. Using masking and exposure to ultraviolet light, we produce a pattern on the top and bottom of the copper layer. The copper is removed at the marked positions by etching and the channels are opened using a kapton solvent. Coupled with a drift electrode above and a readout electrode below the GEM layer, the device acts as a highly performing micro-amplifier. Figure 4.50 (left) shows the electric-field and equipotential lines. The advantage of this detector is that amplification and detection are decoupled allowing an operation of the readout at zero potential. The gain depends on the hole size and lies typically between a few hundred and 2000. The GEM foils are typically characterized by three parameters,

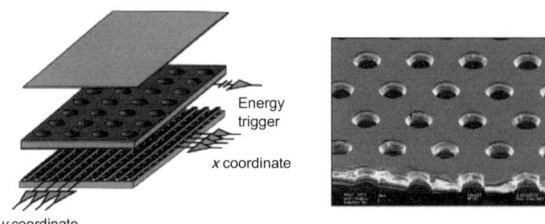

Fig. 4.48 Left: Schematic layout of a GEM detector. Reprinted with kind permission from [124], © 2016, Elsevier. All rights reserved. Right: Electron microscope picture of a GEM detector. Reprinted with kind permission from [125], © 2002, Elsevier. All rights reserved

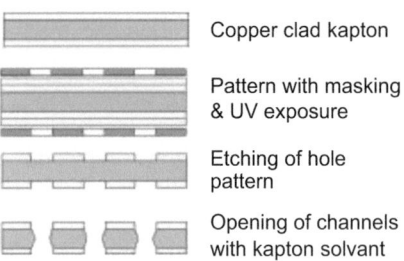

Fig. 4.49 Production of a GEM detector: a sheet of copper-clad kapton (top) is masked and exposed to ultraviolet light producing patterns on the top and bottom of the copper layer (center top). With etching the copper is removed at the marked positions (center bottom) and channels are opened using a kapton solvent (bottom). Reprinted with kind permission from [124], © 2016, Elsevier. All rights reserved

4.7 Gas Electron Multiplier

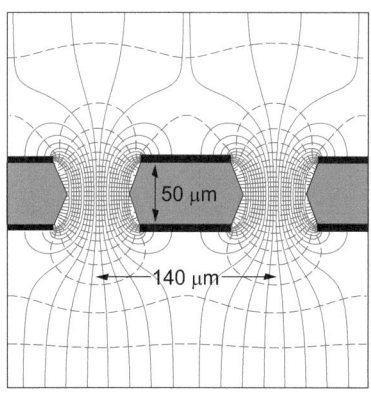

Fig. 4.50 Left: Electric-field configuration in a GEM detector depicting electric-field lines (solid curves) and equipotential lines (dashed curves) [56]. The top part is the conversion and drift region, the GEM layer is the amplification region and the bottom part is the transfer region. Reprinted under CC-BY-NC-4.0 Licence with kind permission from [56], © 2024, the Particle Data Group LBNL. Right: Avalanche formation of two primary electrons in the hole of a GEM detector [127]. Reprinted with kind permission from [10], © 1992, CERN. All rights reserved

the hole pitch (h_p), the hole diameter at the foil surface (h_2) and the hole diameter at foil center (h_1). Due to the etching process, the holes are not cylindrical but conical.

Figure 4.50 (right) illustrates the avalanche formation of two primary electrons in the hole. Primary electrons produced by a traversing charged particle drift towards the GEM, where they encounter a very high electric field and start to avalanche. The avalanche of electrons moves to the readout plane, where it is recorded. Assuming parallel plate geometry, the gain is obtained from the first Townsend coefficient α_I and the distance between the electrodes d_{el} by

$$A_{GEM} \approx \exp(\alpha_I d_{el}). \tag{4.42}$$

The Townsend coefficient can be approximated by (3.16). Figure 4.51 (left, right) shows measurements of $\ln(A_{GEM})/(P \cdot d_{el})$ versus the electric field inside the hole divided by the pressure for pure argon, an Ar-CO_2 (70:30) gas mixture and methane in comparison to the calculated reduced Townsend coefficient α_I/P assuming parallel plate geometry.

Figure 4.52 shows the gain of GEM detectors as a function of the potential across the hole for different geometries. Using an Ar-CO_2 (70:30) gas mixture, the electric drift field in these measurements is set to 1 kV/cm. The highest gains of $A_{GEM} > 1000$ are achieved with small holes ($h_p/h_2/h_1$)=(140/45/30) and (140/60/30). The corresponding gains for large holes (200/140/110) are about 30. Figure 4.53 (left) shows the GEM amplification factor as a function of the metal hole diameter for the same geometries depicted in Fig. 4.52. The gain drops exponentially with the hole diameter. At hole diameters below 70 μm a saturation is observed. This is probably caused by a loss of charge due to diffusion. The pitch plays no role in the gain characteristics but restricts the values of the drift field for full collection. To increase

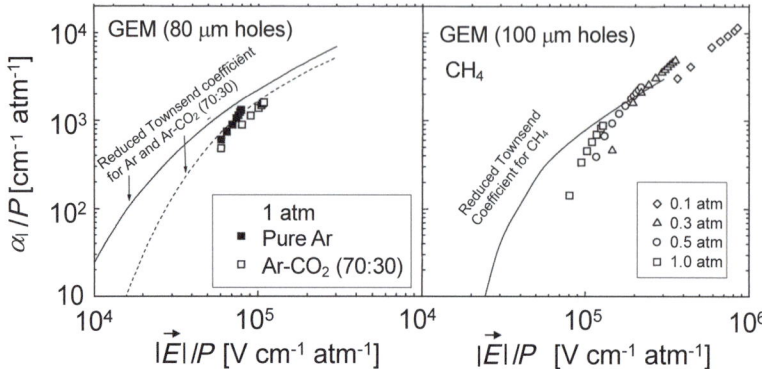

Fig. 4.51 Measurements of $\ln(A_{GEM})/(P \cdot d_{el})$ versus $|\vec{E}|/P$ in comparison to calculated reduced Townsend coefficients α_I/P for pure argon and Ar-CO_2 (70:30) (left) and CH_4 (right) at different pressures. Both reprinted with kind permission from [128], © 1999, Elsevier. All rights reserved

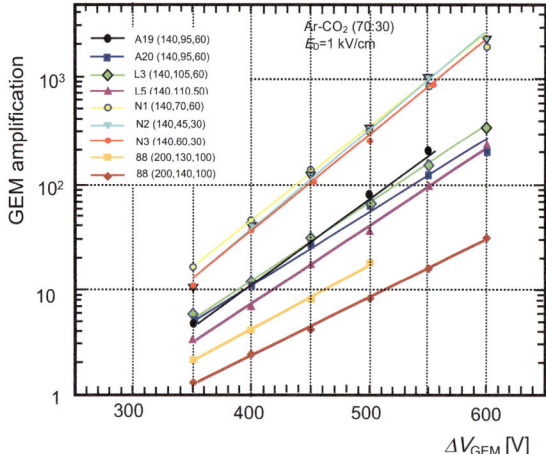

Fig. 4.52 Amplification factor of GEM detectors as a function of applied voltage for different GEM geometries: 140/95/60 (black solid points and blue solid squares), 140/105/60 (green solid diamonds), 140/110/50 (magenta solid upward triangles), 140/70/60 (yellow open circles), 140/45/30 (cyan solid downward triangles), 140/0/30 (red solid points), 200/130/110 (orange squares) and 200/140/110 (brown diamonds). The notation corresponds to the pitch/the diameter of the hole/the diameter of the hole at the center in units of μm. The drift electric field is $E_{drift} = 1$ kV/cm in an Ar-CO_2 (70:30) gas mixture. Reprinted with kind permission from [129], © 1998, CERN. All rights reserved

4.7 Gas Electron Multiplier

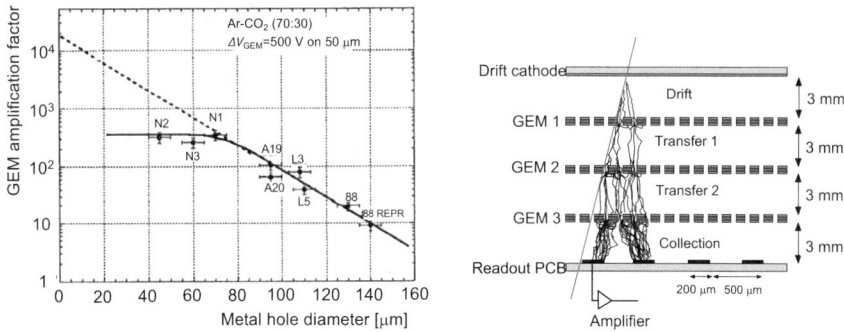

Fig. 4.53 Left: Amplification factor of the same GEM detectors as a function of the metal hole diameter. Reprinted with kind permission from [128], © 1999, CERN. All rights reserved. Right: Schematic view of a triple GEM used in the CMS experiment. The avalanche formation in the three GEMs is clearly visible. Reprinted with kind permission from [134], © 2020, IOP Publishing. All rights reserved

the gain several GEM layers may be combined. Figure 4.53 (right) shows a schematic layout of a triple GEM detector. It consists of a drift region, two transfer regions and a collection region. Electrons produced in the drift region start amplification in the first GEM foil. They transfer to the second GEM foil where amplification continues. The final amplification then occurs in the third GEM. Thus, higher amplifications are achieved than in a single GEM detector. Figure 4.54 (left) shows the gain of a triple GEM detector as a function of the applied voltage for different gas mixtures. In an Ar-CO_2 (70:30) gas mixture gains above 10^5 are achieved with voltages above 1,200 V distributed over the three GEM sections. An effective gain of 10^4 is achieved with an Ar-CO_2-C_4H_{10} (65:28:7) gas mixture around 1 kV.

Another important issue is the ion backflow. In a single GEM, the ion backflow fraction is close to the ratio of the drift field to the induction field. In a triple GEM a suitable sharing of the gains can substantially reduce its value. Figure 4.54 (right) shows the ion backflow fraction as a function of the gain in a double logarithmic scale for two drift field strengths and for different gas mixtures. The ion backflow fraction drops with increasing gain. A reduction of the drift field by a factor of five reduces the ion backflow by an order of magnitude. At a drift field of 0.5 kV/cm and low gains, the Ar-CF_4 (90:10) gas mixture shows a lower ion backflow than the Ar-CH_4 (90:10) gas mixture. At higher gains, however, the ion backflow for both gas mixtures becomes similar. For example, at a gain of 5×10^4 the ion back flow is less than 2%.

Several experiments use triple GEM detectors, including COMPASS[13] [130, 131], TOTEM[14] [132, 133], LHCb and CMS. For the readout of the ALICE TPC, the wire chambers are replaced by four-layer GEM detectors [137]. The GEM foils can be

---

[13] COMPASS is a fixed-target experiment at the SPS at CERN.
[14] TOTEM is located on both sides of the CMS detector and measures total, elastic and diffractive cross sections.

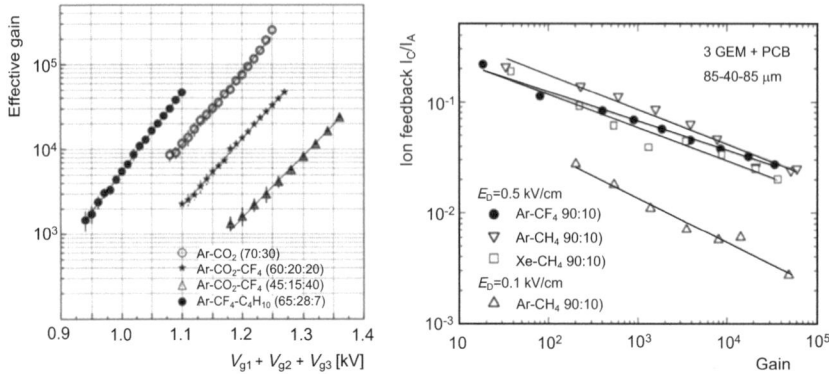

Fig. 4.54 Left: Effective gain of a triple GEM as a function of the applied voltage ($V_{g1} + V_{g2} + V_{g3}$) in the three GEM sections for different gas mixtures, Ar-CO_2 (70:30) [open circles], Ar-CO_2-CF_4 (60:20:20) [stars], Ar-CO_2-CF_4 (45:15:40) [triangles] and Ar-CF_4-C_4H_{10} (65:28:7) [solid points]. Reprinted with kind permission from [135], © 2004, Elsevier. All rights reserved. Right: The ion backflow fraction of a triple GEM with 85-40–85 μm hole sizes in the three layers. Data are shown for gas mixtures of Ar-CF_4 (90:10) [solid points], Ar-CH_4 (90:10) [downward triangles], Xe-CH_4 (90:10) [open squares], all for an electric drift field of $E_D = 0.5$ kV/cm and Ar-CH_4 (90:10) [upward triangles] for an electric drift field of $E_D = 0.1$ kV/cm. Reprinted with kind permission from [136], © 2003, Elsevier. All rights reserved

formed also into cylindrical shape as shown in Fig. 4.55 (left). This technology is used in the BoNus[15] TPC [138] and the upgraded four-layer KLOE tracker [139]. Further details on GEM detectors are given in reference [124]. We focus here on the LHCb and COMPASS chambers as an example and show some properties of the CMS GEM detectors.

The LHCb experiment used 12 GEM chambers in the innermost region of the first muon station because of their performances and radiation hardness [28, 140]. Each chamber consisted of two triple GEM detectors superimposed. The sensitive area was 20 cm × 24 cm. The three GEM foils were sandwiched between anode and cathode planes. A GEM was made by a 5 μm thick kapton foil that was copper clad on each side and was perforated with a high surface density of holes. Each hole had a bi-conical structure with an external (internal) diameter of 70 μm (50 μm) and a pitch of 140 μm. The foils were stretched and were attached to fiber glass frames that defined the gaps. In order to limit the damage in case of discharge, one side of the GEM foil was divided into six sectors, 33 mm × 240 mm in size with a separation of 200 μm between sectors. The anode electrodes were segmented into 192 pads of size 1 cm × 2.5 cm. The detectors were read out on two sides. The chambers were operated with a gas mixture of Ar-CO_2-CF_4 (45:15:40) since a time resolution of 3 ns was measured instead of 10 ns for the Ar-CO_2 (70:30) gas mixture. The typical voltage difference between the two copper sides was 350 V to 500 V yielding electric

---

[15] BoNus is an experiment at the Thomas Jefferson Laboratory that studies the neutron structure in electron-neutron interactions.

4.7 Gas Electron Multiplier

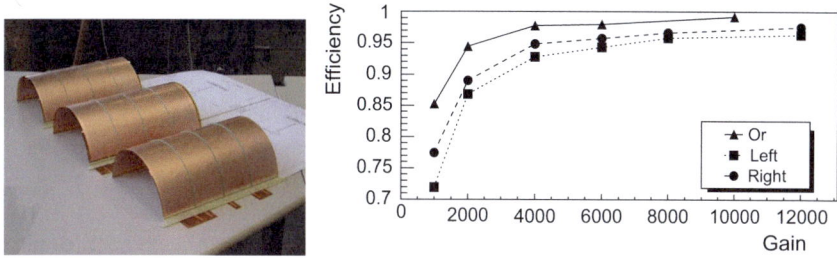

Fig. 4.55 Left: Cylindrical GEM electrodes. Reprinted with kind permission from [124], © 2016, Elsevier. All rights reserved. Right: The efficiency versus gain in a 20 ns time window for a chamber consisting of two triple GEM detectors. The terms left and right refer to one and the other triple GEM. Reprinted with kind permission from [141], © 2007, Elsevier. All rights reserved

fields up to 100 kV/cm inside the holes. Figure 4.55 (right) shows the efficiency as a function of gain for a chamber consisting of two triple GEM detectors. For a gain of 4,000 an efficiency over 98% was achieved if both sides of the chambers were read out.

The Compass experiment at CERN, which was built to study the spin structure in the proton and perform hadron spectroscopy, was the first experiment to use a triple GEM tracking detector [130,131]. The three GEM foils consisted of 50 μm thick kapton sheets covering an area of 31 cm × 31 cm with 5 μm copper cladding on each side. The holes were 70 μm in diameter and had a pitch of 140 μm. The three GEM foils were separated from each other and from the two-dimensional readout board by 2 mm spacers. The readout anode consisted of two sets of perpendicular strips placed at 400 μm pitch. The drift electrode was separated from the top GEM foil by a 3 mm spacer. The electric field for the drift region was 2.49 kV/cm while that across each GEM foil was 3.73 kV/cm. The gas mixture consisted of Ar-CO_2 (70:30). The amplification of the three-layer GEM was 9,800. The GEM detectors were mounted in pairs onto a large-area tracking detector with one detector rotated by 45° with respect to the other. This way, four projections (X,Y and U,V) were measured for each track crossing both GEM detectors of one station. The average efficiency was 97.2% in a single dimension and 95.6% in two dimensions. The position resolution in each plane was $\sigma_{x,y} = 71.6$ μm. If large-angle tracks were discarded the resolution improved to $\sigma_{x,y} = 69.6$ μm.

The CMS experiment has started to upgrade the inner region of the first two layers of the forward muon system using triple GEM detectors [142]. The first layer will have 144 chambers. The GEM foils are 50 μm thick copper-cladded polyimide foils with holes etched in a hexagonal pattern. The holes have a diameter of 70 μm separated by a 140 μm pitch resulting in a distance of 70 μm between two consecutive holes. The chambers are operated with an Ar-CO_2 (70:30) gas mixture. The electric field in the hole is 60 kV/cm. The efficiency is expected to be larger than 97%.

4.8 Micro Mesh Gaseous Structure

To overcome limitations of MWPCs, the micro-strip gas chamber was developed. However, one limitation of MSCGs is that the gain does not exceed 10^4 because the electric field cannot be increased further due to the electric breakdown of the insulator. One new approach proposed by Giomataris [143] is the Micro-Mesh Gaseous Structure (Micromegas), which overcomes this problem. Figure 4.56 (left) shows a schematic layout of a Micromegas detector, which is a two-stage parallel plate avalanche chamber. Figure 4.56 (right) shows a cross section through a Micromegas detector [144]. A drift electrode is placed 3–5 mm above a micro-mesh, which in turn is positioned 50–130 μm above the anode plane consisting of 9 μm thick and 200–400 μm wide gold-plated copper strips that are printed on a 1 mm substrate. Using vacuum deposition, thinner strips can be produced. If needed, both metal deposition techniques are applicable on a 50 μm thick Kapton substrate. The strip pitch is 50–100 μm larger than the strip width. Each readout strip is connected to a low-noise charge-sensitive amplifier ($\sim$ 4 V/pC) with a feedback capacitor. With high-precision polymide spacers, a precise gap can be maintained. The micro-mesh, a 3 μm thick metallic grid with 17 μm openings every 25 μm, is made of nickel using the electroforming technique [145]. With use of this high-resolution emulsion process, a high precision is ascertained to better than 1 μm. The transparency of the micro-mesh was measured to be 45%. The drift electrode consisted of a 100 μm thick nickel mesh having a transparency of 80%. The gas consisted of Ar or Ne with 5–10% admixtures of CO_2, CH_4, C_4H_{10} or DME. A window is built in the center of the chamber allowing to place radioactive sources that are necessary for testing. A newer production technique is based on a woven wire mesh [146].

While readout strips are at ground potential, the micro-mesh and the drift electrode are at different high voltage. The electric field $\vec{E}_D$ in the drift region is 1–5 kV/cm, while the electric field $\vec{E}_A$ in the amplification region is 40–100 kV/cm. The ratio $|\vec{E}_A|/|\vec{E}_D|$ is tuned to large values to achieve optimal operation of this detector. It is also required to catch the positive ions produced in the amplification gap. The high electric field in the amplification region is necessary to quickly collect the positive ions on the micro-mesh. Figure 4.56 (right) also shows the ionization processes of a straight and an inclined charged track traversing the Micromegas detector. The primary electrons from the ionization processes of the straight track are closely spaced. They drift to the amplification region and produce avalanches that are collected by the resistive strips and induce signals on the readout strips. Typically, only one or two strips record the signal. For the inclined tracks the electrons are more spread out. So more readout strips record the signal.

Figure 4.57 (left) shows the electric-field configuration in the drift region and in the amplification region. Due to the two different magnitudes of $|\vec{E}_D| = 1$–5 kV/cm and $|\vec{E}_A| = 30$–50 kV/cm, a funnel effect occurs as illustrated in Fig. 4.57 (right). The electric-field lines are compressed between drift region and the amplification region. Due to the funnel effect, the signal S_1 in the drift region will be projected into the signal S_2 on the anode plane. The ratio S_1/S_2 is approximately given by the ratio $|\vec{E}_A|/|\vec{E}_D|$. For large values of $S_1/S_2 \sim 25/1$, where S_2 is small with respect

4.8 Micro Mesh Gaseous Structure

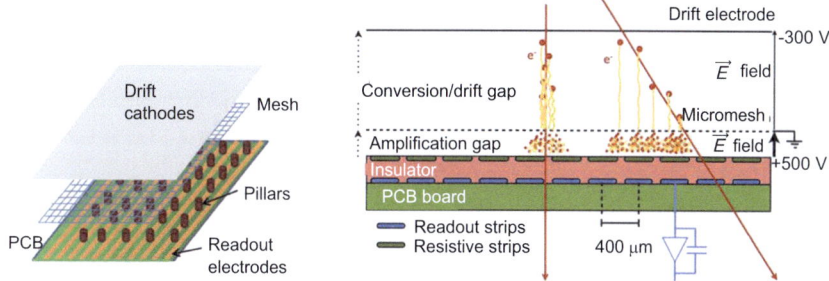

Fig. 4.56 Left: Schematic layout of a Micromegas detector (see text). Reprinted with kind permission from [147], © 2004, Elsevier. All rights reserved. Right: Cross section through a Micromegas detector showing the drift electrode, the micro-mesh, the amplification gap and the readout electrodes. The production of primary ions in the drift region and the avalanche formation in the amplification region is shown for a straight track and an inclined track. Reprinted with kind permission from [148], © 2017, Elsevier. All rights reserved

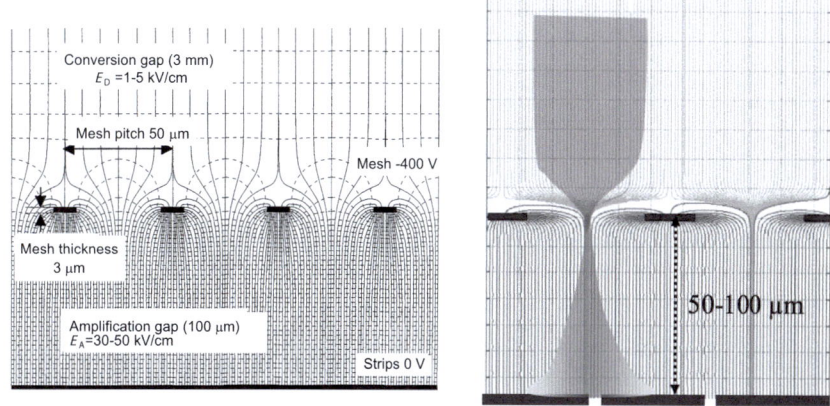

Fig. 4.57 Left: Micromegas electric-field map showing the 3 mm thick conversion gap, the 3 μm thick mesh and the 100 μm thick amplification gap. The electric field in the drift region is $|\vec{E}_D| = 1$–5 kV/cm and that in the amplification region is $|\vec{E}_A| = 30$–50 kV/cm. Reprinted with kind permission from [149], © 1999, Elsevier. All rights reserved. Right: The funnel effect in a Micromegas detector. Charge produced in the drift region is transmitted through a narrow funnel into the amplification regions where due to diffusion a broadening occurs before collection by the resistive strips. Reprinted under CC-BY-4.0 Licence from [150], © 2013, M. Titov

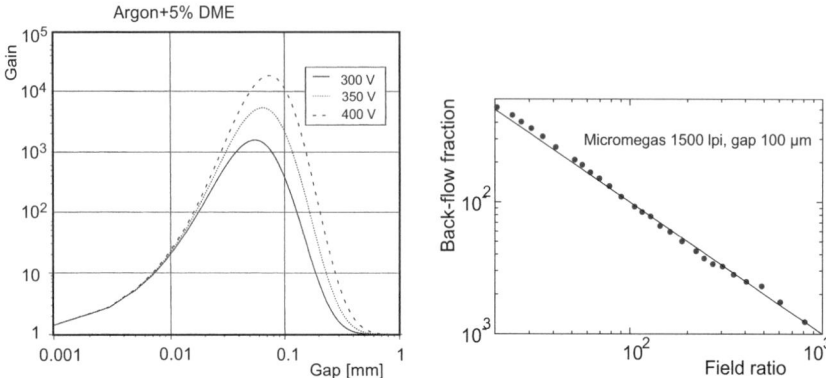

Fig. 4.58 Left: Gain as a function of the gap distance between the micromesh and the anode plane in an Ar-DME (95:5) gas mixture for three different voltages. The optimal gap size is 50–100 μm. Reprinted with kind permission from [149], © 1999, Elsevier. All rights reserved. Right: Ion backflow fraction as a function of the field ratio for a Micromegas detector with 1500 lines per inch and a 100 μm gap size. Reprinted with kind permission from [147], © 2004, Elsevier. All rights reserved

to the avalanche cloud size, it is unlikely that the ions will follow back the field lines into the drift region. Most ions get collected by the mesh.

Figure 4.58 (left) shows the gain as a function of the gap between the micro-mesh and the anode plane in an Ar-DME (95:5) gas mixture for different voltages. The optimal gap size is 50–100 μm. Micromegas detectors operating with such gap sizes are not sensitive to defects in flatness and to variations of the gas pressure. Thus, they have a much larger dynamic range of operation before breakdown. For gap sizes of 50 - 100 μm, signals are fast. Similar gap sizes have been found for a He-isobutane (92:8) gas mixture [144]. Another important parameter is the ion back flow, which drops with a larger $|\vec{E}_A|/|\vec{E}_D|$ ratio. Figure 4.58 (right) shows the ion back flow fraction as a function of the field ratio $|\vec{E}_A|/|\vec{E}_D|$ for an Ar-CH_4 (93:7) gas mixture. The Micromegas detector has a 100 μm gap and uses a mesh of 1500 lines per inch (lip). The ion back flow is independent of magnetic fields [147].

In the narrow amplification gap the electron avalanche is typically collected by one strip. Thus, the spatial resolution is determined by the strip pitch. In order to spread the electron avalanche among several strips two modifications can be done. The first one is to introduce a pre-amplification in the drift gap. This is accomplished by adding a second high-voltage grid in the drift gap as shown in Fig 4.59 (left) [151]. With a potential of 1.52 kV, the electrons are accelerated sufficiently to produce several secondary electrons before reaching the amplification region. With a conversion gap of 3 mm the charge distribution is sufficiently broadened when reaching the amplification region. The avalanche formation in the amplification region broadens the charge distribution further, which is recorded by multiple cathode strips.

The second procedure consists of using a resistive readout. To make a Micromegas detector resistive the pads are covered by a 200 μm insulating layer acting as the capacitance and then a 50 μm kapton with a thin Diamond-Like Carbon layer provid-

4.8 Micro Mesh Gaseous Structure

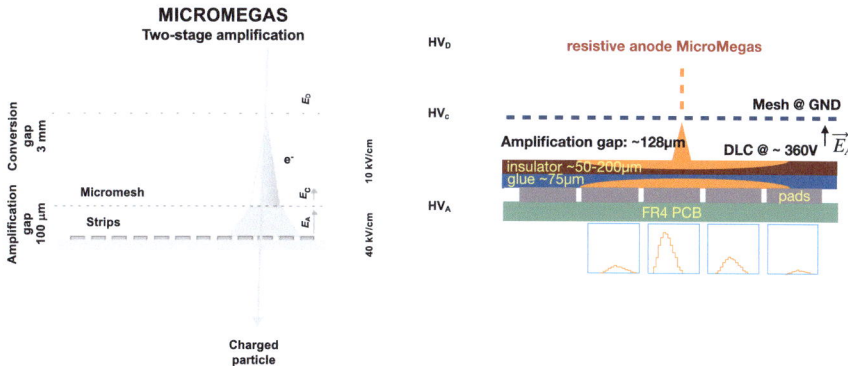

Fig. 4.59 Left: Micromegas detector with preamplification. Here, a high-voltage plane is introduced into the drift region to produce secondary electrons. Since these are more spread out than the primary electrons, the amplification in the amplification gap yields a more spread out charge distribution that induces signals on more readout strips. Reprinted with kind permission from [151], © 2002, Elsevier. All rights reserved. Right: Layout of the resistive Micromegas detectors. The bottom plots show the signal distributions. The electron distribution (light grey) is spread across several pads. With the COG method excellent spatial resolutions are achieved. Reprinted with kind permission from [152], © 2020, IOP Publishing. All rights reserved

ing a designed resistivity of 2.5 MΩ/square.[16] Figure 4.59 (right) shows the layout of a resistive Micromegas detector. The electron avalanche in resistive Micromegas detectors is spread over several strips. Thus, the position is obtained from the COG method yielding better spatial resolutions than for a single strip.

Micromegas detectors may encounter sparking. Though sparking does not damage the chambers it leads to increased dead times. Figure 4.60 shows the efficiency and spark probability as a function of amplification for a Micromegas detector operating with a Ne-isobutane gas mixture (95:5). For a gain above 7,000, the efficiency reaches a plateau above 98% independent whether pre-amplification is used or not. The spark probability is rather low. Without a preamplifier the spark rate is less than 10^{-6} for a gain of 10,000. With a preamplifier the spark rate is further reduced by more than two orders of magnitude. Furthermore, dead zones are potentially very small and the chambers are robust with respect to aging. So, Micromegas detectors are stable devices that are easy to build at low costs. They operate at high efficiencies at high rates. Figure 4.61 (left) shows a comparison of the relative gain versus rate for an MWPC, an MSCG and a Micromegas detector. While the efficiency of MWPCs already starts to drop at rates above 10^4 mm^{-2}s^{-1} and that of MSGCs drops above 10^6 mm^{-2}s^{-1}, Micromegas detectors are still 100% efficient at rates around 10^{10} mm^{-2}s^{-1}. Thus, they are superior to the other two tracking chambers. They also provide ionization measurements.

---

[16] This is the sheet resistance, which is used when measuring a layer or thin film of a semi-conductive material.

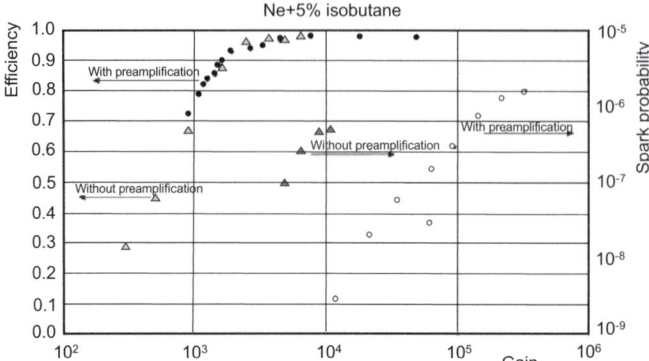

Fig. 4.60 Efficiency (solid points and triangles on the left-hand side of the plot) and spark probability (open points and triangles on the right-hand side of the plot) of a Micromegas detector as a function of the total gain for a Ne-isobutane gas mixture (95:5). Measurements are shown with pre-amplification (points) and without pre-amplification (triangles). Reprinted with kind permission from [151], © 2002, Elsevier. All rights reserved

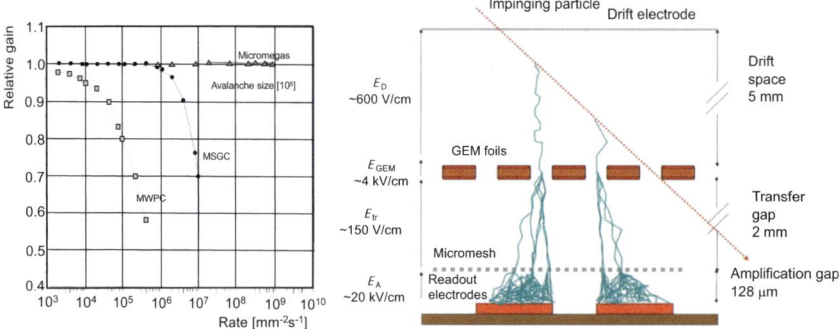

Fig. 4.61 Left: Comparison of the relative gain for an MWPC (squares), an MSGC (points) and a Micromegas detector (triangles) as a function of exposed event rate [154]. Reprinted with kind permission from [154], © 1998, Elsevier. All rights reserved. Right: Schematic layout of a Micromegas detector with a GEM foil in the drift gap [155]. The drift region is 5 mm, followed by a 2 mm transfer gap and a 128 μm amplification gap. The drift field is 600 V/cm and the electric field in the GEM is 4 kV/cm. In the transfer region the electric field is 150 V/cm increasing to 20 kV/cm in the amplification gap. Reprinted under CC-BY-3.0 Licence from [155], © 2013, F. Thibaud

Resistive Micromegas detectors have excellent spatial resolution. In his first prototype, Charpak measured the spatial resolution with two detectors placed back-to-back [149]. Since the two measurements on the track are so close, multiple scattering is negligible. Assuming that both Micromegas have the same resolution, the standard deviation of the position difference measurement divided by $\sqrt{2}$ yields the position resolution of one Micromegas detector. For a detector with a 3 mm drift gap and a 100 μm amplification gap in an Ar-DME (90:10) gas mixture, Charpak measured a spatial resolution of the order of 65 μm independent of the gain of the chamber.

4.8 Micro Mesh Gaseous Structure

Simulation studies indicate that the resolution is mainly determined by the transverse diffusion in the gas. For a 3 mm drift, the measurements are consistent with a transverse diffusion of 200 μm. Using mixtures with CF_4 instead of cyclohexane reduces the diffusion coefficient by a factor of three and increases the number of primary electrons by a factor of 3–4. Simulations indicate that for this geometry a resolution of 10 μm is expected. For drift fields of 0.4 kV/cm and 2.7 kV/cm in a CF_4-iC_4H_{10} (80:20) gas mixture, measurements yielded position resolutions of 18 μm and 11 μm, respectively [153]. In fact, Charpak built Micromegas prototypes at Saclay that consisted of nickel-electroformed micromesh cathodes with square holes of 39 μm × 39 μm at a 50.8 μm pitch with 500 lines per inch (LPI). The anodes consisted of copper strips on a PCB with a pitch of 100 μm. The drift gap was 3 mm and the amplification region was 100 μm thick. All gaps were maintained with precise spacers. The chambers were operated with a CF_4-isobutane gas mixture (80:2). For a drift electric field of $E_D = 1.5$ kV/cm, the measured spatial resolution was 14 ± 3 μm [153].

Micromegas detectors are used in several experiments, for example, in the COMPASS tracker, the T2K TPCs, in digital hadron calorimeters (DHCAL) and ATLAS New Small Wheels. Their properties are summarized in Table 4.6 and photographs of a 1.8 m × 0.8 m plane with 12 detectors for the T2K TPC, a 1 m × 1 m plane with six detectors for a DHCAL prototype and a 1 m × 2 m plane of a Micromegas prototype for the ATLAS Small Wheel are shown in Fig. 4.62 (left, center, right), respectively. We present some more details of the COMPASS and the ATLAS Micromegas.

The COMPASS experiment was the first high-energy physics experiment to use Micromegas detectors for tracking. The 12 Micromegas detectors had an active area of 40 cm × 40 cm with a blind disk of 5 cm diameter around the beam direction [156, 157]. The chambers were arranged in three sets of X, Y, U (45°) and V (−45°) layers. The drift gap was 2.5 mm and the amplification gap was 100 μm. While the drift electric field was 1 kV/cm, the amplification electric field was 50 kV/cm. The chambers were operated with a Ne-C_2H_6-CF_4 (80:10:10) gas mixture at a typical gain of 6,000. The readout strips had a pitch of 360 μm in the central part of the detector (512 strips), which was increased to 420 μm in the outer part (2 × 256 strips). The chambers performed well. The detection efficiency was above 98%. The spatial resolution was 70 μm for inner strips and 78 μs for outer strips. The time resolution was measured to 9.3 ns. However, an issue of discharges remained. To reduce the rate by a factor of 10–100 one solution was to add a GEM foil into the drift region.

The upgraded COMPASS experiment uses 12 layers of Micromegas detectors arranged into three stations with four layers each [164]. The active area is 40 cm × 40 cm around the beam pipe with a hole 5 cm in diameter cut out around the beam pipe. To reduce the spark probability a GEM foil is inserted into the drift gap as shown in Fig. 4.61 (right). The anode plane is divided into three zones, a central zone with 512 strips placed at a 360 μm pitch and two outer zones, each with 256 strips placed at a 420 μm pitch. The detectors have a parallel plate electrode structure with a volume separated into a 5 mm conversion gap, a 2 mm transfer gap and a 128 μm amplification gap. The electric fields in the drift region, across the

Table 4.6 Examples of Micromegas detectors showing the task, amplification mesh structure, drift mesh structure, diffusion gap size, amplification gap size, active area, total area, gain, gas mixture, gas fraction, electric field in the drift region $|\vec{E}_D|$, electric field in the amplification region $|E_A|$, readout strip pitch, readout pad pitch, efficiency and spatial resolution in $r\phi$. Note that LPI stands for lines per inch. [‡]Four active detector layers are combined into a quadruplet. [†]The spatial resolution depends on the electron drift distance in the TPC, which is affected by diffusion. The two values correspond to the small and maximum drift distance, respectively. [§]The first value is for inner strips, the second for outer strips; the old detector COMPASS measured 65 μm

Experiment	COMPASS	T2K	DHCAL	ATLAS
Reference	[157,158]	[159,160]	[161,162]	[163]
Task	Tracking	TPC readout	Hadron energy	Muon tracker[‡]
Amplification mesh [LPI]	500	400	230	325
Diffusion gap [mm]	3.2	–	3	5
Amplification gap [μm]	100	128	128	128
Active area [cm × cm]	40 × 40	96 × 96	33 × 50	
Total area [m^2]	0.32 × 12	9	1	1,280
Gain	6,000	500–2,000	4×10^4	7,000
Gas mixture	Ne-C_2H_6-CF_4	Ar-CF_4-iC_4H_{10}	Ar-CF_4-iC_4H_{10}	Ar-CO_2-iC_4H_{10}
Fraction [%]	(80:10:10)	(95:3:2)	(95:3:2)	(93:5:2)
$|\vec{E}_D|/|\vec{E}_A|$ [kV/cm]	1/50	0.2/25–28.1	0.3/28.9	0.6/40.6
Strip pitch [μm]	360/420[§§]	–	–	& 425–450
Pad pitch [mm × mm]	–	9.8 × 7.0	10 × 10	–
Efficiency [%]	>98	≈ 100	≈100	>90
$\sigma_{r\phi}$ [μm]	70/78[§]	300–700[†]	–	109

GEM layer and in the amplification region are 0.6 kV/cm, 4 kV/cm and 20 kV/cm, respectively. The micromesh consists of a 5 μm thick non ferromagnetic copper grid foil. The chambers are run with the same gas mixture as before. The GEM foil reduces the spark probability as shown in Fig. 4.63 (left), which decreases with a larger electric field across the GEM gap. For a gap voltage of 1260 V the spark probability is already reduced by more than two orders of magnitude with respect to the Micromegas detector without the GEM foil. The mean efficiency of the chamber pixels is better than 97%. The spatial resolution of the pixels (strips) is < 87 μm (72 μm) and the time resolution for both is about 9 ns.

The ATLAS New Small Wheels consists of 16 detector planes, eight detection layers of small-strip Thin Gap Chambers for the trigger (discussed in Sect. 4.14)

4.8 Micro Mesh Gaseous Structure

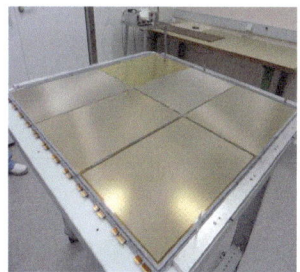
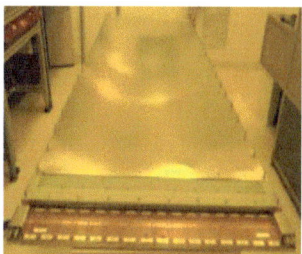

Fig. 4.62 Left: Photograph of a T2K-TPC Micromegas readout plane that consists of 12 detectors spanning an area of 1.8 m × 0.8 m. Reprinted under CC-BY Licence from [165], © 2011, A. Delbart. Center: Photograph of a 1 m × 1 m Micromegas prototype for reading out a digital hadron calorimeter. Reprinted under CC-BY Licence from [161], © 2011, C. Adloff. Right: Photograph of a 1 m × 2 m Micromegas prototype for the ATLAS New Small Wheel. Reprinted with kind permission from [166], © 2023, ATLAS Collaboration. All rights reserved

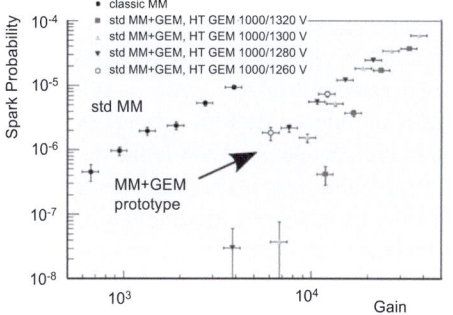
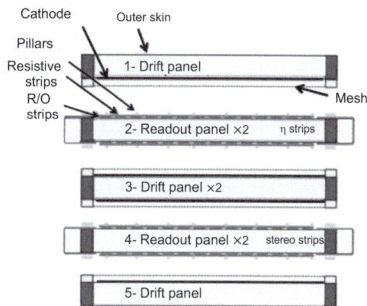

Fig. 4.63 Left: Spark probability of a Micromegas detector with and without a GEM foil as a function of gain for different potentials across the GEM, 1,320 V (squares), 1,300 V (upward triangles), 1,280 V (downward triangles) and 1,260 V (open circles). Reprinted with kind permission from [167], © 2010, the COMPASS Collaboration. All rights reserved. Right: Expanded view of an ATLAS quadruplet Micromegas detector used in the New Small Wheel. The top panel houses the first Micromegas detector with the drift region (light grey), cathode and the micromesh. The second panel, which houses the readout (dark dashes) and resistive strips (grey dashes) of the first Micromegas is just separated by pillars (light grey dots) from the first panel. For illustrative purpose, the panels are pulled apart. The bottom part of this panel holds the readout and resistive strips of the second Micromegas, which follows on the top of the third panel. The bottom side of this panel houses the third Micromegas whose readout is placed on the top side of the fourth panel. The bottom side houses the strips of the fourth Micromegas, which is placed on the fifth panel. Reprinted with kind permission from [163], © 2019, ATLAS Collaboration. All rights reserved

and eight resistive Micromegas layers for precision muon tracking [163]. The Micromegas layers use four axial layers and four U and V stereo layers with angles of $\pm 1.5°$, respectively. Four layers are combined into a quadruplet as shown in Fig. 4.63 (right). The chambers have a 5 mm conversion gap and a 128 μm amplification gap. The electric fields in the two regions are 0.6 kV/cm and 44.5 kV/cm, respectively. The resistive strips have a resistivity of 10 to 20 MΩ/cm. The copper readout strips have a width of 300 μm with a pitch of 425 μm or 450 μm. The mesh is supported by 128 μm high pillars, which guarantee the uniformity of the amplification gap. The cathode is also a PCB having a copper surface. The chambers operate with an Ar-CO_2 (93:7) gas mixture. The drift velocity is $v_d^e = 5$ cm/μs leading to a maximum drift time of 100 ns. For an electric field of $|\vec{E}_A| = 4.5$ kV/cm the gain is 20,000 yielding 100% efficiency. A spacial resolution of 90 μm has been measured.

4.9 Resistive Plate Chambers

Resistive plate chambers (RPCs) were first proposed by Santonico and Sardelli [168]. Figure 4.64 shows the layout of a generic single-gap RPC, which consists of two parallel plates made of a semi-conductor material like painted glass or bakelite covered with linseed oil. The semi-conductor has high resistivity of $\rho_{sc} = 10^{11} - 10^{13}$ Ω-cm. A 2 mm thick gas gap between the plates is introduced by PVC spacers or fishing lines. Graphite layers on the outside of the semiconductor plates define electrodes. The top layer is at ground potential while the bottom layer is set to high voltage. The chamber is read out by x and y strips that are separated from the graphite electrodes via an insulator. Instead of strips also readout pads are used. The charge on the strips or pads is produced via capacitive coupling. While the x strips record positive signals the y strips see negative signals. The RPC is enclosed in a copper sheet providing a

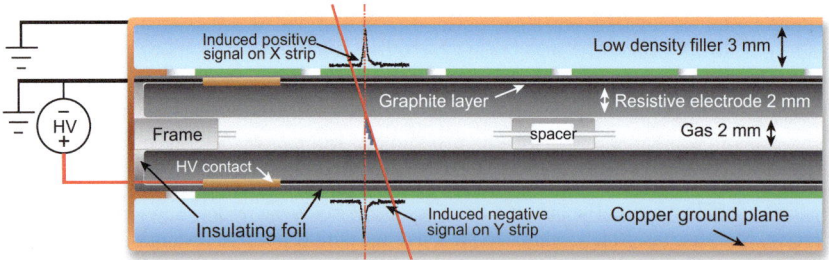

Fig. 4.64 Schematic cross section of a generic single-gap resistive plate chamber. A frame and spacers or fishing lines define a 2 mm gas gap between 2 mm thick resistive electrodes. The electrodes are connected to high voltage, where the cathode is on top and the anode on the bottom. An insulating foil decouples the readout strips from the high voltage. Strips on the top (bottom) run in the x (y) direction. The strips are covered by a 3 mm thick layer of low-density filler and a copper ground plane. A traversing charged particle induces a positive (negative) signal on the x (y) strips. Reprinted under CC-BY-NC-4.0 Licence with kind permission from [56], © 2024, the Particle Data Group LBNL

4.9 Resistive Plate Chambers

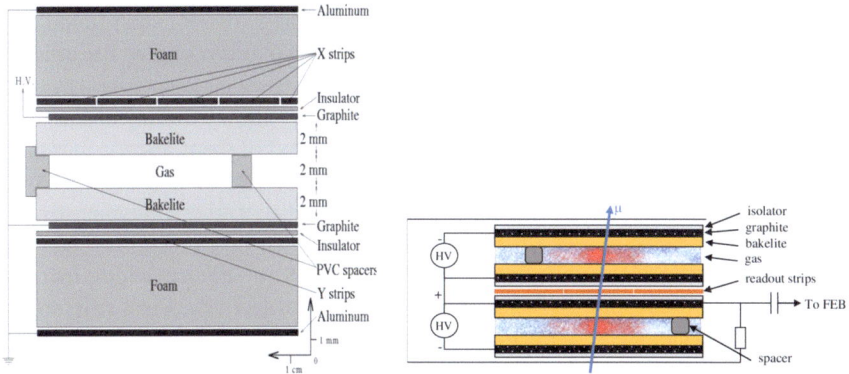

Fig. 4.65 Left: Schematic layout of a single-layer resistive plate chamber used in the BABAR experiment. Reprinted with kind permission from [169], © 2003, Elsevier. All rights reserved. Right: Schematic layout of the CMS double-gap RPC. The readout strips at ground potential lie between the two gaps [170]. Reprinted with kind permission from [170], © 2020, IOP Publishing. All rights reserved

shield, which is insulated from the readout strips by a low-density filler like foam. Note that these chambers use no wires in the sensitive region. Large chambers can be built. Typical sizes are 2 m length and 1 m width. The potential across the gas gap is 8 kV or even higher. Most RPCs use either argon-based gases with admixtures of freon and isobutane or $C_2H_2F_4$-based gases with admixtures of C_4H_{10} and SF_6. For example, the ATLAS RPCs use $C_2H_2F_4$-C_4H_{10}-SF_6 (94.7:5.0:0.3).

The BABAR experiment had instrumented the iron flux return with 19 single-layer planar RPCs in the barrel and 18 single-layer planar RPCs in each endcap [77]. Figure 4.65 (left) shows the layout of the planar RPCs. The readout strips run both parallel (ϕ strips) and orthogonal to the beam direction (z strips). The layers in the barrel were arranged into hexagonal sectors. A barrel layer was segmented into three modules of rectangular shape. In addition, a cylindrical double layer of RPCs was inserted between the electromagnetic calorimeter and the coil with a total of 32 curved modules. So, there were 374 RPC modules in the barrel and 216 RPC modules in each endcap. The endcap modules had trapezoidal-shaped RPCs in the top, center and bottom regions. The two center modules had an additional circular cutout for beam elements. All single-layer RPCs had a 2 mm gas gap while the double-layer RPCs had two 2 mm gas gaps. Each barrel planar layer had 96 ϕ (z) strips running parallel (perpendicular) to the beam direction. For all layers the z strip pitch was 2.8 cm while the ϕ strip pitch varied between 2 cm and 3.5 cm depending on the layer position. Besides ϕ and z strips, the cylindrical double layers had two planes of strips rotated by 30° in addition to reduce ambiguities. An endcap layer had 64 horizontal strips with 2.8 cm pitch in each region and between 55 and 80 vertical strips with 3.8 cm pitch depending on the position in the layer. The RPC electrodes were high-resistivity ($10^{11} - 10^{12}$ Ω-cm) bakelite plates painted with graphite. Two pick-up strip planes, placed on both sides of the chamber provided two-dimensional readout. The chambers used a gas mixture of Ar-$C_2H_2F_4$-C_4H_{10} (60.6:34.9:4.5).

The two graphite surfaces were connected to high voltage (8 kV) and ground, and protected by an insulating mylar film. The position resolutions was of the order of 1 cm.

Other multi-gap devices have been built as well. The CMS experiment uses two-gap RPCs. Figure 4.65 (right) depicts a schematic layout of them. Each 2 mm thick gap is sandwiched between two 2 mm thick layers of bakelite having a resistivity of $1 - 2 \times 10^{10}$ Ω-cm. The readout strips kept at ground potential and being parallel to the beam direction are located between the two gaps. In the barrel 480 rectangular RPCs are installed, most of which have a length of 2.455 m. The strip widths increase from the inner stations to the outer ones to preserve projectivity. Each strip covers $0.13°$ in the azimuth angle. The high voltage across each gap is 12 kV. The chamber gas consists of $C_2H_2F_4$-iC_4H_{10}-SF_6 (96.2:3.5:0.3). Further details are given in reference [171]. Some experiments use even higher-gap RPCs. For example, the ALICE experiment uses a ten-gap chamber in the time-of-flight counters [172].

If a charged particle passes through the gas, an electron from an electron-ion pair produces an avalanche with high amplification. Depending on the electric-field strength the electron produces an avalanche (proportional mode) or a streamer (streamer mode). Thus, Resistive Plate Chambers can be operated in the proportional (avalanche) or in the streamer mode. For example, the *BABAR* planar RPCs were operated in the streamer mode. Figure 4.66 (left) shows the RPC efficiency versus particle rate for operation in the proportional and streamer modes. In the avalanche mode, the chamber is more efficient at higher rates than in the streamer mode. Figure 4.66 (right) shows the RPC efficiency versus high voltage for operation in proportional and streamer modes. For operation in the proportional mode high efficiency is obtained at lower high voltages. Increasing the high voltage keeps the efficiency of the chamber high. However, the contribution from proportional signals to streamer signals changes.

Resistive Plate Chambers are simple, cheap, large area detectors with high efficiency. Thus, they are used in multipurpose detectors, typically in muon systems,

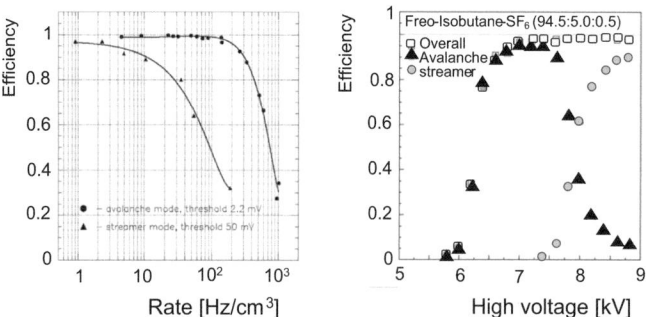

Fig. 4.66 Left: Efficiency of RPCs as a function of event rate for avalanche mode (points) and streamer mode (triangles). Reprinted with kind permission from [173], © 2004, Elsevier. All rights reserved. Right: Efficiency of RPCs as a function of the applied high voltage for avalanche mode (triangles), streamer mode (points) and their combination (squares). Reprinted with kind permission from [174], © 2005, World Scientific. All rights reserved

4.9 Resistive Plate Chambers

where large areas need to be instrumented. Besides *BABAR* also Belle [176] used RPCs in the segmented muon systems, where two RPCs with two readout planes were merged into one superlayer unit. We already mentioned the RPC systems in the ATLAS [180, 181] and CMS [171] muon systems as well as the usage in the ALICE time-of-flight system [185]. The ALICE experiment also uses RPCs in the muon system [172]. Table 4.7 summarizes characteristics of RPCs in some multipurpose detectors. We selected a few examples out of the many RPCs used in experiments [186–191]. Most RPCs are single-gap devices. The main gas component is $C_2H_2F_4$. Some are operated in the streamer mode, some in the proportional mode. Since most of the RPCs are used in muon systems for the trigger, the spatial resolution was not optimized. However, good timing resolution and high efficiencies are important for the trigger, which RPCs have. Figure 4.67 (left) shows the time resolution of the ATLAS RPCs in η and ϕ directions yielding $\sigma_t^\eta = 1.44$ ns and $\sigma_t^\phi = 1.64$ ns. Time resolutions measured in the CMS experiment are 1–2 ns. Figure 4.67 (right) shows the muon efficiency in the CMS experiment as a function of the effective high voltage for different background rates. For a low background rate, an efficiency > 95% is achieved around 7.2 kV. For a background rate of 2.3 kHz the high voltage has to be increased to >7.5 kV to obtain an efficiency of > 95%. The CMS experiment will replace RPCs in the inner rings of station 3 and 4 in the endcaps [170]. In the new chambers both the thickness of the electrode and that of the gas gap will be decreased to 1.4 mm. This reduces the recovery time and thus increases the rate capability in the high-luminosity running of the LHC. In addition, a smaller gap reduces the cluster size for readout strips with a smaller strip pitch (<10 mm). Also, the chambers will operate at lower high voltage.

In *BABAR*, the RPCs developed problems [193]. Dark currents increased and efficiencies dropped down to 40%. This was caused by the formation of linseed oil stalagmites. Three basic conditions have to be met enable the formation of stalagmites:

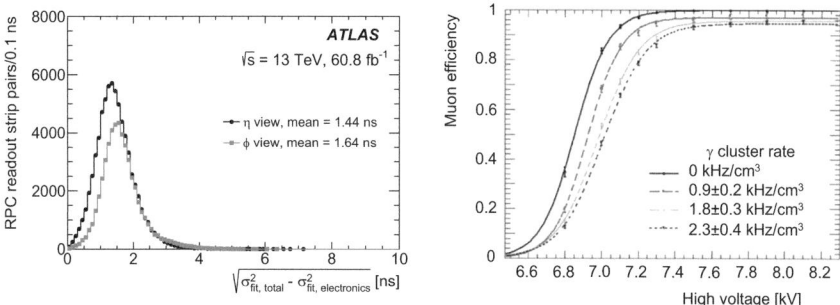

Fig. 4.67 Left: Intrinsic ATLAS RPC time resolution in the η and ϕ directions. Reprinted with kind permission from [192], © 2021, ATLAS Collaboration. All rights reserved. Right: Muon efficiency in the CMS experiment as a function of the effective high voltage measured at different background rates, 0 kHz (solid curve), 0.9 kHz (dashed curve), 1.8 kHz (dash-dotted curve) and 2.3 kHz (dotted curve). Reprinted with kind permission from [170], © 2020, IOP Publishing. All rights reserved

Table 4.7 Examples of resistive plate chambers in recent experiments at B-factories and at the LHC. Listed are the detector system, total area, number of layers (barrel/endcaps), operation mode, number of gaps per chamber, gap size, type of semiconductor electrode, high voltage, bulk resistivity ρ_b, surface resistivity ρ_{surf}, gas mixture, gas fractions, readout pitch in the barrel (ϕ/z or η/ϕ), readout pitch in the endcaps (x/y or ϕ/θ)), efficiency per layer, position resolutions $\sigma_{r\phi}(\sigma_\eta)$, $\sigma_{rz}(\sigma_\phi)$ and time resolution σ_t. In CMS strips run parallel to the beam in the barrel and in radial direction in the endcaps. [†]There are two one-gap chambers. *In layer 2 chambers in the center have 3 gaps. *This is for the barrel. For the endcaps it is a factor of ten higher. [‡]The efficiency deteriorated over time. [§]The first value refers to η strips and the second to ϕ strips. [#]The first value is for clusters with one strip and the second for clusters with two strips

Experiment	BABAR	Belle	ALICE	ATLAS	CMS	ALICE
Reference	[77, 175]	[176, 177]	[172, 178, 179]	[180–182]	[171, 183]	[184, 185]
System	muon	muon	muon	muon trigger	muon	TOF
Area [m^2]	2,000	2,200	143	3,650	2,953	141
# Layers b/e	19/18	15/14	-/4	2 × 3/-	6/4	1/-
Mode	Streamer	Streamer	Streamer	Prop	Prop	Prop
# gaps	1	2[†]	1	1	2 (3)*	10
Gap [mm]	2	1.9	2	2	2	0.25
Electrode	Bakelite	Glass	Bakelite	Bakelite	Bakelite	Glass
HV [kV]	8.0	8.0	10.3	9.35/9.55	8.5–9.0	12–13
ρ_b [Ω-cm]	10^{11}–10^{13}	10^{12}	3×10^9	10^{10}	$1 - 2 \times 10^{10}$	10^{13}
ρ_{surf} [MΩ/square]	0.1	1–10*	–	0.1	–	few
Gas mixture	Ar-$C_2H_2F_4$-iC_4H_{10}	Ar-$C_2H_2F_4$-C_4H_{10}	SF_6-$C_2H_2F_4$-C_4H_{10}	$C_2H_2F_4$-C_4H_{10}-SF_6	$C_2H_2F_4$-C_4H_{10}-SF_6	$C_2H_2F_4$-C_4H_{10}-SF_6
Composition [%]	60.6:34.9:4.5	30:62:8	0.3:89.7:10	94.7:5:0.3	96.2:3.5:0.3	90:5:5
b: RO strip	ϕ/z	ϕ/z	–	η/ϕ	ϕ	–
[cm]	2.0–3.3/3.8	5/5	–	2.3–3.5/same	2.3–4.1/2.0–3.6	–
e: RO strip	x/y	ϕ/θ	x/y	–	ϕ	–
[cm]	2.8/3.8	1.9–4.7/3.6	1,2,4/same	–	1.95–3.63	–
Pad RO [cm^2]	–	–	–	–	–	3.5 × 2.5
Efficiency [%]	94.8[‡]	≥ 98	95	98.5	96–97	99
σ_ϕ (σ_η) [mm]	10	11–17	2.1–5.9	(3.7, 7.3)[#]	b: 8.1–13.2	–
σ_z (σ_ϕ) [mm]	10	11–17	2.1–5.9	(5.8, 6.3)[#]	e: 8.6–12.8	–
σ_t [ns]	1.3	few	2	1.46, 1.56[§]	≤ 2	0.056

- Sufficient linseed oil on the bakelite surface.
- Elevated temperature to soften the uncured oil film so oil molecules become movable.
- A high electric field on the surface to help pull the softened oil film away from the electrode and reach the opposite electrode.

Since the oil stalagmites could not be removed from the chambers without opening them, three procedures were tried out in a test chamber to reduce the impact of the stalagmites:

- Flow of an N_2-O_2 (60:40) gas mixture.
- Flow of dry air.
- Placing the chamber in a metal box maintained at 10% of relative humidity by flushing it with dry nitrogen gas and flowing pure argon gas through the chamber.

After some short-term improvements the efficiency dropped again. Thus, *BABAR* replaced poorly working RPCs in the barrel with limited streamer tubes. Some of the Belle RPCs developed problems showing increased dark currents and lower efficiencies [194]. This was tracked down to high levels of water vapor that was coming through the plastic tubing. After replacing the plastic tubes with copper tubes the chambers returned to normal behavior.

4.10 Limited Streamer Tubes

Limited streamer tubes (LST) were first developed by Iarocci and collaborators at Frascati [195]. Thus, they are also called "Iarocci Tubes". The basic structure is an eight-cell extruded PVC open profile with a cross section of (1–3 cm) × (1–3 cm) and a length of 1–10 m. Figure 4.68 shows an exploded view (left) and a cross section (right) of a single-layer LST. In the center, we see the extruded PVC open profile with anode wires. On top we have a ground plate covered with ϕ strips and on the bottom a ground plate with z strips. The tube walls are coated with graphite having a resistivity of $10 - 10^2$ kΩ/cm. The center of each cell holds an anode wire that is typically 100 μm in diameter. The cathode strips run both parallel and perpendicular to the wires. The whole system is inserted into an uncoated PVC box. A high voltage of 4.5–5.0 kV is applied. The gas typically consists of an Ar-CO_2-iC_4H_{10} mixture, where CO_2 is the dominant component. Figure 4.69 (left, right) shows the construction of a *BABAR* LST chamber and a close-up view of the readout side, respectively.

A charged particle produces electron-ion pairs in the gas volume. The electrons drift towards the anode wire and start to avalanche in the vicinity of the wire. Due to the high electric field, some avalanche clusters grow large and dense enough to excite neighboring ions and create a streamer. The streamer grows throughout the cluster and the high charge density affects the sensitivity of the wire in a region around the streamer. This region of the wire will be unable to produce new streamers until the ions have moved away. Atac and collaborators have photographed limited streamers in an Ar-ethane (50:50) gas mixture [199]. They found that limited streamers are filaments that are 150 μm to 200 μm wide extending along the electric-field lines. The filaments grow with increasing high voltage to a few millimeters from the anode wire. Similar streamer sizes are expected for other gases.

In the proportional mode, the measured charge of the anode wire is proportional to the charge of the primary ionization. In the limited streamer mode, this is different.

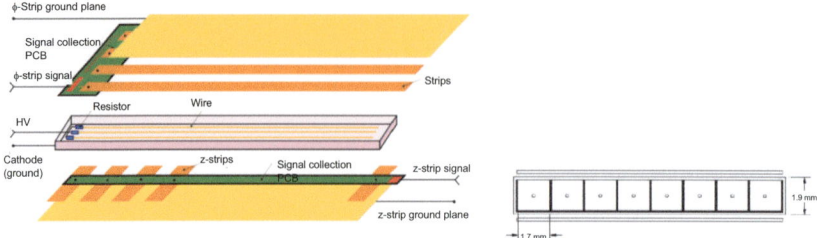

Fig. 4.68 Left: Exploded view of a limited streamer tube [196] showing the extruded PVC profiles hosting the anode wires, a top ground plate with ϕ strips and a bottom ground plate with z strips. Reprinted with kind permission from [196], © 2003, BABAR Collaboration. All rights reserved. Right: Cross section through the BABAR single-layer LST [197] showing the eight cells with the anode wire in the center and the top lid. The cell size is 17 mm × 15 mm. Reprinted with kind permission from [197], © 2005, SLAC. All rights reserved

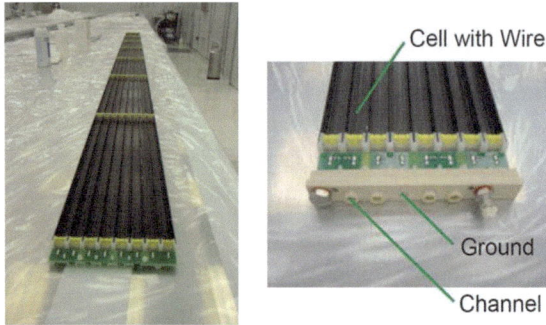

Fig. 4.69 Left: Construction of a BABAR LST chamber. Reprinted with kind permission from [198], © 2005, BABAR Collaboration. All rights reserved. Right: Close-up view of a BABAR LST module near the readout side. Reprinted with kind permission from [78], © 2013, BABAR Collaboration. All rights reserved

The discharge is nearly independent of the primary ionization. It occurs on a section of the anode wire. It has been shown that the discharge takes place over a small region of the wire and does not extend the entire distance between the anode and the cathode [199]. Therefore, it is different from a Geiger-Müller discharge in which the entire cell would be discharged. The limited streamer mechanism is observed in chambers using thick wires (>40 μm) and heavily quenched gases such as Ar-CO_2-iC_4H_{10}.

Limited Streamer Tubes have several advantages:

(i) A large anode signal of $\sim$50 mV/50 Ω load, which permits the usage of simple readout electronics.
(ii) A small dead zone (30 − 300 μs · cm).
(iii) Easy fabrication and reliable performance due to the thick anode wire.
(iv) No need for preamplifiers since signals are large (30 pC).
(v) Reasonably fast pulses with a rise time of the order of 30 ns.

4.10 Limited Streamer Tubes

(vi) Low cost detector allowing to cover large areas.
(vii) High efficiencies, typically >95%.

In the proportional mode the charge is approximately given by [17],

$$Q(V) = Q_0 \exp\left\{\kappa_p \sqrt{\frac{V_0}{V_{\text{tp}}}} \left(\sqrt{\frac{V_0}{V_{\text{tp}}}} - 1\right)\right\} \simeq Q_0 \exp\left(\kappa_p \frac{V_0}{V_{\text{tp}}}\right), \quad (4.43)$$

where Q_0 is the initial ionization charge, κ_p is a constant depending on the gas mixture, temperature and pressure, V_0 is the applied voltage and V_{tp} is the threshold for operation in the proportional mode. In the streamer mode the relation simplifies to

$$Q(V) = \beta_{ls}(V_0 - V_{\text{ts}}), \quad (4.44)$$

where V_{ts} is the threshold for operation in the streamer mode and β_{ls} is another constant. Note that the anode charge Q depends only weakly on the input charge. To minimize crosstalk between adjacent strips, the termination resistor in the high-voltage board circuit was tuned to $R_T = 150\ \Omega$. The efficiency for x-strips is larger than 90% and for y-strips it is about 90%. Figure 4.70 (left) illustrates that the linearity in (4.44) holds for higher voltages. Another important characteristics of an LST is that the slope $\beta_{ls} = dQ/dV$ is independent of the gas composition [200]. Furthermore, for different wire diameters d_w the slopes and anode voltages satisfy the relation,

$$\beta_1 V_1 = \beta_2 V_2, \quad (4.45)$$

or

$$Q - Q_c = \beta_i V_i, \quad (4.46)$$

where Q_c is a constant independent of the wire diameter. Figure 4.70 (right) shows that the data follow the linear behavior of the charge on $V_0 - V_{\text{th}}$. The anode voltages for different wire diameters and the same outer diameter d_{out}, satisfy the relation

$$\frac{V_1}{V_2} = \xi_{12} = \frac{\log(d_{\text{out}}/d_2)}{\log(d_{\text{out}}/d_1)}. \quad (4.47)$$

Figure 4.71 (left) depicts the voltage ratio as a function of ξ_{12} demonstrating that the linear relation is correct. The positive ions produced near the anode wire influence the electric field and thus shape the avalanche formation. The number of electrons increases exponentially,

$$dN_e = N_e \alpha_I(r) dr, \quad (4.48)$$

where dN_e is the number of electrons produced in the radial segment dr and $\alpha_I(r)$ is the first Townsend coefficient, which was defined in (3.16). The total number of electrons in the avalanche is given by

$$N_e = N_{0,e} \exp \int_{d_w/2}^{r_0} \alpha_I(r) dr, \quad (4.49)$$

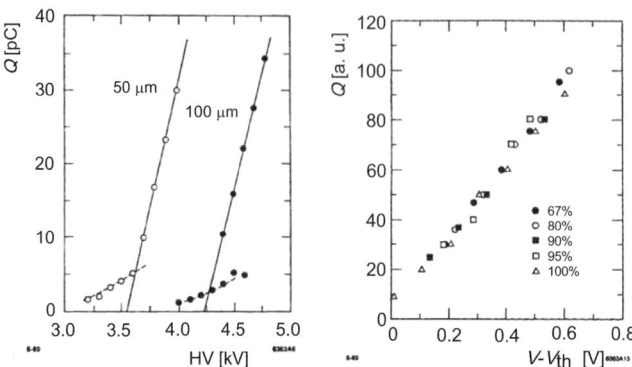

Fig. 4.70 Left: Anode charge versus anode voltage for a chamber with 50 μm diameter wires (open points) and 100 μm diameter wires (solid points) measured in reference. The low-charge points are measurements in the proportional mode while the high-charge points are those in the streamer mode. Note that the charge in the proportional modes has an exponential dependence on the voltage while that in the streamer mode has a linear voltage dependence. Reprinted with kind permission from [201], © 1989, Elsevier. All rights reserved. Right: The collected charge versus $V - V_{th}$ (V_{th} is threshold voltage) for wires with 70 μm diameter in an 8 mm × 8 mm cell for different Ar-CO_2 gas mixtures, with CO_2 contents of 67% (solid points), 80% (open circles), 90% (solid squares), 95% (open squares) and 100% (triangles). Reprinted with kind permission from [200], © 1989, Elsevier. All rights reserved

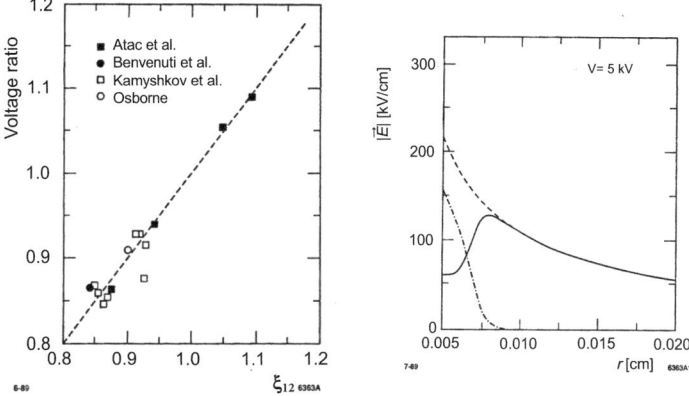

Fig. 4.71 Left: The voltage ratio versus ξ_{12} for data from Osborne [202] (open circles), Benvenuti [201] (solid points), Kamyshkov [203] (open squares) and Atac [199] (solid squares). Right: Electric-field strength as a function of the distance from the wire for an anode voltage of 5 kV showing the total electric field (solid line), that of the positive ions $|\vec{E}_{ion}(r)|$ (dash-dotted curve) and that of the wire $|\vec{E}_w(r)|$ (dashed curve). Both reprinted with kind permission from [200], © 1989, Elsevier. All rights reserved

4.10 Limited Streamer Tubes

where the integral extends from the wire radius to the radius of primary ionization and $N_{0,e}$ is the number of primary electrons. So the electron avalanche is highly non-linear but is proportional to the number of primary electrons. Let us look at an example. For argon gas, $c_1^{Ar} = 14$/cm/Torr and $c_2^{Ar} = 180$ V/cm/Torr while for CO_2 we get $c_1^{CO_2} = 20$/cm/Torr and $c_2^{CO_2} = 466$ V/cm/Torr. Thus, for a wire of 100 μm in diameter held at 4.75 kV the electric field is $|\vec{E}| = 2.1 \times 10^5$ V/cm yielding Townsend coefficients of 5.6×10^3/cm for argon and 2.8×10^3/cm for CO_2. Most of the multiplication occurs near the wire. This causes a large space charge effect of slowly moving ions reducing the electric field

$$|\vec{E}_{\text{total}}| = \frac{V}{r \ln(d_{\text{out}}/d_1)} - |\vec{E}_{\text{ion}}(r)|, \qquad (4.50)$$

where $|\vec{E}_{\text{ion}}(r)|$ is the field of the positive ions. The electric field of the wire $|\vec{E}_w(r)|$ becomes distorted by the positive ions. The total electric field near the anode becomes quasi constant as shown in Fig. 4.71 (right). As the voltage increases, the distortion increases but the value of the total electric field does not change.

Most LSTs are used in muon systems or in hadron calorimeters. Thus, the spatial resolution is not crucial and has not been pushed. Typical resolutions are of the order of few mm to more than 10 mm. D'Agostini and collaborators built a prototype chamber at CERN to study the spatial resolution [204]. The LST chamber had a sensitive area of about 40 cm × 40 cm built from five basic modules. A basic module consisted of eight 9 mm × 9 mm cells, each of which had an anode wire 100 μm in diameter placed in its center. The walls of the eight-cell single extruded PVC profile were coated with graphite having a resistivity of 150–200 kΩ/square. The chambers were operated with an argon-isobutane (33:67) gas mixture at 4.8 kV. For the signal readout, the chambers used two planes of strips. Strips parallel to the wires had a pitch of 0.5 cm while those perpendicular to the wires had a pitch of 1 cm. In the direction perpendicular to the wires the spatial resolution was 350–500 μm and that parallel to the wires was 600 μm. The reason for having a worse resolution for the smaller pitch strips was allocated to higher noise.

As examples, we list properties of the SLD, *BABAR*, OPAL, DELPHI, ZEUS, H1 and MACRO LSTs in Table 4.8. Like RPCs, LSTs have a wide range of applications for large systems like tracking, muon detection and hadron calorimeters. The main gas component is CO_2, butane or pentane operating at high voltages typically above 4 kV. For high reliability, the diameter of the sense wires is typically 100 μm. Efficiencies are high and spatial resolution are less important. We give some more details of the SLD and *BABAR* LSTs.

The warm-iron calorimeter of the SLD detector, which served as a tail catcher of hadron showers and as a muon identifier, used 10,000 LSTs in the iron flux return yoke of the SLD magnet. The LSTs consisted of single layers sandwiched between two electrode sheets. Their lengths varied between 1.9 m to 8.6 m. The LSTs covered an area of 4,500 m². The tube walls were covered with graphite with resistivities between 50 kΩ/square and 2 MΩ/square. In addition to the sense wires, the tubes were equipped with 0.8 cm wide copper cathode strips (x–strips) on the bottom

Table 4.8 Examples of limited streamer tubes in different experiments showing the detector subsystem, number of LST cells N_{cell}, number of cells per extrusion N_{cell}/ext, profile material, wire material, cell size, cell length ℓ_{LST}, wire diameter d_w, pitch of strips running parallel to anode wires x_{str}, pitch of strips running perpendicular to anode wires y_{str}, pad size, anode high voltage, resistivity of cathode coating ρ_c, gas mixture, gas fractions, efficiency, spatial resolution in $r\phi$ or x and spatial resolution in z or y. All Cu-Be wires have a silver wire coating. Note that b (e) stands for barrel (endcap), w (s) stands for wire (strip) and h (v) stands for horizontal (vertical). [‡]Properties are for the endcap muon detectors, which have two double layers. A double layer contains one layer with horizontal wires and one with vertical wires. [†]The resolution is determined from the pitch divided by $\sqrt{12}$. [⋆]The hadron calorimeter has a tower structure; so, the pad sizes are given by $\Delta\theta \times \Delta\phi$, which correspond to pad sizes of about 20 cm × 30 cm. In the endcaps the polar angle bin is $\Delta\theta = 2.62°$. [∗]The wire pitch is 10 mm. [§]The first value refers to the barrel and the second to the endcaps. [#]The strips are at a stereo angle of 26.5°. [&]Two of the 63 planes have an efficiency of 60%. [§§]The resolution of the pads is 10 cm

Exp.	SLD	*BABAR*	OPAL	DELPHI	ZEUS	H1	MACRO
Ref.	[201]	[78,206]	[207,208]	[209,210]	[211,212]	[63,213]	[214,215]
System	Muon	Muon	Muon [‡]	HCAL	Muon	Muon	Tracking
N_{cell}	40,336	9,096	42,496	19,032	43,392	103,000	49,536
N_{cell}/ext	8	7,8	8	8	8	8	8
Material	PVC	PVC	PVC	PVC	Noryl	Luranyl	PVC
Wire	Cu-Be	Au-W	Cu-Be	Cu-Be	Cu-Be	Cu-Be	Cu-Be
Cell [cm^2]	0.9 × 0.9[∗]	1.5 × 1.7	0.9 × 0.9[∗]	0.9 × 0.9[∗]	0.9 × 0.9[∗]	1.0 × 1.0[∗]	2.9 × 2.7
ℓ_{LST} [m]	1.9–8.6	3.75	6	0.4–4.1	0.7–10.1	0.4–2.4	12
d_w [μm]	100	100	100	80	100	100	100
x_{str} (ϕ) [mm]	8	w: 17	8	8	–	–	31+1.5[#]
y_{str} (z) [mm]	40/20[§]	(35+2)	8	–	13+2	17+3	55+5
Pad size [cm^2]	20 × 20	–	–	$\Delta\phi$: 3.75°	–	b: 40 × 50	–
	–	–	–	$\Delta\theta$: 2.96°[⋆]	–	e: 28 × 28	–
HV [kV]	4.75	5.5	4.3	3.92	4.8	4.5	4.23
ρ_c [$\frac{k\Omega}{square}$]	50–2,000	100–600	5,000	50& 150	100	10–30 & 10^4	< 1kΩ
Gas mixture	Ar-CO$_2$-iC$_4$H$_{10}$	Ar-CO$_2$-iC$_4$H$_{10}$	Ar-iC$_4$H$_{10}$ –	Ar-CO$_2$-iC$_4$H$_{10}$	Ar-iC$_4$H$_{10}$ -	Ar-CO$_2$-iC$_4$H$_{10}$	He-C$_5$H$_{12}$ –
Fraction [%]	(2.5:88:9.5)	(3:89:8)	(25:75)	(10:60:30)	(25:75)	(2.5:88:9.5)	(73:27)
Efficiency [%]	90	88	96.7	>95	80–90	80 ± 10[&]	95
σ_x (σ_ϕ) [mm]	2.3–3.5	w: 4.9[†]	h: 3	$\Delta\phi$ 1.08°[†]	w: 2.5	w: 3–4	w: 10
σ_y (σ_z) [mm]	11.5–17/5.8–8.6	(10.7)[†]	v: 1	$\Delta\theta$ 0.85°[†]	s: 0.94	s: 10–15[§§]	s: 0.2°

4.11 Drift Tubes

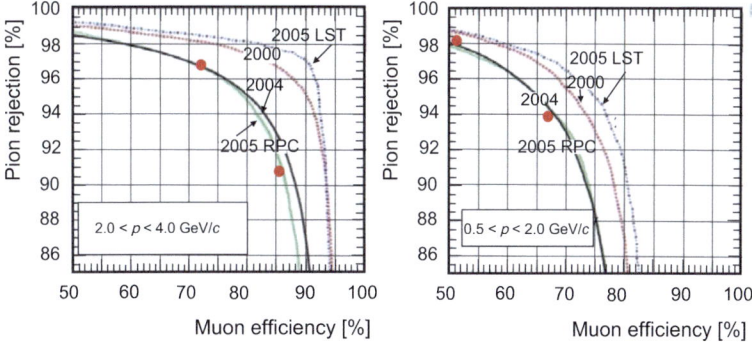

Fig. 4.72 Left: Pion rejection versus muon efficiency for high-momentum muons of the BABAR limited streamer tubes (upper dotted curve) compared to that of the resistive plate chambers in 2000 (lower dotted curve), 2004 (solid black curve) and 2005 (solid green curve). The red point points represent cut-based selections made in 2000. Right: The same curves for low-momentum muons. Both reprinted with kind permission from [218], © 2005, SLAC. All rights reserved

running parallel to the wires and cathode pads (20 cm × 20 cm) on top of the tube. Some chambers also had 4 cm wide copper strips (y–strips) that ran perpendicular to the anode wires. Signals were induced via capacitive coupling similarly as in the RPCs. The LSTs were operated with an Ar-CO_2-iC_4H_{10} (2.5:88:9.5) gas mixture at a high voltage of ∼ 4.7 kV [201].

As mentioned before the BABAR experiment replaced all RPCs in the barrel with single-layer LSTs made from extruded PVCs. The cell size was 1.5 × 1.7 cm². The inner surface was coated with graphite providing a resistivity between 0.2 to 1 MΩ/square. In addition, ϕ- and z- cathodes strips were implemented both having a width of 38.5 mm. The ϕ-strips ran parallel and the z-strips perpendicular to the anode wires. The sense wires were connected to 5.5 kV while the cathode was kept at ground. The LSTs operated with an Ar-iC_4H_{10}-CO_2 (3:8:89) gas mixture. The efficiency was above 95%. Figure 4.72 (left, right) displays the performance of the BABAR LSTs with respect to that of the BABAR RPCs. The LSTs showed a higher efficiency at a higher pion rejection than the RPCs even before degradation. For low-momentum muons, the LSTs performed better than the RPCs.

4.11 Drift Tubes

A drift tube consists of a few-meter-long aluminum pipe that has a diameter of a few centimeters. In its center an anode wire is positioned with a precision of better than 20 μm. The anode wire is on a high potential while the tube is grounded. The counter is operated with an Ar-CO_2 gas mixture. The tubes are glued together to form layers. Typically, three or four layers are combined to form a chamber. Thus, large-area chambers can be built easily at low costs. Since they have excellent spatial resolution, they are used as large-area tracking detectors in muon systems as in the ATLAS and CMS experiments, which we discuss briefly below. Figure 4.73 (left) shows

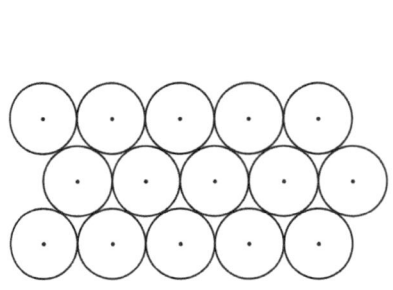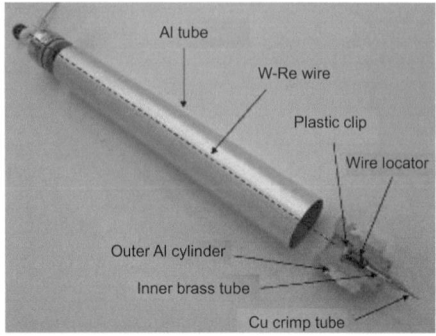

Fig. 4.73 Left: Schematic layout of a three-layer monitored drift tube chamber. Reprinted with kind permission from [42], © 2014, F. Sauli. All rights reserved. Right: Photograph of a monitored drift tube. Reprinted with kind permission from [219], © 2007, Elsevier. All rights reserved

a schematic layout of a monitored drift tube (MDT) chamber and Fig. 4.73 (right) shows a photograph of an individual MDT counter. Besides cylindrical drift tubes also square tubes are used. In the preparations for the GEM (Gammas, Electrons, Muons) detector [205] at the Superconducting Super Collider (SSC) in Texas,[17] studies with drift tubes operating in the streamer mode were conducted. The aluminum tubes had a square cross section of 22.8 mm × 22.8 mm in a 4 × 4 layout. The anode wires in the center of each cell consisted of gold-plated tungsten wires 100 μm in diameter. Readout strips 2.5 cm wide were placed orthogonal to the wires. The chamber was tested with different gas mixtures. For an Ar-C_4H_{10} (25:75) gas mixture at 5.4 kV a spatial resolution of 55 μm was measured in a muon beam [216]. Figure 4.74 (left) shows the drift-time-to-drift-distance relation in this prototype, which is nearly linear. Figure 4.74 (right) shows the spatial resolution as a function of the drift distance. The dependence of the spatial resolution on the drift distance is given by (4.33). For muons the resolution was about 50 μm except in the vicinity of the wire.

The ATLAS muon system uses monitored drift tubes (MDTs) for precision tracking of muons from 10 GeV to 1,000 GeV [217]. The MDT chambers are assembled in three nested cylindrical barrel stations and three main coaxial disks in each endcap. A fourth annulus of endcap chambers covers a limited pseudo-rapidity region spanning from the barrel to the endcap. The 354,000 drift tubes are arranged in 1,172 MDT chambers covering an area of 5,000 m^2. They are constructed with pairs of close-packed multilayers of 3 cm diameter cylindrical aluminum drift tubes. Their lengths vary from 1 m for the inner stations close to the beam up to 6 m for the outer stations. Each multilayer comprises of either three or four single planes, depending on the barrel (endcap) chamber radial (z-axis) position. Each plane contains from 12 to 72 tubes. In the center of each tube, a 50 μm diameter gold-plated tungsten-rhenium sense wire is positioned with an accuracy of <20 μm. Each tube operates

---

[17] The project was cancelled in 1993.

4.11 Drift Tubes

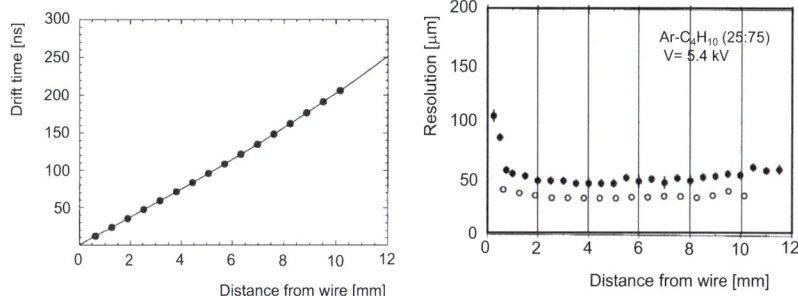

Fig. 4.74 Left: The drift-time-to-drift-distance relation in the drift tube prototype for the GEM detector showing laser data (points) and a fit (solid curve). Right: The spatial resolution as a function of the drift distance from the wire in the drift tube prototype for 500 GeV/c muons (solid points) and laser data (open circles). Both reprinted with kind permission from [216], © 1994, Elsevier. All rights reserved

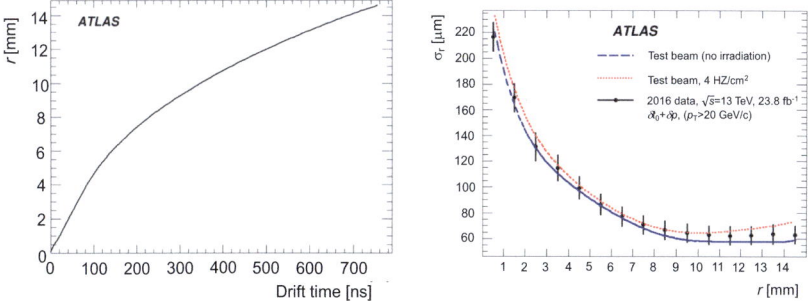

Fig. 4.75 Left: Drift-distance-to-drift-time relation in the ATLAS Monitored Drift Tube. Right: Spatial resolution as a function of drift distance (points with error bars) for the ATLAS MDTs in comparison to that of test beam data before (solid line) and after (dotted line) irradiation. Both reprinted under CC-BY-3.0 Licence from [217], © 2019, ATLAS Collaboration

with an Ar-CO_2 (93:7) gas mixtures at a pressure of 3 bar. The anode wire is at a high voltage of 3.08 kV yielding a gain of 2×10^4.

Figure 4.75 (left) shows the drift-distance-to-drift-time relation, which is not linear but may be approximated by a piecewise linear dependence. The slopes decrease with increasing drift time. Figure 4.75 (right) shows the spatial resolution as a function of the drift distance, which is around 200 μm at 1 mm drift distance from the wire improving to ∼60 μm for distances above 9 mm. The average spatial resolution of a single tube in the barrel is $81.7 \pm 0.2_{stat} \pm 2.2_{sys}$ μm after time-slewing correction at low background. In the endcap the resolution is 100 μm. At a background rate of 500 Hz/cm^2, the single-tube spatial resolution deteriorates to 104 μm. The tube efficiency is ∼100% for nearly the entire track impact radius. The accuracy in momentum measurement is 3% at 100 GeV/c increasing to 10% at 1,000 GeV/c over a pseudo-rapidity range of $|\eta| \leq 2.7$. The dead time is 790 ns.

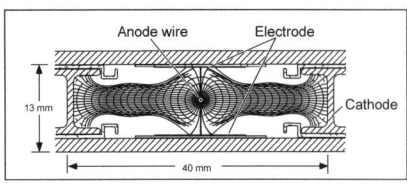 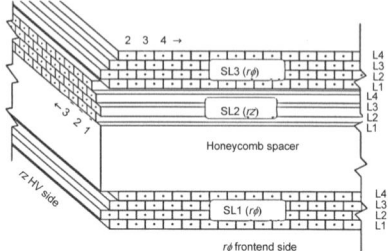

Fig. 4.76 Left: Schematic layout of a CMS drift tube. The electric-field lines and isochrones are also shown. Right: Schematic layout of a CMS drift tube chamber showing three superlayers that consist of four layers each. Superlayer 1 is separated from the other two superlayers by a honeycomb spacer. Both reprinted with kind permission from [220], © 2010, IOP Publishing. All rights reserved

For the high-luminosity upgrade, ATLAS will build new MDTs. The tube diameter is reduced to 15 mm. The chambers operate with the same gas mixture and pressure at 2.73 kV. Each chamber consists of 624 aluminum drift-tubes 2.2 m in length arranged in two multi-layers. Each multilayer is composed of 4 layers with 78 tubes yielding a total width of 1.2 m. Each drift-tube is a multi-component device and must satisfy three requirements. First, the sense wires must be placed in the center of the tube with high precision (better than 20 μm). Second, the wires must be tensioned to 3.5 ± 0.15 N. Third, they must be crimped correctly to the feedthrough.

The CMS experiment uses drift tubes in the muon system in the barrel [221]. The muon barrel consists of five wheels. Each wheel has 12 sectors each covering 30° in azimuth angle. Each sector has four stations equipped with RPCs and drift tubes. In the top and bottom sectors the outer stations are split into two distinct chambers. The drift tubes have a square profile with a cross section of 4 cm $\times$ 1.4 cm. In its center a stainless steel wire 50 μm in diameter is placed. The wire lengths vary between 2 m and 4 m. The cell is made from aluminum I-beams on the sides serving as cathode and aluminum plates on the top and bottom as shown in Fig. 4.76 (left). Copper electrodes are placed on the plates above and below the wire. A drift tube chamber is made of three super layers, each made by four layers of rectangular drift cells staggered by half a cell as shown in Fig. 4.76 (right). The wires in the innermost and outermost superlayers are parallel to the beam line and provide the track measurement in the magnetic bending plane. In superlayer 2, the wires are perpendicular to the beam direction and measure the track position in the z direction. The barrel uses a total of 250 drift tube chambers arranged in four layers containing 172,000 individual cells. The drift tubes are filled with an Ar-CO_2 (85:15) gas mixture. While anode wires are kept at a potential of +3.6 kV, electrodes on the plates are set at +1.8 kV and cathodes are set at -1.2 kV. The single-hit spatial resolution is about 200 μm while resolutions on the segment position and angle in the bending plane are 70 μm and 0.5 mrad, respectively [220].

4.12 Straw Tube Chambers

Straw Tube trackers are similar to drift tube trackers. They are built from straw tubes whose diameters are typically less than 1 cm. A sense wire is placed in its center. Many straws are combined into chambers. Before the era of silicon detectors straw tubes have been used frequently as vertex chambers. Table 4.9 lists properties of straw tube chambers in the HRS, MAC, Mark II, Mark III, AMY and ATLAS experiments. Most of these chambers have 4–6 layers and thin anodes wires of $15 - 30\,\mu m$ diameter. The closest chambers are about 4 cm from the interaction point. The typical gas mixture is argon with ethane. The chambers are very efficient and position resolutions are 50–100 μm. The best resolution is achieved in the MAC chamber yielding 45 μm. The ATLAS transition radiation tracker is somewhat different and will be discussed below after giving some more details about the Mark III straw tube chamber.

Table 4.9 Examples of straw tube chambers used in different experiments as vertex detectors. We show the inner and outer radius r_i, r_o, the wire length ℓ_{wire}, straw tube thickness d_{tube}, sense wire diameter d_w, wire material, number of straws N_{straws}, number of layers N_{layer}, high voltage, gas mixture, gas fraction, gas pressure (P), efficiency and spatial $r\phi$ resolution. [†] PC stands for polycarbonate. [§] The first value refers to wires in layers one and two, while the second refers to wires in layers three and four. [#] The first value refers to the barrel and the second to the endcaps. If only one value is shown it refers to both barrel and endcaps

Experiment	HRS	MAC	Mark II	Mark III	AMY	ATLAS[#]
Reference	[222]	[223]	[224]	[225]	[226]	[180,227]
Material	Al-mylar	Al-mylar	Al-mylar	Al-mylar	Al-PC[†]	Al-polymide
r_i [cm]	9.1	3.56	9.49	5.42	11.8	55.4/61.5
r_o [cm]	11.5	9.04	14.76	13.0	14.7	108.2/110.6
ℓ_{wire} [cm]	40.6–45.7	43.2	75	83.8	57.15	144/37
d_{tube} [mm]	13.82	6.9	8.0	7.8	5.28–5.94	4
d_w [μm]	20	30	20	50	16.3, 15[§]	31
Wire	Au-W	Au-W	Au-W	Au-W	Au-W	Au-W
N_{straw}	352	324	552	640	576	5.044×10^6 4.915×10^6
N_{layers}	4	6	6	12	4	73/160
HV [kV]	1.65	3.9	1.9	3.9	1.75–1.8	1.53
Gas Mixture	Ar-C_2H_6	Ar-CO_2 CH_4	Ar-C_2H_6	Ar-C_2H_6	Ar-C_2H_6	Xe-O_2-CO_2
Fraction [%]	(75:25)	(49.5:49.5:1)	(50:50)	(50:50)	(50:50)	(70:3:27)
P [atm]	1	4	1	3	1.45	1
Efficiency [%]	92	93.9	93		95–97	96
$\sigma_{r\phi}$ [μm]	100	45	90	49	80	130

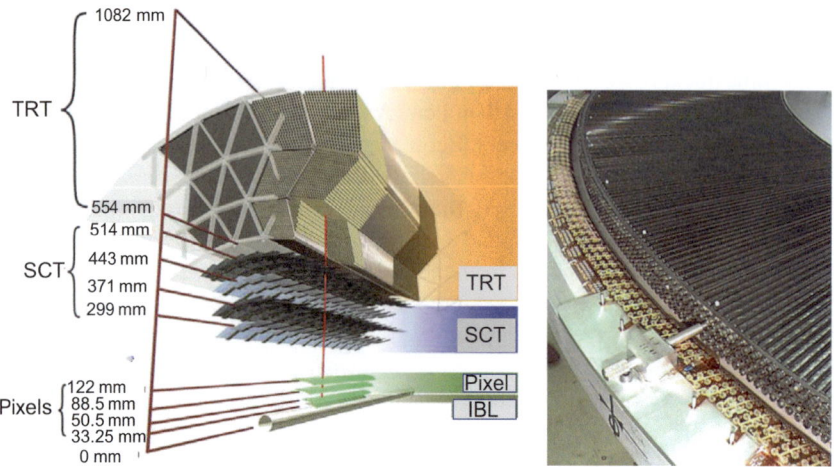

Fig. 4.77 Left: Schematic layout of the ATLAS Inner Detector in the barrel consisting of the Pixel detector, SCT and TRT. Right: Photograph of a four-plane TRT endcap wheel during assembly. Both reprinted with kind permission from [180], © 2008, IOP Publishing. All rights reserved

The Mark III experiment [225], for example, used a 12-layer straw tube array as vertex chamber with eight axial layers and four stereo layers. It was placed at radii between 5.4 cm and 13 cm. The chamber consisted of 640 straws, 8 mm in diameter and 84 cm in length. They were operated with an Ar-C_2H_6 (50:50) gas mixture at 3 atmospheres. The maximum drift distance was 4 mm corresponding to a drift time of 80 ns. The sense wires consisted of gold-plated tungsten wires, 50 μm in diameter. The chambers were operated with a high voltage of 3.9 kV yielding a gain of few $\times$ 10^5. The chamber worked rather well. The overall spatial resolution was 49 μm, which was close to the spatial resolution of 45 μm achieved by the MAC straw chamber. Note that this is still more than an order of magnitude worse than spatial resolutions achieved with silicon vertex detectors that are discussed in Chap. 6.

The ATLAS Transition Radiation Tracker (TRT) depicted in Fig. 4.77 (left) is a straw-tube tracker, which is used for tracking and for electron identification via transition radiation measurements [180]. The straw tubes are 4 mm in diameter. The walls are wound kapton foils reinforced with thin carbon fibers. In the center of each tube a gold-plated tungsten wire 31 μm in diameter is placed. While sense wires are kept at ground potential, the high voltage (+1.5 kV) is placed on the straw tube walls. The tubes operate with a Xe-CO_2-O_2 (70:27:3) gas mixture. Figure 4.77 (right) shows a photograph of a four-plane TRT endcap wheel during assembly. The TRT barrel contains 5,044,224 straw tubes arranged in 73 layers interleaved with fibers running parallel to the beam axis. They are 156 cm in length and extend from 55.4 cm to 108.2 cm in radial direction covering a pseudo-rapidity range of $|\eta| < 1$. The sense wires are read out on both ends. The endcaps have 0.4 m long straws that run perpendicular to the beam direction covering the pseudo-rapidity range from $0.8 < |\eta| < 2.7$. Each endcap consists of 2,457,600 straw tubes (4 mm in diameter,

31 μm diameter sense wire in the center) arranged in 160 layers, which are read out at the outer end. To use the TRT for tracking, we need to know the t_0 for each straw, which is the offset between the start of the readout and the arrival of particles, and the time-to-distance relation. The latter is modeled well with a third-order polynomial. The measured spatial resolution in $r\phi$ is 110 μm.

4.13 Cathode Strip Chambers

A Cathode Strip Chamber (CSC) is a multi-wire proportional chamber that has cathode strips perpendicular to the beam direction that are read out [39]. However, the CSC technology has some advantages with respect to drift chamber technology, such as performance stability and inherent mechanical precision. Furthermore, low-cost detectors can be produced allowing to cover large areas. Since typically charge is induced on several strips, an excellent spatial resolution is achieved via the center-of-gravity method.

In the CMS muon system each endcap has 270 CSCs arranged on four disks (ME1, ME2, ME3, ME4) [228]. The trapezoidal-shaped chambers come in six different sizes, the largest being 3.4 m × 1.5 m. They cover an overall area of more than 1,000 m^2. Figure 4.78 (left) shows the layout of a CSC trapezoidal wedge. All chambers have six wire planes interleaved between seven cathode planes. The wires run at approximately constant η with a pitch of 3.2 mm while the precise ϕ measurement is made by radial strips whose widths vary from 2.5 mm at the inner edge to 15.9 mm at the outer edge. Figure 4.78 (top right) shows the avalanche formation near an anode wire coming from a passing muon in the x direction. Charge is induced on the cathode strips that run perpendicular to the anode wires as shown in the y view in Fig. 4.78 (bottom right). The total number of wires exceeds 2.0 million while the number of cathode strips is approximately 235,000. All strips are instrumented, whereas ten wires are ganged together yielding about 235,000 readout channels.

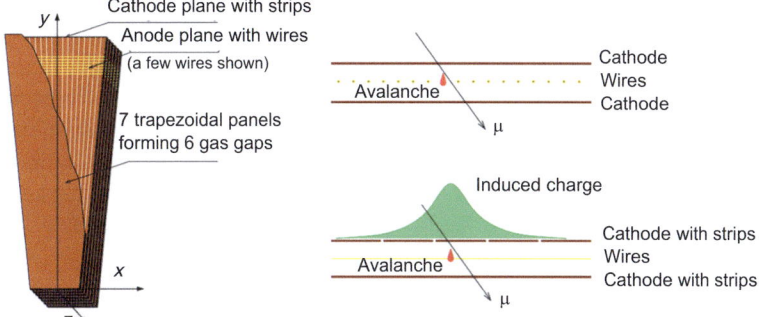

Fig. 4.78 Left: Schematic layout of a CMS cathode strip chamber trapezoidal wedge. Top right: A muon passing through one CSC layer produces an avalanche near the anode wire (x-view). Bottom right: Charge induced on cathode strips (y-view). All reprinted under CC-BY-NC-ND-4.0 Licence from [229], © 2013, B. M. Joshi

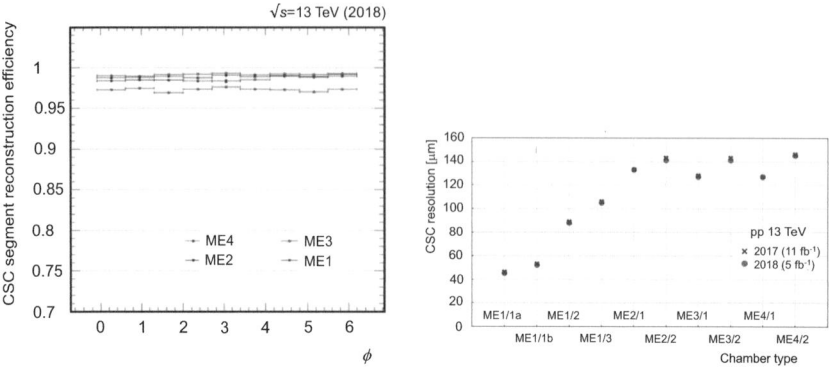

Fig. 4.79 Left: Efficiency of the different cathode strip chamber types in the CMS endcap. The efficiencies have been determined with two different reconstruction algorithms. Right: Spatial resolutions of the CSCs in the four disks in the CMS endcaps. Both reprinted with kind permission from [231], © 2020, IOP Publishing. All rights reserved

The CSCs combine both the precision and trigger functions in one device. They can operate at high rates, in large and non-uniform magnetic fields and require no tight gas, temperature or pressure controls. The chambers are operated with an Ar-CO_2-CF_4 (40:50:10) gas mixture at a high voltage of 3.4 kV. Each chamber has an efficiency >92% providing a track finding efficiency of >99%. Figure 4.79 (left) shows the hit efficiency in the different types of CMS CSCs. The highest efficiency of >98% is obtained with the Road Usage algorithm [230]. Figure 4.79 (right) shows the spatial resolution of the different types of CMS CSCs. In the first disk, the spatial resolutions are around 45 μm increasing to 125-140 μm in disks three and four.

ATLAS uses CSCs in the innermost station of the endcap muon system covering pseudo-rapidities of $2.2 < |\eta| < 2.8$, where particle fluxes are highest [232]. The 32 CSCs are arranged in two rings of eight chambers each. A chamber consists of two identical modules, each having four wire planes. The configuration is equivalent to the one in the MDT system but having finer granularity. The chambers have trapezoidal shape and come in two sizes. The small chambers have a minimum width of 40.3 cm and a maximum width of 73.4 cm with a radial extent of 108.7 cm. The large chambers have dimensions of 61.0 cm, 114.0 cm and 104.0 cm, respectively. The wire pitch is 2.54 mm while the cathode pitch is 5.08 mm. The anode cathode separation is 2.54 mm. The wires are gold-plated tungsten wires with 30 μm diameter. The chambers are operated with an Ar-CO_2-CF_4 (30:50:20) gas mixture at 2.6 kV. The single-point spatial resolution is 60 μm while that of a chamber (four points) is 40 μm. The first pulse of the four planes has a timing resolution of 3.5 ns.

4.14 Thin Gap Chambers

Thin Gap Chambers (TGCs) are also multi-wire chambers consisting of a row of parallel anode wires sandwiched between two cathode planes as shown in Fig. 4.80

4.14 Thin Gap Chambers

(left) [233]. The cathodes are coated with a highly resistive material like graphite. Readout strips or pads are capacitively coupled to the cathodes as depicted in Fig. 4.80 (right). The anode wires are gold-plated tungsten wires with diameters between 20 μm and 100 μm. The anode wire pitch is $\sim$ 2 mm. The anode-cathode gap is 1.0–1.5 mm. Electrons liberated in an ionization process produce a signal on anode wires and induce signals on the readout strips or pads. These chambers have some nice properties. They produces fast signals with a rise time of less than 10 ns and are essentially 100% efficient. Using an operation voltage of 3.5–4.0 kV, they can take particle rates of 300k/cm^2/s. With operation in high multiplication mode the Landau tail becomes small. Thin-Gap Chambers are insensitive to magnetic fields. They can be constructed with high accuracy at moderate costs.

The OPAL experiment used TGCs for electromagnetic presampling in the endcaps [208, 236], which consisted of 32 chambers arranged in 16 sectors, where each sector had one large and one small trapezoidal chamber. While the small chamber was perpendicular to the beam, the large chamber was inclined by 18° with respect to the plane perpendicular to the beam. The cathode planes were 200 μm thick G10 sheets covered with resistive carbon paint separated by 3.2 mm. The anode plane in the center consisted of 69 (432) gold-plated tungsten wires 50 μm in diameter spaced by 2 mm for small (large) chambers. The small (large) chamber had 22 (32) strips placed on one G10 outer surface and 10 (16) pads placed on the other G10 outer surface. The chambers ran with a gas mixture of CO_2-C_6H_{12} at a high voltage of 3.5–3.7 kV. For minimum ionizing particles, the intrinsic resolution was 2.17 mm for single wire hits and 4.62 mm for single strip hits. If the charge was shared by two or more wires or strips the spatial resolution improved. For low-momentum tracks, however, the spatial resolution was worsened by a multiple scattering.

The ATLAS experiment uses TGCs in the muon endcaps, which provide two functions: first, to trigger on muons and second to measure the azimuthal coordinate [180, 237]. The TGCs are arranged in one doublet layer before the toroid magnet and in three layers consisting of one triplet layer before and two doublet layers behind

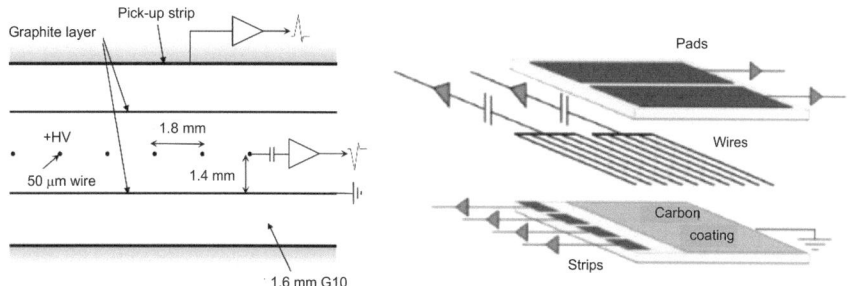

Fig. 4.80 Left: Schematic layout of a thin-gap chamber. Reprinted with kind permission from [234], © 2003, ATLAS Collaboration. All rights reserved. Right: Exploded view of a small thin-gap chamber. Reprinted with kind permission from [235], © 2013, ATLAS Collaboration. All rights reserved

the Big Wheel [238].[18] Each TGC has a trapezoidal shape with a typical size of 1.3 m for the longer base and 1.3 m in height. Each chamber consists of two 1.6 mm thick G-10 plates separated by a 2.8 mm wide gap. The inner walls are coated with graphite. The resistivity is 0.5–2.0 MΩ/square. Individual chambers in a doublet or triplet are separated from each other by a 20 mm honeycomb stiffener. In the center of each chamber, a plane of 50 μm thick gold-plated tungsten anode wires with 1.8 mm pitch is placed that run in the azimuthal direction to measure the radial coordinate. The alignment of sense wires in consecutive layers is staggered to optimize the position resolution and lift the left-right ambiguity. Each sense wire, connected to the high voltage of 2.9 kV, is read out via a decoupling capacitor connected to a preamplifier. Two chambers in a triplet or doublet have 32 copper strips each running along the radial direction to measure the azimuthal coordinate. The gas gap is 2.8 mm. The strip pitch varies between 14.6 mm and 49.1 mm. Anode wires have lengths between 39 cm and 167 cm, while the strip lengths lie between 104 cm and 216 cm. Between 6 and 31 sense wires are grouped together for readout as a function of pseudo-rapidity.

The chambers operate with a CO_2-C_4H_8 (55:45) gas mixture at a pressure of one bar and a high voltage of 2.9 kV. The gas gain is 3×10^5. The highly quenching gas prevents the creation of streamers at high gain. Low-energy neutrons produce pulse heights that are just a factor of 30 above those of minimum ionizing particles. The efficiency is larger than 99%. The long anode wires are supported to reduce sagging. A total of 3,588 TGCs with 318,000 channels is used in the ATLAS detector. The high electric field around the anode wire and the small wire pitch yield good time resolutions of 4 ns. Within a time window of 25 ns more than 99% of the signal is recorded. The spatial resolutions are 2–6 mm in the radial direction and 3–7 mm in the azimuthal direction.

For Run 3, ATLAS has upgraded the Small Wheels in the endcaps [235] with New Small Wheels. Figure 4.81 shows a photograph of a completed New Small Wheel and Fig. 4.82 (left) displays a schematic layout. Each New Small Wheel is made from eight large and eight small pie-shaped sectors. A sector consist of two Micromegas detectors sandwiched between two small-strip Thin Gap Chambers (sTGC) as shown in Fig. 4.82 (right). Each of the four chambers consist of four layers. The cell geometry, electric field and gas mixture are the same as those of the ATLAS TGCs in the endcaps. The cathode planes of inner (outer) chambers consist of a graphite-epoxy mixture with a surface resistivity of 100 (200) kΩ/square that are sprayed onto 100 (200) μm thick G10 sheets. One outer G10 surface is covered with copper strips that run perpendicular to the sense wires while the other outer G10 surface is covered with copper pads that are used for fast trigger purposes. Both, the Cu-strips and Cu-pads are on ground potential and are read out. The strips have a 3.2 mm pitch (2.7 mm strip width 0.5 mm gap) that is much smaller than that of the ATLAS TGCs. The pad size is about 8.7 cm $\times$ 8.7 cm. Five anode wires are ganged together for the readout. The four chambers are mounted together in a honeycomb structure. Positioning of the strip planes is required to better than

---

[18] This is the second tracking station behind the toroid in each endcap consisting of MDTs.

4.14 Thin Gap Chambers

Fig. 4.81 Photograph of an ATLAS New Small Wheel. Reprinted with kind permission from [239], © 2013, ATLAS Collaboration. All rights reserved

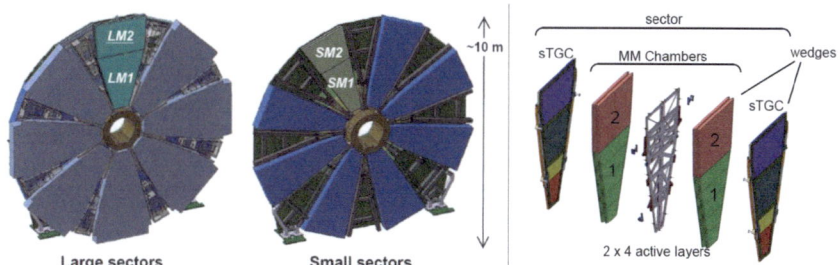

Fig. 4.82 Left: Schematic view of the ATLAS New Small Wheel that is built from large sectors on one side and small sectors on the opposite side. Right: Layout of a New Small Wheel wedge consisting of one four-layer sTGC wedge, two quadruplet Micromegas (MM) and another four-layer sTGC. Both reprinted with kind permission from [240], © 2020, IOP Publishing. All rights reserved

40 μm. The chambers are operated with the same gas mixture and high voltage as the TGCs. Figure 4.83 (left) shows the spatial resolution measured with a four-layer prototype in a CERN test beam with 32 GeV/c pions in six different runs. The average over different runs and layers for perpendicular tracks yields a spatial resolution of $\sigma = 44 \pm 4$ μm. About 95% of all events are contained within a 25 ns time window. Figure 4.83 (right) shows the intrinsic strip spatial resolution of a production module versus the anode high voltage. At the nominal operation voltage of 2.9 kV, a spatial

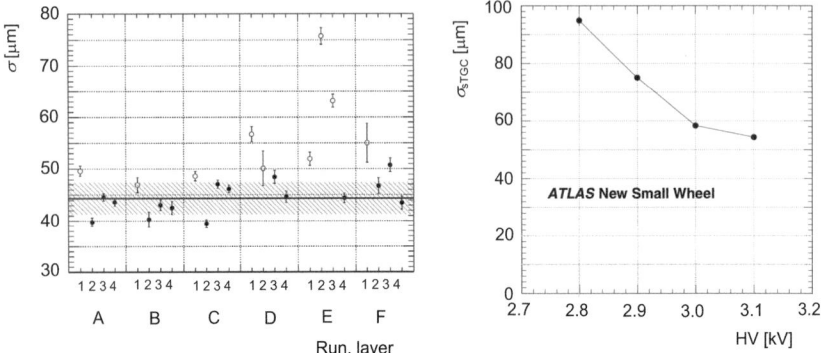

Fig. 4.83 Left: Spatial resolution per layer measured in six different runs in a beam test with a four-layer prototype of sTGCs. The solid points show runs with no expected degradation. The line is a fit to all layers in all runs. Reprinted under CC-BY-4.0 Licence from [241], © 2016, A. Abusleme. Right: Spatial resolution as a function of the anode high voltage for a new Small Wheel production module. Reprinted with kind permission from [240], © 2020, IOP Publishing. All rights reserved

resolution of about 75 μm is measured. Though this is worse than the measured resolutions with the prototype, it is smaller than the design requirement of 100 μm.

4.15 Bubble Chambers

The bubble chamber was invented by Glaser in 1952 [242] for which he was awarded the Nobel Prize in Physics in 1960. A bubble chamber is a volume filled with a liquefied gas with a pressure below but near the boiling point (T_b). After a charged particle passes through the chamber, the volume is expanded to exceed the boiling point. Due to the heat from recombining ions, bubbles are formed along the particle trajectory. A camera takes pictures of the trajectories, which are scanned by people to look for interesting processes. Using high-grain films, resolutions of a few μm were achieved.

At CERN, the Gargamelle bubble chamber was designed to detect neutrino interactions [243]. It operated with 12 m^3 of freon in a muon neutrino beam from 1970 to 1978. In 1973, the discovery of neutral currents was announced shortly after its prediction [244]. In its time of operation, about 83,000 neutrino events were observed of which 102 were neutral-current events. Figure 4.84 (left) shows a photograph of a neutral current interaction, which represents the process $\nu_\mu e^- \rightarrow Z^0 \rightarrow \nu_\mu e^-$. An incoming energetic neutrino interacts with an electron in the atom via a virtual Z^0 boson. The electron is kicked out and radiates photons that are converted into e^+e^- pairs. The neutrino leaves the bubble chamber (on the top of the picture). Gargamelle terminated operation after it developed a crack in 1978. An even larger bubble chamber, the Big European Bubble Chamber (BEBC), started operation inside a 3.5 T superconducting magnet at the Proton Synchrotron in 1973 [245]. In 1977 it was moved to the Super Proton Synchrotron where it operated till 1984. The chamber

4.15 Bubble Chambers

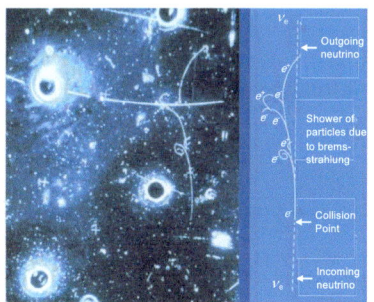
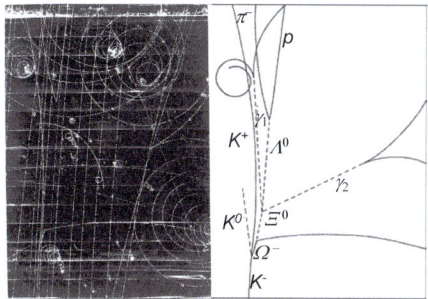

Fig. 4.84 Left: Observation of neutral currents in the Gargamelle bubble chamber showing the photograph on the left-hand side and an interpretation of the particles on the right-hand side [249]. The neutrino enters from the bottom (dashed line). At the collision point an electron is kicked out (solid line). The electron first moves in the direction of the neutrino. It radiates off photons (dotted lines) that convert into e^+e^- pairs. Reprinted with kind permission from [10], © 1992, CERN. All rights reserved. Right: Observation of the Ω^- in the Brookhaven bubble chamber showing the photograph on the left-hand side and an interpretation of the particles on the right-hand side. The Ω^- cascades down via a $\Xi^0\pi^-$, and $\Lambda^0\pi^0$ to a proton and a π^-. Reprinted under CC-BY-NC-ND 3.0 Deed Licence from [250], © 2011, BNL

had a volume of 35 m³ of hydrogen, deuterium, or hydrogen-neon gas mixture. It produced 6.3 million photographs during its operation.

The 80-inch bubble chamber at Brookhaven discovered the Ω^- baryon in 1964 [246]. This confirmed the approach by Gell-Mann, who classified mesons and baryons into multiplets in the Eightfold Way [247,248]. The Ω^- is a bound state of three strange quarks that had not been seen before. Gell-Mann predicted its quantum numbers including mass, spin and strangeness.[19] So, the discovery was a triumph for the Eightfold Way, marking the beginning of a wider acceptance for the concept of quarks. Figure 4.84 (right) shows a photograph taken by the Brookhaven bubble chamber. The Ω^- was produced in a K^-p interaction. Besides the Ω^-, a K^0 and a K^+ are produced to conserve the strangeness quantum number in strong interactions. The Ω^- then cascades down: $\Omega^- \to \Xi^0\pi^-$, $\Xi^0 \to \Lambda\pi^0(\to 2\gamma)$ and $\Lambda \to p\pi^-$.

Despite the superb position resolution, the bubble chambers has several disadvantages. The formation of bubbles cannot be read out electronically. One has to rely on photographs, which are two-dimensional. Scanning photographs is time consuming and cannot compete with high-statistics measurements. The procedure is not well-suited to measure short-lived particles, since the super-heated phase must be ready at the right time. The bubble chamber has a considerable deadtime ($\sim$1 ms). Thus, it is not suited to operate at higher rates since the efficiency rapidly drops with an increased event rate. Furthermore, it is not possible to trigger on specific events.

---

[19] Strangeness is a quantum number that refers to the number of strange quarks in the meson or baryon.

4.16 Other Detectors

In past fixed-target experiments other tracking detectors were used, which include

- Streamer chambers that produce visible streamers along the particle trajectory caused by discharge of the ionized gas [251].
- Spark chambers that produce a discharge of the ionized gas, which is photographed [252,253].
- Flash chambers that produce a large discharge that is recorded [252,254].
- Emulsion detectors using photographic plates that are exchanged and developed from time-to time [255,256].

We will not discuss these detectors in further detail. The interested reader can get more information in older textbooks [257,258]. Except for the emulsion plates, all other detectors listed above are no longer used since it is difficult to read them out electronically, which is important for high efficiencies at high counting rates in multi-purpose detectors. Like a bubble chamber, emulsion plates have a very high position resolution determined by the grain size of the plate. A charged particle passing through the emulsion leaves an imprint. Since these detectors cannot be exchanged frequently, they are only used for rare processes. After development of the film, the process becomes visible. The analysis, however, is cumbersome as each plate needs to be inspected visually and may contain more than one event. The development of semiconductor detectors discussed in Chap. 6 have revolutionized the measurement of particle positions.

4.17 Position Resolution of Tracking Detectors

Table 4.10 shows a comparison of spatial resolutions of the different tracking detectors we discussed or we mentioned. Large drift chambers, jet drift chambers and TPCs provide good $r\phi$ resolution of the order of 100 μm. Small drift and jet drift chambers yield $r\phi$ resolutions of 60 μm. Time resolutions depend on the drift length and drift velocity. For $v_d^e = 1$ cm/μs, a variation in drift length of 10 μm corresponds to 1 ns. For faster drift velocities, larger drift lengths correspond to 1 ns. Thus, time resolutions are typically a few ns. These detectors were used as tracking detectors. The MWPCs, which also have reasonable spatial resolution, have been used mostly in TPC endcaps, muons systems and TOF counters. Both GEM and Micromegas detectors reach excellent position resolutions. The GEM and MicroMegas detectors are used for tracking in several experiments as well as for the readout of TPCs. Both RPCs and LSTs are mainly used in muon detectors and hadron calorimeters, where excellent spatial resolution is not too important. The poor spatial resolutions measured in different experiments are due to large strip and pad pitches to save readout electronics. In the ATLAS experiment for example, RPCs are used to trigger on muons. So excellent spatial resolution is not relevant but time resolution is important, which is of the order of 1 ns. Other trigger chambers are TGCs, whose

4.17 Position Resolution of Tracking Detectors

Table 4.10 Comparison of average position resolutions in tracking detectors. §Resolution for cylindrical detectors is typically in $r\phi$ and z while for planar detectors it is in x and y. †Value in parentheses is from SLD. *This depends on the gas mixture; spatial resolutions of 11 μm were achieved with a CF$_4$-iC4H$_{10}$ (80:20) gas mixture at $|\vec{E}| = 2.7$ kV/cm in a prototype. ‡Here the resolutions are for ϕ and η

Detector	$\sigma_{r\phi}$ or σ_x [μm]§	σ_z or σ_y [μm]§	reference
Small MWPC	150	35–45	[41]
Planar MWPC	150 – 300	150 – 300	[27]
Cylindrical MWPC	179	440	[27,32]
Small DC	60	60	[60]
Planar DC	55–250	160	[61,63]
Cylindrical DC	90–150	100–150	[74,77]
Jet DC	135–170	0.6–1% ·ℓ_{wire}	[90,92]
Small Jet DC	40–55	1% ·ℓ_{wire}	[87,98]
Mini-jet vertex chamber	20–25	35 (at 560 kV/cm)	[99]
TPC	150–180	160–900	[36,100,110]
GEM	70	70	[131,259]
Micromegas	70–110 (11–14)*	–	[153,158,260]
RPC	2,000–11,000	2,000–11,000	[180]
LST	2,300- 5,000	900–11,000	[208,212]
MDT	70–200	–	[180]
Straw chamber	45–130	–	[222,223]
CSC‡	40–140		[180]
TGC	2,000–6,000	3,000–7,000	[180]
sTGC	44, 75		[240]
Bubble chamber	Few	–	[242,261]
Emulsion	1	1	[255,256]
Si strip	2–7	Few	[262]
Si pixel	2	2	[262]

spatial resolution is quite large as well. For the ATLAS run 3, upgrade of the new Small Wheel new sTGCs providing triggers also yield excellent spatial resolution. Both, drift tubes and CGCs provide excellent spatial resolution and are used for precise muon tracking. Despite their excellent spatial resolution, Bubble chambers have been phased out since they cannot be read out electronically and they are too slow to record short-lived particles. Similarly, streamer chambers, flash chambers and spark chambers are no longer in use since they are not competitive as well. Emulsions have been used in neutrino experiments because of their excellent spatial resolution. Since event rates are low due to the small neutrino cross section, the exchange of the photographic plates on a few days basis was fine. With the development of solid-state detectors spatial resolution were improved drastically, so that secondary vertices could be reconstructed. It first started with silicon strip sensors. Then due to increased occupancies, silicon pixel sensor were produced. We discuss these detectors in great detail in Chap. 6.

Exercises

4.1 Consider a big MWPC, which has one plane of wires in the x-direction, spaced every 3 mm. The distance to the frame is also 3 mm. The wires have a diameter of 25 μm. The wires are tensioned such that sagging may be ignored. After charged particles have passed through the detector, you observe the following patterns of charges: a) 11 μC on wire 9, 72 μC on wire 10 and 20 μC on wire 11; b) 51 μC on wire 10; c) 43 μC on wire 9 and 59 μC on wire 10. Determine the y-coordinates of the passing tracks and their uncertainties. Assume that the noise contribution is 1 μC. Hint: to determine the uncertainties on the charge measurements, assume that the MWPC has a gain of 10^7 and that the signal is further amplified in a shaping amplifier with a gain of 3×10^4. What systematic errors have we ignored?

4.2 Consider the same chamber as in problem 4.1. To measure the x coordinate, you instrument both sides of the wires. On side A you see charges of 42 μC and 60 μC for wires 9 and 10. On side B you see 28 μC and 24 μC, respectively. Determine the x position and its uncertainty.

4.3 Determine the RMS of a box distribution with width Δz. In silicon strip detectors one measures the position of particle tracks. Assume a detector that has an array of parallel strips; each strip is 40 μm wide and is separated from its neighbor by 10 μm. If only one strip records a signal, what position resolution do you measure?

4.4 a) Determine the electron drift velocity in Ar gas under normal conditions. There is an electric field present along the y-direction with $\vec{E} = (0, E_y, 0)$ and $E_y = 1$ kV/cm as well as a magnetic field along the z-direction with $\vec{B} = (0, 0, B_z)$ and $B_z = 1$ T. The time between collisions is about 5 ps. How big is the Lorentz angle in degrees? What can I do to reduce the Lorentz angle? b) Change the electric field such that it only has a z component $\vec{E} = (0, 0, E_z)$. Determine the drift velocity and Lorentz angle in this case?

4.5 A drift chamber filled with Ar gas sits in a magnetic field of 1.5 T. The electrons need to drift a distance of 15 cm with a constant velocity of 2.2 cm/μs. The electrons collide with the gas atoms every 10^{-11} s. The mean free path is 5.66×10^{-5} cm. The electric field is parallel to the magnetic field. What is the expected position resolution? What resolution do you obtain if you switch off the magnetic field?

4.6 Consider a cylindrical straw chamber with a radius of $r_a = 1$ cm filled with Ar gas at normal conditions. In the center a 50 μm thick wire is positioned. Electric fields of +100 V/cm and +10 kV/cm are applied to the wire. Signals recorded on the wire are read out. Assume that an e-ion pair is produced at a position $r_0 = 0.5 r_a$, where r_a is the radius of the straw. If the electron would not loose any energy by collisions what would the maximal kinetic energy be for the two field configurations when it hits the wire. The observed pulse has an electron and an ion contribution. What are the relative contributions from electrons and ions for the two field configurations?

4.7 Consider an MWPC with $a_w = 6$ mm, $d_w = 10$ μm and $s_w = 2$ mm. The chamber operates with a gain of 10^6. The wire diameter is known to 2%, the anode cathode gap has an uncertainty of 80 μm and the wire spacing is precise to 50 μm. What are the uncertainties on the gain?

4.8 The endplates of a TPC are instrumented with MWPCs. The chambers are operated with an $Ar - CH_4$ (90:10) gas mixture at NTP in a 1.5 T solenoidal magnetic field. The electric field is 125V/cm. What is the drift velocity? After 1 m drift how big is the transverse diffusion with and without the magnetic field?

4.9 A GEM detector uses 50 μm thick Kapton sheets with 80 μm holes separated with a pitch of 140 μm. The detector is operated with pure argon gas at NTP. The electric field across the hole is $|\vec{E}| = 80$ kV/cm. How large is the first Townsend coefficient and how large is the amplification? How do your results compare to the measurement shown in Fig. 4.51?

4.10 Consider a limited streamer tube that is operated in an Ar-isobutane (25:75) gas mixture. The sense wires are 100 μm in diameter. The threshold voltage is $V_{ts} = 4.25$ kV. The slope is 70 pC/kV. What charge will be measured at operating voltages of 4.5 kV and 4.9 kV? For 50 μm diameter sense wires the threshold voltage is reduced to 3.55 kV. What charge will be measured at 4.0 kV? The chamber may be operated also in the proportional mode. For the 100 μm wire the threshold voltage is $V_{tp} = 3.5$ kV. What charge will be measured at 4.5 kV? The initial charge is 0.4 pC and the constant is $\kappa_p = 16$. Compare the charge measurements at 4.0 kV for the two modes of operation.

References

1. C.T.R. Wilson, Proc. Roy. Soc. Lond. A **87**-595, 277 (1912)
2. C.D. Anderson, Science **76**, 238 (1932)

3. C.D. Anderson, S.H. Neddermeyer, Phys. Rev. **50**, 263 (1936)
4. G.D. Rochester, C.C. Butler, Nature **160**, 855 (1947)
5. V.A. Mikhailov, V.N. Roinishvili, AEC-TR-5417 (1962)
6. G.E. Chikovani et al., Phys. Lett. **6**, 254 (1963)
7. M. Conversi, A. Gozzini, Nuo. Cim. **2**, 189 (1955)
8. M. Conversi, L. Federici, Nucl. Instrum. Meth. **151**, 93 (1978)
9. G. Charpak et al., Nucl. Instrum. Meth. **62**, 262 (1968)
10. G. Charpak, https://physicsmasterclasses.org/exercises/hands-on-cern/hoc_v21en/
11. S.K. Mitra, SLAC **108** UC-37 (1969)
12. F. Sauli, *Gaseous radiation detectors* (Cambridge University Press, Cambridge, UK, 2014)
13. G. Charpak, D. Rahm, H. Steiner, Nucl. Instrum. Meth. **80**, 13 (1970)
14. P. Morse, H. Feshbach, *Methods of theoretical Physics* (Mc Graw Hill, New York, 1953)
15. H. Buchholz, *Elektrische und magnetische Potentialfelder* (Springer, Berlin, 1957)
16. G.A. Erskine, Nucl. Instrum. Meth. **105**, 565 (1972)
17. F. Sauli, CERN report CERN **77**-09, 1 (1977)
18. H.G. Fischer et al., *Proc (Int* (Dubna, JINR report D, Meeting on Proportional and Drift Chambers, 1975), pp.13–9164
19. G. Charpak, Ann. Rev. Nuclear Sci. **20**, 195 (1970)
20. M. Breidenbach, F. Sauli, R. Tirler, Nucl. Instrum. Meth. **108**, 23 (1973)
21. L.E. Price, Nucl. Instrum. Meth. **112**, 507 (1973)
22. G. Charpak et al., Nucl. Instrum. Meth. **97**, 377 (1971)
23. R. Bouclier et al., Nucl. Instrum. Meth. **88** (1970) 149. Nucl. Instrum. Meth. **88**, 149 (1970)
24. P. Schilly et al., Nucl. Instrum. Meth. **91**, 221 (1971)
25. C. Trippe, CERN NP Internal Report **69**-18 (1969)
26. R. Bouclier et al., Nucl. Instrum. Meth. **115**, 235 (1974)
27. G. Aiken et al., SLAC-PUB-**2642**, (1980)
28. The LHCb Collaboration (A. Augusto Alves et al., JINST **3**, S08005 (2008)
29. M. de Palma et al., Nucl. Instrum. Meth. **217**, 135 (1983)
30. S. Özkorucuklu, SELEX experiment, FERMILAB-THESIS-2000-33 (2001)
31. K. Ackerstaff et al., Nucl. Instrum. Meth. A **417**, 230 (1998)
32. H.J. Behrend, Comp. Phys. Comm. **22**, 365 (1981); The Cello Collaboration (M. J. Schachter et al.), Phys. Scripta **23**, 610 (1981)
33. D. Bloch et al., Nucl. Instrum. Meth. A **273**, 847 (1988)
34. The ALICE Collaboration (A. Gallas et al.), Nucl. Instrum. Meth. A **581**, 402 (2007)
35. The ALEPH Collaboration (D. Decamp et al.), Nucl. Instrum. Meth. A **294**, 121 (1990)
36. C. Brand et al., Nucl. Instrum. Meth. A **283**, 567 (1989)
37. The ALICE Collaboration (A. di Mauro et al.), Nucl. Instrum. Meth. A **433**, 190 (1999)
38. A. Breskin et al., Nucl. Instrum. Meth. **143**, 29 (1977)
39. G. Charpak et al., Nucl. Instrum. Meth. **167**, 455 (1979)
40. M. Atac et al., Nucl. Instrum. Meth. **176**, 1 (1980)
41. G. Charpak et al., Nucl. Instrum. Meth. **148**, 471 (1978)
42. F. Sauli, LBNL Symposium, Berkeley, May2-3 (2014)
43. R. Turchetta, Nucl. Instrum. Meth. **335**, 44 (1993)
44. The LASS Collaboration (D. Aston et al.), SLAC-0298 (1987)
45. The LHCb TDR, CERN-LHCC-2003-030 (2003); A.A. Alves Jr. et al., arXiv:1211.1346v2 [physics.ins-det] (2013)
46. A. Andreev et al., Nucl. Instrum. Meth. A **465**, 482 (2001)
47. J. Va'vra, Wire Aging, talk at DESY workshop, https://desy.de/~agingw/trans/ps/vavra.pdf (2003)
48. O. Ullaland, The OMEGA and SFMD experience in intense beams, talk at the Workshop on Radiation Damage to Wire Chambers. LBL- **21170**, 107 (1986)
49. I. Juricic, LBL- **21170**, 141 (1986)
50. M. Binkley et al., Nucl. Instrum. Meth. A **515**, 53 (2003)
51. L. Malter, Phys. Rev. **50**, 48 (1936)

52. F. Sauli, talk on Gas Detectors, KEK, March 14, (2009)
53. R. Openshaw et al., Nucl. Instrum. Meth. A **307**, 298 (1991)
54. G.S. Oehrlein et al., J. Vac. Sci. Tech. A **12**, 333 (1994)
55. A.H. Walenta, J. Heintze, B. Schürlein, Nucl. Instrum. Meth. **92**, 373 (1971)
56. The Particle Data Group (S. Navas et al.), Phys. Rev D. **110**, 030001 (2024)
57. A. Breskin et al., Nucl. Instrum. Meth. **124**, 189 (1975)
58. G. Marel et al., Nucl. Instrum. Meth. **141**, 43 (1977)
59. M. Holder et al., Nucl. Instrum. Meth. **148**, 235 (1977)
60. D. Bettoni et al., Nucl. Instrum. Meth. A **252**, 272 (1986)
61. B. Adeva et al., Nucl. Instrum. Meth. A **289**, 35 (1990)
62. Y. Peng, Report RX-1231, AMSTERDAM (1988)
63. The H1 Collaboration (I. Abt et al.), Nucl. Instrum. Meth. A **386**, 348 (1997); DESY-93-103 (1993)
64. J.M. Bailey, Nucl. Instrum. Meth. A **323**, 184 (192)
65. The H1 Collaboration (I. Abt et al.), DESY-93-103 (1993)
66. P.J. Laycock et al., JINST **7**, T08003 (2012)
67. H. Grässler et al., Nucl. Instrum. Meth. A **283**, 459 (1989)
68. W. Farr et al., Nucl. Instrum. Meth. **154**, 175 (1978)
69. D. Bernstein et al., Nucl. Instrum. Meth. A **226**, 301 (1984)
70. J. Roehrig et al., Nucl. Instrum. Meth. A **226**, 319 (1984)
71. The TASSO Collaboration (R. Brandelik et al.), Phys. Lett. B **83**, 261 (1979)
72. H. Boerner et al., Nucl. Instrum. Meth. **176**, 151 (1980)
73. The ARGUS Collaboration (H. Albrecht et al.), Nucl. Instrum. Meth. A **275**, 1 (1989)
74. SLD Design Report, SLAC **273** (1985)
75. J. Fero et al., Nucl. Instrum. Meth. A **367**, 111 (1995)
76. The CLEO II Collaboration (Y. Kubota et al.), Nucl. Instrum. Meth. A **320**, 66 (1992)
77. The BABAR Collaboration (B. Aubert et al.), Nucl. Instrum. Meth. A **479**, 1 (2002)
78. BABAR Collaboration (B. Aubert et al.), Nucl. Instrum. Meth. A **729**, 615 (2013)
79. The Belle II Collaboration (T. Abe et al.), Belle II TDR arXiv:1011.0352, (2010); (N. Taniguchi et al.), JINST **12**-06, C06014 (2017)
80. T.V. Dong et al., Nucl. Instrum. Meth. A **930**, 132 (2019)
81. The CDF Collaboration (A. Affolder et al.), Nucl. Instrum. Meth. A **526**, 249 (2003)
82. T. LeCompte and H. T. Diehl, Ann. Rev. Nucl. Part. Sci. **50**, 71 (2000); R. Blair et al., Technical Design Report, FERMILAB-PUB-96-390-E (1996)
83. W.D. Li et al., Int. J. Mod. Phys. A **24**-S1, 9 (2009)
84. M. Moritsu et al., PoS EPS-HEP **2019**, 128 (2020)
85. G.F. Tassielli et al., JINST **15**-09, C09051 (2020)
86. M. Hauschild, https://indico.cern.ch/event/996326/contributions/4200962/attachments/2191650/3704305/dEdx.pdf (2021)
87. W. Blum, L. Rolandi, *Particle Detection with Drift Chambers* (Springer Verlag, Berlin, 1994)
88. The BABAR Collaboration (G. Sciolla et al.), Nucl. Instrum. Meth. A **419**, 310 (1998)
89. The BABAR Collaboration (B. Aubert et al.), BABAR photos SLAC (2000); https://phas.ubc.ca/~hearty/BaBar-photos/BaBar-photos.html;
90. The JADE Collaboration (H. Drumm et al.), Nucl. Instrum. Meth. **176**, 333 (1980)
91. G. Abrams et al., Nucl. Instrum. Meth. A **281**-7, 55 (1989)
92. J. Bürger et al., Nucl. Instrum. Meth. A **279**, 217 (1989)
93. H.M. Fischer et al., Nucl. Instrum. Meth. A **283**, 492 (1989)
94. O. Biebel et al., Nucl. Instrum. Meth. A **323**, 169 (1992)
95. J. Heintze, Nucl. Instrum. Meth. A **196**, 293 (1982)
96. G. Hanson et al., Nucl. Instrum. Meth. A **252**, 343 (1986)
97. R.D. Heuer, A. Wagner, Nucl. Instrum. Meth. A **265**, 11 (1988)
98. J.P. Alexander et al., Nucl. Instrum. Meth. A **283**, 519 (1989)
99. M. Roney, Nucl. Instrum. Meth. A **279**, 236 (1989)

100. D. R. Nygren and J. N. Marx, Phys. Today **31**-N10 (1978); H. Aihara et al., IEEE Trans. Nucl. Scienc. NS-**30**, No. 1 (1983)
101. D. Fancher et al., Nucl. Instrum. Meth. **161**, 383 (1979)
102. P. Christiansen, The ALICE Time Projection Chamber, lecture at Lund University, https://www.hep.lu.se/staff/christiansen/teaching/spring_2013/the_alice_tpc.pdf (2013)
103. S.R. Amendolia et al., Nucl. Instrum. Meth. A **239**, 192 (1985)
104. The TOPAZ TPC Group (T. Kamae et al.), Nucl. Instrum. Meth. A **252**, 423 (1986)
105. R. Itoh et al., IEEE Trans. Nucl. Sci. **34**, 533 (1987)
106. A. Shirahashi et al., IEEE Trans. Nucl. Sci. **35**, 1 (1988); K. Fujii et al., Nucl. Instrum. Meth. A **264**, 297 (1988)
107. W. Blum, Nucl. Instrum. Meth. A **225**, 557 (1984)
108. W.B. Atwood et al., Nucl. Instrum. Meth. A **306**, 446 (1991)
109. The ALEPH Collaboration (D. Buskulic et al.), Nucl. Instrum. Meth. A **360**, 481 (1995)
110. J. Alme et al., Nucl. Instrum. Meth. A **622**, 316 (2010)
111. The XENON1T Collaboration, https://xenonexperiment.org/
112. The LZ Collaboration (D. S. Akerib et al.), Nucl. Instrum. Meth. A **953**, 163047 (2019)
113. The DUNE Collaboration (A. Falcone et al.), Nucl. Instrum. Meth. A **1041**, 167216 (2022)
114. ALEPH photographs, https://aleph.web.cern.ch/aleph/aleph/newpub/detector.html (1989)
115. The ALICE TPC Collaboration (M. M. Aggarwal et al.), Nucl. Instrum. Meth. A **903**, 215 (2018)
116. H.J. Hilke, Rep. Prog. Phys. **73**, 116201 (2010)
117. J. Alme et al., Nucl. Instrum. Meth. A **622**, 316 (2010)
118. A. Oed, Nucl. Instrum. Meth. A **263**, 351 (1988)
119. The ALEPH Collaboration, https://aleph.web.cern.ch/aleph/dali/normal/dc015768_005906_960307_1622.gif_3jet (1996)
120. T. Francke and V. Peskov, Innovative Applications and Developments of Micro-Pattern Gaseous Detectors, IGI Global (2004)
121. L. Alunni et al., Nucl. Instrum. Meth. A **348**, 344 (1994)
122. F. Angelini et al., Nucl. Instrum. Meth. A **360**, 22 (1995)
123. R. Bouclier, Nucl. Instrum. Meth. A **332**, 100 (1993)
124. F. Sauli, Nucl. Instrum. Meth. A **805**, 2 (2016)
125. C. Altunbasa et al., Nucl. Instrum. Meth. A **490**, 177 (2002)
126. F. Sauli, Nucl. Instrum. Meth. A **386**, 531 (1997)
127. Cern Courier, Micropattern detectors promise a big future, March issue, p15 (2001)
128. A. Buzulutskov et al., Nucl. Instrum. Meth. A **433**, 471 (1999)
129. J. Benlloch et al., CERN-PPE/97-146 (1997)
130. C. Altunbas et al., Nucl. Instrum. Meth. A **490**, 177 (2002)
131. B. Ketzer et al., Nucl. Instrum. Meth. A **535**, 314 (2004)
132. G. Anelli et al., JINST **3**, S08007 (2008)
133. S. Lami et al., Nucl. Phys. Proc. Suppl. **172**, 231 (2007)
134. The CMS Muon Group (M. Abbas et al.), JINST **15**-10, P10013 (2020)
135. M. Alfonsi et al., Nucl. Instrum. Meth. A **518**, 106 (2004)
136. A. Bondar et al., Nucl. Instrum. Meth. A **496**, 325 (2003)
137. The ALICE Collaboration (D. Miśkowiec et al.), J. Phys. Conf. Ser. **1561**- 1, 012017 (2020)
138. H. Fenker et al., Nucl. Instrum. Meth. A **592**, 273 (2008)
139. A. Balla et al., Nucl. Instrum. Meth. A **732**, 221 (2013)
140. F. Archilli et al., JINST **8**, P02022 (2013)
141. M. Alfonsi et al., Nucl. Instrum. Meth. A **581**, 283 (2007)
142. The CMS Collaboration (M. Bianco et al.), JINST **15**, C09045 (2020)
143. Y. Giomataris et al., Nucl. Instrum. Meth. A **376**, 29 (1996)
144. G. Charpak et al., Nucl. Instrum. Meth. A **478**, 26 (2002)
145. M.I. Ismail, Journal of Applied Electrochemestry **9**, 407 (1979)
146. I. Giomataris et al., Nucl. Instrum. Meth. A **560**, 405 (2006)
147. P. Colas, I. Giomatarisa, V. Lepeltier, Nucl. Instrum. Meth. A **535**, 226 (2004)

148. F. Kuger, Nucl. Instrum. Meth. A **845**, 248 (2017)
149. G. Barouch et al., Nucl. Instrum. Meth. A **423**, 32 (1999)
150. M. Titov, arXiv:1308.3047 [physics.ins-det] (2013)
151. A. Delbart et al., Nucl. Instrum. Meth. A **478**, 205 (2002)
152. C. Jesus-Valls, JINST **15**-08, C08016 (2020)
153. J. Derré et al., Nucl. Instrum. Meth. A **459**, 523 (2001)
154. Y. Giomataris et al., Nucl. Instrum. Meth. A **419**, 239 (1998)
155. F. Thibaud et al., JINST **9**, C02005 (2014)
156. The COMPASS Collaboration (D. Thers et al.), Nucl. Instrum. Meth. A **469**, 133 (2001)
157. F. Kunne et al., Nucl. Phys. A **721**, 1087 (2003)
158. C. Bernet et al., Nucl. Instrum. Meth. A **536**, 61 (2004)
159. The T2K TPC Collaboration (N. Abgrall et al.), Nucl. Instrum. Meth. A **637**, 25 (2010)
160. The T2K TPC Collaboration (J. Beucher et al.), NSS/MIC 2008 / RTSD 2008, 3384 (2008)
161. C. Adloff et al., J. Phys. Conf. Ser. **293**, 012078 (2011)
162. C. Adloff et al., Nucl. Instrum. Meth. **763**, 221 (2014)
163. I. Manthos et al., AIP Conf. Proc. **2075**-1, 080010 (2019). (arXiv:1901.03160 [phys.ins.det] (2021))
164. D. Neyret et al., JINST **7**, C03006 (2012)
165. A. Delbart et al., J. Phys. Conf. Ser. **308**, 012017 (2011)
166. R. de Oliveira, MicroPattern Gaseous Detectors, talk at TIPP2023, https://indico.tlabs.ac.za/event/112/contributions/3303/attachments/1158/1568/TIPP%202023.pdf (2023)
167. The COMPASS Collaboration (F. Gautheron et al.), COMPASS II Proposal, SPSC-P-340, CERN-SPSC-2010-014 (2010)
168. R. Santonico, R. Cardarelli, Nucl. Instrum. Meth. A **187**, 377 (1981)
169. F. Anulli et al., Nucl. Instrum. Meth. A **515**, 322 (2003)
170. P. Kumari et al., JINST **15**-11, C11012 (2020)
171. The CMS Collaboration (S. Chatrchyan et al.), JINST **3**, S08004 (2008)
172. The ALICE Collaboration (K. Aamodt et al.), JINST **3**, S08002 (2008)
173. V. Ammosov et al., Nucl. Instrum. Meth. A **533**, 130 (2004)
174. G. Drake et al., Int. J. Mod. Phys. A **20**, 3830 (2005)
175. F. Anulli et al., Nucl. Instrum. Meth. A **539**, 155 (2005); Nucl. Instrum. Meth. A **552**, 275 (2005)
176. The Belle Collaboration (A. Abashian et al.), Nucl. Instrum. Meth. A **479**, 117 (2002)
177. A. Abashian et al., Nucl. Instrum. Meth. A **449**, 112 (2000)
178. R. Arnaldi et al., Nucl. Instrum. Meth. A **490**, 51 (2002); M. Gagliardi et al., Nucl. Instrum. Meth. A **661**, S45 (2012)
179. A. Ferretti, JINST **14**-06, C06011 (2018)
180. The ATLAS Collaboration (G. Aad et al.), JINST **3**, S08003 (2008)
181. The ATLAS Collaboration (G. Aad et al.), JINST **15**-09, P09015 (2020)
182. M. Sessa, PoS EPS-HEP2021, 750 (2022); The ATLAS Collaboration (G. Aad et al.), JINST **16**-07, P07029 (2021)
183. The CMS Collaboration (P. Paolucci et al.), JINST **8**, P04005 (2012); ibid (S. Chatrchyan et al.), JINST **5**, T03017 (2010)
184. The ALICE Collaboration (N. Jacazio et al.), PoS LHCP2018, 232 (2018); A. Alici, Nuovo Cim. C **027**, 403 (2004)
185. A.N. Akindinov et al., Nucl. Instrum. Meth. A **533**, 178 (2004)
186. The PHENIX Collaboration (K. Adcox et al.), Nucl. Instrum. Meth. A **499**, 469 (2003)
187. The FOPI Collaboration (J. Ritman et al.), Nucl. Phys. B Proc. Suppl. **44**, 708 (1995)
188. The HADES Collaboration (K. Zeitelhack et al.), Nucl. Instrum. Meth. A **433**, 201 (1999)
189. The HARP Collaboration (M. H. Catanesi et al.), Nucl. Instrum. Meth. A **571**, 527 (2007)
190. ARGO-YBJ Collaboration (B. Bartoli et al.), Astropart. Phys. **67**, 47 (2015)
191. A. Bertolin et al., Nucl. Instrum. Meth. A **602**, 631 (2009)
192. The ATLAS Collaboration (O. Kortner et al.), JINST **16**-06, C06001 (2021)
193. F. Anulli et al., Nucl. Instrum. Meth. A **508**, 128 (2003)

194. J.G. Wang et al., Nucl. Instrum. Meth. A **508**, 133 (2003)
195. G. Battistoni et al., Nucl. Instrum. Meth. **164**, 57 (1979)
196. R. Kass, talk on Key LST Project goals OSU (2003)
197. W. Menges, SLAC-PUB-12080 (2005)
198. B. Fulsom, SLAC graduate student seminar, SLAC (2005)
199. M. Atac, Nucl. Instrum. Meth. A **200**, 345 (1982)
200. F. Taylor, Nucl. Instrum. Meth. **289**, 283 (1989)
201. A.V. Benvenuti et al., Nucl. Instrum. Meth. A **290**, 353 (1989)
202. L.S. Osborne, eConf C **851031**, 026 (1985)
203. Y. Kamyshkov et al., Nucl. Instrum. Meth. A **257**, 125 (1987)
204. G. D'Agostini et al., Nucl. Instrum. Meth. A **252**, 431 (1986)
205. S. Bhadra et al., Nucl. Instrum. Meth. A **268**, 92 (1987)
206. G. Cibinetto et al., Int. J. Mod. Phys. A **20**, 3834 (2005)
207. G. Artusi et al., Nucl. Instrum. Meth. A **279**, 523 (1989); G. T. J. Arnison et al., Nucl. Instrum. Meth. A **94**, 431 (1990);
208. The OPAL Collaboration (K. Ahmet et al.), Nucl. Instrum. Meth. A **305**, 275 (1991)
209. The DELPHI Collaboration (P. Aarnio et al.), Nucl. Instrum. Meth. A **303**, 233 (1991)
210. I. Ajinenko et al., IEEE Trans. Nucl. Sci. **43**, 1751 (1996)
211. I. Kudla et al., Nucl. Instrum. Meth. A **300**, 480 (1991)
212. G. Abbiendi et al., Nucl. Instrum. Meth. A **333**, 342 (1993)
213. The H1 Collaboration (I. Abt et al.), Nucl. Instrum. Meth. A **386**, 348 (1997)
214. The MACRO Collaboration (M. Ambrosio et al.), Nucl. Instrum. Meth. A **486**, 663 (2002)
215. The MACRO Collaboration (S. Ahlen et al.), Nucl. Instrum. Meth. A **324**, 337 (1993); The MACRO Collaboration (M. Ambrosio et al.), Astropart. Phys. **7**, 109 (1997)
216. A. Korytov et al., Nucl. Instrum. Meth. A **338**, 375 (1994)
217. The ATLAS Collaboration (G. Aad et al.), JINST **14**-09, P09011 (2019)
218. M. Andreotti, SLAC-PUB-12205 (2005)
219. C. Adorisio et al., Nucl. Instrum. Meth. A **575**, 532 (2007)
220. The CMS Collaboration (S. Chatrchyan et al.), JINST **5**, T03016 (2010)
221. The CMS Collaboration (G. L. Bayatian et al.), CMS TDR, CERN/LHCC **7**-32 (1997)
222. P.S. Bahringer et al., Nucl. Instrum. Meth. A **254**, 542 (1987)
223. W.W. Ash et al., Nucl. Instrum. Meth. A **261**, 399 (1987)
224. W.T. Ford et al., Nucl. Instrum. Meth. A **255**, 486 (1987)
225. J. Adler et al., Nucl. Instrum. Meth. A **276**, 42 (1989)
226. M. Frautschi et al., Nucl. Instrum. Meth. A **307**, 52 (1991)
227. The ATLAS Collaboration (M. Aaboud et al.), JINST **12**, P05002 (2017)
228. D. Acosta et al., Nucl. Instrum. Meth. A **453**, 182 (2000)
229. The CMS Muon Group (B. M. Joshi et al.), PoS LHCP2018, 074 (2018)
230. M. Paneva, EPJ Web of Conferences **214**, 02014 (2019)
231. N. Manganelli, JINST **15**, C03047 (2020)
232. The ATLAS Collaboration (G. Aad et al.), ATLAS muon spectrometer: Technical Design Report, LHCC-**97**-22 (1997)
233. S. Majewski et al., Nucl. Instrum. Meth. A **217**, 265 (1983)
234. Y. Gernitzky et al., ATLAS note ATL-MUON-2003-005 (2003)
235. The ATLAS Collaboration (G. Aad et al.), New Small Wheel TDR, https://cds.cern.ch/record/1552862/files/ATLAS-TDR-020.pdf (2013)
236. C. Beard et al., Nucl. Instrum. Meth. A **286**, 117 (1990)
237. The ATLAS Collaboration (G. Aad et al.), ATLAS new Small Wheel TDR, CERN-LHCC-**2013**-006 (2013)
238. The ATLAS Collaboration (G. Aad et al.), Eur. Phys. J. C **75**, 120 (2015)
239. The ATLAS Collaboration, https://atlas.cern/updates/news/NSW-final-slice
240. The ATLAS Muon Collaboration (D. Pudzha et al.), JINST **15**, C09064 (2020)
241. A. Abusleme et al., Nucl. Instrum. Meth. A **817**, 85 (2016)
242. D. Glaser, Phys. Rev. **87**, 665 (1952)

References

243. P. Queru, Conf. Proc. C **700610V1**, 460 (1970)
244. F.J. Hasert et al., Phys. Lett. B **46**, 138 (1973)
245. H.P. Reinhard, Conf. Proc. C **730508**, 87 (1973)
246. G.S. Abrams et al., Phys. Rev. Lett. **13**, 670 (1964)
247. M. Gell-Mann, CTSL-20, TID-**12608** (1961)
248. M. Gell-Mann, Phys. Rev. Lett. **12**, 155 (1964)
249. http://cds.cern.ch/record/2033326
250. http://it-spots.de/2011/10/brookhaven-national-laboratory-flickr-3/
251. G.E. Chikovani et al., Phys. Lett **6**, 254 (1963)
252. M. Conversi, A. Gozzini, Nuovo Ci. **2**, 189 (1955)
253. M.C. Alkofer, Thiemig München (1969)
254. M. Conversi, I. Federici, Nucl. Instrum. Meth. A **151**, 93 (1978)
255. S. Kinoshito, Proc. Roy. Soc. A **83**, 432 (1910)
256. R. Reiganum, Z. Physik **12**, 1076 (1911)
257. K. Kleinknecht, Detectiors for Particle Radiation, Cambridge University Press, 2nd ed. (1998)
258. W.R. Leo, *Techniques for Nuclear and Particle Physics Experiments* (Springer Verlag, Berlin, 1987)
259. R.K. Carnegie et al., Nucl. Instrum. Meth. A **538**, 372 (2005)
260. L. Sohl et al., JINST **15**, C04053 (2020)
261. D.A. Glaser, Handbuch der Physik **45**, 314 (1958)
262. See Chapter 6

Open Access This chapter is licensed under the terms of the Creative Commons Attribution 4.0 International License (http://creativecommons.org/licenses/by/4.0/), which permits use, sharing, adaptation, distribution and reproduction in any medium or format, as long as you give appropriate credit to the original author(s) and the source, provide a link to the Creative Commons license and indicate if changes were made.

The images or other third party material in this chapter are included in the chapter's Creative Commons license, unless indicated otherwise in a credit line to the material. If material is not included in the chapter's Creative Commons license and your intended use is not permitted by statutory regulation or exceeds the permitted use, you will need to obtain permission directly from the copyright holder.

Momentum Measurements 5

The trajectories of charged particles in a tracking detector are straight lines in the absence of a magnetic field. They become helices in the presence of a magnetic field, where the motion in the plane orthogonal to the magnetic field is circular and that parallel to the magnetic field is linear. The magnitude of the momentum is proportional to the bending radius. Since tracking detectors provide rather precise trajectory position measurements, also the momentum components are precisely measured. Precise charged-particle momenta are important kinematic observables in high-energy physics experiments since they play a crucial role, for example, in determining invariant masses of decaying particles, missing momenta and the jet energy.

5.1 Deflections in Magnetic Fields

In a fixed target experiment, an incident beam is sent to a stationary target. The beam direction is along the z axis. Secondary particles produced in an interaction are typically concentrated within a cone around the incident beam direction since due to the Lorentz boost in the z direction their transverse momenta are typically rather small, so $p_x, p_y \ll p_z$. In order to measure the momentum, we place a magnetic dipole field $\vec{B}$ along the y direction, $\vec{B} = (0, B_y, 0)$. The charged track is deflected by the Lorentz force along a circular orbit with radius r_t in the xy plane. To compute the momentum, we equalize the Lorentz force $\vec{F}_L$ to the centripetal force $\vec{F}_c$,

$$|\vec{F}_L| = |\vec{F}_c| = \frac{pv}{r_t} = q_e v B_y, \tag{5.1}$$

© The Author(s) 2025
G. Eigen, *Detectors in High-Energy Physics Experiments*, Graduate Texts in Physics,
https://doi.org/10.1007/978-3-031-67336-8_5

where $p = |\vec{p}|$, $v = |\vec{v}|$ is the velocity and $B_y = |\vec{B}|$. The radius is given by

$$r_t = \frac{p}{q_e B_y}. \tag{5.2}$$

Thus, the momentum is determined by $p = 0.3 r_t B_y$, where p, B_y and r_t are given in units of [GeV/c], [T] and [m], respectively.[1] If the magnetic field is active on length L_t,[2] the angular deflection is given by

$$2 \sin\frac{\theta}{2} = \frac{L_t}{r_t} = -\frac{q_e B_y L_t}{p}, \tag{5.3}$$

as shown in Fig. 5.1 (left), where θ is the angle between the momenta of incident and deflected particle. The change in transverse momentum is given by

$$\Delta p = p \cdot \sin\theta \simeq -q_e B_y L_t = -q_e \int_0^{L_t} B_y dz, \tag{5.4}$$

in good approximation for small deflection angles.

For an inhomogeneous $\vec{B}$ field we need to replace $B_y L_t$ by $\int_0^{L_t} B_y dz$. For example for $B_y L_t = 1$ Tm, the momentum change is $\Delta p_x \sim 0.3$ GeV/c. In this approximation, the knowledge of θ and B_y yields a measurement of the momentum transverse

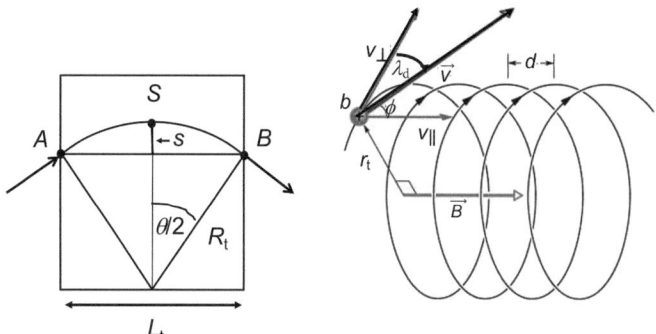

Fig. 5.1 Left: Deflection of a charged particle in a magnetic field on length L_t. The particle enters the magnetic field at point A and leaves it at point B. The path in the magnetic field is a circular segment with radius r_t. The deflection of point S with respect to the line connecting points A and B is called sagitta s. Right: The helical trajectory of a charged particle in the magnetic field, showing the direction of the magnetic field, parallel and transverse velocity components and the dip angle λ_d. Reprinted with kind permission from [1], © 2016, A. Andreazza. All rights reserved

---

[1] The factor 0.3 results from $c/10^9$.
[2] This is the track length in the plane orthogonal to the magnetic field since the distance along the magnetic field remains unaffected.

to the magnetic field. If θ_1 is the angle of the incoming particle and θ_2 is that of the outgoing particle particle,[3] we have

$$\sin\theta = \sin(\theta_2 - \theta_1) \simeq \sin\theta_2 - \sin\theta_1. \tag{5.5}$$

If the magnetized volume is evacuated and multiple scattering in tracking chambers can be neglected, the momentum measurement error solely comes from the error in the position measurement σ_x, yielding

$$\frac{\sigma_p}{p} = \frac{2p}{\Delta p_x}\frac{\sigma_x}{h}, \tag{5.6}$$

where h is the lever arm for the angle measurements before and after the magnet. For example, for $L_t|\vec{B}| = 2$ Tm, a position resolution of $\sigma_x = 100$ μm, a lever arm of $h = 2$ m and a momentum of $p = 100$ GeV/c we get $\sigma_p/p \simeq 1.7\%$ or $\sigma_p/p^2 \simeq 1.7 \times 10^{-4}$ c/GeV.

5.2 Particle Motion in a Magnetic Field

The equation of motion of a charged particle in a magnetic field is

$$\frac{d\vec{p}}{dt} = q_e c \vec{\beta} \times \vec{B}, \tag{5.7}$$

where p is the particle momentum and $\vec{\beta}$ is its relative velocity. The momentum is related to the particle energy by $\vec{p}c = E\vec{\beta}$, yielding

$$\frac{d\vec{p}}{dt} = \frac{1}{c}\frac{dE}{dt}\vec{\beta} + \frac{E}{c}\frac{d\vec{\beta}}{dt} = q_e c \vec{\beta} \times \vec{B}, \tag{5.8}$$

Since the magnetic field does not perform work on the particle, the energy is a constant of motion,

$$\frac{dE}{dt} = 0. \tag{5.9}$$

So, the final equation of motion is

$$\frac{d\vec{\beta}}{dt} = \vec{\beta} \times \vec{\omega}_v, \tag{5.10}$$

where

$$\vec{\omega}_v = \frac{q_e \vec{B}}{E}c^2. \tag{5.11}$$

---

[3] Please note that both angles θ_1 and θ_2 are small.

If $\vec{\beta}$ is normal to $\vec{B}$, the trajectory is a circle with a revolution period of $T_{\rm rp} = 2\pi/\omega_v$. Using $\vec{p} = m\gamma\vec{\beta}$ and $\vec{\beta}c = d\vec{r}_t/dt$, we can rewrite the equation of motion

$$m\gamma \frac{d^2\vec{r}_t}{dt^2} = q_e \frac{d\vec{r}_t}{dt} \times \vec{B}. \tag{5.12}$$

Using the length along the path $d\ell = c\beta dt$ we get

$$m\beta\gamma c \frac{d^2\vec{r}_t}{d\ell^2} = q_e \frac{d\vec{r}_t}{d\ell} \times \vec{B}, \tag{5.13}$$

which leads to the differential equation

$$\frac{d^2\vec{r}_t}{d\ell^2} = \frac{q_e}{p} \frac{d\vec{r}_t}{d\ell} \vec{B}. \tag{5.14}$$

5.2.1 The Particle Trajectory

For an inhomogeneous magnetic field, B depends on the path length ℓ and (5.14) needs to be solved numerically. For a homogeneous field, the trajectory is a helix that can be described in a parametric form,

$$x(\ell) = x_0 + r_t \left[\cos\left(\Phi_0 + \frac{h_p \ell \cos \lambda_d}{r_t}\right) - \cos\Phi_0 \right],$$

$$y(\ell) = y_0 + r_t \left[\sin\left(\Phi_0 + \frac{h_p \ell \cos \lambda_d}{r_t}\right) - \sin\Phi_0 \right],$$

$$z(\ell) = z_0 + \ell \sin \lambda_d, \tag{5.15}$$

where $h_p = \pm 1$ is the sense of rotation on the helix. Figure 5.1 (right) shows the helical trajectory. The projection of the helix onto the xy plane is a circle,

$$(x(\ell) - x_0 + r_t \cos\Phi_0)^2 + (y(\ell) - y_0 + r_t \sin\Phi_0)^2 = r_t^2. \tag{5.16}$$

The points x_0, y_0 and z_0 are the starting points for $\ell = 0$. Note that point z_0 is the longitudinal impact parameter. The points x_0 and y_0 are usually expressed in terms of the transverse impact parameter d_0 and the azimuth angle Φ_0,

$$x_0 = d_0 h_p \cos\Phi_0,$$
$$y_0 = d_0 h_p \sin\Phi_0. \tag{5.17}$$

Figure 5.2 (left) shows the track parameters. Near the origin, the track is specified by the transverse and longitudinal impact parameters, which are respectively the closest distance of approach in the xy and rz planes of a track to the origin. Note that the

5.2 Particle Motion in a Magnetic Field

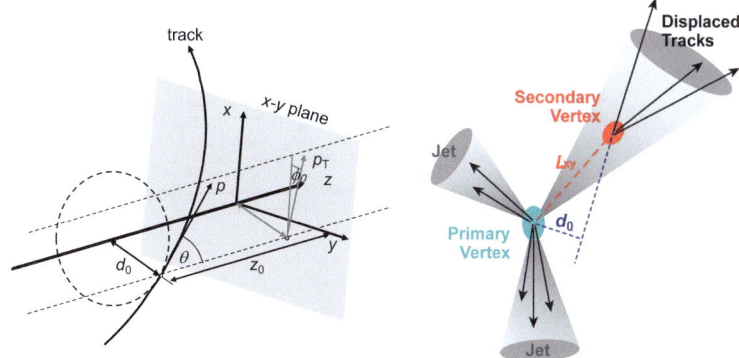

Fig. 5.2 Left: Definition of the tracking parameters, the transverse impact parameter d_0, the longitudinal impact parameter z_0, the track momentum p, the transverse momentum component p_T, the azimuth angle Φ_0 and the polar angle θ. Right: Three-dimensional view of the primary and secondary vertex showing the transverse impact parameter [2]. The parameter L_{xy} is the distance of the secondary vertex from the primary vertex. Reprinted with kind permission from [2], © 2011, N. Wermes. All rights reserved

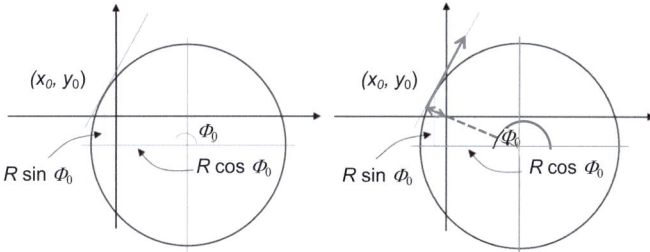

Fig. 5.3 Left: Projection of the helix onto the xy plane for an arbitrary definition of the starting point. Right: Projection of the helix onto the xy plane for defining the start values x_0 and y_0 at $\ell = 0$. Both reprinted with kind permission from [1], © 2016, A. Andreazza. All rights reserved

impact parameters can be positive or negative. Thus, we have added the parameter $(h_p = \pm 1)$.

Impact parameters are important tools for determining secondary and tertiary vertices of charm and beauty decays as illustrated in Fig 5.2 (right) if the full vertex is not reconstructed. Figure 5.3 (left) depicts the starting parameters x_0 and y_0 as well as the parameter Φ_0, which determines the slope of the tangent to the circle at $\ell = 0$.

The projection onto the rz plane is a linear motion,

$$z = z_0 + r \cdot \tan \lambda_d. \tag{5.18}$$

Note that the linear motion is decoupled from the circular motion.

Though the starting point is arbitrary, we typically define it in terms of the impact parameters as depicted in Fig. 5.3 (right). Since the track is reconstructed in a region

$\ell \ll r_t$, we can perform a Taylor expansion in terms of ℓ/r_t, yielding the high-p_T parabolic approximation,

$$\begin{aligned} x(\ell) &= x_0 - h_p \ell \cos \lambda_d \sin \Phi_0 - \frac{1}{2} \frac{\ell^2 \cos^2 \lambda_d}{r_t} \cos \Phi_0, \\ y(\ell) &= y_0 + h_p \ell \cos \lambda_d \cos \Phi_0 - \frac{1}{2} \frac{\ell^2 \cos^2 \lambda_d}{r_t} \sin \Phi_0, \\ z(\ell) &= z_0 + \ell \sin \lambda_d. \end{aligned} \quad (5.19)$$

Introducing the scattering angle $\theta = \pi/2 - \lambda_d$, the curvature $\kappa_t = h_p/r_t$, $\cos \phi_0 = h_p \sin \Phi_0$ and $\sin \phi_0 = -h_p \cos \Phi_0$, we get

$$\begin{aligned} x(\ell) &= -d_0 \sin \phi_0 + \ell \sin \theta \cos \phi_0 + \frac{1}{2} \kappa_t \ell^2 \sin^2 \theta \sin \phi_0, \\ y(\ell) &= d_0 \cos \phi_0 + \ell \sin \theta \sin \phi_0 - \frac{1}{2} \kappa_t \ell^2 \sin^2 \theta \cos \phi_0, \\ z(\ell) &= z_0 + \ell \cos \theta. \end{aligned} \quad (5.20)$$

However, due to multiple scattering, dE/dx losses, magnetic-field inhomogeneities tracks are not simple helices. These effects are implemented in the Kalman track fit discussed in Sect. 5.2.4. The momentum of the track is given by

$$\begin{aligned} p_x &= p_T \cos(\phi_0), \\ p_y &= p_T \sin(\phi_0), \\ p_z &= p_T \tan(\lambda_d). \end{aligned} \quad (5.21)$$

5.2.2 Impact Parameter Determination

As discussed in Appendix B, we can easily extract the longitudinal impact parameter from a χ^2 distribution. We measure track points in n layers in the rz direction. The z position is measured in a vertex detector with resolutions σ_i. We define the sums by $A = \sum_i^n r_i/\sigma_i^2$, $B = \sum_i^n 1/\sigma_i^2$, $C = \sum_i^n z_i/\sigma_i^2$, $D = \sum_i^n r_i^2/\sigma_i^2$ and $E = \sum_i^n r_i z_i/\sigma_i^2$. Then

$$z_0 = \frac{C \cdot D - A \cdot E}{D \cdot B - A^2} \quad (5.22)$$

The determination of the transverse impact parameter is more complex since the circle is non linear. The five track parameters are typically determined using a Kalman fitter as discussed below.

5.2.3 Position Resolution and Impact Parameter Resolution

Using Appendix B we can determine the longitudinal impact parameter resolution,

$$\sigma_{z0} = \frac{B}{B \cdot D - A^2} \tag{5.23}$$

The resolution of the transverse impact parameter resolution is obtained from the Kalman fitter.

Multiple scattering degrades both the transverse and longitudinal impact parameter resolutions by adding a momentum-dependent term,

$$\sigma_{d'_0} = \sigma_{d_0} \oplus \frac{b}{p_T \sin^{1/2}\theta},$$
$$\sigma_{z'_0} = \sigma_{z_0} \oplus \frac{b}{p_T \sin^{3/2}\theta}, \tag{5.24}$$

where b is given by [1]

$$b = 13.6 \text{ MeV}\left(\sqrt{\frac{x}{X_0}}\left(1 + 0.038 \log \frac{x}{X_0}\right)\right), \tag{5.25}$$

x is the thickness of the detector material and X_0 is the radiation length. This is typically also done in the Kalman fit.

5.2.4 Track Finding and Reconstruction

A charged track is characterized by five parameters,

- d_0, the distance of closest approach from a given point P in the x-y plane,
- z_0, the distance of closest approach from the z axis,
- ϕ_0, the azimuth angle of the track at the distance of closest approach,
- $\tan \lambda_{d,0} = \tan(\pi/2 - \theta_0)$,
- $\kappa_t = 1/r_t$, the track curvature.

Note that in LHC experiments instead of $\lambda_{d,0}$ the pseudo-rapidity η_0 is used. To determine the track parameters many experiments use the so-called Kalman fitter, which is an optimal recursive fitter algorithm that incorporates all available information and measurements to estimate the track parameters regardless of their precision. First, it exploits the information on noise, measurement errors as well as uncertainties in dynamic models such as multiple scattering. Second, it is optimal in determining an estimate of the output parameters, providing a conditional probability density that is extracted from the given data sample. A Kalman fitter determines the probability density for systems that can be described by a linear model and in which the measurement noise is white (i.e. uniformly distributed) and Gaussian. Though these assumption seem to be rather restrictive, many physical systems can be described by a linear model. Note that Kalman fitter exist also for nonlinear systems but they are used only if the linear model is inadequate. For track fitting, the linear approach works well.

Let us consider a simple one-dimensional example. We measure the position of a trajectory $x(t)$ as a function of time in an MWPC. Using the information from the anode wires, we measure the position $x_1 = x(t_1)$ at time t_1 with an error of $\sigma(x_1) = \sigma_1$. From the cathode strips, we get a second independent measurement $x_2 = x(t_2)$ at time $t_2 \simeq t_1$ with an error σ_2. The weighted mean is

$$\langle x(t_2) \rangle = \frac{x_1 \cdot w_1 + x_2 \cdot w_2}{w_1 + w_2}, \tag{5.26}$$

where the weights are given by $w_1 = 1/\sigma_1^2$ and $w_2 = 1/\sigma_2^2$. The variance is $\sigma^2(t_2) = (w_1 + w_2)^{-1}$. For simplicity, we assume that the magnetic field is zero. The motion of the particle is given by $dx(t)/dt = v + v_{\text{noise}}$, where v is the velocity of the particle and v_{noise} is a noise term with mean zero and uncertainty σ_{noise}. At time $\tilde{t}_3$, the expected position is $\tilde{x}_3 = \langle x(\tilde{t}_3) \rangle = \langle x(t_2) \rangle + v(\tilde{t}_3 - t_2)$ with expected variance $\tilde{\sigma}_3^2 = \sigma_2^2 + \sigma_{\text{noise}}^2 \cdot (\tilde{t}_3 - t_2)^2$ yielding a weight $\tilde{w}_3 = 1/\tilde{\sigma}_3^2$. It is important to note that in this prediction x_1 and x_2 and the corresponding variances and weights do not enter explicitly but just through $\tilde{x}_3 = \langle x(\tilde{t}_3) \rangle$. Now, a measurement at time t_3 yields a new position x_3 with standard deviation σ_3 and weight w_3. Using the linear extrapolation and the latest measurement, yields

$$\langle x(t_3) \rangle = \frac{\tilde{w}_3 \cdot \tilde{x}_3 + w_3 x_3}{\tilde{w}_3 + w_3}, \tag{5.27}$$

where the corresponding weight is $w(t_3) = \tilde{w}_3 + w_3$ and the variance is $\sigma^2(t_3) = 1/w(t_3)$. The basic principle of the Kalman fitter is to take the weighted average whenever new measurements become available. This reduces the variance and increases the weight. If other uncertainties affect the position measurement, the appropriate variances have to be added.

In a typical detector, we have a solenoidal field. Thus, a particle trajectory is a helix, which is a combination of a circular motion in the transverse plane and a linear motion in the longitudinal direction. The helix is specified by five parameters listed above. However, due to dE/dx losses and multiple scattering the trajectory is not a perfect helix. But both effects are accounted for in the fit.

For the Kalman fitter, we represent the state vector by a five-component vector $\vec{x}(r_t)$ where r_t is the radius of the helix. Instead of a simple variance we have a 5×5 covariance matrix $\Sigma^2(r_t)$ that takes correlations between the track parameters into account. The weight matrix $\Omega(r_t)$ is the inverse of the covariance matrix. Since the model here is not linear, we perform a Taylor expansion and use the linear first-order term. This works reasonably if the fit parameters are relatively small. So, we write the state vector as $\langle \vec{x}(r_t) \rangle = \langle \vec{x}_{\text{ref}}(r_t) \rangle + \langle \delta \vec{x}(r_t) \rangle$ where the first term is the reference

5.2 Particle Motion in a Magnetic Field

and the last term is a correction. The transport matrix $\mathcal{T}(r', r_t)$ propagates $\langle \delta\vec{x}(r_t) \rangle$ to $\langle \delta\vec{x}(r') \rangle$

$$\langle \delta\vec{x}(r') \rangle \sim \mathcal{T}(r', r_t) \cdot \langle \delta\vec{x}(r_t) \rangle. \tag{5.28}$$

The associated variance transforms as

$$\Sigma^2_{\delta x}(r') \sim \mathcal{T}(r', r_t) \cdot \Sigma^2_{\delta x}(r_t) \cdot \mathcal{T}^T(r', r_t), \tag{5.29}$$

where $\mathcal{T}^T$ is the transposed matrix. Please note that we do not fit track parameters any longer but their deviation from the reference. The first-order transport matrix consists of ratios of derivates of the parameters at positions r_t and r'. The measurements that are incorporated into the Kalman fitter have to be expressed in terms of deviations from the reference as well.

The fit starts at point r_1. We assume $\langle \delta(\tilde{r}_1) \rangle = \vec{0}$. We start with measurement 1.

$$\langle \delta\vec{x}_m(r_1) \rangle = \langle \vec{x}_m(r_1) \rangle - \langle \vec{x}_{ref}(r_1) \rangle. \tag{5.30}$$

The weight is $\Omega_{\delta\vec{x}_m}(r_1) = 1/\Sigma^2_{\delta\vec{x}_m}(r_1)$. So,

$$\langle \delta\vec{x}(r_1) \rangle = \frac{\Omega_{\delta\vec{x}}(\tilde{r}_1) \cdot \langle \delta\vec{x}(\tilde{r}_1) \rangle + \Omega_{\delta\vec{x}_m}(r_1) \cdot \langle \delta\vec{x}_m(r_1) \rangle}{\Omega_{\delta\vec{x}}(\tilde{r}_1) + \Omega_{\delta\vec{x}_m}(r_1)}, \tag{5.31}$$

and the new weight is $\Omega_{\delta\vec{x}}(r_1) = \Omega_{\delta\vec{x}}(\tilde{r}_1) + \Omega_{\delta\vec{x}_m}(r_1)$ with $\Omega_{\delta\vec{x}}(\tilde{r}_1) = 1/\Sigma^2_{\delta\vec{x}}(\tilde{r}_1)$. Note that this procedure is very similar to that of the one-dimensional case. Before the best estimate can be propagated to the radius of the next measurement, multiple scattering and dE/dX effects for the previous measurement have to be added. Multiple scattering is symmetric. Thus, it does not change the mean values of the parameters but increases the errors on some parameters. So, $\Sigma^2_{\delta\vec{x}}(r'_i) = \Sigma^2_{\delta\vec{x}}(r'_{\tilde{r}_i}) + \Sigma^2_{MS}$, where the last term is the covariance matrix for multiple scattering. The dE/dx loss affects the curvature and increases the error. The uncertainty $\sigma_{dE/dx}$ is small with respect to the uncertainty on the curvature and is neglected. The other measurements at this point are included one by one in a similar way. Other points are incorporated in a similar way using the information from previous points. The Kalman fitter is a recursive method. Exact knowledge of the starting point is very important. The fit results are validated with Monte Carlo samples. For track fitting, typically the fit works best from the outside to the inside to obtain the smallest errors on the d_0, z_0 and ϕ_0 as illustrated in Fig. 5.4 (left). Figure 5.4 (right) shows the properties of a forward fit. With each step the error becomes smaller. The best estimate of the track parameters is obtained for positions outside the vertex detector. This is useful for extrapolations into the drift chamber. In order to get good track parameters at each point, we can combine the two fits as shown in Fig. 5.5.

$$\langle \delta\vec{x}(r_t) \rangle = \frac{\Omega_{\delta\vec{x}_F}(r_t) \cdot \langle \delta\vec{x}_F(r_t) \rangle + \Omega_{\delta\vec{x}_B}(r_t) \cdot \langle \delta\vec{x}_B(r_t) \rangle}{\Omega_{\delta\vec{x}_F}(r_t) + \Omega_{\delta\vec{x}_B}(r_t)}, \tag{5.32}$$

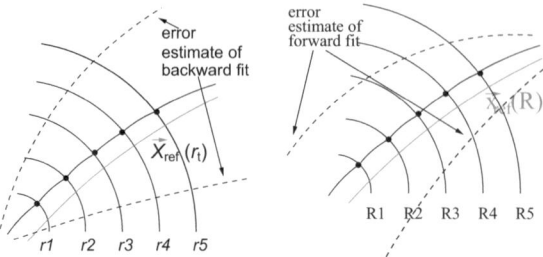

Fig. 5.4 Left: Illustration of a track fit that starts from the outside inwards. Right: Illustration of a track fit that starts from the inside outwards. Both reprinted with kind permission from [3], © 2003, S. Menzemer. All rights reserved

Fig. 5.5 The weighted mean of the forward and backward fit yields the best estimate at every radius. Reprinted with kind permission from [3], © 2003, S. Menzemer. All rights reserved

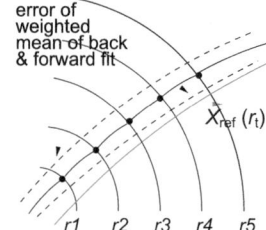

with

$$\Omega_{\delta\vec{x}}(r_t) = \Omega_{\delta\vec{x}_F}(r_t) + \Omega_{\delta\vec{x}_B}(r_t), \qquad (5.33)$$

where F and B denote the corresponding positions and weights for the forward and backwards fits.

5.3 Magnet Shapes for Fixed-Target Experiments

In fixed-target experiments, typically dipole fields are used. The dipole can be arranged in a shape of an H-magnet (Fig. 5.6 left), which has symmetric flux return yokes or a C-magnet (Fig. 5.6 center) that has an asymmetric flux return yoke yielding a less uniform magnetic field. Figure 5.6 (right) shows the magnetic-field lines of a dipole field. The two coils are separated by gap h_d. The field is uniform. Figure 5.7 shows photographs of a typical dipole magnet (left) and the LHCb magnet (right). The two sets of coils and the gap h_d are clearly visible. The LHCb magnet is an H-type dipole magnet with a bending power of 4 Tm [4].

The amount of iron needed since the magnetic flux return is correlated with the desired field strength. The ratio of iron volume to the volume of magnetized air gap is approximately given by

$$\frac{V_{\text{Fe}}}{V_{\text{mag}}} \simeq \left(2 + \frac{|\vec{B}|}{B_s}\right)\frac{|\vec{B}|}{B_s}, \qquad (5.34)$$

5.3 Magnet Shapes for Fixed-Target Experiments

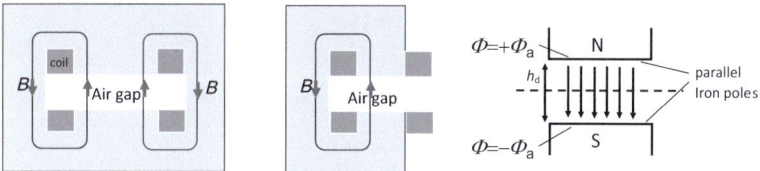

Fig. 5.6 Left: H magnet with an air gap. The grey boxes show the arrangement of the coils. Center: C magnet with an air gap. Right: Magnetic-field lines in a dipole magnet

Fig. 5.7 Left: Photograph of a dipole magnet. The two set of coils, which are clearly visible (brown ribbons), run parallel to each other. Reprinted with kind permission from [5], © 2024, Everson Tesla company. All rights reserved. Right: Photograph of the LHCb dipole magnet. At the entrance and exit the two sets of coils are bent upwards and downwards. Reprinted under CC-BY-4.0 Licence from [6], © 2024, CERN

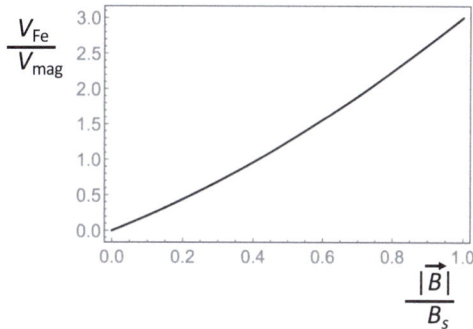

Fig. 5.8 Left: The ratio of volume of iron to volume of magnetization as a function of the ratio of magnetic-field strength $|\vec{B}|$ to the maximum magnetic-field strength B_s

where $|\vec{B}|$ is the actual magnetic-field strength and B_s is the saturation field strength, which is the maximum magnetic-field strength. For example, for a solenoid it is given by (5.46). For a ratio of $|\vec{B}|/B_s \sim 1$, we get a ratio of volume of iron to volume of magnetization of $V_{Fe}/V_{mag} \sim 3$ as shown in Fig. 5.8. For a ratio of $|\vec{B}|/B_s \sim 1/2$ to $1/3$, we get $V_{Fe}/V_{mag} \sim 5/4$ to $7/9$, which is a more realistic configuration.

5.3.1 Properties of Dipole Magnets

In a dipole magnet we have a constant field,

$$B_0 = cons. \tag{5.35}$$

So, we define the potential by

$$\Phi(x, z) = B_0 z. \tag{5.36}$$

The equipotential line $\Phi(x, z) = \Phi_a = const$ is a line parallel to the x-axis at distance z. Figure 5.6 (right) shows a schematic layout of a dipole field consisting of two parallel iron poles that are separated by $h_d = 2z$. The $\vec{B}$ field is determined by the current I flowing through the coil with N_{wind} windings in gap h_d.

$$B_0 = \mu_0 \frac{N_{\text{wind}} I}{h_d} \quad \text{with } \mu_r = 1. \tag{5.37}$$

The dipole strength is given by

$$\frac{1}{r_t} = \frac{q_e}{p B_0} = \frac{q_e \mu_0}{p} \frac{N_{\text{wind}} I}{h_d}. \tag{5.38}$$

5.4 Momentum Measurements in Fixed-Target Experiments

In iron-core magnets the momentum resolution is limited by multiple scattering (MS) of muons in the iron and the measurement error of the track (M). The multiple scattering contributions yields a mean transverse momentum change

$$\sigma p_t^{\text{MS}} = 21 \text{ MeV/c} \sqrt{\frac{L_t}{X_0}}, \tag{5.39}$$

where X_0 is the radiation length. This leads to a momentum error due to multiple scattering in the x direction of[4]

$$\left(\frac{\sigma_p}{p}\right)^{\text{MS}} = -\left(\frac{\sigma_{p_x^{\text{MS}}}}{\sigma_{p_x}}\right) = \frac{15 \text{ MeV/c} \sqrt{L_t/X_0}}{q_e \int B_y dz} = 0.05 \frac{1}{B_y \sqrt{X_0 L_t}}. \tag{5.40}$$

For a 1.5 T magnetic field in iron this becomes

$$\left(\frac{\sigma_p}{p}\right)^{\text{MS}} = 0.25 \frac{1}{\sqrt{L_t}}. \tag{5.41}$$

---

[4] Note that the factor of 15 results from $21/\sqrt{2}$.

5.4 Momentum Measurements in Fixed-Target Experiments

For example for $L_t = 5$ m, the multiple scattering contribution to the momentum resolution is $(\sigma_p/p)^{MS} \simeq 11.2\%$. Let us now turn to the track measurement errors. If the position is measured at 3 equidistant points along the projected (transverse) track length L_t in iron, the sagitta s_x of the circular orbit is given by

$$s_x = R_t - R_t \cos\frac{\theta}{2} \simeq \frac{R_t \theta^2}{8}. \tag{5.42}$$

Since $R_t = p/(0.3 \cdot |\vec{B}|)$, we get $\theta \simeq L_t/R_t \simeq 0.3 \cdot |\vec{B}| \cdot L_t/p$ and in turn

$$s_x = 0.3\frac{|\vec{B}|L_t^2}{8p}. \tag{5.43}$$

Typically, we use SI units, which gives the magnetic fields in T, the distance L_t in m, the energy in GeV and the momentum in GeV/c. The sagitta s_x determined from 3 measurements as shown in Fig. 5.1 has a precision of $\sigma_s = \sqrt{\frac{3}{2}}\sigma_x$, yielding the measurement contribution to the momentum resolution of

$$\left(\frac{\sigma_p}{p}\right)^M = \frac{\sigma_s}{s_x} = \frac{\sigma_s 8p\sqrt{3/2}}{0.3|\vec{B}|L_t^2}. \tag{5.44}$$

If the track is measured at N equidistant points, the position measurement contribution to the momentum resolution becomes [7],

$$\left(\frac{\sigma_p}{p}\right)^M = \frac{\sigma_s p}{0.3|\vec{B}|L_t^2}\sqrt{\frac{720}{N+4}}. \tag{5.45}$$

For example, for a 100 GeV/c track, $|\vec{B}| = 1.5$ T, $L_t = 5$ m, $\sigma_s = 1.5 \times 10^{-3}$ m and N = 6 we obtain $(\sigma_p/p)^M = 11.3\%$. For this example, the error contributions from multiple scattering and position measurement are similar. Note that the contributions from multiple scattering and position measurement depend on length L_t with different dependences while the dependence on the magnetic-field strength is the same. Figure 5.9 show the momentum resolution of muons measured with an iron core magnet for $|\vec{B}| = 1.5$ T and different lengths $L_t = 5$ m (left), $L_t = 10$ m (center) and $L_t = 20$ m (right). For small lengths L_t the track measurement error is the dominant component of the momentum resolution at higher momenta as depicted in Fig. 5.9 (left). For large lengths L_t, multiple scattering becomes the dominant component over the entire momentum range as shown in Fig. 5.9 (right). If measurements are taken at equidistant positions along the track, the momentum error decreases as $L_t^{-5/2}$ since the field integral and lever arm of the angle measurement each increase with L_t, while the number of measurements along the track (N) increases as $\sqrt{L_t}$. The momentum error from multiple scattering decreases as $L_t^{-1/2}$.

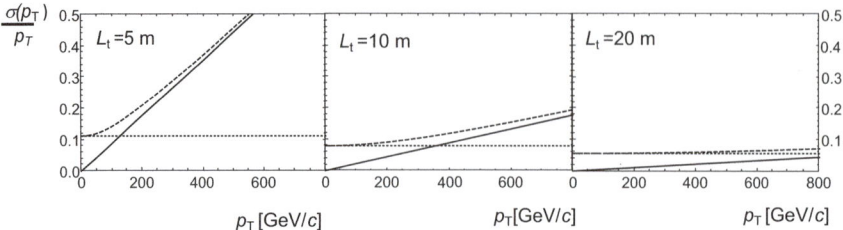

Fig. 5.9 Total muon momentum resolution in an iron-based muon spectrometer for $|\vec{B}| = 1.5$ T and lengths of $L_t = 5$ m (left), $L_t = 10$ m (center) and $L_t = 20$ m (right) shown by solid curves. Contributions from measurement errors and from multiple scattering errors are shown by dashed curves and dotted curves, respectively

5.5 Magnet Shapes for Storage Ring Experiments

In most storage ring experiments the laboratory and center-of-mass (CM) systems coincide. Thus, particles produced in an interaction can move along any direction. In order to minimize losses near the beam directions, the detector needs to cover as much of the total solid angle as possible. The most common magnets used in such detectors are solenoids.

5.5.1 Solenoidal Magnet

The solenoidal magnetic $\vec{B}$ field is parallel to the beam axis. This set-up is used in e^+e^-, $p\bar{p}$, pp and ep storage ring as well as in e^+e^- linear-collider experiments. Figure 5.10 (left) shows the coil arrangement and the magnetic-field lines

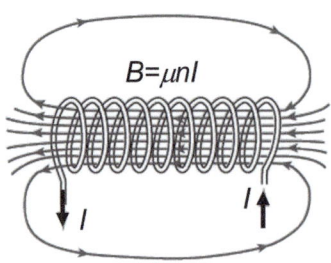

Fig. 5.10 Left: Configuration of a solenoidal magnetic field. The field lines are parallel to the beam direction. The $\vec{B}$ field is produced by the current flowing though the coils. Reprinted with kind permission from [8], © 2019, C. R. Navi GSU. All rights reserved. Right: A photograph of the ATLAS barrel solenoid. Reprinted with kind permission from [9], © 2008, IOP Publishing. All rights reserved

5.5 Magnet Shapes for Storage Ring Experiments

Table 5.1 Relative permeability for some materials

Medium	μ_r
Bi	0.999834
Cu	0.999994
Pt	1.000265
Carbon steel	100
Ni	100–600
Ferrite (Ni Zn)	16–640
Stainless steel	750–950
Fe (99.8% pure)	5,000
Permalloy	8,000
Co-Fe	18,000
Nanopem	80,000
Fe (99.95% puree annealed in H)	200,000

of a solenoid. The coils are arranged in a spiral configuration through which the current I flows. With N_{coil} turns per length ℓ_{coil}, the magnetic-field strength is given by

$$|\vec{B}| = \frac{\mu_0 \mu_r N_{\text{coil}} I}{\ell_{\text{coil}}}, \tag{5.46}$$

where $\mu_0 \mu_r$ is the magnetic permeability. For many materials, $\mu_r \sim 1$. Table 5.1 lists the relative permeability for various materials, in particular those with high μ_r. For 99.8% pure iron, $\mu_r(Fe) = 5,000$. So, the magnetic-field strength $|\vec{B}|$ is proportional to the current flowing through the coils and the number of coils per unit length. Beam particles are not affected by the $\vec{B}$ field while particles produced in an interaction typically carry transverse momentum and thus are deflected in the $\vec{B}$ field. Table 5.2 list properties of solenoid magnets used in different high-energy physics experiments.

As an example let us look at the ATLAS coils. Figure 5.10 (right) shows the ATLAS central solenoid coil. Its dimensions consist of an inner radius of $r_i = 1.23$ m, an outer radius of $r_o = 1.28$ m and a length of $\ell_{\text{coil}} = 5.8$ m. The coils have 1,173 turns/coil transmitting a current of 7.73 kA. The central magnetic field in the bore is 2 T and the peak field in the windings is 2.6 T. The conductor is made of Al:Cu:NbTi in a ratio of 15.6:0.9:1.0. Its size is 30 mm×4.25 mm and 10 km in total length. The critical current is 20.4 kA at 4.2 K and 5 T. The system is cooled to 4.5 K with a temperature margin of 2.7 K. The cool-down time is eight days.

Table 5.2 Examples of solenoidal magnetic fields in high-energy physics experiments presenting the magnetic-field strength $|\vec{B}|$, coil radius r_{coil} and length ℓ_{coil}, inactive material in front of the calorimeter (X/X_0), the momentum resolution where the first term is the measurement contribution and the second term is the multiple scattering contribution and reference. †This resolution is for the drift chamber alone. *The calorimeter is inside the coil, so there is no significant amount of inactive material in front of the calorimeter

Experiment	$|\vec{B}|$ [T]	r_{coil} [m]	ℓ_{coil}	X/X_0	σ_{p_T}/p_T [%]	Ref.
TOPAZ	1.2	1.45	5.4	0.7	1.5 $p_T \oplus$ 1.6	[10,11]
VENUS	0.75	1.5	5.07	0.84	0.5 $p_T \oplus$ 1.29†	[10,12]
AMY	3.0	1.29	3	*	0.9 $p_T^\dagger$	[10,13]
SLD	0.6	2.79	6.8	*	0.5 $p_T \oplus$ 1.0†	[14,15]
ALEPH	1.5	2.75	7.0	2.0	0.06 $p_T \oplus$ 0.5	[10,16]
DELPHI	1.2	2.8	7.4	1.7	0.057 p_T	[10,17]
L3	0.5	5.93	11.9	*	2 p_T	[18]
OPAL	0.435	2.18	6.3	2.1	0.15 $p_T \oplus$ 2†	[19,20]
CDF II	1.5	1.5	5.07	0.84	0.1 $p_T \oplus$ 0.7	[10,21,22]
D0	2.0	0.6	2.73	0.9	0.2 p_T	[10,23]
CLEO II	1.5	1.55	3.8	*	0.11 $p_T \oplus$ 0.67	[10,24]
BABAR	1.5	1.5	3.46	*	0.13 $p_T \oplus$ 0.45	[10,25]
Belle	1.5	1.8	4.0	*	0.2 $p_T \oplus$ 0.29	[10,26]
BES III	1.0	1.475	3.5	*	0.5 at 1 GeV/c	[10]
ZEUS	1.8	1.5	2.85	0.9	0.58 $p_T \oplus$ 0.65 $\oplus$ 0.14/p_T	[10,27]
H1	1.2	2.8	5.75	1.8	0.2 $p_T \oplus$ 1.5	[10,28]
ATLAS	2.0	1.25	5.3	0.66	0.012 $p_T \oplus$ 1.2	[10,29]
CMS	4.0	6.0	125	*	0.0086 $p_T \oplus$ 2.6	[10,29]

5.5.2 Toroidal Magnet

A toroid is a donut-shaped magnetic coil as shown in Fig. 5.11 (left). The current flows in connected loops around the donut-shaped coil producing a magnetic field in the center of the coil perpendicular to the loop as shown in the figure. Toroidal magnetic fields are typically used in inductors and transformers. In high-energy physics toroid magnets have been used in experiments operating at a hadron collider to measure muon momenta. Since in toroid magnets the current is parallel to the beam axis, they can be used only in hadron colliders. In e^+e^- colliders the beams would be distorted so much that they would be lost quickly. The magnetic field lies in azimuth direction so that deflections are in rz plane. The inner conducting cylinder causes multiple scattering of tracks coming from the interaction region. The momentum resolution is deteriorated and beam particles penetrate the detector in the field free region. The magnetic field is given by

$$|\vec{B}| = \mu_0 \mu_r n_{tor} I, \tag{5.47}$$

5.5 Magnet Shapes for Storage Ring Experiments

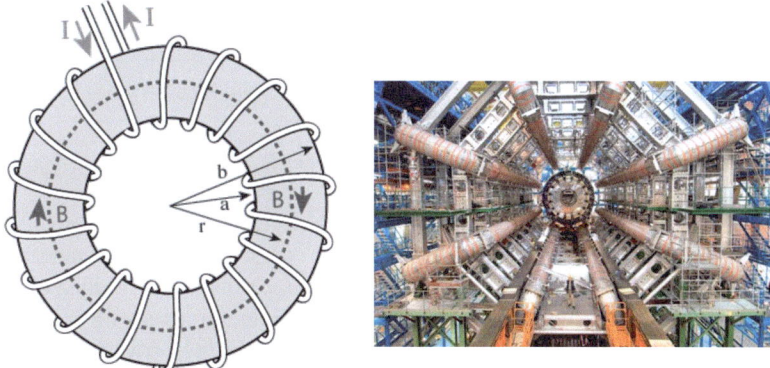

Fig. 5.11 Left: Configuration of a toroidal magnetic field. Reprinted with kind permission from [30], © 2019, C. R. Navi GSU. All rights reserved. Right: Photograph of the ATLAS toroid magnetic system in the barrel. Reprinted with kind permission from [9], © 2008, IOP Publishing. All rights reserved

where $n_{tor} = N_{tor}/(2\pi r_\tau)$, N_{tor} is the number of turns and r_τ is the radius of the torus. We take the ATLAS barrel toroid as an example. The eight coils have an inner radius of $r_i = 4.7$ m, an outer radius of $r_o = 10.05$ m and a length of $\ell_{tor} = 25.3$ m. The coils have 120 turns/coil, transmitting a current of 20.5 kA. The central magnetic field in the bore is 2.5 T while the peak field in the windings is 3.9 T. The conductor is made of Al:Cu:NbTi in a ratio of 28:1.3:1.0. Its size is 57 mm×12 mm and 56 km in total length. The critical current is 58 kA at 4.2 K and 5 T. The system is cooled to 4.5 K with a temperature margin of 1.9 K. The cool-down time is five weeks. Figure 5.11 (right) shows a photograph of the ATLAS barrel toroid system.

In each endcap, the eight coils have an inner radius of $r_i = 0.825$ m, an outer radius of $r_o = 5.35$ m and a length of $l = 5.0$ m. The coils have 116 turns/coil, transmitting a current of 20.5 kA. The typical magnetic field in the bore is 3.5 T with a peak field in the windings of 4.1 T. The field integral is 4–8 Tm. The conductor is made of Al:Cu:NbTi in a ratio of 19.1.3:1.0. Its size is 41 mm×12 mm and 13 km in total length. The critical current is 50 kA at 4.2 K and 5 T. The system is cooled to 4.5 K with a temperature margin of 1.9 K. The cool-down time is four weeks.

5.5.3 Dipole Magnet with Two Compensator Magnets

In a hadron collider a dipole magnet with two compensator fields has been used to steer beams in the interaction region. Figure 5.12 (left) shows the schematic setup of the dipole magnet. The compensator magnets on each side are necessary to correct for the beam distortions after the dipole magnet. The field integral of the three magnets are adjusted appropriately to keep beam particles in orbit. The advantage is a homogeneous field around the interaction region. The momentum resolution is best for secondary particles emitted forward or backward relative to the beam axis. However, the momentum of particles at 90° with respect to the beam axis cannot be measured. Due to the beam deflections, which will produce synchrotron radiation,

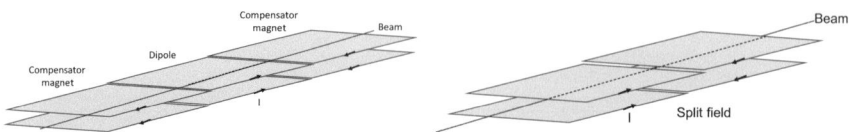

Fig. 5.12 Left: Dipole magnet for the interaction region in a hadron collider. Compensator fields need to be installed on both sides. Right: Split-field dipole magnet

this magnet arrangement is not well-suited for e^+e^- machines. For example, this magnetic-field configuration was used in the UA1 experiment [31], which operated at the CERN Super Proton Synchrotron collider, colliding protons with anti-protons at beam energies of 315 GeV [32]. The 0.7 T magnetic field covered the central region of the detector. The compensator magnets were located in the endcaps.

5.5.4 Split Field Dipole Magnet

In hadron colliders, also split field dipole magnets were used. They provide the most precise momentum measurement in the forward and backward directions. Figure 5.12 (right) shows a schematic view. However, particles emitted transverse to beam direction have a complicated track reconstruction as they pass through an inhomogeneous magnetic field.

5.6 Momentum Measurements in Solenoids

Most colliding beam experiments use solenoids to measure momenta of particles produced in an interaction. Thus, we focus here on solenoidal fields. Since we encounter cylindrical geometry, we use cylindrical coordinates (r, ϕ, z), where r is the radius, ϕ the azimuth angle and z is the coordinate along the beam direction. The magnetic field is given by $\vec{B} = (0, 0, B_z)$. If $\sigma_{r\phi}$ is the measurement error in the $r\phi$ plane, the transverse momentum p_T, which is the momentum component in the $r\phi$ plane, is measured with an error of

$$\left(\frac{\sigma_{p_T}}{p_T}\right)^M = \frac{\sigma_{r\phi} p_T}{0.3|\vec{B}|L_t^2}\sqrt{\frac{720}{N+4}}, \tag{5.48}$$

where N is the number of measured points along the track at uniform spacing. As before, the transverse momentum has units of [GeV/c], the magnetic field comes in units of [T] and the transverse length in units of [m]. This parametrization is a good description for $|\eta| < 1.9$. For larger values, we need to apply a θ correction

$$\left(\frac{\sigma_{p_T}}{p}\right)^M_\theta = \left(\frac{\sigma_{p_T}}{p_T}\right)^M \frac{1}{\tan^2 \theta}. \tag{5.49}$$

5.6 Momentum Measurements in Solenoids

The contribution from multiple scattering is

$$\left(\frac{\sigma_{p_zT}}{p_T}\right)^{MS} = \frac{0.05}{|\vec{B}|L_t}\sqrt{\frac{1.43L_t}{X_0}}. \tag{5.50}$$

Let us consider an example. A 2 GeV/c kaon traverses a drift chamber under a polar angle of 60° that has $N = 40$ measurements in a chamber filled with an Ar-methane gas (0.8:0.2) at normal conditions. The transverse length is $L_t = 0.6$ m. The positions are measured with a resolution of $\sigma_{r\phi} = 200\,\mu$m. The momentum resolution from the measurement contribution is 0.5% and that from multiple scattering is 0.34%, yielding a combined momentum error of 0.6%. In the *BABAR* experiment the momentum resolution was $\sigma_{p_T}/p_T = 0.13\% \cdot p_T \oplus 0.45\%$.

Note that the momentum resolution is directly correlated with the position resolution in the $r\phi$ plane. Thus, it is important to achieve small values of $\sigma_{r\phi}$ in order to obtain low momentum resolutions. Furthermore, we should minimize the inactive material in the tracking detector to minimize multiple scattering. The momentum resolution further improves with the radial coverage of the tracking detector as $1/L_t^2$, the magnetic-field strength as $1/|\vec{B}|$ and the number of measurement points as $1/\sqrt{N}$. Thus, an increase in the tracking distance improves the momentum resolution more than a similar increase in the magnetic-field strength or number of measurement points. For example, increasing L_t by 10% improves the resolution by 20%, while a 10% increase in the magnetic-field strength and the number of measuring points provides a 10% and 5% improvement of the momentum resolution, respectively.

To determine the momentum resolution of muons in the muon system, we need to include a third term that represents the energy loss in the calorimeters,

$$\left(\frac{\sigma_{p_T}}{p_T}\right)^{cal} = \frac{\sigma_{E_{loss}}}{p_T}. \tag{5.51}$$

The total momentum resolution is then

$$\left(\frac{\sigma_p}{p}\right)^{tot} = \left(\frac{\sigma_{p_T}}{p_T}\right)^{M} \oplus \left(\frac{\sigma_{p_T}}{p_T}\right)^{MS} \oplus \left(\frac{\sigma_{p_T}}{p_T}\right)^{cal}. \tag{5.52}$$

In the ATLAS experiment the momentum resolution of muons measured in the Inner Tracker is $\sigma_{p_T}/p_T = 0.042\% \cdot p_T \oplus 1.6\%$. If the muon detectors are included the momentum resolution becomes $\sigma_{p_T}/p_T = 0.017\% \cdot p_T \oplus 3.3\% \oplus 250\%/p_T$, where the momentum is given in units of [GeV/c]. The total momentum is reconstructed from the polar angle θ and the transverse momentum p_T, yielding

$$p = \frac{p_T}{\sin\theta}. \tag{5.53}$$

Figure 5.13 (left, right) shows the charged-particle momentum resolutions as a function of the transverse momentum. We have determined this for a gaseous tracking detector using Ar-CH$_4$ gas (80:20) at normal conditions, a magnetic field strength of

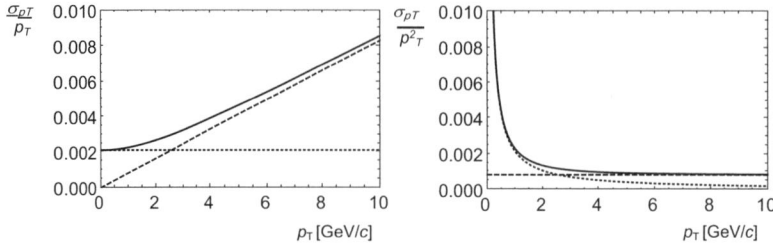

Fig. 5.13 Momentum resolution σ_{p_T}/p_T (left) and σ_{p_T}/p_T^2 (right) as a function of transverse momentum for the measurement contribution (dashed curves), the multiple scattering contribution (dotted curves) and the total momentum resolution (solid curves)

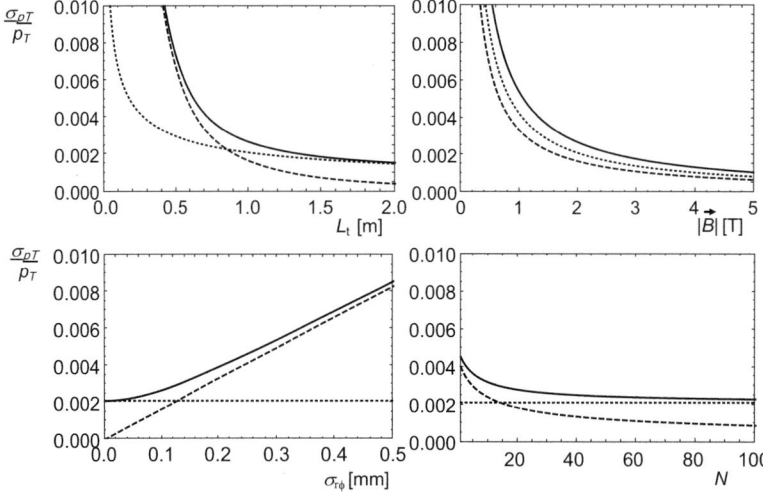

Fig. 5.14 Charged particle momentum resolution as a function of L_t (top left), $|\vec{B}|$ (top right), $\sigma_{r\phi}$ (bottom left) and N (bottom right). Shown are the measurement contribution (dashed curves), the multiple scattering contribution (dotted curves) and the total momentum resolution (solid curves)

2 T, a position resolution of 100 μm and 25 measurement points. Figure 5.14 shows the momentum resolutions as a function of effective length of the tracking detector in the magnetic field L_t (top left), magnetic-field strength $|\vec{B}|$ (top right), position resolution $\sigma_{r\phi}$ (bottom left) and number of measurements N (bottom right). The transverse momentum is fixed at 2 GeV/c. The other fixed parameters are the same.

Figure 5.15 (left) shows σ_{p_T}/p_T as a function of the transverse momentum in the BABAR experiment. The multiple scattering component is about 0.5%. Above 1 GeV/c the measurement component dominates. Figure 5.15 shows the ATLAS muon momentum resolutions σ_{p_T}/p_T in the barrel as a function of the transverse momentum based on the Inner Detector only (center) and a combination of Inner Detector and Muon system (right).

For the measurement of the polar angle θ, again we have contributions from the track measurement and from multiple scattering that are added in quadrature. The

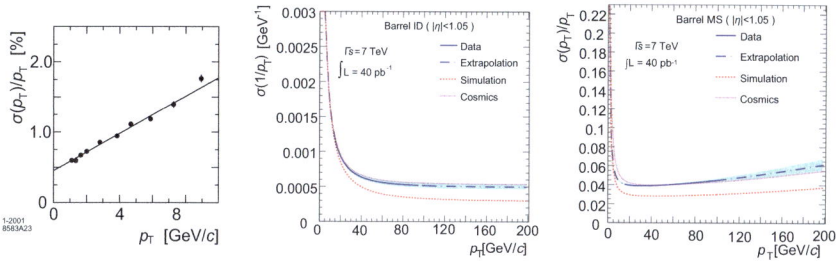

Fig. 5.15 Left: Charged particle transverse momentum resolution in the *BABAR* experiment as a function of transverse momentum. Reprinted with kind permission from [25], © 1978, Elsevier. All rights reserved. Center: The ATLAS muon momentum resolution σ_{p_T}/p_T^2 using the Inner Detector only. Right: The ATLAS muon momentum resolution σ_{p_T}/p_T using the Inner Detector and Muon system. The curves correspond to cosmic muons (dotted curves), pp data at 7 TeV (solid curves), extrapolations (dash-dotted curves) and simulation (dashed curves). Both reprinted with kind permission from [33], © 2011, the ATLAS Collaboration. All rights reserved

measurement term of the angular resolution is given by

$$(\sigma_\theta)^M = \frac{\sigma_z}{L_t}\sqrt{\frac{12(N_z-1)}{N_z(N_z+1)}}, \tag{5.54}$$

where σ_z is the position resolution in the z direction and N_z is the number of z measurements. The error improves as $1/L_t$ and $1/\sqrt{N_z}$. The error due to multiple scattering is given by

$$(\sigma_\theta)^{MS} = \frac{0.015}{\sqrt{3}\,p}\sqrt{\frac{L_t}{X_0}}. \tag{5.55}$$

The multiple scattering contribution decreases with momentum as $1/p$ and with radiation length as $1/\sqrt{X_0}$ while it increases with $\sqrt{L_t}$.

Exercises

5.1 Consider a colliding beam experiment that is covered with a symmetric detector. A pion of momentum p = 1 GeV/c is produced at the origin in the direction $\vec{u} = (0.6124, 0.6124, 0.5)$. The pion traverses a cylindrical drift chamber that has an inner radius of 20 cm, an outer radius of 80 cm and a total length of 2 m. The chamber is placed inside a solenoidal magnetic field (in z direction) of a strength $|\vec{B}| = 1.5\,\text{T}$. The chamber is filled with Ar gas plus some small additional contributions that we can neglect ($X_0 = 19.55\,\text{g/cm}^2$). The pressure chosen corresponds to a density of $\rho = 1.8\,\text{mg/cm}^3$. The drift chamber is laid out to produce 50 measurements in radial direction, each giving a spatial resolution of $\sigma_{r\phi} = 180\,\mu\text{m}$. What momentum resolution is expected for the pion? How big are the contributions from the measurements and from multiple scattering components? How does the momentum resolution change if the pion is emitted in the direction $\vec{u} = (0.4478, 0.0, 0.8942)$. How could you improve the momentum resolution? (Hint: make a sketch of the track in the chamber.)

5.2 Consider a 1 GeV/c particle that is produced in an interaction and travels in a 1.5 T magnetic field. Plot the $x(\ell)$ and $y(\ell)$ dependence versus ℓ and the correlation $y(\ell)$ versus $x(\ell)$. For what value of ℓ do we get a full circle. What are the circle midpoints?

5.3 Consider a cylindrical drift chamber that has 40 layers placed at radial positions between 20 cm and 80 cm. Inside the drift chamber is a vertex detector. The chamber is operated in a magnetic field of 1.5 T. What is the minimum momentum for a track to exit the chamber at the outer chamber radius? What is the minimum momentum for a track to leave the vertex detector that has an outer radius of 18 cm?

5.4 We make 16 rz measurements in the axial wire planes of a drift chamber, which are summarized the following table. Determine the longitudinal impact parameter and its resolution and check if the χ^2 of the fit is consistent with the fit function.

Table 5.3 The measured r_i and z_i positions with their resolutions σ_i in a cylindrical drift chamber

r_i [cm]	26.04	27.14	28.24	29.34	42.27	43.37	44.47	45.57
z_i [cm]	15.10	15.77	16.4	17.03	24.54	25.15	25.79	26.42
σ_i [cm]	0.021	0.020	0.021	0.019	0.02	0.019	0.019	0.019
r_i [cm]	58.54	59.64	60.74	61.84	74.72	75.82	76.92	78.02
z_i [cm]	33.88	34.54	35.15	35.69	43.22	43.85	44.51	45.13
σ_i [cm]	0.021	0.019	0.019	0.019	0.02	0.019	0.019	0.02

5.5 We make five rz measurements in a silicon strip detector which are summarized the following table. Determine the longitudinal impact parameter and its resolution and check if the χ^2 of the fit is consistent with the fit function.

Table 5.4 The measured r_i and z_i positions with their resolutions σ_i in a silicon strip detector

r_i [mm]	32	40	54	127	144
z_i [mm]	5.6	7.12	9.57	22.41	25.48
σ_i [mm]	0.015	0.016	0.017	0.031	0.035

5.6 Determine the lifetime of the following decay measurements listed in the table below. The errors are given by $\sigma_i = 1/\sqrt{N_i}$. What is the uncertainty of the lifetime. Calculate the χ^2 to determine if this is a good fit.

Table 5.5 Measured activity of a radioactive source at 14 time intervals

$t_i [10^{-12}$ s]	0.1	0.2	0.3	0.4	0.5	0.6	0.7
N_0 [decays/s]	7,150	4,950	3,100	2,540	1,780	1,290	860
$t_i [10^{-12}$ s]	0.8	0.9	1.0	1.1	1.2	1.3	1.4
N_0 [decays/s]	650	420	335	240	140	120	78

5.7 (a) A 4 m long detector uses a stainless steel superconducting coil to produce a 1.5 T magnetic field. What current flows through the 1000 coils? (b) A dipole magnet has a gap of 20 cm. The magnet is 50 cm wide and 1 m deep. What is the iron thickness to achieve an 80% average magnetic field?

5.8 A 10 GeV/c muon traverses a 1 m thick iron wall that is located in a 1.2 T homogeneous magnetic field. Four measurements are performed with spatial resolutions of 120 μm. What are the contributions of the momentum resolution from multiple scattering and the measurements?

5.9 A cylindrical drift chamber measures 40 space points in a 1.5 T magnetic field with a resolution of 150 μm. The chamber is located in the radial space between 20 cm and 80 cm. The chamber is operated with a He − C_4H_{10} (80:20) gas mixture. The radiation length is 807 m. Determine the resolution contribution from measurements and multiple scattering for a 1 GeV/c pion.

5.10 Determine the angular-resolution contributions from the measurements and multiple scattering components for a 0.8 GeV/c kaon in a cylindrical drift chamber that is located in a 1 T magnetic field. The chamber operated with an Ar − CH_4 − iC_4H_{10} (88.2:9.8:2) gas mixture at NTP has 159 space point measurements and is located in a radial space from 25.5 cm and 183.5 cm. The spatial resolution is 135 μm. The radiation lengths are $X_0(\text{Ar}) = 19.55$ g/cm^2, $X_0(\text{CH}_4) = 46.47$ g/cm^2 and $X_0(C_4H_{10}) = 45.23$ g/cm^2.

References

1. A. Andreazza, Tracking systems, Lecture at the University of Milano, http://www2.fisica.unimi.it/andreazz/AA_TrackingSystems.pdf (2017)
2. L. Reuen, dissertation at University of Bonn (2011)
3. S. Menzemer, PhD thesis, University of Karlsruhe (2003)
4. The LHCb Collaboration (A. Alves et al.), JINST **3**, S08005 (2008)
5. The Everson Tesla Company, https://www.eversontesla.com/projects
6. The LHCb Collaboration, https://cds.cern.ch/record/808276
7. R.L. Glückstern, Nucl. Instrum. Meth. **24**, 384 (1963)
8. C.R. Navi, http://hyperphysics.phy-astr.gsu.edu/hbase/magnetic/solenoid.html
9. The ATLAS Collaboration (G. Aad et al.), JINST **3**, S08003 (2008)
10. The Particle Data Group (S. Navas et al.), Phys. Rev D. **110**, 030001 (2024)
11. The TOPZ Collaboration (Y. Inoue et al.), Eur. Phys. J. C **18**, 273 (2000)
12. A. Arai et al., Nucl. Instrum. Meth. A **254**, 317 (1987); K. Okada, KEK-Preprint-88-5 (1988)

13. The AMY Collaboration (H. Sagawa et al.), Phys. Rev. Lett. **60**, 93 (1988); K. Ueno, UR-1050, ER13065-528 (1988)
14. M.D. Hildreth et al., IEEE Trans. Nucl. Sci. **42**, 452 (1995)
15. M. Breidenbach, IEEE Trans. Nucl. Sci. **33**, 46 (1986)
16. The ALEPH Collaboration (D. Buskulic et al.), Nucl. Instrum. Meth. A **360**, 481 (1995)
17. The DELPHI Collaboration (P. Abreu et al.), Nucl. Instrum. Meth. A **378**, 57 (1995)
18. B. Adeva et al., Nucl. Instrum. Meth. A **289**, 35 (1990); F. Beissel et al., Nucl. Instrum. Meth. A **332**, 33 (1993)
19. The OPAL Collaboration (K. Ahmet et al.), Nucl. Instrum. Meth. A **305**, 275 (1991)
20. O. Biebel et al., Nucl. Instrum. Meth. A **232**, 169 (1992)
21. The CDF Collaboration (D. Acosta et al.), Phys. Rev. D **71**, 052003 (2005)
22. R.G.C. Oldeman, eConf C **0304052**, FO005 (2003)
23. The D0 Collaboration (S. Abachi et al.), FERMILAB-PUB-96-357-E (1996)
24. The CLEO II Collaboration (Y. Kubota et al.), Nucl. Instrum. Meth. A **320**, 66 (1992)
25. The *BABAR* Collaboration (B. Aubert et al.), Nucl. Instrum. Meth. A **479**, 1 (2002)
26. The Belle Collaboration (A. Abashian et al.), Nucl. Instrum. Meth. A **479**, 117 (2002)
27. D. Bailey, R. Hall-Wilton, Nucl. Instrum. Meth. A **515**, 37 (2003)
28. The H1 Collaboration (V. Andreev et al.), Eur. Phys. J. C **80**-12, 1189 (2020)
29. Y.M.T. Zeng, https://twiki.cern.ch/twiki/pub/Main/ShihChiehHsuSemiconductorTracker/momentum_resolution_hw1.pdf
30. C.R. Navi, https://www.pinterest.com/pin/590112357394312019/
31. The UA1 Collaboation (M. Barranco-Luque et al.), Nucl. Instrum. Meth. **176**, 255 (1980)
32. Proton antiproton collider, R. Schmidt, Particle accelerators **50**, 47 (1995)
33. The ATLAS Collaboration (G. Aad et al.), ATLAS-CONF-2011-046 (2011)

Open Access This chapter is licensed under the terms of the Creative Commons Attribution 4.0 International License (http://creativecommons.org/licenses/by/4.0/), which permits use, sharing, adaptation, distribution and reproduction in any medium or format, as long as you give appropriate credit to the original author(s) and the source, provide a link to the Creative Commons license and indicate if changes were made.

The images or other third party material in this chapter are included in the chapter's Creative Commons license, unless indicated otherwise in a credit line to the material. If material is not included in the chapter's Creative Commons license and your intended use is not permitted by statutory regulation or exceeds the permitted use, you will need to obtain permission directly from the copyright holder.

Detectors for Vertex Measurements 6

In this chapter, we present detectors that are especially designed to measure decay vertices of heavy-flavored particles with high precision. Most of them are based on doped silicon by combining p-type and n-type silicon in a pn junction whose principle of operation we discussed in Sect. 3.3.1. A traversing charged particle produces copious electron-hole pairs via energy loss in the depletion zone of the pn junction. The liberated charge is collected. Since diffusion is small due to short paths the particle trajectory is precisely determined.

The first solid-state vertex detectors consisted of silicon microstrip sensors, which are still used in many experiments today. We discuss their development and show different examples. The SLD experiment followed a different approach. They built a vertex detector with charge-coupled devices, which have superb position resolution but have a rather slow readout rate. At the Stanford Linear Collider (SLC), bunch-crossing and event rates were sufficiently low to permit the usage of CCDs in SLD. At circular e^+e^- and hadron colliders, however, bunch-crossing and event rates are much higher so that CCDs are inefficient. They are still used today in astroparticle physics experiments. The higher occupancies near the beam pipe triggered the development of pixel sensors. First detectors were already developed for LEP II experiments. The demands of the high-luminosity LHC experiments and for Belle II, where radiation hardness becomes an issue, have prompted further developments including monolithic active pixel sensors, depleted p-channel field effect transistors, low-gain avalanche detectors and three-dimensional detectors, which will be presented after discussing radiation hardness issues of silicon sensors. We conclude the chapter with a discussion of diamond detectors.

6.1 Silicon Microstrip Detectors

The basis of silicon microstrip detectors is mono-crystalline silicon. Therefore, we start with a discussion of the fabrication of this material before presenting the layout of a microstrip sensor. Different crystal growing techniques are discussed in the Springer handbook of crystal growth [1]. Here, we present two techniques that play an important role for silicon production.

6.1.1 Fabrication of Mono-crystalline Silicon

One method to grow large mono-crystalline silicon ingots is based on the Czochralski growing method [2]. Figure 6.1 (left) shows a schematic concept of the Czochralski method. One starts with high-purity silicon that is melted at temperatures above the silicon melting temperature of 1414 °C in a quartz crucible. At this stage, one can add impurities like boron or arsenic to produce p-type or n-type silicon of the desired doping concentration. One dips a seed crystal mounted precisely on a rod

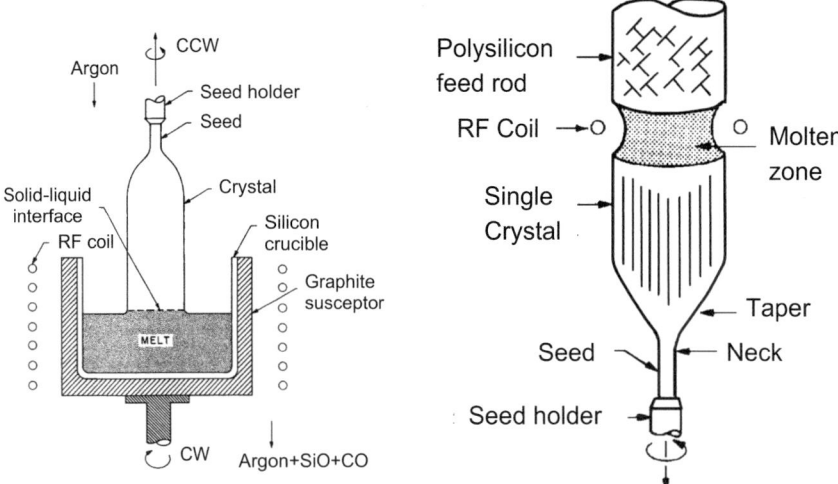

Fig. 6.1 Left: Schematic setup of the Czochralski (Cz) crystal-growing method. A SiO_2 crucible is filled with molten silicon. The crucible surrounded by an RF coil is rotated clockwise. A seed crystal is mounted on a rod that rotates counter clockwise. The seed crystal is slowly pulled upwards moving the grown crystal out of the melt. Right: Schematic setup of the float-zone crystal-growing method. A high-purity polycrystalline silicon rod is held face-to-face with a mono-crystalline seed crystal. At the interface there is an RF coil that partially melts the rod. The seed crystal rotating counter clockwise touches the molten zone. A necking process is done to establish a dislocation-free crystal. The neck is allowed to increase in diameter. It is tapered reaching the desired diameter of the crystal on the upper side. The molten silicon crystallizes on the seed crystal, which is then pulled slowly downward. Since the molten zone is only in contact with the formed crystal, FZ crystals achieve higher purity and higher resistivity than crystals grown with the Czochralski method. Both reprinted with kind permission from [3], © 1999, M. Moll. All rights reserved

6.1 Silicon Microstrip Detectors

Fig. 6.2 Photograph of silicon ingots and silicon wafers. Reprinted with kind permission from [4], © 2018, SUMCO Corporation. All rights reserved

into the molten silicon by slowly lowering the rod till the seed crystal is submerged. In addition, the rod rotates counter clockwise while the crucible rotates clockwise. By adjusting temperature and pulling speed the crystal diameter is first brought down to several millimeters. This helps to eliminate dislocations generated by the seed-melt contact shock. Then the diameter is widened to the full crystal size. The quartz crucible gradually dissolves in the process releasing large amounts of oxygen into the melt of which 99% is lost as SiO gas. Some of the SiO interacts with the graphite producing SiC that re-enters the melt. With a set of RF coils, a precisely controlled temperature gradient is placed along the crucible. At the surface of the molten silicon the temperature is cooler than the silicon melting point. Thus, the silicon crystal starts to grow around the seed crystal. The rod is pulled out at a speed of about 25 mm per hour. Due to the rotation, a cylindrically-shaped ingot is obtained. The ingot length can be up to two meters. The diameter has been increased from six inches to eight inches. Figure 6.2 shows silicon ingots and wafers that have been sliced off from ingots.

Another technique is the float-zone (FZ) method [5], which is based on the zone-melting principle as shown in Fig. 6.1 (right). The process starts with a high-purity poly-crystalline silicon rod and a mono-crystalline seed crystal placed face-to-face in a vertical position in vacuum or in an inert-gas atmosphere. Both are rotated and partially melted using a radio frequency field. The seed crystal is brought in contact with the melt at the tip of the rod. The bottom of the crucible is funnel-shaped, so that the crystal first grows at a small diameter before it is increased to the desired size for steady-state growth. The reason is to produce a dislocation-free crystal. The molten zone is slowly moved upwards so that the crystal can continue to grow from the bottom as a single crystal and the material gets simultaneously purified. The crystals are doped with gaseous phosphine (PH_3) or diborane (B_2H_6) to produce n-type or p-type silicon, respectively. Since the silicon is not in contact with anything besides the ambient gas, higher-purity and higher resistivity sensors are obtained than with the Czochralski method. Impurities can be further reduced by performing multi-zone refining. While phosphorus evaporates from the melt at a fairly high rate, boron has an equilibrium segregation coefficient that is close to the effective segregation coefficient. Therefore, it is easier to produce more homogeneous p-type silicon with this method than n-type silicon. On the other hand, high-resistivity p-type silicon can be obtained from poly-silicon only with low boron concentrations. Note that silicon produced with the float zone method shows microscopic dopant inhomogeneities or dopant striations compared to silicon produced with the Czochralski method since due to the small melting zone more temperature fluctuations, remelting phenomena and dopant segregations occur. The uniformity is better for p-type silicon than for n-type silicon. High-resistivity, very homogenous n-type silicon can be produced via neutron transmutation doping of high-purity p-type silicon [6].

Epitaxial layers are thin crystalline layers with a well-defined orientation with respect to the substrate they are grown on. Different methods exist to produce them [7]. One of them is the vapor phase epitaxy in which a low-resistivity substrate wafer acts as a seed crystal onto which silicon is deposited from a compound like silicon tetrachloride. With hydrogen gas, solid silicon is produced and HCl gas. The reaction temperature is around 1200 °C. The growth rate is 1 μm/min. Adding gases like diborane or phosphine, p-type or n-type doping can be achieved.

6.1.2 Layout of a Silicon Microstrip Detector

Charged particles and photons of sufficient energy generate electron-hole (e-h) pairs in silicon. The goal of any silicon detector is to collect these charges separately and amplify them. In silicon, 3.62 eV are necessary to generate an e-h pair. To be able to detect a signal, the generated charge has to be much larger than the noise. Note that photons and charged particles interact with silicon in different ways. Below 1 MeV, photo-absorption is the dominant process. Thus, photons typically deposit their entire energy in a tiny volume creating a localized charge cloud. Fast charged particles typically traverse the silicon as minimum ionizing particles (MIPs) loosing energy in silicon via dE/dx. The mean energy loss is 3.87 MeV/cm. The most probable energy is about 70% of that. Since the energy for producing an electron-hole pair is 3.62 eV in silicon, a traversing charged particle generates 80 e-h pairs per 1 μm. So, a 300 μm thick detectors liberates about 24,000 e-h pairs. The electrons and holes are separated by a strong electric field that runs across a fully depleted microstrip detector, which is based on n-type silicon. It consists of a p^+n junction[1] that uses acceptor concentrations of $N_A = $ few $\times\ 10^{15}$ – few $\times\ 10^{17}$ cm^{-3} for the p^+ region and donor concentrations of $N_D = 1$–5×10^{12} cm^{-3} for the n-type bulk [8]. The resistivity is larger than 2 kΩ-cm. The typical operation voltage is $V_b = 100$–200 V.

---

[1] The p^+ notation indicates that the p-layer is highly doped. Correspondingly, n^+ indicates high n doping.

6.1 Silicon Microstrip Detectors

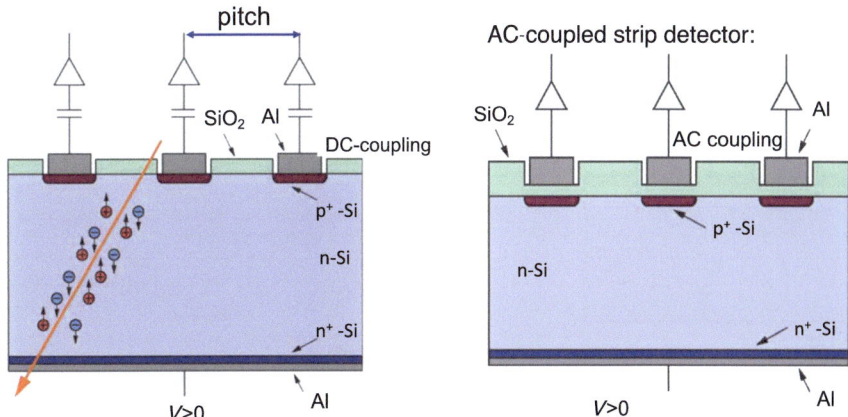

Fig. 6.3 Left: Generic layout of an n-type microstrip detector in which the readout strip is DC coupled to the preamplifier. As indicated the strip pitch is the distance between two neighboring strip centers. Right: Generic layout of an n-type microstrip detector in which the readout strip is AC coupled to the preamplifier. Both reprinted with kind permission from [8], © 2010, M. Krammer. All rights reserved

The backplane is made from an n^+ layer to improve the ohmic contact. The outer coating consists of an aluminum metallization.

The electrons drift towards the n^+ plane while the holes drift towards the p^+ strips, where they are recorded. Figure 6.3 (left) depicts a layout for reading out strips with DC coupling. Here, the aluminum readout strip is directly coupled to the p^+ strip. The current produced by a traversing charged particle is typically converted into a voltage pulse via a resistor. The signal is decoupled from the preamplifier with an external capacitor. This setup has the disadvantage that leakage currents are recorded as well. Figure 6.3 (right) shows a sensor layout for an AC-coupled strip readout. Here, a thin layer of SiO_2 insulates the aluminum readout strip from the p^+ strip. This introduces an internal capacitor that prevents leakage currents from being detected while the signal is recorded.

6.1.3 Noise, Current Flow and Collection Time

The main source of noise results from statistical fluctuations in the number of carriers, leading to changes in the conductivity. This effect is temperature dependent. Figure 6.4 (left) shows an alternate circuit diagram of a silicon microstrip detector. The noise is specified as the mean equivalent noise charge, $\overline{ENC}$. The most important noise contributions result from four sources [9]:

- The leakage current ($\overline{ENC}_L = \frac{e}{2}\sqrt{(I_{dt}t_f/q_e)}$).
- The detector capacity ($\overline{ENC}_C = a_C + b_C C_d$).

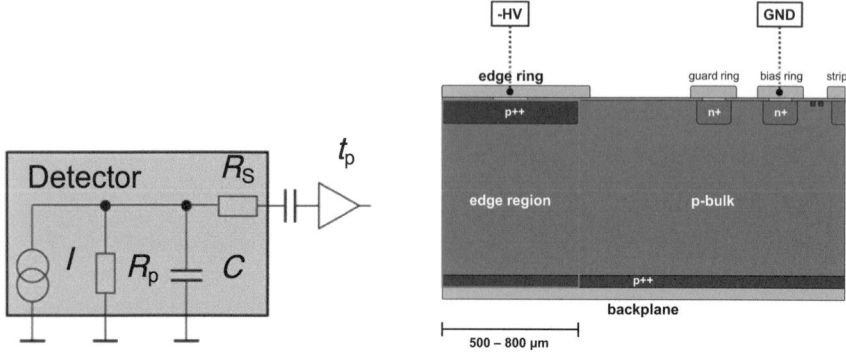

Fig. 6.4 Left: Alternate circuit diagram of a silicon detector. Reprinted with kind permission from [8], © 2010, M. Krammer. All rights reserved. Right: Cross section through the *n-in-p* sensor with front-side biasing. The sensor has three guard rings and an edge ring. Reprinted under CC-BY-3.0 Licence from [10], © 2018, M. Baselga

- The detector parallel resistor $\overline{ENC}_{R_p} = \frac{e}{q_e}\sqrt{\frac{k_B T t_f}{2 R_p}}$.
- The detector serial resistor $\overline{ENC}_{R_s} = 0.395 C_d \sqrt{(R_s/t_f)}$,

where e is the Euler number, q_e is the charge of the carrier, I_d is the dark current, t_f is the integration time, R_p is the parallel resistor, R_s is the serial resistor and C_d is the detector capacitance. The parameters a_C and b_C are determined by the parameters of the preamplifier, typically $a_C \sim 160$ e$^-$ and $b_C \sim 12$ e$^-$/pF. The overall equivalent noise charge is the quadratic sum of all contributions. Note that the detector capacity is typically the dominant noise source. For optimized electronics that matches the detector capacitance C_d with an amplifier input capacitance C_a, we obtain an equivalent noise charge squared of

$$\overline{ENC}^2 = \frac{8 k_B T C_d}{f_T \tau_f}, \tag{6.1}$$

where k_B is the Boltzmann constant, T is the temperature, τ_f is the time constant of the filter and f_T is the frequency of the amplifier for unity gain. For typical values of $f_T = 1$ GHz and $\tau_f = 100$ ns, we estimate $\overline{ENC} = 1.13 \times 10^2 \sqrt{C_d}$ [rms e$^-$] with C_d in units of pF. So for $C_d = 1$ pF, we obtain $\overline{ENC} \simeq 113$ e$^-$, while for $C_d = 100$ pF we get $\overline{ENC} \simeq 1130$ e$^-$. Thus, the noise is small compared to the 24,000 *e-h* pairs. The average current flow squared is given by

$$\langle I^2 \rangle = 2 q_e I_d \Delta B, \tag{6.2}$$

where ΔB is the band width.

Note that standard microstrip sensors are biased from the backside. For LHC applications, frontside biasing was designed as discussed in Sect. 6.2.2. Figure 6.4 (right) shows a cross section of a silicon strip sensor with frontside biasing. The bias voltage is inserted through the edge ring. In addition, there is a bias ring that is kept

6.1 Silicon Microstrip Detectors

on ground potential and a guard ring that is usually set to low voltage. For stable operations guard rings are essential. They typically surround the entire sensor and are kept at a fixed potential, for example at ground potential.

The tracking accuracy is also limited by multiple scattering. Thus, it does not help to make detectors too thick. Since the radiation length of silicon is $X_0^{Si} = 9$ cm, we determine a thickness of $0.0032 X_0$ for a 300 μm thick sensor, which corresponds to that of a 50 cm thick Ar-CO_2 (50:50) detector. The measurement precision is also affected by diffusion. Ionized particles that are initially localized within 1 μm will broaden to 30 μm (FWHM) after a drift distance $d_{drift} = 1$ mm. If the strip pitch is small enough, the charge may be collected by more than one strip, which yields a better position resolution than that of a single strip. Collection times of the signal are fast. For an electric field of $|\vec{E}| = 2 \times 10^4$ V/cm, an electron mobility of $\mu_e = 1,450$ cm^2/(Vs) and a 0.3 mm thickness, we estimate a collection time t_c of

$$t_c = \frac{L_{path}}{v_d^e} = \frac{L_{path}}{\mu_e |\vec{E}|} = 1 \text{ ns}. \tag{6.3}$$

The electron drift velocity can be parameterized in terms of electric field and temperature by [11],

$$v_d^e(\text{Si}) = \frac{1.42 \times 10^9 |\vec{E}|/T^{2.42}}{\left(1 + \left(|\vec{E}|/(1.01 T^{1.55})\right)^{0.0257 T^{0.66}}\right)^{-(0.0257 T^{0.66})}} \text{ cm/s}, \tag{6.4}$$

where $|\vec{E}|$ is the electric field in [V/cm] and T is the temperature in Kelvin. Similarly, we get for holes,

$$v_d^h(\text{Si}) = \frac{1.31 \times 10^8 |\vec{E}|/T^{2.2}}{\left(1 + \left(|\vec{E}|/(1.24 T^{1.68})\right)^{0.46 T^{0.17}}\right)^{-(0.46 T^{0.17})}} \text{ cm/s}. \tag{6.5}$$

Figure 6.5 shows the electron and hole drift velocities as a function of the electric field at NTP (left) and temperature for $|\vec{E}| = 1$ kV/cm (right). At temperatures above 50 K, the electron drift velocity is larger than the hole drift velocity.

The electron and hole mobilities depend on temperature and the doping concentrations [12],

$$\mu_e = 88 \left(\frac{T}{300}\right)^{-0.57} + \frac{7.4 \times 10^8 / T^{2.33}}{1 + 0.88 \left[\frac{N_D}{1.26 \times 10^{17} (T/300)^{2.4}}\right] (T/300)^{-0.146}} \text{ cm}^2/(\text{V} \cdot \text{s}),$$

$$\mu_h = 54.3 \left(\frac{T}{300}\right)^{-0.57} + \frac{1.36 \times 10^8 / T^{2.33}}{1 + 0.88 \left[\frac{N_A}{2.35 \times 10^{17} (T/300)^{2.4}}\right] (T/300)^{-0.146}} \text{ cm}^2/(\text{V} \cdot \text{s}), \tag{6.6}$$

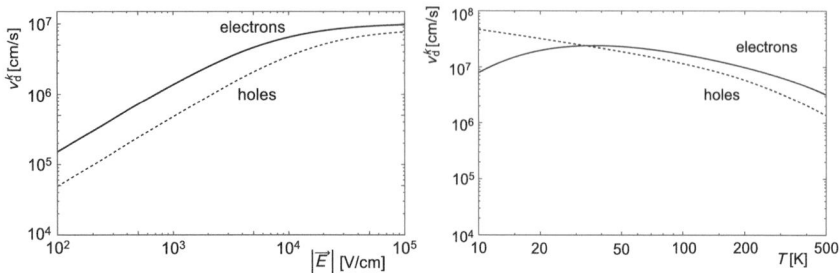

Fig. 6.5 Left: The drift velocity of electrons (solid) and holes (dashed) in silicon as a function of the electric field for NTP. Right: The drift velocity of electrons (solid) and holes (dashed) in silicon as a function of temperature for NTP

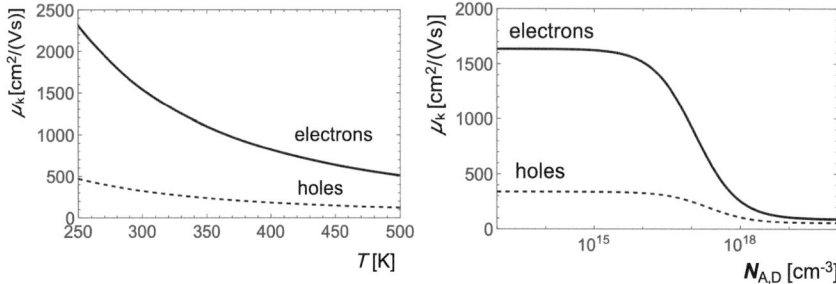

Fig. 6.6 Left: The mobility of electrons (solid) and holes (dashed) in silicon as a function of temperature for $N_{D,A} = 10^{16} \text{cm}^{-3}$. Right: The mobility of electrons (solid) and holes (dashed) in silicon at room temperature as a function of the doping concentration

where N_D and N_A are the concentrations of donors and acceptors in silicon, respectively. We see that the mobilities have a temperature dependence for doping concentrations below $10^{18}/\text{cm}^3$. Figure 6.6 shows the electron and hole mobilities as a function of temperature (left) and doping concentration (right). The dependence on the doping concentration is effective between $10^{16}/\text{cm}^3$ and $10^{18}/\text{cm}^3$.

We characterize sensors by measuring the current-voltage dependence called IV curves. For example, defects and radiation damage on sensors will modify the IV curves by showing higher currents compared to undamaged sensors. Therefore, IV curves are a useful tool in examining sensors. Figure 6.7 (left) shows IV curves for a backward-biased device when the guard ring is left on a floating potential [13]. For some strips, the leakage current increases rapidly already at low reverse bias voltages. Keeping the guard ring on a fixed potential can reduce the leakage currents significantly. Figure 6.7 (right) shows the same IV curves as in Fig. 6.7 (left) when the guard ring is grounded. Even for a reverse bias voltage of around −300 V, the leakage currents of all strips remain low. These results underline the importance of guard rings.

6.1 Silicon Microstrip Detectors

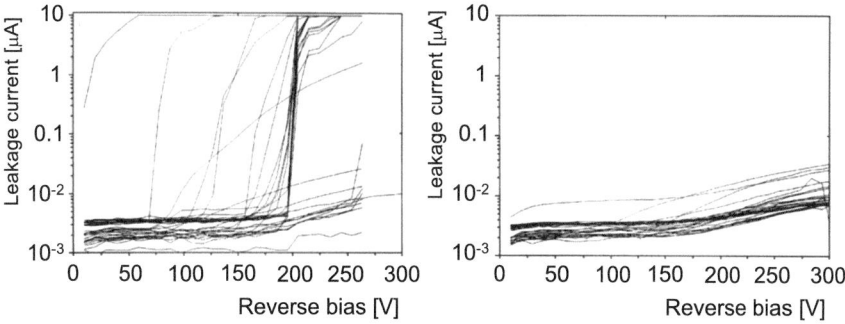

Fig. 6.7 Left: The IV curves of the strips when the guard ring is left on a floating potential. Right: The IV curves of the strips when the guard ring is grounded. Both reprinted with kind permission from [13], © 2005, IOP Publishing. All rights reserved

1. Start with an *n*-type silicon wafer.

2. Coat wafer with an SiO_2 layer via passivation.

3. Open windows by etching the SiO_2 away at specified places.

4. Implant boron ions from the top and arsenic ions from the bottom.

5. Anneal ions from the SiO_2 layer.

6. Perform aluminum metalization on the *p*-side.

7. Remove aluminum on the SiO_2 layer.

8. Perform aluminum metalization on the *n*-side.

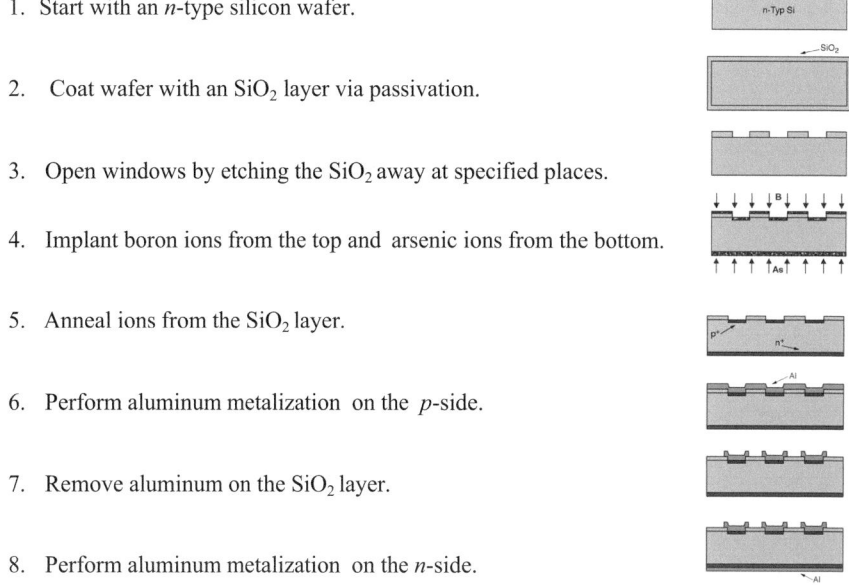

Fig. 6.8 Fabrication of microstrip detectors. Reprinted with kind permission from [15], © 1984, Elsevier. All rights reserved

6.1.4 Fabrication of a Silicon Microstrip Detector

The first silicon strip detector was built for the NA11 experiment by physicists from TUM, MPI Munich and CERN [14]. The detector was fabricated in several steps as shown in Fig. 6.8 [15].

The *n*-type silicon bulk typically has a concentration of $N_D = 1-5 \times 10^{12}$ cm^{-3}. The SiO_2 layer produced via passivation is about 200 nm thick. The p^+ strips are produced via boron implantation and have concentrations of $N_A = 5 \times 10^{16} - 3 \times$

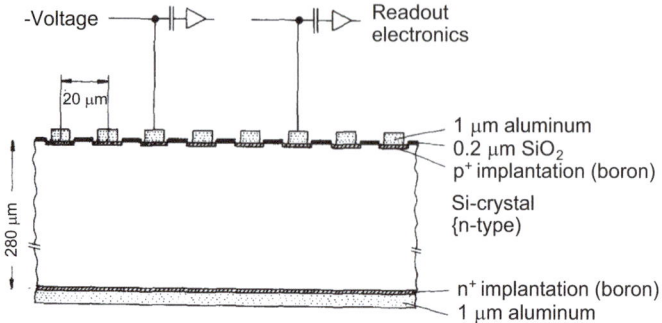

Fig. 6.9 Cross section of the first CERN-Munich microstrip detector with capacitive charge division. The sensor consisted of 280 μm thick n-type silicon. The 12 μm wide strips were placed at a 20 μm pitch. Every third strip was read out with capacitive coupling. The high voltage of -160 V was supplied just to those strips that were read out. Reprinted with kind permission from [14], © 1983, Elsevier. All rights reserved

Fig. 6.10 Photograph of a completed microstrip detector. Reprinted with kind permission from [14], © 1983, Elsevier. All rights reserved

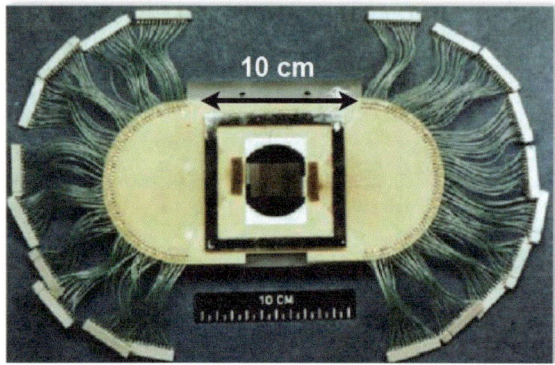

10^{19} cm^{-3}. The n^+ layer is produced by implantation of arsenic ions yielding a concentration of $N_D = 5 \times 10^{15} - 10^{18}$ cm^{-3}.

6.1.5 The First Microstrip Detector

Figure 6.9 shows a cross section through the first microstrip detector, which consisted of a 24 mm × 36 mm rectangular array housing 1,200 strips. The detector thickness was 280 μm made of n-type silicon with a resistivity of $\rho = 2$ kΩ-cm. The bottom was covered with an n^+ implantation and a 1 μm thick Aluminum layer. The top housed 36 mm long and 12 μm wide p^+ strips separated by 20 μm pitch. The strips covered by a 1 μm thick aluminum contact were insulated from each other by SiO$_2$. To save electronics only every third strip was read out with capacitive coupling. The detector was operated with a reverse bias voltage of -160 V. Figure 6.10 shows a mounted microstrip detector used in the NA11 experiment [14, 16].

Fig. 6.11 Top: Charge distribution of the track position with respect to the center of the strip versus distance with no magnetic field for $V_b = -120\,V$ (left) and $V_b = -200\,V$ (right) [17]. Bottom: Charge distribution of the track position with respect to the center of the strip versus distance for $V_b = -120\,V$ in a magnetic field of a $+1.68\,T$ (left) and $-1.68\,T$ (right). All reprinted with kind permission from [17], © 1983, Elsevier. All rights reserved

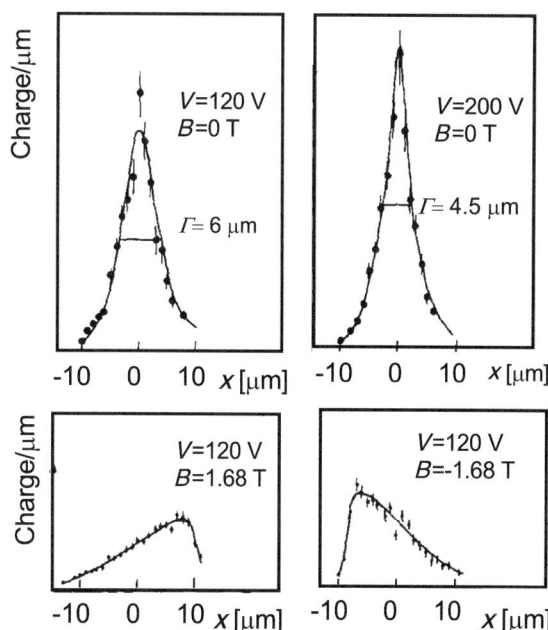

A traversing charged particle produces about 22,400 electron–hole pairs in the 280 µm thick silicon that are collected at electrodes within 10 ns. The electrons move to the n^+ side while the holes move to the p^+ strips. The intermediate strips were kept at the potential of the readout strip. The impedance between the readout strips was much greater than the input impedance of the electronics in order to avoid cross talk. The interstrip capacitance was much greater than the strip-to-ground capacitance. The charge collected at intermediate strips could be divided among the neighboring readout strips. The position of the traversing charged particle was obtained by determining the center-of-gravity of the collected charges.

Figure 6.11 (top) shows the charge distributions as a function of the position measured in the absence of a magnetic field. Exploiting capacitive charge division between adjacent read-out strips, a position resolution (FWHM) of $\Gamma = 6.0$ µm was measured for reading out every third strip at a bias voltage of 120 V (top left). The position resolution improved to $\Gamma = 4.5$ µm for a bias voltage of 200 V (top right) [17]. Furthermore, a two-particle separation of 120 µm was obtained. With limited electronics counting rates of 10^6 counts/s were achieved. In high magnetic fields, however, the position resolution deteriorated. Figure 6.11 (bottom) shows the charge distributions in the presence of a $+1.68\,T$ (bottom left) and $-1.68\,T$ magnetic field (bottom right). The distributions have become non Gaussian and are not centered at zero any longer. For a magnetic field of $|\vec{B}| = 1.68\,T$ that is parallel to the readout strips, the measured coordinate is displaced due to the Lorentz force by ~ 10 µm on average. In addition, the FWHM of the charge distribution is increased from 6.0 to 12 µm. If the magnetic field is reversed to $-1.68\,T$, the x coordinate is shifted by ~ -10 µm with the same width. Fluctuations from δ rays are of the order of 1 µm.

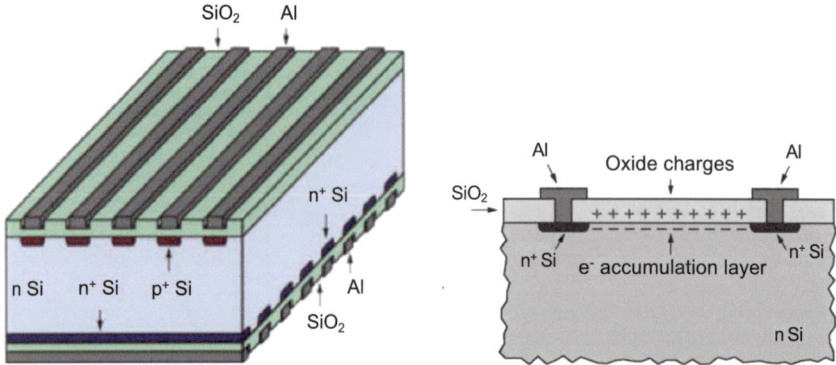

Fig. 6.12 Left: Three-dimensional layout of a double-sided microstrip sensor. The holes drift to the p^+ strips while the electrons drift to the n^+ strips. Right: Cross section through a double-sided microstrip sensor in the vicinity of the n^+ strips. In the SiO_2 region, positive ions accumulate that cause an electron accumulation in the n bulk below. Since the spread occurs over many strips, no position measurement is possible. Both reprinted with kind permission from [8], © 2010, M. Krammer. All rights reserved

The position resolution could be improved if every strip was read out. Calculations show that a resolution of $\Gamma = 2.8$ μm could be achievable in the absence of a magnetic field and if the electronics were placed directly on the wafer [17].

6.1.6 Double-Sided Microstrip Detectors

Single-sided strip detectors measure only one coordinate. A measurement of the other coordinate requires a second detector. Double-sided strip detectors measure two coordinates in one sensor, either ϕ and z or x and y. Figure 6.12 (left) shows the schematic layout of a double-sided microstrip sensor. In n-type detectors, the n^+ backside becomes segmented. The n^+ strips run orthogonal to the p^+ strips. So, holes drift to the p^+ strips while electrons drift to the n^+ strips. With respect to single-sided microstrip sensors, double-sided microstrip sensors have a disadvantage. Since production, handling and testing are complex, they are more expensive. Furthermore, we encounter a problem with the n^+ segmentation. Positive oxide charges are produced in the Si-SiO_2 interface that are static as shown in Fig. 6.12 (right). They attract electrons, which accumulate underneath the oxide layer. Thus, the n^+ strips are no longer isolated from each other and in turn no position measurement is possible. To solve the problem, we need to interrupt the accumulation layer with p^+ stops, p^+ spray or field plates.

For example, p^+ implants between the n^+ strips interrupt the electron accumulation layer as illustrated in Fig. 6.13 (top left). With p^+ stops, the interstrip resistance reaches GΩ levels again. Another option is to do p^+ doping in a layer just below the SiO_2 layer over the whole surface, called p^+-spray as illustrated in Fig. 6.13 (top right). This disrupts the electron accumulation layer. Typically, manufacturers use a combination of these two methods. The accumulation of electrons below Si-SiO_2

6.1 Silicon Microstrip Detectors

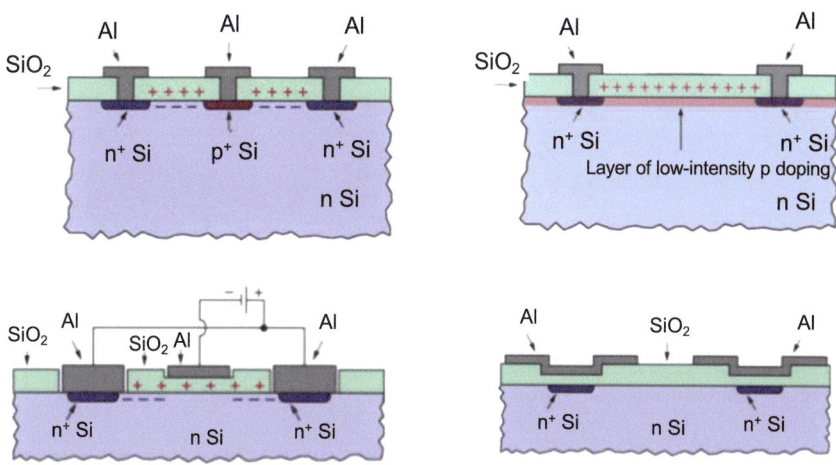

Fig. 6.13 Top left: Insertion of p^+ stops between n^+ strips, which disrupt the accumulation of electrons below the positive ions in the SiO_2 layer. Top right: p doping as a layer over the whole surface also called p spray that prevents the accumulation of electrons. Bottom left: Field plate at negative potential interrupts the accumulation layer of electrons. Bottom right: Extended metal strips on top of the n^+ strips that also prevent the accumulation of electrons below the SiO_2 layer. All reprinted with kind permission from [8], © 2010, M. Krammer. All rights reserved

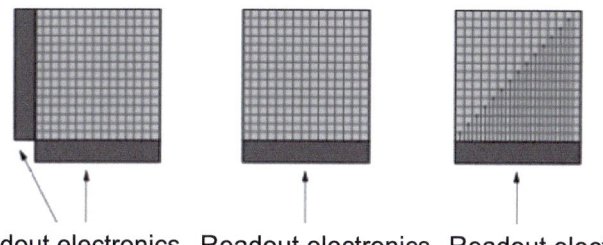

Fig. 6.14 Left: Readout of perpendicular strips on two sides. Center: Preferred readout on one side. Right: Readout on one side by introducing a second metal layer in which the lines run perpendicular to the strips connecting each strip with the electronics. All reprinted with kind permission from [8, 18], © 2010, M. Krammer. All rights reserved

interface can be disrupted also by placing a metal or a metal-oxide semiconductor (MOS) structure at negative potential with respect to the n^+ strips. Figure 6.13 (bottom left) shows this configuration. Instead of placing a separate metal strip between two n^+ strips for AC-coupled double-sided microstrips, the aluminum readout strips can be extended on both sides as illustrated in Fig. 6.13 bottom (right).

Placing the readout electronics of double-sided strips directly at the strip ends requires the instrumentation of two sides as illustrated on Fig. 6.14 (left). This is not practical since the electronics of strips perpendicular to the z directions would interfere with those of the neighboring sensor. Thus, a readout on one side is preferred as indicated in Fig. 6.14 (middle). We accomplish this by introducing a second metal layer in which the lines run perpendicular to the strips connecting each strip with the

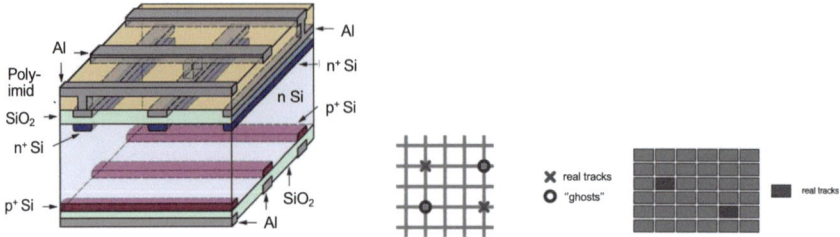

Fig. 6.15 Left: Layout of the electronic readout of double-sided microstrip sensors with metal lines running perpendicular to the n^+ strips. Reprinted with kind permission from [8], © 2010, M. Krammer. All rights reserved. Center: Observation of ghost hits for two charged tracks passing through a double-sided microstrip sensor. Right: Unambiguous hit recording in pixel sensors. Both latter plots reprinted with kind permission from [18], © 2010, M. Krammer. All rights reserved

electronics. This is sketched in Fig. 6.14 (right). The electronics typically consists of printed circuit boards or of a hybrid readout system.

Figure 6.15 (left) shows double-sided microstrip sensors in three dimensions. The two metal planes are insulated from each other via a layer of polyamide or SiO_2. Double-sided microstrip sensors in the barrel (endcaps) measure the ϕ (x) and z (y) positions of a traversing charged track. If two charged tracks pass through the sensor, the measured position is no longer unambiguous. Figure 6.15 (middle) illustrates this. Besides two real hits originating from the two charged particles also two ghost hits appear. This causes a problem at high occupancies. For N hits, the number of ghost hits is $N^2 - N$. The solution is to use pixel sensors that we will discuss in Sect. 6.4. Figure 6.15 (right) shows the hit pattern for pixel detectors in which the recorded hits are unambiguous.

The standard double-sided microstrip detectors consist of n^+np^+ structures. However, it is also possible to build n^+pp^+ sensors. As we will see in Sect. 6.5, under high irradiation n^+np^+ structures convert into n^+pp^+ structures. These two configuration have different characteristics. Figure 6.16 shows a comparison of n^+np^+ and n^+pp^+ configurations for non-depleted and fully depleted situations. For the n^+np^+ arrangement, the depletion layer grows from the bottom as a function of the applied reverse bias voltage until it reaches full depletion. For the n^+pp^+ arrangement, the depletion layer grows from the top as a function of the applied reverse bias voltage until it reaches full depletion. If the reverse bias voltage is sufficiently large, both configurations yield fully depleted detectors.

6.1.7 Position Resolution and Impact Parameter Resolution

Semiconductor detectors are typically positioned close to the beam pipe to measure space points of particle trajectories near the interaction region. Since modern colliding beam experiments typically use solenoidal magnetic fields, the trajectories are helices in the absence of multiple scattering, energy loss and magnetic inhomogeneities (see Chap. 5).

6.1 Silicon Microstrip Detectors

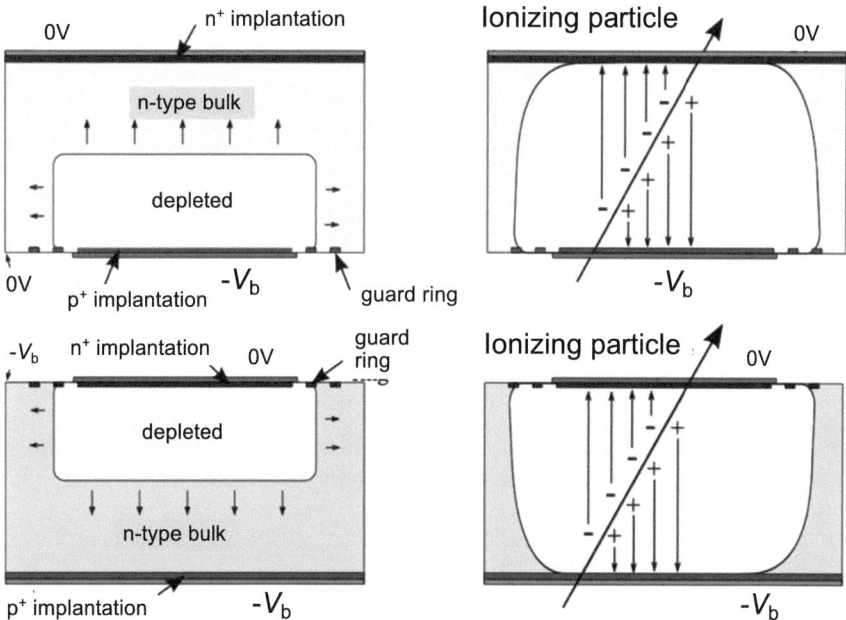

Fig. 6.16 Top left: An n^+np^+ sensor, where the entire sensor is not fully depleted. The depletion layer starts from the bottom. Top right: A n^+np^+ sensor, where the entire sensor is fully depleted. Bottom left: An n^+pp^+ sensor, where the entire sensor is not fully depleted. The depletion layer starts from the top. Bottom right: An n^+pp^+ sensor, where the entire sensor is fully depleted. All reprinted with kind permission from [19], © 2001, K. Dette. All rights reserved

For a single strip, the position resolution is

$$\sigma_x = \frac{x_p}{\sqrt{12}}, \tag{6.7}$$

where x_p is the pitch. For clusters with two or more hits, we use the center-of-gravity method (see (4.24)). The resolution improves like

$$\sigma_x \propto \frac{x_p}{S/N}, \tag{6.8}$$

where S/N is the signal-to-noise ratio. So for increasing S/N ratios, the spatial resolution improves. Figure 6.17 (left) shows spatial resolution as a function of the S/N ratio for a 50 μm readout pitch and a 25 μm readout pitch. For an S/N ratio of 20 and a readout pitch 25 μm a spatial resolution of 2.5 μm is achieved. Note that diffusion broadens the spatial resolution. The broadening depends on the drift length. Energy losses of the charged particle affect the curvature of the path. This needs to be corrected for. Typically, the fit is not just a simple χ^2 fit but uses a Kalman fit, where energy losses, multiple scattering and other effects are included step-by-step. Figure 6.17 (right) shows the energy loss distribution for 310 MeV/c pions and 310 MeV/c protons. While the pions are close to the minimum, the protons

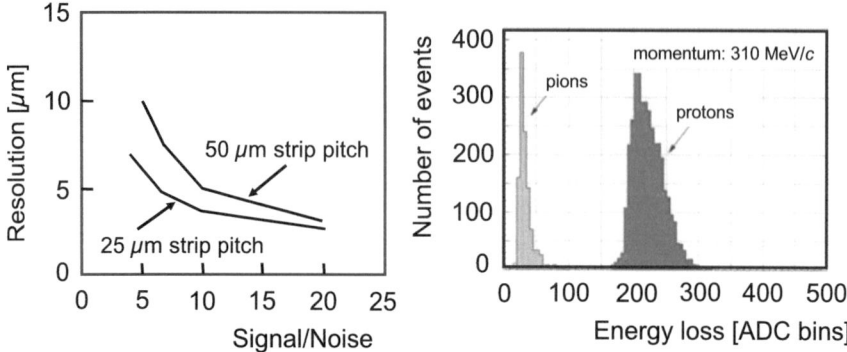

Fig. 6.17 Left: Spatial resolution as a function of the signal-to-noise ratio for a 25 μm and a 50 μm pixel readout pitch. Reprinted with kind permission from [20], © 1992, DELPHI Collaboration. All rights reserved. Right: Energy loss distribution for 310 MeV/c pions and 310 MeV/c protons. Reprinted with kind permission from [8], © 2010, M. Krammer. All rights reserved

are far below and, therefore, loose much more energy. Both distributions reveal the asymmetric Landau-like shape. The resolutions are similar (about 10%).

6.2 Recently Used Microstrip Detectors

All major multipurpose detectors have incorporated silicon microstrip detectors. We list here recent major experiments using microstrip detectors to emphasize their importance. We discuss the *BABAR* and ATLAS microstrip detectors as examples in more detail in the next subsection.

- At the LEP collider (CERN): all four LEP detectors used microstrip sensors. The Delphi experiment [21] used three concentric layers with single-sided and double-sided sensors. The ALEPH [22], L3 [23] and OPAL [24] experiment used two concentric layers with double-sided silicon microstrip sensors.
- At the Tevatron (Fermilab): The main experiments CDF [25] and D0 [26] had vertex detectors based on silicon microstrip sensors. The CDF experiment used seven concentric layers in the barrel, where the first layer was placed at a radius of 1.5 cm and the seventh layer at a radius of 28 cm. The D0 experiments used five concentric layers in the barrel all located within a 10 cm radius around the beam direction.
- At the *B*-factories: The CLEO II experiment [27] used two concentric layers with double-sided microstrip sensors. In the Belle experiment [28], four concentric layers with double-sided microstrip sensors were used that were arranged within the radial space of 2.0 and 8.8 cm. This was the second vertex detector since the first one deteriorated due to radiation damage. In the *BABAR* experiment [29], the vertex detector consisted of five layers with double-sided silicon microstrip sensors as discussed below that survived the entire running period.

- At the HERA collider (DESY): The ZEUS detector [30] used three layers in the barrel with two single-sided sensors in each layer providing $r\phi$ and rz readout, respectively. In the forward region, four wheels were placed, each having two layers. The H1 experiment [31] used two layers with double-sided silicon microstrip sensors.
- At the RHIC collider (Brookhaven): The PHENIX [32] experiment used four layers of pixels detectors in the barrel and and four layers of single-sided silicon microstrips in each endcap. The STAR experiment [33] used one layer of double-sided silicon microstrip detectors as the fourth layer of the silicon vertex detector in the barrel, where the first three layers consisted of silicon drift detectors.
- At the LHC (CERN): The ATLAS SemiConductor Tracker (SCT) [34] has four concentric barrel layers and nine disks in each endcap using two single-sided microstrip sensors as discussed below. The CMS experiment [35] has four layers in the inner barrel, six layers in the outer barrel, three disks in the inner endcap and nine disks in the outer endcap. Each layer uses single-sided microstrips. The first two layers in each region and in disk 5 in the outer endcap have a second microstrip layer rotated by 100 mrad. The LHCb Trigger Tracker used four layers of microstrip detectors with a 183 µm pitch. In the long shutdown 2 it was replaced by the upstream tracker (UT) that consists of four planes equipped with silicon microstrip detectors [36]. While the first and fourth layer are vertical, the two middle layers are tilted by $\pm 5°$. Sensors around the beam pipe have 1024 strips with a pitch of 95 µm. They are followed by a ring of sensors with 1024 strip at a 95 µm pitch. All other sensors have 512 strips with a pitch of 190 µm. The ALICE experiment [37] used two layers with double-sided silicon strip sensors as outer layers in the vertex detector. The entire vertex detector has been replaced in the long shutdown 2 with seven layers containing pixel sensors [38] as discussed in Sect. 6.7.2.

6.2.1 The *BABAR* Silicon Vertex Tracker

Figure 6.18 shows the layout of the *BABAR* silicon vertex tracker (SVT) in the $r\phi$ (left) and rz (right) views. The SVT consisted of five layers of 300 µm-thick, double-sided silicon strips built from six different wafer sizes ranging from $r\phi \times rz = 42$ mm $\times$ 43 mm to 53 mm $\times$ 68 mm [39]. Along the z direction wafers were combined into half modules. The $r\phi$ strips in the same half module were electrically connected to form a single readout strip. In the inner (outer) layer the strips in a half module covered a length of 140 mm (240 mm) in the z direction. The $r\phi$ strips from different wafers in a half-module were connected via microbonds. The strips of two half modules, which form a module, were connected to the front-end electronics on opposite sides.

In each of the first three layers six modules were arranged around the beam pipe at radii of 3.3, 4.0 and 5.9 cm. The $r\phi$ strips ran parallel to the beam, while rz strips ran orthogonal to the $r\phi$ strips. The two outer layers were arch-shaped to minimize

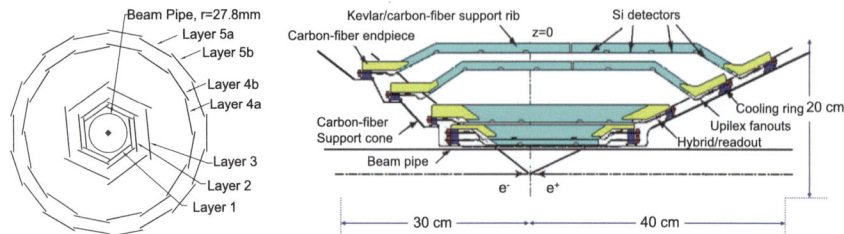

Fig. 6.18 Left: Layout of the $B\!A\!B\!A\!R$ SVT in the $r\phi$ view. Right: Layout of the $B\!A\!B\!A\!R$ SVT in the rz view. Both reprinted with kind permission from [29], © 2002, Elsevier. All rights reserved

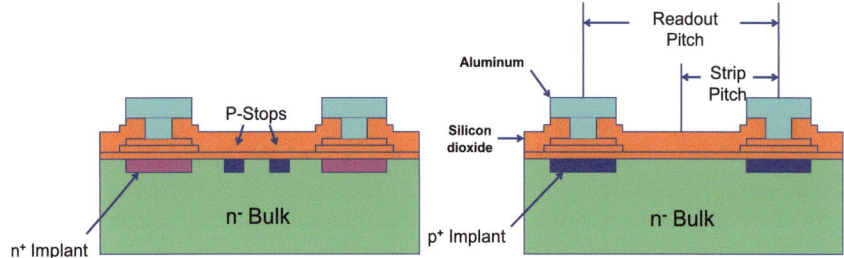

Fig. 6.19 Left: Readout of the silicon microstrip sensors on the n^+ side, which was used for reading out the rz strips. The p^+ stops between the n^+ stops prevent the accumulation of electrons as discussed in Sect. 6.1.6. Right: Readout of the silicon microstrip sensors on the p^+ side, which was used for reading out the $r\phi$ side. Both reprinted with kind permission from [40], © 1995, $B\!A\!B\!A\!R$ Collaboration. All rights reserved

the amount of silicon needed to cover the solid angle. They were built with 16 and 18 modules that were positioned at radii of 12.7 and 14.4 cm, respectively. The arches on each side reduced the radii to 9.1 and 11.4 cm, respectively. In the arches trapezoidal sensors were used. The modules in the three inner layers were tilted by $5°$ to allow for an overlap of adjacent modules in the $r\phi$ direction. To avoid gaps in the two outer layers, they were subdivided into sublayers as shown in Fig. 6.18 (left). The strip pitch in $r\phi$ was 50 μm (55 μm) in layers 1, 4 and 5 (2 and 3). In the trapezoidal sensors the pitch was reduced to 41 μm. The strip pitch in the rz plane was 50 μm in layers 1–3 and 105 μm in layers 4 and 5. To keep the number of readout channels low, the readout pitch was typically twice as large except for some strips in the $r\phi$ view in layers 1 and 2, where each strip was read out. Intermediate strips, which were not read out, were kept on a floating potential to achieve the required position resolution. Typical depletion voltages were in the range of 25–35 V. The strips on the p^+ and n^+ sides were biased via polysilicon resistors to ascertain the required radiation hardness in order to keep the voltage drop across the resistor and the parallel noise as low as possible. Leakage currents were 50 nA/cm^2.

Figure 6.19 (left) shows the readout of the silicon microstrip sensors on the n^+ side. The bulk n^- silicon has a resistivity of 4.8 kΩ-cm. The electron built-up at the Si-SiO_2 interface is stopped by the insertion of two p^+ stops between two n^+ strips. Figure 6.19 (right) shows the readout of the silicon microstrip sensors on the p^+

6.2 Recently Used Microstrip Detectors

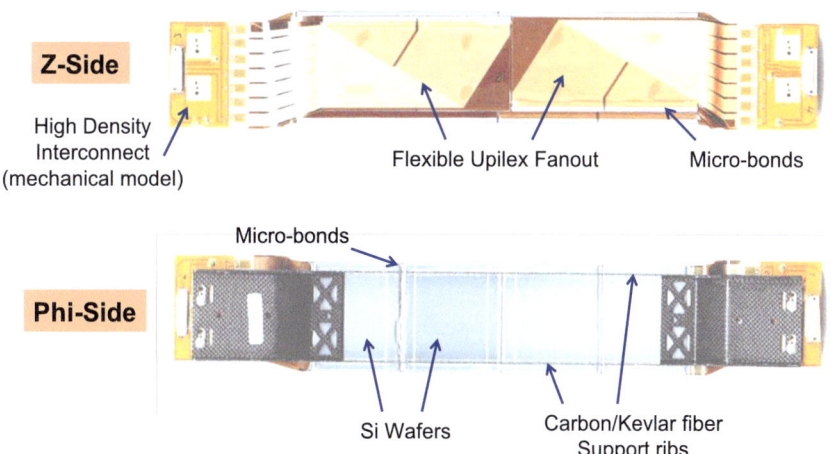

Fig. 6.20 Mounting of the silicon wafers on the carbon/kevlar fiber support ribs on the rz side (top) and on the $r\phi$ side (bottom). The $r\phi$ strips of adjacent wafers and rz strips are connected by microbonds to the fanout. The two half modules are visible. The $r\phi$ strips of adjacent neighbor wafers in each half module are connected and are read out at each end. Both reprinted with kind permission from [41], © 2005, C. Campagnari. All rights reserved

side. All strips are internally AC-coupled to the frontend electronics via integrated decoupling capacitors. The silicon wafers are mounted on Carbon/Kevlar fiber support ribs. Figure 6.20 shows the mounting and readout of the silicon wafers on the rz (left) and $r\phi$ (right) side. On the rz side, the strips are coupled via microbonds to a flexible upilex fanout that runs them under 45° towards the center before running them parallel to the $r\phi$ strips to the readout side. The material budget of the flexible upilex fanout amounts to less than $0.03\% \cdot X_0$. Its capacitance is 0.52 pF/cm. Figure 6.21 shows a photograph of the completed *BABAR* SVT. The silicon wafers in the outer layer of each half module are visible. The black ribbons are part of the support structure. The silicon in layer 1 covers an area of 457 cm^2 increasing up to 2,089 cm^2 in layer 5. The coverage in polar angle is $-0.95 < \cos\theta_{CM} < 0.87$, where θ_{CM} is the polar angle in the center-of-mass frame. The SVT hit efficiencies typically are larger than 95%.

Figure 6.22 shows the single hit resolution as a function of the track's incident angle for $r\phi$ strips (left) and rz strips (right). The observed single-hit resolutions are consistent with the design specifications of 10–15 μm for the three inner layers and 40 μm for the two outer layers. The asymmetry in the $r\phi$ view relative to normal incidence is introduced by the magnetic field and the 5° tilt of the modules in the three inner layers. The reduced number of measurements in the $r\phi$ view of the outer two layers is caused by smaller ϕ coverage of each module (see Fig. 6.18 left). The SVT performed reliably during its ten-year operation. Radiation from beam backgrounds caused bulk damage of the silicon sensors and a shift in the depletion voltage. This affected the position resolution only modestly. Though an increase in noise was observed due to larger leakage currents, it was much smaller

Fig. 6.21 Photograph of the completed *BABAR* SVT. Reprinted with kind permission from [29], © 2002, Elsevier. All rights reserved

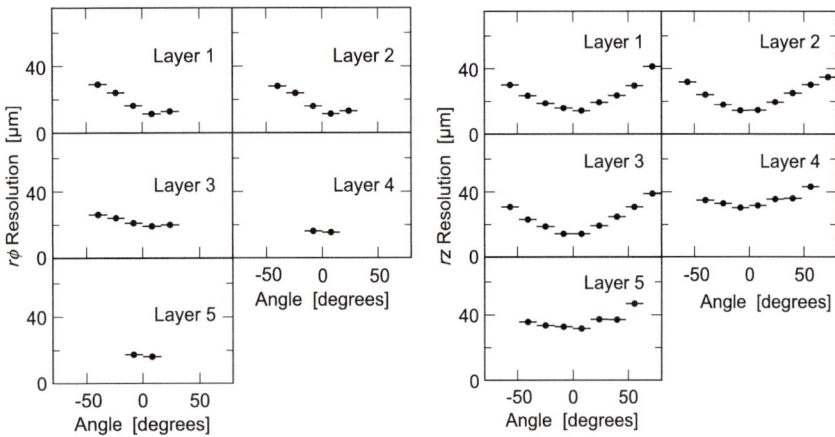

Fig. 6.22 Single hit resolution of the *BABAR* SVT in the $r\phi$ view (left) and rz view (right) as a function of the track incident angle in layers 1 to 5. Both reprinted with kind permission from [29], © 2002, Elsevier. All rights reserved

than the electronic noise. Figure 6.23 shows the *BABAR* transverse and longitudinal impact parameter resolutions, which are extracted using the entire tracking system. The longitudinal impact parameter resolution is slightly worse than the transverse impact parameter resolution. At a momentum of 1 GeV/c the resolutions of the transverse and longitudinal impact parameters are $\sigma_{d_0} = 50$ μm and $\sigma_{z_0} = 67$ μm, respectively. For tracks with momenta above 3 GeV/c they improve to $\sigma_{d_0} = 23$ μm and $\sigma_{z_0} = 29$ μm, respectively.

6.2 Recently Used Microstrip Detectors

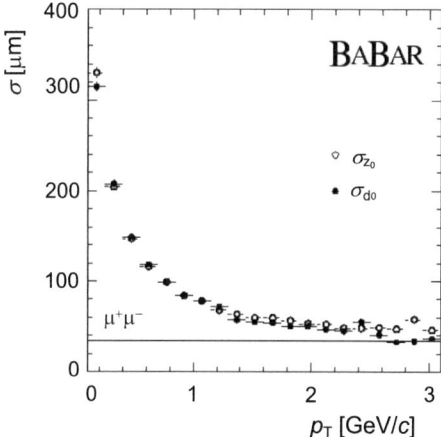

Fig. 6.23 The BABAR impact parameter resolution in $r\phi$ (black points) and in rz (open circles). Reprinted with kind permission from [29], © 2002, Elsevier. All rights reserved

6.2.2 The ATLAS Semi-conductor Tracker

Figure 6.24 shows the schematic layout of the ATLAS SCT, which consists of four concentric barrel layers at radii of 54.8, 71.0, 85.4 and 99.6 cm as well as nine wheels in each endcap [34]. The barrel length is 1.53 m. The two endcaps extend the length to 5.6 m. The outer diameter of the endcap wheels is 1.04 m. The figure

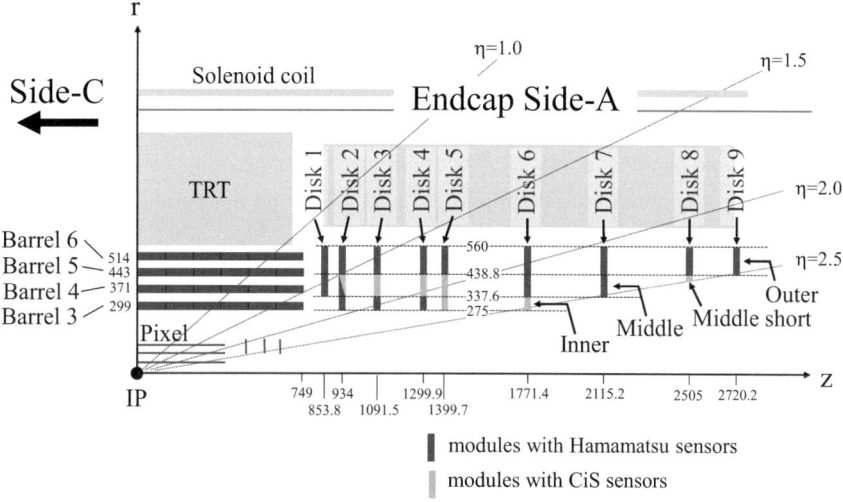

Fig. 6.24 Schematic layout of one quadrant of the ATLAS SCT in the barrel and one endcap. The barrel consists of four concentric cylindrical layers while each endcap has nine disks. The microstrip sensors come from Hamamatsu (black) and CiS (light grey). Reprinted under CC-BY-3.0 Licence from [42], © 2014, ATLAS Collaboration

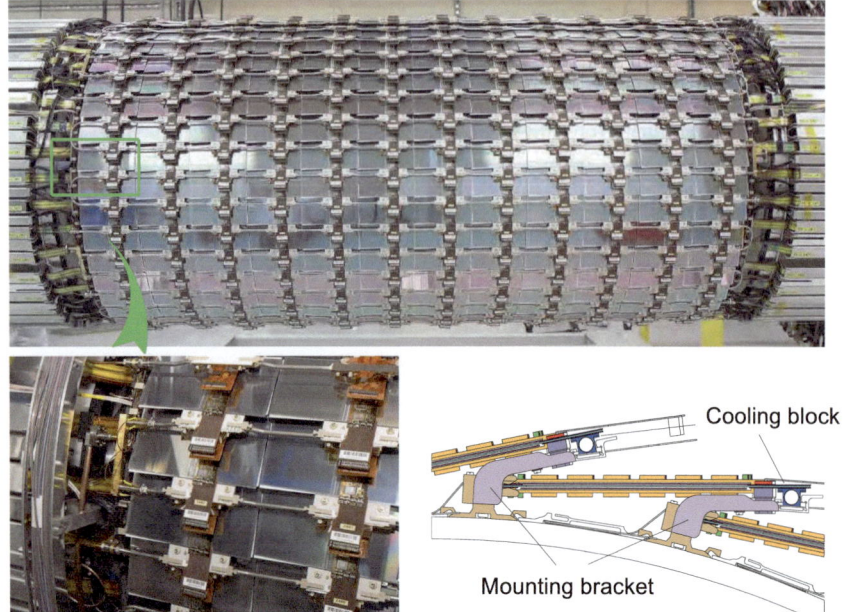

Fig. 6.25 Top: Photograph of the completed barrel SCT before insertion into the ATLAS detector. Bottom left: Enlarged view of barrel SCT modules mounted on the support cylinder together with the module services including signal and power cables as well as cooling tubes. Bottom right: Schematic view of the mounting brackets, which are attached to the barrel SCT cylinders and which hold the modules and cooling pipes. All reprinted with kind permission from [34], © 2008, IOP Publishing. All rights reserved

also shows the coverage in pseudo-rapidity. The ATLAS SCT[2] is built from 4,088 modules of which 2,112 modules are placed in the barrel layers and the remaining ones are mounted in the endcap wheels. Figure 6.25 (top) shows a photograph of the completed barrel SCT. Figure 6.25 (bottom) shows an enlarged view of barrel SCT modules mounted on the support cylinder (left) and a schematic view of the mounting brackets (right).

Figure 6.26 shows a schematic layout (left) and a photograph (right) of a barrel module, which consists of four 285 µm thick, single-sided square p^+n silicon microstrip sensors (64 mm in length) [43].[3] Two pairs of sensors are glued back-to-back to the baseboard. The sensors on the lower side are tilted by 40 mrad with respect to the sensors on the upper side to achieve readout in two dimensions via stereo layers. The strips in each pair of sensors are connected yielding a total strip length of 128 mm. The sensors consist of p^+nn^+ silicon. The strips are AC-coupled and have a readout pitch of 80 µm.

---

[2] The SCT uses sensors from Hamamatsu and CiS (Institut für Mikrosensorik, Erfurt).
[3] These are also called *p-in-n* sensors, where p^+ are the strips and n is the bulk. Note that after an irradiation with a fluence of $\approx 2 \times 10^{13}$ cm^{-2}, the n-type bulk material becomes p-type.

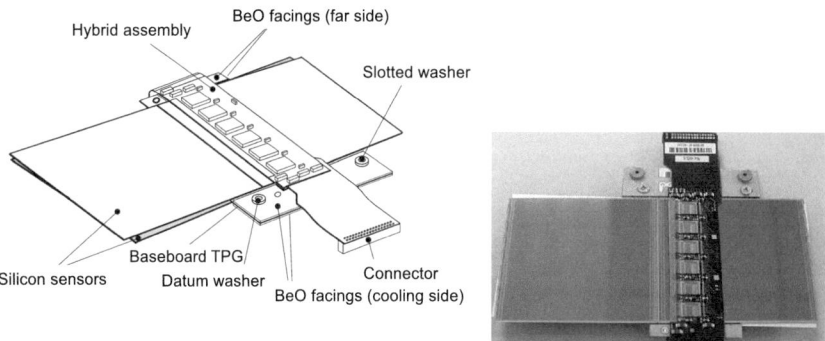

Fig. 6.26 Left: Schematic layout of an ATLAS SCT module, which holds four silicon sensors. The bottom sensors are tilted by 40 mrad with respect to the two top sensors. The read-out electronics uses a hybrid assembly. Right: Photograph of an ATLAS SCT module. Both reprinted with kind permission from [34], © 2010, IOP Publishing. All rights reserved

The two layers of microstrip sensors are fixed to a base board and are read out with hybrids that are mounted on top of the upper layer and on the bottom of the lower layer. The hybrids are attached in the center of the module. The hybrid is a multi-layer design consisting of copper/polyimide flex with a carbon-carbon bridge for strength. It has four active layers in total that contain connections for signal circuits, ground planes and a high-voltage plane. The copper/polyimide flex is 279 μm thick. The carbon-carbon bridge is 0.3 mm thick with 0.5 mm thick legs at both ends and has good thermal conductivity, a high Young's modulus and low radiation lengths. Since the microstrip pitch is larger than the pitch of the readout pads (47 μm), a pitch adaptor is required. Wire bonds first provide connection from the pads to the pitch adaptor and then other wire bonds connect the pitch adaptor to the microstrips. The ASICS[4] are attached to the hybrid with conductive glue. In the $r\phi$ view, the modules in the four layers are tilted with respect to the surface of the support structure by $11.0°$, $11.0°$, $11.25°$ and $11.50°$, respectively.

In the two endcaps, we use four different types of modules depending on the radial position, which are all tapered to account for the radial geometry of the wheels. The strip pitch varies from 57 μm at the innermost sensors to 90 μm at the outermost ones. The strip lengths vary from 54 to 61 mm. Innermost modules use just one sensor. The hybrids are attached at the outer edge. The silicon in the SCT covers an area of 61 m^2 using 15,392 silicon wafers with 6.3×10^6 readout channels. The reverse bias voltage for full depletion of the silicon wafer is about 100 V. To achieve an efficiency above 98%, we need an operation voltage of about 150 V. The SCT modules are operated at temperatures of -7 to $+6\,°C$.

The ATLAS SCT performs rather well. In the barrel, hit efficiencies for standalone SCT hits are above 99.8% while in the endcaps they are above 99.7%. The simulation agrees well with the measurements. Figure 6.27 (left) shows the cluster size as a function of the track incident angle. The minimum is shifted from zero due to

---

[4] ASIC stands for application specific integrated circuit.

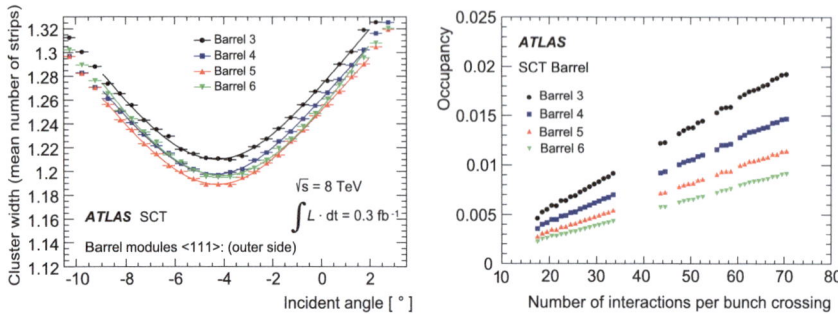

Fig. 6.27 Left: Cluster-size as a function of the particle's incident angle for each SCT barrel. The displacement of the minimum from zero yields a measurement of the Lorentz angle α_L. Right: Mean occupancy of barrel layers as a number of interactions per bunch crossing in minimum-bias pp data. Both reprinted under CC-BY-3.0 Licence from [42], © 2014, ATLAS Collaboration

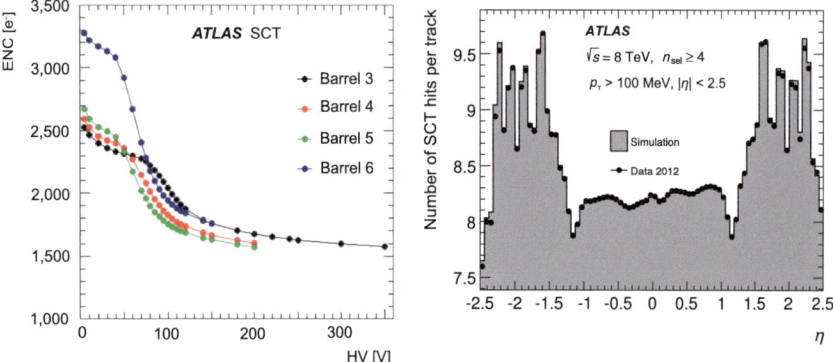

Fig. 6.28 Left: The $\overline{ENC}$ in the barrel layers as a function of the reverse bias voltage for barrel 3 (black points), barrel 4 (red points), barrel 5 (green points) and barrel 6 (blue points) [44]. Reprinted under CC-BY-NC-ND-4.0 Licence from [44], © 2020, S. Hirose. Right: Average number of SCT hits on a track versus pseudo-rapidity for $\sqrt{s} = 8$ TeV proton data. Data are black points and the histogram shows the simulation. Reprinted under CC-BY-3.0 Licence from [42], © 2014, ATLAS Collaboration

the Lorentz angle. This shows that the Lorentz angle is $\alpha_L \simeq 4.3 \pm 0.2°$. At the minimum, about 80% of the clusters have a single hit and only 20% have two hits. This changes when we move away from the minimum. The two-hit clusters increase to about 30% when we move by $\pm 6°$. Figure 6.27 (right) shows the occupancy as a function of number of collision per bunch crossing. The level is still rather small, so the SCT performance is not impacted by a large number of interactions per bunch crossing. Figure 6.28 (left) shows the noise in the barrel layers as a function of the reverse bias voltage. At the operation voltage, the $\overline{ENC}$ is around 1,600–1,750 $rms\ e^-$. Figure 6.28 (right) shows the average number of SCT hits on a track as a function of pseudo-rapidity. In the barrel region, the average number of SCT hits is above eight. The observed shape is well represented in simulations. In the endcaps, the average number of SCT hits is around nine.

Though the SCT layers are built with high precision, additional alignment of individual modules is necessary to achieve the required precision. In the direction perpendicular to the strips, the SCT modules must be aligned with a precision of 12 μm. We use a χ^2 method that minimizes the residual to fitted tracks from pp collision events. For each module we have six alignment parameters, three translations and three rotations. The alignment is performed sequentially. First, the barrel and endcaps are aligned. Next, individual layers are aligned and finally the positions of individual modules are optimized. At the first step, the number of degrees of freedom is 24 while at the last step it increases to 24,528 for all 4,088 SCT modules. It is important to note that we encounter two types of misalignments. In the first type, a misalignment changes the χ^2 distribution of the residuals to a track. In the second type, this is not the case. However, we will encounter a systematic shift in the measured value of the momentum. We have to use different techniques to spot these misalignments and fix them. This can be achieved by using tracks from the decay products of resonances like the J/ψ and the Z bosons. The single hit resolution in the $r\phi$ view is 17 μm. The two-particle resolution is about 120 μm. Further details are given in [42,45]. The impact parameter resolutions are shown in Sect. 6.4.2.

Figure 6.29 shows a cosmic muon that traverses the three ATLAS pixel layers and the four SCT layers on each side.[5] All hits are clearly visible. The muon has

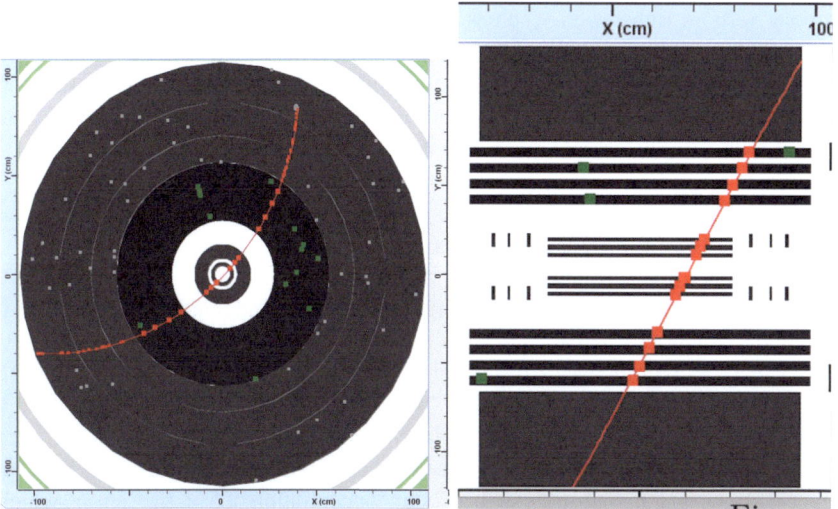

Fig. 6.29 A cosmic muon bent under the force of the ATLAS solenoidal field in the $r\phi$ view (left) and rz view (right). Reprinted with kind permission from [60], © 2008, ATLAS Collaboration. All rights reserved

---

[5] The pixel layers are discussed in Sect. 6.4. The Insertable B Layer was not installed here.

sufficient low momentum so that it is bent by the solenoidal magnetic field in the $r\phi$ plane. All eight SCT hits and six pixel hits are well lined up.

6.2.3 Examples of Microstrip Detectors

For comparison, we have compiled properties of some silicon microstrip detectors used in recent multipurpose experiments, which are listed in Table 6.1. Except for the LEP experiments, most experiments use 4–5 layers of microstrip sensors, placed outside the beam pipe close to interaction region. In the LHC experiments, microstrip detectors follow the pixel detectors. The microstrip sensors are typically around 300 μm thick. Most experiments use double-sided sensors or two single-sided sensors with a small stereo angle. The typical strip pitch is 50–100 μm. The single hit resolution depends on the pitch. Best resolutions are 5–10 μm.

6.2.4 The Power of Microstrip Detectors

Silicon microstrip detectors have changed the determination of vertex finding considerably. This is illustrated in Fig. 6.30 which shows an ALEPH event display in which the vertex detector information was used (top) and was not used (bottom). Using the information from the vertex detector allows the reconstruction of secondary and tertiary vertices rather well whereas without this information the tracks from the TPC point in random directions near the primary vertex region. So, the introduction of silicon microstrip detectors lead to a considerable improvement in determining secondary and tertiary vertices. Thus, all modern colliding beam experiments use silicon microstrip detectors close to the interaction region.

6.3 Charge-Coupled Devices

Charged-Couple Devices (CCD) were invented by Boyle and Smith at Bell Labs [61]. Figure 6.31 shows a schematic layout of a charge-coupled device and Fig. 6.32 (left) shows its cross section. The CCD consists of an array of potential wells that form an array of pixels. In the x direction, p^+ stops form pixel boundaries while in the y direction potentials create the pixel boundaries. The CCD is fabricated on a ∼20 μm thick epitaxial p-type silicon layer with a donor concentration of 5×10^{14} cm^{-3}. This layer is grown on a p^+ substrate with a donor concentration of 5×10^{18} cm^{-3}, which is inactive from point of view of particle detection. The surface of the epitaxial silicon is converted to a 1 μm thick n-type silicon with a donor concentration of 10^{16} cm^{-3}. Above a thin oxide layer, transparent electrodes are located called imaging Gates, which are insulated from one another and from the substrate. In the y direction, each pixels is controlled by three ϕ Gates, ϕ_1, ϕ_2 and ϕ_3. In the initial condition, we keep $\phi_2 = 10$ V and $\phi_1 = \phi_3 = 0$ V. This creates a matrix of potential wells with minima near the upper interface defined by the ϕ_2 Gates in the y direction and the

6.3 Charge-Coupled Devices

Table 6.1 Properties of microstrip detectors in different experiments showing the number of barrel layers B_{layers}, number of endcap layers EC_{layers}, sensor type (single-sided, ss, or double-sided, ds), stereo angle α_{stereo}, sensor thickness d_{Si}, radius for barrel layers r_{Si}, z position of endcap disks z_{Si}, number of channels N_{chan}, high voltage, p-side readout pitch s_p, n-side readout pitch s_n, $r\phi$ resolution $\sigma_{r\phi}$ and rz resolution σ_z. †This value corresponds to the number of channels in the barrel; the total number is 792,576. §The first number is for the barrel, the second for the endcap disks; D0 uses also double-sided-double metal sensors in the barrel whose s_n pitch is 153.5 μm. §§ This is the pitch divided by $\sqrt{12}$. &These values result from pitch/$\sqrt{12}$. ‡The 80 μm refers to the barrel and the range to the endcap disks. *Here, a second single-sided layer is attached in layers 1, 2, 5 and 6 in the barrel and in the endcap rings 1, 2, 4, 5 and 8 under a 100 mrad stereo angle. *The first values refer to layers 1–3 and the second to layers 4 and 5. #In the barrel, the pitch is 80 μm (layers 1,2), 120 μm (layers 3,4), 183 μm (layers 5–8) and 122 μm (layers 9, 10). In the endcap first three disks, the pitch varies from 100 μm to 141 μm and 97–184 μm in the remaining disks. $The first range is for the inner barrel and the second for the outer barrel

Exp.	DELPHI VTX [46,47]	ALEPH SVD [48]	L3 SMD [49]	OPAL μVTX3 [50]	CDF II SVX II [51]	D0 SMT [52]	CLEO III Si3 [53]
B_{layers}	3	2	2	2	5	4	4
EC_{layers}	Other	–	–	–	–	12+4	–
Sensor	ss+ds	ds	ds	ds	ds	ss+ds	ds
α_{stereo}	–	–	($\pm 2°$)	–	($\pm 1.2°$)	$b: \pm 2°$	–
d_{Si} [μm]	290-320	300	300	250	300	300	300
r_{Si} [cm]	6.3–10.8	6.3, 11	6, 8	6.05, 7.38	2.45–10.6	2.7–9.4	2.5–10.1
N_{chan}	149,504	95,088	73,728	65,502	405,504	387,072†	
HV [V]	60–95	20–62	<100	<40	45–70	20–60	100–110
s_p [μm]	44–176	50	50	50	60–65	50/50–80§	50
s_n [μm]	42–150	100	150/200	100	60–141	62.5/62.5§	100
$\sigma_{r\phi}$ [μm]	9–13	10	6.1	5	9–17	14/14–23§§	11
σ_{rz} [μm]	11–70	15	20	12	17–41&	18/18§§	24

(continued)

narrow p^+ implants in x direction. Typically, a CCD has 576×385 such pixels in a sensitive area of 12.5×8.5 mm². So, the pixel size is 22 μm $\times$ 22 μm. Note that the depletion region is only 5–8 μm from the surface of the n-type layer plus epitaxial layer.

Figure 6.32 (right) shows the electrostatic potential in the CCD substrate when all Gate voltages are held at 0 V. Electrons generated in the depletion region diffuse to the n channel. Their drift in the x direction is limited by p^+ stops. In the y direction, the spread among several channels improves accuracy by using the center-of-gravity method. Application of a low voltage to the ϕ_2 Gates freezes the charges inside the pixels. At room temperature, the potential well is rapidly filled with electrons. Since this effect is reduced at low temperatures, we need appropriate cooling. Thus, CCDs

Table 6.1 (continued)

Exp.	BABAR SVT [29]	Belle SVD2 [54]	Belle II SVD [55]	ATLAS SCT [34,56]	CMS SST [35,57]	LHCb TT [36,58]	ALICE SSD [59]
B_{layers}	5	4	4	2×4	4+6	–	2
EC_{layers}	–	–	–	2×9	$2 \times (3+9)$	4	–
Sensor	ds	ds	ds	ss	ss	ss	ds
α_{stereo}	–	–	–	40 mrad	100 mrad*	$\pm 5°$	35 mrad
d_{Si} [μm]	300	300	300–320	285	320, 500	500	300
r_{Si} [cm]	3.2–14.4	2.0–8.8	3.8–14	25.1–54.9	20–116		37.8&38.4 42.8&43.4
$|z_{Si}|$ [cm]	–	–	–	81–280	0–282	–	–
N_{chan}	$\approx$150,000	110,592	243,456	6.3×10^6	9.3×10^6	143,360	2.72×10^6
HV [V]	~35–45	75	<120	150–350	< 500	300	20–90
s_p [μm]	50–110	50, 65	50, 75	80/57–94‡	80–184#	183	95
s_n [μm]	100–210	73, 75	160, 240	–	–	–	
$\sigma_{r\phi}$ [μm]	10–30, 15*	12	2–5	16	17–28, 23–40$	52.6	20
σ_{rz} [μm]	10–30, 35–45*	19	7-27	580	–	–	830

require a cryostat using a bath of liquid nitrogen. The operation temperature is around 180–190 K.

Figure 6.33 (left) shows the temperature dependence of the pair creation energy W_{Si} in silicon. Note that at room temperature a MIP produces more charge than at lower temperature. Figure 6.33 (right) shows the paths of electrons produced in the epitaxial layer of the CCD. In the depletion region, the electrons drift directly to the anode. In the undepleted region, electrons first move randomly by diffusion until they reach the depletion region, where they move directly towards the anode. Electrons are typically produced in the epitaxial layer within a 1 μm wide column. Due to diffusion, the image will be broadened. Typically, around 1,600 electrons are produced. The uncertainty is much smaller than expected from Poisson statistics due to a small Fano factor. The noise is of the order of 100 electrons.

The readout of a CCD is a complex procedure as illustrated in Fig. 6.34. To shift the charge by one pixel in the y direction, we need a three-phase procedure. In the first phase, the voltage on the ϕ_2 Gate is shifted to the ϕ_3 Gate, while $\phi_1 = \phi_2 = 0$ V. In the second step, the voltage is shifted to the ϕ_1 Gate while $\phi_2 = \phi_3 = 0$ V. In the last phase, the voltage is restored to the ϕ_2 Gate. Now, all charges of the CCD are shifted by one pixel in the y direction. The charge in the bottom row has been shifted

6.3 Charge-Coupled Devices

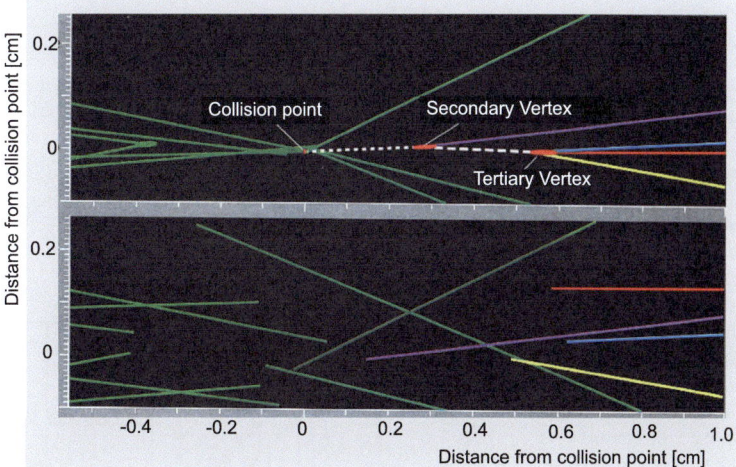

Fig. 6.30 An ALEPH event display of a B meson decay near the interaction region [62]. Using tracking information from the vertex detector clearly shows the collision point, a secondary vertex and a tertiary vertex (top). Without the vertex detector information the tracks near the interaction region point in random directions (bottom). Reprinted with kind permission from [62], © 1995, A. Litke. All rights reserved

to the R_1, R_2 and R_3 registers that are located below the bottom row as shown in Fig. 6.31.

If the R registers are set to $R_1 = R_3 = 0$ V and $R_2 = 10$ V, the charges will be located in the region of R_2. Now we apply the three-phase procedure to the R Gates. After one cycle, the charges in the R registers will shift by one pixel in the x direction. The charge in the outer-right R_2 Gate will be send to the readout. We have to repeat the procedure 385 times until all charges in the R Gates have been read out. Then, we apply the three-phase procedures again to the ϕ Gates to shift the next row to the R Gates and in turn go to the R Gates to read them out one-by-one. This procedure is repeated until all charges of the CCD are collected. The readout speed is about 25 images per second.

When a charged particle ionizes the depletion region, the initial column of ionization electrons is localized to a 1 μm width. The electrons start drifting towards the region of minimal potential, which is part of the depletion region closest to the ϕ Gates. They are collected within 30 ns, during which the width increases to 10 μm due to transverse diffusion. The diffusion is stopped in the x direction by the p^+ implants. The lateral diffusion in the y direction would continue if all ϕ Gates were at same potential. Therefore, the ϕ Gates are polarized as soon as the accelerator is active and the potential matrix is permanent. Events are clocked out continuously at a rate of 3 MHz to the R register. The clearing time is a few 100 μs. Having a second device back-to-back with first one imposes a precise geometrical relation onto the track. The large number of pixels compensates for lack of time resolution. At a rate of 10^5/s there is no problem in tagging beam tracks superimposed on a good event. Please note that CCDs have a high sensitivity. The MIP generates 1,300 electrons

Fig. 6.31 Schematic layout of a Charged-Coupled Device. A 25 μm thick epitaxial layer is placed on a p^+ substrate. In the y direction, the pixel boundaries are defined by the ϕ_1, ϕ_2, ϕ_3 gates while in the x direction they are defined by the p^+ stops. With changing voltage on the ϕ registers, charge can be moved in the y direction. At the bottom of the CCD, there are the R_1, R_2, R_3 registers that are needed to move the charge in the x direction. The black square denotes one pixel. Reprinted with kind permission from [63], © 1981, Elsevier. All rights reserved

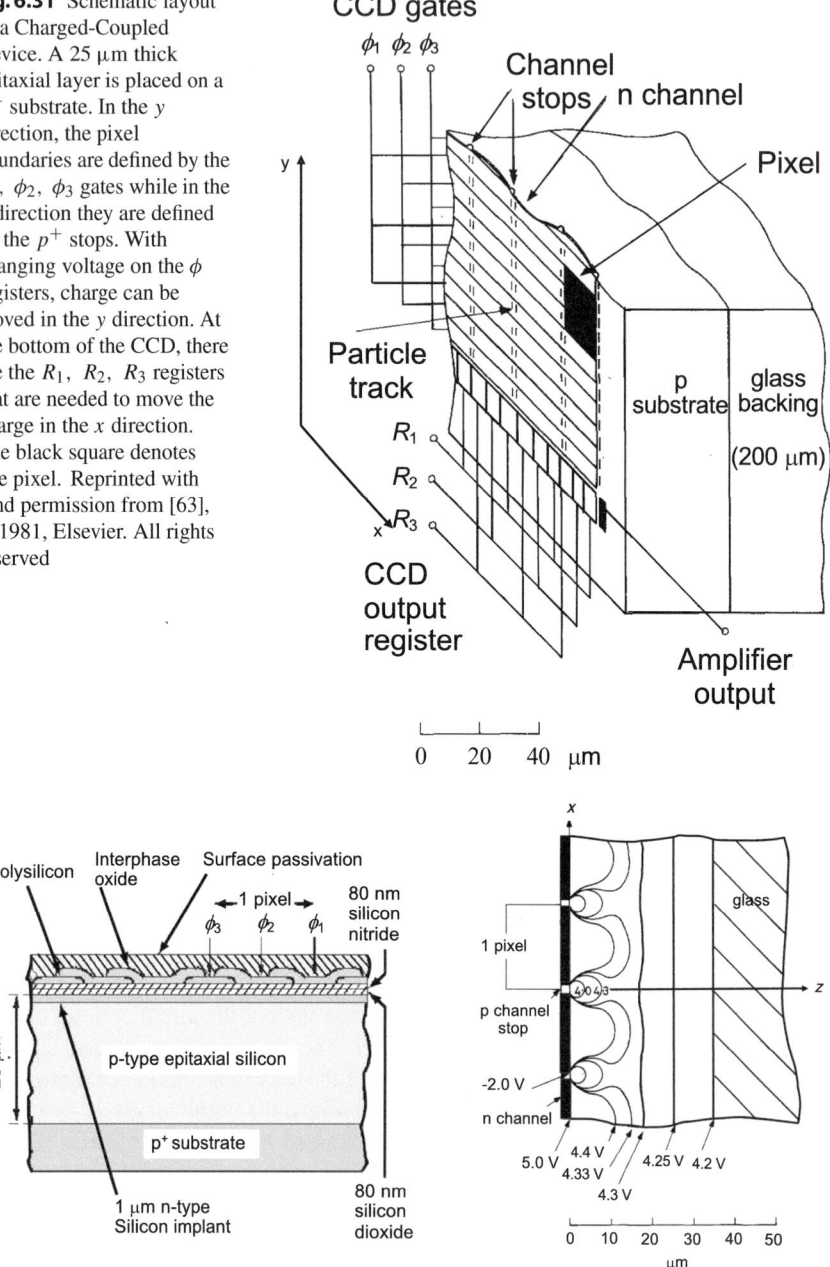

Fig. 6.32 Left: Cross section of a CCD in the yz direction. Reprinted with kind permission from [64], © 2001, J. E. Brau. All rights reserved. Right: Map of the electrostatic potentials in the CCD along the x direction when all Gate voltages are held at 0 V [63]. The p stops are clearly visible. Reprinted with kind permission from [63], © 1981, Elsevier. All rights reserved

6.3 Charge-Coupled Devices

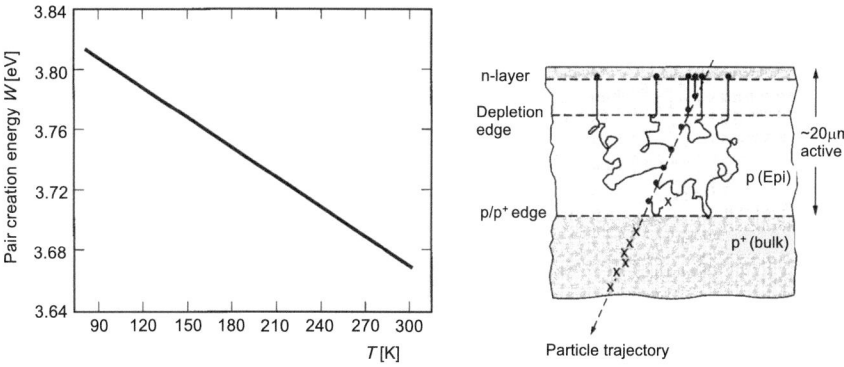

Fig. 6.33 Left: The pair-creation energy as a function of temperature for silicon. Right: Charge collection in a charge-coupled device. The edge of the depleted region and that of the epitaxial layer are indicated. Both reprinted with kind permission from [65], © 1998, AIP Publishing. All rights reserved

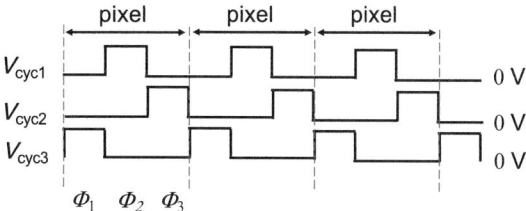

Fig. 6.34 The three-phase readout system of a CCD showing the clock voltage (left) and the charge location (right). In phase 1 (top) we have $\phi_1 = \phi_3 = 0$ V and $\phi_2 = -10$ V. In phase 2 (center) the potential is shifted to ϕ_3 while $\phi_2 = \phi_1 = 0$ V. In phase 3 the potential is shifted to ϕ_1 while $\phi_2 = \phi_3 = 0$ V. The next shift places the potential on ϕ_2, which is identical to phase 1. The charge has been moved by one pixel

on top of a noise of only a few tens of rms electrons. The pulse height distribution from a MIP is $(98 \pm 2)\%$ efficient.

6.3.1 The SLD VXD3 Vertex Detector

The SLD experiment used CCDs for the vertex detector. The VXD3 detector, which was the second full vertex detector starting operation in 1996, had three layers housing 307 million pixels with a size of 20 μm × 20 μm. Figure 6.35 shows the layout of the VXD3 detector in the rz (left) and $r\phi$ (right) views. The three layers were located at radii of 28, 38 and 48 mm covering $\cos\theta = 0.9$ in the electron forward direction and $\cos\theta = 0.85$ in the positron forward direction. In the $r\phi$ view, neighboring sensors were overlapping. The measured point resolutions were $\sigma_{r\phi} = 4.3$ μm in the $r\phi$ view and $\sigma_{rz} = 4.4$ μm in the rz view. The position resolution could have been improved to $\simeq 2$ μm if high-resistivity silicon sensors had been used. The two-track resolution was 40 μm in space with a 2% overlap. This allowed for an excellent separation of

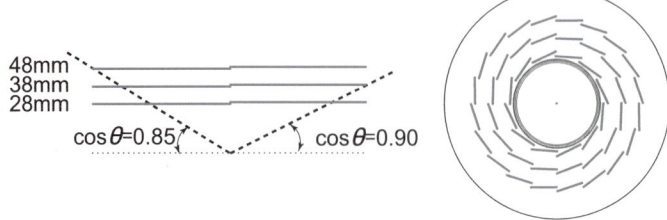

Fig. 6.35 Layout of the SLD VXD3 detector in the rz view (left) and in the $r\phi$ view (right). The CCDs are slightly tilted to provide a small overlap in $r\phi$. Reprinted with kind permission from [66], © 1997, Elsevier. All rights reserved

tracks in a jet. For the SLD VDX3, both the efficiency and purity were $\approx 100\%$ for polar angles up to $|\cos\theta| = 0.9$.

The CCDs were mounted on a kapton flex stiffened by a beryllium substrate. Figure 6.36 (left) shows the layout. On the left-hand side, the CCD is mounted on the top of the kapton flex while on the right-hand-side it is mounted on the bottom. This allowed for a small overlap at the junction. Figure 6.35 (right) shows a photograph of one completed half of the SLD VXD3 detector. The three layers are clearly visible. The CCDs were operated at a temperature of 190 K. This kept the noise at a few 10 rms electrons.

The SLD VXD3 achieved excellent impact parameter resolutions. Figure 6.37 shows the transverse (left) and longitudinal (right) impact parameter resolutions as a function of momentum with respect to that of the SLD VXD2 detector. Both distributions are determined using the full tracking system, though the precision measurements from the vertex detector play a key role in obtaining high-precision impact

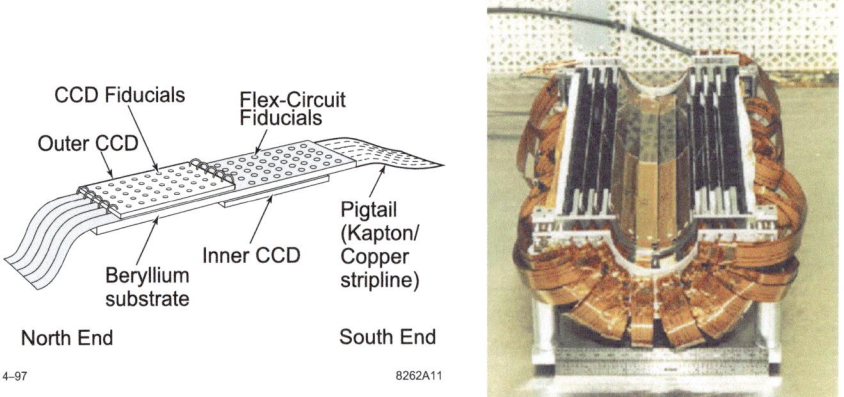

Fig. 6.36 Left: A kapton flex with two mounted CCDs. On the left-hand side, the CCD sits at the top while on the right-hand side it sits on the bottom allowing for an overlap region at the boundary. Reprinted with kind permission from [66], © 1997, Elsevier. All rights reserved. Right: A photograph of one half of the SLD VXD3 vertex detector. Reprinted with kind permission from [64], © 2001, J. E. Brau. All rights reserved

6.3 Charge-Coupled Devices

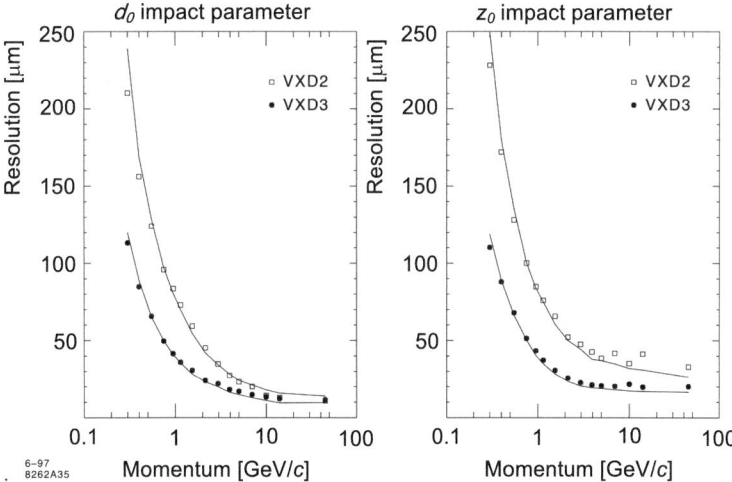

Fig. 6.37 Left: Transverse impact parameter resolution of VXD3 with respect to that of VXD2 as a function of momentum. Reprinted with kind permission from [66], © 1997, Elsevier. All rights reserved. Right: Longitudinal impact parameter resolution of VXD3 with respect to that of VXD2 as a function of momentum. The points are measurements and the curves are simulations. Reprinted with kind permission from [68], © 1998, Elsevier. All rights reserved

parameters. At higher momenta, the $r\phi$ impact parameter resolution approached 10 μm while the rz impact parameter resolution was less than 20 μm. For a polar angle range $|\cos\theta| \leq 0.7$, VXD3 achieved transverse and longitudinal impact parameter resolutions [67] of

$$\sigma_{d_0} = 7.8 \oplus \frac{33}{p \sin^{3/2}\theta},$$
$$\sigma_{z_0} = 9.7 \oplus \frac{33}{p \sin^{3/2}\theta}. \quad (6.9)$$

Note that the second term results from multiple scattering. Figure 6.38 (top) shows a Z^0 decay in the $r\phi$ view (left) and rz view (right), while Fig. 6.38 (bottom) shows an expanded rz view. This illustrates the power of the SLD VXD3 vertex detector in the reconstruction of charged-particle tracks near the primary vertex. The primary vertex is clearly visible as well as a secondary vertex on each side of the primary vertex. Apart from SLD, CCDs are used in astrophysics as infrared cameras. For example, many supernova explosions have been observed by taking pictures of the night sky with CCDs [69].

6.4 Pixel Detectors

We have seen that at high rates the occupancy in silicon strip detectors becomes a problem (see Fig. 6.15 center). Hits may not be associated to tracks unambiguously.

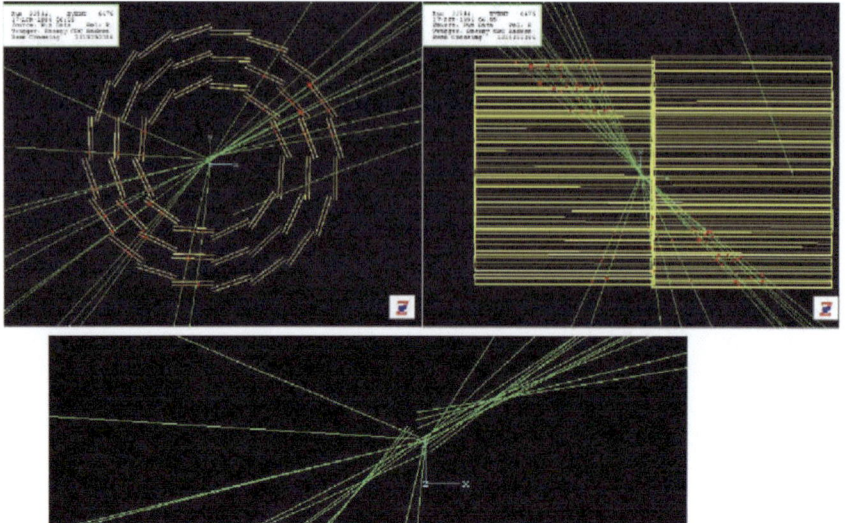

Fig. 6.38 Top: Event display in the $r\phi$ (left) and rz (right) views. Bottom: Close-up view of the rz plane. Clearly visible are the primary vertex and a secondary vertex on each sides of the primary vertex. Reprinted with kind permission from [64], © 2001, J. E. Brau. All rights reserved

This is particularly a problem at hadron colliders operating at high luminosity because of the large number of interactions per beam crossing and beam backgrounds. Using pixels instead of strips solves the problem. However, CCDs are too slow for operation at the LHC. Thus, the solution is to make physically separated silicon pixels, which provide both $r\phi$ and rz measurements. The next question is how to read them out. One way is a so-called hybrid design in which the silicon pixel sensor is coupled to a readout chip via an indium bump bond. Figure 6.39 (left) shows the schematic layout of a hybrid pixel detector. The top layer shows physically separated silicon pixels, which are soldered to the bottom layer of the readout electronics board. Figure 6.39 (center) shows the bump bond in a schematic view. Figure 6.39 (right) shows a photograph of bump bonds. The pixel sizes are limited by the number of readout channels. For example, the ATLAS experiment uses 50 μm pixels in one dimension. Since the pixel area is typically rather small, the detector capacitance is also small yielding a signal-to-noise ratio of better than 150:1. Besides taking high occupancies, pixel detectors have an excellent two-track resolution. Figure 6.40 shows photographs of a CMS pixel sensor (left) and the readout chip (right).

The heavy-ion experiment WA97 at CERN was one of the first to use pixel detectors [72]. Concerning colliding-beam experiments, DELPHI at LEP was the first to implement pixel sensors into the vertex detector [73]. At the LHC, all experiments use silicon pixel detectors. In the following subsections we present the pixel detectors in the WA97, DELPHI, ATLAS and CMS experiments.

6.4 Pixel Detectors

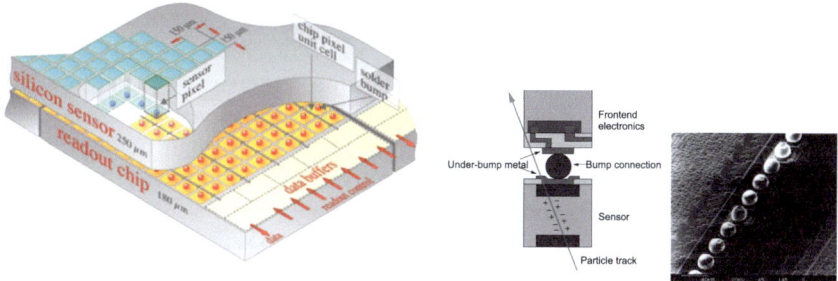

Fig. 6.39 Left: Three-dimensional schematic layout of a pixel detector. The silicon sensor is on the top, followed by a layer of bump bonds (red spheres) and the readout chip on the bottom. Reprinted with kind permission from [70], © 2012, CMS Collaboration. All rights reserved. Center: Schematic layout of a bump bond [8]. Here, the sensor lies on the bottom and the frontend electronics is on the top. The gray solid circle shows the bump bond that connects the aluminum strip on the sensor with the contact electrode of the readout chip. Reprinted with kind permission from [8], © 2010, M. Krammer. All rights reserved. Right: Photograph of bump bonds [71]. Reprinted with kind permission from [71], © 2001, Springer. All rights reserved

Fig. 6.40 Left: Photograph of a silicon pixel sensor used by the CMS vertex detector. Right: Photograph of the readout chip for the pixel sensor shown on the left-hand side. Both reprinted with kind permission from [74], © 2002, CMS Collaboration. All rights reserved

6.4.1 Pixel Detectors in the WA97 and DELPHI Experiments

The WA97 experiment used 5×10^5 pixels that were arranged in two staggered layers covering an area of 30 cm^2. The 300 µm thick pixels had an active area of $r\phi \times rz = 75 \, \mu\text{m} \times 500 \, \mu\text{m}$ and used a DC-coupled readout. They consisted of p^+ implants in an n-type substrate having a resistivity of 10 and 5 kΩ-cm, respectively. The p^+-side metallization was directly coupled to the p^+ implant, thus requiring no poly-resistors or large-area capacitors. The backside was connected to ohmic n^+ implants and an aluminum layer. Applying a positive voltage to the back side produced a partially depleted sensor. Using a reverse bias voltage of 40–70 V, the depletion was maximum near the readout electrodes and vanished at the backside. To suppress noise, only pixels above a high threshold were read out. An ionization event produced a signal of $S = 5,000$ electrons compared to a noise of 170 electrons plus a non-uniformity of 170 electrons. The efficiency was about 98.0%. The single-pixel position resolution was 22 µm in $r\phi$ and 144 µm in rz. For a two-pixel cluster, the position resolution improved to 7.8 µm in $r\phi$.

The DELPHI experiment used pixel sensors arranged into two layers at both ends of the three-layer barrel microstrip detectors (Closer Layer, Inner Layer and Outer Layer). Figure 6.41 (left) shows a schematic layout of one quadrant of the DELPHI silicon tracker near the barrel end. The inner pixel layer was inserted between the Closer Layer and Inner Layer of the microstrip detector under an angle of 12° with respect to the z axis. The outer pixel layer covered the gap between the Closer Layer and Outer Layer. It was inclined by 32° with respect to the z axis. Outside the pixel detectors, there were two layers of ministrip detectors followed by electronics for the different subdetectors.

Each pixel layer consisted of 38 sector-shaped modules, where a module contained 8,064 pixels that were arranged in ten regions of 24×24 pixels and six regions of 24×18 pixels in order to match a 380 µm thick readout chip that was bump-bonded to each sensor. The readout electronics was integrated into the chip. In the $r\phi$ view, the modules overlapped by 35%. They were held by a 3 mm thick carbon fiber structure and were precisely mounted onto aluminum rings that housed cooling pipes [75]. The pixel size was 330 µm $\times$ 330 µm. The sensor thickness ranged between 290 and 320 µm. The pixel detector comprised of 1.23×10^6 pixels, covering an area of 0.15 m^2. For readout, the pixels were DC coupled. The detector operated at a threshold of about 9,000 electrons per pixel yielding an efficiency above 99% [75]. For a single-pixel cluster, the resolution was expected to be $\sigma_{r\phi} = \sigma_{rz} = 95$ µm. For a two-pixel cluster the position resolution improved to $\sigma_{r\phi} = 40$ µm. The measured cluster resolutions were $\sigma_{r\phi} = 99$ µm (105 µm) and $\sigma_{rz} = 100$ µm (76 µm) for the inner (outer) pixel layers, respectively, which are consistent with expectations. Figure 6.41 (right) shows a photograph of the upgraded DELPHI silicon vertex detector in the forward region. The sensors of the outer pixel layers are clearly visible.

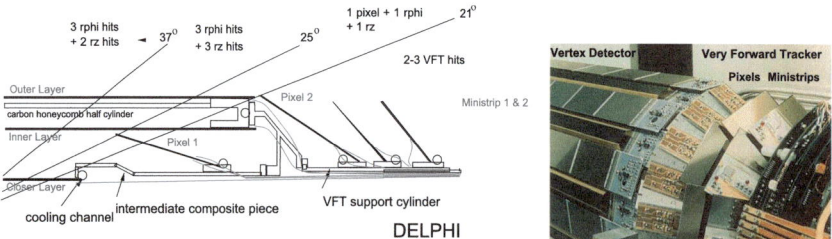

Fig. 6.41 Left: Schematic layout of one quadrant of the DELPHI silicon tracker near the barrel end showing three layers of microstrip detectors in the barrel called Closer, Inner and Outer layers, two layers of pixel sensors and two layers of ministrip detectors. The electronics for readout and services for the different subdetectors was located behind the last layer of ministrip detectors. The open circles show cooling pipes. The polar angle coverage is indicated. Note that the vertex detector covers polar angles down to $|\theta| = 10.5°$. Right: Photograph of the upgraded DELPHI silicon vertex detector in the forward region. Some pixel modules in the outer layer are visible. Both reprinted with kind permission from [76], © 1998, Elsevier. All rights reserved

6.4.2 Pixel Detectors in the ATLAS Experiment

The ATLAS pixel detector consists of four concentric layers in the barrel located inside the SCT and three disks in each endcap as shown in Fig. 6.42. Figure 6.43 (left) depicts the $r\phi$ projection of the barrel region. The four cylinders are placed at radii of 3.1, 5.05, 8.85 and 12.25 cm. The insertable B layer (IBL), which was installed in 2014, is located directly outside the beam pipe. The three outer layers use identical support structures called staves that also provide cooling. They are equipped with 22, 38 and 52 staves, respectively. All staves are inclined in azimuth angle at 20° providing overlap between adjacent staves in the azimuth angle. A stave holds 13 pixel modules yielding a length of 80.1 cm in the z direction. Staves are combined into half shells, which are mounted on the beam pipe into cylinders. Each endcap is equipped with three disks at locations of $|z| = \pm 49.5$ cm, ± 58.0 cm and ± 65 cm [34,77]. The modules are identical to those in the barrel. Each disk consists of eight sectors with six modules each yielding 288 modules containing 13×10^6 pixels.

Figure 6.43 (right) shows interactions of charged hadrons in detector components in the vicinity of the collision point. This "X-ray" picture shows the location of the beam pipe, the IBL and the B layer. The individual staves are clearly visible. For $\eta = 0$, the inactive material in the IBL is about $1.88\% \cdot X_0$ [80] and that in the other barrel layers $2.81\% \cdot X_0$ [34]. The pixel sensors in the three outer layers and the endcap disks are qualified to withstand a total dose of 500 kGy or about 10^{15} (1 MeV) n_{eq} cm^{-2} while the planar sensors in the IBL are qualified for a radiation hardness of 5×10^{15} (1 MeV) n_{eq} cm^{-2}. For more details on radiation hardness see Sect. 6.5.

Figure 6.44 (left) shows the layout of an ATLAS pixel module. It consists of the sensor that is bump bonded to 16 frontend (FE) electronic chips on the bottom layer. The top layer consists of a fine-pitch, double-sided, flexible printed circuit board (flex hybrid) that provides services and routes signals. It hosts the module controller chip (MCC), the power supply, HV guard rings, a negative-temperature coefficient (NTC)

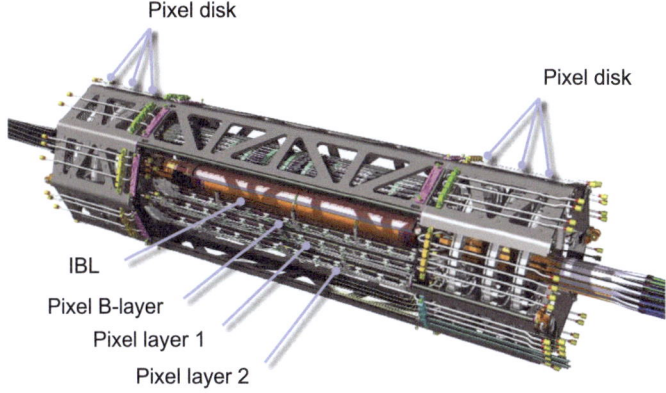

Fig. 6.42 Left: Three-dimensional schematic layout of the ATLAS pixel detector [78]. Reprinted under CC-BY-4.0 Licence from [78], © 2016, ATLAS Collaboration

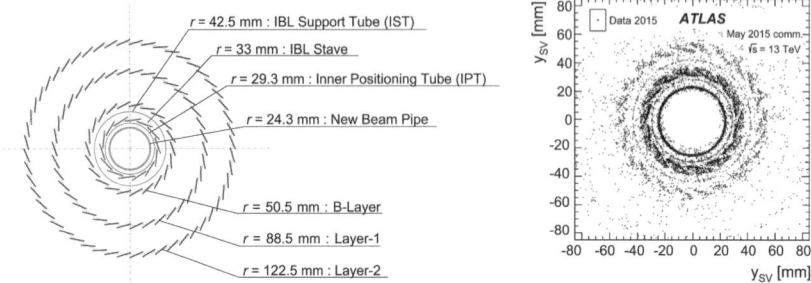

Fig. 6.43 Left: The $r\phi$ view of the ATLAS barrel pixel detector. From the inside out we see the IBL, B layer, layer 1 and layer 2. The 20° inclination of the staves is clearly visible. Reprinted under CC-BY-4.0 Licence from [78], © 2016, ATLAS Collaboration. Right: Interactions of hadron tracks in detector components near the collision point. Clearly visible are the beam pipe, the IBL and the B layer. Reprinted under CC-BY-NC-ND-4.0 Licence from [79], © 2020, ATLAS Collaboration

thermistor and for barrel modules another flexible foil (barrel pigtail) with transfer cables to a connector that provides the link to electrical services via microcables. For endcap modules, the microcables are directly attached without pigtails. The FE chips are connected to the flex hybrid and MCC, which transmits and receives digital signals. The backside of the flex hybrid is glued to the high-voltage side of the sensor using a multiple solder mask layer. The NTCs provide a remote monitoring of the module temperature allowing for a fast power off in case of overheating. The pixel sensor has an active area of 16.4 mm × 60.8 mm holding 47,232 pixels (256 μm thick) that are all read out. The typical pixel pitch is 50 μm × 400 μm. To ensure full coverage in regions between front-end chips, about 11% of the sensor pixels have a pitch of 50 μm × 600 μm. To ensure full depletion, the pixels are operated with a reverse bias voltage of 150–600 V. Table 6.2 summarizes a few properties of the ATLAS pixel detector, which has over 92 million readout channels covering an area of 1.9 m^2.

Figure 6.44 (right) shows the layout of a planar pixel sensor in the ATLAS Pixel detector near the edge. The pixels are made of so-called n^+-in-n type silicon, which

Table 6.2 Properties of the ATLAS pixel detector listing the pixel thickness, pixel size, number of staves, number of modules, total number of pixels and total area covered by silicon pixels. Reprinted with kind permission from [82], © 2013, ATLAS Collaboration. All rights reserved

Pixel detector	Thickness [μm]	Pixel size [μm × μm]	# staves # sectors (EC)	# modules	# pixels [10^6]	area [m^2]
IBL	200 & 230	50 × 250	14	280	12.04	0.15
3 outer layers	250	50 × 400 (600)	112	1,456	67.092	1.45
2 × 3 endcap disks	250	50 × 400 (600)	48	288	13.271	0.28

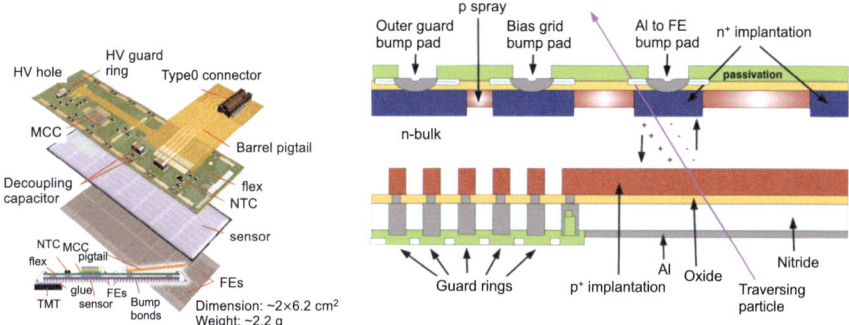

Fig. 6.44 Left top: Layout of a pixel module showing the frontend electronics plane (below), bump bonded to the pixel plane (center) and the service plane on top. Left bottom: Side view of the pixel module showing the bump bonds, barrel pigtails, the sensor, the MCC, NTC, Thermal Management Tile (TMT), FE electronics and the flex. Reprinted with kind permission from [77], © 2008, IOP Publishing. All rights reserved. Right: Layout of the ATLAS pixel sensor. The bulk is an n-type silicon (white area). On the top n^+ pads (blue rectangles) are positioned, which collect the electrons. They are connected by bump pads (grey semi-cylinders) to the readout chip connected to the frontend electronics. With p spray (shiny brown rectangles) between the n^+ pixels, the accumulation of electrons on the oxide layer is prevented. The bottom side has a p^+ layer (dark brown band) that is segmented into several strips on the edge of the sensor (left-hand side). These p^+ strips are coupled to aluminum strips providing guard rings. The outermost n^+ pads serve as guards bumps. Another n^+ pad close to the edge is used to supply the reverse bias. Reprinted with kind permission from [81], © 2013, T. Wittig. All rights reserved

consist of n^+ strips on an n bulk. The n^+ pixels are separated by p-spray. So, the electrons produced in the silicon by ionization of a traversing charged particle are collected at the n^+ pixels. At the pixel edge a bias grid bump pad and an outer guard bump pad are placed. On the bottom several p^+ guard rings stabilize the sensor. The idea of this layout is that due to radiation the n bulk will convert to p bulk after exposure to a fluence of 2×10^{13} cm^{-2}. The detector is still operable as illustrated in Fig. 6.16. We will discuss radiation issues in Sect. 6.5.

Figure 6.45 (left) shows a layout of the ATLAS IBL, which is positioned at a radius of 3.35 cm from the beam line. The IBL has 14 staves, each containing 20 modules, which are equipped to 75% with 200 μm thick planar sensors and to 25% with 230 μm thick 3D sensors (see reference Sect. 6.7.5) located at both ends. The length in the z direction is 64 cm. For all sensors, the pixel size is reduced to 50 μm × 250 μm due to an enhanced occupancy near the beam line. The planar sensors also consist of n^+-in-n type silicon while the 3D sensors use n^+-in-p type silicon. The staves are tilted in ϕ by 14° covering a pseudo-rapidity of $|\eta| \leq 2.9$. The IBL is operated at a temperature of -35 °C whereas the three outer barrel layers and the endcap disks are operated at -20 °C. Figure 6.45 (right) shows a three-dimensional layout of the IBL.

To extract the signal, typically the charge is integrated over some time interval. ATLAS measures the time-over-threshold (ToT), which is expected to have a rather linear dependence on the collected charge. Figure 6.46 (left) illustrates the principle. The top left plot shows the preamplifier signal for two different-size pulses. The

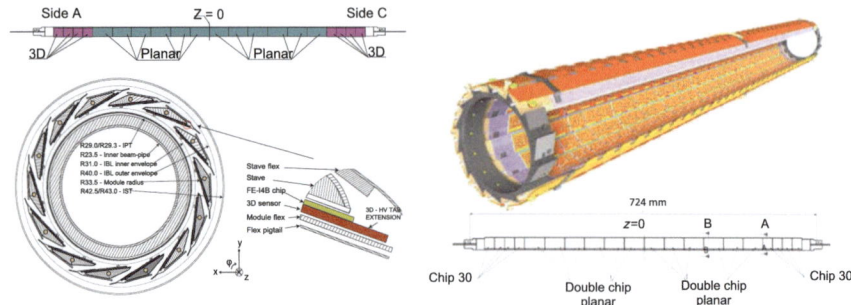

Fig. 6.45 Left: Layout of the IBL detector; (top) the longitudinal layout of planar and 3D modules on a stave; (bottom left) the $r\phi$ view depicting the beam pipe, the inner positioning tube (IPT), the staves of the IBL detector and the inner support tube (IST), as viewed from the C-side; (bottom right) expanded $r\phi$ view from the corner of a 3D module fixed to the stave. Reprinted under CC-BY-4.0 Licence from [80], © 2018, ATLAS Collaboration. Right: Three-dimensional layout of the Insertable B Layer (top) and projection to the rz plane (bottom). Reprinted under CC-BY-4.0 Licence from [78], © 2016, ATLAS Collaboration

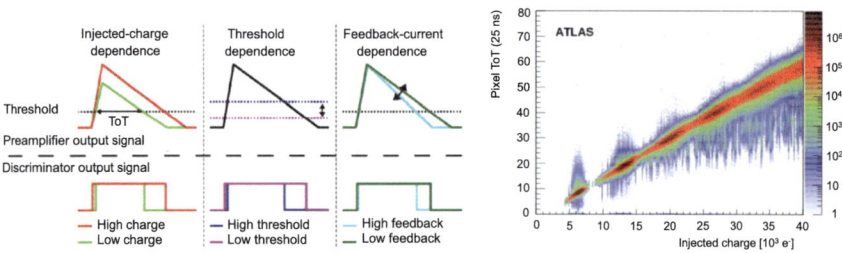

Fig. 6.46 Left: Correlation between preamplifier output signal (top) and the digital discriminator output signal (bottom) for the time-over-threshold signal extraction showing the dependence on the injected charge (left), dependence on threshold (center) and dependence on the feedback current (right). Reprinted with kind permission from [83], © 2007, A. Dobos. All rights reserved. Right: Measured correlation between ToT and injected charge. Reprinted under CC-BY-NC-ND-4.0 Licence from [84], © 2009, ATLAS Collaboration

corresponding digital discriminator output signals shown on the bottom left plot have different lengths depending on the time the input signal is above a fixed threshold. This dependence is rather linear. The length of the output signal also depends on the height of the threshold and the feedback current. The effect of the threshold height is illustrated in the top and bottom center plots. Low (high) thresholds yielding long (short) output signals provide a larger (smaller) dynamic range while being more (less) sensitive to noise. So, there is typically an optimum. The effect on the feedback currents is illustrated in the top and bottom right plots. Higher feed-back currents yield shorter output pulses and thus a lower dynamic range. Figure 6.46 (right) shows the time-over-threshold as a function of injected charge confirming the linear dependence for charges between 5,000 to 40,000 electrons.

Since the electric and magnetic fields are not parallel, the drift velocity has a Lorentz angle with respect to the electric field. For the 2 T solenoidal field, the Lorentz angle amounts to 200–220 mrad in the four pixel layers. Thus, if the track angle differs

6.4 Pixel Detectors

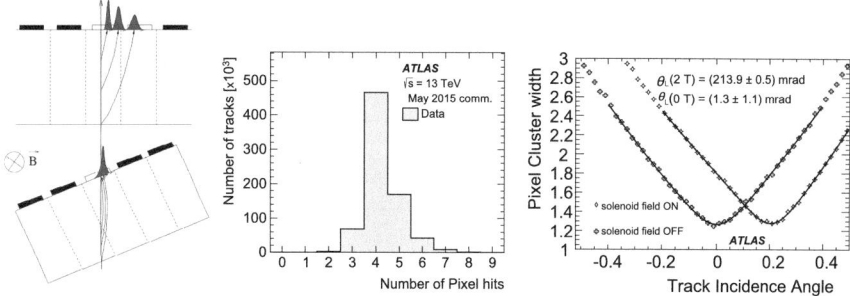

Fig. 6.47 Left top: Charged track traversing a pixel under a different angle than the Lorentz angle [85]. The produced electrons are collected by more than one pixel improving the position resolution to $\sigma_{r\phi} = 12\,\mu$m. Left bottom: Charged track traversing a pixel under the Lorentz angle. Here, typically only one pixels collects the charge. So the resolution in $r\phi$ is 14 μm and in rz is 115 μm. Reprinted with kind permission from [85], © 1989, Elsevier. All rights reserved. Center: The number of pixel clusters per beam crossing in barrel layer 1. Reprinted under CC-BY-4.0 Licence with kind permission from [86], © 2015, ATLAS Collaboration. All rights reserved. Right: The average number of pixels as a function of pseudo-rapidity with and without a 2 T magnetic field. Due to the Lorentz angle the distribution is shifted by 214 mrad [84]. Reprinted under CC-BY-NC-ND-4.0 Licence from [84], © 2009, ATLAS Collaboration

from the Lorentz angle, typically more than one pixel records the signal, improving the position resolution. Figure 6.47 (left) illustrates this. On the top plot, the incoming charged track has an angle different from the Lorentz angle. Here, at least two pixels collect the charges. On the bottom plot, the angle of the incoming charged track is close to the Lorentz angle, which results in a single pixel hit. Figure 6.47 (center) shows the number of pixel hits per beam crossing produced by a traversing track. As expected, the typical number of pixels is four. Figure 6.47 (right) shows the cluster width as a function of pseudo-rapidity with and without magnetic field. We see that for the 2 T magnetic field the track incidence angle is shifted by about ≃214 mrad.

The pixel detector works rather well. In run 1, the efficiency of finding a hit on a reconstructed track was larger than 99% for pixel detectors in the three barrel layers and the endcap disks 1 and 2. For disks 3, it was 1–2% lower due to a few bad modules. For the IBL, the corresponding efficiency is larger than 98% for the full p_T spectrum and $> 99\%$ for high p_T tracks. Figure 6.48 (left) shows the efficiency for planar pixel hits associated with a reconstructed track as a function of luminosity. Both the IBL and the B layers are affected by radiation. In the IBL, the high voltage is raised over time yielding a recovery of the efficiency. In the B-layer the threshold was lowered leading to an efficiency recovery. Figure 6.48 (right) shows the charge collection for planar IBL modules as a function of luminosity for different bias voltages. The collected charge drops for increased luminosity. This can be compensated by increasing the bias voltage. The IBL is operated with a threshold of 1, 500 e having a mean noise below 150 rms electrons. The standard pixels in the other layers and disks have a mean noise of 180 rms electrons. They are operated with a threshold of 3,500 electrons. The dispersion is about 40 e. For the long pixels the noise increases to 300 rms electrons. A minimum ionizing particle with $p_T > 0.1$ GeV/c produces about 20,000 rms electrons. About 80% of the pixel clusters are single-hit clusters for which the position

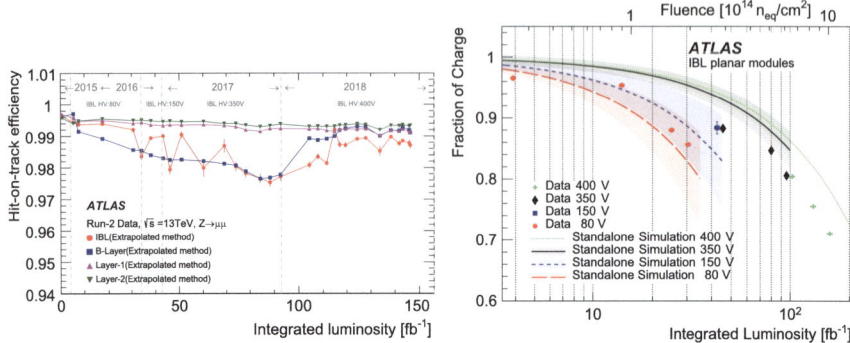

Fig. 6.48 Left: The efficiency for planar pixel hits associated with a reconstructed track in the ATLAS experiment for the IBL (red points), B-layer (blue squares), Layer 1 (magenta upward pointing triangles) and Layer 2 (black downward pointing triangles) as a function of integrated luminosity during Run-2 pp collisions at 13 TeV CM energies. The measurements were obtained from $Z \to \mu^+\mu^-$ events over a period of about three years. Note that the high voltage was increased from 80 to 150 V to 350 and 400 V in the three-year period to adjust for radiation effects. On the B-layer, the threshold was lowered in 2018 to recover the efficiency. The rattling structure in IBL observed since 2016 is probably due to a single problematic module. Reprinted with kind permission from [87], © 2020, ATLAS Collaboration. All rights reserved. Right: The charge collection efficiency for IBL planar modules as a function of integrated luminosity showing measurements and simulations for different bias voltages, 80 V (red points), 150 V (blue squares), 350 V (black diamonds) and 400 V (green crosses). Reprinted with kind permission from [88], © 2020, IOP Publishing. All rights reserved

resolution is about the pitch divided by $\sqrt{12}$. Figure 6.49 shows the spatial resolution of IBL hits associated with reconstructed particle tracks as a function of integrated luminosity. The spatial resolutions in the IBL are $\sigma_{r\phi} = (10.0 \pm 0.1)$ μm in $r\phi$ and $\sigma_{rz} = (66.5 \pm 0.8)$ μm in rz, respectively [90]. Note that they are slightly better than what we expect for single clusters. For the other layers and disks the spatial resolutions are around $\sigma_{r\phi} = 10.0$ μm in $r\phi$ and $\sigma_{rz} = 110$ μm in rz.

Figure 6.50 (top) shows the measured transverse impact parameter resolutions obtained using the full tracking system (σ_{d_0}) as functions of transverse momentum for $0 < |\eta| < 0.2$ (left) and pseudo-rapidity for 0.4 GeV/$c < p_T < 0.5$ GeV/c (right). Inclusion of the IBL yields considerable improvements of σ_{d_0} particularly for transverse momenta below 2 GeV/c. For 0.4 GeV/$c < p_T < 0.5$ GeV/c the improvement is almost a factor of two. At low p_T, the transverse impact parameter resolution is 140 μm decreasing to 30 μm at high p_T. Figure 6.50 (bottom) shows the corresponding distributions for the longitudinal impact parameter resolution (σ_{z_0}). Here, the improvement is again about a factor two at low p_T and still 50% at $p_T = 20$ GeV/c. With the IBL, we achieve $\sigma_{rz} = 220$ μm at low p_T decreasing to 75 μm at high p_T.

6.4 Pixel Detectors

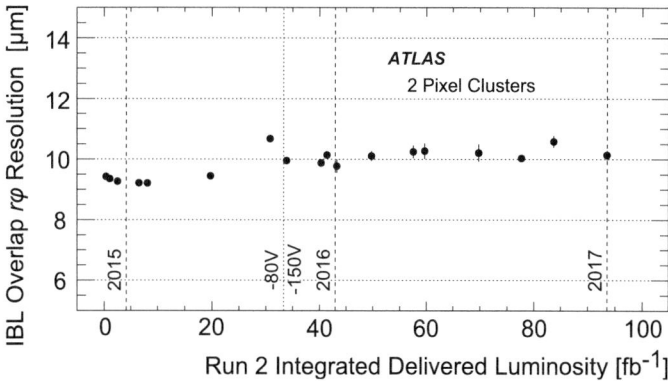

Fig. 6.49 Spatial resolution of IBL hits associated with reconstructed particle tracks in di-jet events as a function of integrated luminosity. Reprinted with kind permission from [88], © 2020, IOP Publishing. All rights reserved

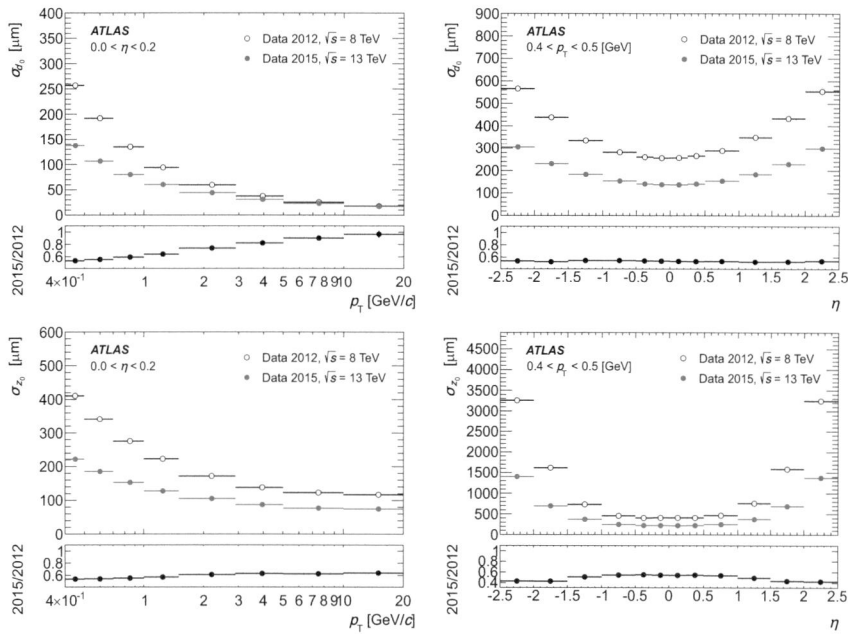

Fig. 6.50 The transverse (top) and longitudinal (bottom) impact parameter resolutions in the ATLAS detector as a function of transverse momentum for $0.0 < |\eta| < 0.2$ (left) and as a function of pseudo-rapidity for $0.4 \text{ GeV}/c < p_T < 0.5 \text{ GeV}/c$ (right). The plots show data without (open points) and with (black points) the IBL included. Reprinted under CC-BY-NC-ND-4.0 Licence from [89], © 2020, ATLAS Collaboration

6.4.3 Pixel Detectors in the CMS Experiment

Originally, the CMS pixel detector had three barrel layers and two disks in each endcap. It was upgraded in 2014 to four barrel layers and three disks in each endcap [91] housing 123.5 million channels. Figure 6.51 shows schematic layouts of the present CMS pixel detector in the rz view (left) and $r\phi$ view (right) in comparison to the previous pixel detector [92]. In the barrel four barrel layers concentric with the beam pipe are positioned at radii of 2.9, 6.8, 10.9 and 16.0 cm. In addition, three disks are placed in each endcap at $z = \pm 29.1$ cm, $z = \pm 39.6$ cm and $z = \pm 51.6$ cm. Figure 6.52 (left) shows the $r\phi$ projection of the barrel layers. Figure 6.52 (right) shows a close-up view of a few modules in one $r\phi$ layer. It should be noted that neighboring modules have a large overlap. The cooling pipes are positioned such that they cool two modules. One module type is used to build the barrel layers and endcap disks. The latter consist of an inner ring (45 mm $< r <$ 110 mm) and an outer ring (96 mm $< r <$ 161 mm).

A module consists of one silicon sensor having an active area of 16.2 mm × 64.8 mm that is bump-bonded to 2 × 8 readout chips. The silicon sensors in the barrel are 285 μm thick while the readout chip is 180 μm thick except for modules in layer 1, where it is thinned to 75 μm. Since the pixel size is 100 μm × 150 μm, a module contains 66,560 pixels. The pixels are also n^+-in-n type silicon sensors as in the ATLAS experiment. For barrel modules, CMS also uses p spray to eliminate the electron accumulation under the oxide layer. The sensors in the endcaps are 300 μm thick. Each pixel is surrounded by an individual p stop, which has a hole on one side.

In the barrel, eight modules along the z-direction are combined into a facet, which is held by a carbon fiber structure. Facets are combined into half cylinders that can be mounted around the beam pipe. The four barrel layers need 12, 28, 44 and 60 facets, respectively. In the six endcap disks, 672 pixel modules are used. The modules are mounted radially on a light-weight structure called blade. Blades are combined into half disks containing 56 modules of which 44 modules are arranged in an outer ring and 22 modules are arranged in an inner ring. The pixel sensors are operated at reverse

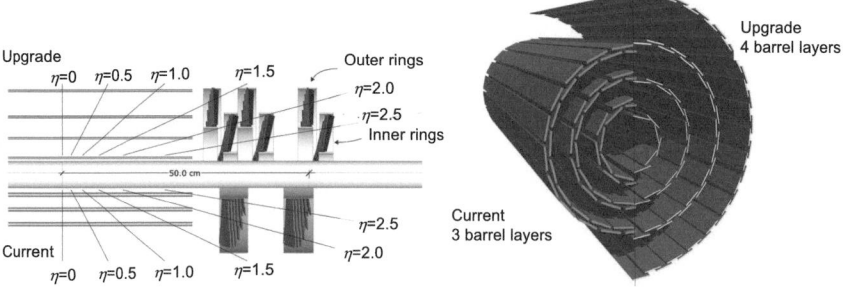

Fig. 6.51 Left: Schematic layout of the upgraded CMS pixel detector (top) in the longitudinal direction in comparison to the old pixel detector (bottom). Right: Schematic layout of the upgraded CMS pixel detector (right) in the transverse oblique view in comparison to the old pixel detector (left). Both reprinted with kind permission from [93], © 2013, CMS Collaboration. All rights reserved

6.4 Pixel Detectors

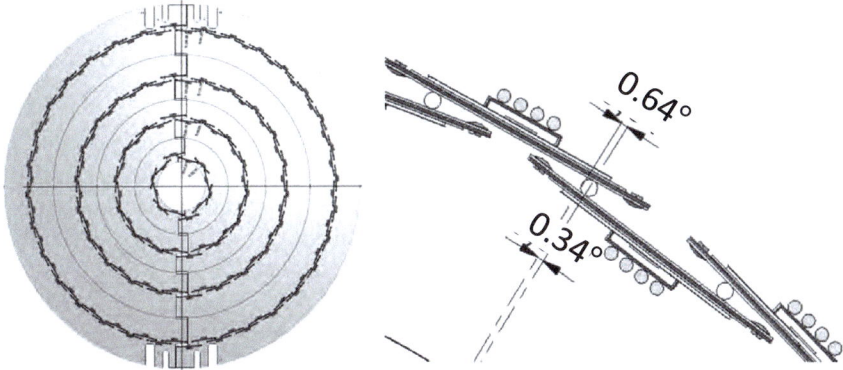

Fig. 6.52 Left: $r\phi$ projection of the CMS barrel pixel detector. Right: Close-up view of CMS modules in a layer showing the large overlap. Cooling pipes (open circles) and cables (grey circles) are visible. Both reprinted with kind permission from [93], © 2013, CMS Collaboration. All rights reserved

bias voltage of 150 V. In the 3.8 T magnetic field the Lorenz angle is 27° producing typically two-pixel $r\phi$ clusters, improving the position resolution considerably with respect to single hit resolution. Note that the Lorentz angle for electrons is about three times larger than that for the holes. The pixel detector is kept at a temperature below $-4\,°C$. The upgraded CMS pixel detector has coverage of $|\eta| \leq 2.5$.

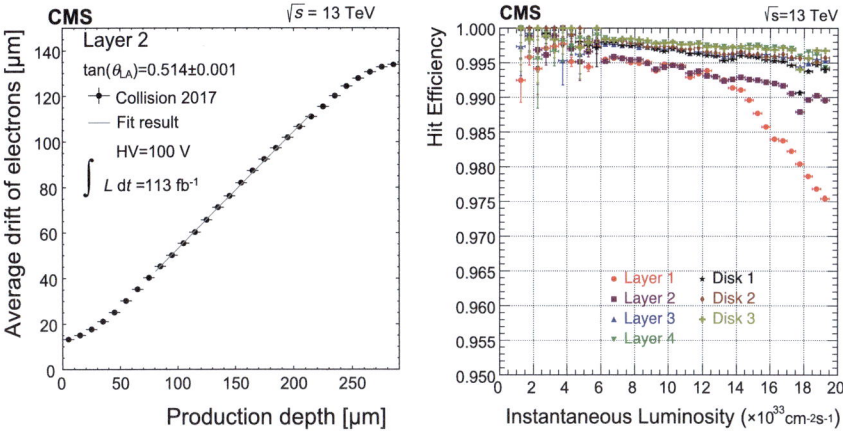

Fig. 6.53 Left: The average drift of electrons as a function of depth at which the charge was generated. The data are for barrel layer 2. The slope of the linear fit (solid curve) is interpreted as the tangent of the Lorentz angle. Right: The cluster hit efficiency as a function of instantaneous luminosity for layer 1 (red points), layer 2 (violet squares), layer 3 (blue upward triangles), layer 4 (green downward triangles), disk 1 (black stars), disk 2 (brown diamonds) and disk 3 (light-green crosses). Note that the efficiency in layer 1 decreases with higher instantaneous luminosity. This effect is also visible in layer 2. Both reprinted under CC-BY-4.0 Licence from [91], © 2018, CMS Collaboration

A minimum-ionizing track produces about 21,000 electrons in the pixel. The thresholds have been set to 2,200 e in layer 1, 1400 e in layers 2–4 and 1,500 e in the endcap disks. For these thresholds, the number of noise hits was rather low. Figure 6.53 (left) shows the average drift of electrons in the 3.8 T CMS magnetic field as a function of the depth at which they are produced. For intermediate depths, the correlation is linear. A fit gives a slope of (0.514 ± 0.001), which represents $\tan \alpha_L$ yielding $\alpha_L = 27°$. Figure 6.53 (right) shows the cluster hit efficiency as a function of instantaneous luminosity. Except for layer 1, the hit efficiency is typically above 99%. For layer 1, the efficiency is above 99% at low luminosity. At $\mathcal{L} = 2 \times 10^{34}$, it is still 97.5%.

The position resolution is measured with the so-called triplet method [94] in which a high p_T track is measured by the entire tracking detector to determine the radius of the helix. Now, two hits in pixel layers 2 and 4 are sufficient to specify the track and predict the position in layer 3. We can compare the predicted position with the measured position and calculate the residuals, which are fitted with a Student's t function [95] to estimate their width. In order to subtract the effect of the two other layer measurements, the width of the fit is divided by $\sqrt{(3/2)}$ to get the intrinsic pixel resolution, where we assumed that the intrinsic resolutions of the three layers are similar. The resolution is measured separately for $r\phi$ and rz planes. Figure 6.54 (left) shows as an example the pixel hit residuals in the $r\phi$ plane for barrel layer 3 and Fig. 6.54 (right) shows the corresponding distribution in the rz plane [91]. The measured hit resolutions are $\sigma_{r\phi} = (9.48 \pm 0.02)\,\mu m$ and $\sigma_{rz} = (22.23 \pm 0.04)\,\mu m$, respectively. Figure 6.55 (left) shows the transverse impact parameter resolution as a function of transverse momentum for the entire pseudo-rapidity range and for pseudo-rapidities in the barrel region. For transverse momenta larger than 7 GeV/c, the transverse impact parameter resolution is 20 μm. Figure 6.55 (right) shows the transverse impact parameter resolution as a function of pseudo-rapidity, which is obtained using the entire tracking system. In the barrel region we get $\sigma_{d_0} = 40$–$50\,\mu m$. In the endcap for large values of $|\eta|$ the transverse impact parameter resolution rises up to 200 μm.

6.5 Radiation Hardness of Silicon Pixel Sensors

For the LHC vertex detectors, radiation hardness of silicon sensors is an issue. Figure 6.56 (left) shows the expected level of radiation as a function of the radius from the beam pipe for the ATLAS experiment. In the region of the ATLAS pixel and silicon microstrip detectors, radiation levels are rather high. In the innermost pixel layer, the radiation dose reaches $\Phi_{eq} = 3 \times 10^{14}(1\text{MeV})\,n_{eq}/cm^2$,[6] while that in the innermost microstrip layer is still $\Phi_{eq} = 2 \times 10^{13}$ (1 MeV) n_{eq}/cm^2. High doses of proton, gamma and neutron radiation may cause temporary or permanent damage of the semiconductor structure since particles interacting with crystalline silicon can introduce reversible and irreversible changes in the sili-

---

[6] Φ_{eq} stands for 1 MeV equivalent neutron fluence.

6.5 Radiation Hardness of Silicon Pixel Sensors

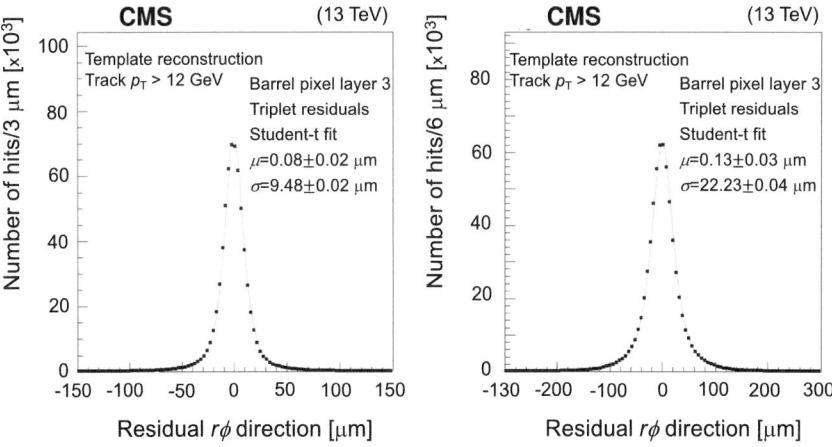

Fig. 6.54 Pixel hit residuals in the $r\phi$ plane (left) and those in the rz plane (right) in the CMS barrel pixel layer 3. The distributions are fitted with a single Student's t-fit. Reprinted under CC-BY-4.0 Licence from [91], © 2018, CMS Collaboration

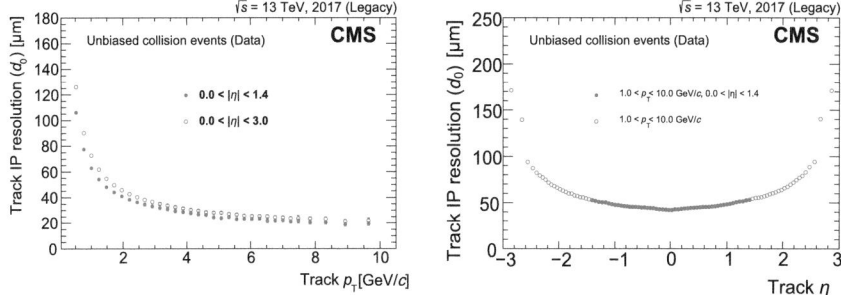

Fig. 6.55 Left: The transverse impact parameter resolution as a function of p_T in unbiased collision data for $|\eta| < 1.4$ (points) and $|\eta| < 3.0$ (open circles). Right: The transverse impact parameter resolution as a function of η in unbiased collision data for $1.0 < p_T < 10.0$ GeV/c. The error bars take into account the limited number of analyzed events and include the systematic uncertainty due to the finite bin width in the fitted distributions. Both reprinted with kind permission from [96], © 2022, CMS Collaboration. All rights reserved

con. Sizable changes take place at the silicon surface barriers if irradiation levels exceed 10^{14} (1 MeV) n_{eq} cm^{-2} for electrons, 10^{13} (1 MeV) n_{eq} cm^{-2} for protons, 10^{11} (1 MeV) n_{eq} cm^{-2} for α particles, 3×10^{8} (1 MeV) n_{eq} cm^{-2} for fission products, and $> 10^{4}$ Gray for X-rays. We distinguish between damage on the surface (surface damage) and that in the detector bulk (bulk damage). Surface damage is mostly caused by photons while bulk damage is created by hadrons. A traversing hadron may knock off a silicon atom from the lattice as shown in Fig. 6.56 (right). The silicon atom can move through the crystal as an interstitial atom. The vacancy-interstitial system is also called Frenkel pair. The threshold energy is about 25 eV. Low-energy charged hadrons typically create point defects while neutrons and high-energy charged hadrons create cluster defects.

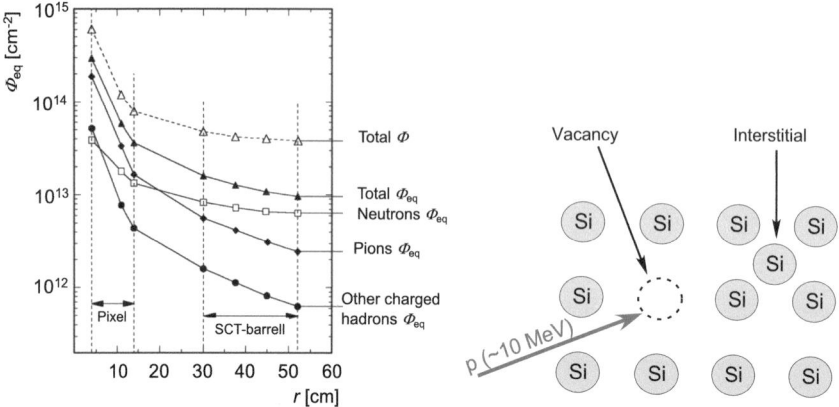

Fig. 6.56 Left: Expected particle fluences for different particle species in the ATLAS detector as a function of the distance from the beam in one year of operation (10^7 s) at high luminosity of $10^{34}/(\text{cm}^2 \cdot \text{s})$ assuming an inelastic pp interaction cross section of 80 mb. Reprinted with kind permission from [3], © 1999, M. Moll. All rights reserved. Right: The Frenkel defect in which a silicon atom is dislocated from its lattice position. At the LHC this occurs via an interaction of the silicon atom with a proton

Defects in the silicon lattice impact the behavior of a silicon particle detector. Vacancies and interstitial atoms can react to form additional defects or anneal. For example, a vacancy can react with other vacancies or impurities like carbon, oxygen or phosphorus [97]. The defects may cause four different effects as illustrated in Fig. 6.57. First, we may observe increased leakage currents. This is caused by defects that occupy levels in the energy gap near the middle allowing to transfer charge carriers from the valence band to the conduction band. Second, space charges may be created that happens both for donors and acceptors. The generated space charge will change the effective doping concentration N_{eff}. Third, defects with energy levels near the middle of the gap energy may create traps for electrons or for holes. Depending on the trapping time, the trapped particles may be lost for the readout. Fourth, according to the inter-center charge transfer model, combinations of the different defects in so-called defect clusters lead to enhanced effects.

Figure 6.58 shows the locations of several point defects in the energy gap. Figure 6.59 illustrates examples of possible point and cluster defects in the silicon crystal structure. Note that defects are not stable but move through the crystal lattice.

The energy loss that leads to damage of the crystal structure is called non-ionizing energy loss (NIEL). The assumption that the created damage scales linearly with the energy lost in the displacement of atoms is called NIEL scaling. The primary knock-on atom carries most of the energy of the primary particle and thus creates further damage in the silicon lattice, which depends on the recoil energy E_R but not on the particle type and interaction process. The recoiling atoms excite and ionize the silicon sensor. They undergo elastic and inelastic interactions, which may cause further displaced silicon atoms. Using the Thomas-Fermi model of the atoms,

6.5 Radiation Hardness of Silicon Pixel Sensors

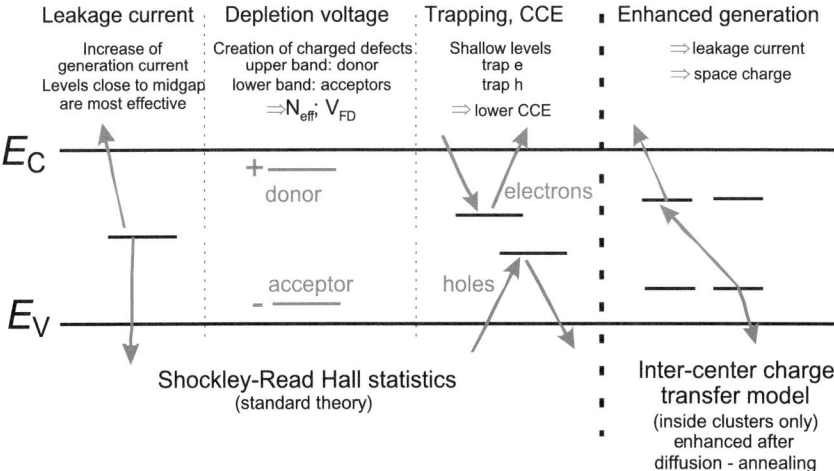

Fig. 6.57 Different defect level locations and their effects [98], which all occur in the forbidden gap. Mid-gap levels (left) increase the leakage current and decrease the charge carrier lifetime. Space charge created by charged donors in the upper half or acceptors in the lower half of the band gap can affect the doping concentration (center left). Deep levels, which can trap electrons or holes, may have trapping times much longer than the time to record the signal (center right) and thus the charge is lost reducing the charge collection efficiency (CCE). According to the inter-center charge transfer model, combinations of the different defects in defect clusters additionally enhance the effects (right). Reprinted with kind permission from [36], © 1985, Springer. All rights reserved

Lindhard *et al.* developed a treatment of these atomic interactions to calculate the energy-displacement damage function,

$$D(E) = \sum_k \sigma_k(E) \int_{E_{th}}^{E_R^{max}} dE_R \, P_k(E, E_R) f_{part}(E_R), \tag{6.10}$$

where the sum runs over all reactions, $\sigma_k(E)$ is the corresponding cross section, $P_k(E, E_R)$ is the probability for a neutron with energy E to produce a recoil with energy E_R through reaction k and $f_{part}(E_R)$ is the partition function derived by Lindhard [99]. Figure 6.60 shows the displacement damage cross section $D(E)$ normalized to 95 MeV mb for neutrons as a function of kinetic energy for different particle types in a double-logarithmic scale. The displacement damage cross section has a strong energy dependence. It is largest for protons decreasing rapidly with energy. For 100 eV protons, it is 10^4 dropping below 1 for 10 GeV protons. For neutrons, $D(E)$ has a minimum of $\simeq 10^{-5}$ for an energy around 100 eV. For higher energies, $D(E)$ increases up to 1 around 1 GeV showing a rich resonant structure before decreasing again. At lower energies, $D(E)$ increases slowly. For pions, $D(E)$ rises to a value around 1 at $E = 100$ MeV before dropping again. Electrons have a much smaller displacement damage cross section, which after a slow rise saturates around 0.1. For an energy of $E = 9$ GeV, the displacement damage cross sections are 0.446, 0.508, 0.379 and 0.1 for neutrons, proton, pions and electrons, respectively.

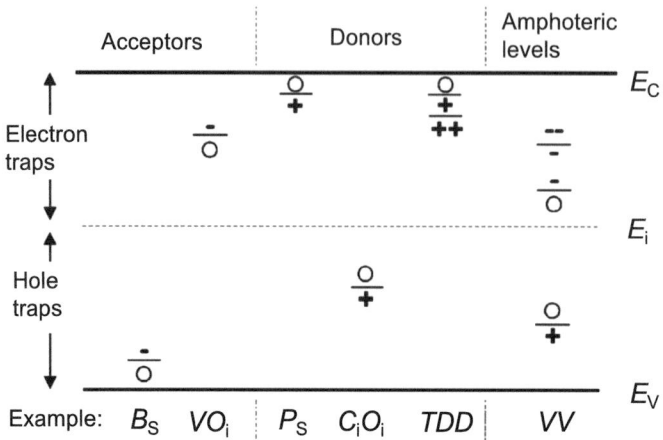

Fig. 6.58 Location of different point defects in the energy gap. The symbols (+, o, −) indicate the charge of the impurity. Shown are states of a boron interstitial (B_s), vacancy-oxygen complex (VO_i), phosphorus interstitial (P_s), carbon-oxygen complex (C_iO_i), a thermal double donor (TDD) and amphoteric di-vacancy (VV). Electron traps are displayed at the top and hole traps on the bottom. While the boron interstitial, carbon-oxygen complex and amphoteric di-vacancy are hole traps all the other are electron traps. Note that acceptors are defects that are negatively charged when occupied with an electron while donors are defects that are neutral when occupied with an electron. If the Fermi level is located above the defect level, acceptors are negatively charged and donors are neutral. If the Fermi level lies below the defect level, acceptors are neutral and donors are positive. Reprinted with kind permission from [3], © 1999, M. Moll. All rights reserved

Note, a 1 MeV neutron has $D(E) = 0.9 \pm 0.03$ [100], while a 1 MeV proton has a value of 50. A value of around one is found for 100 MeV protons.

The damage inflicted by a particle with a specified energy E can be scaled to the damage caused by a 1 MeV neutron. The scaling factor is called the hardness factor

$$\kappa_E = \frac{\int D(E)\Phi(E)dE}{D(E_n = 1\text{ MeV}) \cdot \int \Phi(E)dE} = \frac{\Phi_{eq}}{\Phi}, \quad (6.11)$$

where $\Phi = \int \Phi(E)dE$ is the irradiation fluence and $D(E_n = 1\text{ MeV})$ is set to 95 MeV mb to ensure independence of different calculations from binning effects [101]. The equivalent 1 MeV neutron fluence Φ_{eq} can be calculated by

$$\Phi_{eq} = \kappa_E \Phi = \kappa_E \int \Phi(E)dE. \quad (6.12)$$

Before irradiation, leakage currents are typically below 1 nA. They increase rapidly after irradiation. Figure 6.61 (left) shows the change in leakage current ΔI normalized to the volume of the sensor V_o as a function of the fluence Φ_{eq} for various silicon sensors produced by various process technologies. The current was measured after annealing the sensor for 80 min at 60 °C. All sensors show the same

6.5 Radiation Hardness of Silicon Pixel Sensors

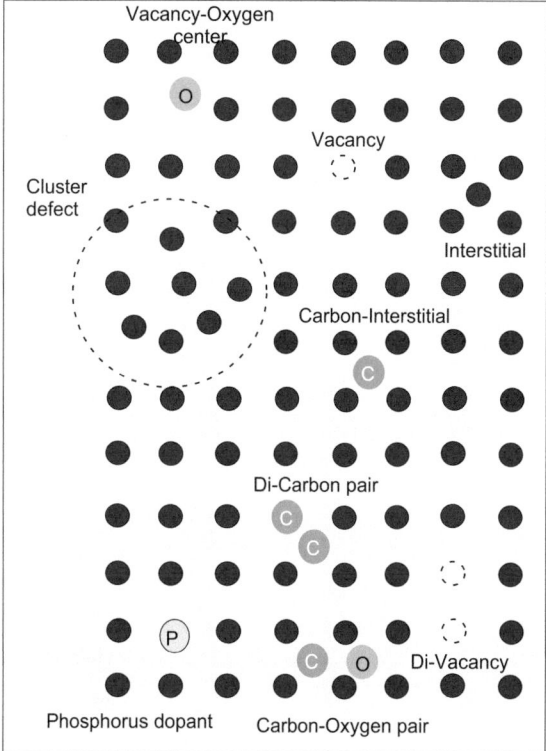

Fig. 6.59 Possible defects in the silicon crystal structure, including a vacancy, an interstitial silicon atom, a cluster defect, a vacancy-oxygen center, a di-vacancy, a carbon interstitial, a phosphorus dopant, a di-carbon pair and a carbon-oxygen pair

linear behavior in the double logarithmic plot,

$$\frac{\Delta I}{V_o} = \alpha_{fl} \Phi_{eq}. \tag{6.13}$$

The slope called the current-related damage rate is $\alpha_{fl} = (3.99 \pm 0.03) \times 10^{-17}$ A/cm. The fluence range was from $10^{11}/\text{cm}^2$ to $5 \times 10^{14}/\text{cm}^2$. It looks like that the different silicon sensors all show the same behavior with respect to radiation damage.

After irradiation, the sensors show annealing effects. Keeping a sensor at a fixed temperature as a function of time reduces the current-related damage rate. Figure 6.61 (right) shows the time dependence of α_{fl} in a semi-logarithmic plot for times up to 10^6 min. The distributions are fitted with the function

$$\alpha_{fl}(t) = \alpha_0 + \alpha_1 \cdot \exp\left(\frac{t}{t_1}\right) + \beta_1 \cdot \log\left(\frac{t}{t_0}\right), \tag{6.14}$$

where $\alpha_0, \alpha_1, \beta_1$ and t_1 are the fit parameters and t_0 is fixed to 1 min. The fitted parameters α_1 and β_1 are rather similar for the different temperature data. For $T_1 = 21\,°\text{C}$,

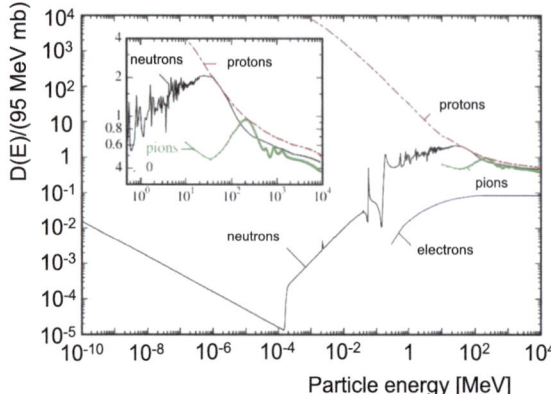

Fig. 6.60 Displacement damage functions $D(E)$ (black curve) normalized to 95 MeV mb for neutrons in a double-logarithmic scale using data from [102–106]. In addition to the neutrons, $D(E)$ is also shown for protons (red dash-dotted curve), pions (green curve) and electrons (blue solid curve). The inset shows an expanded view in the energy range of $1-10^4$ MeV. Reprinted with kind permission from [107], © 2003, Elsevier. All rights reserved

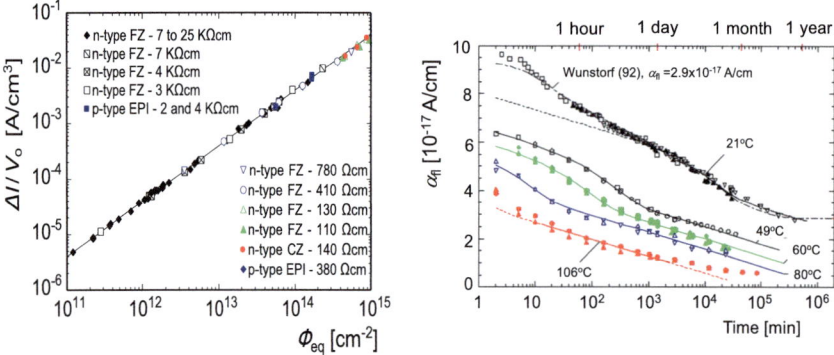

Fig. 6.61 Left: Leakage current as a function of fluence for silicon detectors produced by various process technologies (FZ, Cz, EPI) from different silicon materials (n-type or p-type) with different resistivity. The measurements were conducted at a temperature of 21 °C. Right: The current-related damage rate α_fl as a function of cumulated annealing time at different temperatures (21° to 106°) for different samples. At each temperature point at least one type-inverted and one not-type-inverted sample was used. The solid curves represent fits using (6.14). The dash-dotted curve shows a simulation based on a sum of exponential functions. Both reprinted with kind permission from [3], © 1999, M. Moll. All rights reserved

6.5 Radiation Hardness of Silicon Pixel Sensors

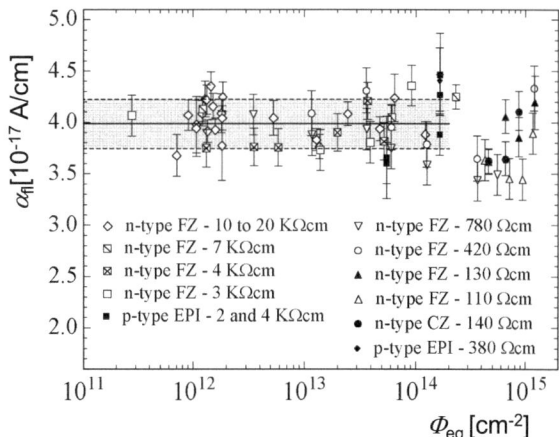

Fig. 6.62 The damage rate α_{fl} measured after 80 min at $60\,^\circ\text{C}$ as a function of the fluence for different samples. The shaded region corresponds to the average value $\alpha_{\text{fl}} = (3.99 \pm 0.24) \times 10^{-17}$ A/cm. Samples without guard rings are shown by filled symbols. Reprinted with kind permission from [3], © 1999, M. Moll. All rights reserved

we get $\alpha_1(T_1) = 1.23 \times 10^{-17}$ A/cm and $\beta_1(T_1) = 3.29 \times 10^{-18}$ A/cm while for $T_2 = 106\,^\circ\text{C}$ the parameters decrease slightly to $\alpha_1(T_2) = 1.13 \times 10^{-17}$ A/cm and $\beta_1(T_2) = 2.97 \times 10^{-18}$ A/cm. However, parameter α_0 drops by a factor of $\simeq 2$ from 7.07×10^{-17} A/cm at T_1 to 3.38×10^{-17} A/cm at T_2. The time constant t_1 is very temperature dependent. At T_1, we get $t_1 = 1.4 \times 10^4$ min decreasing to 9 min at $T = 80\,^\circ\text{C}$. We note that at larger times the annealing rate becomes the same, independent of temperature and after several months at room temperature a saturation value is reached [3].

It looks like that consistent values of α_{fl} are measured at the same fluence for different production processes. Measurements show that α_{fl} is independent of the hadron radiation source and of the silicon sensor, depending on temperature and annealing time. At fixed temperature and fixed annealing time, we measure the same value of α_{fl}, independent of fluence as shown in Fig. 6.62. At $T = 60\,^\circ\text{C}$ and after 80 min annealing, we measure $\alpha_{\text{fl}} = (2.99 \pm 0.24) \times 10^{-17}$ A/cm. The damage rate is basically independent of the production process and material type. Thus, α_{fl} is a very good damage parameter for measuring the hardness factor of high-energy particles and radiation fields. The increased leakage currents after hadron irradiation are mainly due to the formation of cluster defects. Irradiation with gammas from a ^{60}Co source produces point defects in the bulk and damages on the surface. So, we do not observe the same scaling behavior. For example, we see no leakage current annealing at room temperature.

Defects create space charge, thus changing the effective doping concentration and in turn the electric field and depletion voltage V_{dep}. Figure 6.63 (left) shows the depletion voltage as a function of the fluence. With increasing fluence, the depletion voltage drops at $\Phi = 2 \times 10^{12}/\text{cm}^2$, rising again at higher fluences. This represents the type inversion. The effective doping concentration n_{eff} changes under irradiation due to donor removal and generation of acceptor-like states. In contrast to what we found for the time dependence of α_{fl}, the change in the effective doping concentration ΔN_{eff} due to radiation shows a complex time behavior as illustrated in Fig. 6.63 (right). With respect to its value before irradiation, ΔN_{eff} can be expressed as a sum of three contributions

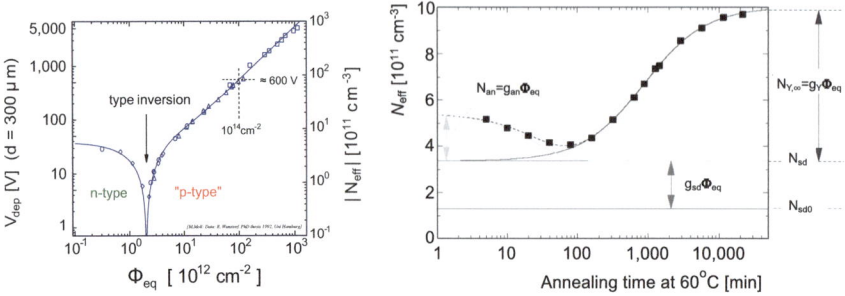

Fig. 6.63 Left: Depletion voltage as a function of the fluence. For a supply voltage of 600 V, a 300 µm thick sensor can be fully depleted up to a fluence of $\Phi_{eq} = 10^{14}$ (1 MeV) n_{eq}/cm^2 [108]. Type inversion occurs here at a fluence of 2×10^{12} (1 MeV) n_{eq}/cm^2. Reprinted with kind permission from [108], © 1992, R. Wunstorf, All rights reserved. Right: Annealing behavior of radiation-induced change in the effective doping concentration $\Delta N_{\mathrm{eff}}(\Phi_{eq}, t(T_a))$ at 60 °C as a function of annealing time after an irradiation dose of 1.4×10^{13} (1 MeV) $n_{eq}\,cm^{-2}$. After 100 min annealing time the effective doping $\Delta N_{\mathrm{eff}}(\Phi_{eq}, t(T_a))$ increases again. Reprinted with kind permission from [3], © 1999, M. Moll. All rights reserved

$$\Delta N_{\mathrm{eff}}(\Phi_{eq}, t(T_a)) = N_{\mathrm{an}}(\Phi_{eq}, t(T_a)) + N_{\mathrm{sd}}(\Phi_{eq}) + N_{\mathrm{Y}}(\Phi_{eq}, t(T_a)), \quad (6.15)$$

where $N_{\mathrm{an}}(\Phi_{eq}, t(T_a))$ is the annealing component, $N_{\mathrm{sd}}(\Phi_{eq})$ is the stable damage contribution that consists of the incomplete donor removal and stable damage generation while $N_{\mathrm{Y}}(\Phi_{eq}, t(T_a))$ is the so-called reverse annealing contribution, which is zero at $t = 0$; at room temperature it saturates at a fluence of $N_{\mathrm{Y},\infty}$ for very large times with a time constant of 500 days. Note that time t depends on the annealing temperature T_a. For more details, see reference [3].

As indicated in Fig. 6.63 (right), all three components are proportional to Φ_{eq},[7]

$$N_{\mathrm{an}}(\Phi_{eq}, t(T_a)) = g_{\mathrm{an}} \Phi_{eq} \exp\left(-\frac{t(T_a)}{\tau_{\mathrm{an}}}\right) \text{ with } g_{\mathrm{an}} = (1.81 \pm 0.14) \times 10^{-2}\,cm^{-1},$$

$$N_{\mathrm{sd}}(\Phi_{eq}) = g_{\mathrm{sd}} \Phi_{eq} + N_{\mathrm{sd0}}\left(1 - \exp(-c_{\mathrm{dr}} \Phi_{eq})\right) \text{ with } g_{\mathrm{sd}} = (1.49 \pm 0.04) \times 10^{-2}\,cm^{-1},$$

$$N_{\mathrm{Y}}(\Phi_{eq}, t(T_a)) = g_{\mathrm{Y}} \Phi_{eq}\left(1 - \frac{1}{1 + t(T_a)/\tau_{\mathrm{Y}}}\right) \text{ with } g_{\mathrm{Y}} = (5.16 \pm 0.09) \times 10^{-2}\,cm^{-1}, \quad (6.16)$$

where τ_{an} is the longest decay time constant, c_{dr} is the initial dopant removal constant, $N_{\mathrm{sd0}}/N_{\mathrm{eff},0}$ is the fraction of initial dopant removal, which depends strongly on the oxygen concentration, and τ_{Y} is another time constant. The dopant removal constant c_{dr} is anti-correlated with the removable initial donor concentration N_{sd0}. Thus, the product is a constant. A fit to data with standard oxygen concentrations yields $N_{\mathrm{sd0}} \cdot c_{\mathrm{dr}} = (7.5 \pm 0.6) \times 10^{-2}/cm$. The weighted average of the product of the initial dopant removal constant and donor concentration before irradiation

---

[7] Note that we considered the second-order process for $N_{\mathrm{Y}}(\Phi_{eq}, t(T_a))$. The results for the first-order process are given in [3].

6.5 Radiation Hardness of Silicon Pixel Sensors

yields $N_{\text{eff},0} \cdot c_{\text{dr}} = (10.9 \pm 0.8) \times 10^{-2}$/cm. The annealing times are temperature dependent. For example at 60 °C annealing temperature, the annealing times yield $\tau_Y = (24.1 \pm 2.3)$ min and $\tau_Y = (1060 \pm 110)$ min. At room temperature they are considerably longer. The change in the effective doping concentration first decreases with annealing time before increasing again. While the component $N_{\text{sd}}(\Phi_{\text{eq}})$ is time independent, $N_{\text{an}}(\Phi_{\text{eq}}, t(T_a))$ drops with the $t(T_a)$ to zero around 100 min and $N_Y(\Phi_{\text{eq}}, t(T_a))$ increases with $t(T_a)$ saturating around 5000 min.

The initial doping concentrations of phosphorus or boron are typically of the order of $N_{eff} \simeq 10^{12}$/cm^3. They are changed during irradiation creating additional donors and acceptors. If more donors than acceptors are created, the positive space charge will increase and in turn the depletion voltage increases. On the other hand, if more acceptors than donors are created, negative space charge will build up cancelling the positive space charge from the initial donor concentration. At some fluence, the effective space charge will be zero. Increased irradiation will increase the negative space charge such that the material behaves like p-type silicon. Such a device is called type-inverted. The depletion voltage at 40 V, which is proportional to the effective doping concentration, first decreases continuously with increasing fluence to $\Phi_{\text{eq}} = 2 \times 10^{12}$ (1 MeV) n_{eq}/cm^2 and then rises again due to an increase in the absolute effective space charge density N_{eff} (see Fig. 6.63 left). Note that the rise is linear at fluences well above the type inversion point. This is what happens in n-type silicon. A vacancy can combine with a phosphorus atom or an oxygen atom to form a vacancy-phosphorus (V-P) or vacancy-oxygen (V-O) complex. The V-P complex is electrically neutral and, therefore, does not contribute to the doping concentration any longer.

Let us look at the type-inversion in more detail. Figure 6.64 shows the depletion layer formation of n-type silicon before irradiation (left), effective doping concentration N_{eff} as a function of the fluence for standard and oxygen-rich silicon irradiated with protons, pions and neutrons (center) and the depletion layer formation of n-type silicon after irradiation (right). For low levels of radiation, the depletion zone is formed from the bottom (see Fig. 6.64 (left)). For low fluence, the doping concentration drops, becoming zero around $N_{\text{eff}}(\Phi_{\text{eq}}, t(T_a)) = 3 \times 10^{13}$ (1 MeV) n_{eq}/cm^2. Here, type-inversion takes place. For increasing fluence, the doping concentration rises again. For pions and protons, the rise is larger in standard silicon than in oxygen-rich silicon. This is not the case for neutrons, which show a large rise for both materials. Generally, pions produce a slightly higher rise than protons. After type-inversion, the depletion zone is formed from the top (see Fig. 6.64 (right)). With sufficiently large reverse bias voltage, the type-inverted silicon still works. The effect of donor removal depends on the energy, particle type and resistivity of the n-bulk silicon. Note, that for p-type silicon the behavior is just vice versa. Nice detailed discussions on radiation damage in silicon detectors are given in the thesis of Moll [3].

In the SCT, for example, ATLAS uses p-in-n (p^+nn^+) silicon sensors. Before irradiation, the depletion layer starts at the bottom as shown in Fig. 6.64 (left). After radiation, the n-type silicon converts into p-type silicon. So, the structure now is p^+pn^+ in which the depletion layer grows from the top as depicted in Fig. 6.64 (right). The reverse bias voltage needs to be increased to obtain a fully depleted sensor. The device is cooled to limit the reverse annealing of the sensor.

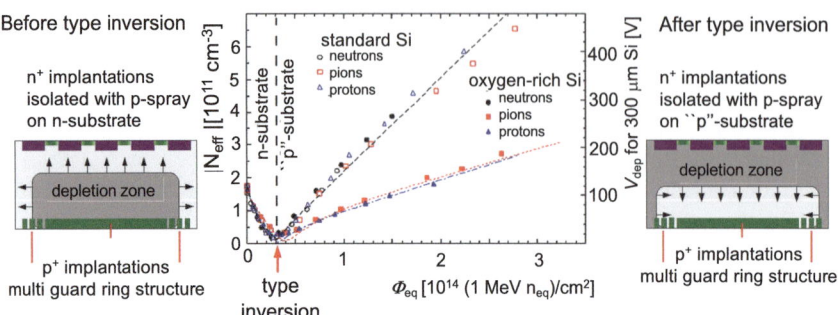

Fig. 6.64 Left: Depletion layer formation in n-type silicon before type inversion. The depletion zone grows from the bottom with increasing reverse bias voltage. Center: The doping concentration as a function of fluence in standard silicon (open symbols) and oxygen-rich silicon (filled symbols) for protons (blue triangles), neutrons (black points) and pions (red squares). The expected dependence is shown for protons (blue dash-dotted line), pions (red dotted line) and neutrons (black dashed line). Right: Depletion layer formation in "p"-type silicon after type inversion. The depletion zone now grows from the top with increasing reverse bias voltage. All reprinted with kind permission from [83], © 2007, A. Dobos. All rights reserved

Electrons and holes produced by an ionizing particle may be captured by deep defects before they reach the detecting electrodes. The effective charge collection efficiency $Q^{e,h}(t)/Q_0^{e,h}(t)$ deteriorates due to trapping

$$\frac{Q^{e,h}(t)}{Q_0^{e,h}} = \exp\left(-\frac{t}{\tau_{\text{eff}}^{e,h}}\right), \qquad (6.17)$$

where $\tau_{\text{eff}}^{e,h}$ is the effective trapping time for electrons and holes that is inversely proportional to the concentration of defect N_i capturing electrons or holes,

$$\frac{1}{\tau_{\text{eff}}^{e,h}} = \sum_i N_i (1 - P_i^{e,h}) \sigma_i^{e,h} v_d^{e,h}, \qquad (6.18)$$

where $P_i^{e,h}$ is the probability that the defect is already occupied, $\sigma_i^{e,h}$ is the capture cross section of the defect for electrons or holes and $v_d^{e,h}$ is the drift velocity for electrons or holes. The trapping time, which is related to the mean free path of the electron or hole in the irradiated sensor, depends on the concentration of the defects and thus on the radiation dose. Figure 6.65 (left) shows $1/\tau_{\text{eff}}^{e,h}$ as a function of the radiation dose.

The effective trapping time is slightly smaller for electrons than for holes. Both decrease with Φ_{eq}^{-1}. So, for short trapping times the mean free path is small and the probability that the electron or hole is trapped and thus cannot be detected increases. For a fluence of $2 \times 10^{14}/\text{cm}^2$, the trapping time is $\tau_{\text{eff}}^e = 10$ ns decreasing to 2 ns at a fluence of $10^{15}/\text{cm}^2$.

For 300 µm thick non-irradiated p-in-n sensors, a signal typically produces 20,000–25,000 electrons. This changes for radiation doses above

6.5 Radiation Hardness of Silicon Pixel Sensors

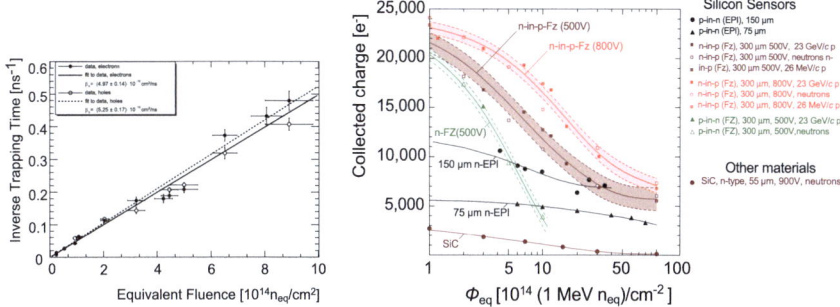

Fig. 6.65 Left: The inverse trapping time for electrons (solid points with error bars) and holes (open points with error bars) as a function of the fluence. Superimposed are fits for electrons (dashed curve) and holes (solid curve) [109]. Reprinted with kind permission from [109], © 2004, O. Krasel, All rights reserved. Right: The measured signal charge as a function of fluence for different FZ sensors, *p-in-n* (green triangles with green band), *n-in-p* at $V_b = 500$ V (brown squares with brown band), *n-in-p* at $V_b = 800$ V (red squares with red band) as well as epitaxial *p-in-n* sensors (black points and black triangles) and a SiC sensor (lower brown points). Reprinted with kind permission from [110], © 2009, M. Moll. All rights reserved

$\Phi_{eq} = 10^{14}$ (1 MeV)n_{eq}/cm^2 as Fig. 6.65 (right) shows. The charge loss occurs more rapidly with increasing fluence for 300 μm thick *p-in-n* sensors than for a 300 μm thick *n-in-p* sensors. For higher operation voltage, the decrease is reduced as the 800 V data show. For the 75 μm and 150 μm thick epitaxial *p-in-n* sensors, the signal charge for non-irradiated sensors is reduced by factors of four and two, respectively. However, the relative charge loss with increasing fluence is smaller so that at $\Phi_{eq} \simeq 10^{15}$/cm^2 ~ 5,000 and ~ 8000 electrons are observed, respectively, which is higher than that for the 300 μm thick *p-in-n* sensor. At a fluence of $\Phi_{eq} \simeq 2 \times 10^{15}$/cm^2, the signal yield of the 150 μm thick epitaxial sensor becomes comparable to that of the 300 μm thick *n-in-p* sensor at 500 V. For the LHC phase-2 upgrade, *n-in-p* sensors will be used that undergo a type change and have a higher signal charge production at high irradiation levels.

Besides bulk damage, we also encounter surface damage, which is typically caused by photon irradiation. The surface of a silicon sensor consists of silicon dioxide. In addition, there may be a layer of silicon nitride. The damage typically occurs in the silicon dioxide or at the interface of the silicon dioxide and the silicon bulk. Figure 6.66 (left) illustrates the impact of the surface damage on a segmented p^+n sensor schematically. Photon radiation damages the silicon dioxide in the following way. For photon energies below 1 MeV, the typical interaction process is the photoelectric effect or Compton scattering. The secondary electrons produce *e-h* pairs. The average energy for *e-h* production is 18 eV. Some fraction of the *e-h* pairs annihilate via recombination. Some electrons and holes respectively may be captured by electron traps or hole traps in the silicon dioxide. These are called oxide-trapped charges, which can be positive or negative depending on the type of defect. Due to their higher mobility, the remaining electrons leave the SiO$_2$ quickly. The remaining holes will move via shallow traps in the SiO$_2$ via so-called polaron hopping. Some fraction of the holes may be captured by deep traps in the oxide bulk or near the Si-

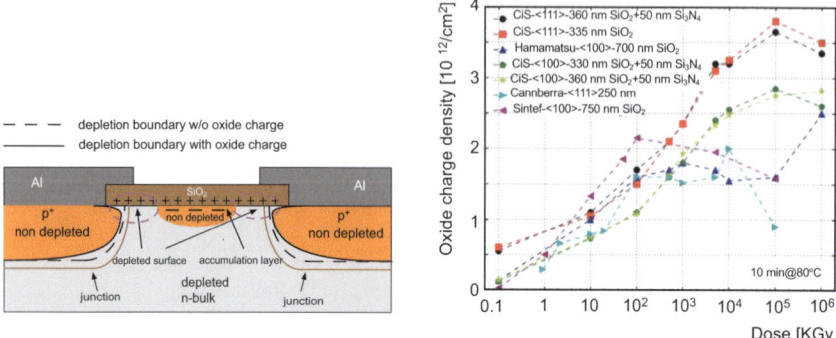

Fig. 6.66 Left: Effects of oxide charge in a p^+n sensor. Positive oxide charge moves the depletion boundary, increases the width of the accumulation layer and decreases the width of the depleted region at the Si-SiO$_2$ interface. Reprinted with kind permission from [112], © 2012, IOP Publishing. All rights reserved. Right: The fixed oxide-charge density N_{ox} as a function of the gamma radiation dose scaled to 20 °C after annealing at 80 °C for ten minutes. Samples are from CIS (solid points and squares), Hamamatsu (upward triangles), Canberra (right-pointing triangles) and Sintef (left-pointing). Reprinted with kind permission from [113], © 2013, Elsevier. All rights reserved

SiO$_2$ interface forming fixed positive charges. In addition, reactions between holes and hydrogen-containing defects or dopant complexes at the interface can lead to the formation of interface traps, which have energy levels distributed throughout the silicon band gap. The details depend on oxide thickness, electrical field, dose rate, crystal orientation and fabrication technology [111].

The positive oxide charge changes the electric field near the junction such that the depletion width is reduced and the depletion boundary shows a larger curvature with respect to the case without oxide charges. This produces a locally high electric field and a reduced breakdown voltage. Furthermore, an electron accumulation layer forms between the junctions below the oxide, which prevents a full depletion of the surface and which affects the inter-pixel capacitance and the inter-pixel resistance that in turn may cause charge losses. Figure 6.66 (right) shows the measured oxide charge concentration as a function of radiation dose for different sensors from different manufacturers. The measurements were performed after a 10 min annealing period at 80 °C. For most sensors, the oxide charge density increases with higher irradiation dose. At 20 °C, it takes about three years to anneal 50% of the oxide charges.

At the Si-SiO$_2$ interface, dangling bonds of the atoms create energy levels in the band gap near its center. The effects of these interface traps are two-fold. First, they can be positively or negatively charged depending on the trap type (donor or acceptor), the energy level and the Fermi energy; they contribute to the effective oxide charge. Second, due to surface recombination at the depleted surface, they are responsible for the surface leakage current

$$I_s = 0.5 q_e v_{\text{rec}} N_i A_s, \tag{6.19}$$

where q_e is the elementary charge, N_i is the carrier density, v_{rec} is the surface recombination velocity and A_s is the depleted surface. The recombination velocity is given

6.5 Radiation Hardness of Silicon Pixel Sensors

by the capture cross section σ_{cap}, the thermal velocity of the carriers $u_{e,h}$ and the interface trap density N_{it}

$$v_{rec} = \sigma_{cp} u_{e,h} N_{it}. \qquad (6.20)$$

For example, before irradiation the surface recombination velocity is 8 cm/s and the fixed oxide density is $1 \times 10^{11}/\text{cm}^2$. At a radiation dose of 1 MGy, the surface recombination velocity is 7.5×10^3 cm/s and the fixed oxide density is $2.1 \times 10^{12}/\text{cm}^2$, while at 100 MGy the surface recombination velocity increases to 1.1×10^4 cm/s and the fixed oxide density becomes $2.9 \times 10^{12}/\text{cm}^2$ [111].

Figure 6.67 shows the leakage current in the three ATLAS Pixel barrel layers as a function of luminosity in runs 1 and 2, which constitute an average over a representative sample of modules in azimuth angle and pseudo-rapidity. The leakage current has increased a little less than linearly with luminosity each year. This increase indicates radiation damage. We observe larger drops in the leakage current at run begin each year and smaller drops during the year, which are due to annealing effects of the pixel sensors. The B layer, which was closest to the beam before 2014, exhibits the largest leakage current, exceeding 4 mA/cm^3 at the end of 2018 at a luminosity of $\sim$190 fb^{-1}, while layer 2 just had 1 mA/cm^3. Predictions of the leakage currents with the Hamburg model [3] are consistent with measurements. Note that the bias voltage has been increased every year. Figure 6.68 shows leakage currents in the ATLAS IBL detector, which is closest to the beam, as a function of Run 2 luminosity. The leakage currents averaged over the azimuth angle with similar z values show a similar behavior as those of the other pixel barrel layers. We observe the annealing effect at run begin every year and during the year. The leakage currents increase a little less than linearly with luminosity. For sensors closer to the interaction region,

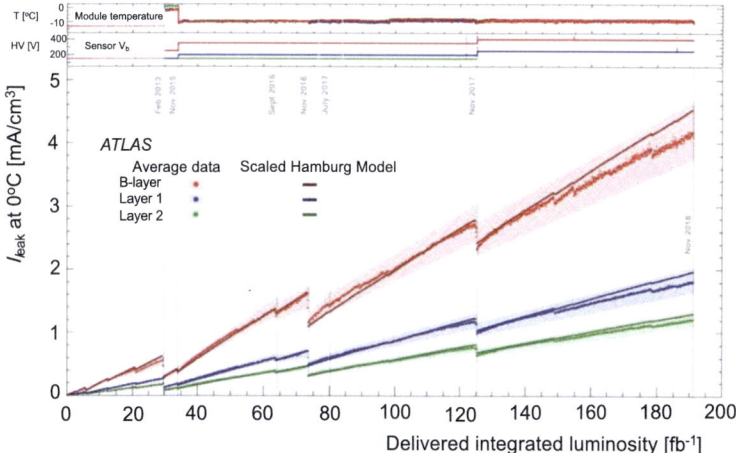

Fig. 6.67 The leakage current in the ATLAS Pixel barrel layers as a function of luminosity for layer 2 (green points), layer 1 (blue points) and the B layer (red points). In addition to the data, predictions of the Hamburg model are shown. The bands represent the errors. The steps in the distribution result from annealing effects when the LHC was not running. Reprinted under CC-BY-4.0 Licence from [114], © 2021, ATLAS Collaboration

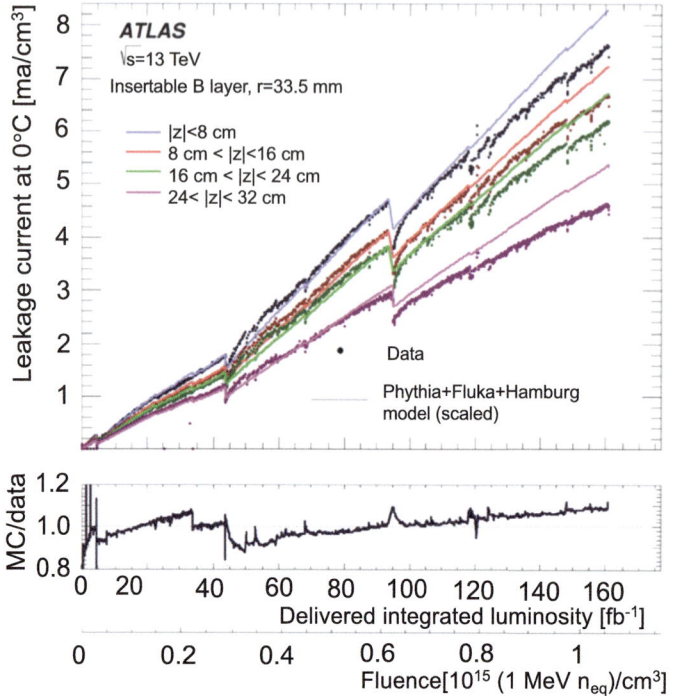

Fig. 6.68 Leakage currents in the ATLAS IBL detector as a function of luminosity at different z positions for planar and 3D sensors. With increasing z position, the leakage current decreases. The solid curves show simulations. Reprinted under CC-BY-4.0 Licence from [114], © 2021, ATLAS Collaboration

the leakage at the end of Run 2 approached 8 mA/cm^3. The CMS experiment sees a similar effect in their pixel detector. The 3D detectors see much lower leakage currents than the planar sensors, which approached $\sim$3.5 mA/cm^3 at the end of 2018. The high voltage of the planar sensors was changed every year from 80–150 V in 2016, to 300 V in 2017 and to 400 V in 2018. The high voltage of the 3D sensors was 20 V in the beginning and was raised to 40 V in 2017.

6.6 Vertex Detectors for the LHC Phase II Upgrade

Both the ATLAS and CMS experiments will replace their entire inner detectors for operation at the high-luminosity LHC. In the mid 2020s, the LHC machine will be upgraded to run at an instantaneous luminosity of 7.5×10^{34} cm^{-2} s^{-1}, yielding at least a total luminosity of $3,000$ fb^{-1} around 2040. Figure 6.69 (left) shows the new baseline layout of the inner detector in the ATLAS experiment. The inner tracking detector will be an all-silicon device consisting of a barrel with five pixel layers and four microstrip layers plus two endcaps, each with nine pixel disks and six microstrip disks. The coverage in pseudo-rapidity extends to $|\eta| < 4.0$. Figure 6.69

6.7 New Structures

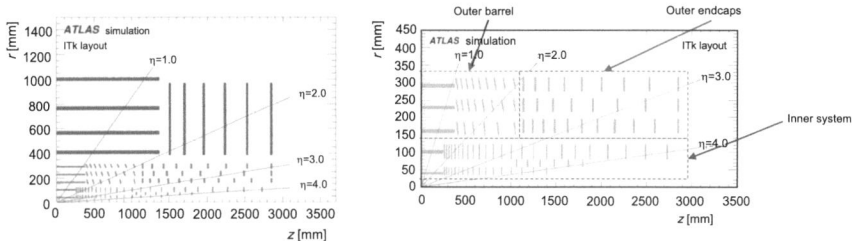

Fig. 6.69 Left: Schematic layout of the ATLAS inner tracker. Right: Enlarged view of the ATLAS pixel region. Reprinted with kind permission from [115], © 2017, ATLAS Collaboration. All rights reserved

(right) shows a close-up view in the pixel region. The pixel detector has five billion channels covering an area of 13 m^2 and the microstrip detector has 50 million channels covering an area of 160 m^2 of silicon. For the high luminosity, the expected number of interactions per bunch crossing will be around 200. This layout provides a minimum of 13 hits in the barrel and nine hits in the endcap in the forward direction. The design of the inner tracker (ITk) is based on the following requirements. The ITk has to be radiation hard up to 10 MGy and 10^{16} (1 MeV) n_{eq}/cm^2. The track efficiency should be better than 99% for muons and better than 85% for electrons and pions with misidentification rates smaller than 10^{-5}. The occupancy is smaller than 1% and the readout rate at most 4 MHz. The system should be robust against a 15% loss of channels. The readout bandwidth should be less than 5.2 Gb/s for each frontend chip. The material budget should be 1.5–2% · X_0 per layer.

To operate in the high-irradiation environments, n^+pp^+ sensors are the baseline for both the ATLAS and CMS upgrades. In the ATLAS experiment, layer 0 will be equipped with 3D sensors (see Sect. 6.7.5) having pixel sizes of 50 μm × 50 μm or 25 μm × 150 μm. In layer 1, pixels will be 100 μm thick and have a size of 50 μm × 50 μm or 25 μm × 150 μm. In layers 2–4, they will be 150 μm thick and have a size of 50 μm × 50 μm. For the microstrip detector all strips are 320 μm thick. In the barrel, the strip pitch is 75 μm. Short strips have a length of 24.1 mm while long strips are 48.3 mm. In the endcaps, the pitch varies from 69.9 to 80.7 μm and lengths vary between 15.1 to 60.2 mm. The sensors are rotated by 20 mrad.

6.7 New Structures

The pixel detectors we have discussed so far are hybrid designs in which pixels are bump-bonded to the readout chips. Even though hybrid pixel detectors have achieved very good performance in many experiments, they have some limitations. For example, the pixel size is limited by the bump bonding technology to 50 μm. The bump bond adds a non-negligible amount of inactive material and increases production costs. Furthermore, sensors can be tested only after bump bonding. Thus, new detector concepts have been developed to remove the bump bonding and integrate the readout into the silicon detector directly. This concept is realized in the silicon drift chamber, monolithic active pixel sensor, depleted p-channel field effect transistor, low-gain avalanche detector, silicon on insulator design, and 3D silicon detectors, which we discuss below.

6.7.1 The Silicon Drift Chamber

Figure 6.70 (left) shows a schematic layout of a silicon drift chamber that was proposed first by Gatti and Rehak [116, 117]. On an n-type silicon wafer, parallel p^+ strips are implanted on the top and bottom planes of the sensor. This produces two depleted regions, one on the top and one on the bottom of the wafer while the central region remains undepleted. On the top left-hand edge n^+ pads are implanted. By applying reverse bias voltage, the sensor becomes fully depleted. Figure 6.70 (right) shows the potential inside the detector. It has a parabolic shape with a minimum along the center. A traversing charged particle liberates electron-hole pairs. The electrons drift along the center to the n^+ pads. Each n^+ pad is read out. The electronics are placed on the detector. The position along the strips is obtained from the n^+ pads while the position orthogonal to the strips is obtained from the drift time. The resolution is $\sigma = 5\,\mu$m.

The CERES experiment at CERN was the first to use this detection concept [119]. Furthermore, silicon drift detectors have been used in other experiments, for example the STAR experiment at RHIC (Brookhaven) [120, 121] and the ALICE experiment at CERN [122].

The ALICE experiment used two layers of silicon drift detectors in their vertex detector for Run 1 and Run 2 [123]. They contained 260 modules, each of which had a sensitive area of 70.17×75.26 mm^2 that was divided into two drift regions by a central cathode kept at -1800 V. Thus, the electrons produced by a traversing charged particle drifted along the $r\phi$ direction away from the cathode. On both sides of the cathodes on the top and the bottom of the chamber 291 p^+ cathode strips

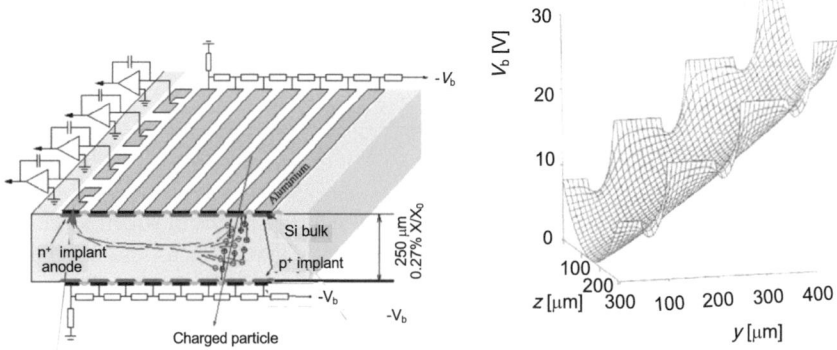

Fig. 6.70 Left: Three-dimensional schematic layout of a silicon drift chamber [116–118]. Right: Three-dimensional view of the potential inside the silicon drift chamber versus drift distance. Both reprinted with kind permission from [116], © 1984, Elsevier. All rights reserved

parallel to the cathode were implanted at a pitch of 120 μm fully depleting the detector volume. The electric field was ~500 V/cm. A second voltage of −40 V was used to keep the bias voltage in the collecting region independent of the drift voltage guiding the electrons towards the collecting anodes. At each end of the two drift regions, 256 anode pads with 294 μm pitch were implanted. The drift velocity was ~6.5 μm/ns. The achieved resolutions were $\sigma_{r\phi} = 35$ μm and $\sigma_{rz} = 25$ μm. For Run 3, the vertex detector was upgraded as discussed in the next subsection.

6.7.2 Monolithic Active Pixel Sensors

In a Monolithic Active Pixel Sensor (MAPS), the silicon sensor and the electronics are integrated on the same substrate. MAPS were originally developed for cost-efficient optical imaging and became successful in digital photography. Their main economical advantage is that the sensor and electronics can be produced on one cheap monolithic CMOS chip[8] [124], whereas in more traditional devices like CCDs, the sensor and the readout units are on two separate chips [125].

Figure 6.71 (left) shows a cross section through a MAPS device. The basic material is a CMOS image sensor. A 10–15 μm thick p-type epitaxial layer is implanted between a p^{++} substrate and a p^+-well layer including circuitry. Into the p-well layer, an n-well collection electrode is placed. Thus, the sensor only has a small depletion layer directly under the n-well electrode, where the charge collection occurs. The p-well regions have n-channel MOSFETs built-in consisting of two n^{++} pads sandwiching a polysilicon pad. Electron-hole pairs created in the substrate typically recombine. Electrons produced in an ionization process in the epitaxial layer move via diffusion until they reach the depletion layer under the n-well diode. Now they drift along the electric-field lines until they reach the n^+ pad.

Figure 6.71 (right) shows the doping concentration as a function of the wafer depth. The n^{++} diode has a concentration of $5 \times 10^{20}/\text{cm}^3$ dropping to $10^{17}/\text{cm}^3$ at the beginning of n^+-well region and to ~$5 \times 10^{13}/\text{cm}^3$ at the end of n^+-well region. In the p^+-well region, the concentration changes from $10^{17}/\text{cm}^3$ to $10^{15}/\text{cm}^3$, which is the concentration of the epitaxial layer. In the p^{++} substrate the concentration rises to $10^{19}/\text{cm}^3$ [125]. These cheap, thin and highly granular sensors have low noise of $< 10\ rms\ e^-$ and a signal charge of a few 100 e^-. A single-point resolution of 1–2 μm and a detection efficiency of 100% was measured in a 4 cm² large sensor carrying 10^6 pixels [126, 127]. The advantage of MAPS with respect to other silicon pixel detector technologies is the excellent spatial resolution, low material budget, reliable particle detection and low price. The initially modest tolerance to radiation was dramatically improved during the last two decades.

---

[8] Complementary metal-oxide-semiconductor or CMOS is a type of metaloxide-semiconductor field-effect transistor (MOSFET) fabrication process that uses complementary and symmetrical pairs of p-type and n-type MOSFETs for logic functions.

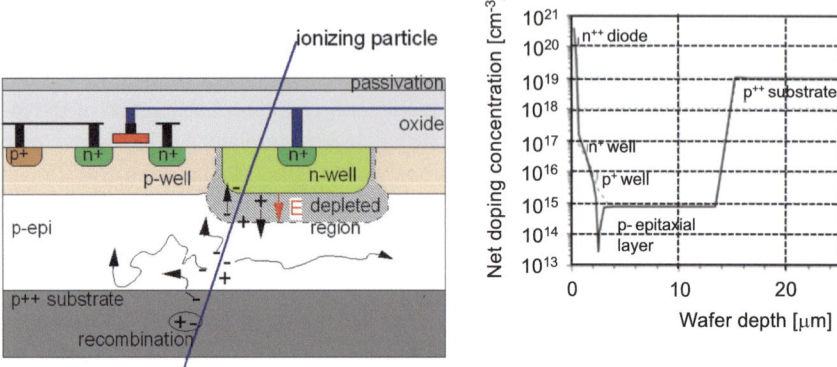

Fig. 6.71 Left: Cross section of a MAPS device. A 10–15 μm thick p-type epitaxial layer is placed between a p^{++} substrate and a p well. An n^+-well collection electrode is placed into the p-well layer. In the n-well region an n^+ pad is placed that is connected to the readout. The p well region has n-channel MOSFETs built-in consisting of two n^+ pads sandwiching a polysilicon pad. The depletion layer is localized just under the n well. An ionizing particle produces e-h pairs in the epitaxial layer. The electrons move via diffusion (zigzag paths) through the epitaxial layer until they reach the depletion region. Reprinted with kind permission from [128], © 2019, D. Bortoletto. All rights reserved. Right: The doping concentration across the sensor depth. Reprinted with kind permission from [129], © 2019, D. Contarato. All rights reserved

As discussed in Sect. 6.5, bulk defects are caused by irradiation of hadrons and surface effects introduced into the SiO_2 that result from irradiation with photons. The tolerance of the sensors to non-ionizing radiation is limited mostly by two radiation effects: first, the radiation-induced reduction of the lifetime for trapping signal electrons before they are collected and second the radiation-induced increase of leakage currents. The first effect dominates in sensors that have a small depletion layer. If the active layer is fully depleted, the charge collection efficiency remains high up to fluences of 10^{15} (1 MeV) n_{eq}/cm^2. The leakage current, however, increases with the depleted volume to high values. This in turn produces a shot noise that leads to an accelerated clearing of the pixel signal in the pre-amplifier. Appropriate cooling may solve both issues.

To limit effects from surface damage, one introduces a p^+ guard ring around the n-well diode that is grounded. This creates radiation-hardened MAPS in which the leakage current after irradiation with 20 krad γs from a ^{60}Co source is nearly the same as that before irradiation, in particular at temperatures near 0 °C [125]. The charge collection efficiency is improved by reducing the pixel size.

Figure 6.72 (left) shows the noise of a standard and a radiation-hardened MAPS device as a function of integration time for non-irradiated samples and samples after 1 Mrad of γ irradiation from a ^{60}Co source at a temperature of 10 °C. For the non-irradiated samples, the standard sensor has an $\overline{ENC}$ of ten rms electrons for a short integration time of < 1 ms while the radiation-hardened sample has an $\overline{ENC}$ of about 12 rms electrons. For 80 ms integration time, the $\overline{ENC}$ increases to 20 and 25 rms electrons, respectively. After 1 Mrad irradiation with γ's from a ^{60}Co source, the radiation-hardened device looks much better. At an integration time

6.7 New Structures

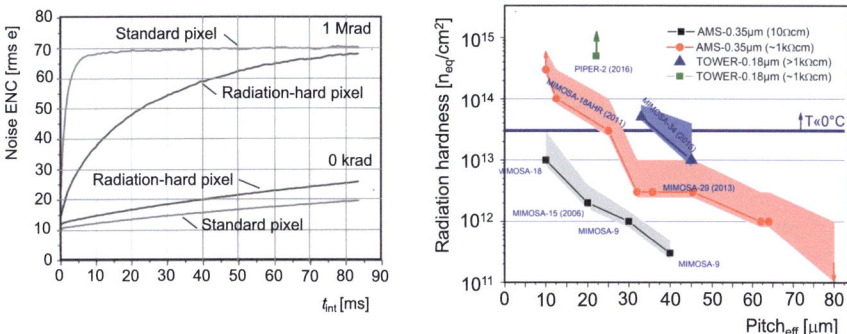

Fig. 6.72 Left: Noise of a standard and radiation-hardened MAPS device as a function of integration time for non-irradiated samples and samples irradiated with 1 Mrad of γs from a ^{60}Co source at a temperature of 10 °C. Right: Radiation tolerance of MAPS from the MIMOSA series for different resistivities of the epitaxial layer and 0.35 and 0.18 μm CMOS pixel pitches as a function of the sensor pixel pitch. Sensors were cooled to temperatures below 0 °C to control leakage currents. Both reprinted with kind permission from [125], © 2019, IOP Publishing. All rights reserved

of < 1 ms, the standard sample has $\overline{ENC} \simeq 30\ rms$ electrons while the radiation-hardened sample has $\overline{ENC} \simeq 16\ rms$ electrons. With longer integration times the noise rises much faster in the standard sample saturating at $\overline{ENC} \simeq 70\ rms$ electrons at 20 ms integration time whereas the radiation-hardened sample has a noise of $\overline{ENC} \simeq 50\ rms$ electrons. It increases to 70 rms electrons for integration times above 80 ms. Using a short integration time, the radiation-hardened sample has a sufficiently low noise to work well. Figure 6.72 (right) shows the radiation tolerance of MAPS from the MIMOSA[9] series for different resistivities of the epitaxial layer and two CMOS pixel pitches as a function of the sensor pixel pitch. The radiation tolerance is higher for small-pitch, high-resistivity sensors. For example, a high-resistivity MAPS sensor with a pixel pitch of 10 μm and a 0.35 μm CMOS pixel pitch, can tolerate a fluence of $\Phi_{eq} \simeq 3 \times 10^{14}$ (1 MeV) n_{eq}/cm^2. A similar performance is reached with a high-resistivity MAPS sensor having a 20 μm pixel pitch and a 0.18 μm CMOS pixel pitch. The performance of different MIMOSA MAPS is presented in [125].

The STAR experiment at Brookhaven has successfully used MAPS (MIMOSA28) fabricated with the 0.35 μm twin-well technology,[10] which were placed in two cylindrical layers at 2.8 and 8.0 cm. The pixel size was 20.7 μm × 20.7 μm. The sensors had a thickness of 50 μm and a rate capability of 10^6 Hz/cm^2. The single-hit resolution was 3.7 μm. The active area had 0.9 million pixels. The material budget was $0.39\% \cdot X_0$. They tolerated a radiation dose of 10^{13} (1 MeV) n_{eq}/cm^2. The transverse and longitudinal resolutions for pions were measured to be about 30 μm at a transverse momentum of 1 GeV/c dropping to 20 μm at $p_T = 2$ GeV/c [130]. Figure 6.73 (left) illustrates the twin-well technology.

---

[9] The MIMOSA sensors were produced by the IPHC Strasbourg.

[10] These sensors have a p-well and an n-well transistor.

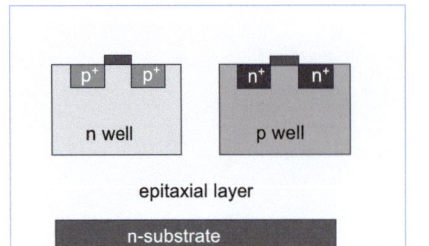
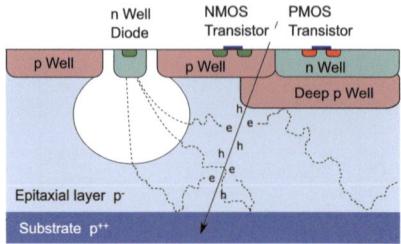

Fig. 6.73 Left: Illustration of the twin-well technology. An n-well and a p-well transistor is implemented into the epitaxial layer. Right: Schematic view of the well-structure used for ALPIDE sensors. The electrons produced in the epitaxial layer move via diffusion through the epitaxial layer until they reach the depletion region, when they are accelerated to the n-well diode. Both reprinted with kind permission from [131], © 2014, ALICE Collaboration. All rights reserved

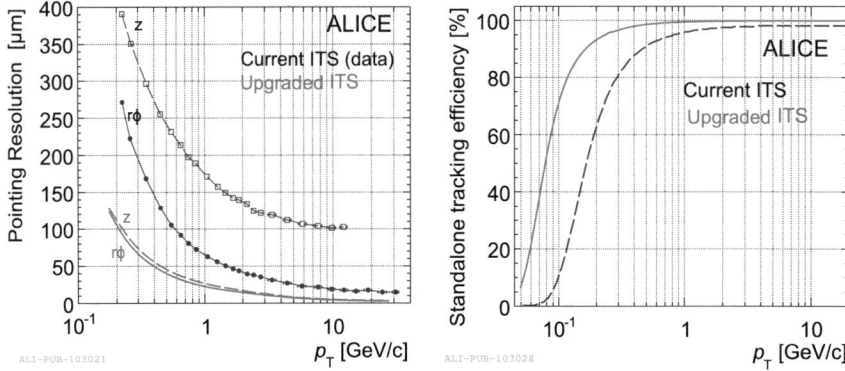

Fig. 6.74 Left: The expected pointing resolution of the new ALICE inner tracker (lower solid and dashed curves) with respect to the measured pointing resolution of the present inner tracker (upper open and solid points) as a function of p_T in the $r\phi$ and rz planes. Right: The expected tracking efficiency of the new ALICE inner tracker (solid curve) with respect to that of the present inner tracker (dashed curve) as a function of p_T. Both reprinted under CC-BY-NC-ND-4.0 Licence from [133], © 2020, G. Contin

The inner tracker in the ALICE experiment (ITS) has been upgraded with a seven-layer vertex tracker using the next generation MAPS with the $0.18\,\mu$m CMOS process technology called ALPIDE that was developed at CERN and is illustrated in Fig. 6.73 (right) [131]. The new vertex tracker is arranged in an inner barrel, a center barrel and an outer barrel covering an area of 10 m^2 with 12.5×10^9 pixels. The inner barrel comprises of three 27 cm long layers at radii of 2.3, 3.1 and 3.9 cm. Two 84 cm long center layers are placed at radii of 24 and 30 cm and two 148 cm long outer layers are located at radii of 42 and 48 cm. The epitaxial layer is made from high-resistivity p-type silicon (1kΩ-cm). The pixels have a size of 29 μm $\times$ 27 μm. In the inner layer, they are 50 μm thick increasing to 100 μm in the center and outer layers. The signal is integrated within 20 μs. The sensor is laid out to tolerate a dose of 10^{14} (1 MeV) n_{eq}/cm^2 and 0.7 Mrad of γ s. Figure 6.74 shows the expected track pointing resolution (left) and tracking efficiency (right) of the new ALICE inner tracker as a function of the transverse momentum. Both the track pointing resolution and the tracking efficiency are considerably improved compared to those of the present inner tracker.

Furthermore, the BES III experiment is considering a three-layer inner tracker based on the MAPS chip MIMOSA28 that was used in the STAR experiment [132]. In linear-collider experiments, MAPS are also considered for the inner layers of tracking detectors.

6.7.3 DEpleted P-Channel Field Effect Transistors

Another monolithic pixel detector is the DEpleted P-channel Field Effect Transistor (DEPFET) sensor first proposed by Kemmer and Lutz [134]. It integrates a MOSFET or junction field effect transistor (JFET) onto a high-ohmic lightly-doped n^- detector substrate. Figure 6.75 shows cross sections through a DEPFET pixel sensor in the x-y (left) and z-y (right) views. The MOSFET sensor consisting of two p^+ stops (source and drain) separated by an n^+ implant on a p channel (external Gate) sits on top of the n^- substrate that is about 75 μm thick. Directly underneath the external Gate at a depth of a few μm, another n stop is implanted called internal Gate. This produces a local potential minimum at the position of the internal Gate. The backside of the n^- bulk is covered by a p^+ layer. Since the front side has p^+ implants, we have a similar layout as in a silicon drift chamber. Thus, the depletion proceeds sideways. Figure 6.76 illustrates this. With low bias voltage, the depletion starts both at the bottom p^+ layer and at the top p^+ implant. With increasing bias voltage the two depletion zones move towards each other and then increase sideways. With

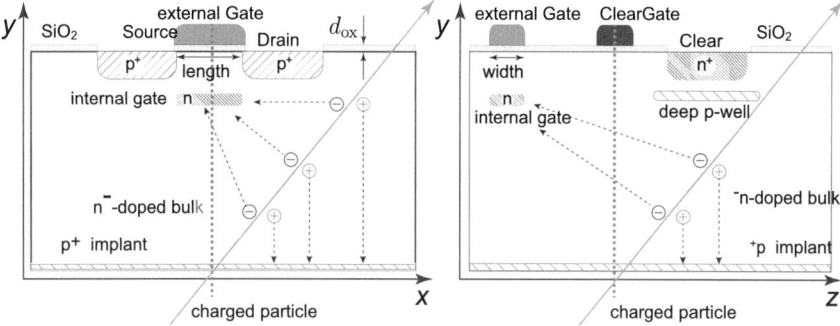

Fig. 6.75 Cross section of a DEPFET pixel sensor in the x-y-view (left) and the z-y view (right), which consists of a MOSFET transistor (source p^+ implant, drain p^+ implant, external n^+ Gate and a polysilicon ClearGate) that is integrated onto a sidewards-depleted lightly doped n^- silicon bulk. The bottom of the sensor is covered with a p^+ layer. Underneath the external Gate another n-doped implant is placed that serves as an internal Gate. A charged particle penetrating the sensor creates electron-hole pairs. While the holes drift towards the backside p^+ implants, the electrons accumulate in the internal Gate, where they modulate the source-drain current. An n^+-doped Clear implant is placed in each pixel to provide a pixel reset. Both reprinted with kind permission from [135], © 2017, C. Kiesling. All rights reserved

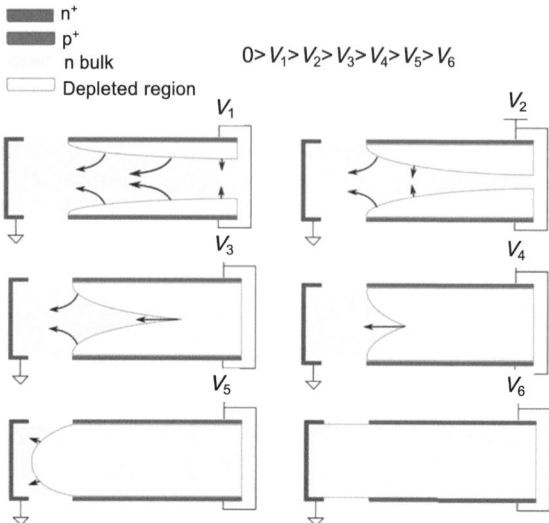

Fig. 6.76 Sideways depletion procedure in a DEPFET. Top left: After applying voltage V_1, the depletion zone forms from the top and from the bottom. Top right: Increasing the voltage to V_2, the depletion zone expands to the center. Center left: After a further increase to V_3, the depletion zone continues to grow towards the left-hand side. Center right: After a voltage increase to V_4, the depletion zone is complete under the p^+ region. Bottom left: At voltage V_5, the depletion zone extends into the n^+ region. Bottom right: Increasing the voltage to V_6, yields a fully depleted sensor. All reprinted with kind permission from [136], © 2013, N. Wermes. All rights reserved

sufficient bias voltage, the entire n^- bulk is depleted. Concerning the transistor, the drain is at ground, while the source is at a potential V_s. The amount of current I_d between the Source implant and the Drain implant is adjusted by the Gate voltage V_g. Applying a sufficient negative V_g relative to V_s, forms a conductive channel between the Source implant and the Drain implant below the external Gate, allowing holes to move between the Source and the Drain. Applying a sufficient positive voltage V_g relative to V_s on the other hand, switches off the p^+ channel MOSFET. So, here no current flows between the Source implant and the Drain implant. The transistor becomes disabled when V_g exceeds the threshold voltage V_{th}.

A charged particle traversing the sensor creates electron-hole pairs in the fully depleted n^- silicon substrate. While holes drift to the p^+ contact on the backside, the electrons accumulate at the internal Gate changing its potential. This leads to a modulation of the drain current I_d, where the amount depends on the number of accumulated electrons. The n^+ Clear implant is needed to remove the accumulated charge in the internal Gate, so the pixel becomes ready again to record the next event. This is achieved by forming a conductive path between the internal Gate and the Clear implant. Applying a positive voltage of $\simeq +15$ V with respect to the source voltage yields a punch-through at the internal Gate since the electrons move to the most positive potential. In addition, the p well underneath the Clear implant prevents electrons from moving back from the Clear region into the internal Gate. The ClearGate, which consists of a polysilicon structure, has the following function.

6.7 New Structures

In the charge collection mode, the ClearGate forms a potential barrier between the internal Gate and the Clear region. One possibility is to use capacitive coupling for the ClearGate. Applying positive voltage to the ClearGate, decreases the potential barrier between the internal Gate and the Clear region, permitting the electrons to move to the Clear region. The readout is non-destructive and can be repeated many times.

The accumulated charge Q_g in the internal gate produces

$$I_d = g_q \cdot Q_g, \tag{6.21}$$

where the internal amplification g_q is

$$g_q = \frac{\partial I_d}{\partial Q_g}. \tag{6.22}$$

The units are nA per unit charge. The charge Q_g also leads to a change in the Gate Source voltage V_g,

$$\partial V_g = f_{\text{gate}} \frac{\partial Q_g}{C_{\text{ox}}}, \tag{6.23}$$

where f_{gate} is a reduction factor that accounts for couplings of the internal Gate to other regions and C_{ox} is the oxide capacitance per unit area. The general relations of the electrical properties of the DEPFET are given in the textbook by Lutz [71]. Since the source-drain current is much more influenced by the Gate voltage than the bulk voltage, we consider the latter constant and write the Source-Drain current of a DEPFET in a simplified form [135],

$$I_{d,\text{sat}} = -\frac{1}{2} \frac{w_{\text{ch}}}{\ell_{\text{ch}}} C_{\text{ox}} \mu_h \left(f_{\text{gate}} \frac{Q_g}{w_{\text{ch}} \ell_{\text{ch}} C_{\text{ox}}} + V_g - V_{\text{th}} \right)^2, \tag{6.24}$$

where w_{ch} and ℓ_{ch} are width and length of the internal Gate and μ_h is the hole mobility. The charge amplification g_q and the transconductance $g_m = \partial I_{d,\text{sat}}/\partial V_{g,\text{eff}}$ are related[11]. Using (6.24) we get

$$g_m = -\frac{w_{\text{ch}}}{\ell_{\text{ch}}} C_{\text{ox}} \mu_h \left(f_{\text{gate}} \frac{Q_g}{w_{\text{ch}} \ell_{\text{ch}} C_{\text{ox}}} + V_g - V_{\text{th}} \right) = \sqrt{\frac{2 w_{\text{ch}} C_{\text{ox}} \mu_h}{\ell_{\text{ch}}}} \sqrt{-I_{d,\text{sat}}}, \tag{6.25}$$

and

$$g_q = -\frac{f_{\text{gate}} \mu_h}{\ell_{\text{ch}}^2} \left(f_{\text{gate}} \frac{Q_g}{w_{\text{ch}} \ell_{\text{ch}} C_{\text{ox}}} + V_g - V_{\text{th}} \right) = f_{\text{gate}} \sqrt{\frac{2 \mu_h}{w_{\text{ch}} \ell_{\text{ch}}^3 C_{\text{ox}}}} \sqrt{-I_{d,\text{sat}}}. \tag{6.26}$$

---

[11] $V_{g,\text{eff}}$ is a short for the voltage in the parenthesis in (6.24).

Thus,

$$g_q = g_m \frac{f_{\text{gate}}}{w_{\text{ch}} \ell_{\text{ch}} C_{\text{ox}}}. \tag{6.27}$$

The oxide capacitance C_{ox} per unit area is given by

$$C_{\text{ox}} = \frac{\epsilon_{\text{ox}} \epsilon_0}{d_{\text{ox}}}, \tag{6.28}$$

where d_{ox} and ϵ_{ox} are thickness and dielectric constant of the SiO_2 layer (see Fig. 6.75). The capacitance C_{ox} of DEPFET pixels is tiny. For example, for a typical pixel with dimensions $w_{\text{ch}} = 20$ μm, $\ell_{\text{ch}} = 5$ μm, $d_{\text{ox}} = 40$ nm and $\epsilon_{\text{ox}} = 3.85$, we get $C_{\text{ox}} w_{\text{ch}} \ell_{\text{ch}} = 0.085$ fF. Since I_d is variable, g_q is also variable while the other parameters are fixed by the manufacturer. Note that the drain current of a MOSFET transistor is independent of V_d. Thus, a change in the Gate voltage is necessary to influence the drain current during operation. Furthermore, the usage of the fully depleted bulk gives large signals for low noise. Typically, devices have high detection efficiencies and perform well with low power consumption. Since a DEPFET provides simultaneous detection and amplification of a signal, it is a very attractive device for low-noise applications in high-energy physics. Thin detectors of 50 μm may be produced that are operated with low power consumption. Since the capacitance of the internal Gate can be made small, the noise should be rather low.

The $\overline{ENC}$ of a DEPFET sensor is given by [137],

$$\overline{ENC} = \sqrt{\alpha_{\text{therm}} \frac{8}{3} k_B T \frac{g_m}{g_q^2} \frac{1}{\tau_{\text{sh}}} + q_e I_{\text{leak}} \Delta t_f + 2\pi a_f C_{\text{eff}}^2}, \tag{6.29}$$

where $k_B T$ is the thermal energy, α_{therm} is a constant of order unity depending on the exact shaping of the amplifier, τ_{sh} is the shaping time of the amplifier, I_{leak} is the DEPFET leakage current, Δt_f is readout time, a_f is a normalization parameter and C_{eff} is the effective capacitance. The first term represents the thermal noise, which decreases with the shaping time, the second term is the shot noise, which increases with the readout time, and the last term is the $1/f$ noise. At a bandwidth of 50 Mhz, the noise is about 40 rms electrons dominated by the thermal noise contribution. For example, for a 50 μm thick sensor a minimum ionizing particle produces about 4000 electrons yielding a signal-to-noise ratio of 40 : 1.

The Belle II experiment uses two concentric layers of DEPFET pixel sensors (PXD) placed at radii of 1.4 and 2.2 cm inside the four layer silicon vertex detector. The silicon in the two layers covers an area of 266.5 cm^2 containing 7.7 million pixels. Figure 6.77 (left) shows them in a three-dimensional layout and Fig. 6.77 (right) shows a transverse view. The inner layer is 9 cm long while the outer layer has a length of 12.3 cm. The pixel modules in both layers are tilted to provide an overlap in $r\phi$ for adjacent modules. The two layers are built from 40 half ladders, 16 half ladders in the inner layer and 24 half ladders in the outer layer. Two half

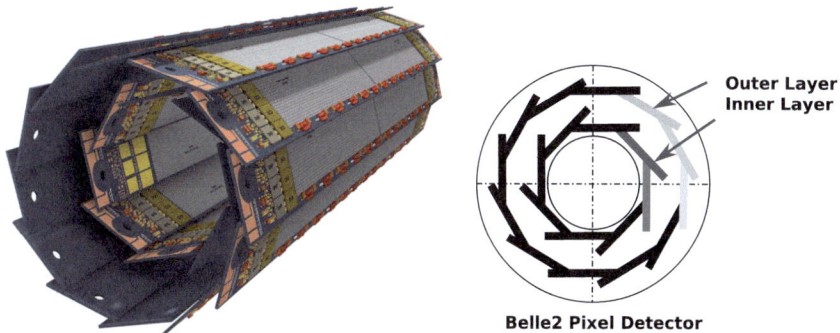

Fig. 6.77 Left: Three-dimensional layout of the Belle II pixel detector. Reprinted with kind permission from [135], © 2017, C. Kiesling. All rights reserved. Right: Transverse view of the Belle II pixel detector [138]. Reprinted under CC-BY-NC-ND-4.0 Licence from [138], © 2016, D. Levit

ladders are glued together to form a ladder. In August 2023 a complete improved PXD2 replaced the incomplete first PXD installed in 2018.

Figure 6.78 (left center) depicts a half ladder, which holds 768 × 250 pixels. In the inner (outer) layer, the pixels dimensions are 55 µm × 50 µm (70 µm × 50 µm) for the 256 inner pixels on each side of the interaction region. The pixel size increases to 60 µm × 50 µm (85 µm × 50 µm) for the remaining 512 outer pixels on each side. The analog and digital electronics is located at the ends of each half ladder away from the interaction region. The pixel sensors are thinned to 75 µm to reduce the inactive material. They are held by a 525 µm thick and 2 mm wide frame at one side that provides a self-supporting structure. As shown in Fig. 6.78 (center right) the frame is etched at two positions to reduce the inactive material but keep the necessary mechanical stiffness. The inactive material of the pixel detector amounts to $0.21 \cdot X_0$. Both, the sensors and ASICs are radiation hard. Figure 6.78 (left top/bottom) shows the electrical configuration for a switched-on and a switched-off pixel.

Figure 6.78 (right) shows a photograph of PXD2. We see the DEPFET matrix in the center, the End-Of-Stave (EOS) outside the acceptance located on each side of the active area, which hosts eight ASICs, namely the Drain Current Digitizers (DCDs) for signal digitization and Data Handling Processors, as well as a cooling block. Alongside the sensitive area are six Switcher ASICs that are responsible for steering the rolling shutter readout of the module within 20 µs integration time. A single PXD module produces a power of 9 W that has to be removed, where 7.2 W comes from the DCDs. The cooling is accomplished by a two-phase CO_2 cooling system.

The DEPFET signals are read out row-wise by four drain current digitizer ASICs in the EOS at the outer end of a half ladder. Rows of 250 pixels are multiplexed to one readout row. So, four drain current digitizer chips are sufficient to read out 1,000 channels. Thus, only 192 readout steps are necessary for reading out an entire matrix. In the default state the DEPFET pixels are switched off since the transistor consumes almost no power but the pixel is still sensitive to ionizing particles. Electrons generated by ionization in the depleted bulk accumulate in the internal Gate.

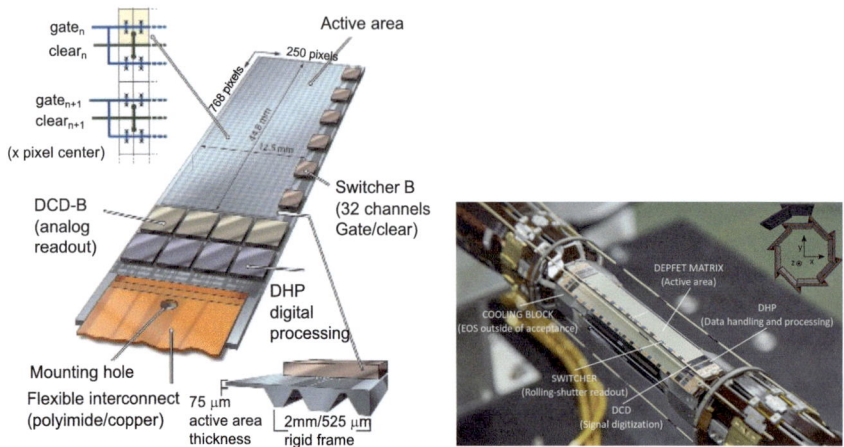

Fig. 6.78 Electrical configuration for a switched-on and a switched-off pixel (left top/bottom), a half-ladder of the Belle II DEPFET detector (left center) and cross section through a Belle II DEPFET sensor (left-right). All reprinted with kind permission from [139], © 2019, IOP Publishing. All rights reserved. Right: Photograph of the Belle II PXD2 detector. Reprinted under CC-BY-NC-ND-4.0 Licence from [140], © 2016, G. Giakoustidis

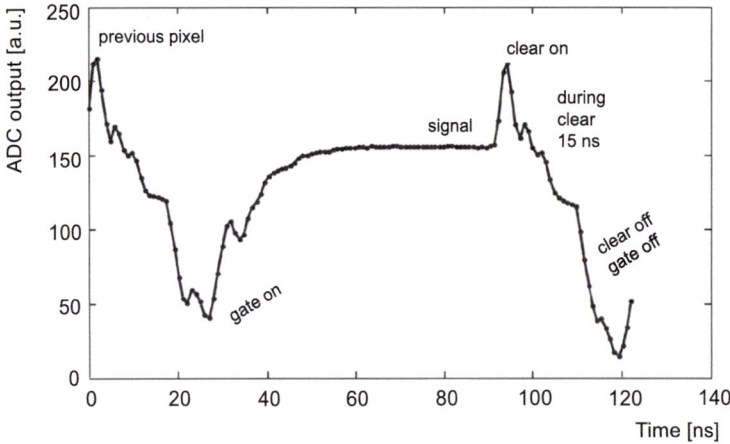

Fig. 6.79 The drain current during the DEPFET readout cycle (see text). Reprinted under CC-BY-NC-ND-4.0 Licence from [138], © 2016, D. Levit

A pixel is only switched on for the readout. Figure 6.79 shows the DEPFET readout cycle. It takes about 30 ns to clear the Gate from the previous pixel readout. Turning on the Gate on the next pixel, the signal current increases. After about 70 ns the Source-Drain current, composed of a pedestal current defined by the external Gate and a signal current, proportional to the charge in the internal Gate, is recorded and digitized. By applying a voltage to the Clear contact the charge in the internal Gate is removed. After 30 ns the Gate and the Clear are switched off and the system is ready to read out the next pixel. The current is measured again to obtain a pedestal

6.7 New Structures

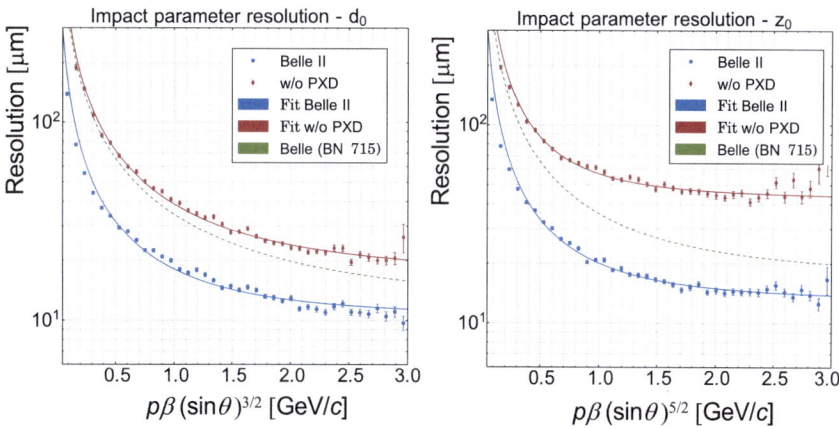

Fig. 6.80 Left: The expected transverse impact parameter resolution as a function of $p\beta \sin^{3/2}\theta$ for the Belle II detector without the pixel detector (red points) and with the pixel detector included (blue squares). The solid curves show the corresponding fits. The dashed line shows the result for the Belle detector. Right: The corresponding expected longitudinal impact parameter resolutions as a function of $p\beta \sin^{5/2}\theta$. Both reprinted with kind permission from [142], © 2015, C. Kiesling. All rights reserved

current measurement, which is subtracted from the previous current measurement to obtain the signal contribution. Since the digitization of one pixel takes about 100 ns, it takes 20 μs to read out the entire matrix. The next step involves the data handling process, which performs a pedestal subtraction, common-mode correction and zero-suppression.

The performance of the DEPFET modules was tested in a DESY test beam. The average efficiency is above 99%. The noise is rather low yielding signal-to-noise ratios of 25–57. The impact parameter resolutions were determined using the head-tail position finding algorithm [141]. Figure 6.80 (left) shows the expected Belle II transverse impact parameter resolution as a function of $p\beta \sin^{3/2}\theta$. A fit of the data points to the function

$$\sqrt{\left(\sigma_{d_0}^{M}\right)^2 + \left(\frac{\sigma_{d_0}^{MS}}{p\beta(\sin\theta)^{3/2}}\right)^2} \tag{6.30}$$

yields $\sigma_{d_0}^{M} = (10.3 \pm 0.1)$ μm for the measurement term and $\sigma_{d_0}^{MS} = (14.9 \pm 0.1)$ GeV/c μm for the multiple-scattering contribution. Figure 6.80 (right) shows the expected longitudinal impact parameter resolution as a function of $p\beta \sin^{5/2}\theta$. A fit of the data points to the function

$$\sqrt{\left(\sigma_{z_0}^{M}\right)^2 + \left(\frac{\sigma_{z_0}^{MS}}{p\beta(\sin\theta)^{5/2}}\right)^2} \tag{6.31}$$

yields $\sigma_{z_0}^M = (12.9 \pm 0.1)$ μm for the measurement term and $\sigma_{z_0}^{MS} = (15.4 \pm 0.1)$ GeV/c μm for the multiple-scattering contribution. These impact parameter resolutions are improved with respect to the Belle results. Without the DEPFETs the impact parameter resolutions are considerably worse.

6.7.4 Low-Gain Avalanche Detectors

Low-Gain Avalanche Detectors (LGAD) are thin solid-state devices with an intrinsic gain of typically 5 to 50. This new concept was pioneered by the RD50 detector collaboration [143] with the Centre National de Microelectronica in Barcelona [144] and the Bruno Kessler Institute in Trento (FBK) [145]. Figure 6.81 (left) shows a schematic layout of an LGAD. The sensor has a thickness of 20–60 μm that is fully depleted. Pad sizes are 50 μm × 50 μm to 5 mm × 5 mm. On a high-resistivity p substrate (10 kΩ-cm), one implants highly doped n^{++} electrodes on one side and a p^+ anode plane on the other side. Below the n^+ cathode, one inserts a moderately doped thin p^+ layer, which acts as a multiplication region of the primary charges. The shallow n-type electrode (~1 μm) has a doping concentration of 1×10^{19} cm^{-3}, whereas the p-type multiplication implant has a concentration of 1×10^{16} cm^{-3} and is somewhat deeper (~4 μm). At the n^+p junction a high electric field is produced, which is proportional to the square root of the p-type doping density and the square root of the reverse bias voltage. The voltage needed to deplete the gain layer, V_{dep}, is given by

$$V_{\text{dep}} = \frac{q_e \rho_g}{2\epsilon_0 \epsilon_r} w_g^2 \tag{6.32}$$

where ρ_g and w_g respectively are doping density and thickness of the gain layer. Since w_g is fixed, V_{dep} is proportional to the doping concentration in the gain layer. Besides boron doping also gallium doping has been tested.

Figure 6.81 (right) shows the time evolution of the signal pulse of an LGAD sensor that was illuminated from the backside with red light. The early part of the signal is determined by drifting electrons in the bulk. At about 7.5 ns, the electrons enter the n^+ region and start amplification. This lasts for about 11 ns after which the signal starts to decline. The nearly linear decline is caused by drifting holes that ends around 26 ns. The remaining rapidly declining tail is caused by diffusion and electronics. Figure 6.82 (left) shows the gain as a function of V_b for several different LGAD sensors. For $V_b \simeq 150\text{--}200$ V, the typical gain is $G = 10\text{--}15$. Figure 6.82 (right) shows the most probable charge as a function of V_b for a non-irradiated sensor and for irradiated sensors with doses $1 \times 10^{14} - 1 \times 10^{16}$ (1 MeV) $n_{\text{eq}}/\text{cm}^2$. With increasing dose, the collected charge declines rapidly. The non-irradiated sample produces at 0.5 kV more than 250,000 e^-, while the sample irradiated to $\Phi_{\text{eq}} = 2 \times 10^{15}$ (1 MeV) $n_{\text{eq}}/\text{cm}^2$ yields around 20,000 e^- at 1.5 kV. The samples have been annealed for 80 min at 60 °C. The measurements were performed at a temperature of -20 °C. Note that basically no charge is measured at $V_{\text{mr}} = 30$ V, which is the voltage needed to deplete the p^+ multiplication layer for non-irradiated W8 sensors.

6.7 New Structures

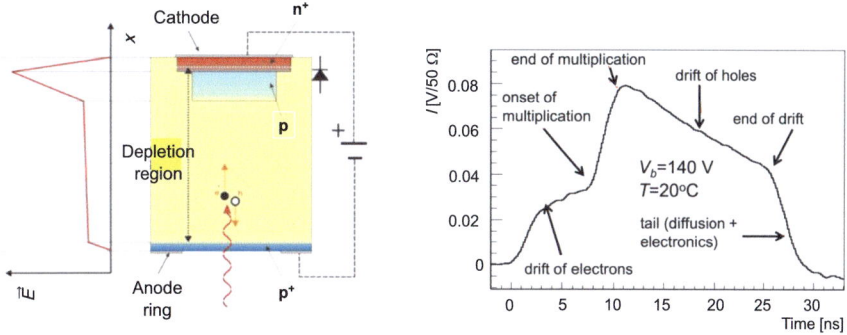

Fig. 6.81 Left: Schematic layout of an LGAD and the electric-field distribution across the sensor. Reprinted with kind permission from [146], © 2017, Elsevier. All rights reserved. Right: The measured current in an LGAD sensor as a function of time produced by illumination of the sensor's back side with short pulses of red light. The different stages of the signal evolution are labeled: (1) drift of electrons, (2) onset of electron multiplication, (3) end of electron multiplication, (4) drift of holes, (5) end of drift of holes, (6) tail from diffusion and electronics. Reprinted with kind permission from [147], © 2015, IOP Publishing. All rights reserved

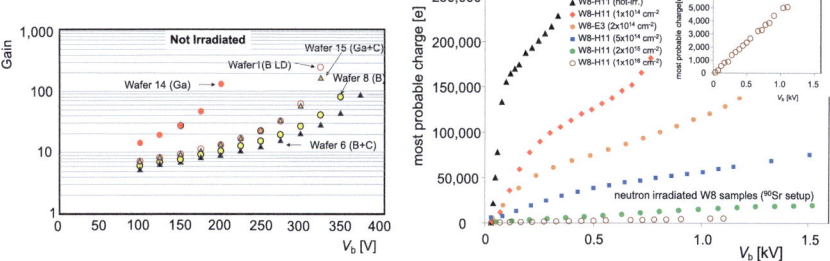

Fig. 6.82 Left: The gain of different LGADs as a function of V_b for different types of doping, boron (yellow points), gallium (red points), gallium plus carbon (orange triangles), boron low diffusion (brown circles) as well as boron plus carbon (black triangles). The LGADs were produced by FBK. The doping material is shown in parentheses. Reprinted with kind permission from [148], © 2019, Elsevier. All rights reserved. Right: The most probable charge of the minimum ionization peak of electrons from a ^{90}Sr source as a function of V_b for the unradiated LGAD sensor from FBK (black triangles) and after neutron irradiation with $\Phi_{eq} = 10^{14}$ (red diamonds), 2×10^{14} (orange points), 5×10^{14} (blue squares), 2×10^{15} (green points) and 10^{16} (1 MeV) n_{eq}/cm^2 (yellow circles). The inset shows an enlargement for $\Phi_{eq} = 1 \times 10^{16}$ (1 MeV) n_{eq}/cm^2 in the low bias voltage region. Reprinted with kind permission from [147], © 2015, IOP Publishing. All rights reserved

The electric field in the multiplication layer becomes sufficiently high to start charge multiplication. As seen in Fig. 6.81 (right), there is a steep rise in the expected charge for voltages above V_{mr}, because the bulk regions starts to deplete and, therefore, generates new electrons that become available for multiplication. Once the bulk is completely depleted, the charge rises moderately since additional voltage is equally distributed over the entire thickness. So, $V_b = 300$ V is needed to increase the electric field in the multiplication layer by 1 V/μm. The voltage V_{mr} decreases exponentially with fluence, dropping faster for proton irradiation than for neutron irradiation. For example for neutron irradiation with $\Phi_{eq} = 10^{15}$ (1 MeV) n_{eq}/cm^2, $V_{mr} \simeq 15$ V. At $V_{mr} = 30$ V, the leakage of the non-irradiated W8 sensor is less than 10 µA. It increases rapidly with V_b yielding about 25 µA at 300 V.

Figure 6.83 (left) shows the leakage current versus the reverse bias voltage for an LGAD sensor before irradiation and for neutron irradiation with fluences from 10^{14} to 10^{16} (1 MeV) n_{eq}/cm^2. The measurements were performed at a temperature of $-20\,^\circ$C. The leakage current increases rapidly with bias voltage. For a fluence of $\Phi_{eq} = 10^{16}$ (1 MeV $n_{eq})/cm^2$ and a bias voltage of 500 V the leakage current is 300 µA increasing to ~500 µA at 1,000 V. For a fluence of $\Phi_{eq} = 2 \times 10^{15}$ (1 MeV) n_{eq}/cm^2 and a bias voltage of 100 V the leakage current is 200 µA. The change in leakage current is given by

$$\Delta I_{leak} = M_I(\Phi_{eq})\alpha_{leak}\Phi_{eq}V_b, \qquad (6.33)$$

where $M_I(\Phi_{eq})$ is the current multiplication factor and α_{leak} is the leakage current damage constant, which is 2.9×10^{-17} A/cm for protons and 4.0×10^{-17} A/cm for neutrons. For a 50 µm thick sensor with a pad size of 1.0 mm × 1.0 mm, a gain of 10 and a fluence of 2×10^{15} (1 MeV) n_{eq}/cm^2 neutron irradiation, we calculate a $\Delta I = 40$ µA. For a 100 µm × 100 µm pixel size the leakage current is reduced to 400 nA. The temperature dependence of the leakage current is given by [150],

$$I_{leak}(T) \propto T^2 \exp\left[-\frac{1.21\ eV}{2k_B T}\right]. \qquad (6.34)$$

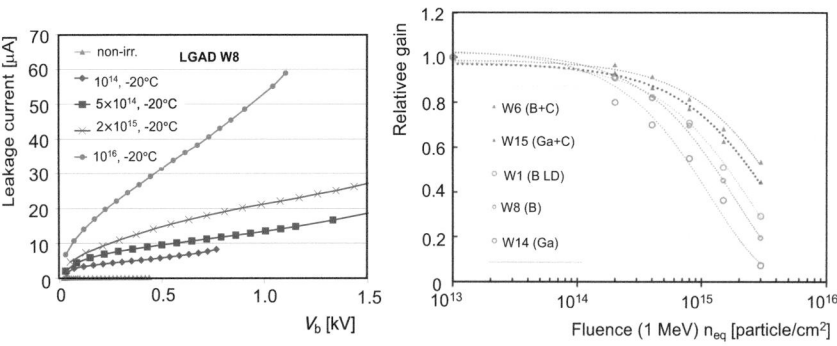

Fig. 6.83 Left: Leakage current versus V_b for an FBK LGAD W8 sensor before irradiation (triangles), irradiation with $\Phi_{eq} = 10^{14}$ (diamonds), 5×10^{14} (squares), 2×10^{15} (crosses) and 10^{16} (1 MeV) n_{eq}/cm^2 (points). Measurements were performed at a temperature of $-20\,^\circ$C. Reprinted with kind permission from [147], © 2015, IOP Publishing. All rights reserved. Right: The gain of FBK LGAD sensors with respect to the gain at $\Phi_{eq} = 10^{13}$ (1 MeV) n_{eq}/cm^2 as a function of fluence for different doping materials: boron and carbon (upper triangles and upper dotted curve), gallium and carbon (lower triangles and thick dotted curve), boron with low diffusion (upper large circles and associated curve), boron with high diffusion (small circles and associated curve) and gallium (lower circles and lower curve). The points are measurements and the curves are fits. Reprinted with kind permission from [149], © 2019, M. Moll. All rights reserved

6.7 New Structures

Reducing the temperature from 25 to 17 °C cuts the leakage current by a factor of two. If we measure the leakage current at room temperature T_r, we can estimate it at any temperature T,

$$I_{\text{leak}}(T) = I_{\text{leak}}(T_r)\frac{T^2}{T_r^2} \exp\left[-\frac{1.21 \text{ eV}(T_r - T)}{2k_B T T_r}\right]. \quad (6.35)$$

Figure 6.83 (right) shows the relative gain for LGAD sensors with different p-type dopings with respect to that at $\Phi_{\text{eq}} = 10^{13}$ (1 MeV) $n_{\text{eq}}/\text{cm}^2$ as a function of fluence for neutron irradiation. The gain reduction increases with fluence and is larger for neutron irradiation than for proton irradiation. The reason is an initial acceptor removal, which is worse for gallium-doped LGADs than for boron-doped LGADs. The addition of carbon reduces the effect both for boron and gallium doping. The radiation effects can be partly compensated by increasing the bias voltage.

Thin LGAD detectors have excellent time resolution and, therefore, are considered for timing detectors at the high-luminosity LHC upgrade, where a time resolution of less than 50 ps is necessary. Figure 6.84 (left) shows the measured time resolution as a function of the thickness of the depletion layer. For a 50 μm thick depletion layer a resolution of 20 ps has been achieved. Figure 6.84 (right) shows the measured time resolution as a function of the gain. All measurements were conducted at a temperature of −20 °C. The gain needs to be above 10 to achieve the best time resolution. Furthermore, the standard LGAD fill factor needs to be improved, which is defined as the ratio of the gain area over the total area. The traditional gain isolation design has a rather large inactive area and thus, a small fill factor. In particular for smaller pixel pitches below 100 μm, this becomes an issue. To achieve good time resolutions, the signal must be homogeneous. This implies sensors to have no large inactive areas and to have a homogeneous weighting field. However, a pixel-border termination is necessary to host all structures controlling the electric field but it needs to be reduced. Several new approaches have been proposed that are shown in Fig. 6.85. The first approach shown in Fig. 6.85 (left) consists of a reduction of the inactive area by trench isolation. The original inactive zone of 66 μm is reduced to 11 μm. The fill factor is increased to 0.6 for a pixel pitch of 50 μm. The second approach depicted in Fig. 6.85 (center) consists of an AC-coupled readout of the LGAD. An SiO_2 layer is inserted between the n^+ zone and the readout pad. The third approach shown in Fig. 6.85 (right) consists of an inverse LGAD setup in which the amplification is done on the back side and the front side is instrumented with several p^+ strips. So here, the n-in-p structure is changed to a p-in-p structure.

The time resolution has several contributions that are added in quadrature,

$$\sigma_t = \sigma_{\text{Jitter}} \oplus \sigma_{\text{Time-Walk}} \oplus \sigma_{\text{Landau-Noise}} \oplus \sigma_{\text{Distortion}} \oplus \sigma_{\text{TDC}}, \quad (6.36)$$

where σ_{jitter} is the contribution from the time jitter, i.e. variations in the time measurements, $\sigma_{\text{Time-Walk}}$ is the contribution caused by variations in the arrival time due to the size of the signal, $\sigma_{\text{Landau-Noise}}$ results from a non-uniform energy deposition

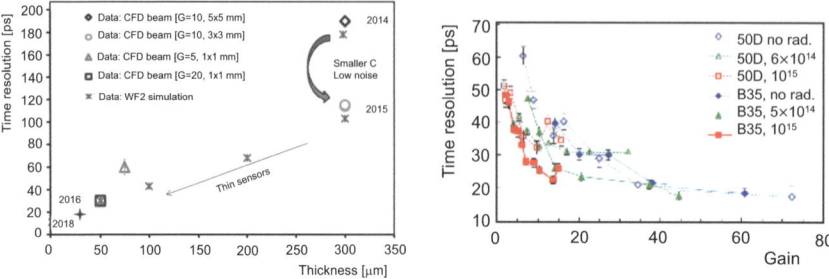

Fig. 6.84 Left: Time resolution as a function of the sensor thickness for LGAD sensors with pad sizes of 5 mm × 5 mm operated with gain 10 (diamond), 3 mm × 3 mm operated with gain 10 (point), 1 mm × 1 mm operated with gain 5 (triangle,) 1 mm × 1 mm operated with gain 20 (square) and 1 mm × 1 mm operated with gain 20 (star). Measurements were conducted in a test beam except for the 1 mm × 1 mm sensor operated with gain 5 for which the time resolution was measured with a β source. In addition, simulated time resolutions are shown (crosses). Right: Time resolution as a function of gain for a 50/35 μm thick sensor before irradiation (open/solid blue diamonds), after irradiation with $\Phi_{eq} = 6/5 \times 10^{14}$ (open/solid green triangles) and with $\Phi_{eq} = 10^{15}$ (1 MeV) n_{eq}/cm^2 (open/solid red squares). All measurements were performed at $-20\,°C$. Both reprinted with kind permission from [151], © 2018, H. Sadrozinski. All rights reserved

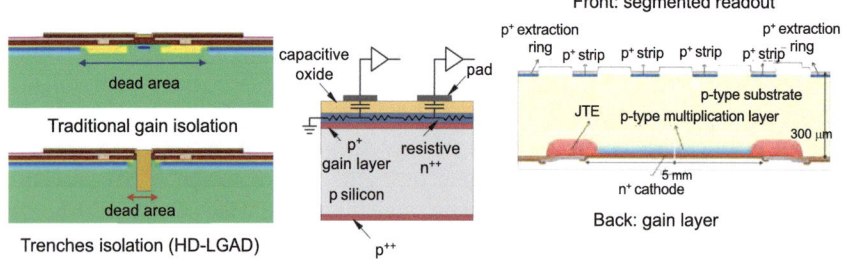

Fig. 6.85 Left: Reduction of the inactive area using a smaller trench isolation. Center: Use of AC-coupled readout [149]. Both reprinted with kind permission from [149], © 2019, M. Moll. All rights reserved. Right: Placing the gain layer on the back plane. Reprinted with kind permission from [152], © 2016, Elsevier. All rights reserved

of the energy loss of a traversing charged particles, $\sigma_{Distortion}$ accounts for the uncertainty caused by the non-uniform drift velocity and the weighting field and σ_{TDC} accounts for the TDC binning of the Time-to-Digital Converter (TDC).

The contribution σ_{jitter} is caused by the time the comparator fires, which is affected by electronic noise and typically is the dominant component. It can be parameterized in terms of the signal-to-noise ratio S/N and the signal rise time t_{rise},

$$\sigma_{jitter} = \frac{N}{dV/dt} \sim \frac{t_{rise}}{S/N}, \quad (6.37)$$

where dV/dt is pulse slope. To reduce σ_{jitter}, we need to maximize the pulse slope by recording fast and large signals. The signal size depends on the gain. Thus, the

Fig. 6.86 Time resolution of LGAD detectors from Hamamatsu (HPK) and FBK as a function of the radiation dose Φ_{eq} [151, 153–156]. The Hamamatsu sensors are 50 μm (diamonds) and 35 μm (squares and triangles) thick while the FBK sensor (open and solid points) has a thickness of 60 μm. Reprinted with kind permission from [151], © 2018, H. Sadrozinski. All rights reserved

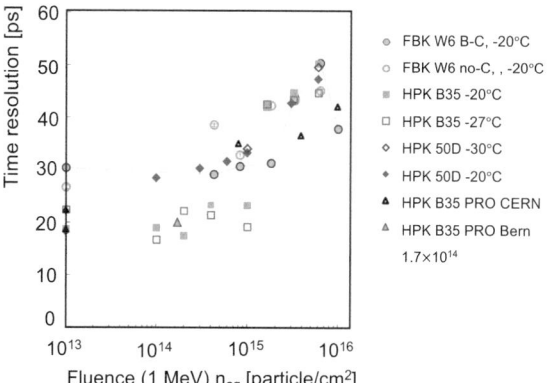

jitter contribution decreases inversely with the gain. The contribution $\sigma_{\text{Time-Walk}}$, which can be reduced by usage of a constant fraction discriminator is given by

$$\sigma_{\text{Time-Walk}} = \left[\frac{V_{\text{th}}}{S/t_{\text{rise}}}\right] \propto \left[\frac{N}{dV/dt}\right]_{\text{RMS}}, \quad (6.38)$$

where V_{th} is the constant fraction discriminator threshold or the ToT threshold. The error due to time walk can be reduced to 10 ps. The contribution $\sigma_{\text{Distortion}}$ accounts for the uncertainty caused by the non-uniform drift velocity and weighting field as stated by the Ramo-Shockley's theorem [157], stating that the current induced by a charge carrier is proportional to its electric charge, drift velocity and weighting field. We obtain a constant drift velocity of the carriers in the sensor volume by producing an electric field that is sufficiently high so that carriers move with saturated velocity. We may also reduce the weighting field difference caused by electrode geometry by using electrodes with sizes similar to the pitch and being much larger than the sensor thickness. The contribution σ_{TDC} accounts for the TDC binning that is typically less than 10 ps and is negligible. The best time resolutions with LGAD sensors are around 20 ps [158]. Figure 6.86 shows the time resolution as a function of radiation dose for the different sensors. The time resolution becomes worse with increased radiation dose. For a fluence of $\Phi_{eq} = 6 \times 10^{15}$ (1 MeV) n_{eq}/cm^2, the time resolution increases from around 20 ps for the non-irradiated sensor to about 35 ps for the irradiated sensor if cooled to $-30\,°\text{C}$. For a fixed temperature, the time resolution seems to increase logarithmically with fluence.

6.7.5 3D Silicon Detectors

In order to produce sensors that tolerate higher radiation doses, the drift distances have to be shortened. In traditional planar sensors, the electrodes are implanted on the top and/or bottom of the sensor. Some electrons and holes have to drift through the entire sensor before they are recorded. So, for these the probability is much

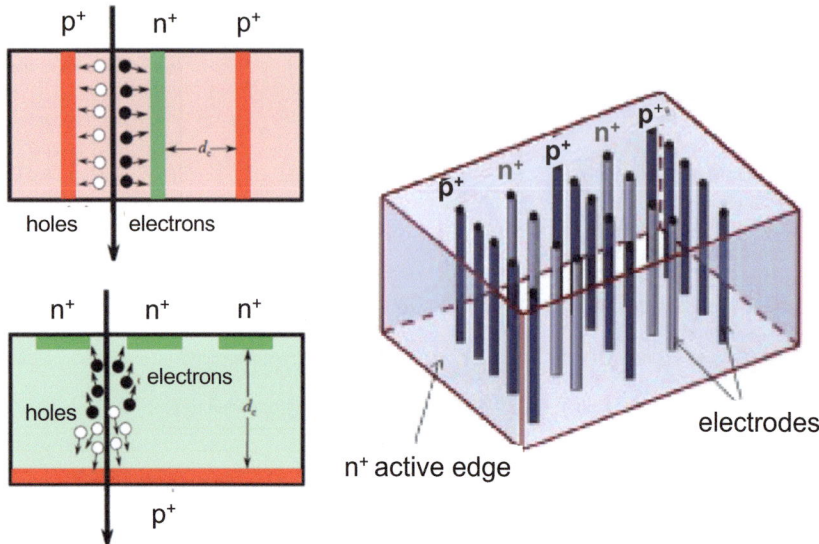

Fig. 6.87 Left: Principle of a 3D sensor (top) in comparison to that of a planar sensor (bottom). The drift directions of electrons and holes are indicated, which are much shorter in a 3D sensor than in a planar sensor. Reprinted with kind permission from [159], © 2016, M. Moll. All rights reserved. Right: Three-dimensional view of a 3D sensor. Alternating rows of p^+ and n^+ rods are used as electrodes. Reprinted with kind permission from [160], © 2013, Elsevier. All rights reserved

higher to be caught in a defect trap. In a three-dimensional (3D) detector the electrodes are placed inside the detector by alternating p^+-type and n^+-type rods [161]. Figure 6.87 (top left) illustrates this in comparison to the layout of a planar sensor displayed in Fig. 6.87 (bottom left). In the 3D sensor, holes and electrons have rather short drift distances. Figure 6.87 (right) depicts the arrangement of the rods in a three-dimensional view. In a 3D sensor, alternating rows of n^+ and p^+ rods are implemented into a high-resistance (>5 kΩ-cm) p bulk. In a single-sided 3D detector, just the n^+ rods are read out while in a double-sided 3D detector both n^+ and p^+ rods could be read out. Single-sided sensors have been produced by the Stanford Nanofabrication Facility [162] and the Norwegian SINTEF MiNaLab [163], while double-sided sensors have been made by the Italian research institute Fondazione Bruno Kessler (FBK) [145] and the Spanish research institution Centro Nacional de Microelectrónica (CNM) [144]. The distance d_c between the n^+ rod, which is kept on high voltage, and the p^+ rod, which is kept on ground potential, is typically much smaller than that of a conventional planar sensor. Thus, also a smaller depletion voltage can be used. In addition, 3D detectors have a $\sim$100% fill factor and have short drift distances. However, sensors need to be cooled at $-30\,°$C to achieve a high detection efficiency after radiation.

Figure 6.88 (top) shows two different schematic layouts of double-sided 3D detectors, a sensor made by FBK (top left) and one made by CNM (top right), which have been used in the ATLAS IBL. The first step is to create column-like holes into

6.7 New Structures

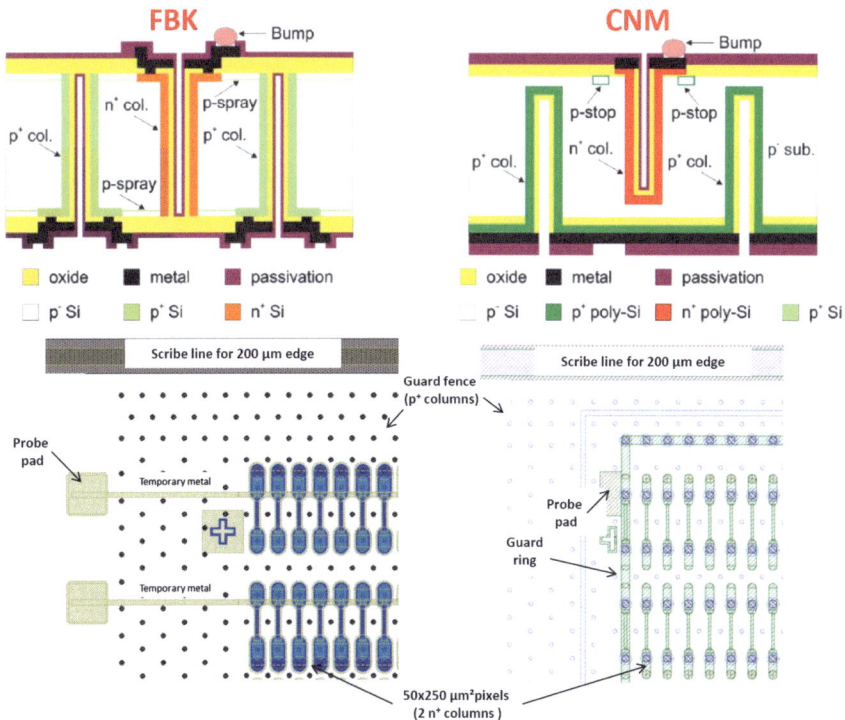

Fig. 6.88 Top left: Schematic side view of a double-sided 3D sensor fabricated by FBK. Top right: Schematic side view of a double-sided 3D sensor fabricated by CNM. For both designs n^+ holes are etched from the top and p^+ holes from the bottom. For the FBK sensors, the holes are etched completely though, while for CNM sensors the etching stops about 20 μm before the end. Bottom left: Top view of the FBK sensor. Bottom right: Top view of the CNM sensor. The FBK sensors are shielded by a guard fence of p^+ columns whereas the CNM sensors use a guard ring. The pixel size is 50 μm × 250 μm in both designs. All reprinted under CC-BY-NC-ND-4.0 Licence from [165], © 2015, J. Lange

the p^- bulk via etching. This is achieved with the deep reactive-ion-etching technique [164]. Initially, there was an issue with yielding smooth rims since this is important for avoiding disruption in the electric field. The n^+ and p^+ columns are etched from opposite sides, since this does not require a support wafer and the bias voltage can be applied at the back sides like for planar sensors. In the FBK sensors, the columns are etched completely through the sensor, while in CNM sensors the etching stops about 20 μm before penetrating the other side (see Fig. 6.88 (top right)). To prevent the build-up of electrons on the SiO_2 between the n^+ and p^+ channels, FBK uses the p-spray technique, while CNM uses the p stop technique. Without the support wafer, no fully active edges can be produced. Thus, FBK uses a guard fence of p^+ rods, while CNM uses a guard fence plus a 3D guard ring. These techniques also allow for very small insensitive edges of ∼10 μm or ∼150 μm, respectively [166,167]. Another difference is the on-wafer-selection method. For the FBK sensors, a pad is coupled by temporary strips to the rods, while for CNM sensors the pad is coupled to the guard ring as shown in Fig. 6.88 (bottom), which displays the top layout of the FBK sensor (bottom left) and CNM sensor (bottom right).

In the ATLAS IBL detector, 25% of the pixels are 3D sensors placed at the outer part of the barrel. The double-sided pixel sensor consists of a 230 μm thick bulk that is made of p-type silicon holding 26,880 pixels on an area of 2.00 cm × 1.68 cm. Thus, the pixel size, which is defined by six p^+ rods surrounding two n^+ rods, is 250 μm × 50 μm. The distance between an n^+ and a p^+ electrode is $d_c = 67$ μm. An edge fence of 200 μm width has been implemented, introducing a small inactive area. The pixels are read out with a dedicated readout chip, ATLAS FE14 [168]. These sensors have a lower capacitance and the number of trapped charges is reduced, thus providing a much better radiation resistance. Studies have shown that a good signal-to-noise ratio can be maintained up to a fluence of $\Phi_{\text{eq}} = 1.4 \times 10^{16}$ (1 MeV) $n_{\text{eq}}/\text{cm}^2$. The signal produces more than 16, 000 e^-.

Based on the successful operation of the 3D sensors in the ATLAS IBL detector, they have been selected for the ATLAS Forward Proton detector (AFP) [167, 169, 170] and the CMS Precision Proton Spectrometer (PPS) [171]. The pixel size for PPS will be 100 μm × 150 μm, while that of AFP has the same size as the IBL pixels. For the high-luminosity LHC upgrade, both ATLAS and CMS consider 3D pixel detectors as well. To cope with higher particle densities near the interaction region, different cell sizes of 50 μm × 50 μm and 25 μm × 100 μm have been studied. The first cell has four p^+ rods surrounding one n^+ cell, yielding $d_c = 35$ μm, while the second cell has six p^+ rods surrounding two n^+ rods, yielding a slightly smaller inter-electrode distance of $d_c = 28$ μm.

For operations at the High-Luminosity LHC, the production of thinner sensors is desired. Besides a reduction of the rod diameters, other advantages are a reduced sensor capacitance, smaller leakage currents, smaller cluster sizes at large pseudo-rapidity, which provides a better two-particle separation in a high-particle density, less trapping due to the smaller electrode distance and smaller multiple scattering, improving the vertex resolution. Disadvantages, however, are that an ionizing particle deposits less charge at normal incidence, which may be compensated by reducing the threshold, as well as higher production and assembly complexity, increasing costs. For the double-sided technology, the thickness cannot be reduced to less than 200 μm, while for the single-sided technology we expect a sensor thickness of 75–150 μm, depending on the manufacturer. However, thin single-sided wafers need an additional support structure. One possibility is the utilization of Si-Si wafer-bonding techniques in which the thin sensitive high-resistivity substrate is bonded to a low-resistivity support wafer [42, 172–174]. The principle is illustrated in Fig. 6.89 (left). Another technology is to produce sensors with the silicon-on-insulator (SOI) technique [174], which is illustrated in Fig. 6.89 (right) and is discussed in the next subsection. For both technologies, the holes are etched from the front side. For n^+ rods, they end just before they reach the support structure, while the p^+ rods go all the way through. Thus, the bias voltage to the p^+ rods can be supplied from the back side, simplifying the assembly process. Figure 6.90 shows an X-ray picture of a 3D

6.7 New Structures

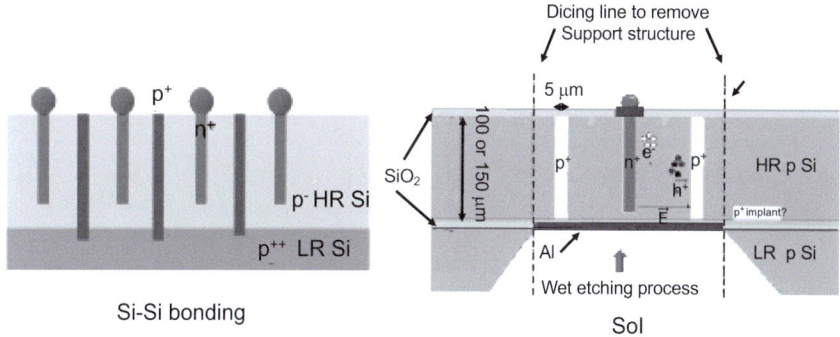

Fig. 6.89 Left: Layout of thin 3D sensors produced with the Si-Si wafer bonding. Right: Layout of thin 3D sensors produced with the silicon-on-insulator technique. Both reprinted under CC-BY-NC-ND-4.0 Licence from [165], © 2016, J. Lange

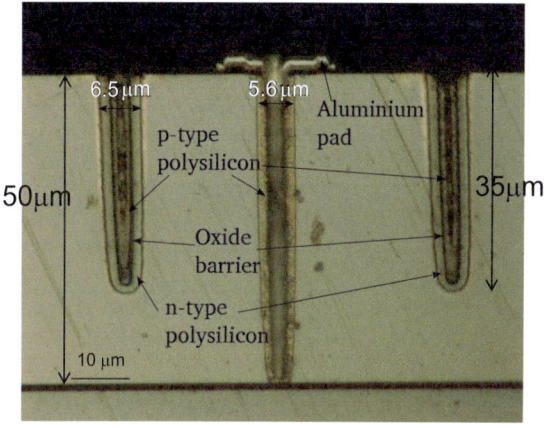

Fig. 6.90 An X-ray picture of a 3D sensor. Reprinted with kind permission from [175], © 2015, G. Pellegrini. All rights reserved

sensor, which illustrates the complex structure. The walls of the holes are smooth, which is important for homogeneous electric fields.

For the High-Luminosity LHC operation, a new readout chip was produced, the RD53A ASIC [173]. The chip was bump-bonded to 150 µm thick CNM sensor. The tin-silver bumps atop the aluminum pixels covering the n^+ rods were melted via thermo-compression. Two assembled modules with pixel sizes of 50 µm × 50 µm and 25 µm × 100 µm were irradiated uniformly with protons at a dose of 5×10^{15} (1 MeV) n_{eq}/cm^2 without powering the ASIC. Figure 6.91 (left, right) shows the IV curves for the two modules before and after irradiation. Both modules were measured with a linear and differential frontend readout. For the non-irradiated 50 µm × 50 µm pixel sensor, the leakage current is rather small but starts to increase rapidly at voltages larger than 10 V. For the 25 µm × 100 µm pixel sensor, the increase starts above 55 V. After irradiation the leakage currents are significantly

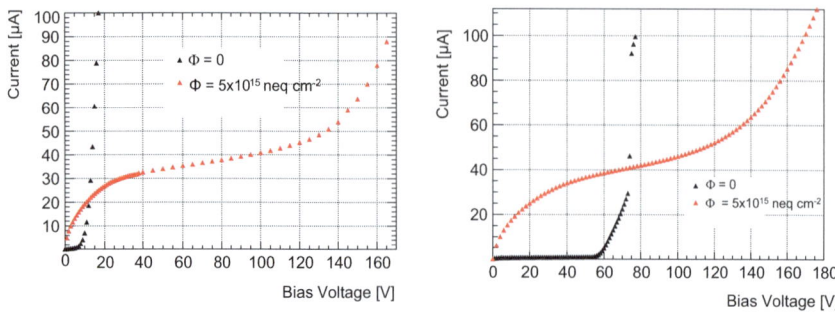

Fig. 6.91 Left: Leakage current as a function of the bias voltage for the 50 μm × 50 μm sensor. Right: Leakage current as a function of the bias voltage for the 25 μm × 100 μm sensor. Measurements were performed before irradiation (black triangles) at 20 °C and after a radiation dose of $\Phi_{eq} = 5 \times 10^{15}$ (1 MeV) n_{eq}/cm^2 (red triangles) at -25 °C. Both reprinted with kind permission from [176], © 2019, IOP Publishing. All rights reserved

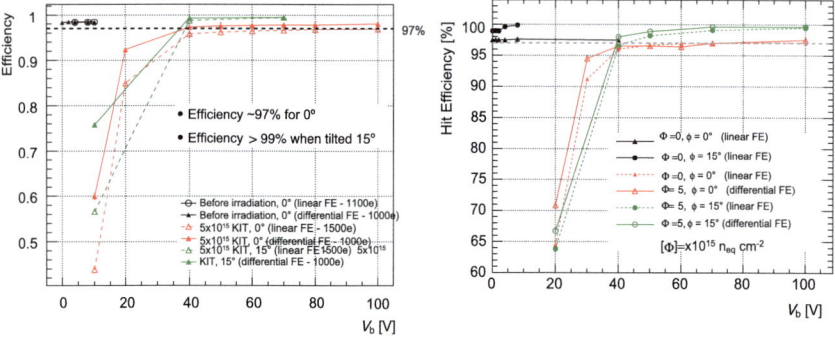

Fig. 6.92 Left: Hit efficiency as a function of the bias voltage for a 50 μm × 50 μm sensor before irradiation (open circles and black triangles) and after irradiation with $\Phi_{eq} = 5 \times 10^{15}$ (1 MeV) n_{eq}/cm^2 for 0° (red triangles) and 15° tilt positions (green triangles), using linear (open symbols) and differential readout (solid symbols). Right: Similar measurements for a 25 μm × 100 μm sensor. Though the color code the same, solid (open) symbols show linear (differential) readout. The black dashed lines indicate the ATLAS benchmark hit efficiency of 97%. Both reprinted with kind permission from [176], © 2019, IOP Publishing. All rights reserved

higher but the detectors are still operable. After a fast rise to 30 μA for bias voltage around ~30 V, the leakage currents almost level off until $V_b = 140$ V when they start to rise again rapidly. Figure 6.92 (left, right) shows the hit efficiencies as a function of bias voltage for the two sensors before and after irradiation for both linear frontend and differential frontend readout. For the irradiated samples, the hit efficiency is also measured for modules tilted at 15°. For both irradiated samples, a hit efficiency of 97% is reached for bias voltages above 70 V. For the tilted samples the hit efficiency is above 99% before and after irradiation. The reason is that particles are not passing any longer through inactive rods.

In 3D sensors, the time resolution is determined mainly by jitter and time-walk contributions. The latter is dominated by the hit resolution. For a 300 μm thick sensor with a pixel size of 50 μm × 50 μm a time resolution of about 75 ps was measured at 27 °C for a bias voltage of 50 V. Simulations show that the time resolution improves for smaller cell sizes, larger bias voltages and cooled sensors. For example, cooling the 50 μm × 50 μm sensor to −20 °C, reduces the time resolution to 45 ps. If, in addition, the bias voltage is increased to 100 V, a time resolution of 35 ps is predicted [177].

6.7.6 Silicon-on-Insulator Sensors

Another monolithic pixel detector consists of a silicon-on-insulator (SoI) device shown in Fig. 6.93 (left). A high-resistivity (>1 kΩ-cm) 50–500 μm thick n-type silicon sensor that is fully depleted is used for particle detection. It is separated from a low-resistivity (<10 Ω-cm) silicon layer (800 nm) in which the readout is accomplished by a layer of SiO_2 (∼200 nm) called buried oxide (BOX). The backside is covered with an n^+-type layer and an aluminum metallization. In the bulk under the SiO_2 p^+, pixels or strips are implanted, which consist of p^+ electrodes embedded in a doped region called buried p (n)-well under the BOX [178] and which are connected to NMOS (n-p-n) and PMOS (p-n-p) transistors, implemented in the low-resistivity silicon. While the bias voltage is supplied on the back side, the transistors on the front side are connected to the readout. So, here the holes are recorded. The pixel size is 36 μm × 36 μm. This configuration provides both high-speed readout circuits and a thick depletion layer in a monolithic device in which the sensor is fully depleted. The bias voltage is below 100 V and depends on the radiation dose.

In sensors without the buried p well, the so-called back-Gate effect causes problems [181]. Here, the transistor threshold is changed when applying the detector bias voltage. When the back-Gate voltage is increased, the threshold voltage of NMOS transistors are decreased and that of PMOS transistors are increased. The circuit eventually stops working due to an excessive back-Gate voltage. To circumvent this back-Gate effect, p^+ implants are inserted into the substrate and are connected to

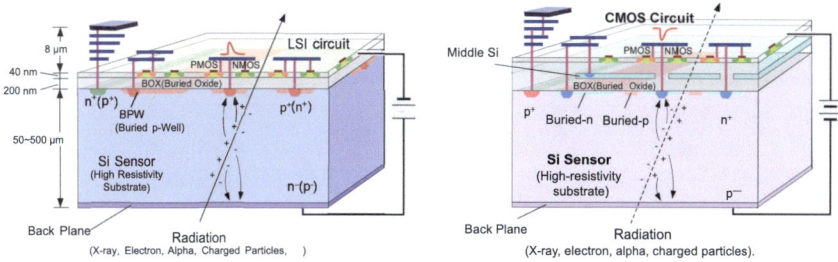

Fig. 6.93 Left: Schematic layout of a silicon-on-insulator device. Reprinted with kind permission from [179], © 2016, Elsevier. All rights reserved. Right: Schematic layout of a double SoI device. Reprinted with kind permission from [180], © 2016, Elsevier. All rights reserved

ground potential to keep the back-Gate potential sufficiently low. The buried p-well connected to the pixel below the BOX layer provides a protecting layer. This simple modification in the process resolves the back-Gate effect by shielding the sensor's electric field. However, the pixel capacitance increases as well as the parasitic coupling between the electronics and the pixel.

Another solution is the double silicon-on-insulator protection that is discussed below (D-SoI). It contains an additional thin silicon layer inside the BOX layer, which removes the drawback of increased pixel capacitance. In addition, the crosstalk between electronic circuit and the sensor is suppressed and the total ionization damage effect is reduced. Figure 6.93 (right) shows the layout of a double SoI device. Here, the high-resistance bulk is made from p^--type silicon to increase the charge collection efficiency. Two buried n-type silicon layers are inserted. A top layer with a thickness of 145 nm functions as the standard SoI layer, which embeds the n^+ electrode. A middle layer with a thickness of 145 nm acts as a shielding layer between the pixels and the transistors. In the top layer the n^+ electrodes are separated by p^+ stops to prevent the built-up of positive charges in the SiO_2 layer. So, here the electrons are recorded. By applying a fixed voltage to the highly-doped buried p well, it acts as an electrostatic shield and thus reduces the capacitive coupling between the sensor and circuit layers. The step-like buried n-wells help the signal charge to drift towards the readout electrodes without touching the interface between the sensor and SiO_2 layers as illustrated in Fig. 6.94 (left), which depicts the potential of a double SoI device. Therefore, the charge collection efficiency is expected to be higher than that of single SoI devices. In this layout, the cross talk is reduced by a factor 20 with respect to that of a single SoI device. The electrons drift towards the minimum, where they are stored until they are recorded.

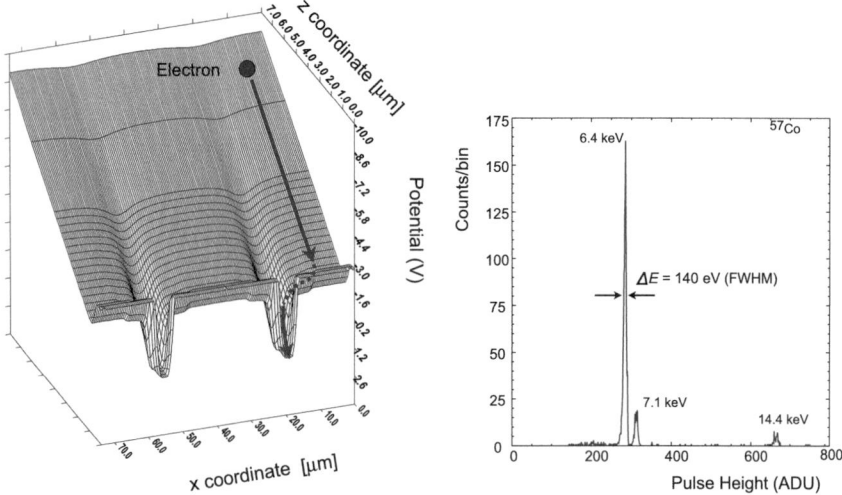

Fig. 6.94 Left: Potential in a double SoI device. The electrons are trapped in the end in one of the minima. Right: Pulse height measured with a ^{57}Co source. The 6.4 keV line has an excellent energy resolution of $\sigma_E = 59$ eV. Both reprinted under CC-BY-4.0 Licence from [178], © 2018, T. Tanaka

The double SoI pixel detector depicted in Fig. 6.93 (right) has a thickness of 200 μm hosting 48 × 48 pixels with a pixel size of 36 μm × 36 μm. The middle layer compensates the electric field that is generated by radiation trapped holes. Thus, it can be used in high-radiation environments. The signal-to-noise ratio typically is $S/N > 100$ for bias voltages of $V > 60$ V. The device is excellent for measuring X-rays. Figure 6.94 (right) shows the energy spectrum measured with a ^{57}Co source. The energy resolution is FWHM=140 e^- at a 6.4 keV photon. A position resolution of 5 μm has been measured. After irradiation with 100 kGy, measured I-V curves resemble those obtained at the pre-irradiation level if the potential on the middle silicon layer is set to -5 V [178].

Thus, the double SoI pixel detector has many good features, such as fine segmentation, low material thickness, good radiation hardness, and low power consumption. In addition, lower cost compared with hybrid pixel detector can be expected. The pixel circuit has a charge-sensitive amplifier followed by a correlated double sampling circuit. The signal is further processed by the peripheral readout circuit composed of a column amplifier and an output buffer. There is a sequential readout of one row of pixels The readout takes about 6 μs, which is much faster than that of a CCD. The double SoI devices are considered for X-ray detectors in experiments in astronomical satellites. In addition, R&D is ongoing to use them in detectors at future linear colliders.

6.8 Diamond Detectors

Since a diamond has a large resistivity, it can be used as a solid-state ionization chamber. Diamonds are produced by the Chemical Vapour Deposition (CVD) technique [182]. Both single crystals and poly-crystalline structures are produced. Single crystals are grown on a high-pressure high-temperature diamond substrate forming a perfect diamond lattice. Polycrystalline structures are grown on a non-diamond substrate. Thus, small crystal grains start forming on the substrate, which are in random orientations. They grow, where the larger ones typically grow faster and eventually terminate the growth of the smaller ones. The average grain size increases across the poly-crystalline thickness from the substrate side to the growth side. With growth thickness, the charge collection properties also improve.

Figure 6.95 shows a schematic layout of a diamond detector. We take a diamond substrate and sandwich it between electrodes, which may consist of a set of strips or pads with a typical thickness of a few hundred μm. The sensor is operated with a high bias voltage and the electrons are recorded by reading out the anode. A traversing charged particle produces on average 36 e-h pairs per μm. The liberated charges drift across the diamond due to the applied electric field. In a 500 μm thick detector, we get 18,000 electrons that produce a detectable electric signal. The gap energy is 5.4 eV. So, at room temperature the thermal noise is low. Figure 6.96 (left) shows the average energy loss of a charged particle in diamonds with respect to that in silicon

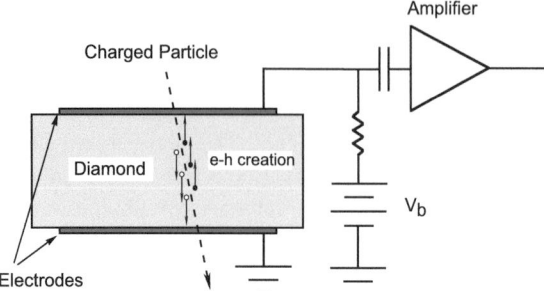

Fig. 6.95 Schematic layout of a diamond detector [183]. Reprinted with kind permission from [183], © 2008, IOP Publishing. All rights reserved

before irradiation. Silicon produces about a factor of two more e-h pairs. Since there may be traps in the diamond, we often use the so-called charge collection distance to characterize the material, which corresponds to the average distance the e-h pairs move apart. In thin diamond sensors, the charge collection distance is limited by the sensor dimension, while in thick diamond sensors it approaches the sum of mean free paths of electrons and holes.

Since single crystal diamonds exhibit not much trapping, the charge collection distance corresponds to the detector thickness. In poly-crystalline samples, the charge collection distance is limited by trapping. It helps to grow thick wafers (>1 mm) and polish off the substrate side that contains more small grains since the trapping in large grains is reduced. Top quality poly-crystalline material can be grown in six inch wafers with a charge collection distance of 300 μm in a 500 μm tick sensor with an electric field of 2 V/μm. Though single crystals have superb quality, they are limited in size to 1 cm × 1 cm and are expensive.

To produce a trap, an atom has to be moved out of the crystal structure. The energy needed is called the threshold displacement energy. For diamonds, it is ∼42 eV per atom compared to ∼26 eV per atom for silicon. Thus, we expect diamond sensors

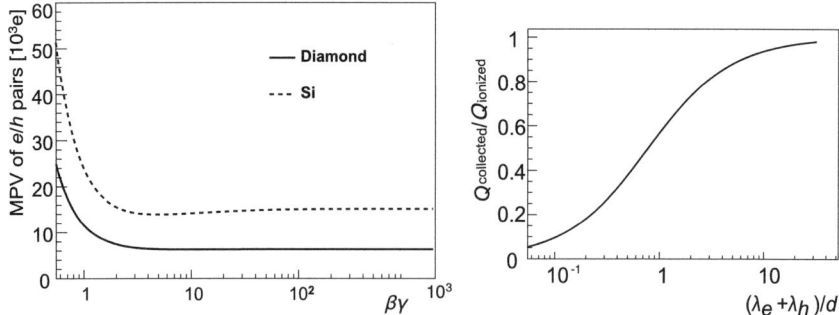

Fig. 6.96 Left: Most probable energy loss as a function of $\beta\gamma$ for diamonds in comparison to that for silicon. Right: Ratio of the collected charge to ionized charge as a function of the mean free paths of electrons and holes with respect to the thickness of the sensor. Both reprinted with kind permission from [185], © 2012, IOP Publishing. All rights reserved

6.8 Diamond Detectors

to be more radiation hard than silicon sensors. In diamond detectors, irradiation causes no increase in leakage currents. However, charge can be trapped as in silicon. The cross sections for the capture of electrons and holes can differ by orders of magnitude. If the energy of the trapped charge is sufficiently far from the energy of the conduction band (deep-level traps), de-trapping times can be large so that the charge is lost for signal collection. These traps permanently capture a charge carrier, which passivates the trap. This is the basis of the so-called pumping procedure in which diamonds are exposed to ionizing radiation in order to fill the traps and thus passivize them. The pumping procedure is accomplished with a strong ^{90}Sr source before performing measurements. In a hadron collider, this is achieved by the beam backgrounds. In addition, we have shallow traps. These remain active since they capture the charge just for short time [184].

Traps can be present already in non-irradiated diamond sensors due to lattice imperfections. We expect this to occur more often in poly-crystalline diamond samples, where charge also may be trapped at grain boundaries. The radiation-induced traps also occur in single crystal diamond. The mean free path $\lambda_{e+h} = \lambda_e + \lambda_h$ is inversely proportional to the fluence,

$$\frac{1}{\lambda_{e+h}} = \frac{1}{\lambda^0_{e+h}} + k_{e,h}\Phi_{\text{eq}}, \tag{6.39}$$

where λ^0_{e+h} is the mean free path before irradiation, Φ_{eq} is the fluence and $k_{e,h}$ is the radiation damage constant. Solving for λ_{e+h} yields

$$\lambda_{e+h} = \frac{\lambda^0_{e+h}}{1 + \lambda^0_{e+h} \cdot k_{e,h}\Phi_{\text{eq}}}. \tag{6.40}$$

For example for irradiation with 25 MeV and 24 GeV protons, the radiation damage constants are $k_{e,h} = 3.02^{+0.42}_{-0.36}$ and $k_{e,h} = 0.69^{+0.14}_{-0.17}$, respectively. For thick sensors, the charge collection distance is the sum of the mean free paths for electrons and holes, where $\lambda_e \sim \lambda_h$ since the collected signal is quasi independent of the electric-field direction. In thinner detectors, the measured charge collection distance is given by

$$d_c = \frac{Q_{\text{collected}}}{Q_{\text{ionized}}} = \frac{\lambda_{e+h}}{d_{\text{sen}}}\left[1 - \frac{\lambda_{e+h}}{d_{\text{sen}}}\left(1 - \exp\left(-\frac{d_{\text{sen}}}{\lambda_{e+h}}\right)\right)\right], \tag{6.41}$$

where d_{sen} is the thickness of the sensor. Figure 6.96 (right) shows the ratio of collected charge to ionized charge as a function of $\lambda_{e+h}/d_{\text{sen}}$. To achieve a collection efficiency above 90%, the sum of the mean free paths for electrons and holes have to be at least ten times larger than the detector thickness [185].

Strong irradiation in diamonds causes a loss of signal charge due to trapping and a reduction of λ_{e+h}. Figure 6.97 shows the mean free paths $\lambda_e + \lambda_h$ as a function of fluence for a single crystal diamond sample, two poly-crystalline diamond samples and a silicon sensor after irradiation with 25 MeV (left) and 24 GeV (right) protons. The samples were operated with the same electric-field strength of 1.8–2.2 V/μm,

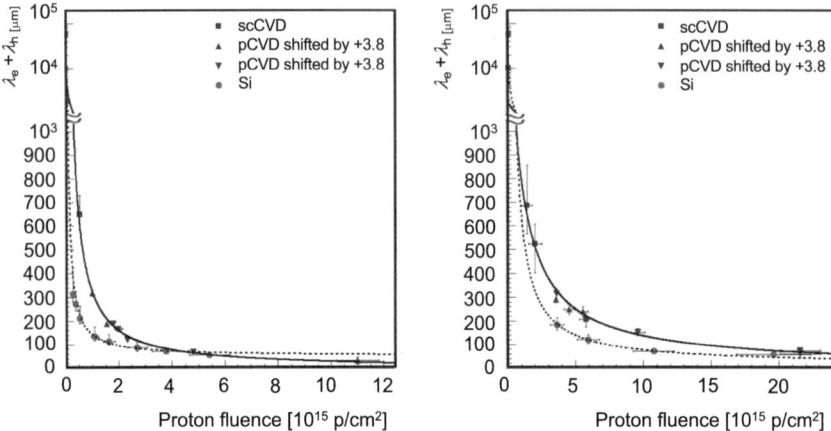

Fig. 6.97 Mean free paths $\lambda_e + \lambda_h$ as a function of fluence for 25 MeV proton irradiation (left) and for 24 GeV proton irradiation (right) for silicon (red points), single-crystal diamond (blue squares) and two poly-crystalline diamond samples (blue triangles). For irradiation with 25 MeV protons the poly-crystalline diamond samples are shifted to the right by $1.0 \times 10^{15}\,p/\text{cm}^2$ and $1.8 \times 10^{15}\,p/\text{cm}^2$ while for irradiation with 24 GeV/c protons they are shifted to the right by $3.8 \times 10^{15}\,p/\text{cm}^2$. Reprinted with kind permission from [185], © 2012, IOP Publishing. All rights reserved

corresponding to a bias voltage of 900 V for the 500 μm thick diamond sensors and 600 V for the 310 μm thick silicon n-in-p sensors. All samples were fully depleted. This is the maximum bias voltage considered for operation at the LHC. The diamond samples have a strip pitch of 80 μm.

The effect of irradiation with low-energy protons is more severe than that with high-energy protons. A fit to the functional form in (6.41) is performed to extract the damage constant. Note that the shifts for the poly-crystalline samples become less important at high irradiation doses since the changes in the mean-free path become tiny. For 24 GeV/c protons the diamond sensors perform better than silicon sensors at high fluences.

Presently, diamond detectors are used in high-energy physics experiments for monitoring beam conditions and for measuring the luminosity. The first diamond detectors were used in the BABAR [186] and CDF [187] experiments. At the LHC, the ATLAS [188] and CMS [189] experiments use them for both tasks. Beam conditions monitoring is also accomplished with diamond detectors by the LHCb [190] experiment. The ATLAS beam conditions monitor (BCM) has four detector modules on each side of the interaction region placed symmetrically around the beam pipe that are mounted on the ATLAS Pixel detector support structure at an angle of 45° with respect to the beam pipe. The detectors are placed at $|\eta| = 4.21$ at a radius of 5.5 cm. Here, the radiation dose is harsh, expecting to amount to 10^{15} particles per cm^2 and an ionization dose of 0.5 MGy in ten years of ATLAS operation. So, both sensors and electronics have to be laid out to survive the radiation dose. The BCM

6.8 Diamond Detectors

detectors should provide signal pulses with a rise time of ∼1 ns, a width of ∼3 ns and a restoration time to the baseline of about 10 ns. In addition, the BCM detectors should be able to record MIPs.

The sensors consist of polycrystalline diamond samples that are operated at room temperature. They are 500 μm thick with an 8 cm × 8 cm readout matrix. They are operated at a bias voltage of 1 kV. The leakage current is less than 100 pA and the charge collection distance is 220 μm. Two diamond sensors are combined to increase the signal amplitude, where the adjacent electrodes of the two sensors are set to ground potential and are connected to the amplifier, while the outer electrodes are connected to the bias voltage. In both sensors, the drift field is 2 V/μm. The detectors are mounted on an Al_2O_3 ceramic board. The front-end amplifier consists of a two-stage current amplifier. Each stage has an amplification of 20 db with a noise of less 0.9 db. The S/N ratio is ∼7–7.5.

Besides planar sensors, 3D sensors are produced as well. The holes are drilled from the backside using a Ti-Sapphire femto-second laser, which can induce a phase transition of the diamond lattice into resistive material consisting of diamond-like carbon, amorphous carbon and graphitic material such that continuous electrodes can be produced. The laser operates at a wavelength of 800 nm. The laser beam spot size is about 4 μm. The 3D sensors are 500 μm thick. Pixel sizes of 100 μm × 100 μm and 150 μm × 150 μm have been produced. The electrodes have a resistivity of (45 ± 11.8) kΩ. The column production efficiency was about 92%.

In 2014, the ATLAS experiment upgraded the BCM by installing one of the first diamond pixel detectors called diamond beam monitors (DBM). The goal is to measure the instantaneous bunch-by-bunch luminosity and the bunch-by-bunch position of the beam spot. The diamond beam monitor comprises of eight telescopes, four on each side of the interaction region. However, just six of the telescopes are based on poly-crystalline diamond sensors and the other two use silicon sensors as a reference. The diamond sensors have a thickness of 500 μm and a pixel size of 150 μm × 150 μm. Each of the telescopes has three detector planes to add tracking capability to the existing precise time-of-flight (ToF) measurements of the original eight pad detectors of the BCM. The readout uses the same chip as that of the ATLAS IBL. The mean of the measured pulse height is 13,500 electrons, which is much higher than 6,900 electrons measured for a diamond strip detector.

The system increases the segmentation and spatial resolution of the beam monitor. Due to the projective geometry pointing to the interaction region, it can distinguish particles coming from collisions and those from background. Figure 6.98 shows the longitudinal (left) and transverse (right) closest distance-of-approach to the interaction point. The detectors clearly distinguish between signal events and backgrounds.

In 2016, 3D diamond detectors with a pixel size of 100 μm × 100 μm were produced that had 1,199 working cells. The column production efficiency was 99%. Furthermore, strip detectors with 25 μm strips and a 50 μm pitch width were manufactured on each diamond. For the non-irradiated single-crystal diamonds, measurements reported a spatial resolution of (7.6 ± 1.0) μm, while for non-irradiated poly-crystalline diamonds a spatial resolution of (15.0 ± 1.5) μm was measured. A possible explanation for this difference may be the effect of grain boundaries, which

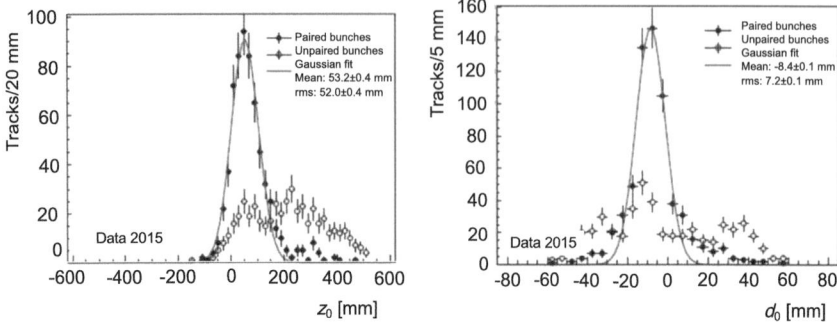

Fig. 6.98 Longitudinal (left) and transverse (right) closest distance-of-approach to the interaction point. The data are for paired (solid points) and unpaired bunches (open points). The distributions are obtained from the reconstruction of tracks from three modules using the initial alignment. Reprinted with kind permission from [191], © 2017, RD42 Collaboration. All rights reserved

allows the charge to move laterally. After irradiation with 24 GeV/c protons at a fluence of $18.0 \times 10^{15}\,p/\text{cm}^2$, the spatial resolution of the poly-crystalline diamonds improved to $(7.6 \pm 0.8)\,\mu\text{m}$, the same spatial resolution as that of single crystals. For irradiation with 800 MeV/c protons at a fluence of $12.6 \times 10^{15}\,p/\text{cm}^2$, a resolution of $(8.6 \pm 0.9)\,\mu\text{m}$ was measured. For irradiated single-crystal diamonds, the spatial resolution was basically independent of the fluence.

Exercises

6.1 Consider a 300 µm thick silicon strip detector. The strips are 150 mm long and 20 µm wide at a 30 µm pitch. The detector is operated at room temperature. The strips are read out with an amplifier that has $f_T = 1$ GHz and a time constant of 100 ns. What is the mean equivalent noise charge on a strip? If you cool the detector to $-40\,°\text{C}$, how much does the noise improve?

6.2 Using the same parameters as in the previous problem calculate the $\overline{ENC}$ of the dark current ($I_d = 1$ nA), that of the serial resistor ($R_s = 50\,\Omega$) and that of the parallel resistor $R_p = 2$ MΩ).

6.3 An electron produced at the top of a silicon strip detector has to traverse the entire thickness of 300 µm. For a drift velocity of $v_d^e = 1.4 \times 10^6$ cm/s at NTP, how much time does the electron travel? How large is the spread due to diffusion at room temperature if the electron mobility is $\mu_e^{Si} = 1,400\,\text{cm}^2/(V \cdot s)$? In a 3D detector, the typical distance between a p^+ and an n^+ electrode is 67 µm. How fast is the collection time for the same drift velocity? How large is the diffusion spread?

6.4 Plot the depletion layer thickness as a function of the reverse bias voltage for a silicon detector that has acceptor and donor densities of $N_A = 10^{15}/\text{cm}^3$ and $N_D = 3 \times 10^{12}/\text{cm}^3$. First, you need to calculate the contact potential V_0. What reverse bias voltage is needed to fully deplete a 300 µm thick sensor?

6.5 Consider a CCD that has a 20 μm p-type epitaxial layer, where the surface 1 μm layer is converted to n-type. The detector is operated at a temperature of 190 K. The ϕ wells have potentials of $\phi_1 = 0$ V, $\phi_2 = 10$ V and $\phi_3 = 0$ V. The depletion layer is 5 μm. How large is the drift velocity in the depletion region? How fast are these electrons collected in the ϕ_2 well? How large is the diffusion coefficient? How fast are the electrons collected in the undepleted region? How many electrons are produced in the epitaxial layer? How large is the spread due to diffusion?

6.6 The ATLAS barrel pixel detector has 67 million channels arranged in 46,080 modules. Each module has an active surface of 6.08×1.64 cm^2. The sensor thickness is 250 μm. The average damage rate is $\alpha_\text{fl} = 3.99 \times 10^{-17}$ A/cm. How large is the leakage current ΔI for a fluence of 10^{15}/cm^2? Compute the amplitudes of the three contributions to the effective doping concentration, N_an, N_sd and N_Y for annealing at 60 °C. Plot the time dependence for each component. Plot the time dependence of α_fl for 21 °C.

6.7 In the ALICE silicon drift chamber, the electrons are drifted in an electric field of 460 V/cm at a maximum length of 35 mm at room temperature. How large is the drift velocity? What is the maximum drift time? The resolution of the electron cloud arriving at the anode is given by $\sigma^2 = 2D_f \cdot t + \sigma_0^2$, where $\sigma_0 = 200$ ns. Calculate D_f and the spatial resolution.

6.8 Consider a DEPFET similar to that in Belle II. The width and length of the transistor are $w_\text{ch} = 20$ μm and $l_\text{ch} = 5$ μm, respectively. The dielectric constant in the 0.1 μm thick SiO$_2$ layer is $\epsilon_{\text{SiO}_2} = 3.85$. The gate voltage is $V_\text{g} = 18$ V and the threshold voltage is $V_\text{th} = 3$ V. We assume a gate reduction factor of $f_\text{gate} = 0.8$ and an input signal of 10 fC. Calculate the hole mobility at room temperature (20 °C) for a doping concentration of 10^{15}/cm^3. How large is capacitance of the DEPFET pixel C_ox? Further compute the drain source current $I_\text{d,sat}$, the charge amplification g_m and the transconductance g_q.

6.9 Consider a Low-Gain Avalanche Detector that has a 4 μm thick gain layer. How large is the depletion voltage for a doping concentration of 10^{16}/cm^3? Plot the depletion voltage versus the doping concentration. Plot the depletion voltage versus gain layer for a doping concentration of 10^{16}/cm^3.

6.10 For an LGAD sensor with 100 μm × 100 μm pixels that have a 50 μm depth, plot the leakage current as a function of fluence. The leakage current damage constant is $\alpha_\text{leak} = 4 \times 10^{-17}$ A/cm. The gain is 15. The measurements were performed at $T = -20$ °C. What is the leakage current at room temperature?

References

1. *Springer Handbook of Crystal Growth*, ed. by G. Dhanaraj et al. (Springer, 2010)
2. J. Czochralski, Metalle. Z. Phys. Chem. **92**, 219 (1918)
3. M. Moll, dissertation at Hamburg University, Hamburg DESY-THESIS-1999-040 (1999)
4. https://www.exportersindia.com/vritra-technologies/monocrystalline-silicon-wafer-5132632.htm

5. H.C. Theuerer, US patent, 3,060, 123 (1962)
6. G. Bertolini, A. Coche, *Semiconductor Detectors* (North-Holland Publishing Company, Amsterdam, 1968)
7. S.M. Sze, *Semiconductor Devices, Physics and Technology* (Wiley, 1985)
8. M. Krammer, Halbleiterdetektoren, Lectures (Vienna, 2010)
9. M. Krammer, F. Hartmann, Silicon detectors, https://indico.cern.ch/event/124392/contributions/1339904/attachments/74582/106976/IntroSilicon.pdf (2011)
10. M. Baselga et al., JINST **13**, P11007 (2018)
11. C. Canali et al., IEEE Trans. Electron Device **22**-11, 1045 (1974)
12. N.D. Arora, J.R. Hauser, D.J. Roulston, ampx.tugraz.at/ hadley/psd/L4/mobility.php
13. V. Mishra, V.D. Srivastava, S.K. Kataria, Pramana J. Phys. **65**(2), 259 (2005)
14. B. Hyams et al., Nucl. Instrum. Meth. **205**, 99 (1983)
15. J. Kemmer, Nucl. Instrum. Meth. A **226**, 89 (1984)
16. J. Kemmer et al., Proc. Silicon Detectors for HEP **195** (1981)
17. E. Belau et al., Nucl. Instrum. Meth. **214**, 253 (1983)
18. T. Bergauer, Lecture on Silicon Detectors in High-Energy Physics, HEPHY Vienna, April (2015)
19. K. Dette, Diploma Thesis, Technische Universität Dortmund (2013)
20. A. Peisert, DELPHI 92-143 MVX 2, CERN (1992)
21. The DELPHI Collaboration (P.A. Aarnio et al.), Nucl. Instrum. Meth. A **303**, 233 (1990)
22. The ALEPH Collaboration (D. Buskulic et al.), Nucl. Instrum. Meth. A **360**, 481 (1995)
23. The L3 Collaboration (B. Adeva et al.), Nucl. Instrum. Meth. A **289**, 35 (1990)
24. The OPAL Collaboration (K. Ahmet et al.), Nucl. Instrum. Meth. A **305**, 275 (1991)
25. The CDF Collaboration (F. Abe et al.), Nucl. Instrum. Meth A **271**, 387 (1988)
26. The D0 Collaboration (V.M. Abazov et al.), Nucl. Instrum. Meth. A **565**, 463 (2006)
27. The CLEO II Collaboration (Y. Kubota et al.), Nucl. Instrum. Meth. A **320**, 66 (1992)
28. The Belle II Collaboration (A. Abashian et al.), Nucl. Instrum. Meth. A **479**, 117 (2002)
29. The BABAR Collaboration (B. Aubert et al.), Nucl. Instrum. Meth. A **479**, 1 (2002)
30. The ZEUS Collaboration (G. Abbiendi et al.), Nucl. Instrum. Meth. A **333**, 342 (1993)
31. The H1 Collaboration (D. Pitzl et al.), Nucl. Instrum. Meth. A **454**, 334 (2000)
32. The PHENIX Collaboration (C. Aidala et al.), Nucl. Instrum. Meth. A **755**, 44 (2014)
33. The STAR Collaboration (Y.V. Fisyak et al.), Nucl. Instrum. Meth. A **549**, 27 (2005)
34. The ATLAS Collaboration (G. Aad et al.), JINST **3**, S08003 (2008)
35. The CMS Collaboration (S. Chatrchyan et al.), JINST **3**, S08004 (2008)
36. The LHCb Collaboration (R. Aaij et al.), Int. J. Mod. Phys. A **30**-07, 1530022 (2015); LHCb Tracker Upgrade Technical Design Report, Tech. Rep. CERN-LHCC-2014-001, CERN, Geneva, (2014)
37. The ALICE Collaboration (P. Kuijer et al.), Nucl. Instrum. Meth. A **447**, 251 (2000)
38. The ALICE Collaboration, https://home.cern/press/2022/ALICE-upgrades-LS2 (2022)
39. The BABAR Collaboration (J. Richman et al.), Nucl. Instrum. Meth. A **409**, 219 (1998); The BABAR Collaboration (C. Bozzi et al.), Nucl. Instrum. Meth. A **435**, 25 (1999)
40. The BABAR Collaboration (D. Boutigny et al.), BABAR Technical Design Report, SLAC-R-95-457 (1995)
41. C. Campagnari, https://slideplayer.com/slide/4948486/ (2005)
42. The ATLAS Collaboration (G. Aad et al.), JINST **9**, P08009 (2014)
43. A. Ahmad et al., Nucl. Instrum. Meth. A **578**, 98 (2007)
44. The ATLAS SCT group (S. Hirose et al.), PoS Vertex2019, 005 (2020)
45. The ATLAS Collaboration (C. S. Hsu et al.), Nucl. Phys. Proc. Suppl. **215**, 92 (2011)
46. P. Chochula et al., Nucl. Instrum. Meth. A **412**, 304 (1998)
47. A. Andreazza et al., Nucl. Instrum. Meth. A **367**, 198 (1995)
48. D. Creanza et al., Nucl. Instrum. Meth. A **409**, 157 (1998)
49. M. Acciarri et al., Nucl. Phys. B **44**, 296 (1995); Nucl. Instrum. Meth. A **360**, 103 (1995)
50. S. Anderson et al., Nucl. Instrum. Meth. A **403**, 326 (1998); S. de Jong, Nucl. Instrum. Meth. A **386**, 23 (1997)

References

51. The CDF Collaboration (R. Blair et al.), The CDF-II detector: Technical design report (1996); D. Bortoletto, Nucl. Instrum. Meth. A **386**, 87 (1997)
52. The D0 Collaboration (S. N. Ahmed et al.), Nucl. Instrum. Meth. A **634**, 8 (2010)
53. R. Kass et al., Nucl. Instrum. Meth. A **501**, 32 (2003)
54. Z. Natkaniec et al., Nucl. Instrum. Meth. A **560**, 1 (2006)
55. The Belle II Collaboration (T. Abe et al.), arXiv:1011.0352 (2010)
56. G. Barone, Nucl. Instrum. Meth. A **732**, 57 (2013)
57. Y. Gotra, Nucl. Instrum. Meth. A **652**, 680 (2011)
58. The LHCb Silicon Tracker group (M. Tobin et al.), Nucl. Instrum. Meth. A **831**, 174 (2016)
59. The ALICE Collaboration (K. Aamodt et al.), JINST **3**, S08002 (2008)
60. The ATLAS Collaboration, https://twiki.cern.ch/twiki/pub/AtlasPublic/ApprovedPlotsED/curvedcosmic6.png
61. W.S. Boyle, G.E. Smith, Bell Labs Bell Syst. Tech. J. **49**, 587 (1970)
62. A. Litke, A. Schwarz, Sci. Am. **272**-5, 76 (1995)
63. C.J.S. Damerell et al., Nucl. Instrum. Meth. **185**, 33 (1981)
64. J.E. Brau, CCD Vertex Detectors, IEEE NPSS Technology Lectures at Snowmass 2001, https://pages.uoregon.edu/jimbrau/LC/ccd-tutorial.pdf
65. C.J.S. Damerell, Rev. Sci. Instrum. **69**-4, 1549 (1998)
66. The SLD Collaboration (K. Abe et al.), Nucl. Instrum. Meth. A **400**, 287 (1997)
67. The SLD Collaboration (K. Abe et al.), Nucl. Instrum. Meth. A **447**, 90 (1999)
68. N.B. Sinev et al., Nucl. Instrum. Meth. A **409**, 243 (1998)
69. A.G. Riess et al., Astrophys. J. **504**, 935 (1998)
70. https://cms.cern/detector/identifying-tracks/silicon-pixels
71. G. Lutz, *Semiconductor Radiation Detectors*, 2nd edn. (Springer, 2001)
72. The WA97 Collaboration (F. Antinori et al.), Nucl. Instrum. Meth. A **360**, 91 (1995)
73. K.H. Becks et al., DELPHI note 96-56 MVX14 (1996)
74. W. Erdmann, The DMILL Readout Chip for the CMS Pixel Detector, talk at Int. Pixel 2002 Workshop, https://www.slac.stanford.edu/econf/C020909/weslide.pdf (2002)
75. K.H. Becks et al., Nucl. Instrum. Meth. A **395**, 398 (1997); ibid A **409**, 229 (1998)
76. The DELPHI Collaboration (P. Chochula et al.), Nucl. Instrum. Meth A **412**, 304 (1998)
77. The ATLAS Collaboration (G. Aad et al.), JINST **3**, P07007 (2008)
78. The ATLAS Collaboration (M. Backhaus et al.), Nucl. Instrum. Meth. A **831**, 65 (2016)
79. K. Potamianos, PoS EPS-HEP2015, 261 arXiv:1608.07850 [physics.ins-det] (2015)
80. The ATLAS IBL Collaboration (B. Abbott et al.), JINST **13**-05, T05008 (2018)
81. T. Wittig, PhD Thesis. TU Dortmund (2013)
82. The ATLAS Collaboration (G. Aad et al.), New Small Wheel TDR, https://cds.cern.ch/record/1552862/files/ATLAS-TDR-020.pdf (2013)
83. D.A. Dobos, Dissertation at the University of Dortmund (2007)
84. M. Keil, Proc. Sci., PoS 001 (2009)
85. I. Gorelov et al., Nucl. Instrum. Meth. A **481**, 204 (2002)
86. ATLAS public Pixel plots, https://atlas.web.cern.ch/Atlas/GROUPS/PHYSICS/PLOTS/IDTR-2015-007/
87. ATLAS approved plots, https://atlas.web.cern.ch/Atlas/GROUPS/PHYSICS/PLOTS/PIX-2020-003/
88. P. Sabatini, JINST **15**-02, C02039 (2020)
89. The ATLAS Collaboration (M.P. Giordani et al.), PoS ICHEP2016, 254 (2016)
90. The ATLAS Collaboration (G. Aad et al.), ATLAS Note ATL-INDET-PUB-2016-001 (2016)
91. The CMS Tracker Collaboration (W. Adam et al.), JINST **16**-02, P02027 (2021)
92. The CMS Collaboration (A. Dominguez et al.), CMS-TDR-011 (2012)
93. The CMS Collaboration (D.A. Matzner et al.), CMS Technical Design Report for the Pixel Detector Upgrade (2012)
94. A. Burgmeier, Dissertation at the Karlsruhe Institute of Technology (2014)
95. F.R. Helmert, Z. Math. U. Physik **20**, 300–3 (1875)
96. The CMS Collaboration (W. Adam et al.), Performance note CMS DP-2020/032 (2022)

97. M. Huhtinen, Nucl. Instrum. Meth. A **491**, 194 (2002)
98. F. Hartmann, *Evolution in Silicon Sensor Technology in Particle Physics* (Springer, 2017)
99. J. Lindhard et al., Mat. Fys. Med. **33** No 10, 2 (1963)
100. D. Zontar, Dissertation at the University of Ljubljana (1998)
101. ASTM E722-85. Standard practice for characterizing neutron fluence spectra in terms of an equivalent monoerergetic neutron fluence for radiation hardness testing of electronics. ASTM E772-93 (revision) (1993)
102. P. Griffin et al., Unpublished but available from G. Lindstrom, Hamburg (gunnar@sesam.desy.de) (1996)
103. A. Yu. Konobeyev et al., J. Nucl. Mater. **186**, 117 (1992)
104. M. Huhtinen, P. Aarino, HU-SEFT R 1993-02 (1993)
105. M. Huhtinen, P. Aarnio, Nucl. Instrum. Meth. A **335**, 580 (1993)
106. G.P. Summers et al., IEEE **40**, 1372 (1993)
107. G. Lindström, Nucl. Instrum. Meth. A **512**, 30 (2003)
108. R. Wunstorf, PhD thesis, Hamburg University, Hamburg (1992)
109. O. Krasel, PhD thesis Dortmund. https://doi.org/10.17877/DE290R-14839 (2004)
110. M. Moll, CERN Detector Seminar (2009)
111. R. Klanner et al., Nucl. Instrum. Meth. A **730**, 2 (2013)
112. J. Schwandt et al., JINST **7**, C01006 (2012)
113. R. Klanner et al., Nucl. Instrum. Meth. A **732**, 117 (2013)
114. The ATLAS Collaboration (G. Aad et al.), JINST **16**, P08025 (2021)
115. The ATLAS Collaboration (J.D. Bossio Sola et al.), ATLAS Inner Tracker Pixel Detector TDR (2017)
116. E. Gatti et al., Nucl. Instrum. Meth. A **226**, 129 (1984)
117. W. Chen et al., Report SSC-PC-015 (1989)
118. G. Gramegna et al., IEEE Trans. Nucl. Sci. **42**, 1497 (1995)
119. U. Faschingbauer et al., Nucl. Instrum. Meth. A **377**, 362 (1996)
120. The STAR SVT Collaboration (R. Bellwied et al.), Nucl. Instrum. Meth. A **377**, 387 (1996)
121. The STAR SVT Collaboration (J. Takahashi et al.), Nucl. Instrum. Meth. A **439**, 497 (2000)
122. The ALICE Collaboration (J. Schuhkraft et al.), ALICE-INT-1993-27, (1993)
123. B. Alessandro et al., JINST **5**, P04004 (2010)
124. https://www.tf.uni-kiel.de/matwis/amat/semitech_en/kap_5/backbone/r5_1_5.html
125. M. Deveaux, JINST **14**-11, R11001 (2019)
126. G. Deptuch et al., Nucl. Instrum. Meth. A **511**, 240 (2003)
127. D. Contarato et al., Nucl. Instrum. Meth. A **565**, 119 (2006)
128. D. Bortoletto, CERN summer lectures, https://indico.cern.ch/event/387979/
129. D. Contarato, Dissertation at the University of Hamburg (2005)
130. G. Contin, Nucl. Instrum. Meth. A **907**, 60 (2018)
131. The ALICE Collaboration (S. Acharya et al.), Technical Design Report for the Upgrade of the ALICE Inner Tracking System, J. Phys. G: Nucl. Part. Phys. **41**, 087002 (2014)
132. M.Y. Dong et al., Nucl. Instrum. Meth. A **924**, 287 (2019)
133. G. Contin, PoS Vertex 2019, arXiv:2001.03042 [physics-ins-det] (2020)
134. J. Kemmer, G. Lutz, Nucl. Instrum. Meth. A **253**, 365 (1987)
135. F. Müller, Dissertation at the Ludwig Maximilian University Munich (2017)
136. M. Lamarenko, Dissertation at the University of Bonn (2013)
137. H.G. Moser et al., PoS VERTEX2007, 022 (2007)
138. D. Levit, PoS BORMIO2016 **029** (2016)
139. P. Kodys, JINST **10**, C02037 (2015)
140. G. Giakoustidis et al., PoS Pixel **2022**, 005 (2023)
141. R. Turchetta, Nucl. Instrum. Meth. A **335**, 44 (1993)
142. A. Moll, Dissertation at the Ludwig Maximilian University Munich (2015)
143. The RD 50 Collaboration, https://rd50.web.cern.ch/
144. http://www.cnm.es/
145. Fondazione Bruno Kessler, https://www.fbk.eu/en/

146. G. Kramberger et al., Nucl. Instrum. Meth. A **891**, 68 (2017)
147. G. Kramberger et al., JINST **10**, P07006 (2015)
148. M. Ferrero et al., Nucl. Instrum. Meth. A **919**, 16 (2019)
149. M. Moll, Talk at LHCC open session, https://indico.cern.ch/event/835603/contributions/3504560/attachments/1905621/3147133/190911-RD50-Statusreport-LHCC_a.pdf (2019)
150. A. Chilingarov, Generation current temperature scaling. http://rd50.web.cern.ch/rd50/doc/Internal/rd50-2011-001-I-T-scaling.pdf. Version: 2011. RD50 Technical Note RD50-2011-01 (2011)
151. H.F.W. Sadrozinski, Talk at Trento 2018 Meeting, https://indico.cern.ch/event/666427/contributions/2881734/attachments/1604247/2544544/1_Trento-2018-HFWS-2.pdf (2018)
152. G. Pellegrini et al., Nucl. Instrum. Meth. A **831**, 24 (2016)
153. Z. Galloway et al., https://archiv.org/abs/1707.04961 (2017)
154. S.M. Mazza et al., JINST **15**-04, T04008 (2018)
155. Y. Zhao et al., Nucl. Instrum. Meth. A **924**, 387 (2019)
156. G. Casse, Talk at 27th Int. WS on Vertex Detectors, https://indico.cern.ch/event/710050/ (2018)
157. W. Shockley, J. Appl. Phys., **9**, 635 (1938); S. Ramo, Proc. I.R.E., 584 (1939)
158. Y. Yang et al., arXiv:1912.13211 [physics.ins-det], 13pp (2019)
159. M. Moll, Talk at LHCC open session, https://indico.cern.ch/event/527359/contributions/2158543/attachments/1278623/1898336/160525-RD50-Status_2016.pdf (2016)
160. M. Obertino et al., Nucl. Instrum. Meth. A **730**, 33 (2013)
161. S.I. Parker, C.J. Kenney, J. Segal, Nucl. Instrum. Meth. A **395**, 328 (1997)
162. The Stanford Nanofabrication Facility, https://snfexfab.stanford.edu/
163. The Sintef Company, https://www.sintef.no/en/all-laboratories/minalab/
164. B. Wu et al., J. Appl. Phys. **108**, 051101 (2010)
165. J. Lange, PoS VERTEX2015, 026 (2015)
166. M. Povoli et al., JINST **8**, C11022 (2013)
167. J. Lange, et al., JINST **10**, C03031 (2015)
168. M. Garcia-Sciveres et al., Nucl. Instrum. Meth. A **636**, 155 (2011)
169. The ATLAS Collaboration (L. Adamcyk et al.), Technical Design Report for the ATLAS Forward Proton Detector, CERN-LHCC-2015-009, The ATLAS-TDR-024 (2015)
170. The ATLAS Collaboration (G. Aad et al.), ATLAS Note ATL-Phys-PUB-2015-18 (2015)
171. The CMS-TOTEM Collaboration (M. Albrow et al.), CERN-LHCC-2014-021, TOTEM-TDR-003, CMS-TDR-13 (2014)
172. The CMS Collaboration (B. Vormwald et al.), JINST **14**, C07008 (2019)
173. G. Pellegrini, M. Manna, D. Quirion, Nucl. Instrum. Meth. A **924**, 69 (2019)
174. The CMS Collaboration (J. Sonneveld et al.), arXiv:1807.08987 [ins-det] (2019)
175. G. Pellegrini, Talk at the 27th RD50 Workshop, CERN (2015)
176. S. Terzo et al., JINST **14**-06, P06005 (2019)
177. G. Kramberger et al., Nucl. Instrum. Meth. A **934**, 26 (2019)
178. T. Tanaka et al., arXiv:1812.05803 [astro-ph.IM] (2018)
179. M. Yamada et al., Nucl. Instrum. Meth. A **831**, 309 (2016)
180. M. Asano et al., Nucl. Instrum. Meth. A **831**, 315 (2016)
181. R. Ichimiya et al., https://doi.org/10.5170/CERN-2009-006.68 (2009)
182. H.F. Sterling, R.C.G. Swann, *Solid State Electronics*, vol. 8, p. 663 (Pergamon Press, 1965)
183. V. Cindro et al., JINST **3**, P02004 (2008)
184. W. Adam et al., Nucl. Instrum. Meth. A **476**, 686 (2002)
185. J.W. Tsung et al., JINST **7**, P09009 (2012)
186. The BABAR Collaboration (A. Edwards et al.), Nucl. Instrum. Meth. A **552**, 176 (2005)
187. The CDF Collaboration (R. Eusebi et al.), IEEE Nuclear Science Symp. Conf. Record N **18**-2, 709 (2006); The CDF Collaboration (P. Dong et al.), IEEE Trans. Nucl. Sci. 55328 **55**, 328 (2008)
188. The ATLAS Collaboration (A. Gorisek et al.), Nucl. Instrum. Meth. A **572**, 67 (2007); The ATLAS Collaboration (V. Cindro et al.), JINST **3**, P02004 (2008)

189. The CMS Collaboration (A. Bell et al.), IEEE Nuclear Science Symp. Conf. Record N30-242, 2322 (2008); The CMS Collaboration (E. Bartz et al.), Nucl. Phys. B **197**, 171 (2009)
190. The LHCb Collaboration (M. Domke et al.), IEEE Nuclear Sci. Symp. Conf. Record N **58**-6, 3306 (2008)
191. The RD42 Collaboration (A. Alexopoulos et al.), PoS (Vertex 2016) 027 (2016)

Open Access This chapter is licensed under the terms of the Creative Commons Attribution 4.0 International License (http://creativecommons.org/licenses/by/4.0/), which permits use, sharing, adaptation, distribution and reproduction in any medium or format, as long as you give appropriate credit to the original author(s) and the source, provide a link to the Creative Commons license and indicate if changes were made.

The images or other third party material in this chapter are included in the chapter's Creative Commons license, unless indicated otherwise in a credit line to the material. If material is not included in the chapter's Creative Commons license and your intended use is not permitted by statutory regulation or exceeds the permitted use, you will need to obtain permission directly from the copyright holder.

Detectors for Energy Measurements 7

In this chapter, we discuss measurements of electromagnetic and hadronic energy and present various sub-detectors to measure these. We discuss new concepts, one that is based on particle flow and another that combines measurements of scintillation light and Cherenkov light. We start with summarizing some characteristics of electromagnetic showers.

7.1 Characteristics of the Electron-Photon Shower

We have seen in Sect. 2.6.4 that high-energy photons and electrons produce a shower of photons, electrons and positrons via e^+e^- pair creation and bremsstrahlung. The development of the electromagnetic cascade was illustrated in Fig. 2.41 (left). The process stops when the photon energy is too low for an e^+e^- pair creation and the energy of electrons and positrons approaches the critical energy. We discussed a simplified shower model that determined the shower maximum at $t_{max} = X_{max}/X_0 = \ln(E_0/E_c)/\ln 2$ with the number of particles $n_{max} = E_0/E_c$, where E_0 is the initial energy, E_c is the critical energy and X_{max} is the distance of the shower maximum. For $t_r = x/X_0 > t_{max}$, the particles still loose energy, photons via Compton scattering and photoabsorption and $e^{\pm}$ via ionization and excitation. Nearly stopped positrons annihilate with an electron from the atomic shell into two 511 keV photons. So, these particles travel several radiation lengths before they are stopped.

The integrated path length of all positrons and electrons with energy E_c is

$$\ell_{tr} = \frac{2}{3}X_0 \sum_{\nu=1}^{n} 2^{\nu} + \frac{2}{3}s_0 n_{max} = \left(\frac{4}{3}X_0 + \frac{2}{3}s_0\right)\frac{E_0}{E_c}, \quad (7.1)$$

where s_0 is the range of electrons with energies $E \leq E_c$ and n_{max} is the number of particles at shower maximum. Thus, the path length ℓ_{tr} is proportional to primary energy E_0. The exact proportionality factor depends on whether the e^- and e^+ also can be detected through deceleration from the critical energy until they are stopped. This is usually not the case since a minimum energy is required for detection. We can extract the track length from the longitudinal shower distribution introduced in the next section or from simulations since the most exact calculation of the electromagnetic shower development is obtained with the EGS4 program [1].

Simulations confirm our earlier findings of the electron-photon shower from the simple model.

1. The number of particles at shower maximum (n_{max}) is proportional to primary energy E_0.
2. The total track length ℓ_{tr} of electrons and positrons is proportional to E_0.
3. The depth at which shower maximum (t_{max}) occurs increases logarithmically with E_0.

A more realistic shower model distinguishes between electrons and photons and gives the following expression for the shower maximum [2],

$$t_{max} = \log\left(\frac{E_0}{E_c}\right) + t_e, \tag{7.2}$$

where the offset is $t_e = -0.5$ for electrons and $t_e = 0.5$ for photons. So, shower maximum of a 1 GeV photon is shifted by 1 X_0 with respect to that of a 1 GeV electron. Let us consider an example. For a 1 GeV photon in a NaI crystal, the critical energy is $E_c = 12.5$ MeV and the radiation length is $X_0 = 2.59$ cm, yielding $t_{max} = 4.9$, $n_{max} = 80$ particles and $X_{max} = 12.6$ cm.

7.1.1 Longitudinal Shower Profile

Figure 7.1 shows the longitudinal shower profile of 6 GeV electrons in different materials. When plotted in terms of radiation lengths, the shower profiles for different materials look rather similar. With increasing Z, the position of the shower maximum shifts to higher values since the critical energy becomes smaller.

The longitudinal shower profile is parameterized by

$$\frac{dE}{dt_r} = E_0 C_n t_r^\alpha \exp(-\beta t_r), \tag{7.3}$$

where $t_r = x/X_0$, $\beta \approx 0.5$, $\alpha = \beta t_{max}$ and $C_n = \beta^{\alpha+1}/\Gamma(\alpha+1)$. Figure 7.2 (left) shows simulated longitudinal shower profiles for different electron energies in copper using the EGS4 generator [1]. We see that the shower maximum increases logarithmically with the electron energy.

7.1 Characteristics of the Electron-Photon Shower

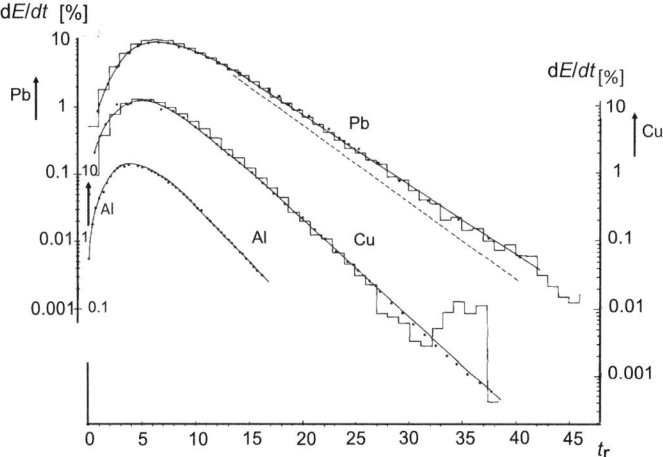

Fig. 7.1 Measured (curves) and simulated (histograms) longitudinal $E^{-1}\mathrm{d}E/\mathrm{d}t_r$ distribution as a function of $t_r = x/X_0$ for a 6 GeV electron in Pb, Cu and Al. The left-hand scale is for Pb and Al, while the right-hand scale is for Cu. The dashed line indicates the slope for $X_0 = 0.581$ cm. Reprinted with kind permission from [3], © 1970, Elsevier. All rights reserved

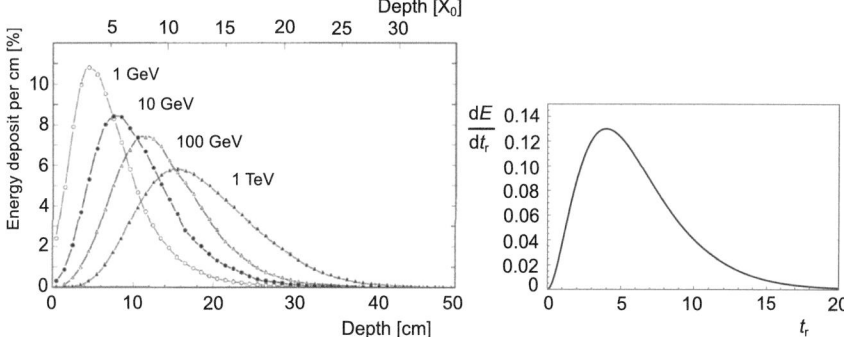

Fig. 7.2 Left: Simulated longitudinal shower profiles as a function of shower depth in copper for electron energies of 1 GeV (open circles, light grey curve), 10 GeV (solid points, black curve), 100 GeV (open triangles, grey curve) and 1 TeV (solid triangles, dark gray curve) [4]. Reprinted with kind permission from [4], © 2019, R. Wigmans. All rights reserved. Right: The $\mathrm{d}E/\mathrm{d}t_r$ distribution for 1 GeV photons in lead as a function of t_r

For 1 GeV photons in a Pb converter, we obtain the approximate form

$$\frac{\mathrm{d}E}{\mathrm{d}t_r} = 0.06 E_0 t_r^2 \exp(-0.5 t_r). \tag{7.4}$$

Since for lead $E_c = 6.9$ MeV and $X_0 = 0.56$ cm, the shower maximum is at $t_{\max} \approx 5.5$ as depicted in Fig. 7.2 (right). We see that the longitudinal energy profile drops exponentially and extends to several radiation lengths beyond the shower maximum.

For a CsI(Tl) calorimeter, we get $t_{max} \approx 5.0$. The length for 95% containment in units of radiation length is approximately given by

$$t_{95} = t_{max} + 9.6 + 0.08 \cdot Z, \qquad (7.5)$$

where Z is the atomic number. For a 1 GeV photon in lead, we get $t_{95} = 21.6$ whereas in CsI(Tl) we obtain $t_{95} = 18.9$, corresponding to 12.1 and 35.1 cm, respectively. For most $25 X_0$ thick calorimeters, the longitudinal shower leakage is less than 1% for electron energies up to 300 GeV [2].

7.1.2 Transverse Shower Profile

Another distribution is the transverse shower shape. Multiple scattering of low-energy electrons and positrons and Compton scattering of photons lead to a broadening of the radial shower shape. Figure 7.3 (left) shows the measured transverse energy distribution $E^{-1} dE/d\alpha_R$ for copper, lead and aluminum as a function $\alpha_R = R_{sh}/R_M$, where R_{sh} is the radius and R_M is the Moliére radius,

$$R_M = \sqrt{\frac{4\pi}{\alpha}} (m_e c^2) \frac{X_0}{E_c} = 21 \text{ MeV} \frac{X_0}{E_c}. \qquad (7.6)$$

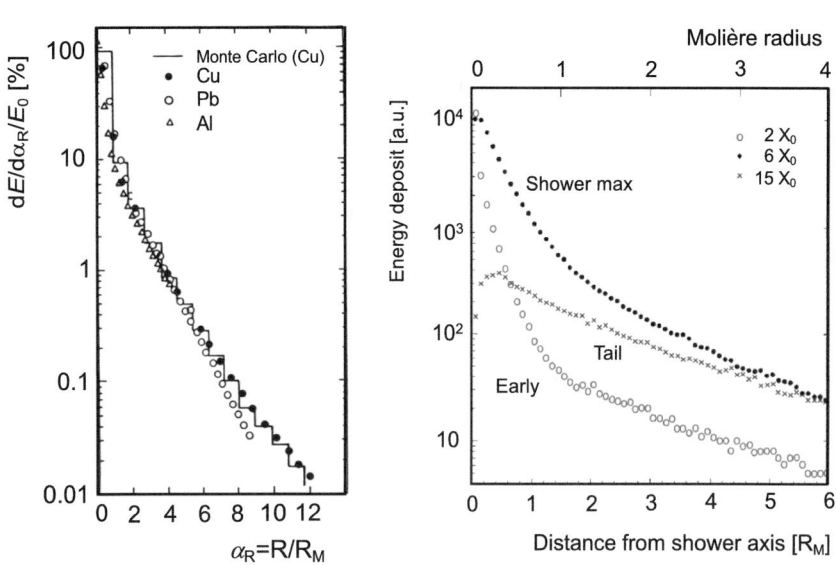

Fig. 7.3 Left: Measured transverse energy distribution $E^{-1} dE/d\alpha_R$ as a function of α_R in cylindrical intervals around the shower axis. Data are for aluminum (triangles), copper (solid points) and lead (open circles), while the histogram shows the simulation for copper. Reprinted with kind permission from [3], © 1970, Elsevier. All rights reserved. Right: Simulated transverse energy distributions of 10 GeV electrons in copper for 2 X_0 (open points), 6 X_0 (solid points) and 15 X_0 (crosses). Reprinted with kind permission from [4], © 2019, R. Wigmans. All rights reserved

7.1 Characteristics of the Electron-Photon Shower

The transverse shower shape parameterized in terms of Moliére radii is nearly independent of the material. Small deviations are visible at large α_R values. Inside 1 R_M, 90% of the electromagnetic shower is contained, while a 99% shower containment requires 3 R_M. To explore the transverse shower development, transverse shower profiles have been generated after 2 X_0, 6 X_0 and 15 X_0 for 10 GeV electrons in copper using the EGS4 simulation program. Figure 7.3 (right) shows the resulting distributions as a function of the distance from the shower axis (α_R). In the early development, the shower is rather narrow. At shower maximum most of the energy is deposited in the narrow core. For distances $t_r > t_{max}$, the transverse shower starts to broaden. At 15 X_0, the core has diminished and only a large tail is visible. In high-Z absorbers, we can parameterize the lateral energy distribution after shower maximum by

$$\frac{dE}{dR_{sh}} = a \exp(-\alpha_R) + b \exp(-\frac{R_{sh}}{\lambda_{min}}), \qquad (7.7)$$

where a, b and λ_{min} are free parameters. Note that λ_{min} introduces a slope different from α_R, describing the fall-off of the tail.

Figure 7.4 (left) shows the shower evolution in three dimensions for 6 GeV electrons in lead. This clearly illustrates the lateral shower spreading after shower maximum.

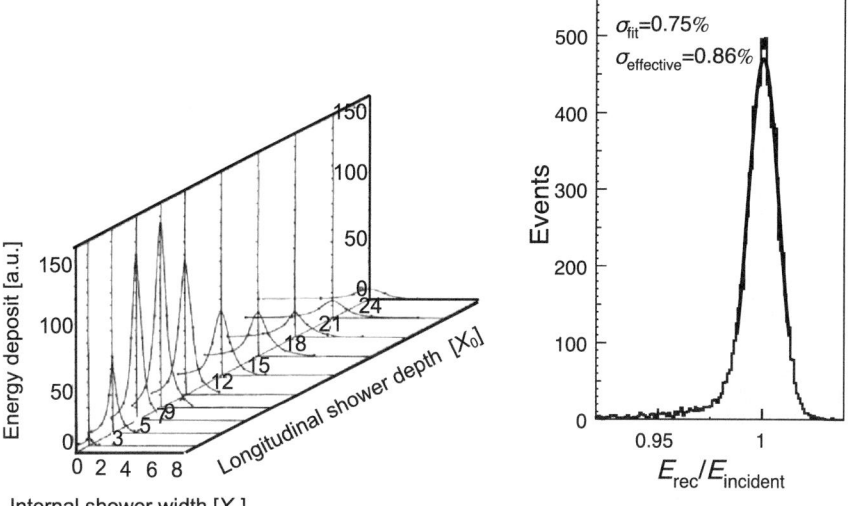

Fig. 7.4 Left: Three-dimensional shower development of a 6 GeV/c electron in lead [5]. Reprinted with kind permission from [5], © 1993, Springer. All rights reserved. Right: The measured photon energy spectrum normalized to the incident photon energy for PbWO$_4$ crystals. Reprinted with kind permission from [6], © 2004, Elsevier. All rights reserved

7.1.3 Energy Reconstruction

An electromagnetic shower is extended both longitudinally and laterally. While all calorimeters basically have lateral segmentation, the longitudinal segmentation is usually limited. We will see different examples in the following sections. A calorimeter is segmented into many cells. In a crystal or lead glass calorimeter, a cell consists of a single crystal or lead glass block read out with a photodetector. In a homogeneous liquid krypton (xenon) calorimeter or a sampling calorimeter, a cell consists of a readout pad or tile. Each cell has to be calibrated. First calibrations are performed in the laboratory with radioactive sources, cosmic muons, LED systems and lasers. Electronics is calibrated with well defined calibration signals. After moving the calorimeter into the experiment, calibrations need to be continued. Different experiments use different methods. In the *BABAR* experiment a fluorine source was used, which emitted a mono-energetic photon line at 6.13 MeV. By sending the fluorine through pipes in front of the calorimeter between data runs, all crystals were calibrated as often as needed. The photon is produced through the process $^{19}F + n \rightarrow\ ^{16}N + \alpha$ succeeded by $^{16}N \rightarrow\ ^{16}O^* e^- \nu_e$ and $O^* \rightarrow\ ^{16}O + \gamma$ (6.13 MeV). Radiative Bhabhas ($e^+e^- \rightarrow e^+e^-\gamma$) and processes like $\pi^0 \rightarrow \gamma\gamma$ and $\eta \rightarrow \gamma\gamma$ provide other calibration sources, since the energy of the three particles or the two photons is fixed by kinematics. In the CMS experiment, a laser system was installed to provide independent calibrations. Furthermore, physics processes such as $Z \rightarrow e^+e^-$ and $J/\psi \rightarrow e^+e^-$ decays are also used.

To define a cluster, we typically start with a cell that contains the largest amount of energy, called the seed cell. In the CMS PbWO$_4$ crystal calorimeter, about 70% of the photon energy is contained in the seed crystal and 96% of the energy in the seed crystal plus the eight surrounding neighbor crystals. We then add all energy above threshold of the eight nearest neighbors. The threshold is set to reduce the electronic noise. Typical values are around 1 MeV as in the *BABAR* experiment. Then, we extend the search to the second nearest neighbors and add energies above a threshold. So, a cluster is formed from all connected cells having energy above a fixed threshold. To get the energy from the signal amplitude in a cell, we have to apply a number of calibrations and corrections. For example, in the CMS PbWO$_4$ calorimeter, the energy of a cluster is determined by

$$E_{\text{EM}} = F_{\text{EM}} \Big[G(\eta) \sum_i S_i(t) C_i A_i H_{i,\eta} + E_{\text{pre}} \Big], \tag{7.8}$$

where A_i is the signal amplitude in the ith crystal and $G(\eta)$ is a conversion factor from ADC bins to the GeV scale that depends on pseudo-rapidity (1.38). The intercalibration factor C_i accounts for differences in light yield and photodetector responses in each cell in order to equalize the channel response. This is typically achieved using $\pi^0 \rightarrow \gamma\gamma$ or $Z \rightarrow e^+e^-$ decays. The precision in the barrel is < 0.3%. Since the crystal transparency may change with time, we need to correct for this effect by $S_i(t)$. The factor F_{EM} is an electron- or photon-specific multivariate correction of the shower that accounts for imperfect clustering as well as geometry and material

7.1 Characteristics of the Electron-Photon Shower

effects. The factor is different for photons and electrons. The factor $H_{i,\eta}$ corrects for the η-scale to equalize the η ring response with respect to the simulation.[1] Finally, E_{pre} is the energy measured in a pre-shower detector. Many calorimeters like *BABAR* in which the amount of inactive material in front of the calorimeter is small do not have a dedicated detector to measure this energy. Otherwise, the energy reconstruction is based on similar algorithms.

Figure 7.4 (right) shows the measured photon energy spectrum normalized to the incident energy, which was recorded with the CMS PbWO$_4$ crystals. The distribution, which is measured by summing 25 crystals, peaks at one and shows a tail at the lower energy side. The distribution is typically fitted with a Crystal Ball function, which consists of a Gaussian plus an exponential tail on the lower-energy side [7].

7.1.4 Cluster Shapes

Electromagnetic showers have characteristic shapes. There are different shape variables that provide discrimination of electromagnetic showers from background including neutral hadron showers. They become important in the reconstruction of photons. In an electromagnetic shower, most of the energy is contained in a single cell, the seed cell and the surrounding cells carry the rest. Thus, the energy ratios

$$S_1 S_9 = \frac{S1}{S9} = \frac{E_1}{\sum_{i=1}^{9} E_i},$$

$$S_9 S_{25} = \frac{S9}{S25} = \frac{\sum_{i=1}^{9} E_i}{\sum_{i=1}^{25} E_i} \qquad (7.9)$$

provide useful information. The sum in S_9 runs over the seed cell and all other eight cells that surround it. In S_{25} the energy of all cells in the second ring around the seed cell is added to S_9 as well. For electromagnetic showers, we typically have $0.4 < S_1 S_9 < 1$ and $S_9 S_{25} > 0.9$. Note that the exact shapes of these distributions depend on the cell size and the shower energy.

Another shape variable is the second angular moment,

$$TP2 = \frac{\sum_i E_i \cdot [(\theta_i - \theta_0)^2 + (\phi_i - \phi_0)^2]}{\sum_i E_i}, \qquad (7.10)$$

where E_i is energy deposited in the ith cell and θ_i and ϕ_i are polar and azimuth angle coordinates of the ith cell, respectively. The sum runs over all energies in the cluster. The parameters θ_0 and ϕ_0 are the coordinates of the cluster centroid. The lateral moment is defined by,

---

[1] The η-dependent correction factors are characteristic for hadron colliders. At e^+e^- machines the correction factors may be θ dependent.

$$LAT = \frac{\sum_{i=3}^{n} E_i \cdot r_i^2}{\sum_{i=3}^{n} E_i \cdot r_i^2 + E_1 \cdot r_0^2 + E_2 \cdot r_0^2}, \qquad (7.11)$$

where r_i is the radial position of the i cell with respect to the shower center, r_0 is the average distance between two cells, which is ~ 5 cm in the *BABAR* experiment. The energies are ordered as $E_1 > E_2 > \ldots\ldots > E_n$. The sums runs over all crystals in a shower except for the two cells with the highest energies. Two additional shape variables are the absolute values of the Zernike moments $Z(2, 0)$ and $Z(4, 2)$,

$$Z(2,0) = \sum_{r_i < a_0}^{n} \frac{E_i}{E} \left[2(r_i/a_0)^2 - 1 \right]$$
$$Z(4,2) = \sum_{r_i < a_0}^{n} \frac{E_i}{E} \left[4(r_i/a_0)^4 - 3(r_i/a_0)^2 \right] \exp\left[-2\phi_i\right], \qquad (7.12)$$

where $a_0 \simeq 15$ cm in *BABAR*. Figures 7.5 show these simulated distributions for signal and background in the *BABAR* experiment. The background consist of neutral hadron showers and split offs of hadronic events. We see that these distributions are different for true photons and background. The S_9 distribution for real photons lies between 0.3 and 0.98, while background shows a peak at one. The S_{25} distribution lies above 0.9 with a peak at one, where most of the background is located. The second angular moment has small values (<0.003). A large faction of the background is at zero. The lateral moment takes values between 0.05 and 0.5, while most of the background is located at zero. The Zernike moments $Z(2, 0)$ and $Z(4, 2)$ lie between 0.6 and 0.98 and 0.01 and 0.2, respectively, where background populate bins one and zero. Since many distributions have large overlaps between signal and background, they are typically combined into a multivariate analysis like neural networks [9] or boosted decision trees [10] to maximize the separation between photons and background.

7.1.5 Electromagnetic Energy Resolution

One important question is how well can we measure the electromagnetic energy. The intrinsic energy resolution of an ideal calorimeter, i.e. a calorimeter of infinite size that has no deterioration due to inefficiencies in signal collection and mechanical non-uniformities, is basically limited by statistical fluctuations of the track length ℓ_{tr}. Since the shower development is a stochastic process and the track length is proportional to the number of track segments, the intrinsic energy resolution is

$$\sigma_E \propto \sqrt{\ell_{\text{tr}}}. \qquad (7.13)$$

Since $\ell_{\text{tr}} \propto E_0$ we get

$$\frac{\sigma_E}{E} \propto \frac{1}{\sqrt{\ell_{\text{tr}}}} \propto \frac{1}{\sqrt{E_0}}. \qquad (7.14)$$

In homogeneous calorimeters, intrinsic fluctuations are small since the energy deposited in the active medium by the electron or photon does not fluctuate event-by-event. Therefore, the intrinsic energy resolution can be better than that given in (7.13)

7.1 Characteristics of the Electron-Photon Shower

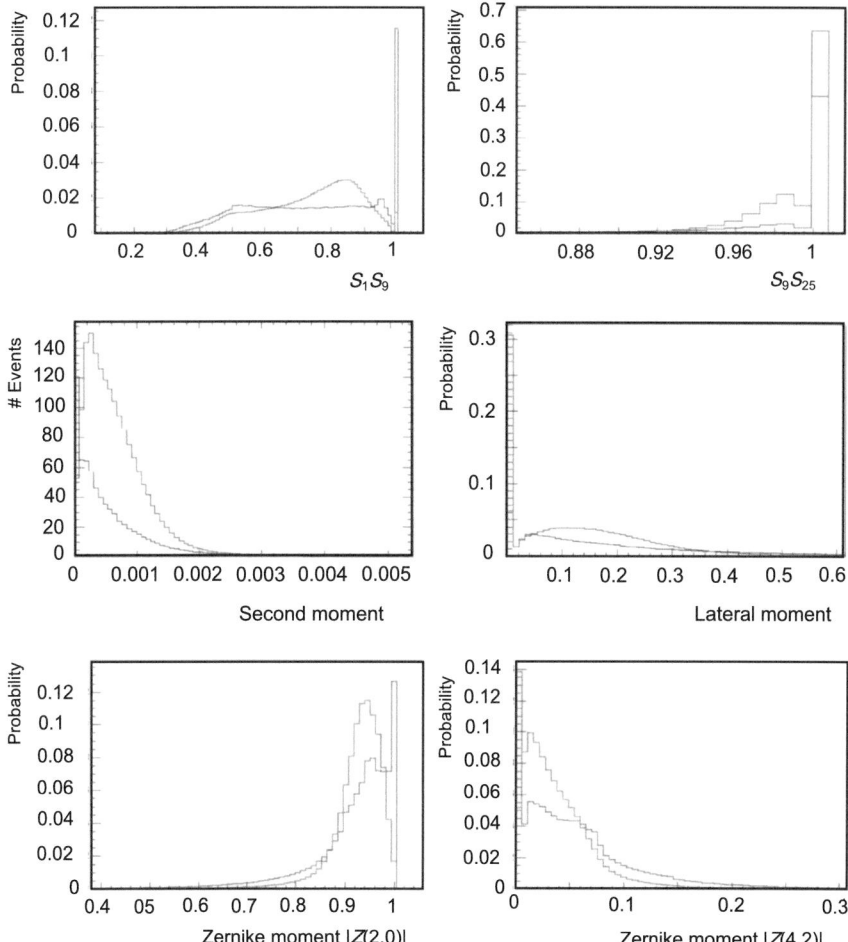

Fig. 7.5 Top left: The $S_1 S_9$ distribution. Top right: The $S_9 S_{25}$ distribution. Center left: The second angular moment. Center right: The lateral distribution. Bottom left: The Zernike moment $Z(2, 0)$. Bottom right: The Zernike moment $Z(4, 2)$. Shown are simulations in the BABAR CsI(Tl) crystal calorimeter for true photons (red curves) produced in generic $b\bar{b}$ events and for background (grey curves). All reprinted with kind permission from [8], © 2006, D. Bard. All rights reserved

by a Fano factor [11]. Particularly, in semiconductor, noble-gas and noble-liquid calorimeters, the Fano factor improves the intrinsic energy resolution considerably. In general, in homogeneous calorimeters the stochastic term is of the order of a few percent in units of $1/\sqrt{E[\text{GeV}]}$.

In a real calorimeter, we encounter several other effects that degrade the energy resolution. Homogeneous calorimeters are read out with photodetectors. While PhotoMultiplier Tubes (PMTs) have negligible electronic noise due to high-gain multiplication of the signal, this is different for some solid-state devices, where preamplifiers become necessary. The noise contribution is relevant at low energies.

Other effects include the uniformity of light production in the crystal, intercalibration effects with other crystals, the linearity of the crystal with energy and leakage effects. Thus, the energy resolution of a real homogeneous calorimeter can be written as

$$\left(\frac{\sigma_E}{E}\right) = \sqrt{\frac{\sigma_{\text{noise}}^2}{E^2} + \frac{a^2}{E} + b(E)^2}, \quad (7.15)$$

where σ_{noise}/E is the electronic noise term, $a/\sqrt{E}$ is the stochastic term and b is a constant term apart from energy leakage out the rear. Note that the three terms have been added in quadrature. The constant term consists of several contributions,

$$b(E)^2 = +\tilde{b}^2 + \sigma_{\text{FL}}^2 + \sigma_{\text{SL}}^2 + \sigma_{\text{RL}}^2 f(E)^2 + \sigma_{\text{NC}}^2, \quad (7.16)$$

where $\tilde{b}$ is a term for intercalibration effects and inhomogeneities, σ_{FL} is the front leakage, σ_{SL} is the side leakage, σ_{RL} is the rear leakage and σ_{NC} is the nuclear counter effect.[2] The function $f(E)$ is simply an energy expansion that accounts for rear leakage in calorimeters that are not sufficiently deep,

$$f(E) = 1 + a_1 E + a_2 E^2 + \cdots \quad (7.17)$$

The parameters a_1 and a_2 need to be determined from a fit to b(E). For a sufficiently long calorimeter, $f(E) \simeq 1$. For small rear leakage, all energy-independent terms can be represented by one constant b. Figure 7.6 (left) qualitatively illustrates this shape. If the calorimeter is sufficiently deep, we can ignore the rear leakage and the

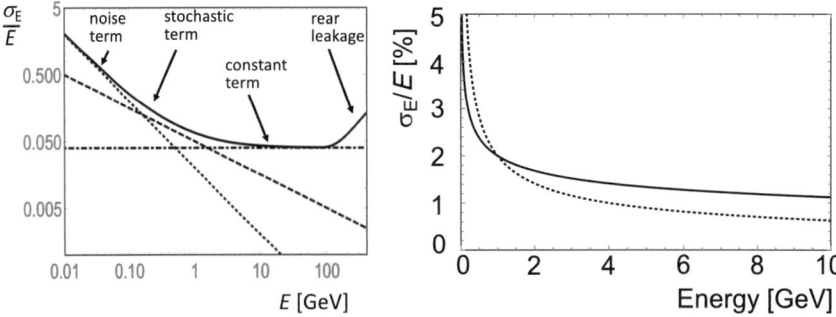

Fig. 7.6 Left: Energy resolution of a crystal calorimeter as a function of energy (solid curve) showing the noise term (dotted curve), the stochastic term (dashed curve), the constant term (dash-dotted curve) and energy leakage at the rear. Right: Comparison of $1/\sqrt{E}$ (dashed curve) and $1/\sqrt[4]{E}$ (solid curve) dependence in the stochastic term

---

[2] In this the case, the charged particle passes the photodetector and creates a signal.

energy resolution of homogeneous shower counters discussed in the next section can be written in a compact form

$$\left(\frac{\sigma_E}{E}\right) = \frac{\sigma_{\text{noise}}}{E} \oplus \left(\frac{a}{\sqrt{E}}\right) \oplus b, \tag{7.18}$$

where we have introduced the notation $\sqrt{(\sigma_1^2 + \sigma_2^2)} = \sigma_1 \oplus \sigma_2$. For lead glass and liquid noble calorimeters, the noise term is negligible and is set to zero. As already stated, the $1/\sqrt{E}$ dependence of the stochastic term results from the fact that the energy measurement is a statistical process. However, for NaI(Tl) and CsI(Tl) a deviation from the $1/\sqrt{E}$ dependence is observed. The data are better described by a $1/\sqrt[4]{E}$ energy dependence. The reason is not fully understood, but both NaI(Tl) and CsI(Tl) are hygroscopic, which affects the surface uniformity, which in turn affects the wave guide properties. At higher light densities, part of the light is not transported to the photodetector. Figure 7.6 (right) shows the $1/\sqrt{E}$ energy dependence of the stochastic term with respect to the $1/\sqrt[4]{E}$ dependence. At 1 GeV both resolutions are equal. At higher energies, the $1/\sqrt{E}$ dependence yields lower values, while below 1 GeV the $1/\sqrt[4]{E}$ term yields lower values. At 5 GeV, the $1/\sqrt[4]{E}$ term yields a 1.49 times higher resolution than the $1/\sqrt{E}$ term.

7.1.6 Noise

The noise term is important for readout with photodiodes and avalanche photodiodes at low energies. The observed noise has three contributions:

- parallel noise, which results from all voltage sources,
- serial noise, which results from all current sources,
- 1/f noise.

The noise is expressed in terms of equivalent noise charge ($\overline{ENC}$). Figure 7.7 (left) shows an equivalent circuit for the noise evaluation. Each electronic component is a noise source, where $\langle i^2 \rangle$ ($\langle e^2 \rangle$) indicates a noise current (voltage) source. The bias resistor contributes to the parallel noise, while the serial resistor contributes to the serial noise.[3] Note that these noise sources are so-called "white" noise sources, which means that they are frequency independent,

$$\langle i_{nd}^2 \rangle = 2q_e I_d \tag{7.19}$$
$$\langle i_{nb}^2 \rangle = \frac{4k_B T}{R_b} \tag{7.20}$$
$$\langle e_{ns}^2 \rangle = 4k_B T R_s, \tag{7.21}$$

---

[3] Generally, resistors shunting the input act as noise current sources and resistors in series with the input act as noise voltage sources.

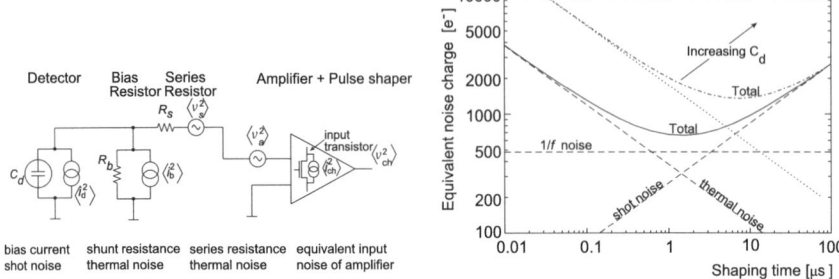

Fig. 7.7 Left: Equivalent circuit for the noise evaluation. The current noise sources are the detector noise current (i_{nd}), the dark current from the bias resistor (i_{nb}) and the dark current from the amplifier (i_{na}). The noise voltage sources are the equivalent noise from the serial resistor (e_{ns}), the equivalent input noise source from the amplifier (e_{na}) and that from the shaper (e_{sh}). The electronic components are the detector capacitance (C_d), the bias resistor (R_b) and the serial resistor (R_s). Right: The equivalent noise charge as a function of the shaping time for the total $\overline{ENC}$ (solid curve), the current-based or parallel noise, the voltage-based, thermal or serial noise and the $1/f$ noise. For increased voltage noise (e.g. increase in C_d) the curves are shifted along the diagonal (dash-dotted curve). Reprinted under CC-BY-NC-4.0 Licence with kind permission from [12], © 2024, the Particle Data Group LBNL

where I_d is the detector bias current, R_b is the bias resistor and R_s is the serial resistor. In addition, there may be a noise contribution from the preamplifier whose parameters e_{na} and i_{na} are typically of order nV/$\sqrt{\text{Hz}}$ and pA/$\sqrt{\text{Hz}}$.

Trapping and detrapping processes in resistors, dielectrics and semiconductors can introduce additional fluctuations whose noise power frequently exhibits a 1/f spectrum. The spectral density of the $1/f$ noise voltage is

$$\langle e_{nf}^2 \rangle = \frac{A_f}{f}, \tag{7.22}$$

where the noise coefficient A_f is device specific and is of order 10^{-10}–10^{-12} V^2. For a capacitive sensor and negligible amplifier noise the $(\overline{ENC})$ squared is given by

$$\overline{ENC}^2 = F_i \tau_{sh} \left[2q_e I_d + \frac{4k_B T}{R_b} \right] + \frac{F_v \cdot C_d^2}{\tau_{sh}} \left[4k_B T R_s \right] + F_{vf} \frac{A_f}{f} [C_d^2], \tag{7.23}$$

where C_d is the detector capacitance, τ_{sh} is the shaping time, $k_B T$ is the thermal energy and f is the frequency. The form factors F_i, F_v and F_{vf} depend on the shape of the pulse $P_{sh}(t)$ from the shaper. The first two form factors are given by

$$F_i = \tfrac{1}{2\tau_{sh}} \int_{-\infty}^{\infty} P_{sh}(t) dt \tag{7.24}$$
$$F_v = \tfrac{\tau_{sh}}{2} \int_{-\infty}^{\infty} \left[\tfrac{dP_{sh}(t)}{dt} \right]^2 dt.$$

For example, a pulse shaper formed by a single differentiator and integrator with equal time constants has $F_i = F_v = 0.9$ and $F_{vf} = 4$, independent of the shaping time constant. This $\overline{ENC}$ represents the circuit in Fig. 7.7 (left). If a circuit has more capacitances, then C_d has to be replaced by the sum of all capacitances.

The three terms in (7.23) are the parallel noise, serial noise and $1/f$ noise, respectively. The contribution from the detector current (I_d) is also called shot noise. The unit of $\overline{ENC}$ is rms electrons. The parallel noise is the current-based noise, which is independent of the detector capacitance and increases linearly with the shaping time τ_{sh}. The serial noise is a voltage noise, which decreases with shaping time as $1/\tau_{sh}$ and increases with the detector capacitance as C_d^2. The $1/f$ noise is independent of the shaping time and increases with C_d^2. Note that pulse shapers can be designed to reduce the effect of current noise, for example to mitigate radiation damage. An increase in pulse symmetry causes a decrease of F_i and an increase of F_v. For a shaper with one CR differentiator and four cascaded integrations we get 0.45 and 1, respectively.

Figure 7.7 (right) shows the time dependence of the equivalent noise charge. Since the serial noise decreases while the parallel noise increases, there is an optimal shaping time indicated by the minimum. For example, for a detector capacitance of 10 pF, the $1/f$ noise is 12.5 rms electrons. For $R_p = 1$ MΩ, $I_d = 1$ nA and room temperature, the parallel noise is 24 rms electrons at 1 ns. For a serial resistor of 50 Ω the serial noise at room temperature is 1,705 rms electrons at 1 ns. The minimum lies at $\tau = 70$ ns reducing the noise to 204 rms electrons.

7.1.7 Timing

The time resolution of measuring photons in a calorimeter is given by

$$\sigma_t = \sigma_{\text{phot}} \oplus \sigma_{\text{clock}} \oplus \sigma_{\text{elec}} \oplus \sigma_{\text{digi}} \oplus \sigma_{\text{noise}}, \quad (7.25)$$

where σ_{photo} is the contribution of recording the photon energy, σ_{clock} is the time resolution of the clock, σ_{elec} is the contribution of the electronics, σ_{digi} is the contribution of the digitization and σ_{noise} is the error from the noise contribution. The error from photon statistics is

$$\sigma_{\text{phot}} \propto \sqrt{\frac{\tau_{rt} \tau_{ft}}{N_{pe}}}, \quad (7.26)$$

where τ_{rt} is the signal rise time, τ_{ft} is the signal fall time and N_{pe} is the number of observed photoelectrons. The latter depends on the photodetection efficiency of the photodetector. Present test beam results yield $\sigma_{\text{phot}} \simeq 25$ ps [13]. The uncertainty in the clock is $\sigma_{\text{clock}} \simeq 15$ ps, while the electronics and digitization contribute with $\sigma_{\text{elec}} \simeq 8$ ps and $\sigma_{\text{digi}} \simeq 7$ ps, respectively. The noise increases with luminosity. At the end of the LHC High-Luminosity running after 3,000 fb^{-1}, the noise term is expected to be $\sigma_{\text{noise}} \simeq 50$ ps. The expected time resolution for new or upgraded detectors is of the order of 30–50 ps. The goal, however, is to improve this to 10 ps.

Fig. 7.8 Time resolution of electrons measured with a single CMS PbWO$_4$ crystal, using radiation hard electronics. Reprinted with kind permission from [14], © 2020, Elsevier. All rights reserved

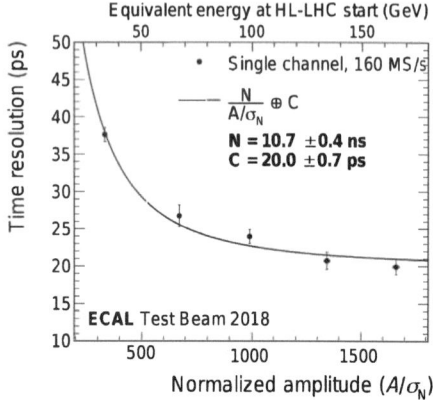

In another study, CMS has tested the time performance of PbWO$_4$ crystals [14]. They read out an array of crystals with multi-channel plates (see Sect. 8.2). The time resolution can be parameterized by

$$\sigma_t = \frac{N}{A/\sigma_N} \oplus C, \tag{7.27}$$

where N, C are fit parameters and A/σ_N is the normalized amplitude. They used new electronics that is foreseen for the High-Luminosity LHC running. It is radiation hard, has dual-gain trans-impedance amplifiers and a sampling ADCs with lossless data compression. With 160 MHz sampling, they obtained an improved time performance. Figure 7.8 shows the result. A fit yields $N = (10.7 \pm 0.4)$ ns and $C = (20.0 \pm 0.4)$ ps. So, for high amplitudes or large energies a time resolution of 20 ps is feasible.

7.1.8 Radiation Hardness

Most inorganic scintillating crystals are not very radiation hard. For example, both NaI(Tl) and CsI(Tl) crystal already show radiation effects after 10 krad as presented in Fig. 7.9. The performance depends on impurities introduced in the growing process and thus is dependent on the manufacturer. Hilger and Kharkov crystals show the largest light losses, which are up to 20% in the endcaps at 1 krad, while Shanghai crystals show the lowest light losses amounting to less than 2% in the barrel at 700 rad and 5–7% at 1,5 krad. The St. Gobain crystals lie in between. Pure CsI crystals perform somewhat better, being operable to around 1 Mrad. Similarly, BGO and BaF$_2$ are fine up to 1 Mrad. The most radiation-tolerant crystals are PbWO$_4$ and LYSO (LSO), which work fine above 100 Mrad.

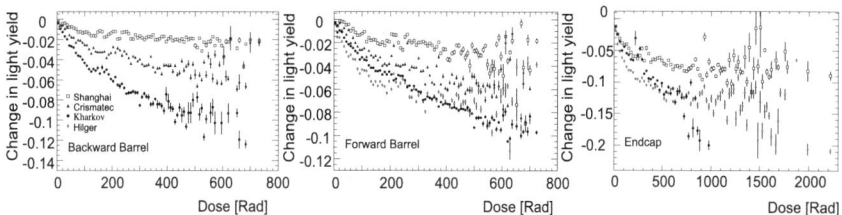

Fig. 7.9 Changes in light yield of the $BABAR$ electromagnetic calorimeter crystals as a function of accumulated radiation dose, showing crystals from different manufacturers, from Shanghai (squares), St Gobain (triangles), Kharkov (solid points) and Hilger (diamonds) in the backward barrel (left), forward barrel (center) and forward endcap (right). All reprinted with kind permission from [15], © 2013, Elsevier. All rights reserved

7.2 Homogeneous Shower Counters

Electromagnetic calorimeters are based on two different concepts. The first concept is based on homogeneous shower counters in which the total energy of the electromagnetic shower is measured in the active material. The second concept is based on sampling shower counters in which only a fraction of the shower energy is recorded. In this section, we discuss homogeneous shower counters. We can subdivide them further based on the active absorber, which consist of

1. Inorganic scintillating crystals.
2. Pb glass blocks.
3. Liquid noble gases.
4. Semiconductors.

We start with a discussion of inorganic scintillating crystals.

7.2.1 Inorganic Scintillating Crystals

We discussed the scintillation mechanism of inorganic scintillating crystals in Sect. 2.8.3. The typical crystal growing methods consist of the Czochralski [16] and Bridgman [17] methods.[4] There are many different scintillating crystals. For example, NaI(Tl), CsI(Tl), BaF_2, BGO, $PbWO_4$ and LSO/LYSO have been or are used in electromagnetic calorimeters in high-energy physics experiments. Figure 7.10 shows excitation spectra, photo-luminescence spectra and transmission curves of eight inorganic scintillators as a function of wavelength. The excitations of these crystals all

---

[4] The Czochralski method was discussed in Sect. 6.1.1. The Bridgman method has a crucible set inside an oven. The crucible is filled with a melt and a seed crystal is placed at the bottom. The rotating crucible is slowly lowered into the lower part of the oven, which has lower temperature. So, the crystal grows at the bottom and is pushed down.

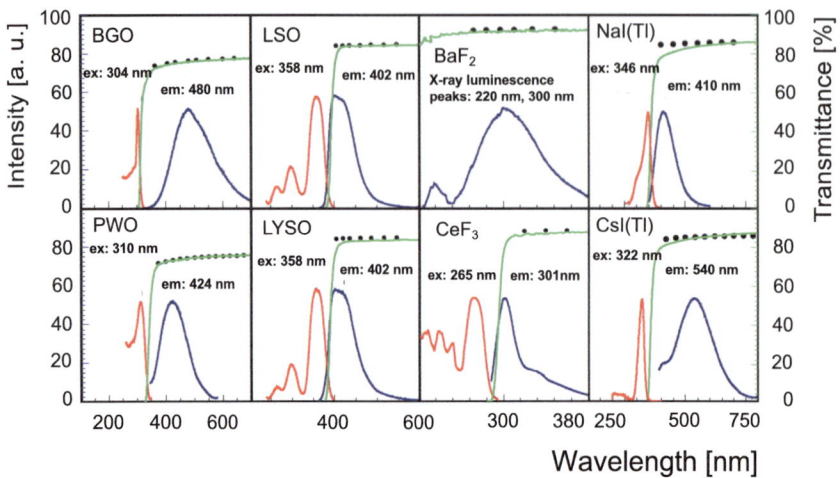

Fig. 7.10 Measured excitation spectra (red), photo-luminescence spectra (blue) and optical transmittance (green curve) of inorganic crystals as a function of wavelength; top row: BGO, LSO, BaF$_2$ and NaI(Tl); bottom row: PbWO$_4$, LYSO, CeF$_3$ and CsI(Tl). The left scale is for photo-luminescence (excitation) spectra, while the right scale is for transmittance curves. The black dots represent the theoretical limit of the transmittance. Reprinted with kind permission from [33], © 2006, R. Y. Zhu. All rights reserved

lie in the UV region. The photo-luminescence spectra lie in the UV region for BaF$_2$ and CeF$_3$ and in the blue region for the other crystal, except for CsI(Tl) that emits in the green-light region. All the crystals are transparent in the visible light spectrum. The transmission curves are above 80%, except for PbWO$_4$ and BaF$_2$, where the transmission curves are 75 and 90%, respectively.

Table 7.1 shows properties of the most commonly used inorganic crystals. Figure 7.11 shows photographs of CsI(Tl), PbWO$_4$, BGO and LYSO crystals. While surfaces of the latter three crystals are clear, that of CsI(Tl) looks foggy. This is caused by the hygroscopic properties of the crystal. Therefore, CsI crystals have to be operated in an inert atmosphere. Note that PbWO$_4$ is the fastest crystal, while CsI(Tl) has the highest light yield and an emission wavelength in the green spectrum that is well matched to the spectral sensitivity of silicon-based solid state detectors. Concerning radiation length, BGO and LYSO have the smallest. Both also have small Molière radii. The crystals with highest radiation hardness are PbWO$_4$ and LSO/LYSO.

Figure 7.12 (left) shows the measured energy spectrum of a ^{22}Na source in a 30 cm CsI(Tl) crystal read out with an avalanche photodiode. The peaks of the 511 and 1275 keV photon lines are clearly visible. Below each peak, the Compton edge is also visible as a shoulder. The measured resolutions are $\sigma_E/E = 9.3\%$ for the 511 keV line and $\sigma_E/E = 5.3\%$ for the 1275 keV line. Figure 7.12 (right) shows the measured energy spectrum of ^{22}Na in a 4 cm long LYSO crystal read out with a silicon photomultiplier. Again, the two lines and their Compton edges are clearly

7.2 Homogeneous Shower Counters

Table 7.1 Properties of different inorganic scintillators. Apart from the effective atomic number Z_{eff}, melting point T_{melt} and density ρ, we list the radiation length X_0, Moliére radius R_M, critical energy E_c, nuclear interaction length λ_I, energy loss at the minimum dE/dx, absorption length λ_{abs}, refractive index n, maximum emission wavelength λ_{scint}, decay time τ_d, light yield LY, dLY/dT, ratio of fast and slow components, radiation hardness, hygroscopic properties and experiments that use the crystal [12]. The element in parenthesis indicates the doping material of the scintillator.
[†] The KLOE experiment uses LYSO crystals, which are LSO crystals in which Lutecium atoms are replaced by Yttrium (amount between 5 and 70%). The properties of LYSO crystals depend on the amount of Yttrium and are similar as those of LSO

	NaI(Tl)	CsI(Tl)	CsI	BGO	PbWO$_4$	BaF$_2$	LSO(Ce)
Z_{eff}	32	54	54	19	31	25	54
T_{melt} [K]	933.2	894.2	894.2	1,317	1,396	1,641	2,050
ρ [g/cm^3]	3.667	4.51	4.51	7.13	8.30	4.893	7.40
X_0 [cm]	2.588	1.86	1.86	1.118	0.8903	2.026	1.143
R_M [cm]	4.105	3.531	3.531	2.259	1.959	3.117	2.07
E_c [MeV]	12.94	10.8	10.8	10.14	9.31	13.78	11.32
λ_I [cm]	42.16	38.04	38.04	22.32	20.27	30.46	20.64
$\frac{dE}{dx}$ [$\frac{\text{MeV}}{\text{cm}}$]	4.785	5.605	5.605	8.918	10.2	6.374	9.648
λ_{abs} [cm]	41.4	37.0	37.0	21.8	18.0	29.9	21
n	1.775	1.787	1.787	2.15	2.2	1.474	1.82
λ_{scint} [nm]	410	560	420 (310)	480	560 (420)	300 (220)	420
τ_d [ns]	230	680, 3340	35, 6	300	10, 30	0.9, 630	40
LY [γ/MeV]	40,000	54,000	2,500	8,200	100	11,350	27,000
$\frac{dLY}{dT}$ [%/°C]	~ 0	0.3	−0.6	−1.6	−1.9	−2.0	−0.3
Fast:slow	–	(64:36)	(29:61)	–	(20:80)	(12:88)	–
Rad. h. [Mrad]	0.01?	0.01	0.01–1	0.1–1	100	1	100
Hygroscopic	Strongly	Slightly	Slightly	No	No	No	No
Experiment	Crystal Ball	*BABAR*	KTeV	L3	CMS	TAPS	PET
	CUSB	Belle (II)[†]	Mu2e		ALICE		KLOE[†]
		CLEO II/III			PANDA		CMS BTL
		BES III					
Reference	[18, 19]	[20–25]	[26, 27]	[28]	[29]	[30]	[31, 32]

visible. The energy resolution of the 511 keV line is $\sigma_E/E = 7.0\%$. The resolution from photon statistics without any losses is expected to be around 5% at 511 keV.

7.2.2 Examples of Crystal Calorimeters

Several experiments have chosen scintillating inorganic crystals in their electromagnetic calorimeters. These include NaI(Tl) for Crystal Ball [19] and CUSB [18], CsI(Tl) for CLEO II [20], Crystal Barrel [21], *BABAR* [22], Belle/Belle II [23, 24]

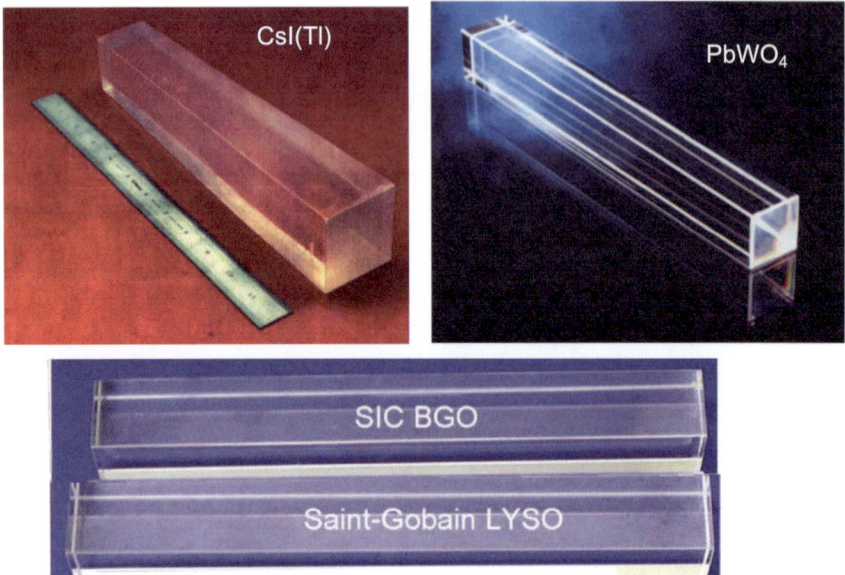

Fig. 7.11 Top left: Photograph of 30 cm long CsI(Tl) crystal. Reprinted with kind permission from [34], © 2002, CLEO II Collaboration. All rights reserved. Top right: Photograph of a 26 X_0 long PbWO$_4$ crystal. Reprinted with kind permission from [35], © 2012, ALICE Collaboration. All rights reserved. Center: Photograph of a large BGO crystal [36]. Bottom: Photograph of a large LYSO crystal. Both reprinted with kind permission from [36], © 2023, R. Y. Zhu. All rights reserved

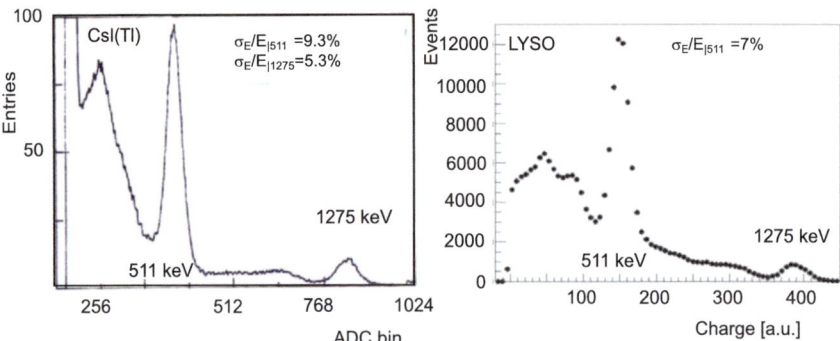

Fig. 7.12 Left: The ^{22}Na spectrum measured with a 30 cm long CsI(Tl) crystal read out with an avalanche photodiode. In addition to the 511 and 1275 keV photon peaks, the Compton edges are also visible. The energy resolutions of the two lines are 9 and 5.3%, respectively. Right: The ^{22}Na spectrum measured with a 4 cm long LYSO crystal, read out with a silicon photomultiplier. In addition to the 511 and 1275 keV peaks, the Compton edges are also visible. The energy resolution of the 511 keV line is 7%

7.2 Homogeneous Shower Counters

and BES III [25], pure CsI for KTeV [26] and Mu2e [27], BGO for L3 [28], BaF$_2$ for TAPS [29] and PbWO$_4$ for CMS [30]. As examples, we present here the *BABAR* CsI(Tl), KTeV pure CsI and CMS PbWO$_4$ calorimeters.

7.2.2.1 The *BABAR* CsI(Tl) Electromagnetic Calorimeter

Figure 7.13 (left) shows a longitudinal cross section of the *BABAR* electromagnetic calorimeter (EMC) indicating the arrangement of the 56 crystal rings. The crystals have trapezoidal shapes. They are arranged in projective geometry but with a small offset, so a photon from the interaction point cannot escape detection through the boundary between two adjacent crystals. The crystal lengths vary from 16.0 X_0 in the backward region to 17.5 X_0 in the forward region of the barrel continuing into the endcap except for the innermost ring, where the length is reduced to 16.5 X_0. Figure 7.13 (right) shows a schematic view of a wrapped CsI(Tl) crystal with the front-end readout package mounted on the rear face. Each crystal is wrapped in two layers of white TYVEK paper, a diffuse reflector,[5] plus an aluminum foil that acts as an RF shield. Note that there is an air gap between the Tyvec wrapping and the CsI crystal. This ascertains that light trapped inside the crystal (for angles $\theta > \theta_{\text{abs}}$) will be transported via total reflection to the rear of the crystal, where two photodiodes are mounted. Most of the light emitted under an angle $\theta < \theta_{\text{abs}}$ is lost. Some of it may be recovered by the highly efficient diffuse reflector. To approximately match refractive indices, the photodetector is coupled to the crystal typically by glue, optical grease or a silastic silicone cookie. In *BABAR*, the photodiodes are glued to a piece of plastic that is attached to the crystal. The two photodiodes are read out separately. The signal

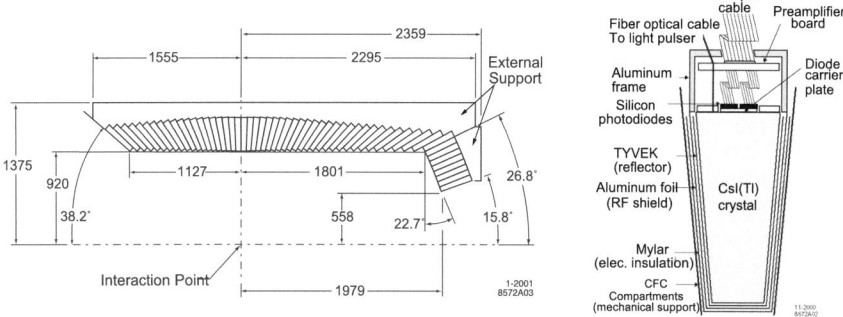

Fig. 7.13 Left: Longitudinal cross section of the top half of the *BABAR* projective electromagnetic calorimeter (EMC), indicating the arrangement of the 56 crystal rings. The detector is axially symmetric around the z axis. All dimensions are given in units of [mm]. Right: Schematic view of a wrapped *BABAR* CsI(Tl) crystal and the frontend readout package mounted on the rear face. The tapered, trapezoidal carbon fiber composite compartment is open at the front. Both reprinted with kind permission from [22], © 2002, Elsevier. All rights reserved

---

[5] In white reflectors, the light is emitted under a different angle than the incident angle whereas in mirrors outgoing angle equals incoming angle.

is amplified in a low-noise, charge-sensitive preamplifier with gains of one and 32. The signals are sent to the minicrate located on the back support structure, where they are digitized. Each crystal is inserted into a trapezoidal carbon fiber composite compartment that is open at the front.

Three crystals in ϕ and seven crystals in θ are mounted into a module that consists of a tapered trapezoidal carbon fiber composite compartment. The crystal weight is held from the back by an aluminum support structure. The modules are mounted on an aluminum support cylinder. Figure 7.14 (center) shows the barrel support structure onto which the individual modules are mounted. Figure 7.14 (left) shows an enlarged view of the electronics mini-crates, which house the amplification and digitization boards, the input/output boards and a cooling system. Figure 7.14 (right) shows the details of a module. Figure 7.15 shows a photograph of the inner barrel cylinder displaying the wrapped front face of each crystal in the barrel.

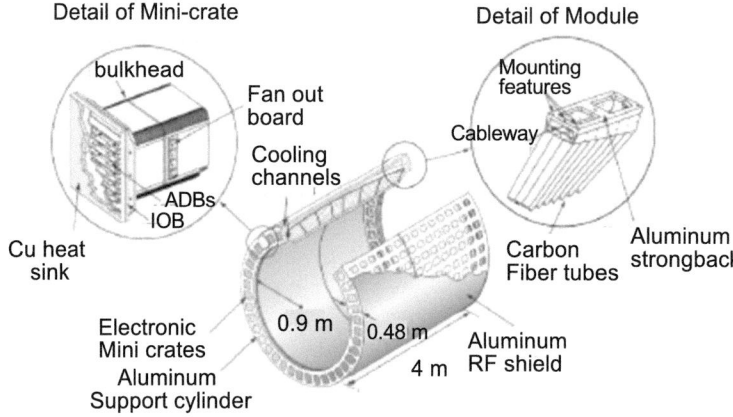

Fig. 7.14 The $BABAR$ calorimeter barrel support structure (center) with details of the module (right) and the electronics creates (left). Reprinted with kind permission from [22], © 2002, Elsevier. All rights reserved

Fig. 7.15 Left: Photograph of the $BABAR$ inner barrel cylinder showing the installation of the CsI(Tl) crystals. Reprinted with kind permission from [37], © 2019, C. Jessop. All rights reserved

7.2 Homogeneous Shower Counters

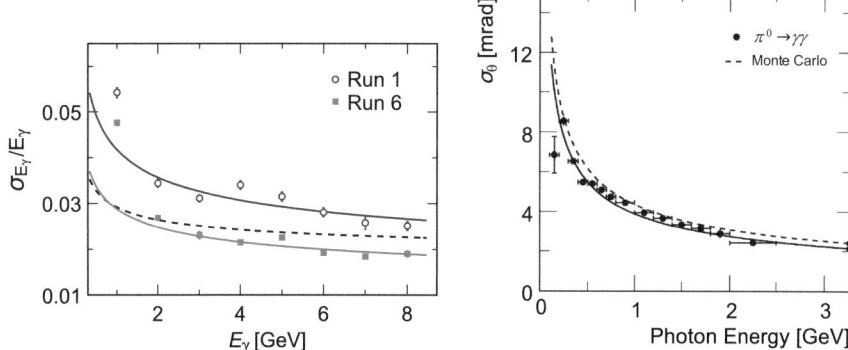

Fig. 7.16 Left: Measured energy resolution of the *BABAR* electromagnetic calorimeter as a function of energy, using radiative Bhabhas in Run 1 (open points) and Run 6 (solid squares) [15]. The dashed curve shows the parameterized energy resolution determined in 2001 that was based on photons from π^0 decays and $e^{\pm}$ from radiative Bhabha scattering. Right: Measured angular resolution of the *BABAR* electromagnetic calorimeter as a function of energy using $\pi^0 \to \gamma\gamma$ decays (black points). The solid curve shows the fit and the dashed curve depicts the simulation. Both reprinted with kind permission from [22], © 2002, Elsevier. All rights reserved

The energy resolution has been measured with radiative Bhabhas. Figure 7.16 (left) shows the energy resolution as a function of energy measured with the *BABAR* electromagnetic calorimeter at the beginning and end of the experiment. The improvement, which is clearly visible, results from improved intercalibrations and refined correction factors. The fit yields a resolution of

$$\frac{\sigma_E}{E} = \frac{2.7\%}{\sqrt[4]{E}} \oplus 1.0\%. \quad (7.28)$$

Below 1 GeV, the simulations are in good agreement with the data. For higher energies, they are too optimistic. Figure 7.16 (right) shows the corresponding angular resolution. The fit to the data yields

$$\sigma_\theta = \left(\frac{3.87 \pm 0.07}{\sqrt{E}} \oplus 0 \pm 0.04\right) \text{mrad}. \quad (7.29)$$

The simulations are in good agreement with the data.

7.2.2.2 The KTeV Pure CsI Electromagnetic Calorimeter

The KTeV CsI calorimeter consist of a 0.95 m × 0.95 m wall of pure CsI crystals that are stacked just with paper between crystals [39] to minimize systematic effects in the energy measurement. In the inner region the crystals are blocks with dimensions of 2.5 × 2.5 cm × 50 cm increasing to 5 × 5 cm × 50 cm in the outer part, corresponding to 27 radiation lengths. Figure 7.17 (left) shows a photograph of the backside of the KTeV CsI calorimeter. The crystals are read out with low-gain highly linear PMTs. The average light yield is 20 photoelectrons per MeV. The

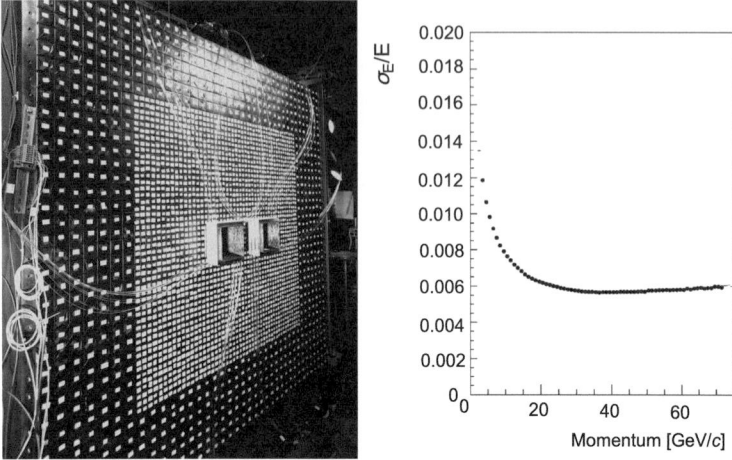

Fig. 7.17 Left: Photograph of the KTeV pure CsI calorimeter [38]. In the center there are two holes for the beam particles to pass. Reprinted with kind permission from [38], © 1998, KTEV Collaboration. All rights reserved. Right: The measured energy resolution of electrons from $K_L^0 \rightarrow \pi^\pm e^\mp \nu$ decays [26]. Reprinted with kind permission from [26], © 2001, Elsevier. All rights reserved

typical photon energy is between 10 and 40 GeV. The electron energy can go up to 70 GeV. The electron energy resolution is depicted in Fig. 7.17 (right), yielding $\sigma_E/E = 2\%/\sqrt{E} \oplus 0.5\%$. The position resolution for the fine-segmented crystals was 1.2 and 2.4 mm for the larger ones. The KTeV crystals showed no degradation after around 20 krad of irradiation.

7.2.2.3 The CMS PbWO$_4$ Electromagnetic Calorimeter

Figure 7.18 shows a schematic view of one quarter of the CMS electromagnetic calorimeter. The crystals are arranged in projective geometry in the barrel and end-

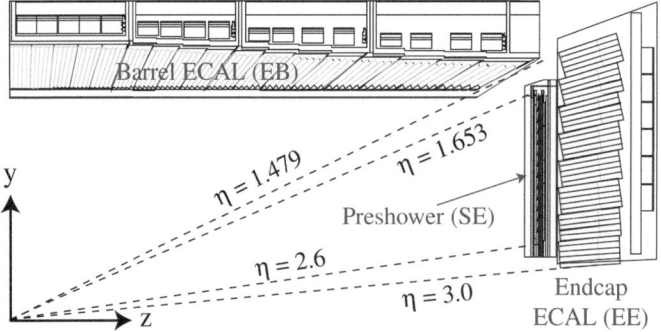

Fig. 7.18 Schematic view of one quarter of the CMS electromagnetic calorimeter showing the nearly projective geometry of the PbWO$_4$ crystals. Reprinted with kind permission from [40], © 2005, CMS Collaboration. All rights reserved

7.2 Homogeneous Shower Counters

 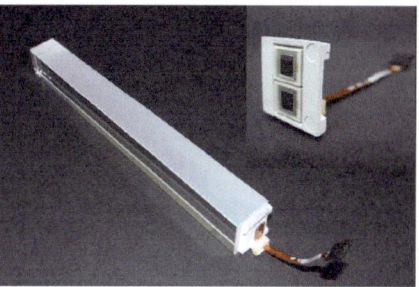

Fig. 7.19 Left: Photograph of the inner side of CMS barrel electromagnetic calorimeter. Reprinted with kind permission from [41], © 1998, CMS Collaboration. All rights reserved. Right: Photograph of a wrapped PbWO$_4$ crystal read out by two avalanche photodiodes. All reprinted with kind permission from [42], © 2008, IOP Publishing. All rights reserved

caps. There is a small crack in pseudo-rapidity where photons are not detected. Figure 7.19 (left) shows a photograph of the inner side of the CMS barrel electromagnetic calorimeter. The barrel holds 61,000 crystal arranged into 36 modules, 18 in each half of the detector covering a pseudo-rapidity range of $-1.48 < \eta < 1.48$. The barrel crystals are tapered, where the exact size depends on η. The crystals are about 23 cm long and have a front face cross section of 2.2 cm × 2.3 cm that extends to 2.5 cm × 2.6 cm at the rear. They are held in a glass fiber structure. Lead tungstate has a very short radiation length of 0.89 cm and a Molière radius of 2.2 cm. Figure 7.19 (right) shows a photograph of a wrapped PbWO$_4$ crystal and the two readout APDs.

Each endcap holds 7,300 22 cm long crystals, which have a front cross section of 2.86 cm × 2.86 cm extending to 3.0 cm × 3.0 cm at the rear. While barrel crystals are 25.8 X_0, endcap crystals are 1 X_0 shorter because of the presence of a pre-shower detector, which consists of two orthogonal layers of silicon strip detectors and two layers of lead planes contributing an extra 2.8 X_0 in front of each calorimeter endcap. The principal axes of the crystals are not pointing to the interaction point. They are offset by 3° in both the θ and ϕ directions. The barrel crystals are read out by two avalanche photodiodes, while for the readout of endcap crystals vacuum phototriodes are used.

The PbWO$_4$ calorimeter performs quite well. For a threshold of $p_T = 12.5$ GeV/c, the electron efficiency saturates around 95% for transverse momenta above 40 GeV/c and all pseudo-rapidities except for the barrel-endcap boundary. Figure 7.20 (left) shows the energy resolution of electrons as a function of energy measured in test beam with radiation-hard electronics. A fit to the data yields

$$\frac{\sigma_E}{E} = 0.37\% \oplus \frac{2.9\%}{\sqrt{E}} \oplus \frac{510 \text{ MeV}}{E}. \qquad (7.30)$$

The position resolution was also measured. Figure 7.20 (right) shows the result in the x direction. The positions were specified by hodoscopes whose 150 μm resolution has been subtracted. For showers larger than 80 GeV, positions were measured to better than 0.8 mm. Similar results were obtained in the y direction.

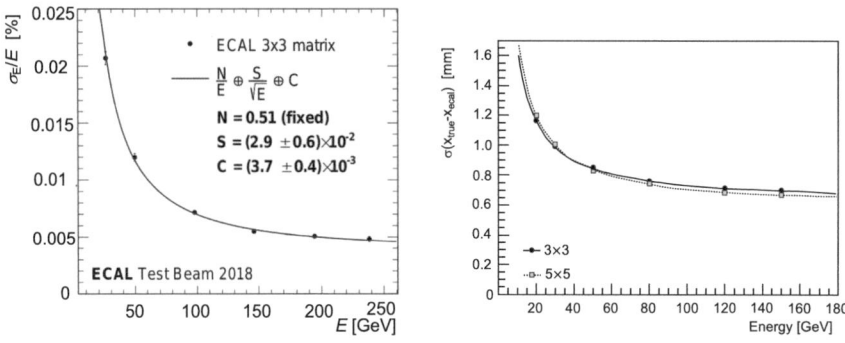

Fig. 7.20 Left: Electron energy resolution as a function of energy measured in a test beam, where the energy of 3 × 3 CMS crystals is summed, showing data (points with error bars) and a fit to $N/E \oplus S/\sqrt{E} \oplus C$ (solid line) that yields $N = 0.51 \, (fixed)$, $S = (2.9 \pm 0.6) \times 10^{-2}$ and $C = (3.7 \pm 0.4) \times 10^{-3}$. Reprinted with kind permission from [14], © 2020, Elsevier. All rights reserved. Right: Position resolution of CMS crystals in the x direction as a function of energy. Reprinted under CC-BY-4.0 Licence from [43], © 2009, CMS Collaboration

7.2.3 Beam Radiation Monitor

In colliding beam experiments, beam-related backgrounds are produced. They have to be kept sufficiently small in order not to damage detector components and provide serious backgrounds to the data analysis. Thus, they need to be monitored. One kind of background is beam radiation. For example, at the DAΦne collider photons in the 100 keV energy range need to be measured. My group in Bergen has been asked to build a beam radiation monitor.

Figure 7.21 (left) shows a schematic layout of a beam radiation monitor we designed and built in Bergen to measure and record beam radiation photons in the 100 keV energy range. Eight counters consisting of a LYSO crystal each that is read out with a SiPM is arranged in a ring around the beam pipe. Each crystal has dimensions of 0.3 cm × 0.3 cm × 4 cm and is wrapped in 3 layers of reflective ESR film from 3M. Each crystal is attached to the SiPM with radiation-hard glue. The SiPM is an experimental device from KETEK having an area of 3 mm × 3 mm with 12100 20 μm × 20 μm pixels. The signal is amplified with a charge-sensitive preamplifier, which has a gain of 8.5. The signal is digitized by a 14-bit Caen ADC that is controlled by Labview. We have tested all eight crystals with various radioactive sources, including ^{133}Ba, ^{57}Co, ^{22}Na, ^{137}Cs and ^{60}Co measuring the linearity and energy resolution.

We showed the energy spectrum of ^{22}Na in Fig. 7.12 (right). Figure 7.21 (right) depicts the ^{133}Ba spectrum, which has a broad structure around 356 keV and a clearly resolved line at 81 keV. Figure 7.22 (left), showing the measured energy as a function of the nominal energy in the range of 0 to 1.33 MeV, illustrates that the counters have a linear energy response at least up to 1.33 MeV. Using all well-measured peaks, we determine the energy resolution depicted in Fig. 7.22 (right) as a function $1/\sqrt{E}$. The data are consistent with the expected $1/\sqrt{E}$ dependence.

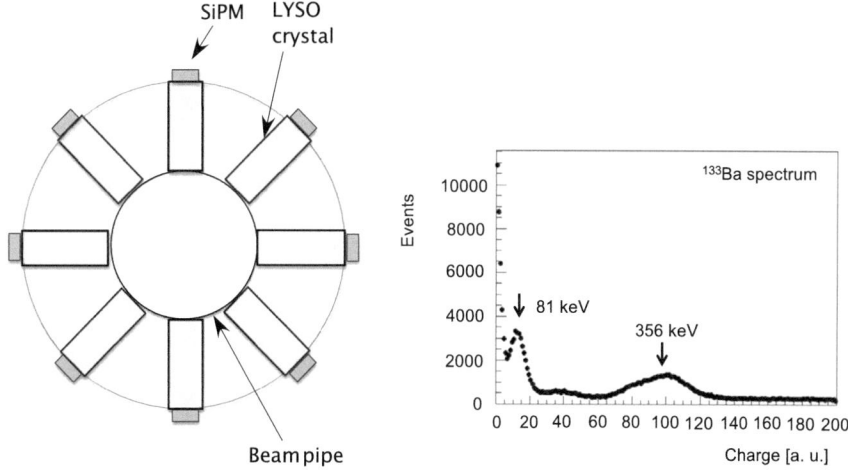

Fig. 7.21 Left: Schematic layout of a beam radiation monitor consisting of eight LYSO crystals read out by SiPMs. Right: The ^{133}Ba spectrum measured with a 0.3 cm × 0.3 cm × 4 cm LYSO crystal read out with a 3 mm × 3 mm SiPM

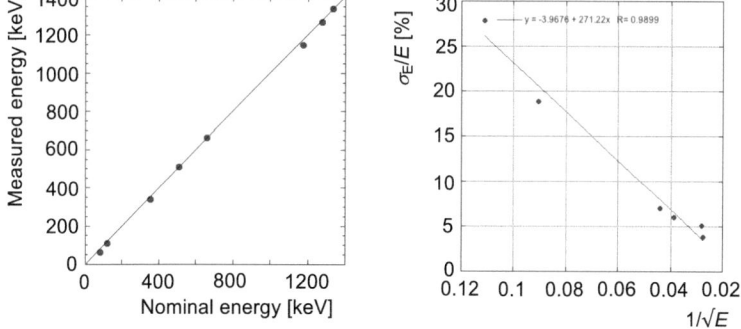

Fig. 7.22 Left: Measured energy versus nominal energy showing good linearity. Right: Measured energy resolution as a function of $1/\sqrt{E}$

7.2.3.1 Comparison of Homogeneous Electromagnetic Calorimeters

Table 7.2 summarizes properties of homogeneous shower counters that have been used in different experiments. The energy resolution of inorganic crystals and liquid krypton calorimeters yield the best energy resolution in the stochastic term, which is $2-3\%/\sqrt{E}$. In lead glass calorimeters, it is about a factor of two worse. As already mentioned, for NaI(Tl) and CsI(Tl) calorimeters, the stochastic term deviates from the $E^{-1/2}$ dependence. For these hygroscopic crystals, we encounter non-uniformities in the light collection due to surface losses on the crystal faces, leakage and calibration errors, which are due to the hygroscopic properties of these crystals. If calorimeters are too short, the energy resolution may be affected also by shower leakage. Note that for the KTEV calorimeter, the constant is zero.

Table 7.2 Absorber, experiment, depth, energy resolution, starting date and reference of homogeneous shower counters [12]. The absorber consists of inorganic crystals, lead glass and LKr. The energy resolution refers to the barrel. The lead glass and LKr detectors are discussed in the next subsections. The energy is in units of GeV. ‡The first value refers to the barrel, the second to the endcaps. §The first value refers to top and bottom rows, the second to the central rows

Absorber	Experiment	Depth	σ_E/E	Date	Ref.
NaI(Tl)	Crystal Ball	16 X_0	$\frac{2.8\%}{\sqrt[4]{E}}$	1983	[44]
BGO	L3	21 X_0	$\frac{1.6\%}{\sqrt{E}} \oplus 0.35\%$	1993	[45]
CsI	KTeV	27 X_0	$\frac{1.45\%}{E^{0.32}} \oplus \frac{0.7\%}{\sqrt{E}} \oplus \frac{0.7\%}{E}$	1996	[46,47]
CsI(Tl)	CLEO II	16.2 X_0	$\frac{0.35}{E^{0.75}}\% + 1.9\% - 0.1\% \cdot E$	1991	[20]
CsI(Tl)	BABAR	(16–17.5) X_0	$\frac{2.7\%}{\sqrt[4]{E}} \oplus 1\%$	1999	[15]
CsI(Tl)	Belle	16X_0	$\frac{0.81\%}{\sqrt[4]{E}} \oplus 1.18\% \oplus \frac{0.66\%}{E}$	1998	[23]
CsI(Tl)	BES III	15.1 X_0	$\frac{2.3\%}{\sqrt{E}} \oplus 1\%$	2010	[25]
PbWO$_4$	CMS	26/25 $X_0^{\ddagger}$	$\frac{2.9\%}{\sqrt{E}} \oplus 0.37\% \oplus \frac{0.51\%}{E}$	1997	[14]
PbWO$_4$	ALICE	20.2 X_0	$\frac{3.58\%}{\sqrt{E}} \oplus 1.12\% \oplus \frac{1.3\%}{E}$	2008	[48]
Lead glass	SLAC BC 72/73	10.5, 21§ X_0	$\frac{4.8\%}{\sqrt{E}} \oplus 0.84\%$	1981	[49]
Lead glass	SHMS	18.2 X_0	$\frac{3.75\%}{\sqrt{E}} \oplus 1.64\% \oplus \frac{1.96\%}{E}$	2010	[50]
Lead glass	OPAL	24.6 X_0	$\frac{6.3\%}{\sqrt{E}} \oplus 0.2\%$	1990	[51]
Lead glass	K2K	20. X_0	$\frac{7.9\%}{\sqrt{E}}$	1999	[52]
LKr	NA48	27 X_0	$\frac{3.2\%}{\sqrt{E}} \oplus 0.42\% \oplus \frac{9\%}{E}$	1998	[53]

7.2.4 Lead Glass Calorimeters

Lead glass is an amorphous solid consisting to about 80% of SiO_2 and PbO with additives of 10% sodium or potassium, 10% calcium and 1% other components like iron, tin and fluorine. Thus, different types of lead glass have been produced, which have different properties as shown in Table 7.3.

Lead glass detectors record Cherenkov light produced by showering electrons and positrons. The transparent lead glass blocks serve as waveguides. Photons are not detected directly. However, they typically convert with probability of 54% to an e^+e^- pair after one radiation length, which produces Cherenkov radiation. The number of Cherenkov photons is limited and for photons below 1.022 MeV some energy may be lost. Thus, the energy resolution is limited by the limited light yield. A computation yields

$$\frac{\sigma_E}{E} = \sqrt{0.006^2 + \frac{0.032}{\xi_p E \text{ [GeV]}}}, \quad (7.31)$$

7.2 Homogeneous Shower Counters

Table 7.3 Examples of lead glass types used in high-energy physics experiments showing the density ρ, the index of refraction n, the radiation length X_0 and the producer

Material	ρ [g/cm³]	n	X_0 [cm]	Producer	Ref.
SF2	3.6	1.62	3.06	Schott	[49]
SF5	4.08	1.673	2.36	Schott	[54]
SF6	5.18	1.8	1.7	Schott	[55]
SF57	5.54	1.847	1.5	Schott	[51]
TF1	3.86	1.65	2.5	CERN	[56]
TF101	3.86	1.65	2.8	CERN	[56]
F8	3.6	1.62	3.1	Lytkarino Optical Glass Factory	[56]
CEREN 25	4.06	1.708	2.51	Corning	[51]

where ξ_p is the ratio of the photocathode area to the counter back face. An actual measurement of 208 blocks of size 38 mm × 36 mm × 420 mm gave

$$\frac{\sigma_E}{E} = \sqrt{0.012^2 + \frac{0.053}{E\,[\text{GeV}]}}. \tag{7.32}$$

From this we determine $\xi_p = 0.35$ in agreement with expectation.

Lead glass counters have a long history in high-energy physics, since lead glass is cheap and easy to handle. The detectors typically consisted of walls built with lead glass blocks in fixed-target experiments at CERN [57–59], Fermilab [54] and SLAC [49]. Further, they have been used in electromagnetic calorimeters in many other experiments later, for example in OPAL [51], SELEX [56], COMPASS [60], HERMES [61], K2K [62] and detectors at CEBAF [50]. Typical lead glass block sizes are (4–15) cm in width and height and (35–65) cm in depth. The Cherenkov light was recorded with UV-sensitive PMTs. Energy resolutions varied around 5–10%/$\sqrt{E}\oplus$1–2%. As an example, we discuss the OPAL calorimeter.

7.2.4.1 The OPAL Calorimeter

The OPAL experiment at LEP used new heavy lead glass blocks for the electromagnetic calorimeter consisting of 9,440 blocks in the barrel (SF57) and 1,132 blocks in each endcap (CEREN 25) [51]. Figure 7.23 (left) shows a photograph of a lead glass block read out by a photomultiplier tube. Figure 7.23 (right) shows a photograph of a quarter of the lead glass barrel calorimeter. Both barrel and endcaps have an angular coverage of $|\cos\theta| < 0.98$. The blocks are quasi pointing to the interaction point. In ϕ, 80 identical blocks are used. In θ, 16 different sizes are needed to ensure quasi pointing blocks. In the barrel, the block size is of the order of ~ 10 cm $\times \sim 10$ cm $\times 37$ cm, corresponding to a total length of 24.6 X_0. The lead glass blocks in the endcaps come in different sizes, which have 22 X_0 on average. Each lead glass block is wrapped with a black 70 μm thick sheet of vinyl fluoride

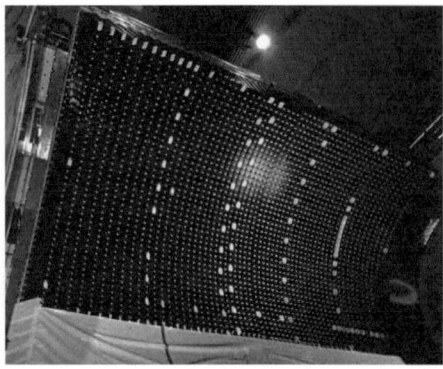

Fig. 7.23 Left: Photograph of an OPAL lead glass block. Reprinted with kind permission from [63], © 2017, OPAL Collaboration. All rights reserved. Right: Photograph of a quarter of the OPAL lead glass barrel calorimeter. Reprinted with kind permission from [64], © 2018, OPAL Collaboration. All rights reserved

laminated on each side with a polyester film. The inner surface of the sheet is coated with aluminum for efficient light reflection. The block is coupled via a 4 cm or 6 cm light guide to a 3-inch photomultiplier tube in the barrel and a phototriode in the endcaps. A stainless steel flange holds the lead glass block from the back. An optical fiber coupled to a Xenon flash lamp is inserted through one side of the flange. The blocks were calibrated with 50 GeV electrons. The gain was monitored with the Xe lamp. Since detectors in front of the calorimeter contribute to about $2X_0$ of material in the barrel becoming worse in the endcaps, OPAL installed pre-samplers. In the barrel they consisted of limited streamer tubes, while in the endcaps they used MWPCs.

The position resolution in the lead glass improves exponentially from 11 mm at 6 GeV to about 4 mm at 50 GeV. The energy response is linear. The energy resolution without any material in front of the lead glass blocks in the barrel is

$$\frac{\sigma_E}{E} = \frac{6.3\%}{\sqrt{E\,[\text{GeV}]}} \oplus 0.2\%. \tag{7.33}$$

Including the inactive material in front of the calorimeter, deteriorates the energy resolution at 6 GeV by a factor of two. However, including the information from the pre-sampler about 50% of the degradation can be recovered. The energy resolution in the endcaps was

$$\frac{\sigma_E}{E} = \frac{5.0\%}{\sqrt{E\,[\text{GeV}]}}. \tag{7.34}$$

For electrons inclined at 15°, the position in the endcaps was measured with an accuracy of 8–14 mm at 6 GeV.

7.2.5 Liquid Noble Gases

Liquid noble gases such as liquid argon (LAr), liquid krypton (LKr) and liquid xenon (LXe) are homogeneous materials that can be used for electromagnetic calorimeters. We discussed their properties in Chap. 2. The NA48 experiment at CERN, which studied CP violation in the decay $K_L^0 \to \pi\pi$, used a tank of liquid krypton for their electromagnetic calorimeter [53].

7.2.5.1 The NA48 Liquid Krypton Calorimeter

The NA48 liquid krypton calorimeter depicted in Fig. 7.24 (left) has an octagonal shape with a cross section of 5.5 m^2 and a 1.25 m depth along the beam direction corresponding to 27 X_0. The volume houses 10 m^3 of liquid krypton at a temperature of 120 K in a cryostat. The beam of undecayed neutral particles moves through a vacuum pipe of 8 cm radius in the center of the calorimeter. Figure 7.24 (center) shows a schematic view of one quarter of the calorimeter. Between the front plate and the back plate several spacer plates are placed as depicted in the figure. Outer rods hold the structure together. Thin copper-beryllium ribbons of dimensions 40 μm × 18 mm × 127 cm, running from front to back of the tank in longitudinal direction, form a tower-structured readout. There are 13,212 readout channels, each having a cell size of $x \times y = 2$ cm $\times$ 2 cm. They consist of an anode operated at a high voltage of 3 kV sandwiched between two cathodes kept at ground potential. In the horizontal direction, electrodes are separated by 1 cm thick gaps, while in vertical direction the separation is 2 mm. Figure 7.24 (right) shows a close-up view of the ribbons near a spacer plate, which guides the ribbons in an accordion geometry with a ±48 mrad zig-zag angle in the horizontal direction. The ionization electrons drift towards the anode, producing a current that is measured within 70 ns. Note that in this calorimeter only charged particles are recorded. By injecting calibration pulses into the calorimeter, the response uniformity and stability of the electronics is

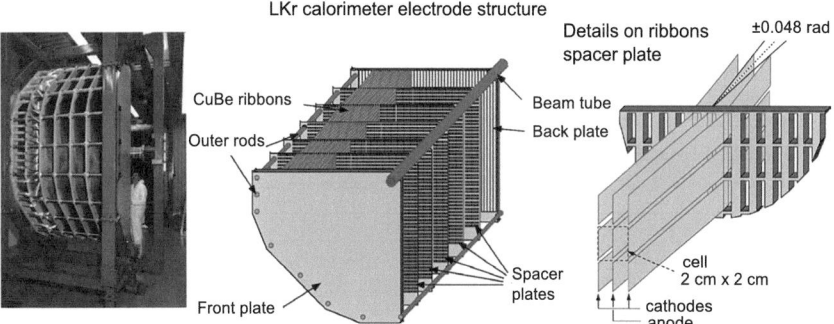

Fig. 7.24 Left: Photograph of the NA48 LKr calorimeter. Reprinted with kind permission from [65], © 2001, NA48 Collaboration. All rights reserved. Center: Schematic layout of one quadrant of the NA48 LKr calorimeter [66]. Right: Close-up view of the ribbon feedthrough at a spacer plate. The dashed square shows the cell size of the NA48 LKr calorimeter. Both reprinted with kind permission from [66], © 2000, NA48 Collaboration. All rights reserved

permanently checked. Furthermore, physics processes are exploited. As said before, the oxygen level needs to be kept at a very low level. At the top of the cryostat, the krypton evaporates continuously. The gas is taken out and is purified before it is recondensed and returned to the cryostat. The krypton purity is such that the lifetime of secondary electrons exceeds 100 µs.

This quasi homogeneous calorimeter has a Molière radius of 6.1 cm. The position resolution was measured in a test beam with electrons by comparing the center-of-gravity of the shower over a 3 × 3 grid corrected for its non-linearity, with the position reconstructed from the two drift chambers placed upstream of the calorimeter. The position resolutions in x and y are identical, yielding

$$\sigma_{x,\,y} = \frac{4.2}{\sqrt{E}} \oplus 0.6 \,[\text{mm}]. \quad (7.35)$$

So, the position resolution is better than 1 mm above 25 GeV. The time resolution for 50 GeV electrons is $\sigma_t = 262.8$ ps. The energy dependence of the time resolution is parameterized by

$$\sigma_t = \frac{2.5}{\sqrt{E}} \,[\text{ns}]. \quad (7.36)$$

Concerning the linearity of the shower reconstruction, Fig. 7.25 (left) shows the linearity of the calorimeter as a function of energy. The non-linearity across the entire energy range is less than 0.1% and is in good agreement with expectation. Figure 7.25 (right) shows the energy resolution as a function of energy. At high energies, the resolution approaches 0.6%. A fit to the data yields [68],

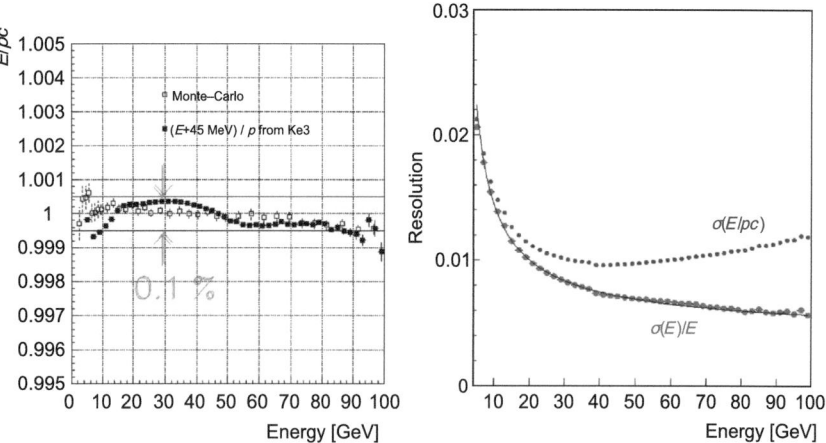

Fig. 7.25 Left: Linearity of the LKr calorimeter measured by the ratio of energy divided by momentum as a function of energy for data (solid points) and simulation (open circles). Right: Energy resolution as a function of energy showing σ_E/E (lower points) and $\sigma(E/pc)$ (upper points). Both reprinted with kind permission from [67], © 2007, Elsevier. All rights reserved

7.2 Homogeneous Shower Counters

$$\frac{\sigma_E}{E} = \frac{(3.2 \pm 0.2)\%}{\sqrt{E}} \oplus (0.42 \pm 0.05)\% \oplus \frac{(9 \pm 1)\%}{E}. \quad (7.37)$$

7.2.5.2 Liquid Xenon Calorimeters

Many experiments use liquid Xenon in their experiments because it has a higher atomic number, a shorter radiation length, a smaller Moliére radius and a smaller gap energy. We will mention here a few experiments briefly.

The MEG experiment at PSI [69], which has been built to look for charged-lepton flavor violation in the decay $\mu \to e\gamma$, uses a LXe calorimeter that has a volume of 900 liters and covers 10% of the solid angle. Figure 7.26 (left) shows a schematic layout of the calorimeter in the $r\phi$ direction. The MEG calorimeter focuses on detecting the scintillation light. The production mechanism was discussed in Sect. 2.8.2. The scintillation light is read out with 846 2-inch PMTs. The MEG experiment needs to measure a 52.83 MeV photon from the decay $\mu^+ \to e^+\gamma$. The achieved energy resolution is 2.4% for shallow events and 1.7% for deep events. The position resolution is 5 mm and the time resolution is 67 ps. In the upgraded MEG II experiment, the readout with PMTs in the inner wall is replaced by 3 mm × 3 mm silicon photomultipliers. Figure 7.26 (right) depicts a photograph of the MEG II inner calorimeter faces, showing mounted silicon photomultipliers and PMTs. The energy resolution is improved to 1.1 and 0.7% for shallow and deep events, respectively. Position and time resolutions are improved to 2 mm and 50 ps, respectively.

Most experiments in the search for dark matter in the form of weakly interacting massive particles (WIMP) use LXe detectors. The XENON1T experiment is a third-generation experiment to search for WIMPs. It uses a so-called dual-phase TPC [70] illustrated in Fig. 7.27 (left). A tank is filled with liquid xenon except for the top region that is filled with xenon gas. To record the scintillation light PMTs are installed on the top and bottom faces. The prompt scintillation light (S1) produced in the LXe is recorded by the 248 bottom and top PMTs that have a diameter of 76.2 mm.

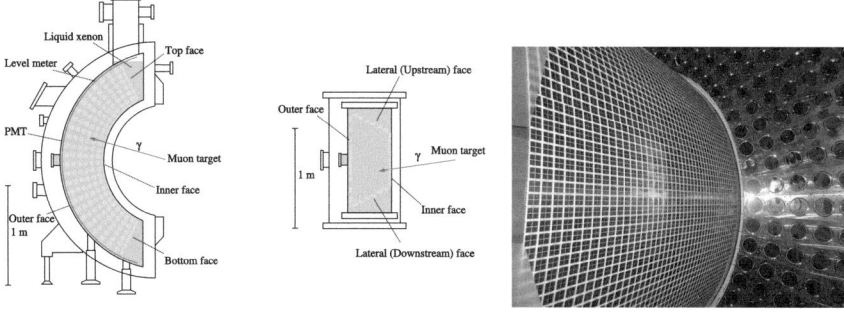

Fig. 7.26 Left: Schematic layout of the MEG LXe calorimeter in the $r\phi$ direction [69]. Center: Lateral view of the MEG LXe calorimeter. Both reprinted under CC-BY-NC-ND-4.0 Licence from [69], © 2014, MEG Collaboration. Right: Photograph of the inside of the MEG II calorimeter after silicon photomultipliers and PMTs were mounted. Reprinted under CC-BY-4.0 with kind permission from [71], © 2013, MEG2 Collaboration

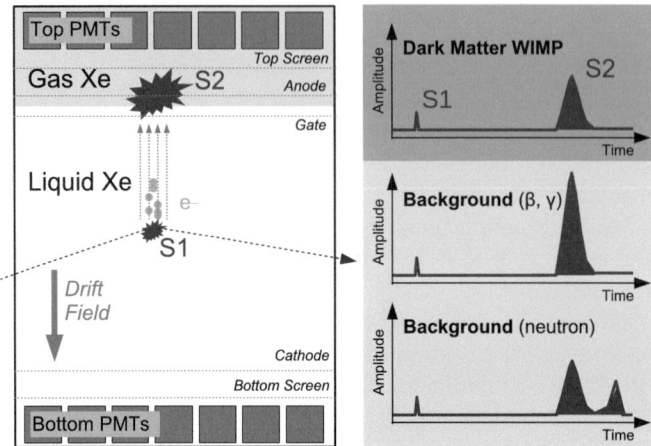

Fig. 7.27 Left: Principle of a dual LXe TPC operation. Right: Signal for a WIMP (top), backgrounds from β and γ events (center) and background from neutrons (bottom). Both reprinted under CC-BY-4.0 Licence with kind permission from [70], © 2017, XENON1T Collaboration

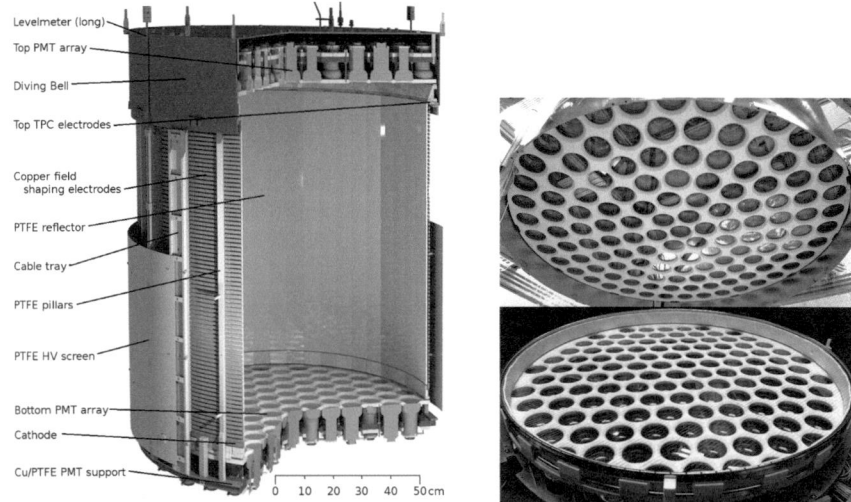

Fig. 7.28 Left: Schematic layout of the XENON1T detector. Right: Photographs of the top and bottom readout planes of the XENON1T detector. The holes hold the photomultipliers. Both reprinted under CC-BY-4.0 Licence with kind permission from [70], © 2017, XENON1T Collaboration

The same PMTs also record the delayed signal (S2) that is created by proportional scintillation induced by ionization electrons in the gas phase. A uniform electric field is produced by electrodes on the wall. By measuring the time difference between S1 and S2 and the S2 signal pattern, a three-dimensional space point of the interaction is achieved. Background signals are rejected by the charge-to-light ratio (S2/S1) and the scatter multiplicity as indicated in Fig. 7.27 (right).

Figure 7.28 (left) shows a schematic view of the cylindrical XENON1T TPC [70], which has a length of 97 cm and a diameter of 96 cm. It contains an active LXe target of 2.0 t in which both scintillation and ionization signals can be detected. The photodetectors are located at the bottom and top of the chamber as depicted in Fig. 7.28 (right). The chamber is built from low-radioactivity materials. To provide a homogeneous electric field, the TPC is surrounded by 74 electrodes with a cross section of 10 mm × 5 mm. The electric field is provided by the gate at the top on ground potential ($z = 0$ mm) and the negatively biased cathode of -12 kV at the bottom ($z = -969$ mm). This field is used to move the ionization electrons away from the interaction site, drifting them to the liquid gas interface. A second electric field is built up in the gas region on the top between the gate and the positively biased anode at 4 kV ($z = 5$ mm). Its purpose is to accelerate them sufficiently in the gas phase, so they excite and ionize the gas atoms. This generates the secondary scintillation signal (S2), which is proportional to the number of extracted electrons [70]. The position of the initial interaction, as well as the scatter multiplicity, can be reconstructed in three dimensions from the position and number of S2 signals observed by the top photosensors and the S1–S2 time difference. The ratio S2/S1 can be employed for electronic recoil background rejection, with typically >99.5% discrimination at 50% signal acceptance. In addition, there are two screening electrodes on the top (63 mm) and on the bottom (-1017 mm) in front of the PMTs, which can be biased to minimize the electric field in front of the photocathodes.

The top screen and the anode are made of 178 μm thick stainless steel wires with a pitch of 10.2 and 3.5 mm, respectively. The anode is made from 127 μm thick stainless steel wires with a pitch of 3.5 mm. The cathode and bottom screen use 216 μm thick gold-plated tungsten wires with a pitch of 7.75 mm. The top and bottom PMT arrays are equipped with 127 and 121 PMTs, having a diameter of 76.2 mm, respectively. The TPC is surrounded by a water Cherenkov detector to veto both muons and muon-induced neutrons. The PMTs are calibrated with blue LED light. In addition, ^{228}Th and ^{137}Cs gamma sources and a ^{241}Am source can be inserted into the cryostat to calibrate the electronics and nuclear recoils, respectively,

Describing the observed scintillation and charge signals respectively by Gaussian distributions, gives a light yield of (8.02 ± 0.06) photoelectrons (pe)/keV and a charge yield of (198.3 ± 2.3) pe/keV at an energy of 41.5 keV. The photon detection efficiency is $(12.5 \pm 0.6)\%$, while the electron extraction efficiency is consistent with the calculation of 96%.

Dual-phase LXe TPCs are also used in the ZEPLIN [72], LUX [73] and LUX-ZEPLIN [74] dark matter search experiments. The XMASS dark matter search experiment uses a single-phase TPC [75]. The EXO experiment that looks for neutrinoless double β decay uses a 0.2 t LXE TPC [76]. Other neutrinoless double beta experiments also rely on LXe detectors [77].

7.2.5.3 Semiconductor Calorimeters

In homogeneous semiconductor calorimeters, the energy is measured with silicon or germanium crystals. The latter are rarely used in high-energy physics experiments because they are unpractical for large system. They find application as photon detectors in nuclear physics experiments [78] since they have excellent energy resolution. This results from a large number of produced electron-hole pairs and small Fano factors. For example, a 1 MeV photon measured with a germanium detector produces 3.3×10^5 electron-hole pairs. For a Fano factor of 0.13 the expected energy resolution is 0.63 keV, which is agreement with a measurement of 0.55 keV. For comparison, the energy resolution of a small CsI(Tl) crystal was about 25 keV [79], which is about a factor of 50 worse. Silicon is used in sampling shower detectors, which we discuss in Sect. 7.3.4.3.

7.3 Sampling Shower Detectors

In the homogeneous shower detectors, the energy degradation and the energy detection are accomplished in the same medium. In sampling calorimeters, the fluctuations of the energy degradation and the energy measurement are separated in alternating layers of passive absorbers and active media. Figure 7.29 (left) illustrates this in a schematic layout of a sampling calorimeter. The choices for passive absorbers are plates made of Fe, Cu, W, Pb or U. They are placed on alternating ground- HV-ground potentials. The active media consist of gas mixtures, liquid noble gases, plastic scintillators or silicon. Properties of gases were discussed in Sect. 2.1.6, those of liquid noble gases were given in Sect. 2.8.2, properties and operation principles of plastic scintillators were presented in Sect. 2.8.4 and those of silicon detectors in Sect. 3.3.1 and Chap. 6.

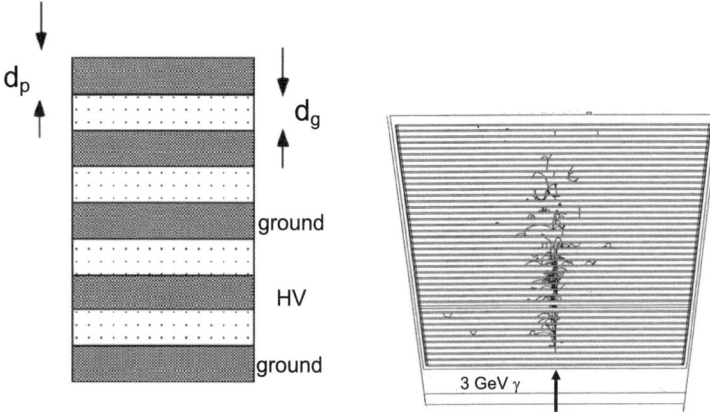

Fig. 7.29 Left: Schematic layout of a sampling calorimeter, which consists of alternating layers of absorber plates (grey, thickness d_p) and active media (light region, thickness d_g). Right: Simulation of a 3 GeV photon in a sampling calorimeter using the EGS4 program. Reprinted with kind permission from [80], © 1985, Springer. All rights reserved

7.3 Sampling Shower Detectors

Sampling calorimeters do not measure the entire energy of a particle but just part of it called sampling fraction f_{samp},

$$f_{samp} = \frac{\text{energy deposited in the active medium}}{\text{total energy deposited in the calorimeter}}. \quad (7.38)$$

The sampling fraction is typically obtained from the minimum energy loss in the active and passive media. For example, in the D0 uranium liquid-argon electromagnetic calorimeter, which used 3 mm thick uranium plates in the barrel, 1 mm thick G10 boards and 2×2.3 mm liquid-argon gaps, the minimum ionizing energy losses were $dE/dx = 6.147$ MeV/layer, 0.4 MeV/layer and 0.968 MeV/layer, respectively. Thus, the sampling fraction was $f_{samp} = 12.9\%$. The energy loss by ionization is smaller for electrons than for muons (see Fig. 2.21).

Sampling calorimeters permit the construction of rather compact devices with optimization for specific experimental requirements:

- e^--π discrimination,
- longitudinal shower profile,
- good angular measurements,
- good position measurements.

The thickness d_p of the absorber plates ranges from a fraction of a X_0 for electromagnetic energy measurements to a few X_0 for hadron energy measurements (see Sect. 7.4). The disadvantage is that only a fraction of total energy of the electromagnetic shower is detected in the active planes d_g. So, here we encounter sampling fluctuations that affect the energy resolution. However, the electromagnetic shower development is well understood, so simulations provide realistic descriptions. Figure 7.29 (right) shows a simulated 3 GeV photon shower in a sampling calorimeter. The curl-up of low-energy electrons and positrons is clearly visible.

The layouts of sampling calorimeters depend on the active medium. Figures 7.30 illustrate this. To collect the light from scintillators, we can use light guides coupled to the photodetector (top left) or a wavelength shifter coupled to the photodetector (top right). In case of liquid argon as active medium, the charge deposited on the absorber plates is collected (bottom left), while the readout with gaseous detectors is illustrated in Fig. 7.30 (bottom right). A fifth configuration (not-shown) consists of tungsten absorber plates and silicon chips as active material.

7.3.1 Properties of Sampling Calorimeters

The sampling fluctuations have been studied rather carefully. They depend on the characteristics of both the passive absorber plates and the active medium, in particular on the thickness and density as they contribute to the total sampling fluctuations. The intrinsic sampling fluctuations express the statistical fluctuations in the number of e^+e^- pairs traversing the actives signal planes.

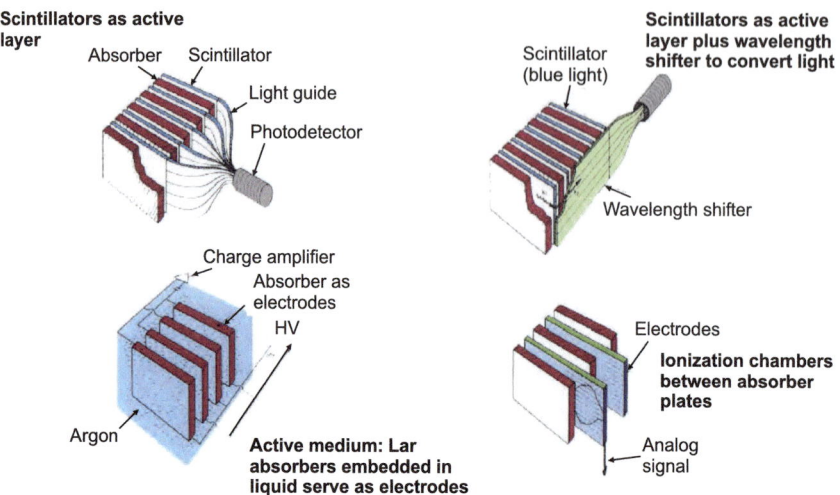

Fig. 7.30 Different active readout schemes of sampling calorimeters, with a plastic scintillator coupled to a light guide read out with a photodetector (top left), with a plastic scintillator coupled to a wavelength shifter that is attached to a photodetector (top right), with liquid argon, where the charge is collected from the absorber plates (bottom left) and with ionization chambers between absorber plates that collect the charge (bottom right). Reprinted with kind permission from [80], © 1985, Springer. All rights reserved

The sampling fluctuations can be estimated by Rossi's approximation B, which is based on the following assumptions:

1. The cross section for ionization is energy independent. The energy loss is $dE/dx = -E_c/X_0$ where E_c is the critical energy and X_0 is the radiation length.
2. Multiple scattering is neglected and the electromagnetic shower is treated in one dimension.
3. Compton scattering is neglected.

The number of crossings, N_r, in a sampling calorimeter for a cutoff energy $E_{\text{cut}} = 0$ is

$$N_r = \frac{\ell_{\text{tr}}}{d_s} = \frac{\text{total track length}}{\text{distance between active plates}}, \qquad (7.39)$$

where the sampling thickness $d_s = d_p + d_g$ is the plate thickness plus the gap depth. Since $\ell_{\text{tr}} = E_0/E_c \cdot X_0$,

$$N_r = \frac{E_0}{E_c} \frac{X_0}{d_s} = \frac{E_0}{\Delta E}, \qquad (7.40)$$

where E_0 is the initial energy and ΔE is the energy loss per unit cell. The contribution to the energy resolution is

$$\left(\frac{\sigma_E}{E}\right)_{\text{sampling}} = \frac{\sigma_{N_r}}{N_r} = \frac{1}{\sqrt{N_r}} = 3.2\% \left[\frac{\Delta E \text{ MeV}}{E \text{ [GeV]}}\right]^{1/2}. \qquad (7.41)$$

7.3 Sampling Shower Detectors

This expression is actually the lower bound on the sampling fluctuation. There are several effects that increase the sampling fluctuations:

1. Tracks originate from pair-produced particles, which means that the number of independent gap crossings would be only $N_r/2$ for a totally correlated production.
2. Rossi's approximation B ignores multiple scattering, which increases the effective distance $d'_s = d_s/\langle \cos\theta \rangle$, where the characteristic multiple scattering angle θ is given by $\langle \cos\theta \rangle \simeq \cos(21 \text{ MeV}/(E_c \pi))$.
3. For an energy cut-off $E_{\text{cut}} \neq 0$, we get the effective track length $\ell_{\text{tr}} = F(\xi_k)t_r \leq t_r$, where

$$F(\xi_k) = \left[1 + \xi_k \log \frac{\xi_k}{1.526}\right] \exp \xi_k, \quad (7.42)$$

and

$$\xi_k = 4.58 \frac{Z}{A} \frac{E_{\text{cut}}}{E_c}. \quad (7.43)$$

Figure 7.31 shows $F(\xi_k)$ as a function of E_{cut}/E_c. For small values of E_{cut}/E_c, simulations are in good agreement with Rossi's approximation B. For larger values of E_{cut}/E_c, calculations for air and lead as well as a simulation of a lead sampling device deviate from Rossi's approximation B. Thus,

$$\left(\frac{\sigma_E}{E}\right)_{\text{sampling}} \geq 3.2\% \left[\frac{\Delta E \text{ [MeV]}}{F(\xi) \cos(\frac{21 \text{ MeV}}{E_c \pi}) E \text{ [GeV]}}\right]^{1/2}. \quad (7.44)$$

Note that in this formula the energy loss has to be inserted in units of [MeV], while the energy is given in units of [GeV]. This parametrization does not include additional effects due to Landau fluctuations of the energy depositions in the active signal plane, which are estimated to be

$$\left(\frac{\sigma_E}{E}\right)_{\text{Landau}} = \left[\frac{3}{\sqrt{N_r}} \cdot \log(1.3 \times 10^4 \delta)\right], \quad (7.45)$$

where δ [MeV] is the energy loss per active detector plane. Such additional fluctuations are small for $\delta \sim$ few MeV in a few mm of scintillator, but they may become compatible to the intrinsic sampling fluctuations for thin detectors such as gaseous detectors where $\delta \sim$ few keV.

Finally, there is an error source that also depends on the density of the medium, yielding path length fluctuations. Here, low-energy electrons may be multiple-scattered into the plane of the active medium and then travel distances much larger than the gap thickness in the gaseous detector planes. Thus, these electrons deposit more energy than under perpendicular traversal. This effect is less significant in dense active layers because the range of low-energy electrons is comparable to the thickness of these layers. In dense detector planes, multiple scattering also tends to reduce this effect more than in light absorbers.

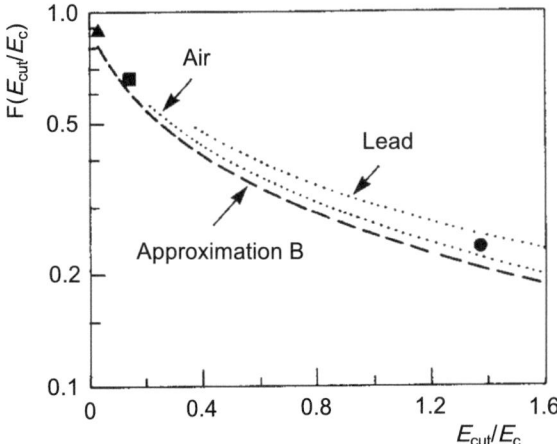

Fig. 7.31 Fraction $F(E_{\text{cut}}/E_c)$ of the total track length as a function of E_{cut}/E_c seen on average in a fully contained electromagnetic shower, where E_{cut} is the cut-off energy. Shown are Rossi's approximation B (dashed line), calculations for air and lead (dotted lines) and simulations for lead glass (triangles) and lead sampling devices (square and point). All reprinted with kind permission from [81], © 1981, IOP Publishing. All rights reserved

7.3.2 Energy Resolution of Electromagnetic Sampling Calorimeters

The total stochastic energy resolution of a sampling calorimeter is given by

$$\left(\frac{\sigma_E}{E}\right)_{\text{stochastic}} = \sqrt{\left(\frac{\sigma_E}{E}\right)^2_{\text{sampling}} + \left(\frac{\sigma_E}{E}\right)^2_{\text{Landau}} + \left(\frac{\sigma_E}{E}\right)^2_{\text{path length}}}. \quad (7.46)$$

Figure 7.32 (left) shows the contributions from sampling, path length and Landau fluctuations to the total energy resolution from a Pb-MWPC sampling calorimeter in which the largest uncertainties result from sampling and path length. In addition, contributions from the electronic noise (1/E) and a constant term may have to be included.

Table 7.4 lists the observed and measured energy resolutions, the individual contributions from sampling, Landau and path length as well as the factors $1/\sqrt{F(\xi_k)}$ and $1/\sqrt{\langle\cos\theta\rangle}$ for different sampling calorimeters. In sampling calorimeters with plastic scintillators as active medium, the sampling term dominates. The path length adds a small correction, while the Landau term is negligible. In the aluminum-scintillator calorimeter the energy resolution is poor due to the light absorber. The resolution improves with increasing Z. For the copper-scintillator calorimeter, the energy resolution is 13%, improving to 11% for the uranium-scintillator calorimeter. The best energy resolution is obtained with the iron-liquid-argon calorimeter. Here, the contribution from the path length is larger than that from sampling. Both the W-Si and Pb-gas calorimeters achieve poor energy resolutions. In the W-Si device, the contributions from sampling and path length are both large and similar in size. For

7.3 Sampling Shower Detectors

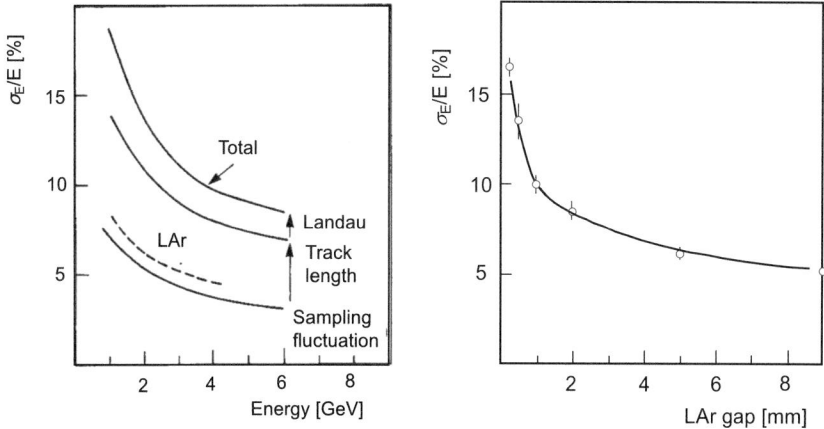

Fig. 7.32 Left: Contributions of sampling, path length and Landau fluctuations to the total energy resolution as a function of energy for a lead-MWPC sampling calorimeter. At 1 GeV the sampling fluctuations are 7%, while the other two contributions are about 12%, yielding a total energy resolution of $18\%/\sqrt{E}$. Reprinted with kind permission from [82], © 1978, Elsevier. All rights reserved. Right: Energy resolution σ_E/E of 1 GeV electrons as a function of the gap thickness for an Fe-LAr sampling calorimeter. Reprinted with kind permission from [80], © 1985, Springer. All rights reserved

Table 7.4 Properties of different electromagnetic sampling calorimeters including measured and estimated energy resolutions, contributions from sampling, Landau fluctuations and path lengths as well as cut-off energies and the correction factors $1/\sqrt{F(\xi_k)}$ and $1/\sqrt{\langle\cos\theta\rangle}$. Reprinted with kind permission from [80], © 1985, Springer. All rights reserved

Device	Al/scint.	FE/LAr	Cu/scint.	W/Si	Pb/Ar/CO_2	U/scint.
Passive/active [mm]	89/30	1.5/2.0	5/2.5	7.0/0.2	2.0/10.0 (NTP)	1.6/2.5
Reference	[83]	[82,84]	[82,85]	[82,86]	[82,87]	[82,88]
σ_E/E at 1GeV [%]	20	7.5	13.0	25.0	$\leq$ 20.0	11.0
E_{cut} [MeV]	3.0	0.7	0.7	0.7	$\leq$ 0.6	0.7
$F(\xi)^{-0.5}$	1.16	1.10	1.10	1.18	1.18	1.20
$\langle\cos\theta\rangle^{-0.5}$	1.00	1.03	1.03	1.27	1.36	1.51
$(\sigma_E/E)_{sample}$ [%]	23	4.8	9.2	19.1	8.2	10.6
$(\sigma_E/E)_{Landau}$ [%]	3.8	1.0	1.0	4.5	8.70	1.0
$(\sigma_E/E)_{path\ length}$ [%]		5.7	6.0	17.5	13.0	6
$(\sigma_E/E)_{estimated}$ [%]	23	7.5	10.0	25.9	17.7	12.2

the Pb-gas device, the path length contribution dominates followed by the sampling contribution.

Figure 7.32 (right) shows energy resolution σ_E/E of 1 GeV electrons as a function of the gap thickness for a Fe-LAr sampling calorimeter. For small gap sizes, the energy resolution becomes worse since the number of measurements in the gap is rather small so that Landau fluctuations become more dominant. The fastest improvements we get by increasing the gap to 2 mm. An increase from 2 to 5 mm improves the resolution from ~ 8 to $\sim 6\%$. A too large gap increases the calorimeter radius and in turn costs. Thus, most sampling calorimeters use gaps of the order of 2–3 mm.

Figure 7.33 (left) shows the energy resolutions of an ATLAS electromagnetic barrel liquid-argon calorimeter module measured in an electron test beam. The ATLAS electromagnetic sampling calorimeter is described in Sect. 7.3.4.2. The lead absorber is 1.5 mm thick and the argon gap is 2 mm. The measured energy resolution is

$$\frac{\sigma_E}{E} = \frac{(10.1 \pm 0.1)\%}{\sqrt{E}} \oplus (0.17 \pm 0.04)\%. \tag{7.47}$$

This satisfies the requirement that the stochastic term is $10\%/\sqrt{E}$ and the constant term is $<0.7\%$. Note that above 100 GeV, the energy resolution is better than 1%. Figure 7.33 (right) shows the energy resolution σ_E/E of a lead-scintillator calorimeter prototype as a function of the electron energy consisting of nine modules in comparison to that of a lead glass and lead liquid-argon calorimeter [90]. The lead-scintillator shower counter consists of an alternating stack of 130 layers of lead plates and scintillator plates. The 40 cm thick module correspond to 12.5 X_0.

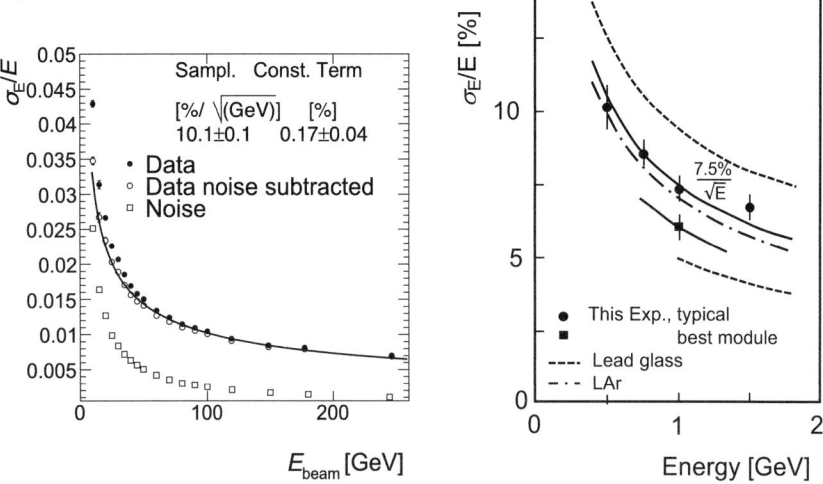

Fig. 7.33 Left: Energy resolution σ_E/E of the ATLAS LAr electromagnetic calorimeter barrel module studied in a test beam showing data (solid points), noise (open squares) and noise-corrected data (open circles). The sampling term is $\sigma_E/E_{\text{sampling}} = (10.1 \pm 0.11)\%/\sqrt{E}$, while the constant term is $(0.17 \pm 0.04)\%$. Reprinted with kind permission from [89], © 2006, Elsevier. All rights reserved. Right: Energy resolution σ_E/E as a function of energy for a lead-scintillator sampling calorimeter (points with error bars and solid curve). In comparison, energy resolutions of a lead glass array (upper dashed curve), an optimized limited area of lead-glass blocks (lower dashed curve) and a large liquid-argon calorimeter (dash-dotted curve) are shown. Reprinted with kind permission from [90], © 1976, Elsevier. All rights reserved

The lead plates have dimensions of 0.1 cm × 10 cm × 10 cm, the scintillator plates have a size of 0.5 cm × 10 cm × 10 cm. On one side the scintillators are coupled to a 5 mm thick BBQ wavelength shifter bar that is read out by a photomultiplier tube. The module is enclosed in a steel box. The stochastic term in the energy resolution is $7.5\%/\sqrt{E}$.

7.3.3 Position Resolution of Electromagnetic Sampling Calorimeters

The impact point of positrons, electrons or photons on an array of shower counters can be obtained by measuring the lateral energy distribution in the shower. The precision of the position information increases with the number of cells hit by showering particles and decreases with increasing cell size. The position resolution is best if energy is shared equally between two adjacent cells. The position resolution depends on the cell size [81],

$$\sigma_y = \sigma_y^0 \exp(d/d_0), \quad (7.48)$$

where d_0 is a reference distance. The energy dependence of the resolution σ_y^0 is given by

$$\sigma_y^0 = \frac{15}{E}\left(\ln^4 E + E/4\right)^{1/2} mm,$$

where E is given in units of [GeV].

Figure 7.34 (left) shows the spatial resolution versus strip width of an iron scintillator calorimeter. For the final calorimeter, a strip width of 5 cm was chosen. For a Pb-scintillator sandwich counter with 10 cm × 10 cm lateral dimensions, a spatial resolution of $\sigma_x = 1.1$ cm/$\sqrt{E}$ was obtained. Figure 7.34 (right) shows the position resolution for the PHENIX lead scintillator sampling calorimeter as a function of the electron energy [91]. The calorimeter has a tower structure. Each tower consists of 66 layers of 1.5 mm thick lead plates and 4 mm thick scintillator plates whose lateral dimension are 5.25 cm × 5.25 cm. It is read out with 36 wavelength-shifting fibers that are coupled to a photomultiplier tube. The position resolution is parameterized by $\sigma_r = 1.4$ mm $+ 5.9$ mm/$\sqrt{E}$. So at 100 GeV, the spatial resolution is 2 mm. The spatial resolution improves with $1/\sqrt{E}$. Often, the angular resolution is given, which is correlated with the position resolution.

7.3.4 Examples of Electromagnetic Sampling Calorimeters

The energy resolution of sampling calorimeters is much worse than that of homogeneous shower counters. While sampling calorimeters provide longitudinal segmentation, most homogeneous shower counters like crystal calorimeters cannot since crystals are not segmented longitudinally. Typically, sampling calorimeters also provide finer transverse shower segmentation yielding better angular resolutions than

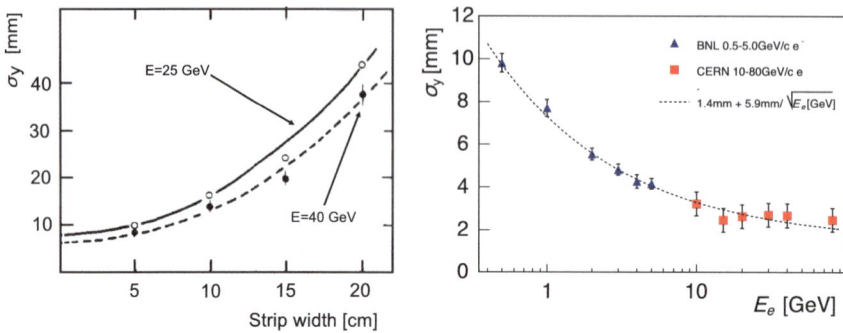

Fig. 7.34 Left: Spatial resolution of an iron-scintillator calorimeter versus strip width measured by the IHEP-IISN-LAPP Collaboration at 25 GeV (circles) and 40 GeV (points). The curves are fits with an exponential function listed in (7.48) with $d_0 = 10$ cm. Reprinted with kind permission from [81], © 1981, Springer. All rights reserved. Right: Position resolution of a lead scintillator calorimeter prototype as a function of the electron energy for low-energy data taken at Brookhaven National Laboratory (triangles) and for high-energy electrons taken at CERN (squares). A fit yields $\sigma_r = 1.4\,\text{mm} + 5.9\,\text{mm}/\sqrt{E}$ (dashed line). Reprinted under CC-BY-NC-ND-4.0 Licence from [91], © 2002, PHENIX Collaboration

crystal calorimeters. They are cheaper and, therefore, have been used in many experiments. The Mark III shower counter used 2.8 mm thick lead plates stacked with a 12.7 mm gap filled with Argon-methane (80:2) gas. The measured energy resolution was $17\%/\sqrt{E}$. The position resolution was 7 mrad in ϕ and 2.7 cm in z, corresponding to 20 mrad in θ. Willis introduced sampling calorimeters based on liquid argon [92]. He built a large prototype using 200 1.5 mm thick steel plates stacked with 2 mm gaps that were filled with liquid argon [93]. For 7 GeV/c muons, he measured a resolution of 11.4%. The Mark II experiment at SLAC was one of the first to use this new technology with a lead absorber, which was arranged into 2 mm thick lead plates followed by 2 mm thick lead strips separated by 3 mm gaps filled with liquid argon. Besides an aluminum trigger layer with 1.6 mm thick and 3.8 mm wide strips, the calorimeter contained 18 lead strip layers organized into six sections. Three of them measured the ϕ coordinate, two the θ coordinate and the sixth was rotated by 45° to resolve ambiguities. The calorimeter was behind the solenoid that was $1X_0$ thick. With the $1X_0$ aluminum, the energy resolution was about $14\%/\sqrt{E}$ improving to $10.8\%/\sqrt{E}$ without the aluminum. Also the SLD detector at SLAC used lead liquid-argon sampling calorimeters, we discuss below. At the LHC, the ATLAS experiment uses an iron-lead liquid-argon sampling calorimeter, which we discuss as well. We summarize the performance of several sampling calorimeters in Table 7.5.

7.3.4.1 The SLD Liquid-Argon Electromagnetic Calorimeter

The SLD lead liquid-argon sampling calorimeter consisted of a cylindrical barrel and two endcaps covering 98% of the total solid angle. The barrel was arranged into 144 modules, 48 modules in the azimuthal angle and three modules in the rz direction (center and end modules) covering the polar angle down to 33°. There were another

7.3 Sampling Shower Detectors

Table 7.5 Properties of different electromagnetic sampling calorimeters in the barrel listing the sampling fraction f_{samp}, total thickness, energy resolution σ_E/E, spatial σ_x or angular resolution $\sigma_\phi/\sigma_\theta$, experiment, starting date and reference. The energy is given in units of [GeV]. The energy resolution of the Mark II calorimeter is without $1X_0$ of aluminum. *This is for perpendicular tracks; for other polar angles a $\sin\theta$-dependent term needs to be added. †This is the spatial resolution in Layer 1; the noise term of $1\%/E$ was omitted. ‡There is only an endcap lead liquid-argon calorimeter

f_{samp}	Type	Total thickn.	σ_E/E [%] barrel	σ_x [mm] or $(\sigma_\phi/\sigma_\theta)$ [mrad]	Exp.	Date	Ref.
0.45	Pb-scint.	$12.5X_0$	$\frac{6.5}{\sqrt{E}} \oplus 7.2$	(13) at 5 GeV	Argus	1980	[94]
0.19	Pb-scint.	$18X_0$	$\frac{13.5}{\sqrt{E \times \sin\theta}}$	2 at 50 GeV	CDF	1988	[95]
0.32	Pb-scint.	$15X_0$	$\frac{7.58}{\sqrt{E}} \oplus 3.57$	–	E865	1992	[96]
0.24	Pb-fiber	$15X_0$	$\frac{5.7}{\sqrt{E}}$	–	KLOE	1995	[97]
0.30	Pb-scint.	18	$\frac{8.1}{\sqrt{E}} \oplus 2.1$	$\frac{5.7}{\sqrt{E}} \oplus 1.55$*	PHENIX	2002	[98]
0.07	U-scint.	$21X_0$	$\frac{18}{\sqrt{E}} \oplus 1$	$\frac{15.2}{\sqrt{E}} \oplus 1.5$	ZEUS	1988	[99,100]
0.017	Si-W	$19.8X_0$	$\frac{20}{\sqrt{E}}$	–	SLD	1980	[101]
0.015	Si-W	$24X_0$	$\frac{35}{\sqrt{E}}$	–	DELPHI	1990	[102]
0.20	Pb-LAr	$14X_0$	$\frac{11.5-12}{\sqrt{E}}$	(3.6–8.0)	Mark II	1978	[103]
0.20	Pb-LAr‡	$13.5X_0$	$\frac{10-12}{\sqrt{E}}$	(6/11)@ 1 GeV	TASSO	1980	[104]
0.33	Pb-LAr	$21X_0$	$\frac{10.1}{\sqrt{E}}$	(3–5)	CELLO	1980	[105]
0.185	Pb-LAr	$21X_0$	$\frac{15}{\sqrt{E}}$	–	SLD	1990	[106]
0.125	Pb-LAr	$20X_0$	$\frac{12}{\sqrt{E}} \oplus 0.6 \oplus \frac{32.5}{E}$	–	H1	1998	[107,108]
0.15	Pb-LAr	$22.3X_0$	$\frac{10.1}{\sqrt{E}} \oplus 0.17$	$\phi: \frac{3.87}{\sqrt{E}} \oplus 0.19$ $\theta: \frac{4.7}{\sqrt{E}} \oplus 0.21$	ATLAS	2010	[89,109]
0.073	U-LAr	$20.5X_0$	$\frac{15.7}{\sqrt{E}} \oplus 0.3 \oplus \frac{40}{E}$	0.8–1.2	D0	1993	[110]
–	Pb-PWC	$12X_0$	$\frac{17.5}{\sqrt{E}}$	(7)	Mark III	1977	[111]
–	Pb-TPC	$18X_0$	$\frac{32}{\sqrt{E}} \oplus 4.2$	(1.7/1.0)	DELPHI	1990	[112]

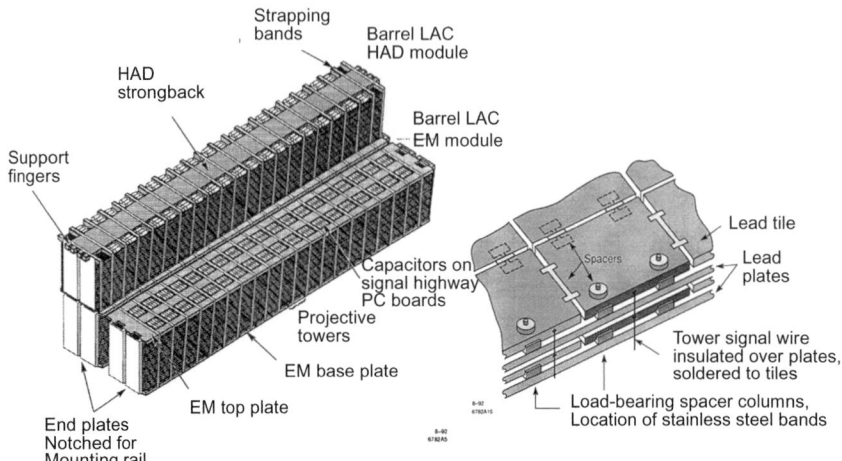

Fig. 7.35 Left: Layout of an SLD electromagnetic and hadronic barrel liquid-argon calorimeter module. Right: Detailed view of an electromagnetic calorimeter tile showing the different types of spacers. All reprinted with kind permission from [113], © 1990, Elsevier. All rights reserved

144 hadronic models on top that are discussed in the next section. Figure 7.35 (left) shows the schematic layout of a barrel module. The 16 wedge-shaped endcap modules extended the coverage down to 8°. The modules were constructed as parallel plate liquid-argon ionization chambers. The absorber structure consisted of alternate planes of large lead sheets and segmented lead tiles, with liquid-argon filling gaps between the planes. The tiles in successive planes were arranged transversely and connected radially to form projective towers. The lead plates were held on ground potential, while the tiles were placed at high voltage (2 kV) serving as charge collecting electrodes. The use of lead as both the high-Z absorber and the electrode structure permitted a compact electromagnetic calorimeter. Figure 7.35 (right) shows the layout of a tile layer. Lead plates and tiles were 2 mm thick separated by 2.75 mm gaps. Sixteen different specially designed spacers provided well-defined gaps between tiles and plates.

In the radial direction, 28 layers were stacked on top of an aluminum base plate corresponding to 21 X_0. The first eight layers were connected forming EM1 and the latter 20 layers were connected to EM2. The tile size in ϕ and θ increased appropriately to yield projective geometry. The tiles provided a tower structure. The ϕ direction was covered with 192 towers and the z direction with 68 towers. The tiles in the layers of EM1 and EM2 were soldered to two separate traces on capton strips that ran on the outer side to the frontend electronics. For a minimum ionizing particle, the sampling fraction was 18%. An aluminum plate on the outer side served as a strong back that carried the weight. It contained bays that housed the frontend electronics, which consisted of charge-sensitive preamplifiers, analog storage and multiplexing circuits, analog-to-digital converters, and associated power, control, calibration, readout, and monitoring devices. The modules were compressed and were held together by stainless steel bands to be operable under cryogenic temperatures. Figure 7.36 shows a

7.3 Sampling Shower Detectors

Fig. 7.36 Photograph of a stacked and compressed SLD electromagnetic liquid-argon calorimeter barrel module, showing the tower structure and kapton readout strips (personal photograph)

photograph of a stacked and compressed SLD electromagnetic-calorimeter module before attaching the stainless steel bands.

The endcaps extended the angular coverage to $8° < \theta < 35°$ and $145° < \theta < 172°$. Each endcap was built from 16 wedge-shaped modules, which were functionally identical to the barrel modules with a similar tower geometry. The tiles had trapezoidal shapes. A MIP signal produced 200,000 (500,000) electrons in EM1 (EM2) in comparison to a noise of 4,000 (5,000–6,000) electrons. The oxygen contamination was lower than 0.7 ppm. For 45 GeV/c Bhabhas, the measured energy resolution in the barrel was 2.3% leading to a stochastic term of $15\%/\sqrt{E}$ in the energy resolution. In the endcaps, it was worse due to inactive material in front of the calorimeter.

7.3.4.2 The ATLAS Liquid-Argon Electromagnetic Calorimeter

Figure 7.37 shows the calorimeters in the ATLAS experiment. The electromagnetic calorimeter consists of an iron-lead liquid-argon sampling calorimeter with a barrel section (22 X_0), two endcaps (24 X_0) and two forward calorimeters [114]. We focus here on the barrel calorimeter, which is built from two half-sections. A half-barrel consists of 16 modules and covers the pseudo-rapidity range of $0 < |\eta| < 1.475$. The absorbers plates and readout electrodes have an accordion shape, which provides complete ϕ symmetry without any azimuthal cracks. The lead thickness in the absorber plates has been optimized as a function of η in terms of the energy resolution. A half barrel contains 1,024 absorber plates interleaved with readout electrodes.

Figure 7.38 shows the layout of a barrel module, which is segmented into three layers in depths, a 4.3 X_0-deep Layer 1, a 16 X_0-deep Layer 2 and a 2 X_0-deep Layer 3. In addition, a 1.7 X_0-thick lead liquid-argon pre-sampler is installed in front of Layer 1 to allow for energy corrections of photons and electrons that interacted before entering the calorimeter. Layer 1 is finely segmented into $\Delta\eta \times \Delta\phi = 0.0031 \times 0.098$ strips. The middle layer collects the bulk of the energy. Its segmentation is

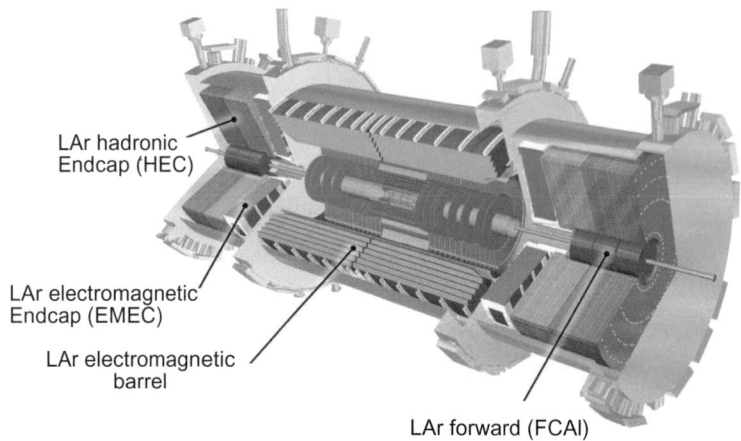

Fig. 7.37 Schematic view of the calorimeters in the ATLAS experiment. Reprinted under CC-BY-4.0 Licence with kind permission from [115], © 2010, ATLAS Collaboration

$\Delta\eta \times \Delta\phi = 0.025 \times 0.0245$. The third layer just collects the tail of the shower. Its segmentation is increased to $\Delta\eta \times \Delta\phi = 0.05 \times 0.025$. The pre-sampler is located in the same cryostat as the barrel. It is made of 64 identical azimuthal sectors, each 3.1 m long and 0.28 m wide. The coverage in $\Delta\eta \times \Delta\phi$ is 1.52×0.2. Figure 7.38 also illustrates the trigger tower size of $\Delta\eta \times \Delta\phi = 0.1 \times 0.0982$.

Figure 7.39 (left) shows the layout of the absorber plates and read out electrodes. The absorber layers consist of lead plates that are sandwiched between two 0.2 mm thick stainless-steel sheets to provide mechanical strength. In the barrel, the lead plates are 1.53 mm thick for $\eta < 0.8$, decreasing to 1.13 mm for $0.8 < \eta < 1.4$. The 2 mm gap between the absorber layers and readout layers is filled with liquid argon. The readout electrodes consist of three conductive copper layers separated by insulating polyimide sheets. The two outer layers are at a potential of 2 kV, while the inner layer is used for reading out the signal via capacitive coupling. The segmentation of the calorimeter in pseudo-rapidity and in depth is obtained by etched patterns on the different layers. In ϕ, electrodes are ganged together. This yields an accordion-shaped tower structure shown by the photograph in Fig. 7.39 (top right). Figure 7.39 (bottom right) shows a simulated electromagnetic shower in the ATLAS liquid-argon barrel.

A charged particle traversing a liquid-argon gap produces a current that has a short rise time of the order of 1 ns and a linear decay determined by the maximum drift time, which is given by

$$t_d = \frac{d_g}{|\vec{v}_d^{\,e}|}, \qquad (7.49)$$

where d_g is the LAr gap and v_d^e is the electron drift velocity. The current has a triangular shape as that of an ionization chamber,

$$I(t) = I_0\left(1 - \frac{t}{t_d}\right) \quad \text{for } 0 < t < t_d, \qquad (7.50)$$

7.3 Sampling Shower Detectors

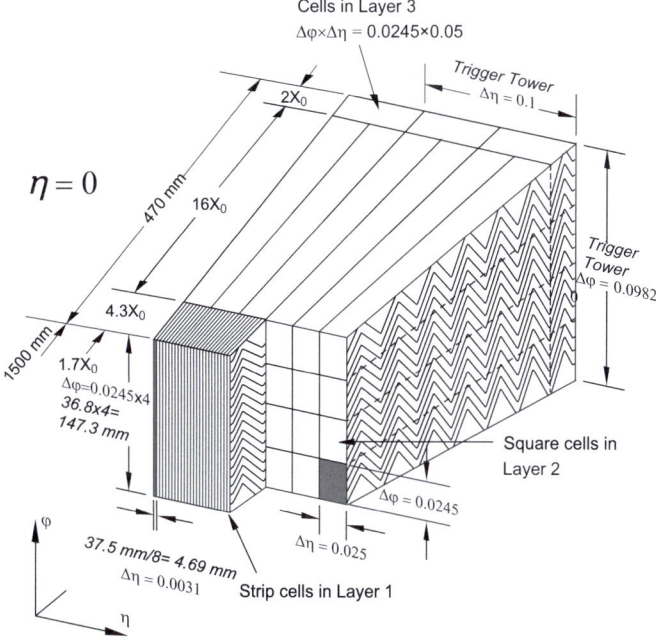

Fig. 7.38 Layout of a barrel liquid-argon calorimeter module showing the three different layers and the accordion shape of the electrodes as well as the granularity in $\Delta\eta$ and $\Delta\phi$. Reprinted from [116]. Public domain

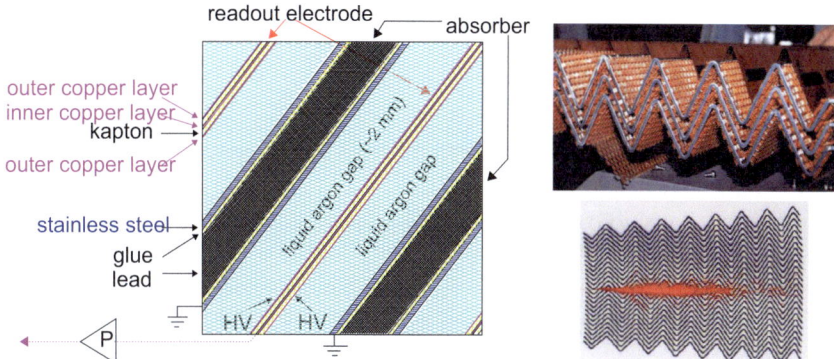

Fig. 7.39 Left: Layout of the absorber plates and the readout electrodes. The lead absorber plates are sandwiched between two stainless steel shims that are glued onto the plates and are placed on ground potential. A 2 mm liquid-argon gap separates the absorber plates from the copper readout electrodes (inner copper layer) that are sandwiched between two kapton layers and two outer copper layers, which are placed on high voltage. Reprinted under CC-BY-4.0 Licence with kind permission from [117], © 2010, ATLAS Collaboration. Top right: Photograph of the accordion-shaped electrodes. Reprinted with kind permission from [118], © 2006, ATLAS Collaboration. Bottom right: Simulated electromagnetic shower in the accordion-shaped readout structure. Reprinted from [119]. Public domain

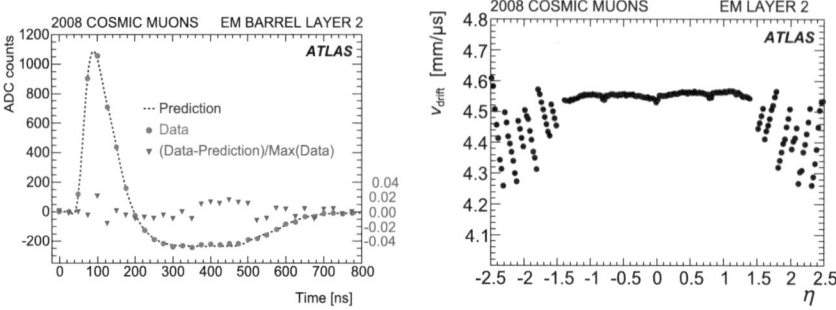

Fig. 7.40 Left: Typical single ionization pulse in a cell in barrel Layer 2 for the prediction (dotted line) and measured points (grey points). The triangles show the deviation from the prediction normalized to the maximum value (right-hand scale). Right: Drift velocity as a function of pseudo-rapidity for the ATLAS barrel liquid-argon calorimeter Layer 2 extracted from drift time measurements. Both reprinted under CC-BY-4.0 Licence with kind permission from [117], © 2010, ATLAS Collaboration

where I_0 is the current at $t = 0$ given by $I_0 = \rho_e v_d^e$ with ρ_e being the negative linear charge density along the direction perpendicular to the readout electrode. With the appropriate readout electronics, the triangular pulse is transformed into a positive spike followed by a uniform undershoot whose length is given by the drift time. Figure 7.40 (left) shows a typical single ionization pulse in a cell in barrel Layer 2, comparing measurements with cosmic muons to the expected shape. Note that in running conditions only five points spaced by 25 ns are taken.

The drift velocity is related to the electric field

$$v_d^e \propto \left[\frac{V}{d_g}\right]^{\tilde{\alpha}}, \qquad (7.51)$$

where $\tilde{\alpha} = 0.3^{+0.04}_{-0.02}$ was extracted from different measurements. For a 2 mm gap and a 2 kV potential, the drift velocity is $|\vec{v}_d^{e0}| = (4.61 \pm 0.07)$ mm/μs. Figure 7.40 (right) shows the drift velocity measured with cosmic muons as a function of pseudo-rapidity. In the barrel, the measured drift velocity is basically uniform around 4.6 mm/μs with a small modulation of about 1% that reflects 1% variations in the absorber thickness with η. In the endcaps, we see six sawtooth distributions in each endcap that result from the fine granularity of the high-voltage distribution. Thus in the energy reconstruction, corrections are applied by normalizing the response of each strip in pseudo-rapidity to the response of the strip in the center of the HV sector, using the power law dependence. Beside these modulations, we notice that the average drift velocity in the endcaps is smaller than that in the barrel and that the measured drift velocity averaged over an HV sector somewhat diminishes with increasing pseudo-rapidity. The total noise is 10–30 MeV in the barrel and a factor of ten higher in the endcaps.

7.3 Sampling Shower Detectors

The total energy deposited in an LAr cell is obtained from [120],

$$E_{\text{cell}} = F_E F_I \frac{1}{f_{\text{gain}}} G_{\text{cell}} \sum_{j=1}^{N_{\text{samples}}} a_j (S_j - S_{\text{ped}}), \tag{7.52}$$

where F_E is a conversion factor from μA to MeV, F_I is a conversion factor from ADC bins to μA, f_{gain} corrects the gain factor G_{cell} obtained with the calibration pulses to adapt it to physics-induced signals, G_{cell} is the cell gain, the S_j are the samples of the shaped ionization signal digitized in the selected electronic gain, measured in ADC counts, S_{ped} is the pedestal and the a_j weights are obtained from the predicted shape of the ionization pulse and the noise autocorrelation accounting for both the electronic and the pile-up components. We typically take the ADC counts in five samples spaced by 25 ns. We have shown the energy resolution in Fig. 7.33 (left).

The spatial resolution of the ATLAS liquid-argon barrel and endcap calorimeter prototypes were tested in a test beam. Figure 7.41 (left) shows the η resolution in the barrel in the front and middle compartments as a function of pseudo-rapidity. The measurements show reasonable agreement with the simulation. The resolution degrades somewhat with large η. For the front and middle compartments, position resolutions of 240 and 540 μm were extracted, respectively. Figure 7.41 (right) shows the resolution variation within a cell. The resolution is better at the border of a cell than in its center, since energy sharing among cells works better. Figure 7.42 (left) shows the η resolution in the barrel in the front and middle compartments as a function of the beam energy. The η resolution in the front compartment is about a factor of two lower than that in the middle compartment. The reason is a much finer segmentation in the front part. The data points in the barrel are fitted with

$$\sigma_\eta = (0.04 \pm 0.088) \times 10^{-3} \oplus \frac{(1.91 \pm 0.22) \times 10^{-3}}{\sqrt{E}} \oplus \frac{(10.3 \pm 1.3) \times 10^{-3}}{E} \quad \text{(front)},$$

$$\sigma_\eta = (0.12 \pm 0.050) \times 10^{-3} \oplus \frac{(5.05 \pm 0.20) \times 10^{-3}}{\sqrt{E}} \oplus \frac{(6.5 \pm 11.4) \times 10^{-3}}{E} \quad \text{(middle).} \tag{7.53}$$

The constant term is rather small and the noise term is only relevant at low energies. In the endcaps, the η resolution is worse, in particular the stochastic term [121].

The time resolution of the calorimeter is rather good. Performing periodic studies with collision data, guarantees that the entire liquid-argon calorimeter is uniform and aligned in time. The timing alignment for all readout channels is better than 1 ns. Using electrons from $W^- \to e\bar{\nu}_e$ decays, show that the time resolution in the middle layer is 300 ps for large energy deposits. Figure 7.42 (right) shows the single channel time resolution as a function of the energy deposited in a cell and reconstructed in the high-gain mode in the middle layer. The time resolution is

$$\sigma_t = \left(\frac{1.85}{E} \oplus 0.3\right) \text{ns}. \tag{7.54}$$

The noise term becomes negligible for energies above 25 GeV. In analyses in particular at the high-luminosity LHC, timing will play an essential role in reducing backgrounds from pile up.

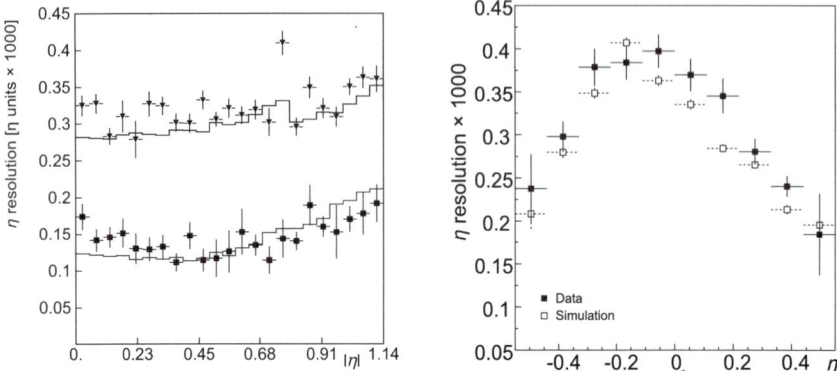

Fig. 7.41 Left: The η resolution as a function of pseudo-rapidity for $\phi = 0.26$ rad and $E = 245$ GeV for the front compartment (squares) and the middle compartment (triangles) of the barrel. The histograms show the Monte Carlo predictions. Note that the values on the ordinate have to be divided by 10^3. Right: Variation of the η resolution across a cell, $\eta = 0.69$ and $\phi = 0.28$ rad, for $E = 245$ GeV data (solid points) and simulation (open squares). Both reprinted with kind permission from [121], © 2005, Elsevier. All rights reserved

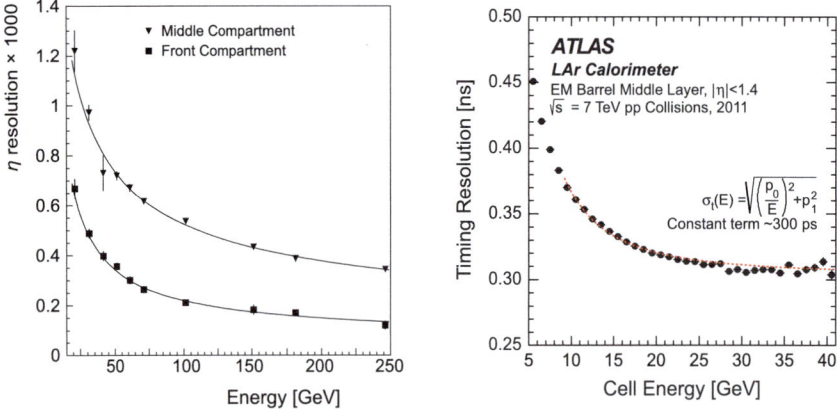

Fig. 7.42 Left: The η resolution as a function of the beam energy in the ATLAS liquid-argon barrel calorimeter measured in the front compartment (squares) and the middle compartment (triangles). The solid lines represent fits described in the text. Note that the values on the ordinate have to be divided by 10^3. Reprinted with kind permission from [121], © 2005, Elsevier. All rights reserved. Right: Time resolution of electrons in the middle layer as a function of the cell energy in Layer 2 reconstructed in the high-gain mode. Reprinted under CC-BY-NC-ND-4.0 Licence from [122], © 2013, ATLAS Collaboration

7.3.4.3 Silicon Tungsten Luminosity Monitors

The SLD and DELPHI experiments used silicon-tungsten calorimeters for their luminosity monitors [123, 124]. The fine transverse segmentation of the silicon permits the determinations of track directions. This is important for separating signal electrons coming from e^+e^- interactions from synchrotron radiation backgrounds produced in the final focus. We discuss the SLD LMSAT luminosity monitor here as an example, which covered an angular region of 23–60 mrad. The region 68–100 mrad was covered by another silicon tungsten calorimeter, MASC. The LMSAT consisted of 23 W plates, which were 3.5 mm thick and were stacked with a 4.5 mm gap into which the silicon layers were inserted. The 23 layers were split into four different configurations to approximate projective geometry. The silicon pads were arranged in r and ϕ in six rings with 160 pads. In the first configuration, the inner (outer) radius was 25 mm (85 mm). The silicon pads were 300 µm thick, high-resistivity n-type sensors with p-type ion implantation. The pad sizes in the inner radius were about 10 mm × 12 mm increasing to 15 mm × 18 mm in the outer ring. For the three other configurations, the pad sizes increased radially accordingly to obtain a projective tower. The pads in each tower in the first six layers were connected. Similarly, those in the remaining 17 layers were connected yielding 320 towers in total in two longitudinal sections. The detectors were operated with a reverse bias voltage of 75 V. They worked reliably. The measured energy resolution for 45 GeV electrons was $\sigma_E/E = 3\%$ [125]. The angular resolutions were $\sigma_\theta = 0.35$ mrad and $\sigma_\phi = 0.92°$. Silicon-tungsten calorimeters are considered in connection with particle flow as the electromagnetic calorimeter in linear-collider experiments. We discuss them in more detail in Sect. 7.5.1.4.

7.3.4.4 Comparison of Electromagnetic Sampling Calorimeter

Table 7.5 summarizes properties of different electromagnetic sampling calorimeters. Lead-scintillator sampling calorimeters yield the best energy resolution, which is still a factor of three to six worse than the energy resolution of CsI(Tl) crystal calorimeters. For the lead scintillating-fiber sampling calorimeter by KLOE, the factor is just 2.5, while for lead liquid-argon sampling calorimeters, the energy resolution is worse by factors of five or more. For lead-gas and silicon tungsten calorimeters, the energy resolution with respect to CsI(Tl) calorimeters is higher by factors of 7 to 14.

7.4 Hadron Shower Measurements

Conceptually, the energy measurement of hadronic showers is analogous to that of electromagnetic cascades. However, a detailed understanding of hadronic cascades is complicated due to the complexity and variety of hadronic processes. No simple analytical description of hadronic showers exists, but elementary processes are well understood. A typical feature of hadronic interactions is multiple particle production with transverse momenta of $\langle p_T \rangle \simeq 0.35$ GeV/c. These particles are mainly $\pi^\pm$ and π^0s. To a lesser amount also charged and neutral kaons are produced as well as other

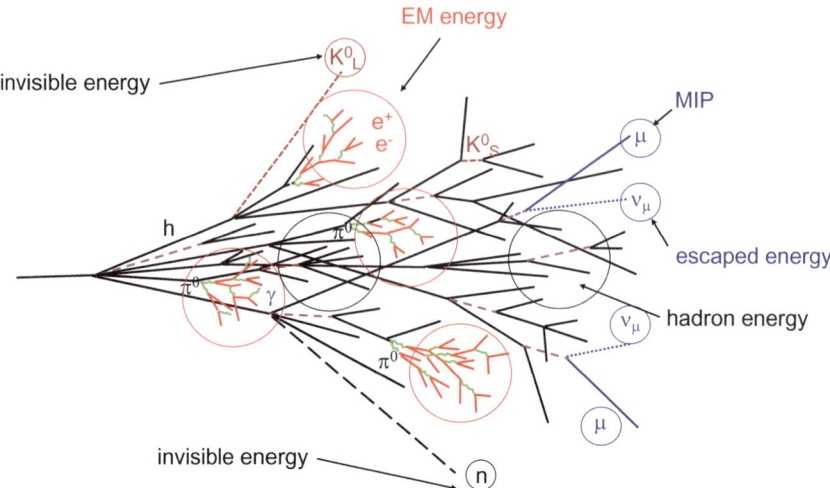

Fig. 7.43 Example of a hadron shower that reveals different kinds of energy losses: electromagnetic energy in form of photons and $e^{\pm}$, invisible energy in form of neutrons and K_Ls, escaped energy in form of neutrinos and pure hadronic energy

light mesons and baryons of which many decay before producing other hadronic interactions. Sometimes, also heavier particle such as charm or beauty hadrons may be produced. The mass of created particles depends on the momentum of the incoming hadron. About half the energy is used up in this process. The remaining energy is carried off by fast, forward going (so-called leading) particles. The multiplicity of secondary particles, which are mostly charged and neutral pions and nucleons, is only weakly energy dependent. Figure 7.43 shows an example of a hadronic shower that consists of several components. Produced π^0s decay to photons, which evolve as electromagnetic showers. Produced neutrons and neutrinos escape detection and produce invisible energy. Energetic muons will penetrate the calorimeter and leave only minimum ionization losses leading to escaped energy. Charged hadrons, antineutrons and K_L^0s will be stopped and deposit hadronic energy.

The interaction of a hadron in matter is described by the total hadronic cross section $\sigma_{\text{tot}}(pA)$, which has both elastic $\sigma_{\text{el}}(pA)$ and inelastic contributions $\sigma_{\text{inel}}(pA)$ after neglecting diffractive processes. We can parameterize $\sigma_{\text{tot}}(pA)$ in terms of the total proton-proton cross section $\sigma_{\text{tot}}(pp)$,

$$\sigma_{\text{tot}}(pA) = \sigma_{\text{tot}}(pp)\tilde{A}^{2/3}, \tag{7.55}$$

where $\tilde{A}$ is the number of protons and neutrons in the nucleus and is related to the atomic weight by $A = \tilde{A} \cdot (m_C)/12$ with units [g/mol]. The $\tilde{A}^{2/3}$ dependence results since the hadronic cross section is proportional to the radius of the nucleus r_A squared, where $r_A = r_f \cdot \tilde{A}^{1/3}$ and $r_f = 1.2 \times 10^{-13}$ cm is the Fermi radius. Figure 7.44 shows the total, elastic and inelastic pp cross section as a function of the laboratory momentum and $\sqrt{s}$. In the $\sqrt{s} = 2 - 100$ GeV energy range, the pn

7.4 Hadron Shower Measurements

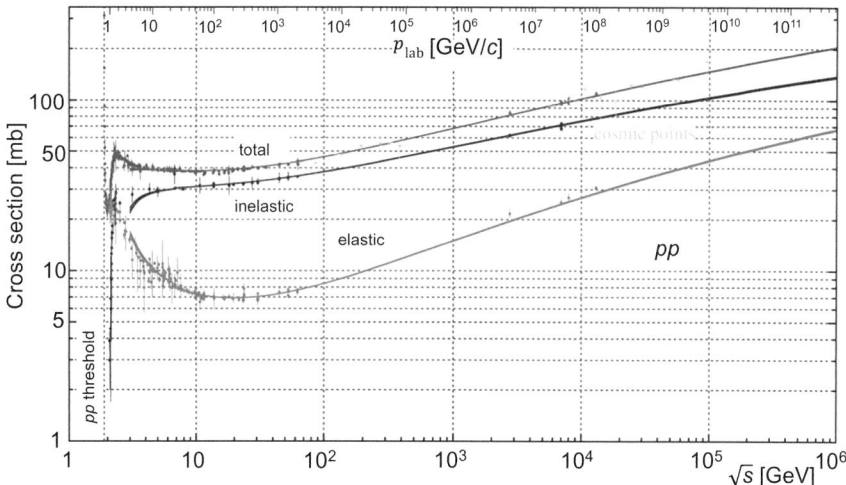

Fig. 7.44 Total, elastic and inelastic pp cross section as a function of the laboratory momentum p_{lab} and $\sqrt{s}$. Reprinted under CC-BY-NC-4.0 Licence with kind permission from [12], © 2024, the Particle Data Group LBNL

cross section is similar. The $\pi^{\pm}p$ and $K^{\pm}p$ cross sections are respectively 30 and 50% lower.

The mean free path between two nuclear interactions is given by the interaction length λ_I,

$$\lambda_I = \frac{1}{\sigma_{\text{tot}} \cdot \mathbf{N}} = \frac{A}{\sigma_{\text{tot}}(pp)\tilde{A}^{2/3} N_A \rho} \propto A^{1/3}. \quad (7.56)$$

Taking out the density, the interaction length is $\lambda_I \sim 40 A^{1/3}$ g/cm^2, since the pp total cross section is 40 mb in the 5–100 GeV energy region. The probability that a particle has not interacted within length x is

$$P = \exp(-x/\lambda_I). \quad (7.57)$$

In addition, we have the hadronic absorption length $\lambda_{\text{abs}} = A/(\sigma_{\text{inel}} \cdot N_A)$, which is determined by the inelastic cross section. Since the inelastic cross section is smaller than the total cross section, we have $\lambda_{\text{abs}} > \lambda_I$.

7.4.1 Characteristics of Hadron Showers

Table 7.6 summarizes characteristic properties of hadronic showers. There are two specific effects that turn out to limit energy resolution of hadron calorimeters.

Table 7.6 Characteristic properties of hadronic cascades listing properties of different reactions, the influence on the energy resolution, the characteristic time scales and characteristic lengths. Reprinted with kind permission from [80], © 1985, Springer. All rights reserved

Reaction	Properties	Influence on energy resolution	Charac. time [s]	Characteristic length [g/cm^2]
Hadron production	$N_{particle} \simeq A^{0.1} \log s$	π^0/π^+ ratio	10^{-22}	Absorption length
	Inelasticity $\simeq 1/2$	Binding energy loss		$\lambda_I \sim 35 A^{1/3}$
Nuclear	$E_{Evaporation} \sim 10\%$	Binding energy loss	10^{-18}–10^{-13}	
De-excitation	$E_{Binding} \sim 10\%$ Fast neutrons $\sim 40\%$ Fast protons $\sim 40\%$	Poor or different response to n, charged particles and γ's		Fast n, $\lambda_n \simeq 100$ Fast p, $\lambda_p \simeq 20$
Pion and muon decays	Fractional energy of μ's and ν's $\simeq 5\%$	Loss of ν's	10^{-8}–10^{-6}	$\gg \lambda_I$
Decays of b and c quarks produced in multi TeV cascades	Fractional energy of μ's, and ν's at percent level	Loss of ν's Tails in resolution function	10^{-12}–10^{-10}	$\ll \lambda_I$

1. A considerable part of secondary particles are π^0s, which will propagate electromagnetically without further nuclear interactions. The average fraction of hadronic energy converted into π^0s is $f_{\pi^0} \sim 0.1 \log E$ [GeV] for energies in the range (few GeV $\leq E \leq$ several 100 GeV). The size of the π^0 component is largely determined by the production in the first interaction and by event-by-event fluctuations about the average.
2. Secondary charged particles deposit energy via ionization and excitation. They also interact with nuclei producing evaporation neutrons, spallation protons and neutrons, and heavier spallation fragments. The charged collision products and γ-rays from the prompt de-excitation of highly excited nuclei produce detectable ionization, while the recoiling nuclei generate little or no detectable signal. Neutrons lose kinetic energy in elastic collisions. They become thermic on a time scale of several microseconds and are captured producing more γ-rays, which typically are outside the acceptance gate of the electronics. Note that between endothermic spallation losses, nuclear recoils and late neutron capture, a significant fraction of the hadronic energy (20–40%, depending on the absorber and energy of the incident particle) is used to overcome nuclear binding energies and, therefore, is lost.

Both processes are intimately correlated. For a given entering hadron these processes may lead to a very different shower decomposition with a very different detectable response. This feature imposes limitations on the performance of hadron calorimeters. Figure 7.45 shows a 500 GeV proton shower in copper simulated with GEANT4. We clearly see the multiple particle production and the electromagnetic showers. While electromagnetic showers scale in radiation lengths, hadronic showers scale in nuclear interaction lengths. Electromagnetic showers are much more compact than hadron showers in particular in high-Z absorbers. Figure 7.46 (left) shows the Z dependence of the radiation length and interaction length. The radiation length decreases much more rapidly than the interaction length. Figure 7.46 (right) shows the ratio λ_I/X_0 as a function of the atomic number. The dependence is nearly linear.

In the mid 1970s, considerable insight was gained from very detailed Monte Carlo calculations, which aimed to simulate full nuclear and particle physics aspects of hadronic cascades that were based on measured cross sections of elementary processes. Besides multiple-particle production, quarks had been discovered in e^+e^- collisions and as jets in hadron interactions. In 1972 after studying protons, Ranft made an energy-dependent model what fraction of the hadron energy is used for the formation of electromagnetic showers, charged-particle production and compensation of nuclear binding energy plus production of nuclear fragments as shown in Fig. 7.47 (left) [127]. At 3 GeV, 50% of the energy is taken by nuclear fragments and binding energy effects, 30% for charged-particle production and 20% for electromagnetic-shower formation. This becomes rather different at 100 GeV, where 50% is taken for electromagnetic shower formation, 30% for nuclear fragments and binding energy effects and 20% for charged-particle production.

In 1974 after studying pion interactions, Baroncelli came up with a different model that is depicted in Fig. 7.47 (center) [128]. At 3 GeV, about 40% of the energy is taken by nuclear fragment production and compensation for binding energy

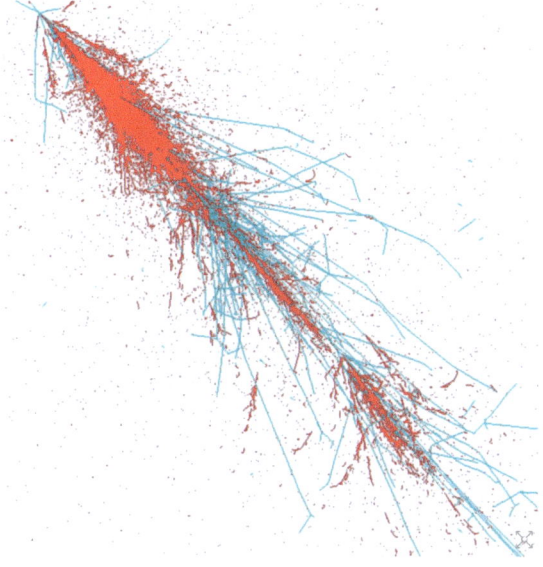

Fig. 7.45 GEANT4 simulation of a 500 GeV proton in copper. The red points indicate electromagnetic energy deposits and the blue points show hadronic energy deposits. Reprinted under CC-BY-4.0 Licence with kind permission from [126], © 2018, IOP Publishing

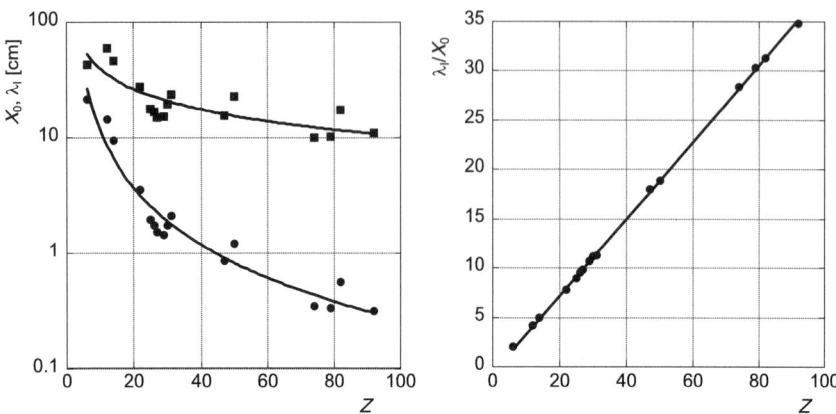

Fig. 7.46 Left: The Z-dependence of the radiation length (lower curve) and the nuclear interaction length (upper curve). Right: The Z-dependence of the ratio λ_I/X_0

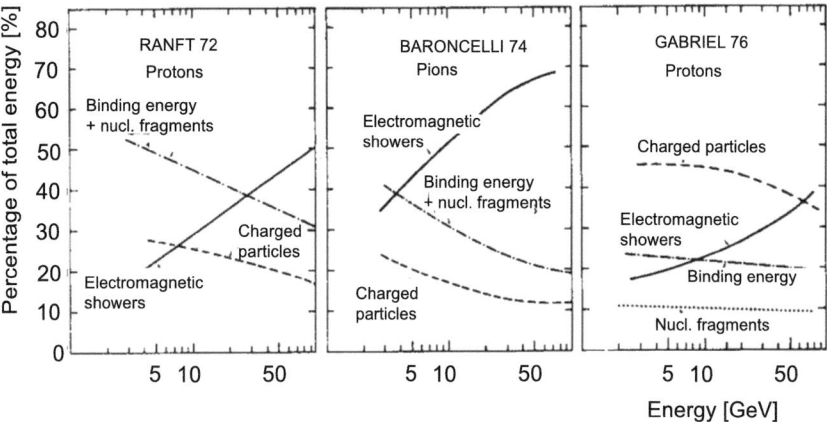

Fig. 7.47 The energy distribution among four processes, the formation of electromagnetic showers, charged-particle production, production of nuclear fragments and compensation of nuclear binding energy effects as a function of energy for the Ranft model (left) [127], the Baroncelli model (center) [128] and the Gabriel model (right) [129]. Reprinted with kind permission from [80], © 1985, Springer. All rights reserved

effects, 35% by formation of electromagnetic showers and 25% by charged-particle production. At 100 GeV, the electromagnetic shower formation increases to 70%, while compensation for nuclear binding effects plus nuclear fragment production and charged particle production decreases to 20 and 10%, respectively.

In 1976, Gabriel formulated another model based on the study of protons that is depicted in Fig. 7.47 (right) [129]. He separated the compensation for nuclear binding effects and nuclear fragment production. At 3 GeV, charged-particle production dominates with 47%, followed by 25% for compensation of binding energy effects, 18% for electromagnetic shower formation and 10% for nuclear fragments production. At 100 GeV, the contribution from charged particle production decreases to

7.4 Hadron Shower Measurements

30%, while that from electromagnetic shower formation increases to 40%. Compensation for nuclear binding energy effects contributes at 20% and nuclear fragment production is 10%.

Since the 1970s, more experimental data became available and model building continued. One important step was the development of the Lund model [130], which is based on the fragmentation of a string and makes predictions for the hadronization of quarks. It is part of a much more sophisticated Monte Carlo program called GEANT. The present version is GEANT4 [131], which contains the EGS4 program for simulations of electromagnetic showers. It further uses different hadronic shower models including Gheisha [132] and Fluka [133]. It also uses libraries for neutron interactions below 20 MeV. The Gheisha and Fluka models are based on our present knowledge and experience, including many known processes, symmetries, and inputs from Quantum Chromodynamics (QCD), which is the theory of strong interactions. The programs are rather complex including many processes.

In Fluka, the hadron-nucleon interactions are divided into three different regions.

- Low energy regime (<100 MeV): here, pp and np cross sections are large and πp and πn interactions are related by isospin.
- Intermediate energy regime (0.1–5 GeV): here various resonances are taken into account.
- High energy regime (>5 GeV): Results of the dual parton model are used [134]. The energy is sufficiently high to break the string producing $q\bar{q}$ jets.

In Gheisha, parameterizations and extrapolations of cross sections are performed. By now, GEANT4 includes 28 physics lists that perform simulations of certain processes in different energy regimes. Some were developed in the context with physics at LHC detectors.

7.4.1.1 GEANT4 Physics Lists

We focus here on the most frequently used physics lists.

- Low- and High-Energy Parameterization model (LEP, HEP) [135].
- The Quark-Gluon String model (QGSP) for high-energy hadrons with $E > 12$ GeV [136,137].
- The Fritiof model (FTFP) for energies $E > 10$ GeV [138].
- BInary Cascade model (BIC) for energies $E < 10$ GeV [139].
- The Bertini cascade model (BERT) for energies $E < 10$ GeV [140].
- High-Precision neutron model (HP) for energies $E < 20$ MeV [141].
- PRE-COmpound model (PRECO) for energies $E < 150$ MeV [142].
- Variations of the standard EM package (EMV) [143].
- CHiral Invariant Phase Space (CHIPS), a three-dimensional low-energy parton model [144].
- Modeling of the radiation environment with Mokhov's code MARS [145].

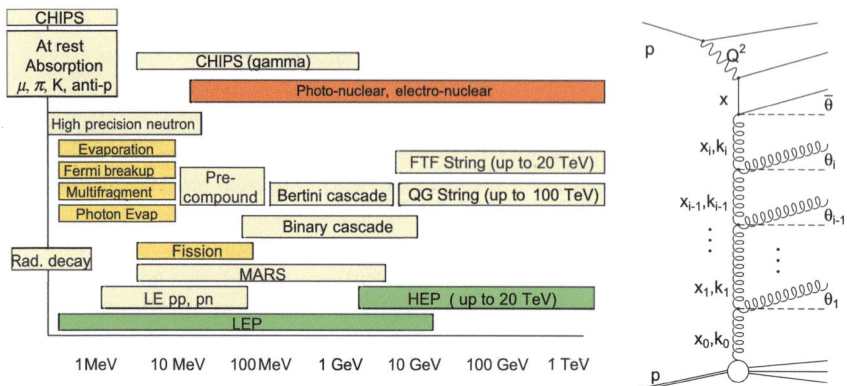

Fig. 7.48 Left: Most frequently used physics lists in GEANT4 as a function of energy. Reprinted from [147]. Public domain. Right: Feynman diagram for the formation of a jet. Reprinted with kind permission from [148], © 2005, G. Flucke. All rights reserved

For the simulation of hadron showers of different energies several models are often combined [146].

Figure 7.48 (left) shows the range of physics lists in GEANT4 as a function of energy. This illustrates in which energy range the physics list is most effective. Figure 7.48 (right) shows the generation of a jet in QCD. The LHEP model is the fastest of all physics lists. It is not the most precise but contains standard electromagnetic processes, providing a good description of electromagnetic showers. The QSG model combined with the Bertini cascade (QSGP-BERT) is the most recommended physics list for high-energy physics. It is used by the ATLAS experiment and contains all standard electromagnetic processes using the Bertini cascade for hadrons of energy below 10 GeV and the QSGP model for high energies above 20 GeV. The QGSP-BERT-EMV model is the same as QSGP-BERT but uses electromagnetic processes tuned for better CPU performance. However, the increase in speed comes with a slight decrease in precision of describing electromagnetic processes. This physics list is used by the CMS experiment. Another modification is the QGSP-BERT-HP model, which is the same as QGSP-BERT but includes a high-precision neutron model that describes neutrons below 20 MeV. It is significantly slower than QGSP-BERT, when the full thermal cross sections are used. It can be accelerated by turning off thermal scattering. This model is used for radiation protection and shielding applications. Another useful physics list is QSGP-BIC, which uses the binary cascade, pre-compound and various de-excitation models for hadrons. It incorporates all electromagnetic processes and is recommended for applications at energies below 200 MeV. The QSGP-BIC-HP model is the same as QGSP-BIC but it incorporates high-precision neutron modeling for neutrons below 20 MeV. Thus, this physics list is rather useful for radiation protection, shielding and medical applications. The FTFP list uses a different string model than QGSP that is based on the Fritof model for describing string excitations and fragmentation.

7.4.1.2 Performance of GEANT4 Physics Lists

Figures 7.49 show the reconstructed energies for different energy hadrons (5–300 GeV) measured with a hadron calorimeter prototype in a test beam in comparison to predictions from the LHEP and QGSP physics lists. At 5 and 7 GeV, the simulations yield somewhat narrower showers than the data. For the other energies, the simulations yield consistent shower shapes. For the 300 GeV shower, the LHEP list yields a better description of the data than the QGSP list.

7.4.1.3 Longitudinal Shower Profiles

The average longitudinal and lateral shower distributions are useful estimates of the characteristic dimensions for near complete shower containment. The average longitudinal distribution exhibits a scaling in units of the interaction length λ_I.

Figure 7.50 shows the longitudinal shower development in different materials for 15 and 100 GeV hadrons as a function of λ_I. Parametrized in terms of λ_I, the longitudinal shower distributions for different materials are rather similar. The shapes resemble those of electromagnetic showers parameterized in terms of radiation lengths. The shower maximum increases logarithmically with energy. The shower distributions are measured from the shower vertex and are more peaked than those measured with respect to the face of the calorimeter. For aluminum and marble, we encounter a special situation, since $X_0^{Al} \simeq X_0^{marble}$ and $\lambda_I^{AL} \simeq \lambda_I^{marble}$. The figure also displays the lateral shower development as a function of λ_I, showing the shower

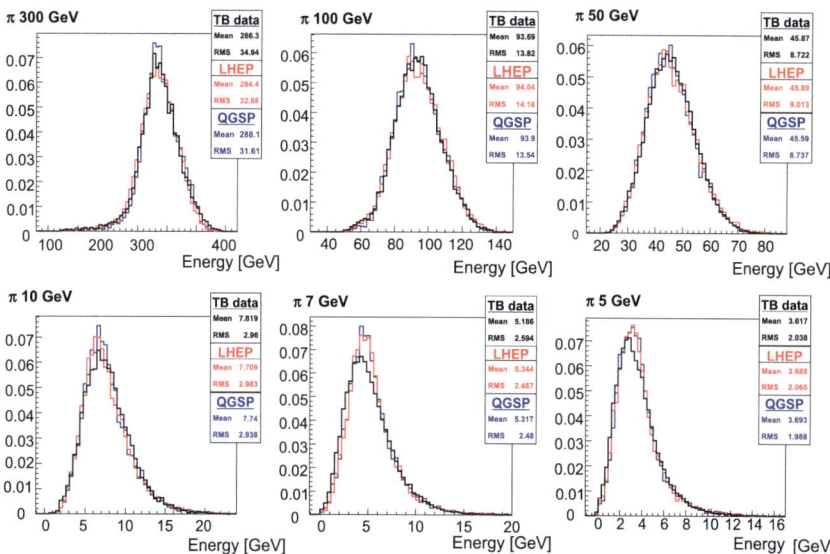

Fig. 7.49 Reconstructed hadron shower in a test beam at different energies, 300 GeV (top left), 100 GeV (top center), 50 GeV (top right), 10 GeV (bottom left), 7 GeV (bottom center) and 5 GeV (bottom right) for data (black histograms), the LHEP physics list (red histograms) and the QGSP physics list (blue histograms). All reprinted with kind permission from [149], © 2007, AIP Publishing. All rights reserved

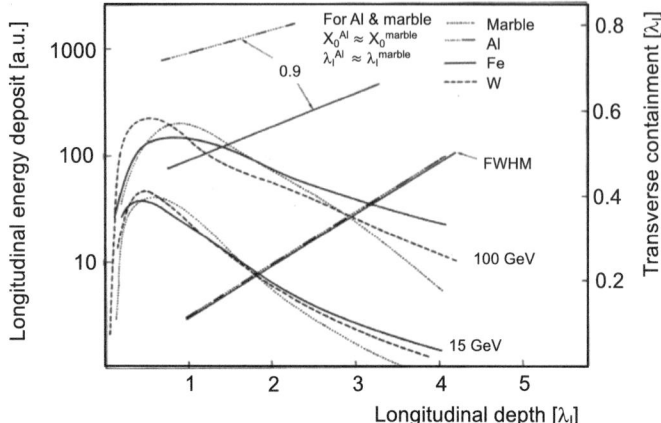

Fig. 7.50 Longitudinal shower distribution (left scale) as a function of interaction lengths λ_I for 15 and 100 GeV hadrons in marble (dotted) [150], aluminum (dash-dotted), iron (solid) [151] and tungsten (dashed) [152] showing approximate scaling in λ_I. Corresponding lateral shower distributions (straight lines) as a function of λ_I (right scale). While the lateral shower distribution for the core scales with λ_I, the radius of 90% lateral containment does not [153]. Reprinted with kind permission from [80], © 1985, Springer. All rights reserved

core and the 90% shower containment. Concerning the lateral distributions, scaling is only observed for the narrow core (FWHM).

Figure 7.51 (left) shows the measured average longitudinal shower profile of the CMS hadron calorimeter prototype for a 300 GeV pion in comparison to LHEP and QGSP predictions. The LHEP physics list provides a rather good description of the

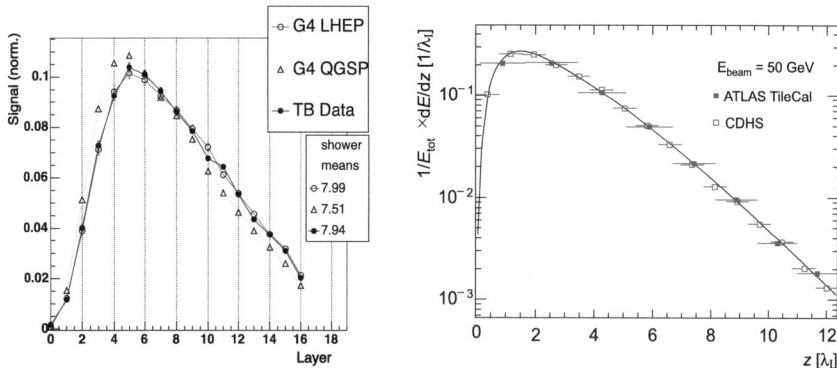

Fig. 7.51 Left: Average longitudinal shower distribution of the CMS hadron calorimeter prototype for a 300 GeV/c pion. In addition to the data, the predictions of the LHEP and QGSP list are depicted. Reprinted with kind permission from [149], © 2007, AIP Publishing. All rights reserved. Right: Average longitudinal shower distribution of 50 GeV charged pions in the CDHS magnetized iron-scintillator sandwich calorimeter measured in test beam calibration runs [151] and in the ATLAS tile calorimeter. The measurements are performed at the front of the calorimeter and are convolved with the first interaction distance. Reprinted with kind permission from [154], © 2010, Elsevier. All rights reserved

7.4 Hadron Shower Measurements

longitudinal shower distribution, whereas the QGSP list yields a too narrow shower. Note that individual distributions have very large variations. Figure 7.51 (right) shows the average longitudinal distributions for 50 GeV pions in the CDHS magnetized iron-scintillator sandwich calorimeter [151] and in the ATLAS tile calorimeter [154]. The shower shapes are identical for the two different calorimeters. It should be noted that proton-induced cascades are somewhat shorter and broader than pion-induced cascades [154].

7.4.1.4 Transverse Shower Profiles

Figure 7.52 (left) shows the lateral shower profile for 5–150 GeV energy pions as a function of the radial depth in a lead scintillating-fiber calorimeter. The shower profile consists of a narrow core that contains most of the energy and a broad halo. The electromagnetic part of the shower is located in the core. Note that different shower profile parameterizations exist. One of them consists of a Gaussian for the core plus an exponential function for the halo. Thus, the energy loss per circular tower area (A_{sh}) is

$$\frac{dE}{dA_{sh}} = \left[\frac{b_1}{r}\exp\left(-\frac{r}{\lambda_1}\right) + \frac{b_2}{r}\exp\left(-\frac{r^2}{\lambda_2^2}\right)\right]/E, \tag{7.58}$$

where λ_1 is the scale length of the halo, λ_2 is that of the core and b_1, b_2 are the fractions of the halo and core contributions, respectively. All these parameters are determined from a fit to the data distribution. The scale of the core is 3.6–4.1 cm for energies of 150 to 9.7 GeV, while that of the halo is around 14.3 cm. Figure 7.52 (right)

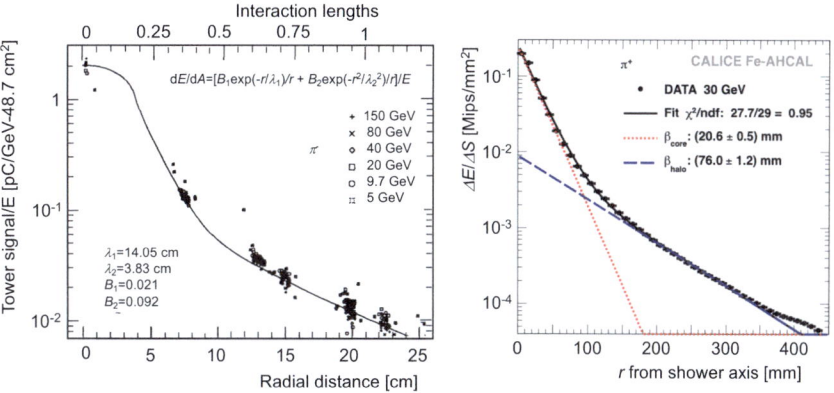

Fig. 7.52 Left: Lateral shower profile for 5 to 150 GeV pions in a lead scintillating-fiber calorimeter. Reprinted with kind permission from [155], © 1992, Elsevier. All rights reserved. Right: Radial shower profile for 30 GeV pions in a lead scintillator-tile calorimeter [156]. The variable ΔS is the area of a ring with radial thickness Δr, $\Delta S = 2\pi r \Delta r$. The fit function is $\Delta E/\Delta A = A_{core}\exp(-\beta_{core}) + A_{halo}\exp(-\beta_{halo})$, where A_{core}, A_{halo} are scaling factors and $\beta_{core} < \beta_{halo}$ are slope parameters. Reprinted under CC-BY-3.0 Licence from [156], © 2016, CALICE Collaboration

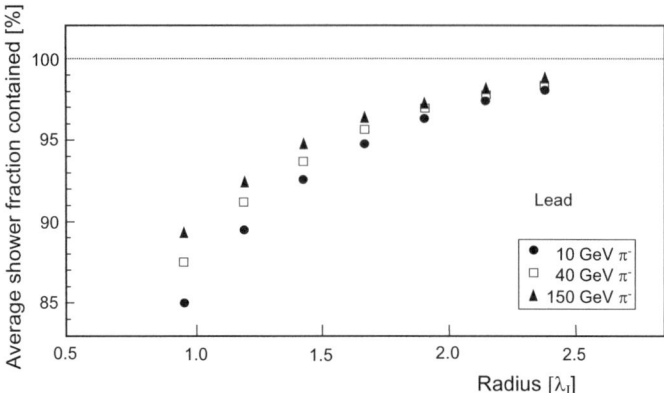

Fig. 7.53 Average shower containment fraction as a function of the radial distance for 10 to 150 GeV pions in a lead scintillating-fiber calorimeter. Reprinted with kind permission from [157], © 2019, Springer. All rights reserved

shows the lateral shower profile of 30 GeV pions measured in an iron scintillator-tile sampling calorimeter. Again we see the narrow core and the broad halo. Here, the data are fitted with two exponential functions. Like electromagnetic showers, the lateral distributions depend on the longitudinal depth. The core of the shower is rather narrow; the FWHM is ~ 0.1–$0.5\,\lambda_I$ increasing with depth. The halo particles may extend to a considerable distance from the shower axis. For 90% shower containment, we typically need a cylinder of radius $r \sim 1\lambda_I$. For 99% containment, $\sim 2.5\lambda_I$ are necessary for 10 to 150 GeV pions as shown in Fig. 7.53. Figure 7.54 shows a three-dimensional plot of the shower development in the longitudinal and radial directions for 300 GeV protons in a lead scintillator calorimeter prototype.

7.4.1.5 Shower Parameterization
The experimental data are consistent with following parameterization:

1. The shower maximum (measured from face of calorimeter) can be parameterized by

$$I_{\max}(\lambda_I) \simeq 0.2 \log E \text{ [GeV]} + 0.7. \tag{7.59}$$

Due to the smaller ratio X_0/λ_I for high Z materials the shower maximum occurs at smaller values than for low-Z materials.

2. The longitudinal dimension for a 95% shower containment measured from the front face is

$$I_{0.95}(\lambda_I) = I_{\max}(\lambda_I) + 2.5\lambda_{\text{att}}, \tag{7.60}$$

where λ_{att} describes the exponential decay of the shower beyond $I_{\max}$ increasing with energy approximately as

$$\lambda_{\text{att}} \simeq \lambda_I \cdot (E \text{ [GeV]})^{0.13}. \tag{7.61}$$

7.4 Hadron Shower Measurements

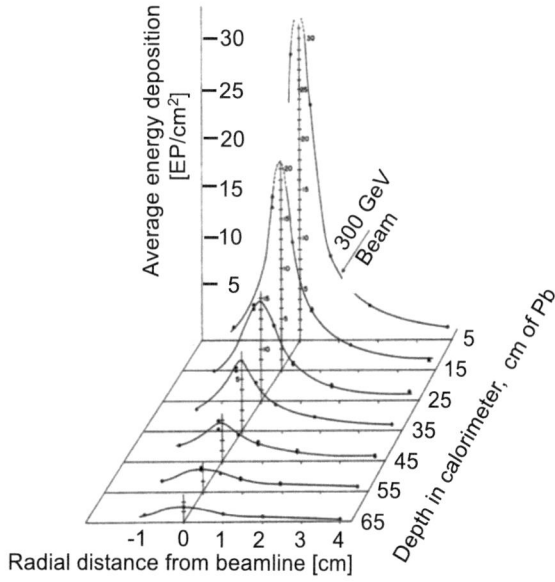

Fig. 7.54 Three-dimensional shower profile in the longitudinal and radial directions for 300 GeV protons in a lead scintillator-tile calorimeter prototype. The acronym EP stands for equivalent particle. The energy deposition of 1 EP corresponds to 1.65 MeV in a 1 cm thick scintillator. Reprinted with kind permission from [158], © 1981, Elsevier. All rights reserved

For high-Z materials the dependence is weaker. Note that $L_{0.95}(\lambda_I)$ describes data in the energy range of few GeV $\leq E \leq$ few 100 GeV within 10%.

3. The transverse radii r of a cylinder with 95% shower containment is $R_{0.95} \leq 1.5\,\lambda_I$. We observe no scaling with λ_I. In high Z-materials $R_{0.95}$ is smaller.
4. Since the longitudinal shower shapes of electromagnetic and hadron showers are similar, we can use the following parameterization,

$$dE/dz = K\left[w(t_r)^a \exp(-b \cdot t_r) + (1-w)(\ell_I)^c \exp(-d \cdot \ell_I)\right], \quad (7.62)$$

where the first term represents the electromagnetic shower contribution with weight w starting at the shower origin and the second part represents the hadronic contribution starting at the shower origin. Note that $t_r = z/X_0$ and $\ell_I = z/\lambda_I$. The scale factor K, the parameters a, b, c, d and the weight w are determined from a fit to the data. They reflect the logarithmic energy dependence.

The curve in Fig. 7.51 (right), for example, is obtained from a fit with the function given in (7.62).

7.4.1.6 Crude Shower Model

In order to simulate fluctuations in a hadron shower we may apply the following crude model.

- We randomly vary the depth of shower origin.
- We smear the incident particle energy to simulate the calorimeter energy resolution.
- We randomly vary the length of the shower by scaling the values of t_r and ℓ_I.

The total depth needed for a near complete absorption (95%) of a 350 GeV pion is 8 λ_I. The depth increases only logarithmically with energy.

7.4.1.7 Calorimeter Response to Pions

Lets us now investigate the response of hadron calorimeters to 10 GeV pions. Conceptually, fluctuations are dominated by the nature of the first inelastic interaction. Figure 7.55 (left) shows the response of a generic hadron calorimeter. On average, the pion produces a certain number of neutral pions, charged pions, nuclear fragments, slow protons and neutrons. Neutral pions decay nearly 100% to two photons. So one extreme is that no π^0s are produced. Here, the shower is purely hadronic. The average hadron energy deposit is depicted by the lower energy distribution. The other extreme is that mostly π^0s are produced. In this case, the deposited energy is that of electrons and photons depicted by the upper energy distribution. For a typical hadron shower, we have a combination of the two effects, which is different event-by-event. Figure 7.55 (center) shows the response of an iron liquid-argon sampling calorimeter to the pion in a logarithmic scale. Here, we observe a broad energy distribution, which is a superposition of the hadron energy deposition and that of electromagnetic showers. The contribution of the electromagnetic shower is visible as a shoulder on the upper part of the energy deposit. In addition, a peak from penetrating muons is visible as well as a noise peak at $E = 0$ GeV. Figure 7.55 (right) shows the response of a uranium-scintillator sampling calorimeter. Nuclear losses effectively are compensated, leading to a response that is nearly equal for $\pi^\pm$ and π^0s. We see two separate peaks for the hadronic and electromagnetic energy deposits, which display Gaussian response. The MIP peak from muons is also clearly visible, which can be used to set the energy scale.

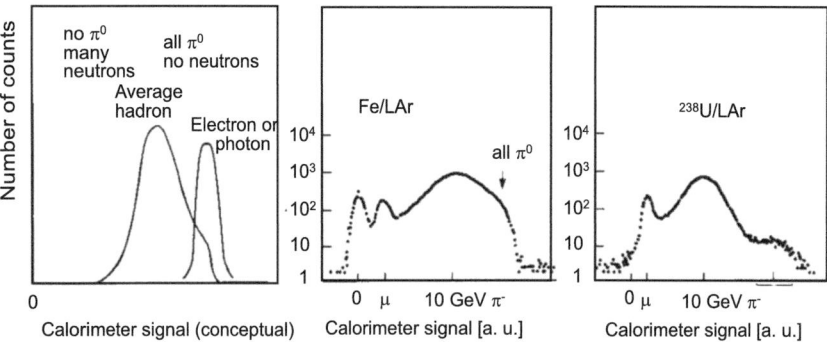

Fig. 7.55 Left: Energy spectrum in a generic hadron calorimeter. Center: Energy spectrum in an iron liquid-argon calorimeter. Right: Energy spectrum in a uranium liquid-argon calorimeter. All reprinted with kind permission from [80], © 1985, Springer. All rights reserved

7.4.2 Intrinsic Energy Resolution

We saw that we encounter large variations in the shower composition from one hadron shower to the next. In addition to charged pions, we have π^0s, slow neutrons, muons and neutrinos, which have rather different detection characteristics and are the principal limitations to the energy resolution. These fluctuations have been found to be large. For example, at 1 GeV they amount to about 50%. This is in strong contrast to electromagnetic showers. Available experimental evidence indicates that the intrinsic hadronic energy resolution is,

$$\left(\frac{\sigma_E}{E}\right)_{\text{intrinsic}} = \frac{45\%}{\sqrt{E \text{ [GeV]}}}, \tag{7.63}$$

which holds for materials from aluminum to lead. The only exception is ^{238}U, where the intrinsic energy resolution is,

$$\left(\frac{\sigma_E}{E}\right)_{\text{intrinsic}}^{\text{U}} = \frac{22\%}{\sqrt{E \text{ [GeV]}}}. \tag{7.64}$$

As we will see the improvement in intrinsic hadron energy resolution originates from the measurement of extra energy produced by nuclear effects, which is more prominent in uranium than in other absorbers. The level of nuclear effects and level of invisible energy is sensitively measured by comparing the response to electrons and hadrons at the same available energy in different calorimeters. All hadron calorimeters except those using ^{238}U show a visible energy of $\sim$70% relative to electrons, increasing slowly with higher energies because of the increasing electromagnetic component. For energies below $E = 1.5$ GeV, the nature of the hadronic cascade changes. Here, the energy is reduced by ionization alone. The hadronic response approaches that of muons. In this limit, all calorimeters give the same response. In the next section, we examine the response to electrons and hadrons in hadron calorimeters in more detail.

7.4.2.1 Response to Electrons and Hadrons

Let us look at the energy flow in a hadronic cascade illustrated in Fig. 7.56. An incoming hadron with energy E transfers part of its energy to π^0s and uses the rest for hadronic interactions. Thus, the energy is split to $E = E_{\pi^0} + E_h$, where $E_{\pi^0} = f_{\pi^0} E$ and $E_h = (1 - f_{\pi^0})E$. The π^0s create an electromagnetic component, while E_h evolves hadronically. The calorimeter samples each component separately, but we measure the combination $E^{\text{vis}} = E^{\text{vis}}_{\pi^0} + E^{\text{vis}}_h$, where $E^{\text{vis}}_{\pi^0} = e E_{\pi^0}$ and $E^{\text{vis}}_h = h E_h$, where e and h are the electromagnetic and hadronic efficiencies, respectively.[6]

In each high-energy interaction of the hadron shower, about 25% of the energy is lost via π^0 production from further hadronic activity. Let us call $\langle f_{\pi^0} \rangle$ the mean π^0

---

[6] Typically, efficiencies are denoted by ϵ. However, it is tradition to call electron and hadron efficiencies e and h.

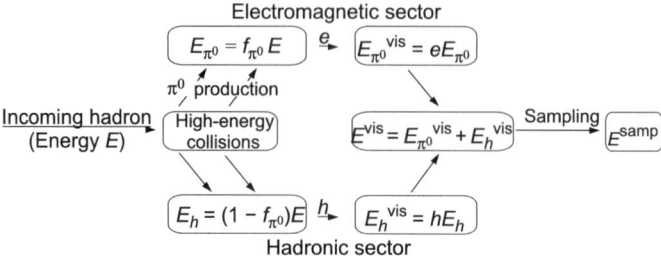

Fig. 7.56 Illustration of how the energy of an incoming hadron is transferred between electromagnetic and hadronic components. Details are described in the text. Reprinted with kind permission from [159], © 2007, Elsevier. All rights reserved

Fig. 7.57 Simulated energy deposits of the hadronic energy and electromagnetic energy of 30 GeV negative pions in lead corrected for albedo losses. The distribution was generated with an older version of FLUKA, which was missing the recording of de-excitation photons. Reprinted with kind permission from [159], © 2007, Elsevier. All rights reserved

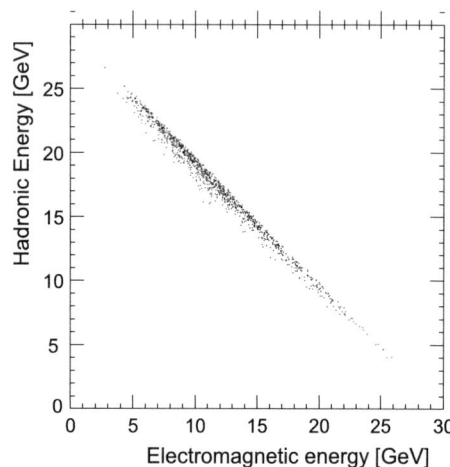

fraction. Then, the mean hadronic fraction is $\langle f_h \rangle = 1 - \langle f_{\pi^0} \rangle$, which scales with energy as [160]

$$\langle f_h \rangle = \left(\frac{E}{E_0}\right)^{(m-1)}, \tag{7.65}$$

where $m \approx 0.83$ with some small Z dependence and $E_0 \approx 1$ GeV for pions and 2.6 GeV for protons, again with some Z dependence. Physically, m is related to the average number of secondary particles and E_0 is the energy scale for which multiple pion production becomes significant. Both parameters need to be determined experimentally for each calorimeter. The electromagnetic and hadronic components of a hadronic cascade are highly correlated as illustrated in Fig. 7.57.[7] The distribution shows that the fluctuations on an event-by-event basis are very large.

---

[7] Note that in this FLUKA simulation γs from nuclear de-excitations have not been accounted for, which result from slow-neutron captures. Thus, the hadron energy is underestimated for some events.

7.4 Hadron Shower Measurements

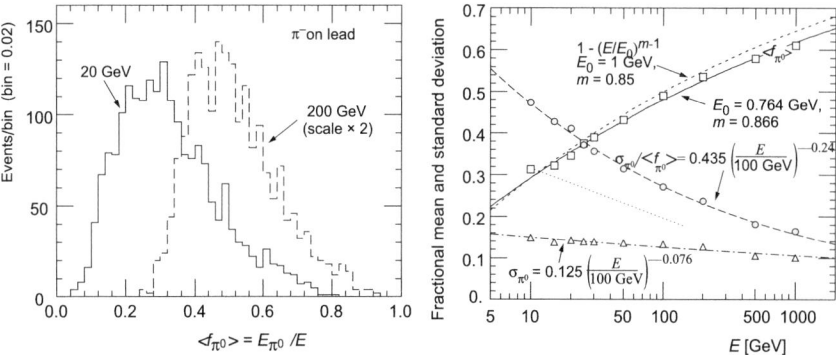

Fig. 7.58 Left: FLUKA simulations of π^0 energy deposits for 20 and 200 GeV π^- mesons interacting in lead. Reprinted with kind permission from [159], © 2007, Elsevier. All rights reserved. Right: Energy dependence of the average value $\langle f_{\pi^0} \rangle$ (open squares), the standard deviation (σ_{π^0}) of the f_{π^0} distribution (triangles) and $\sigma_{\pi^0}/\langle f_{\pi^0} \rangle$ (circles) [159]. The short-dashed curve is for $1 - (E/E_0)^{(m-1)}$ with m = 0.85, while the solid curve shows the fit to the data with m = 0.866. The dash-dotted curve is a fit with $\sigma_{\pi^0} = 0.125 \, (E/100 \text{ GeV})^{-0.076}$ and the long-dashed curve is a fit for $\sigma_{\pi^0}/\langle f_{\pi^0} \rangle = 0.435 \, (E/100 \text{ GeV})^{-0.24}$. The dotted curve represents a fit to the SPACAL data given in [161]. Reprinted with kind permission from [161], © 1991, Elsevier. All rights reserved

Figure 7.58 (left) shows the distribution of π^0s for 20 and 200 GeV pions in lead. The average π^0 fraction increases with energy as indicated in (7.65). Figure 7.58 (right) shows energy dependence of the average value $\langle f_{\pi^0} \rangle$, the standard deviation (σ_{π^0}) of the f_{π^0} distribution and $\sigma_{\pi^0}/\langle f_{\pi^0} \rangle$. An electromagnetic shower originating from an electron or π^0-decay photons produces a visible signal in the calorimeter with efficiency **e**. Most of the ionization is by electrons and positrons with energies below the critical energy. The response R_e is linear in the incident energy,

$$R_e = e E_e. \qquad (7.66)$$

The visible signal produced by hadron interactions also comes predominantly from low-energy ionizing particles whose spectra and relative abundance are independent of the incident hadron energy. Many mechanisms contribute, including endothermic nuclear spallation. Also neutrons play an especially significant role [159]. The sum of all the hadronic energy deposit mechanisms (excluding showers by π^0 decay photons) produces an observable signal with efficiency **h**. In most cases $h/e \leq 1$. The response to the pion is

$$\begin{aligned} R_\pi &= e \langle f_{\pi^0} \rangle E + h \langle f_h \rangle E \\ &= eE[1 - (1 - h/e)\langle f_h \rangle]. \end{aligned} \qquad (7.67)$$

The electromagnetic energy and purely hadronic energy contributions in a pion shower are highly correlated as we saw in Fig. 7.57.

In case of an incident pion the response relative to that of an electron is

$$R_\pi/R_e \equiv \pi/e = [1 - (1 - h/e\langle f_h\rangle)]$$
$$= [1 - (1 - h/e)(E/E_0)^{(m-1)}] \equiv 1 - aE^{(m-1)}, \quad (7.68)$$

where we have used (7.65). The parameter a stands for the additional factors. From this we get

$$\frac{e}{\pi} = \frac{1}{1 - (1 - h/e)(E/E_0)^{m-1}}. \quad (7.69)$$

For example, for the H1 SPACAL [161], $m = 0.788$, $a = 0.141$ and $h/e = 0.836$. For the CDF endplug hadron calorimeter, we get $m = 0.865$, $a = 0.244$ and $h/e = 0.756$ [162]. Note that we cannot determine h/e as a function of energy from a measurement of π/e without any other information or assumptions on E_0. Wigmans uses a different parameterization that yields similar results [163],

$$\frac{e}{\pi} = \frac{e/h}{1 + (e/h - 1) \cdot 0.11 \ln E}. \quad (7.70)$$

We could also relate e/π to e/h yielding

$$\frac{e}{\pi} = \frac{e}{h} \frac{1}{1 + \langle f_{\pi^0}\rangle(e/h - 1)}. \quad (7.71)$$

So only if $e/h = 1$, then $e/\pi = 1$.

Figure 7.59 shows the e/h ratio as a function of energy in different hadron sampling calorimeters. For hadron sampling calorimeters that do not use uranium, the e/h ratio is 0.7 for energies around 0.2 GeV rising rapidly to 1.4 for energies above 1 GeV. Only for sampling calorimeters with a uranium absorber, the e/h ratio levels

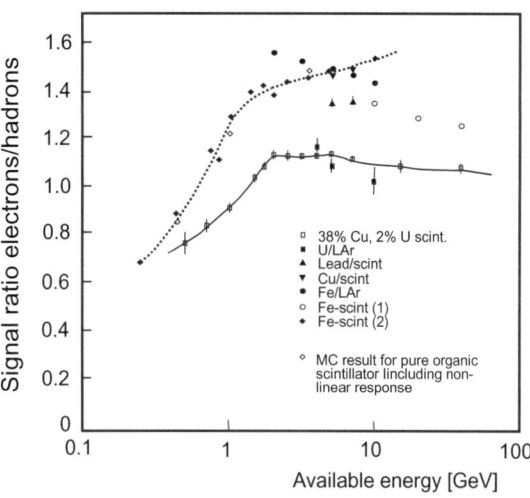

Fig. 7.59 The ratio of electromagnetic energy to hadronic energy response e/h as a function of energy for different hadron sampling calorimeters, Cu-U-scintillator (open squares) [164], U-LAr (solid squares) [84], Cu-scintillator (downward triangles) [85], Fe-LAr (solid points) [84], Fe-scintillator (1) (open points) [165] and Fe-scintillator (2) (diamonds) [166]. Reprinted with kind permission from [80], © 1985, Springer. All rights reserved

7.4 Hadron Shower Measurements

off around one. The response of a calorimeter to hadrons and electrons expressed in terms of the **e/h** ratio is a sensitive probe of the level of nuclear effects. We see that typical values for most materials are $e/h \sim 1.4$. Ideally, we want **e/h** = 1. A hadron-to-electron suppression is strongly correlated with large fluctuations in the electromagnetic component. Thus, the intrinsic energy resolution is $\sigma_E/E = 0.45/\sqrt{E}$ unless event-by-event fluctuations in the electromagnetic component are corrected for, which applies likewise to homogeneous and to sampling calorimeters.

For assumptions of the energy dependence of **h/e**, we can predict the π/e energy dependence. Figure 7.60 (left) shows the energy dependence of π/e response expected for several values of **h/e** under the assumption that $E_0 = 1$ GeV. For uranium-based calorimeters, $\pi/e = 1$ since **h/e** = 1. So, they are compensating. For lead-scintillator and iron-scintillator calorimeters, $\pi/e < 1$, which is under-compensating. Calorimeters with $\pi/e > 1$ are over-compensating. Note that $f_{\pi^0} \to 1$ if $E \to \infty$. In a 10^{10} GeV induced proton air shower $f_{\pi^0} = 0.98$.

In this calculation, photons produced from slow neutrons are not accounted for. If $\langle f_\gamma \rangle$ is the fraction of the hadronic energy deposited by nuclear γs observed within the electronic time window, $\langle f_\gamma \rangle \langle f_h \rangle$ of the incident energy is detected with efficiency **e**. The total electromagnetic fraction is $\langle f_{\pi^0} \rangle + \langle f_h \rangle \langle f_\gamma \rangle$. The remaining $(1 - \langle f_\gamma \rangle)\langle f_h \rangle$ is detected with a redefined hadron detection efficiency **h'**. So the modified equation (7.67) is

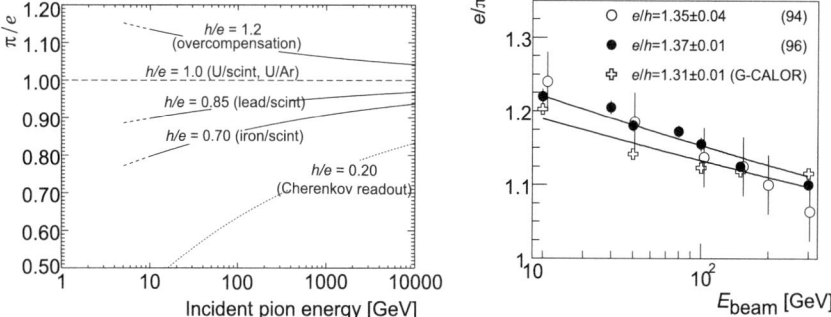

Fig. 7.60 Left: The π/e ratio as a function of energy expected for several values of **h/e** under the assumption that $E_0 = 1$ GeV. Note that for almost all calorimeters **h/e** < 1 (under-compensating). The exact value of **h/e** depends on the actual configuration. The lower dotted curve for **h/e** ≪ 1 should be applicable to a calorimeter with Cherenkov readout. The power-law description is not expected to work well below $\simeq$10 GeV, but nonetheless seems adequate at 5 GeV. Reprinted with kind permission from [159], © 2007, Elsevier. All rights reserved. Right: The e/π ratio as a function of energy for 1994 (circles) and 1996 (triangles) test beam data and a GEANT simulation (crosses). A fit to the data using (7.70) yields **e/h** = 1.35 and **e/h** = 1.37 for the two data sets, respectively. The simulation uses **e/h** = 1.31. Reprinted with kind permission from [168], © 2000, Elsevier. All rights reserved

$$\begin{aligned}\pi &= (\langle f_{\pi^0}\rangle + \langle f_h\rangle\langle f_\gamma\rangle) + (\boldsymbol{h'}/\boldsymbol{e})\langle f_h\rangle(1-\langle f_\gamma\rangle)\\ &= 1-(1-\boldsymbol{h'}/\boldsymbol{e})(1-\langle f_\gamma\rangle)\langle f_h\rangle\\ &\approx 1-(1-\boldsymbol{h'}/\boldsymbol{e})(1-\langle f_\gamma\rangle)(E/E_0)^{(m-1)}\\ &\equiv 1-a'E^{(m-1)},\end{aligned} \quad (7.72)$$

where a' is the product of all factors. Energy deposits by nuclear γs can account for 10–20% of the total energy deposit in materials such as iron, and it can be even higher in high-Z materials [167].

The response for an incoming pion and incoming proton is different if $e/h \neq 1$. We get

$$\begin{aligned}\frac{\pi}{e} &= 1-\bigl(1-h/e\bigr)\langle f_{\pi^-}\rangle \quad \text{and}\\ \frac{p}{e} &= 1-\bigl(1-h/e\bigr)\langle f_p\rangle,\end{aligned} \quad (7.73)$$

where $\langle f_{\pi^-}\rangle$ ($\langle f_p\rangle$) are the mean charged pion (proton) fractions that have an energy dependence of $(E/E_{0\pi^-})^{m-1}$ and $(E/E_{0p})^{m-1}$, respectively. While $E_{0\pi^-}$ (1 GeV) and E_{0p} (2.6 GeV) are different, m and h/e are the same. By rearrangement we get

$$\frac{\langle f_{\pi^-}\rangle}{\langle f_p\rangle} = \frac{1-\pi/e}{1-p/e}. \quad (7.74)$$

Figure 7.60 (right) shows the energy dependence of the e/π ratio for ATLAS Tile-Cal test beam data. The e/h ratios are larger than one, decreasing slightly with energy. The e/h ratios obtained from fits are similar for the two data sets. We present the energy resolution and linearity of these data in Sect. 7.4.2.5. Figure 7.61 (left) shows the electromagnetic shower fraction in pions as a function of the pion energy for a copper-quartz-fiber calorimeter and a spaghetti calorimeter. The electromagnetic shower fraction typically increases with the hadron energy.

7.4.2.2 Compensation

Calorimeters in which $e/h \neq 1$ show some characteristic features.

- The signal shape of a hadron at fixed energy is not Gaussian.
- Fluctuations in $\langle f_{\pi^0}\rangle$ yield an additional contribution to the energy resolution.
- With increasing energy, σ_E/E does not improve as $1/\sqrt{E}$.
- The observed calorimeter signal is not proportional to the hadron energy.
- The measured e/π ratio is energy dependent.

Note that all these effects have been seen experimentally [165, 170, 171] and are reproducible in simulations. Willis suggested that compensation might be possible [84].

7.4 Hadron Shower Measurements

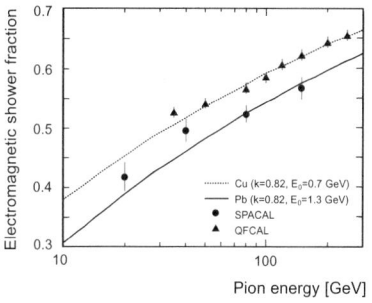
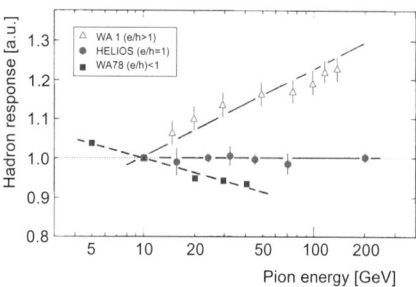

Fig. 7.61 Left: The electromagnetic shower fraction as a function of pion energy in a copper-quartz-fiber calorimeter (triangles) [169] and a spaghetti calorimeter (points) [161]. The dashed and solid curves are fits with the function $1 - (E/E_0)^{(m-1)}$, using $E_0 = 0.7$ GeV, $m = 0.82$ and $E_0 = 1.3$ GeV, $m = 0.82$, respectively. Reprinted with kind permission from [169], © 1997, Elsevier. All rights reserved. Right: The energy dependence of the pion response in three calorimeters with different *e/h* values, for the under-compensating WA1 experiment (triangles) [165], the compensating HELIOS experiment (points) [170] and the over-compensating WA78 experiment (squares) [171]. All data are normalized to the results at 10 GeV. Reprinted with kind permission from [172], © 1991, R. Wigmans. All rights reserved

The response of a hadron sampling calorimeter to a showering hadron is a complex process. The calorimeter signal is basically determined by the numerous very soft particles from the last stages of the shower development. For example, for electromagnetic showers in lead or uranium sampling calorimeters about 40% of the energy is deposited by ionization of electrons softer than 1 MeV [173]. Based on measurements of pion signals in lead-scintillator sampling calorimeters, we know that particles between two consecutive layers nearly show no correlation. Those particles that form the signal traverse only a small fraction of a nuclear interaction length. Thus, the last stages of the shower development must be understood in great detail, i.e. processes at the nuclear and atomic levels. These are triggered by soft protons and neutrons from nuclear reactions. The saturation properties of the active material for densely ionizing particles are of crucial importance because most protons are non-relativistic.

The level of visible compensation in hadron showers is expected to be affected both by the choice of the passive absorber, the choice of the active medium and the ratio of plate thickness to gap thickness. To achieve compensation, we need to equalize the response of the calorimeter to electrons and hadrons. We can either decrease the electron response or boost the hadron response. Note that in a high-Z radiator the electron response will be decreased. Motivated very much by the work of Brau [174,175], Gabriel [176], Brückmann [177], and Wigmans [163,173], several groups explored a variety of compensation mechanisms and built calorimeters, which were very nearly compensating. The degree of compensation was sensitive to the acceptance gate width, and so could be somewhat further tuned.

Figure 7.61 (right) shows the response to hadrons in the WA1, HELIOS and WA78 calorimeters. The WA1 experiment used an iron-scintillator calorimeter having *e/h* > 1. The HELIOS experiment used a compensating uranium-scintillator calorimeter, while the WA78 experiment also used a uranium-scintillator calorimeter having *e/h* < 1. While for the HELIOS experiment the hadronic response is energy independent, it increases in the WA1 calorimeter and decreases in the WA78 calorimeter.

In uranium-scintillator calorimeters, the hadronic component can be boosted. Due to nuclear break-up, we get neutron-induced fissions yielding about 40 fissions/GeV, which altogether liberate about 10 GeV of fission energy. To compensate the nuclear deficit, only a small fraction (300–400 MeV) needs to be detected. This can be accomplished either by detecting the few MeV γ component or by detecting fission neutrons liberated in the fission process. It depends on the active sampler, which component and what fraction is measured. To determine the characteristics of energy depositions in uranium, we have to distinguish five very different types of processes illustrated in Fig. 7.62,

1. Incident or internally created photons; $e^{\pm}$ and π^0s produce electromagnetic showers.
2. Muons and stable charged hadrons deposit energy via dE/dx losses.
3. Neutrons in the MeV range might interact by elastic collisions or nuclear interactions.
4. Fission processes of uranium nuclei might be directly induced by high-energy hadrons or by medium-energy neutrons.
5. Excited fission products or evaporation from the residual nuclei might emit prompt photons, while delayed photons are emitted in neutron capture processes.

We focus here on sampling calorimeters using plastic scintillators or LAr. The first two processes occur in all absorbers and do not provide a new source of energy deposit. Thus, we focus on the latter three processes since neutrons and photons originate in spallation, evaporation and fission processes of uranium, which may deposit additional energy.

Figure 7.63 illustrates the production of neutrons in fission processes. There are two different production mechanisms, either by spallation of a high-Z nucleus or by evaporation from highly excited fission products. Very few high-energy neutrons are produced, which will move through the calorimeter and eventually start another spallation process. Most neutrons are produced with energies between 0.1 and 10 MeV. So, they have typical mean free paths of 2–5 cm, thus traversing several absorber layers before interacting. In calorimeters using scintillators as active medium, the neutrons deliver almost their entire kinetic energy in a few interactions to recoiling protons, which loose energy by ionization losses. The conversion efficiency is determined by Birks' formula (2.197). This energy conversion mechanism is not observed in liquid argon. The neutrons typically perform elastic scattering and vanish into the absorber.

Figure 7.64 shows uranium and hydrogen cross sections for neutron-induced reactions. For energies below 1 MeV, neutrons are captured and produce de-excitation photons, while above 1 MeV they trigger fission in which the fission products are highly excited. Although most of the binding energy released by fission shows up

7.4 Hadron Shower Measurements

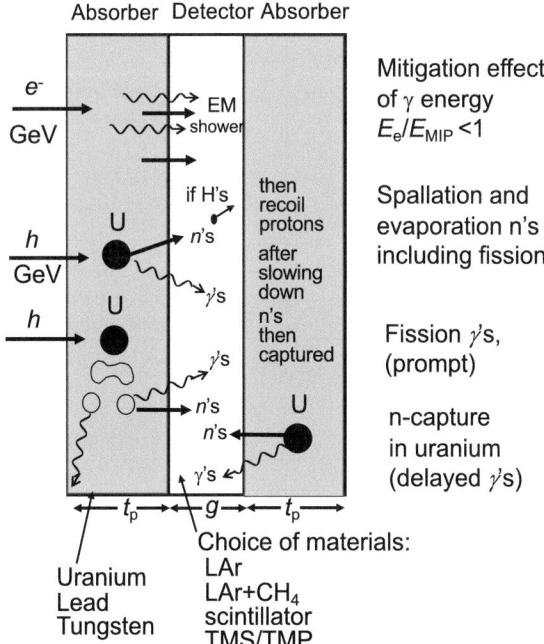

Fig. 7.62 Schematic visualization of important processes that are involved in the physics of compensating sampling calorimeters. Electrons and photons entering the absorber that can be uranium, lead or tungsten produce an electromagnetic shower that is detected in the detector planes, which may consist of liquid argon or scintillator. Muons and charged hadrons typically loose energy via dE/dx, which is recorded in the detector. A charged hadron can interact with uranium emitting neutrons and photons. Spallation and evaporation neutrons are prompt having energies in the MeV range. The charged hadron may cause fission in which prompt photons and neutrons are emitted also. The neutrons are typically slowed down in the absorber. A slow neutron may be captured by a uranium atom emitting a photon delayed in time. Reprinted with kind permission from [177], © 1988, Elsevier. All rights reserved

as kinetic energy of the fission products, it does not contribute to the signal because due to the short range of fission products (10 μm) this energy is dissipated inside the uranium plates. In addition to neutrons, energy is released in form of de-excitation γ-rays. Since the detection efficiency for them is low, they also do not contribute much new energy to the signal. However, the capture of neutrons produces prompt fission γ-rays. The cross section for the (n, γ) process becomes large in the range of neutron energies of several keV (resonance region) and very large at even lower energies (eV to thermal energies) as indicated in Fig. 7.64. Thus, we get delayed contributions from the detection of the capture γ-rays. The detection of the fission γ-rays is similar to that of the de-excitation γ-rays.

A 10 GeV pion releases on average 222 neutrons. Only a small fraction is necessary to achieve *e/h* = 1. Figure 7.65 (left) shows the visible energy and *e/h* ratio as a function of integration time in a uranium scintillator calorimeter. For an integration time of about 200 ns, an *e/h* ratio of one can be achieved. The extra energy from

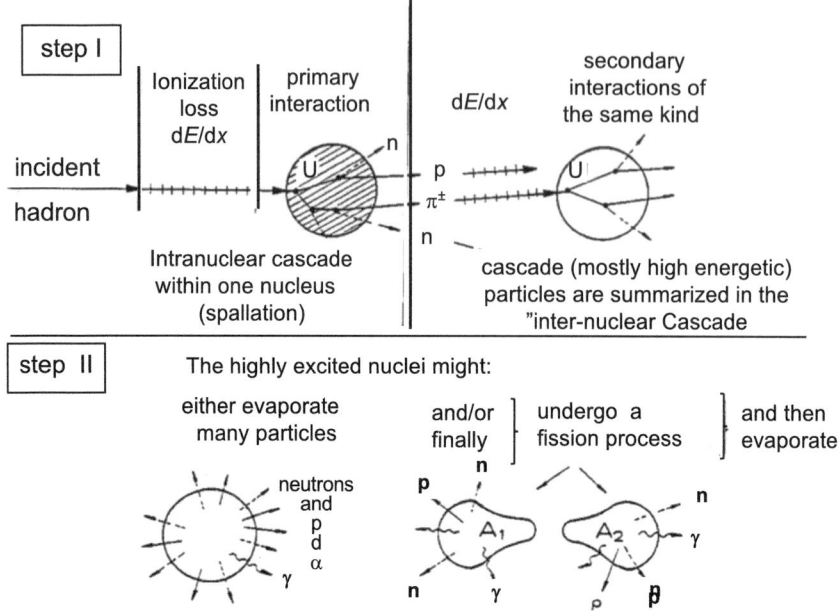

Fig. 7.63 Top: An incident hadron creates an internuclear cascade that releases a few high-energy spallation products including neutrons, which are able to start another internuclear cascade. Bottom: The highly excited nuclei left behind undergo de-excitation. They either evaporate many particles or undergo a fission process before evaporating many particles. Reprinted with kind permission from [177], © 1988, Elsevier. All rights reserved

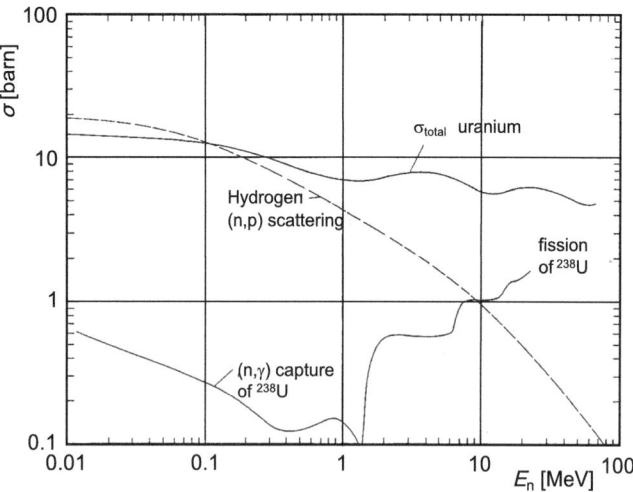

Fig. 7.64 Uranium and hydrogen cross sections for neutron-induced reactions. Reprinted with kind permission from [177], © 1988, Elsevier. All rights reserved

7.4 Hadron Shower Measurements

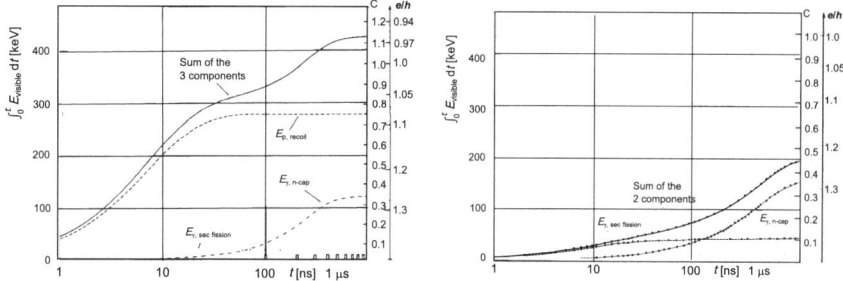

Fig. 7.65 Left: The visible energy as a function of integration time in a uranium scintillator sampling calorimeter. Right: The visible energy as a function of integration time in a uranium liquid-argon sampling calorimeter. The overall compensation C and e/h ratio are shown on the right-hand scale. Both reprinted with kind permission from [177], © 1988, Elsevier. All rights reserved

neutron and γ capture as well as fission γ rays compensates the electromagnetic response. For liquid argon, the contribution of compensation is only 50% with respect to that in scintillators as shown in Fig. 7.65 (right).

The experimentally measurable ratio **e/h** can be written as,

$$e/h = \frac{e/MIP}{h_i/MIP + C_{\text{tot}}(e/MIP - h_i/MIP)}, \tag{7.75}$$

where **MIP** denotes the efficiency for an energy deposit by a minimum ionizing muon. As before, **e** denotes the efficiency of the energy deposit by a pure electromagnetic shower and h_i denotes the efficiency of an energy deposit of the ionizing part of the hadronic shower, which contains both hadronic and electromagnetic components. The quantity C_{tot} characterizes the degree of compensation,

$$C_{\text{tot}} = C_{\text{recoil}} + C_{\gamma,\ 1-\text{fission}} + C_{\gamma,\ 2-\text{fission}} + C_{\text{delayed }\gamma}, \tag{7.76}$$

where C_{recoil} is the contribution from recoil protons, $C_{\gamma,\ 1-\text{fission}}$ is that from prompt fissions, $C_{\gamma,\ 2-\text{fission}}$ is that from secondary fissions and $C_{\text{delayed }\gamma}$ is that from n capture. The normalization is chosen such that $C_{\text{tot}} = 1$ corresponds to full compensation (**e/h** = 1.) For a U-LAr, we get measured values of **e/MIP** = 0.63 and h_i/MIP = 0.41, while for an U-scintillator calorimeter we get **e/MIP** = 0.60 and h_i/MIP = 0.41. For U-LAr sampling calorimeters the neutrons have zero recoil momentum, while for U-scintillator devices the recoil momentum is 91 MeV. Thus, compensation for the U-LAr calorimeter is $C_{\text{tot}} = 48\%$, while $C_{\text{tot}} = 89\%$ for the U-scintillator device, yielding **e/h** ratios of 1.22 and 1.04, respectively.

Figure 7.66 shows the energy deposits of an incoming proton as a function of energy in iron (left) and uranium (right). Besides the electromagnetic energy deposit, which is higher in iron than in uranium and increases with energy, the hadron component can be split into energy deposits of relativistic hadrons, spallation protons, neutrons with energies below 50 MeV and binding and recoil energy effects of nuclear fragments. The energy deposits for the two latter processes are higher in uranium

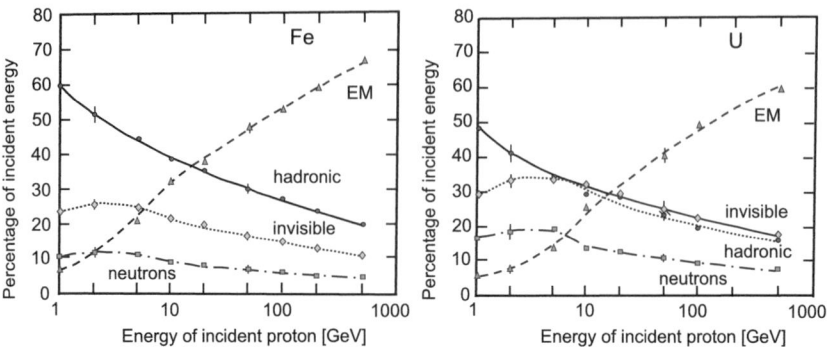

Fig. 7.66 Energy deposits of an incoming proton in iron (left) and in uranium (right) via electromagnetic energy from π^0's + γ's (blue triangles), hadrons like p, $\pi^\pm$ and $\mu^\pm$ (red points), binding energy plus recoil energy of heavy nuclei plus ν energy and electrons from μ decay (orange diamonds) and neutrons with energy less than 50 MeV (green squares) as a function of energy in lead (left) and in uranium (right) [178]

than in lead. While most of the energy of the last process is invisible, the neutrons produce γ rays that are detected and reduce the *e/h* ratio.

Figure 7.67 (left) shows the simulated *e/MIP* ratios as a function of Z for a uranium-liquid-argon calorimeter and a uranium-scintillator calorimeter based on polymethyl methacryate (PMMA), a different solid solvent compared to the standard polysterene. The *e/MIP* ratios decrease with Z, being lower for the U-PMMA calorimeter than for the U-LAr calorimeter. Figure 7.67 (right) shows the *e/MIP* ratio as a function of the uranium plate thickness for readout with PMMA and liquid argon. We see a saturation for 6 mm plate thickness. The *e/MIP* ratio is higher in liquid argon than in PMMA detectors.

Figure 7.68 (left) shows calculated *e/h* curves as a function of the ratio of absorber plate thickness to readout layer thickness R_d for uranium calorimeters, using different readout materials, a plastic scintillator based on PMMA, a plastic scintillator based on polystyrene (SCSN-38), silicon, liquid argon and tetramethylpentane (TMP). For liquid argon and plastic scintillators, also experimental data are displayed. While the plate thickness is varied in R_d, the gap size of the active material is fixed to 2.5 mm for PMMA, SCSN-38, LAr and TMP and 0.4 mm for silicon. For LAr and silicon, two curves are shown for 0.1 and 1 μs gate time lengths. The short gate time yields larger *e/h* values. For $R_d > 2$, the *e/h* curves for LAr become uniform. This indicates that recoil protons do not lose energy in the liquid argon. They perform elastic scattering and disappear into the uranium. This is quite different for the two scintillators and for TMP (C_9H_{20}) for which *e/h* continues to decrease rapidly with increasing R_d. The amount of energy loss that is seen by the detector is given by Birks' law. Since TMP contains a very large fraction of free protons, the contribution of recoil protons is increased. To make predictions, we need to know the saturation and recombination properties of TMP. Since no measurements on the suppression of the signal from soft protons exist, the TMP curves have been calculated assuming that recombination effects are similar as those for liquid argon or PMMA. Thus, two curves were

7.4 Hadron Shower Measurements

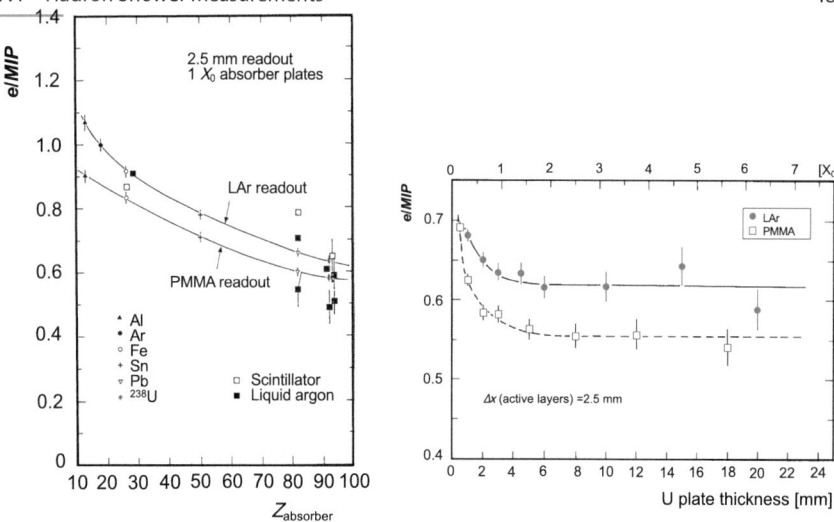

Fig. 7.67 Left: Simulated *e/MIP* ratio as a function of the absorber atomic number for a uranium liquid-argon calorimeter (upper curve) and a uranium scintillator calorimeter (lower curve) using polymethyl methacrylate (PMMA) as solid solvent. The points are measurements. Right: Simulated *e/MIP* ratios as a function of the uranium plate thickness for a uranium liquid-argon calorimeter (dark points) and a uranium scintillator calorimeter using PMMA (open squares). Both reprinted with kind permission from [173], © 1987, Elsevier. All rights reserved

calculated using Birks' constants for liquid argon ($k_{\text{Birks}} = 0.0045$ g/MeV · cm^2) and for PMMA ($k_{\text{Birks}} = 0.00978$ g/MeV · cm^2). To achieve *e/h* = 1, we need large sampling fractions or thin plates. The curves for the two scintillators are slightly different because of the different chemical compositions and number of free protons. For PMMA, we reach *e/h* = 1 at $R_d \approx 1$, while for SCSN-38 this occurs at $R_d \approx 0.8$.

The curves for silicon show a similar behavior as those for liquid argon though the saturation sets in at higher R_d values. The reason for the shift results from the different thickness of the silicon. We used a 0.4 mm silicon thickness compared to a 2.5 mm scintillator thickness. This means that the uranium plates are 4 mm at $R_d = 10$, while in case of the scintillator we used 5 mm thick plates for $R_d = 2$. The plateau values for *e/h* are 1.09 and 0.97 for a 100 ns and 1 µs gate, respectively. So with the appropriate gate width tuning, we can achieve compensation in silicon-uranium calorimeters. It should be noted that the uncertainties in these calculations concern details of the shower evolution, which affect the absolute values of *e/h* but not the R_d dependence. Furthermore, efficient neutron detection is important for efficient hadron signal detection. These neutrons can travel long distances, especially also in lateral directions. Since many measurements were conducted with calorimeter prototypes, neutron leakage may have been underestimated yielding overestimated *e/h* values.

Figure 7.68 (right) shows the corresponding calculated *e/h* curves for lead-based calorimeters. The additional curves arise by considering pure lead and a lead-cadmium alloy. The curves for plastic scintillators use a gate width of 100 ns. We

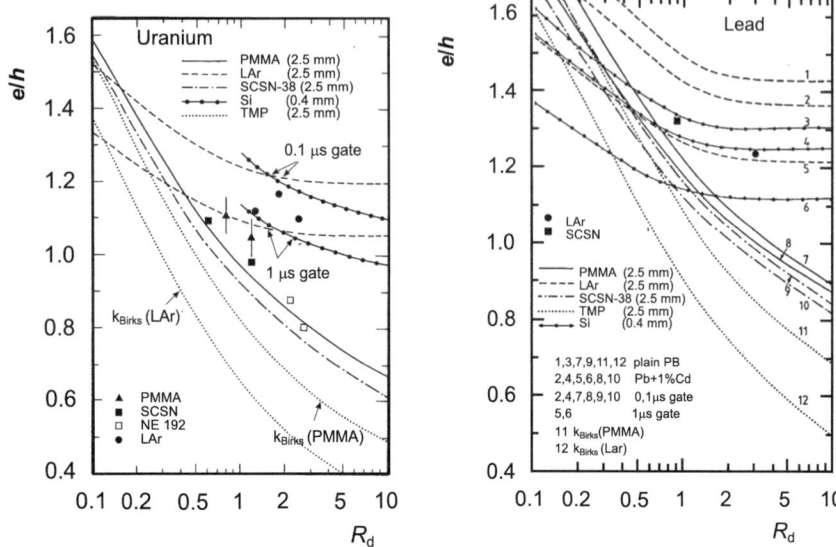

Fig. 7.68 Curves and measurements of the *e/h* ratio as a function of the ratio R_d of absorber thickness d_p to readout layer thickness d_g for uranium (left) and lead (right) calorimeters. The readout layers consist of PMMA (solid curve and triangles [164,179]), liquid argon (dashed curves and points [180,181]), standard scintillator SCSN (dash-dotted curves and squares [182,183]), silicon (solid curves with dots), TMP (dotted curves) and NE 102 scintillator (open square). For the lead-silicon curve, R_d actually runs from 1 to 100, clearly showing saturation above 20. Error bars on measurements are expected to be at least 5%. For the lead absorber predictions are given for pure lead and lead with 1% cadmium. Most predictions are for a gate of 0.1 µs except for two lead +1% Cd devices with Si and LAr that use a gate of 1 µs. Reprinted with kind permission from [173], © 1987, Elsevier. All rights reserved

notice a similar tendency as for uranium-based calorimeters except that *e/h* is shifted to higher values. As before, the decrease of *e/h* with increasing R_d is due to the contribution of recoil protons, while the increase in *e/h* values is due to the absence of fission neutron effects. Since the cross section for the capture of thermal neutrons in lead is more than a factor of ten smaller than that in uranium, the mean free path increases up to 180 cm. So, neutron capture is negligible in lead unless we consider alloys with materials that have a large neutron capture cross section. Using cadmium, we just need a small amount to capture thermal neutrons. In addition, 9 MeV of energy is released per captured neutron. Compared to pure lead, 20% more neutrons are captured in the lead-cadmium alloy as the two curves for scintillators show. Also for Pb-scintillator calorimeters, compensation can be achieved. However, the R_d values are four to five times larger than those for uranium. For the lead-cadmium alloy, we get a 20% reduction in R_d values. For lead-liquid-argon and lead silicon, compensation cannot be achieved while it is possible for TMP.

Figure 7.69 shows the energy loss by spallation protons in the active medium with respect to that in the passive medium for uranium, lead and steel absorbers combined with PMMA scintillator or liquid argon. This is a measure for the calorimeter signal.

7.4 Hadron Shower Measurements

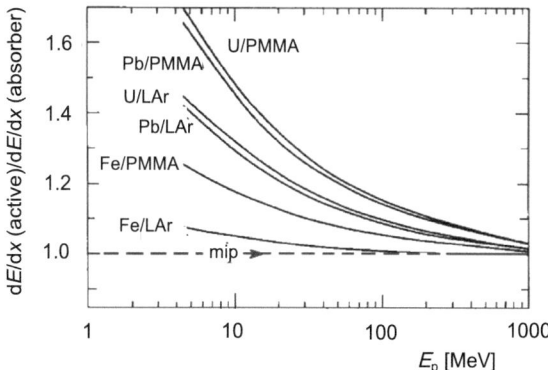

Fig. 7.69 The calorimeter response to spallation protons showing the energy loss in the active medium with respect to that in the passive medium as a function of proton energy for uranium, lead and steel absorbers combined with PMMA scintillator or liquid argon as active medium, where both energy losses are normalized to their minimum value. Reprinted with kind permission from [173], © 1987, Elsevier. All rights reserved

The energy loss is largest for uranium or lead combined with PMMA. The corresponding values for uranium or lead combined with liquid argon are about 15% lower. Note that for uranium absorbers the dE/dx ratio is always slightly higher than that for lead absorbers. For steel calorimeters, the dE/dx losses in the active medium with respect to that in the steel are considerably lower, both for PMMA and liquid argon. So based on the above knowledge, several hadron sampling calorimeters were built that achieved nearly compensation. Table 7.7 summarizes their properties.

In some non-uranium calorimeters, hardware compensation is possible by the appropriate choice of the active readout material and the thickness of plates and active gap. However, we may apply software compensation procedures to non-compensating calorimeters. The charge recorded in a calorimeter cell offers the possibility of identifying the underlying showering process. Typically, electromagnetic showers deposit locally more charge than hadron showers. In a given cell the energy of the deposited charge has a non-linear dependence. So, accounting for both the *e/h* ratio and the tagging efficiency for charge deposits from electromagnetic particles will effectively compensate for the different electron and hadron responses and thus improve the energy resolution.

The total energy of a hadron is given by

$$E = \sum_{i=1}^{N_{EM}} a_{EM}^i Q_i^e + \sum_{j=1}^{N_{Had}} a_{Had}^j Q_j^h, \qquad (7.77)$$

where Q_i^e is the charge measured in cell i in the electromagnetic calorimeter, Q_j^h is the charge measured in cell j in the hadron calorimeter, a_{EM}^i and a_{Had}^j are the scaling parameters in the EM and hadron calorimeter cells, respectively. Note that inactive material has been corrected for. The calibration constants are typically used as starting

Table 7.7 Examples of nearly compensating hadron sampling calorimeters listing the calorimeter, absorber (thickness), active medium (thickness), energy resolution and *e/h* ratio. [†]Helios and D0 quote the e/π ratio. [‡]The first value refers to the horizontal and the second to the vertical direction

Calorimeter	Absorber (th. [mm])	Active layer [mm]	σ_E/E [%]	*e/h* (at p [GeVc])
Akesson et al. [164]	U(3) U/Cu(3/5)	scint. (2.5)	$\frac{36}{\sqrt{E}}$	1.11 (1–10)
HELIOS [170]	U (3)	scint. (2.5)	$\frac{(33.7\pm1.2)}{\sqrt{E}}$	0.984 ± 0.006[†] (8–200)
Drews et al. [184]	Pb (10)	scint. (2.5)	$\frac{(43.5\pm1.0)}{\sqrt{E}}$	1.10±0.01 (30)
Drews et al. [184]	U (3.2)	scint. (3.0)	$\frac{(35.8\pm1.0)}{\sqrt{E}}$	1.02±0.01 (30)
WA80 low I [185]	U (3)	scint. (3.0)	$\frac{33}{\sqrt{E}} \oplus 1.3$	1.12 (135)
WA80 high I [185]			$\frac{67}{\sqrt{E}} \oplus 2$	
ZEUS [186]	U (3.3)	scint. (2.6)	$\frac{34}{\sqrt{E}}$	~ 1 for (> 3)
SPACAL [161]	Pb ($^{2.22}_{1.92}$)[‡]	scint. fibers (1)	$\frac{(33.4\pm0.4)}{\sqrt{E}} \oplus (2.2\pm0.1)$	1.15 ± 0.02 (< 80)
D0 [187]	U (6)	LAr (2 × 2.3)	$\frac{(41\pm4)}{\sqrt{E}} \oplus (3.2\pm0.4)$	1.11[†] (10)

values for a least-square fit by minimizing the energy resolution. This takes intercalibration effects between cells into account. For non-compensating calorimeters the energy resolution can be improved by weighting methods, which provide some kind of "compensation". Several weighting methods exist. We use the following ansatz for the weighting function [188],

$$E(Q_k)/Q_k = C_1 \exp(-C_2 Q_k/V_k) + C_3, \quad (7.78)$$

where k is the index for a cell in the electromagnetic calorimeter or the hadron calorimeter and V_k is the volume of that cell. The parameters C_k are optimized separately for the electromagnetic and hadronic calorimeter parts in two χ^2 fits minimizing the difference between the reconstructed and deposited energy. Since a jet has a larger fraction of electromagnetic energy, the distributions for jets and pions differ leading to a significantly different e/π ratio. For the jet reconstruction H1 uses energy-dependent weighting parameters,

$$\begin{aligned} C_1 &= \alpha_1 \cdot \exp(-\alpha_2 \cdot E_{jet}) + \alpha_3 + \alpha_4 \cdot E_{jet} \\ C_2 &= \alpha_1 \cdot \exp(-\alpha_2 \cdot E_{jet}) + \alpha_3 \\ C_3 &= \alpha_1 \cdot \exp(-\alpha_2 \cdot E_{jet}) + \alpha_3 + \alpha_4 \cdot E_{jet}, \end{aligned} \quad (7.79)$$

where the new weighting parameters $\alpha_1, \alpha_2, \alpha_3$ and α_4 are determined from χ^2 fit separately for cells in the electromagnetic and hadron calorimeters (see [188]). Applying this to the energy range from 50 to 250 GeV and entire angular range improves the energy resolution to $55\%/\sqrt{E}$. The energy resolution of single pions is about $50\%/\sqrt{E}$ [189].

7.4.2.3 Instrumental Effects

Most hadronic shower counters are sampling calorimeters. Preferentially, rather dense passive absorbers are used to reduce the linear dimension of detector. Sampling fluctuations of statistical origin analogous to those of the electromagnetic sampling fluctuations may contribute to the energy resolution, but for the sampling of hadron showers we do not have a similar detailed description. Several energy resolution measurements are consistent with a sampling contribution of [80],

$$\left(\frac{\sigma_E}{E}\right)_{\text{hadronic sampling}} = \frac{9\%\sqrt{\Delta E \text{ [MeV]}}}{\sqrt{E \text{ [GeV]}}}. \quad (7.80)$$

where ΔE is the energy loss per unit sampling for minimum ionizing particles. A detailed study of a Pb-scintillator and a U-scintillator calorimeters yielded a slightly higher value of $\frac{11.5\%\sqrt{\Delta E \text{ [MeV]}}}{\sqrt{E \text{ [GeV]}}}$ [184]. Note that the hadronic sampling fluctuations are approximately twice as large as the electromagnetic sampling fluctuations. The sampling in hadron calorimeters can be made small relative to large intrinsic component, so that the energy resolution does not need to be sacrificed in hadron sampling calorimeters.

7.4.2.4 Contributions to the Energy Resolution

Table 7.8 shows a comparison of contributions to the total energy resolution in electromagnetic and hadronic showers. Both the intrinsic shower fluctuations and sampling fluctuation are much larger in hadron calorimeters than those in electromagnetic calorimeters. Thus, electromagnetic showers can be measured much more precisely than hadron showers. In addition, all properties of electromagnetic showers can be simulated well with the EGS4 program. For hadron showers, this is much more difficult.

7.4.2.5 Measured Hadronic Energy Resolution

The energy resolution also depends on the *e/h* ratio. Figure 7.70 shows the energy resolution $\sigma_E/\sqrt{E}$ as a function of the uranium absorber thickness for 2.5 mm (left) and 5 mm (right) thick PMMA scintillator tiles. For a fixed PMMA thickness, the energy resolution has a minimum at a specific plate thickness. Here, *e/h* = 1 and only the sampling fluctuations and the intrinsic energy resolution contribute to the total energy resolution. The energy-dependent terms are zero. For *e/h* < (>)1, the calorimeter is overcompensating (undercompensating) and the energy resolution becomes worse. It will not scale with energy as $1/\sqrt{E}$. Similar plots exist for liquid argon, silicon and TMP [173].

Table 7.8 Comparison of contributions to the total energy resolution in electromagnetic and hadronic showers. The individual components are added in quadrature to yield the total energy resolution. Reprinted with kind permission from [80], © 1985, Springer. All rights reserved

Mechanism	Electromagnetic showers	Hadronic showers
Intrinsic shower fluctuations	Track length fluctuations $\sigma_E/E \geq 0.5\%/\sqrt{E}$ [GeV]	Fluctuations in energy loss $\sigma_E/E \simeq 45\%/\sqrt{E}$ [GeV] Scaling weaker than $1/\sqrt{E}$ for high E With compensation for nuclear effects $\sigma_E/E \simeq 22\%/\sqrt{E}$ [GeV]
Sampling fluctuations	$\sigma_E/E \simeq 4\%\sqrt{E\,[\text{MeV}]}/E\,[\text{GeV}]$ Nature of readout may augment sampling fluctuations	$\sigma_E/E \simeq 9\%\sqrt{E\,[\text{MeV}]}/E\,[\text{GeV}]$
Instrumental effects	Noise and pedestal width: $\sigma_E/E \sim 1/E$ Determine minimum detectable signal Limits low-energy performance Calibration error and non-uniformities $\sigma_E/E \sim$ constant, therefore limits high-energy performance	
Incomplete shower containment	$\sigma_E/E \sim E^{-\alpha}$ with $\alpha < 1/2$ For leakage fraction $\geq$ few %, non-linear response and non-Gaussian tail	

Figure 7.71 (left) shows the energy resolution $\sigma_E/\sqrt{E}$ as a function of energy for the WA1(CDHS) iron-scintillator calorimeter, the HELIOS uranium-scintillator calorimeter and the WA78 uranium-scintillator calorimeter. For the HELIOS calorimeter, where **e/h** = 1, an energy resolution of $35\%/\sqrt{E}$ was measured with the $1/\sqrt{E}$ scaling up to 100 GeV. The measured energy resolution in the under-compensating WA1 calorimeter only exhibit $1/\sqrt{E}$ scaling after software compensation. For the over-compensating WA78 calorimeter, the energy resolution increases with energy. This underlines again that in calorimeters with **e/h** $\neq$ 1 the energy resolution does not scale with $1/\sqrt{E}$.

The ZEUS Collaboration measured the intrinsic energy resolution in a compensating uranium scintillator sampling-calorimeter prototype, having 3.2 mm thick uranium plates and 3 mm thick scintillator plates to be $(20.4 \pm 2.4)\%/\sqrt{E}$, which is consistent with the quoted value of $22\%/\sqrt{E}$. For the lead scintillator calorimeter prototype that used 100 mm thick lead plates and 2.5 mm thick scintillator plates they measured an intrinsic energy resolution of $(13 \pm 4.7)\%/\sqrt{E}$ [184, 191]. This difference indicates that in principle a better intrinsic energy resolution can be achieved with lead rather than with uranium calorimeters. The reason is the following. How much extra energy is recorded depends on the degree of correlation between nuclear binding energy losses and the neutron kinetic energy. This correlation is higher in lead

7.4 Hadron Shower Measurements

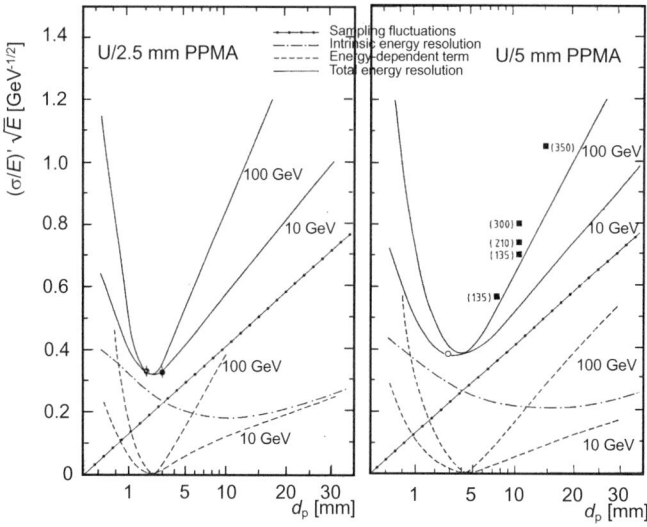

Fig. 7.70 The total energy resolution $\sigma_E/\sqrt{E}$ of 10 and 100 GeV hadrons (solid curves) as a function of uranium plate thickness for 2.5 mm (left) and 5 mm thick PMMA scintillator (right). In addition, the contributions from the intrinsic energy resolution (dash-dotted curves), sampling fluctuations (dotted curves) and energy dependent terms (dashed curves) are displayed as well as measurements (solid points [164, 179], solid squares [182, 183] and open points [173]). Note that the energy-dependent terms arise, when $e/h \neq 1$. At the minimum, where $e/h = 1$, they vanish. Reprinted with kind permission from [173], © 1987, Elsevier. All rights reserved

than in uranium because in uranium many neutrons originate from fission processes, which are less correlated with nuclear binding energy losses. Thus, the fluctuations are smaller in lead than in uranium. In addition to the intrinsic energy resolution, however, the contribution from sampling fluctuations is relevant for the total energy resolution. For uranium the sampling fluctuations amount to $(31.1 \pm 0.9)\%/\sqrt{E}$, while for lead they are $(41.2 \pm 0.9)\%/\sqrt{E}$. Thus, the total energy resolution, which is the quadratic sum of these two contributions plus possibly additional effects, is lower for uranium calorimeters than for lead calorimeters.

Figure 7.71 (right) shows the total energy resolution $\sigma_E/\sqrt{E}$ as a function of energy for the ZEUS uranium scintillator sampling-calorimete prototype, which illustrates that compensation works rather well, since the energy resolution scales as $1/\sqrt{E}$. The ZEUS experiment used uranium scintillator for both the electromagnetic and hadron calorimeters. The layout is briefly discussed in Sect. 7.4.4.2.

Many hadron sampling calorimeters have been based on steel as the absorber material, with active layers consisting of plastic scintillator, liquid argon or wire chambers. None of these under-compensating ($e/h > 1$) calorimeters have achieved energy resolutions better than $\simeq 50\%/\sqrt{E}$. In addition, we observe deviations from the $1/\sqrt{E}$ scaling at higher energies. The H1 experiment operated a steel liquid-argon calorimeter, where the steel absorber plates had a thickness of 18 mm while the liquid-argon gap was 2×2.4 mm [192]. Using energy-depending weighting, the energy resolution of single pions was $\sigma_E/E = \frac{46.1\%}{\sqrt{E}} \oplus 2.6\% \oplus \frac{73\%}{E}$ (with E in units

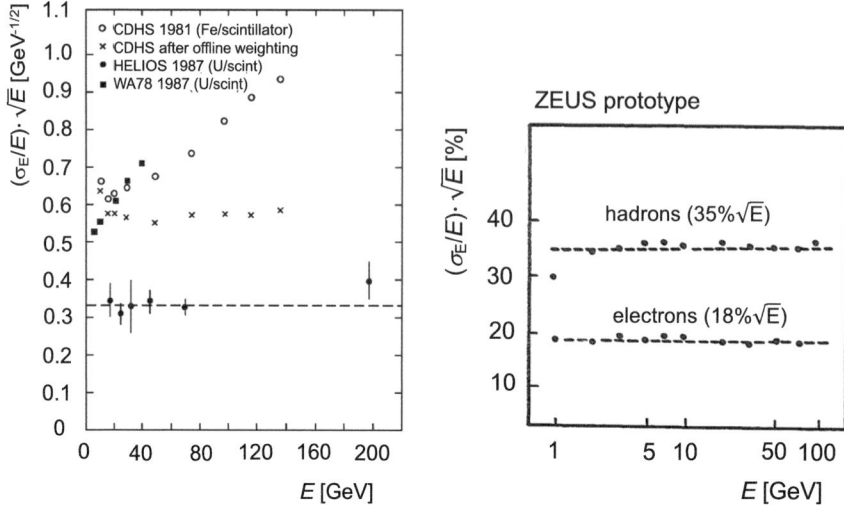

Fig. 7.71 Left: The energy resolution $\sigma_E/\sqrt{E}$ as a function of energy for the WA1 (CDHS) iron scintillator calorimeter before weighting (open circles) and after weighting (crosses [165]), the compensating HELIOS uranium scintillator calorimeter (solid points [170]) and the WA78 uranium scintillator calorimeter (solid squares [171]). The response to pions for these calorimeters was shown in Fig. 7.61. Reprinted with kind permission from [172], © 1991, R. Wigmans. All rights reserved. Right: Energy resolution $\sigma_E/\sqrt{E}$ as a function of energy for the ZEUS uranium scintillator sampling-calorimeter prototype. Reprinted with kind permission from [190], © 1990, Elsevier. All rights reserved

of GeV). For uranium liquid-argon calorimeters, which have *e/h* values closer to one, deviations from the $1/\sqrt{E}$ scaling are smaller. The energy resolution obtained by the D0 experiments amounts to about $60\%/\sqrt{E}$ at 10 GeV [193]. Calorimeter prototype studies with uranium liquid argon for the SLD experiment achieved a similar energy resolution [180].

At the LHC, the ATLAS experiment uses a steel scintillator calorimeter in the barrel while CMS uses a steel-copper scintillator shower detector. The ATLAS calorimeter is briefly discussed below. Figure 7.72 (left) shows the energy resolution σ_E/E measured in an ATLAS iron scintillator prototype as a function $1/\sqrt{E}$ in two sets of test beam data. The energies are reconstructed using two approaches, the benchmark procedure and the cell correction [168]. The benchmark procedure consists of two steps to reconstruct the nominal beam energy. First, the energy E_0 of a pion is obtained as a sum four terms, the energy in the electromagnetic calorimeter, the charge deposited in the hadron calorimeter, a term accounting for energy loss in the cryostat and a correction term. The latter three terms are multiplied by parameters that are obtained from a fit minimizing the fractional energy resolution of 300 GeV pions. In the second step, the energy E_0 is rescaled to the nominal beam energy. The second approach consists of a cell-weighting procedure. The total energy is reconstructed by correcting the energy in each cell of both the liquid-argon calorimeter and the tile calorimeter by a factor that is a function of the energy in each cell and of the beam energy. The energy loss in the cryostat is also considered. The weights are parametrized for each beam energy by a constant and a $1/E_{cell}$ term for both

7.4 Hadron Shower Measurements

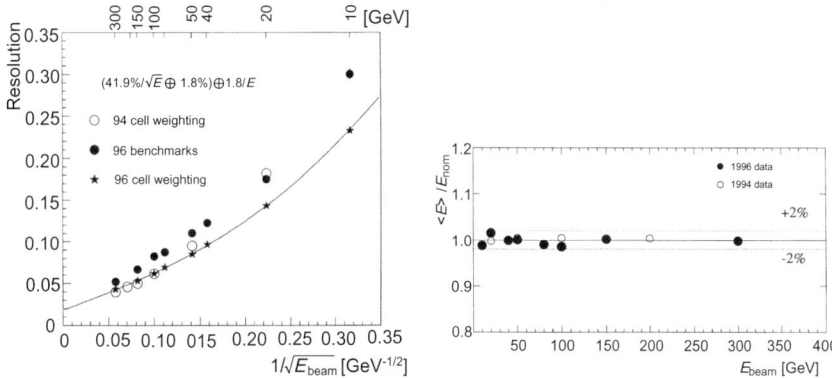

Fig. 7.72 Left: The energy resolution σ_E/E of the ATLAS iron scintillator tile calorimeter prototype as a function of $1/\sqrt{E}$ for 1994 and 1996 data with different weighting procedures: the 1994 data with cell weighting (circles), 1996 data with cell weighting (stars) and 1996 data with benchmark weighting (points). Right: The ratio of reconstructed energy and nominal energy of the ATLAS iron scintillator tile calorimeter prototype for 1994 data (circles) and 1996 data (points) as a function of energy after weighting. Both reprinted with kind permission from [168], © 2000, Elsevier. All rights reserved

calorimeters. The four parameters for eight beam energies are obtained from a fit. With benchmark weighting, the energy resolution is

$$\frac{\sigma_E}{E} = \frac{59.5\%}{\sqrt{E}} \oplus 1.8\% \oplus \frac{2.1\,\text{GeV}}{E}, \tag{7.81}$$

where the energy is given in units of [GeV]. This is considerably worse than for cell weighting,

$$\frac{\sigma_E}{E} = \frac{41.9\%}{\sqrt{E}} \oplus 1.8\% \oplus \frac{1.8\,\text{GeV}}{E}. \tag{7.82}$$

Figure 7.72 (right) shows the ratio of reconstructed energy to nominal energy for two data sets as a function of energy after cell weighting. Both data sets are linear within ±2%.

7.4.2.6 Jet Energy Resolution

So far, we focused on measurements of individual hadrons. However, the task of hadron calorimeters is primarily the measurement of jet energies. Jets arise from the hadronization of quarks and gluons and involve many particles: charged hadrons, neutral hadrons, photons, electrons and muons. They are distributed in a cone around a leading particle. Two important jet characteristics are the energy linearity and the energy resolution, which play a crucial role in the determination of the missing transverse energy ($E_{T,\text{miss}}$) and the missing transverse energy resolution $\sigma_{E_{T,\text{miss}}}$. In hadron collider experiments, the jet energy is typically determined by adding the energy of cells contained in a cone with half angle $\Delta R = \sqrt{(\Delta \eta)^2 + (\Delta \phi)^2}$

whose axis is centered on a seed cell with energy above a predefined threshold. The jet energy resolution is affected by the algorithms that are used to define jets, which involve effects like the cone radius and lateral segmentation. Furthermore, fluctuations in the particle content of jets due to differing fragmentations from one jet to another, fluctuations in the underlying event and fluctuations in energy pileup in high-luminosity hadron colliders affect the jet energy resolution. Jet reconstruction algorithms play an important role, since omitted energy deposits or extra energy deposits affect the resolution. In addition, we have the *e/h* effects, which come from two sources, the composition of the jet and the amount of electromagnetic showers produced in the interaction of the hadrons in the jet. Both effects determine the jet energy resolution.

Figure 7.73 (left) shows as an example the jet energy resolution for a copper and a uranium calorimeter as a function of the jet energy. The copper calorimeter has $e/\pi = 1.48$ and a hadron energy resolution of $\sigma_E/E = 48\%/\sqrt{E}$, while the uranium calorimeter has $e/\pi = 1.11$ and an energy resolution of $\sigma_E/E = 37\%/\sqrt{E}$. We note that at low energies the jet energy resolution is dominated by the very non-linear response of the calorimeter to low-momentum particles, which is similar for both calorimeters. Therefore, they show similar jet energy resolutions. At very high energies, the performance is dominated by the relative electron-to-hadron response. Thus, the energy resolution of the uranium calorimeter decreases with higher energies whereas the copper calorimeter levels off.

As a second example, we consider the jet energy resolution in the ATLAS experiment. Figure 7.73 (right) shows the jet energy resolution as a function of the jet transverse momentum for two jet algorithms. The first one is the topological cluster defined at the electromagnetic energy scale [195], while the second uses new ideas about particle flow [196], which will be discussed in detail in Sect. 7.5.1. Both algorithms use the same absolute jet energy scale (JES) corrections. The latter algorithm provides a better jet energy resolution at low energies.

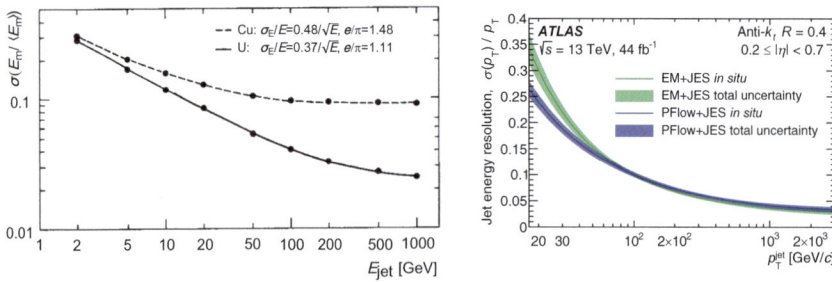

Fig. 7.73 Left: Jet energy resolution for a copper calorimeter (points plus dashed curve) and a uranium calorimeter (points plus solid curve). Reprinted with kind permission from [80], © 1985, Springer. All rights reserved. Right: The ATLAS jet energy resolution as a function of the jet transverse momentum for two jet algorithms, the fully calibrated PFlow+JES jets (blue curve and band) and EM+JES jets (green curve and band). Reprinted under CC-BY-4.0 Licence with kind permission from [194], © 2020, ATLAS Collaboration

7.4.3 Spatial Resolution

Similarly as electromagnetic showers, hadron showers consist of a narrow core surrounded by a halo of particles, which extends to several times the dimension of the core. The measurement of the spatial resolution of the impact point may be parameterized approximately by

$$\sigma_{\text{vertex}}[\text{cm}] = \frac{\langle \lambda_I \rangle}{4\sqrt{E\ [\text{GeV}]}}. \tag{7.83}$$

In compact calorimeters, where $\langle \lambda_I \rangle \leq 20$ cm, we achieve a spatial resolution of a few cm at 1 GeV. For transverse segmentation, we obtain the following dependence

$$\sigma_{\text{vertex}}[\text{cm}] = \sigma^0_{\text{vertex}} \exp(2d_I), \tag{7.84}$$

where σ^0_{vertex} is the intrinsic vertex resolution in the absence of instrumental effects due to finite segmentation and d_I is the segmentation in units of λ_I. Note that improvements become rather modest if the lateral segmentation is increased beyond $d_I \sim 0.1$. Limitations in the angular resolution stem from fluctuations in the $\pi^\pm/\pi^0$ composition of the hadron calorimeter.

As an example, for the ATLAS tile calorimeter prototype the polar angular resolution was extracted from a fit yielding [168]

$$\sigma_\theta = \left(\frac{160.50 \pm 1.48}{\sqrt{E}} \oplus (8.15 \pm 0.20)\right) \text{mrad}. \tag{7.85}$$

The non-pointing geometry in the test beam measurement yields a degradation of the resolution. For the azimuth angle, which has projective geometry, the fit gave

$$\sigma_\phi = \left(\frac{68.17 \pm 0.75}{\sqrt{E}} \oplus (0.91 \pm 0.11)\right) \text{mrad}. \tag{7.86}$$

In the CHARM ν-detector, for example, these effects were minimized by choosing marble as passive absorber, where three radiation length correspond to one interaction length. The angular resolution of the CHARM detector was,

$$\sigma(\theta)_{\text{hadron}} \approx \frac{160}{\sqrt{E\ [\text{GeV}]}} \oplus \frac{560}{E\ [\text{GeV}]} \ [\text{mrad}]. \tag{7.87}$$

7.4.4 Examples of Hadron Calorimeters

We give brief descriptions of three hadron sampling calorimeters, the SLD lead liquid-argon sampling calorimeter, the ZEUS uranium scintillator sampling calorimeter and the ATLAS steel scintillator tile sampling calorimeter. As before, we focus mainly on the barrel region.

7.4.4.1 The SLD Lead Liquid-Argon Sampling Calorimeter

Like the SLD electromagnetic barrel calorimeter, the hadron barrel calorimeter also consists of 144 modules arranged in 48 modules in ϕ and three in z [197]. They are placed on top of the electromagnetic modules as illustrated in Fig. 7.35. Thus, both the electromagnetic and hadronic modules share the same cryostat. In depth, a hadronic module is segmented into two sections with 13 cells each. A cell consists of a 6 mm thick lead plate followed by a 2.75 mm thick argon gap, a 6 mm thick lead tile followed by a 2.75 thick argon. The dE/dx sampling fraction is 7%. The 26 cells amount to two interaction lengths. The electromagnetic calorimeter provides another 0.84 λ_I. The tile size increases with radius to provide projective geometry. To avoid that photons can escape through the cracks without detection the projective pointing to the primary vertex is slightly offset. The liquid-argon calorimeter records 85% of the hadron shower. The remaining energy is recorded by a 5.2 λ_I deep tail catcher consisting of an iron-LST sampling calorimeter that also serves as a muon detector [198].

The hadronic modules are stacked in the reverse order than the electromagnetic modules starting with the strong back that holds the weight. While for electromagnetic modules the load is transferred to the lead through the spacers and plates keeping tiles mechanically floating, the load for hadronic modules is transferred through spacers and tiles keeping the plates floating. To achieve the projective tower geometry in hadronic modules 16 different tile sizes and 12 varieties of spacers have been used. The electrical signal of the 13 layers are ganged together into towers. In the barrel, there are 5,952 hadronic towers. The endcaps hold another 1,536 towers. The tiles are operated with a high voltage of 2 kV, while the plates are kept on ground potential. They are also ganged together at eight locations. Figure 7.74 shows a photograph of a completely stacked SLD hadronic module. The hadronic modules are tensioned to 400 pounds to provide a compressive force of 2g before they are tied together with stainless steel bands.

We had performed studies of different prototypes [199] using lead and uranium absorbers. For an 11 GeV hadron, the e/μ and e/h ratios for the uranium liquid-argon calorimeter yielded 0.51 and 1.19, respectively. The corresponding values for the lead liquid-argon calorimeter were 0.54 and 1.24. This illustrates that the neutrons are not efficiently detected in liquid argon. The energy resolution for the stochastic term was $60\%/\sqrt{E}$.

7.4.4.2 The ZEUS Uranium Scintillator Tile Sampling Calorimeter

The ZEUS uranium scintillator sampling calorimeter consisted of three main components, a forward calorimeter ($2.2° < \theta < 39.9°$), the barrel calorimeter ($36.7° < \theta < 129.1°$) and the backward calorimeter ($128.1° < \theta < 176.5°$). In depth, the forward and barrel calorimeters were segmented into a 1 λ_I thick electromagnetic section and two hadronic sections. In the barrel they were 2×2.1 λ_I thick increasing to 2×3.1 λ_I in the forward calorimeter. In the rear calorimeter there was just the electromagnetic section plus one 3.1 λ_I thick hadronic section.

7.4 Hadron Shower Measurements

Fig. 7.74 Photograph of a completely stacked SLD hadronic module (personal photograph). The tiles are clearly visible

The barrel calorimeter was constructed from 32 identical wedge-shaped modules, each covering an angle of 11.25°. Figure 7.75 (left) shows the layout of a barrel module. The inner radius was 1.22 m and the outer radius was 2.29 m. All modules had projective geometry in ϕ but were tilted by 2.5° to avoid projective module boundaries. While electromagnetic modules were also projective in θ, hadronic modules were not. At the front face, electromagnetic tiles had a size of 5 cm × 20 cm. The front size of the hadronic tiles was 20 cm × 20 cm. The length of the barrel calorimeter was 3.3 m. The uranium plates were 3.3 mm thick, corresponding to one radiation length. The electromagnetic section had 36 layers, while the two hadronic sections used 2 × 69 layers. The 2.6 mm thick scintillator tiles were read out by 2 mm thick wavelength-shifting fibers that took out the signal to the back region, where they were recorded by PMTs as shown in Fig. 7.75 (right).

To achieve $e/h = 1$, the plate thickness, scintillator thickness and the gate time needed to be optimized. Figure 7.76 (left) shows the e/h ratio as a function of the ratio of plate thickness to scintillator thickness, R_d. For the ZEUS calorimeter the ratio was 1.3 yielding $e/h \approx 1$. Figure 7.76 (right) shows the e/h ratio as a function of energy. Except for the 1 GeV point, e/h is close to one or exactly one. As shown below, the ZEUS hadron calorimeter shows excellent compensation and achieves an excellent energy resolution for hadron showers (see Fig. 7.71 (right)).

7.4.4.3 The ATLAS Steel Scintillator Tile Sampling Calorimeter

The ATLAS Tile Calorimeter (TileCal) is a hadron sampling calorimeter that is arranged into one barrel ($-1 < \eta < 1$) and two extended barrel parts ($0.8 < |\eta| < 1.6$) consisting of a stack of steel plates and 460,000 scintillator tiles. Radially, the

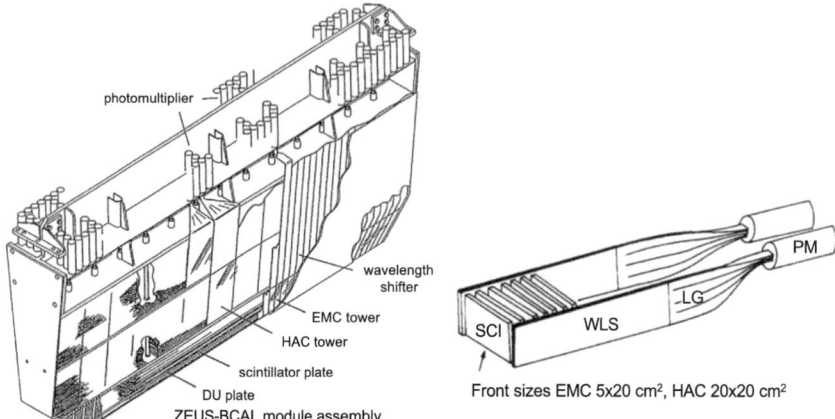

Fig. 7.75 Left: Layout of a ZEUS barrel calorimeter module consisting of an electromagnetic section and two hadronic sections. The tiles are read out via wavelength-shifting fibers that run towards the rear where PMTs are mounted. Reprinted with kind permission from [200], © 1993, Elsevier. All rights reserved. Right: Readout of the scintillator tiles with wavelength-shifting bars on each side coupled via light guides to PMTs. For the electromagnetic part the tiles were 5 cm × 20 cm while for the hadron part they increased to 20 cm × 20 cm. Reprinted with kind permission from [201], © 1987, Elsevier. All rights reserved

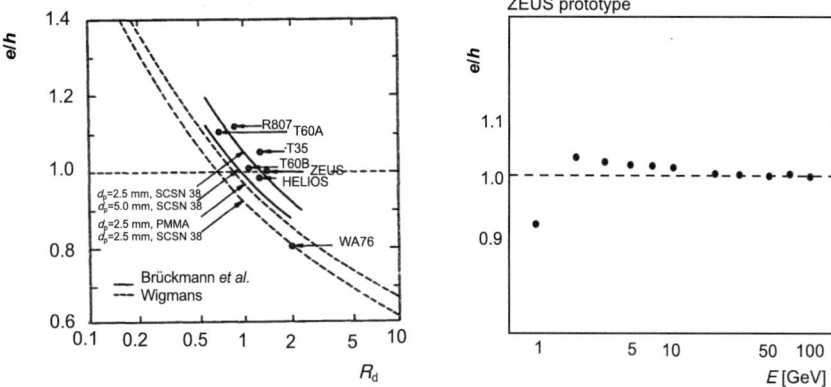

Fig. 7.76 Left: The *e/h* ratio as a function of the ratio of plate thickness to scintillator thickness, R_d, for two ZEUS prototypes, TA50A [202] and T50B [203], the WA87 experiment [204], the R807 experiment [164], and the Helios experiment [170]. In addition, simulations from Brückmann [177] and Wigmans [173] are plotted. Right: Measured *e/h* ratios as a function of energy for the ZEUS uranium scintillator prototype. Both reprinted with kind permission from [205], © 1991, Elsevier. All rights reserved

calorimeter extends from 2.28 to 4.25 m, providing a depth of 7.4 λ_I for particles emitted at $\theta = 90°$. Each section is constructed from 64 modules, where a module covers an azimuth angle $\Delta\phi = 5.2°$. The scintillator is located in pockets of the steel structure and is read out with wavelength-shifting fibers that are coupled to PMTs located inside the outer support girder of the calorimeter support structure [218, 219].

7.4 Hadron Shower Measurements

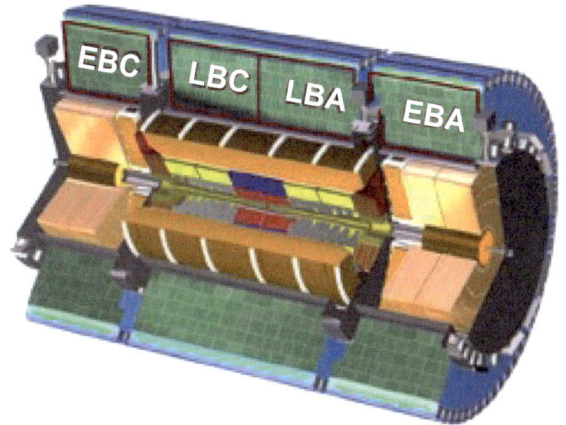

Fig. 7.77 Three-dimensional view of the ATLAS hadron calorimeter [206]. The LBA and LBC sections are long barrels and the EBA and EBC sections are the extended barrels. Reprinted with kind permission from [206], © 2016, Elsevier. All rights reserved

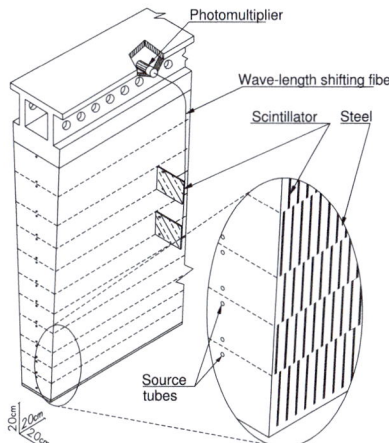 

Fig. 7.78 Left: Layout of an ATLAS hadronic submodule showing the pockets in the steel laminations into which scintillator tiles are inserted. Each tile has two holes through which calibration source tubes run allowing to calibrate all tiles with a ^{137}Cs source. The enlarged inset shows the arrangement of alternating scintillator tiles and steel tiles in the radial direction. The submodule covers $\Delta\phi = 5.2°$ and $\Delta\eta = 0.086$. The readout of the tiles with a wavelength-shifting fiber coupled to a photomultiplier tube is also illustrated. Note that the coordinate system shows r (pointing to the top), ϕ (pointing to the right) and z (pointing to the back). Reprinted under CC-BY-4.0 with kind permission from [207], © 2010, ATLAS Collaboration. Right: Photograph of a module showing the distribution of wavelength-shifting fibers. Reprinted with kind permission from [208], © 2008, the ATLAS Collaboration. All rights reserved

Figure 7.77 shows a three-dimensional view of the hadron calorimeter. A barrel module is built from 19 submodules, while that of an extended barrel is built from nine submodules. Each module is held by a strong back placed at the outer radius. In radial direction, there are three segments, layers 1–3, layers 4–9 and layers 10–11. Figure 7.78 (left) shows a schematic view of a submodule. The tiles are stacked along the z direction. This is clearly visible in the close-up view. Each layer has two 8 mm holes along the z directions for inserting tubes for running through a ^{137}Cs calibration source. The scintillator tiles consist of 1.5% of p-Terphenyl and 0.04% of POPOP mixed into polysterene. The three different trapezoidal sizes were produced in molds. In the radial direction, tile sizes are 97 mm for layers 1–3, 127 mm for layers 4–6, 147 mm for layers 7–9 and 187 mm for layers 10–11. The tiles are read out by 1.0 mm thick double-clad Y11 fibers that run on both sides on each tile in a groove. Both sides are read out and the two fibers are connected to the same photomultiplier tube. Figure 7.78 (right) shows a photograph of a barrel module, illustrating the wavelength-shifting fiber routing. The energy resolution and energy linearity were shown in Fig. 7.72 (left, right).

7.4.5 Properties of Different Hadron Sampling Calorimeters

Table 7.9 summarizes properties of diff..erent hadron sampling calorimeters. The best energy resolution is obtained with the ZEUS uranium-scintillator sampling calorimeter, yielding $35\%/\sqrt{E}$.

Table 7.9 Examples of several hadron sampling calorimeters in the barrel region, showing the cell thickness of the absorber and active medium, the total thickness of the hadron calorimeter in nuclear interaction lengths d_{HCAL}, the energy resolution, the experiment and the starting date. The total thickness refers to the hadron calorimeter only. The electromagnetic calorimeter typically adds another interaction length. ‡Extracted from 30.5 cm thickness. § E_T is the transverse energy. *This excludes the coil. With the coil material included the energy resolution is worsened to $38\%/\sqrt{E} \oplus 1\%$. §§This plate thickness is for the fine hadronic section; for the coarse hadronic section the plate thickness is 46.5 mm

Type	Cell thickness $t_r^{\text{abs}}/t_r^{\text{act}}$ [mm]	d_{HCAL} [λ_I]	σ_E/E barrel	Experiment	Date
HCA: Fe-scint.	25–50/10	4.7	$\frac{50\%}{\sqrt{E_T}} \oplus 3\%$ §	CDF [209]	1990
BCAL: U-scint.	3.3/2.6	5	$\frac{35\%}{\sqrt{E}} \oplus 1\%$ *	ZEUS [210]	1988
TileCal: Fe-scint.	5+4/3	7.4	$\frac{(46.7\pm 2.2)\%}{\sqrt{E}} \oplus (5\pm 0.6)\%$	ATLAS [114, 211]	2009
HB: Brass-scint.	50.5&56.5/3.7	5.8	$\frac{84.7\%}{\sqrt{E}} \oplus 7.4\%$	CMS [212]	2009
HAC: Fe-LAr	19/2 × 2.4	4.7-8	$\frac{50\%}{\sqrt{E}} \oplus 2\% \oplus \frac{90\%}{E}$	H1 [213]	1998
LAC: Pb-LAr	6/2.75	2.0	$\frac{60\%}{\sqrt{E}}$	SLD [106, 113]	1993
FH: U-LAr	6§§/2.3	3.3–4.7	$\frac{47\%}{\sqrt{E}} \oplus 4.5\%$	D0 [214]	1993
Fe-gas	50/22	7.2	$\frac{85\%}{\sqrt{E}}$	ALEPH [215]	1990
HAC: Fe-gas	50/18	6.6	$\frac{112\%}{\sqrt{E}} \oplus 21\%$	DELPHI [112]	1990
Fe-gas	100/25	8	$\frac{120\%}{\sqrt{E}}$	OPAL [216]	1990
U-gas	5/5	2.8‡	$\frac{61\%}{\sqrt{E}} \oplus 3\%$	L3 [217]	1990

7.5 New Calorimeter Concepts

To improve the performance of calorimeters at the international linear collider, new calorimeter ideas have been proposed.

7.5.1 Calorimeters Based on Particle Flow

By studying one-event displays in the ALEPH experiments, Videau had the idea to perform particle tracking in calorimeters since some jets have rather distinct signatures [220]. He proposed to design future calorimeters in an appropriate way to optimize particle tracking in jets such that the jet energy resolution and in turn the two-jet mass resolution can be improved significantly to separate W^+W^- events from Z^0Z^0 and Z^0H events. The standard jet energy resolution at the LEP experiments was $60\%/\sqrt{E}$, which is not sufficient to separate W^+W^- from Z^0Z^0 events as illustrated in Fig. 7.79 (left). If we improve the jet energy resolution to $30\%/\sqrt{E}$, a clear separation of W^+W^- from Z^0Z^0 events is achievable as shown in Fig. 7.79 (right). This concept is of interest for experiments at the international linear collider (ILC), which are designed to perform precision measurements of Higgs decays. A $30\%/\sqrt{E}$ jet energy resolution will also separate Higgs events from Z^0 events.

7.5.1.1 The Concept of Particle Flow
The basic idea of particle flow consists of classifying particles in a jet into three categories.

1. All charged particles including hadrons, $e^{\pm}$s and muons, which amount to 65% of all particles.

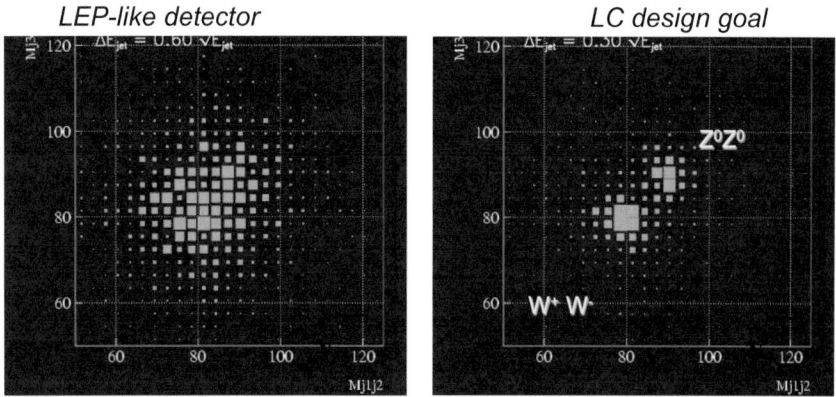

Fig. 7.79 The correlation of one di-jet invariant mass $m_{j_1 j_2}$ versus the other di-jet invariant mass $m_{j_3 j_4}$ for a jet energy resolution of $60\%/\sqrt{E}$ (left) and $30\%/\sqrt{E}$. Reprinted with kind permission from [221], © 2002, J. C. Brient. All rights reserved

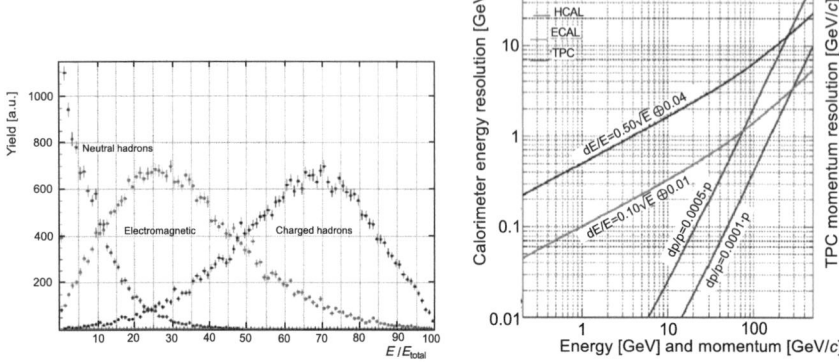

Fig. 7.80 Left: Energy-dependent distributions of neutral hadrons, electromagnetic showers and charged hadrons in a jet as a function of E/E_total [222]. Reprinted with kind permission from [222], © 2004, S. Magill. All rights reserved. Right: Energy resolution as a function of energy for photons in the electromagnetic calorimeter, for neutral hadrons in the hadron calorimeter (left-hand scale) and for the momentum resolution of charged hadrons measured in the tracker (right-hand scale) [223]. Reprinted with kind permission from [223], © 2004, DESY. All rights reserved

2. Photons that amount to 25% of all particles;
3. Neutral hadrons that amount to the remaining 10% of all particles.

Figure 7.80 (left) shows the contribution of these three categories to the total energy and Fig. 7.80 (right) presents a comparison of the energy or momentum resolution of the different particle types. For charged particles, the momentum measurement in a tracking detector provides the most precise measurement. Neglecting multiple scattering, the momentum resolution is $\sigma_p/p = 0.0001\ p$ to $0.0005\ p$. For photons, the energy is measured in an electromagnetic sampling calorimeter with a typical energy resolution of $\sigma_E/E = 0.15/\sqrt{E} \oplus 0.01$. Neutral hadrons are detected in the hadron calorimeter, which has a typical energy resolution of $\sigma_E/E = 0.50/\sqrt{E} \oplus 0.04$ after weighting.

We saw in Fig. 1.4 that the different particles leave distinct "finger prints" in the subdetectors. We use these to select charged tracks, photons and neutral hadrons.

7.5.1.2 Jet Energy Resolution

The jet energy is given by

$$E_\text{jet} = E_\text{ch} + E_\text{ph} + E_\text{nh}, \quad (7.88)$$

where E_ch is the summed energy of all charged particles using the momentum measurements assuming a pion hypothesis, E_ph is the electromagnetic energy of all photons and E_nh is the energy of all neutral hadrons. The variance of the jet energy is given by

7.5 New Calorimeter Concepts

$$\sigma^2_{E_{\text{jet}}} = \sigma^2_{E_{\text{ch}}} + \sigma^2_{E_{\text{ph}}} + \sigma^2_{E_{\text{nh}}} + \sigma^2_{E_{\text{conf}}}, \quad (7.89)$$

where we have added a confusion term $\sigma_{E_{\text{conf}}}$, which accounts for wrong energy allocations in the different categories within the jet and energy that was assigned to a different jet. The energy deposit of charged particles, electromagnetic showers and neutral hadrons in a jet are typically close to each other. So in assigning energy deposits to a specific shower, we have to make assumptions that are based on topological patterns and simulations. We certainly make incorrect assignments and we need a term in the energy resolution that accounts for that. We can parameterize the different terms with the anticipated resolutions and express them in terms of jet energy resolution, assuming a 100 GeV jet. For the charged-particle contribution, we get

$$\sigma^2_{E_{\text{ch}}} \approx (10^{-4})^2 \sum \frac{E^4_{\text{ch}}}{\text{GeV}^2} \approx (0.0036 \text{ GeV})^2 \left(\frac{E_{\text{jet}}}{10 \text{ GeV}}\right)^4, \quad (7.90)$$

where 10^{-4} is the typical momentum resolution. We set $\sum E_{\text{ch}} = 0.65 E_{\text{jet}}$. For an electromagnetic calorimeter with an energy resolution of $15\%/\sqrt{E}$, the contribution of photons to the jet is 25% on average yielding

$$\sigma^2_{E_{\text{ph}}} \approx (0.15)^2 \sum E_{\text{ph}} \cdot \text{GeV} \approx (0.75 \text{ GeV})^2 \left(\frac{E_{\text{jet}}}{100 \text{ GeV}}\right). \quad (7.91)$$

In a hadron calorimeter with weighting procedures, an energy resolution around $50\%/\sqrt{E}$ can be achieved. So, the 10% neutral hadrons in the jet contributes with

$$\sigma^2_{E_{\text{nh}}} \approx (0.50)^2 \sum E_{\text{neutral hadrons}} \cdot \text{GeV} \approx (1.6 \text{ GeV})^2 \left(\frac{E_{\text{jet}}}{100 \text{ GeV}}\right). \quad (7.92)$$

For jet energies of the order of 100 GeV or less, the contribution of the charged particles is negligible. Thus, we get

$$\sigma^2_{E_{\text{jet}}} = (0.18)^2 E_{\text{jet}} \cdot \text{GeV} + \sigma^2_{E_{\text{conf}}} \approx (0.30)^2 (E_{\text{jet}} \cdot \text{GeV}). \quad (7.93)$$

Note that $\sigma^2_{E_{\text{conf}}}$ is the largest term being $\sim 24\%$.

7.5.1.3 Calorimeter System Design

To achieve this jet energy resolution, we need highly granular calorimeters. We have to identify and measure each jet energy component as well as possible. First, we allocate all charged particles in the jet and compute their combined momenta. We then eliminate the associated energy deposits in the electromagnetic and hadronic calorimeters. Next, we identify all photons in the electromagnetic calorimeter and tails in the hadron calorimeter adding up their energies. Then, we eliminate the associated energy deposits in the electromagnetic and hadron calorimeters. Now, only energy deposits from neutral hadrons should be left in the electromagnetic and hadron calorimeters, which we sum up. We then determine the energy resolutions for

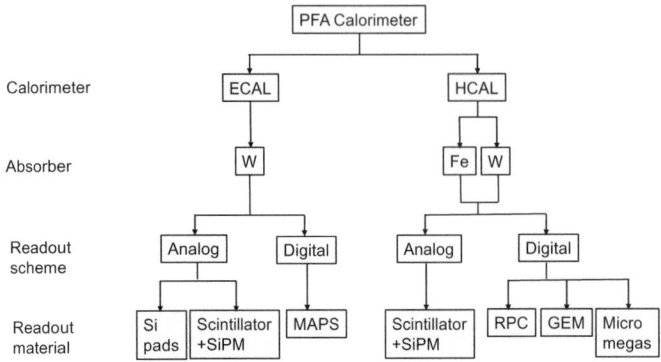

Fig. 7.81 Different technology options for particle flow calorimeters

each component. To follow charged particles through the calorimeter and associate hits, high granularity both in transverse and longitudinal directions is necessary for both the electromagnetic and hadron calorimeters. Different options exist for the electromagnetic and hadron calorimeters that are summarized in Fig. 7.81. For the electromagnetic calorimeter, the leading option is a silicon-tungsten detector. An alternative analog option is to use plastic scintillator tiles or strips instead of the silicon pads or make a hybrid device. For the digital option MAPS are used. For the hadron calorimeter, one option is an analog steel-scintillator sampling calorimeter. An alternative option is a digital or semi-digital steel-RPC/GEM/Micromegas device. As absorber also tungsten is considered. So for the first type, the active layer consists of scintillator tiles while for the second type it consists of gas detectors like RPCs, GEMS or Micromegas that are inserted between the absorber plates. Since the hadron calorimeter typically is not sufficiently deep to contain high-energy hadrons, we need a tail catcher and muon detector in addition. Note that the two experiments SiD and ILD, planned for the International Linear collider, will use highly granular electromagnetic and hadronic calorimeters. We discuss the silicon-tungsten ECAL and the analog steel-scintillator hadron calorimeter in more detail.

7.5.1.4 The Silicon Tungsten Electromagnetic Calorimeter

The leading option consists of a silicon tungsten sampling calorimeter, which uses tungsten absorber plates interleaved with layers of 10 mm × 10 mm silicon pads. Figure 7.82 shows the layout of the central slab (left) and that of the silicon-tungsten electromagnetic-calorimeter prototype (right). The active area is 180 mm × 180 mm. It is segmented into nine silicon wafers, each having 36 500 μm thick, 10 mm × 10 mm pads. There are 30 active silicon diode layers and 30 tungsten plates, which come in three different thicknesses, 1.4, 2.8 and 4.2 mm. Half of the absorber plates are wrapped in carbon fiber and are stacked into a rigid structure with gaps. The sensors are glued onto printed circuit boards (PCB). They transfer their signals to the very frontend electronics, which are mounted on the same boards outside the detector volume. Two sensors planes sandwiching an absorber plate covered with carbon fiber are combined into a middle slab and a bottom slab, which are

7.5 New Calorimeter Concepts

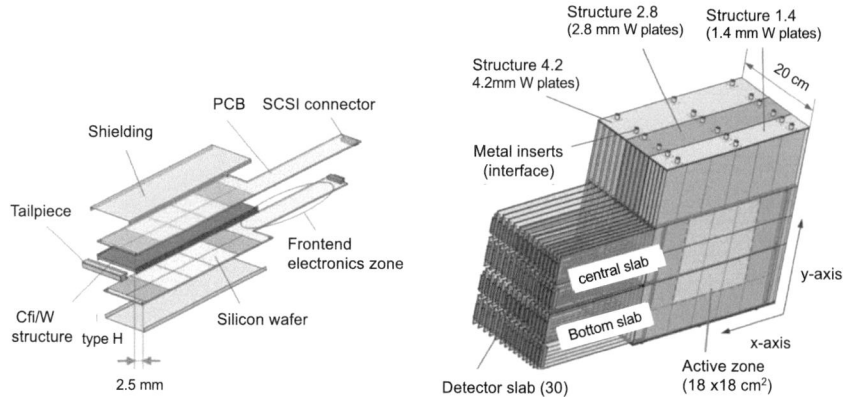

Fig. 7.82 Left: Schematic layout of the central slabs holding two layers of silicon pads connected to the very frontend electronics, sandwiching an absorber plate. Right: Layout of an electromagnetic calorimeter prototype based on the concept of particle flow. The tungsten absorber comes in three thicknesses of 1.4, 2.8 and 4.2 mm. The silicon detectors are inserted into the gaps. Both reprinted with kind permission from [224], © 2011, Elsevier. All rights reserved

inserted into the gaps. The active planes are covered on the outer sides by shields. The very frontend electronics provides pre-amplification. The PCBs are connected via cables to VME readout electronics, which provide digitization and readout. While the tungsten is kept at ground potential the silicon pads are operated at a voltage of 200 V.

In test beams at CERN and Fermilab, the prototype was extensively studied, using electrons in the energy range of 6 and 45 GeV. Figure 7.83 (left) shows the measured energy as a function of the beam energy. We see that the calorimeter exhibits excellent linearity. Figure 7.83 (right) shows the energy resolution as a function of $1/\sqrt{E}$. A fit to the data yields [224],

$$\frac{\sigma_E}{E} = \left[\frac{16.5 \pm 0.1}{\sqrt{E\,[\text{GeV}]}} \oplus 1.1 \pm 0.1 \right]\%. \tag{7.94}$$

The silicon tungsten electromagnetic calorimeter is the baseline designs for detectors planned at the International Linear Collider in Japan. For example, the original baseline design of the SiD detector was a 30-layer calorimeter with 2.5 mm thick W plates in the first 20 layers and 5 mm thick W plates in the last 10 layers amounting to 26 X_0. There was a 1.25 mm gap between plates, which housed the silicon pads and the readout electronics. The silicon sensors used 6-inch wafers holding 1024 hexagonal silicon pads with an area of 13 mm^2. In the other detector, ILD, the silicon tungsten calorimeter has 30 layers amounting to 24 X_0. The silicon pads have an area of 5 mm × 5 mm.

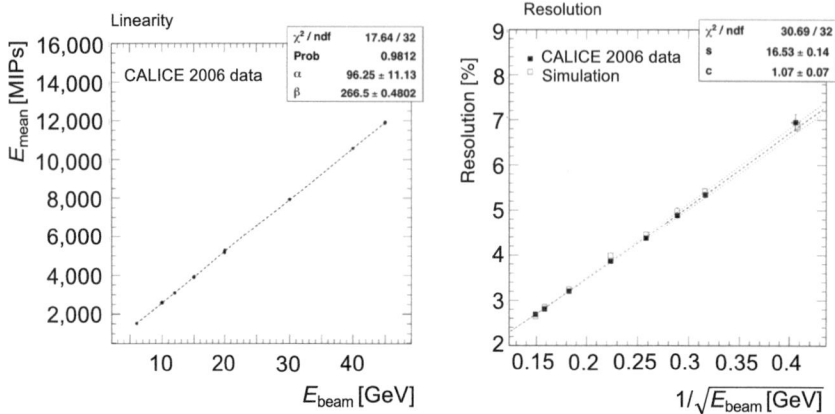

Fig. 7.83 Left: Measured energy of the Si-W electromagnetic-calorimeter prototype in units of MIPs as a function of the beam energy. Reprinted with kind permission from [225], © 2008, Elsevier. All rights reserved. Right: Measured energy resolution σ_E/E of the Si-W electromagnetic-calorimeter prototype as a function $1/\sqrt{E}$ in units of GeV. Reprinted with kind permission from [224], © 2011, Elsevier. All rights reserved

For the SiD electromagnetic calorimeter a new design has been proposed by replacing the silicon pads by MAPS that will be used also in the tracker. The pixel size is 25 µm × 100 µm and pixels will be read out digitally. The calorimeter has 20 layers of 2.243 mm tungsten plus a 1 mm sampling gap and 10 layers of 4.486 mm tungsten with a 1 mm sampling gap yielding 27 X_0. The Moliere radius is 14 mm. With power cycling the produced heat can be handled without an active cooling system. Simulations have shown that an energy resolution of about $10\%/\sqrt{E}$ can be achieved, which is substantially better than that of the silicon tungsten calorimeter. The pixels also have good time resolution.

7.5.1.5 The Analog Hadron Calorimeter

One of the options for a highly granular hadron calorimeter is the analog steel scintillator tile sampling calorimeter. We have built prototypes consisting of 38 layers of 2 cm thick steel plates interleaved with scintillator planes. In the first prototype, the scintillator was 5 mm thick read out with wavelength-shifting fibers coupled to SiPMs. In the second prototype, the SiPM mounted on the readout board was inserted into a dimple in the 3 mm thick tile of dimension 3 cm × 3 cm. So, here we have removed the wavelength-shifting fiber. Four readout boards (each 36 cm × 36 cm in size) are needed per plane housing 576 SiPMs. Figure 7.84 (left) shows cells on the readout board with an unwrapped tile, a wrapped tile and the bare board with the SiPM visible. Figure 7.84 (right) shows a layer with four completed readout boards. For example, the baseline design for the analog hadron calorimeter in the ILD experiment consists of a stack of 48 layers of steel-scintillator tiles. The absorber is effectively 2 cm thick, while the scintillator has a 3 mm thickness. This

7.5 New Calorimeter Concepts

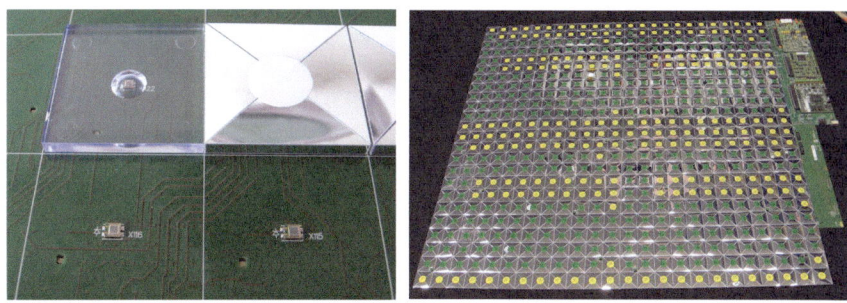

Fig. 7.84 Left: Readout board with an unwrapped and a wrapped tile (top row) and bare board with the silicon photomultiplier (bottom row). Right: A layer holding four readout boards with wrapped scintillator tiles. On the upper right-hand side, interfaces with the LED calibration system, power supply and data acquisition are visible. Both reprinted under CC-BY-3.0 Licence from [226], © 2018, CALICE Collaboration

amounts to about six nuclear interaction lengths. The scintillator pads have an area of 3 cm × 3 cm.

We have performed detailed studies with the electromagnetic and hadron calorimeter prototypes plus a tail catcher in test beams, using pions in the energy range of 6 to 80 GeV, where the steel scintillating-tile calorimeter is linear. With software compensation, we reach quite a good energy resolution as shown in Fig. 7.85 (right). For comparison, the results from the first prototype are overlaid. For the standard reconstruction both prototypes yield the same energy resolution of $\sigma_E/E = 57.6\%/\sqrt{E} \oplus 1.6\%$. For software weighting the energy resolution in the first prototype was improved to

$$\frac{\sigma_E}{E} = \left[\frac{(45.8 \pm 0.3)\%}{\sqrt{E \text{ GeV}}} \oplus (1.6 \pm 0.3)\right]. \tag{7.95}$$

For the second prototype the energy resolution is slightly larger.

Figure 7.86 (top) shows the measured longitudinal shower shapes for 8 (left), 18 (center) and 80 GeV (right) incoming pions in comparison to a simulation using the physics list FTFP-BERT. Simulations with other FTFP models, QGSP models, LHEP and CHIP were also performed [230]. Figure 7.86 (bottom) shows the corresponding measured lateral shower shapes for 8, 18 and 80 GeV incoming pions. None of the simulations reproduces all measured distributions well. At low values of z the reconstructed energies in the data are higher than those in the simulation with FTFP-BERT. Shower maximum is at the right place and for large z values the simulation is reasonable. For the radial distributions we find that the simulation yields too small reconstructed energies at large r values while for small r values the simulation agrees with the data.

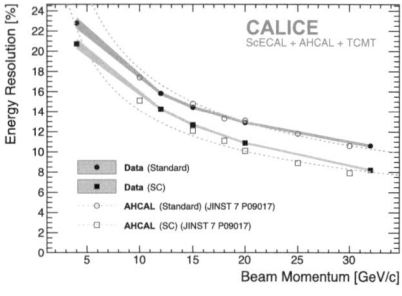

Fig. 7.85 Left: Photograph of the analog hadron calorimeter prototype at CERN. Reprinted from [227], © 2018, J. Kvasnicka. All rights reserved. Right: Energy resolution of single pions with standard (solid points) and software compensation (solid squares) reconstruction including energy in the electromagnetic calorimeter in front and in the tail catcher in the back. For comparison, corresponding results from the first prototype are shown (open circles and squares) [228]. Reprinted under CC-BY-3.0 Licence from [229], © 2018, CALICE Collaboration

7.5.1.6 Study of Particle Flow

In a test beam, the concept of particle flow cannot be studied directly since the beam typically consists of a single particle and not of jets. In particular, a systematic study requires a neutral particle and a charged particle that are separated at a well determined distance from each other, which can be modified over a certain range. With the test beam data, however, we can perform a trick to simulate this dependence. We select a hadronic shower of a given energy and then overlay another hadronic shower at a selected distance. We use the standard shower-finding algorithm to reconstruct two hadronic showers. Then, we examine the two showers and determine the energy that is assigned to the wrong shower. With this technique, pure pion samples were selected by rejecting electron and proton showers. In addition, the two showers were required to be fully contained in the electromagnetic and hadron calorimeter prototypes.

To study how well the concept of particle flow works, we consider two examples, a 10 GeV neutral hadron separated from either a 10 GeV pion or a 30 GeV pion. We consider separations between 5 and 30 cm. We plot the difference between recovered and measured energy in data and compare them to simulations. For a separation of $\Delta z = 5$ cm a fair amount of energy is assigned wrongly for the 10 GeV pion. For the 30 GeV pion, the wrong energy assignment is even worse because a second peak at 7 GeV is produced. Figure 7.87 (left) illustrates this for the 30 GeV pion. For a separation of $\Delta z = 30$ cm, the pion and the neutral hadron are well separated for both pion energies. Figure 7.87 (right) shows this for the 30 GeV pion.

Figure 7.88 (left) shows the mean value of the energy difference between recovered and measured energies as a function of the separation between a 10 GeV neutral hadron and 10/30 GeV pions. The simulation results for the physics lists (LHEP

7.5 New Calorimeter Concepts

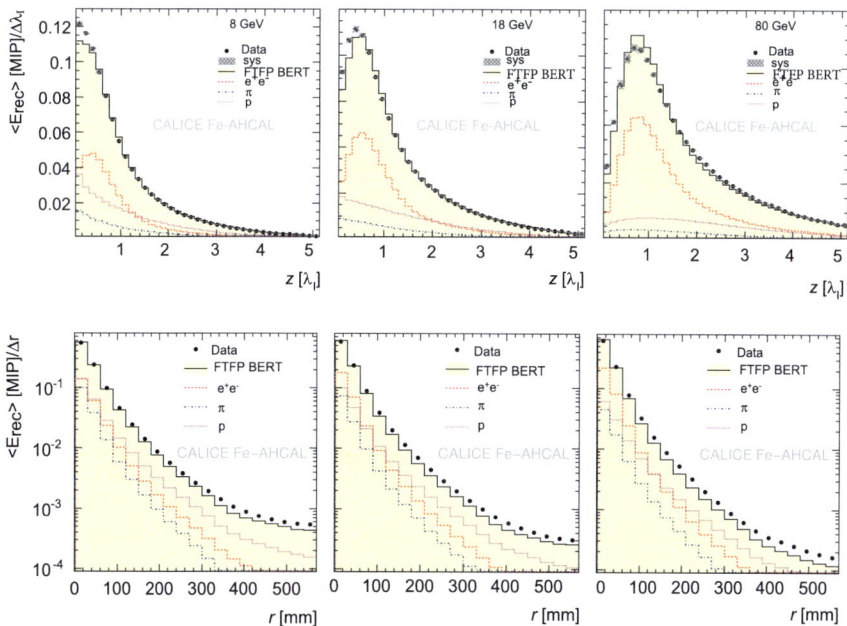

Fig. 7.86 Top: Longitudinal shower distributions for 8 GeV (left), 18 GeV (center) and 80 GeV (right) pions. Bottom: Lateral shower distributions for 8 GeV (left), 18 GeV (center) and 80 GeV (right) pions. Black points show data and black histograms show a simulation with the FTFP-BERT physics list. In addition, simulations for e^+e^- (red dotted histogram), pions (blue dash-dotted histogram) and protons (magenta histogram) are depicted. All reprinted under CC-BY-3.0 Licence from [230], © 2013, CALICE Collaboration

and QGSP-BERT) are shown overlaid. While LHEP is too optimistic, QGSP-BERT describes the data well. The mean value of the difference between recovered and measured energies decreases rapidly with increasing separation Δz from several GeV at $\Delta z = 5$ cm to 0 (-400 MeV) at $\Delta z = 30$ cm for the 10 (30) GeV pion. The rms, however, is rather large, being ~ 3 (~ 5) GeV at $\Delta z = 5$ cm separation and ~ 1 (~ 2.2) GeV at $\Delta z = 30$ cm separation for the 10 (30) GeV pion. The probability for recovering the energy within three standard deviations increases from $\sim 85\%$ ($\sim 50\%$) at $\Delta z = 5$ cm separation to 100% ($\sim 90\%$) at $\Delta z = 30$ cm separation. Figure 7.88 (right) illustrates this.

We can look also what separation is obtained in the electromagnetic calorimeter prototype. Figure 7.89 (left) shows a double particle event in the SiW electromagnetic calorimeter prototype. The two particles are separated by 5 cm. We see good two-particle separation due to the good position resolutions in x and y shown in Fig. 7.89 (right) as a function of the energy after the "S-curve" correction [232]. At 45 GeV,

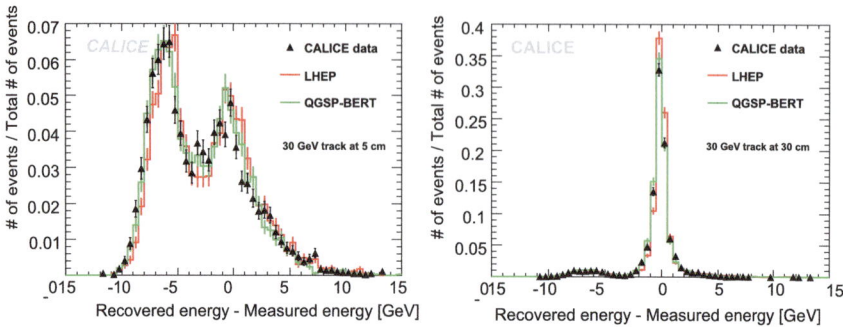

Fig. 7.87 Difference between recovered and measured energy for a 10 GeV neutral hadron separated from a 30 GeV pion at $\Delta z = 5$ cm (left) and at $\Delta z = 30$ cm (right) for data and simulations with LHEP (red curves) and QGSP-BERT (green curves) physics lists. Both reprinted with kind permission from [231], © 2011, IOP Publishing. All rights reserved

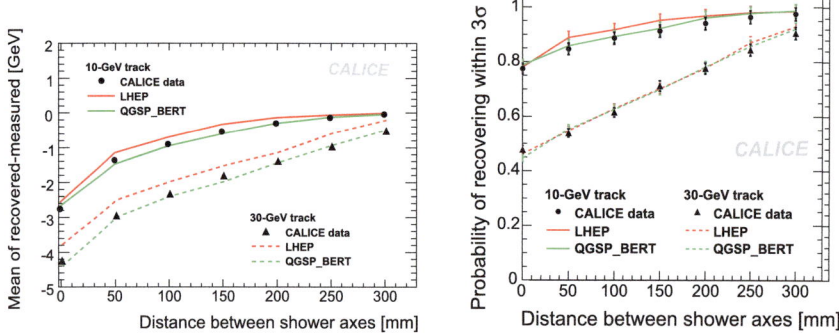

Fig. 7.88 Probability of recovering the energy of 10 GeV neutral hadrons within two (left) and three (right) standard deviations versus the distance from 10 GeV (circles and continuous lines) and 30 GeV (triangles and dashed lines) charged hadrons for beam data (black) and simulations with LHEP (red histograms) and QGSP-BERT (green histograms) physics lists. Both reprinted with kind permission from [231], © 2011, IOP Publishing. All rights reserved

the position resolutions approach 1 mm. The fits to the data yield

$$\sigma_x = \left(\frac{15.7 \pm 0.2}{E} \oplus \frac{0. \pm 0.4}{\sqrt{E}} \oplus 1.04 \pm 0.01 \right) \text{mm} \tag{7.96}$$

$$\sigma_y = \left(\frac{15.3 \pm 0.5}{E} \oplus \frac{1.4 \pm 0.6}{\sqrt{E}} \oplus 0.94 \pm 0.02 \right) \text{mm}. \tag{7.97}$$

The "S-curve" corrections yield similar resolutions in the x and y directions.

7.5 New Calorimeter Concepts

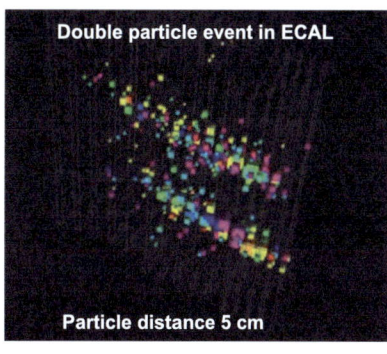
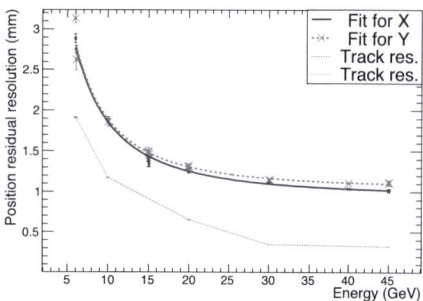

Fig. 7.89 Left: A double-particle event, where the two particles are separated by 5 cm. Reprinted with kind permission from [233], © 2019, J. C. Brient. All rights reserved. Right: Position resolutions σ_x (points) and σ_y (crosses) as a function of the beam energy. The solid and dotted curves are fits to the data. The light dotted curves show the tracking resolutions obtained from simulations, which consist of a constant term added in quadrature with a multiple scattering term. The data have been corrected for the so-called "S-curve" effect. Reprinted from [232], © 2010, CALICE Collaboration. All rights reserved

7.5.1.7 The CMS HGCAL Calorimeter

The CMS collaboration decided to upgrade their endcap calorimeters with a High-Granularity Calorimeter [234], a sampling calorimeter consisting of a 26 radiation lengths (one interaction length) thick electromagnetic calorimeter (CE-E) followed by a 3.5 interaction lengths thick forward (FH) hadron calorimeter and a 5.7 interaction lengths thick backward (BH) hadron calorimeter (CE-H). Figure 7.90 (left) shows a schematic view of half an HGCAL endcap. The 28-layer electromagnetic calorimeter uses 1 cm^2 hexagonal silicon pads as active medium and copper, tungsten-copper and lead plates as absorber material. Figure 7.90 (right) shows the layout of an electromagnetic cassette, which consists of two cassettes mounted on each side of a copper cooling plate and lead plates sandwiched between steel shims that provide a rigid structure. A module is the assembly of the silicon pads that are glued to gold-plated polyimide foils, which in turn are glued to the copper-tungsten base plates and a PCB servicing the front-end ASICs. The signals from the silicon will be routed to the ASICs via wire bonds through holes in the PCB. The copper cooling plate contains thin pipes for CO$_2$ flow to cool the detector to $-30°$C to mitigate the effects of radiation damage to the silicon. Motherboard PCBs are connected to the modules, containing data-concentrator ASICs as well as optical links. The electromagnetic section is 30 cm deep.

The hadron calorimeters have 24 layers, 12 in each section, extending to 1.5 m. The absorber consists of steel plates. In the forward part, the plates are 35 mm thick, increasing to 68 mm in the backward part. In the innermost region of the hadronic calorimeters, where the radiation is expected to be very high (up to 10^{16}(1 MeV) n_{eq}/cm^2), silicon pads are used. In the outermost region, plastic scintillator tiles read out by SiPMs are mounted directly onto the tiles. Unlike the electromagnetic section in which the absorber plates are part of the cassettes, the hadronic active layers are inserted between full disks of steel. As shown in Fig. 7.90 (right),

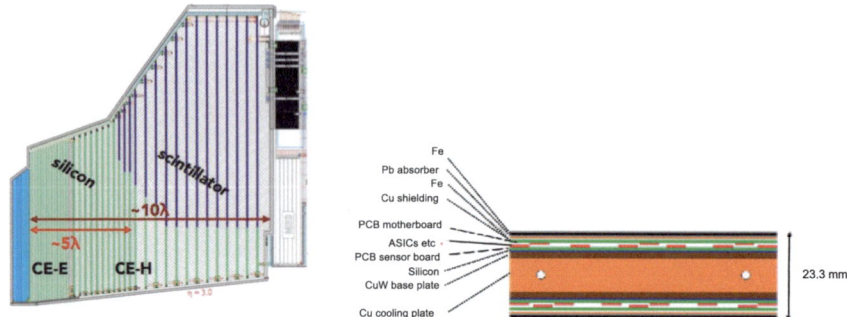

Fig. 7.90 Left: Schematic view of the High-Granularity Calorimeter design. Reprinted under CC-BY-3.0 Licence from [235], © 2018, CALICE Collaboration. Right: Layout of electromagnetic calorimeter cassette. From outside in, lead plates are sandwiched between steel shims followed by a copper shielding. Next, a PCB motherboard (green) follows holding ASICs (red) and a PCB sensor board with ASICs that read out the silicon pads (blue), which sit on top of the copper-tungsten base plate (brown) and a copper cooling plate (orange). Both reprinted from [236], © 2017, CMS Collaboration

the front layers contain only silicon modules, whereas rear layers are mixtures of silicon at the innermost regions and scintillator tiles in the outer regions. The full calorimeter is placed in the same cold volume at $-30°C$. This calorimeter design provides the high transverse and longitudinal granularity needed for particle flow reconstruction algorithms.

The thickness of the silicon sensors is optimized for regions of different radiation levels coming in 120, 200 and 300 μm thickness. For the 120 μm sensors, p-type epitaxial on a handle wafer is the baseline substrate material. For the thicker sensors, physically thinned p-type float zone silicon wafers are chosen. For p-type sensors, the risk of charge-up effects on the Si-SiO$_2$ interface is reduced by introducing additional p-stops, which isolate the frontside n-implants. The 120 μm sensors have a cell size of 0.52 cm^2, while the 200 and 300 μm sensors have cell sizes of 1.12 cm^2. The small-cell sensors are more radiation hard. They can take fluences of $> 2.5 \times 10^{15}$(1 MeV) n_{eq}/cm^2, while the 200 μm (300 μm) thick sensors can take fluences of 0.5-2.5×10^{15}($0.1 - 0.5 \times 10^{15}$)(1 MeV) n_{eq}/cm^2, respectively. The 120 μm sensors have very low leakage currents (0.1 μA) at full depletion at 1000 V. The signal-to-noise ratio is 7.5–10.

One of the goals in HGCAL is the usage of timing information in the shower reconstruction. In a test beam, it was shown that a resolution of 20 ps with 100% efficiency is achievable for photons with $p_T > 2$ GeV/c if only cells within 2 cm of the shower axis are considered. For K_L^0 with $p_T > 5$ GeV/c, a resolution of 30 ps with 90% efficiency was achieved [237]. Figure 7.91 shows the jet energy resolution of the high-granularity endcap calorimeter of the CMS experiment before and after software compensation. The stochastic term after compensation is $(65.4 \pm 0.8)\%/\sqrt{E}$.

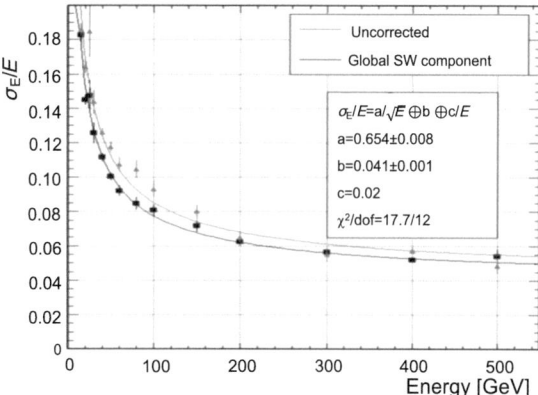

Fig. 7.91 The jet energy resolution of the HGCAL before (light grey points and grey curve) and after (black points and black line) software compensation. Reprinted under CC-BY-3.0 Licence from [238], © 2017, CMS Collaboration

7.5.2 Dual Readout Calorimetry

Another approach is the method of dual readout. Here, the idea originally proposed by Mockett [239] is to measure separately the scintillation light and the Cherenkov light since they have rather different **h/e** values. This provides another compensation procedure. First studies were performed with orange scintillators using orange filters for the scintillation light and blue filters for the Cherenkov light [240]. In the late 1990's, the DREAM collaboration implemented the dual readout into their calorimeter prototype. Figure 7.92 (top left) shows a photograph of the backside of the DREAM calorimeter prototype. Figure 7.92 (bottom left) shows a photograph of the fibers coming out of the back of the calorimeter. Copper tubes inserted into the absorber are filled with both scintillating fibers and quartz fibers as depicted in Fig. 7.92 (bottom right). Calling the Cherenkov signal C and the scintillation signal S, we find for each event,

$$C = [f_{\pi^0} + (h/e)_C (1 - f_{\pi^0})]/E,$$
$$S = [f_{\pi^0} + (h/e)_S (1 - f_{\pi^0})]/E, \qquad (7.98)$$

where $(h/e)_C$ is the ratio of response to hadron and that to electromagnetic showers for the Cherenkov signal and $(h/e)_S$ is the same quantity for the scintillation signal. Figure 7.92 (right) shows a scatter plot of C/E versus S/E. The two responses scatter around a line segment. With increasing energy the distribution moves upward along the line segment and becomes tighter. Note that (7.98) is linear in both $1/E$ and f_{π^0}. Thus, we can solve them to obtain estimators for the corrected energy and f_{π^0} for each event. Both quantities are subject to resolution effects, but contributions due to fluctuations in f_{π^0} are eliminated. The solution for the corrected energy is

$$E = \frac{\xi S - C}{\xi - 1}, \qquad (7.99)$$

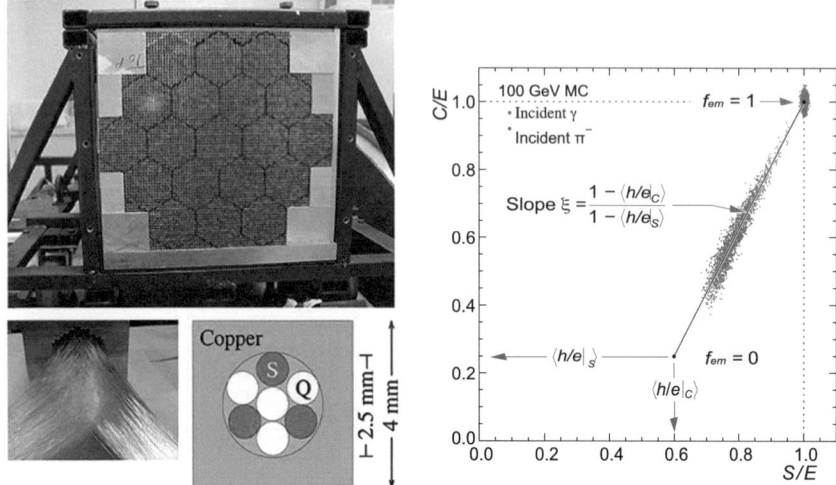

Fig. 7.92 Top left: Photograph of the backside of the dual calorimeter prototype. Bottom left: Photograph of the scintillating and quartz fibers coming out of the back of the prototype. Bottom right: Layout of scintillating fibers (S) and quartz fibers (Q) in a copper tube. All reprinted with kind permission from [241], © 2017, AIP Publishing. All rights reserved. Right: Scatter plot of Cherenkov signal C/E versus scintillation signal S/E in a dual calorimeter for simulated photons (black points) and pions (grey points). Electromagnetic clusters gather around $C/E = 1$ and $S/E = 1$. Reprinted under CC-BY-NC-4.0 Licence with kind permission from [12], © 2024, Particle Data Group LBNL

where

$$\xi = \frac{1 - (h/e)_C}{1 - (h/e)_S}, \qquad (7.100)$$

which is the energy-independent correlation slope of C/E versus S/E shown in Fig. 7.92. The slope ξ is determined either from a fit of the correlation in Fig. 7.92 (right) or by measuring the π/e ratio as a function of E.

For optimized resolution, the scintillator readout of the calorimeter must be as compensating as possible. In this case, the slope is much larger than unity. The energy resolution σ_E/E in an ideal calorimeter can be represented by

$$\frac{\sigma_E}{E} = \frac{\alpha_1(E)}{\sqrt{E}} \oplus [1 - (h/e)\sigma_{f_{\pi^0}}], \qquad (7.101)$$

where $\sigma_{f_{\pi^0}}$ is the error of f_{π^0}. The coefficient α_1 is expected to have a small energy dependence. For example, the sampling variance contribution to α_1 is $(\pi/e)E$ rather than E. The resolution $\sigma_{f_{\pi^0}}$ decreases with increasing energy. Usually, the energy resolution $(\sigma_E/E)^2$ versus $1/E$ is well described by a straight line with a slope and an intercept, called constant term. Precise data indicate that we need to account for significant additional effects:

- Incomplete corrections for leakage, differences in light collection efficiency and electronics calibration.
- Readout transducer shot noise (usually photoelectron statistics), plus electronic noise.
- Sampling fluctuations. As discussed before, only a small part of the energy deposit takes place in the scintillator and this fraction is affected by large fluctuations, which depend on the sensor-absorber ratio.
- Intrinsic fluctuations. Since ionization can be produced in many ways in a hadron shower, we encounter different detection efficiencies, which are subject to stochastic fluctuations. In particular, a very large fraction of the hadronic energy is invisible. The lost fraction depends on the readout being larger for Cherenkov light than for scintillation light.

So far, the concept of dual readout has not been used in any experiment.

7.6 Concluding Remarks

In this chapter, we have given an overview of electromagnetic and hadronic energy measurements and discussed the different detector technologies that accomplish this. For more information, we refer the interested reader to the following review articles [2, 241–243] and books [4, 157].

Exercises

7.1 Plot the longitudinal shower profile for 10 GeV/c electrons in iron. Where is the shower maximum? How large is t_{95}? How large is the Moliére radius?

7.2 You want to detect photons from J/ψ decays with a NaI(Tl) crystal calorimeter. NaI(Tl) has a radiation length of $X_0 = 9.5$ g/cm^2 and a density of $\rho = 3.67$ g/cm^3. The photon energies range between 0.05 and 1.55 GeV. Where is the shower maximum and how much does it vary between 0.05 and 1.55 GeV? How long do the crystals have to be to contain 98% of the shower? Hint: numerical integration of the energy distribution yields pairs of $(X_0, E) = (1, 0.006)$, (1.5, 0.021), (2, 0.050), (2.5, 0.094), (3, 0.151), (3.5, 0.222), (4, 0.301), (4.5, 0.388), (5, 0.478), (5.5, 0.570), (6, 0.661), (6.5, 0.750), (7, 0.843), (7.5, 0.915), (8, 0.988), (8.5, 1.056), (9, 1.118), (9.5, 1.174), (10, 1.224), (10.5, 1.269), (11, 1.308), (11.5, 1.342), (12, 1.373), (12.5, 1.399), (13, 1.422), (13.5, 1.441), (14, 1.458), (14.5, 1.472), (15, 1.484), (15.5, 1.495), (16, 1.504), (16.5, 1.512), (17, 1.518), (17.5, 1.523), (18, 1.528), (18,5, 1.532), (19, 1.535), (19.5, 1.537), (20, 1.539), (20.5, 1.541), (21, 1.543), (22, 1.545), (24, 1.548), (25, 1.549), (30, 1.550), (35, 1.55). Determine the length more precisely by doing the integral yourself.

7.3 An electromagnetic sampling calorimeter is used to measure the energy of e^-. It consists of 2.0 mm thick Pb plates stacked with 2 mm gaps that are filled with LAr. The calorimeter has a depth of 48 cm. What is the expected energy resolution of a 10 GeV e^-, if the contribution of the path length fluctuations is 6% at 1 GeV. Use the particle data table to obtain other relevant quantities.

7.4 For a hadron calorimeter 2 cm thick Pb plates are stacked with a 2.75 mm gap filled with LAr. Since the calorimeter is undercompensating, the intrinsic energy resolution is $\sigma_E/E = 45\%/\sqrt{E}$. What is the expected energy resolution of a 10 GeV pion, at what position do we expect the shower maximum and how long does the calorimeter need to be to contain 95% of the energy?

7.5 A CsI crystal is a scintillator used to detect electromagnetic showers. It has a radiation length of 1.86 cm. Assume you have photons up to an energy of 5 GeV/c. How long do you have to make the crystals to contain on average all its energy except for a 2% residue? Assume that $t_{98} = 2 \times t_{max}$. How many particles are produced here?

7.6 A U-scintillator calorimeter is used to measure the energy of 10 GeV photons. The U plates are 1.6 mm thick and the scintillator is 2.5 mm thick. The total energy resolution has contributions from sampling fluctuations, the Landau fluctuations and the path length fluctuations. In a calibration run with 1 GeV electrons, you have measured $\sigma_{\text{sampling}} = 106$ MeV, $\sigma_{\text{Landau}} = 10$ MeV and $\sigma_{\text{path-length}} = 60$ MeV. In addition, there is a constant term in the energy resolution of 2%. What energy resolution would you expect at 10 GeV. At what energies does the constant term become dominant?

7.7 Compute the interaction length for a 100 GeV/c proton in a lead-gas sampling calorimeter. What is the probability that the proton has not interacted within 10 cm? Where is shower maximum and the 95% containment of the longitudinal shower?

7.8 Consider pions entering a Pb LAr calorimeter. For $m = 0.85$ and $E_0 = 0.95$ GeV, calculate *e/h*. Plot e/π as a function of the pion energy using both the power-law parameterization and Wigmans' parameterization. How big is the difference? Plot e/π as a function *e/h* for a 200 GeV pion.

7.9 (a) A steel scintillating-fiber calorimeter prototype is used to study the performance of hadrons. At an energy of 100 GeV the hadron-to-electron responses are $\pi/e = 0.859$ and $p/e = 0.828$. The other parameters are $m = 0.82$ and $e/h = 0.667$. Plot π/e and p/e as a function of the beam energy up to 1000 GeV. What is the p/π ratio? How does it vary with energy?

(b) Consider three hadron calorimeters using uranium as absorber and pure LAr, LAr-CH$_4$ and scintillator as active medium. For the LAr combinations the U plates and active gaps are are 6.6 mm and 2×2.4 mm thick, while for the scintillator layout they are 3 and 2.5 mm thick. For the three calorimeters $h_i/MIP = 0.41$, while $e/MIP = 0.63$ for the LAr configurations and $e/MIP = 0.60$ for the scintillator layout. The degree of compensation is $C_{\text{tot}} = 48$, 97 and 89%, respectively. What are the *e/h* ratios?

7.10 A 38 layer steel-scintillator tile calorimeter is used to study hadron showers. The steel plates have a thickness of d = 16.7 mm (layers 1–3), d = 17.4 mm (layers 4–24) and d = 17.6 mm (layers 25–38). In addition, the 5 mm thick scintillators are covered by 2 mm thick steel shims. At the end, another 20 mm thick steel plate is positioned. Determine the total thickness relevant for energy measurements in units of pion interaction lengths and radiation lengths. The steel is a composite of Fe (0.98), Mn (0.015), and Cu (0.05). The scintillator is made of polysterene. The pion interaction lengths (λ_π), radiation lengths (X_0) and densities (ρ) of the materials are: Fe ($\lambda_\pi = 160.8$ g/cm^2, $X_0 = 13.84$ g/cm^2, $\rho = 7.874$ g/cm^3), Mn ($\lambda_\pi = 160.2$ g/cm^2, $X_0 = 14.64$ g/cm^2, $\rho = 7.44$ g/cm^3), Cu ($\lambda_\pi = 165.9$ g/cm^2, $X_0 = 12.86$ g/cm^2, $\rho = 8.96$ g/cm^3), and polysterene ($\lambda_\pi = 113.7$ g/cm^2, $X_0 = 43.79$ g/cm^2, $\rho = 1.06$ g/cm^3). How much does the pion interaction length increase if a Si-W calorimeter is placed in front having 30 layers (10 layers each of 1.4, 2.8 and 4.2 mm thick W and 30 layers of 0.5 mm thick Si)? The parameters for these materials are: W ($\lambda_\pi = 218.7$ g/cm^2, $X_0 = 6.78$ g/cm^2, $\rho = 19.3$ g/cm^3) and Si ($\lambda_\pi = 137.7$ g/cm^2, $X_0 = 21.82$ g/cm^2, $\rho = 2.33$ g/cm^3).

References

1. R.L. Ford, W.R. Nelson, SLAC-210, UC-32 (1978). W.R. Nelson, H. Hirayama, D. Rogers, SLAC-R-0265 (1985); W.R. Nelson et al., SLAC-PUB-6625-Rev. (1997)
2. C.W. Fabjan, F. Gianotti, Rev. Mod. Phys. **75**, 1243 (2003)
3. G. Bathow et al., Nucl. Phys. B **20**, 592 (1970)
4. R. Wigmans, *Calorimetry, Energy Measurements in Particle Physics* (Oxford Science Publications, New York, 2000); 2ed (Oxford University, 2017)
5. C. Grupen, Teilchendetektoren, BI Wissenschaftsverlag (1993)
6. J. Nysten, Nucl. Instrum. Meth. A **534**, 194 (2004)
7. M. Oreglia, Ph.D. thesis, SLAC Report SLAC-R-236 (1980)
8. D. Bard, PhD thesis at the University of Edinburgh (2006)
9. J.J. Hopfield, Proc. Natl. Acad. Sci. U.S.A. **79**(8), 2554 (1982)
10. L. Breiman, Mach. Learn. **24**, 123 (1996); I. Narsky, arXiv:physics/0507157 (2005)
11. U. Fano, Phys. Rev. **72**, 26 (1947)
12. The Particle Data Group (S. Navas et al.), Phys. Rev D. **110**, 030001 (2024)
13. The CMS Collaboration (M. Malberti et al.), JINST **15**, C04014 (2020)
14. The CMS Collaboration (A. Benaglia et al.), JINST **9**, C02008 (2014); The CMS Collaboration (F. Ferri et al.), Nucl. Instrum. Meth. A **958**, 162159 (2020)
15. The BABAR Collaboration (B. Aubert et al.), Nucl. Instrum. Meth. A **729**, 615 (2013)
16. J. Czochralski, Metalle. Z. Phys. Chem. **92**, 219 (1918)
17. P.W. Bridgman, Proc. Am. Acad. Arts Sci. **60** (6), 305 (1925)
18. S. Herb, Proceedings of 15th Rencontres de Moriond, 31 (1980)
19. E.D. Bloom, Lepton Photon 79, SLAC-PUB-2425, (1979)
20. Y. Kubota et al., Nucl. Instrum. Meth. A **320**, 66 (1992)
21. E. Aker et al., Nucl. Instrum. Meth. A **321**, 69 (1992)

22. The BABAR Collaboration (B. Aubert et al.), Nucl. Instrum. Meth. A **479**, 1 (2002)
23. The Belle Collaboration (A. Abashian et al.), Nucl. Instrum. Meth. A **479**, 117 (2002)
24. B. Shwarz et al., PoS PhotoDet2015, 051 (2015)
25. A. Zhemchugov, Nucl. Phys. B Proc. Suppl. **189**, 353 (2009); The BES III Collaboration (D.M. Asner et al.), Int. J. Mod. Phys. A **24**, S1 (2009)
26. The KTeV Collaboration (V. Prasad et al.), Nucl. Instrum. Meth. A **461**, 341 (2001)
27. N. Atanov et al., JINST **15**-09, C09035 (2020)
28. R.L. Sumner et al., Nucl. Instrum. Meth. A **265**, 252 (1988)
29. F.M. Marques et al., Nucl. Instrum. Meth. A **417**, 137 (1995)
30. M. Ryan, Nucl. Instrum. Meth. A **598**, 217 (2009)
31. R.Y. Zhu, SNOWMASS-2005-ALCPG0705 (2005)
32. M. Cordelli et al., Nucl. Instrum. Meth. A **718**, 81 (2013)
33. R. Mao, L. Zhang, R.Y. Zhu, IEEE NSS/MIC **2007**, 2285 (2007)
34. Photograph of CLEO CsI(Tl) crystal; https://www.classe.cornell.edu/public/lab-info/cc.html
35. F. Yang et al., IEEE NSS/MIC **2012**, 1681 (2012)
36. R.Y. Zhu, Personal Communication (2023)
37. C. Jessop, https://www3.nd.edu/~cjessop/research/BaBar/BaBar.html
38. The KTeV Collaboration, http://hep.uchicago.edu/ktev/ktev.html
39. R.E. Ray, Conf. Proc. C 940925, 110 (1994)
40. L. Carminati, P. Meridiani, Frascati Physics Series Vol. XXXVIII, 115 (2004)
41. C. Jessop, https://www3.nd.edu/~cjessop/research/CMS/CMS.html
42. The CMS Collaboration (S. Chatrchyan et al.), JINST **3**, S08004 (2008)
43. The CMS ECAL Collaboration (R. Arcidiacono et al.), J. Phys. Conf. Ser. **160**, 012048 (2009)
44. R. Partridge et al., Phys. Rev. Lett. **44**, 712 (1980)
45. Y. Karyotakis, Proc of 1994 Beijing Calorimetry Symp., LAPP-EXP-95-02 (1995)
46. A. Roodman, PRINT-96-313 (CHICAGO), 1684 (1996)
47. A. Roodman, Conf. Proc. C **971109**, 89 (1997)
48. The ALICE Collaboration (D.V. Aleksandrov et al.), Nucl. Instrum. Meth. A **550**, 169 (2005)
49. J. Brau et al., Nucl. Instrum. Meth. **196**, 403 (1982)
50. H. Mkrtchyan et al., Nucl. Instrum. Meth. A **719**, 85 (2013)
51. The OPAL Collaboration (K. Ahmet et al.), Nucl. Instrum. Meth. A **305**, 275 (1991)
52. S. Kawabata et al., Nucl. Instrum. Meth. A **270**, 11 (1988)
53. NA48 Collaboration (M. Jeitler), Nucl. Instrum. Meth. A **478**, 404 (2002)
54. J.A. Appel et al., Nucl. Instrum. Meth. A **127**, 495 (1975)
55. K. Ogawa et al., Jap. J. Appl. Phys. **23**, 897 (1984); https://www.schott.com/shop/advanced-optics/en/Optical-Glass/SF6/c/glass-SF6
56. M.Y. Balatz et al., Nucl. Instrum. Meth. A **545**, 114 (2005)
57. M. Holder et al., Nucl. Instrum. Meth. A **108**, 541 (1973)
58. J.S. Beale et al., Nucl. Instrum. Meth. A **117**, 501 (1974)
59. F. Dydak et al., Nucl. Instrum. Meth. A **137**, 427 (1976)
60. A.V. Dolgopolov et al., Nucl. Instrum. Meth. A **420**, 20 (1999)
61. H. Avakian et al., Nucl. Instrum. Meth. A **417**, 69 (1998)
62. The K2K Collaboration (S.H. Ahn et al.), Phys. Lett. B **511**, 178 (2001); Phys. Rev. D **74**, 072003 (2006)
63. S. Masciocchi, Electromagnetic and hadronic calorimeters, 39th Heidelberg Physics Graduate Days, Heidelberg (2017)
64. OPAL Collaboration, https://pixels.com/featured/opal-detector-e-m-calorimeter-cernscience-photo-library.html
65. NA48 Collaboration, https://palesti2.web.cern.ch/na48_page.htm (2001)

References

66. G. Unal, Frascati Phys. Ser. **21**, 361 (2001)
67. V. Fanti et al., Nucl. Instrum. Meth. A **574**, 433 (2007)
68. The NA48 Collaboration (M. Jeitler et al.), Nucl. Instrum. Meth. A **494**, 373 (2002)
69. The MEG II Collaboration (R. Sawada et al.), PoS TIPP 2014, 033 (2014)
70. The XENON1T Collaboration (E. Aprile et al.), Eur. Phys. J. C **77**-12, 881 (2017)
71. The MEG II Collaboration (A.M. Baldini et al.), Eur. Phys. J. C **78**- 5, 380 (2018)
72. H. Araujo et al., Nucl. Instrum. Meth. A **604**, 41 (2009)
73. The LUX Collaboration (D. Akerib et al.), Nucl. Instrum. Meth. A **704**, 111 (2013)
74. The LZ Collaboration (D. Akerib et al.), Nucl. Instrum. Meth. A **953**, 1910.09124 (2019)
75. The XMASS Collaboration (A. Minamino et al.), Nucl. Instrum. Meth. A **623**, 448 (2010)
76. R. Neilson et al., Nucl. Instrum. Meth. A **608**, 68 (2009)
77. M. Miyajima, S. Sasaki, H. Tawara, IEEE Trans. Nucl. Sci. **41**, 835 (1994)
78. G.F. Knoll, *Radiation Detection and Measurements* (Wiley, New York, 1989)
79. G. Eigen, D. Hitlin, CALT-68-1838, SLAC-BABAR-NOTE-099 (1992)
80. C. Fabjan, N.A.T.O. Sci. Ser. B **128**, 281 (1985)
81. U. Amaldi, Phys. Scripta **23**, 409 (1981)
82. H.G. Fischer, Nucl. Instrum. Meth. **156**, 81 (1978)
83. A.N. Diddens et al., Nucl. Instrum. Meth. **178**, 27 (1980)
84. C.W. Fabjan et al., Nucl. Instrum. Meth. **141**, 61 (1977)
85. O. Botner, Phys. Scripta **23**, 555 (1981)
86. G. Barbiellini et al., Nucl. Instrum. Meth. **235**, 53 (1985)
87. J.A. Appel, Fermilab FN-380 (1982)
88. R. Carosi et al., Nucl. Instrum. Meth. **219**, 311 (1984)
89. The ATLAS Collaboration (M. Aharrouche et al.), Nucl. Instrum. Meth. A **568**, 601 (2006)
90. W. Hofmann et al., Nucl. Instrum. Meth. **163**, 77 (1979)
91. T.C. Awes et al., arXiv:nucl-ex/0202009 (2002)
92. W.J. Willis, BNL-17070, BNL-CRISP-72-46 (1972)
93. W.J. Willis, V. Radeka, Nucl. Instrum. Meth. **120**, 221 (1974)
94. The ARGUS Collaboration (H. Albrecht et al.), Nucl. Instrum. Meth. A **275**, 1 (1989)
95. The CDF Collaboration (L. Balka et al.), Nucl. Instrum. Meth. A **267**, 272 (1988)
96. R. Appel et al., Nucl. Instrum. Meth. A **479**, 349 (2002)
97. M. Adinolfi et al., Nucl. Instrum. Meth. A **494**, 326 (2002)
98. L. Aphecetche et al., Nucl. Instrum. Meth. A **499**, 521 (2003)
99. I.M. Gregor, IEEE NSS/MIC **2007**, 249 (2007)
100. A. Bernstein et al., Nucl. Instrum. Meth. A **336**, 23 (1993)
101. S.C. Berridge et al., IEEE Trans. Nucl. Sci. **39**, 1242 (1992)
102. S. Almehed et al., Nucl. Instrum. Meth. A **305**, 320 (1991)
103. G. S. Abrams et al., IEEE Trans. Nucl. Sci. **25**, 309 (1978); *ibid* **27**, 59 (1980)
104. V. Kandansky et al., Phys. Scripta **23**, 680 (1981)
105. H.J. Behrend et al., Phys. Scripta **23**, 610 (1981)
106. The SLD Collaboration (E. Vella et al.), Texas Calorim. HEP, 192 (1993)
107. The H1 Calorimeter Group (B. Andrieu et al.), Nucl. Instrum. Meth. A **350**, 57 (1994)
108. B. Andrieu et al., Nucl. Instrum. Meth. A **336**, 460 (1991)
109. The RD3 Collaboration (D.M. Gingrich et al.), Nucl. Instrum. Meth. A **364**, 290 (1995)
110. The D0 Collaboration (S. Abachi et al.), Nucl. Instrum. Meth. A **338**, 185 (1994)
111. W. Toki et al., Nucl. Instrum. Meth. A **219**, 479 (1983)
112. The DELPHI Collaboration (P. Abreu et al.), Nucl. Instrum. Meth. A **378**, 57 (1996); The DELPHI HAC Collaboration (I. Ajinenko et al.), IEEE Trans. Nuc. Sci. **43**-3, (1996)
113. D. Axen et al., Nucl. Instrum. Meth. A **328**, 472 (1990)

114. The ATLAS Collaboration (G. Aad et al.), JINST **3**, S08003 (2008)
115. The ATLAS Collaboration (G. Aad et al.), Eur. Phys. J. C **70**, 723 (2010)
116. The ATLAS LAr Collaboration (O.B. Abdinov et al.), Liquid-Argon Calorimeter, Technical Design Report, CERN/LHCC 96-41 (1996)
117. The ATLAS Collaboration (G. Aad et al.), Eur. Phys. J. C **70**, 755 (2010)
118. I. Riu, The ATLAS Liquid Argon Calorimeters, talk at LHC days in Split, https://indico.cern.ch/event/6428/contributions/2067801/attachments/1010504/1437745/ATLAS-ECAL-Split.pdf (2006)
119. ATLAS Collaboration, https://conferences.fnal.gov/lp2003/forthepublic/detectors/index.html
120. The ATLAS Colaboration (G. Aad et al.), Eur. Phys. J. C **74**, 3071 (2014)
121. J. Colas et al., Nucl. Instrum. Meth. A **550**, 96 (2005)
122. L. Aperio Bella, Nucl. Instrum. Meth. A **718**, 60 (2013)
123. S. Berridge et al., IEEE Trans. Nucl. Sci. **37**, 1191 (1990)
124. The DELPHI Collaboration (S. Almehed et al.), DELPHI-92-77 Dallas PHYS 188 (1992)
125. S.L. White, SLAC-R-684 (1995)
126. https://physicsworld.com/a/where-the-energy-goes/ (2018)
127. J. Ranft, Particle Accelerator **3**, 129 (1972)
128. A. Baroncelli, Nucl. Instrum. Meth. A **118**, 45 (1974)
129. T.A. Gabriel, Nucl. Instrum. Meth. A **134**, 271 (1976)
130. B. Andersson et al., Phys. Rept. **97**, 31 (1983)
131. S. Agostinelli et al., Nucl. Instrum. Meth. A **506**, 250 (2003)
132. H. Fesefeld, Tech. Rep. PITHA85-02, Aachen, Germany (1985)
133. A. Fasso et al., CERN 2005-10 (2005); G. Battistoni et al., Annals of Nuclear Energy. **82**, 10 (2014)
134. A. Capella et al., Phys. Rep. **236** (1993)
135. D.H. Wright et al., AIP Conf. Proc. **896**, 1 (2007)
136. A.B. Kaidalov, Phys. Lett. B **116**, 459 (1982); A.B. Kaidalov, K.A. Ter-Martirosian, Phys. Lett. B **117**, 247 (1982)
137. G. Folger, J.P. Wellisch, nucl-th/0306007 (2003); S. Amelin et al., Phys. Rev. Lett. **67**, 1523 (1991); N.S. Amelin et al., Nucl. Phys. A **544**, 463c (1992); L.V. Bravina et al., Nucl. Phys. A **566**, 461c (1994); L.V. Bravina et al., Phys. Lett. B **344**, 49 (1995)
138. B. Andersson et al., Nucl. Phys. B **281**, 289 (1987); B. Andersson et al., Z. Phys. C **57**, 485 (1993)
139. G. Folger, V.N. Ivanchenko, J.P. Wellisch, E. P. J. C **21**, 407 (2004)
140. A. Heikkinen, N. Stepanov, J.P. Wellisch, eConf C 0303241, MOMT008 (2003)
141. T. Koi, SATIF-11, 85 (2012)
142. M. Blann, Phys. Rev. C **54**, 1341 (1996)
143. H. Burkhardt et al., IEEE NSS/MIC, 1907 (2004)
144. https://indico.cern.ch/event/14946/contributions/190670/attachments/149855/212275/G4ReviewCHIPS.pdf
145. https://lss.fnal.gov/archive/test-fn/1000/fermilab-fn-1058-apc.pdf
146. The CALICE Collaboration (C. Adloff et al.), JINST **8**, P07005 (2013)
147. https://llr.in2p3.fr/activites/physique/atf2/Geant4_hadronic_models.html; Aatos Heikkinen, 2nd Finnish Geant4 Workshop, Helsinki (2005)
148. G. Flucke, Thesis (2005). https://doi.org/10.3204/DESY-THESIS-2005-006
149. S. Piperov, A.I.P. Conf. Proc. **896**-1, 195 (2007)
150. M. Jonker et al., Nucl. Instrum. Meth. A **200**, 183 (1982)
151. M. Holder et al., Nucl. Instrum. Meth. A **151**, 69 (1978)
152. D.I. Cheshire et al., Nucl. Instrum. Meth. A **141**, 219 (1977)

153. B. Friend et al., Nucl. Instrum. Meth. A **136**, 505 (1977)
154. P. Adragna et al., Nucl. Instrum. Meth. A **615**, 158 (2010)
155. D. Acosta et al., Nucl. Instrum. Meth. A **316**, 184 (1992)
156. The CALICE Collaboration (G. Eigen et al.), JINST **11**-06, P06013 (2016)
157. L.M. Livan, R. Wigmans, *Calorimetry for Collider Physics* (Springer, 2019)
158. P.J. Gollon et al., Nucl. Instrum. Meth. **189**, 387 (1981)
159. D.E. Groom, Nucl. Instrum. Meth. A **572**, 633 (2007); Nucl. Instrum. Meth. A **593**, 628 (erratum) (2008)
160. T.A. Gabriel et al., Nucl. Instrum. Meth. A **338**, 336 (1994)
161. D. Acosta et al., Nucl. Instrum. Meth. A **308**, 481 (1991)
162. The CDF Collaboration (J.B. Liu et al.), Calorimetry in high-energy physics, 237 (1997)
163. R. Wigmans et al., Nucl. Instrum. Meth. A **265**, 273 (1988)
164. T. Akesson et al., Nucl. Instrum. Meth. A **241**, 17 (1985)
165. H. Abramowicz et al., Nucl. Instrum. Meth. A **180**, 429 (1981)
166. A. Beer et al., Nucl. Instrum. Meth. A **224**, 360 (1984)
167. B. Aubert, et al., Frascati Physics Series, 31 (2001)
168. S. Akhmadaliev et al., Nucl. Instrum Meth. A **449**, 461 (2000)
169. N. Akchurin et al., Nucl. Instrum. Meth. A **399**, 202 (1997)
170. T. Akesson et al., Nucl. Instrum. Meth. A **262**, 243 (1987)
171. T. de Vincenzi et al., Nucl. Instrum. Meth. A **243**, 348 (1986)
172. R. Wigmans, CERN-PPE/91-39 (1991)
173. R. Wigmans et al., Nucl. Instrum. Meth. A **259**, 389 (1987)
174. J.E. Brau et al., Nucl. Instrum. Meth. A **238**, 190 (1989)
175. J.E. Brau, T.A. Gabriel, Nucl. Instrum. Meth. A **275**, 489 (1985)
176. T.A. Gabriel et al., IEEE Trans. Nucl. Sci. **32**, 697 (1985)
177. H. Brückmann, B. Anders, U. Behrens, Nucl. Instrum. Meth. A **263**, 136 (1988)
178. J. Brau, Personal Communication
179. F. Corriveau et al., CERN HELlOS, Int. Note 145 (1986)
180. D. Hitlin, Proc. Workshop on Compensated Calorimetry CALT-68-1305 (1985)
181. B. Cox, Uranium liquid-argon calorimetry, FERMILAB - Conference 86/14-E (1986)
182. G. D'Agostini et al., Nucl. Instrum. Meth. A **274**, 134 (1988)
183. C. Daum et al., DESY, Hamburg, ZEUS Int. Note 86/13 (1986)
184. G. Drews et al., Nucl. Instrum. Meth. A **290**, 335 (1990)
185. G.R. Young et al., Nucl. Instrum. Meth. A **279**, 503 (1989)
186. A. Andresen et al., Nucl. Instrum. Meth. A **290**, 95 (1990)
187. The D0 Collaboration (S. Abachi et al.), Nucl. Instrum. Meth. A **338**, 185 (1994)
188. H.P. Wellisch et al., MPI PhE/94-03, H1, 02/94-346 (1994)
189. The H1 Collaboration (B. Andrieu et al.), Nucl. Instrum. Meth. A **336**, 499 (1993)
190. The ZEUS Calorimeter Group (U. Behrens et al.), Nucl. Instrum. Meth. A **289**, 115 (1990)
191. H. Tiecke et al., Nucl. Instrum. Meth. A **277**, 42 (1989)
192. V. Korbel, Nucl. Instrum. Meth. A **263**, 70 (1988); B. Andrieu et al., ibid A **336**, 460 (1993)
193. M. Abolins et al., Nucl. Instrum. Meth. A **280**, 36 (1989)
194. The ATLAS Collaboration (G. Aad et al.), Eur. Phys. J. C **81** 8, 689 (2021)
195. The ATLAS Collaboration (G. Aad et al.), Eur. Phys. J. C **77**, 490 (2017)
196. The ATLAS Collaboration (G. Aad et al.), Eur. Phys. J. C **77**, 466 (2017)
197. D. Axen et al., Nucl. Instrum. Meth. A **328**, 472 (1993)
198. A.C. Benvenuti et al., Nucl. Instrum. Meth. A **276**, 94 (1989); ibid A **89**, 463 (1990)
199. R. Dubois et al., IEEE Trans. Nucl. Sci. **33**, 194 (1986)
200. A.C. Caldwell, Nucl. Instrum. Meth. A **330**, 389 (1993)
201. E. Hilger, Nucl. Instrum. Meth. A **257**, 488 (1987)
202. U. Behrens et al., Nucl. Instrum. Meth. A **289**, 112 (1990)
203. G. d'Agostini et al., Nucl. Instrum. Meth. A **274**, 134 (1989)
204. M.G. Catanesi et al., Nucl. Instrum. Meth. A **260**, 43 (1987)

205. E. Ros, Nucl. Phys. B Proc. Suppl. **23**, 51 (1991)
206. The ATLAS Collaboration (L.C. Alberich et al.), Nucl. Instrum. Meth. A **824**, 12 (2016)
207. The ATLAS Collaboration (G. Aad et al.), Eur. Phys. J. C **70**, 1193 (2010)
208. The ATLAS Collaboration, https://www.fsp103-atlas.de/e17619/e17623/
209. The CDF Collaboration (P.A. Movilla Fernandez et al.), AIP Conf. Proc. **867** 1, 487 (2006)
210. The ZEUS Collaboration (A. Bernstein et al.), Nucl. Instrum. Meth. A **336**, 23 (1993)
211. M. Volpi, M. Cavalli-Sforza, ATL-TILECAL-PUB-2009-006 (2008)
212. The CMS Collaboration (S. Chatrchyan et al.), JINST **5**, T03012 (2009)
213. The H1 Collaboration, (I. Abt et al.), Nucl. Instrum. Meth. A **386**, 310 (1997)
214. The D0 Collaboration (D. Buchholz et al.), Eur. Phys. J. C **33**, S984 (2004)
215. The ALEPH Collaboration (D. Buskulic et al.), Nucl. Instrum. Meth. A **360**, 481 (1995)
216. The OPAL Collaboration (K. Ahmet et al.), Nucl. Instrum. Meth. A **305**, 275 (1991)
217. C. Chen et al., Nucl. Instrum. Meth. A **272**, 713 (1988); L3 Technical Design Report, LEPC-P-4. (1983)
218. The ATLAS Collaboration (J. Abdallah et al.), JINST **8**, T11001 (2013)
219. The ATLAS Collaboration (J. Abdallah et al.), JINST **8**, P01005 (2013)
220. H. Videau, Energy flow or particle flow, Int. Cof. on ILC, p 105 (2004)
221. J.C. Brient, H. Videau, eConf C 010630, E3047, arXiv:hep-ex/0202004 (2002)
222. S. Magill, Talk at ALCPG Workshop, Victoria (2004)
223. A. White, Talk at DESY PRC, https://www.desy.de/f/prc/talks_open/meet_57/PRC57-CalicePRC_Report-open.pdf
224. The CALICE Collaboration (D. Jeans et al.), Nucl. Instrum. Meth. A **628**, 324 (2011)
225. The CALICE Collaboration (C. Adloff et al.), Nucl. Instrum. Meth. A **608**, 372 (2008)
226. F. Sefkow et al., J. Phys. Conf. Ser. **1162**, 1 (2019)
227. K. Krüger, Talk at AHCAL Meeting DESY; https://agenda.linearcollider.org/event/8082/ (2018)
228. The CALICE Collaboration (C. Adloff et al.), JINST **7**, P09017 (2012)
229. The CALICE Collaboration (J. Repond et al.), JINST **13**, P12022 (2018)
230. The CALICE Collaboration (C. Adloff et al.), JINST **8**, P07005 (2013)
231. The CALICE Collaboration (C. Adloff et al.), JINST **6**, P07005 (2011)
232. The CALICE Collaboration (C. Adloff et al.), CALICE Analysis Note CAN-2010-04 (2010)
233. E. Garutti, Talk at ILC Workshop 2006, https://agenda.linearcollider.org/event/1049/timetable/?view=lcc
234. The CMS Collaboration (A.M. Magnan et al.), JINST **12**-01, C01042 (2017)
235. N. Akchurin et al., JINST **13**-10, P10023 (2018)
236. The CMS Collaboration (C. Seez et al.), The Phase-2 Upgrade of the CMS endcap calorimeter, TDR (2017)
237. The CMS Collaboration (A. Lobanov et al.), JINST **15**-07, C07003 (2020)
238. The CMS Collaboration (F. Chlebana et al.), J. Phys. Conf. Ser. **928**-1, 012027 (2017)
239. P. Mockett, SLAC-267, 335 (1983)
240. D.R. Winn, W.A. Worstell, IEEE TNS NS **36**, 334 (1989)
241. S. Lee, M. Livian, R. Wigmans, Rev. Mod. Phys. **90**-2, 025002 (2018)
242. F. Sefkow et al., Rev. Mod. Phys. **88**, 015003 (2016)
243. E. Aprile, T. Doke, Rev. Mod. Phys. **82**, 2053 (2010)

Open Access This chapter is licensed under the terms of the Creative Commons Attribution 4.0 International License (http://creativecommons.org/licenses/by/4.0/), which permits use, sharing, adaptation, distribution and reproduction in any medium or format, as long as you give appropriate credit to the original author(s) and the source, provide a link to the Creative Commons license and indicate if changes were made.

The images or other third party material in this chapter are included in the chapter's Creative Commons license, unless indicated otherwise in a credit line to the material. If material is not included in the chapter's Creative Commons license and your intended use is not permitted by statutory regulation or exceeds the permitted use, you will need to obtain permission directly from the copyright holder.

Photodetectors, Plastic Scintillators and Time Measurements

8

In this chapter, we discuss different types of photodetectors, plastic scintillators and properties of time measurements. The most commonly used instruments for time measurements consist of fast scintillators read out by photomultiplier tubes (PMT). Though silicon-based devices, such as the SiPM, are used more often today for time measurements, photomultiplier tubes are still important photodetectors since they record photons with high efficiency and excellent time resolution. They are used in Cherenkov detectors, trigger counters and calorimeters. Another photodetector is the Micro-Channel Plate, which presently achieves the best time resolutions. Plastic scintillators are used in sampling calorimeters, trigger counters and time-of-flight detectors.

8.1 The Photomultiplier

The first PMT was built by RCA in 1934 [1]. Since then various photomultiplier tubes have been produced by many companies for various applications. The theory and properties of PMT operation are discussed in references [2,3]. The principle of a photomultiplier tube illustrated in Fig. 8.1 is to convert optical photons into electrons and to amplify them. A photocathode converts an incoming photon into an electron. A high electric field between the photocathode and the first dynode accelerates the electron. When it hits the first dynode, it has sufficient kinetic energy to knock out three to five electrons, which are further accelerated towards the second dynode, where each electron knocks out at least two new electrons. This process is repeated at each succeeding dynode. Photomultipliers typically have 12 to 16 dynodes, yielding amplifications of 10^5 to 10^8. The entire configuration is encapsulated in a glass tube under vacuum. The photocathode is evaporated onto a glass window, which can be made of quartz to be sensitive to UV photons.

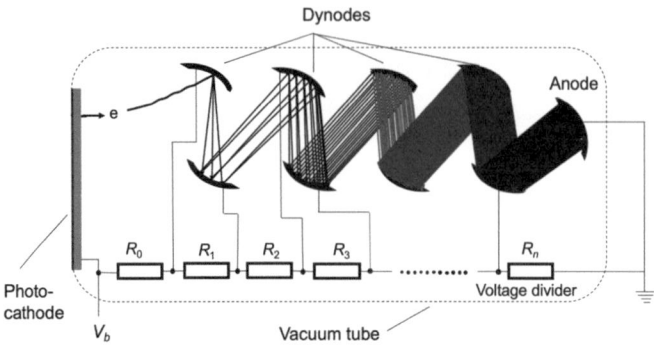

Fig. 8.1 Principle layout of a photomultiplier (see text)

Note that the photocathode is a semiconductor. So, we can apply the band model. When a photon strikes the photocathode, it will transfer its energy onto an electron in the valence band and excite it into the conduction band. The excited electron diffuses towards the photocathode surface. If the diffused electrons have sufficient energy to overcome the vacuum level barrier, they are emitted into the vacuum as photoelectrons. The ratio of liberated photoelectrons to impinging photons is called the quantum efficiency (ϵ_{QE}), i.e. the number of carriers generated per photon.

8.1.1 Photomultiplier Layout

Figure 8.2 (left) shows the layout of a photomultiplier tube. The system is housed in a glass tube operated under vacuum. Shapes are either cylinders or cuboids. The top face is sealed with a glass or quartz window that holds the photocathode on the

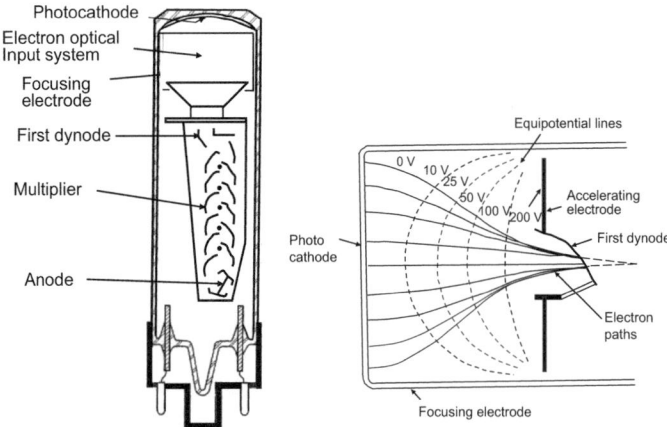

Fig. 8.2 Left: Schematic layout of a photomultiplier tube. Right: Example of the electric-field configuration between photocathode and first dynode, showing equipotential lines (dashed curves) and electron flight paths (solid curves). Both reprinted with kind permissions from [4], © 1970, N.V. Philips' Gloeilampenfabrieken. All rights reserved

inner side. The bottom face is closed with a socket that holds the high voltage and signal connectors. Below the photocathode, a focusing electrode is placed that guides the photoelectrons towards the first dynode, which is followed by the multiplication system, consisting of dynodes and the anode. Figure 8.2 (right) shows the electric field configuration between the photocathode and the first dynode. The electric field is raised from 0 to 200 V such that the electrons gain sufficient kinetic energy before they hit the first dynode. Dashed lines show different equipotentials, while solid lines represent the electron paths. The focusing electrode is arranged such that all photoelectrons reach the first dynode.

Photocathodes are made from combinations of alkali elements (Na, K, Cs) and group V elements (usually Sb). Table 8.1 shows some characteristics of commonly used photocathodes. Typical quantum efficiencies of multi-alkali photocathodes lie around 20–25%. The maximum quantum efficiency of $\epsilon_{QE} = 35$–43% is achieved with bi-alkali or GaAsP photocathodes. Figure 8.3 shows the photocathode radiant sensitivity as a function of the photon wavelength for several photocathodes using alkali, bi-alkali and multi-alkali materials. The dependence of the quantum efficiency as a function of the photon wavelength is indicated. Typically, photocathodes are laid out to work at wavelengths above 400 nm. One reason is that these PMTs use regular borosilicate windows that block UV light. However, the spectral sensitivity of photocathodes can be extended to wavelengths below 200 nm. These PMTs need special UV-transparent windows. The sensitivity in the infrared region can be increased with modified cathode layouts such as the photocathodes 501 and 502 K.

Some newer photocathode materials are based on GaAs as shown in Fig. 8.4 (left). For GaAsP photocathodes, the quantum efficiency reaches values around 40% in the visible region. For GaAs photocathodes, the sensitivity ranges from about 300 nm to 1000 nm with ϵ_{QE} up to 15%. The InP/InGaAsP photocathode extends the wavelength range to 1400 nm. However, the quantum efficiency is only 2%. Figure 8.4 (right) shows the spectral transmittance of photomultiplier windows. Borosilicate is transparent to about 300 nm. The UV-transmitting glass works to about 180 nm. Synthetic silica and sapphire reach 160 and 140 nm, respectively. The most optimal material is MgF_2, which is transparent to wavelengths of about 110 nm.

Photocathodes are operated both in transmission and reflection modes. The operation principles are illustrated in Fig. 8.5 (left, center). In the transmission mode, the photon enters through a glass window and produces an electron in the photocathode that is accelerated into the vacuum tube. In the reflection mode the photon enters from the vacuum and produces an electron in the photocathode, which is reflected back into the vacuum. The quantum efficiency in the reflection mode is a factor of 1.5–2 higher than that in the transmission mode. For head-on photomultiplier tubes that are typically used in particle physics detectors, photocathodes work in the transmission mode.

Typically, dynodes are made of CuBeO(Cs) or GaP except for the first dynode for which a material with a high secondary electron emission coefficient is used, such as Ag-MgO(Cs) or SbKCs. Figure 8.5 (right) shows the secondary emission coefficient δ_{dyn} of different dynode materials as a function of incident electron energy. For 100 eV incident electrons, these materials produce three to five secondary electrons.

Table 8.1 Examples of photocathodes in transmission mode listing the cathode type, photocathode composition, the window material, accepted wavelength range λ_{acc}, the radiant sensitivity at a given wavelength λ_1 and the maximum quantum efficiency ϵ_{QE} at wavelength λ_{max}. †High-temperature bi-alkali. ‡Low-noise bi-alkali (Sb-Na-K). *Super bi-alkali. ◇Ultra bi-alkali. *Extended green bi-alkali photocathode. ⊙Extended red multi-alkali photocathode. ∧Extended red GaAsP (Cs) photocathode. Reprinted with kind permission from [5], © 2017, Hamamatsu Photonics. All rights reserved

Cathode (curve code)	Photocathode composition	Window material	λ_{acc} [nm]	rad. sens. at (λ_1) [mA/W] (nm)	ϵ_{QE} at (λ_{max}) [%] (nm)
100M	Cs-I	MgF$_2$	115–200	14 (140)	13 (130)
200S	Cs-Te	Silica	160–320	29 (240)	16 (210)
200M	Cs-Te	MgF$_2$	115–320	29 (240)	17 (200)
201S	Cs-Te	Silica	160–320	31 (240)	17 (210)
400K	K$_2$-Cs-Sb	Borosilicate	300–650	88 (420)	27 (390)
400U	K$_2$-Cs-Sb	UV	185–650	88 (420)	27 (390)
400S	K$_2$-Cs-Sb	Silica	160–650	88 (420)	27 (390)
401K	Bi-alkali†	Borosilicate	300–650	51 (375)	17 (375)
402K (S_{10})	Bi-alkali‡	Borosilicate	300–650	54 (375)	18 (375)
440K	Bi-alkali*	Borosilicate	300–650	110 (400)	35 (350)
441K	Bi-alkali ◇	Borosilicate	300–650	130 (400)	43 (350)
442K	Bi-alkali*	Borosilicate	230–700	110 (400)	35 (350)
443K	Bi-alkali◇	Borosilicate	230–700	130 (400)	43 (350)
444K	Bi-alkali*	Borosilicate	300–700	127 (420)	40 (380)
500K (S_{20})	Multi-alkali	Borosilicate	300–850	64 (420)	20 (375)
500U	Multi-alkali	UV	185–850	64 (420)	25 (280)
500S	Multi-alkali	Silica	160–850	64 (420)	25 (280)
501K (S_{25})	Multi-alkali⊙	Borosilicate	300–900	40 (600)	8 (580)
502K	Multi-alkali	Borosilicate	300–900	69 (420)	20 (390)
600K	GaAsP (Cs)	Borosilicate	280–720	180 (550–650)	40 (480–530)
601K	GaAs (Cs)	Borosilicate∧	280–820	160 (550–650)	36 (480–530)
602K	GaAs (Cs)	Borosilicate	370–920	85 (750–850)	12 (600–750)
700K (S_1)	Ag-O-Cs	Borosilicate	400–1,200	2.2 (800)	0.36 (740)
900S	In P/InGaAsP (Cs)	Silica	950–1,200	18 (1,100)	2.0 (1,000–1,100)
901S	In P/INGaAs (Cs)	Silica	950–1,700	24 (1,500)	2.0 (1,000–1,550)

8.1 The Photomultiplier 529

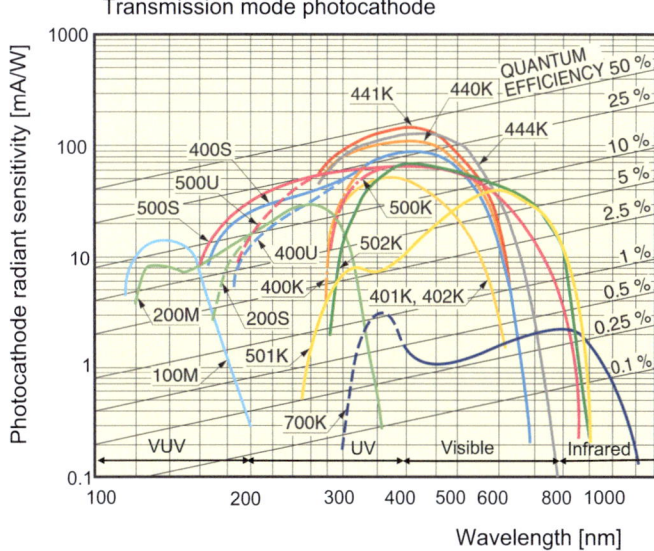

Fig. 8.3 The photocathode radiant sensitivity as a function of the photon wavelength in transmission mode for presently used PMTs [5]. The quantum efficiency as a function of photon wavelength is indicated. The labels are the same as those in Table 8.1 and mean: 100 is Cs-I, 200 is Cs-Te, 400 is bi-alkali, 500 is multi-alkali, and 700 is AgO Cs. The letters indicate the window material, K is borosilicate, M is MgF_2, S is quartz, and U is UV glass. Reprinted with kind permission from [5], © 2017, Hamamatsu Photonics. All rights reserved

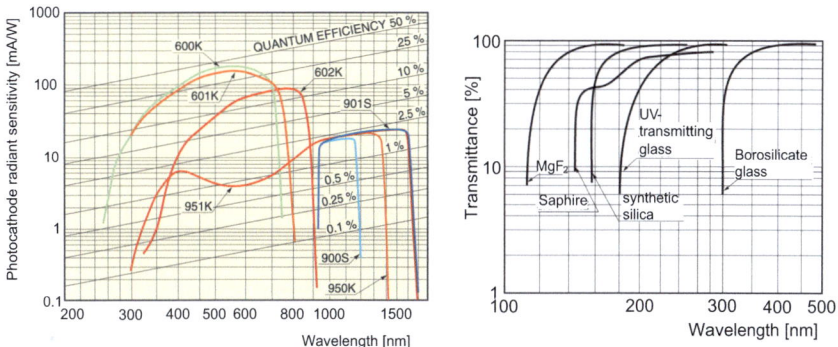

Fig. 8.4 Left: The photocathode radiant sensitivity of GaAs-based materials as a function of wavelength. Right: The transmittance of different photocathode windows as a function of wavelength [5]. Both reprinted with kind permission from [5], © 2017, Hamamatsu Photonics. All rights reserved

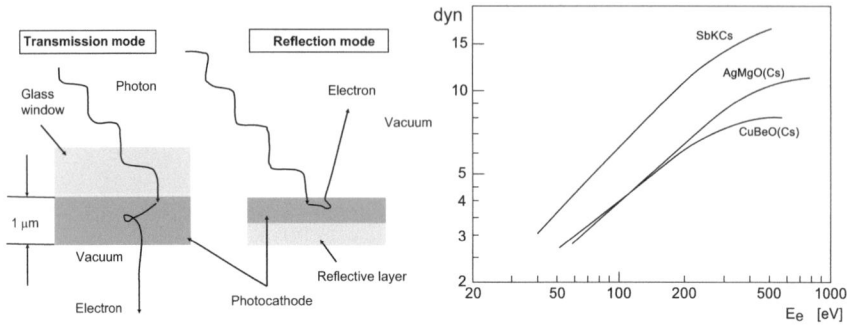

Fig. 8.5 Left: Photocathode in transmission mode. Center: Photocathode in reflection mode. Right: Secondary emission coefficient δ_{dyn} as functions of incident electron energy for SbKCs, AgMgO(Cs) and CuBeO(Cs). Reprinted with kind permission from [6], © 2013, V. Taillandier. All rights reserved

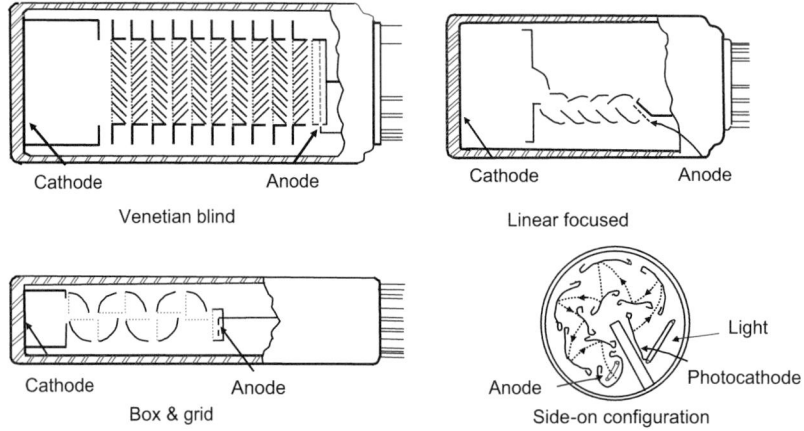

Fig. 8.6 Four different dynode configurations. Top left: Venetian blind configuration in which sets of parallel dynodes are used at each stage to minimize the path of the secondary electrons emitted from the first dynode. Top right: Linear focussed configuration. Bottom left: The box and grid configuration. Bottom right: The side-on configuration in which all dynodes are arranged in a single plane to reduce the size of the photomultiplier tube. All reprinted from [7], © 1970, EMI Coporation. All rights reserved

8.1.2 Dynode Configurations

There is a variety of different dynode arrangements. Figure 8.6 show four examples, the Venetian blind (top left), linear focussed (top right), box and grid (bottom left) and a side-on configuration (bottom right). Table 8.2 lists various properties of PMTs with different dynode configurations displayed in Fig. 8.6. The potential difference between adjacent dynodes is 100–200 V. To achieve a high electron emission coefficient in the first dynode, the potential difference between the photocathode and the first dynode is highest (200 V). The high voltage is applied via a voltage divider (see Fig. 8.8), which is an electronic configuration with selected resistors. The dynode

8.1 The Photomultiplier

Table 8.2 Properties of photomultiplier tubes with different dynode configurations listing rise time, fall time, pulse width, electron transit time and spread of the electron transit time. Reprinted with kind permission from [5], © 2017, Hamamatsu Photonics. All rights reserved

Dynode type	Rise time [ns]	Fall time [ns]	Pulse width [ns]	Electron transit time [ns]	Time spread ns
Linear-focused	0.7–3.0	1–10	1.3–5	16–50	0.37–1.1
Circular-cage	3.4	10	7	31	3.6
Box-and-grid	≤7	25	13–20	57–70	<10
Venetian blind	≤7	25	25	60	<10
Fine mesh	2.5–2.7	4–6	5	15	<0.45
Metal channel	0.65–1.5	1–3	1.5–3	4.7–8.8	0.4

configuration also depends on whether the photomultiplier is used for high gain or for time measurements. In fact, many photomultipliers have two outputs, one at the anode and a second at dynodes 9 to 12. The dynode configuration affects the deviation from linearity at high currents induced by high counting rates. Figure 8.7 (left) shows the deviation from linearity for the four configurations displayed in Fig. 8.6. The box and grid configuration shows a 2% deviation from linearity at a peak anode current of 1 mA. For the Venetian-blind configuration with the standard voltage divider the 2% deviation from linearity occurs at ∼2 mA. Using the Venetian-blind configuration with a high-current voltage divider increases the linearity range. Here, the 2% deviation from linearity is pushed to currents of 4 mA. The linear focussed configuration

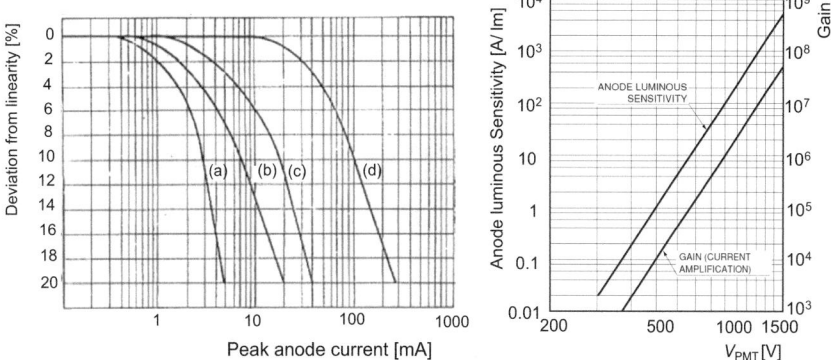

Fig. 8.7 Left: Deviations from a linear response as a function of the anode current for **a** the box and grid configuration, **b** the Venetian blind configuration with a standard voltage divider, **c** the Venetian blind configuration with a high-current voltage divider and **d** the linear focussed configuration with a high-current voltage divider. Reprinted from [7], © 1970, EMI Corporation. All rights reserved. Right: Anode luminous sensitivity and gain versus applied voltage V_{PM} for a typical PMT in double-logarithmic scale. Reprinted with kind permission from [5], © 2017, Hamamatsu Photonics. All rights reserved

with a high-current voltage divider increases the linearity range substantially. The 2% deviation from linearity is pushed to 30 mA.

8.1.3 Photomultiplier Properties

The gain of a PMT is the ratio of anode current I_a to the photocathode input current I_{cath},

$$G_{PM} = I_a/I_{cath}. \tag{8.1}$$

The gain can be expressed as a product of the individual gains at each dynode

$$G_{PM} = \prod_{i=1}^{n} g_i, \tag{8.2}$$

where g_i is the gain at dynode i. An energetic electron from the photocathode hits the surface of the first dynode and liberates electrons from the valence band to the conduction band. For a bi-alkali photocathode illuminated with light in the 400–430 nm wavelength range the kinetic E_e energy of the photoelectrons lies between zero and 1.8 eV with a peak at 1.2 eV. The electrons move towards the dynode-vacuum interface, where some escape as secondary electrons. The procedure occurs at each dynode. The gains g_i are a function of the electron kinetic energy and thus depends on the voltage V_i applied between dynode i and dynode $i + 1$ yielding

$$g_i = k_i V_i^{\alpha_{dyn}}, \tag{8.3}$$

where k_i and α_{dyn} are dynode-specific parameters. The voltage is typically supplied via a voltage divider (see Fig. 8.8), which consists of a chain of resistors R_i placed between the dynodes. If V_{PM} is the voltage between cathode and anode, $R_{tot} = \sum_{i=1}^{n} R_i$ is the total resistor and $f_i = R_i/R_{tot}$, the voltage at dynode i is $V_i = f_i V_{PM}$. Thus, the gain can be parameterized as [8],

$$G_{PM} = \prod_{i=1}^{n} k_i \left(f_i V_{PM} \right)^{\alpha_{dyn}}. \tag{8.4}$$

For example for the Hamamatsu R11410-20 PMT, $k_{dyn} = 0.167$ if all k_i are equal and $\alpha_{dyn} = 0.674$ [8]. The gain is 5×10^6 (6×10^5) at $V_{PM} = 1500$ (1200) V. Apart from V_{PM}, the gain typically depends on the number of dynodes and ranges from a few 10^5 to few 10^8. Figure 8.7 (right) shows the gain versus applied voltage for a typical Hamamatsu PMT.

The voltage at each dynode is provided by a voltage divider. Figure 8.8 (left) shows an example of a typical voltage divider. High voltage is applied between anode and cathode by a high-voltage power supply The voltage divider consisting of a chain of resistors sets specific voltages at each dynode. Typically, the resistors R_2–R_{10} are

8.1 The Photomultiplier

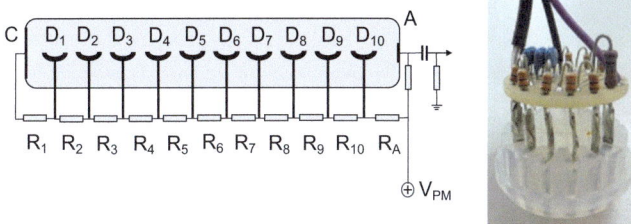

Fig. 8.8 Left: Schematic layout of a ten-stage voltage divider. Typically, the resistors R_2-R_{10} are equal. The photocathode is on ground potential and the anode on high voltage. Right: Photograph of a voltage divider. Reprinted with kind permission from [5], © 2017, Hamamatsu Photonics. All rights reserved

equal whereas R_1 is much larger to increase the kinetic energy of the photoelectrons hitting the first dynode. Figure 8.8 (right) shows a photograph of a voltage divider.

Without illumination, a PMT still produces a current, called dark current, which originates from different sources. There are thermionic emissions from the dynodes and cathode, leakage currents, ionization phenomena, light phenomena and contaminations from radioactive elements. The thermal noise is the main source whose contribution is given by Richardson's equation,

$$J_\mathrm{d} = a_d (k_B T)^2 \exp\left(-\frac{q_e \phi_f}{k_B T}\right), \tag{8.5}$$

where J_d is the emission current density, $a_d = 1.2$ A mm^{-2} K^{-2} is a constant, ϕ_f is the work function for photocathodes [9] and $k_B T$ is the thermal energy. The work function times unit charge ranges from 1.3 to 1.12 eV, yielding dark currents of 10^{-15}–10^{-11} A at the cathode, while the anode dark currents increase with the gain. The signal currents at the anode are much larger, ranging from μA to mA. Note that residual gases that are left or are formed in the PMT may produce detectable currents. At room temperature, the dark current is of the order of a few hundred pA. The gas atoms may be ionized by electrons. Having the opposite charge, the ions may be accelerated back towards the cathode or a dynode, where they may create new electrons. This effect is called after-pulsing. The after-pulse signal is slightly time delayed (30 to 60 ns), depending on where it originates, a dynode or the photocathode. Furthermore, the photoemission and secondary emission processes are affected by statistical fluctuations leading to statistical noise. The number of photoelectrons and number of secondary electrons emitted from the dynodes will fluctuate. This noise is usually called shot noise or noise due to the Schottky Effect. The fluctuations in the photocathode and the multiplication system can be calculated assuming Poisson statistics. For constant illumination in time interval $\Delta \tau$, the *rms* of first contribution is given by

$$\langle \Delta I^2 \rangle = q_e I_\mathrm{cath} / \Delta \tau, \tag{8.6}$$

where I_cath is the cathode current. The contribution from dynodes involve the statistical nature of secondary emissions as well as differences in electron transit times, non-

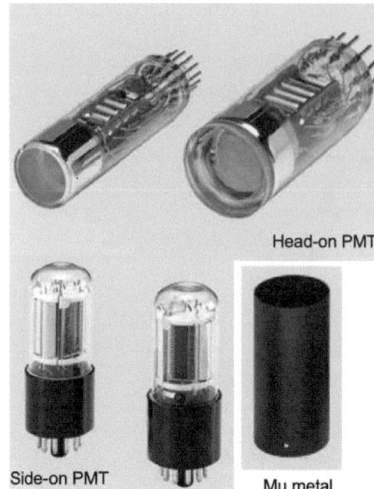
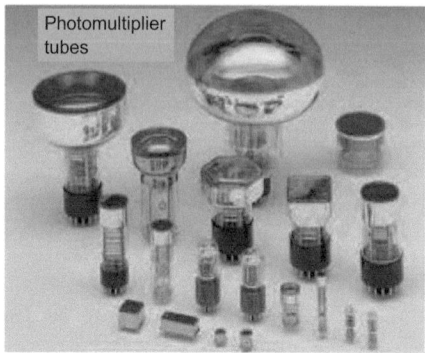

Fig. 8.9 Top left: Head-on PMTs. Bottom left: Side-on PMTs. Both reprinted with kind permission from [25], © 2017, Hamamatsu Photonics. All rights reserved. Bottom center: A μ-metal shield [26]. Note that μ-metals are used to shield the PMT from the earth's magnetic field (personal photograph). Right: Different PMTs from Hamamatsu. Reprinted with kind permission from [25], © 2017, Hamamatsu Photonics. All rights reserved

uniformities in the secondary emission factor over the dynodes, and other effects. Note that the latter contribution accounts at most 10%. The signal-to-noise ratio (S/N) is given by

$$\frac{S}{N} = \frac{I_{\text{cath}}}{\sqrt{2q_e \Delta B \cdot \delta/(\delta - 1) \cdot (I_{\text{cath}} + 2I_d) + N_{\text{amp}}^2}}, \quad (8.7)$$

where I_{cath} is the cathode current ([A]), ΔB is the bandwidth ([Hz]), δ is the secondary emission ratio, I_d is the cathode equivalent dark current ([A]) and N_{amp} is the amplifier noise that is typically small and is ignored. The cathode current can be written as

$$I_{\text{cath}} = q_e \epsilon_{\text{QE}} \frac{P_{\text{opt}}}{\hbar \omega}, \quad (8.8)$$

where P_{opt} is the optical power ([W]) and $\hbar \omega$ is the photon energy. We can rewrite this in terms of the anode current $I_a = I_{\text{cath}} \cdot G_{\text{PM}}$,

$$\frac{S}{N} = \frac{I_a}{\sqrt{2q_e \Delta B \cdot G_{\text{PM}} \cdot \delta/(\delta - 1) \cdot (I_a + 2I_{\text{da}})}}, \quad (8.9)$$

where I_{da} is the dark current at the anode and G_{PM} is the gain.

Many different photomultiplier tubes have been manufactured depending on its specific application. Figure 8.9 shows photographs of several PMTs and Table 8.3

8.1 The Photomultiplier

Table 8.3 Properties of some photomultipliers listing the PMT type, reference, experiment and system, photocathode diameter (size), number of dynodes, window type, photocathode material, dynode material, high voltage V_{PM} between anode and cathode, anode layout, quantum efficiency, gain, signal rise time τ_r, signal transit time τ_t and jitter on the transit time for multiple electrons $\Delta\tau_t$ [FWHM]. Note that HPK stands for Hamamatsu Photonics and HyperK stands for Hyper-Kamiokande. [†]This is a fine mesh PMT that can be operated in magnetic fields of 1 T, where the gain is reduced to 1.8×10^4. [‡]Super bi-alkali photocathode

PMT	Amperex	RCA	HPK	ETL	HPK	HPK	HPK
Type	XP2020	8854	R5900	9125	R12860	R11265	R5505[†]
Reference	[10]	[11,12]	[13,14]	[15–17]	[18,19]	[20,21]	[22,23]
Experiment	Mark II	DELCO	ATLAS	*BABAR*	HyperK	LHCb	H1
system	TOF	Cherenkov	CAL	DIRC	Cherenkov	RICH	fiber Cal.
$\varnothing_C$ [mm]	44	114		29	508		25
$\square_C$ [mm^2]			15.8 × 15.8			23 × 23	
Stages	12	14	10	11	10	12	15
Window	Quartz	UV glass	Borosilic.	Borosilic.	Borosilic.	UV glass	UV glass
Photocath.	Bi-alkali	Bi-alkali	multi-alk.	Bi-alkali	Bi-alkali	Bi-alkali[‡]	Bi-alkali
Dynode	CuBe	GaP/BeO	Metal	SbCs	Box&grid or MCP	Metal	Fine mesh
V_{PM} [kV]	2.2	2.5	0.9	0.9–1.4	2.0	0.9	1.7–1.9
Anode	Single	Single	16	Single	Single	64	Single
ϵ_{QE}	0.25	0.27	0.164	0.28	0.3	0.35	0.23
Gain [10^6]	>30	350	1	17	10	1.2	0.5
τ_r [ns]	1.5	3.2	0.6	4.5	6.0	1.3	1.5
τ_t [ns]	28	–	7.4	33	95	5.8	5.6
$\Delta\tau_t$ [ns]	0.25	–	0.33	4	2.4	0.27	0.35

shows properties of some selected PMTs. Properties of many more PMTs are listed in reference [24].

Typically, the anode is connected to ground via a 50 Ω resistor that leads to the formation of 2 to 200 mV pulses. The pulse rise time (measured from 10 to 90%) is 0.5–6 ns. The total transit time inside the photomultiplier from cathode to anode is 5–100 ns. The time spread (also called time jitter) is caused by different times the photoelectrons need to travel from the photocathode to the anode or readout dynode. There are two effects that contribute:

- A variation in the photoelectron velocities since the kinetic energies typically range from 0 to 1.8 eV with a peak at 1.2 eV for photons with wavelengths of 400 < λ < 430 nm. Thus, for an electric-field strength of $|\vec{E}| = 150$ V/cm between photocathode and the first dynode the time difference for photoelectrons at rest and with kinetic energy $E_e = 1.2$ eV is $\delta t_1 = \sqrt{(2 m_e E_e)/(q_e |\vec{E}|)} \sim 0.2$ ns.
- Different path lengths that depend mainly on the diameter of the cathode. For a diameter of $\varnothing = 44$ mm, we get transition times of $\delta t_2 = 0.25 - 0.7$ ns.

The latter contribution is the ultimate limitation in the time resolution. The diameters of photocathodes typically range from 5–125 mm though the largest tube has

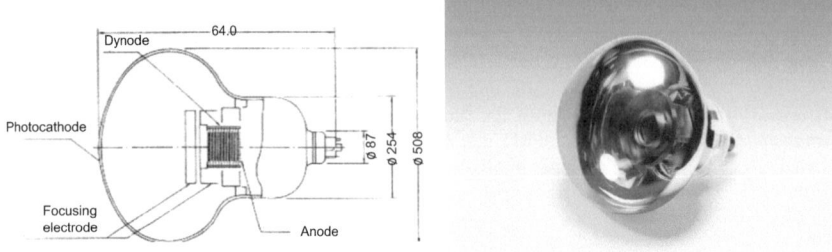

Fig. 8.10 Left: Schematic layout of a Hamamatsu 20-inch phototube. Reprinted with kind permission from [27], © 1983, Elsevier. All rights reserved. Right: Photograph of a 20-inch photomultiplier tube used in the T2K experiment in the water Cherenkov detector. Reprinted with kind permission from [28], © 2017, Hamamatsu Photonics. All rights reserved

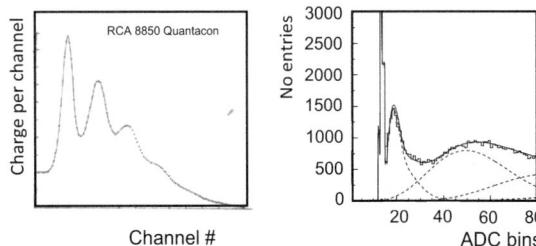

Fig. 8.11 Left: Single-photoelectron (pe) spectrum of the RCA 8850 photomultiplier. Note that the pedestal is suppressed and that visible peaks correspond to the 1-pe, 2-pe and 3-pe peaks. Reprinted with kind permission from [29], © 1973, Elsevier. All rights reserved. Right: Photoelectron spectrum of a Hamamatsu R5600 photomultiplier operated at 900 V that is used in the ATLAS Tile Calorimeter. Besides the pedestal only the single-photoelectron peak is resolved. The dashed curves show the single, two and three photoelectrons peaks, where the latter two are not resolved. Reprinted with kind permission from [30], © 1999, the ATLAS Collaboration. All rights reserved

a diameter of 508 mm. This is the R12860 PMT from Hamamatsu, which is used in water Cherenkov detectors like the HyperKamiokande detector. It is the successor of the R1449 PMT used in the SuperKamiokande experiment. The tube has a rise time of 6 ns and the variations in transit time are 2.4 ns. Figure 8.10 shows the layout (left) and a photograph (right) of a 20-inch photomultiplier. The smallest PMTs have a size of 1 mm × 3 mm. In high-energy physics experiments, typically 1-inch and 2-inch tubes are used.

Some photomultipliers have sufficient gain to observe single photoelectrons. Figure 8.11 (left) shows the photoelectron spectrum of an RCA 8850 photomultiplier. We clearly see the single-photoelectron, two-photoelectron and three-photoelectron peaks. The pedestal is suppressed. Figure 8.11 (right) shows the photoelectron spectrum of a Hamamatsu R5600 photomultiplier operated at a high voltage of 900 V. Besides the pedestal only the single-photoelectron peak is visible, while the 2-pe and 3-pe peaks are not resolved.

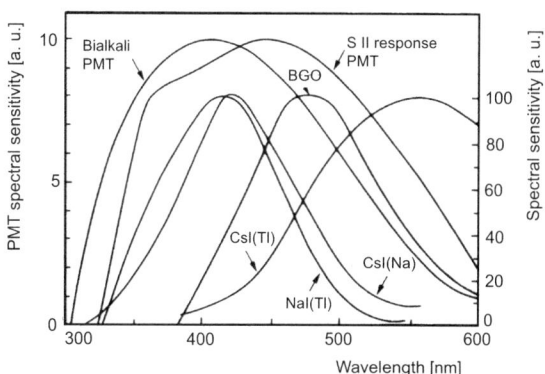

Fig. 8.12 Emission spectra of the inorganic crystals NaI(Tl), CsI(Na), BGO and CsI(Tl) and the spectral sensitivities of bi-alkali and S11 photocathodes. Reprinted from the Harshaw Scintillation Phosphor Catalog [31], © 1975, Harshaw Chemical Company. All rights reserved

Excellent time resolutions were achieved with the Phillips XP2020UR photomultiplier, yielding $\sigma_t = 90$ ps with a ^{60}Co source [32]. Photomultipliers are used further to read out the emission spectra of scintillating crystals. This works only if the emission spectrum of the crystals matches the spectral sensitivity of the PMT photocathode. Figure 8.12 shows the emission spectra of some commonly used scintillating crystals in comparison to the spectral sensitivity of different photocathodes as a function wavelength. The sensitivity of the bi-alkali photocathode matches the emission spectra of NaI(Tl) and CsI(Na) crystals rather well, whereas the sensitivity of the S11 photocathode matches the emission spectrum of BGO crystals well. For CsI(Tl,) only the lower part of the emission spectrum is covered.

8.1.4 Fine-Mesh Photomultiplier Tube

Standard photomultiplier tubes are very sensitive to magnetic fields. The electrons emitted from the photocathode are deflected in the magnetic field and miss hitting the first dynode. In addition, the electron paths in the dynode is deteriorated such that the multiplication is perturbed. So, standard PMTs do not work in magnetic fields. Since even the earth's magnetic field has an impact, PMTs are typically covered with a μ-metal shims, which shields off low magnetic fields from the photomultiplier. To enable operation in magnetic fields, the fine-mesh photomultiplier tube was developed. Figure 8.13 (left) shows a schematic layout of a fine-mesh PMT. Dynodes consist of a fine mesh instead of a metal sheet as illustrated in Fig. 8.13 (right). The mesh diameter is about 4 μm with a pitch of 15 μm. There are also coarse-mesh detectors with larger diameters and a pitch of 1 mm. The dynodes are closely stacked in 15 to 19 layers. For example, for 19 dynode layers the distance cathode-anode is 20 mm. When an electron hits the upper part of the fine-mesh dynode, secondary electrons are emitted upward from the fine-mesh dynode. These electrons return to the same dynode and pass through it to reach to next fine-mesh layer. This process is repeated at each layer up to the last dynode layer before they reach the anode. The amplification is 10^6–10^8. If a magnetic field is present with its axis parallel to

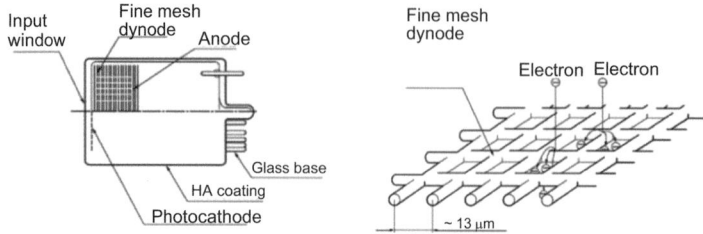

Fig. 8.13 Left: Schematic layout of a fine-mesh photomultiplier tube. Right: Details of a fine-mesh dynode. Reprinted with kind permission from [35], © 2011, Springer. All rights reserved

the PMT axis, the Lorentz force causes the electrons to spiral. The spiral diameter depends on the strength of the magnetic field. When the magnetic field is increased, the number of electrons that move upwards and return to the same dynode also increases leading to a gain reduction. The fine-mesh PMTs work in magnetic fields up to about 2 T.

The photodetector diameters range from 1-inch to 3-inch tubes. The typical operation is at a high voltage of $-2,000$ V, which can be increased to $-2,800$ V. The fine-mesh PMT accepts an order of magnitude higher anode peak currents than a linear-focussed PMT. Let us look at the performance of the Hamamatsu 1-inch PMT R3432 and the 2-inch PMT R2490 [33]. Figure 8.14 (left, center left) shows the gain versus magnetic-field strength for both PMTs, respectively. Both PMTs have a borosilicate window and a bi-alkali photocathode. The sensitive wavelength range is 300–650 nm. The maximum sensitivity is at 420 nm. The nominal operation voltage is 2,000 and 2,500 V, using 15 and 16 dynodes, respectively. Without a magnetic field the gains are 6×10^5 and 10^8. At 1 T the gains are reduced by factors of 10 and 100, respectively. At 2 T the gain loss is even more dramatic but the PMTs still work. Figure 8.14 (center right) shows the gain as a function of the angle between the photomultiplier axis and the direction of the magnetic field for the 1-inch tube for different magnetic-field strengths. For angles up to 35° the gain increases slowly

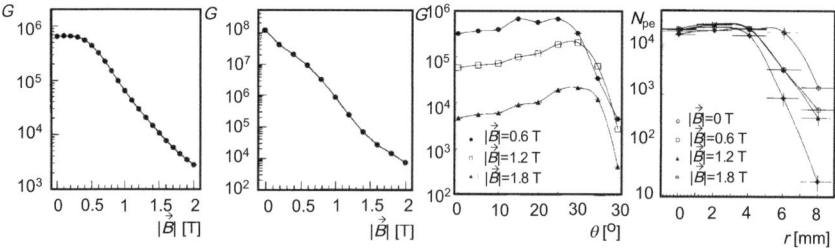

Fig. 8.14 Left: Gain versus magnetic-field strength for the R3432 PMT. Center left: Gain versus magnetic-field strength for the R2490 PMT. Center right: Gain versus the angle between the magnetic field direction and the photomultiplier axis for the R3432 PMT for magnetic fields of 0.6 T (solid points), 1.2 T (squares) and 1.8 T (triangles). Right: Number of photoelectrons versus the distance of the impact point from the photomultiplier axis for the R3432 PMT for no magnetic field (open circle), 0.6 T (squares), 1.2 T (triangles) and 1.8 T (diamonds). All reprinted with kind permission from [33], © 1994, Elsevier. All rights reserved

before dropping off rapidly at higher angles. Figure 8.14 (right) shows the number of photoelectrons for the 1-inch tube as a function of the distance of the impact point from the photomultiplier axis for different magnetic-field strengths. For distances below 4 mm the photomultiplier shows the same response independent of the magnetic fields up to 1.8 T. For larger distances the number photoelectrons starts to drop, which is faster for higher $|\vec{B}|$ fields. Similar results were achieved with the 2-inch PMT R5924-70 from Hamamatsu [34]. The time resolution does not change for $|\vec{B}|$ fields up to 1.8 T.

8.1.5 Photomultipliers in Subdetector Systems

Photomultipliers have been and are still used in large subdetector systems including time-of-flight counters, Cherenkov detectors and calorimeters. The Mark II experiment at SLAC used 96 XP2222 PMTs in a time-of-flight system (see Sect. 9.2) that consisted of 48 scintillator bars read out on both ends [36]. The average time resolution was $\sigma_t = 221$ ps. Another large system was the standoff box in the BABAR DIRC system (see Sect. 9.3.5) that housed 10,752 photomultiplier tubes (ETL model 9125), recording the Cherenkov photons emitted in quartz bars [37]. The PMTs having a diameter of 29 mm were closely packed and were placed inside a water volume. They had a bi-alkali photocathode that cuts off ultra violet light around 300 nm. The quantum efficiency in the UV region was about 25%. The PMT had an 11-stage amplification. The after-pulse rate was 1%. The dark current at a temperature of 20 °C was typically 0.2 nA. The operation voltage was $V_0 = 800 - 1400$ V, yielding a gain of $(1 - 2) \times 10^7$. Figure 8.15 (left) shows the arrangement of PMTs in the DIRC standoff box.

The water Cherenkov detector in the new HyperKamiokande neutrino experiment uses about 50,000 20-inch R12860 PMTs for reading out Cherenkov light produced in the water by a charged lepton [38]. The predecessor SuperKamiokande had 11,129 20-inch PMTs. The properties of the R12860 PMT are listed in Table 8.3. The single-photon detection efficiency is 30%. The new tube has improved the collection efficiency from 70 to 90%. Figure 8.15 (right) shows a photograph of the SuperKamiokande detector. The PMTs covering the walls are clearly visible. More details are given in Chap. 11. In the LHCb experiments, the Cherenkov photons produced in the two RICH systems are read out with multiple-anode PMTs. In RICH 1 and the inner region of RICH 2, 2,800 R13742 PMTs, a customized version of the R11265 1-inch tube, are used. The properties of this PMT are listed in Table 8.3. The outer part of RICH 2 is equipped with 400 2-inch R13743 photodetectors. These are a customized version of the R12699 PMT [39].

Another large system using photomultipliers is the ATLAS hadron calorimeter (TileCal, see Sect. 7.4.4.3), which uses Hamamatsu R5900 photomultiplier tubes for reading out the green Y11 fibers, collecting light from plastic scintillator tiles. The PMTs have a bi-alkali photocathode that is deposited on a borosilicate window, giving a spectral response of 300–900 nm. The PMT has a completely metallic housing and a 10-stage amplification. The maximum response is at 420 nm. The PMT has a square

Fig. 8.15 Left: Photograph of the photomultiplier tubes in the BABAR DIRC standoff box (personal photograph). Right: Photograph of the arrangement of photomultipliers in the water Cherenkov detector of SuperKamiokande [41]. Reprinted with kind permission from [41], © 1996, SuperKamiokande Collaboration. All rights reserved

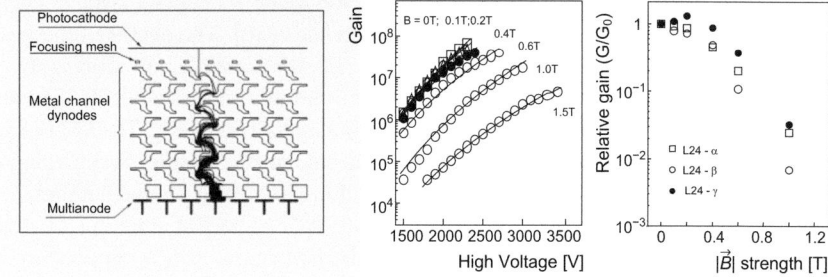

Fig. 8.16 Left: The layout of metal channel dynodes in a multi-anode photomultiplier tube [42]. Reprinted with kind permission from [42], © 2017, Hamamatsu Photonics. All rights reserved. Center: The gain of a multi-anode photomultiplier tube as a function of applied high voltage for different magnetic-field strengths from no field up to 1.5 T. Right: The relative gain as a function of the magnetic-field strength for three different photodetector configurations, Hamamatsu L14α, L14β and L14γ. The L14α PMT had 24 stages of fine mesh dynodes with 1 mm spacing in a square-shaped package (39 mm × 39 mm) and 24 anodes, each of which had a size of 26.5 mm × 0.8 mm. In the L14β PMT the spacing between the photocathode and first dynode was reduced from 2.5–3 mm to 1 mm. For the L14γ PMT the dynode mesh size was reduced from 2,000 lines per inch with a 12.5 μm pitch to 2,500 lines per inch with a 9 μm pitch. Both reprinted with kind permission from [43], © 2002, Elsevier. All rights reserved

shape with dimensions of 30 × 30 mm^2 of which 18 × 18 mm^2 is the sensitive area. With μ-metal shielding, magnetic fields up to 20 Gauss are shielded. A total of 43 fibers is bundled and coupled to one PMT, yielding a gain of 10^7 at 900 V operation voltage [40].

8.1.6 Multi-anode Photomultiplier Tube

In a multi-anode photomultiplier tube, the anode is segmented into several pads. The dynodes called metal channel dynodes are arranged such that a particular amplification path leads to one anode pad as illustrated in Fig. 8.16 (left). Conventional photodetectors (compared to Micro-Channel Plates) have at most an array of 8 × 8 anodes pads. Electrons emitted from the photocathode are focussed onto the first dynode stage. From there the path is fixed. This dynode arrangement has a very low cross talk[1] because the photoelectrons emitted from the photocathode are directed onto the first dynode by the focusing mesh and then move to the second dynode, third dynode, and so on to the last dynode and finally to the anode, while being multiplied with a minimum spatial spread in the secondary electron flow. The gain loss in magnetic fields is similar to that of fine-mesh photomultiplier tubes. Figure 8.16 (center) shows the gain of a multi-anode PMT as a function of applied high voltage for different magnetic fields. For low magnetic-field strength ($|\vec{B}| < 0.2$ T), a gain of few $\times 10^7$ is obtained with $V = 2, 200$ V. At a magnetic-field strength of 1.5 T, the voltage has to be increased to 3,500 V to achieve a gain of 3–4 $\times 10^6$. Figure 8.16 (right) shows the relative gain as a function of the magnetic-field strength for fixed high voltage. At a magnetic-field strength of 1 T, the relative gain has dropped to about 3%. The time resolution is unaffected by the magnetic field. For example, these photodetectors are used in the LHCb RICH systems [44].

8.1.7 Vacuum Phototriodes

A vacuum phototriode is a fine-mesh tube consisting of a photocathode, a mesh anode and a reflective dynode. The mesh anode is located 45 mm from the photocathode, while the dynode is separated from the mesh anode by 2–3 mm. Typically, the photocathode is grounded, the anode is operated at ≈1 kV and the dynode at ≈800 V. A large fraction of the photoelectrons liberated from the photocathode pass through the anode mesh and hit the dynode, where they produce secondary electrons. With a high-gain dynode, the secondary emission factor can be as high as 20. The secondary electrons are attracted to the anode mesh, where a substantial fraction is captured. In the absence of a magnetic field, the effective gain of the vacuum phototriode is around 12. This is about halved in a 4 T magnetic field. The photodetector is radiation hard and, therefore, very suitable for regions near the beam in LHC experiments. For example, the CMS experiment uses vacuum phototriodes (50 mm long and 22 mm in diameter) in the endcap electromagnetic calorimeter to read out the lead tungstate crystals [45]. Furthermore, vacuum phototriodes with a diameter of 75 mm had been used in the OPAL experiment to read out the lead glass block in the endcap electromagnetic calorimeter [46].

---

[1] We speak about cross talk if a signal produced in one channel also creates a signal in a neighboring channel.

8.2 Micro-channel Plates

A micro-channel plate (MCP) consists of an array of 10^4–10^7 parallel channels with diameters of 6–25 μm and a length that is 40 to 100 times larger [47]. Figure 8.17 (left) shows the layout of a micro-channel plate. The channels are holes in Pb-glass. The channel wall is covered with a highly resistive coating that permits the application of a constant potential gradient between the two ends of the channel. The total resistance between the two electrodes is 1 GΩ. The resistive layer is covered with a film that has a high efficiency for secondary electron emission. Thus, the channel coating acts like a continuous dynode. The MCP is operated at high vacuum ($\sim 10^{-6}$ Torr). Figure 8.17 (right) illustrates the electron multiplication along a micro-channel. To prevent positive ions to be accelerated in the reverse direction, the electron channels are shaped such that the ions hit the wall before gaining sufficient kinetic energy to produce secondary electrons. In addition, an aluminum film is placed between the photocathode and the first micro-channel plate. To increase the gain, multiple micro-channel plates can be placed in sequence. The gain of a micro-channel plate array is 10^4–10^7. The upper value is caused by saturation effects. Figure 8.18 (left) shows the setup of two coupled micro-channel plate detectors in the so-called "Chevron" arrangement. The two micro-channel detectors are placed under an angle of $15°$ with respect to each other to limit the ion back-flow. The anodes can be segmented into 2×2, 4×4, 8×8 and 64×64 arrays. Figure 8.18 (right) shows a photograph of the new PHOTEK MCP MAPMT253 with 64×64 multi-channel anodes as an example.

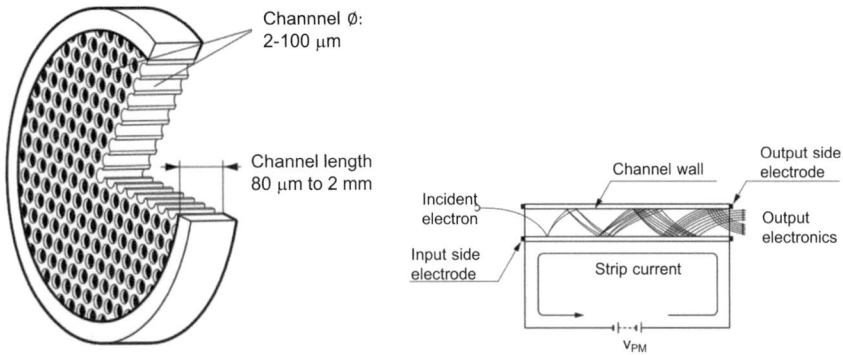

Fig. 8.17 Left: Schematic layout of a micro-channel plate detector. Right: Principle of the electron multiplication in a micro-channel plate. Both reprinted with kind permission from [48], © 2017, Del Mar Photonics Inc.. All rights reserved

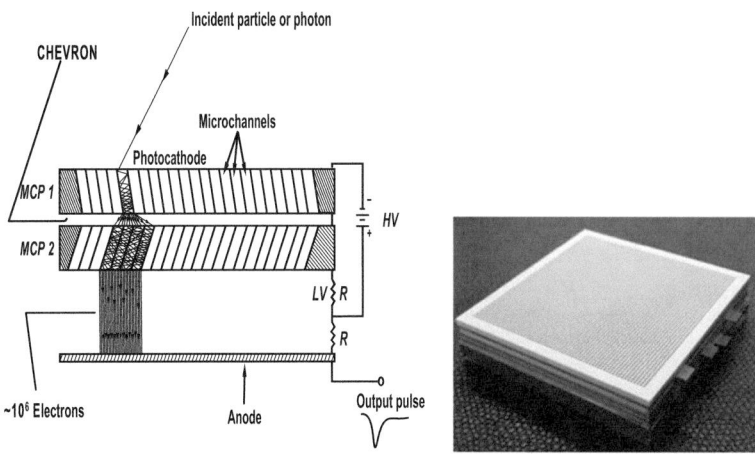

Fig. 8.18 Schematic layout of the two-layer MCP-PMT Chevron design. Reprinted with kind permission from [47], © 1979, Elsevier. All rights reserved. Right: Photograph of the 64 × 64 pixel PHOTEK MAPMT253. Reprinted under CC-BY-4.0 Licence from [49], © 2021, LHCb Collaboration

8.2.1 Detection Properties

The detection efficiency is a product of three components, the quantum efficiency ϵ_{QM}, the collection efficiency ϵ_c and the geometric efficiency ϵ_{geo},

$$\epsilon_{det} = \epsilon_{QM} \cdot \epsilon_c \cdot \epsilon_{geo}. \tag{8.10}$$

The collection efficiency is the ratio of photoelectrons emitted from the photocathode that reach the anode to all photoelectrons produced in the photocathode. In other words, it is the observed charge at the anode divided by the gain and the charge produced in the photocathode. Some photoelectrons are lost due to recoil from the MCP-PMT entrance. In newer devices most of them are recaptured such that collection efficiencies reach 90–95%. The geometric efficiency is the ratio of active area to total area. Typical values range between between 61 and 81%. In some newer MCP-PMTs the three efficiencies have been optimized to achieve a detection efficiency of ∼30%.

The gain depends on the thickness of the secondary electron emission layer and on the coating material. The gain is about a factor of two higher for MgO than for Al_2O_3, increasing with the coating thickness as depicted in Fig. 8.19 (left). Typical coating thicknesses are 10–20 nm. The gain is simply given by

$$G_{MCP} = \delta_{se}^n, \tag{8.11}$$

where δ_{se} is the number of secondary electrons that are produced by one photoelectron and n is the number of stages. Assuming that the secondary electrons are emitted

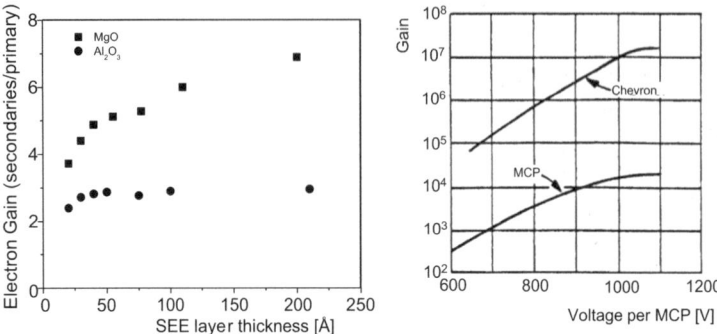

Fig. 8.19 Left: Gain as a function of the secondary electron emission (SEE) layer thickness for MgO and Al$_2$O$_3$. Reprinted with kind permission from [51], © 2012, Elsevier. All rights reserved. Right: Measured gain of a single-stage straight-channel MCP-PMT in comparison to that of a Chevron MCP-PMT as a function of high voltage. Reprinted with kind permission from [47], © 1979, Elsevier. All rights reserved

normal to the wall, we can express the gain of a single-stage MCP-PMT as a function of the channel voltage by V [47,50],

$$G_{\text{MCP}} = \left(\frac{c_m V}{4\alpha_{\text{ld}}^2 V_0}\right)^{4\alpha_{\text{ld}}^2 \frac{V_0}{V}}, \qquad (8.12)$$

where α_{ld} is the ratio of the detector length to its diameter, $c_m \sim 0.2$ is a proportionality constant and $q_e V_0$ is the initial energy of an emitted secondary electron, typically 1 eV. With increasing voltage the secondary electron yield also increases. At the same time the number of collisions within a channel must decrease yielding a maximum gain at voltage V_{max}. Note that the calculation is somewhat idealistic since the emission of secondary electrons under angles that are not orthogonal to the walls is neglected. Figure 8.19 (right) shows the measured gain as a function of applied high voltage per stage. The curves rise exponentially before leveling off around 900–1,000 V. For a single-stage MCP-PMT, the typical gain is 10^3–10^4, while for a two-stage CHEVRON device it is 10^6–10^7. The calculated gain in (8.12) plotted as a function of α_{ld} also has a maximum whose position increases with high voltage.

Emission of thermal electrons from the photocathode produces a so-called dark rate. Though it is temperature dependent, it is typically rather small at room temperature. Figure 8.20 (left) shows the time distribution of a PHOTONIS MCP-PMT in the time window $-10\,\mu\text{s} < t < 1\,\mu\text{s}$. Besides the prominent peak from photons of a laser pulse at 100 ns a small constant distribution of dark hits (200 events per bin) is visible. This produces a dark current of 2–5 nA in comparison to the signal current of a few μA. If we zoom in to the time interval of 80–580 ns, we notice additional peaks as shown in Fig. 8.20 (center). This is called after-pulsing. It is caused by feed-back ions of the residual gas that hit the photocathode and release an electron, which is amplified in the MCP-PMT. The after-pulse hits typically arrive later than the signal. To quantify the after-pulse rate, we determine the ratio of hits in the time window 115–550 and 98–115 ns. The after-pulse rate is typically 1% or less. By zooming

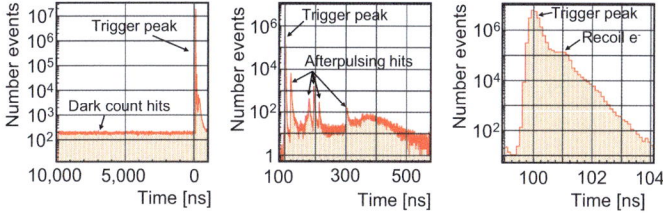

Fig. 8.20 Left: Time distribution of a PHOTONIS XP85012 MCP-PMT in the region $-10\,\mu s$ to $1\,\mu s$ showing dark counts before the signal peak from the laser pulse. The dark hits are uniform. Center: Same time distribution zoomed to the time interval of 80–580 ns. Besides the signal peak several after-pulses are visible. Right: Same time distribution zoomed further to time interval of 98–108 ns. Besides the signal peak a shoulder from recoil electrons is visible. All reprinted with kind permission from [52], © 2018, IOP Publishing. All rights reserved

into the time distribution further to the time interval 98–108 ns (see Fig. 8.20 right), we notice a shoulder on the signal peak. This is caused by electrons that recoil from the input surface of the first MCP-PMT layer. The scattered electrons produce a later and spatially shifted avalanche when they reach the MCP-PMT. For single-channel MCP-PMTs the anode collects the avalanche charge. A certain number of micro-channels is matched to one anode pad. However, the charge distribution may have a larger spread such that charge is recorded in neighboring anode pads. This is called cross talk.

8.2.2 Timing Measurements

Micro-channel plates have been designed for a rapid transit time and in turn for small time jitter. So time resolution is an important property. There is a crude estimate for the time resolution [53],

$$\sigma_{\text{time}} = \frac{\sigma_{\text{noise}}}{(dS/dt)_{\text{threshold}}} \simeq \frac{t_{\text{rise}}}{S/N}, \quad (8.13)$$

where S is the signal amplitude, $N = \sigma_{\text{noise}}$ is the noise contribution and t_{rise} is the signal rise time. So, for a typical $t_{\text{rise}} = 200$ ps and a $S/N = 10$ we expect a time resolution of $\sigma_{\text{time}} = 20$ ps. To achieve the same time resolution for a slower detector having a rise time of 2 ns we need $S/N = 100$. The observed time resolution of a detector has many contributions,

$$\sigma_{\text{tot}} = \sigma_{\text{el}} \oplus \frac{\sigma_{\text{chr}}}{\sqrt{N_{pe}}} \oplus \frac{\sigma_{\text{TTS}}}{\sqrt{N_{pe}}} \oplus \sigma_{\text{track}} \oplus \sigma_{t_0} \oplus \ldots, \quad (8.14)$$

where σ_{el} is a contribution from electronics, σ_{chr} is an error due chromatic effects, σ_{TTS} is transition time spread, σ_{track} is a tracking error contribution and σ_{t_0} is the uncertainty in the start time. There may be other contributions like time walk, cross-talk effects in multi-pixel detectors, baseline ringing in multi-pixel detectors affecting later arriving pulses, charge sharing in multi-pixel detectors, etc. For MCP-PMTs the transition time spread is the most interesting contribution. For a single photoelectron the time resolution of the MCP-PMT is given by the spread in transit time. If several photoelectrons are detected the time resolution improves as $\sigma_{\text{TTS}}/\sqrt{N_{pe}}$.

8.2.3 Examples of MCP-PMTs

One of the first devices was the F4129 by ITT [54]. The photodetector consisted of three micro-channel plates in cascade. Each plate had 500 μm long channels with a diameter of 12 μm. The S-10 photocathode had a diameter of 18 mm. Each micro-channel plate housed 2.25×10^6 channels. Here, the aluminum film between the photocathode and the first micro-channel plate was 7 nm thick. Operating at a high voltage of 2.5 kV with a ∼20% quantum efficiency, the micro-channel plate had a gain of 1.6×10^6. The signal had a rise time of 0.35 ns and a transition time of 2.5 ns. The time jitter was $\sigma_{\text{TTS}} = 53$ ps.

Another early device was the Hamamatsu micro-channel plate R1564U, which consisted of two micro-channel plates placed in series [55,56]. The micro-channels of the two detectors were arranged under an angle of 15° to prevent backward drifts of positive ions. In addition, a 13 nm thick aluminum film was placed between the photocathode and the first micro-channel plate. The photocathode had a diameter of 18 mm. Each micro-channel had a diameter of 12 μm and a length of ≈500 μm. Each stage housed 2.25×10^6 channels, which were operated at a high voltage of 2.9 kV yielding a gain of 1×10^6 and a quantum efficiency of 15%. The rise time was 0.24 ns and the transition time was 0.18 ns with a 85 ps time jitter. Hamamatsu also developed three-stage MCP-PMTs with channel diameters of 12 and 25 μm that had a gain of $G_{\text{MCP}} = 10^7$. Since then many MCP-PMTs have been built by different companies. Table 8.4 summarizes the properties of nine different micro-channel plate PMTs from different manufacturers as examples.

8.2.4 Lifetime Issues and Performance in Magnetic Fields

The first MCP-PMTs had a problem with the lifetime, mainly caused by the positive-ion feedback. Ions may be produced in collisions of secondary electrons with the residual gas or MCP-PMT material. These ions are accelerated backwards into the high electric field. They could damage the photocathode leading to a deterioration of the quantum efficiency. To diminish the effect of quantum efficiency deterioration, the micro-channels are placed under an angle so that the ions hit the wall. In addition, three steps in the MCP-PMT production were applied. First, a ceramic block and an aluminum layer were introduced to block gas and ions from reaching the photocathode. This was already done in the MCP-PMTs. We call these conventional MCP-PMTs. Second, an ALD coating was added to suppress outgassing from the MCP-PMT. Such MCP-PMTs are called ALD MCP-PMTs. In addition, one can also fine-tune the secondary electron emission and increase it. Third, some pro-

8.2 Micro-channel Plates

Table 8.4 Properties of some Microchannel Plate detectors listing the detector size (disk diameter/square detector area), sensitive area (disk diameter/square detector area), photocathode material, quantum efficiency ϵ_{QE}, collection efficiency ϵ_c, number of stages, channel diameter, length over diameter ratio α_{ld}, typical voltage V_{MCP} between anode and cathode, anode type (single or multiple), anode configuration, gain, rise time t_{rise}, transition time resolution σ_{TTS} and usage of an atomic layer deposition (ALD) film. [†]This refers to the Microchannel Plate voltage. [‡]From reference [57]. [§]From reference [58]

MCP-PMT	ITT	HPK	HPK	BINP	Burle	HPK	HPK	Photonis	Photek
Type	F4129	R1564U	R3809U-50-11X	MCP-PMT	85001 501	R10754 X	R13266 -07-M64	XP85112	MAPMT 253
Reference	[59]	[59,60]	[61,62]	[62, 63]	[62,64]	[65]	[57,66]	[67]	[68]
$\varnothing_{MCP}/a_{MCP}$ [mm]/[mm×mm]			45	31	71 × 71	28 × 28	61 × 61	59 × 59	60 × 60
$\varnothing_{sen}/a_{sen}$ [mm]/[mm×mm]	18	18	11	18	50 × 50	22 × 22	51 × 51	53 × 53	53 × 53
Photo-cathode	S20	Bi-alkali	Multi-alkali	Multi-alkali	Bi-alkali	Multi-alkali	Multi-alkali	Bi-alkali	Bi-alkali
ϵ_{QE}	0.2	0.15	0.26	0.18	0.24	0.28	0.18–0.24	0.22	0.21
ϵ_c			38[‡]	45[‡]	29[‡]	76 ± 8[§]	60[‡]	63 ± 6[§]	83 ± 8[§]
# stages	3	2	2	2	2	2	2	2	2
$\varnothing_{ch}$ [μm]	12	12	6	8	25	10	10	10	15
α_{ld}	40	40	40	40	40	40	40	60	
V_{MCP} [kV]	2.5[†]	3.4	3.6	3.2	2.5	2.2	2.4	2.03	2.7
Anode	Single	Single	Single	Single	2 × 2	4 × 4	8 × 8	8 × 8	64 × 64
Gain [10^6]	1.6	0.5	0.2	1.0	0.6	1.0	1.0	1.0	1.0
t_{rise} [ns]	0.35	0.27	0.16	0.3	0.3	0.195	0.24	0.6	<0.175
σ_{TTS}^{rms} [ps]	85	38	30	30	60	31	30	120	<40
ALD film	0	0	1	0	0	1	0–2	0–2	1

cesses were applied in the production to reduce the residual gas on the MCP-PMT. Such devices are called life-extended ALD MCP-PMTs. Figure 8.21 (left) shows the relative quantum efficiency as a function of accumulated output charge density for a conventional, an ALD and a life-extended ALD MCP-PMT. The output charge density, where the relative quantum efficiency drops to 80%, is called "lifetime" of the MCP-PMT. Figure 8.21 (right) shows the measured output charge densities for several conventional, ALD and life-extended ALD MCP-PMTs. Conventional MCP-PMTs have an average output charge density of 1.1 C/cm^2, where the maximum value lies around 1.8 C/cm^2. For ALD MCP-PMTs the average output charge density is 10.5 C/cm^2. However, there are two detectors that reach 16 C/cm^2 and 26 C/cm^2. The measured output charge densities for life-extended ALD MCP-PMTs lie above 16 C/cm^2, where the maximum value is about 33 C/cm^2.

The MCP-PMT gain is affected by the magnetic field [69]. Figure 8.22 shows the gain of the MCP-PMT R10754 as a function of the magnetic-field strength for four tilt angles (left) and four rotation angles (right). For different tilt angles the gain first increases with $|\vec{B}|$ producing a peak around 0.5–0.6 T, where the gain is about twice as high as that at $|\vec{B}| = 0$ T. At higher values of $|\vec{B}|$ the gain drops again. For smaller

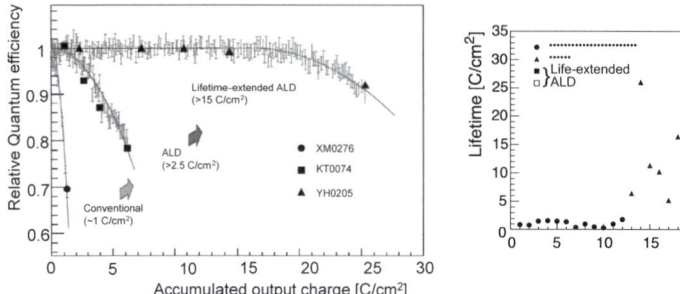

Fig. 8.21 Left: The relative quantum efficiency as a function of accumulated output charge for conventional MCP-PMTs (solid points), atomic layer deposition MCP-PMTs (squares) and life-extended ALD MCP-PMTs (triangles). Right: Lifetime of 12 conventional (points), 8 ALD (triangles) and 8 life-extended ALD (squares) MCP-PMT samples. Solid squares show measured lifetimes while open squares show lower bounds on the lifetime. Both reprinted with kind permission from [70], © 2017, Elsevier. All rights reserved

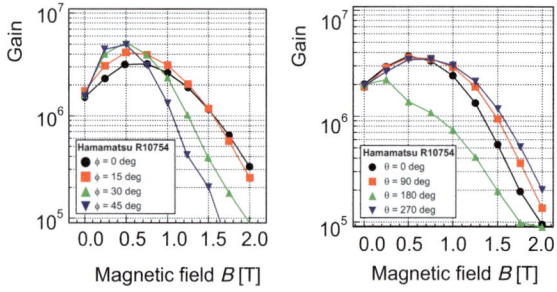

Fig. 8.22 Left: Gain as a function of magnetic-field strength of the Hamamatsu MCP-PMT R10754 for a rotation angle of 90° and different tilt angles, 0° (points), 15° (squares), 30° (upward triangles) and 45° (downward triangles). Right: Gain as a function of magnetic-field strength of the Hamamatsu MCP-PMT R10754 for a tilt angle of 15° and different rotation angles, 0° (points), 90° (squares), 180° upward triangles) and 270° (downward triangles). Both reprinted with kind permission from [69], © 2011, Elsevier. All rights reserved

tilt angles ($\leq 15°$) the gain at $|\vec{B}| = 2$ T is about an order of magnitude lower than that at $|\vec{B}| = 0$ T. A similar magnetic-field dependence is seen for different rotation angles. Except for 180°, the gain exhibits a similar magnetic-field dependence as that for tilt angles. So, for this MCP-PMT operation under magnetic fields up to 2 T looks fine. Other MCP-PMTs show similar behavior.

8.2.5 Applications of MCP-PMTs

Micro-Channel Plates are used in time-of-flight and Cherenkov detectors of new and upgraded experiments. For example, in the Belle II experiment the time-of-propagation system, called TOP counter, uses MPC-PMTs to record Cherenkov photons produced by charged tracks in quartz bars with excellent time resolution to separate kaons from pions for momenta up to 2 GeV [71] (see Sect. 9.3.6). Some other experiments that plan to use MCP-PMTs are the LHCb TORCH detector [39] (see Sect. 9.3.6), Panda [72], the mini TimeCube neutrino experiment [73] and the ALICE experiment [74]. In the JUNO neutrino experiment, 20,000 20-inch MCP PMTs are considered. Prototypes were designed at IHEP [75]. The photodetectors have a quantum efficiency of 26%, a rise time of 1.4 ns and a ∼5 ns transit time spread. The gain is 1×10^7 and the detection efficiency is improved to about 30% [76]. We will discuss the layout of the TOP counter and TORCH detector in more detail in the next chapter.

8.3 Silicon-Based Photodetectors

Silicon-based photodetectors have become important devices in high-energy physics experiments due to compact arrangements, no sensitivity to magnetic fields and good timing properties. This section covers the PIN photodiode, the avalanche photodiode, the silicon photomultiplier and the hybrid phototube. We follow the discussion by Sze [77].

8.3.1 PIN Diodes

A PIN photodiode consists of a highly doped n-type silicon and highly doped p-type silicon that are coupled via a wide intrinsic semiconductor region of high-resistivity low-doped n-type silicon as depicted in Fig. 8.23 (left). Figure 8.23 shows the symbol for a diode circuit (left) and a photograph of a 10 mm × 10 mm sensor (right). Figure 8.24 (left) displays an equivalent circuit of a PIN photodiode. The PIN photodiode works exactly like a pn junction. It is operated in the reverse-bias mode. The depletion region is larger than that of a pn junction and its temperature

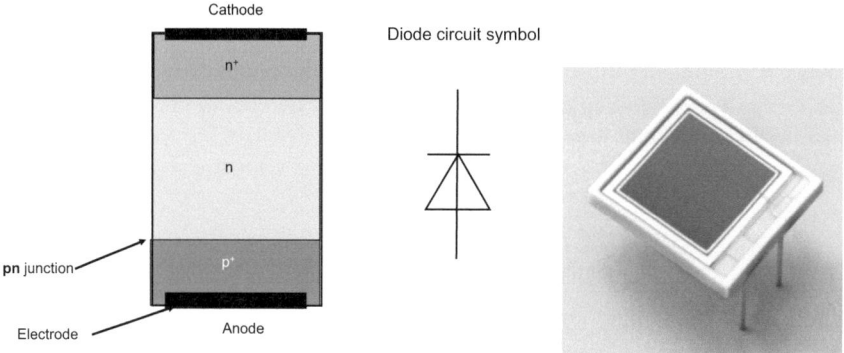

Fig. 8.23 Left: Schematic layout of a PIN photodiode. Photons are absorbed in the n (intrinsic) region. Center: Symbol for a diode circuit. Right: Photograph of a PIN photodiode (personal photograph)

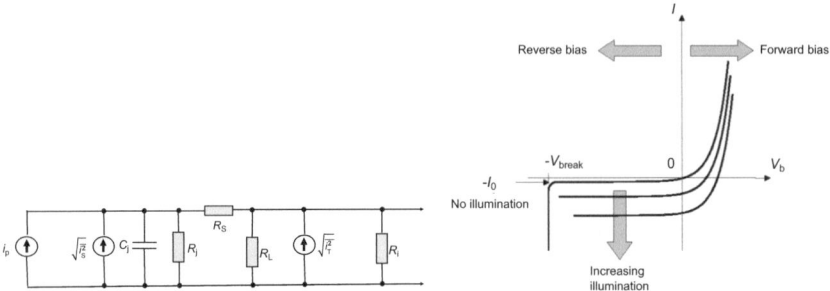

Fig. 8.24 Left: Equivalent circuit of a PIN photodiode. The signal current i_p, the junction capacitance C_j, the junction resistance R_j, the series resistance R_s and the average shot noise $\sqrt{\langle i_s^2 \rangle}$ are associated with the photodiode. The other quantities are the external load resistor R_L, the input resistor R_i and the average thermal noise $\sqrt{\langle i_T^2 \rangle}$. Right: The I-V curve of a PIN photodiode. Reprinted with kind permission from [78], © 2010, Springer. All rights reserved

sensitivity is lower. Most of the photons that enter the PIN photodiode from the p side or i side are absorbed in the intrinsic region. An electron in the valence band is lifted into the conduction band. The electron is recorded on the anode. The photodiode has no intrinsic gain. To observe sizable signals many photons have to impinge simultaneously. The noise is sufficiently low to permit operation at room temperature. Figure 8.24 (right) depicts the I-V curve of a PIN photodiode. In the forward bias direction the current rises steeply with the bias voltage V_b. In the reverse direction the current remains very small until the break-down voltage V_{break} is reached after which it grows extremely fast.

8.3.1.1 General Considerations

The performance of a photodetector is characterized by the quantum efficiency, response time and sensitivity.[2] Under illumination the sensor generates a number of carriers n_0 in a unit volume by a given photon flux at time $t = 0$. At a later time the number of carriers is reduced by recombinations yielding $n_{\text{carrier}}(t) = n_0 \exp(-t/\tau_c)$, where τ_c is the carrier life time. For a steady flow of photons with energy $\hbar\omega$ impinging uniformly on the surface of the photodiode (PD) with area $A_{\text{pin}} = \ell_{\text{pin}} w_{\text{pin}}$, length ℓ_{pin} and width w_{pin}, the total number of photons arriving at the PD surface per unit time is given by [77],

$$N_{\text{ph}} = \frac{P_{\text{opt}}}{\hbar\omega}, \qquad (8.15)$$

where P_{opt} is the incident optical power. For photons in the 800 nm to 400 nm range, their energy ranges from 1.55 to 3.1 eV, respectively. For VUV photons with

---

[2] For avalanche photodiodes, silicon photomultipliers and hybrid phototubes the gain is another characteristic quantity.

8.3 Silicon-Based Photodetectors

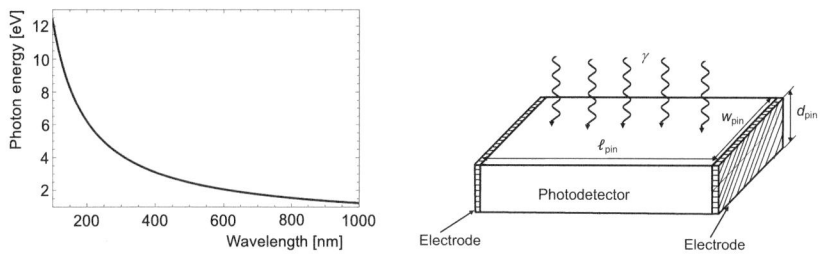

Fig. 8.25 Left: Photon energy as a function of wavelength for photons in the UV, optical and infrared region. Right: Schematic diagram for the illumination of a silicon photodetector

wavelengths of <200 nm, the photon energy is 6.2 eV. For comparison, the thermal energy at room temperature (20 °C) is 0.025 eV. Figure 8.25 (left) shows the photon energy as a function of wavelength in the 100 to 1000 nm range.

Figure 8.25 (right) illustrates the illumination of a PIN photodiode. At steady state the carrier generation rate has to be equal to the recombination rate. If the sensor thickness d_pin is much larger than the light penetration depth, the total steady-state generation rate is

$$R_{cg} = \frac{N_c}{\tau_c} = \frac{\epsilon_\text{QE}(P_\text{opt}/\hbar\omega)}{\ell_\text{pin} w_\text{pin} d_\text{pin}}, \tag{8.16}$$

where ϵ_QE is the quantum efficiency and N_c is the carrier density.

The current between the anode and cathode is

$$I_p = \sigma_\text{con}|\vec{E}|w_\text{pin}d_\text{pin} = q_e\mu_c N_c|\vec{E}|w_\text{pin}d_\text{pin} = q_e N_c|\vec{v}_d^e|w_\text{pin}d_\text{pin}, \tag{8.17}$$

where σ_con is the conductivity, $|\vec{E}|$ is the magnitude of the electric field, μ_c is the carrier mobility and $|\vec{v}_d^e|$ is the carrier drift velocity. Since

$$N_c = \frac{\epsilon_\text{QE}}{w_\text{pin}\ell_\text{pin}d_\text{pin}}\frac{P_\text{opt}}{\hbar\omega}\tau_c, \tag{8.18}$$

we obtain

$$I_p = q_e\epsilon_\text{QE}\left(\frac{P_\text{opt}}{\hbar\omega}\right)\left(\frac{v_d^e \tau_c}{\ell_\text{pin}}\right). \tag{8.19}$$

The primary photon current is

$$I_\text{ph} = q_e\epsilon_\text{QE}\frac{P_\text{opt}}{\hbar\omega}. \tag{8.20}$$

For a CsI(Tl) crystal, the photon yield is 54,000 photons per MeV of which about 34,500 are impinging on the photodiode within 680 ns. For a quantum efficiency of $\epsilon_\text{QE} = 0.84$, we calculate a signal current of $I_\text{ph} = 6.8$ nA.

The gain G_{PIN} is the ratio of the two currents

$$G_{PIN} = \frac{I_p}{I_{ph}} = \frac{|\vec{v}_d^e|\tau_c}{\ell_{pin}} = \frac{\tau_c}{t_r}, \qquad (8.21)$$

where $t_r = \ell_{pin}/|\vec{v}_d^e|$. Thus, the gain depends on the ratio of the carrier lifetime and transition time. For a PIN photodiode they are equal, yielding $G_{PIN} = 1$. If the lifetime is rather long, the gain can become large. However, for a PIN photodiode $t_r = \tau_c$.

An intensity-modulated optical signal is given by

$$P(\omega_{mod}) = P_{opt}(1 + m_{mod} \exp(i\omega_{mod} t)), \qquad (8.22)$$

where m_{mod} is the modulation index and ω_{mod} is the modulation frequency. For the modulated signal, the *rms* optical power is $m_{mod} P_{opt}/\sqrt{2}$ and the *rms* signal current can be expressed by

$$i_p \approx \frac{q_e m_{mod} \epsilon_{QE}}{\sqrt{2}} \frac{P_{opt}}{\hbar\omega} \left(\frac{\tau_c}{t_r}\right) \frac{1}{\sqrt{(1 + \omega_{mod}^2 \tau_c^2)}}. \qquad (8.23)$$

Since for a PIN photodiode $\omega_{mod} \tau_c \ll 1$, the *rms* signal current becomes

$$i_p = \frac{q_e \epsilon_{QE} m_{mod}}{\sqrt{2}} \frac{P_{opt}}{\hbar\omega}. \qquad (8.24)$$

The average shot noise is given by

$$\langle i_s^2 \rangle = 2q_e(I_p + I_b + I_d)\Delta B, \qquad (8.25)$$

where I_b is the current resulting from background radiation, I_d is the dark current originating from thermal generation of *e-h* pairs in the depletion region and ΔB is the bandwidth. The average thermal noise is given by

$$\langle i_T^2 \rangle = 4k_B T \frac{1}{R_{eq}} \Delta B, \qquad (8.26)$$

where

$$\frac{1}{R_{eq}} = \frac{1}{R_j} + \frac{1}{R_L} + \frac{1}{R_i} \qquad (8.27)$$

and R_i, R_j and R_L are the parallel resistors shown in Fig. 8.24 (left).

The quantum efficiency can be parameterized by

$$\epsilon_{QE} = (1 - \epsilon_{ref})\left(1 - \frac{\exp[-\alpha_{abs} \cdot w_{pn}]}{1 + \alpha_{abs}\sqrt{D_p \tau_p}}\right), \qquad (8.28)$$

8.3 Silicon-Based Photodetectors

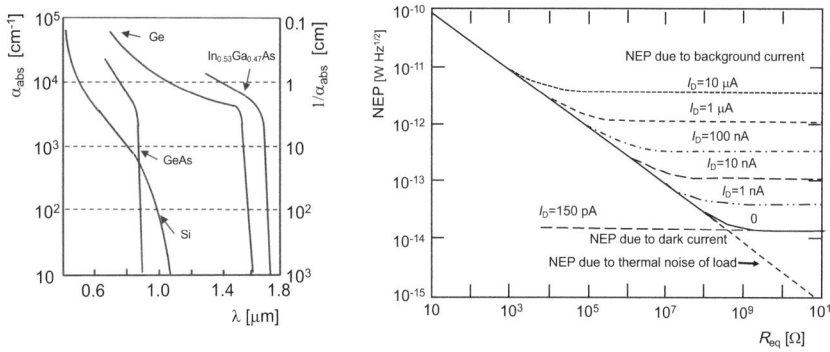

Fig. 8.26 Left: The absorption coefficient as a function of wavelength for different semiconductor materials [79]. Reprinted under CC-BY-4.0 Licence from [79], © 2018, Z. R. Szczepaniak. Right: The noise-equivalent power for a PIN photodiode as a function of R_{eq} at $T = 20°$ and a photon energy of 2 eV/cm^2 for a dark current of $I_d = 120$ nA, quantum efficiency of 0.75 and different background currents I_b (from zero nA to 10 μA).

where α_{abs} is the absorption coefficient, w_{pn} is the depletion layer width, ϵ_{ref} is the reflection coefficient, D_p is the diffusion coefficient for holes and τ_p is the lifetime of excess carriers. In silicon, the wavelength-dependent absorption coefficient ranges from $\alpha_{abs} = 10^6$ cm^{-1} at $\lambda = 300$ nm to 10 at $\lambda = 1100$ nm as displayed in Fig. 8.26 (left). So, let us look at an example. For $\epsilon_{ref} = 0.1$, $\alpha_{abs} = 100$ cm^{-1}, $D_p = 12$ cm^2/s, $\tau_p = 1$ ns and $w_{pn} = 300$ μm, we get a quantum efficiency of $\epsilon_{QE} = 0.85$. With increasing α_{abs} the quantum efficiency increases towards $1 - \epsilon_{ref}$.

8.3.1.2 Signal-to-Noise Ratio

For a 100% modulated signal with average power P_{opt}, we get a signal-to-noise ratio (S/N) of

$$(S/N)_{power} = \frac{i_p^2}{(\langle i_S^2 \rangle + \langle i_T^2 \rangle)} = \frac{\frac{1}{2}(q_e \epsilon_{QE} P_{opt}/\hbar\omega)^2}{2q_e(I_p + I_b + I_d)\Delta B + 4k_B T \Delta B/R_{eq}}. \quad (8.29)$$

The minimum optical power required to achieve a given S/N is

$$(P_{opt})_{min} = \frac{2\hbar\omega \Delta B}{\epsilon_{QE}} \left(\frac{S}{N}\right)\left\{1 + \left[1 + \frac{I_{eq}}{q_e \Delta B (S/N)}\right]^{1/2}\right\}, \quad (8.30)$$

where

$$I_{eq} = I_b + I_d + \frac{2k_B T}{q_e R_{eq}}. \quad (8.31)$$

In the limit $I_{eq}/(q_e \Delta B(S/N)) \ll 1$, the minimum optical power is given by the quantum noise associated with the optical signal. In the opposite limit $I_{eq}/(q_e \Delta B (S/N)) \gg 1$, the background radiation and/or thermal noise of the equivalent resistor become dominant. Under this condition the noise-equivalent power (NEP) is defined

Table 8.5 Properties of some silicon photodiodes (PIN) listing the sensitive area a_{sens}, reverse bias voltage V_{APD}, sensor capacitance at some voltage C_{APD}, dark current I_d, spectral range, wavelength for maximum quantum efficiency, radiant sensitivity, quantum efficiency ϵ_{QE}, noise-equivalent power (NEP), depletion width and applications

Property	HPK	HPK	HPK	Excelitas
Type	S1723-06	S2744-08	S3204-09	VTH2110
Reference	[81]	[82]	[83,84]	[85]
a_{sens} [mm^2]	10 × 10	10 × 20	28 × 28	5 × 5
V_{APD} [V]	≤50	≤100	≤100	≤100
C_{APD} [pF]	70 (30 V)	85 (70 V)	130 (70 V)	30 (20 V)
I_d [nA]	5 (30 V)	3 (70 V)	6 (70 V)	2 (20 V)
Range [nm]	320–1,060	340–1,100	340–1,100	400–1,100
Peak λ_{peak} [nm]	900	960	960	940
Radiant sensitivity [A/W]	0.6 (λ_{peak}) 0.4 (633 nm)	0.66 (λ_{peak}) 0.36 (540 nm)	0.66 (λ_{peak}) 0.41 (540 nm)	0.70 (λ_{peak}) 0.4 (620 nm)
ϵ_{QE} [%]		85 (540 nm)	85 (540 nm)	
NEP [10^{-13} W/Hz$^{1/2}$]	2 (30 V)	0.47 (70 V)	0.66 (70 V)	
Depletion width [μm]	300	300	300	>90
Application	CLEO II	BABAR, Belle, BES III	CsI R&D	CTA R&D

by the *rms* optical power $(P_{\text{opt}})_{\text{min}}$ for an area of 1 cm × 1 cm with $S/N = 1$ and $\Delta B = 1$ Hz yielding

$$NEP = \sqrt{2}\frac{\hbar\omega}{\epsilon_{\text{QE}}}\left(\frac{I_{\text{eq}}}{q_e}\right)^{1/2} \quad [\text{W}/\text{s}^{1/2}]. \tag{8.32}$$

To improve the sensitivity of a photodiode, we need to increase ϵ_{QE} and R_{eq}, while decreasing I_b and I_d. Figure 8.26 (right) shows the noise-equivalent power as a function of R_{eq} for a typical PIN photodiode [77]. From the figure it is evident that we must use high values of R_{eq} to achieve an NEP limited by dark current or by background current shot noise. For a background current of $I_b = 10$ μA, the NEP value levels off around 2×10^{-12}. This decreases to around 10^{-14} for zero background current. The dashed line shows the NEP value due to the thermal noise of the load. Another important quantity is the equivalent noise charge $\overline{ENC}$, which we introduced in (7.23).

8.3.1.3 Properties of PIN Diodes and Applications

One common application of PIN photodiodes is the readout of inorganic scintillating crystals in electromagnetic calorimeters. Table 8.5 lists properties of some selected silicon PIN diodes. As we saw in Sect. 7.2.2.1, the BABAR CsI calorimeter was read out with two PIN photodiodes that were coupled via a plastic plate to the CsI crystals [80]. The PIN photodiodes (S-2744-08 from Hamamatsu) have a capacitance of 105 pF at the depletion voltage of 70 V. The dark current is less than 5 nA. For

8.3 Silicon-Based Photodetectors

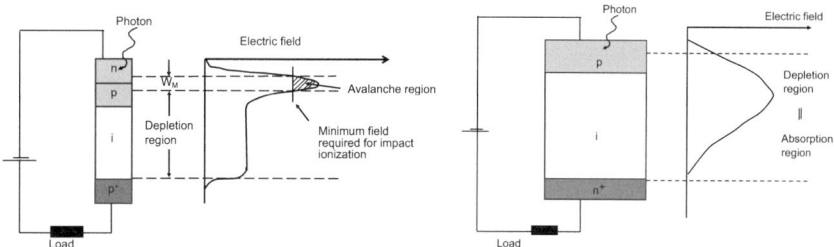

Fig. 8.27 Schematic layout of an APD (left) and a PIN diode (right) and electric fields across the sensors

the CsI(Tl) scintillation light, the quantum efficiency is 85%. The same PDs read out the CsI crystals of the Belle and BES III calorimeters, while S1723-06 PDs read out the CLEO II CsI crystals. In R&D studies of CsI(Tl) crystal for the BABAR electromagnetic calorimeter S3204-05 PDs were used. The VTH2110 PDs from Excelitas company were studied by CTA. Besides single sensors also arrays are produced.

8.3.2 Avalanche PhotoDiodes

An Avalanche Photodiode (APD) consists of a highly doped epitaxial pn junction that sits next to a nearly intrinsic region, a lightly doped p region, and a p^+ region. Figure 8.27 (left) shows a schematic view of an APD and the electric field configuration across the sensor, while Fig. 8.27 (right) displays that of a PIN photodiode for comparison. In the APD at the pn junction boundary, a metallurgical junction may be inserted. In this case the pn junction is moderately doped. Instead of the metallurgical junction also a highly-doped epitaxial p-type or n-type layer could be inserted. The sensor is operated in the reverse bias mode. The electric field is constant between the p^+ edge and the p layer of the pn junction, while it increases across the pn junction. The high-field region W_M is active over a typical range of 2 μm. The intrinsic region is the depletion region. In some devices, the region is not fully depleted. Typically, photons enter from the p^+ side. In the intrinsic region they produce electron-hole pairs. In the low-field regions, electrons and holes get separated. The electrons move towards the pn^+ layer. In the high electric field, the electrons gain kinetic energy, producing more electron-hole pairs. The gain depends on the applied reverse bias voltage. In the PIN photodiode, the electric field runs from the p side across the intrinsic region to the n^+ side.

In Fig. 8.27 (left) the photons enter from the n side. This type is called APD with reverse structure.[3] Besides silicon, APDs use other materials including ger-

---

[3] The APD used by CMS shown in Fig. 8.28 has standard structure, where photons enter from the p side.

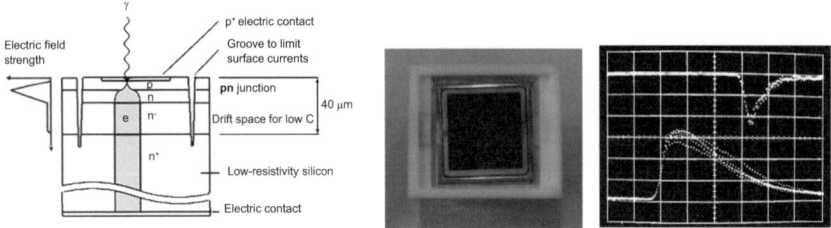

Fig. 8.28 Left: Schematic layout of the APD S8148 from Hamamatsu. Note that grooves surround the amplification region to reduce surface currents. The figure also illustrates the avalanche formation at the *pn* junction. Reprinted with kind permission from [91], © 2004, Elsevier. All rights reserved. Center: Photograph of the Hamamatsu APD S8148. Reprinted with kind permission from [92], © 2005, R. Scheuermann. All rights reserved. Right: Signal pulse of the Advanced Photonix APX-200-DUV APD (top pulse) in comparison to that of an XP2020 PMT (bottom pulse)

manium [86], InGaAs [87], Gallium nitride [88] and HgCdTe [89]. The latter two devices are used for operation with UV photons and infrared photons, respectively.

Figure 8.28 (left) shows the structure of the APD S8148 used in the CMS electromagnetic calorimeter. Here, an epitaxial layer is grown on a low-resistivity n^+-type silicon [90]. With appropriate ion implantation, the p^+-pn-n^- structure is produced. The *pn* junction depth is set at ≈5 μm to minimize the sensitivity to ionizing radiation and to maximize the absorption of the blue light from the crystal. The introduction of the n^- layer leads to an increase of the depletion region decreasing both the capacitance and the sensitivity of the gain to changes in the applied bias voltage. The 30 μm deep grooves introduced on both sides of the *pn* junction help to reduce the surface currents. This becomes particularly important after exposure to high radiation doses. As shown in the figure, the photon is converted to an *e-h* pair in the 2 μm thick p^{++} layer. The electrons drift in the 6 μm thick high electric-field region, where they are accelerated and multiplied. Then, they drift in the low doped n^- region towards the n^{++} layer, where they are collected. Figure 8.28 (center) shows a photograph of the S8148 APD. A typical APD signal is depicted in Fig. 8.28 (right). It has a fast rise time and a decay time that is much faster than that of a PMT signal.

8.3.2.1 Operation Principle of APDs

Figure 8.29 illustrates the principle of electron multiplication in an APD.

1. A photon impinges on the APD and creates an *e-h* pair in the depletion (B, A) layer by lifting an electron from the valence band into the conduction band. The hole (A) moves in the valence band.
2. In the high electric field, the electron (B) is accelerated gaining sufficient energy. At (C) it produces a new *e-h* pair (E, F) loosing the binding energy E_B (D).
3. The newly generated carriers drift in opposite directions (E, F) being accelerated in the electric field.
4. The accelerated hole may gain sufficient energy (G) to produce an *e-h* pair as well (H, I).

8.3 Silicon-Based Photodetectors

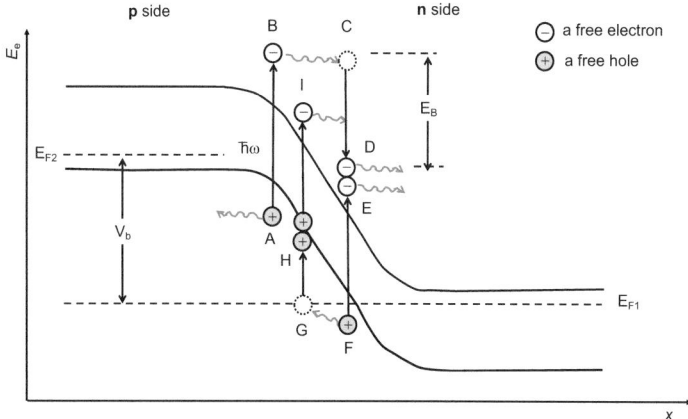

Fig. 8.29 Illustration of the charge multiplication in an APD. On the p^+ side, the Fermi energy E_{F2} and bands are shifted by the bias voltage V_b from the Fermi energy E_{F1} on the n side (see text for explanation)

These processes continue until holes and electrons leave the high-field region. The gain is strongly dependent on the bias voltage. To attain stable conditions, we typically operate APDs at gains below 100.

8.3.2.2 Signal-to-Noise Ratio

The equivalent circuit of an APD is similar to that of the PIN diode. The signal current, background current and dark current are multiplied by the avalanche gain G_{APD} [77]. The multiplied *rms* signal photocurrent becomes

$$i_p = q_e \epsilon_{QE} m_{\text{mod}} G_{APD} \frac{P_{\text{opt}}}{\sqrt{2}\hbar\omega}. \tag{8.33}$$

Similarly for the modulated optical signal, the rms signal current is given by (8.24) and the noise currents I_d and I_b are multiplied by the gain. The mean-square shot noise is also modified by the gain by,

$$\langle i_s^2 \rangle = 2q_e(I_p + I_b + I_d)\langle G_{APD}^2 F(G_{APD})\rangle \Delta B, \tag{8.34}$$

where the signal currents, background radiation current and dark current are the same as those for the PIN photodiode. Since $\langle G_{APD}^2 \rangle$ is the mean-square value of the APD internal gain and the noise factor is $F(G_{APD}) = \langle G_{APD}^2 \rangle / \langle G_{APD} \rangle^2$, which is a measure of the increase in the shot noise with respect to a noiseless multiplier. The thermal noise is the same as that of a photodiode (8.26).

For a 100% modulated signal with average power P_{opt}, we get a signal-to-noise ratio (S/N) of

$$(S/N)_{\text{power}} = \frac{\frac{1}{2}(q_e \epsilon_{QE} P_{\text{opt}}/\hbar\omega)^2 G_{APD}^2}{2q_e(I_p + I_b + I_d)\Delta B \cdot F(G_{APD})G_{APD}^2 + 4k_B T \Delta B/R_{\text{eq}}}. \tag{8.35}$$

The last term in the denominator is also called Johnson noise. For low voltage, the Johnson noise dominates. Thus, the S/N ratio increases as G_{APD}^2. Increasing V_b also increases the shot noise. Once the shot noise exceeds the Johnson noise, S/N starts to drop.

When the shot noise equals the Johnson noise, we obtain the minimum detectable power of

$$(P_{opt})_{min} = \frac{2\hbar\omega}{\epsilon_{QE}}\left(\frac{S}{N}\right)\left\{1 + \left[1 + \frac{I_{eq}}{q_e \Delta B \cdot F^2(G_{APD})(S/N)}\right]^{1/2}\right\}, \quad (8.36)$$

where

$$I_{eq} = (I_b + I_d)F(G_{APD}) + \frac{2k_B T}{q_e R_{eq} G_{APD}^2}. \quad (8.37)$$

In the limit $I_{eq}/(q_e \Delta B(S/N)) \ll 1$, the minimum optical power is given by the quantum noise associated with the optical signal. In the opposite limit $I_{eq}/(q_e \Delta B (S/N)) \gg 1$, the background radiation and/or thermal noise of the equivalent resistor becomes dominant. Under this condition the noise-equivalent power (NEP) is defined similarly as that in (8.32) with the additional factor $1/F(G_{APD})$ yielding

$$NEP = \sqrt{2}\frac{\hbar\omega}{\epsilon_{QE}}\left(\frac{I_{eq}}{q_e F^2(G_{APD})}\right)^{1/2} \quad [W/s^{1/2}]. \quad (8.38)$$

To improve the sensitivity of an avalanche photodiode, we need to increase ϵ_{QE} and R_{eq} while decreasing I_b and I_d. Figure 8.30 (left) shows the quantum efficiency of APDs as a function of the photon wavelength for non-irradiated and irradiated detectors. The quantum efficiency is rather high in the visible spectrum and does not change much after irradiation.

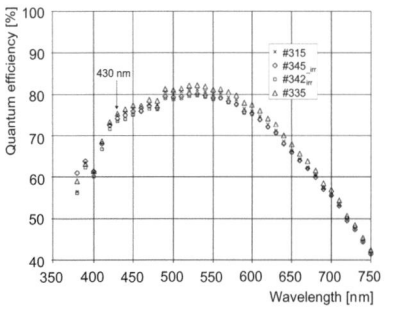
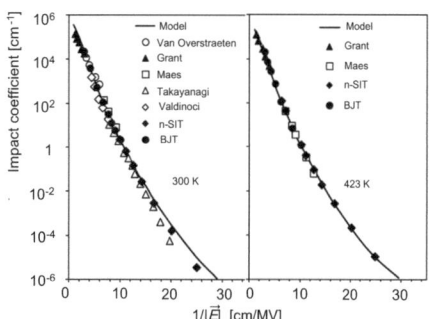

Fig. 8.30 Left: Quantum efficiency as a function of the photon wavelength for two unradiated and two irradiated sensors, where the radiation dose was $2 \times 10^{13} (1\ \text{MeV})\ n_{eq}/\text{cm}^2$. Reprinted with kind permission from [90], © 2000, Elsevier. All rights reserved. Right: Measured electron impact ionization coefficient as a function of $1/|\vec{E}|$ at 300 K (center) and 423 K (right) versus model prediction. Other data come from [94–98]. The solid line represents the new impact ionization model. Reprinted with kind permission from [93], © 2023, S. Reggiani. All rights reserved

8.3 Silicon-Based Photodetectors

8.3.2.3 APD Gain

The low-frequency avalanche gain for electrons is given by,

$$G_{APD} = \left(1 - \int_0^{w_{pn}} \alpha_n \exp\left[-\int_0^x (\alpha_n - \alpha_p) dx'\right] dx\right)^{-1}, \quad (8.39)$$

where w_{pn} is the width of the depletion layer and α_n and α_p are the electron and hole impact ionization coefficients, respectively. For position-independent impact ionization coefficients, the gain of electrons injected into the high $\vec{E}$ field region at $x = 0$ is

$$G_{APD} = \frac{(1 - k_{eff}) \exp(\alpha_n \cdot w_{pn}(1 - k_{eff}))}{1 - k_{eff} \exp(\alpha_n \cdot w_{pn}(1 - k_{eff}))}, \quad (8.40)$$

where $k_{eff} = \alpha_p / \alpha_n$. Figure 8.30 shows the electron impact ionization coefficient as a function of $1/|\vec{E}|$ for 300 K (left) and 423 K (right). For small electric field strengths, we get small α_n increasing exponentially with $|\vec{E}|$. For example, at 300 K $\alpha_n = 10^{-4}$ cm^{-1} for $|\vec{E}| = 44$ kV/cm, $\alpha_n = 1$ cm^{-1} for $|\vec{E}| = 83$ kV/cm and $\alpha_n = 10^3$ cm^{-1} for $|\vec{E}| = 200$ kV/cm [93]. The hole impact ionization coefficient is typically smaller. For equal ionization coefficients, the gain is

$$G_{APD} = \frac{1}{1 - \alpha_n w_{np}}. \quad (8.41)$$

We can also plot the gain as a function of bias voltage as depicted in Fig. 8.31 (left). For a bias voltage $V_b < 50$ V, the gain is one. Thus, a gain measurement at $V_b = 20$ V is used as a reference point. With increasing V_b the gain starts to increase rapidly. The avalanche multiplication depends also on the mean free path of the electrons between two collisions. Since λ_e depends on temperature, the gain changes with temperature. Figure 8.31 (right) shows the gain versus temperature dependence, which shows a nearly linear decrease with increasing temperature. We see a gain change from 52.4 to 51 for an increase of temperature from 16 to 17 °C. Figure 8.32 shows $dG_{APD}/(dV \cdot G_{APD})$ (left) and $dG_{APD}/(dT \cdot G_{APD})$ (right) as a function of gain. At

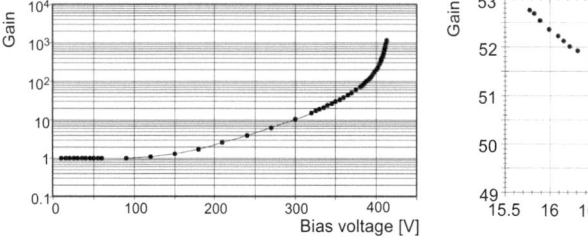
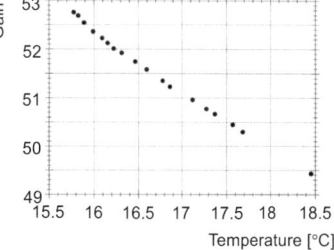

Fig. 8.31 Left: APD gain versus reverse bias voltage. Reprinted with kind permission from [90], © 2000, Elsevier. All rights reserved. Right: APD gain dependence on temperature. Reprinted with kind permission from [99], © 1999, Elsevier. All rights reserved

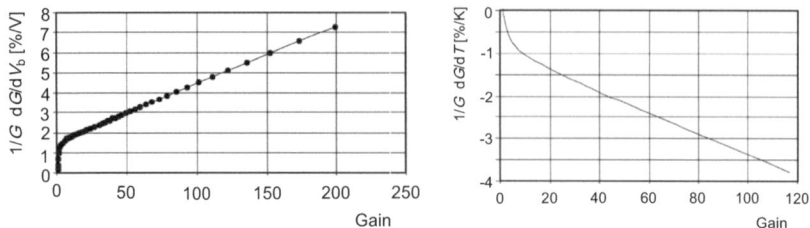

Fig. 8.32 Left: APD gain variation $dG_{APD}/dV_b \cdot 1/G_{APD}$ versus gain. Right: APD gain variation $dG_{APD}/dT \cdot 1/G_{APD}$ versus gain. Both reprinted with kind permission from [90], © 2000, Elsevier. All rights reserved

a gain of 50, $dG_{APD}/(dV \cdot G_{APD}) = 3\%/V$ and $dG_{APD}/(dT \cdot G_{APD}) = -1.7\%/K$, while at a gain of 100 the gain changes become $4.5\%/V$ and $-3.4\%/K$, respectively.

8.3.2.4 Excess Noise Factor

The excess noise factor $F(G_{APD})$ originates from the statistical nature of the avalanche process. It accounts for the fact that the statistical noise on the APD current exceeds that expected from a noiseless multiplier on the basis of Poisson statistics (shot noise) alone. It is a function of k_{eff}. The electron (hole) impact ionization coefficients range from 1–10^5/cm. For high gain, we can express the excess noise factor by

$$F(G_{APD}) = k_{eff} \cdot G_{APD} + \left(2 - \frac{1}{G_{APD}}\right)(1 - k_{eff}). \qquad (8.42)$$

Figure 8.33 (left) shows the excess noise factor as a function of the APD gain. There are two special cases, $k_{eff} = 1$ and $k_{eff} \to 0$ for which $F(G_{APD}) = G_{APD}$ and $F(G_{APD}) \to 2$.

Figure 8.33 (right) depicts the dark current as a function of the bias voltage. At a gain of 50, the dark current is about 10 nA before irradiation. After irradiation with 24 MeV protons at a fluence of 2×10^{13} n/cm^2 the dark current increases to 4 µA.

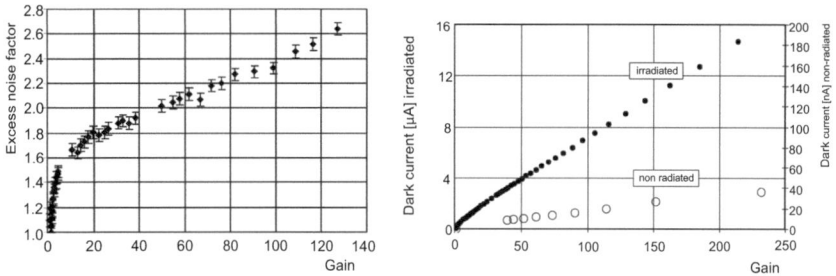

Fig. 8.33 Left: The excessive noise factor as a function of gain. Right: The dark current as a function gain. Both reprinted with kind permission from [90], © 2000, Elsevier. All rights reserved

8.3.2.5 Nuclear Counter Effect

A charged particle passing through the APD produces *e-h* pairs along its path. This effect is called Nuclear Counter Effect (NCE). The number of *e-h* pairs generated by a traversing charged track is given by [100]

$$\frac{dn}{dx} = \frac{dE}{dx} \rho_{\text{apd}} \frac{1}{E_{e,h}}, \quad (8.43)$$

where dE/dx is the energy loss of the charged particle, ρ_{apd} is the density of the silicon layer involved and $E_{e,h}$ is energy of the *e-h* pair. We expect dn/dx to be of the order of 80–100 *e-h* pairs per 1 μm [101]. Note that only electrons that are created before the high-field region and holes that are produced after the high-field region are amplified. This can be expressed in terms of an effective thickness of the APD d_{eff}, which is related to the real thickness of a silicon PIN photodiode d_{PIN} by

$$d_{\text{eff}} = \frac{d_{\text{PIN}}}{Q_{\text{PIN}}} \cdot \frac{Q_{\text{APD}}}{G_{\text{APD}}}, \quad (8.44)$$

where G_{APD} is the APD gain, Q_{APD} and Q_{PIN} are the charges collected by the APD and PIN photodiodes, respectively. The value of d_{eff} depends on the thickness of the p^+ layer. So, a minimum ionizing particle (MIP) traversing the APD will create a signal of energy

$$E_{\text{NCE}} = \frac{dn}{dx} \frac{d_{\text{eff}}}{N_{pe}}, \quad (8.45)$$

where N_{pe} is the number of photoelectrons produced by a 1 MeV photon that is observed by two APDs reading out the PbWO$_4$ crystal in the CMS calorimeter and $d_{\text{eff}} = 5$ μm is the effective thickness of the APD.

A muon deposits about 280 MeV within the crystal. Scintillation light recorded by two APDs yields 1120 photoelectrons. If a muon crosses the APD, about 500 photoelectrons are produced. So,

$$E_{\text{NCE}} = \frac{500}{1120} 280 \approx 125 \text{ MeV}. \quad (8.46)$$

The nuclear counter effect worsens the energy measurements of charged particles. However except for electrons, the energy measurement of charged particles is just used for E/p measurements to separate electrons from charged pions (see Sect. 9.5). From studies, CMS showed that 0.015% of the charged pions with momenta of 5 to 30 GeV produce an $E/p > 0.95$. A signal is also produced by a neutron penetrating the APD.

Table 8.6 Properties of some selected avalanche photodiodes listing the sensitive area a_{sens}, reverse bias voltage V_{APD}, gain G_{max}, capacitance C_{APD}, dark current I_d, wavelength range λ_{min}, quantum efficiency ϵ_{QE}, gain change with respect to voltage $dG_{APD}/dV(1/G_{APD})$, gain change with respect to temperature $dG_{APD}/dT(1/G_{APD})$, excess noise factor (ENF) and rise time t_{rise}. [†]In the visible region $\epsilon_{QE} = 59\%$. [§§]They give a temperature coefficient of 0.88 V/°C. [‡]This is for $G_{APD} = 50$

Property	HPK	HPK	API	EG&G	First Sensor
Type	S8148	S8664-55	200-DUV-01	C30626E	AD800
Reference	[102, 103]	[104]	[83, 105]	[99, 106]	[107]
a_{sens} [mm^2]	5 × 5	5 × 5	201	5 × 5	0.5
V_{APD} [V]	380	<400	1,750–2,050	320–420	120–180
G_{APD}^{max}	>1,000	50	200	20–120	<200
C_{APD} [pF]	80	80 (300 V)	130[§§]	33	2.8
I_d [nA]	<10	5	<600[§§]	50–120	1
λ range	300–1,000	320–1,000	120–1,000	300–800	300–1,000
ϵ_{QE} [%]	72 (430 nm)	70 (420 nm)	95 (120 nm)[†]	88 (580 nm)	80 (500 nm)
$\frac{dG_{APD}}{G_{APD}}dV$ [%/V]	3.3[‡]	3.2	2.8[‡]	2.0[‡]	
$\frac{dG_{APD}}{G_{APD}dT}$ [%/°C]	−2.3[‡]	−2.5[‡]		−3.5[‡]	
ENF (at G)	2[‡]	0.2[‡]		2.7[‡]	2.0
t_{rise} [ns]	<2	1		3.3	1

8.3.2.6 Properties of Avalanche Photodiodes and Applications

Avalanche photodiodes are typically made of silicon. Other materials are InGaAs. One main area of application is the light collection of inorganic scintillating crystals with APDs. Research and development work for the *BABAR* CsI(Tl) calorimeter included studies with APD readout. Since CsI(Tl) is producing sufficient light, we found no advantage of APD over PIN photodiode readout. As already mentioned in the last chapter, the CMS experiment is using two APDs to read out the PbWO$_4$ crystals in the barrel electromagnetic calorimeter [108, 109]. For stability reasons their gain is limited to $G_{APD} = 50$. Table 8.6 summarizes properties of some commonly used silicon APDs.

8.3.3 Silicon PhotoMultipliers

The silicon photomultiplier (SiPM)[4] is a pixilated APD operated in the Geiger mode.

8.3.3.1 Operation Principle

Figure 8.34 (left) shows a schematic layout of a SiPM. On a p^+ substrate a few μm thick epitaxial p^- layer is grown onto which a pixellated p^+n^+ junction is created by ion implantation. The n^+ pixel is fed through the SiO$_2$ layer and is coupled through a

---

[4] SiPMs are also called Multi-Pixel Photon Counters (MPPC), Pixelated Photon Detector (PPD), or Geiger Avalanche Photo Detector (G-APD).

8.3 Silicon-Based Photodetectors

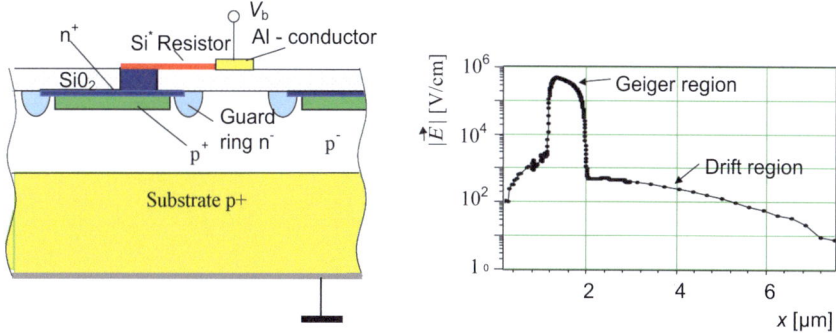

Fig. 8.34 Left: Schematic layout of a silicon photomultiplier pixel. The high electric field is created in about a 1 μm thick layer at the n^+ and p^+ boundary. At the pixel border, n^- type guard rings are introduced to prevent avalanche breakdown in this region. The p^- region serves as the drift region increasing the sensitivity for red light and reducing the pixel capacity. Right: Electric field versus pixel depth. At the 1 to 2 μm depth, the electric field increases by almost three orders of magnitude. Both reprinted with kind permission from [110], © 2001, E. Popova. All rights reserved

silicon resistor (order 1 MΩ) to the aluminum readout trace. Each pixel is surrounded by n^- guard rings. Across the p^+ region in the p^+n^+ junction, which is 1 μm thick, we apply a high electric field of 3–5×10^5 V/cm. The n^- region is depleted and provides a drift region. Figure 8.34 (right) shows the electric-field configuration across the SiPM. The electrons from e-h pairs created in the p^- region drift towards the high electric-field region, where they are amplified. A single electron reaching the high electric-field region triggers a Geiger avalanche. A SiPM has a few-hundred to few-ten-thousand pixels. Each pixel acts as a separate detector. The dynamic range is given by the number of pixels. Since a fired pixel is not sensitive to accept another signal during the recovery time, a sufficient number of pixels is needed to measure a variable like energy well. For a compact readout of many channels single detectors are also arranged into arrays.

8.3.3.2 Detection Properties

Figure 8.35 (left) shows a schematic three-dimensional view of a SiPM. Typically, the total sensitive area comes from 1 mm × 1 mm to 6 mm × 6 mm. An enlargement of a pixel indicates the location of a high-ohmic resistor and the aluminum traces. Figure 8.35 (center) depicts a photograph of a SiPM and Fig. 8.35 (right) shows a close-up view of pixels. Though pixel sizes of 10 μm × 10 μm have been produced, typical pixel sizes are 25 μm × 25 μm or 50 μm × 50 μm. For example, a 1 mm × 1 mm detector with 25 μm × 25 μm pixels holds 1600 pixels. The reverse bias voltage V_b is typically 10–15% above the break-down voltage, yielding $V_b = 25$ to 60 V. The pixel capacitance for 25 μm pixels is 53 pF. The SiPM gain depends on the reverse-bias voltage ranging from a few $\times 10^5$ to few $\times 10^6$. Due to the high gain, individual photoelectrons are visible. The charge of a fired pixel Q_{pix} is given by

$$Q_{\text{pix}} = C_{\text{pix}}(V_b - V_{\text{break}}), \tag{8.47}$$

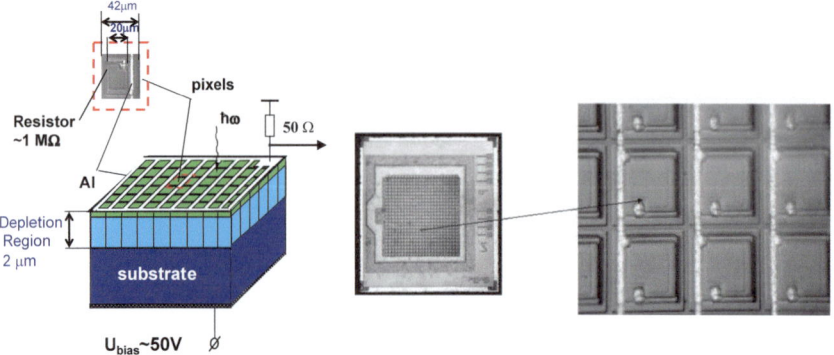

Fig. 8.35 Left: Schematic view of a silicon photomultiplier in three dimensions. Reprinted with permissions from [111], © 2011, World Scientific. All rights reserved. Center: Photograph of a SiPM. Right: Photograph of individual pixels. The shiny traces and blobs are aluminum contacts. The dark traces are the silicon resistors. Both reprinted with kind permission from [112], © 2004, E. Popova. All rights reserved

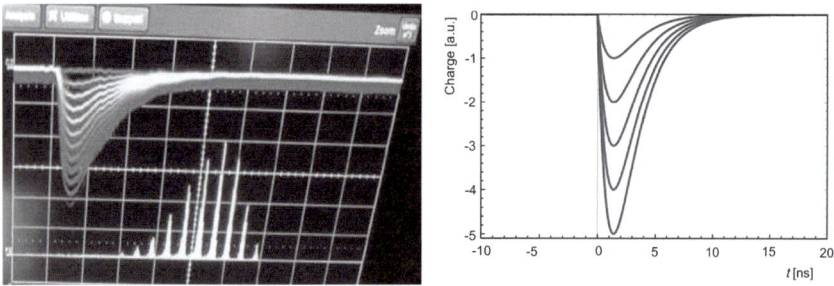

Fig. 8.36 Left: Silicon photomultiplier waveform displayed on a digital oscilloscope. Right: Calculated waveforms as a function of time using (8.48)

where C_{pix} is the pixel capacitance and V_{break} is the break-down voltage. The voltage difference $V_b - V_{\text{break}}$ is often called over-voltage. The recovery time from quenching an avalanche is less than 100 ns/pixel. Silicon photomultipliers are insensitive to magnetic fields, permitting usage in solenoidal fields in multipurpose detectors. Due to the Geiger-mode operation, the nuclear counter effect is negligible for SiPMs since only one out of the total number of pixels is affected, which may have been included in the readout anyhow.

The Geiger-mode operation permits the observation of individual photoelectrons. Figure 8.36 (left) shows the waveform of a Hamamatsu third-generation SiPM. Individual photoelectrons are clearly visible by eye. The inset also shows the extracted photoelectron spectrum. The waveform without after-pulsing can be parameterized by the following function [117],

$$Q(t) = -A\left[\exp\left(-\frac{t-t_0}{\tau_r}\right) - \exp\left(-\frac{t-t_0}{\tau_d}\right)\right]\Theta(t-t_0), \qquad (8.48)$$

8.3 Silicon-Based Photodetectors

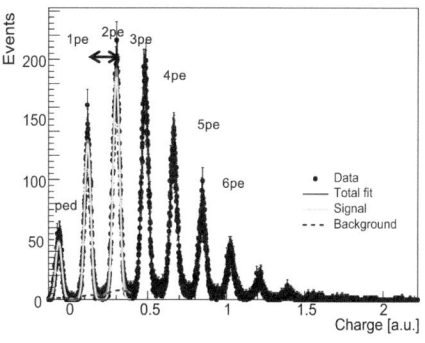
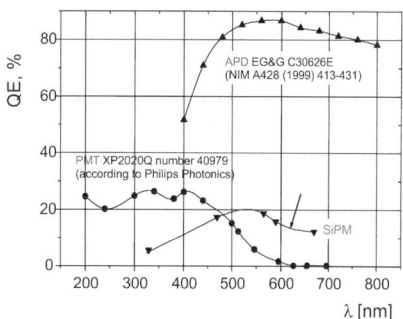

Fig. 8.37 Left: Single photoelectron spectrum of a SiPM. In addition to the pedestal, eight photoelectron peaks are visible. Right: Quantum efficiency as a function of wavelength for an APD (upward triangles), a photomultiplier (points) and a SiPM (downward triangles). Reprinted with kind permission from [113], © 2003, Elsevier. All rights reserved

where A is an amplitude, t_0 is the start time of the signal, τ_r is the rise time, τ_d is the decay time, $\Theta(t)$ is the Heaviside function, which is zero for $t < 0$ and one for $t > 0$. The amplitude can be expressed as

$$A = Q_0 \left(b^{\frac{1}{1-b}} - b^{\frac{1}{1/b-1}} \right), \tag{8.49}$$

where $Q_{\text{tot}} = n_{\text{pix}} Q_0$ is the charge for n_{pix} photoelectrons and $b = \frac{\tau_r}{\tau_d}$. Figure 8.36 (right) shows calculated waveforms using (8.48) for up to five photoelectrons. Here, the decay is represented by one exponential function. Depending on the internal SiPM structure, a second exponential may be needed to fit the measured waveform.

Figure 8.37 (left) displays a reconstructed photoelectron spectrum after integrating the waveform over a time interval, which is determined by the time difference of the start of the signal and the time when the waveform returns to the baseline. This time difference is of the order of 60 ns. The photoelectron spectrum reveals the pedestal and eight well-separated photoelectron peaks. Thus, SiPMs can be used for photon counting. The SiPM gain, which increases with the reverse bias voltage, can be simply obtained from the charge difference of two adjacent photoelectron peaks or the charge difference between the first photoelectron peak and the pedestal.

The SiPM photon detection efficiency is given by

$$\epsilon_{\text{ph}} = \epsilon_{\text{QE}} \cdot \epsilon_{\text{geo}} \cdot P_{\text{g}}, \tag{8.50}$$

where ϵ_{QE} is the quantum efficiency, ϵ_{geo} is the fill factor, i.e. the ratio of sensitive area to total area, and P_{g} is the probability to trigger a Geiger avalanche. Figure 8.37 (right) shows a comparison of quantum efficiencies of an APD, a photomultiplier and a SiPM. Though the APD has the highest quantum efficiency, in the range of 500 to 700 nm the quantum efficiency of a SiPM is higher than that of a photomultiplier. The fill factor is improved in newer SiPM devices. Expected improvements yield $\epsilon_{\text{geo}} \sim 0.7$. The SiPM gives a fast signal pulse yielding good time measurements. The rise time is about 1 ns. The measured time resolution is about 50 ps [113]

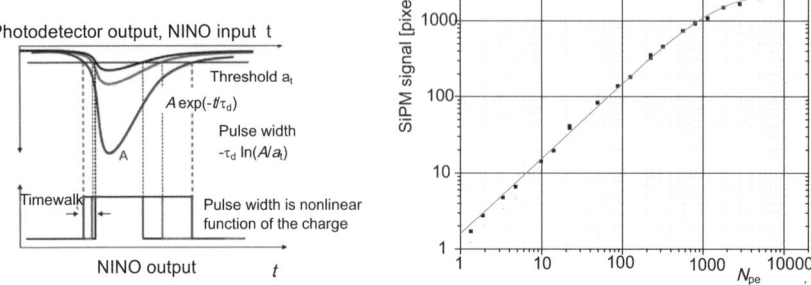

Fig. 8.38 Left: The time development of SiPM signals. The start time depends on the signal height introducing a time walk since t_0 is typically defined by the time the signal crosses a threshold. NINO is an ultrafast discriminator amplifier developed at CERN [114] that employs the time-over-threshold technique. Reprinted under CC-BY-SA-4.0 Licence from [115], © 2018, S. Gundacker. Right: The SiPM signal as a function of the number of incoming photons in a double-logarithmic scale. The solid curve is a fit based on the distribution given in (8.51). Reprinted with kind permission from [116], © 2004, Elsevier. All rights reserved

including the laser pulse width and contribution of the electronics. Figure 8.38 (left) illustrates the introduction of time-walk in time measurements. Depending on the SiPM amplitude, the threshold that starts the time measurement occurs at different times. The start time is earlier for large amplitudes than for small amplitudes. This time-walk affects the time resolution and needs to be corrected for unless a constant fraction discriminator or the Time-over-Threshold method are used (see Sect. 9.2) is used.

Figure 8.38 (right) shows the number of fired pixels as a function of the number of photons that impinge on the SiPM with 1,064 pixels. The distribution saturates around 2,000 indicating that during the measurement each pixels fires twice on average. The SiPM is linear up to 300–400 photons. In order to increase this range either a larger-area SiPM or one with more pixels needs to be used. The measured charge is given by

$$Q(N_{ph}) \propto N_{pix}\left[1 - \exp\left(-\frac{N_{ph} \cdot \epsilon_{ph}}{N_{pix}}\right)\right], \quad (8.51)$$

where N_{pix} is the number of pixels in the SiPM and N_{ph} is the number of fired pixels.

8.3.3.3 Properties of SiPMs

Table 8.7 summarizes properties of some selected SiPMs. Gain stabilization tests have used the Hamamatsu sensors S13360 and the KETEk sensors PM33. The Hamamatsu S14160 SiPMs were studied in the readout of plastic scintillator tiles. The MEPhI SiPMs were used in an analog hadron calorimeter prototype. The FBK RGB-HD sensors were studied by performing readout measurements of LYSO and Ce:GAAG (Cerium-doped gadolinium aluminum gallium garnet) crystals and radiation hardness tests. The Hamamatsu S13170 SiPMs have a high photon detection efficiency in the VUV region and are suited for liquid noble gas detectors.

8.3 Silicon-Based Photodetectors

Table 8.7 Properties of some selected silicon photomultipliers listing the sensitive area a_{sens}, pixel size, number of pixels, wavelength sensitivity λ, wavelength for peak sensitivity λ_{peak}, breakdown voltage V_{break}, operating voltage V_{SiPM}, gain G_{SiPM}, capacitance C_{SiPM}, dark count rate N_d, photon detection efficiency, ϵ_{ph}, temperature coefficient $\Delta T/V_{SiPM}$, probability for cross talk $P_{crosstalk}$ and fill factor ϵ_{geo}. [†]Sensor works highly efficient in the VUV region

Property	HPK	HPK	HPK[†]	MEPhI	FBK	KETEK
Type	S13360	S14160	S13370		RGB-HD	PM33
Reference	[118]	[119]	[120]	[121,122]	[123]	[124]
a_{sens} [mm^2]	1.3×1.3 3×3	1.3 × 1.3	3 × 3	1.1 × 1.1	1 × 1	3 × 3
Pixel size [μm^2]	25 × 25	15 × 15	50 × 50	32 × 32	25 × 25	25 × 25
# pixels	2,668 14,400	7,284	3,600	1,156	1,600	13,408
λ range [nm]	320–900 270–900	290–900	120–900	390–800	260–900	340–900
λ_{peak} [nm]	450	460	500	700	410	430
V_{break} [V]	53 ± 5	38 ± 3	53 ± 5	47	28.5	25
V_{SiPM} [V]	V_{break} + 5	V_{break} + 4	V_{break} + 4	V_{break} + 3	V_{break} + 6.5	V_{break} + 5
G_{SiPM}	7 × 10^5	3.6 × 10^5	2.55 × 10^6	1 × 10^6	4 × 10^6	1.7 × 10^6
C_{SiPM} [pF]	60 320	100	320	58	270	790
N_d [kHz]	70 400	120	1,000	1,000–2,000	480	500 (5 V)
ϵ_{ph} (λ_{peak}) [%]	25	32	35	20	33	34
$\frac{\Delta T}{V_{SiPM}}$ [mV/°C]	54	34	54	58	27	18
$P_{crosstalk}$ [%]	1	<1	3	70		30
ϵ_{geo} [%]	47	49	60	<70	21	

8.3.3.4 Noise, Cross Talk and After-Pulsing

The thermal noise in the SiPM is able to trigger a Geiger mode discharge. This is called dark count. The rate depends on temperature and over-voltage. Figure 8.39 (left) shows the number of dark counts as a function of the deposited charge for a Hamamatsu S13360-2050 SiPM. The measurements were performed at an over-voltage of 3 V and at room temperature. We clearly see 1-pe, 2-pe and 3-pe peaks. By requiring a minimum charge we can reduce the dark counts. For $Q > 0.214$ pC we remove the 1-pe contribution. Figure 8.39 (right) shows the dark count rate as a function of over-voltage for different charge thresholds. At an over-voltage of 3 V and room temperature the dark count rate for a 0.2 pC threshold is about 150 kH decreasing to about 600 Hz for a threshold of 0.6 pC. Another set of measurements for six SiPMs from Hamamatsu and Novel Device Laboratory is depicted in Fig. 8.40 (left). The Hamamatsu SiPMs have a much lower dark-count rate than the NDL detectors. For example, for the S13360-1325 SiPM the dark-count rate at 5 V over-voltage is around 100 kHz compared to 3 MHz for the NDL 3030C detector.

A fired pixel is blind to new photons until the avalanche is broken off and the pixels is reset. In the Geiger discharge photons are produced at a rate of $3 \times 10^{-5}/e^-$ with wavelengths $\lambda < 1100$ nm. Low-energy photons have long absorption lengths (order 1 mm at $\lambda = 1100$ nm) and may produce photoelectrons in neighboring pixels. This

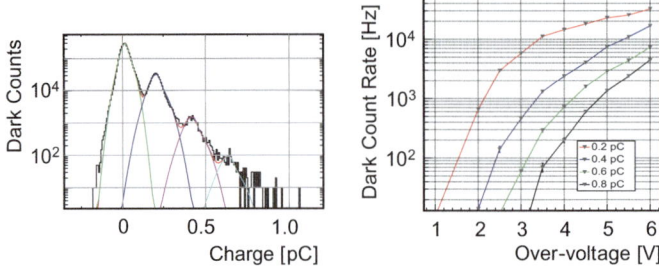

Fig. 8.39 Left: Dark counts as a function of measured charge. Right: Dark count rate as a function of over-voltage for four minimum charge thresholds, 0.2 pC (red line), 0.4 pC (blue line), 0.6 pC (green line) and 0.8 pC (black line). Both reprinted under CC-BY-4.0 Licence from [125], © 2018, M. Jangra

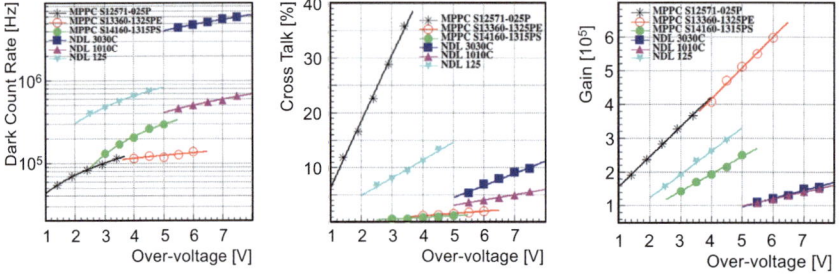

Fig. 8.40 Left: Dark count rate versus over-voltage. Center: Cross talk versus over-voltage. Right: Gain versus over-voltage. Results are presented for Hamamatsu SiPMs S12751-025P (black stars), S13360-1325 (red open circles), S14160-1315 (green solid circles) and Novel Device Laboratory SiPMs 3030C (blue squares), 1010C (magenta upward triangles) and 125 (cyan downward triangles). All reprinted with kind permission from [126], © 2020, Elsevier. All rights reserved

effect is called cross talk. Figure 8.40 (center) shows the cross talk of the six SiPMs depicted in Fig. 8.40 (left) as a function of over-voltage. For these we also show the gain as a function of over-voltage in Fig. 8.40 (right). The prompt component can be suppressed by implementing trenches around the pixels, while the delayed component can be suppressed by an additional *pn* junction [127]. For example, for the S13360-1325 and S14160-1315 SiPMs the cross-talk probability is 1% at an over-voltage of 5 V and room temperature. However, many commercially available SiPMs still have a cross talk probabilities of 5–10% at typical over-voltages. The NDL SiPMs 3030C and 1010C belong to this group. However, their gain is only 1×10^5 at an over-voltage of 5 V. The Hamamatsu S12571 SiPM, which has a similar gain-versus-voltage dependence as the S13360-1325, has a much larger cross talk probability since it has no trenches. At an over-voltage of 3 V and a gain of 3.2×10^5 it is already 28%.

Another effect is after-pulsing, which is caused by delayed Geiger discharges in the same pixel in which the primary Geiger discharge occurred. The cause is that charges, which are trapped by states in the silicon band gap and which are released

8.3 Silicon-Based Photodetectors

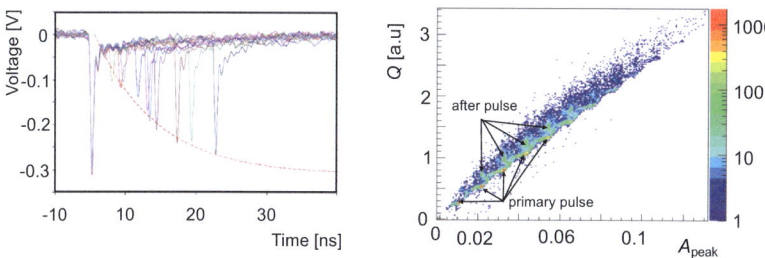

Fig. 8.41 Left: A SiPM signal with after-pulses. The dashed line shows the voltage during the recovery after the initial pulse. Reprinted with kind permission from [121], © 2009, Elsevier. All rights reserved. Right: Integrated charge of the signal pulse versus peak value of the signal amplitude. With appropriate requirements after-pulse-free amplitudes can be clearly separated from amplitudes containing after-pulses

after some time, trigger a new Geiger discharge. Figure 8.41 (left) shows the time dependence of the SiPM amplitude. Besides the initial signal several after-pulses are visible, which come delayed with respect to the initial pulse and have a smaller amplitude since the pixel voltage is not yet completely restored. Most after-pulses come soon after the initial signal with a decay time of 18 ns. Some after-pulses have a longer decay time of 90 ns. The decay times decrease with higher temperature. The after-pulse probability is proportional to the number of electrons that grows as overvoltage squared. Cross talk and after-pulses increase the excess noise factor, which is closer to one than that of APDs. Figure 8.41 (right) shows the integrated charge of the SiPM signal versus the peak value of the pulse amplitude. Since after-pulses occur at delayed times, the peak value of the amplitude is free of after-pulses. The integrated charge, however, is affected by after-pulses since we integrate the pulse until it reaches the baseline again. So, the scatter plot allows us to separate signals without after-pulses from those with after-pulses. In Fig. 8.41 (right) we see the 1-pe, 2-pe, ... 7-pe narrow clusters without after-pulsing, which lie on the lower side of the plot. Those with after-pulsing are broader and lie on the upper side of the plot. We can introduce a separation line between the two distributions and just keep charges below the line to get an after-pulse-free sample.

The pixel recovery time depends on the pixel capacitance and the quenching resistor. For 100 ps recovery time of 25 μm pixels we need a 3 $M\Omega$ quenching resistor. For small recovery times a pixel can fire more than once particularly during a long signal pulse. So the response depends on the shape of the light pulse making calibration procedures more complicated. For many SiPM types the pixel recovery time is 1 μs. The SiPM dead time, however, is much smaller due to the large number of pixels. Therefore, SiPMs can accept high counting rates.

8.3.3.5 Gain Stabilization

The gain of SiPMs increases with bias voltage and decreases with temperature, similarly as for APDs. Figure 8.42 shows the SiPM gain as a function of bias voltage (left) and temperature (right). For stable operations, we need to have constant gain.

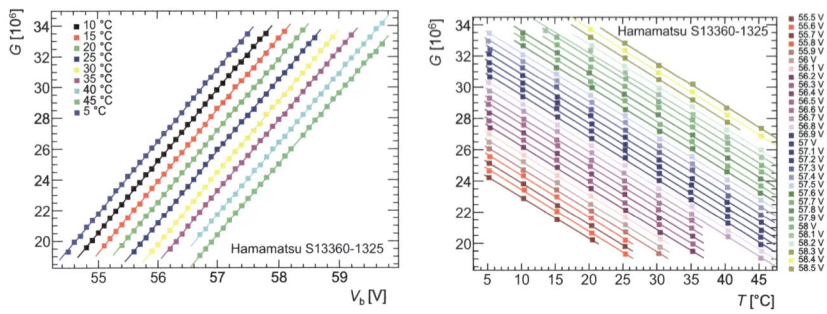

Fig. 8.42 Left: The gain as a function of V_b for a Hamamatsu SiPM (S13360) measured at different temperatures. Right: The gain as a function of temperature for the same SiPM measured for different bias voltages

We can achieve this, for example, by readjusting the bias voltage. For stable gain, dG_{SiPM}/dT needs to be temperature independent, implying we need to determine

$$\frac{dV_b(T)}{dT} = -\frac{dG_{SiPM}(V_b, T)/dT}{dG_{SiPM}(V_b, T)/dV}. \tag{8.52}$$

The gain can be expanded into a Taylor series,

$$G_{SiPM}(V_b, T) = G_{SiPM}^0 + \frac{dG_{SiPM}(V_b, T)}{dV_b}(V_b - V_0) + \frac{dG_{SiPM}(V_b, T)}{dT}(T - T_0) + \mathcal{O}(V_b^2, T^2), \tag{8.53}$$

where $G_{SiPM}^0 = G_{SiPM}(V_0, T_0)$, $T_0 = 25\,°C$ and $V_0 = V_b(25\,°C)$. We can express the derivatives in terms of polynomials

$$\frac{dG_{SiPM}(V_b, T)}{dT} = a + b \cdot V_b(T) + \mathcal{O}(V_b^2),$$
$$\frac{dG_{SiPM}(V_b, T)}{dV_b} = c + d \cdot T + \mathcal{O}(T^2), \tag{8.54}$$

where the parameter a is the gain change with respect to temperature and c is that with respect to the bias voltage. The parameters b and d are small linear corrections due to bias voltage and temperature variations, respectively. So,

$$\frac{dV_b(T)}{dT} = \frac{a + b \cdot V_b(T)}{c + d \cdot T}. \tag{8.55}$$

If both derivates are given by first-order polynomials, we obtain the solution

$$V_b(T) = -\frac{a}{b} + \frac{K_b}{\left(c + d \cdot T\right)^{b/d}}, \tag{8.56}$$

8.3 Silicon-Based Photodetectors

where K_b is an integration constant. For $d = 0$ this simplifies to

$$V_b(T) = -\frac{a}{b} + K_b \cdot \exp\left(\frac{-b \cdot T}{c}\right). \tag{8.57}$$

If in addition $b = 0$, we obtain

$$V_b(T) = -\frac{a}{c} \cdot T + K_b. \tag{8.58}$$

Finally, for $b = 0$ and $d \neq 0$ the solution becomes

$$V_b(T) = K_b - \frac{a \cdot \log(c + d \cdot T)}{d}. \tag{8.59}$$

If the derivatives are constant, $V_b(T)$ is exactly linear with T. Figure 8.43 (left) shows the calculated voltage-versus-temperature dependence of a Hamamatsu SiPM S13360. The dependence is rather linear though parameters b and d are not zero.

By readjusting V_b by $dV_b(T)/dT$ if T changes, the gain can be kept stable. We define gain stability as a dG_{SiPM}/dT change of less than $\pm 0.5\%$ in the temperature range 20 to 30 °C. Gain stability was tested with 30 different SiPMs in a temperature range of 0 to 50 °C [128]. For S13360 SiPMs, we measured $dV_b/dT = (57.8 \pm 0.1)$ mV/°C as shown in Fig. 8.43 (right). Figure 8.44 (left) shows the gain changes of four S13360 SiPMs as a function of temperature, which are all stabilized with a single value of $dV_b/dT = 57.8$ mV/°C. All four SiPMs satisfy the requirement of varying less than $\pm 0.5\%$ in the temperature range 20 to 30 °C. In this study, we also measured the breakdown voltage as a function of temperature, which increases with a slope of $0.058 \cdot T$ in units of °C. Figure 8.44 (right) shows the breakdown voltage as a function of temperature for one of the S13360 SiPMs. At T = 25° the breakdown voltage is $V_{break} = (51.463 \pm 0.004)$ V.

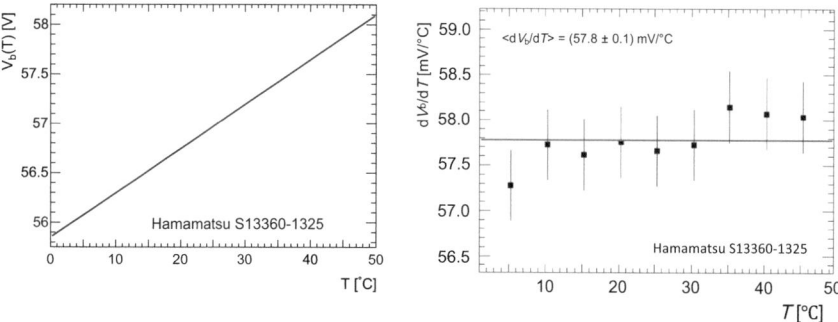

Fig. 8.43 Left: Calculated dependence of the bias voltage versus temperature for the Hamamatsu SiPM (S13360). Right: Measurement of dV_b/dT (points with error bars) at different temperatures for the same Hamamatsu SiPM with a linear fit overlaid

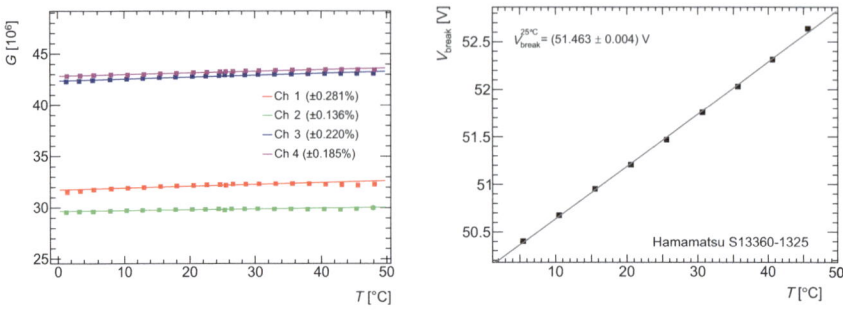

Fig. 8.44 Left: The gain as a function of temperature for four Hamamatsu SiPMs (S13360) after stabilization with one fixed correction dV_b/dT. All SiPMs satisfy the requirement of $\pm 0.5\%$ stability within the 20 to 30 °C temperature range. Right: The breakdown voltage as a function of temperature for a Hamamatsu SiPM (LCT4)

8.3.3.6 Applications of SiPMs

In addition to single SiPMs, also SiPM arrays are produced. Standard arrays come in 2×2, 4×4 or 8×8 matrices. The pitch is 25 µm or 50 µm. Figure 8.45 (left) shows the especially designed readout board of a 8×8 SiPM array that is used to perform readout studies of the ATLAS TileCal fibers. An enlarged photograph of the SiPM array with 64 1.3 mm × 1.3 mm SiPM sensors is depicted in Fig. 8.45 (center). These new SiPM arrays provide a compact readout scheme. They are used also in other experiments. For example, Fig. 8.45 (right) shows a 128-channel SiPM array that is used to read out the scintillating fibers in the LHCb SciFi tracker [129].

Due to the compact arrangement and low high-voltage operations, SiPMs have become rather popular being used or being considered in many applications. We give some examples here. Analog hadron calorimeters considered at e^+e^- linear colliders use plastic scintillator tiles that are read out with SiPMs [116, 130]. The Scintillating-Fiber Tracker built by LHCb uses SiPM arrays to read out the fibers (see Sect. 8.8 [131]). Also, the new CMS HGCal uses SiPMs to read out scintillators [132]. Silicon photomultipliers are well suited to read out LYSO crystals [133] and pure

Fig. 8.45 Left: The readout board of a 8×8 SiPM array located in the center (personal photograph). Center: A close-up photograph of the 8×8 SiPM array, housing 64 1.3 mm × 1.3 mm SiPM sensors (personal photograph). Right: A 128-channel SiPM array used for the LHCb Scintillating-Fiber Tracker. Reprinted with kind permission from [129], © 2017, IOP Publishing. All rights reserved

8.3 Silicon-Based Photodetectors

CsI crystals [134]. Furthermore, they are considered for gamma-ray telescopes [135], neutron monitors in the CMS experiment [136], time-of-flight counters in the PANDA experiment [137] and medical applications such as positron-emission tomography (PET) detectors [138].

8.3.4 Digital Silicon Photomultipliers

The SiPM is a digital device. So instead of using an analog readout, the digital character can be utilized directly in the readout as the Philips company has done. Each pixel is equipped with its own active quenching element, recharge element for the recovery of the sensitive state after breakdown and a one-bit on-chip ADC in the form of a CMOS inverter. This active quenching and recharging in each pixel improves the detector's recovery time and, in turn, the dead time of the detection system by about an order of magnitude. In addition, the power consumption is reduced due to the exclusion of high-resistive elements in the readout. This means that a photon, which triggers a microcell, produces a digital pulse. The output of a digital SiPM is the sum of all of digital pulses triggered by the incoming photons having energy information as well as timing information. While an analog SiPM requires dedicated readout circuitries, which adds electronic noise to the signal, the digital SiPM digitizes the signal at the microcell level within the SiPM [139,140]. Furthermore, each microcell has its own control circuit, so microcells with high dark-count rates can be disabled. The digital SiPM does not use analog signal processing, producing faster and more accurate photon counts with extremely well-defined timing in reference to the first photon detection. Each microcell has integrated electronics on the chip including digital summation, trigger network, TDC and pixel controller electronics.

Thus, digital SiPMs work in the counting mode providing both the total number of detected photons and the arrival time related to the time of the signal. The digital SiPMs are primarily used for medical detectors such as positron emission tomography. Here, two key measurements are needed, the number of photons and the precise time of the arrival of the photons, which the digital SiPMs provide.

8.3.5 Visible Light Photon Counters

A Visible-Light Photon Counter (VLPC) is a solid state photomultiplier that was designed to convert single photons into a few$\times 10^4$ electrons with high quantum efficiency (95%) [141]. Figure 8.46 (left top) shows a schematic layout of a VLPC, which consists of several silicon epitaxial layers grown on a doped Si substrate. The gain and drift layers are highly doped with arsenic and slightly counter-doped with boron. At operating temperatures, the As dopants in the gain and drift layers form a populated impurity band of 54 meV below the conduction band. The total thickness is 30 μm. The device is operated at low bias voltage and at very low temperature, typically between 6.0–7.5 V and 5.5–7.0 K, respectively. Figure 8.46 (left bottom) shows the electric field across the VLPC. At the operating bias voltage field-assisted

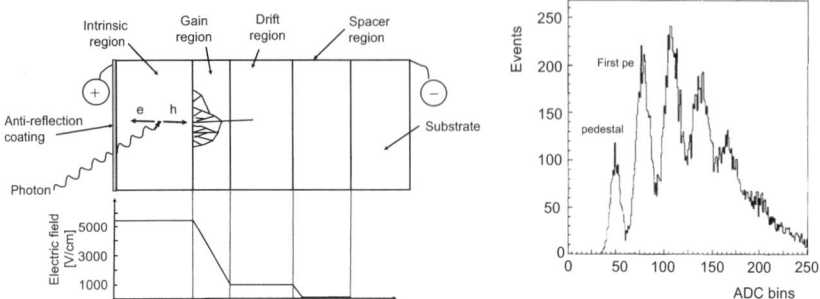

Fig. 8.46 Left top: Schematic layout of a VLPC (see text for details). Left bottom: Electric-field configuration in a VLPC. Right: Photoelectron spectrum of a VLPC measured with a red LED at $V_b = 6.5$ V and $T = 6.5$ K. Both reprinted with kind permission from [142], © 1997, Elsevier. All rights reserved

thermal ionization of donors leads to a stable bias current flowing through the sensor. This sets up a region of constant electric field that defines the drift layer. This is crucial for achieving a high quantum efficiency.

A photon converts in the intrinsic region into an electron-hole pair. The hole accelerates into the drift layer, liberating an electron with high probability via collisions with a neutral donor atom. This secondary electron moves back into the gain layer, where it starts an electron avalanche inside the gain layer. The avalanche produces a pulse at the anode containing tens of thousands of electrons. Long-wavelength photons may propagate beyond the intrinsic layer before they ionize an atom. In this case the primary electron will produce the avalanche. Since the impurity band gap is only 54 meV, the electron avalanche develops with very small gain dispersion. Note that the gain, quantum efficiency and dark rate depend on the layer thickness and donor concentrations. Figure 8.46 (right) shows a typical photoelectron spectrum measured with a VLPV. Single photon counting is clearly visible. The gain is obtained in a similar way as for a SiPM, taking the charge difference between the first photoelectron peak and the pedestal. At $V_b = 7.2$ V and $T = 7.0$ K measurements yield a gain of few $\times 10^4$, an rms time jitter of $\sigma_{\text{jitter}} = 102$ ps and a dark count rate of 30 kHz.

The D0 experiment used VLPCs to read out the scintillating-fiber tracker [142]. A disadvantage is the operation at rather low temperatures. In addition, the single photoelectron peaks are much broader than those of SiPMs. So, with the development of PIN diode, APDs and SiPMs, which work at room temperature VLPCs became less attractive.

8.3.6 Hybrid PhotoDiodes

In a Hybrid PhotoDiode (HPD), the virtues of high gain in a PMT are combined with that of excellent position resolution and pixilation in a silicon photodiode. Figure 8.47 (left) shows the schematic layout of an HPD, which consists of a glass tube under

8.3 Silicon-Based Photodetectors

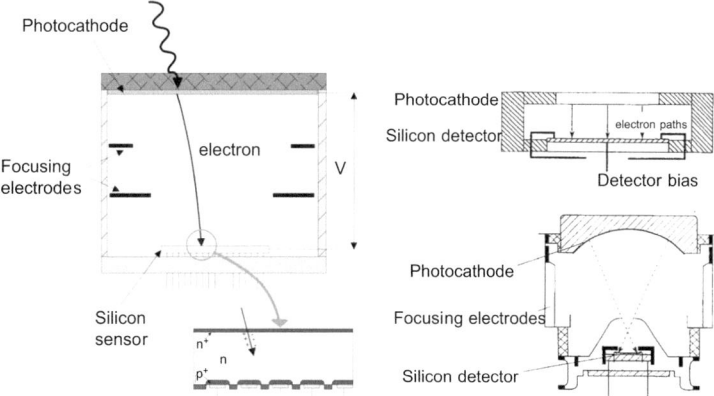

Fig. 8.47 Left: Schematic layout of an HPD (top) and details of the pixellated silicon sensor (bottom). Top right: Schematic layout of a proximity-focussed HPD. Bottom right: Schematic layout of a cross focussed HPD. All reprinted with kind permission from [143], © 1970, Elsevier. All rights reserved

vacuum that has a UV-extended borosilicate window covered with a bi-alkali photocathode, focussing electrodes and a pixellated silicon sensor. A high electric field of 20 kV is applied over a few centimeters. The electrons gain kinetic energies of 20 keV before they hit the silicon detector, where focussing electrodes guide their paths. Since the energy to create an e-h pair is 3.6 eV in silicon, about 5,600 e-h pairs are produced. Figure 8.47 (top right) shows the schematic layout of proximity-focused HPD.[5] Here, an electric field of 15 kV is applied over a 1.5 mm gap. The silicon is pixellated providing some position information. This rather compact layout needs no focussing electrodes. Figure 8.47 (bottom right) shows the schematic layout of a cross-focussed HPD. Here, photoelectrons emitted from the left-hand side of the photocathode are recorded on the right-hand side of the photodetector and vice versa.

Figure 8.48 displays photographs of a DEP HPD with 80 mm diameter (left) and a PAD HPD with 127 mm diameter (right). The quantum efficiency is typically the same as that of photomultipliers. However, using GaAsP photocathodes provides a considerable improvement. Figure 8.49 (left) shows the quantum efficiency as a function of wavelength. In the green light spectrum, the quantum efficiency is 55%, which is more than a factor of two higher than that of standard photocathodes. There is a correlation between the cathode position and the position in the silicon sensor as illustrated in Fig. 8.49 (right). The layout produces a linear correlation over the full active diameter. A fit to the measurements yields a slope of $m = -2.7$. Thus, by segmenting the silicon sensor position information can be obtained. Hybrid photodiodes also resolve individual photoelectron peaks. Figure 8.50 (left) shows the

---

[5] The photoelectrons produced in the photocathode are channeled by a strong electric field parallel to the tube axis into a close-by silicon detector.

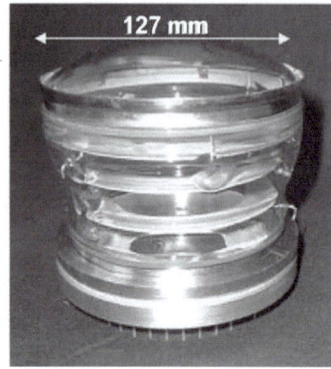

Fig. 8.48 Left: Photograph of a pixel HPD from the DEP company. Reprinted with kind permission from [144], © 2002, LHCb Collaboration. All rights reserved. Right: Photograph of a Pad HPD built at CERN. Reprinted with kind permission from [145], © 2002, Elsevier. All rights reserved

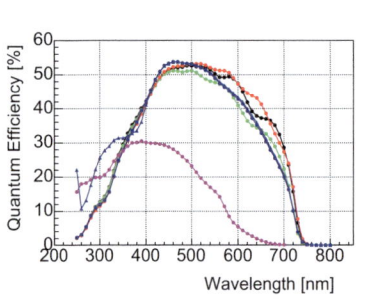
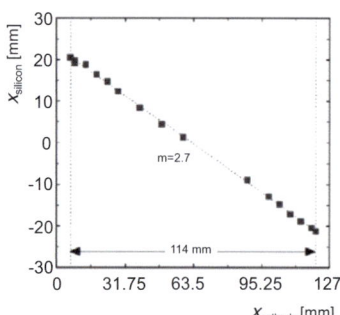

Fig. 8.49 Left: The quantum efficiency of HPDs with a GaAsP photocathode (four upper curves) in comparison to that of a photomultiplier (lower curve). Reprinted under CC-BY-SA-4.0 Licence from [148], © 2007, T. Saito. Right: Correlation of the position in the photocathode with the position in the silicon sensor. Reprinted with kind permission from [150], © 2000, Elsevier. All rights reserved

photoelectron spectrum of an HPD. Apart from the individual photoelectron peaks, a shoulder is visible on the right-hand side of each photoelectron peak. This is caused by electrons that back-scatter from the silicon into the vacuum. The probability is about 18% for 13 keV electrons. The shoulders can be represented by additional peaks. The spectrum can be parameterized in terms of photoelectron peaks and back-scattered electrons yielding a complex parametrization [146].

Figure 8.50 (right) shows the normalized signal as a function of the magnetic-field strength. For a zero degree tilt angle of the photodetector symmetry axis with respect to the magnetic field, the gain is not affected by the magnetic field. This changes for tilt angles of 30°. The gain decreases with increasing magnetic-field strength. For $|\vec{B}| \geq 0.5$ T the relative gain saturates at a gain of 50%. This shows that HPDs may be used in multi-purpose detectors without significant loss if tilt angles are small.

8.3 Silicon-Based Photodetectors

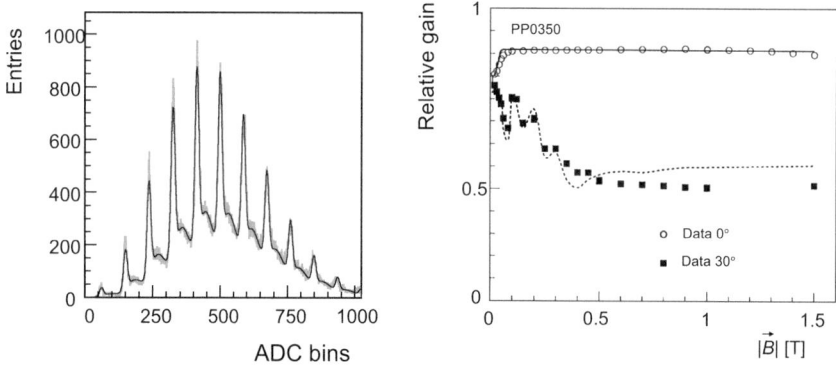

Fig. 8.50 Left: Measured photoelectron spectrum showing individual photoelectron peaks. In addition, shoulders on the right-hand side of each photoelectron peak are visible, which result from backscattered electrons. Reprinted with kind permission from [146], © 2002, Elsevier. All rights reserved. Right: The normalized signal as a function of the magnetic-field strength for tilt angles of 0° (open circles) and 30° (squares) of the photodetector symmetry axis with respect to the magnetic-field. The solid and dashed lines are calculations. Reprinted with kind permission from [151], © 1999, Elsevier. All rights reserved

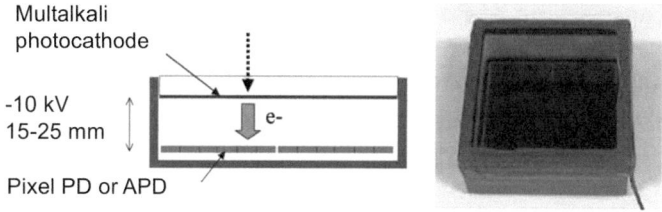

Fig. 8.51 Schematic layout (left) and photograph (right) of an HAPD used in the Belle II aerogel RICH counter [153]. Reprinted with kind permission from [153], © 2010, Elsevier. All rights reserved

For example, HPDs are used as Cherenkov detectors. The LHCb experiment uses HPDs in their Cherenkov detectors (RICH) [147]. The MAGIC II experiment decided to use HPDs with a GaAsP photocathode for the Cherenkov telescope [148, 149]. For the Belle II experiment, a new HPD with 144 multi-anodes was developed to read out the aerogel ring-imaging Cherenkov counter. Instead of a plain silicon photodiode, they use an avalanche photodiode with a gain of about 40 increasing the total gain up to 10^5. Figure 8.51 shows a schematic layout (left) and a photograph (right) of this HAPD. The detector holds 12×12 pixels on a 72 mm $\times$ 72 mm array [152]. The photodetector works well in a 1.5 T magnetic field and tolerates radiation up to 10^{12} (1 MeV) n_{eq}/cm^2. A photoelectron generated in the photocathode generates 1,800 electron-hole pairs in the APD after acceleration in a 6 kV electric field. A reverse bias voltage of 350 V of the APD generates about 40 electron-hole pairs. The quantum efficiency at 400 nm is 32.2%. For the readout of the 248 aerogel tiles in the aerogel ring-imaging Cherenkov detector, 420 HAPDS are necessary covering an area of 3.5 m^2.

8.4 Gaseous Photon Detectors

A Gaseous Photon Detector (GPD) is able to detect photons and produce an avalanche in a high field region. Charge multiplication and collection processes are similar to those in MWPCs, Micromegas, or GEMs, which we discussed in Chap. 4. Detectors need a photosensitive material for converting the photons into electrons that are then collected. One possibility is to use CsI or gases like tetrakis dimethyl-amine ethylene (TMAE) or tri-ethyl-amine (TEA), which respectively have photoionization work functions of 5.3 and 7.5 eV. Thus, they are suited to detect deep UV photons. Since devices like GEMs offer sub-mm spatial resolution, GPDs are often used as position-sensitive photon detectors. They can be made into flat panels to cover large areas ($\mathcal{O}(1\ m^2)$), can operate in high magnetic fields, and are relatively inexpensive. Many of the ring-imaging Cherenkov (RICH) detectors have used GPDs for the detection of Cherenkov light (see Sect. 9.3.3). Special care must be taken to suppress the ion-feedback and photon-feedback processes in GPDs. It is also important to maintain high purity of the gas as tiny traces of oxygen or water can significantly degrade the detection efficiency.

8.5 Comparison of Photodetectors

Table 8.8 shows a comparison of the properties of the different photodetectors we have discussed.

Table 8.8 Properties of some photodetectors used in particle physics experiments including the wavelength range λ, quantum efficiency ϵ_{QE} times collection efficiency (ϵ_c), typical gain, rise time t_r, rms time resolution σ_t for single photons, detector area, dark counting rate per area ν_{dark} and operation voltage (adapted from [154]). †These devices often come in multi-anode configurations. In such cases, area and noise are to be considered on a "per readout-channel" basis. ‡The APD does not resolve single photoelectrons. *Hybrid APDs (HAPDs) may add an avalanche multiplication step to boost the gain by a factor of 50. †Gaseous photon detectors (GPD). Reprinted under CC-BY-NC-4.0 Licence with kind permission from [154], © 2024, the Particle Data Group LBNL

Type	λ [nm]	$\epsilon_{QE}\epsilon_c$	Gain	t_r [ns]	σ_t [ps]	Area [mm^2]	ν_{dark} [Hz/mm^2]	HV [kV]
PMT†	115–1,700	0.15–0.25	10^5–10^7	0.7–10	~200	10–10^5	10^{-2}–10^2	0.5–3.0
MCP†	115–650	0.01–0.10	10^3–10^7	0.15–0.3	~20	1–10^4	1–10	0.5–3.5
APD‡	300–1,700	~0.7	10–10^5	$\mathcal{O}(1)$	1	1–10^3	$\mathcal{O}(10^7)$	0.4–1.4
SiPM	125–1,000	0.15–0.4	10^5–10^7	~1	50	1–36	10^4–10^5	0.03–0.07
HPD†	115–850	0.1–0.3	10^3–10^4	$\mathcal{O}(1)$	~1000	10–10^5	10–100	20
HAPD†,*	115–850	0.1–0.3	10^4–10^5	$\mathcal{O}(1)$	30	10–10^5	~1	10
GPD†	115–500	0.15–0.3	10^3–10^6	$\mathcal{O}(0.1)$	~100	$\mathcal{O}(10)$	~1	0.3–2.

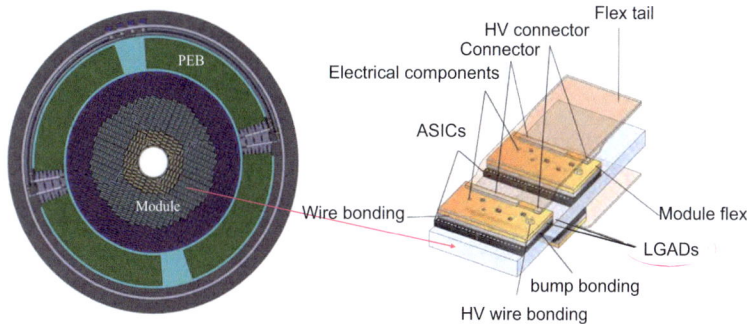

Fig. 8.52 Left: Schematic layout of an ATLAS timing detector disk. The active area (modules) is located in the inner disk, while the Peripheral Electronics Board (PEB) is located on the outside. Right: Schematic layout of a module. Both reprinted with kind permission from [155], © 2020, ATLAS Collaboration. All rights reserved

8.6 Timing Detectors at LHC Experiments

For the operation at the High-Luminosity LHC, the increase in pile-up becomes an issue. Fast timing provides a powerful tool for pile-up suppression. Thus, ATLAS and CMS will install special timing detectors.

The ATLAS experiment plans to install a high-granularity timing detector for the high-luminosity upgrade in the gap between the barrel and the endcaps [155]. It consists of two disks, one placed on each side of the barrel. The disks are equipped with low-gain avalanche silicon diodes covering an active region of $2.4 < |\eta| < 4.0$ and 12 cm $<$ r $<$ 64 cm. They tolerate a radiation of 2.5×10^{15} (1 MeV) n_{eq}/cm. Each disk has LGAD sensors on both sides that are bump-bonded to the ASICs in the center containing 8,034 modules as shown in Fig. 8.52 (left). A module depicted in Fig. 8.52 (right) has two bump-bonded sensors plus the ASIC glued and wire-bonded to a module flex. The 50 μm thick LGAD sensors are arranged in arrays of 15 × 15 pads, where each pad is 1.3 mm × 1.3 mm. The LGADs are operated with a gain of 20. The anticipated timing resolution is 35 ps per hit and 30 ps per track at the beginning of the high-luminosity LHC operation increasing to 70 ps per hit and 50 ps per track after a luminosity of 4,000 fb^{-1}. Figure 8.53 shows the time resolution of eight different LGAD sensors measured at $-30\,°C$ with electrons from a ^{90}Sr source at a fluence of $\Phi = 2.5 \times 10^{15}$(1 MeV) n_{eq} cm^{-1} [156]. The three sensors with boron-carbon doping yield the required time resolution of 30-50 ps at lower bias voltage.

The CMS experiment will install MIP timing detectors between the tracker and the calorimeters [157] consisting of a Barrel Timing Layer (BTL) and Endcap Timing Layers (ETL). The technology choice is driven by radiation hardness requirements. In the barrel it consists of two thin layers of LYSO(Ce) crystals read out by silicon photomultiplier arrays that detect charged particles in the region $|\eta| < 1.45$ and $p_T >$

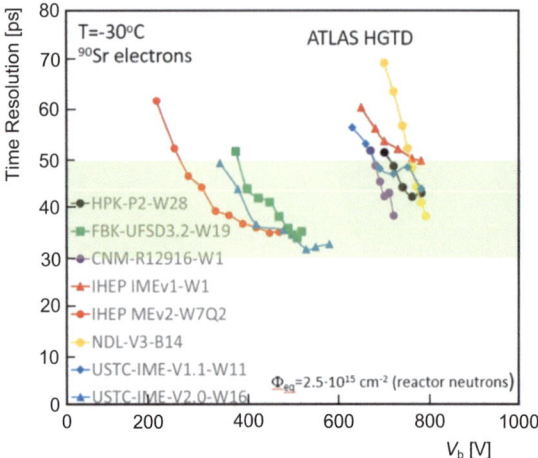

Fig. 8.53 Time resolution of different LGAD sensors measured at $-30\,°C$ with electrons from a ^{90}Sr source at a fluence of $\Phi_{eq} = 2.5 \times 10^{15}(1\,\text{MeV})n_{eq}\,\text{cm}^{-1}$. The eight sensors come from Hamamatsu (black points), FBK (green squares), CNM (brown points), IHEP-IME[†] (red triangles and red points), NDL[‡] (orange points), and USTC-IME[§] (blue diamonds and blue triangles). The FBK, IHEP MEv2 and USTC-IMEV2.0 use boron-carbon doping. The shaded region is the targeted time resolution. Note that HGTD stands for High-Granularity Timing Detector. [†]IHEP-IME is the Institute for High-Energy Physics and Institute for Micro Electronics in Bejing. [‡]NDL is Novel Device Laboratory of Bejing Normal University. [§]USTC is the University of Science and Technology of China. Reprinted with kind permission from [156], © 2023, ATLAS Collaboration. All rights reserved

0.7 GeV/c. The active area covers 38 m² and consists of 332,000 readout channels. The SiPMs tolerate a fluence of 2×10^{14} (1 MeV) n_{eq}/cm, which is expected after 3,000 fb^{-1}. In the endcaps, the two layers consist of LGAD detectors that have to tolerate a fluence of 2×10^{15} (1 MeV) n_{eq}/cm. They cover a pseudo-rapidity region of $1.6 < |\eta| < 3.0$. The active area is 14 m² and consists of 8.5 million readout channels.

The barrel uses 175,000 LYSO crystals of dimension 54.7 mm × 3.2 mm × 3.75 mm. The crystals are read out on each side with SiPM arrays having 25 μm pixels whose photodetection efficiency is 30-50%. First, two layers of 16 bars each are arranged into modules (Fig. 8.54 left top). Next, 3 × 8 modules are combined into readout units (Fig. 8.54 left center). Then, six readout units are combined into trays (Fig. 8.54 left bottom). Finally, 72 trays (2 × 36) make the BTL. Figure 8.54 (right) shows the arrangement of the trays inside the tracker support tube. The detector is cooled to $-35\,°C$.

The time resolution in the barrel timing detector is given by

$$\sigma_t^{\text{BTL}} = \sigma_t^{\text{clock}} \oplus \sigma_t^{\text{digi}} \oplus \sigma_t^{\text{elec}} \oplus \sigma_t^{\text{photo}} \oplus \sigma_t^{\text{DCR}}, \quad (8.60)$$

where σ_t^{clock} is the error of the clock digitization (15 ps), σ_t^{digi} is the error of digitization (7 ps), σ_t^{elec} is the error of the electronics (8 ps), σ_t^{photo} is the photostatistic error

8.6 Timing Detectors at LHC Experiments

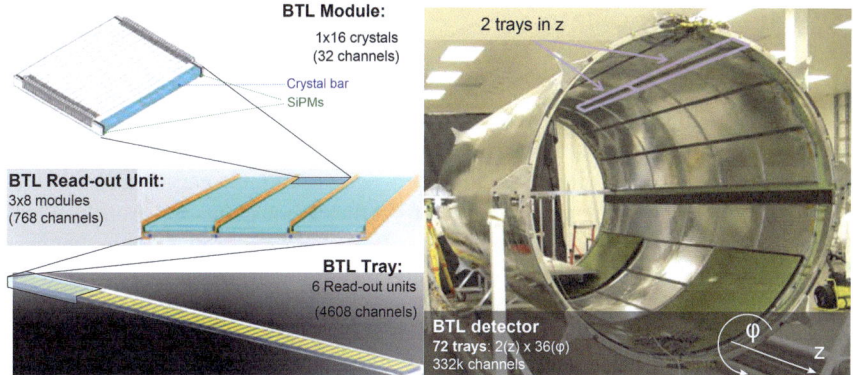

Fig. 8.54 Schematic layout of the CMS Barrel Timing Layer showing modules (left top), readout units (left center) and trays (left bottom). The arrangement of the trays inside the tracker support tube is indicated by purple rectangles (right). Reprinted with kind permission from [157], © 2020, IOP Publishing. All rights reserved

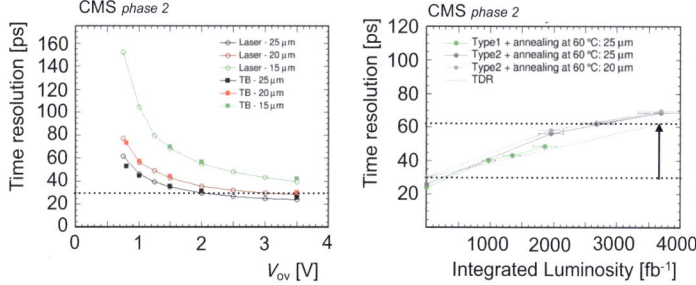

Fig. 8.55 Left: Time resolution of the CMS timing detector in the barrel as a function of SiPM over-voltage for pixel sizes of 15 μm (green squares and green circles), 20 μm (red circles and red squares) and 25 μm (black circles and black squares) measured with a laser and in a test beam. Right: Time resolution of the CMS timing detector in the barrel as a function of luminosity. About 12% of the sensors will be exposed to a fluence of 1×10^{15} (1 MeV) n_{eq}/cm. Both reprinted with kind permission from [159], © 2023, CMS Collaboration. All rights reserved

term (25–30 ps) and σ_t^{DCR} is the noise term due to dark counts. At startup the noise term is negligible. After a luminosity of 3,000 fb^{-1} it rises to 50 ps. Figure 8.55 (left) shows the time resolution of the CMS timing detector in the barrel as a function of SiPM over-voltage for different pixel sizes measured with a laser and in a test beam. Figure 8.55 (right) shows the time resolution of the CMS timing detector in the barrel as a function of luminosity. At startup a resolution of 30 ps is targeted, which will increase to 65 ps at the end of operation.

For the two planes in the endcaps 50 μm thick planar LGADs are used that are operated with a gain of 10–30. They are arranged in a 16 × 16 array with pad sizes of 1.3 mm × 1.3 mm. The sensors are bump-bonded to the readout ASICs. Modules are assembled as two double-sided disks on each endcap. There are 8.5 million readout channels. The targeted resolution is 50 ps per hit and 35 ps per track.

8.7 Plastic Scintillators

Frequently, photodetectors are coupled to plastic scintillators, which are organic scintillators built up from condensed or linked benzene-like structure [158]. A key element of the molecular structure is the presence of extended groupings of conjugated double bonds that form typically between unsaturated carbon atoms. Thus, only two or three of the four valence electrons of the carbon atoms are strongly bound within the molecule. These electrons occupy the so-called σ-orbital that lie between the atoms they bind. The remaining valence electrons are delocalized within the molecule. So, they are not associated with a particular atom occupying the so-called π-molecular orbitals as illustrated in Fig. 8.56. They are above and below the molecular plane containing the σ-orbitals. We discussed the scintillation process in Sect. 2.8.4. Transitions between the π electronic states produce luminescence observed in the scintillation process. The light produced by the primary fluor is typically emitted in the UV to blue spectrum and is absorbed right away in most organic materials, which have absorptions lengths of a few mm for UV light but are transparent in the visible light range.

Table 8.9 lists some properties of the most commonly used organic scintillators. Figure 8.57 shows the chemical structures of these primary fluors. The table lists the wavelength where the maximum emission occurs. The emission is typically in the UV to blue region. These primary fluors also have an absorption spectrum, which lies in

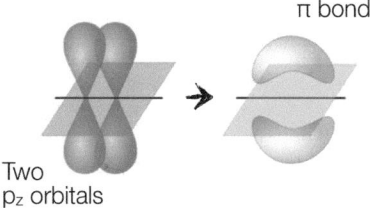

Fig. 8.56 Formation of π orbitals in the molecule [160]. Reprinted under CC-BY-SA-3.0 Licence from [160], © 2009, J. V. Vladsinger

Table 8.9 Properties of most commonly used organic scintillators listing the name, the chemical formula, emission wavelength range λ, wavelength of maximum emission λ_{max}, decay time τ_d and relative light yield (LY) with respect to that of anthracene. [†]Organic crystal. [‡]Liquid in plastic

Scintillator	Formula	λ [nm]	λ_{max} [nm]	τ_d [ns]	rel. LY	Reference
Naphtalene[†]	$C_{10}H_8$	310–380	325	96	0.24	[161, 162]
Anthracene[†]	$C_{14}H_{10}$	375–455	400	30	1.0	[161, 163]
p-Terphenyl (PTP)[‡]	$C_{18}H_{14}$	310–390	338	3.3	0.28	[161, 164]
PBD[‡]	$C_{20}H_{14}N_2O$	340–440	365	1.2	0.7	[161, 165]
Trans-Stilbene[†]	$C_{14}H_{12}$	320–420	345	4.3	0.5	[166, 167]
2,5-diphenyloxazole (PPO)[‡]	$C_{15}H_{11}NO$	330–450	355	1.6		[168, 169]

8.7 Plastic Scintillators

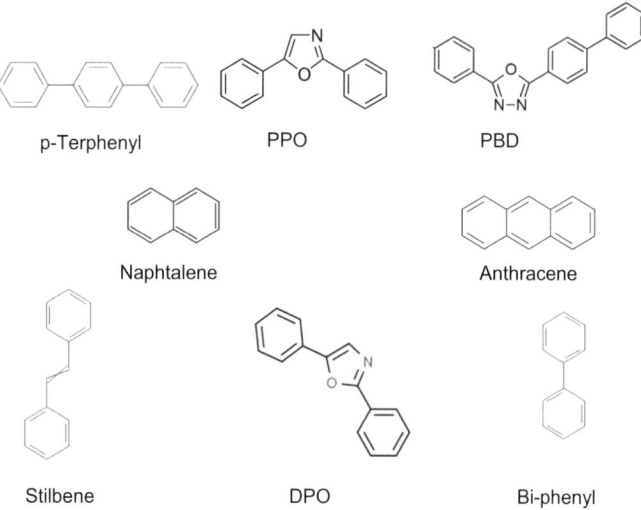

Fig. 8.57 Chemical structure of different organic scintillators. Top left: p-Terphenyl. Top center: 2,5-Diphenyloxazole (PPO). Top right: 2-phenyl-5(4-biphenyl-1,3,4-oxadizole) (PBD). Center left: Naphtalene. Center right: Anthracene. Bottom left: Stilbene. Bottom center: Diphenyl phosphine oxide (DPO). Bottom right: Bi-phenyl. The hexagons represent benzine rings

the UV to VUV region. Figure 8.58 (left) shows the absorption and emission spectra of anthracene, which partially overlap. The emission spectrum lies in violet-blue region and has a range of about 80 nm. Figure 8.58 (right) shows the absorption and emission spectra of bPBD and PPO, which are rather similar. The emission spectrum of PPO extends a little more into the UV region. Note that their light yield is lower than that of NaI(Tl) or CsI(Tl). Anthracene is the organic scintillator with the highest light yield, while PBD is the fastest. Often several organic scintillators are mixed. Table 8.10 shows examples of presently commercially available organic scintillators from St Gobain and Eljen. All scintillators are polymerized in polyvinyltoluene, have a density of 1.023 g/cm^3 and have a refractive index of $n = 1.58$.

The extraction of a visible light yield signal becomes possible only by introducing a second fluorescent dye, called wavelength shifter, which converts the UV light into visible light as discussed in the next section. We choose this dye such that its absorption band is matched to emission spectrum of primary fluor, while its emission spectrum should match the spectral sensitivity of a second fluor or that of the photodetector. The two active components are either dissolved into suitable organic liquids or are mixed with a monomer, which is capable of polymerization. The resulting polymer can be cast into the desired shape for practical use. Liquid solvents comprise of bexylene, benzine (C_6H_6), toluene ($C_6H_5 \cdot CH_3$), phenylcyclohexane, xylene ($C_6H_4 \cdot (CH_3)_2$), triethylbenzene and decalin. Organic materials that polymerize include polysterene, polyvinyltoluene and polyphenylbenzine. Note that polysterene also absorbs light in the UV region and reemits it in the blue region.

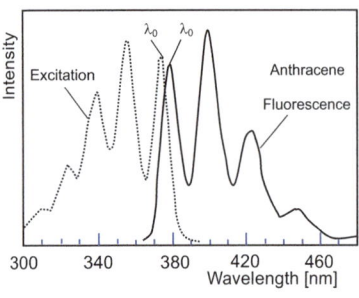 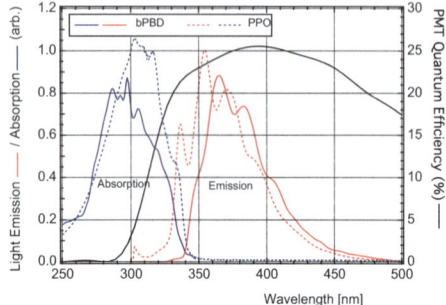

Fig. 8.58 Left: Absorption (dotted curve) and emission spectrum (solid curve) of anthracene. Reprinted with kind permission from [163], © 1991, American Chemical Society Publishing. All rights reserved. Right: Absorption (blue wavelength spectra) and emission (red wavelength spectra) of bPBD (solid lines) and PPO (dashed lines) [170]. The broad black curve shows the quantum efficiency of a typical PMT. Reprinted with kind permission from [170], © 2012, MiniBooNE Collaboration. All rights reserved

Table 8.10 Properties of common commercial organic scintillators listing the name, light yield with respect to anthracene (LY), number of photons per MeV N_γ, wavelength of maximum emission λ_{max}, light attenuation length λ_{att}, rise time t_r, decay time τ_d and pulse width Δt. These scintillators are produced by Saint Gobain [171] and Eljen [172]

Scintillator	LY [%]	N_γ	λ_{max} [nm]	λ_{att} [cm]	t_r [ns]	τ_d [ns]	Δt FWHM [ns]
BC400	65	10,000	423	160	0.9	2.4	2.7
BC404	68	10,400	408	140	0.7	1.8	2.2
BC408	64	10,000	425	210	0.9	2.1	2.5
BC412	60	9,200	434	210	1.0	3.3	4.2
BC416	38	5,850	434	210	–	4.0	5.3
BC420	64	10,000	391	140	0.5	1.5	1.2
EJ 200	64	10,000	425	380	0.9	2.1	2.5
EJ 204	68	10,400	408	160	0.7	1.8	2.2
EJ 208	60	9,200	435	400	1.0	3.3	4.2
EJ 212	65	10,000	423	250	0.9	2.4	2.7
EJ 230	64	9,700	391	120	0.5	1.5	1.3

8.7.1 Wavelength Shifters

In addition to the primary fluor, a secondary and maybe a tertiary fluor is added to shift the light into the visible spectrum, where the photodetector has a high quantum efficiency. Figure 8.59 illustrates the principle of wavelength shifting. The absorption spectrum of the secondary fluor is matched to the emission spectrum of the primary fluor, which typically emits light in the UV spectrum. So most of the light emitted by the primary fluor will be shifted to a higher wavelength. Another fluor may be added whose absorption spectrum matches the emission spectrum of the secondary fluor.

8.7 Plastic Scintillators

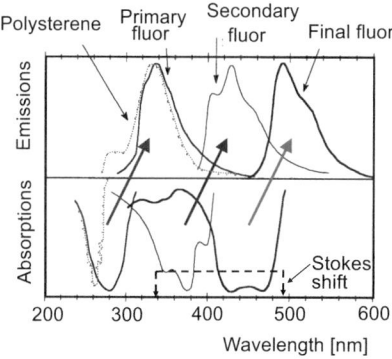

Fig. 8.59 Illustration of wavelength shifting in a polystyrene plastic scintillator that is doped with several fluors [173–175]. The primary fluor (buthyl PBD) has an absorption spectrum around 280 nm to absorb UV light shifting it to 340 nm, which lies in the absorption region of the secondary fluor (BDB). Polystyrene (dotted curve) itself absorbs UV light around 260 nm in a narrower absorption spectrum, while the emission spectrum is close to that of the primary fluor. The secondary fluor shifts the light to 430 nm, which overlaps with the absorption spectrum of the final fluor (Y7 from Kuraray) whose emission spectrum lies in the green-light region, which matches the absorption spectrum of silicon photodetectors. Reprinted with kind permission from [173], © 2001, C. Joram. All rights reserved

The shift in wavelength from the absorption maximum to the emission maximum is called Stokes' shift as illustrated in Fig. 8.60 (left). Figure 8.60 (right) shows the explanation of the Stokes' shift. The secondary fluor is excited into a higher vibrational state. The time scale is 10^{-14} s. The higher excited vibrational state first moves to the lowest excited vibrational state within 10^{-12} s. The de-excitation to a vibrational ground states occurs at a time scale of 10^{-8} s. Since there are many vibrational states a continuous emission spectrum results. Note that primary and secondary fluors are transparent to visible light. Table 8.11 list properties of the most common wavelength shifters, where the Kuraray YS fibers are new developments that have lower decay times. Figure 8.61 shows the chemical structure of bis-MSB, 3HF, POPOP, BBQ and TPB.

Figure 8.62 (left) shows the absorption and emission spectra of the primary fluor bi-phenyl and the wavelength shifter POPOP. Bi-phenyl absorbs light around 250 nm and emits it in the 300–400 nm region, which overlaps with the absorption spectrum of POPOP. The light emitted from POPOP spans from the blue to green optical region. Self-absorption of POPOP is rather small. Figure 8.62 (right) shows the absorption and emission spectra of BBQ in comparison to the emission spectra of PBD and POPOP. The broad BBQ absorption spectrum covers the PBD emission spectrum rather well, while it only covers part of the POPOP emission spectrum. Figure 8.63 shows the absorption length of light emitted by POPOP in the BBQ-doped wavelength shifter bar (with 90 mg/L) as a function of thickness. The absorption length is $\lambda_{abs} = (5.2 \pm 0.2)$ mm. We see that for a thickness of 15 mm more than 90% of the light is absorbed.

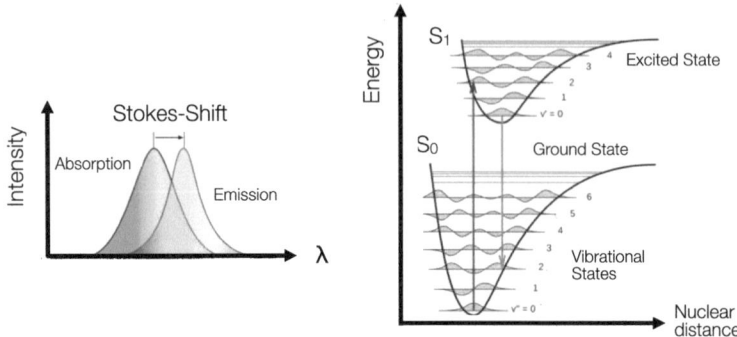

Fig. 8.60 Left: Definition of the Stokes' shift. Reprinted under CC-BY-SA-4.0 Licence from [176], © 2024, Wikipedia. Right: The Franck-Condon principle that explains the Stokes' shift [177]. A fluor has a ground state E_0 with several vibrational states denoted by V''. Similarly, the first excited state E_1 also has several vibrational states denoted by V'. The absorption of a photon typically happens in the lowest vibrational state of E_0, populating a vibrational state in E_1 (upward going arrow). The excited state first moves to the lowest vibrational state before emitting a photon that transfers it into a vibrational state of E_0 (downward going arrow). The lower-energy emitted photon has a higher wavelength than the original photon. Note that in nuclear coordinates the ground state and first excited state are shifted favoring in the figure transitions between $V' = 0$ and $V'' = 2$. Which transitions occur depends on the overlap of the wave functions in the vibrational states. Reprinted under CC-BY-SA-3.0 Licence from [177], © 1997, J. Franck

Table 8.11 Properties of wavelength shifters listing the name, chemical formula, wavelength for maximum absorption λ_{abs}, wavelength for peak emission $\lambda_{emission}$, decay time and attenuation length

Scintillator	Formula	λ_{abs} [nm]	$\lambda_{emission}$ [nm]	Decay time [ns]	Att. length [m]	Reference
POPOP	$C_{24}H_{16}N_2O_2$	359	420	1.6		[161]
bis-MSB	$C_{24}H_{22}$	350	420	1.2		[161]
BBQ		300, 382–430	480	11.5	1.3	[178, 179]
TPB	$C_{28}H_{22}$	90–250	440	1.8		[180]
Kuraray Y7		439	490		>2.8	[181]
Kuraray Y8		455	511		>3.0	[181]
Kuraray Y11	$C_8H_{17}N_4OBr$	430	476	6.9	>3.5	[181, 182]
Kuraray B2		375	437		>3.5	[181]
Kuraray B3		351	450		>4.0	[181]
Kuraray O2		535	550		>1.5	[181]
Kuraray R3		577	610		>2.0	[181]
Kuraray YS1		395	454	2.7	>3.5	[182, 183]
Kuraray YS2		422	474	3.2	>3.5	[182, 183]
Kuraray YS4		420	470	1.4	>3.0	[182, 183]
Kuraray YS6		414	462	1.3	>3.0	[182, 183]
SCSF-3HF	$C_3H_{12}F_3N$	340	530	7.0	>4.5	[184]

8.7 Plastic Scintillators

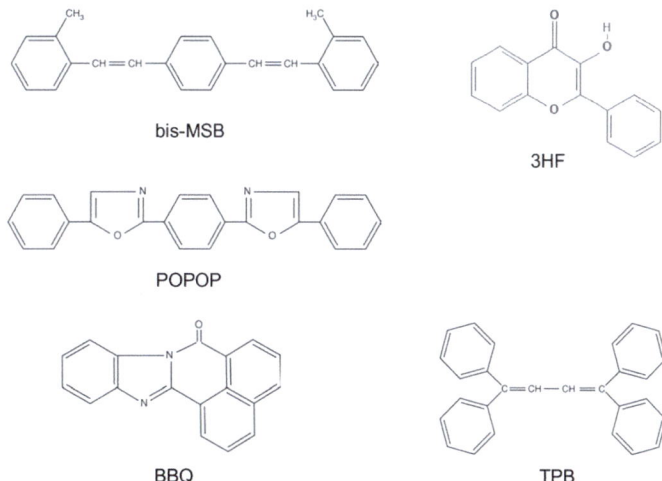

Fig. 8.61 Chemical structure of different wavelength shifters. Top left: 1,4-bis(2-methylstyryl)benzene (bis MSB). Top right: 3-hydroxyflavone (3HF). Center: 1,4-bis(5-phenyloxazol-2-yl) benzene (POPOP). Bottom left: BBQ. Bottom right: Tetraphenyl butadiene (TPB)

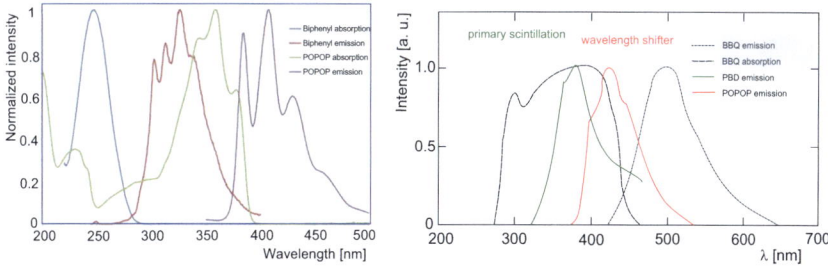

Fig. 8.62 Left: Absorption and emission spectra of the primary fluor bi-phenyl (blue and red curves) and the wavelength shifter POPOP (green and violet curves). Reprinted with kind permission from [186], © 2014, IOP Publishing. All rights reserved. Right: Absorption (black dashed curve) and emission spectra (black dotted curve) of BBQ with the emission spectra of PBD (red curve) and POPOP (green curve) overlaid. Reprinted with kind permission from [161], © 1989, Springer. All rights reserved

Figure 8.64 (left) shows the absorption and emission spectra of the Kuraray fibers Y7, Y8 and Y11, which shift light from the blue to the green-yellow optical region. The fibers have a small overlap with their absorption spectra yielding a small amount of self-absorption. The Y11 fiber covers lower wavelengths and is frequently used in subdetectors. Figure 8.64 (right) shows the absorption and emission spectra of the new Kuraray fibers YS1, YS2, YS4 and YS6 in comparison to those of the Y11 fiber. The absorption spectra of the new fibers lie lower than that of the Y11 fiber. In turn, the shifted light is also at slightly lower wavelengths. However, their decay times (1.3–3.2 ns) are considerably lower than 6.9 ns.

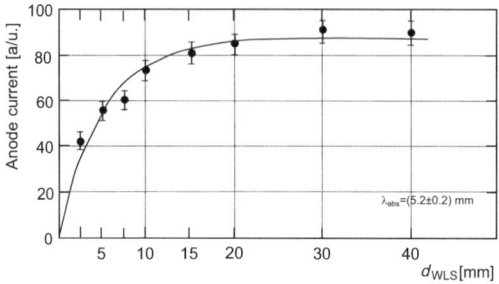

Fig. 8.63 The measured absorption of scintillation light in a BBQ-doped acrylic piece (concentration of 90 mg/L) as a function of the wavelength shifter thickness. Reprinted with kind permission from [187], © 1981, Elsevier. All rights reserved

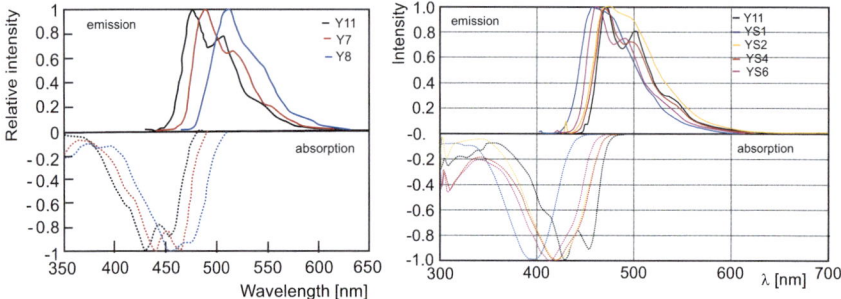

Fig. 8.64 Left: Absorption (dotted) and emission spectra (solid) of the wavelength-shifting fibers Y11 (black), Y8 (blue) and Y7 (brown) from Kuraray. Right: Absorption (dotted) and emission spectra (solid) of the new wavelength-shifting fibers from Kuraray, YS1 (blue), YS2 (orange), YS4 (brown) and YS6 (magenta) in comparison to those of Y11 (black)

As we saw, Cherenkov light is preferentially produced in the UV spectrum. So, to read out Cherenkov detectors with PMTs, the PMTs need to be equipped with a UV-sensitive window, such as a fused-silica window. However, in particular for large tubes these are much more expensive than PMTs with regular borosilicate glass windows. So an idea evolved to cover standard glass windows with wavelength shifters [185]. Studies were conducted with PTP and PBD in polystyrene and paraloid as a binder. Several 5-inch PMTs were tested. The PMT window was simply dumped into the dissolved scintillator. The gain in the number of observed photoelectrons with respect to the uncovered PMT was between 1.4 and 1.85. Figure 8.65 (left) shows the cathode sensitivity of a PMT coated with PTP-polystyrene or PTP-paraloid as a function of wavelength. The increased sensitivity of the PTP-paraloid configuration in the 200–240 nm region is due to a higher transmission of paraloid in this region as shown in Fig. 8.65 (right).

8.7 Plastic Scintillators

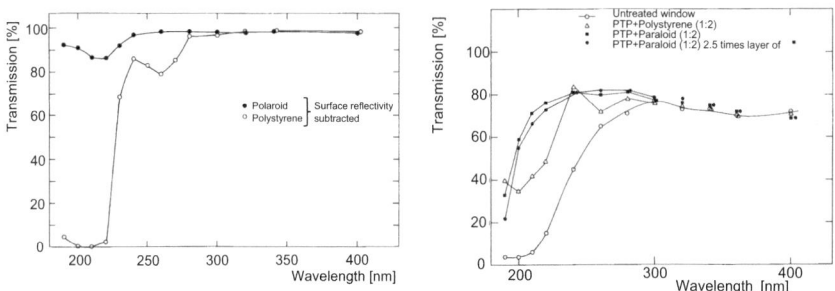

Fig. 8.65 Left: Cathode sensitivity of PMT RCA 8854 coated with PTP-polystyrene or PTP-paraloid as a function of wavelength. Curves are normalized to similar measurements with a quartz-window tube. Right: Transmission of thin layers of paraloid and polystyrene as a function of wavelength. Both reprinted with kind permission from [185], © 1979, Elsevier. All rights reserved

8.7.2 Usage of Plastic Scintillators

Organic scintillators are used in many sub-detectors, including trigger counters, time-of-flight systems (see Sect. 9.2) and sampling calorimeters. Instead of rigid scintillator plates, often scintillating fibers are used nowadays, in particular in calorimeters and tracking detectors. In Sect. 8.8, we discuss the Scintillating-Fiber Tracker that was built by the LHCb collaboration. For large detectors such as calorimeters or time-of-flight counters, where the scintillators are several meters long, it is important to have uniform light yield over the entire length of the scintillator. Figure 8.66 (left) shows the light output of an organic scintillator using 3% Naphtalene and 1% PBB as primary fluor and 0.01% bis-MSB as wavelength shifter dissolved in plexiglas at the near end (top) and at the far end (bottom). Without a filter, the huge light yield below 430 nm visible at the near end is lost at the far end. It is a typical feature of plastic scintillators that the absorption length for short wavelengths is small. To make the light yield more uniform, we insert a yellow filter in front of the photocathode that absorbs preferentially light at short wavelengths [188]. With the filter, the light spectrum at the near and far end are nearly equalized. However, the filter reduces the overall light output.

Figure 8.66 (top right) shows the attenuation of a large block (180 cm × 15 cm × 0.5 cm) with and without a yellow filter at the photocathode and with the backend blackened or covered by a reflector. Note that the configuration with a yellow filter and the far end blackened basically shows an exponential intensity reduction. This is not achieved without the filter or with a reflector at the far end. A total of 6,000 scintillators was produced in the configuration with a yellow filter the rear end blackened. The measured attenuation lengths are depicted in Fig. 8.66 (bottom right) yielding an average of $\lambda_{att} = 210$ cm with a standard deviation of $\sigma_{\lambda_{att}} = 18$ cm.

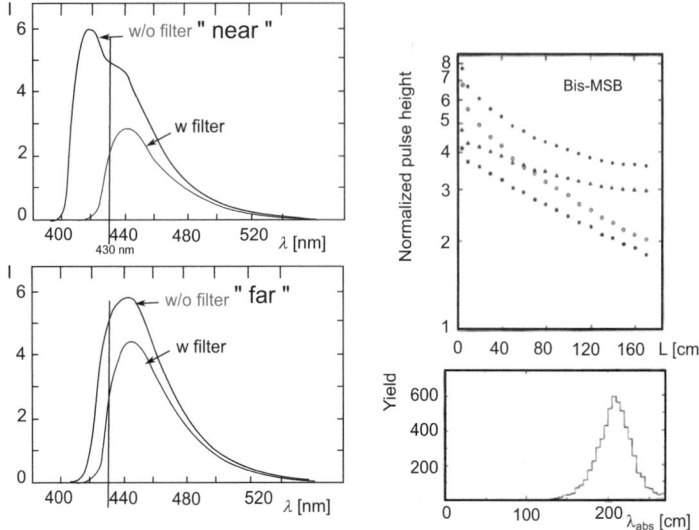

Fig. 8.66 Left top: Light yield of a plastic scintillator (dimension: 180 cm × 15 cm × 0.5 cm) as a function of wavelength measured 10 cm from the photomultiplier tube with and without a yellow filter placed in front of the photocathode. Left bottom: Light yield of the same scintillator as a function of wavelength measured 170 cm from the photomultiplier tube with and without a yellow filter placed in front of the photocathode. Right top: Light attenuation as a function of the distance from the PMT for the same plastic scintillator for different readout configurations, without a filter and far end reflecting (black solid points), without a filter and far end blackened (open points), with a filter and far end reflecting (black triangles) and with a filter and far end blackened (black squares). All reprinted with kind permission from [188], © 1982, Elsevier. All rights reserved. Right bottom: Absorption length of 6,000 bis-MSB strips. Reprinted with kind permission from [161], © 1989, Springer. All rights reserved

8.7.3 Collection of Scintillation Light and Light Guides

To transport the light produced from large scintillator plates or scintillator bars to the photodetector, we use adiabatic light guides. The light guide either couples directly to the scintillator or to a wavelength-shifting bar that is coupled to the scintillator via an air gap. Both scintillators and wavelength-shifting bars have long attenuation lengths for visible light, so light can travel long distances without large attenuation. The light typically propagates via multiple total reflections at the surface of the scintillator or wavelength-shifting bar. The total reflection angle $\alpha_g = \arcsin(1/n)$ is defined with respect to the normal to the scintillator/wavelength shifter plane, where n is the refractive index of the solid or liquid solvent. Polystyrene has $n = 1.581$ yielding $\alpha_g = 39°$, while plexiglas has $n = 1.49$ yielding $\alpha_g = 42°$.

The back face of a scintillator or wavelength shifter that typically has a rectangular profile with area F is imaged onto the area f of a photodetector. According to Liouville's theorem, which states that the phase space of the light rays consisting of three position dimension and three directions, is constant. Since f is smaller than F and the directions cannot be increased, the collected light is smaller than the ratio

8.7 Plastic Scintillators

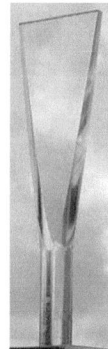

Fig. 8.67 Left: Wedge-shaped light guide. Right: Bent light guide. Both reprinted with kind permission from [189], © 2011, M. Krammer. All rights reserved

f/F. Figure 8.67 (left) shows a wedge-shaped light guide and Fig. 8.67 (right) shows a bent light guide as examples.

The time resolution of scintillator bars has three contributions:
1. Fluctuations in transit time in the PMT.
2. Variations of light paths in the scintillator light guide.
3. Fluctuations in light production due to the decay times of dyes implying to use fast dyes.

For long scintillators (order of 2 m), the contribution (2) becomes dominant. The best achieved time resolutions are 200 ps for long scintillators. Figure 8.68 (left) shows a comparison of the *rms* resolution in small and large scintillators. Short scintillators achieve time resolutions of 25–100 ps, while long scintillators achieve 200–300 ns. In addition of using the scintillator directly as a wave guide, we can attach a wavelength shifting bar to one side of the scintillator via an air gap. Figure 8.68 (right) illustrates the principle of light collection from scintillator bars with wavelength shifting bars. So, here the light leaves the scintillator and enters the wavelength shifter, where the light is absorbed and reemitted at higher wavelength. The reemitted light travels in the wavelength shifter until it reaches the photodetector. Note that the air gap between the wavelength shifter and the scintillator maintains the total reflection conditions in the wavelength shifter. A suitable wavelength shifter for the readout of light from a scintillator that has POPOP or bis-MSB admixtures is BBQ, which has a maximum emission wavelength of 480 nm. A concentration of 90 mg/L BBQ in a plexiglass is sufficient to achieve high yields.

8.7.4 Winston Cone

Another possibility to focus light from a scintillator or wavelength shifter to a photodetector is a Winston cone whose working principle is illustrated in Fig. 8.69 (left). A Winston cone has a parabolic surface with an entrance face diameter d_1 and an

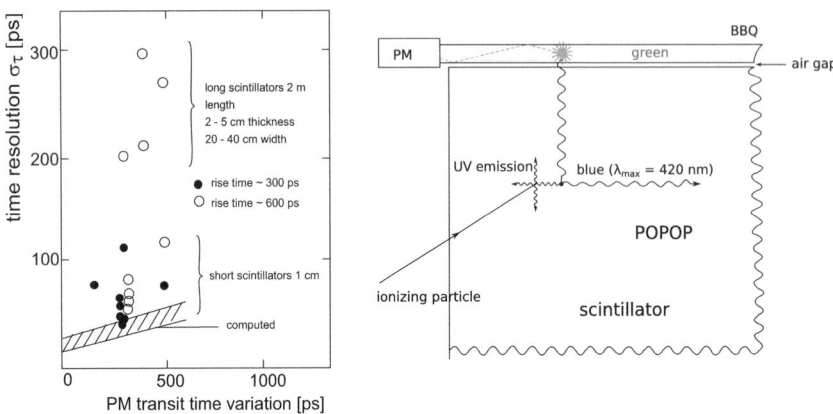

Fig. 8.68 Left: Measurements of the *rms* time resolution σ_t as a function of the transit time spread in the PMTs for short scintillators and for 2 m long scintillators. Solid points (open circles) denote signal rise times of 300 ns (600 ns). The calculated resolutions (shaded region) use rise times of 100-400 ps and decay times of 1500 ps. Reprinted with kind permission from [190], © 1981, IOP Publishing. All rights reserved. Right: Principle of light collection from a scintillator block with wavelength shifting bars. Reprinted with kind permission from [191], © 1978, B. Barish

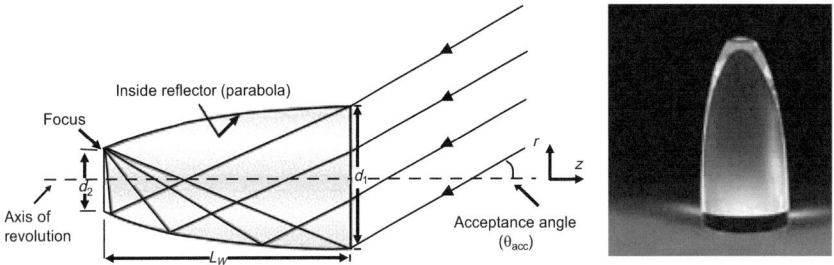

Fig. 8.69 Left: Principle of light focussing in a Winston cone. Reprinted under CC-BY-4.0-Deed Licence with permissions from [192], © 2016, EDP Sciences. All rights reserved. Right: Photograph of a Winston cone. Reprinted with permissions from [193], © 2024, Edmund Optics. All rights reserved

exit face diameter d_2. Its length is L_w. Light entering the cone under angle $\theta < \theta_{max}$ is reflected on the parabolic surface to exit the cone on point $r \leq d_2/2$. For θ_{max}, this exit point is the focus point labelled in the figure. The maximum acceptance angle is determined by

$$\sin\theta_{max} = \frac{d_2}{d_1}. \tag{8.61}$$

The length is also determined by

$$L_w \left(\frac{d_1 + d_2}{2}\right) \cot\theta. \tag{8.62}$$

For light entering the Winston under $-\theta_{\max}$, the reflection occurs on the upper parabolic surface focussing the light onto the point on the opposite side of the symmetry axis. Figure 8.69 (right) shows a photograph of a Winston cone. The typical usage is focusing scintillation light and Cherenkov light from cosmic rays.

8.7.5 Wavelength-Shifting Fibers

Wavelength-shifting fibers consist of a core made of Polystyrene or Polyvinyl toluene including a primary scintillator and wavelength shifter (refractive index n_0). The core is surrounded by one or two layers of thin material with refractive indices $n_2 < n_1 < n_0$. Figure 8.70 displays a schematic view (left) and a cross section (right) of a double-clad fiber. Unclad fibers have the problem that an air gap needs to be maintained along the entire fiber in order to prevent a destruction of the total reflection. In the double-clad fiber, photons with angles $\alpha_1 > \alpha_{g1}$ and $\alpha_2 > \alpha_{g2}$ experience total reflection at the boundaries n_0-n_1 and n_1-n_2, respectively. The fiber diameter is of the order of 1 mm. Typically, only a small fraction of the emitted light is transported to the fiber end.

In the previous chapter, we discussed the ATLAS Tile Calorimeter, which uses green wavelength-shifting fibers to read out the scintillator tiles. The CHORUS experiment at CERN used scintillating fibers depicted in Fig. 8.71 (left) for reading out the electromagnetic calorimeter. The fibers were arranged in parallel in a plane. In order to attach them to a photodetector they were bundled at one end. The MINOS experiment at Fermilab also used 1 mm green fibers to read out the plastic scintillator strips that were sandwiched between steel plates. Figure 8.71 (right) shows bundles of these fibers.

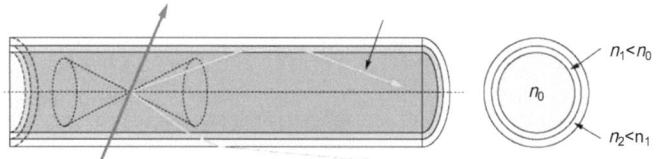

Fig. 8.70 Left: Schematic view of a double-clad scintillating fiber consisting of a core fiber with refractive index n_0, a first layer with refractive index n_1 and a second layer with refractive index n_2. The arrow shows a charged particle crossing the fiber. Light emitted under angles larger than the total reflection angles for $\alpha_1 > \alpha_{g1}$ with $\sin \alpha_{g1} = n_1/n_0$ and $\alpha_2 > \alpha_{g2}$ with $\sin \alpha_{g2} = n_2/n_1$ will be totally reflected (see upper light grey arrow) while light emitted under angles $\alpha_1 > \alpha_{g1}$ and $\alpha_2 < \alpha_{g2}$ may leave the fiber. Right: Cross section of a double-clad scintillating fiber. Both reprinted with kind permission from [189], © 2011, M. Krammer. All rights reserved

 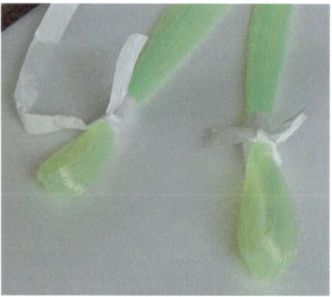

Fig. 8.71 Left: Scintillating fibers of the electromagnetic calorimeter of the CHORUS experiment at CERN. Right: Scintillating fibers of the MINOS experiment at Fermilab. Both reprinted with kind permission from [189], © 2011, M. Krammer. All rights reserved

8.8 The LHCb Scintillating-Fiber Tracker

The LHCb experiment has built the Scintillating-Fiber Tracker (SciFi) consisting of three tracking stations that are separated by 4.55 m from each other. The total depth is 18.18 m. Each tracking station consists of an axial layer X, two stereo layers U, V followed by another axial layer X. The stereo angles of the U,V layers are $\mp 5°$, respectively. Each layer is made of six layers of scintillating fibers. The dimensions are 2×3 m in the X direction and 2×2.4 m in the y direction. Figure 8.72 shows a schematic three-dimensional layout of the SciFi. The chambers are built from modules that have a size of 0.5 m × 4.8 m. Each module comprises of eight

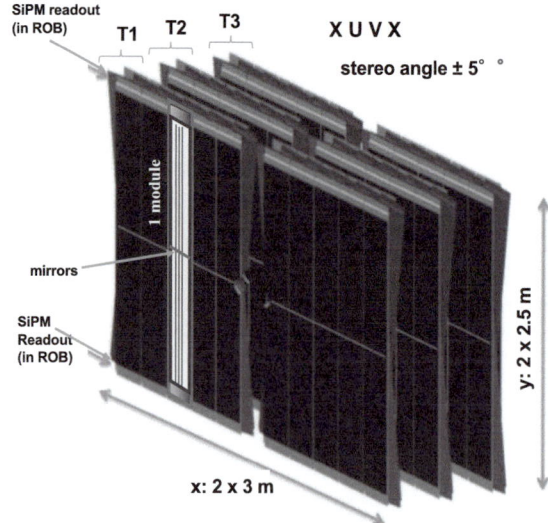

Fig. 8.72 Schematic layout of the LHCb scintillating-fiber tracker that consists of three tracking stations, each having four layers (X, U, V, X), where U, V are stereo layers with stereo angles of $\mp 5°$ [194]. Reprinted with kind permission from [194], © 2014, LHCb Collaboration. All rights reserved

8.8 The LHCb Scintillating-Fiber Tracker

scintillating-fiber mats with a length of 2.4 m as the active detector material. The 13 cm wide fiber mats consist of six layers of densely packed blue-emitting scintillating fibers. The scintillation light is detected with arrays of multi-channel SiPMs, which are cooled to $-40\,°C$ to minimize the expected high dark noise from neutron radiation. The readout of 524,000 channels occurs through custom-designed front-end electronics.

The fibers start in the center plane ($y = 0$) and run to the top and bottom. The fiber ends at $y = 0$ are covered with a mirror. A total of 11,000 km fibers is used in the detector covering an area of 342 m^2. The fibers, 250 μm in diameter, are double cladded providing a 5.4% internal reflection efficiency. They are read out at the top and bottom with arrays of SiPMs having 50 μm pixels. The decay constant of the scintillation light is 2.4 ns. The emission spectrum with a peak at 450 nm extends from 400 to 600 nm. The SiPM arrays are operated with an over-voltage of 3.5 V. The photon detection efficiency is 45% at 450 nm.

Figure 8.73 (left) shows a photograph of the six layers of scintillating fibers and Fig. 8.73 (right) depicts the mapping of the fibers onto the plane of SiPM arrays. A charged track passing through the fibers will produce photons that are transmitted onto the SiPM array. The yellow-shaped pixels show hits producing the cluster shown on top right. Dedicated electronics is used to amplify and digitize the signal and save it on disk. Note that fibers near the beam see a lot of radiation. Thus, the photon yield produced in the center plane is about half of that produced near the SiPM array. Figure 8.74 shows the relative light yield as a function of the fiber length from the readout side for different configurations. After irradiation, the light yield at the center plane for a time selection of 25 ns and a mirror placed at the center plane is reduced by 50%. Note that a non-irradiated fiber without the mirror and no time selection has a light yield of less than 40% at the center plane. The mirror increases the light yield near the SiPM array by 15% with respect to the fiber without a mirror. Near the center plane the mirror increases the light yield to 70%.

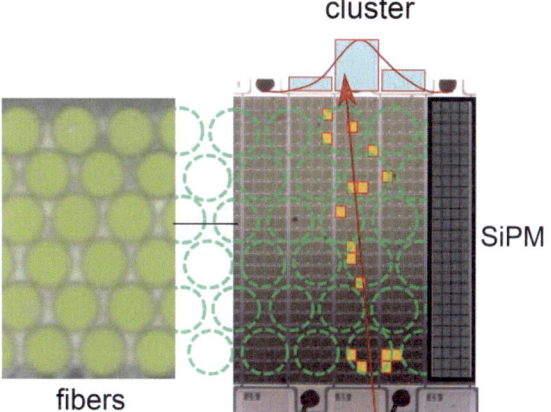

Fig. 8.73 Left: Photograph of the six layers of scintillating fibers. Right: Mapping of the fibers onto the SiPM arrays. The right arrow shows a penetrating charged track. The yellow-shaded pixels show hits produced by the charged track. The cluster shape is depicted on the top right. Both reprinted with kind permission from [195], © 2020, IOP Publishing. All rights reserved

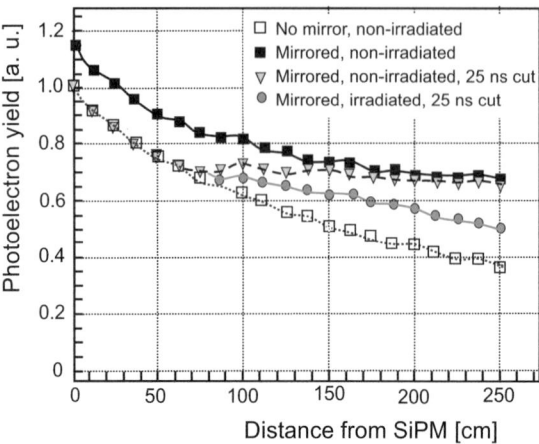

Fig. 8.74 Light yield as a function of the distance from the SiPM for a non-irradiated fiber with no mirror and no integration time selection (open squares with dashed curve), a non-irradiated fiber with a mirror in the center plane and no integration time selection (squares with solid black curve), a non-irradiated fiber with a mirror in the center plane and a 25 ns integration time selection (triangles with solid curve) and an irradiated fiber with a mirror in the center plane and a 25 ns integration time selection (grey points with solid curve). Reprinted with kind permission from [196], © 2014, LHCb Collaboration. All rights reserved

The new tracker adds less than 1% of a radiation length. It is expected to be more than 99% efficient and radiation tolerant. The single hit spatial resolution is $< 100\,\mu\mathrm{m}$. It is designed to run at 40 MHz and its performance needs to last for an integrated luminosity of 50 fb^{-1}.

Exercises

8.1 Design a voltage divider for a 16-stage photomultiplier that is operated with a voltage of $+1700$ V. Each dynode should see the same potential difference, except for the first dynode that is a factor of two larger. What is the dark current at 20 °C? How large is the gain if the first dynode produces four electrons and each other dynode produces on average 2.4 electrons per incoming electrons?

8.2 Consider a microchannel plate that is 500 μm in length and has 15 μm diameter channels. The incoming electrons have an energy of 1.2 eV. Assume that 100 electrons were produced. Plot the gain voltage-dependence for $c_m = 0.2\,\mathrm{V}^{-1/2}$. What is the gain at 1 kV if 100 electrons enter the MCP-PMT? For which ratio of length/diameter is the gain maximum? Now consider a Chevron design. How large is the gain?

8.3 Consider a 1 cm × 1 cm PIN diode that is 300 μm thick. Compute the quantum efficiency for 500 nm light that has an absorption coefficient of $\alpha_{\mathrm{abs}} = 10^4/\mathrm{cm}$. Assume that about 90% of the photons are absorbed in the PIN diode, which is fully depleted. The diffusion coefficient is $D_p = 12\,\mathrm{cm}^2/\mathrm{s}$ and the lifetime of the excess carriers is 1 ns. In a LYSO crystal 27,000 photons per MeV are produced within 40 ns. How large are the primary photon current and the optical power? Check that the gain is one. Note that the signal is not modulated.

8.4 Consider the same PIN photodiode as in the previous problem. For a dark current of $I_d = 150$ nA, a background current of $I_B = 10$ nA and a bandwidth of 1 Hz, calculate the average current-based noise and the average thermal noise at room temperature. Assume that $R_i = 100$ MΩ, $R_j = 100$ kΩ and $R_L = 1$ MΩ. What is the $(S/N)_{\text{power}}$?

8.5 Consider a 1 cm × 1 cm Avalanche Photodiode that is 300 μm thick. Determine the gain for $\alpha_n = 0.04$ and $\alpha_p = 0.1 * \alpha_n$. What is the electric-field strength? Plot the gain versus α_n.

8.6 For the APD in the previous problem compute the S/N ratio. The quantum efficiency is $\epsilon_{\text{QE}} = 0.84$. Assume $P_{\text{opt}} = 1.14 \times 10^5$ MeV/s and an energy of 1 MeV. Measurements occur at room temperature and the bandwidth is 100 kHz. The equivalent resistor is obtained from $R_i = 100$ MΩ, $R_j = 100$ kΩ and $R_L = 1$ MΩ. Use $k_{\text{eff}} = 0.1$ and $k_{\text{eff}} = 1$. Calculate $F(G)$ and determine the noise-equivalent power.

8.7 Consider a SiPM at room temperature with 20 μm × 20 μm pixels. The depletion layer is 4 μm. Calculate the pixel capacitance and the pixel charge for an over-voltage of 5 V. Plot the dependence of the charge as a function of over-voltage. For $Q_0 = 6\, Q_{\text{pix}}$ plot the charge as a function of time. The rise time is $\tau_r = 1$ ns and $\tau_d = 5$ ns. How large is the amplitude? Where is the minimum?

8.8 For a SiPM, we measured the gain-voltage and gain-temperature dependence, yielding $dG/dV = c + d \cdot T$ and $dG/dT = a + b \cdot V$ with $a = -52{,}090$, b $= 1{,}378$, c $= 384{,}766$ and d $=2{,}265$. Plot dV/dT as a function of temperature. How large is the voltage difference from $T = 5\,°C$ to $T = 45\,°C$?

8.9 Consider a scintillating tile of dimensions 3 cm × 3 cm × 0.5 cm that is read out with a green wavelength-shifting fiber 1 mm in diameter. The fiber sits in a groove having a shape of a quarter circle with a radius of 2.8 cm and is attached to a SiPM on one side and has a mirror on the other side. The scintillator has a refractive index of 1.59. The fiber is single clad with refractive indices of 1.59 and 1.49. The attenuation lengths in the tile and that in the fiber are 3.7 m. The absorption length of blue light in the fiber is 5 mm. The quantum efficiency of the SiPM is 0.75. A charged particle looses 2.05 MeV/cm energy in the tile. The produced photons are reflected inside the tile. List all the efficiencies and estimate them. Make a model for trapping light in the fiber.

8.10 A cylindrical double-clad scintillating fiber has a diameter of 250 μm and is 4.8 m long. One end is covered with a mirror, the other end is attached to a photodetector. The refractive indices from inside out are $n_0 = 1.59, n_1 = 1.49$ and $n_2 = 1.42$. The cladding layers are each 3.75 μm thick. A charged particle looses 2.052 MeV/cm energy on average. The attenuation length is 3.7 m. How much light is totally reflected inside the fiber? Assume that the light is produced 10 cm from the mirror end. How much light reaches the other end if it is reflected at maximum angle and if it is parallel to the fiber?

References

1. H. Iams, B. Salzberg, Proc. IRE **23**, 55 (1935)
2. S.O. Flyckt, C. Marmonier, *Photomultiplier Tubes* (PHOTONIS Company, Brive, France, 2002)
3. R.W. Engstrom, *RCA Photomultiplier Handbook* (1980)
4. J.M. Schonkeren, *Photomultipliers*, Philips Application Book Series, ed. by H. Kater, L.J. Thompson (Philips Eindhoven, 1970)
5. Hamamatsu Photonics, https://www.hamamatsu.com/content/dam/hamamatsu-photonics/sites/documents/99_SALES_LIBRARY/etd/PMT_handbook_v4E.pdf
6. V. Taillandier, PhD thesis, Leicester University (2013)
7. EMI photomultiplier catalogue 1979, EMI Industrial Electronics Ltd; https://frank.pocnet.net/other/EMI/EMI_Photomultiplier_Tubes_1970.pdf
8. D.Y. Akimov et al., J. Phys.: Conf. Ser. **798**, 012211 (2017)
9. V.P. Beguchev, I.A. Shefova, A.L. Musatov, J. Phys. D: Appl. Phys. **26**, 1499 (1993)
10. Philips, https://wwwusers.ts.infn.it/rui/univ/Acquisizione_Dati/Manuals/PhilipsXP2020.pdf
11. W.E. Slater et al., Nucl. Instrum. Meth. **154**, 223 (1978)
12. C.C. Lo, B. Leskovar, IEEE Trans. NS **29**, 183 (1982)
13. Hamamatsu Photonics, https://www.hamamatsu.com/eu/en/product/optical-sensors/pmt/pmt_tube-alone/metal-package-type/R5900U-01-L16.html
14. G. Montarou et al., Report PCCF-RI-9705 (1997)
15. https://et-enterprises.com/images/data_sheets/9125B.pdf
16. I. Adam et al., Nucl. Instrum. Meth. A **433**, 121 (1999)
17. P. Bourgeois et al., Nucl. Instrum. Meth. A **442**, 105 (2000)
18. Hamamatsu Photonics, https://www.hamamatsu.com/content/dam/hamamatsu-photonics/sites/documents/99_SALES_LIBRARY/etd/LARGE_AREA_PMT_TPMH1376E.pdf
19. C. Bronner et al., J. Phys.: Conf. Ser. **1468**, 012237 (2020)
20. Hamamatsu Photonics, https://www.hamamatsu.com/content/dam/hamamatsu-photonics/sites/documents/99_SALES_LIBRARY/etd/R11265U_H11934_TPMH1336E.pdf
21. M. Calvi et al., LHCb-PUB-2014-043 (2014)
22. Hamamatsu Photonics, https://datasheetspdf.com/datasheet-pdf/353391/R5505.html
23. R.D. Appuhn et al., Nucl. Instrum. Meth. A **386**, 397 (1997)
24. Photonis Corp., https://hallcweb.jlab.org/DocDB/0008/000809/001/PhotonisCatalog.pdf
25. Hamamatsu Photonics, https://www.hamamatsu.com/eu/en/product/optical-sensors/pmt/pmt_tube-alone/index.html
26. https://www.emi-shielding.net/magnetic-shields-gallery/
27. H. Kume et al., Nucl. Instrum. Meth. **205**, 443 (1983)
28. Hamamatsu Photonics, https://www.hamamatsu.com/eu/en/why-hamamatsu/academic-projects/exploring-neutrinos/index.html
29. H. Houtermans, Nucl. Instrum. Meth. **112**, 121 (1973)
30. S. Tokar et al., ATLAS TileCal-99-005 (1999)
31. Harshaw catalog, file:///Users/gxe/Downloads/Harshaw-3.pdf
32. M. Moszynski, Nucl. Instrum. Meth. A **324**, 269 (1993)
33. J. Janoth et al., Nucl. Instrum. Meth. A **350**, 221 (1993)
34. V. Sulkosky et al., Nucl. Instrum. Meth. A **827**, 137 (2016)
35. H. Dho, V. Kuznetsov, W. Kim, Korean Phys. Soc. **61**-1, 49 (2011)
36. G.S. Abrams et al., Nucl. Instrum. Meth. A **281**, 55 (1989)
37. J. Schwiening et al., Nucl. Instrum. Meth. A **553**, 317 (2005)
38. K. Nakamura, Int. J. Mod. Phys. A **18**, 4053 (2003)
39. M. Fiorini, PoS(EPS-HEP2017), 494 (2017)
40. ATLAS tile calorimeter TDR, CERN-LHCC-96-42 (1996)
41. Superkamiokande, https://www-sk.icrr.u-tokyo.ac.jp/en/sk/experience/gallery/
42. https://slideplayer.com/slide/7107770/; Hamamatsu photodetector catalogue
43. T. Hokuue et al., Nucl. Instrum. Meth. A **494**, 436 (2002)

44. J. Bibby et al., Nucl. Instrum. Meth. A **546**, 93 (2005)
45. K.W. Bell et al., IEEE Trans. NS **51**, 2284 (2004)
46. R.M. Brown et al., IEEE Trans. Nucl. Sci. NS-**32**, 736 (1985)
47. J.L. Wiza, Nucl. Instrum. Meth. **162**, 587 (1979)
48. Del Mar Photonics, http://www.dmphotonics.com/MCP_MCPImageIntensifiers/microchannel_plates.htm
49. R. Gao et al., JINST **17**, C05015 (2022)
50. P. Schagen, vol. 1 (New York, 1974)
51. A.U. Mane et al., Phys. Proc. **37**, 722 (2012)
52. A. Lehmann et al., JINST **13**-02, C02010 (2018)
53. J. Va'vra, arXiv:1906.11322 [physics.ins-det] (2020)
54. C.C. Lo, B. Leskovar, IEEE Trans. Nucl. Sci. NS-**28**-1 (1981)
55. B. Leskovar, T.T. Shimizu, IEEE Trans. Nucl. Sci. NS-**34**-1 (1987)
56. I. Yamazaki et al., Rev. Sci. Instrum. **56**, 1187 (1985)
57. Hamamatsu Photonics, https://mosphys.ru/indico/event/1/contributions/6/attachments/38/63/Lehmann_Talk_MCPs.pdf (2018)
58. K. Inami, https://indico.cern.ch/event/999817/contributions/4253050/attachments/2240234/3798061/20210506MCP_inami_s.pdf (2021)
59. B. Leskovar, LBL-26321 report (1989)
60. B. Leskovar, T.T. Shimizu, IEEE NS-**34**-1, 427 (1987)
61. Hamamatsu Photonics, https://www.hamamatsu.com/eu/en/product/optical-sensors/pmt/pmt_tube-alone/mcp-pmt/R3809U-50.html
62. M. Akatsu et al., Nucl. Instrum. Meth. A **528**, 763 (2004)
63. Y. Barnyakov et al., Nucl. Instrum. Meth. A **567**, 17 (2006)
64. K. Kosev et al., Nucl. Instrum. Meth. A **624**, 641 (2010)
65. Hamamatsu Photonics, https://www.hamamatsu.com/content/dam/hamamatsu-photonics/sites/documents/99_SALES_LIBRARY/etd/R10754_TPMH1364E.pdf
66. The PANDA Collaboration (F. Davi et al.), Technical Design Report, arXiv:1912.12638 [physics.ins-det] (2020)
67. M. Pfaffinger et al., https://eic.jlab.org/dirc/images/c/c6/Photonis_XP85112_SpecificationSheet.pdf; https://indico.gsi.de/event/6949/contributions/31421/attachments/22529/28278/20180306_PANDA_Darmstadt_Pfaffinger.pdf (2017)
68. Photek, https://www.photek.com/wp-content/uploads/2021/10/PH-DS006-PhotomultiplierTube-Rev05-Oct21.pdf; https://www.photek.com/pdf/datasheets/detectors/DS034-Auratek-MAPMT253-Detector-Datasheet.pdf
69. A. Lehmann et al., Nucl. Instrum. Meth. A **639**, 144 (2011)
70. K. Matsuoka et al., Nucl. Instrum. Meth. A **876**, 93 (2017)
71. The Belle II TOP Group (K. Suzuki et al.), Nucl. Instrum. Meth. A **876**, 252 (2017)
72. The PANDA Cherenkov Group (A. Lehmann et al.), JINST **9**, C02009 (2014)
73. V.A. Li et al., AIP Adv. **8**, 095003 (2018)
74. The ALICE Collaboration (Y.A. Melikyan et al.), Nucl. Instrum. Meth. A **952**, 161689 (2020)
75. Y. Chang et al., Nucl. Instrum. Meth. A **824**, 143 (2016)
76. L. Ren et al., JPS Conf. Proc. **27**, 011014 (2019)
77. S.M. Sze, *Physics of Semiconductor Devices* (Wiley, 1981)
78. M. Razeghi, *Technology of Quantum Devices* (Springer, 2010)
79. Z.R. Szczepaniak, B.A. Galwas, JTIT **3**, 86 (2001)
80. The *BABAR* Collaboration (B. Aubert et al.), Nucl. Instrum. Meth. A **479**, 1 (2002)
81. Hamamatsu Photonics, https://www.datasheets360.com/pdf/6680401822795397410
82. Hamamatsu Photonics, https://www.hamamatsu.com/content/dam/hamamatsu-photonics/sites/documents/99_SALES_LIBRARY/ssd/s2744-08_etc_kpin1049e.pdf
83. G. Eigen, D.G. Hitlin, Int. WS on Heavy Scint., Caltech Report CALT-68-1836, SLAC-BABAR-NOTE-098 (1992)
84. Hamamatsu Photonics, https://www.hamamatsu.com/content/dam/hamamatsu-photonics/sites/documents/99_SALES_LIBRARY/ssd/s3204-08_etc_kpin1051e.pdf

85. Excelitas technology, https://www.excelitas.com/product/vth2110-si-pd-chip-form-25-mm2
86. GPD optoelectronics, https://www.gpd-ir.com/germanium-avalanche-photodiodes/
87. W.T. Tsang (ed.), *Semiconductors and Semimetals*, vol. 22, Part D "Photodetectors" (Academic Press, 1985)
88. T.T.T. Pham, J. Semicond. Tech. Sci. **18**(5) (2018)
89. G.L. Hansen, J.L. Schmit, J. App. Phys. **53**, 7099 (1982); ibid **54**, 1639 (1983)
90. K. Deiters et al., Nucl. Instrum. Meth. A **453**, 223 (2000)
91. I. Britvitch et al., Nucl. Instrum. Meth. A **535**, 523 (2004)
92. A. Stoykov, R. Scheuermann, https://indico.hep.caltech.edu/event/11/attachments/38/51/apd_intro_jra.pdf
93. S. Reggiani, Personal Communication (2023)
94. R. Van Overstraeten, H. De Man, Solid State Electron. **13**, 583 (1970)
95. W.N. Grant, Solid State Electron. **16**, 583 (1973)
96. H. Maes, K. De Mayer, R. Van Overstraeten, Solid State Electron. **33**, 705 (1990)
97. I. Takayanagi, K. Matsumoto, J. Nakamura, J. Appl. Phys. **72**-5 (1992)
98. M. Valdinoci et al., Proc. SISPAD 27 (1999)
99. A. Karar, Y. Musienko, J.Ch. Vanel, Nucl. Instrum. Meth. A **428**, 413 (1999)
100. M. Dittmar, D. Zürcher, CERN-CMS-NOTE-1999-040 (1999)
101. The CMS Collaboration (G.L. Bayatian et al.), CERN/LHCC 97-33, CMS TDR 4 (1997)
102. J. Grahl et al., Nucl. Instrum. Meth. A **504**, 44 (2003)
103. D. Renker, Nucl. Instrum. Meth. A **486**, 164 (2002)
104. Hamamatsu Photonics, https://www.hamamatsu.com/content/dam/hamamatsu-photonics/sites/documents/99_SALES_LIBRARY/ssd/s8664_series_kapd1012e.pdf
105. Advanced Photonix incoporated, https://www.advancedphotonix.com/wp-content/uploads/2016/07/Deep-UV-Large-Area-Avalanche-Photodiodes-LAAPD-V1.16.pdf
106. J.P. Peigneux et al., Nucl. Instrum. Meth. A **378**, 410 (1996)
107. First Sensor, https://www.first-sensor.com/cms/upload/datasheets/AD800-11_TO_3001356.pdf
108. M. Ryan, Nucl. Instrum. Meth. A **598**, 217 (2009)
109. The CMS ECAL Collaboration (R. Arcidiacono et al.), J. Phys. Conf. Ser. **160**, 012048 (2009)
110. P. Buzhan et al., ICFA Instrum. Bull. **23**, 28 (2001)
111. P. Buzhan et al., Particle Physics, 7th Int. Conf. Adv. Tech. part. phys., 717 (2002)
112. J. Barrel, talk, https://www.slideserve.com/javen/study-of-silicon-photomultipliers-powerpoint-ppt-presentation (2004)
113. P. Buzhan et al., Nucl. Instrum. Meth. A **504**, 48 (2003)
114. F. Anghinolfi et al., IEEE Trans. Nucl. Sci. **51**-5, 1974 (2004)
115. S. Gundacker et al., PoS Photo Det **2012**, 016 (2012)
116. V. Andreev et al., Nucl. Instrum. Meth. A **540**, 368 (2004)
117. F. Acerbi, S. Gundacker, Nucl. Instrum. Meth. A **926**, 16 (2019)
118. Hamamatsu Photonics, https://www.hamamatsu.com/content/dam/hamamatsu-photonics/sites/documents/99_SALES_LIBRARY/ssd/s13360_series_kapd1052e.pdf
119. Hamamatsu Photonics, https://www.hamamatsu.com/content/dam/hamamatsu-photonics/sites/documents/99_SALES_LIBRARY/ssd/s14160-1310ps_etc_kapd1070e.pdf
120. Hamamatsu Photonics, https://hamamatsu.su/files/uploads/pdf/3_mppc/s13370_vuv4-mppc_b_(1).pdf
121. M. Danilov, Nucl. Instrum. Meth. A **604**, 183 (2009)
122. The CALICE Collaboration (C. Adloff et al.), JINST **5**, P05004 (2010)
123. N. Serra et al., JINST **8**, P03019 (2013)
124. KETEK, https://www.ketek.net/wp-content/uploads/2017/01/KETEK-PM3325-EB-PM3350-EB-Datasheet.pdf
125. M. Jangra et al., J. Phys. Conf. Ser. **2374**(1), 012125 (2022)
126. J. Jiang et al., Nucl. Instrum. Meth. A **980**, 164481 (2020)
127. R. Mirzoyan et al., NDIP'08, Aix-les-Bains (2008)
128. G. Eigen et al., JINST **14**-05, P05006 (2019)

129. The LHCb Collaboration (R. Greim), JINST **12**-02, C02053 (2017)
130. M. Danilov, Nucl. Instrum. Meth. A **453**, 223 (2000)
131. A. Massafferri et al., JINST **15**-08, C08006 (2020)
132. The CMS Collaboration (A.M. Magnan et al.), JINST **12**-01, C01042 (2017)
133. N. D'Ascenzo et al., JINST **15**, C07006 (2020)
134. N. Atanov et al., Nucl. Instrum. Meth. A **989**, 164967 (2019)
135. The CTA consortium and SST-1M consortium (E.J. Schioppa et al.), JINST **11**-01, C01038 (2016)
136. E. Popova et al., JPS Conf. Proc. **27**, 012018 (2019)
137. N. Kratochwil et al., Springer Proc. Phys. **212**, 283 (2018)
138. P. Lecoq, S. Gundacker, EPJ Plus **136**, 292 (2021). J. CH
139. J.Y. Yeom et al., Phys. Med. Biol. **58**, 1207 (2013)
140. T. Frach et al., IEEE Nucl. Sci. Symp. Conf. Rec. (NSS/MIC) (2009)
141. M.D. Petroff, M.G. Stapelbroek, W.A. Kleinhans, Appl. Phys. Lett. **51**-6, 408 (1987); M.D. Petroff, M. Atac, IEEE Trans Nucl. Sci. **36**-1 (1989); IEEE Trans. Nucl. Sci. **36**, 158 (1989) 158
142. M.R. Wayne, Nucl. Instrum. Meth. A **387**, 278 (1997)
143. C. Joram, Nucl. Phys. B Proc. Suppl. **78**, 407 (1999)
144. C. Matteuzzi, particle Identification Talk, https://speakapp.link/to/LDxQFg (2002)
145. A. Braem et al., Nucl. Instrum. Meth. A **478**, 400 (2002)
146. K. Hoepfner, A. Skiba, C. Hensel, Nucl. Instrum. Meth. A **483**, 747 (2002)
147. M. Adinolfi et al., Eur. Phys. J. C **73**, 2431 (2013)
148. T. Saito et al., PoS PD07, 041 (2006)
149. R. Orito et al., arXiv: 0907.0865 [astro-ph.IM] (2009)
150. A. Braem et al., Nucl. Instrum. Meth. A **442**, 128 (2000)
151. N. Kanaya et al., Nucl. Instrum. Meth. A **421**, 512 (1999)
152. I. Adachi et al., PoS PD-**07**, 035 (2006)
153. I. Adachi et al., Nucl. Instrum. Meth. A **623**, 285 (2010)
154. The Particle Data Group (S. Navas et al.), Phys. Rev D. **110**, 030001 (2024)
155. The ATLAS Collaboration (M.F. Daneri et al.), CERN-LHCC-2020-007 (2020)
156. The ATLAS HGTD group (S. Ali et al.), Talk at TIPP Conference (2023)
157. The CMS Collaboration (M. Malberti et al.), JINST **15**, C04014 (2020)
158. F.D. Brooks, Nucl. Instrum. Meth. **162**, 477 (1979)
159. The CMS Collaboration (P. Meridiani), Talk at TIPP Conference (2023)
160. https://en.wikipedia.org/wiki/File:Pi-Bond.svg
161. K. Kleinknecht, *Detectors for Particle Radiation*, 2nd edn. (Cambridge University Press, 1998)
162. S. Sau et al., J. Incl, Phen. Macrocyc. Chem. **48**, 173 (2004)
163. C.M. Byron, T.C. Werner, J. Chem. Ed. **68**, 433 (1991)
164. U. Akgun et al., IEEE (2008). https://doi.org/10.1109/NSSMIC.2008.4774796, 2228
165. The MiniBoone Collaboration (R. Dharmapalan et al.), arXiv:1210.2296 [hep-ex] (2012)
166. N.Z. Galunov, O.A. Tarasenko, V.A. Tarasov, Func. Mat. **20**-03 (2013)
167. https://omlc.org/spectra/PhotochemCAD/html/110.html; https://inradoptics.com/pdfs/Inrad_AN_Stilbene.pdf, https://chemistry.stackexchange.com/questions/77410/is-it-possible-for-a-substance-to-absorb-a-longer-wavelength-of-em-wave-and-emit
168. https://omlc.org/spectra/PhotochemCAD/html/020.html
169. A. Wieczorek, Dissertation, Krakow (2017)
170. The MiniBooNE Collaboration (R. Dharmapalan et al.), Letter of Intent, arXiv:1210.2296 [hep-ex] (2012)
171. St Gobain, https://www.crystals.saint-gobain.com/products/plastic-scintillators; https://www.luxiumsolutions.com/radiation-detection-scintillators/plastic-scintillators/bc400-bc404-bc408-bc412-bc416
172. Eljen, https://eljentechnology.com/products/plastic-scintillators
173. C. Joram, Summer Student Lectures at CERN (2001)
174. C. Zorn, *Instrumentation in High-Energy Physics* (World Scientific, 1992)

175. T. Kamon et al., Nucl. Instrum. Meth. A **213**, 261 (1983)
176. https://en.wikipedia.org/wiki/Stokes_shift
177. https://en.wikipedia.org/wiki/Franck%E2%80%93Condon_principle
178. L.A. Allemand et al., Nucl. Instrum. Meth. **164**, 93 (1979)
179. M. Bourdinaud, J.C. Thevenin, Phys. Scr. **23**, 534 (1981)
180. R. Francini et al., JINST **8**, C09010 (2013)
181. http://kuraraypsf.jp/psf/ws.html
182. http://kuraraypsf.jp/pdf/YSSeries_230926.pdf
183. S. Kodama et al., Prog. Theor. Exp. Phys. **2024**-5, (2024)
184. Kuraray, http://kuraraypsf.jp/psf/sf.html
185. G. Eigen, E. Lorenz, Nucl. Instrum. Meth. **166**, 165 (1979)
186. M. Hamel, A.M. Frelin-Labalme, S. Normand, EPL **106**, 52001 (2014)
187. P. Klasen et al., Nucl. Instrum. Meth. **185**, 67 (1981)
188. F. Klawonn et al., Nucl. Instrum. Meth. **195**, 483 (1982)
189. M. Krammer, Lectures at the IPM School, Teheran, https://www.oeaw.ac.at/fileadmin/Institute/HEPHY/PDF/ausbildung/praktikum/VO-5-Scintillators.pdf (2011)
190. P.J. Carlson, Phys. Scripta **23**, 393 (1981)
191. B. Barish et al., FERMILAB-PUB-77-180-E (1978)
192. A. Segal, Renew. Energy Environ. Sustain. **1**, 1 (2016)
193. https://www.edmundoptics.com/p/25deg-434mm-output-dia-compound-parabolic-concentrator/17843/
194. Ch. Joram, https://slidetodoc.com/lhcb-sci-fi-the-new-fibre-tracker-for/ (2014)
195. The LHCb Collaboration (A. Massafferri), JINST **15**-08, C08006 (2020)
196. The LHCb Tracker Upgrade TDR, CERN/LHCC 2014-001 (2014)

Open Access This chapter is licensed under the terms of the Creative Commons Attribution 4.0 International License (http://creativecommons.org/licenses/by/4.0/), which permits use, sharing, adaptation, distribution and reproduction in any medium or format, as long as you give appropriate credit to the original author(s) and the source, provide a link to the Creative Commons license and indicate if changes were made.

The images or other third party material in this chapter are included in the chapter's Creative Commons license, unless indicated otherwise in a credit line to the material. If material is not included in the chapter's Creative Commons license and your intended use is not permitted by statutory regulation or exceeds the permitted use, you will need to obtain permission directly from the copyright holder.

ns and Devices

This chapter focusses on measurement techniques and subdetectors for identifying charged particles and neutrons. We saw in Chap. 5 that the transverse momentum is obtained from the radius of the spiral track and the magnitude of the magnetic field. Now by measuring the velocity β, we can determine the particle mass m since

$$m = \frac{p_T}{\beta \gamma c}, \tag{9.1}$$

where p_T is the magnitude of the transverse momentum $\vec{p}_T$. So,

$$\frac{dm}{m} = \gamma^2 \frac{d\beta}{\beta} + \frac{dp_T}{p_T}. \tag{9.2}$$

The velocity appears in the energy loss of charged particles in matter, particle time-of-flight measurements, Cherenkov radiation and transition radiation. We will discuss these processes in detail below. While time-of-flight, Cherenkov radiation and transition radiation measurements require special detectors, dE/dx energy losses come for free in basically all experiments that have tracking and vertex detectors. The most precise velocity measurement is done in Cherenkov detectors yielding accuracies in the velocity measurement down to $\Delta\beta/\beta = 10^{-7}$. At the present energy scale, transition radiation provides electron/pion separation, while the other three techniques provide pion/kaon separation in certain momentum regions.

In neutrino and astrophysical experiments, Cherenkov detectors are used. In neutrino experiments, these are typically water Cherenkov detectors, which consist of a large tank of water whose walls are covered with large photomultiplier tubes. The goal is to detect a neutrino interaction produced inside the water tank. The neutrino will produce an electron or a muon, which emits Cherenkov radiation that is recorded by PMTs. On the outside of the tank veto counters are mounted that detect the background from incoming particles such as cosmic muons. The signature of the signal is an event produced and contained in the water tank. The presently largest water Cherenkov detector is Superkamiokande [1], which uses 40,000 tons of water and about 11,000 PMTs. It will be succeeded by Hyperkamiokande, which will use 260,000 tons of water and about 50,000 large PMTs. We showed a photograph of Superkamiokande in Fig. 8.15 (right) and give some more details in Chap. 11. The Sudbury Neutrino Observatory (SNO) is another neutrino experiment that used heavy water [2]. This allowed them to measure charged-current interactions, neutral-current interactions and elastic scattering processes separately, which helped to solve the solar neutrino problem. The Ice Cube experiment in Antarctica utilizes the ice as radiator [3]. It consists of a large array of PMTs chains inserted into the ice. We discuss it briefly in Chap. 11. Astrophysical experiments like MAGIC [4], HESS [5] and the Cherenkov Telescope Array (CTA) [6] detect the Cherenkov light produced by charged particle in cosmic showers. Except for the CTA array briefly presented in Chap. 11, we will not discuss these other experiments here.

9.1 Energy Loss Measurements

The energy loss of a charged particle in matter depends on the velocity β as discussed in Sect. 2.1.1. For a heavy charged particle, the Bethe-Bloch formula yields the stopping power dE/dx (2.13), while for electrons and positrons (2.54) and (2.55) provide the energy loss. Thus, we can extract β and in turn the particle mass from dE/dx measurements if we know the particle momentum. We saw that the stopping power for a specific material plotted as a function of $\beta\gamma$ provides a unique function that is valid for all heavy charged particles. If we plot the stopping power as a function of momentum as depicted in Fig. 9.1 (left), the distributions for different particles split up. Note that the height of the plateau in the relativistic rise depends on the gas pressure as illustrated in Fig. 9.1 (right). It is substantially higher for low gas pressures than for high gas pressures. As Fig. 9.1 (left) shows, two particle species

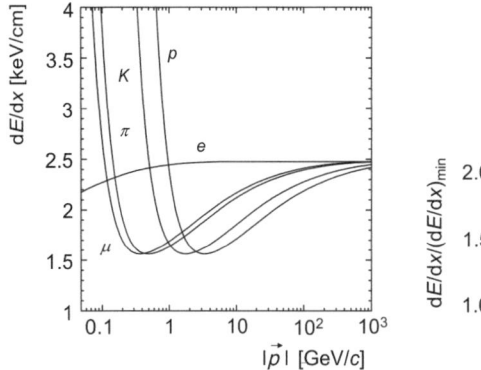
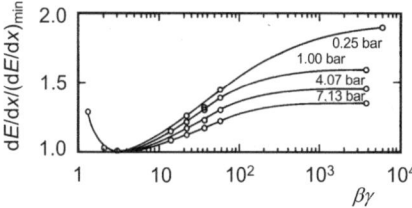

Fig. 9.1 Left: The stopping power as a function of momentum for different particles types. Reprinted with kind permission from [7], © 2021, M. Hauschild. All rights reserved. Right: The stopping power as a function of $\beta\gamma$ for an Ar-CH$_4$ (90:10) gas mixture at different gas pressures. All reprinted with kind permission from [8], © 1979, Elsevier. All rights reserved

9.1 Energy Loss Measurements

are well separated in some momentum regions. We divide the energy loss distribution into three momentum regions,

- Low momenta, where the stopping power drops like $1/\beta^{1.4-1.7}$.
- The cross over region around the minimum ionization.
- Higher momenta where the stopping power shows the relativistic rise.

Note that for many experiments π/K and K/p separations are of greatest interest.

The most common method to determine the stopping power is to measure the charge in each layer of the tracking chamber. We saw, however, that the instantaneous energy loss is typically described by a Landau-like distribution, which has a long tail on the high-energy side. In addition, primary electrons with sufficient kinetic energy can produce secondary electrons. If they are recorded in the same time window as the primary electrons, they yield an increased charge measurement. In order to determine the most probable value, we typically discard 30% of the measurements with the largest dE/dx values. For a sufficient number of measurements, the peak position provides a good approximation of the most probable energy loss. We calculate the weights for each particle hypothesis,

$$w(e, \mu, \pi, K, p) = \left[\frac{(dE/dx)_{\text{meas}} - (dE/dx)_{(e,\mu,\pi,K,p)\text{pred}}}{\sigma_{dE/dx}^{\text{meas}}} \right], \quad (9.3)$$

where $(dE/dx)_{\text{meas}}$ is the measured energy loss, $(dE/dx)_{(e,\mu,\pi,K,p)\text{pred}}$ is the predicted value for a given mass hypothesis and $\sigma_{dE/dx}^{\text{meas}}$ is the measured resolution of the truncated energy distribution. This techniques was pioneered by the LBL TPC experiment at PEP [9] and then used in many other experiments.[1] Figure 9.2 (left) shows the energy loss distribution of the LBL TPC experiment for electrons, muons, pions, kaons, protons and deuterons as a function of momentum, which clearly shows their separation in certain momentum ranges. This idea was first doubted by many physicists. But at that time many new things were tried at SLAC and were successful. In the LBL TPC muons could be separated from pions up to 300 MeV/c, while π/K separation worked up to 700 MeV/c. After that many experiments adopted the energy loss measurements in their wire chambers and vertex detectors. Figure 9.2 (right) shows the energy loss distributions for electrons, muons, pions, kaons, protons and deuterons as a function of momentum in the *BABAR* drift chamber [10]. The individual bands are clearly visible and provide good particle separation in certain momentum regions. Figure 9.3 (left) shows the corresponding distributions for the ALICE TPC [11]. For the *BABAR* drift chamber, the upper bound for a 3σ π/K separation lies at 0.7 GeV/c. Similar upper bounds hold for the CLEO II drift chamber [12] and the DELPHI TPC [13], while for the ALICE TPC the upper bound has increased to 1.1 GeV/c.

---

[1] Another method consists of discarding all dE/dx measurements that are larger than the most probable value plus $1.5\sigma_{dE/dx}$. A third technique is to fit the data with a Landau distribution to determine the peak position and rms.

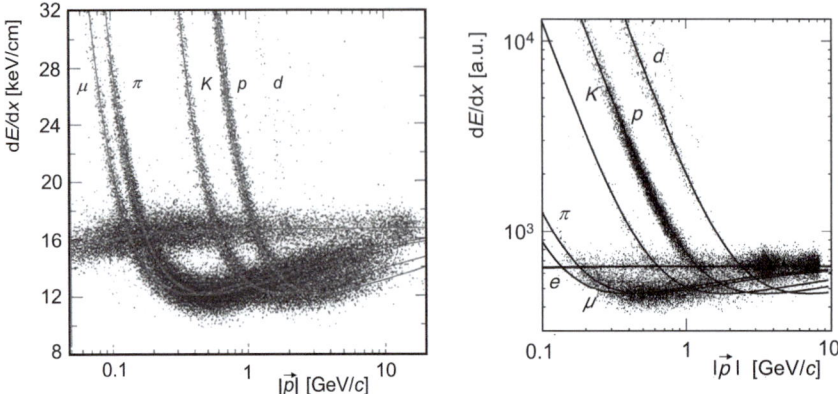

Fig. 9.2 Left: The energy loss distribution as a function of momentum for different particles measured in the TPC built by Nygren [9]. Reprinted under CC-BY-NC-4.0 Licence with kind permission from [9], © 1998, Particle Data Group LBNL. Right: The energy loss distribution as a function of momentum measured in the BABAR drift chamber. Reprinted with kind permission from [10], © 2002, Elsevier. All rights reserved

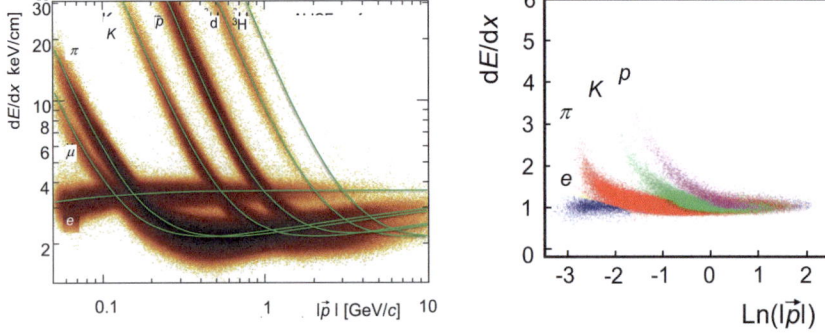

Fig. 9.3 Left: The energy loss distribution as a function of momentum measured in the ALICE TPC. Next to the deuteron band, also the tritium ion band is visible. Reprinted under CC-BY-NC-4.0 Licence with kind permission from [14], © 2024, Particle Data Group LBNL. Right: The energy loss distributions as a function of $\ln(p)$ for electrons (blue points), pions (red points), kaons (green points) and protons (magenta points) measured in the BABAR SVT. Reprinted with kind permission from [15], © 2013, Elsevier. All rights reserved

The dE/dx energy loss can be measured in vertex detectors as well. Figure 9.3 (right) shows the dE/dx curves in the BABAR SVT. For low-momentum tracks, the π/K separation works well. This adds measurement points to higher-momentum tracks. For low-momentum tracks, which curve in magnetic fields and may not reach the tracking detector or pass through the first few layers, the vertex detector may provide important particle identification. Both ATLAS and CMS use dE/dx measurements in the pixel detector.

The truncated $dE/dx|_{\text{trunc}}$ resolution of gaseous detectors depends on the number of samples N_{samp}, the thickness of the sampling layer d_{samp} and on the gas pressure

9.1 Energy Loss Measurements

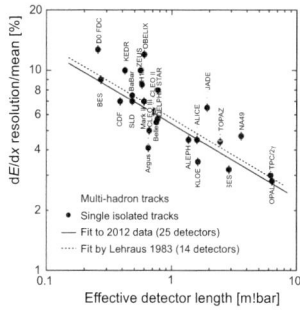
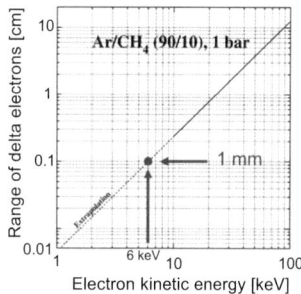

Fig. 9.4 Left: The correlation between dE/dx resolution and track length in a double logarithmic plot displaying single isolated tracks (solid points), multi hadronic tracks (open circles), the original Lehraus fit to 14 detectors (dashed line) [19] and a fit to 25 detectors (solid line). Note that the measurements are not corrected for gas parameters. Right: Range of δ-electrons as a function of their kinetic energy. Both reprinted with kind permission from [7], © 2021, M. Hauschild. All rights reserved

P. For pure argon, an empirical formula was developed [16] from the PAI model [17],

$$\sigma_{dE/dx}/(dE/dx)|_{\text{trunc}} = 0.41 N_{\text{meas}}^{-0.46}(d_{\text{samp}} \cdot P)^{-0.32}. \quad (9.4)$$

For comparison with other gases, an extension has been suggested,

$$\sigma_{dE/dx}/(dE/dx)|_{\text{trunc}} = 0.345 N_{\text{meas}}^{-0.46}(A_{\text{gas}} d_{\text{samp}} \cdot P)^{-0.32}, \quad (9.5)$$

with

$$A_{\text{gas}} = 6.83 N_e P S_{\text{pow}}/(\beta^2 dE/dx)|_{\text{trunc}}), \quad (9.6)$$

where

$$S_{\text{pow}} = \frac{(dE/dx)|_{\text{trunc,A}} - (dE/dx)|_{\text{trunc,B}}}{(\sigma_{dE/dx}|_{\text{trunc,A}} + \sigma_{dE/dx}|_{\text{trunc,B}})/2} \quad (9.7)$$

is the separation power and N_e is the mean number of electrons per molecule. For noble gases (He, Ne, Ar, Kr, Xe) we get $A_{\text{gas}} = 0.32, 0.50, 0.62, 0.65$ and 0.70, respectively. Due to the truncation of the Landau-like distribution, the $1/\sqrt{N_{\text{meas}}}$ improvement of the dE/dx resolution is modified. At nominal pressure, typical values are $\sigma_{dE/dx}/(dE/dx) = 4.5$–$7.5\%$, for a sampling thickness of $d_{\text{samp}} = 0.4$–1.5 cm, an appropriate detector length and $N_{\text{meas}} = 40$–300 for gaseous detectors. Due to the high gas pressure of 8.5 bar, PEP-4/9 TPC achieved an unprecedented resolution of 3% [18]. The number of measurements and the effective cell thickness is correlated with the effective track length. So, Lehraus looked at the correlation of the dE/dx resolution and detector length L_{det} for 14 detectors [19]. Figure 9.4 (left) illustrates this correlation for 25 detectors showing that dE/dx resolution improves with increasing effective detector length. The correlation would be even better if the measurements were corrected for gas parameters. A fit to 14 detectors gave $\sigma_{dE/dx}/(dE/dx) = 0.057(L_{\text{det}})^{-0.37}$ and a fit to the 25 detectors gives,

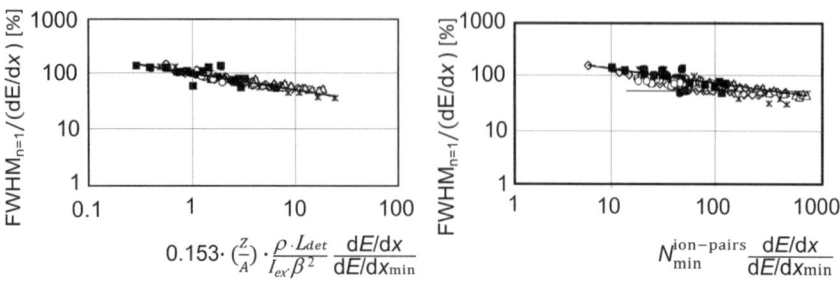

Fig. 9.5 Left: The relative resolution $FWHM_{n=1}/(dE/dx)$ as a function of the electron density of the medium corrected for the relativistic rise. The fit is given by (9.9) [20]. Right: The relative resolution $FWHM_{n=1}/(dE/dx)$ as a function of primary ions. The points result from Allison-Cobb calculations for He-based gas mixtures (open circles), data from large experiments (black squares) and other test results (open triangles and crosses). Both reprinted with kind permission from [20], © 2000, Elsevier. All rights reserved

$$\frac{\sigma_{dE/dx}}{dE/dx} = 0.054(L_{det})^{-0.37}. \tag{9.8}$$

These results are similar yielding the same slope. So, for a 1 m long detector the dE/dx resolution is 5.4%. However, most important is the separation power S_{pow}. Concerning the gas pressure, the optimal separation power is achieved with 3–4 bar.

Figure 9.5 shows the relative resolution $FWHM_{n=1}/(dE/dx)$ for a single sample corrected for gas properties and the relativistic rise as a function of ζ'_e (left) and the number of primary ions, $N_{min}^{ion-pairs} \cdot \frac{dE/dx}{(dE/dx)_{min}}$ (right). Here, $\zeta'_e = \zeta_e \cdot \frac{dE/dx}{(dE/dx)_{min}}$, where $\zeta_e = 0.153 \cdot \frac{Z}{A} \cdot \frac{\rho \cdot L_{det}}{I_{ex} \cdot \beta^2}$ is the electron density. We see that the spread around the fit in both distributions is much smaller than that in Fig. 9.4 (left). The fit function in Fig. 9.5 (left) is parameterized by

$$FWHM_{n=1} = 97.448 \cdot \zeta'^{-0.31633}_e. \tag{9.9}$$

The resolution of n samples shows the scaling

$$\sigma_n = \sigma_{n=1} \cdot n^{-0.4}. \tag{9.10}$$

The determination of the charge requires to sum up all energy deposits in a cluster. Typically, the full energy loss is not recorded as detectors measure the restricted energy loss discussed in Chap. 2. Some of the primary electrons have sufficient energy to fly and produce secondary electrons. So, they may not be included in the energy clusters of the track. There is a correlation of the electron energy and the range as illustrated in Fig. 9.4 (right) that shows the kinetic energy of of δ-rays as a function of their kinetic energy in an Ar-CH_4 (90:10) gas mixture at a pressure of 1 bar. An introduced energy threshold defines the maximum energy of electrons that are associated with a track. Defining a typical range of 1 mm corresponds to a kinetic energy threshold of 6 keV. Please note that the energy threshold modifies the

9.1 Energy Loss Measurements

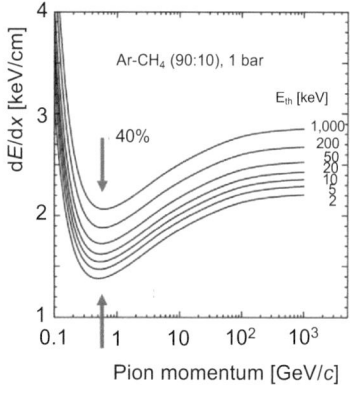
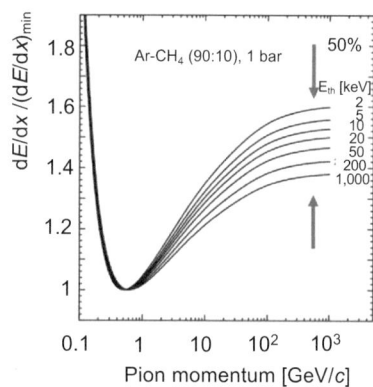

Fig. 9.6 Left: The energy loss distribution dE/dx as a function of pion momentum for different energy thresholds. Right: The energy loss distribution $(dE/dx)/(dE/dx_{min})$ as a function of pion momentum for different energy thresholds. Both plots are for an Ar-CH$_4$ gas mixture at a pressure of 1 bar. Both reprinted with kind permission from [7], © 2021, M. Hauschild. All rights reserved

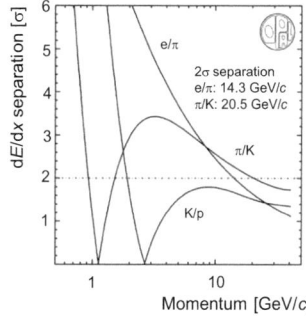
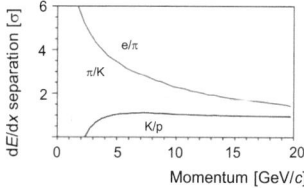

Fig. 9.7 Left: Particle separation in the OPAL jet chamber that is operated at a pressure of 4 bar as a function of momentum. Right: Particle separation in the ALEPH TPC that is operated at a pressure of 1 bar as a function of momentum. Both reprinted with kind permission from [7], © 2021, M. Hauschild. All rights reserved

energy loss distributions, which is illustrated in Fig. 9.6 (left), showing the stopping power of pions as a function momentum for different energy thresholds from 2 keV to 1 MeV. For high thresholds, the ionization curves increase by about 40% with respect to those with low threshold. If we normalize each ionization curve to its minimum energy loss, we obtain the curves depicted in Fig. 9.6 (right). While the minimum is basically at the same energy loss, the relativistic rise at 1000 GeV decreases by 50% for a 1 MeV threshold with respect to a 2 keV threshold.

In the momentum region around the pion and kaon ionization minimum, the π/K separation is limited. In the relativistic-rise part π/K separations of 2σ or larger are possible. The OPAL jet chamber achieved excellent π/K separation abilities above the minimum ionization. Figure 9.7 (left) shows the e-π, π/K and K/p separation curves as a function of momentum in the OPAL jet chamber. The π/K separation is

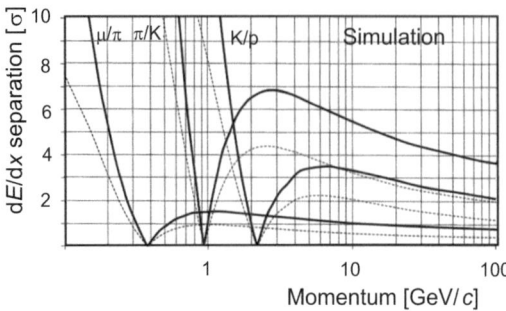

Fig. 9.8 Separation of two particle types as a function of momentum for the dE/dx technique (dashed lines) and the cluster counting technique dN/dx (solid lines). Shown are the μ-π, π/K and K/p separations. Reprinted with kind permission from [42], © 2019, Elsevier. All rights reserved

above 2σ for momenta $1.6 < p_T < 20.5$ GeV/c. Even e-π separation of $>2\sigma$ works up to 14.3 GeV/c. This is similar in the ALEPH TPC depicted in Fig. 9.7 (right). Note that the CLEO II experiment also used the 2σ π/K separation for momenta above 2 GeV/c for identifying kaons.

Instead of measuring the charge of a cluster, we could measure the number of clusters along a track. This has the advantage that it is not affected by Landau fluctuations. First ideas came from Davidenko in 1969 [40]. Detailed studies were performed already in the mid 1990s by Malamud, Breskin and Chechik [41]. In test beam experiments, cluster counting works well since conditions are fully under control. But this method has not been used in large-scale detectors. Studies of the performance of a drift chamber operated with a He-CH$_4$ (90:10) gas mixture have been performed in simulations [42]. The chamber extends from 0.3 to 2.0 m, has 112 coaxial layers and is operated in a 2 T magnetic field. Figure 9.8 shows the separation of two particle types as a function of momentum using dE/dx and cluster counting dN/dx. The cluster counting technique outperforms the dE/dx technique by a factor of two. A π/K separation of $>3\sigma$ is found in the entire momentum region except for the small region of 0.85 GeV/c to $\sim$1.0 GeV/c. The dE/dx resolution was assumed to be 4.5% and the cluster counting efficiency is 80%. Table 9.1 lists examples of the dE/dx resolution in tracking detectors along with some other properties.

9.2 Time-of-Flight Measurements

Another technique for determining particle velocities consists of measuring the particle flight time. For a particle with known momentum $p = |\vec{p}|$, which traverses a path length L_{path} in time t, we can determine its mass by

$$m = \frac{p}{c}\sqrt{\frac{c^2 t^2}{L_{\text{path}}^2} - 1}. \tag{9.11}$$

9.2 Time-of-Flight Measurements

Table 9.1 Examples of central tracking systems, gas mixtures, gas pressures, number of measurements, thickness of the sampling layers d_{samp} and measured dE/dx resolutions. The dE/dx resolution corresponds to that of single tracks. The values in parentheses correspond to that of dense tracks. [†]Sense wires and pads. [§]The first value is for Bhabhas and the second for minimum-ionizing particles. [*]The first value refers to layers 1–8 and the second to the outer layers

Exp.	Gas mixture	P [bar]	N_{meas}	d_{samp} [mm]	$\sigma_{dE/dx}/dE/dx$ [%]	Reference
ALEPH TPC	Ar-CH_4 (91:9)	1	338[†]	4	4.5	[21]
ALICE TPC	Ne-CO_2 (90:10)	1	159	7.5, 10,15	4.5 (5.0)	[11,22]
ARGUS DC	C_3H_8–Methylal	1	36	18.0	4.1 (4.4)	[23,24]
ATLAS Pixel	Silicon	–	≥ 5	–	9.5	[25]
BABAR DCH	He-iC_4H_{10} (80:20)	1	40	12	7.5	[10]
Belle CDC	He-C_2H_6 (50:50)	1	47	16	5.7, 7.0[§]	[26]
BES III MDC	He-C_3H_8 (60:40)	1	43	12, 16.2 [*]	6	[27]
CDF COT	Ar-C_2H_6-C_2H_6O (49.6:49.6:0.8)	1	32	12	7.0	[7]
CLEO II DR	Ar-C_2H_6 (50:50)	1	51	14	6.2 (7.1)	[12]
CMS silicon	Silicon	–	≥ 10	–	6.0	[28]
DELPHI TPC	Ar-CH_4 (80:20)	1	192	4	5.7 (6.2)	[29]
D0 FDC	Ar-CH_4-CO_2 (93:4:3)	1	32	8	12.7	[30]
H1 CTC	Ar-C_2H_6 (50:50)	1	56	10	10.0	[31]
JADE JET	Ar-CH_4-C_4H_{10} (88.7:8.5:2.8)	4	48	10	6.5 (7.2)	[32]
KEDR JET	DME (100)	1	42	10	10.0	[33]
KLOE DC	He-C_4H_{10} (90:10)	1	58	28	3.5	[34]
Mark II drift	Ar-CO_2-CH_4 (89:10:1)	1	72	8.33	7.0	[35]
OPAL JET	Ar-CH_4-iC_4H_{10} (88.2:9.8:2)	4	159	10	2.8 (3.2)	[36]
SLD CDC	CO_2-Ar-iC_4H_{10} (75:21:4)	1	80	6	7.0	[37]

(continued)

Table 9.1 (continued)

Exp.	Gas mixture	P [bar]	N_{meas}	d_{samp} [mm]	$\sigma_{dE/dx}/dE/dx$ [%]	Reference
TOPAZ TPC	Ar-CH$_4$ (90:10)	3.5	175	4	4.4 (4.6)	[38]
TPC/2γ TPC	Ar-CH$_4$ (80:20)	8.5	183	4	3.0	[18,39]
ZEUS CTC	Ar-CO$_2$-C$_2$H$_6$ (90:8:2)	1	72	8	8.5	[7]

The uncertainty in the mass measurement is given by,

$$\frac{dm}{m} = \frac{dp}{p} + \gamma^2 \left[\frac{dt}{t} + \frac{dL_{path}}{L_{path}} \right]. \quad (9.12)$$

Momenta are measured rather accurately, so typically $dp/p < 1\%$. The path length is determined precisely as well, $dL_{path}/L_{path} < 10^{-3}$. Thus, dm/m is mainly determined by the time resolution since we typically have $\gamma \gg 1$.

The time difference for two particles with masses m_1 and m_2 with velocities β_1 and β_2 for a flight path L_{path} is given by,

$$\Delta t_{tof} = \frac{L_{path}}{\beta_2 c} - \frac{L_{path}}{\beta_1 c} = \frac{L_{path}}{c} \left[\sqrt{1 + \frac{m_2^2 c^2}{p^2}} - \sqrt{1 + \frac{m_1^2 c^2}{p^2}} \right]. \quad (9.13)$$

For $p^2 \gg m_2^2 c^2$, this can be approximated by

$$\Delta t_{tof} \approx \left(m_2^2 - m_1^2 \right) \frac{c \cdot L_{path}}{2 p^2}. \quad (9.14)$$

Typically, we require that $\Delta t_{tof} \geq 3\sigma_t$, where σ_t is the time resolution. The TOF measurement yields $\Delta t_{tof} = n_{sig} \sigma_t$, where n_{sig} is the number of standard deviations. From this we can determine the momentum threshold,

$$p_{th} = \sqrt{\left(\frac{m_2^2 c^2 \Delta t_{tof}^2 L_{path}^2}{D_1} + \frac{m_1^2 c^2 \Delta t_{tof}^2 L_{path}^2}{D_1} + \frac{2 \cdot \sqrt{c^2 \Delta t_{tof}^2 L_{path}^4 \left[c^2 \cdot m_2^2 m_1^2 \Delta t_{tof}^2 + L_{path}^2 (m_2^2 - m_1^2)^2 \right]}}{D_1} \right)} \quad (9.15)$$

where

$$D_1 = c^2 \cdot \Delta t_{tof}^4 - 4 \Delta t_{tof}^2 L_{path}^2. \quad (9.16)$$

Plugging in masses in units of [GeV/c^2] yields the momentum in units of [GeV/c]. Figure 9.9 (left) shows the momentum threshold as a function of the flight time for a detector located 1.5 m from the interaction region. Figure 9.9 (right) shows the momentum threshold as a function the radial distance of the TOF counter from the interaction region for 50 ps time resolution.

9.2 Time-of-Flight Measurements

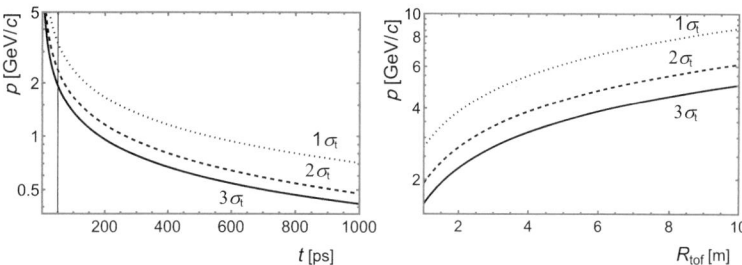

Fig. 9.9 Left: Momentum threshold as a function of measured flight time Δt_{tof} for π/K separation at a $3\sigma_t$ (solid curve), $2\sigma_t$ (dashed curve) and $1\sigma_t$ level (dotted curve). The TOF counter is located 1.5 m from the interaction region. Right: Momentum threshold as a function of the radial distance of the TOF counter R_{tof} for π/K separation at a $3\sigma_t$ (solid curve), $2\sigma_t$ (dashed curve) and $1\sigma_t$ level (dotted curve). The time resolution is 50 ps

Time-of-flight counters consist of thin plastic scintillator bars placed at a large distance of 1.0 to 1.5 m from the primary interaction vertex. The scintillators are read out by photodetectors at each end, which used to be photomultiplier tubes. In newer designs, micro-channel plates and silicon-based solid state detectors are used also. To achieve good time resolution, both the scintillator and the photodetector have to be fast. As we saw in the previous chapter, a fast scintillator has a decay time of the order of 1 ns, while fast photodetectors have a rise time of a few ns and transit time resolutions of 50–100 ps. Micro-channel plates have a rise time of a few hundred picoseconds and a transit time of 30–40 ps. The photodetector output signal is read out via a constant-fraction discriminator, which provides a stop signal of a TDC that was started by beam-crossing signal. Figure 9.10 (left) illustrates the operation mode of a regular discriminator. The recorded time is defined by the signal crossing a certain threshold, which depends on the signal size. A large signal crosses the threshold earlier than a weak signal. This introduces an additional error in the time measurement called time jitter. On the other hand, a constant-fraction

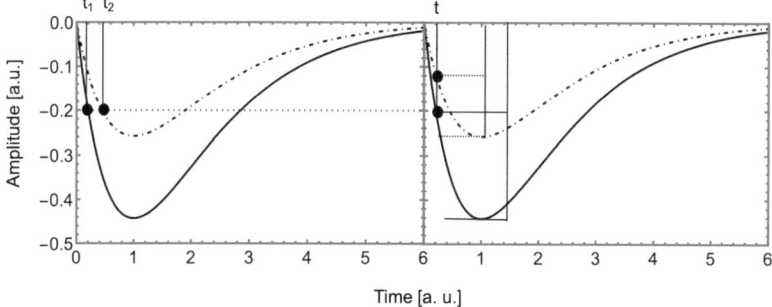

Fig. 9.10 Left: Stopping time defined by a signal pulse in a regular discriminator. For a fixed threshold (dotted curve) a large signal (solid curve) crosses the threshold at t_1, while a small signal (dash-dotted curve) crosses at t_2. This introduces a time jitter $\Delta t = t_2 - t_1$ that broadens the time resolution. Right: Stopping time defined in a constant-fraction discriminator. The stopping time t is now independent of the signal size

discriminator removes this error source as depicted in Fig. 9.10 (right). The threshold is varied such that the 90% value of the signal amplitude always occurs at the same time. Without a constant-fraction discriminator the time resolution is worsened by the time jitter. In present experiments the time jitter may be removed in software if waveform digitizing electronics is used. Many experiments use the time-over-threshold technique discussed in the previous chapter. An example is the ALICE TOF system [43].

The measurement of the flight time involves several steps. A specific event trigger provides the start signal. The charged particle moves through the detector. When it traverses the scintillator bar, it produces scintillation photons that travel in both directions in the scintillator, which serves as a waveguide. Photons with angles larger than α_g are totally reflected in the scintillator. The photon path length also depends on the angle. It is maximum for angle α_g and minimum for 90°. At the bar end, the photons enter the photodetector and produce the stop signal in the time measurement. Since we read out the left-hand and right-hand ends separately, we compute a weighted time average

$$t_{\text{tof}} = \frac{A_L t_L + A_R t_R}{A_L + A_R}, \quad (9.17)$$

where A_L, t_L and A_R, t_R are the pulse heights and times measured in the left-hand and right-hand counters, respectively. The time measurements is typically done with TDCs.[2] The time difference is

$$\Delta t = t_{tof} - t_0, \quad (9.18)$$

where t_0 is the start time of the trigger TDC. The time resolution has the following contributions,

$$\sigma_t^2 = \frac{A_L^2}{(A_L + A_R)^2}(\sigma_L^2 + \sigma_{el}^2) + \frac{A_R^2}{(A_L + A_R)^2}(\sigma_R^2 + \sigma_{el}^2) + \sigma_0^2 + \sigma_w^2, \quad (9.19)$$

where σ_L, and σ_R are the expected time resolutions for the left-hand, and right-hand counters that result from photon statistics and the transit time spread of the PMT, σ_0 is the resolution in the start signal, σ_{el} is the resolution of the electronics including a possible time walk of discriminator non-linearities and σ_w accounts for additional time smearing effects. Typically, the time resolutions of the two counters are identical $\sigma_L = \sigma_R$ and the intrinsic time resolution scales with the amplitude as $1/\sqrt{A}$ due to photon statistics. If σ_{ps} is the expected time resolution when the particle passes through the center of the counter with observed pulse height A_C, then $\sigma_{L,R} = \sqrt{A_C/A_{L,R}}\sigma_{ps}$ and the time resolution becomes [44]

$$\sigma_t^2 = \frac{A_C}{A_L + A_R}\sigma_{ps}^2 + \frac{A_L^2 + A_R^2}{(A_L + A_R)^2}\sigma_{TDC}^2 + \sigma_0^2 + \sigma_w^2. \quad (9.20)$$

---

[2] A TDC is a Time-to-Digital Converter, which is triggered by a start and stop signal allowing to determine time differences.

9.2 Time-of-Flight Measurements

Table 9.2 Examples of barrel time-of-flight counters in different experiments, listing the detector type, radius R_{tof} and length L_{tof} of the time-of-flight system, the time resolution and the momentum threshold for 3σ π/K separation at dip angles $\lambda_d = 0°$ and $\lambda_d^{\max}$ in the barrel. The latter dip angle refers to the longitudinal edges of the counter. [†] The lower value corresponds to the center of the counter, while the higher value corresponds to the ends of the counter

Experiment	Detector type	R_{tof} [m]	L_{tof} [m]	Hadron σ_t [ps]	p_{thr} at 3σ π/K sep. ($\lambda_d = 0°$) & $\lambda_d^{\max}$ [GeV/c]	Reference
Mark III	scint.	1.2	3.2	189	0.86 & 1.1	[45]
Mark II	scint.	1.524	3.0	221	0.9 & 1.07	[48]
ARGUS	scint.	1.0	2.18	220	0.72 & 0.89	[49]
CLEO II	scint.	0.98	2.79	154	0.86 & 1.15	[12]
OPAL	scint.	2.36	6.84	280–350[†]	1.0 & 1.2	[50]
Belle	scint.	1.2	2.55	100	1.2 & 1.46	[51]
BES III	scint.	0.86	2.30	100	1.02 & 1.32	[52]
CDF II	scint.	1.4	2.79	120	1.18 & 1.42	[53,54]
STAR	MRPC	2.1	4.2	130–140	1.35–1.4 & 1.6–1.68	[55,56]
ALICE	MRPC	3.7	7.41	56	2.87 & 3.42	[47]

From (9.14), we can determine the separation between two particles in units of σ,

$$N_\sigma = \left(\frac{m_2^2 - m_1^2}{\sigma_t}\right) \frac{c \cdot L_{\text{path}}}{2 p^2}. \quad (9.21)$$

Many colliding beam experiments have used time-of-flight systems. Table 9.2 lists properties of some ToF counters used in colliding beam experiments. Figure 9.11 shows the ToF resolutions for Bhabhas (left) and hadrons (center) in the Mark III experiment measuring time resolutions of $\sigma_t = 171$ ps and $\sigma_t = 189$ ps, respectively. These were typical time resolutions of experiments at that time. In the Mark II experiment for example, the single counter time resolution was 221 ps on average, yielding a 2σ π/K separation up to 1.1 GeV/c [48]. In the CLEO II experiment, the measured time resolution was improved to $\sigma_t = 139$ ps for Bhabhas and $\sigma_t = 154$ ps for hadrons, setting the 2σ π/K separation around 1.35 GeV. The CDF II experiment added a ToF system for b tagging and particle identification of specific B meson decays. The counters were placed behind the drift chamber at $R_{\text{tof}} = 1.4$ m. With a measured time resolution of $\sigma_t = 120$ ps, they achieved a 2σ π/K separation up to 1.46 GeV/c. To achieve a 3σ π/K separation at higher momenta, we need to increase R_{tof} or decrease σ_t. With thin counters and fast PMT's, we achieve resolutions of σ_t of 50 ps. For example, the time-of-flight counter of the ALICE experiment achieves a time resolution of 56 ps using large-area resistive plate chambers as depicted in Fig. 9.11 (right). For $R_{\text{tof}} = 3.7$ m, this increases the momentum threshold to around 3 GeV/c.

Figure 9.12 (left) shows the distribution of $1/\beta_{\text{meas}}$ versus momentum in the STAR experiment at Brookhaven [57]. The length is $R_{\text{tof}} = 2$ m and the time resolution is $\sigma_t < 100$ ps. The 3σ π/K separation is observed for momenta up to ~1.6 GeV/c

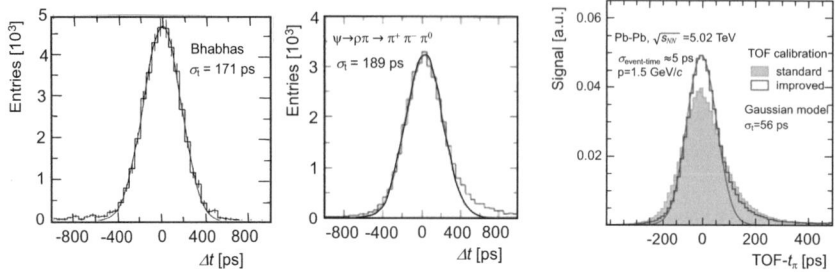

Fig. 9.11 Left: Time resolution of the Mark III time-of-flight counter for Bhabhas (histogram) with a fit overlaid (solid line) yielding $\sigma_t = 171$ ps. All reprinted with kind permission from [45], © 1984, Elsevier. All rights reserved. Center: Time resolution of the Mark III time-of-flight counter for hadrons (histogram) with a fit overlaid (solid line) yielding $\sigma_t = 189$ ps. Reprinted with kind permission from [46], © 1981, Mark III Collaboration. All rights reserved. Right: Time resolution of the ALICE time-of-flight counter for raw data (grey histogram) and for data after corrections (black histogram) with a fit overlaid (solid line) yielding $\sigma_t = 56$ ps. Reprinted with kind permission from [47], © 2018, IOP Publishing. All rights reserved

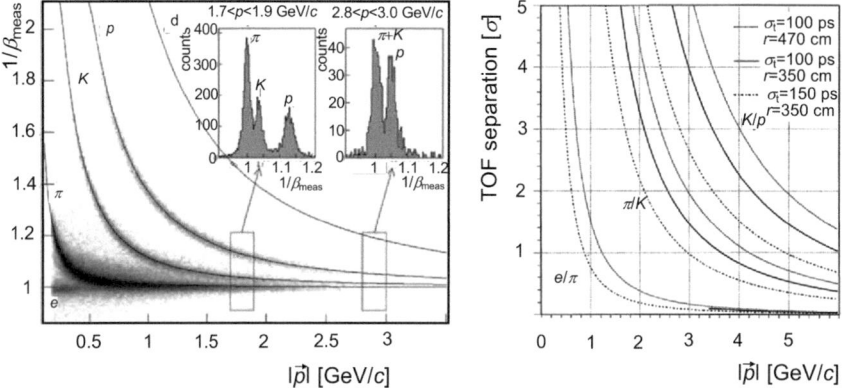

Fig. 9.12 Left: Measurement of $1/\beta$ versus momentum in the STAR experiment. The insets show the $1/\beta$ projections for momenta around 1.8 and 2.8 GeV/c. Reprinted with kind permission from [57], © 2003, Elsevier. All rights reserved. Right: The e/π, π/K and K/p separation curves for different configurations, $R_{\text{tof}} = 470$ cm and $\sigma_t = 100$ ps (dotted line), $R_{\text{tof}} = 350$ cm and $\sigma_t = 100$ ps (solid line), and $R_{\text{tof}} = 350$ cm and $\sigma_t = 150$ ps (dash-dotted line). Reprinted with kind permission from [58], © 1999, Elsevier. All rights reserved

and the 3σ K/p separation goes up to nearly 3 GeV/c. Figure 9.12 (right) shows the curves for e/π, π/K and K/p separation as a function of momentum for three configurations, R_{tof}=470 cm and $\sigma_t = 100$ ps, R_{tof}=350 cm and $\sigma_t = 100$ ps and R_{tof}=350 cm and $\sigma_t = 150$ ps. For 350 cm and 100 ps time resolution the 3σ time resolution is around 2.2 GeV/c. Even electrons can be separated from pions up to 0.8 GeV/c.

9.3 Cherenkov Counters

The third technique of measuring the velocity of a charged particle is via the Cherenkov effect. A review on Cherenkov counters is given in [59].

We saw in Sect. 2.4 that Cherenkov radiation occurs if the particle velocity is $\beta > 1/n$, where n is the refractive index of the medium Thus, one way of distinguishing between different particles is to look if at a given momentum Cherenkov radiation occurred or not. Detectors based on this yes-no decision are called threshold counters. Choosing an appropriate refractive index n, we can do π/K separation in a certain momentum region. Near the threshold, the threshold velocity is $\beta_s = 1/n$ and the number of detected photons is modified by the threshold factor

$$\gamma = \left(1 - \beta^2\right)^{-\frac{1}{2}} > \left(1 - \frac{1}{n^2}\right)^{-\frac{1}{2}} = \gamma_s. \tag{9.22}$$

Table 9.3 shows the β_s and γ_s threshold factors and the Cherenkov angles for different radiators. The refractive indices and threshold factors of gases depend on the pressure. For pressures that are not too high, the following approximation holds to high accuracy,

$$(n - 1) = (n_0 - 1)P, \tag{9.23}$$

where n_0 is the refractive index at normal pressure (1 atmosphere) and P is the pressure in atmospheres. Besides the threshold we can measure also the number of Cherenkov photons N_γ and the Cherenkov angle as discussed below. For particles

Table 9.3 Refractive indices, γ_s threshold factor, velocity threshold β_s and maximum Cherenkov angle θ_c^{max} for different radiators. [†]Depends on pressure and temperature

Material	n	γ_s factor	β_s	θ_c^{max}
Diamond	2.419	1.098	0.419	65.58°
ZnS	2.355	1.105	0.425	64.87°
Lead fluorite	1.767	1.21	0.566	55.52°
Glass	1.75–1.46	1.219–1.372	0.571–0.685	55.15°–46.77°
Plastic scintillator	1.58	1.292	0.633	50.73°
LiF	1.392	1.438	0.718	44.08°
Plexiglass	1.495	1.345	0.669	48.02°
Fused silica	1.4–1.55	1.309–1.429	0.645–0.714	49.82–44.42°
Water	1.333	1.5119	0.750	41.41°
C_6F_{14}	1.2515	1.663	0.799	36.96°
LAr	1.233	1.709	0.811	35.80°
Silica Aerogel	1.1–1.0006	2.400–28.88	0.9091–0.9994	24.62°–1.98°
Gases (Pentane, He)[†]	1.0017–1.000033	17.17–123.1	0.9983–0.99997	3.339°–0.4655°

with momentum p, the velocity difference between two particles with masses m_1 and m_2 is given by

$$\left(\frac{\Delta\beta}{\beta}\right)_{m_1 m_2} = \frac{(m_2^2 - m_1^2)c^2}{2p^2}. \tag{9.24}$$

9.3.1 Threshold Cherenkov Counters

Threshold Cherenkov counters ideally show no signal if the particle momentum is below threshold and a signal if it is above threshold. The threshold momentum for a particle with mass m is given by

$$p_{th} = \frac{mc}{\sqrt{n^2 - 1}}. \tag{9.25}$$

However, the detection of a signal depends on the efficiency and is affected by various background sources, including detector/electronic noise, non-Cherenkov light sources, extra tracks, interactions of the track in the detector, particle decays and δ rays, which may mimic a signal. To improve the signal sensitivity, we can measure the number of recorded photoelectrons N_{pe} in addition. The number of produced Cherenkov photons is given by the integral of (2.100) over the accepted photon energies E_{max} and E_{min},

$$N_\gamma = \frac{z^2 \alpha}{\hbar c} [E_{max} - E_{min}] L_r \sin^2 \theta_c = 2\pi z^2 \alpha \left[\frac{1}{\lambda_{min}} - \frac{1}{\lambda_{max}}\right] L_r \sin^2 \theta_c. \tag{9.26}$$

where L_r is the radiator length, λ_{min} is the minimum-detectable photon wavelength and λ_{max} is the maximum-detectable photon wavelength. For a photon detector that is sensitive in the wavelength range 220 nm $\leq \lambda \leq$ 700 nm, we observe 1,430 photons while in the range of 400 nm $\leq \lambda \leq$ 700 nm we only see 490 photons.

The number of photoelectrons may be parameterized by

$$N_{pe} = N_0 L_r \sin^2 \theta_c, \tag{9.27}$$

where

$$N_0 = \frac{\alpha z^2}{\hbar c} \epsilon_{det} \int \epsilon_{coll}(E_\gamma) dE_\gamma = 370 \cdot \epsilon_{det} \int \epsilon_{coll}(E_\gamma) dE_\gamma \tag{9.28}$$

is a photodetector-dependent parameter that specifies how many photoelectrons are produced per centimeter, z is the charge of the particle, $\epsilon_{det} \approx 0.9$ is the track-independent photodetector efficiency and $\epsilon_{coll}(E)$ is the photon collection efficiency that varies for different tracks. Note that N_γ and N_{pe} both depend on the photodetector efficiency and photon collection efficiency. Photomultipliers with a fused silica window and a bi-alkali photocathode have $N_0 = 100$–150 cm^{-1}, while

9.3 Cherenkov Counters

PMTs with a glass window and lower-quantum-efficiency window have $N_0 = 50$–60 cm^{-1} [60,61]. Since for gas Cherenkov counters the Cherenkov angle is small, $\sin^2 \theta_c$ term may be approximated by $\theta_c^2/2$. A PMT with a bi-alkali photocathode, has $\int \epsilon_{\text{coll}}(E_\gamma) dE_\gamma = 0.3$ yielding $N_0 = 100$ cm^{-1}.

If a threshold counter detects the lighter particle with mass m_1 at an average Cherenkov angle $\langle \theta_c \rangle$, while the heavier particle has no Cherenkov radiation $\theta_c = 0°$, we have

$$\cos \langle \theta_c \rangle = (n \cdot \beta_1)^{-1}, \tag{9.29}$$

and[3]

$$n = 1/\beta_2. \tag{9.30}$$

For small angles, we can expand the cosine term,

$$1 - \frac{1}{2}\theta_c^2 = \frac{\beta_2}{\beta_1} \tag{9.31}$$

from which we get

$$\langle \theta_c^2 \rangle = 2\left(1 - \frac{\beta_2}{\beta_1}\right) = 2\frac{\Delta \beta}{\beta}, \tag{9.32}$$

where we used $\Delta \beta = \beta_1 - \beta_2$ and $\beta \equiv \beta_1$. From this we get

$$\langle \theta_c^2 \rangle = \frac{(m_2^2 - m_1^2)c^2}{p_T^2}. \tag{9.33}$$

So for a PMT with a bi-alkali photocathode, we get

$$N_{pe} = 100 \text{ cm}^{-1} L_r \frac{(m_2^2 - m_1^2)c^2}{p_T^2}. \tag{9.34}$$

The resolution of a threshold counter is given by

$$\frac{\sigma_\beta}{\beta} \approx \frac{\tan^2 \theta_c}{2\sqrt{N_{pe}}}. \tag{9.35}$$

A 2 cm thick quartz bar read out with a PMT having a UV-sensitive window yields a resolution of 4.0% for $N_{pe} = 200$.

The use of five threshold counters with different refractive indices provides particle identification (π, K, p) over a wide momentum range. Table 9.4 shows the properties of the five threshold counters. Note that such an arrangement is even hard for a fixed-target experiment. The five counters have been selected in such a way that three pion thresholds approximately match three kaon thresholds and that four kaon thresholds approximately match four proton thresholds.

---

[3] In general $n \geq 1/\beta_2$, where the equal sign holds if particle 2 approaches the threshold.

Table 9.4 Properties of five threshold Cherenkov counters including the refractive index n, radiator material, radiator length L_r, counter length L_{count}, photoelectron yield, pion threshold p_π, kaon threshold p_K and proton threshold p_p. The refractive indices have been chosen such that the kaon thresholds of counters A, B, C approximately correspond to the pion thresholds of counters C, D, E. Similarly, the proton thresholds of counters A, B, C, D approximately correspond to the kaon thresholds of counters B, C, D, E and the proton thresholds of counters A,B also correspond to the pion thresholds of counters D, E. Reprinted with kind permission from [62], © 1981, IOP Publishing. All rights reserved

Counter	n	Radiator	L_r [cm]	L_{count} [cm]	Yield [p.e.]	p_π [GeV/c]	p_K [GeV/c]	p_p [GeV/c]
A	1.022	Aerogel	20	50–100	5–6	0.66	2.34	4.45
B	1.006	Aerogel	25	50–100	3–4	1.27	4.51	8.55
C	1.00177	Neopentane	30	50	≈10	2.34	8.32	15.8
D	1.00049	N_2O-CO_2 or Freon 14	100	≈120	≈10	4.44	15.8	30.0
E	1.000135	Ar-Ne or H_2	185	≈200	≈5	8.46	30.1	57.1

Figure 9.14 shows the threshold momenta for π, K, p for different combinations of the five counters specified in Table 9.4. If we only use the output of counters A and B, we have the following picture: for the combination $A = yes$ and $B = yes$, we identify a pion in the momentum range 0.66 to 4.5 GeV/c; for the combination $A = yes$ and $B = no$, we clearly identify a kaon in the momentum range of 2.34 to 4.5 GeV/c since for larger momenta counter B will turn on for kaons and the kaon is not distinguishable from a pion; for the combination $A = no$ and $B = no$, we clearly identify a proton in the momentum range 2.34 to 4.5 GeV/c. Concerning protons, combination $A = yes$ and $B = no$ extends the range from 4.5 to 8.5 GeV/c. If we add counter C, we extend the pion identification to 8.5 GeV/c for the combination $A = yes$, $B = yes$ and $C = yes$; a kaon is distinguished from a pion by the combination $A = yes$, $B = yes$ and $C = no$ in the momentum range 2.34 to 8.5 GeV/c; for a proton we have $A = yes$, $B = no$ and $C = no$ in the momentum range 2.34 to 8.5 GeV/c and $A = yes$, $B = yes$ and $C = no$ for 8.5 to 15.8 GeV/c. In a similar fashion, counter D and E can be added or other three counter combination can be selected.

In the design of the *BABAR* experiment, two silica aerogel threshold counters were considered with refractive indices of 1.06 and 1.0087. Figure 9.15 (right) shows a photograph of a block of silica aerogel. The block sizes were 10 cm × 8 cm × 4.0 cm and 10 cm × 8 cm × 12.5 cm, respectively. The blocks, housed in aluminum containers, were read out with fine-mesh photomultipliers. A prototype was tested at CERN in a test beam with pions, kaons and protons in the momentum range of 1–10 GeV/c [63]. Figure 9.15 (right) shows the measured thresholds for pions, kaons and protons for the $n = 1.0087$ aerogel. For an efficiency of 99%, the kaon misidentification was about 5%.

The Belle experiment realized a silica aerogel threshold Cherenkov counter (ACC). Due to the asymmetric beam energies and the resulting correlation between particle momenta and polar angle of B meson decay products, the Cherenkov threshold varied as a function of polar angle. Thus, the refractive index was optimized, varying from $n = 1.028$ in the backward part of the barrel to $n = 1.01$ in the forward part of the barrel to $n = 1.03$ in the endcap. The aerogel was read out with a fine-mesh PMT that works well in a 1.5 T magnetic field. During the ten-year operation, Belle achieved a kaon efficiency of up to 90% for a pion misidentification rate of 6% [64,65].

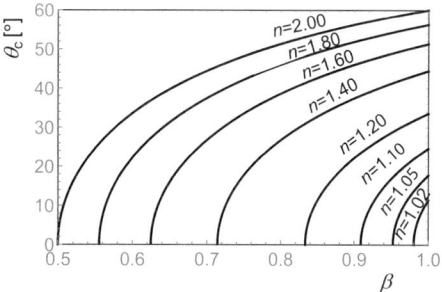

Fig. 9.13 The Cherenkov angles as a function of the relative velocity β of the charged particle for different indices of refraction ($n = 1.02$ to 2.0)

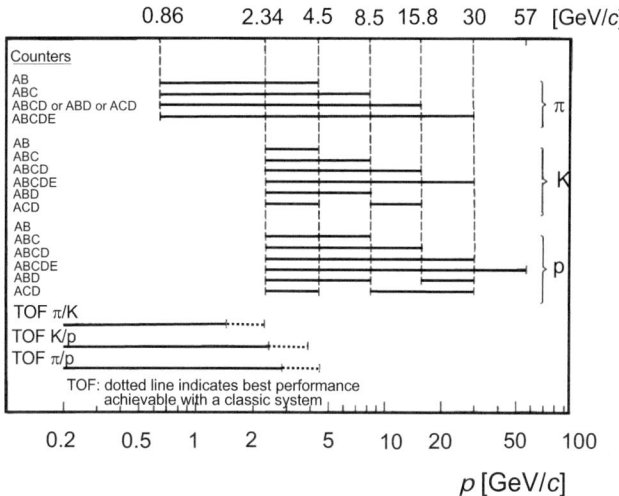

Fig. 9.14 Momentum ranges in which the pion, kaon and proton can be identified positively by a combination of two or more counters listed in Table 9.4. With additional time-of-flight measurements the particle identifications can be extended at low momenta down to 0.2 GeV/c assuming a time resolution of 150 ps and a 3σ separation for flight paths of 3 m (solid line) and 7 m (dashed line). Reprinted with kind permission from [62], © 1981, IOP Publishing. All rights reserved

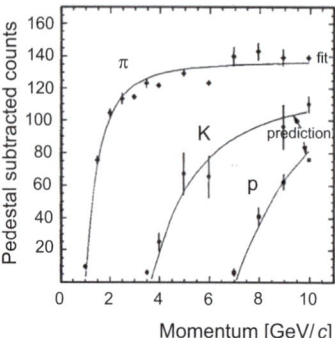

Fig. 9.15 Left: Photograph of a block of silica aerogel. Reprinted under CC-BY-SA-3.0 Licence from [67], © 2024, Wikipedia. Right: Measured threshold curves versus momentum for pions, kaons and protons in $n = 1.0087$ silica aerogel. Reprinted with kind permission from [63], © 1995, Elsevier. All rights reserved

9.3.2 Differential and DISC Cherenkov Counters

Besides detecting the threshold and measuring the number of Cherenkov photons, we can measure also the Cherenkov angle from which we obtain the velocity β for known n. This is done in differential and DISC Cherenkov counters, which have been used in fixed-target experiments. Since they are not practical for colliding-beam experiments, we just discuss them briefly. Figure 9.13 shows the Cherenkov angle as a function of the particle velocity β for various refractive indices. For $n > 1.6$ and very fast particles, the Cherenkov angles can be larger than $50°$. For $n < 1.02$ and very fast particles, Cherenkov angles are small ($\theta_c < 10°$).

The conical emission pattern around a radiating particle can be focussed into a ring-shaped image. If we place an adjustable diaphragm at the focus, the Cherenkov light that is emitted in a small angular range can be transmitted into a PMT. Changing the diaphragm radius allows us to scan through regions of different velocity. Such counters are called differential gas counters. After correcting for the chromatic dispersion in the radiator, we achieve resolutions of $\Delta\beta/\beta \simeq 10^{-7}$. These differential counters are called DISC counters. Though counter lengths are limited to a few meters, π/K separation at several 100 GeV/c is possible in such devices. Instead of changing the radius of the diaphragm, we can vary also the gas pressure and leave the optical system in place.[4]

Figure 9.16 (left) shows the layout of a differential Cherenkov counter with a fixed diaphragm [66]. The particle beam enters from the left-hand side. The Cherenkov light is emitted in a small cone around the beam. At the end of the counter, mirror M1 reflects the light back onto mirror M2, which sends it onto a 1 mrad annular iris that blocks out all light except for Cherenkov photons with angles between 0.7 and 1 mrad. Between M2 and the iris a shutter is located, which can be closed remotely. After the iris, mirror M3 reflects the light through a window onto a lens

---

[4] Such counters are still used for particle identification in test beams.

9.3 Cherenkov Counters

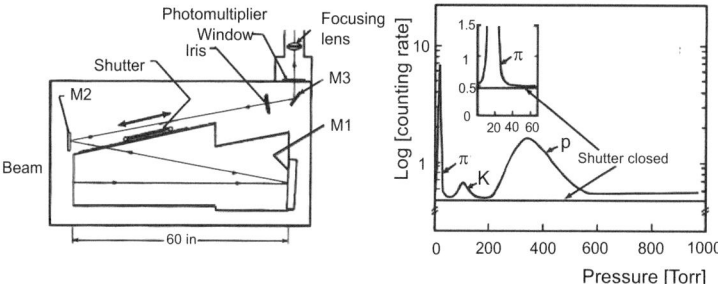

Fig. 9.16 Left: Layout of a differential Cherenkov counter with a fixed diaphragm. By variation of the gas pressure, the refractive index is varied. Right: Counting rate as a function of gas pressure for the differential Cherenkov counter in the left-hand plot, showing peaks for positively charged pions, kaons and protons at a momentum of $p = 165$ GeV/c. The inset depicts an enlargement of the low-momentum region around the pion peak. Both reprinted with kind permission from [66], © 1983, Springer. All rights reserved

that focuses it on the photomultiplier tube. Figure 9.16 (right) shows the counting rate as a function of the gas pressure for 165 GeV/c pions, kaons and protons. Each particle type appears at a specific pressure range, where lower pressures produce low particle masses and high pressure high particle masses.

Figure 9.17 (left) shows the Cherenkov angle for π/K separation as a function of momentum for a threshold counter, two differential counters and five DISC counters [68]. The points show the maximum-observed Cherenkov angle in these Cherenkov detectors. The two differential counters were built at Serpukov, one 5 m and one 10 m long, reaching $\theta_{c,\max} = 23$ and 12 mrad for momenta of 100 and 200 GeV/c, respectively [69]. Serpukov also built a DISC counter that reached

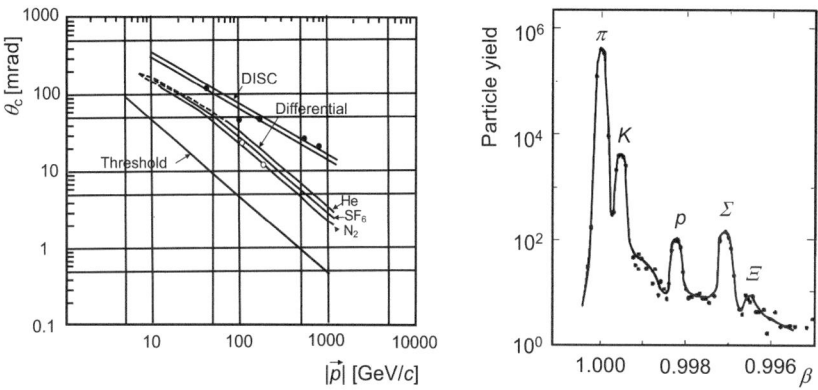

Fig. 9.17 Cherenkov angle as a function of momentum for π/K separation in a threshold counter, two differential counters (open squares) and five DISC counters (solid points). Solid lines show the expected Cherenkov angle momentum dependence for each counter type. Reprinted with kind permission from [61], © 1972, Elsevier. All rights reserved. Right: Pressure curve for a DISC counter used in the CERN 20 GeV/c charged hyperon beam, showing peaks for negatively charged pions, kaons, protons, Σs and Ξs. Reprinted with kind permission from [68], © 1972, Elsevier. All rights reserved

$\theta_{c,\max} = 45$ mrad at a momentum of 100 GeV/c. The Hyperon experiment obtained $\theta_{c,\max} = 120$ mrad at momenta around 30 GeV/c. As expected the DISC counters have the highest sensitivity, while threshold counters have the lowest. We gain a factor of about 10 by changing from a threshold counter to a differential counter and a factor of 30–40 by going to a DISC counter. Figure 9.17 (right) shows pressure curves measured with a DISC counter in the CERN 20 GeV/c charged hyperon beam [68]. In addition to the narrow pion, kaon and antiproton peaks, we see peaks for the hyperons Σ^- and Ξ^-.

9.3.3 Ring-Imaging Cherenkov Counters

Since differential and DISC counters can be used only for particles that are parallel to optical axis of the detector, a velocity measurement for diverging particles from the interaction region (IR) requires a different approach. Here, we need the concept of ring imaging that was first proposed by Roberts [70]. Seguinot and Ypsilantis extended this idea to the first practical large acceptance device [71], illustrated schematically in Fig. 9.18 (left). Many salient features of RICH detectors have been discussed in various reviews [72–77]. A comparison of the performance of RICH detectors to dE/dx and time-of-flight measurements is given in reference [78].

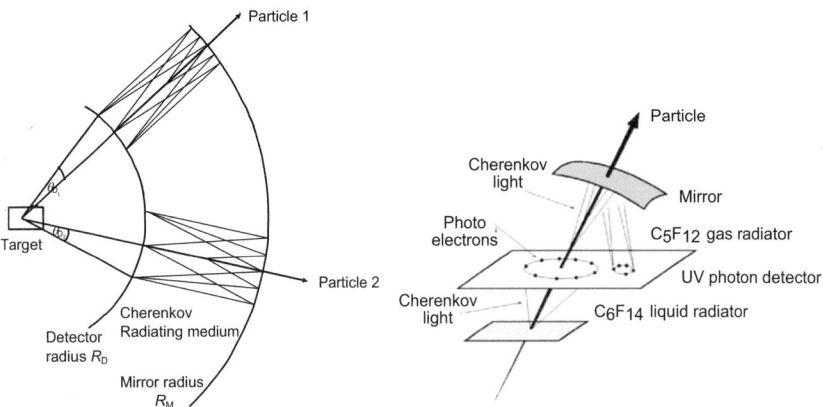

Fig. 9.18 Left: Principle of ring-imaging Cherenkov counters. A mirror is placed at radius R_M and a detector at radius R_D. Here, $R_D = R_M/2$ was chosen. Two particles are shown as examples that produce Cherenkov photons in the radiator in a cone around the particle's trajectory. In the mirror the photons are reflected back onto the detector plane. The projection of the cones onto the depicted plane yields two points for each particle where all photons produced in this projection are collected. This is illustrated in the Figure by three rays as an example. The radius of the cone is defined by the Cherenkov angle. Reprinted with kind permission from [71], © 1977, Elsevier. All rights reserved. Right: Schematic layout of a RICH detector with a liquid and a gas radiator. The liquid radiator uses C_6F_{14} and is placed below a UV-sensitive photon detector. The Cherenkov rings are directly detected. The C_8F_{12} gas radiator is placed above the UV-sensitive photon detector. A mirror placed at twice the detector radius reflects the Cherenkov photons back onto the detector plane. Reprinted with kind permission from [79], © 2017, DELPHI Collaboration. All rights reserved

A spherical mirror of radius R_M centered at the IR focuses the Cherenkov cone produced in a radiator between the sphere with radius R_D and the mirror into a ring-shaped image on the detector sphere. Seguinot and Ypsilantis chose $R_D = 1/2 R_M$ for convenience, but this is not required. Figure 9.18 (right) shows a schematic layout for recording rings from both liquid and gas radiators in the same UV-sensitive detector. The UV photon detector needs to record the Cherenkov photons with high efficiency. This is achieved with a TPC-like chamber containing a photosensitive gas such as TEA, or TMAE that converts UV photons with high efficiency into electrons. The top and bottom sides of the photodetector facing the radiators are covered with fused silica windows that allow UV photons to enter. The converted electrons drift under a strong electric field to the end plates, where they are detected in an MWPC. Figure 9.19 shows a schematic layout of the photon detector.

In the SLD CRID, the TPC gas was C_2H_6 with an admixture of 0.1% TMAE. The top and bottom faces of the TPC had fused silica windows. The ring-shaped photon distribution from the gas entered the TPC after reflection of a spherical mirror through the top window. The ring-shaped photon distribution from the liquid entered the TPC through the bottom window. After photon conversion in the TPC, we obtained two rings of electrons, a small gas ring and a large liquid ring. One end of the chamber was equipped with an MWPC, which had a cathode plane and an anode wire plane, where the wire direction was orthogonal to the electric field. Under the guidance of a strong electric field along the z axis, the electrons drifted towards the MWPC, where they got amplified in the vicinity of the wires. The x-y position was measured in the MWPC, while the z position was determined by the drift time. Since the electrons on

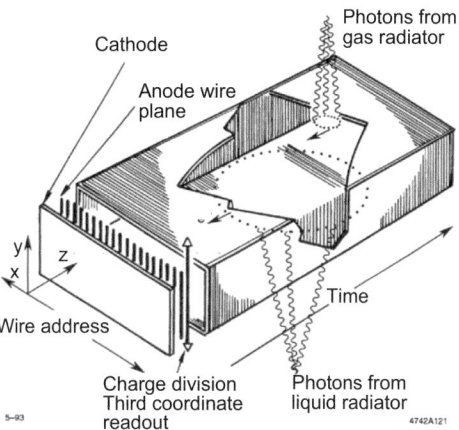

Fig. 9.19 Schematic layout of the TPC-like photodetector for recording converted Cherenkov photon rings. The converted photons from the gas and liquid rings are illustrated. While top and bottom faces are covered with fused silica windows, the left-hand end is covered with an MWPC. The right-hand end is the midplane. The electrons drift along the z direction (arrows) towards the MWPC, where they are recorded. Reprinted with kind permission from [80], © 1993, Elsevier. All rights reserved

a ring have different arrival times, the ring can be reconstructed after detection and θ_c can be extracted. Note that there were two TPCs in the z directions with MWPCs on opposite ends.

If the focal length of the mirror is $R_M/2$, the Cherenkov cones of opening angle $\theta_c = \arccos[1/(\beta n)]$ emitted along the particle's path in the radiator are focused into a ring with radius r_c on the detector sphere. In first approximation, the opening angle θ_D of this ring equals θ_c. For this special geometry, the radius of ring image yields θ_c via

$$\tan\theta_c = \frac{2r_c}{R_M}. \tag{9.36}$$

Since $\beta = 1/(n \cdot \cos\theta_c)$ we get

$$\frac{\Delta\beta}{\beta} = \sqrt{\left[\tan\theta_c^2(\Delta\theta_c)^2 + \left(\frac{\Delta n}{n}\right)^2\right]}. \tag{9.37}$$

Neglecting the error Δn yields the resolution on the velocity,

$$\frac{\sigma_\beta}{\beta} = \tan\theta_c \sigma(\theta_c). \tag{9.38}$$

The resolution of the Cherenkov angle measurement improves as $1/\sqrt{N_{pe}}$,

$$\sigma(\theta_c) = \frac{\langle\sigma(\theta_c^i)\rangle}{\sqrt{N_{pe}}} \oplus C, \tag{9.39}$$

where $\langle\sigma(\theta_c^i)\rangle$ is the average single-photoelectron resolution, N_{pe} is the number of photoelectrons and the parameter C combines several other contributions including correlated terms such as tracking, alignment, and multiple scattering, hit ambiguities, background hits from random sources, as well as hits from other tracks. In addition, physics effects such as decays in flight and interactions in material also limit the performance.

The SLD (CRID) and DELPHI (RICH) particle identification systems were based on this concept. Both experiments used a liquid and a gas radiator. Figure 9.20 (left) shows the layout of a quarter of the SLD barrel CRID. It included 40 liquid radiator trays filled with C_6F_{14} followed by 40 TPCs filled with C_2H_6 plus a 0.1% TMAE admixture and a vessel containing a gas radiator filled with a mixture of C_5F_{12}-N_2 (76:24) and a system of 400 spherical mirrors [81,82]. The top and bottom faces of the TPC were made of fused silica to let the photons from the radiators enter. The photoelectrons drifted in an electric field of 400 V/cm to an MWPC, where its position was reconstructed using a combination of drift time for the z-coordinate, wire clusters for the x-coordinate and charge division for the y-coordinate. They were measured with a precision of $\sigma_x = 1$ mm, $\sigma_y = 2$ mm and $\sigma_z = 1$ mm [81]. The maximum drift length was 1.2 m. The system operated at 35 °C. The MWPCs had 3,720 anode wires, 7,440 amplifiers, 64,000 field shaping electrodes and 6,520

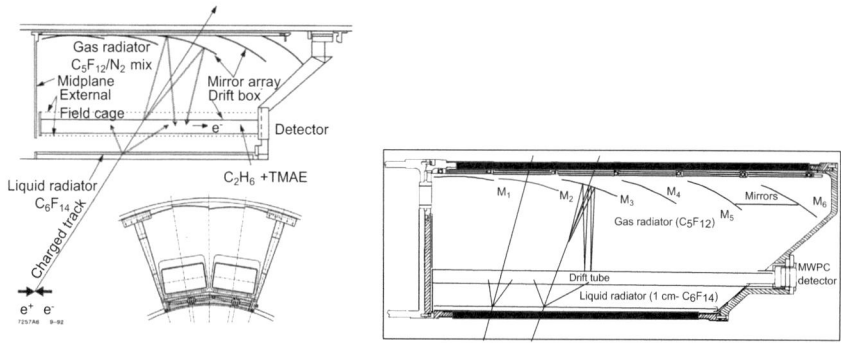

Fig. 9.20 Left: Longitudinal view (top) of one quarter and $r\phi$ side view of two sectors (bottom) of the SLD barrel CRID. Reprinted with kind permission from [80], © 1993, Elsevier. All rights reserved. Right: Longitudinal view of one quarter of the DELPHI barrel RICH. Reprinted with kind permission from [83], © 1991, Elsevier. All rights reserved

corona-preventing field wires. The SLD CRID endcaps consisted of a gas radiator only arranged in ten azimuthal sectors, which were operated with C_4F_{10} gas. Each sector had a TPC with a 28 cm maximum drift length operated with C_2H_6-CO_2 (85:15) plus a 0.1% admixture of TMAE. The Cherenkov light was focused by 60 spherical mirrors. The drift box was kept at a potential of -55 kV corresponding to a uniform electric field of 400 V/cm. The MWPCs in each TPC contained 128 carbon wires with a 7 μm diameter. The typical gain was 3×10^5 [84]. Note that the Malter effect is a particular problem for detectors operating with TMAE, because unlike in other chambers substantial amounts of water or alcohol cannot be added. The gas also attacks most elastomers. In SLD the chosen O-rings worked out fine.

Figure 9.20 (right) shows the layout of a quarter of the DELPHI barrel RICH, which also used a liquid and a gas radiator in the barrel [85]. In the endcaps, DELPHI also used liquid and gas radiators. The liquid radiators consisted of C_6F_{14}, while gas radiators used C_4F_{10} in the barrel and C_5F_{12} in the endcaps. The average pathlength was 45 cm in the barrel and 60 cm in the endcaps. The Cherenkov light was detected with TPC-like photodetectors whose windows consisted of fused quartz plates. In the barrel, the counting gas was a methane-ethane (75:25) mixture plus a small amount of TMAE for UV photon detection. The cylindrical RICH detector was arranged in 2×12 sectors, each covering $30°$ in azimuth angle. Each sector comprised of two liquid radiator trays of 1 cm thickness and two rows of parabolic mirrors that focused the Cherenkov photons produced in the gas radiator onto the photodetector. The complete barrel RICH was operated with a temperature of $40 °C$ to avoid condensation of the C_5F_{12} gas. The drift field in the photon detector was 360 V/cm for drift paths up to 1.5 m. In the gas radiator, the sectors were all connected to generate one gas volume. The endcap RICH detectors were segmented into $30°$ sectors. Five spherical mirrors focused the Cherenkov light generated in the C_4F_{10} gas volume onto the same photodetector consisting of ethane gas plus a small amount of TMAE. The drift field was 1 kV/cm. Since the drift field was orthogonal to the 1.2 T magnetic field, the Lorentz angle was about $50°$.

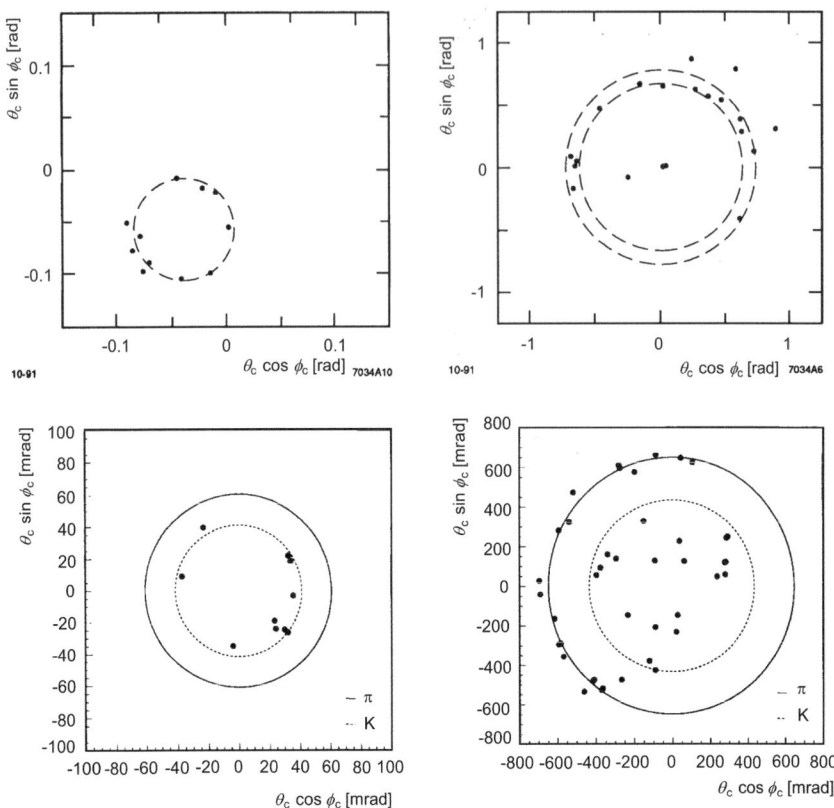

Fig. 9.21 Top left: Reconstructed Cherenkov ring in the SLD gas radiator for a cosmic ray event [87]. The dashed curve shows a ring with a radius 45 mrad. Top right: Reconstructed Cherenkov ring in the SLD liquid radiator for a cosmic ray event. The two dashed circles represent the $\pm\sigma$ range. Both reprinted with kind permission from [87], © 1991, SLAC. All rights reserved. Bottom left: Reconstructed Cherenkov ring in the DELPHI gas radiator for two tracks from Z^0 decay [88]. Bottom right: Reconstructed Cherenkov ring in the DELPHI liquid radiator for two tracks from Z^0 decay. The solid circle shows the expected ring for pions and the dashed circle that for kaons. Both reprinted with kind permission from [88], © 1996, Elsevier. All rights reserved

Figure 9.21 (top) shows reconstructed Cherenkov rings for a cosmic ray event in the SLD gas (left) and liquid (right) radiators. The gas ring had a mean Cherenkov angle of $\theta_c = 57.33$ mrad, while that in the liquid ring was 666.24 mrad. Figure 9.21 (bottom) shows reconstructed Cherenkov rings for two tracks from Z^0 decay in the DELPHI gas (left) and liquid (right) radiators. Note that large rings are often seen as partial ellipses due to geometric and refractive effects. Figure 9.22 shows the Cherenkov angles measured in the gas (left) and liquid (right) radiators of the SLD CRID. The measured resolutions were 4–5 mrad in gas rings and 12–15 mrad in liquid rings. The number of Cherenkov photons were around 10 and 16–17, respectively [86]. The average photon energies were 6.70 and 6.50 eV in the gas and liquid rings for which the detection efficiencies were 17% and 12%, respectively.

9.3 Cherenkov Counters

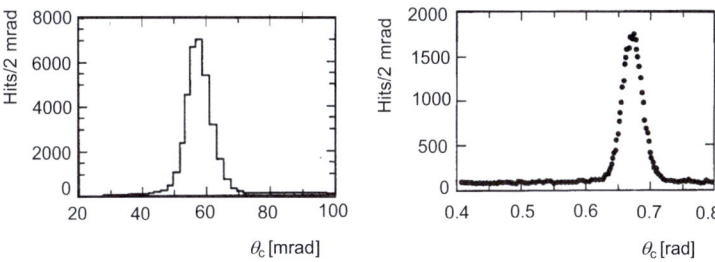

Fig. 9.22 Cherenkov angles measured in the gas radiator (left) and in the liquid (right) radiator of the SLD CRID system. Both reprinted with kind permission from [86], © 1999, Elsevier. All rights reserved

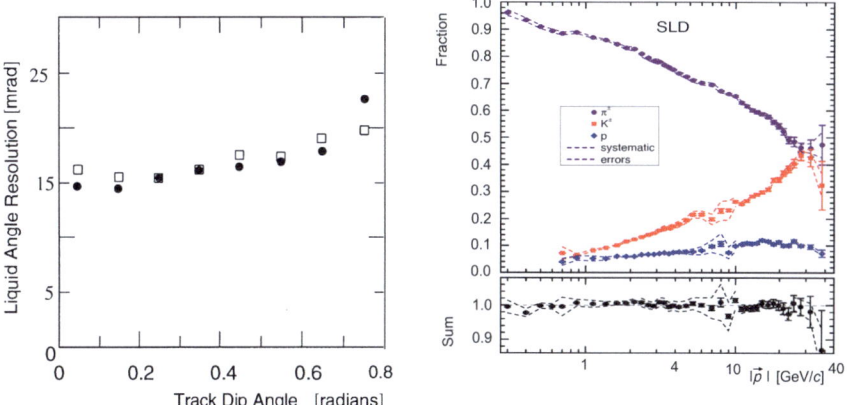

Fig. 9.23 Left: Angular resolution for liquid rings as a function of the dip angle for hadrons (squares) and leptons (solid points) from $e^+e^- \to 6\mu^+\mu^-$, $6e^+e^-$. Right: Particle fractions from Z^0 decays as a function of momentum. Both reprinted with kind permission from [86], © 1999, Elsevier. All rights reserved

The detection efficiencies are determined by the TMAE quantum efficiency and the wavelength-dependent transmission through the quartz window. For liquid rings we have the transmission through the C_6F_{14} liquid while for gas rings we have the transmission through the C_5F_{12}-N_2 gas mixture (87:13) and the mirror reflectivity. In addition come various correction factors. In the DELPHI RICH the corresponding resolutions were 3–5 and 12–15 mrad, while the number of Cherenkov photons were around 10 and 16–17, respectively. Figure 9.23 (left) shows the angular resolution for rings in the liquid radiator as a function of the dip angle. For larger dip angles the resolution increases to values around 20 mrad.

For a particle with $\beta \approx 1$ and momentum p, the separation between two mass hypotheses m_1 and m_2 in units of σs is

$$N_\sigma \approx \frac{|m_2^2 - m_1^2|}{2p^2 \sigma(\theta_c)\sqrt{n^2-1}}. \tag{9.40}$$

Note that the range of momenta over which a particular counter can separate particle species extends from the momentum where the counter operates efficiently as a threshold device to the momentum in the imaging region (9.25). Let us consider two examples. The first example is the C_5F_{12} gas under normal conditions in the SLD gas CRID. The refractive index is $n = 1.00165$. If we increase the pion threshold by 20%, we get a pion momentum of 2.85 GeV/c. Requiring a 3σ π/K separation for a resolution of $\sigma(\theta_c) = 3$ mrad, we get an upper momentum of 14.8 GeV/c. Thus, pions are identified in the momentum range $2.85 \geq p \geq 14.8$ GeV/c at a $\geq 3\sigma$ level. The second example concerns a counter using a fused silica radiator with $n = 1.47$. Here, for a 3 mrad resolution π/K separation at a $\geq 3\sigma$ level works in the momentum range $0.15 \geq p \geq 3.4$ GeV/c. Note that the resolution of a RICH counter is much better than that of a threshold counter by a factor of $\tan \theta_c/(2\sigma(\theta_c))$. The reason is that in a RICH counter we measure the Cherenkov angle directly, while in a threshold counter it is inferred from the number of Cherenkov photons. This factor can be as large as 250. Figure 9.23 (right) shows the particle identification performance of the SLD CRID. Pions (kaons) are identified in the momentum range 0.35 (0.75) GeV/c and 35 (35) GeV/c. Proton identification works in the range 0.75 GeV/c $\leq p \leq$ 45 GeV/c. The DELPHI RICH showed a similar performance.

Figure 9.24 (left) shows DELPHI simulations for the energy loss distribution in the TPC (top) and Cherenkov angle distributions in the liquid RICH (center) and gas RICH (bottom) as a function of momentum. The electron, pion, kaon and proton bands are clearly visible. Misidentification is rather small. Figure 9.24 (right) shows the corresponding distributions in DELPHI data. The liquid RICH provides 3σ π/K separation for momenta up to about 2 GeV/c. The gas RICH extends this to momenta up to 25 GeV/c. Thus, the three PID systems provide 3σ π/K separation over the entire momentum region up to 20.6 GeV/c.

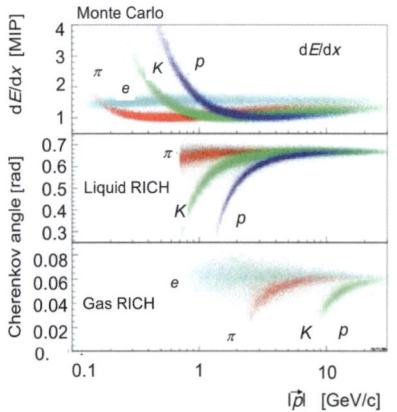
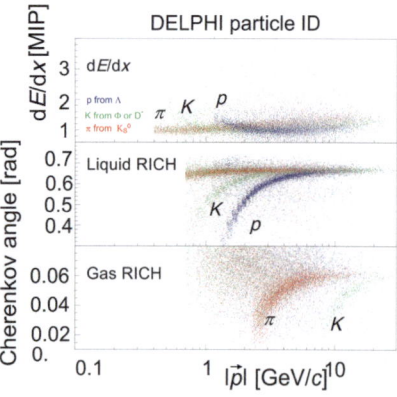

Fig. 9.24 Expected particle identification in the DELPHI experiment showing the dE/dx distribution measured in the TPC and the Cherenkov angles measured in liquid RICH and gas RICH for simulations (left) and for data (right). The electron, pion, kaon and proton bands are clearly visible. Note that pions, kaons and protons were selected from K_S^0, ϕ, D^* and Λ decays via kinematic constraints, respectively. All reprinted with kind permission from [13], © 1996, Elsevier. All rights reserved

9.3.4 Other RICH Detectors

The LHCb experiment at the LHC also uses two RICH detectors. The first RICH combines silica aerogel with a C_4F_{10} gas radiator.[5] Its layout is depicted in Fig. 9.25 (left). The second RICH uses a CF_4 gas radiator. Table 9.5 shows the properties of the radiators in RICH 1 and RICH 2. Note that the number of photoelectrons in the aerogel is rather low. Both RICH detectors are read out with hybrid photodiodes [89, 90], where RICH1 uses 196 HPDs and RICH2 288 HPDS. The photocathode has a high granularity of 2.5 mm × 2.5 mm pads. Photoelectrons produced in a pad are focussed to a specific region of the silicon detector, which is segmented into 1,024 500 μm × 500 μm pixels. The HPDs are sensitive to record single photons. The quantum efficiency of the photocathode is 33% at 270 nm. The HPDs detect Cherenkov photons in the range 200–600 nm. While the anode is grounded, the photocathode is kept at a high potential of −20 kV. The voltage of the first electrode is at −19.7 kV and that of the second electrode is at −16.4 kV. Figure 9.25 (right) shows the measured Cherenkov angle as a function of momentum for the gas radiator in RICH 1. Note that besides the excellent pion/kaon separation also muon and pions are well separated at low momenta.

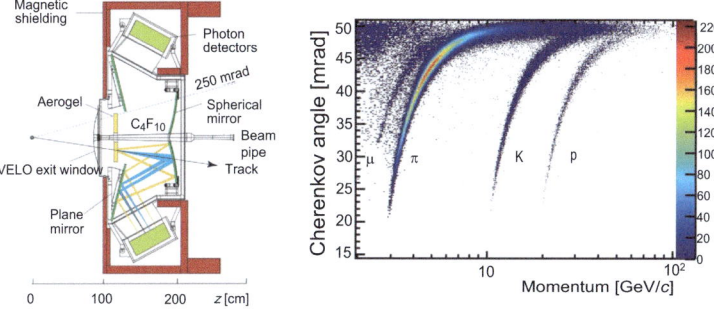

Fig. 9.25 Left: Layout of the LHCb RICH-1 detector. Reprinted with kind permission from [92], © 2001, Elsevier. All rights reserved. Right: The measured Cherenkov angle as a function of momentum for the LHCb RICH-1 C_4F_{10} gas radiator. Reprinted under CC BY 4.0 with kind permission from [89], © 2013, LHCb Collaboration

Table 9.5 Properties of the LHCb RICH radiators, including index of refraction n, maximum Cherenkov angle θ_c^{max}, number of photoelectrons N_{pe} and kaon threshold p_{th}. Reprinted with kind permission from [92], © 2010, Elsevier. All rights reserved

Radiator	n	θ_c^{max} [mrad]	N_{pe}	K p_{th} [GeV/c]
Aerogel	1.03	242	6	2
C_4F_{10}	1.0014	53	24	9
CF_4	1.0005	32	18	15

---

[5] For Run3, the silica aerogel detector has been removed.

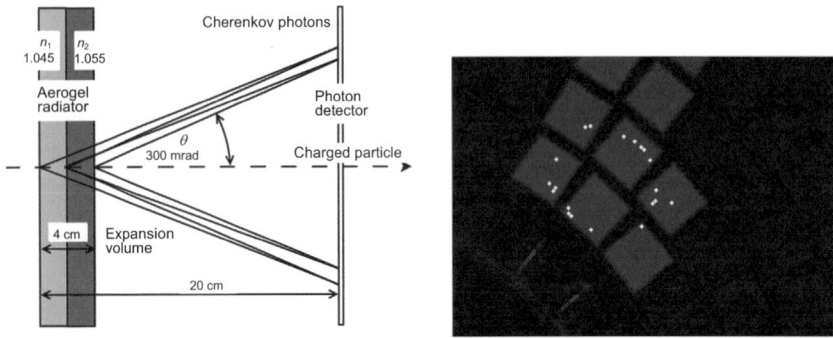

Fig. 9.26 Left: Principle of operation of the Belle II aerogel RICH counter. Right: Observation of a Cherenkov ring produced from a cosmic muon in the Belle II aerogel RICH counter. Both reprinted with kind permission from [93], © 2017, Elsevier. All rights reserved

For run 3 the two RICH detectors were upgraded. The aerogel was removed. The optical system was improved and the HPDs were replaced by two types of 8×8 channel multi-anode photomultiplier tubes (MaPMTs) from Hamamatsu: a 1 inch $\times$ 1 inch version for RICH 1 and the high-occupancy regions of RICH 2 and a 2 inch $\times$ 2 inch version for the low- occupancy regions of RICH 2. The photon detectors and their front-end electronics are combined into modular units called elementary cells [91]. In test beam data an average of 10.6 photon hits was observed consistent with the expected yield of 10.5 photon hits.

The Belle II experiment uses a RICH detector with an aerogel radiator in the forward endcap [93]. It consists of two back-to-back 2 cm thick aerogel blocks with refractive indices of $n_1 = 1.045$ and $n_2 = 1.055$. They are read out with hybrid avalanche photodiodes. Figure 9.26 (left) illustrates the focusing property of the ARICH. The lower-refractive-index detector gives a larger Cherenkov angle than the higher-refractive-index counter. Figure 9.26 (right) shows the detection of a Cherenkov ring. At 4 GeV/c, π/K separation yields a $(93.5 \pm 0.6)\%$ kaon efficiency for a $(10.9 \pm 0.9)\%$ pion misidentification rate [94].

9.3.5 The BABAR DIRC

The BABAR DIRC detector (Detector of Internally Reflected Cherenkov light) is a novel design. Figure 9.27 (left) illustrates the principle. The radiator consists of long rectangular fused-silica bars (n = 1.473). A charged particle traversing the bar emits Cherenkov photons in a cone around the particle's trajectory. The photons propagate along the bar in both directions. The bars also act as wave guides. If the Cherenkov photons have the right angle with respect to the boundary faces, they are totally reflected, otherwise they are typically lost. At one end of the bar a mirror is placed, while the other end of the bar is coupled to the readout system, which uses PMTs. Since PMTs cannot be operated inside the solenoidal magnetic field, the bars were extended through the endcap.

9.3 Cherenkov Counters

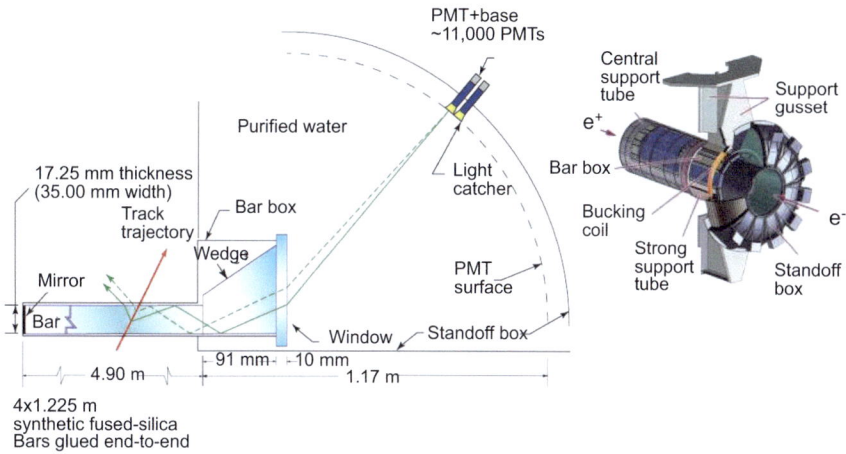

Fig. 9.27 Left: Operation principle of the BABAR DIRC (see text). Right: Three-dimensional schematic view of the DIRC particle identification detector in the BABAR experiment showing the bar box, a strong support tube, the standoff box holding the PMTs and a bucking coil. Both reprinted with kind permission from [95], © 2002, Elsevier. All rights reserved

To measure the different Cherenkov angles with good resolution, the photons exiting at the end of the bar need to be spread out. This is achieved with the principle of a pin hole camera. Here, the pin hole is the end face of the bar. Note that pin hole cameras have an infinite depth of field and the image can be magnified as much as necessary by moving the detector away from the bar end. The angular resolution is limited by the pin hole size and the detector pixel size,

$$\sigma_\alpha = \sqrt{\frac{\frac{t_{x,y}^2}{12} + \sigma_{x,y}^2(\det)}{L_{\text{sob}}^2}}, \qquad (9.41)$$

where $t_{x,y}$ is the size of the bar in the x, y position, $\sigma_{x,y}(det)$ is the resolution of the position measurement in the detector plane and L_{sob} is the distance between the bar end and the detector plane.

Figure 9.27 (right) shows a three-dimensional schematic view of the DIRC. A central support tube housed the 12 modules, each covering an angular region of 30°. A module whose layout is depicted in Fig. 9.28 (left) housed 12 fused-silica bars. Adjacent modules had no overlap. A strong support tube and a bucking coil were installed in front of the standoff box shown by the disk-shaped detector. Each bar was 17 mm thick, 35 mm wide and 4.9 m long and showed excellent transmission properties. This is illustrated in Fig. 9.28 (right), which displays a photograph of the light path in a fused-silica bar. If the light has the correct reflection angle it travels all the way through the 4.9 m long bar. For wavelengths $\lambda > 300$ nm, the light transmission was very good. The photons were detected with 10,752 closely spaced PMTs at the outer spherical detector surface of the stand-off box. A photograph of the PMT wall, which is placed at 1.17 m from the bar end, was shown in Fig. 8.15 (left).

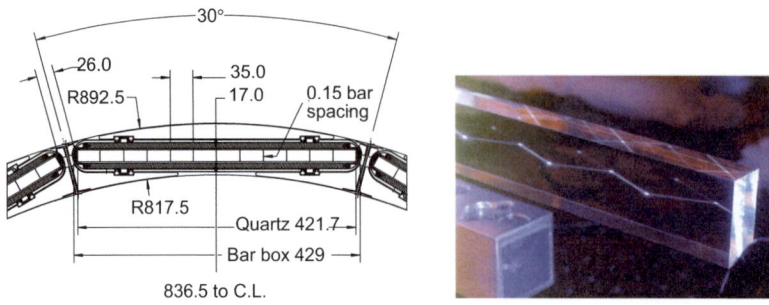

Fig. 9.28 Left: Sideview of a DIRC module. Reprinted with kind permission from [96], © 2004, Elsevier. All rights reserved. Right: Path of UV light in a fused silica bar. Reprinted with kind permission from [97], © 1997, SLAC. All rights reserved

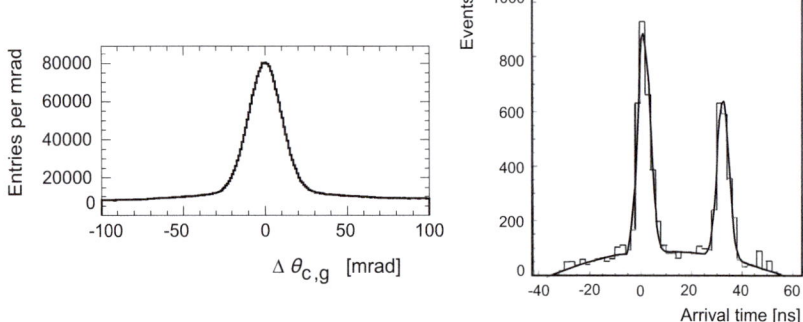

Fig. 9.29 Left: The difference between measured and expected Cherenkov angle for single photons. Reprinted with kind permission from [10], © 2011, Elsevier. All rights reserved. Right: Arrival time of Cherenkov photons at the detector. The first peak corresponds to the detection of direct photons and the second to that of reflected photons. Reprinted with kind permission from [98], © 2020, Elsevier. All rights reserved

As depicted in Fig. 9.27 (left), the volume of the stand-off box was filled with purified water to better match the refractive index of fused silica than with air. The fused-silica bar end was extended with a wedge to guide the photons into the water. Since the refractive indices of fused silica and water are not identical, the refraction angle had to be accounted for. The PMTs measured the positions of Cherenkov ring segments from which the Cherenkov angle was extracted.

Figure 9.29 (left) shows the difference between measured and expected Cherenkov angle for single photons. The observed resolution is 9.6 mrad compared to the expectation of 9.5 mrad, which was dominated by 7 mrad from geometry and optics, 5.4 mrad from chromatic effects and (2–3) mrad from imperfections on the bar surface and angles. In addition to the clear peak, we noticed a 10% background originating from combinatorial background, track overlaps, accelerator backgrounds, δ rays and reflections at the glued bar junctions. Using timing information, this background can be reduced. The time distribution measured for single muons gave a time

9.3 Cherenkov Counters

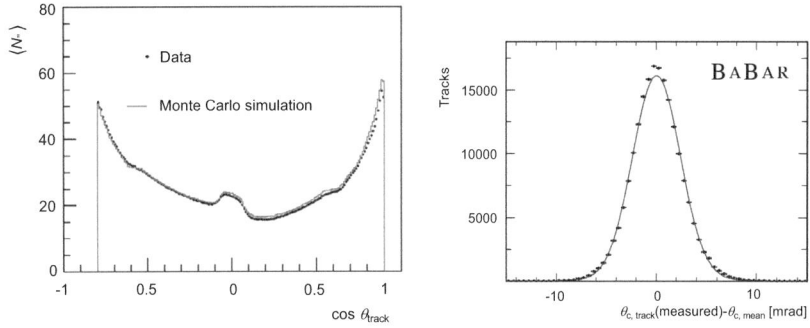

Fig. 9.30 Left: Number of detected photons as a function of $\cos\theta_{track}$ for data (points) and simulation (solid line) in $\mu^+\mu^-$ events. Right: Difference between measured and expected Cherenkov angle per track in $\mu^+\mu^-$ events. Both reprinted with kind permission from [95], © 2002, Elsevier. All rights reserved

resolution of 1.7 ns. Figure 9.29 (right) shows the arrival time of the Cherenkov photon at the detector. The two peaks correspond to the detection of direct and reflected Cherenkov photons, respectively.

The DIRC produces a large number of Cherenkov photons. We saw in Sect. 9.3.1 that the expected number of Cherenkov photons and detected photoelectrons can be obtained from equations (9.26) and (9.27), respectively. Figure 9.30 (left) displays the number of detected Cherenkov photons per track as a function of $\cos\theta_{track}$ of the incident particle in comparison to simulation. We see between 20 and 60 photons in the DIRC with a quite strong angular dependence. Since the higher-momentum tracks tend to be at small angles in the forward direction, this peaking helped the actual track resolution just where it was needed. The actual Cherenkov angular resolution for muons in $\mu^+\mu^-$ events was 2.4 mrad as illustrated in Fig. 9.30 (right). Due to the geometric arrangement of the DIRC, the Cherenkov images in the PMT array are not full rings but circular and elliptical segments. In the *BABAR* experiment, we are most interested in π/K separation. Figure 9.31 (left) shows the kaon identification efficiency and pion misidentification rate as a function of the charged track momentum for $D^0 \to K^-\pi^+$ events. The kaon identification efficiency is larger than 90% for momenta up to 3.2 GeV/c, while the pion misidentification rate is less than 10%. For momenta below 2 GeV/c, the pion misidentification rate drops to $\leq 2\%$. Figure 9.31 (right) shows the π/K separation as a function of of track momentum for $D^0 \to K^-\pi^+$ decays. The $\geq 3\sigma$ π/K separation holds for momenta of 4 GeV/c. In (9.40), we specified the two-particle separation in units of σs. Figure 9.32 shows the observed Cherenkov angles as a function of momentum for pions (left) and kaons (right) that were kinematically identified in the decay $D^{*+} \to D^0\pi^+ \to K^-\pi^+\pi^+$. We observe a clear π/K separation, which is 4.3σ at $p = 3$ GeV/c and 3σ at 4 GeV/c. A few pions are misidentified as kaons, while a few more kaons are misidentified as pions.

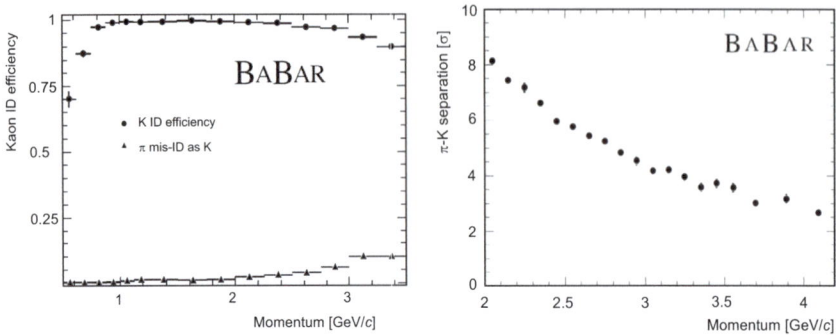

Fig. 9.31 Left: Kaon identification efficiency and rate of pions misidentified as kaons as a function of the charged track momentum in $D^0 \to K^-\pi^+$ decays selected kinematically from an inclusive D^* sample. Reprinted with kind permission from [98], © 2020, Elsevier. All rights reserved. Right: The DIRC π/K separation as a function of track momentum. Reprinted with kind permission from [96], © 2004, Elsevier. All rights reserved

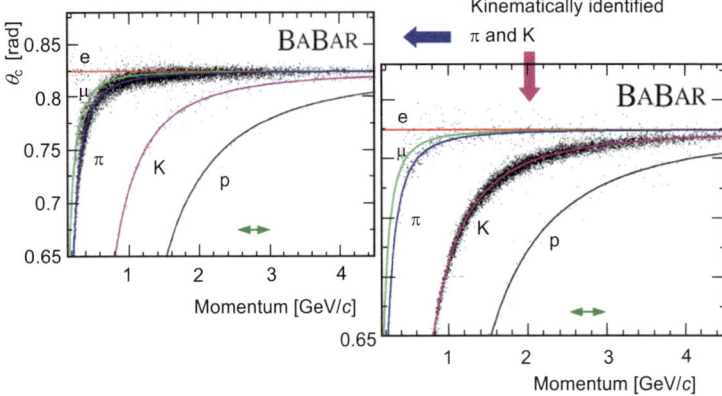

Fig. 9.32 Left: The measured Cherenkov angle for kinematically identified pions from the process $D^0 \to K^-\pi^+$. Right: The measured Cherenkov angle for kinematically identified kaons from the process $D^0 \to K^-\pi^+$. Both reprinted with kind permission from [95], © 2002, Elsevier. All rights reserved

9.3.6 Next Generation DIRC Counters

Next-generation DIRC counters take advantage of the development of new, very fast, pixelated photodetectors such as the multi-anode photomultiplier tubes (MaPMT) or the micro-channel plate PMTs. By utilizing either time imaging or lens/mirror-focused optics, they accomplish precision measurements of the Cherenkov angle. In some cases, they measure the time-of-flight rather precisely correcting for the chromatic dispersion in the radiator. One of these devices is the Belle II time-of-propagation (TOP) counter, which emphasizes precision timing for both Cherenkov imaging and TOF to perform π/K separation of at least three standard deviations for momenta up to 4 GeV/c. The TOP counter consists of sixteen identical modules. Each module is composed of four parts glued together, two fused silica bars of dimensions 125 cm × 45 cm × 2 cm acting as Cherenkov radiator, a mirror located at the forward end of the bars and a 10 cm long prism that couples the bar with an array of microchannel-plate photomultiplier tubes. The expected number of observed Cherenkov photons is 20 [99]. The transition time spread is about 30 ps.

The LHCb experiment is developing the TORCH detector (Timing Of internally Reflected Cherenkov photons), which is a high-precision time-of-flight detector combining timing measurements with DIRC-style reconstruction. The concept is to perform DIRC imaging for individual photons with fast photon detectors. For time resolutions of 10–15 ps it provides efficient π-K separation up to 10 GeV/c [100]. Figure 9.33 shows a schematic layout of the TORCH detector. The detector consists of 18 modules (Fig. 9.33 left) of synthetic fused silica radiator plates. Each module has 1 cm thick, 66 cm wide and 250 cm long plates (Fig. 9.33 center). Each plate is coupled to a focusing block at the outer end. The focusing block has a curved shape such that the same exiting angles in the yz plane θ_z are reflected into the same point on the opposite side where the MCP is mounted (Fig. 9.33 right). The MCPs cover geometric angles $\theta_z = 0.45$ rad to $\theta_z = 0.85$ rad. The flight time difference between a 10 GeV/c pion and a 10 GeV/c kaon over a 10 m distance is 35 ps. In order to achieve a 3σ separation, a time resolution of 10–15 ps per track is necessary. Since a track will produce about 30 detected Cherenkov photons, a time resolution of 70 ps per photon is required. This translates into a 1 mrad angular resolution. LHCb selected MCPs from Photek with a 64×64 anode pixel structure spread over an active area of 53 mm × 53 mm (Fig. 8.18 right). The MCPs have ALD coating and are designed to withstand an integrated charge of 5 C/cm^2. The pixelated anode is connected via an Anisotropic Conductive Film to a readout PCB. To obtain an effective granularity of 8×128 pixels required to achieve the 1 mrad angular precision, charge sharing and pixel grouping is employed, where charge sharing relies on capacitive coupling to

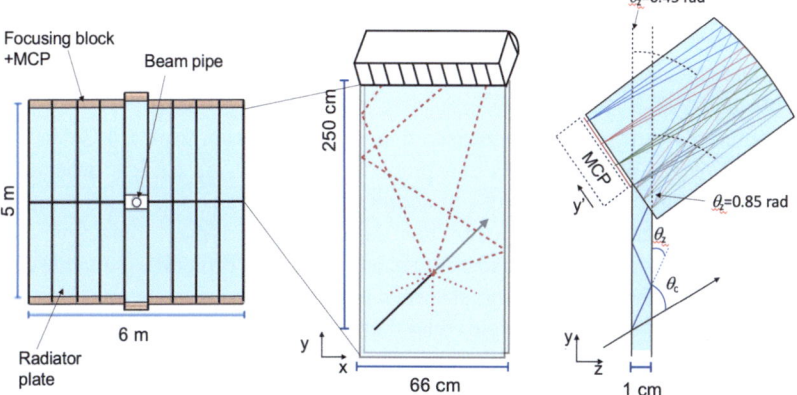

Fig. 9.33 Left: Schematic view of the TORCH detector. Reprinted under CC-BY-NC-ND-4.0 Licence from [103], © 2014, T. Blake. Center: Schematic layout of a TORCH module. Right: Illustration of the focusing of Cherenkov light in a TORCH module. Both reprinted under CC-BY-4.0 Licence from [104], © 2019, N. Harnew

produce a smooth well-defined charge footprint on the anode. More details of these projects are discussed in [98].

The GlueX experiment at Jefferson Lab places four *BABAR* DIRC modules in their upgraded DIRC [101]. The modules are coupled to upgraded optics and readout. The quartz bars run perpendicular to the beam line. So, this is the first application of a DIRC in an endcap. The PANDA experiment at FAIR plans for a barrel DIRC [102]. This is the first DIRC that will use lens focusing. The Panda physicists expect to achieve three standard deviation π/K separation for momenta up to 3.5 GeV/c.

9.4 Transition Radiation Measurements

In Fig. 9.16 (right) we observed a non-zero counting rate at zero pressure. This indicates that radiation is emitted already below the Cherenkov threshold. It can be explained as radiation that is emitted, when particles traverse boundaries between a dense medium and vacuum as well as vacuum and a dense medium. This is transition radiation (TR) discussed in Sect. 2.5. It is emitted, when a charged particle traverses a medium with varying dielectric constants, such as an array of thin foils and air gaps. The radiation is emitted from the interface between the two materials. In Sect. 2.5 we explained how transition radiation is generated phenomenologically in terms of a time-varying dipole.

For a relativistic particle with time dilatation factor $\gamma = E/(mc^2)$, the radiation is concentrated in a cone with half-opening angle $\phi \sim 1/\gamma$. The intensity of transition radiation increases linearly with γ. If we use a periodic arrangement of many foils and gaps as radiator, interference will produce a threshold effect in γ. Counters measuring transition radiation can, therefore, be used to discriminate between particles of different mass but the same momentum via the γ factor. Transition radiation is typically emitted in X-ray region. Thus, actual counters consist of radiators followed by a proportional chamber to detect the X-rays. Since X-ray absorption vanishes like $Z^{3.5}$, we need a Z as low as possible for the foil. We choose Li ($Z = 3$) and Xe gas. Actual devices contain up to 1,000 50 μ thick Li foils. Note that a 1.4 GeV/c electron has a γ factor of 2,740.

Figure 9.34 (left) shows the measured energy distribution from 1.4 GeV/c electrons and pions. Both electrons and pions loose energy via dE/dx producing a Landau distribution. The electrons in addition produce transition radiation quanta, which leave higher energy deposits than those from the typical ionization. Since the impact points of electrons and photons are close, it is difficult to separate them. Furthermore, without the radiator, electrons produce the Landau energy loss, which is slightly shifted to higher energies than that of the pions.

The average energy of the transition radiation quanta depends on the electron momentum and the radiator layout. Figure 9.34 (right) shows the average energy of transition radiation X rays for different radiator configurations followed by a proportional chamber filled with Xe-CO_2 (80:20). In the setup with 1,000 (500) 50 μm thick Li foils spaced at a 500 μm gap, we observe about 30 (25) keV X-rays radiated by 2.5 GeV/c electrons. By reducing the gap to 200 μm decreases

9.4 Transition Radiation Measurements

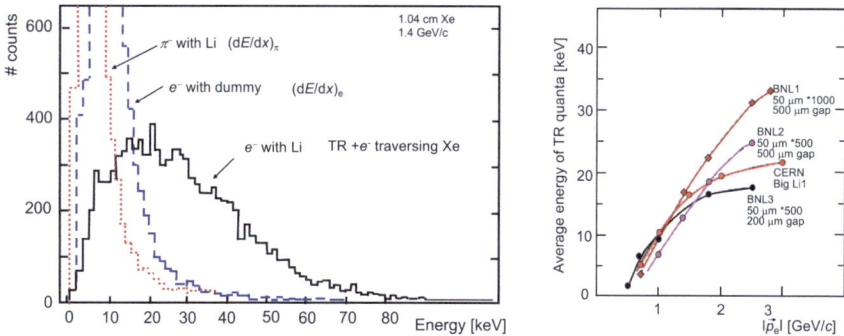

Fig. 9.34 Left: Measured energy distribution of 1.4 GeV/c electrons and 1.4 GeV/c pions in a detector consisting of 1,000 Li foils and a Xe chamber [105]. For electrons, the measured spectrum (solid histogram) is a superposition of transition radiation quanta and ionization losses. Without the foils the electrons only produce the dE/dx energy loss (blue dashed histogram). For pions, the spectrum results from the Landau-shaped ionization loss distribution (red dotted histogram). Reprinted with kind permission from [105], © 1980, IOP Publishing. All rights reserved. Right: Average energy of the transition radiation quanta measured with different radiators and a 12.7 mm thick Xe-CO$_2$ proportional chamber having different configurations: 1,000 50 μm thick Li foils separated at 500 μm (brown squares and curve), 500 50 μm thick Li foils separated at 500 μm (magenta points and curve), 500 50 μm thick Li foils separated at 200 μm (black points and curve) and a device built at CERN (red points and curve) with 650 50 μm thick Li foils separated by 300 μm. Reprinted with kind permission from [106], © 1977, Elsevier. All rights reserved

the X-ray energy to about 15 keV. The electron momentum range of 0.5 GeV/c to 3 GeV/c corresponds to γ factors of 1–6 $\times 10^3$. Typically, transition radiation detectors are usable for electrons with momenta larger than 0.5 GeV/c or pions with $p > 140$ GeV/c. If we need to extend transition radiation detection to $\gamma < 10^3$, we need to detect X-rays in the 1–5 keV energy range.

In the separation of transition radiation signals from pulses caused by the energy loss of charged particles, the upper end of the Landau distribution is disturbing since it overlaps with the TR pulse height. In order to separate the two contributions, we can measure two quantities, the total energy deposited by the transition radiation quanta and the number of ionization clusters along the charged track. The number of clusters is Poisson distributed, whereas the energy loss curve has a very long tail. Figure 9.35 (left) shows the schematic layout of the radiator followed by a drift chamber. The beam enters from the left, passes through the radiator and enters the drift chamber. While clusters produced by dE/dx losses lie near the particle trajectory, the transition radiation clusters are produced away from the axis. They are typically more energetic than those from dE/dx losses. Figure 9.35 (center) shows clusters as a function of the drift time. The charge of a transition radiation cluster is typically larger than that produced by ionization of a charged particle. By selecting a threshold on the measured charge, we can reduce the contamination from δ rays. Furthermore, we can look at the number of measured clusters that are larger for transition radiation quanta than those produced by ionization of charged particles. Figure 9.35 (right) shows the number of clusters (top) and the charge distribution (bottom) measured in the drift chamber. The number of clusters is much larger for electrons than for pions, where

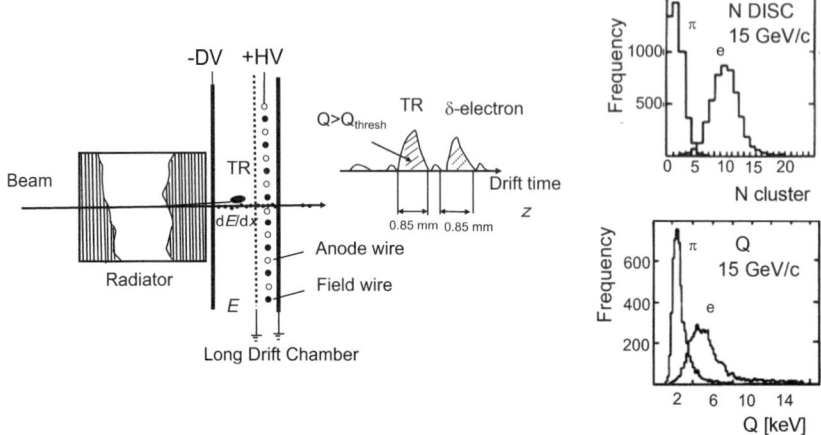

Fig. 9.35 Left: Schematic layout of a transition radiation detector showing the radiator followed by a drift chamber. The deposition of transition radiation quanta and dE/dx energy losses is illustrated. Center: Charge distribution for transition radiation quanta and δ-rays as a function of the drift time. Transition radiation quanta produce more charge than δ electrons. Both reprinted with kind permission from [107], © 1981, Elsevier. All rights reserved. Right top: Observed number of clusters for 15 GeV/c pions and 15 GeV/c electrons [108]. Right bottom: Charge distributions for 15 GeV/c pions and 15 GeV/c electrons. Both reprinted with kind permission from [108], © 1981, Elsevier. All rights reserved

the number of electron clusters is the sum of dE/dx hits and transition radiation quanta. Furthermore, the charge measured for electrons is also larger than that for pions, since the distribution is dominated by the higher-charge transition radiation quanta.

Figure 9.36 (left) shows the pion rejection as a function of electron efficiency after applying selections on the charge and the number of cluster distributions in a detector consisting of 12 radiators, each with 160 35 µm thick foils spaced at 240 µm followed by a proportional chamber. The entire detector has a depth of 66 cm. With a requirement on a minimum charge, we can achieve an 80% electron efficiency for a ~1% pion contamination. Adding a requirement of at least 4 keV on the cluster energy improves the pion rejection to 2×10^{-3} for a 90% electron efficiency. A further requirement on the discriminator threshold of 4 keV reduces the pion misidentification to 8×10^{-4}. For a radiator made of pure carbon fiber with 7 µm thick foils, we achieve a pion rejection of 2×10^{-3} at a 90% electron efficiency. For a detector with 20 µm thick mylar foils and a selection of only charge measurements yields almost the same performance as that of the carbon fiber detector. Note that a minimum charge requirement alone in the fiber detector yields a much worse performance.

We can use these detectors also to perform pion-kaon separation at momenta above 140 GeV/c. Figure 9.36 (right) shows the pion misidentification rate as a function of the kaon efficiency for two setups: setup A using 24 carbon fiber radiators each filled with a Xe chamber having a total length of 132 cm and setup B using 20

9.4 Transition Radiation Measurements

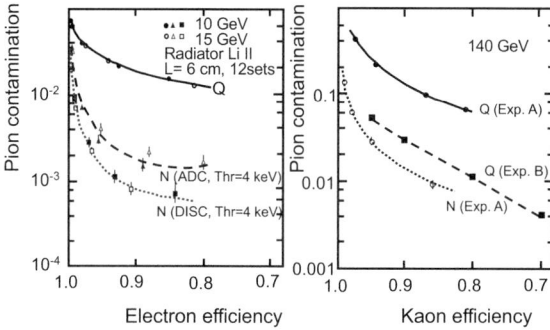

Fig. 9.36 Left: The pion contamination as a function of electron efficiency for 10 GeV/c (solid) and 15 GeV/c (open) particles after requirements on the measured charge (points), cluster energy (triangles) and discriminator threshold (squares). Right: The pion contamination as a function of the kaon efficiency for two experimental setups, selecting a specific charge and number of clusters (see text). Both reprinted with kind permission from [108], © 1981, Elsevier. All rights reserved

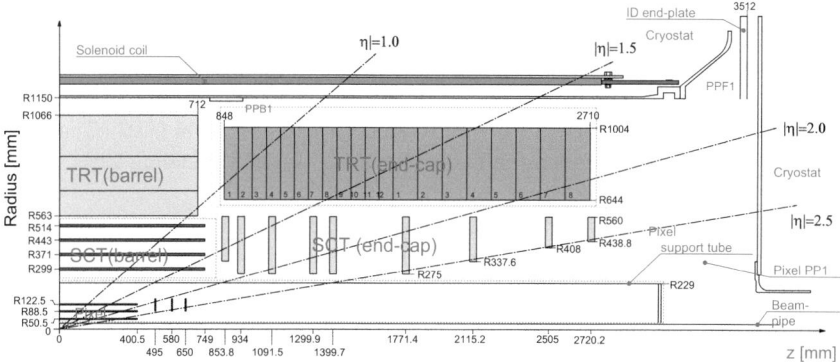

Fig. 9.37 Schematic layout of one quarter of the ATLAS inner detector in the barrel and in an endcap. Reprinted with kind permission from [109], © 2017, IOP Publishing. All rights reserved

radiators made of 5 μm thick mylar foils with one chamber each. The detector has a total length of 147 cm. Just imposing requirements on the measured charge, setup A yields a pion misidentification rate of around 10% at a 90% kaon efficiency, while setup B yields a 3% pion misidentification rate at a 90% kaon efficiency. We can improve this further by adding a requirement on the number of charge clusters. For setup A we achieve a pion misidentification rate of 10^{-2} at an 85% kaon efficiency.

9.4.1 The ATLAS Transition Radiation Tracker

The ATLAS experiment uses a Transition Radiation Tracker (TRT) to identify electrons and to measure track positions. Figure 9.37 shows a schematic layout of one quarter of the ATLAS inner detector. The TRT barrel follows the SCT barrel and the

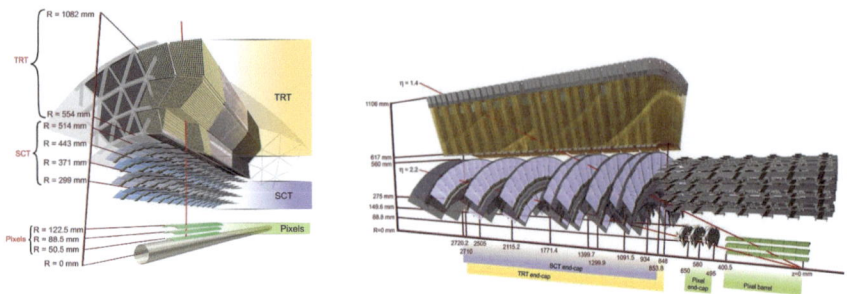

Fig. 9.38 Layout of the ATLAS TRT in the barrel (left) and in the endcaps (right). Reprinted with kind permission from [109], © 2017, IOP Publishing. All rights reserved

TRT endcap covers the entire SCT endcap region. Figure 9.38 (left) shows a three-dimensional view of the inner tracker in the barrel, while Fig. 9.38 (right) shows a three-dimensional view of the inner tracker in the endcaps. The TRT Barrel is divided into 96 modules of three types, arranged in three cylinders of 32 modules of each type. Module types 1, 2, 3 have 19, 24 and 30 layers, respectively. The basic unit is a 144 cm long straw tube with a diameter of 4 mm. A 31 μm thick gold-plated tungsten wire is placed in its center. The tube is filled with a gas mixture of Xe:CO_2:O_2 (70:23:7). The total number of straw tubes in the barrel is 52,544 [110]. The straw cathodes are typically operated at a high voltage of -1.53 kV, yielding a gas gain of 2.5×10^4. The drift velocity is around 5.0 cm/μs.

In addition to the barrel TRT, there are two endcap TRTs in which the straw tubes are arranged in so-called eight-plane wheels, which combine eight layers of straw tubes into one wheel. The straw tubes run perpendicular to the beam direction in the $r\phi$ plane. The same-type straw tubes are used as in the barrel. The length is about 50 cm. Each endcap has 20 eight-plane wheels, where 12 are arranged in wheel type A and eight in wheel type B. Each endcap spans about 1.9 m in the z direction and holds 122,880 straw tubes. The spatial resolution is 118 μm in the barrel and 130 μm in the endcaps.

In the TRT we measure the time when the signal exceeds the threshold and the time when it falls below the threshold in 3.125 ns bins as depicted in Fig. 9.39 (left). The time difference is called time-over-threshold (ToT), which is used to extract the signal. It is corrected for systematic variations along the z coordinate and is divided by the transverse particle trajectory length in the straws. We use two thresholds, a low threshold at 300 eV, which is well above the noise level of 40 eV, for measuring track positions. The high threshold is set at 6 keV for identifying electrons via transition radiation. The average energy loss of a minimum-ionizing particle is 2 keV. If a charged particle or an X ray has created an electron in the straw tube, the electron drifts towards the wire (t_d) where it starts avalanching. The arrival of the first electron will start the ToT and the last electron will stop it. Note that the time-over-threshold method with track-position correction improves particle identification in the TRT

9.4 Transition Radiation Measurements

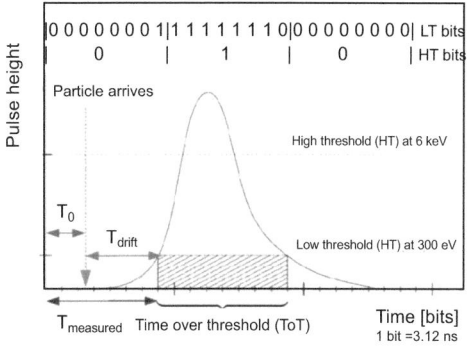
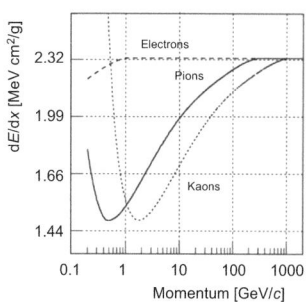

Fig. 9.39 Left: The pulse height as a function of the recorded time. The shaded region defines the time-over-threshold. The measured time t_m consists of the drift time t_d and the offset t_0. The time is digitized in time bins whose values for a low threshold (300 eV) and a high threshold (6 keV) are displayed on the top of the figure. The dotted arrow indicates that a charged particle has created an electron in an ionization process. Reprinted with kind permission from [111], © 2017, IOP Publishing. All rights reserved. Right: Stopping power distributions of electrons, pions and kaons in the ATLAS xenon gas straw tubes. For momenta less than 20 GeV/c, good particle identification based on an accurate measurement of the energy loss is possible. Reprinted with kind permission from [112], © 2001, Elsevier. All rights reserved

at low particle momenta. Figure 9.39 (right) shows the energy loss distributions of electrons, pions and kaons in the ATLAS TRT. For low momenta the energy loss of electrons is substantially larger than that of pions and kaons. Note that for a momentum of 1 GeV/c the electron already has a γ factor of $\sim$2,000.

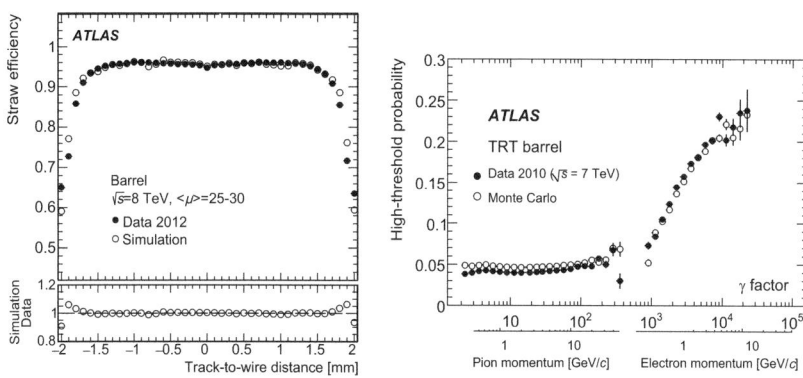

Fig. 9.40 Left: Straw efficiency as a function of the distance from the wire in the barrel showing data (black points) and simulation (open points). Reprinted with kind permission from [109], © 2017, IOP Publishing. All rights reserved. Right: Probability of a TRT high-threshold hit as a function of the γ factor in the barrel showing data (black points) and simulations (open squares). The pion and electron momentum scales are indicated. Reprinted with kind permission from [113], © 2012, Elsevier. All rights reserved

Tracks cross on average ∼35–40 straws but not all of them will record signals. Since the number of hits on a track is important for momentum measurements, the hit reconstruction efficiency is an important parameter. Figure 9.40 (left) shows the efficiency of the straw tubes in the barrel as a function of the drift distance to the wire. For distances less than $|d_{tr}| < 1.5$ mm, the efficiency is above 90%, except for a small efficiency reduction near the wire and the efficiency drop in the vicinity of the tube wall, where the efficiency drops to 65%. The measured efficiency is slightly lower than that obtained in simulations. Similar efficiencies are found for the straw tubes in the endcaps. To produce a single transition radiation photon, requires about 30 interface crossings. The typical photon energy is 5 keV. The photons are produced with an angle of the order $1/\gamma$ relative to the track. Absorption of the photon in the xenon-based gas in the straw tube yields high-amplitude signals. Figure 9.40 (right) shows the probability of a TRT high-threshold hit as a function of the Lorentz factor in the barrel for 7 TeV collision data as well as for simulations. A pure electron sample from photon conversions was used in this region. For $\gamma > 1{,}000$ the Monte Carlo describes the rise of the high-threshold probability rather well.

Figure 9.41 (left) shows the summed time-over-threshold corrected for systematic variations along the z coordinate and divided by the transverse particle trajectory length for pions and electrons in 7 TeV collision data. The pions are clearly separated from the electrons. Note that we have made no momentum selection. The average pion momentum is 1.0 GeV/c, while the average electron momentum is 1.4 GeV/c. We see that 99.6% of the selected particles are electrons. Figure 9.41 (right) shows the high-threshold probability for different particles as a function of the γ factor in test beam data. Note that the event selection imposes tight requirements on tracking and particle species selection. The contamination is estimated to be below 0.1%. The high-threshold probability curve shows no dependence on the particle species as expected, just on the Lorentz factor. Thus, there is no discontinuity in the curve in

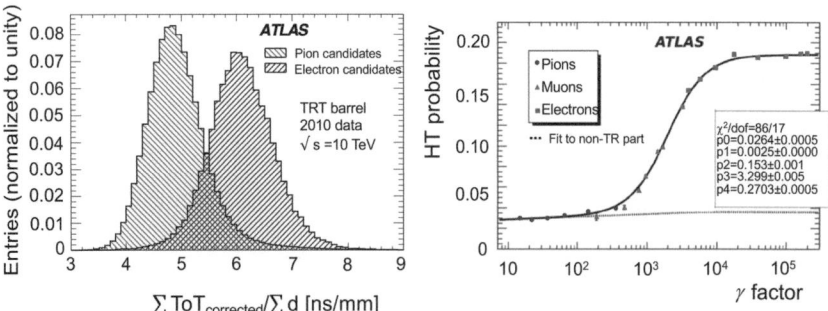

Fig. 9.41 Left: Corrected time-over-threshold divided by the transverse particle trajectory length for pions (left histogram) and electrons (right histogram). Reprinted with kind permission from [114], © 2011, Elsevier. All rights reserved. Right: High-threshold probability as a function of the γ factor for pions (blue points), muons (magenta triangles) and electrons (red squares) in barrel modules using test beam data. The black curve is a fit to the function given in (9.42). The fit parameters are $p_0 = 0.0264 \pm 0.0005$, $p_1 = 0.0025 \pm 0.0000$, $p_2 = 0.153 \pm 0.001$, $p_3 = 3.299 \pm 0.005$ and $p_4 = 0.2703 \pm 0.0005$. Reprinted with kind permission from [115], © 2011, Elsevier. All rights reserved

the regions between different particle species. The solid curve is a fit to the following generic onset function

$$P(\gamma) = p_0 + p_1 \log_{10}(\gamma) + \frac{p_2}{1 + \exp\left[-(\log_{10}(\gamma) - p_3)/p_4\right]}, \quad (9.42)$$

where p_0, p_1, p_2, p_3 and p_4 are fit parameters. This onset function provides a good description of the data. The fit parameters are listed in the caption of Fig. 9.41 (right). The TRT effectively provides e/π separation over the momentum range $1 < p < 150$ GeV/c.

During Run 1, some tubing that supplies the active gas to the detector developed leaks due to mechanical stress and corrosion caused mainly by ozone. The ATLAS Collaboration decided to use a gas mixture based on argon in places where the xenon losses were too high so that the tracking in the TRT was not impacted. Though the number of leaks has increased during Run 2, the TRT track operation works fine. This will be continued until the high-luminosity upgrade, when the entire inner detector will be replaced by a new inner tracker (ItK).

9.5 Electron Identification via the E/p Method

We identify electrons by calculating the ratio of the measured energy to measured momentum. As we saw in Chap. 7, the energy of electrons E_e is measured in an electromagnetic calorimeter, while the momentum p_e is measured in a tracking detector inside a magnetic field (see Chap. 6). Since typically $E_e \gg m_e c^2$, we can neglect the electron mass yielding $E_e/p_e = 1$. The measured E_e/p_e distribution is asymmetric with a peak value below one due to energy losses via radiation. Hadrons on the other hand are minimum-ionizing particles and thus lose a small fixed amount of energy via dE/dx. The exact amount E_h depends on the material, the calorimeter thickness and the hadron momentum, but it is much smaller than p_h. In the momentum range of B factories we observe values around 0.25–0.3. For the production of δ rays larger energies may be observed but the probability is small. Figure 9.42 (left) shows the E/p distribution for electrons and pions measured in the Belle experiment. Since the electrons are not corrected for energy loss due to radiation, the E_e/p_e distribution peaks around 0.98 with an rms of $\simeq 0.02$. The E_h/p_h distribution for pions peaks around 0.24 and has a long high-energy tail, which is caused by low momentum hadrons and δ rays. Figure 9.42 (right) shows the average $\langle E/p \rangle$ as a function of the laboratory momentum both for electrons and hadrons. For low momenta, $\langle E_e/p_e \rangle$ is much smaller than one because the electrons loose energy via interactions and radiation before they reach the electromagnetic calorimeter. However, $\langle E_e/p_e \rangle$ increases with p_{lab}. At 1 GeV/c, it is already above 0.9. The $\langle E_h/p_h \rangle$ distribution for hadron decreases with p_{lab}.

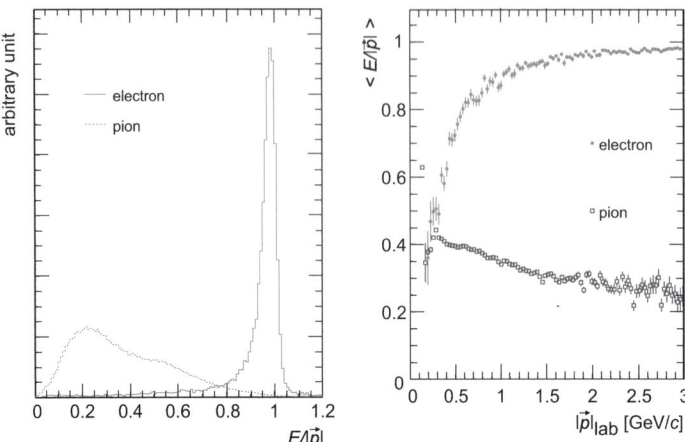

Fig. 9.42 Left: The $E/|\vec{p}|$ distribution for electrons (solid) and pions (dotted) in the Belle experiment. Right: The distribution $\langle E/|\vec{p}|\rangle$ as a function of the laboratory momentum for electrons (solid points) and pions (open squares) in the Belle experiment. Both reprinted with kind permission from [116], © 2002, Elsevier. All rights reserved

Figure 9.43 shows the electron efficiency (left) and pion misidentification probability (right) in the *BABAR* experiment as a function momentum. For momenta above 0.7 GeV/c, the electron efficiency for tight, very tight and likelihood-based electron identification is 93–96% and the pion misidentification probability is less than 0.2%. Below 2 GeV/c, the misidentification probability is 0.1% or less. So, tight electron identification provides excellent electron-pion separation. Note that most high-energy physics experiments measure the momentum of charged particles in a solenoidal magnetic field and they typically have an electromagnetic calorimeter. Thus, they can perform electron identification via E_e/p_e. However, the resolution in E_e/p_e depends on the energy resolution in the electromagnetic calorimeter and

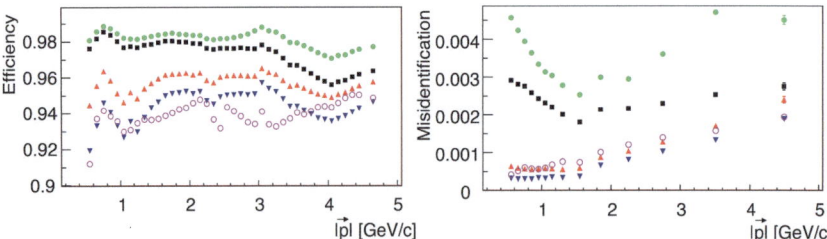

Fig. 9.43 Electron efficiency (left) and pion misidentification probability (right) as a function of momentum for very loose (green points), loose (black squares), tight (red upward-pointing triangles), very tight (blue downward-pointing triangles) and a likelihood-based (magenta open circles) electron identification in the *BABAR* experiment. Both reprinted with kind permission from [10], © 2002, Elsevier. All rights reserved

the momentum resolution, which is typically much smaller. With tighter selection criteria we get purer electron samples at the cost of lower efficiency.

9.6 Muon Identification

Sufficiently energetic muons (>1 GeV/c) penetrate several interaction lengths of iron without being stopped, while hadrons are typically stopped in a few interaction lengths. The exact thickness of the iron depends on the hadron momentum. This property is utilized in muon detectors. Colliding-beam experiments use iron to return the flux of the magnetic-field lines. By placing chambers or scintillators in front of the iron and behind the iron, we can produce a muon detection system. Often, additional chambers or scintillators are placed in between the iron. Most experiments use such muon systems including ATLAS and CMS. Figure 9.44 (left) shows a schematic layout of the ATLAS muon system. The barrel uses three layers of resistive plate chambers for triggering and three layers of muon drift tubes for track measurements. The endcaps use three layers of muon drift tubes and three layers of thin gap chambers. The muon chambers are located inside the toroidal magnetic field. Figure 9.44 (right) shows a schematic layout of the CMS muon system. The barrel uses six layers of RPCs for triggering and four layers of muon drift tubes for track measurements. The endcaps use four layers of RPCs and four layers of cathode strip chambers for track measurements.

The muon systems in the ATLAS and CMS experiments work very well. Figure 9.45 shows the resolution of the $1/p_T$ residual distribution as a function of transverse momentum for stand-alone and combined muons in the ATLAS bar-

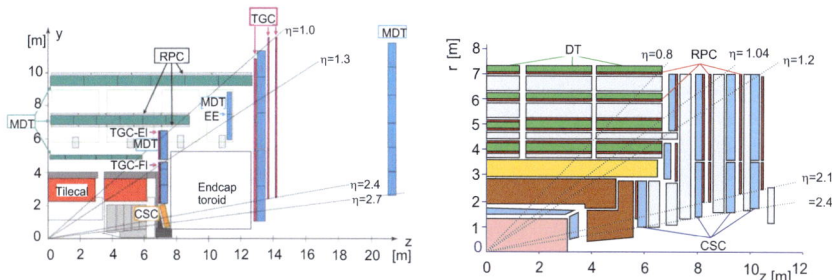

Fig. 9.44 Left: Schematic layout of the ATLAS muon system, consisting of three layers of RPC trigger chambers in the barrel, three layers of TGC trigger chambers in the endcaps, three layer of MDTs in the barrel and endcaps plus one layer of CSCs in the endcap close to the beam. The MDTs and CSCs are used for track measurements. Note that the first TDC layer and first MDT layer in the endcaps were located in the old small wheel, while the remaining TDC layers and the second MDT layer are arranged in the large wheel. Reprinted with kind permission from [117], © 2020, IOP Publishing. All rights reserved. Right: Schematic layout of the CMS muon system [118], consisting of six layers of RPC trigger chambers in the barrel, five layers of RPC trigger chambers in the endcaps, four layers of DTs in the barrel and four layers of CSCs in the endcaps. Reprinted with kind permission from [118], © 2004, Springer. All rights reserved

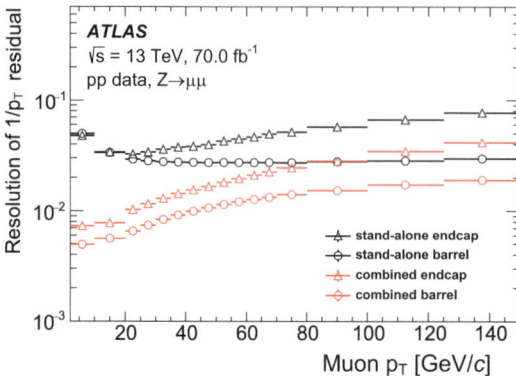

Fig. 9.45 The $1/p_T$ resolution as a function of the transverse momentum for stand-alone muons (black triangles and black points) and combined muons (red triangles and red points) in the barrel (points) and endcap regions (triangles) for $Z \to \mu\mu$ events at 13 TeV pp collision energy in the ATLAS experiment. Reprinted with kind permission from [117], © 2020, IOP Publishing. All rights reserved

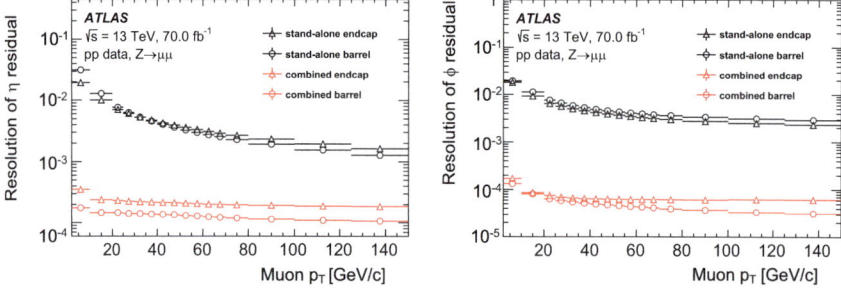

Fig. 9.46 The pseudo-rapidity (left) and azimuth angle (right) resolutions as a function of the transverse momentum for stand-alone muons (black points and black triangles) and combined muons (red points and red triangles) in the barrel (points) and endcap regions (triangles) for $Z \to \mu\mu$ events at 13 TeV pp collision energy in the ATLAS experiment. Reprinted with kind permission from [117], © 2020, IOP Publishing. All rights reserved

rel and endcap muon systems, using 70 fb^{-1} of $Z \to \mu\mu$ data. Stand-alone muons just use the information of the muon chambers, while combined muons combine the tracking information of the muon chambers with those of the inner detector. The best $1/p_T$ resolution is achieved with combined muons in the barrel. Above 20 GeV/c, all resolutions increase with transverse momentum. Figure 9.46 shows the resolutions of the pseudo-rapidity (left) and azimuth angle (right) as a function of transverse momentum. The resolutions improve with increasing transverse momentum. They are best for combined muons by nearly two orders of magnitude with respect to stand-alone muons.

The *BABAR* IFR discussed in Sect. 4.9 served both as muon detector and hadron calorimeter. We observe different patterns for pions and muons. While muons pass

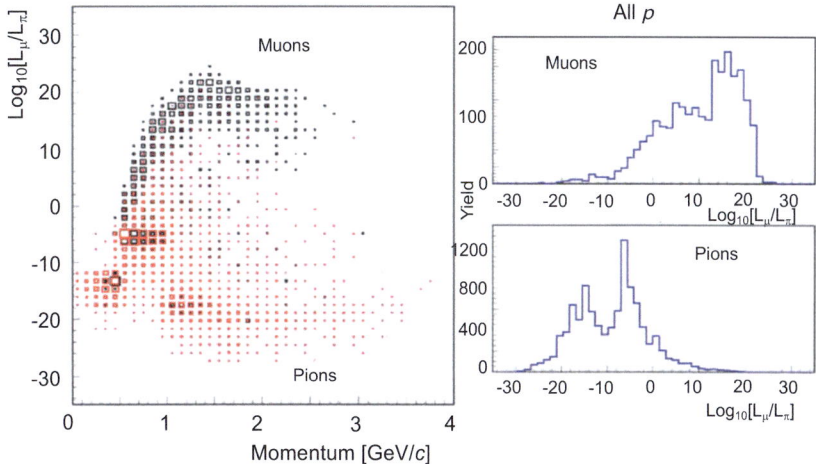

Fig. 9.47 Left: Scatter plot of $\log_{10}(L_\mu/L_\pi)$ versus momentum for pions (red) and muons (black). Top right: The $\log_{10}(L_\mu/L_\pi)$ distribution for muons. Bottom right: The $\log_{10}(L_\mu/L_\pi)$ distribution for pions. All reprinted with kind permission from [119], © 2001, *BABAR* Collaboration. All rights reserved

through the entire iron, hadrons range out, lower-momentum tracks faster than higher-momentum tracks. Figure 9.47 (left) shows a scatter plot of the likelihood ratio ($\log_{10}(L_\mu/L_\pi)$) versus momentum, where L_π denotes the likelihood for identifying a pion and L_μ is the likelihood for identifying a muon. For pions the ratio is typically small, while for muons it is large. Even by eye a clear separation between muons and pions is visible. Figure 9.47 (top right) shows $\log_{10}(L_\mu/L_\pi)$ for muons, while Fig. 9.47 (bottom right) shows the distribution for pions. We can conclude that for pions $\log_{10}(L_\mu/L_\pi)$ is typically below zero while for muons it is above zero. The average muon efficiency for the highest purity was 63% for a 1.2% pion misidentification probability compared to an 80% efficiency after the upgrade. For a higher pion misidentification rate of 5%, the muon average was 89% compared to 95% after the upgrades. This translates to about 6–17% of muons lost due to hardware problems. The IFR is a hadron calorimeter and can be used to identify K^0_Ls, antineutrons and antiprotons for which we observe characteristic shower patterns that are different from those of neutrons and hadrons.

9.7 Neutron Identification

In multipurpose detectors at colliding-beam experiments, we usually do not implement neutron detectors. The neutron may interact in the electromagnetic or hadron calorimeters leaving a neutral shower that is distinguishable from a photon shower by its shape. Typically, we try to remove odd looking showers from the photon samples by evaluating Zernike moments, cluster lateral moments and ratios of energy sums, e.g. $E9/E25$. However, in some nuclear physics experiments some interest

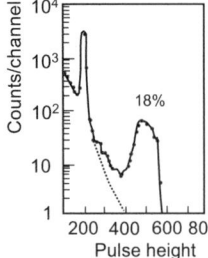
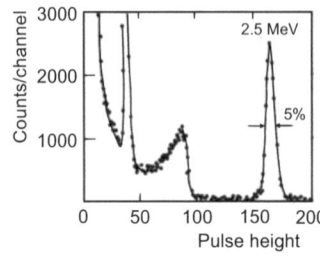

Fig. 9.48 Left: Pulse height distribution from a LiI(Eu) scintillator detecting 5.3 MeV neutrons at 20 °C. All reprinted with kind permission from [120], © 1958, Elsevier. All rights reserved. Right: Pulse height distribution from a ^{3}He counter detecting 2.5 MeV neutrons. Both reprinted with kind permission from [121], © 2017, AIP Publishing. All rights reserved

may exist to identify neutrons. The neutron detectors are based on nuclear interaction like $n + ^6$Li $\rightarrow \alpha + ^3$H, $n + ^{10}$B $\rightarrow \alpha + ^7$Li or $n + ^3$He $\rightarrow p + ^3$H.

Figure 9.48 (left) shows the pulse height spectrum measured with 5.3 MeV neutrons in a LiI crystal doped with Europium. The ionizing α particles and tritium atoms produce scintillation light. The broad peak at the upper spectrum results from α particles and tritium atoms, while the lower narrow peak originates from thermal neutrons and γ-rays. At room temperature, the resolution is about 18%. Cooling the detector to -142 °C reduces the resolution to 10%. For low-energy neutrons, we typically use detectors based on the ^{3}He reaction. Figure 9.48 (right) shows the pulse height spectrum from a ^{3}He counter impinged by 2.5 MeV neutrons. The 2.5 MeV mono-energetic neutron produces a narrow peak with a 5% resolution (FWHM). The peak at low pulse heights originates from thermal neutron background. The peak in the middle results from the recoil distribution, which is expected at 0.75 E_n.

Exercises

9.1 In a 30-layer chamber, energy loss measurements (arbitrary scale) are performed, yielding $(dE/dx, \sigma_{dE/dx})$ = (2.8000, 0.0023293), (0.10000, 0.017700), (−0.20000, 0.018060), (0.0000, 0.017880), (5.1000, 0.038000), (1.1000, 0.014000), (1.5000, 0.012420), (0.30000, 0.017187), (−0.60000, 0.017430), (10.300, 0.011300), (2.4000, 0.0091340), (0.35000, 0.017033), (0.05000, 0.017800), (−0.05000, 0.017955), (3.4000, 0.064800), (20.200, 0.00294), (0.80000, 0.015360), (−0.3000, 0.018040), (−0.010, 0.017900), (0.7500, 0.015560), (2.300, 0.0094567), (0.200, 0.017467), (0.1150, 0.017670), (14.600, 0.0057), (−1.200, 0.013346), (−1.6000, 0.0088720), (17.300, 0.0040440), (3.9000, 0.005500), (0.5500, 0.016300), (1.4000, 0.012800). Plot the distribution. Determine the most probable energy loss. What is the resolution of the truncated distribution? The pion hypothesis is at −0.24, while the kaon hypothesis is at 0.8. Which hypothesis do the data fit better?

Exercises

9.2 A tracking chamber operating with pure argon gas measures the ionization of charged particles at NTP. The chamber of 60 cm length provides 40 measurements. What is the dE/dx resolution? Plot relative dE/dx resolution of the truncated distribution as a function of measurements for d_{samp} and as a function of N_{samp} for argon gas. What resolution would you measure in xenon gas? How much would the resolution improve if you had $d_{samp} = 90$ cm and 80 measurements in xenon gas?

9.3 In a test beam experiment, a hadron beam contains 10 GeV protons and 10 GeV pions. You want to study hadronic shower shapes with a hadron calorimeter. Is it possible to distinguish between pions and protons using a TOF counter, consisting of a plastic scintillator and a photomultiplier? A standard TOF system achieves a time resolution of $\sigma_t = 100$ ps, while the best devices reach $\sigma_t = 50$ ps. How far do you have to place the TOF counter from the calorimeter in order to separate pions from protons at a 3 standard deviation level with the standard counter? If the test beam hall is 200 m long what is the maximum momentum for proton/pion separation at 3 standard deviations with the best TOF counter?

9.4 A time-of-flight counter (length 1 m, height 1 cm) is used to discriminate charged pions from charged kaons. The counter is made from a scintillator that is mixed into polystyrene (n = 1.581) and is parallel to the beam 1 m away. The counter has a time resolution of $\sigma_t = 50$ ps. Which momenta can be separated at a 3σ level? What is the variation in path lengths of the photons produced in the counter?

9.5 In a test beam experiment, a gas Cherenkov counter can be used to separate electrons from pions up to 80 GeV/c. This is useful for comparing the interaction of these particles in a calorimeter. The Cherenkov counter is filled with He gas (25 °C). By varying the gas pressure you can can vary the index of refraction. For standard conditions, the index of refraction is specified in the PDG table. By varying the pressure, you can vary the index of refraction. For what relative pressure would you just observe electrons and no pions in the in the full energy range? For what pressure would 10 GeV pions be visible? Plot the pressure curve between 10 and 80 GeV. Note that the index of refraction for an ideal gas is given by $n - 1 = (n_0 - 1) * \frac{P}{P_0} * \frac{1}{1+\alpha*(T-T_0)}$, where n_0, T_0 and P_0 are refractive index, temperature and pressure at standard conditions (STP) and $\alpha = 0.003/°C$.

9.6 Consider the set of Cherenkov counters: A, aerogel with $n_A = 1.022$; B, aerogel with $n_B = 1.005$; C, neopantene with $n_C = 1.00177$; D, freon 14 with $n_D = 1.00049$; and E, Ar-Ne with $n_E = 1.000135$. By using the combination ABCD or ABCDE what can you say about identifying π, K and p unambiguously in what momentum range? What would you learn from the output of counter A or B alone?

9.7 An experiment uses ring-imaging detectors with a gas radiator ($n_{gas} = 1.00173$) and a liquid radiator ($n_{liquid} = 1.277$) to separate kaons from pions. The liquid ring-imaging counter starts at a radius $r_i = 102$ cm and is 1 cm in depth. It is followed by a 14 cm deep drift region, the 4 cm thick gaseous photon detector, and the 45 cm deep gas radiator. The mirrors at the outer radius are arranged such that all Cherenkov photons produced inside the gas radiator along the track for the same azimuth angle of the Cherenkov cone are focussed into a point in the photon detector. Assume that the ring is produced on average in the middle of the photon detector (around 2 cm in depth). The resolutions in the gas and liquid rings are 4.3 and 13 mrad per photoelectron, respectively. On average 7.9 (12.9) photoelectrons are produced in the gas (liquid) radiator. Is it possible to separate a 10 GeV/c pion from a 10 GeV/c kaon in the two Cherenkov detectors? How big are the rings in each radiator and how many standard deviations are they separated? (Hint: make a sketch of the geometry).

9.8 A multipurpose detector has a tracking system that provides π/K separation at more than 3-σ for momenta below 600 MeV/c. Now we want to extend this with a set of two threshold counters such that we achieve π/K separation at a 3-σ level up to 3.5 GeV/c. Each counter should be sufficiently long to detect at least 10 photoelectrons. The detector lengths should be as small as possible. What materials should we consider? What refractive indices are optimal? Determine the pion and kaon thresholds for this arrangement. How long should the individual radiator lengths be?

9.9 Consider the *BABAR* DIRC counter. A particle penetrating the 17.25 mm thick and 4.9 m long synthetic fused silica (n = 1.473) produces Cherenkov photons, which are transported by total reflection to the Stand-off box. One end is covered with a mirror, the other end ends in the Stand-off box. How many Cherenkov photons does a minimum ionizing particle produce. How many photons reach the Stand-off box? The Cherenkov photons are detected with photomultiplier tubes that have an extended photocathode. If the quantum efficiency is 25% and the lowest detectable wavelength is 300 nm, how many photoelectrons are produced. If the charged particle penetrates through the center of the bar, how long are transition times for particles moving towards the Stand-off box and those moving in the opposite direction?

9.10 Consider a transition radiation tracker that consists of 200 25 μm thick lithium foils separated by 1.5 mm air gaps. Plot the formation zones and the differential transition radiation intensity as a function of θ for 15 keV photons and a γ factor of 10^4. How many photons are emitted for $\hbar\omega/(\gamma\hbar\omega_p) > 0.1$? Plot the differential yield per interface as a function of the photon energy without absorption.

References

1. The Super-Kamiokande Collaboration (S. Fukuda et al.), Nucl. Instrum. Meth. A **501**, 418 (2003)
2. THE SNO Collaboration (J. Roger et al.), Nucl. Instrum. Meth. A **449**, 172 (1999)
3. The Ice Cube Collaboration (A. Achterberg et al.), Astropart. Phys. **26**, 155 (2006)
4. The MAGIC Collaboration (D. Ferenc et al.), Nucl. Instrum. Meth. A **553**, 274 (2005)
5. The HESS Collaboration (G. Vasileiadis et al.), Nucl. Instrum. Meth. A **553**, 268 (2005)

References 653

6. The CTA Collaboration (M. Punch et al.), 44th Rencontres de Moriond on Very High Energy Phenomena in the Universe, 189 (2009)
7. M. Hauschild, Talk at RD51 Workshop, https://indico.cern.ch/event/996326/contributions/4200962/attachments/2191650/3704305/dEdx.pdf (2021)
8. A.H. Walenta et al., Nucl. Instrum. Meth. A **161**, 45 (1979)
9. The Particle Data Group (C. Caso et al.), Eur. Phys. J. C **3**, 1 (1998)
10. The *BABAR* Collaboration (B. Aubert et al.), Nucl. Instrum. Meth. A **479**, 1 (2002)
11. The ALICE TPC Collaboration (W. Yu et al.), Nucl. Instrum. Meth. A **706**, 55 (2013)
12. The CLEO II Collaboration (Y. Kubota et al.), Nucl. Instrum. Meth. A **320**, 66 (1992)
13. The DELPHI Collaboration (P. Abreu et al.), Nucl. Instrum. Meth. A **378**, 57 (1996)
14. The Particle Data Group (S. Navas et al.), Phys. Rev. D. **110**, 030001 (2024)
15. The *BABAR* Collaboration (B. Aubert et al.), Nucl. Instrum. Meth. A **729**, 625 (2013)
16. H.J. Hilke, Rep. Prog. Phys. **73**, 116201 (2010)
17. W.W.H. Allison, J.H. Cobb, Ann. Rev. Part. Sci. **30**, 253 (1980)
18. H. Aihara et al., IEEE Trans. NS-**30**, 63 (1983)
19. I. Lehraus, Nucl. Instrum. Meth. A **217**, 43 (1983)
20. J. Va'vra, Nucl. Instrum. Meth. A **453**, 262 (2000)
21. The ALEPH Collaboration (D. Buskulic et al.), Nucl. Instrum. Meth. A **360**, 481 (1995)
22. The ALICE TPC Collaboration (J. Alme et al.), Nucl. Instrum. Meth. A **622**, 316 (2013)
23. H. Albrecht et al., Nucl. Instrum. Meth. A **275**, 1 (1989)
24. Y. Oku, PhD thesis, University of Lund, LUNDFD6/(NFFL 7024)/ (1985)
25. The ATLAS Collaboration (G. Aad et al.), ATL-ATLAS-CONF-2011-016 (2011)
26. The Belle Collaboration (E. Nakano et al.), Nucl. Instrum. Meth. A **494**, 402 (2002)
27. The BES III Collaboration (M. Ablikim et al.), Nucl. Instrum. Meth. A **614**, 345 (2010)
28. S. Banerjee et al., CMS-NOTE-1999-056 (1999)
29. J. Dahm, M. Elsing, M. Reale, DELPHI 95-48 TRACK 81 (1995)
30. R. Rajagopalan, FERMILAB-THESIS-1992-41 (1992)
31. The H1 Collaboration (I. Abt et al.), Nucl. Instrum. Meth. A **386**, 348 (1997)
32. K. Ambrus, PhD thesis, University of Heidelberg (1986)
33. S.E. Baru et al., Nucl. Instrum. Meth. A **323**, 151 (1992)
34. A. Andryakov et al., Nucl. Instrum. Meth. A **409**, 390 (1998)
35. A. Bojarski et al., Nucl. Instrum. Meth. A **283**, 617 (1989)
36. M. Hauschild, Nucl. Instrum. Meth. A **379**, 436 (1996)
37. M.D. Hildreth et al., IEEE Trans. Nucl. Sci. **42**, 451 (1995)
38. M. Iwasaki et al., Nucl. Instrum. Meth. A **365**, 143 (1995)
39. D.R. Nygren, Phys. Scripta **23**, 584 (1981)
40. A. Davidenko et al., JETP **28**-2, 223 (1969)
41. G. Malamud, A. Breskin, R. Chechik, Nucl. Instrum. Meth. A **372**, 19 (1996)
42. G. Chiarello et al., Nucl. Instrum. Meth. A **936**, 503 (2019)
43. The ALICE Collaboration (F. Carnesecchi et al.), JINST **14**-06, C06023 (2019). TOF Technical Design Report, CERN / LHCC 2000–12 (2000)
44. The CDF Collaboration (F. Abe et al.), FERMILAB-PROPOSAL-0830A, CDF-Note-2573 (1995)
45. D. Bernstein et al., Nucl. Instrum. Meth. **224**, 301 (1984)
46. R. Mozley, SLAC Beam Line **12**, 3 (1981)
47. The ALICE Collaboration (F. Carnesecchi et al.), JINST **14**-06, C06023 (2019); ALICE TDR-8, CERN/LHCC 2000–12 (2000)
48. The Mark II Collaboration (G.S. Abrams et al.), Nucl. Instrum. Meth. A **281**, 55 (1989)
49. The ARGUS Collaboration (H. Albrecht et al.), Nucl. Instrum. Meth. A **275**, 1 (1988)
50. The OPAL Collaboration (K. Ahmet et al.), Nucl. Instrum. Meth. A **305**, 275 (1991)
51. The Belle Collaboration (A. Abashian et al.), Nucl. Instrum. Meth. A **479**, 117 (2002)
52. W.D. Li, Y.J. Mao, Y.F. Wang, Int. J. Mod. Phys. A **24S1**, 9 (2009)
53. The CDF Collaboration (S. Cabrera et al.), Nucl. Instrum. Meth. A **494**, 416 (2002)
54. The CDF Collaboration (R.G.C. Oldeman et al.), eConf C0304052, FO005 (2003)

55. J. Wu et al., J. Phys. G: Nucl. Part. Phys. **34**, S729 (2007)
56. The STAR Collaboration (W.J. Llope), AIP Conf. Proc. **1099**-1, 778 (2009)
57. W.J. Llope et al., Nucl. Instrum. Meth. A **522**, 252 (2004)
58. W. Klempt, Nucl. Instrum. Meth. A **433**, 542 (1999)
59. B. Ratcliff, J. Schwiening, *Handbook of Particle Detection and Imaging*, ed. by C. Grupen, I. Buvat (Springer, 2012)
60. J. Litt, R. Meunier, Ann. Rev. Nucl. Part. Sci. **23**, 1 (1973)
61. M. Benot, J. Litt, R. Meunier, Nucl. Instrum. Meth. **105**, 431 (1972)
62. P. LeComte et al., Phys. Scripta **23**, 377 (1981)
63. G. Eigen, Nucl. Phys. B Proc. Suppl. **44**, 51 (1995)
64. T. Iijima et al., Nucl. Instrum. Meth. A **453**, 321 (2000)
65. E. Nakano et al., Nucl. Instrum. Meth. A **494**, 402 (2002)
66. A. Bodek et al., Z. Phys. C **18**, 289 (1983)
67. https://en.wikipedia.org/wiki/Aerogel
68. J. Badier et al., Phys. Lett. B **39**, 414 (1972)
69. S.P. Denisov et al., Nucl. Instrum. Meth. **92**, 77 (1971)
70. A. Roberts, Nucl. Instrum. Meth. **9**, 55 (1960)
71. J. Seguinot, T. Ypsilantis, Nucl. Instrum. Meth. **142**, 377 (1977)
72. J. Seguinot, T. Ypsilantis, Nucl. Instrum. Meth. A **343**, 1 (1994)
73. T. Ypsilantis, J. Seguinot, Nucl. Instrum. Meth. A **343**, 30 (1994)
74. J. Va'vra, Nucl. Instrum. Meth. **371**, 33 (1996)
75. T. Ypsilantis, J. Seguinot, Nucl. Instrum. Meth. A **433**, 1 (1999)
76. P. Glassel, Nucl. Instrum. Meth. A **433**, 17 (1999)
77. B. Ratcliff, Nucl. Instrum. Meth. A **502**, 211 (2003)
78. B. Ratcliff, Nucl. Instrum. Meth. A **595**, 1 (2008)
79. P. Baillon, Adv. Ser. Direct. High Energy Phys. **27**, 249 (2017)
80. The SLD Collaboration (K. Abe et al.), Nucl. Instr. Meth. A **300**, 501 (1993)
81. The SLD Collaboration (K. Abe et al.), Nucl. Instr. Meth. A **343**, 74 (1993)
82. The SLD Collaboration (K. Abe et al.), IEEE Trans. Nucl. Sci. **45**, 648 (1998)
83. M. Dracos, D. Loukas, Nucl. Instrum. Meth. A **302**, 241 (1991)
84. The SLD Collaboration (K. Abe et al.), IEEE Trans. Nucl. Sci. **42**, 518 (1995)
85. E. Albrecht et al., Nucl. Instr. Meth. A **433**, 47 (1999)
86. The SLD CRID Collaboration (J. Va'vra et al.), Nucl. Instrum. Meth. A **433**, 59 (1999)
87. K. Abe et al., IEEE Trans. Nucl. Sci. **39**, 685 (1991)
88. W. Adam et al., Nucl. Instr. Meth. A **371**, 240 (1996)
89. The LHCb RICH Collaboration (M. Adinolfi et al.), Eur. Phys. J. C **73**, 2431 (2013)
90. The LHCb Collaboration (R. Aaij et al.), Int. J. Mod. Phys. A **30**-07, 1530022 (2015)
91. The LHCb RICH Collaboration (M.P. Blago et al.), Springer Proc. Phys. **213**, 249 (2018)
92. R. Forty, Nucl. Instrum. Meth. A **623**, 294 (2010)
93. R. Pestotnik et al., Nucl. Instrum. Meth. A **876**, 265 (2017)
94. M. Yonenaga et al., PTEP **2020**-9, 093H01 (2020)
95. The *BABAR* Collaboration (D. Leith et al.), Nucl. Instrum. Meth. A **494**, 389 (2002)
96. The *BABAR* DIRC Collaboration (I. Adam et al.), Nucl. Instrum. Meth. A **538**, 281 (2005)
97. The *BABAR* DIRC Collaboration (I. Adam et al.), IEEE Trans. Nucl. Sci. **45**, 450 (1998)
98. B. Ratcliff, J. Va'vra, Nucl. Instrum. Meth. A **970**, 163442 (2020)
99. T. Yonekura et al., PoS TIPP **2014**, 082 (2014)
100. K. Föhl et al., JINST **11–05**, C05020 (2016)
101. F. Barbosa et al., Nucl. Instrum. Meth. A **876**, 69 (2017)
102. The PANDA Collaboration (B. Singh et al.), e-Print: 1710.00684 (2017)
103. T. Blake et al., PoS ICHEP2018, 667 (2018)
104. N. Harnew et al., Nucl. Instrum. Meth. **936**, 595 (2019)
105. C.W. Fabjan, H.G. Fischer, Rep. Prog. Phys. **43**, 003 (1980)
106. J.H. Cobb et al., Nucl. Instrum. Meth. **140**, 413 (1977)
107. T. Ludlam et al., Nucl. Instrum. Meth. **180**, 413 (1981)

References

108. C.W. Fabjan et al., Nucl. Instrum. Meth. **185**, 119 (1981); ibid **216**, 106 (1983)
109. The ATLAS Collaboration (M. Aaboud et al.), JINST **12**-05, P05002 (2017)
110. The ATLAS TRT Collaboration (E. Abat et al.), JINST **3**, P02014 (2008)
111. The ATLAS Collaboration (B. Mindur et al.), Nucl. Instrum. Meth. A **845**, 257 (2017)
112. T. Akesson et al., Nucl. Instrum. Meth. A **474**, 172 (2001)
113. The ATLAS TRT Collaboration (J.M. Stahlman et al.), Phys. Procedia **37**, 506 (2012)
114. The ATLAS TRT Collaboration (A. Alonso et al.), Nucl. Phys. B Proc. Suppl. **215**, 185 (2011)
115. The ATLAS Collaboration (E.B. Klinkby et al.), Nucl. Instrum. Meth. A **706**, 79 (2011)
116. K. Hanagaki et al., Nucl. Instrum. Meth. A **485**, 490 (2002)
117. The ATLAS Collaboration (G. Aad et al.), JINST **15**, P09015 (2020)
118. S. Lacaprara et al., Eur Phys. J. C **34**, s75 (2004)
119. M. Piccolo, (Private Communication)
120. R.B. Murray, Nucl. Instrum. Meth. **2**, 237 (1958)
121. A. Sayres, M. Coppola, Rev. Sci. Instr. **35**, 431 (1964)

Open Access This chapter is licensed under the terms of the Creative Commons Attribution 4.0 International License (http://creativecommons.org/licenses/by/4.0/), which permits use, sharing, adaptation, distribution and reproduction in any medium or format, as long as you give appropriate credit to the original author(s) and the source, provide a link to the Creative Commons license and indicate if changes were made.

The images or other third party material in this chapter are included in the chapter's Creative Commons license, unless indicated otherwise in a credit line to the material. If material is not included in the chapter's Creative Commons license and your intended use is not permitted by statutory regulation or exceeds the permitted use, you will need to obtain permission directly from the copyright holder.

Trigger and Data Acquisition   10

10.1 Overview

The main role of the Trigger and the Data AcQuisition is to select interesting events, process them and save the raw data on a permanent storage. Figure 10.1 illustrates this schematically for the ATLAS experiment. The ATLAS sub-detectors send out trigger signals that are processed in the trigger unit. In parallel, the data signals are send to the data acquisition system. If the trigger processor decides to keep the event, the raw data are sent to storage. For analysis, the raw data are adjusted with updated calibration constants and are reconstructed. In addition, simulated data are produced that follow the same path in the processing as the raw data.

At the LHC, we encounter high cross sections and high luminosity for pp collisions. Figure 10.2 shows the cross section and expected event rates for several physics processes at the LHC design luminosity for a center-of-mass energy of 14 TeV as a function of the jet transverse energy or particle mass. The cross section for $b\bar{b}$ production is two orders of magnitude smaller than the inelastic cross section, which consists mainly of hadronic interactions. Weak processes, like the production of W and Z bosons are another four orders of magnitude lower. For example, the process $H \rightarrow \gamma\gamma$ in which the Higgs boson was discovered at the LHC is another six orders of magnitude smaller.

The measured inelastic cross section at 13 TeV is about 79 mb [2]. So at the design luminosity of 10^{34} cm^{-2}s^{-1}, the inelastic event rate is $79 \times 10^{-27} * 10^{34} = 790$ million events per second. Presently, the instantaneous luminosity is already a factor of 2.3 larger than the design luminosity, yielding a rate of 1.82 GHz. The LHC bunch spacing within a bunch train is 25 ns, yielding a collision rate of up to 40 MHz. Due to the high cross section, multiple interactions occur per bunch crossing. At the present luminosity, we get on average ~ 45 interactions per bunch crossing. In addition, sub-detectors produce hits from machine backgrounds and noise. The experiments have 100 million channels with data sizes of 1–2 MB yielding a data rate

© The Author(s) 2025
G. Eigen, *Detectors in High-Energy Physics Experiments*, Graduate Texts in Physics, https://doi.org/10.1007/978-3-031-67336-8_10

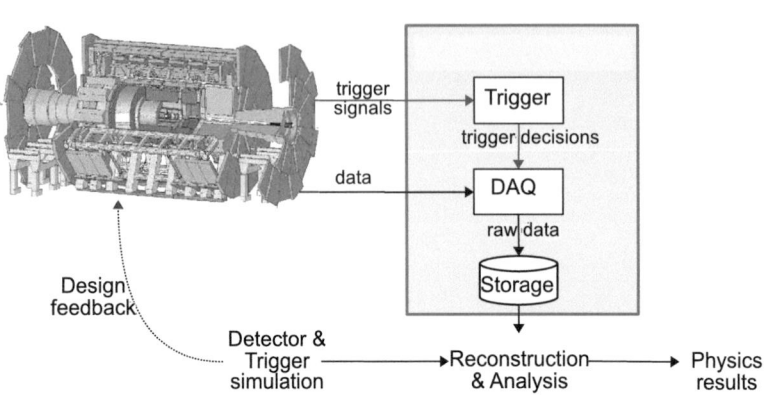

Fig. 10.1 General overview of the trigger and data acquisition system in the ATLAS experiment. Reprinted with kind permission from [1], © 2017, ATLAS Collaboration. All rights reserved

Fig. 10.2 Cross sections and expected event rates for several physics processes at the LHC design luminosity as a function of the jet transverse energy or particle mass in a double-logarithmic scale [3]. The Level 1 output rate in ATLAS, which is the High-Level Trigger input rate, is indicated as well as the maximum High-Level Trigger output rate. Reprinted under CC-BY-3.0 Licence from [3], © 2009, ATLAS Collaboration

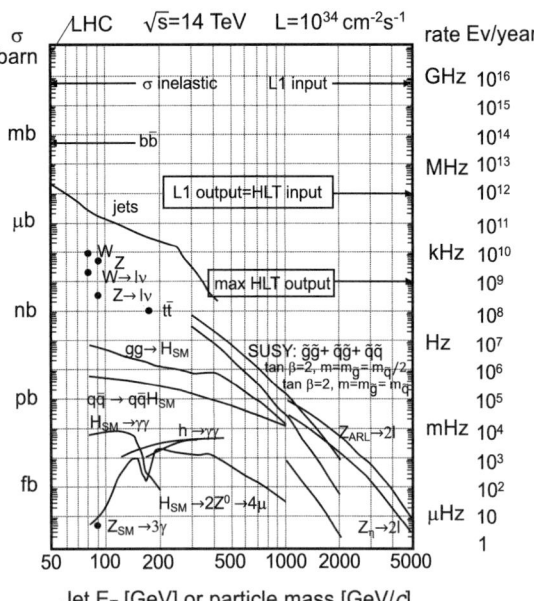

of 60 TB/s. The affordable rate is 1.5 GB/s, corresponding to a rate of 1 kHz. Most of the interactions are standard strong interaction processes as Fig. 10.2 shows. Other event classes are too high as well, like those containing single muons or muon pairs. The rate of interesting processes is many orders of magnitudes lower. For example, the rate for W (Z) production is about 1 kHz while that for Higgs production is about 1 Hz. To satisfy the 1 kHz readout rate, we need to reduce the acceptance of QCD and other high-rate events. To accomplish this, we can introduce scale factors and just record every nth event, or impose minimum transverse momentum requirements

10.1 Overview

known as selecting thresholds. Furthermore, we need to set up a readout system that collects the relevant information from the millions of readout channels.

Figure 1.17 (right) illustrated the readout of a single crystal with a PMT, amplifier and ADC. In todays experiments, the amplifier and ADC are integrated into the frontend (FE) electronics board. A buffer and an integrator may be added. The FE board sends information to the first level trigger system and transfers the data to the data acquisition board. At the LHC, data have to be recorded within the 25 ns bunch crossing. For example, in the LHCb electromagnetic calorimeter, many detector cells have to be read out with PMTs within the 25 ns bunch crossing. To fit the readout into the time frame, the signal is clipped directly after the PMT with a specific circuit as illustrated in Fig. 10.3 (top left). The clipped signal is transported via a 50 Ω cable to the frontend electronics board shown in Fig. 10.3 (top right), which contains the analog chip with a buffer plus an integrator circuit and a 12-bit ADC. Figure 10.3 (bottom left) shows the PMT signal before and after clipping, while Fig. 10.3 (bottom right) depicts the integrated signal. Other sub-detectors are read out in a similar way. However, often the frontend electronics board is rather close to the sub-detector also holding a preamplifier with one or two amplification stages.

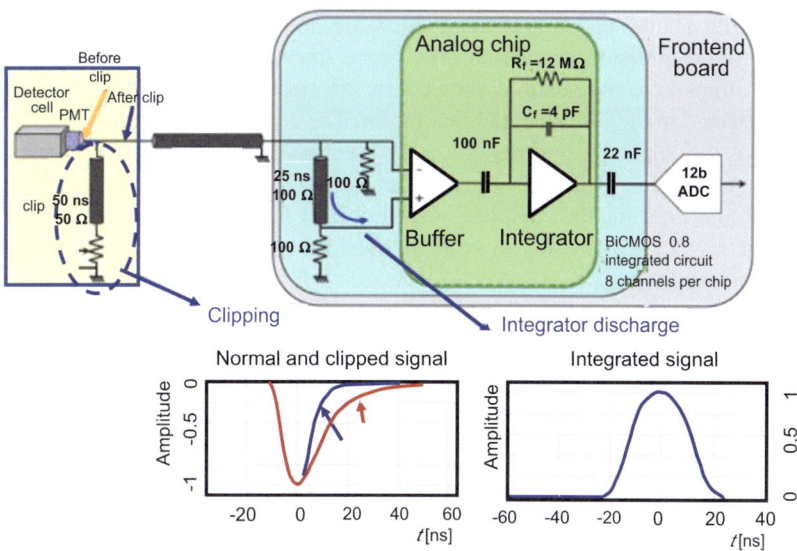

Fig. 10.3 Top: Detector cell read out by a PMT (left) and frontend electronics board (right) [4]. Bottom: Detector signal before and after clipping (left) and the integrated signal (right). Reprinted with kind permission from [4], © 2014, F. Lanni. All rights reserved

10.2 Trigger Layout

In most experiments, the trigger system is typically performed in three stages:

- Level 1 (L1): This is typically a hardware trigger.
- Level 2 (L2): This is a software trigger that may be omitted or integrated into the Level 3 trigger.
- Level 3 (L3): This is a higher-level software trigger.

Note that the L1 hardware trigger typically is a custom-designed synchronous hardware system, while the L2/L3 software triggers are asynchronous, mostly commodity electronics-based systems. They process fast signals and provide timing-based event identifications. To keep the network manageable, they provide a quick rate reduction. A more detailed analysis is performed later in the CPU farm.

In the *BABAR* experiment, for example, the trigger system consisted of an L1 hardware and an L3 software trigger [5]. The L1 trigger configuration was designed for maximum efficiency and background suppression using customized electronics that processed input from the drift chamber, electromagnetic calorimeter and muon system. With a bunch crossing rate of about 4.2 ns, the L1 input rate was continuous. The L1 trigger decision was based on charged tracks in the drift chamber above a certain transverse momentum, showers in the electromagnetic calorimeter and tracks detected in the instrumented flux return. The output rate was set to ≤ 2 kHz, increasing with higher luminosity. For a given collision, the maximum L1 response latency, i.e. the time delay for transmission, was 12 μs. The L3 trigger selected events of interest using the complete event and the L1 trigger information. The rate to storage was 120 Hz in order to avoid an overload of the downstream storage and processing capacity.

Figure 10.4 shows a schematic view of the *BABAR* Level1 trigger and data acquisition system. Each *BABAR* sub-detector sent raw analog signals to the FE board, which were digitized and were transferred to the Online Data Flow (ODF), where event building was done including fast control and timing. In addition, the drift chamber, electromagnetic calorimeter and instrumented flux return sent data to the L1 trigger processor. The ODF also had a link to the L1 trigger, which processed the

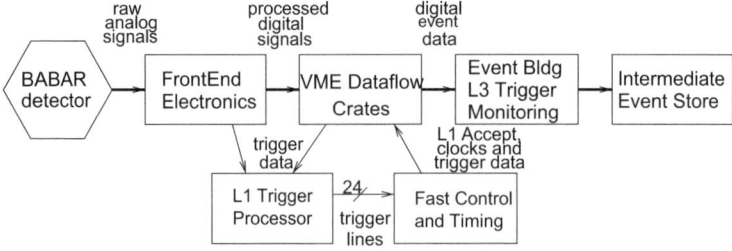

Fig. 10.4 Data acquisition in the *BABAR* experiment. Reprinted with kind permission from [5], © 2002, Elsevier. All rights reserved

10.2 Trigger Layout

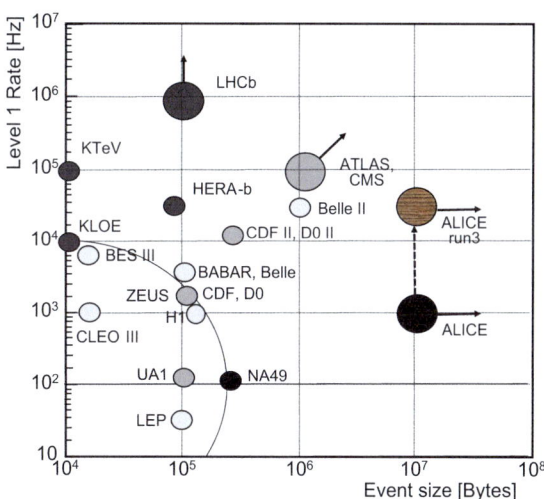

Fig. 10.5 Level-1 trigger rate as a function of the event size showing experiments at e^+e^- and ep colliders (light grey points), hadron colliders (grey points), fixed target experiments (dark grey points) and heavy ion experiments (black/brown points)

data. If the event was accepted, the fast control and timing systems sent an accept signal to ODF, which transferred the data to the L3 level trigger, where the final decision was reached. If L3 accepted the event, it was moved to intermediate storage. Further details are given in [5]. Similarly, the Belle trigger consisted of a Level-1 hardware trigger and a Level-3 software trigger [6].

Figure 10.5 shows the L1 trigger rate as a function of the event size for different experiments. The trigger rate increases with high luminosity and high cross sections, while the event size depends on the physics process and the complexity of the detector (many sub-detectors, many channels). The experiments KLOE, *BABAR*, H1, Zeus, CDF, D0 and NA49 lie on an elliptical boundary in the L1 trigger rate versus event size plane. This indicates a correlation between the L1 trigger rate and event size. For KLOE, the L1 trigger rate was 10 kHz for an event rate of 10 kB, while for NA49 the event size increased to more than 200 kB for an L1 trigger rate of 100 Hz. For *BABAR* and Belle, the event rate was increased due to high luminosity, while the event size did not change. For the LEP experiments, the trigger rate was less than 100 Hz at an event size of 100 kB. Though the event size remained at 10 kB, the L1 trigger rate increased by an order of magnitude. For CDF II and D0 II, the trigger rates increased due to higher beam energy and in turn increased cross section. Furthermore, the upgraded detectors collected more decay channels increasing the event size. Due to improved technology, both the L1 trigger rates and event sizes at LHC are higher. The LHCb experiment has a trigger rate of 1 MHz, but the event size is only 100 kB. The detector records mainly charm and beauty events, which are produced with large cross section and thus high rates but their event size is small due to low multiplicity. This is very different in the ALICE experiment, where the L1 trigger rate is 1 kHz but the event size is 10 MB. The reason is that the ALICE experiments records heavy ion collisions that have lower cross sections but very high multiplicities.

10.3 Trigger Signatures

In Sect. 1.2, we discussed the imprints different particles leave in sub-detectors. The different responses of these particle species are implemented in the trigger definitions. Furthermore, trigger algorithms look for charge, position, p_T and quality of each object. In addition, we also use global algorithms to define missing transverse energy (E_T^{miss}), total transverse energy (E_T) and the total number of objects. The goal is to keep high efficiency for selecting interesting physics events without bias with minimum dead time, while having a large rate reduction for instrumental backgrounds and uninteresting physics events. The system has to be affordable with respect to computing, connectivity and budget. It should have a highly robust system that functions reliably under all experimental conditions. Note that simple and inclusive triggers are preferred.[1] Furthermore, the trigger needs to be flexible to accommodate changing experimental conditions such as an increase in luminosity. In addition to physics triggers, we also need calibration, background and monitoring triggers.

10.3.1 The ATLAS Trigger Systems in Run 1 and Run 2

In Run 1, ATLAS used a three stage trigger system as shown in Fig. 10.6 (top), consisting of a hardware-based Level-1 (L1) trigger and two software-based triggers, the Level 2 (L2) trigger and the Event Filter (EF). The LHC operated with a bunch crossing within a bunch train of 50 ns, yielding a maximum event rate of 20 MHz though the trigger was laid out for a maximum rate of 40 MHz (i.e. 25 ns bunch crossing).

The L1 trigger was built with fast custom-made ASICs and FPGAs[2] to identify basic signatures of interesting physics with high efficiency. The data acquisition system records in parallel the different experimental quantities like charge, hit positions, time, etc, in the tracking detectors, calorimeters and muon systems from the FE electronics, processing only channels with active signals, formatting them and applying calibrations. Trigger decisions are based on multiplicities for local trigger objects, using various p_T thresholds for muons, electromagnetic clusters, as well as narrow and regular jets. For global trigger objects, missing transverse energy and the total scalar transverse energy are included in addition. The muon and calorimeter trigger systems employ simple algorithms to make fast decisions and to define Regions of Interest (RoI), using coarse position information (η, ϕ) from the calorimeter and muon systems in regions of activity. The trigger information is processed in parallel with a latency of < 2.5 μs. When L1 accepts an event, a signal is sent to the data acquisition system and the ROIs are transferred to Level 2 for further processing. In the DAQ, the data from each subsystem are collected in ReadOut Drivers (ROD)

---

[1] An inclusive trigger selects an entire class of decays like a single muon or di-muon events.
[2] An ASIC is an application-specific circuit and an FPGA is a field-programmable gate array.

10.3 Trigger Signatures

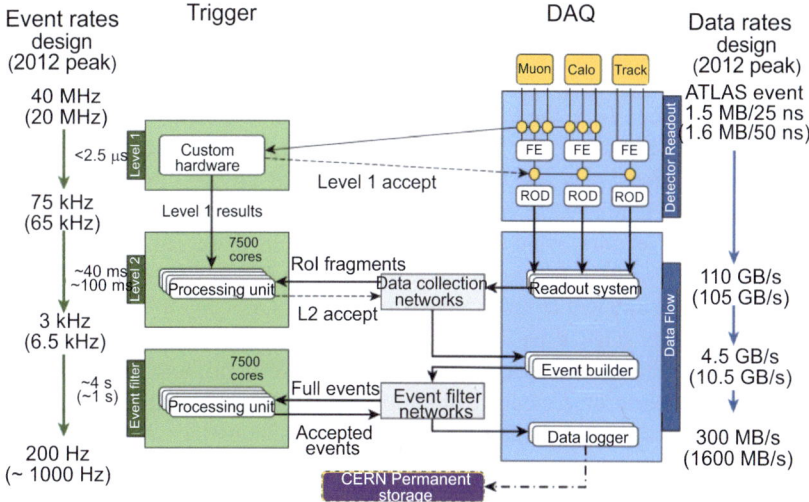

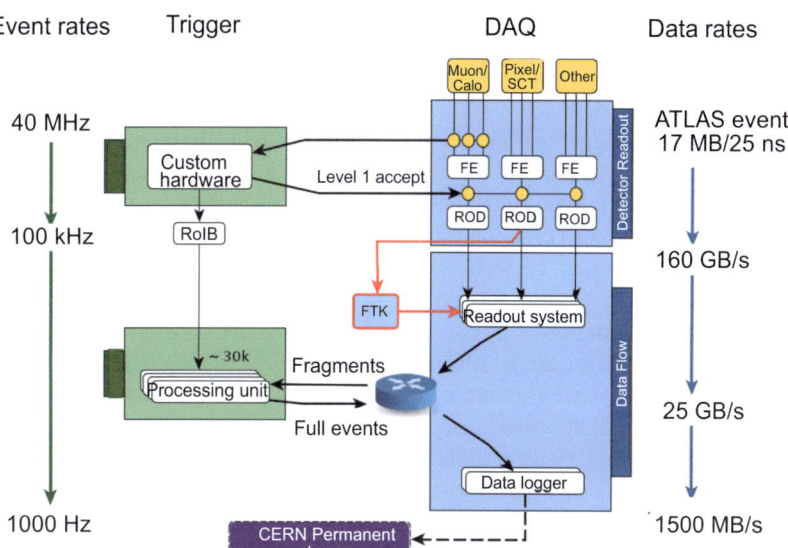

Fig. 10.6 ATLAS trigger and data acquisition in Run 1 (top) and in Run 2 (bottom). In addition, the event rates and data rates are shown at each trigger level. For Run 1, the rates correspond to expected data taking at 25 ns bunch crossing, while those in parentheses refer to the actual condition in Run 1 for 50 ns bunch crossing. Reprinted under CC-BY-NC-ND-4.0 Licence from [7], © 2015, ATLAS Collaboration

that assemble them and transfer them to the ReadOut Buffers (ROB) if an L1 accept is received. The L1 output rate is 65 kHz.

The L2 software trigger processes the data with 7,500 CPU cores, combining information from the different subsystems. For local objects, the information provided is the position (η, ϕ) and the p_T range. In addition, the components of the missing transverse energy vector and the total scalar transverse energy value are given. The L2 trigger processes the RoIs with fine-grain detector data and adds tracking information from the Pixel detector, SCT and TRT. The information from all systems is then combined to form global trigger objects that become candidates for electrons, muons, photons, tau leptons, jets and B-physics objects as well as events with missing transverse energy. If L2 accepts the event, a signal is sent to the Data Collection Network, which transfers the data to the Event Builder. The rate is reduced to about 6.5 kHz, processing the events within 100 ms. About 5–10% of the full event data is sufficient to reach a decision.

The final trigger stage is the Event Filter that also has 7,500 cores for processing. Using the full granularity of the detector and offline-like algorithms, a refined reconstruction can be performed though calibration and alignment constants are not final. Vertex reconstruction, track fitting including bremsstrahlung recovery are possible. For accepted events, the Event Filter Network sends the data to the Data Logger, which transfers them to permanent storage. The Event Filter provides the last step to decide if the event is kept or not. It classifies accepted events and establishes a catalogue of discovery type candidates. The rate for events to tape is $\sim$ 1 kHz, writing data with 1.6 GB/s to storage. Since the fine granularity of the calorimeter is used and information from other subsystems is combined, the processing time increases to about 1 s per event.

For Run 2, the trigger was modified as depicted in Fig. 10.6 (bottom). The Level 2 Trigger and Event Filter have been merged into a single software-based High-Level Trigger (HLT). Major upgrades for L1Calo, L1Muon and the HLT have been implemented. The bunch crossing was reduced to its nominal value of 25 ns, leading to an event rate of up to 40 MHz. The L1 again uses fast electronics and collects coarse information from the calorimeter and the muon spectrometer, processing it with a latency of less than 2.5 μs. The L1 trigger consists of the L1 calorimeter trigger system (L1Calo), the L1 muon trigger system (L1Muon), the Central Trigger Processors (CTP), as well as the new L1 topological trigger modules (L1Topo). As before, data from the calorimeter and muon systems are processed at a rate of 1.7 MB/25 ns. For an L1 accept signal, the readout drivers of the Calorimeters, Muon detectors, Pixel, SCT and other system transfer the data to the readout system and the High-Level Trigger (HLT). The rate is up to 100 kHz, processing events with 160 GB/s. As in Run 1, event building is followed by event filtering. The HLT does the final selection to keep or to toss an event. Fast offline-like algorithms have access to higher resolution information than L1 and run on PC computing clusters on RoIs defined by L1 with an average processing time of 0.2 s per event at a rate of 25 GB/s. The HLT output rate is 1 kHz, writing data to permanent storage at a rate of 1.5 GB/s.

10.3.2 The ATLAS Level 1 Calorimeter Trigger System

The L1 Calorimeter Trigger System processes signals from the electromagnetic and hadronic calorimeters and provides trigger signals to the Central Trigger Processors.

10.3 Trigger Signatures

The pedestal-subtracted energies are processed in a PreProcessor that is a complex system of different modules with many cards containing ASICs and FPGAs and working in parallel. The inputs to the L1 Calorimeter Trigger System are 7,200 so-called Trigger Towers that are formed by analog summation of calorimeter cells across the longitudinal layers in a region of $\Delta\eta \times \Delta\phi = 0.1 \times 0.1$. These signals are sampled in the Pre-Processor Modules (PPMs) at a frequency of 40 MHz and are processed to provide calibrated transverse energy (E_T) that is time-aligned to the correct bunch crossing. The resultant digital data are transmitted to the Cluster Processor and Jet-Energy Processor subsystems in which electron and photon candidates, tau leptons and jets are identified, respectively. The Jet-Energy Processor also provides the total transverse energy and the magnitude of the missing transverse energy E_T^{miss}. The trigger multiplicity information is then sent to the Central Trigger Processors. The L1 Calorimeter Trigger System uses 57 thresholds compared to 28 thresholds in Run 1. These are split into 25 thresholds for jets and forward jets and 16 thresholds each for electromagnetic showers and tau leptons.

For Run 3, the granularity for the L1 Calorimeter Trigger System is increased to cope with increased pile-up levels that degrade the present trigger performance [9]. Figure 10.7 shows the layout of a Super Cell in the LAr calorimeter. With respect to Run 2, the granularity of layers 1 and 2 has been increased in $\Delta\eta$ by a factor of four. The $\Delta\eta \times \Delta\phi$ cell sizes are 0.1×0.1 for the pre-sampler, 0.025×0.1 for the front section, 0.025×1 for the center section and 0.1×0.1 for the back section of the barrel LAr calorimeter. To process the data, new LAr Trigger Digitizer Boards with 12-bit ADCs and LAr Digital Processing Blades have been introduced in the on-detector front-end crates. Ten Super Cells correspond to a Trigger Tower. The shower shape information is used to discriminate electromagnetic showers from jets.

Figure 10.8 shows a schematic view of the upgraded L1 Calorimeter Trigger System architecture for Run 3, which employs feature extractor modules for electrons/photons, jets and global event features. The digitized signal from the ECAL

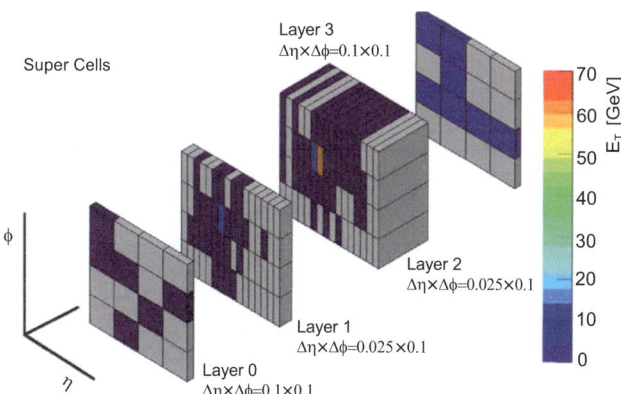

Fig. 10.7 Layout of a Super Cell with the energy deposition of a simulated electron shower. The colored cells show the deposited energy. Reprinted with kind permission from [8], © 2017, ATLAS Collaboration. All rights reserved

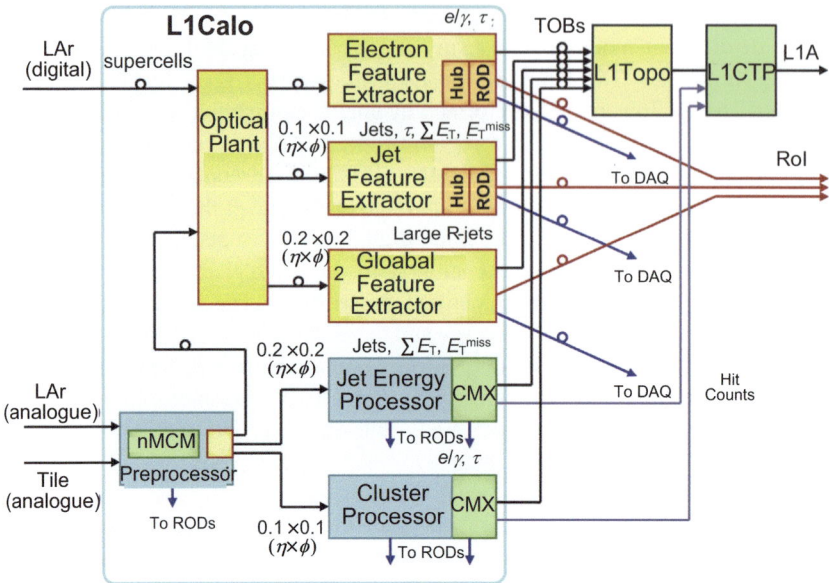

Fig. 10.8 Schematic view of the upgraded L1 Calorimeter Trigger System architecture for Run 3. The η, ϕ binning used in the cluster, jet and global processing are shown. Reprinted under CC-BY-3.0 Licence from [11], © 2016, ATLAS Collaboration

supercells is collected in an optical plant. The analog ECAL and HCAL data are collected in a Pre-Processor. The output is sent to the optical plant, the jet energy processor and the cluster processor. The optical plant sends its output to the Electron Feature Extractor, the Jet Feature Extractor and the Global Feature Extractor. The Electron Feature Extractor searches for $e^{\pm}$, γs and τs. The Jet Feature Extractor looks for jets and τs. In addition, it computes the linear sum of E_T and E_T^{miss}. The Global Feature Extractor looks for large-R jets. The outputs of these three extractors are sent to the level 1 Topological Trigger, the DAQ and to define RoIs for the High-Level Trigger.

Figure 10.9 (left) shows the simulated L1 trigger rate for electrons as a function of the transverse energy. Simulations show that for a 95% trigger efficiency the E_T^{miss} threshold can be reduced to 21.5 GeV in Run-3 to maintain a rate of 20 kHz. Note that the threshold was 28.5 GeV in Run-2.[3] Figure 10.9 (center) shows the simulated L1 trigger rate for jets as a function of the E_T^{miss} threshold with and without pile-up correction. The E_T^{miss} is calculated for all Supercell Towers with

---

[3] The selection in the upgraded calorimeter is based on three shape variables constructed with transverse energy. The first variable R_η is the ratio of transverse energy in an $\eta \times \phi = 3 \times 2$ group of Supercells over 7×2 Supercells in the center region of the calorimeter. The second variable f_3 is the transverse energy in the back region of the calorimeter with respect to that in all three regions. The third variable is $w_{\eta,2}$ is the spread of the shower in the center region of a 3×2 Supercell. For more details see [10].

10.3 Trigger Signatures

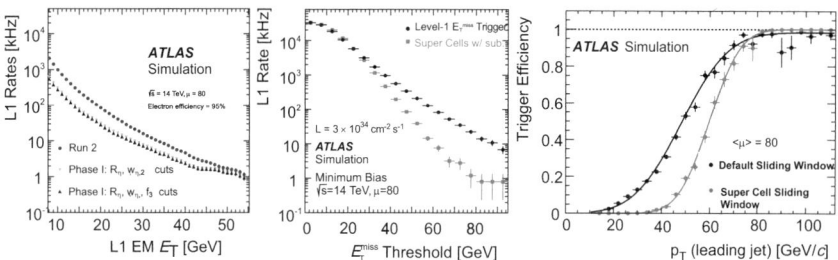

Fig. 10.9 Left: Simulated L1 trigger rate as a function of the calorimeter transverse energy for electrons showing run 2 rates (points) and rates expected for run 3 that result from tighter selections on two (downward triangles) and three shape variables (upward triangles) constructed with transverse energy. In the calorimeter upgrade, the trigger elements have a finer granularity, which allow for more effective selections. Center: Simulated L1 trigger rate as a function of the missing energy threshold showing the L1 missing transverse energy trigger with (squares) and without pile-up subtraction (black points). Right: The L1 trigger efficiency for jets as a function of p_T for the leading jet, comparing the present setup (upper points) to the run 3 upgrade (lower points). All simulations were performed at a center-of-mass energy of 14 TeV, a luminosity of 3×10^{34} cm^{-2}s^{-1} and a pile-up of 80 interactions per beam crossing. All reprinted with kind permission from [9], © 2016, Elsevier. All rights reserved

$|\eta| < 4.9$. For higher E_T^{miss} thresholds, the trigger rate with pile-up correction is significantly reduced. Figure 10.9 (right) shows the trigger efficiency as a function of the transverse momentum of the leading jet for the present setup and for the Run-3 upgrade. For the upgraded system, the turn-on curve is much steeper than for the present setup.

10.3.3 The ATLAS Level 1 Muon Trigger

The inputs to the muon trigger come from the Resistive Plate Chambers in the barrel and the Thin Gap Chambers in the endcaps. Figure 10.10 (left) shows a schematic view of a quarter section of the ATLAS muon system. In the L1 trigger, the RPCs are used in the barrel, while the TGCs are used in the endcaps. Figure 10.10 (right) shows the arrangement of the RPCs in the barrel. Each of the three layers consists of eight large chambers. At the boundaries eight small chambers bridge the gaps.

First, we look for hits in the so-called pivot layer RPC2 in the barrel or TGC3 in the endcaps. Next, we apply predefined trigger roads to look for a two-station, or low-p_T coincidence, in RPC1 or TGC2. The search for a high-p_T trigger starts from a low-p_T trigger candidate, looking again in the trigger roads for a verification of a hit in the third station. If a low-p_T or high-p_T candidate is found, the trigger road becomes a RoI. The transverse momentum is determined from matching the trigger roads predefined for the different p_T values to the hit patterns. The width of the region of interest increased for lower thresholds, taking the bending in the solenoid into account. The strategy is depicted in Fig. 10.10 (left). The typical dimensions of the regions-of-interest are $\Delta\eta \times \Delta\phi = 0.1 \times 0.1$ (0.03×0.03) in the RPCs (TGCs). The geometric coverage of the Level-1 trigger is about 99% in the endcap regions

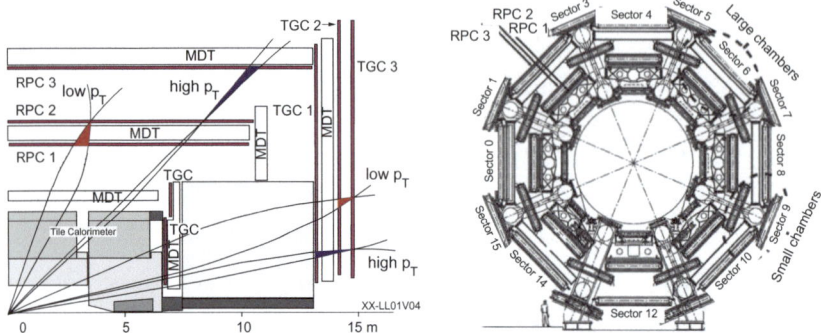

Fig. 10.10 Left: Schematic view of a quarter section of the ATLAS muon system, highlighting the RPCs in the barrel and TGCs in the endcaps used for the Level 1 muon trigger. The first layer of MDT and TGC in the endcaps are mounted in the small wheel, while the second MDT layer and three TGC layers are mounted in the Large wheel. Reprinted with kind permission from [12], © 2013, Elsevier. All rights reserved. Right: The RPC trigger stations in the ATLAS barrel muon detector. Reprinted with kind permission from [13], © 2009, Elsevier. All rights reserved

and about 80% in the barrel region. The acceptance loss in the barrel is caused by the muon support structures, which prevent a full coverage.

To keep the trigger rate acceptable while maintaining high efficiency, false Level-1 trigger signals induced by beam backgrounds were removed by requiring coincidences between chambers installed inside and outside of the magnet system. For Run 2, also hits from the Tile Calorimeter were included. For Run 3, the muon chambers in the endcap's inner most layer was replaced with the New Small Wheel. In addition, new Resistive Plate Chambers have been installed in the $1.0 < |\eta| < 1.3$ region.

10.3.4 The ATLAS Level-1 Topological Trigger

For Run-2, a Topological Trigger was installed at Level-1 to cope with the increased luminosity and higher collision energies. The motivation for such a trigger resulted from the idea to have topological information of different particles available in one processor at the first trigger stage. These triggers allow for a reduction of trigger rates while keeping transverse energy thresholds low. Figure 10.11 shows a schematic layout of the new Level-1 trigger. The Topological Trigger is just added as a separate entity before the Central Trigger. The topological processor receives inputs from the jet/energy processor, the cluster processor and the muon processor. It applies topological algorithms on electrons, γs, τs, jets, energy sums and muon particle candidates, accounting for angular separation ($\Delta R = \sqrt{\Delta \phi^2 + \Delta \eta^2}$) between trigger objects. It performs calculations of masses and transverse masses and determines the interaction hardness expressed as the scalar sum of transverse momenta in jets.

10.3 Trigger Signatures

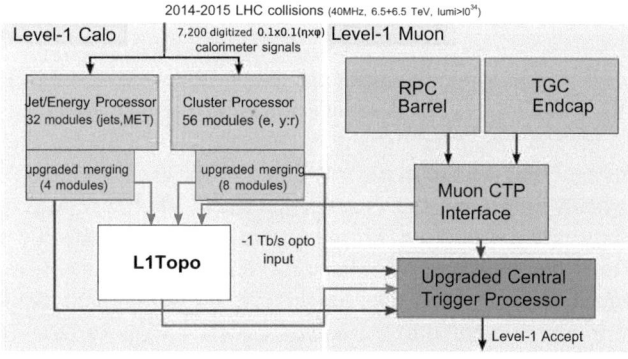

Fig. 10.11 Schematic layout of the Level-1 Trigger for Run 2. The L1 Topo has been added before the Central Trigger. Signals from the L1 Calo upgraded merging modules as well as from the muon interface are fed into the L1 Topo, where topological algorithms are applied. The results are sent to the Central Trigger Processor, which generates the L1 accept. Reprinted under CC-BY-NC-ND-4.0 Licence from [14], © 2016, ATLAS Collaboration

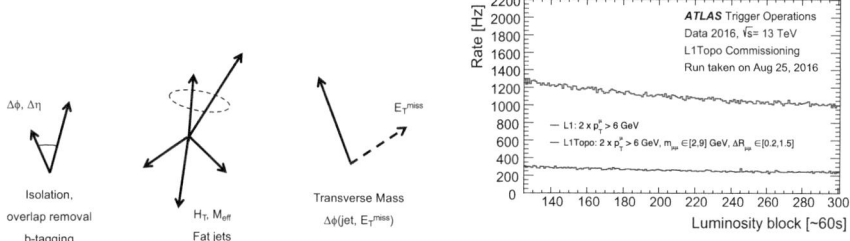

Fig. 10.12 Left: Examples of topologies used for topological trigger decisions: isolation based an angular separation (left), the hardness of the interaction by forming sums of energy and momenta (center) and transverse mass (right). Reprinted under CC-BY-NC-ND-4.0 Licence from [14], © 2016, ATLAS Collaboration. Right: The ATLAS rate of L1 triggers for two muons with $p_T > 6\,\text{GeV}/c$ as a function of luminosity block with (lower histogram) and without (upper histogram) the L1 topological trigger requirement that includes di-muon masses of $2\,\text{GeV}/c^2 < m_{\mu\mu} < 9\,\text{GeV}/c^2$ and an angular separation of $0.2 < \Delta R < 1.5$. One luminosity block is the time interval of data recorded over which the experimental conditions are assumed to be constant (usually 60 s). The average pile up was $\langle \mu \rangle = 46.4$. Note that the rate with the L1 topological requirement is reduced by a factor of four. Reprinted under CC-BY-4.0 Licence from [15], © 2020, ATLAS Collaboration

Figure 10.12 (left) illustrates some examples of the topological quantities, such as angular separation between objects, scalar sums of transverse energy, masses, transverse masses and separation between jets and missing transverse energy. To determine these quantities within 150 ns, 15 different algorithms have been implemented. The final trigger decision is made in the central trigger processor, which obtains inputs from the calorimeter triggers, the muon trigger and the topological trigger. Figure 10.12 (right) shows the event rate versus time of Level-1 triggers without and with topological requirements. The latter yields a rate reduction by more than a factor of five.

10.3.5 The ATLAS High-Level Trigger

During the Level-2 processing, the event data fragments were stored in ROBs, distributed over the ReadOut System. A set of event building nodes collected fragments of accepted events and forwarded them to the Event Filter (EF) farm, where offline-like algorithms analyzed the full event assembled by the event builder. The average rate for events ready for prompt analysis was 400 Hz. In addition, events were stored for delayed analysis at a rate of 200 Hz plus a rate for data generated by calibration triggers. In Run 1, the total rate was about 700 Hz. Events accepted by the Event Filter were sent to a small set of data loggers, which stored them temporarily on disk. Data were then forwarded to the central CERN storage system. The combination of the Level-2 trigger and the Event Filter is called High-Level Trigger. The triggers are based on identifying combinations of candidate physics objects such as electrons, photons, muons, jets, jets with b-flavor tagging (b-jets) as well as global event properties such as missing transverse energy and summed transverse energy. Data are written to inclusive data streams based on the trigger type.

False muon candidates induced by physics backgrounds are rejected by using more precise muon momentum measurements. The muon tracks are reconstructed by adding data from the MDT chambers to obtain a more precise estimate of the track parameters. Then, the muon is combined with a track found in the inner detector. The closest inner-detector track in the η and ϕ planes is selected as the best matching track and the p_T value is refined by taking the weighted average between that of the track in the muon system and the inner detector track.

10.3.6 Trigger Performance in the ATLAS Experiment

In the following plots, we illustrate the ATLAS trigger performance in Run-2. Figure 10.13 shows the L1 muon trigger efficiencies using the tag-and-probe method

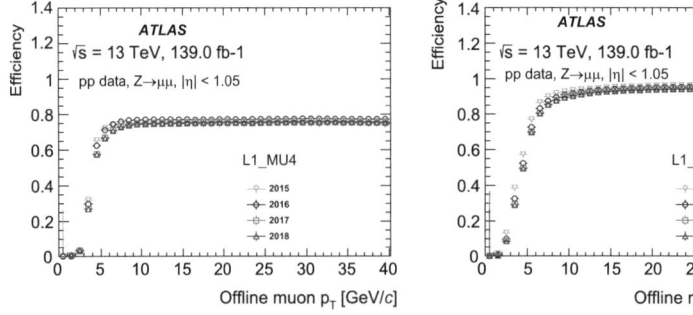

Fig. 10.13 Efficiencies of L1 single-muon triggers for a threshold of 4 GeV/c as a function of the reconstructed muon p_T in the barrel (left) and endcaps (right) in different years: Run 2, 2015 (downward triangles), 2016 (circles), 2017 (squares) and 2018 (upward triangles). The efficiency is measured using the tag-and-probe method in $Z \to \mu^+\mu^-$ events. The error bars show statistical uncertainties only. Reprinted under CC-BY-4.0 Licence from [15], © 2020, ATLAS Collaboration

10.3 Trigger Signatures

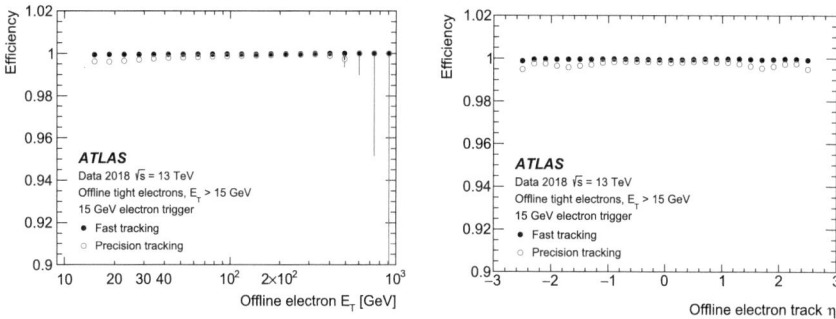

Fig. 10.14 Left: The efficiency of offline tight electrons from a tag and probe analysis of a 15 GeV trigger width as a function of transverse energy for Run 2 $Z^0 \to e^+e^-$ data. Right: The electron efficiency as a function of pseudo-rapidity. Data are shown for fast tracking (solid points) and precision tracking (open points). Both reprinted under CC-BY-4.0 Licence from [16], © 2022, ATLAS Collaboration

for muons from Z^0 decays in the barrel (left) and endcaps (right). Here, we select $\mu^+\mu^-$ pairs from a Z^0 or J/ψ decays. From kinematic reconstruction we can identify the two muons. This procedure provides efficiencies as a function of p_T and η. We select one tightly identified muon as the tag and use the second muon as a probe. We check if the probe muon passed the trigger. The Z-boson events cover the transverse muon momentum range $10\,\text{GeV}/c \leq p_T \leq 100\,\text{GeV}/c$, whereas the J/ψ meson events cover $p_T \leq 10\,\text{GeV}/c$. For transverse momenta above the threshold, the absolute efficiency in the barrel (endcaps) is about 70% (90%).

Offline electrons are selected having at least two pixel hits, an IBL hit if passing through at least one active IBL module, and at least four clusters in the SCT. The pseudo-rapidity is restricted to $|\eta| < 2.5$, for transverse momenta $p_T > 5\,\text{GeV}/c$ and for different transverse energy triggers. The efficiency is obtained from the tag and probe method of $Z^0 \to e^+e^-$ decays. Figure 10.14 shows the efficiency of offline electrons for $E_T > 15$ GeV as a function of the transverse energy (left) and pseudo-rapidity (right). For both tracking methods the efficiency is above 99%.

Jet triggers are used for signal selection in various physics measurements and detector performance studies. Events selected by the single-jet triggers are also used for the calibration of the calorimeter jet energy scale and resolution. Figure 10.15 (left) shows the single-jet trigger efficiencies for L1 as a function of the leading-jet transverse momentum in the barrel for different jet thresholds. After the typical rise above the threshold, the efficiency levels off at 100%. Similar results are obtained for the endcaps. More results, on the HLT trigger efficiencies and multi-jet triggers are found in [17].

The missing transverse energy is used in searches for final states with jets and large missing energy. Such decays, for example, are of great interest in searches for physics beyond the SM, where invisible particles are produced. However, since the rate of hadronic jet production is very large, a mismeasurement of the jet energy is possible even with reasonably good calorimeter resolution, which can mimic a large E_T^{miss}. Pile up, in addition, adds energy to the calorimeter and degrades the E_T^{miss}

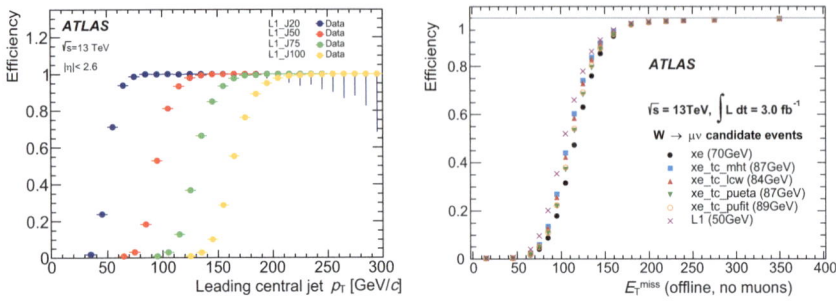

Fig. 10.15 Left: Efficiency of single-jet triggers as a function of offline jet transverse momentum for L1 triggers in the barrel region for jet thresholds of 20 GeV/c (blue points), 50 GeV/c (red points), 75 GeV/c (green points) and 100 GeV/c (yellow points). Right: The E_T^{miss} trigger efficiency as a function of the missing transverse energy for $W \to \mu\nu$ events for L1 triggers with a 50 GeV threshold (crosses) and for a combination of L1 with HLT triggers using different E_T^{miss} algorithms, for cells with a 70 GeV threshold (solid points), jets with an 87 GeV threshold (blue squares), topological clusters with an 84 GeV threshold (red upward triangles), pile-up suppression with an 87 GeV threshold (green downward triangles) and a pile-up fit with an 89 GeV threshold (red open circles). The thresholds for the different algorithms correspond to an approximately equal trigger rate. Both reprinted under CC-BY-4.0 Licence from [17], © 2017, ATLAS Collaboration

resolution. Since the rate is rather large, thresholds need to be adjusted. To ensure the desired broad E_T^{miss}-based physics program, different HLT algorithms based on cells, jets, topological cluster, pile-up suppression algorithm and a pile-up fit were developed. They are explicitly defined in [17]. We illustrate their efficiencies on a sample of $W \to \mu\nu_\mu$ events that was selected by requiring only one muon with $p_T > 25$ GeV/c, a transverse mass $m_T > 50$ GeV/c^2 and a single-muon trigger of > 20 GeV/c. Figure 10.15 (right) shows the E_T^{miss} trigger efficiency as a function of the offline E_T^{miss} for the different HLT algorithms and for the L1 trigger. The trigger efficiencies are all rather similar and reach 100% after the threshold turn on.

10.4 The CMS Trigger System

The CMS experiment uses a hardware Level-1 trigger and a software Level-3 trigger. It is based on FPGA and custom ASIC technology, reducing the LHC 16 MHz collision rate in 2012 to 100 kHz with a latency of less than 4 μs. The High-Level Trigger is implemented in software running on a farm of commercial computers, including over 13,000 CPU cores that reduce the output rate to about 1 kHz. The software consists of a streamlined version of the offline reconstruction algorithms that is optimized to comply with the strict time requirements of the online selection. In 2012, the average processing time was about 200 ms per event. The HLT split the selected events into various non-exclusive streams, with different purposes and event content, where most of the bandwidth was devoted to data for physics analysis. Streams of data with limited event content were also saved for data quality monitoring (DQM), calibration, and trigger studies. The conceptual layout is similar to that in the ATLAS experiment [18].

10.4 The CMS Trigger System

For rising collisions rates, the CMS trigger system has been upgraded. Since the Higgs mass is relatively low, energy thresholds need to be kept low to maintain sensitivity to all Higgs decay products and carry out precision measurements of Higgs parameters. Thus, the following steps have been accomplished.

- A better resolution and precision of trigger objects with respect to ϕ, η, p_T and E_T has been achieved, eventually using the full detector resolution already at Level-1.
- Perform more complex operations at an early level such as the pileup subtraction in the calorimeter trigger.
- More sophisticated trigger algorithms such as complex correlations between different types of trigger data, calculation of invariant masses or transverse masses of pairs of trigger objects have been used.
- A larger number of trigger objects (such as muon or electron candidates) and a larger number of algorithms (trigger paths) to allow for more combinations have been considered.
- Information from additional parts of the CMS detector, in particular from the silicon strip tracker has been incorporated.

For Run 2, the size of the 5×5 crystal trigger tower in the electromagnetic calorimeter remained at $\Delta\phi \times \Delta\eta = 0.087 \times 0.087$. However, more sophisticated algorithms were introduced. In the muon system, CMS used Drift Tubes (DT), Cathode Strip Chambers (CSC) and Resistive Plate Chambers (RPC) in the trigger. In Run 1, the merging happened at a later stage in the Global Muon Trigger. In Run 2, three track finders covering the barrel, the endcaps and the barrel-endcap overlap regions used all signals available in their region, providing a better discrimination between signal and background. The data of these three track finders were merged in the upgraded Global Muon Trigger that forwards them to the Global Trigger.

Figure 10.16 shows a schematic layout of the upgraded Level-1 Trigger, which processes information from the calorimeter and muon triggers. The electromagnetic calorimeter barrel and endcaps (ECAL) as well as the hadronic calorimeters in the barrel, endcaps and forward region send data to the Calo Trigger Layer 1 and Layer 2. The output is transferred to both the Global Trigger and the Global Muon Trigger. The muon detectors CSCs, DTs and RPCs send their data to the Muon Track Finder Layer, where it handles barrel muons, endcap muons and barrel-endcap overlap muons. The output proceeds to the Sorting and Merging Layer and then to the Global Muon Trigger. The output is sent to the Global Trigger.

For the high-luminosity upgrade, the CMS experiment will replace the endcap calorimeters with new radiation-hard calorimeters based on silicon tungsten, which has finer segmentation. The CMS silicon tracker will be replaced with a new system in the barrel and both endcaps, having strip sensors on the outside and a combination of strip and pixel sensors on the inside. The tracking information will be used at a special trigger stage to validate muon and calorimeter objects from the respective systems. An increase in latency to 120 μs makes it possible to include tracking information in the Level-1 Trigger. This will improve the $\pi^0 \to \gamma\gamma$ selection by discriminating against electrons and hadrons from jets, improve the p_T assignment of objects, correct for

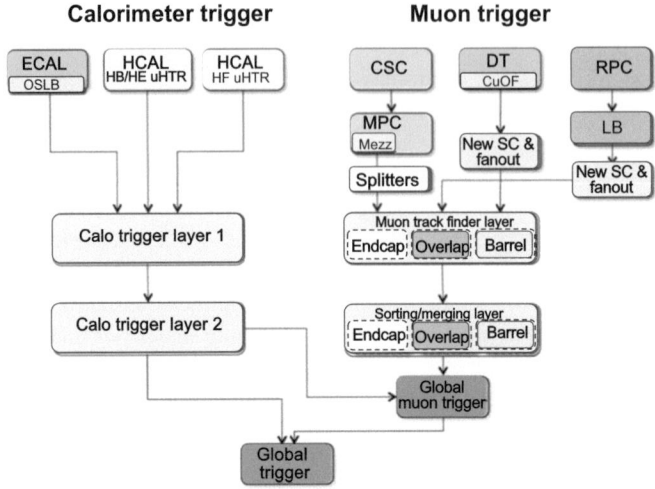

Fig. 10.16 Schematic view of the upgraded CMS Level-1 Trigger showing the calorimeter Level-1 trigger (left) and the muon Level-1 trigger (right). The calorimeter trigger combines information from the electromagnetic calorimeter and the different hadron calorimeters. The muon trigger combines information from the CSCs, DTs and RPCs. The information is fed into a global muon trigger and together with the calorimeter trigger into a global trigger. Reprinted with kind permission from [19], © 2017, IOP Publishing. All rights reserved

pileup in calorimeter objects, determine the isolation of electrons, photons, muons and taus and provide missing transverse-energy calculations for individual proton-proton collisions. In addition to the tracking information of silicon strip detectors, information of the pixel detectors may be useful. The Level-1 trigger also will be upgraded using the principle of particle flow. Figure 10.17 shows the layout of the CMS upgraded Level-1 trigger for Run 3. The Level-1 trigger is divided into nine subsystems that sent information to Trigger Primitive Generators (TPG). A new Correlator Trigger system will match tracks with the Calorimeter and Muon Trigger information, apply intricate object identification algorithms and provide a list of sorted trigger objects to a Global Trigger.

For Run 3, the Global Trigger will have all the logic functionality transferred into one Micro Telecommunications Computing Architecture that is equipped with powerful FPGAs. For the High-Luminosity Upgrade several such modules will run in parallel. The optical input data from calorimeters and muons merged with tracker information will be split to serve all the parallel Global Trigger boards. Using modern powerful FPGAs with built-in Digital Signal Processor will allow the Global Trigger to calculate complex correlations between different trigger objects and physical quantities such as invariant masses or transverse masses of pairs of trigger objects.

For Run-2, the High-Level Trigger hardware consisted of a processor farm, the Event Filter Farm with about 26,000 cores, processing 100 kHz data with a latency of 260 ms. A crucial role consisted of gathering all information from the Silicon Pixel Tracker to perform track reconstruction and the primary vertex finding. The track reconstruction software is basically identical to that used for offline reconstruction.

10.4 The CMS Trigger System

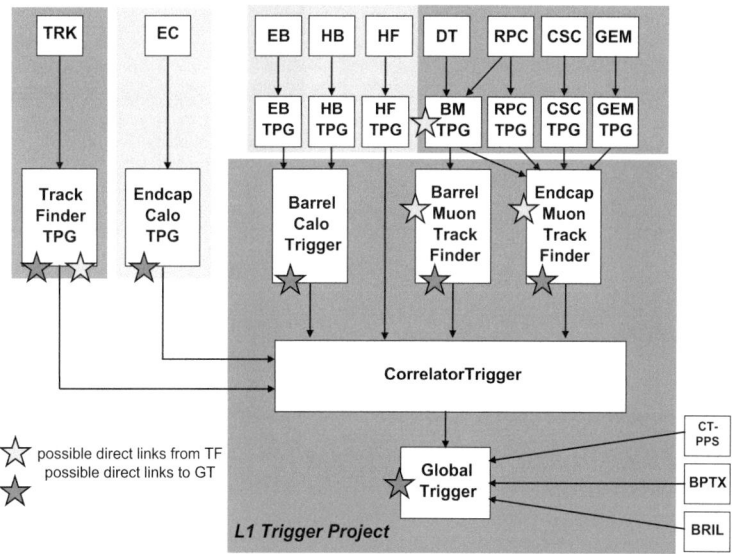

Fig. 10.17 Schematic view of the Level-1 Trigger in the Phase II upgrade. The individual components are Outer Tracking Detector (TRK), the Endcap Calorimeter (EC), the ECAL Barrel (EB), the HCAL Barrel (HB), the HCAL Forward Detector (HF), the Muon Drift Tube Detectors (DT), the Resistive Plate Chambers (RPC), the Cathode Strip Chambers (CSC) and the Gas Electron Multiplier Chambers (GEM). Reprinted under CC-BY-3.0 Licence from [20], © 2019, CMS Collaboration

The track reconstruction procedure is organized into four steps: first, construction of seeds (few pixel hits); second, definition of the initial estimate for trajectory parameters; third, finding of further compatible hits by means of Kalman fitter techniques and track building; fourth, the track fitting marked by quality flag. The CMS trigger performs similarly well as the ATLAS trigger. Figure 10.18 shows as examples the trigger efficiency for electrons (left) and that for the linear scalar sum of transverse energies (right).

For the high-luminosity operation the High-Level Trigger has to cope with increased pileup and significantly increased rates of individual triggers. The new high-granularity endcap calorimeter will comprise of about six million channels. Thus, sophisticated reconstruction at Level-1, with Track Trigger, Particle Flow and pileup mitigation techniques are being developed. At the High-Level Trigger, the deployment of heterogeneous hardware (FPGAs, GPUs), the optimization of the reconstruction algorithms and the adoption of alternative data-taking strategies like scouting are the keys to success. Depending on the pileup, the High-Level Trigger accept rate will increase from 1 kHz at pileup of $\langle \mu \rangle = 60$ to 5 kHz (7.5 kHz) at $\langle \mu \rangle = 140$ (200). The corresponding storage capacity increases from 0.2 PB to 2.7 PB (5.3) PB, respectively.

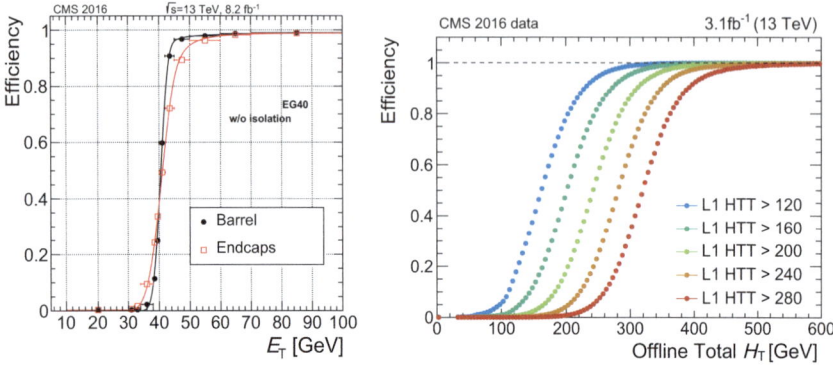

Fig. 10.18 Left: The trigger efficiency for electromagnetic showers in the barrel (solid points) and endcaps (open circles) as a function of the transverse energy. Right: The trigger efficiency as a function of the linear scalar sum of transverse energies H_T for different thresholds varying between 120 GeV and 280 GeV. Both reprinted under CC-BY-NC-ND-4.0 Licence from [21], © 2016, CMS Collaboration

10.5 The LHCb Trigger System

The LHCb experiment originally used a three-level trigger system, a level 0 hardware trigger and two high-level triggers HLT1 and HLT2. The LHCb detector is limited by design in terms of data bandwidth. Instead of the 40 MHz bunch crossing rate, the LHCb detector accepted a rate of 1 MHz in Run 1. The Level 0 hardware trigger looks for high transverse-energy and transverse-momentum signatures. The 1 MHz rate is distributed as 450 kHz for charged hadrons, 400 kHz for single muons and di-muons and 150 kHz for electrons and photons. This selection enters the Software High-Level trigger, which performs partial event reconstruction, finds displaced tracks and vertices and selects di-muons. So at this level, tracking and particle identification information is used. Selected candidates that are sent to disk are a mixture of exclusive and inclusive triggers. The output rate to disk was 5 kHz consisting of 2 kHz of inclusive topologies, 2 kHz of inclusive and exclusive charm events and 1 kHz of muon and di-muon events, writing data with 0.3 GB/s.

For Run 2, the trigger was adjusted to the higher cross section at 13 TeV [22] and HLT1 and HLT2 were combined. Figure 10.19 (left) shows a schematic layout of the LHCb trigger strategy for Run 2. Though the input rate to the High-Level Trigger remained the same, the rate to storage was increased to 12.5 kHz at a data rate of 0.6 BG/s. The HLT performed online detector calibrations and alignments. Furthermore, it performed full offline-like events selections and buffered a fraction of the events to temporary disk for analysis purposes.

Now for Run 3, LHCb has adopted a novel design as depicted in Fig. 10.19 (right). The Trigger consists only of the Software High-Level Trigger, which accepts the full 30 MHz inelastic event rate and performs full event building. The HLT performs the full event reconstruction online and makes exclusive kinematic and geometric

10.5 The LHCb Trigger System

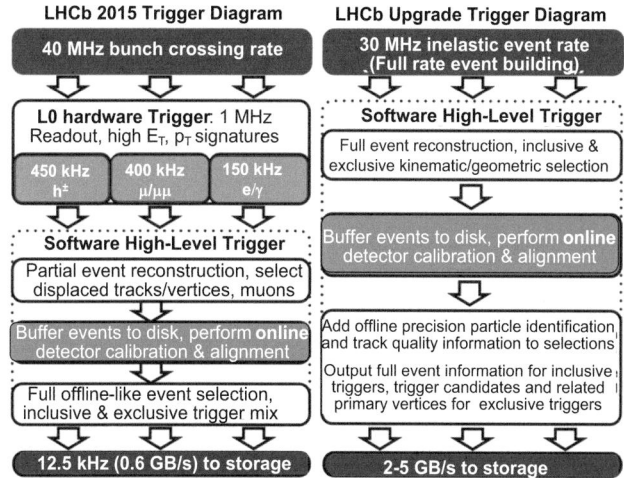

Fig. 10.19 Left: The LHCb trigger strategy for Run 2. Right: The LHCb trigger strategy for Run 3. Both reprinted with kind permission from [23] © 2016, Elsevier. All rights reserved

selections. Online detector calibrations and alignments are done. About 75% of the physics program is written to the TURBO stream, a reduced format that allows for flexible event information and can discard a user-specified fraction of raw and reconstructed data. Selected events are buffered to disk, allowing for fast physics analyses. Before sending the data to storage at a rate of 2–5 GB/s, offline precision particle identification and track quality information is added to the selections. The system outputs the full event information for inclusive triggers, trigger candidates and related primary vertices for exclusive triggers. Presently, LHCb is the first and only hadron collider experiment to deploy a trigger exclusively in software.

Figure 10.20 (left) shows the steps performed in the offline tracking. First, tracks are located in the Vertex Locator (VELO). Tracks with transverse momenta $p_T > 70 \, \text{MeV}/c$ are added and a primary vertex is determined. For all tracks, a full Kalman fit is performed and the particle identification information is added. In the new trigger, the track reconstruction is done in the HLT as illustrated in Fig. 10.20 (right). In addition to VELO tracks, tracks with $p_T > 200 \, \text{MeV}/c$ from the Upstream Tracker (UT) and tracks with $p_T > 500 \, \text{MeV}/c$ from the Forward Tracker are combined to determine the primary vertex. Event requirements are added to keep the output rate below 1 MHz. In addition, muon candidates are identified. Then, a simplified Kalman fit is performed and particle identification information is added. At a rate of 2 GB/s, the LHCb physics program will be limited, while at 5 GB/s a wide beauty and charm program can be performed.

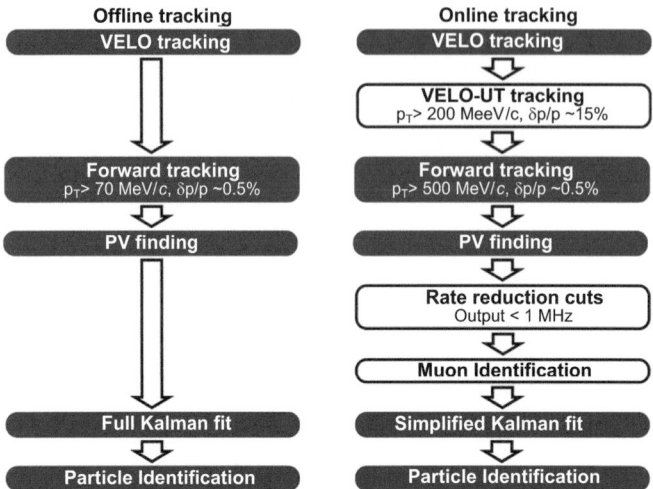

Fig. 10.20 Left: The LHCb offline track reconstruction scheme in Run 2. Right: The LHCb online track reconstruction scheme in Run 3. Both reprinted with kind permission from [23] © 2016, Elsevier. All rights reserved

10.6 New Developments

The future of HEP computing heads to new technologies, advanced computing units, improved software and faster networks. One new device is the Graphical Processing Unit (GPU), which is used in general purpose computing, since it provides a streaming data parallelism. Simple repetitive algorithms can be processed in parallel, which speeds up the processing time considerably. The new technology enables the user to process parallel-computing pipelines on modern graphics cards. These cards have hundreds or thousands of cores and are well-suited for handling parallel algorithms. Thus, GPUs have become of great interest for high-energy physics experiments, in particular for track reconstructions. At the LHC, ATLAS [24], CMS [25], LHCb [26] and ALICE [27] are exploring the use of GPCs in tracking and cluster finding.

The LHCb DAQ and event building infrastructure considers two processing structures. The first one is based on CPU only, implementing both HLT1 and HLT2 using Event Filter Farm CPU servers. Here, both HLT1 and HLT2 are executed on the Event Filter Farm. The second one is hybrid implementing the HLT1 with GPU cards installed in the Event Builder servers, while the CPU-based HLT2 runs on Event Filter Farm. Figure 10.21 illustrates the CPU-only design. The system has to perform various tasks.

- A set of custom-made FPGA cards, called PCIe40, which collect data from the sub-detectors at a rate of 32 Tb/s.

10.6 New Developments

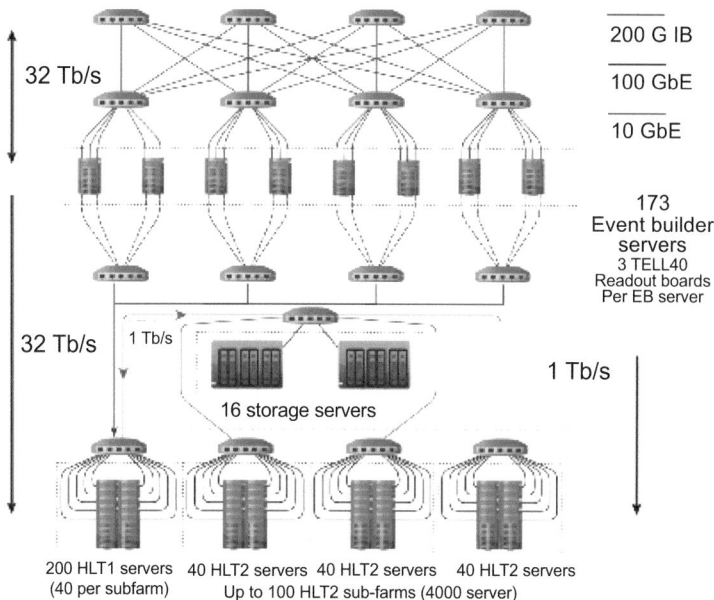

Fig. 10.21 A CPU only design for the LHCb DAQ in Run 3 including the Event Builder, Event Filter Farm and dedicated storage servers for the disk buffer. Reprinted under CC-BY-4.0 Licence from [26], © 2021, LHCb Collaboration

- A set of Event Building servers that host the PCIe40 cards and implement a network protocol to combine the sub-detector data fragments produced in a single LHC bunch crossing. They group $\mathcal{O}(1000)$ of such events into multi-event packets.
- An Event Filter Farm executes HLT1. When LHCb is not taking data, this farm can also receive events from the disk servers and run the HLT2 process on them.
- A high-performance network connects the Event Building servers and HLT1 Event Filter Farm to transmit a data rate of 32 Tb/s.
- An array of disk server buffers the HLT1 output data, while detector alignment and calibrations are performed.
- A second Event Filter Farm receives events from the disk servers once the alignment and calibration constants are available and runs the HLT2 process on them.

Figure 10.22 illustrates the Hybrid architecture. In comparison to the CPU-only design, here GPU cards are added running the HLT1 to the Event Builder servers, reducing the data rate at the output of the Event Builder by a factor of 30-60. Thus, the communication between the Event Builder and the Event Filter Farm can be performed with a lower-bandwidth network. The HLT2 then runs similarly on the Event Filter Farm as in the CPU-only solution. It must be able to run the complete offline-quality reconstruction, using the highest quality alignment and calibration at all times.

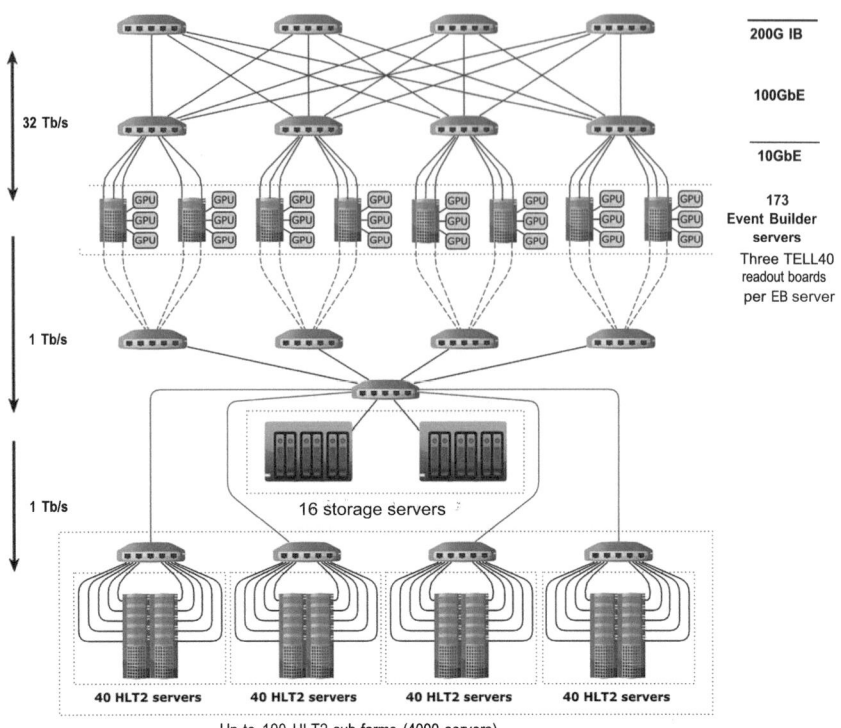

Fig. 10.22 A hybrid design for the LHCb DAQ in Run 3 with GPUs placed in the Event Builder servers. Reprinted under CC-BY-4.0 Licence from [26], © 2021, LHCb Collaboration

Using simulations, LHCb studied both designs. They determined the efficiencies for reconstructing particles in the Vertex Locator, the Vertex Locator and Upstream Tracker, as well as Vertex Locator, Upstream Tracker and Scintillating-Fiber Tracker as a function of momentum and transverse momentum. The performance of both technologies is rather similar. While the CPU-only design is slightly more efficient at low momenta, the hybrid design is slightly more efficient at high momenta. Similar performance was found for the muon identification efficiency and the HLT1 operation. Both designs yield close to identical performance in all aspects. Minor differences are considered to be a matter of tuning, which are expected to be neutral in terms of throughput. The hybrid design, however, is about 20% more cost effective.

In the CMS experiment, GPUs have been tested in two-dimensional cluster reconstruction of the HGCAL [25]. A new CPU-based design and three GPU-based architectures were tested. The CPU-based design needed 203 ms to locate the 2D clusters, while the third GPU-based architecture just took 32 ms. Note that the algorithmic computation just takes 6 ms, while the latency due to data transfer is 20 ms and the structure of array conversion needs another 6 ms. The ATLAS collaboration is also looking into using GPUs for the high-luminosity LHC operation.

Exercises

10.1 Consider the LHC high-luminosity upgrade. The center-of-mass energy will remain at the design energy of $\sqrt{s} = 14$ TeV. The inelastic cross section is 79 mb and the instantaneous luminosity will increase to 7×10^{34} cm^{-1}s^{-1} at the beginning of a fill. How large is the inelastic event rate? To reach a luminosity of 3,000 fb^{-1} in 12 years of running, an annual luminosity of 250 fb^{-1} needs to be achieved. How long does the accelerator need to run each year to reach this luminosity?

10.2 The decays $B_s^0 \to \mu^+\mu^-$ and $B_d^0 \to \mu^+\mu^-$ are of great interest in search for new physics. The cross section for $b\bar{b}$ production is 790 µb. The fraction of B_s and B_d mesons at 14 TeV is 0.1 and 0.4 and the branching fractions are $\mathcal{B}(B_s^0 \to \mu^+\mu^-) = 3.7 \times 10^{-9}$ and $\mathcal{B}(B_d^0 \to \mu^+\mu^-) = 1.06 \times 10^{-10}$. The signal efficiency in ATLAS is 18%. How many events do you expect for each mode?

10.3 What handles do experimenters have to deal with high input rates in the trigger system? Which sub-detectors are typically used in the level L1 trigger? What information is used?

10.4 The LHCb experiment is using a new trigger philosophy by removing the level 1 hardware trigger. Instead they use a software-only trigger that handles a 30 MHz input rate. What are the advantages for such a concept? Discuss if such a concept is feasible for the multipurpose experiments ATLAS and CMS.

10.5 Design a trigger and data acquisition system for a multipurpose detector operating at an e^+e^- collider that operates at a center-of-mass energy of 500 GeV. The detector has a four-layer pixel detector followed by a six-layer silicon strip tracker. A 24 X_0 silicon-tungsten electromagnetic calorimeter and a 3 λ_I steel-scintillating-tile hadron calorimeter lie inside the 1.5 T coil. Outside is a muon system. The detector should be efficient in measuring Higgs boson decays, top quark decays, electroweak decays and search for new physics phenomena.

References

1. A. Negri, https://indico.cern.ch/event/643308/contributions/2610520/attachments/1467722/2536718/isotdaq18.Negri.DaqIntro.pdf
2. The ATLAS Collaboration (H. Stenzel et al.), arXiv:1611.02454 [hep-ex] (2017)
3. The ATLAS Collaboration (G. Aad et al.), Report: SLAC-R-980, CERN-OPEN-2008-020, arXiv:0901.0512 [hep-ex] (2009)
4. F. Lanni, https://indico.cern.ch/event/347919/contributions/819131/attachments/686441/942792/Calorimeter_Electronics_Upgrades.pdf
5. The BABAR Collaboration (B. Aubert et al.), Nucl. Instrum. Meth. A **479**, 1 (2002)
6. The Belle Collaboration (A. Abashian et al.), Nucl. Instrum. Meth. A **479**, 117 (2002)
7. P. Czodrowski, PoS EPS-HEP-**2015**, 273 (2015)
8. The ATLAS Collaboration (G. Aad et al.), ATL-TDR-027, LHCC-2017-018 (2017)
9. The ATLAS Collaboration (S. Yamamoto et al.), Nucl. Instrum. Meth. A **824**, 374 (2016)
10. The ATLAS Collaboration (G. Aad et al.), ATL-TDR-022 á LHCC-2013-017 (2013)
11. The ATLAS Collaboration (R. Schwienhorst et al.), JINST **11**, C01018 (2016)

12. A. Ventura, Nucl. Instrum. Meth. A **732**, 48 (2013)
13. F. Anulli et al., JINST **4**, P04010 (2009)
14. S. Artz, PoS LHCP-**2016**, 200 (2016)
15. The ATLAS Collaboration (G. Aad et al.), JINST **15**-09, P09015 (2020)
16. The ATLAS Collaboration (G. Aad et al.), Eur. Phys. J. C **82**, 206 (2022)
17. The ATLAS Collaboration (G. Aad et al.), Eur. Phys. J. C **77**, 317 (2017)
18. The CMS Collaboration, Technical Design Report: Trigger systems, CERN-LHCC-2000-038 CMS-TDR-6-1 (2000)
19. L. Cadamuro, JINST **12**, C03021 (2017)
20. The CMS Collaboration (A. Zabi et al.), J. Phys. Conf. Ser. **1162**, 012040 (2019)
21. The CMS Collaboration (A. Tapper et al.), PoS ICHEP-**2016**, 242 (2016)
22. The LHCb Collaboration (F. Dordei et al.), EPJ Web Conf. **164**, 01016 (2017)
23. The LHCb Collaboration (A. Dziurda et al.), Nucl. Instrum. Meth. A **824**, 277 (2016)
24. The ATLAS Collaboration (D. Emeliyanov1 et al.), J. Phys. Conf. Ser. **396**, 012018 (2012)
25. The CMS Collaboration (Z. Chen et al.), EPJ Web Conf. **245**, 05005 (2020)
26. The LHCb Collaboration (R. Aaij et al.), arXiv:2105.04031 [physics.ins-det] (2021)
27. The ALICE Collaboration (D. Rohr et al.), EPJ Web Conf. **245**, 10005 (2020)

Open Access This chapter is licensed under the terms of the Creative Commons Attribution 4.0 International License (http://creativecommons.org/licenses/by/4.0/), which permits use, sharing, adaptation, distribution and reproduction in any medium or format, as long as you give appropriate credit to the original author(s) and the source, provide a link to the Creative Commons license and indicate if changes were made.

The images or other third party material in this chapter are included in the chapter's Creative Commons license, unless indicated otherwise in a credit line to the material. If material is not included in the chapter's Creative Commons license and your intended use is not permitted by statutory regulation or exceeds the permitted use, you will need to obtain permission directly from the copyright holder.

Examples of Full Detector Systems

11

In this chapter, we present examples of full detector systems at colliding beam experiments, for non-colliding beam environments and for medical applications. We start with colliding beam experiments presenting the BABAR experiment at PEPII, the Belle II experiment at Super KEKB, the LEP and SLC experiments ALEPH, DELPHI and SLD, the Tevatron experiment CDF, as well as the LHC experiments ATLAS, CMS, LHCb and ALICE. Though we see some similarity among the different colliding beam experiments, there are also some significant differences for some of them. The succeeding section presents the Mu2e, Superkamiokande, IceCube, LUX/LZ and CTA experiments. Finally we present a PET scanner.

11.1 Colliding-Beam Detectors

We start with the BABAR experiment.

11.1.1 The BABAR Detector

The BABAR detector was built to measure CP violation in the B-meson system [1]. The PEP II asymmetric storage ring operated at center-of-mass energies of the Υ resonances, colliding 3.1 GeV positrons with 9.0 GeV electrons. The emphasis on the detector performance focussed on very good vertex resolution, very good momentum measurements, very good π/K separation, very good photon energy resolution and muon efficiency. The experiment operated from 1999 to 2008. In 2001, both BABAR and Belle at KEK discovered CP violation in the B^0-meson system.

Figure 11.1 shows a schematic view of the BABAR detector. Due to the asymmetric beam energies, the detector is asymmetric. Around the beryllium beam pipe,

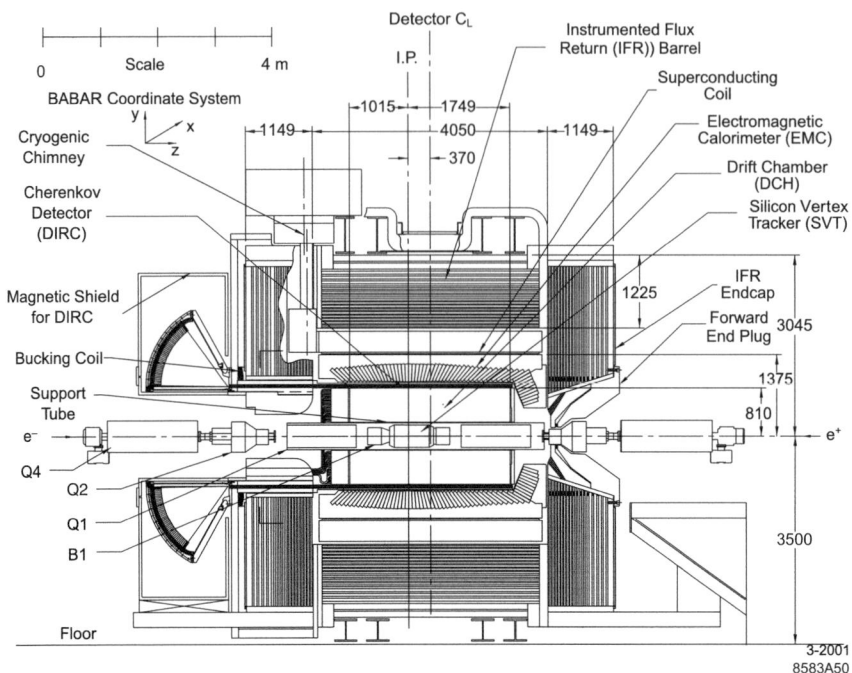

Fig. 11.1 Schematic view of the $B\!A\!B\!A\!R$ detector. Reprinted with kind permission from [5], © 2002, Elsevier. All rights reserved

a five-layer silicon vertex detector is placed, which we discussed in Sect. 6.2.1. It is surrounded by the 30-layer drift chamber we presented in Sect. 4.3.4. Next, follows the DIRC particle identification system that we discussed in Sect. 9.3.5. Before the solenoid that produced the 1.5 T magnetic field, the CsI(Tl) electromagnetic calorimeter discussed in Sect. 7.2.2.1 was positioned consisting of 6,580 CsI(Tl) crystals. The magnetic field was closed with the instrumented flux return consisting of iron plates and chambers. It served as a hadron calorimeter and a muon detector as discussed in Sect. 9.6. Figure 11.2 (left) shows a photograph of the $B\!A\!B\!A\!R$ detector. To illustrate the excellent operation of the detector, Fig. 11.2 (right) shows an event display of $e^+e^- \to B^0 \bar{B}^0$, where one B^0 decays to $J/\psi K_S^0$. The detector sees one electron (top right) one positron (bottom left) and a pion pair (right) that comes from a K_S^0 decay. There are another electron and three charged pions from the $\bar{B}^0$ decay in the event. The open circles are hits in the drift chamber. The green squares show the energy deposit in the electromagnetic calorimeter. The electron and positron leave large showers in the electromagnetic calorimeter while the pions are minimum ionizing.

11.1 Colliding-Beam Detectors

Fig. 11.2 Left: Photograph of the $B\!A\!B\!A\!R$ detector. Reprinted with kind permission from [6], © 2003, $B\!A\!B\!A\!R$ Collaboration. All rights reserved. Right: Event display of $e^+e^- \to B^0\bar{B}^0$, where one B^0 meson decays as $B^0 \to J/\psi K_S^0$ with $J/\psi \to e^+e^-$ and $K_S^0 \to \pi^+\pi^-$. The $\bar{B}^0$ decays semi-leptonically with an electron in the final state. Reprinted with kind permission from [7], © 1999, D. MacFarlane. All rights reserved

11.1.2 The Belle II Detector

The Belle II detector is an upgrade of the original Belle detector [2]. It operates at the asymmetric B factory Super KEKB, which collides 7 GeV electrons with 4 GeV positrons [3]. Since it operates in the same energy range as the $B\!A\!B\!A\!R$ detector, the detector requirements are rather similar. Figure 11.3 shows a schematic view of the Belle II detector. Around the beam pipe, a two-layer silicon pixel detector based on the DEPFET technology is placed at radii of 14 and 22 mm. It is followed by a four-layer doubled-sided silicon strip detector with 320 μm thick rectangular and 300 μm thick trapezoidal shapes positioned at radii of 39, 80, 115 and 140 mm [4]. The pitch on the p strips is 75 μm (50 μm) for large (small) rectangular sensors and 50 μm to 75 μm for trapezoidal sensors. The pitch on the n strips is 240 μm for large and trapezoidal sensors, while it is 160 μm for small rectangular sensors. The expected spatial resolutions are 8.2 μm for p strips and 23.6 μm for n strips.

The 56-layer central drift chamber follows, with 32 axial layers and 24 stereo layers. The inner radius is 16 cm and the outer radius is 113 cm [8]. The longitudinal extensions are from -83 cm to $+159$ cm. The radial cell size is 10 mm for the innermost superlayer (eight layers) and 18.2 mm for the other superlayers. The 14,336 sense wires are 30 μm diameter gold-plated tungsten wires, while the 42,240 field wires are 126 μm diameter aluminum wires. The gas mixture is He-C_2H_6 (50:50), which is the same as that in Belle. The position resolution in $r\phi$ depends on the layer and on the incident angle yielding values in the range 50 to 120 μm, typically being better than 100 μm. The impact parameter resolutions are $\sigma_{d_0} = 126$ μm and

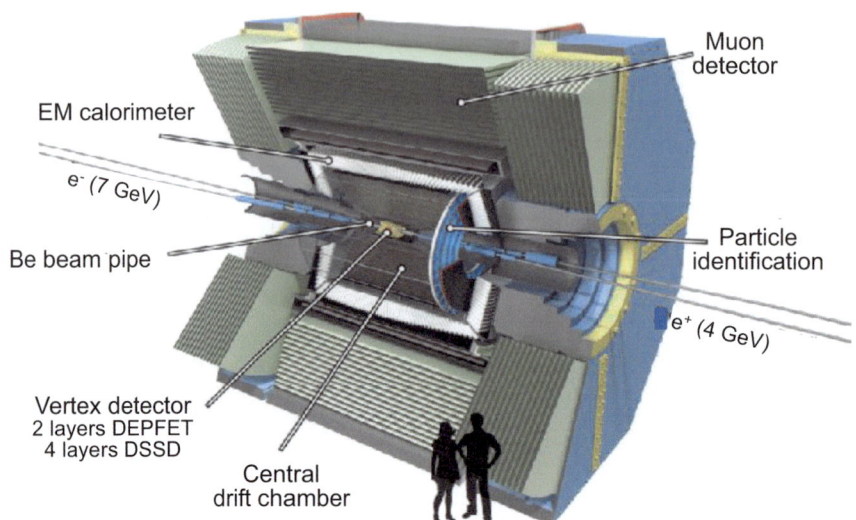

Fig. 11.3 Schematic view of the Belle II detector. Reprinted under CC-BY-4.0 License from [3], © 2017, Belle2 Collaboration

$\sigma_{z_0} = 1.43$ mm. The momentum resolution is $\sigma_{p_T}/p_T = 0.127\% \cdot p_T \oplus 0.321\%$ and the dE/dx resolution is about 5%.

Outside the drift chamber, the time-of-propagation counter is placed at a radius of 120 cm. It consists of 2 cm thick and 275 cm long quartz bars that are arranged in 16 ϕ segments. The quartz bars are read out with 4×4 channel MCP PMTs. Typically, about 20 photoelectrons are recorded with a time resolution of 40 ps. Concerning the π/K separation, they achieve an efficiency larger than 98% with less than a 1% pion misidentification rate for momenta less than 2.5 GeV/c and most polar angles in the barrel. Particle identification in the forward endcap is achieved with an aerogel RICH (ARICH) detector, which we discussed in Sect. 9.3.4.

The electromagnetic CsI crystal calorimeter follows next. It has 6,624 CsI(Tl) crystals in the barrel, 1,152 pure CsI crystals in the forward endcap and 960 pure CsI crystals in the backward endcap. In the barrel, the 37 cm long crystals are placed at a radius of 125 cm. In the backward (forward) endcap, the crystals are located at a position of $z = -102$ cm ($z = 196$ cm). The energy resolution is $\sigma_E/E = 0.066\%/E \oplus 0.81\%/\sqrt[4]{E} \oplus 1.34\%$.

The supeconducting solenoid follows next providing a magnetic field of 1.5 T. The flux return in the barrel is provided by fourteen 5.0 cm thick iron plates that are spaced by 4 cm gaps into which two RPC's are inserted. In the endcaps, there are 14 iron layers in the forward region and 12 layers in the backward region. The gaps are equipped with (7–10) mm $\times$ 40 mm scintillator strips, which are read out with wavelength-shifting fibers and SiPMs. The angular resolution for K_L^0 reconstruction is 10 mrad and the hadron misidentification rate for muons is about 1%.

11.1.3 The ALEPH Detector

The ALEPH experiment (A detector for LEP PHysics) was one of four experiments operating at the LEP storage ring to study the Z^0 boson and search for new physics at higher energies [9]. Figure 11.4 shows a schematic view of the ALEPH detector. A two-layer vertex detector with double-sided silicon strips was placed close to the beam pipe that was surrounded by a small eight-layer inner drift chamber, which measured only $r\phi$ coordinates with 100 µm precision. The z coordinate was obtained from the arrival time of the signal at the wire with a 3 mm precision. This chamber was used also for the level L1 trigger. Next, followed the 4.7 m long time projection chamber that was discussed in Sect. 4.5.2. It covered the radial distance from 0.31 to 1.8 m. The TPC measured 21 three-dimensional space points for each track traversing the inner and outer field cages. The magnetic field was provided by a 1.5 T solenoid yielding a transverse momentum resolution of $\sigma_{p_T}/p_T \sim 0.0015 p_T$. The dE/dx resolution was 4.5%. Between the TPC and the solenoid, the electromagnetic calorimeter was placed, which consisted of a $22 X_0$ thick lead/wire-chamber sampling calorimeter. Energy and position of the electromagnetic showers were measured with 30 mm × 30 mm wide cathode pads, which were connected internally to form towers pointing back to the interaction point. Each tower was read out in three sections in depth. The calorimeter was a highly granular, hermetic detector with 2% cracks in the barrel and 6% cracks in the endcaps. The energy resolution in the barrel was $\sigma_E/E = (16.5 \pm 0.3)\%/\sqrt{E} \oplus (0.3 \pm 0.1)\%$. Outside the solenoid lied the instrumented iron flux return that served both as a hadron calorimeter and muon system. In the barrel, 22 layers of 5 cm thick iron slabs were mounted with

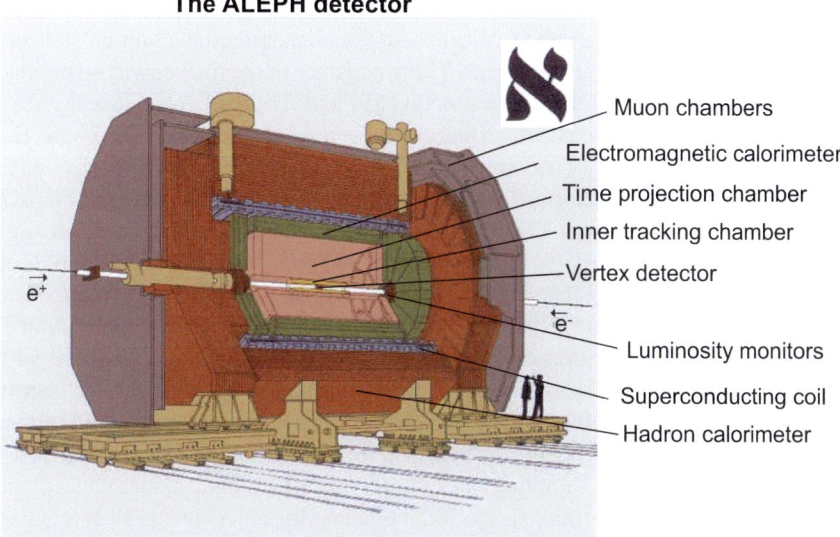

Fig. 11.4 Schematic view of the ALEPH detector. Reprinted with kind permission from [10], © 2001, ALEPH Collaboration. All rights reserved

22 mm gaps between them. A 10 cm thick outer layer was added to obtain 22 gaps. The total thickness was 1.2 m, corresponding to 7.16 λ_I. The gaps were filled with limited streamer tubes made from extruded PVC cells. Each tube had eight cells with a cross limited section of 9 mm × 9 mm and 1 mm thick walls. The surface of each cell was painted with graphite and housed a 100 µm thick wire in the center. Signals were induced on electrodes located on both sides of the eight-cell unit. The hadron calorimeter also exhibited a tower structure that was quasi projective. All towers had the same width in polar angle and covered 3.7° in azimuth angle. There were 2,688 towers that were fully contained in the barrel and 768 towers that overlapped with the endcaps.

Each endcap disk was built from six 60° sector-shaped modules that consisted of 22 steel slabs of 5 cm thickness and a 10 cm thick slab as outer layer except for the overlap region with the barrel, where the number of layers was reduced to 16. This created 22 or 15 gaps into which LSTs were inserted just as in the barrel. To ensure stability of the gaps due to the magnetic forces five iron rods were welded into each gap. The tower structure in the barrel was continued into the endcaps. In total, there were 2,100 towers in the endcaps of which 768 were shared with the barrel. The width in azimuth-angle increased to 7.5° at $\theta = 34°$ and to 15° at $\theta = 18°$. The gas mixture for the hadron calorimeter consisted of Ar-CO_2-isobutane (13:57:30), yielding an energy resolution of $\sigma_E/E = 84\%/\sqrt{E}$. At the outside of the iron, two double layers of LSTs served as muon chambers that were used to obtain two track measurements from each layer. Thus, the chambers had strip electrodes parallel and perpendicular to the wires, which were 4 mm wide with a 10 mm pitch and 10 mm wide with a 12 mm pitch, respectively. The gas in the muon chambers was the same as that in the hadron calorimeter.

ALEPH used a luminosity monitor located at $z = \pm 2.7$ m from the interaction point consisting of a tracker and a calorimeter, which measured the luminosity from the number of recorded Bhabha events. The acceptance of the tracker was 40 mrad to 90 mrad and that of the calorimeter was 45 to 155 mrad. The tracker had nine layers of drift tubes arranged in sectors, which had a cross section of 0.95 mm × 0.95 mm. The azimuth-angle resolution of the entire chamber was 13 mrad, which corresponded to an *rms* of the fitted track position of 200 µm in the radial direction and 2 mm in the azimuth-angle direction. The luminosity calorimeter was a 38-layer lead/wire-chamber sampling calorimeter with a depth of 24.6 X_0. It was arranged into three sections around the beam pipe covering the radial position from 10 to 52 cm. The lead plates in section 1 and 2 were 2.8 mm thick increasing to 5.6 mm in section 3. The total energy and position of the electromagnetic showers were measured with cathode pads of 3 cm × 3 cm. The position resolution was 1.4 mm and the energy resolution was $\sigma_E/E = 20\%/\sqrt{E} \oplus 1.4\%$. A second calorimeter was placed further upstream and downstream behind the beta quadrupoles at $z = \pm 7.8$ m. It was a 22 X_0 thick tungsten-scintillator sampling calorimeter with ten sampling planes plus a plane of vertical silicon strips arranged into four counters. In each counter, all layers had 7 mm thick plates except for layer 1 that had 14 mm thick plates. The scintillator tiles were 3 mm thick and were read out by photomultiplier tubes 1 cm in diameter. The plane of 40 silicon strips was located after the first eight radiation lengths. The strips were 5 cm long and 300 µm thick, having a 0.5 mm pitch. A 2.8 mm thick tungsten plate was used to protect the calorimeter from synchrotron radiation photons that may enter from the back.

11.1.4 The DELPHI Detector

The DELPHI experiment (DEtector with Lepton, Photon and Hadron Identification) was another experiment at the LEP collider [11, 12]. Figure 11.5 shows a schematic view of the DELPHI detector. The three-layer Vertex Detector discussed in Sect. 6.2 lied outside the beryllium beam pipe, followed by the Inner Detector consisting of an inner drift chamber with jet-chamber geometry, providing 24 $r\phi$ points per track and a five-layer MWPC covering polar angles down to 15°. Each layer had 192 wires and 192 circular cathode strips with about 5 mm pitch. The wires provided fast trigger information and resolved left-right ambiguities from the jet section. The strips gave z information that was used also in the trigger. The single-wire precision was 85 µm and the local track element precisions was $\sigma_{r\phi} = 40$ µm and $\sigma_\phi = 0.89$ mrad. The main tracking device was the Time Projection Chamber presented in Sect. 4.5.2. Next, the barrel RICH counters were placed that we discussed in Sect. 9.3.3. The Outer Detector followed, which was built from 24 modules, each consisting of 145 drift tubes arranged in five layers. It was used for fast trigger information in $r\phi$ and rz and to improve the momentum resolution for fast particles. The single–point precision was $\sigma_{r\phi} = 110$ µm and $\sigma_{rz} = 3.5$ cm.

To identify electrons and photons, the High-density Projection Chamber followed. The design was based on the time projection principle for calorimetry. It consisted of 144 modules arranged in six rings, using the lead converter in each module as electric-field cage. The ionization charge of showers and tracks was thus extracted onto a single proportional wire plane at one end of each HPC module. The converter consisted of 41 lead walls spaced by 8 mm gas gaps. The granularity was 4 mm in z and 1° in azimuth angle. The radial sampling was nine-fold. The energy resolution was $\sigma_E/E = 32\%/\sqrt{E} \oplus 4.3\%$. The position resolution was $\sigma_{rz} = 0.13$ cm in the inner ring decreasing to $\sigma_{rz} = 0.22$ cm in the middle rings and $\sigma_{rz} = 0.31$ cm in the outer ring, corresponding to a resolution in the polar angle of 0.6 mrad for 45 GeV electrons. The resolution in the azimuth angle was 3.1 mrad dominated by the uncertainty of track extrapolation from the TPC through the material of the RICH and Outer Detector.

All these subsystems lied inside the superconducting solenoid that provided a magnetic field of 1.23 T. A Time-of-Flight system followed, consisting of 172 two cm thick scintillator bars read out with PMTs on both sides. The TOF counters served as a fast trigger and as a veto for cosmic muons. Further out the barrel hadron calorimeter and the barrel muon chambers were located. The hadron calorimeter was a six-λ_I-thick iron-gas sampling calorimeter made of 20 layers of 5 cm thick iron plates with 2 cm gaps into which limited streamer tubes were inserted. The LSTs consisted of eight cells each having a cross section of 9 mm × 9 mm. One

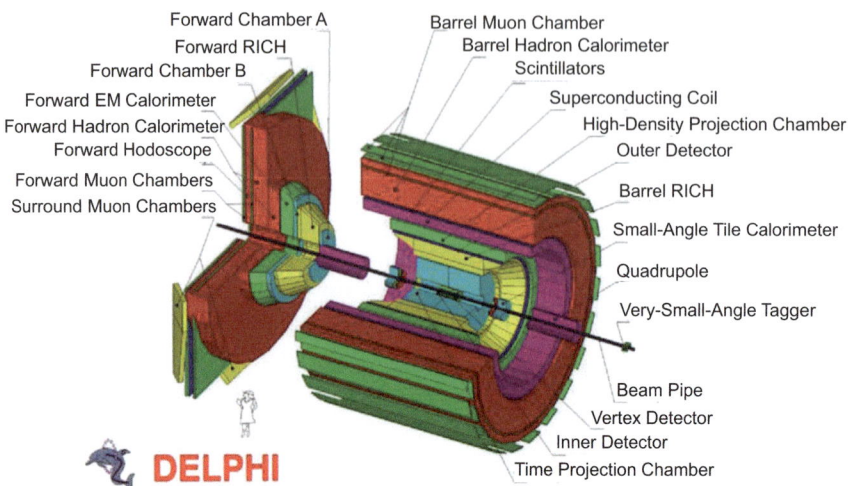

Fig. 11.5 Schematic view of the DELPHI detector showing the barrel and one endcap. Reprinted with kind permission from [13], © 2001, DELPHI Collaboration. All rights reserved

80 μm diameter copper-beryllium anode wire was placed in the center of each tube. The surface resistivity was larger than 50 kΩ/square. The copper-clad readout boards were segmented into pads, which picked up the streamer charges. Pads were shaped to form towers pointing to the interaction point. Each tower covered the angular region of $\Delta\phi \times \Delta\theta = 3.75° \times 2.96°$. The energy resolution was $\sigma_E/E = 120\%/\sqrt{E}$. The barrel muon detector was composed of 2 layers. The first layer contained three staggered drift chamber planes, while the second layer had two staggered planes of drift chambers. The resolution of extrapolated tracks was $\sigma_{r\phi} = 4$ mm and $\sigma_{rz} = 2.5$ cm.

The endcaps consisted of the Forward Chambers A, the forward RICH systems, the Forward Chambers B, forward electromagnetic and hadronic calorimeters, a forward hodoscope and forward muon chambers. The Forward Chambers A were mounted on both ends of the TPC. Each side consisted of three chambers, each with two staggered layers of limited streamer tubes, which were turned with respect to each other by 120°, thus providing 2 × 3 coordinates. The Forward Chambers B consisted of 12 planes of drift chambers. Successive planes were rotated by 120°. The chambers covered the polar angle region $11° \leq \theta \leq 36°$ and $144° \leq \theta \leq 169°$. The x and y coordinates were measured with a precision of 150 μm, yielding resolutions of $\sigma_\theta = 3.5$ mrad and $\sigma_\phi = 4.0$ mrad/$\sin\theta$.

The forward RICH systems consisted of a liquid radiator and a gas radiator. Each arm used 24 time projection chambers. The liquid radiator was the same as that in the barrel RICH, while the gas RICH used C_4F_{10}. The Cherenkov angle resolutions were $\sigma = 2.5$ mrad in the gas and $\sigma = 11.4$ mrad in the liquid radiators. The Forward ElectroMagnetic Calorimeters, FEMC, consisted of two arrays of 4,532 lead glass blocks placed at $z = \pm 2.84$ m from the interaction region. They covered the polar angle regions of $8° < \theta < 35°$ and $145° < \theta < 172°$. The blocks were truncated pyramids with inner (outer) face dimensions of 5.0 cm × 5.0 cm (5.6 cm × 5.6 cm) and 40 cm depths, corresponding to 20 X_0. They were read out by single-stage PMTs designed to operate inside the DELPHI magnetic field. The energy resolution was $\sigma_E/E = 12\%/\sqrt{E} \oplus 3\% \oplus 11\%/E$. The endcap hadron calorimeter had a sampling depth of 19 layers of LSTs, using the same size iron plates as those in the barrel. It also exhibited a tower structure with a segmentation of $\Delta\theta \times \Delta\phi = 3.75° \times 2.62°$.

The forward scintillator hodoscopes improved the muon detection and trigger system efficiencies for beam events and cosmic rays. They were placed between the endcap yoke and the second muon chambers arranged into four quadrants each containing 28 counters. The extruded plastic scintillators, 1 cm thick, 20 cm wide and up to 450 cm long, were mounted with some overlap. They were read out with photomultiplier tubes. The time resolution was 5 ns. Both arms of the forward muon chambers had two planes of chambers, one inside the yoke behind > 85 cm of iron and the second 30 cm further out behind another 20 cm of iron and the forward hodoscopes. Each plane was composed of four quadrants. A quadrant consisted of two orthogonal layers of 22 drift chambers. By fitting muon trajectories the residual distribution over an average of 16 detector layers gave a spatial resolution of $\sigma = 3$ mm.

In the very forward region two additional detectors were used to measure the luminosity, the Small-angle Tile Calorimeter (STIC) and the Very Small Angle Tagger (VSAT). The STIC was a lead-scintillator sampling calorimeter formed by two cylindrical detectors placed on either side of the DELPHI interaction region at a distance of 2.2 m, covering a wider angular region between 29 and 185 mrad in polar angle. The blue light produced in the scintillator was read by wavelength-shifting fibers placed perpendicular to the scintillator planes. The geometry was quasi-projective. The total length of the detector was 27 X_0. Each STIC arm was divided into 10 rings and 16 sectors, giving a segmentation of $r \times \phi = 3$ cm $\times 22.5°$. The energy resolution was $\sigma_E/E = 13.5\%/\sqrt{E} \oplus 1.5\%$.

The VSAT in each arm was composed of two rectangular tungsten-silicon calorimeter stacks, 24 X_0 deep, 5 cm high, 3 cm wide and 10 cm long. The blocks were mounted at $z = \pm 7.7$ m to both horizontal sides of the elliptic beam pipe and were fixed to the support of the superconducting quadrupoles. They covered polar angles between 5 and 7 mrad and a $\pm 45°$ azimuth angle around the horizontal axis. Each stack consisted of 12 tungsten plates, interleaved with full-area silicon detectors (3 cm × 5 cm) for energy measurements. Each full-area detector was monitored. Two silicon planes with 32 vertical strips (1 mm pitch) were inserted behind 5 X_0 and 9 X_0 and one plane with 48 horizontal strips behind 7 X_0. The energy resolution was $\sigma_E/E = 35\%/\sqrt{E}$.

11.1.5 The SLD Detector

The SLD detector was the first experiment that was designed specifically for the operation at the SLAC Linear Collider (SLC) that operated at the CMS energy of the Z^0 boson. The SLC operated with polarized electrons, where the maximum polar-

Fig. 11.6 Schematic layout of one quarter of the SLD detector. Reprinted with kind permission from [16], © 1993, Elsevier. All rights reserved

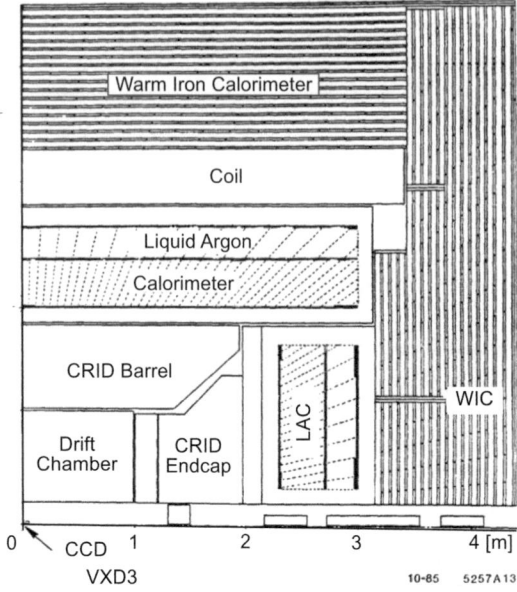

ization was 80% [14]. This allowed for the measurement of the left-right asymmetry, which the LEP experiments could not do [15]. It also compensated for the reduced luminosity of the SLC and made measurements competitive with LEP results.

Figure 11.6 shows a schematic layout of one quarter of SLD. The VXD3 vertex detector discussed in Sect. 6.3.1 surrounded the beam pipe. Next followed the central drift chamber, which had 80 layers arranged in ten superlayers of which six were stereo layers with a stereo angle of 41 mr. The drift chamber covered a radial space from 20 to 100 cm. The sense wires were gold-plated tungsten wires 25 μm in diameter, while the field-shaping wires were gold-plated aluminum wires 150 μm in diameter. The chamber was operated with a gas mixture of CO_2-Ar-isobutane (75:21:4) with a small addition of water. The drift velocity was $v_d^e = 7.9$ μm/ns. In the global drift region, the mean track resolution was 92 μm. The momentum resolution was $\sigma(p_T/p_T) = 0.005 p_T \oplus 0.01$. By combining the vertex detector with the drift chamber, position resolutions of $\sigma_{r\phi} = 10$ μm and $\sigma_{rz} = 36$ μm were achieved. Outside the drift chamber lied the Cherenkov Ring-Imaging Detector that we discussed in Sect. 9.3.3. Next, followed the lead liquid-argon calorimeter with two electromagnetic and two hadronic sections, all located inside the coil as discussed in Sect. 7.3.4.1. The electromagnetic section had a thickness of 22 X_0 measuring electromagnetic showers with a resolution of $15\%/\sqrt{E}$. The hadronic section has a thickness of 2 λ_I. Together with the electromagnetic section this increased to 2.86 λ_I. The hadron energy resolution was $60\%/\sqrt{E}$. The magnetic field of 0.6 T was produced by a conventional aluminum coil. The iron of the flux return was instrumented with limited streamer tubes (Sect. 4.10) serving as a tail catcher and muon system. It increased the depth by over 4 λ_I measuring hadronic energy with a resolution of $75\%/\sqrt{E}$.

11.1 Colliding-Beam Detectors

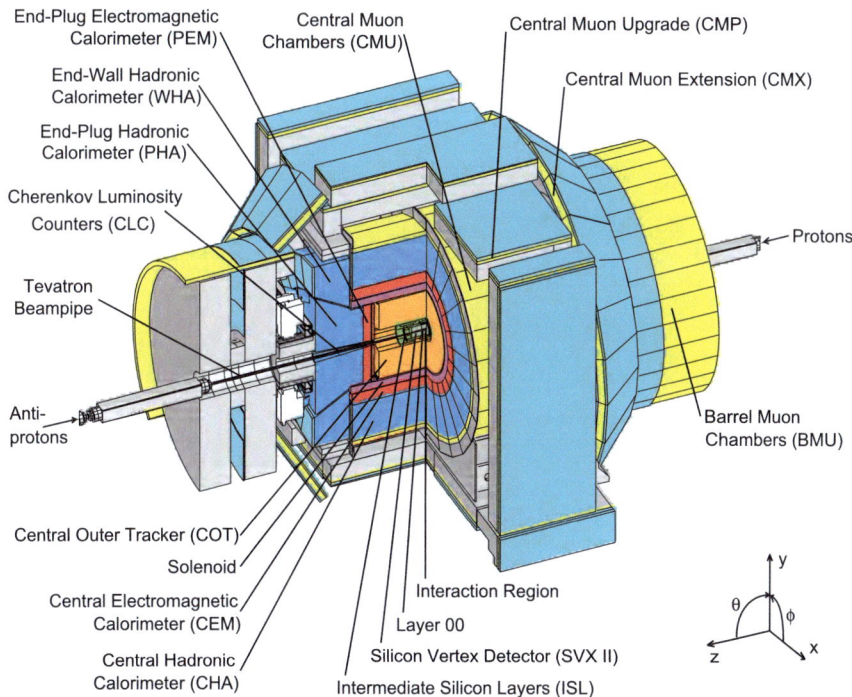

Fig. 11.7 Schematic view of the CDF II detector. Reprinted with kind permission from [17], © 2013, AIP Publishing. All rights reserved

11.1.6 The Collider Detector at Fermilab

The Collider Detector at Fermilab (CDF) discovered the top quark together with D0 and $B_s^0 \bar{B}_s^0$ oscillations [17]. Figure 11.7 shows a schematic view of the upgraded CDF II detector. Outside the beam pipe at a radius of 1.5 cm, 48 ladders of single-sided silicon strips with a readout pitch of 50 µm were mounted directly onto the beam pipe. The position resolution lied in the range of 6–7 µm. Next, came the SVXII double-sided silicon strip detector that covered the radial region from 2.5 cm to 10.6 cm. The vertex detector consisted of 360 ladders that were arranged in five radial layers, six axial sections and twelve 30° ϕ slices. The axial layers had a readout pitch of 60 to 65 µm depending on the layer. Three of the layers had a 90° stereo angle with readout pitches between 125 and 140 µm, while the other two layers had small-angle stereo strips with the same pitch as the corresponding axial layers. The third silicon strip detector (ISL) was placed at a maximum radius of 32 cm, consisting of one or two layers of double-sided silicon strips arranged in 296 ladders, which had a readout pitch between 110 and 146 µm. The three vertex detectors had a length of 1.0 m, providing a full coverage of the luminous region. The three sub-detectors shared the same readout system. The charge collection efficiency was above 99% and the single-hit efficiency was larger than 90%. The transverse impact parameter

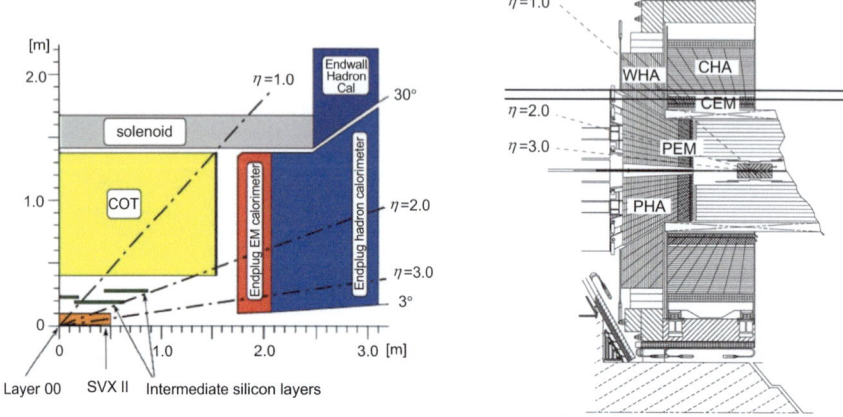

Fig. 11.8 Left: Layout of the tracking system in the upgraded CDF II detector, where COT stands for Central Outer Tracker. Reprinted with kind permission from [18], © 2004, Elsevier. All rights reserved. Right: Schematic layout of the CDF II detector focussing on calorimeters. The acronyms stand for Central EM (CEM), Central HAdronic (CHA), Wall HAdronic (WHA), Plug EM (PEM) and Plug HAdronic (PHA). Reprinted with kind permission from [19], © 2002, CDF II Collaboration. All rights reserved

resolution was 30 µm. The signal-to-noise ratio was larger than ten for both $r\phi$ and rz strips.

Figure 11.8 (left) shows the layout of the CDF tracking systems. The silicon trackers were followed by the Central Outer Tracker (COT), reaching from 43.4 cm to 310 cm in radius. In longitudinal direction, the chamber was 310 cm long. The COT had 30,240 sense wires in 96 layers grouped into eight superlayers. Half of them were axial layers and the other half were stereo layers with an angle of $\pm 2°$. The sense wires were 40 µm diameter gold-plated tungsten wires. Each superlayer was divided into supercells in the azimuth angle, where each supercell had 12 sense wires and a maximum drift distance that was about the same for all supercells. The supercell layout consisted of a wire plane containing sense and field wires and a field shield on either side. The field shields were shared with neighboring cells. The supercell was tilted by 35° with respect to the radial direction to compensate for the Lorentz angle of the drifting electrons in the magnetic field. The electric drift field was 1.9 kV/cm. The chamber ran with a gas mixture of Ar-ethane (50:50) with some small amount of isopropyl added. It provided a maximum drift time of 177 ns on a maximum drift distance of 0.88 cm. The COT had a single-hit resolution of $\sigma_{r\phi} = 140$ µm and an impact parameter resolution of $\sigma_{d_0} = 35$ µm at 2 GeV/c. The measurement term in the momentum resolution yielded $\sigma_{p_T}/p_T^2 = 0.15\%$ $(\text{GeV}/c)^{-1}$. The chamber provided dE/dx measurements that were used for particle identification. The π/K separation at momenta greater than 2 GeV/c was 1.4σ.

Outside the drift chamber resided the time-of-flight system (TOF), which consisted of plastic scintillating bars having a cross section of 4×4 cm^2 and a length of 279 cm. The scintillating bars had an attenuation length of 2.5 m. Since each bar covered 1.6° in ϕ, 216 bars were necessary to cover the entire azimuth-angle region. The scintillator bars were read out with photomultiplier tubes on both ends. The single hit resolution was 110 ps, providing π/K separation of $> 2\sigma$ for momenta smaller than 1.5 GeV/c. Combining the dE/dx measurements with the TOF measurements yielded π/K separation of better than 1.4 σ in the entire momentum range.

The trackers and the TOF system lied insides the superconducting solenoid that provided a magnetic field of 1.41 T. The magnet was surrounded by segmented electromagnetic and hadron calorimeters that were arranged in a quasi projective geometry as shown in Fig. 11.8 (right) [20]. The electromagnetic calorimeters consisted of lead-scintillator sampling calorimeters. In the central region ($|\eta| < 1.1$), the electromagnetic calorimeter had 31 layers of 3.2 mm thick lead plates and 5 mm thick plastic scintillator tiles [21], which were organized in projective towers, each subtending 0.1 units of pseudo-rapidity and 15° in azimuth angle. The tiles of each tower were read out with wavelength-shifting fibers spliced to same-diameter clear fibers, which took out the signal to a photomultiplier tube. Both sides of the tiles were read out with separate PMTs. The calorimeter was 19 X_0 thick, corresponding to 1 λ_I. The energy resolution was $\sigma_E/E = 13.5\%/\sqrt{E} \oplus 2\%$. The forward region was covered by the plug and wall calorimeters, covering the pseudo-rapidity range of $1.10 \leq |\eta| \leq 3.64$ [22]. The electromagnetic part consisted of 24 layers of 4.5 mm lead plates and projective 4.0 mm thick plastic scintillator tiles read out by wavelength-shifting fibers spliced to clear fibers, which were coupled to 16-channel multi-anode PMTs. The total thickness was 21 X_0, corresponding to 1 λ_I. There were 20 towers of 15° in the azimuth-angle direction and 20 towers in pseudo-rapidity. The energy resolution was $\sigma_E/E = 14.4\%/\sqrt{E} \oplus 0.7\%$.

The electromagnetic calorimeter was followed by a hadron calorimeter that was arranged into a central region, a plug region and a wall region. The central hadron calorimeter consisted of 2.5 cm thick iron plates and 10 mm thick scintillator tiles that were read out by wavelength-shifting fibers coupled to PMTs. The thickness was 4.5 λ_I. There were 48 modules with projective geometry, each covering 15° in azimuth angle and 0.1 units in pseudo-rapidity. The towers consisted of 32 layers. The scintillator tiles had average dimensions of 1 cm × 35 cm × 70 cm. The tiles of each tower were read out by wavelength-shifting fibers coupled to two PMTs. The central hadron calorimeter had an energy resolution of $\sigma_E/E = 50\%/\sqrt{E} \oplus 3\%$.

The hadron plug calorimeter was an iron-scintillator sampling calorimeter with 5 cm thick iron plates and 0.6 mm thick scintillator tiles arranged in projective geometry. Wavelength-shifting fibers embedded in each scintillator tile were spliced to clear fibers that took out the light to PMTs. The fibers of a tower were coupled to two PMTs. There were 22 layers in depth yielding 7 λ_I. The energy resolution was $\sigma_E/E = 68\%/\sqrt{E} \oplus 4\%$. The endwall hadron calorimeter had 15 layers of 5 cm thick steel plates and 1.0 cm thick plastic scintillator tiles that were arranged in towers with projective geometry. A tower covered 15° in the azimuth angle and 0.11 units in pseudo-rapidity. The calorimeter had 288 towers with lengths of 35–78 cm in ϕ and widths of 25–40 cm in η. The plastic scintillator tiles were read out with wavelength-shifting fibers. The fibers of each tower were coupled to two PMTs.

The endwall hadron calorimeter was 4.5 λ_I thick and had an energy resolution of $\sigma_E/E = 80\%/\sqrt{E} \oplus 5\%$.

Two muon systems were used in the CDF detector, one in the central region and another one in the endcaps [23]. The central muon system [24] lied outside the central hadron calorimeter at a radius of 3.47 m from the beam axis. In the azimuth angle, it was segmented into 12.6° wedges with a gap of 2.4° between wedges. Each wedge was further segmented into three 4.2° modules. Each module consisted of four layers of four rectangular drift cells. A drift cell was 6.35 cm wide, 2.68 cm high and 226.1 cm long. A stainless steel sense wire was placed in the center of the drift cell, which was read out at both ends. The chamber measured four points along a track with a precision of 250 μm per point. Charge division along the z direction gave an accuracy of 1.2 mm per point. The chambers had an efficiency of 100%.

In the forward and backward endcaps, the muon spectrometer consisted of magnetized steel toroids with drift chamber planes and trigger scintillation counters. In the intermediate region, there was partial coverage for isolated muons. Each spectrometer contained two 1 m thick steel toroids that had a diameter of 7.82 m. Four coils excited the toroids to a magnetic field ranging from 2.0 T at the inner radius to 1.6 T at the outer radius. Three layers of electrode-less drift chambers measured the muon trajectory with an accuracy of 5° in the ϕ direction and 200 μm in the radial direction. Two layers of scintillation counters provided trigger information. The angular region covered by each spectrometer lied between 3° and 16° from the beam line. The momentum resolution was 13%, independent of momentum, for muons with total momentum p above 8 GeV/c. Muons were matched to trajectories in the trackers.

11.1.7 The ATLAS Detector

The ATLAS experiment is one of the two multi-purpose experiments at the Large Hadron Collider (LHC) [25]. Figures 11.9 and 11.10 show a three-dimensional layout and a photograph of the ATLAS detector, respectively. Inside the solenoid that provides a 2 T magnetic field, three tracking detectors are located. In the barrel, the four-layer Pixel Detector is directly mounted onto the beam pipe. Next follows the four-layer SemiConductor Tracker (SCT) and the 73-layer transition radiation tracker (TRT). We have discussed these tracking detectors in the previous sections (Sects. 6.4, 6.2.2 and 9.4.1). In each endcap, we have three Pixel disks, nine SCT disks and 160 straw planes. Figure 11.11 (left) shows an enlarged view of the ATLAS Inner Detector. For the high-luminosity upgrade, the inner detector will be completely rebuilt using silicon pixel and silicon strip detectors.

Outside the superconducting coil, a hybrid calorimeter system is positioned [25]. Figure 11.11 (right) shows an enlarged view of the calorimeter positions. The 22 X_0 deep electromagnetic calorimeter in the barrel is a lead-liquid argon calorimeter with accordion-shaped kapton electrodes and lead absorber plates. The accordion geometry provides complete ϕ symmetry without azimuthal cracks. The calorimeter covers a pseudo-rapidity region, $|\eta| < 1.475$. We have discussed this detector in

11.1 Colliding-Beam Detectors

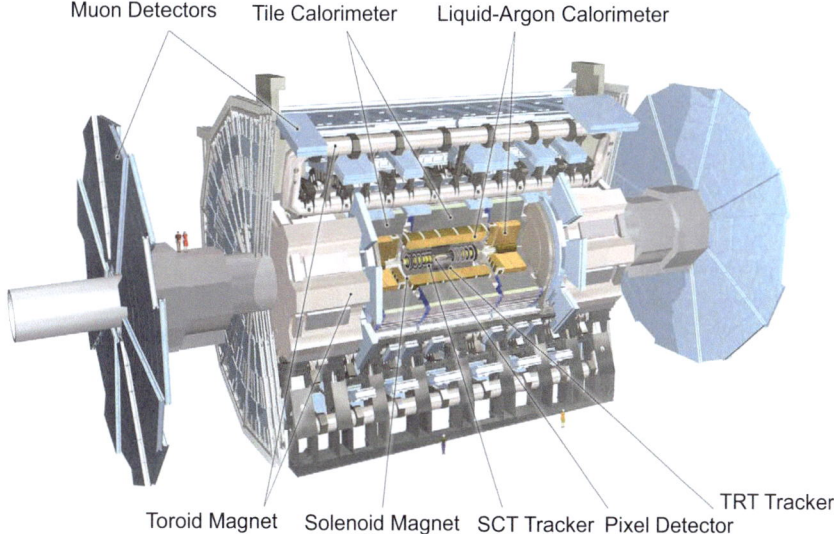

Fig. 11.9 Schematic view of the ATLAS detector. Reprinted with kind permission from [25], © 2008, IOP Publishing. All rights reserved

Fig. 11.10 Photograph of the ATLAS detector. Four of the eight barrel toroid coils are visible. Reprinted from [26], public domain

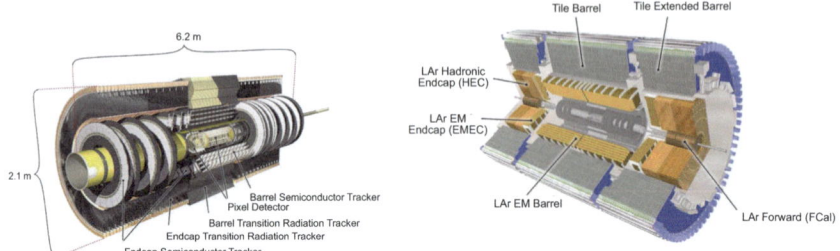

Fig. 11.11 Left: Schematic view of the ATLAS Inner detector. Right: Schematic view of the ATLAS calorimeter system. Both reprinted with kind permission from [25], © 2008, IOP Publishing. All rights reserved

Sect. 7.3.4.2. The 24 X_0 deep electromagnetic endcap calorimeters arranged in two coaxial wheels extend the pseudo-rapidity range to $1.375 < |\eta| < 3.2$. The forward calorimeter increases the η range to 4.9. The endcap calorimeters have two layers except for the region $1.5 < |\eta| < 2.5$, where three layers are used. In front of the endcap, there is also a one-layer pre-sampler that has the same segmentation as the one in the barrel. The segmentation in the endcap calorimeters increases from $\Delta\eta \times \Delta\phi = 0.05 \times 0.1$ to 0.1×0.1 (for details see Table 1.3 in [25]).

In the barrel and extended barrel region, the hadron calorimeter is the steel-scintillator tile sampling calorimeter (TileCal), which is segmented in depth into three layers and 64 azimuthal modules. Radially, the tile calorimeter extends from an inner radius of 2.28 m to an outer radius of 4.25 m. Together with the electromagnetic calorimeter it provides 9.7 λ_I. Two sides of the scintillating tiles are read out with wavelength-shifting fibers that are sent to separate photomultiplier tubes. In pseudo-rapidity, the towers are projective with a small offset. The pseudo-rapidity coverage of the barrel and extended barrel is $\eta < 1.7$. The segmentation is $\Delta\eta \times \Delta\phi = 0.1 \times 0.1$ except for the last layer, where it is increased to 0.2×0.1.

The Hadronic Endcap Calorimeters (HEC) are copper-liquid-argon sampling calorimeters. Each consists of two independent wheels, located directly behind the electromagnetic endcap calorimeters sharing the same LAr cryostats. Each wheel is segmented into 32 identical azimuthal wedge-shaped modules and two sections in depth. The wheels closest to the interaction point are built from 25 mm parallel copper plates, while those further away use 50 mm copper plates. The radius of the wheels extends from 0.475 to 2.03 m, except for the overlap region with the forward calorimeter, where the inner radius is reduced to 0.372 m. The copper plates are interleaved with 8.5 mm LAr gaps. The segmentation in the first disk is $\Delta\eta \times \Delta\phi = 0.1 \times 0.1$ increasing to 0.2×0.1 in the second disk. The HEC has small overlaps with the tile calorimeter and the forward calorimeter.

The forward calorimeter (FCal) is housed inside the endcaps and thus is integrated into the endcap cryostats, since this provides better uniformity and reduces radiation background levels in the muon spectrometer [27]. In order to reduce the amount of neutron albedo in the inner detector cavity, the front face of the FCal is recessed by about 1.2 m with respect to the EM calorimeter front face. The FCal detector is

11.1 Colliding-Beam Detectors

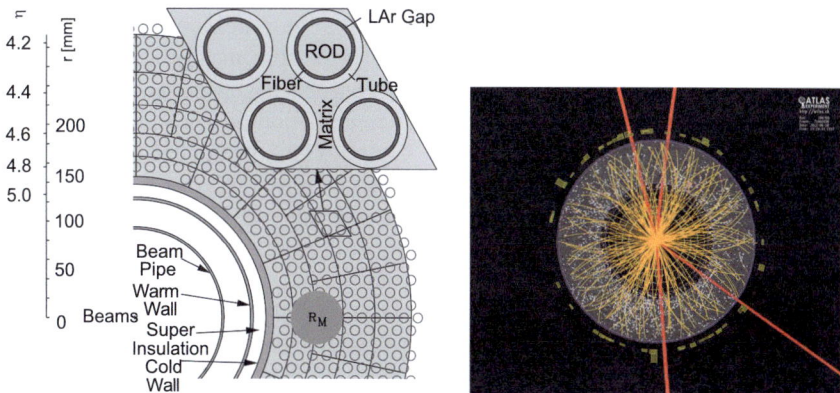

Fig. 11.12 Left: Transverse cross section of the ATLAS FCal detector. The inset shows an enlargement of the tube region. Reprinted with kind permission under CC-BY-4.0 Licence from [28], © 2009, ATLAS Collaboration. Right: ATLAS event display of a Higgs Decay to four muons. The invariant mass of the four muons is m = 124.1 GeV/c^2 without a Z^0 constraint, increasing to m = 125.1 GeV/c^2 with Z^0 mass constraints. Reprinted with kind permission from [29], © 2013, ATLAS Collaboration. All rights reserved

approximately 10 λ_I deep and consists of three sections. The first section, FCAL1, is the electromagnetic calorimeter part, which uses copper absorber plates that are optimized for electromagnetic measurements. The other two sections, FCAL2 and FCAL3, are hadron calorimeters that use tungsten absorber plates for the measurements of hadron energies. The three sections consist of ∼ 44.5 cm long tubes that have a rod in its center. The gap between the rod and the tube, which is filled with liquid argon, is maintained by a fiber that is helically wound around the rod. Figure 11.12 (left) shows a transverse cross section of the FCal detector. Gap sizes vary from 269 µm in FCal1 to 376 µm in FCal2 to 508 µm in FCal3. Such small gaps are necessary to provide fast signal collection and to reduce ion buildup that could alter the electric field across the gap. While the tube is grounded, the rod is set on high voltage, which varies from 250 V for FCal1 to 375 V for FCal2 to 500 V for FCal3. In the electromagnetic calorimeter section, the granularity in x-y in the first module layer is 3.0 cm × 2.6 cm.

Outside the calorimeters ATLAS uses an air-core toroid system that generates an average magnetic field of 0.5 T in the barrel and 1 T in the endcaps. The large barrel toroids cover the range $|\eta| < 1.4$, while small endcap toroids cover the range $1.4 < |\eta| < 2$. We discussed the ATLAS muon system in Sect. 9.6 (see Fig. 9.44 (left)). The toroid magnet system produces a bending power between 2.0 and 7.5 Tm in the barrel and endcaps, respectively. Four different types of detectors are used, two for triggering and two for precision measurements.

Over most of the η-range, tracks are measured by Monitored Drift Tubes (MDT) discussed in Sect. 4.11. The mechanical isolation in the drift tubes of each sense wire from its neighbors guarantees a robust and reliable operation. The MDTs provide 20 measurements per track with a position resolution of 35 µm in the z direction. At large pseudo-rapidities ($2.0 < |\eta| < 2.7$), Cathode Strip Chambers (CSC) are used, which provide higher granularity (see Sect. 4.13). This helps to deal with the demanding rate and background conditions. The CSCs provide four measurements per track with a resolution of 40 μm in r. The stringent requirements on the relative alignment of the muon chamber layers are met by the combination of precision mechanical-assembly techniques and optical-alignment systems both within and between muon chambers. The trigger system covers the pseudo-rapidity range $|\eta| < 2.4$. Resistive Plate Chambers (see Sect. 4.9) are used in the barrel and Thin Gap Chambers in the end-cap regions (see Sect. 4.14). The trigger chambers provide bunch-crossing identification, yield well-defined p_T thresholds and measure the muon coordinate in the direction orthogonal to that determined by the precision-tracking chambers. The RPCs provide six measurements per track with a time resolution of 1.5 ns. The TGCs provide nine measurements per track with a time resolution of 4 ns.

Three smaller detector systems cover the ATLAS forward region. The main function of the first two systems is to determine the luminosity delivered to ATLAS. At ±17 m from the interaction point lies LUCID (LUminosity measurement using Cherenkov Integrating Detector), which detects inelastic p-p scattering in the forward and backward directions and which is the main online relative-luminosity monitor for ATLAS. The second detector is ALFA (Absolute Luminosity For ATLAS), which is located at ±240 m from the interaction point and which consists of scintillating-fiber trackers that are rather close to the beam. The third system is the Zero-Degree Calorimeter (ZDC), which plays a key role in determining the centrality of heavy-ion collisions. It is located at ±140 m from the interaction region. The ZDC modules consist of layers of alternating quartz rods and tungsten plates, which measure neutral particles at pseudo-rapidities $|\eta| \geq 8.2$. One of the greatest achievements at the LHC is the discovery of the Higgs boson by ATLAS and CMS in 2012. The Higgs boson was observed in the four-lepton and two-photon final states. Figure 11.12 (right) shows an ATLAS event display of a Higgs boson decay to $ZZ^* \to \mu^+\mu^-\mu^+\mu^-$.

11.1.8 The CMS Detector

The other multipurpose detector at the LHC is the Compact Muon Solenoid detector (CMS), which is 21.6-m long and has a diameter of 14.6 m [30]. Figure 11.13 shows a schematic three-dimensional view of the CMS detector. Outside the beam pipe, the pixel detector is placed, which consists of three pixel layers in the barrel at radii of 4.4, 7.3 and 10.2 cm and two pixel disks in the endcaps at $|z| = \pm 34.5$ cm and $|z| = \pm 46.5$ cm, providing coverage in pseudo-rapidity of $|\eta| < 2.5$. The pixel cell size was 100 μm × 150 μm. The barrel region contained 40 million pixels, while the encaps had 18 million pixels. The pixels were n-on-n sensors, having a signal-to-noise ratio of typically above 20. For signals that shared two neighboring sensors, the hit resolution was 15–20 μm. This has been upgraded for Run 2 to four layers in the barrel and three disks in each endcap as discussed in Sect. 6.4.3.

The pixel detector is followed by a silicon strip detector with p-in-n type silicon microstrip sensors. Figure 11.14 (left) shows a schematic view of the inner detector. The tracker cylinder is 5.5 m long and 2.4 m in diameter. It consists of ten layers in the

11.1 Colliding-Beam Detectors

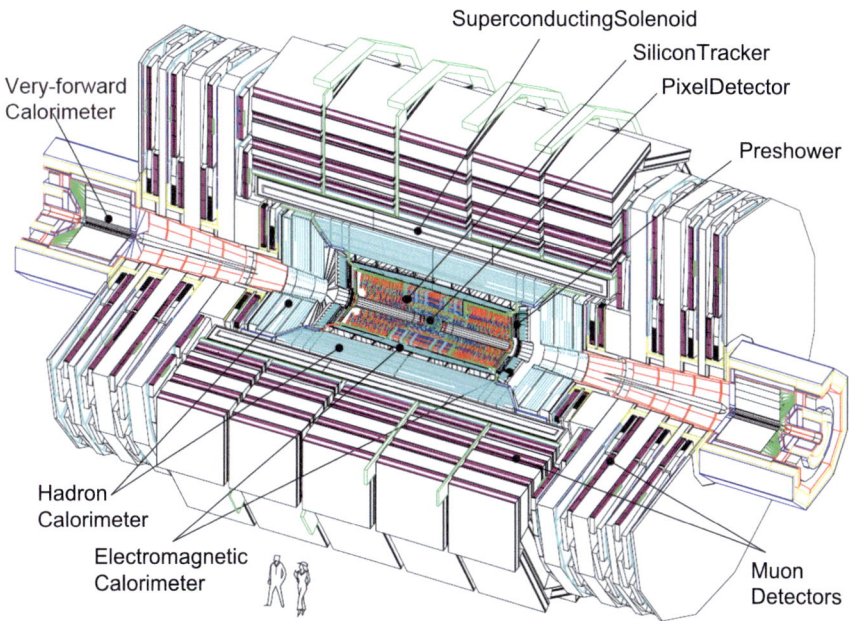

Fig. 11.13 Schematic three-dimensional view of the CMS detector. All reprinted with kind permission from [30], © 2008, IOP Publishing. All rights reserved

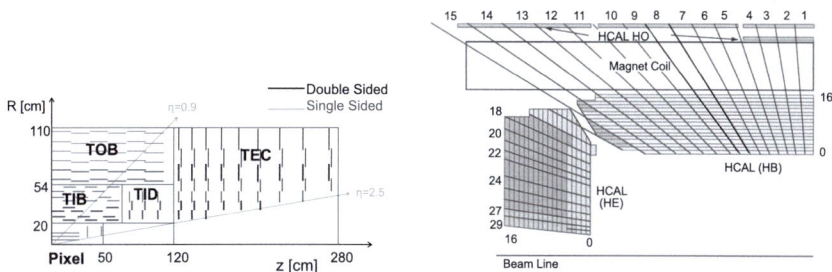

Fig. 11.14 Left: Enlarged view of the CMS silicon tracker. The acronyms stand for Tracker Inner Barrel (TIB), Tracker Outer Barrel (TOB), Tracker Inner Disks (TID) and Tracker End-Cap (TEC) disk. Reprinted under CC-BY-NC-SA-4.0 Licence from [32], © 2011, CMS Collaboration. Right: Tower structure of the CMS hadron calorimeters, showing the barrel (HB), endcap (HE) and part of the outer hadron calorimeter (HO). Reprinted with kind permission from [30], © 2008, IOP Publishing. All rights reserved

barrel and nine disks in each endcap, covering a pseudo-rapidity range of $|\eta| < 2.5$. The four Tracker Inner Barrel layers (TIB) are assembled in shells and three small Tracker Inner Disks (TID) that close the coverage in each endcap. The Tracker Outer Barrel structure (TOB) consists of six concentric layers, closing the tracker towards the calorimeter. The Tracker EndCap modules (TEC) are mounted in seven rings on nine disks. Each disk is built from sixteen 22.5° ϕ sectors. All layers use single-sided silicon strip sensors except for the first two layers in the TIB, the first two rings in the TID and layers 1, 2 and 5 in the TEC, where two single-sided sensors are mounted back-to-back. In the inner region, the sensors are 320 μm thick, while the other sensors are 500 μm thick. The strip pitch is 80 μm (120 μm) on layers 1, 2 (3, 4) in the TIB, leading to a single-point resolution of 23 μm (35 μm). In the TID, the mean pitch varies between 100 and 141 μm. In the TOB, the strip pitch is 183 μm on the first four layers and 122 μm on layers 5 and 6. In the TEC, the radial strip pitch increases from 97 to 184 μm. For non-isolated particles with $1 < p_T < 10$ GeV/c, the transverse momentum resolution is around 1.5% for pseudo-rapidities of $|\eta| < 3$. The transverse impact parameter resolution is $\sigma_{d_0} = 20$–75 μm for $|\eta| < 3.0$, while the longitudinal impact parameter resolution is $\sigma_{z_0} = 45$–150 μm [31]. For the high-luminosity upgrade, the inner detector will be completely rebuilt with radiation-hard sensors. The technology is not yet finalized. In addition, the calorimeter endcap will be rebuilt with a high-granularity silicon tungsten and scintillator tungsten as discussed in Sect. 7.5.1.7.

The 25-X_0-deep electromagnetic calorimeter is placed directly behind the tracker. It consists of 61,200 lead tungstate (PbWO4) crystals mounted in a central barrel and 7,324 crystals in two endcaps, covering a pseudo-rapidity range of $|\eta| < 3.0$. The scintillation light of the $PbWO_4$ is detected by avalanche photodiodes in the barrel region and vacuum phototriodes in the endcaps. A pre-shower system is installed in front of the endcap ECAL for π^0 rejection. The ECAL was discussed in the calorimeter Sect. 7.2.2.3. It is surrounded by a brass/scintillator hadron sampling calorimeter (HCAL) with coverage up to $|\eta| < 3.0$ that is split into two half barrels housing about 70,000 tiles. Each half barrel is built from 18 wedges in the azimuth angle, where each wedge is segmented into four ϕ sectors. In pseudo-rapidity the scintillators are segmented into 16 sectors yielding a segmentation of $\Delta\phi \times \Delta\eta = 0.087 \times 0.087$. In the radial direction there are 17 layers. Figure 11.14 (right) shows the tower structure of the hadron calorimeters. The absorber consists of a 40-mm-thick front steel plate, followed by eight 50.5-mm-thick brass plates, six 56.5-mm-thick brass plates and a 75-mm-thick steel back plate. The scintillator tiles are 3.7 mm thick except for layers zero and 16 where they are 9 mm thick. Wrapped in Tyvek paper with edges painted white, they are arranged in trays that are inserted into the gaps between the absorber plates. The scintillation light from each tile is collected with 0.94-mm-diameter green double-clad wavelength-shifting Y11 fibers that are placed in machined grooves in the scintillator. After exiting the scintillator, the Y11 fibers are spliced to clear double-clad fibers that take the light out to optical connectors at the end of the tray. An optical cable takes the light to an optical decoding unit, which arranges the fibers into readout towers and brings the light to hybrid photodiodes (HPD), which consist of a photocathode held at a potential of 8 kV at a distance of ~ 3.3 mm from a pixelated silicon photodiode. The HPD, which has a gain of approximately 2,000, is segmented into 19 hexagonal 20 mm × 20 mm pixels. Furthermore, for calibration purposes, each tray has 1-mm-diameter stainless steel tubes, called source tubes, which carry ^{137}Cs radioactive sources through the center of each tile. An additional quartz fiber is used to inject UV laser light into the layer 9 tiles. At perpendicular incidence, the total absorber thickness is 5.82 λ_I increasing with polar angle θ as $1/\sin\theta$, yielding 10.6 λ_I at $|\eta| = 1.3$. Note that the electromagnetic calorimeter in front adds another 1.1 λ_I.

The hadron endcap calorimeters cover the pseudo-rapidity region $1.3 < |\eta| < 3.0$ with 19 active layers having a depth of 10 λ_I. The absorber consists of 79 mm thick brass plates spaced by a 9 mm gap. The scintillator tiles are 3.7 mm thick except for layer zero (9 mm) and are wrapped in Tyvek paper. They are mounted in trays that are inserted into the gaps. There are 72 cells in ϕ and 14 cells in r. The granularity of the calorimeters is $\Delta\eta \times \Delta\phi = 0.087 \times 0.087$ for $|\eta| < 1.6$ and $\Delta\eta \times \Delta\phi = 0.17 \times 0.17$ for $|\eta| > 1.6$. Each endcap houses 10,458 tiles that are arranged into 684 trays. The light collection of the tiles is achieved in a similar manner as that in the barrel. The energy resolution is $\sigma_E/E = 105\%/\sqrt{E} \oplus 4\%$.

The barrel calorimeter is complemented by a tail-catcher in the barrel region (HO) ensuring that hadronic showers are sampled with nearly 11 λ_I. The tail catcher is located outside the solenoid. To return the magnetic field, an iron yoke is designed in the form of five 2.536 m wide rings. The hadron tail catcher is placed as the first sensitive layer in each of the five rings, each of which has 12 identical ϕ sectors separated by 75-mm-thick stainless steel beams. Sizes and positions of the tiles in the tail catcher are supposed to map the layers of the barrel hadron calorimeter approximately, in order to produce towers of granularity $\Delta\eta \times \Delta\phi = 0.087 \times 0.087$. Coverage up to a pseudo-rapidity of 5.0 is provided by an iron/quartz-fiber calorimeter, which consists of 5 mm thick steel absorber structures with grooves into which the cladded quartz fibers are inserted. The calorimeter has a two-fold longitudinal segmentation and is built from 18 wedges in the ϕ and 13 towers in $|\eta|$ direction on either side of the interaction point. The Cherenkov light emitted in the quartz fibers is detected by photomultipliers.

The tracker and the barrel calorimeters lie inside a 13-m-long and 6-m-inner-diameter, 4-T superconducting solenoid, providing a large bending power of 12 Tm before the muon system shown schematically in Fig. 11.15. The return field is sufficiently large to saturate 1.5 m iron. This is sufficiently deep to integrate four muon stations arranged in concentric cylinders. In the barrel region, where the neutron-induced background is small, the muon rate is low. The 4-T magnetic field is uniform and mostly contained in the steel yoke. Both the barrel and each endcap have four measurement stations, consisting of drift tubes in the barrel and cathode strip chambers in the endcaps and five to six trigger planes consisting of resistive plate chambers. We discussed all these detectors in Chap. 4. The muon system has a pseudo-rapidity coverage of up to $|\eta| = 2.4$.

The barrel drift tube (DT) chambers reduce the neutron-induced noise. They cover the pseudo-rapidity region $|\eta| < 1.2$ and are organized into four stations interspersed

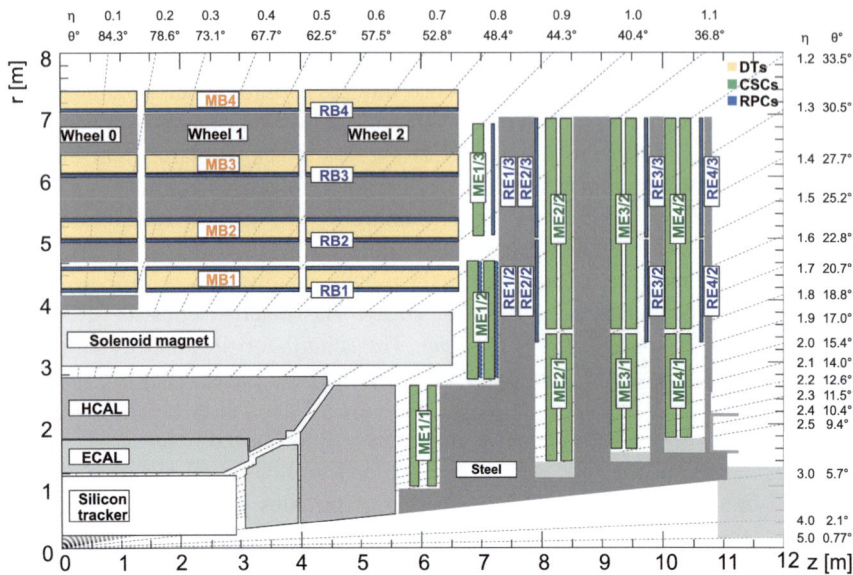

Fig. 11.15 Schematic layout of the CMS muon system. The drift tubes (orange) are arranged in four layers in the barrel (MB1-MB4) with four RPC layers (blue) in front (RB1-RB4). In the endcap there are four layers of cathode strip chambers (green) (ME1/1,2,3 -ME4/1,2) plus four layers of RPCs (RE1/2,3 -RE4/2,3). Reprinted under CC-BY-4.0 Licence from [33], © 2021, ATLAS Collaboration

among the layers of the flux return plates, operating 172,000 sense wires. A drift tube chamber consists of three superlayers that each contains four layers of rectangular drift cells in which consecutive layers are staggered by half a cell. In superlayers 1 and 3, the tubes with 50 μm diameter anode wires in the center run parallel to the beam direction and thus measure $r\phi$ coordinates, while superlayer 2 has tubes and wires running perpendicular to the beam line measuring rz coordinates. The fourth station has no rz superlayer and thus just measures $r\phi$ positions. These chambers, which operate with an Ar-CO_2 (85:15) gas mixture, are the self-triggering and have bunch crossing identification capabilities. The bunch and trigger identification electronics combines the drift time measured in three consecutive layers, applying a mean timer technique. The excellent linearity of the time-to-distance relationship of the cell allows for a 4 ns time resolution. Note that the maximum drift time is 380 ns. The multilayer structure of the chamber improves the spatial resolution of the single cell, reaching $\sigma_{r\phi} \sim 100$ μm.

In the endcap region, where the background originating from thermal neutrons and halo muons is expected to be quite large and the magnetic field is strong and highly inhomogeneous, drift tubes are not suited. Therefore, CMS uses 468 CSCs arranged in four stations in each endcap. The chambers operate with an Ar-CO_2-CF_4 (30:50:20) gas mixture without any hydrogen to reduce neutron-induced noise. The cathode strips of each chamber run radially outwards and provide a precision measurement in the $r\phi$ bending plane. The anode wires run approximately perpendicular to the strips and are also read out in order to provide measurements of η and the beam-crossing time of a muon. The spatial resolution is $\sigma_{r\phi} \sim 100$ to 240 μm for strip measurements, depending on the chamber, while that for wire measurements is $\sigma_{r\phi} \sim 5$ mm. The chambers have a fast response due to the small wire spacing of 3.4 mm.

The six (five) RPCs in the barrel (endcaps) are arranged as shown in Fig. 11.15. They have a double gap read out in the center by a single set of strips, use a $C_2H_2F_4$-iC_4H_{10}-SF_6 (95.2:4.5:0.3) gas mixture and operate at a high voltage of 9.5 kV. The excellent time resolution of 1 ns provides an unambiguous bunch crossing assignment. The RPCs are also used to identify and measure muon tracks at the Level 1 trigger. The spatial resolution in the bending coordinate is determined by the strip pitch and is of the order of $\sigma_{r\phi} \sim 10$ mm.

Due to multiple-scattering in the detector material before the first muon station, the offline muon momentum resolution of the standalone muon system is about 9% for small values of pseudo-rapidity and transverse momenta up to 200 GeV/c. At 1 TeV, the standalone momentum resolution varies between 15% and 40%, depending on η. In the endcap region, there is a plane of RPCs in each of the first 3 stations in order for the trigger to use the coincidences between stations to reduce background, to improve the time resolution for bunch crossing identification and to achieve a good p_T resolution. The RPCs provide a fast, independent, and highly-segmented trigger with a sharp p_T threshold over a large portion of $|\eta| < 1.6$ of the muon system.

The CMS luminosity is measured by using the signals from the forward hadron calorimeter and the Pixel Luminosity Telescope. There are three very forward detectors, two dedicated calorimeters CASTOR and Zero Degree Calorimeter and the tracking detector TOTEM. The Cherenkov-radiation-based CASTOR calorimeter is placed 14.38 m from the interaction region. It consists of tungsten absorber plates and fused silica quartz plates. For the electromagnetic section, the thickness of the tungsten plates and quartz plates is 5.0 and 2.0 mm, respectively. For the hadronic section, the tungsten and quartz plates have thicknesses of 10.0 and 4.0 mm, respectively. The electromagnetic section is 20.1 X_0 deep, while the total calorimeter is 10 λ_I deep. The Zero Degree Calorimeter is located 140 m from the interaction region, consisting of an electromagnetic and hadronic part using tungsten absorbers and quartz fibers. The electromagnetic section is built from 33 layers of 2 mm thick tungsten plates and 33 layers of 0.7 mm diameter quartz fibers, while the hadron section has 24 layers of 15.5 mm thick tungsten plates and 24 layers of 0.7 mm diameter quartz fibers. The total calorimeter is 7.5 λ_I deep.

As mentioned before, the Higgs boson was discovered by ATLAS and CMS in two decay modes. Therefore, we would like to show the status at the end of Run 2. Figure 11.16 (left) shows the CMS two-photon spectrum. At a two-photon mass of 125 GeV/c^2, a bump is seen on top of a large background. After background subtraction, a prominent signal arises. Figure 11.16 (right) shows the ATLAS four-lepton invariant-mass spectrum. Again a prominent peak is seen at 125 GeV/c^2.

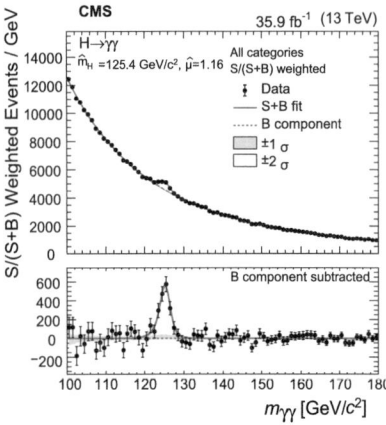
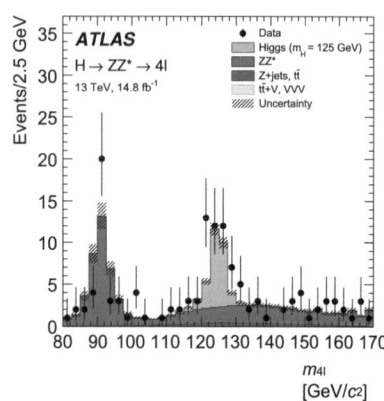

Fig. 11.16 Left: The two-photon invariant-mass spectrum measured by the CMS Collaboration. The top figure shows the observed spectrum (points with error bars) with a signal plus background fit overlaid (solid line). The background is shown by the dotted line. The bottom plot shows the spectrum after background subtraction, revealing a prominent Higgs signal. Reprinted under CC-BY-NC-ND-4.0 Licence from [34], © 2018, CMS Collaboration. Right: The four-lepton invariant-mass spectrum measured by the ATLAS Collaboration for data (points with error bars), signal (grey histogram), ZZ^* events (dark grey histogram), Z plus jets and $t\bar{t}$ events (dark uniform histogram under ZZ^* background) and $t\bar{t}$ plus one vector boson or three vector bosons (light grey histogram). Reprinted under CC-BY-NC-ND-4.0 Licence from [35], © 2016, ATLAS Collaboration

11.1.9 The LHCb Detector

The LHCb detector is a single-arm spectrometer and thus looks more like a fixed-target detector [36]. Figure 11.17 shows a schematic view of the upgraded LHCb experiment. Surrounding the interaction region, the Vertex Locator is positioned (VELO), followed by RICH 1 and the Upstream Tracker (UT). All three detectors are placed before the dipole magnet, which provides a bending power of 4 Tm. After the magnet, the Scintillating Fiber-Tracker (SciFi) and RICH 2 are placed, which are followed by a neutron shield, the electromagnetic calorimeter, the hadron calorimeter and the muon system.

Figure 11.18 (top) shows a schematic layout of the upgraded VELO, which is composed of two halves [37, 38]. During beam injection, the two halves are separated as depicted in Fig. 11.18 (bottom right), while during stable beam conditions they are returned to their nominal operating positions, 5.1 mm from the beam line as shown in Fig. 11.18 (bottom left). Each half consists of a bank of 26 modules, oriented perpendicular to the beam axis. Four silicon pixel sensors are mounted on each module. To provide overlapping acceptance two sensors are mounted on the front and two are placed on the back. Each sensor (43 mm × 15 mm) is bump-bonded to three ASICs. The 200 μm thick silicon pixels have sizes of 55 μm × 55 μm. The sensors are laid out to tolerate a fluence of 8×10^{15} (1 MeV) n_{eq} cm^{-2}. After full irradiation, sensors are operated with a bias voltage of 1 kV, with a hit efficiency exceeding 99%. The sensors and ASICs are cooled at $-20\,^\circ$C using a novel scheme of micro-channel cooling [39].

11.1 Colliding-Beam Detectors

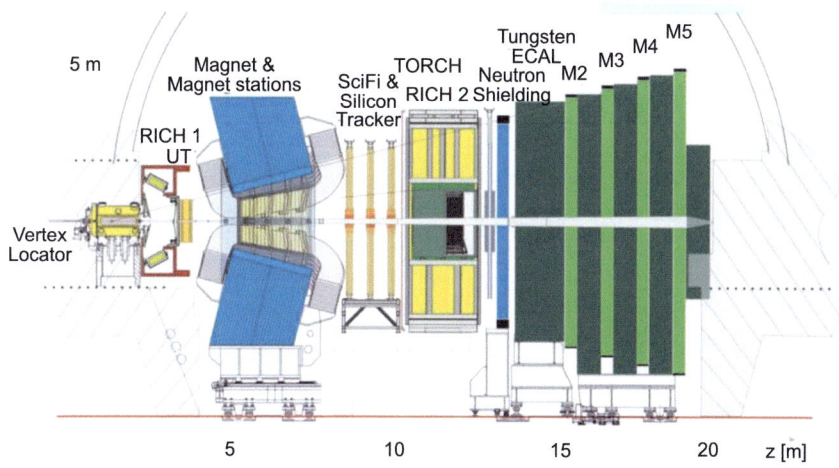

Fig. 11.17 Schematic view of the upgraded LHCb detector. The muon station M1 has been removed for run 3 leaving four muon stations (M2-M5). Reprinted with kind permission from [40], © 2023, Elsevier. All rights reserved

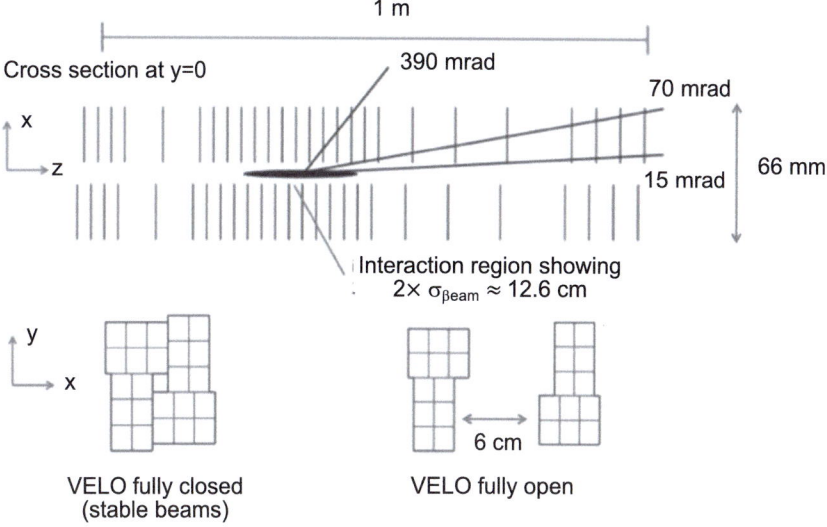

Fig. 11.18 Top: Schematic layout of the upgraded LHCb Vertex Locator. The grey ellipse shows the interaction region. Bottom left: Fully closed VELO. Bottom right: Fully open VELO. All reprinted under CC-BY-3.0 Licence from [37], © 2013, ATLAS Collaboration

The RICH 1 operates with C_4F_{10} gas, providing particle identification up to 40 GeV/c [41]. The Cherenkov light is reflected by a set of spherical mirrors and a set of planar mirrors onto a plane of 64-pixel 1-inch multi-anode PMTs. The Upstream Tracker consists of four planar detection layers made of silicon microstrip detectors [42]. The first and last layer have vertical readout strips, while the two intermediate layers have readout strips rotated by stereo angles of $+5°$ and $-5°$, respectively. Microstrip sensors around the beam pipe have 5 cm long strips with a 95 μm pitch followed by a ring of 10 cm long strips with the same pitch. The sensors use the n-in-p technology for improved radiation hardness. The remaining sensors are 10 cm long strips with a pitch of 190 μm using the p-in-n technology. The sensors are read out by a custom-made radiation-hard ASIC. The detector is operated at a temperature of 5 °C.

The Scintillating-Fiber Tracker, we discussed in Sect. 8.8, sits directly behind the dipole magnet followed by RICH 2, which uses CF_4 gas as radiator providing particle identification up to 100 GeV/c. The Cherenkov light is reflected by two sets of planar mirrors onto a plane of 64-pixel 2-inch multi-anode PMTs. In front of the RICH, the Time-of-Flight Identification with Cherenkov Radiation detector (TORCH) will be installed, which will measure the Cherenkov angle and time-of-flight [43] of charged particles. It consists of a plane of 10 mm thick quartz strips.

The Time-of-flight difference of 35 ps of a charged particle and the time of-propagation difference of 49 ps of photons yield a total time difference of $\sigma_t = 84$ ps. Typically, about 30 Cherenkov photons with an average time resolution of 70 ps are detected, yielding a time resolution of $\sigma_t < 15$ ps [44]. The neutron shield is 12.5 m from the interaction point and covers an area of 3.5 m × 4.0 m. It is followed by a pre-shower detector that consists of a 14 mm thick lead absorber plate sandwiched between two 15 mm thick scintillator planes. The scintillator is read out with wavelength-shifting fibers coupled to 64 multi-anode PMTs.

Figure 11.19 shows a schematic layout of the LHCb calorimeter system. The electromagnetic calorimeter (ECAL) is a scintillator-lead sampling calorimeter. The outer dimensions match projectively those of the tracking system, $\theta_x < 300$ mrad and $\theta_y < 250$ mrad. The inner acceptance is mainly limited by $\theta_{x,y} > 25$ mrad around the beam pipe due to the substantial radiation dose level. The hit density is a steep function of the distance from the beam pipe and varies over the active calorimeter surface by two orders of magnitude. Therefore, the calorimeter is subdivided into inner, middle and outer sections with appropriate cell sizes. The inner section holds 176 modules with nine cells each. Figure 11.20 (left) shows the layout of an inner ECAL module, which is built from alternating layers of 2 mm thick lead tiles and 4 mm thick scintillator tiles, wrapped in 120 μm thick TYVEK paper. The cell size varies from 4.04 cm × 4.04 cm in the inner section to 6.06 cm × 6.06 cm in the middle section and 12.12 cm × 12.12 cm in the outer section. The number of modules increases to 448 with four cells each in the middle section and 2,688 with one cell each in the outer section. In depth, the 66 Pb and 66 scintillator layers form a 42 cm deep stack corresponding to 25 X_0. The Molière radius of the stack is 3.5 cm. Light from the scintillator tiles is absorbed, reemitted and transported by 1.2 mm thick wavelengths-shifting fibers, which traverse the entire module. In order

11.1 Colliding-Beam Detectors

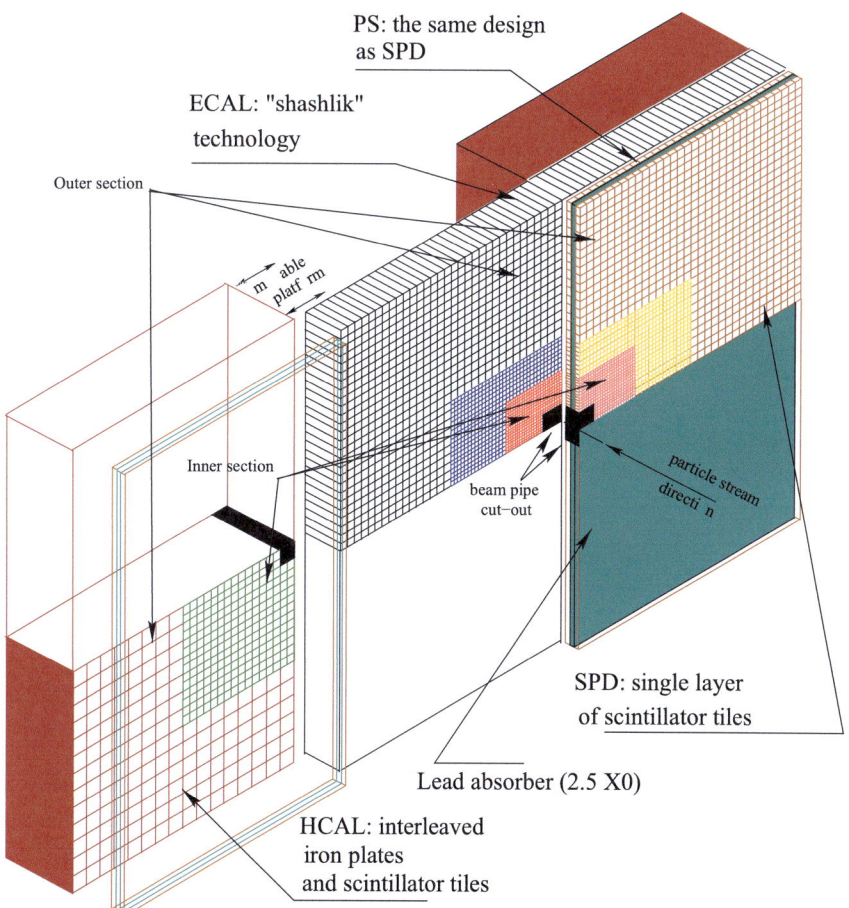

Fig. 11.19 Schematic layout of the LHCb calorimeter system consisting of a scintillator pad detector (SPD, green layer), a preshower detector (PS, red array), an electromagnetic calorimeter (ECAL, black array) and a hadron calorimeter (HCAL, brown stack) arranged in two movable halves. Note that the left HCAL half is moved out. Reprinted under CC-BY-4.0 Licence from [46], © 2020, LHCb Collaboration

to improve the light collection efficiency, the fibers are looped such that each traverses the module twice. The light is read out with PMTs. The electromagnetic calorimeter has an energy resolution of $\sigma_E/E = (145 \pm 13)$ MeV$/E \oplus (9.4 \pm 0.2)\%/\sqrt{E} \oplus (0.83 \pm 0.02)\%$.

The LHCb hadron calorimeter is a 5.6-λ_I-deep steel-scintillator sampling calorimeter. The overall HCAL structure is built as a wall, positioned at a distance of 13.33 m from the interaction region with dimensions of 6.8 m in height, 8.4 m in width and 1.65 m in depth. Together with the ECAL and the presampler, the HCAL has a depth of 6.8 λ_I. The calorimeter is segmented transversely into square cells with a size 13.1 cm in the inner section increasing to 26.2 cm in the outer section. The optical system is designed such that the two different cell sizes can be realized with

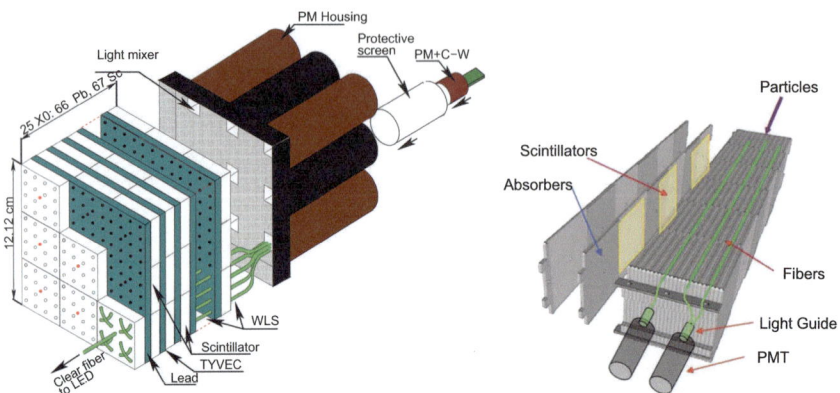

Fig. 11.20 Left: Schematic layout of an LHCb inner ECAL module, showing the scintillator tiles (white squares), lead plates (green layers), wavelength-shifting fibers (green lines) and PMTs (brown cylinders). Note that the fibers run through the tiles and absorber plates. The four fibers that read out a tower are sent to the same PMT. Right: The LHCb cell structure showing the arrangement of absorber plates, scintillator tiles, the fiber routing and connection to the PMTs. The particle enters from the front (purple arrow). Both reprinted under CC-BY-4.0 Licence from [46], © 2020, LHCb Collaboration

an absorber structure that is identical over the whole HCAL. The absorber structure is made from laminated steel plates of 20 mm thickness that are glued together. The 3 mm thick scintillator tiles consist of 1.75% PTP and 0.05% POPOP dissolved in polystyrene.

Like in the ATLAS TileCal, the scintillating tiles are oriented parallel to the beam axis. In the lateral direction they are interspersed with 1 cm of iron plates, whereas in the longitudinal direction the length of tiles and iron spacers corresponds to a hadron interaction length in steel. The light is collected by wavelength-shifting fibers running along the detector towards the back side, where the PMTs are housed. Figure 11.20 (right) shows a schematic view of the hadron cell structure. There are three scintillator tiles arranged in depth that are in optical contact with 1.2 mm thick wavelength-shifting fibers, which run along the tile edges. The WLS fibers are coupled to a small square light mixer just in front of the PMT entrance window. The tile edges are wrapped in such a way that the fiber can be easily inserted between the envelope and the tile edge during the module optics assembly. This reflective TYVEK envelope avoids light crosstalk between adjacent edges of the small tiles, enhances the light collection in the wavelength-shifting fiber and protects the optical reflective surface of the tiles. Each fiber collects light from three scintillator tiles. Readout cells of different sizes are defined by grouping together different sets of fibers onto one PMT that is fixed on the rear side of the sampling structure. The HCAL uses 50,000 fibers, which all have the same length of 1.6 m. The fiber end opposite to the PMT is coated with a layer of reflective aluminum. The hadron calorimeter performs well. The measured energy resolution is $\sigma_E/E = (67 \pm 5)\%/\sqrt{E} \oplus (9 \pm 2)\%$.

The muon system is composed of five muon stations placed behind the HCAL (M1-M5), which are interleaved with 80 cm thick iron absorber plates. Each muon

11.1 Colliding-Beam Detectors

station is divided into four regions, R1 to R4 with increasing distance from the beam axis. Note that for Run 3 the M1 station has been removed and is not discussed further. The linear dimensions of the regions R1, R2, R3, R4 and their segmentations scale in the ratio 1:2:4:8. With this geometry, the particle flux and channel occupancy are expected to be approximately the same for the four regions. Each station consists of 276 MWPCs, which are composed of four gas gaps arranged in two sensitive layers with independent readout. In M2 (M3) in region R1, the pad size is $\Delta x \times \Delta y = 0.63 \text{ cm} \times 3.1 \text{ cm}$ ($0.67 \text{ cm} \times 3.4 \text{ cm}$), increasing in each direction by factors of two, four and eight going to regions R2, R3, R4, respectively. For the chambers in M4 (M5), the pad sizes in R1 are $\Delta x \times \Delta y = 2.9 \text{ cm} \times 3.6 \text{ cm}$ ($3.1 \text{ cm} \times 3.9 \text{ cm}$), again increasing by factors of two, four and eight going to regions R2, R3, R4. All the chambers are segmented into anode wire pads or cathode pads. The MWPCs have four gaps, each 5 mm thick. The anode wires are gold-plated tungsten wires 30 μm in diameter with an anode-cathode spacing of 2.5 mm. The wire pitch is 2.0 mm and the length is 25–31 cm. The MWPCs hold 3×10^6 wires, which are tensioned to 0.7 N. They operate with an Ar-CO_2-CF_4 (40:55:5) gas mixture at a high voltage of 2.5–2.8 kV, yielding a typical gain of 10^5. They have high efficiency (>99%). The position resolution in M3, for example, is $\sigma_x \times \sigma_y = 10 \text{ mm} \times 12 \text{ mm}$ in R1, increasing to 40 mm × 96 mm in R4. The much larger M5 stations have position resolutions of 33 mm × 40 mm in R1, increasing to 150 mm × 180 mm in R4. The time resolution is 3.5 ns.

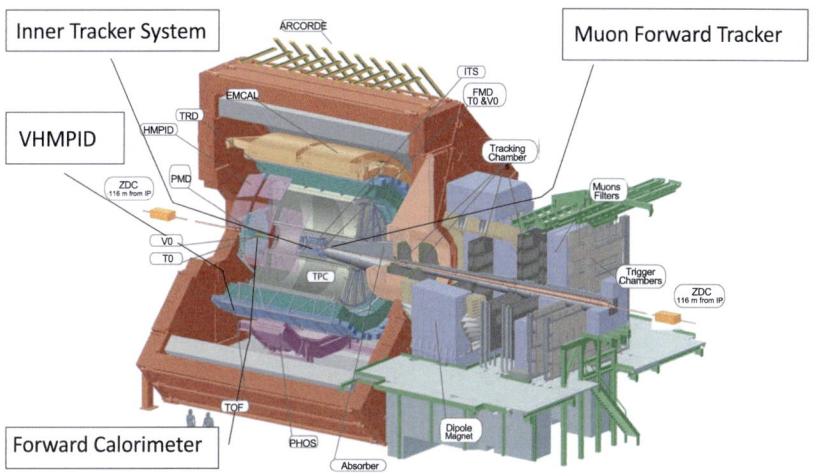

Fig. 11.21 Schematic three-dimensional view of the ALICE detector. Reprinted with kind permission from [47], © 2008, IOP Publishing. All rights reserved

11.1.10 The ALICE Detector

The ALICE experiment is fourth experiment at the LHC and is dedicated to heavy-ion physics [47]. Figure 11.21 shows a schematic three-dimensional view of the ALICE detector, which consists of a central barrel part and a forward muon spectrometer. The central barrel detectors are installed inside a large solenoid magnet (0.5 T)[1] and cover polar angles from 45° to 135°. The Inner Tracking System (ITS) includes six layers of silicon pixel (SPD), silicon drift (SDD) and silicon strip detectors (SSD), which provide vertex reconstruction, tracking and particle identification by dE/dx. The large cylindrical Time Projection Chamber (TPC) is the core tracking detector. The Time-of-Flight (TOF) identifies hadrons at intermediate transverse momenta. Arrays of Ring Imaging Cherenkov detectors (HMPID) extend the hadron identification to high p_T. Electron identification is achieved by the Transition Radiation Detector (TRD). Two electromagnetic calorimeters (PHOS and EMCal) are located inside the solenoid. All detectors except for HMPID, PHOS and EMCal have full azimuth-angle coverage. The forward muon arm consists of several passive absorbers, a large dipole magnet (3 Tm bending), ten planes of muon tracking stations and four planes of triggering chambers. The forward detectors located at small angles include the Photon Multiplicity Detector (PMD), Forward Multiplicity Detector (FMD), Zero Degree Calorimeter (ZDC), V0 detector and T0 detector. These are used for global event characterization and triggering. An array of scintillators on top of the solenoid magnet (ACORDE) is used to trigger on cosmic rays.

The six-layer silicon vertex detector uses silicon pixel and silicon drift detectors in the innermost four layers due to the high particle density. The outer layers are equipped with double-sided silicon micro-strip detectors. The four outer layers have analog readout for independent particle identification via dE/dx in the non-relativistic region, which provides the inner tracking system with stand-alone capability as a low-p_T particle spectrometer. The TPC is the main tracking device, which we discussed in Sect. 4.5.2. For particle identification, ALICE uses a TOF array that is optimized for large acceptance and average momenta, covering the central barrel over an area of 140 m^2 with 160,000 individual cells at a radius of close to 4 m. The HMPID is a single-arm, 10 m^2 array of proximity focusing ring-imaging Cherenkov counters with liquid radiator and solid CsI photocathode evaporated on the segmented cathode of MWPCs. It extends the hadron identification capabilities towards higher momenta in about 10% of the barrel acceptance. The TRD identifies electrons above 1 GeV/c to study production rates of quarkonia and heavy quarks near mid-rapidity. It consists of six layers of Xe-CO$_2$-filled time expansion wire chambers following a composite foam and fiber radiator. In central collisions, a hadron rejection of order 100 is needed to reduce the background from misidentified hadrons below the level of real electrons.

ALICE uses two electromagnetic calorimeters. A small single arm, high-resolution and high-granularity crystal calorimeter is placed at the bottom of the detector, cover-

---

[1] The magnet came from the L3 experiment at LEP.

ing the azimuth angle $220° < \phi < 320°$ and pseudo-rapidity $|\eta| < 0.12$. It consists of three modules with 3,584 detection cells, each arranged in 56 rows of 64 cells. A detection cell consists of 22 mm × 22 mm × 180 mm lead-tungstate crystals, coupled to a 5 mm × 5 mm avalanche photodiode followed by a low-noise preamplifier. The crystal length corresponds to 20 X_0. A set of multiwire chambers in front of the PbWO$_4$ calorimeter acts as a charged particle veto.

On the top is the second electromagnetic calorimeter, which covers $\Delta\phi = 107°$ and $|\eta| < 0.7$. It consists of a lead-scintillator sampling calorimeter, which is read out by 1 mm wavelength-shifting fibers coupled to APDs. The detector is segmented into 12,288 towers, each of which is approximately projective in pseudo-rapidity and azimuth angle with respect to the interaction vertex. Each module contains 2 × 2 towers built up from 76 alternating layers of 1.44 mm thick lead plates and 77 layers of 1.76 mm thick polystyrene-based organic scintillator tiles. The inner layer is a scintillator plane to tag those photons that started to shower before the calorimeter. The towers are grouped into 10 super modules with a cell size of $\Delta\eta \times \Delta\phi = 0.7 \times 20°$ and two super modules with $\Delta\eta \times \Delta\phi = 0.7 \times 7°$. Each tower has a size of $\Delta\eta \times \Delta\phi = 0.0143 \times 0.819°$ and has a thickness of 20.1 X_0. The APDs have an active area of 5 mm × 5 mm. The fiber bundle in a single tower terminates in a 6 mm diameter disk and connects to the APD photosensor through a short light guide/diffuser with a square cross section of 7 mm × 7 mm that tapers slowly down to 4.5 mm × 4.5 mm.

The forward muon arm covers the pseudo-rapidity of $-4 < \eta < 2.4$ and the azimuth-angle range of $2° < \phi < 9°$ to provide good acceptance down to zero transverse momentum. It consists of a 10 λ_I thick composite absorber made of layers with both high-Z and low-Z materials starting 90 cm from the vertex, a large dipole magnet with a 3 Tm bending radius placed outside the L3 magnet and ten planes of very thin, high-granularity, cathode strip tracking stations. A second 7 λ_I thick muon filter at the end of the spectrometer and four planes of resistive plate chambers are used for muon identification and triggering. The spectrometer is shielded by a dense conical absorber tube of about 60 cm outer diameter, which protects the chambers from secondary particles created in the beam pipe.

The event time is measured with very good precision (<25 ps) by the T0 detector, which consists of two sets of 12 Cherenkov counters that are mounted around the beam pipe. They consist of fused silica that are read out with fine mesh photomultiplier tubes. Two arrays of segmented scintillator counters (V0) are used as minimum bias trigger and for rejection of beam-gas backgrounds. They also participates in the measurement of luminosity in pp collisions with a precision of about 10%. An array of 60 large scintillators (ACORDE) on top of the L3 magnet will trigger on cosmic muons for calibration and alignment purposes. The Forward Multiplicity Detector provides multiplicity information over a large fraction of the solid angle. Charged particles are counted in rings of silicon strip detectors located at three different positions along the beam pipe. The Photon Multiplicity Detector measures the multiplicity and spatial distribution of photons event-by-event in the region of $2.3 < \eta < 3.7$. It consists of two planes of gas proportional counters with cellular honeycomb structure, preceded by two lead converter plates of 3 X_0 each. Two sets of two compact calorimeters are used to measure and trigger on the impact parameter of the collision. They are made of tungsten and brass with embedded quartz fibers and are located on both sides in the machine tunnel, about 116 m from the interaction region. In addition, two small electromagnetic calorimeters are installed on one side, 7 m from the vertex, to improve the centrality selection.

11.2 Non-collider Detectors

In addition to detectors at colliders, there are several large detector systems that are built to study dedicated topics in non-collider experiments. Among these are measurements of the anomalous magnetic moment of the muon and electric dipole moments, searches for charged lepton-flavor violation, rare kaon decays, study of neutrino interactions, search for dark matter, search for dark energy, study of cosmic rays, exploration of astro-particle physics and studies of antimatter. For example, the anomalous magnetic moment of the muon is measured in the muon g-2 experiment E989 at Fermilab [48]. A search for charged-lepton flavor violation in muon decays is conducted in the Mu2e experiment [49] at Fermilab, the MEG/MEG2 experiment [50,51] at PSI and the COMET experiment [52] at JPARC. We present, herein, the Mu2e experiment.

There are several neutrino experiments in operation or under construction, Super-Kamiokande [53,54], T2K [55], SNO [56], Juno [57], Noνa [58], KATRIN [59], Ice Cube [60], Hyper-Kamiokande [61], ORCA [62], PINGU [63] and DUNE [64]. We present here the Super-Kamiokande and IceCube detectors. Though some of the neutrino experiments can also look for dark matter, several dedicated experiments have been constructed to search for the Weakly-Interacting Massive Particle (WIMP) including LUX-LZ [65], Xenon 1T, [66] SENSEI at SNOLab [67] and in future DAMIC [68] and others. We present the LUX − LZ experiment.

Furthermore, the Large Synoptic Survey Telescope (LSST) also called Vera C. Rubin observatory is under construction in Chile [69]. With an 8.4 m mirror it is the largest camera ever built for atronomy and astrophysics. The goal is to explore dark energy. Data taking will start in 2025. Several cosmic ray experiments are in operation including the Auger observatory [70], AMS [71], CALET [72], PAMELA [73] and FERMI-LAT [74]. For gamma ray studies, the observatories MAGIC [75], HESS [76] and VERITAS [77] have been in operations. The new project is the Cherenkov Telescope Array (CTA) [78] that we will present. There are also dedicated experiments that look for axions, light dark matter and double beta decay.

11.2.1 The Mu2e Detector

The Mu2e experiments looks for muon-to-electron conversions $\mu N \to eN$ [49,79]. The distinctive signature of the conversion electron is a mono-energetic electron with energy very close to the muon mass, $E_e = 104.96$ MeV. Muons stopped on aluminum have a 39% probability of undergoing a three-body decay when orbiting

11.2 Non-collider Detectors

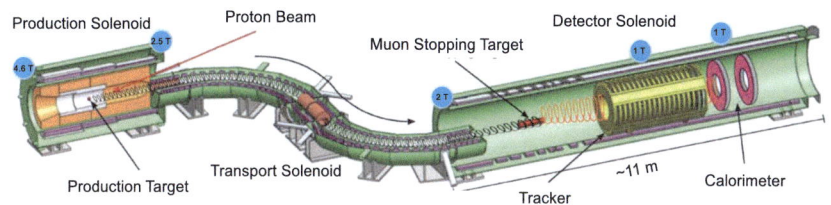

Fig. 11.22 Schematic view of the Mu2e detector. Reprinted with kind permission from [49], © 2015, Mu2e Collaboration. All rights reserved

around the nucleus. The electron spectrum of this decay-in-orbit process substantially differs from that of a free decay, due to the presence of an extended tail resulting from recoil and radiative correction effects that falls rapidly as the electron energy approaches the kinematical endpoint.

Figure 11.22 shows the layout of the Mu2e experiment. An eight GeV/c pulsed proton beam hits a tungsten target. A superconducting solenoid covering the target area produces a gradient field from 2.5 to 4.6 T that acts as a magnetic lens to focus the backward-produced low-energy particles (pions, muons and a small number of antiprotons) into the transport channel. An S-shaped transport solenoid efficiently transfers low-energy, negatively charged particles to the end of the beam line, while allowing a large fraction of pions to decay into muons. Positively and negatively charged particles move in opposite directions while traveling through the curved solenoidal field. A mid-section collimator removes nearly all the positively charged particles. The transfer solenoid ends in the detector solenoid, which encompasses the muon stopping target, the tracker and the calorimeter. The graded field, which reflects backward-produced particles, runs from 2 to 1 T in the upstream region. Approximately 50% of the muon beam whose momentum is $\sim$ 50 MeV/c is stopped in the aluminum target. These muons are captured in an atomic excited state and promptly cascade to the 1S ground state, where 39% decay in orbit and the remaining 61% are captured by the nucleus. Low-energy photons, neutrons and protons are emitted in the nuclear capture process, which make up an irreducible source of accidental activity being the origin of a large neutron fluence on the detection systems. Together with the flash of particles accompanying the beam, the capture process produces the bulk of the ionizing dose observed in the detector system and its electronics.

The Mu2e tracker is the primary device to measure the momentum of the electron and to separate it from background. It consists of nearly 20,000 low-mass straw drift tubes, 5 mm in diameter with 15 μm thick Mylar walls and a 25 μm diameter sense wire in the center. Straws are oriented in a direction transverse to the solenoid axis. The basic element is the panel, where straws are organized in two staggered layers. Six panels are arranged as shown in Fig. 11.23 (left top-left) and two panels rotated by 30° define a station as shown in Fig. 11.23 (left top-right). The 3.2 m long tracker consists of 18 stations as shown in Fig. 11.23 (left bottom). A central hole, 38 cm in diameter, makes the tracker blind to low momentum background particles, primarily from muon decay in orbit. The momentum is reconstructed with an uncertainty of 115 keV/c.

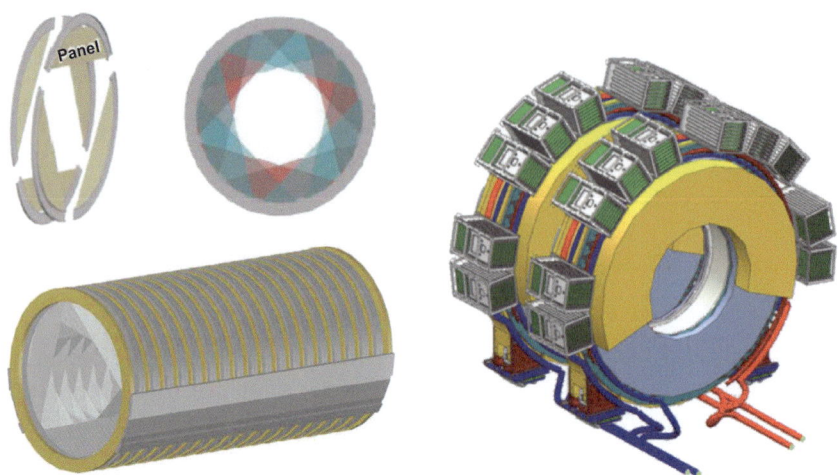

Fig. 11.23 Left top-left: Arrangement of the six panels in the Mu2e tracker [79]. Left top-right: Three consecutive panels are rotated by 30° each [79]. Left-bottom: Three-dimensional view of the Mu2e tracker. All reprinted under CC-BY-4.0 Licence from [79], © 2018, MU2E Collaboration. Right: Schematic view of the Mu2e calorimeter. Reprinted under CC-BY-NC-ND-4.0 Licence from [80], © 2020, MU2E Collaboration

The measured transverse position resolution is $\sigma_{\text{trans}} = (0.133 \pm 0.022)$ mm, while the longitudinal position resolution is $\sigma_{\text{long}} = (42 \pm 1)$ mm.

The crystal calorimeter plays a crucial role in providing particle identification capabilities and a fast online trigger filter, while aiding the track reconstruction capabilities. Figure 11.23 (right) shows the layout of the calorimeter that consists of two disks made by 674 undoped CsI scintillating crystals, each having a size of 3.4 cm × 3.4 cm × 20 cm. Each crystal is read out with two UV-extended SiPMs, having a 6 mm × 6 mm active surface and a 30% photon detection efficiency. A time resolution of 100 ps and an energy resolution of 6.5% was measured for 100 MeV particles. The crystals will receive an ionizing dose of 90 krad and a fluence of 3×10^{12} (1 MeV) $n_{\text{eq}}/\text{cm}^2$ in three years running. The photosensors, being shielded by the crystals, will get a three times smaller dose.

An external veto for cosmic rays surrounds the solenoid since cosmic ray muons are the major background source in Mu2e. Cosmic rays may mimic conversion electrons when interacting with the detector materials. These events occur at a rate of approximately one event per day. Thus the cosmic ray veto has to reject cosmic ray muons with an efficiency of at least 99.99% for cosmic ray tracks while withstanding an intense radiation environment. The cosmic ray veto consists of four staggered scintillator slabs, which are read out with two embedded wavelength-shifting fibers, each one coupled to a 2 mm × 2 mm SiPM. In addition, an extinction monitor detects scattered protons from the production target to evaluate the fraction of out-of-time beam and a stopping target monitor measures the rate and the number of negative muons that are stopped in the target. The experiment is expected to take data in 2025.

11.2 Non-collider Detectors

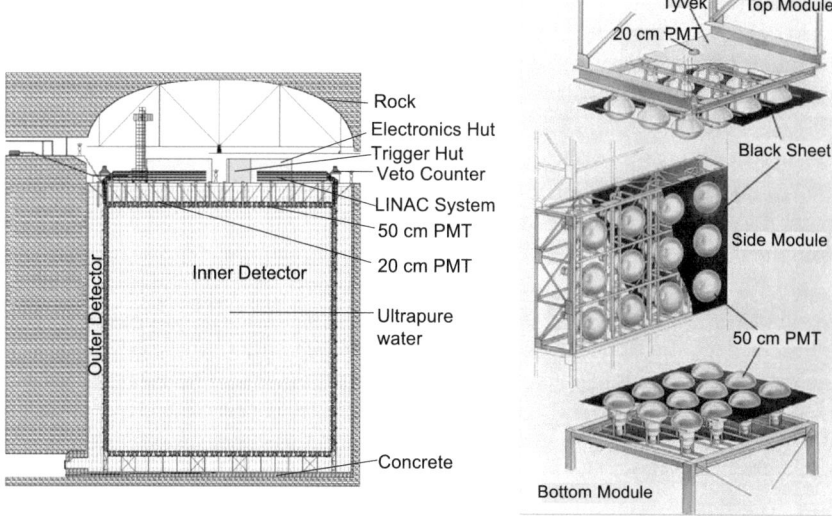

Fig. 11.24 Left: Schematic view of the Super-Kamiokande detector. Right: Schematic view of support structures for the Super-Kamiokande inner detector holding the photomultiplier tubes. The setups for top, side and bottom modules are illustrated. Both reprinted with kind permission from [54], © 2003, Elsevier. All rights reserved

11.2.2 The Super-Kamiokande Detector

The Super-Kamiokande detector is the world's largest water Cherenkov detector located in the Mozumi mine in Japan 1 km under the peak of mount Ikenoyama. The scientific program of the Super-Kamiokande experiment includes searches for proton decays and studies of neutrinos from various sources such as neutrinos from the sun, the atmosphere, supernovae, gamma ray bursts, other astrophysical sources as well as neutrinos from beams. The Super-Kamiokande detector can detect events over a wide range of energy, from 4.5 MeV to over 1 TeV. Neutrino interactions are detected via the Cherenkov light emitted by the charged leptons produced in the interaction. The incoming neutrino enters the detector and interacts $\nu_\ell p \to n\ell^+$. So, a charged lepton is produced, typically a muon or an electron that radiates Cherenkov radiation in the water. The signal is the observation of a charged particle produced in the water, while background consists of charged particles entering the water from the outside.

Figure 11.24 (left) shows a schematic view of the Super-Kamiokande experiment. This water Cherenkov detector consists of a welded stainless-steel tank, 39 m in diameter and 42 m tall. Access to the tank is provided by two hatches on the top and one hatch at the bottom. The tank top houses the electronics hut, which has the equipment for calibrations, water quality monitoring and other facilities. Inside the tank a support structure spaced 2–2.5 m from the wall for housing PMTs is mounted on all walls. There is an array of 11,146 PMTs with a 50 cm diameter facing inside. This part is called the Inner Detector, which contains 32 ktons of water. A photograph of the Inner Detector and the PMTs was shown in Fig. 8.15 (right). Figure 11.24 (right) shows the layout of the support structure with inner PMTs mounted. Another array of 1,885 outward-facing PMTs with a diameter of 20 cm is placed that is optically isolated from the inner detector. This water volume with the PMTs is called the Outer Detector and serves as a veto. Each Outer Detector PMT is attached to a 50 cm × 50 cm acrylic wavelength-shifting plate. If the Inner Detector records an event without a signal from the Outer Detector, a neutrino interaction is rather likely. However, if the Outer Detector sees a signal, a charged particle entered from the outside.

The Inner Detector PMTs are operated with a high voltage of 1.7–2.0 kV, providing a gain of 10^7. The resolution of the transit time is $\sigma_{trans} = 2.16$ ns. Several sources are used for calibrations. To obtain precise knowledge of the absolute energy scale, energy resolution, angular resolution, spatial resolution and detection efficiency for low-energy electrons, Super-Kamiokande uses a LINAC, which provides electrons with energies between 4.9 and 16.1 MeV. The absolute energy scale is cross checked with a ^{16}N source. The water transparency is measured with a dye laser and cosmic rays. The relative gain of the PMTs is measured with light from a Xe lamp that is injected into a scintillator ball via a clear fiber, which is lowered into the water. Stopping muons, electrons from muon decays and π^0 produced in interactions are used for cross checks.

The next step is an upgrade to Hyper-Kamiokande [61]. The tank size will be increased to 68 m in diameter and 71 m in height. The Inner detector will contain 217 ktons of water and is viewed by an array of 40,000 new PMTs, 50 cm in diameter. Hyper-Kamiokande will receive an intense neutrino beam from J-PARC at an off-axis angle of 2.5°. The physics goals of Hyper-Kamiokande are measurements of *CP* violation, searches for proton decay, dark matter and physics beyond the Standard Model.

11.2.3 The Ice Cube Detector

Ice Cube is another neutrino experiment located in Antarctica, which basically consists of PMT arrays submerged into the ice that detect Cherenkov light from traversing charged particles [60]. Figure 11.25 shows a layout of the Ice Cube detector. It is arranged in three parts, the main in-ice array, the smaller densely instrumented central array (DeepCore) and the surface array (IceTop). The in-ice components are comprised of 5,160 digital optical modules on a total of 86 strings with 60 digital optical modules per string. Holes were drilled into the ice using a combination of a firm drill for the top layer of compacted snow and a hot-water drill to continue down into the glacial ice to a depth of 2,450 m below the surface. Into such a hole, 78 strings were deployed with 17-m spacing vertically and approximately 125 m horizontally in a hexagonal grid. The other eight strings are used in DeepCore with smaller spacing. After string installation, the water in each hole eventually refreezes. The cables from all strings are connected to the IceCube Laboratory located in the

11.2 Non-collider Detectors

center of the grid. The in-ice array is comprised of a cubic kilometer of instrumented volume, or approximately one gigaton of glacial ice.

The DeepCore central subarray is equipped with six strings that have higher-quantum-efficiency PMTs, while the other two strings have the standard PMTs used in main array. The six extra strings were deployed more compactly, reducing the horizontal separation to 70 m and the vertical spacing to seven meters. In addition, the digital optical modules were deployed towards the bottom of the detector, where the ice is especially clear. The combination of higher-efficiency PMTs with the reduced string spacing and deeper ice permits a reduction of the energy threshold of DeepCore to 10 GeV. The IceTop array consists of 81 surface stations near each of the IceCube strings. Each IceTop station consists of two ice tanks with two digital optical modules in each tank. IceTop is used mainly for calibration and as a veto of the cosmic ray background for in-ice studies. It is also used for cosmic ray air shower analyses in the higher energy range. IceCube has an uptime of over 99.8%.

Figure 11.26 (left) shows a schematic layout of a digital optical module that consists of a 33 cm deep ocean Benthos sphere made of 1.3 cm thick, low-potassium glass to reduce radioactivity. The sphere is purged with dry nitrogen gas and evacuated to 0.5 atmospheres. This is done to keep the two sphere halves pressed firmly together while being handled before descending as far as 2.5 km into the glacier. The glass is transparent to photons with a wavelength greater than 330 nm. Thus, Cherenkov photons with smaller wavelengths are not detected. The primary IceCube PMT is a 25.4-cm diameter photodetector with a sensitive cathode area of 500 cm^2 in the

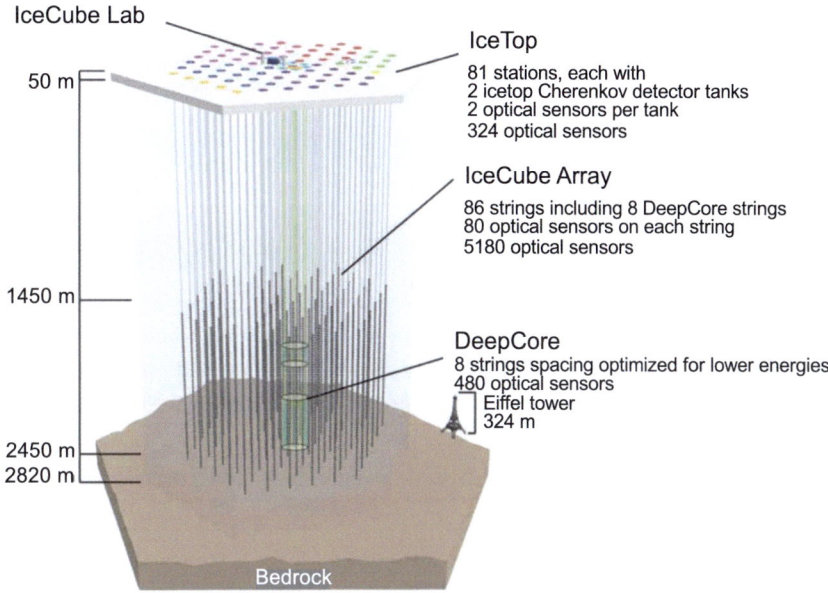

Fig. 11.25 Schematic view of the Ice Cube detector. Reprinted from [81]. Public domain

300 to 550 nm wavelength range with a peak quantum efficiency of 25%. For the special PMTs used in six of the DeepCore strings, the quantum efficiency is increased to 40% at 400 nm photon wavelengths. The PMTs are mounted in the bottom half of the glass sphere facing downward to emphasize the detection of photons from up-going particles. The PMTs are coupled to the glass with optically-transparent silicone gel. A mu-metal grid covers the PMT that shields it from the Earth's magnetic field.

The data acquisition system is controlled by an FPGA located on the Main Board of the digital optical module. The FPGA interfaces with the digitizers, acquires the data, compresses signals and controls communications with the surface. The Main Board is located just above the PMT and amplifies, records, and digitizes the PMT signals, when the trigger conditions are met. The Main Board divides the PMT signal into three paths, the PMT trigger, the PMT signal, and a signal sent to a converter, in order to handle longer signals with a lower sampling. In addition, the high-voltage divider and an LED flasher board are housed. Figure 11.26 (center) shows an event display of a track-like event in the ICECube detector, which may be a muon produced in a neutrino interaction. Figure 11.26 (right) shows a shower-like event. One of the DOMs records a huge energy. This should illustrate the different kind of events IceCube observes.

11.2.4 The LUX-ZEPLIN Detector

The LUX experiment has been searching for WIMPs at the Sanford Underground Research Facility [83]. The WIMPs are dark matter particles that may roam the universe and that are not expected in the Standard Model. The LUX-ZEPLIN (LZ) experiment is next iteration of the LUX experiments, using about a ten-times bigger

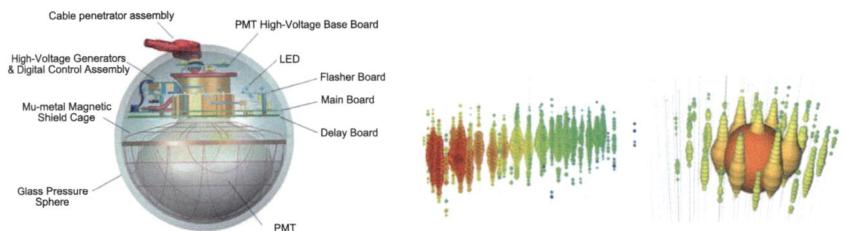

Fig. 11.26 Left: Layout of a Digital Optical Module (DOM) in the IceCube detector. The PMT is housed insides a glass pressure sphere. A mu-metal magnetic shield cage protects the PMT from the Earths magnetic field. The PMT high-voltage base board is located on top of the PMT. The connection to the outside is provided by the cable penetrator assembly. A main electronics board is installed inside the sphere, which contains an LED and flasher board. It also contains the high-voltage generator and digital-control assembly. Reprinted with kind permission from [60], © 2006, Elsevier. All rights reserved. Center: Track-like event in the IceCube detector. Right: Shower-like event in the IceCube detector. Each sphere shows the light observed by a DOM. The size indicates the amount of Cherenkov light a PMT observed, while the color shows the relative time of the photons with respect to each other. Both reprinted under CC-BY-3.0 Licence from [82], © 2017, IceCube Collaboration

11.2 Non-collider Detectors

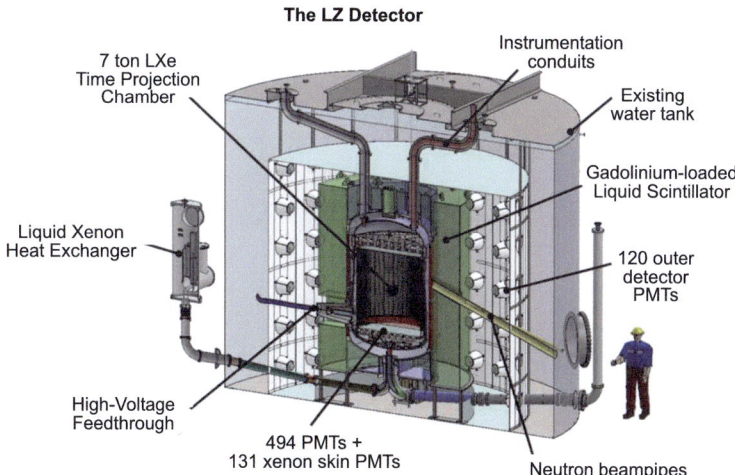

Fig. 11.27 Schematic view of the LUX-ZEPLIN detector. The core consists of the two-phase liquid Xenon time projection chamber. Signals are recorded by 494 three-inch photomultiplier tubes housed on the bottom and top of the cylindrical Xenon vessel. In addition, 113 side skin and 18 dome PMTs are mounted as a veto. The Xenon is surrounded by Gadolinium-loaded liquid scintillator followed by the 120 eight-inch outer detector PMTs that stand in the water tank to veto background. Reprinted with kind permission from [84], © 2017, LUX-ZEPLIN Collaboration. All rights reserved

Xenon tank [65, 84]. Figure 11.27 shows a schematic layout of the LUX-ZEPLIN detector. The core of the experiment is a two-phase xenon time projection chamber, containing about seven fully active tons of liquid xenon. The energy deposited from the particle interaction in the xenon creates a primary scintillation signal and electrons from ionization. The electrons are drifted by an electric field (181 V/cm) to the liquid-gas interface at the top of the detector, enter the gas phase, where they are accelerated by an electric field of 6.0 kV/cm to produce a proportional scintillation signal. Both signals are read out by two arrays of three-inch photomultiplier tubes, 253 PMTs view the TPC from above, and 241 PMTs view from below. The difference in time of arrival between the two signals measures the position of the event in the z direction, while the x, y position is determined from the light pattern observed in the top PMT array. Highly reflective polytetrafluoroethylene forms the inner wall of the TPC to enable efficient light collection of the proportional scintillation signal.

The TPC is housed in a double-wall vacuum-insulated titanium cryostat with a layer of "skin" LXe acting as a high-voltage stand-off. The skin is separately instrumented with PMTs to veto gamma and neutron interactions in this region. The TPC is surrounded on all sides by a hermetic, gadolinium-loaded, liquid scintillator serving as an Outer Detector. The cathode high-voltage connection is made horizontally at the lower left. The organic scintillator is viewed by an array of 120 eight-inch PMTs standing in the water, which provides shielding for the detector.

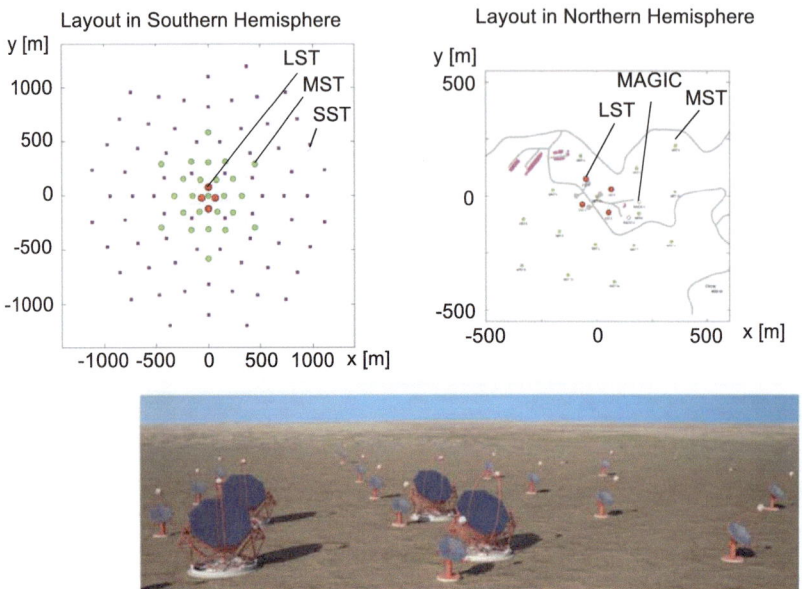

Fig. 11.28 Top left: Layout of the Cherenkov Telescope Array in the southern hemisphere [85]. Top right: Layout of the Cherenkov Telescope Array in the northern hemisphere. The telescopes come in three sizes, large (LST) with a diameter of 23 m, medium (MST) with a diameter of 9–12m and small (SST) with a diameter of 4 m. The array is built around the MAGIC telescope. Both reprinted under CC-BY-4.0 Licence from [85], © 2020, CTA Collaboration. Bottom: Illustration of the CTA experiment [86]. Reprinted under CC-BY-4.0 Licence from [86], © 2015, CTA Collaboration

11.2.5 The Cherenkov Telescope Array

The Cherenkov Telescope Array (CTA) is the largest gamma ray observatory that is under construction [78]. It builds upon the technology of existing gamma ray detectors MAGIC [75], HESS [76] and VERITAS [77] but will be ten times more sensitive and have unprecedented accuracy in the detection of high-energy gamma rays. The principle is to detect Cherenkov radiation that is produced by charged particles that are produced in the atmosphere by interactions of gamma rays. Arrays will be placed at two sites, in La Palma, Spain (northern site) and in Eso, Chile (southern site). Figure 11.28 (top left, top right) shows the layout of the two arrays. Figure 11.28 (bottom) shows an illustration of the experiment. To cover a wide energy range in a cost-effective way, the arrays will contain several different types of telescopes. The array in the northern site consists of four large-size telescopes and 15 medium-size telescopes. The array in the southern site consists of four large-size telescopes, 25 medium-size telescopes and 70 small-size telescopes. Figure 11.29 illustrates the layout of the different-size telescopes. The telescopes are spherical mirrors that focus the reflected light onto a photodetector located in the focus. The mirrors are pixelated into hexagonal pads.

The small-size telescope has a diameter of four meters with a field of view of at least 8°. It is sensitive to the highest-energy gamma rays, which come from our

11.3 Applications in other Fields

Fig. 11.29 Left: Schematic layout of a small-size telescope (left), medium-size telescope (center) and large-size telescope (right) in CTA. Reprinted under CC-BY-4.0 Licence from [86], © 2015, CTA Collaboration. Right: Photograph of the first CTA Large-Size Telescope. Reprinted under CC-BY-NC-ND-4.0 Licence from [87], © 2021, CTA Collaboration

own galaxy that is best observed from the southern hemisphere. The telescopes are spread out over several square kilometers. The photodetectors consist of both PMTs and SiPMs. The sensitive energy range is up to 300 TeV. The medium-size telescope has a diameter of 12 m with a field of view of at least $7°$. The Cherenkov light is recorded with PMTs. The large-size telescopes have a diameter of 23 m and a parabolic shape. They have a field of view of about $4.5°$. The Cherenkov light is recorded with PMTs. The first Large-Sized Telescope started first measurements in La Palma in 2018.

11.3 Applications in other Fields

High-energy physics detection techniques are used in many other fields. We saw some examples in astrophysical telescopes. Another example is the AMS detector at the International Space Station [71]. Modern cameras use CCDs or a CMOS image sensors instead of films, since they are reusable. X-ray detectors use image plates or flat panel detectors. For dose measurements, devices consist of ionization chambers or Geiger counters. Furthermore, particle sub-detectors are used in geophysics and medical physics. We present a few examples in the following.

11.3.1 Geophysics Applications

In geophysics, there is some interest to image the internal structure of active volcanoes. Atmospheric muons are rather useful for this task This radiography is also called muography. The muons are detected by large-area, high-efficiency and high-precision muon detectors. A Japanese group used two sets of position-sensitive emulsion cloud chambers to map the crater region of an active Japanese volcano [88]. Each set consisted of two emulsion cloud chamber planes that were interleaved with 3 mm thick iron plates to shield off 1.31 MeV electrons from ^{40}K found in the rock. The detectors were placed in an underground chamber located at an elevation of 2,256 m from sea level and at a horizontal distance of 1,000 m from the crater center.

The advantage of emulsion cloud chambers is that they do not need any power, which is an advantage in the wilderness.

The favored detector technology consists of plastic scintillator strips. For example, the MU-RAY detector at the Puy de Dôme volcano in France uses plastic scintillator bars with a triangular cross section having a 3.3 cm base and a 1.7 cm height [89,90]. The 107 cm long bars are extruded and are coated with TiO_2. In the bar's center a wavelength-shifting fiber is placed that is coupled to a SiPM. Thirty-two bars are combined into a module. The 32 SiPMs are hosted by a custom-designed Printed Circuit Board. Two modules are assembled side-by-side to form either an X or a Y layer with an area of 1 m × 1 m. Two layers with orthogonal bars form an X-Y plane. The detector consists of three X-Y planes mounted on an aluminum alloy frame that can be rotated to collect data from different regions in the sky. A 3 cm thick iron shield of 1 m^2 area was placed between the first and second X-Y plane to stop muons with energies below 70 MeV. The detector was placed 1.3 km away from the 1,465 m high summit at an altitude of 950 m. There are several other geophysical detectors that use arrays of plastic scintillator strips [91].

A second detector, the TOMUVOL muon telescope, was placed at the Puy de Dôme volcano at the same time as the MU-RAY detector [92]. The detector consists of four layers of glass RPCs that are operated in avalanche mode with a high voltage of 7 kV. The 31 mm thick chambers and their readout electronics are placed in aluminum cassettes for protection. They cover an area of 36.1 cm × 74.6 cm. Each detection layer consists of six cassettes held by a light aluminum frame, covering an area of about 1 m × 1 m. The separation between layers is 33 cm.

11.3.2 Positron Emission Tomography

Positron emission tomography (PET) is an imaging procedure for organs and tissues to show how well they function. A positron emitting drug is inserted into the patient by injection, swallowing or inhalation. The drug accumulates in the organ or tissue of interest in regions of higher chemical activity. The emitted positron annihilates with an electron in the radiating atom producing two back-to-back 511 keV γ quanta. By recording the position and arrival time of both photons, the position of the radiating atom can be traced. The distribution of the radiating atoms provides a three-dimensional image of the higher chemical activity of the organ or tissue.

The PET scanner consists of a cylindrical ring or tube of length L_{PET} that is sufficiently large to insert a patient. Figure 11.30 (left, center) shows a schematic layout of a PET detector. The tube is instrumented with an array of γ-ray detectors, typically consisting of inorganic crystals read out by solid-state photodetectors, providing segmentation in the azimuth-angle ϕ and z directions. Figure 11.30 (right) shows a PET scanner using inorganic crystals. Most PET scanner only record fully absorbed γ quanta and ignore Compton-scattered photons. Various inorganic crystals have been used, including NaI(TL), GSO, BGO, LSO, LYSO and $LaBr_3$. The photon detectors used to be PMTs. Today, they are APDs and SiPMs. The most popular scintillators for new scanners are Cerium-doped LSO and LYSO crystals, which have a light yield of 68% of that of NaI(Tl) crystals. Crystal sizes are typically around 3 mm × 3 mm with lengths of 10–50 mm. For readout, 4 × 4 SiPM arrays are used. Each SiPM has a size of 3 mm × 3 mm holding 3,600 pixels with a 50 μm pitch. The operational voltage is around 70 V yielding a gain of the order of 10^6.

The Philips Gemini ToF PET/CT Scanner is a new high-performance, time-of-flight (TOF) capable, fully 3-dimensional (3D) PET scanner together with a 16-slice Brilliance CT scanner [94] that has incorporated the ideas of LYSO crystals and SiPMs for readout. The crystals with sizes 4 mm × 4 mm × 22 mm, are arranged in flat modules, containing a matrix of 23 × 44 crystals. The spatial resolution is $\sigma_p = 2.0$ mm in the axial and transverse direction. The energy resolution is $\sigma_E/E = 4.9\%$ at 511 keV and the timing resolution is $\sigma_t = 249$ ps.

New PET scanners are built also with liquid xenon. One example is the PETALO concept [95]. The PETALO group has shown that liquid xenon in combination with SiPM-based readout and fast electronics provides a significant improvement for TOF-PETs. They use exclusively the scintillation light of liquid xenon, which provides sufficient information to achieve good energy and spatial resolution. Other groups have explored to read the ionization charge in addition [96,97]. If both signal are combined further improvement is achieved.

The PETALO detector consists of a continuous ring of liquid xenon that is surrounded by a dense array of VUV SiPMs. The ring is located inside a cylindrical vacuum chamber with a bore at room temperature, leaving a cavity where the patient will be placed. In a first arrangement a ring was equipped with 4,512 SiPMs covering an active area of 3 mm × 3 mm at a pitch of 4 mm. The dimensions of this setup were an internal radius of 15 cm, a depth of 3 cm and a length of 6.4 cm. The spatial resolutions were $\sigma_r = 1.17$ mm, $\sigma_z = 0.79$ mm and $\sigma_\phi = 3.6$ mrad in radial, longitudinal and azimuth-angle directions, respectively. The time resolution is $\sigma_{pet} = 15.4$ ps.

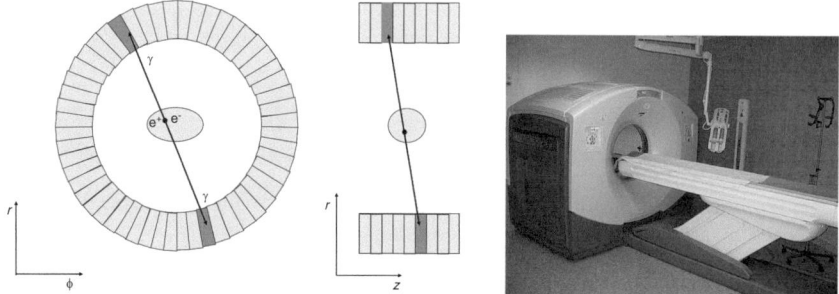

Fig. 11.30 Left: Schematic layout of a PET scanner in the $r\phi$ view. Right: Schematic layout of a PET scanner in the rz view. Right: A three-dimensional view of a PET scanner using inorganic crystals [93]. Reprinted from [93], public domain

11.4 Conclusion

Colliding beam experiments are typically segmented into a barrel region and two endcaps. Around the beam pipe, usually made of beryllium, a vertex detector or detector system is placed, followed by tracking detectors. Some experiments use time-of-flight counters or Cherenkov detectors for π-K separation. All detectors have an electromagnetic calorimeter followed by a hadron calorimeter and a muon detection system. In some experiments, the electromagnetic calorimeter is placed inside the coil and the hadron calorimeter as well. In many experiments, the calorimeters lie outside the coil. These use a pre-sampler to correct for energy losses in the coil. A tail catcher and the muon system lie behind the coil. For luminosity measurements, special detectors (electromagnetic calorimeters in e^+e^- colliders) are placed near the beam pipe at small polar angles. For high luminosities, mini-beta and corrector magnets are positioned close to the interaction region. We have seen that different experiments apply different detector technologies. We have discussed these in the previous chapters.

In the forthcoming years the focus at LHC will be the upgrade of the ATLAS and CMS experiments for the High-Luminosity LHC operation. Furthermore, new detectors are designed for future projects, for example for an e^+e^- Higgs factory, a new hadron collider, new neutrino experiments, charged-lepton flavor violation and dark-matter search experiments. Thus, R&D will continue on many subdetectors, for example on solid-state detectors, tracking detectors, calorimeters as well as read out systems, electronics, and data acquisition as well as data transfer and processing. To proceed in an organized way, the CERN committee ECFA has organized nine detector R&D groups who started their work. Challenges include improved background rejection with better timing resolutions, improved impact parameters resolutions, fast highly-efficient triggers and fast data acquisition. New detectors may have finer segmentation and thus more channels requiring faster data processing and data collection. Maybe the ASICs in individual detector planes will be able to transfer data wireless between layers and to the control room. Trigger decisions may be based purely on software and event reconstruction may be done during data acquisition. There are certainly many fruitful years ahead of us.

Exercises

11.1 Design a multipurpose detector for studying charm and τ decays, which are produced at a symmetric e^+e^- collider with beam energies of 1.5–2.2 GeV. Below the charm threshold, charmonium resonances (J/ψ and $\psi(2S)$) are produced, which decay via strong and electromagnetic interactions, while the ψ'' above the charm threshold decays to pairs of D mesons, (D^+D^- or $D^0\bar{D}^0$). The detector should be optimized to study all decays with high efficiency and good resolution. It should measure charged particles and photons well. It should be able to separate kaons from pions. Make a little sketch of the overall detector and explain the layouts of the sub-detectors. Justify your choices. Give some details about the expected performance of each sub-detectors you selected in terms of coverage, efficiency and resolutions.

11.2 An e^+e^- linear collider is planned to operate at a center-of-mass energy of 250 GeV that should be upgradeable to 500 GeV and maybe even higher energies. The main goal of the detector is to perform precision measurements of various Higgs boson decays to test the predictions of the SM. Another goal is to study top quark decays that are produced in pairs. Top quarks typically decay to b quark and a W^+ boson, where the latter can decay into e^+, μ^+, τ^+ and a neutrino or into two different flavored quarks. So the detector has to detect leptons and jets. The b-jets have a typical lifetime of 1 ps. The detector should also be able to perform lifetime measurements and provide b flavor tagging. Thus, B decay vertices should be measurable with good resolution. Since the beam energies are well defined, full energy momentum conservation is applicable. Make a little sketch of the overall detector and explain the layouts of the sub-detectors. Justify your choices. Give some details about the expected performance of each sub-detectors you selected in terms of coverage, efficiency and resolutions.

11.3 Design a multipurpose detector for studying rare kaon decays. The kaons are produced in $\phi(1020)$ decays, which are produced at a symmetric e^+e^- collider with beam energies of 0.51 GeV. The main decay modes are $\phi(1020) \to K^+K^-$, $\phi(1020) \to K^0_S K^0_L$ and $\phi(1020) \to \pi^+\pi^-\pi^0$. The detector should be able to measure, $K^\pm, K^0_L, \pi^\pm, \pi^0$s, $\mu^\pm$s and $e^\pm$s well in terms of momentum, energy and particle species. Make a little sketch of the overall detector and explain the layouts of the sub-detectors. Justify your choices. Give some details about the expected performance of each sub-detectors you selected in terms of coverage, efficiency and resolutions.

11.4 Design a multipurpose detector for measuring all processes that may be produced in pp collisions at a beam energy of 50 TeV per beam. At this energy new unknown processes may occur. Such processes involve jets, electrons, muons, photons and missing particles, which should be measured with high efficiency and excellent resolution. Leakage should be minimized. Some of these particles may be produced from decays of b-hadrons or charmed hadrons that fly some distance before decaying. So the detector should detect secondary vertices with excellent position resolution. The detector should have excellent energy measurements of electromagnetic showers and hadron showers. It should measure charged particle momenta with good resolution. Make a little sketch of the overall detector and explain the layouts of the sub-detectors. Justify your choices. Give some details about the expected performance of each sub-detectors you selected in terms of coverage, efficiency and resolutions.

11.5 Design a detector system for a long-baseline neutrino experiment, which consists of a near detector and a far detector. The detector should have good sensitivity to measure CP violation in the neutrino sector. Make a little sketch of the overall detector and explain the layouts of the sub-detectors. Justify your choices. Give some details about the expected performance of each sub-detectors you selected in terms of coverage, efficiency and resolutions.

References

1. *BABAR* Collaboration (B. Aubert et al.), Phys. Rev. Lett. **86**, 2515 (2001)
2. The Belle Collaboration (A. Abashian et al.), Nucl. Instrum. Meth. A **479**, 117 (2002)
3. F. Bernlochner et al., EPJ Web Conf. **150**, 00014 (2017). arXiv:1011.0352 [physics.ins-det] (2010)
4. K.R. Nakamura et al., PoS(Vertex 2016), 012 (2016)
5. The *BABAR* Collaboration (B. Aubert et al.), Nucl. Instrum. Meth. A **479**, 1 (2002)
6. *BABAR* Collaboration, http://wnppc10.phys.uvic.ca/general/babar.html
7. D. McFarlane, Exploring Mattar-Antimatter Asymmetries at the B-factories, talk at UC Berkeley, https://slideplayer.com/slide/14623305/ (2006)
8. T.V. Dong et al., Nucl. Instrum. Meth. A **930**, 132 (2019)
9. The ALEPH Collaboration (D. Decamp et al.), Nucl. Instrum. Meth. A **294**, 121 (1990)
10. http://hst-archive.web.cern.ch/archiv/HST2001/detectors/trackdata/ALEPH.htm
11. The DELPHI Collaboration (P. Abreu et al.), Nucl. Instrum. Meth. A **294**, 121 (1991)
12. The DELPHI Collaboration (P. Abreu et al.), Nucl. Instrum. Meth. A **378**, 57 (1996)
13. DELPHI Collaboration, https://physicsmasterclasses.org/exercises/keyhole/en/detectors/DELPHI.html
14. J.E. Clendenin et al., 1993 IEEE Particle Accelerator Conference (PAC 93), 2973 (1993)
15. The SLD Collaboration (K. Abe et al.), Phys. Rev. Lett **70**, 2515 (1993)
16. The SLD Collaboration (D. Axen et al.), Nucl. Instrum. Meth. A **328**, 472 (1993)
17. The CDF Collaboration (T. Aaltonen et al.), Phys. Rev. D **87**, 7 (2013)
18. The CDF Collaboration (A. Affolder et al.), Nucl. Instrum. Meth. A **526**, 249 (2004)
19. The CDF Collaboration (C. Currat et al.), FERMILAB-CONF-02-185-E (2002)
20. M. Mattson, A.I.P. Conf. Proc. **867**(1), 11 (2006)
21. The CDF Collaboration (L. Balka et al.), Nucl. Instrum. Meth. A **267**, 272 (1988)
22. R. Oishi et al., Nucl. Instrum. Meth. A **453**, 227 (2000)
23. The CDF Collaboration (F. Abe et al.), Nucl. Instrum. Meth. A **303**, 233 (1991)
24. G. Ascoli et al., Nucl. Instrum. Meth. A **268**, 33 (1988)
25. The ATLAS Collaboration (G. Aad et al.), JINST **3**, S08003 (2008)
26. ATLAS Collaboration, https://www.pinterest.com/pin/160722280436088788/
27. The ATLAS Collaboration (A. Artamonov et al.), JINST **3**, P02010 (2008)
28. L. Heelan, J. Phys. Conf. Ser. **160**, 012058 (2009)
29. CERN, http://bowshooter.blogspot.com/2013/03/cern-lhc-new-results-indicate-that.html
30. The CMS Collaboration (S. Chatrchyan et al.), JINST **3**, S08004 (2008)
31. The CMS Collaboration (A. Tumasyan et al.), arXiv:2111.08757v2 [physics.ins-det] (2021)
32. G. Sguazzoni, PoS (VERTEX2011), 013 (2011)
33. The CMS Collaboration (A. M. Sirunyan et al.), JINST **16**, P07001 (2021)
34. M. Planer, PoS (EPS-HEP2017), 328 (2017)
35. L. Aperio Bella, PoS (ICHEP2016), 386 (2016)
36. The LHCb Collaboration (A. Augusto Alves Jr et al.), JINST **3**, S08005 (2008)
37. The LHCb Collaboration (I. Bediaga et al.), CERN-LHCC-2013-021 (2013)
38. The LHCb Collaboration (M. Williams et al.), JINST **12**, C01020 (2017)
39. A. Nomerotski et al., JINST **8**, P04004 (2013)
40. The LHCb Collaboration (G. Bencivenni et al.), Nucl. Instrum. Meth. A **1049**, 168075 (2023). https://cerncourier.com/a/lhcbs-momentous-metamorphosis/
41. The LHCb Collaboration (M. Fiorini et al.), Nucl. Instrum. Meth. A **952**, 161688 (2020)
42. The LHCb Collaboration (O. Steinkamp et al.), Nucl. Instrum. Meth. A **831**, 367 (2016)
43. The LHCb Collaboration (S. Easo et al.), Nucl. Instrum. Meth. A **876**, 160 (2017)
44. K. Föhl et al., JINST **11**, C05020 (2016)
45. The LHCb Collaboration (I. Machikhiliyan et al.), J. Phys. Conf. Ser. **160**, 012047 (2009)
46. The LHCb Collaboration (C. Abell'an Beteta et al.), arXiv:2008.11556 [physics.ins-det] (2020)
47. The ALICE Collaboration (K. Aamodt et al.), JINST **3**, S08002 (2008)

References

48. The Muon g-2 Collaboration (J. Grange et al.), arXiv:1501.06858 [physics.ins-det] (2015)
49. The Mu2e Collaboration (L. Bartoszek et al.), Technical Design Report, arXiv:1501.05241 [physics.ins-det] (2015)
50. The MEG Collaboration (P.W. Cattaneo et al.), Nucl. Instrum. Meth. A **623**, 350 (2009)
51. The MEG II Collaboration (A.M. Balidini et al.), Eur. Phys. J. C **78**-5, 380 (2018)
52. The COMET Collaboration (R. Abramishvili et al.), http://comet.kek.jp/Documentsfiles/PAC-TDR-2016/COMET-TDR-2016v2.pdf (2016)
53. Y. Fukuda et al., Phys. Rev. Lett. **81**, 1158 (1998)
54. The Super-Kamiokande Collaboration (S. Fukuda et al.), Nucl. Instrum. Meth. A **501**, 418 (2003)
55. The T2K Collaboration (K. Kaneyuki et al.), Nucl. Phys. B Proc. Suppl. **145**, 178 (2005)
56. Q.R. Ahmad et al., Phys. Rev. Lett. **89**, 011302 (2002)
57. The Juno Collaboration (C. Cderna et al.), Nucl. Instrum. Meth. A **958**, 162183 (2020)
58. D.S. Ayres et al., NOvA Technical Design Report, FERMILAB-DESIGN-2007-01 (2001); P. Adamson et al., Nucl. Instrum. Meth. A **806**, 279 (2016)
59. The KATRIN Collaboration (J. Wolf et al.), Nucl. Instrum. Meth. A **623**, 442 (2010)
60. The Ice Cube Collaboration (A. Achterberg et al.), Astropart. Phys. **26**, 155 (2006)
61. S. Fukasawa, O. Yasuda, Adv. High Energy Phys. **2015**, 820941 (2015)
62. D.N.A. Murphy, J.E. Geach, R.G. Bower, Mon. Not. Roy. Astron. Soc. **420**, 1861 (2012)
63. D.J. Koskinen, Mod. Phys. Lett. A **26**, 2899 (2011)
64. The DUNE Collaboration (B. Abi et al.), JINST **15**, T08008 (2020); arXiv:2002.03005; JINST **15**, T08010 (2020)
65. The LUX-ZEPLIN Collaboration (D.S. Akerib et al.), arXiv: 1509/1509.02910 [physics.ins-det] (2015)
66. The XENON1T Collaboration (E. Aprile et al.), Springer Proc. Phys. **148**, 93 (2013)
67. J. Tiffenberg, Astrophys. Space Sci. Proc. **56**, 137 (2019)
68. The DAMIC Collaboration (A.E. Chavarria et al.), J. Phys. Conf. Ser. **1342**-1, 012057 (2020)
69. J.A. Tyson et al., Nucl. Phys. B Proc. Suppl. **124**, 21 (2003)
70. The Pierre Auger Collaboration (A. Aab et al.), Nucl. Instrum. Meth. A **798**, 172 (2015)
71. The AMS Collaboration (J. Alcaraz et al.), Nucl. Instrum. Meth. A **478**, 119 (2002)
72. S. Torii et al., Nucl. Phys. B Proc. Suppl. **113**, 103 (2002)
73. The PAMELA Collaboration (V. Bonvicini et al.), Nucl. Instrum. Meth. A **461**, 262 (2001)
74. The Fermi-LAT Collaboration (A. Abdo et al.), Ap. J. S. **188**, 405A (2010)
75. The MAGIC Collaboration (D. Ferenc et al.), Nucl. Instrum. Meth. A **553**, 274 (2005)
76. The HESS Collaboration (G. Vasileiadis et al.), Nucl. Instrum. Meth. A **553**, 268 (2005)
77. The VERITAS Collaboration (T.B. Humensky et al.), Phys. Conf. Ser. **60**, 309 (2007)
78. The CTA Collaboration (M. Punch et al.), 44th Rencontres de Moriond on Very High Energy Phenomena in the Universe, 189 (2009)
79. The Mu2e Collaboration (S. Giovannella et al.), EPJ Web Conf. **179**, 01003 (2018)
80. The Mu2e Collaboration (L. Morescalchi et al.), PoS ICHEP20, 759 (2020)
81. The IceCube Detector, https://icecube.wisc.edu/gallery/press/view/1336
82. The IceCube Collaboration (M. Kowalski et al.), J. Phys. Conf. Ser. **888**, 012007 (2017)
83. The LUX Collaboration (D. S. Akerib et al.), Nucl. Instrum. Meth. A **704**, 111 (2003)
84. LUX-ZEPLIN (LZ) Technical Design Report, B.J. Mount et al., arXiv:1703.09144 [physics.ins-det] (2017)
85. The CTA Collaboration (E. Bissaldi et al.), Nuovo Cim. C **42**-6, 244 (2020)
86. A. Carrasco-Casado, J. Sanchez-Pena, R. Vergaz, arXiv:1512.00002 [astro-ph.IM] (2015)
87. The CTA Collaboration (D. Mazin et al.), PoS ICRC2021, 872 (2021)
88. H.K.M. Tanaka et al., Nucl. Instrum. Meth. A **757**, 489 (2007)
89. F. Ambrosino et al., JGR Solid Earth **120**-11, 7290 (2015)
90. A. Anastasio et al., Nucl. Instrum. Meth. A **732**, 423 (2013)
91. J. Marteau et al., Nucl. Instrum. Meth. A **695**, 23 (2012)
92. C. Cârloganu et al., Geosci. Instrum. Method. Data Syst. **2**, 55 (2013)
93. https://en.wikipedia.org/wiki/Positron_emission_tomography

94. S. Surti et al., J. Nucl. Med. **48**-3, 471 (2007)
95. https://indico.cern.ch/event/716539/contributions/3245979/attachments/1799570/2934772/VCI2019_Romo.pdf
96. P. Ferrario, JINST **13**-01, C01044 (2018)
97. L.G. Manzano et al., Nucl. Instrum. Meth. A **787**, 89 (2015)

Open Access This chapter is licensed under the terms of the Creative Commons Attribution 4.0 International License (http://creativecommons.org/licenses/by/4.0/), which permits use, sharing, adaptation, distribution and reproduction in any medium or format, as long as you give appropriate credit to the original author(s) and the source, provide a link to the Creative Commons license and indicate if changes were made.

The images or other third party material in this chapter are included in the chapter's Creative Commons license, unless indicated otherwise in a credit line to the material. If material is not included in the chapter's Creative Commons license and your intended use is not permitted by statutory regulation or exceeds the permitted use, you will need to obtain permission directly from the copyright holder.

Types and Sources of Radiation

We know two main sources of natural radiation.

- Cosmic and solar radiation from outer space, which typically consists of protons, light nuclei, electrons, gamma rays and neutrinos. In addition, we have charged pions and muons that are produced via interactions in the earth's atmosphere.
- Natural radioactivity from nuclei of chemical elements producing α, β and γ radiation as well as neutrons.

In addition, radioactive nuclei can be produced in reactors. So let us go through the individual types of radiation in more detail.

A.1 α Radiation

A helium nucleus penetrates the Coulomb barrier of nucleus $^A_Z X$ by tunneling, leading to the reaction $^A_Z X \rightarrow ^{A-4}_{Z-2} Y + ^4_2 He$. The α particle is mono-energetic in an energy range from 2 to 10 MeV. For example, ^{235}U decays to ^{231}Th with the emission of a 4.68 MeV α particle. The half life is 703.8 million years. The ^{231}Th nucleus is not stable, decaying via various isotopes into lead. Another example is ^{247}Am, which produces α particles with energies of 5.44 and 5.49 MeV with a half life of 433 years.

A.2 β Radiation

In a β decay a nucleus X emits an electron-antineutrino pair via the process $^A_Z X \rightarrow ^A_{Z+1} X + e\bar{\nu}_e$. In this three-body decay, only the electron is observed, since neutrinos and antineutrinos have very weak cross sections and, therefore, escape detection and the recoil of the nucleus is typically not seen. Thus, the electron spectrum is continuous as illustrated in Fig. A.1 (left). In the following we list a few examples.

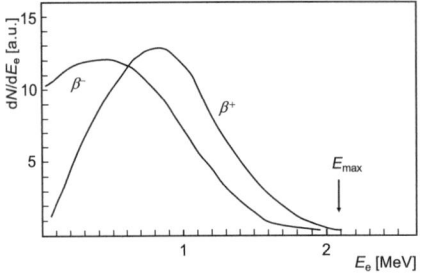 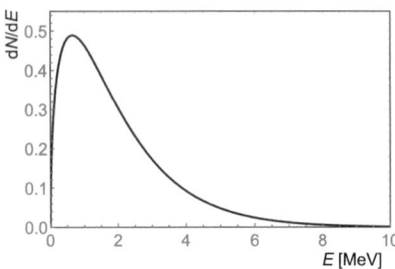

Fig. A.1 Left: Typical energy spectrum of electrons from β^- decay as a function of the kinetic energy in comparison to that of positrons from β^+ decays. Reprinted with kind permission under CC-BY-SA-4.0 Licence from [1], © 2019, C. R. Navi GSU. Right: Neutron energy distribution for ^{237}Cf

- ^{3}H decays to ^{3}He with an endpoint energy $E_{end} = 18.6$ keV and a half life $t_{1/2} = 12.26$ y.
- ^{14}C decays to ^{14}N with $E_{end} = 156$ keV and $t_{1/2} = 5730$ y.
- ^{90}Sr decays to ^{90}Y with $E_{end} = 0.546$ MeV and $t_{1/2} = 28.8$ y; then ^{90}Y decays under β emission to ^{90}Zr with $E_{end} = 2.279$ MeV and $t_{1/2} = 2.67$ d.
- ^{106}Ru decays to ^{106}Rh with $E_{end} = 0.039$ MeV and $t_{1/2} = 371.5$ d; then ^{106}Rh decays under β emission to ^{106}Pd with $E_{end} = 3.546$ MeV and $t_{1/2} = 30.1$ s.

In addition to β decay we also have $^A_Z X \rightarrow ^A_{Z-1} X + e^+ \nu_e$, producing a positron-neutrino pair. The positron annihilates with an atomic electron via $e^+ e^- \rightarrow 2\gamma$, where the photons each have 511 keV energy providing a good γ source. For example, ^{22}Na has a half life of $t_{1/2} = 2.6$ y. Since sodium decays to an excited neon state, there is a coincident 1.275 MeV γ in addition. Another example is ^{64}Cu, which has a half life $t_{1/2} = 12.8$ h.

A.3 γ Radiation

High-energy photons originating from excited nuclei are the source of γ radiation. The radiation is mono-energetic and observable energies vary from tens of keV to a few MeV depending on the excited nucleus. The ^{22}Na isotope mentioned before has a 1.275 MeV line.

Other sources are ^{137}Cs, which has a half life of $t_{1/2} = 5.27$ y and produces a photon with energy $E_\gamma = 0.662$ MeV, and ^{60}Co, which has a half life of $t_{1/2} = 30$ y and produces two monochromatic-energy photons at $E_\gamma = 1.173$ MeV and $E_\gamma = 1.333$ MeV.

A.4 Radiation from the Interaction of Excited Nuclei with the Atomic Shell

Excited nuclei interact with electrons in the atomic shell and can produce either photons or electrons.

- One such process is electron capture in which an electron from the atomic K or L shell is captured via the reaction $p + e \rightarrow n + \nu$, where the neutrino escapes. The new isotope may be in an excited state emitting additional mono-energetic γs. An example is ^{55}Fe, which has two photon lines at $E_\gamma = 5.9$ keV and $E_\gamma = 6.5$ keV and a half life $t_{1/2} = 2.7$ y.
- Another process is internal conversion in which the excitation in the nucleus is transferred to an electron in the K shell, producing mono-energetic electrons with an energy of $E_e = E^* - E_b$, where E^* is the excitation energy and E_b is the binding energy.
- A third process is the production of Auger electrons, where the excitation in the atomic shell is transferred to an electron in an outer shell. The electron has a discrete kinetic energy.

A.5 Neutrons from Spontaneous Fission

Neutrons are produced in spontaneous fission in transuranium elements. They have a continuous energy spectrum that is given by a Maxwellian shape

$$dN/dE_n = \sqrt{E_n} \exp\left(-\frac{E_n}{E_{\text{eff}}}\right), \tag{A.1}$$

where E_n is the neutron energy and E_{eff} is an energy specific to the fission element. For example, for ^{257}Cf $E_{\text{eff}} = 1.3$ MeV. The peak position is located at $E_{\text{eff}}/2$. Figure A.1 (right) shows the neutron energy distribution for ^{237}Cf. Neutron energies can run up to 10 MeV.

A.6 Neutrons from Nuclear Reactions

Neutrons are also produced in nuclear (α, n) and (γ, n) reactions. For example, α particles interacting with ^{9}Be produce an excited ^{13}C, which decays in three ways: $3\alpha + n$, ^{8}Be $+ \alpha + n$ and ^{12}C$^* + n$. The excited C* produces a 4.4 MeV mono-energetic photon. The continuous neutron energy is less than 10 MeV. In ^{210}Am, the yield is 70 neutrons per million α particles, while in ^{242}Cm the yield increases to 106 neutrons. To produce mono-energetic neutrons, we need to look at photon-nucleon interactions. For example, we have the processes $\gamma + ^9$B $\rightarrow ^8$B $+ $n and $\gamma + ^2$H $\rightarrow$ H $+$ n. The yield for (γ, n) reactions is a factor of 10 to 100 lower than that for (α, n) reactions.

A.7 Usage of Radioactive Sources

In particle and nuclear physics, cosmic rays and radioactive sources are useful tools for detector calibrations, system checks, track tracing and performance monitoring. More generally, ^{14}C is used for age determinations [2] since the amount of ^{12}C to ^{14}C has a fixed ratio in a living organism. Once the organism dies, ^{14}C just decays and from the ratio of ^{12}C to ^{14}C we can determine the age due to the known half life of ^{14}C. In the previous sections we have listed several sources. A more complete list of useful radioactive sources is found in [3].

References

1. https://physics.stackexchange.com/questions/462674/spectrum-of-beta-decay
2. J.R. Arnold, W.F. Libby, Science **110**-2869, 678 (1949)
3. The Particle Data Group (S. Navas et al.), Phys. Rev D. **110**, 030001 (2024)

Advanced Statistical Techniques   B

B.1 Introduction

In this appendix, we present some advanced statistical techniques, which are used in data analyses of measurements as well as in physics analyses. We assume the reader to know basic statistics. Excellent reviews are given in [1, 2]. Here, we just summarize a few important properties, list commonly used distributions and settle on a common notation before we start with the least-squares method. For an underlying Gaussian distribution it is equal to the maximum likelihood method we discuss after that. Then, we focus on linear and non-linear fits before turning to Multivariate analyses, where we discuss the Fisher discriminant, neural networks and decision trees. Measured distributions are typically Poisson or Gaussian distributions. If the possible outcome is yes or no and the number of samples is fixed, we encounter a binomial distribution.

B.1.1 Expectation Value, Variance and Covariance

Consider a continuous distribution $P(x)$ in the interval $a \leq x \leq b$. The expectation value is given by,

$$\mu = \int_a^b x P(x) \mathrm{d}x. \tag{B.1}$$

Note that μ denotes the theoretical mean calculated from a theory distribution that is different from the experimental average $\hat{\mu}$ of repeated measurements. The second moment or variance is given by,

$$\sigma^2 = \int_a^b (x - \mu)^2 P(x) \mathrm{d}x. \tag{B.2}$$

The variance is the average squared deviation of x from the mean μ. The square root of the variance is known as the standard deviation σ, measuring the dispersion or width of the distribution $P(x)$, showing how much x fluctuates around μ. For a discrete distribution, the integral is replaced by a sum.

A distribution may depend on several variables $x, y, z, \ldots$, which is described by a multivariate probability density function (pdf) $P(x, y, z, \ldots)$. So, here we have the mean values μ_x, μ_y and μ_z and variances σ_x^2, σ_y^2 and σ_z^2. In addition, we have correlations among the variables, which we characterize by covariances. The covariance between variables x and y is defined by,

$$cov(x, y) = \int_a^b (x - \mu_x)(y - \mu_y) P(x, y) dx dy. \tag{B.3}$$

For a distribution with three variables, we have the covariances $cov(x, y)$, $cov(x, z)$ and $cov(y, z)$. The covariances indicate how correlated the variables are. We define the correlation coefficient

$$\rho(x, y) = \frac{cov(x, y)}{\sigma_x \sigma_y}. \tag{B.4}$$

The correlation coefficient varies between -1 and $+1$. The sign indicates the sense of the correlation, anti-correlated or correlated. If two variables are perfectly (anti)correlated, we have $|\rho(x, y)| = 1$. If two variables are independent, we get $|\rho(x, y)| = 0$. Conversely, if $|\rho(x, y)| = 0$, then we know that x and y are not linearly dependent. However, they could be parabolically dependent, $y = x^2$.

B.1.2 Common Probability Density Distributions

We briefly define the Binomial, Poisson, Gaussian and χ^2 distributions.

B.1.2.1 Binomial Distribution

The binomial distribution accounts for repeated, independent trials of a process in which the outcome of a single trial is dichotomous, for example, yes/no, head/tail, hit/miss. Let us designate the two possible outcomes as success or failure. If the probability of success does not change from one trial to the next, the probability of r successes in N trials is

$$P(r) = \frac{N!}{r!(N-r)!} p^r (1-p)^{N-r}, \tag{B.5}$$

where p is the probability of a success in a single trial and $P(r)$ represents a discrete distribution with mean

$$\mu = \sum_r r P(r) = Np \tag{B.6}$$

and
$$\sigma^2 = \sum_r (r-\mu)^2 P(r) = Np(1-p). \tag{B.7}$$

The binomial distribution is normalized by summing $P(r)$ from $r = 0$ to $r = N$. In the large N limit for not too small p ($N > 30$, $p \geq 0.05$), the binomial distribution may be approximated by a Gaussian distribution. If p is small ($p < 0.05$) but Np is finite, the binomial distribution may be approximated by a Poisson distribution.

B.1.2.2 Poisson Distribution

The Poisson distribution is the limiting case for $p \to 0$ and $N \to \infty$ with $\mu = Np$ finite and is given by

$$P(r) = \frac{\mu^r \exp(-\mu)}{r!} \tag{B.8}$$

is also discrete, describing processes for which the single trial probability of success is very small but in which the number of trials is so large that there is reasonable rate of events. An Example is particle decays. Note that only μ appears, so the knowledge of p and N is not needed. Since μ is the only relevant parameter, the variance is

$$\sigma^2 = \mu. \tag{B.9}$$

This implies that the standard deviation is $\sigma = \sqrt{\mu}$, explaining the square root law in counting experiments. For large μ, the Poisson distribution approaches a Gaussian distribution with $\sigma^2 = \mu$.

B.1.2.3 Gaussian Distribution

The Gaussian distribution plays a central role in science since measurement errors and particularly instrumental errors are generally described by this pdf. Even in cases that are not strictly Gaussian, it provides a good approximation. The Gaussian is a continuous symmetric distribution, with a pdf

$$P(r) = \frac{1}{\sigma\sqrt{(2\pi)}} \exp\left(-\frac{(x-\mu)^2}{2\sigma^2}\right). \tag{B.10}$$

The two parameters μ and σ^2 are the mean and the variance. The full width at half maximum is related to the variance by

$$\text{FWHM} = 2\sigma\sqrt{(2\ln 2)} = 2.35\sigma. \tag{B.11}$$

Since the integral cannot be represented in analytical form, we have to perform a numerical integration. Any Gaussian can be brought into the reduced form by the variable transformation

$$z = \frac{x-\mu}{\sigma}. \tag{B.12}$$

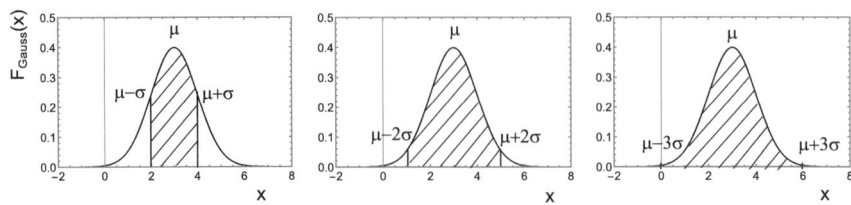

Fig. B.1 Left: The area of $\pm 1\sigma$ interval around the mean. Center: The area of $\pm 2\sigma$ interval around the mean. Right: The area of $\pm 3\sigma$ interval around the mean. Reprinted with kind permission from [2], © 1987, Springer. All rights reserved

The area under a Gaussian between integral values of σ have an important practical application for interpreting measurements errors as depicted in Fig. B.1. A measurement has a 68% probability to lie within 1σ of the mean, a 95% probability to be within 2σ and a 99.7% probability to be within 3σ. This implies that we have a 1/3 probability that the true value lies outside the 1σ limit and a 5% probability for being outside 2σ.

B.1.2.4 χ^2 Distribution

The χ^2 distribution is particularly useful for testing the goodness of fit of a theoretical hypothesis to experimental data. We assume that we have a set of n independent random variables x_i, distributed with Gaussian pdf's that have theoretical means μ_i and standard deviations σ_i.

The sum

$$\chi^2 = \sum_{i=1}^{n} \left(\frac{x_i - \mu_i}{\sigma_i} \right)^2 \tag{B.13}$$

is called χ^2. Since x_i is a random variable, so is χ^2 that follows the distribution

$$P(\chi^2) = \frac{\left(\frac{\chi^2}{2}\right)^{\frac{\nu}{2}-1} \exp\left(\frac{\chi^2}{2}\right)}{2\Gamma\left(\frac{\nu}{2}\right)}. \tag{B.14}$$

Figure B.2 depicts $P(\chi^2)$ for different values of ν. The integer ν is the degree of freedom (dof) and is only parameter in the distribution that can be interpreted as parameter related to the number of independent variables in the sum.

The mean and variance are found to be

$$\mu = \nu \tag{B.15}$$
$$\sigma^2 = 2\nu \tag{B.16}$$

The χ^2 represents a sum of deviations of measurements x_i from their mean μ_i weighted by $1/\sigma_i$. It characterizes the fluctuations in the data x_i. If the x_i are indeed Gaussian distributed with means μ_i and variances σ_i^2, then on average each ratio

Appendix B: Advanced Statistical Techniques

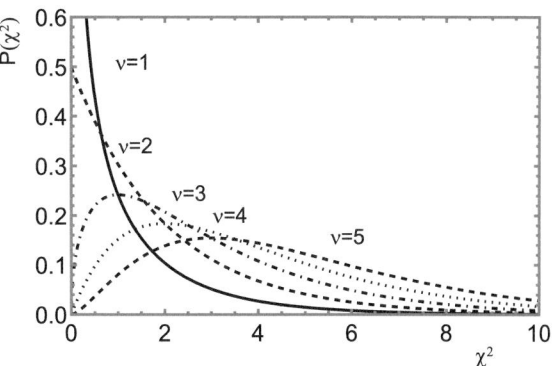

Fig. B.2 The χ^2 distribution for $\nu = 1$ (solid curve), $\nu = 2$ (upper dashed curve), $\nu = 3$ (dash-dotted curve), $\nu = 4$ (dotted curve) and $\nu = 5$ (lower dashed curve)

should be about 1 and $\chi^2 = \nu$. For any given set of x_i, of course, there will be fluctuations of χ^2 from this mean with a pdf given by (B.14). Thus, the χ^2 is useful for hypothesis testing, providing a measure of reasonableness of fluctuations in data around expectation. If an improbable χ^2 is found, we need to question the hypothesis.

B.1.3 A Word on Errors

Measurements in any kind of experiment are always subject to uncertainties or errors. We will argue that the measurement process is in fact a random process described by an abstract probability distribution whose parameters contain the information desired. The results of a measurement are then samples from this distribution, which allow for an estimate of the theoretical parameters. In this view, measurement errors can be viewed as sampling errors. Before following this path, it is necessary to discuss different types of errors.

B.1.3.1 Random or Statistical Errors

Random errors may be handled by the theory of statistics. They arise from an instrumental imprecision and/or from the inherent statistical nature of the phenomena of being observed. Statistically, both uncertainties are treated in the same manner as uncertainties arising from finite sampling of an infinite population of events. The measurement process is a sampling process, where the experimenter attempting to determine the parameters of a distribution too large to measure in its entirety takes finite-size random samples using the sample parameters as estimate of true values. This view point is most easily seen in measurements of statistical processes such as a radioactive decay or two-particle scattering, which are governed by statistical laws of quantum mechanics. The number of disintegrations or scatterings in a given time period is a random variable. We are interested in the mean of theoretical probability distribution.

When measuring the number of decays or scatterings per unit time, we take a sample from this distribution. The variable x takes values $x_1, x_2, \ldots, x_n$. From these data, we can estimate the mean value. Due to the finite sample, we get an uncertainty on the estimate representing our measurement error. Errors arising from measurements of inherently random processes are called statistical errors. Now consider measurements of quantities such as table length or the voltage in an electronic circuit. By taking several readings, we notice that our results will fluctuate. The differing values are the result of many small factors, which are not controlled by the experimenter and which may change between measurements, such as misalignment of ruler, round off errors and others. Such errors are called instrumental errors, including the observer. The magnitude of fluctuations becomes smaller if more effects determining them are well controlled. In the extreme limit, we get a δ function. The pdf for these fluctuations is Gaussian, where the mean of the Gaussian should be equal to true value of the quantity being measured and the standard deviation is proportional to precision of instrument.

B.1.3.2 Systematic Errors

Systematic errors are uncertainties that influence or bias the measurement process. Examples are, the zero position of an electrical instrument, the length and scale of a ruler, the precision of electric and magnetic fields and the precision of the luminosity measurement. If a voltmeter is not correctly zeroed, all measurements will be offset by some constant amount or some factor. Even if the utmost care is taken in zeroing the instrument, we can achieve only a zero within some limits $\pm\delta$, which may be small. This will affect the precision of our measurements. Typically for a systematic error is that it affects all measurements in the same direction though the exact value is usually unknown. This is conceptually different from a random error. There is no consistent treatment of systematic errors, e.g. estimation of magnitude or combination of different systematic errors. Different systematic errors are often approximated and treated as random errors since they are assumed to be independent and are based on approximate Gaussian pdfs. Typically, we distinguish between additive and multiplicative systematic errors.

Additive systematic errors affect signal yields directly. For example, if we determine the signal yield from a fit to signal and background, where the shapes, determined from simulation, are fixed and only the normalization is kept free. Here, we need to vary the shape by changing each parameter that fixes the shape by 1σ. This will modify the signal yield becoming larger or smaller. Multiplicative systematic errors result, if one calculates a new quantity from the measured yield, such as a branching fraction or cross section. Typically, the new quantity depends on several factors each of which has an uncertainty. For example, the total sample is uncertain by the luminosity error. First, we combine all additive and all multiplicative errors separately. At the end we may combine them.

B.1.3.3 Theoretical Non-statistical Errors

These are uncertainties that affect the precision of a prediction since a prediction may be affected by additional parameters whose functional form is not well known. For example, the prediction of a decay rate may be affected by some processes that have not been calculated but may be non negligible. So, often the effect is accounted for by an error estimate. We distinguish these errors from those based measured inputs (systematic errors) and call them theoretical non-statistical errors. Generally, it is not justified to assume that the error is associated with a pdf that is approximately Gaussian. Typically the pdf is unknown. Thus, we need to be careful with interpreting and combining theoretical errors with other systematic errors. There are no consistent treatments of these errors as well.

B.2 Analysis Methods

We have a set of measurements and need to find the underlying theoretical distribution. Since the number of data points is finite, we sample the theoretical distribution. Important is to have a representative unbiased sample, meaning that the experimentalist should not reject data because they do not look right. Then, we minimize the sum of squared deviations from the estimator. This is the best we can do since the true parameters are unknown. In statistics, this procedure is known as estimation. The estimation problem consists of two parts, first determine the best estimate and second determine the uncertainty on the estimate. The most widely accepted method for this task is the least-square method. Before describing this method, we need to define a few terms. Consider a sample $x_1, x_2, \ldots, x_n$ of size n taken from a distribution whose theoretical mean and variance are μ and σ^2, respectively. This is called the sample population. The sample mean is defined by

$$\bar{x} = \frac{1}{n} \sum_{i=1}^{n} x_i \tag{B.17}$$

and corresponds to the arithmetic average of the sample. In the limit of $n \to \infty$, this approaches the theoretical mean,

$$\mu = \lim_{n \to \infty} \frac{1}{n} \sum_{i=1}^{n} x_i \tag{B.18}$$

Similarly, the sample variance denoted by s^2 is given by

$$s^2 = \frac{1}{n} \sum_{i=1}^{n} (x_i - \bar{x})^2. \tag{B.19}$$

It is the average of the squared deviations. In the limit of $n \to \infty$, this approaches the theoretical variance σ^2. In the case of multivariate samples, e.g. (x_1, y_1), $(x_2, y_2), \ldots, (x_n, y_n)$, sample means and variances are calculated as above. In an analogous manner a sample covariance can be calculated by

$$cov(x, y) = \frac{1}{n} \sum_{i=1}^{n} (x_i - \bar{x})(y_i - \bar{y}). \tag{B.20}$$

In the limit of infinite n, this approaches the theoretical covariance.

B.2.1 The Maximum Likelihood Method

The method of maximum likelihood is only applicable if the form of the theoretical distribution is known. For most measurements in physics, this is either a Gaussian or Poisson distribution. To be general, suppose we have a sample of n independent observations $x_1, x_2, \ldots, x_n$ from a theoretical distribution $f(x|\theta)$, where θ is the parameter to be estimated. The method consists of calculating the likelihood function

$$L(\theta|x) = f(x_1|\theta) f(x_2|\theta) \ldots f(x_n|\theta), \tag{B.21}$$

which can be recognized as the probability for observing the sequence of values $x_1, x_2, \ldots, x_n$. Given hypothesis H, $P(R|H)$ is the probability to obtain result R and $L(H|R)$ is the likelihood of hypothesis H given results R. The principle states that this probability is a maximum for the observed values. This indicates that parameter θ must be such that L is a maximum. If L is a regular function, parameter θ can be found by solving

$$\frac{\partial L}{\partial \theta} = 0. \tag{B.22}$$

If there is more than one parameter, then partial derivatives of L with respect to each parameter must be taken, obtaining a system of equations. Depending on the form of L, it may be easier to maximize $\ln(L)$ rather than L, solving

$$\frac{d \ln L}{d\theta} = 0. \tag{B.23}$$

The solution, $\hat{\theta}$, is known as the maximum likelihood estimator for the parameter θ. Note that $\hat{\theta}$ is also a random variable since its a function of the x_i. If a second sample is taken, $\hat{\theta}$ will have a different value. Next we have to determine the error on $\hat{\theta}$, which is determined from the standard deviation of the estimator function. We can calculate it from the likelihood function if we recall that L is just the probability for observing the sampled values $x_1, x_2, \ldots, x_n$, yielding

$$\sigma^2(\hat{\theta}) = \int (\hat{\theta} - \theta)^2 L(\theta|x) dx_1 dx_2 \ldots dx_n. \tag{B.24}$$

This is a general formula and only in special cases has an analytical form. An easier but just approximate form that works in the limit of large numbers is to calculate the inverse second derivate of $\ln(L)$ at the maximum,

$$\sigma^2(\hat{\theta}) = -\left(\frac{d^2 \ln L}{d\theta^2}\right)^{-1} \tag{B.25}$$

In case of more than one parameter we need to determine matrix,

$$U_{ij} = -\frac{\partial^2 \ln L}{\partial \theta_i \partial \theta_j}. \tag{B.26}$$

The diagonal elements of the inverse matrix give approximate variances,

$$\sigma^2(\hat{\theta}_i) = \left(U^{-1}\right)_{ii}. \tag{B.27}$$

Note that we assumed here that the mean value $\hat{\theta}$ is the theoretical θ value, which is a desirable but not essential property for an estimator. It is guaranteed by the maximum likelihood method only for an infinite sample. The variance remains valid for all $\hat{\theta}$ since the error is the deviation from the true mean.

B.2.1.1 Estimator for the Poisson Distribution

Let us consider the Poisson and Gaussian distributions as an examples. The likelihood function for the Poisson distribution is

$$L_{\text{Poisson}}(\mu|x) = \prod_{i=1}^{n} \frac{\mu^{x_i}}{x_i!} \exp(-\mu) = \exp(-n\mu) \prod_{i=1}^{n} \frac{\mu^{x_i}}{x_i!}. \tag{B.28}$$

Taking the logarithm converts the product into a sum,

$$\ln L_{\text{Poisson}} = -n\mu + \sum_{i=1}^{n} x_i \ln \mu - \sum_{i=1}^{n} \ln x_i!. \tag{B.29}$$

Differentiating and setting the result to zero yields,

$$\frac{d \ln L_{\text{Poisson}}}{d\mu} = -n + \frac{1}{\mu} \sum_{i=1}^{n} x_i = 0. \tag{B.30}$$

The solution is

$$\hat{\mu} = \frac{1}{n} \sum_{i=1}^{n} x_i = \bar{x}. \tag{B.31}$$

The variance is obtained from (B.25)

$$\sigma^2(\bar{x}) = \frac{\bar{x}}{n}, \tag{B.32}$$

where we have substituted the estimated value $\hat{\mu}$ for the theoretical μ value.

B.2.1.2 Estimator for the Gaussian Distribution

The likelihood function for the Gaussian distribution is

$$L_{\text{Gauss}} = \prod_{i=1}^{n} \frac{1}{\sigma\sqrt{(2\pi)}} \exp\left(-\frac{(x_i - \mu)^2}{2\sigma^2}\right). \quad (B.33)$$

Taking the logarithms yields,

$$\ln L_{\text{Gauss}} = -\frac{n}{2} \ln(2\pi\sigma^2) - \frac{1}{2} \sum \left[\frac{(x_i - \mu)^2}{\sigma^2}\right]. \quad (B.34)$$

Taking derivatives with respect to μ and σ^2 and setting them to zero, yields

$$\frac{\partial \ln L_{\text{Gauss}}}{\partial \mu} = \sum \frac{x_i - \mu}{\sigma^2} = 0,$$

$$\frac{\partial \ln L_{\text{Gauss}}}{\partial \sigma^2} = -\frac{n}{2\sigma^2} + \frac{1}{2} \sum \frac{(x_i - \mu)^2}{\sigma^2} \frac{1}{\sigma^2} = 0. \quad (B.35)$$

The solutions are

$$\hat{\mu} = \frac{1}{n} \sum x_i = \bar{x}, \quad (B.36)$$

$$\sigma(\bar{x}) = \frac{\sigma}{\sqrt{n}}, \quad (B.37)$$

which represent mean and standard deviation of the mean, respectively. As n increases, the estimate $\bar{x}$ becomes more and more precise. For $n = 1$, $\sigma(x)$ reduces to σ. So for a measuring device, σ represents the instrument precision.

Solving the second equation in (B.35) yields,

$$\hat{\sigma}^2 = \frac{1}{n} \sum_{i=1}^{n} (x_i - \mu)^2 \simeq \frac{1}{n} \sum_{i=1}^{n} (x_i - \bar{x})^2 = s^2. \quad (B.38)$$

For finite values of n, the sample variance turns out to be a biased estimator. It means that the expectation value of s^2 does not equal the true value but is offset by a constant factor. In fact, the expectation value of s^2 is $\sigma^2 - \sigma^2/n = (n-1)\sigma^2/n$. For very large n, s^2 approaches the true value, while for small n σ^2 is underestimated by s^2. A somewhat better estimate involves the factor $n/(n-1)$, yielding

$$\hat{\sigma}^2 = \frac{1}{n-1} \sum_{i=1}^{n} (x_i - \bar{x})^2. \quad (B.39)$$

This result is unbiased and is recommended for estimating the variance. Note that unlike the mean, it is impossible to estimate the standard deviation from one measurement because of $(n-1)$ term in the denominator. The variance of σ^2 becomes

$$\sigma(\hat{\sigma}^2) = \frac{2\sigma^4}{n-1} = \frac{2\hat{\sigma}^4}{n-1}, \tag{B.40}$$

yielding a standard deviation of

$$\sigma(\hat{\sigma}) = \frac{2\sigma}{\sqrt{(2n-1)}} = \frac{2\hat{\sigma}}{\sqrt{(2n-1)}}. \tag{B.41}$$

B.2.1.3 Weighted Mean

So far we have discussed the estimation of the mean and standard deviation from a series of measurements of the same observable with the same instrument. It often occurs that we have to combine two or more measurements of same observable with different errors. Thus, we need to weight each measurement in proportion to its error. The maximum likelihood method allows us to determine the weighting function. From a statistical view point, we have a sample of $x_1, x_2, \ldots, x_n$, where each value comes from a Gaussian distribution with same mean μ but different standard deviation σ_i. The likelihood function is same as that in (B.33) with σ replaced by σ_i. Maximizing these yields, the weighted mean becomes

$$\hat{\mu} = \left(\sum_i \frac{x_i}{\sigma_i^2} \bigg/ \sum_i \frac{1}{\sigma_i^2} \right). \tag{B.42}$$

The weighting factor is $1/\sigma_i^2$ and the error of weighted mean is,

$$\sigma^2(\hat{\mu}) = \left(1 \bigg/ \sum_i \frac{1}{\sigma_i^2} \right). \tag{B.43}$$

If all errors are the same, the weighted mean and variance reduce to the results in (B.36).

B.2.2 The Least-Square Method

Lets us suppose to make measurements at n points x_i of observable $y_i = y(x_i)$ with experimental uncertainties σ_i ($i = 1, 2, \ldots, n$). We want to fit the data to a function $f(x; a_1, a_2, \ldots, a_m)$, where $a_1, a_2, \ldots, a_m$ are unknown parameters to be determined. It is obvious that the number of points must be greater than the number of parameters. The method of least squares states that the best values of a_i are those for which the following sum becomes a minimum,

$$S = \sum_{i=1}^{n} \left(\frac{y_i - f(x_i; a_j)}{\sigma_i} \right)^2 \tag{B.44}$$

This just represents the sum of squared deviations of the data points from the curve $f(x_i)$ weighted by the errors on y_i, respectively. This procedure is often called χ^2 minimization. Strictly speaking, this is not quite correct as the y_i must be Gaussian distributed with a mean $f(x_i; a_j)$ and a variance σ_i^2 for S to be true χ^2. However, for most measurements in physics this is a valid hypothesis most of the time. Note that the least-square method is totally general and requires no knowledge of the parent distribution. If the parent distribution is known, the method of maximum likelihood may also be used. In case of Gaussian errors, this yields identical results. To find values of parameters a_j, we must solve system of equations,

$$\frac{\partial S}{\partial a_j} = 0. \quad (B.45)$$

Depending on the function $f(x; a_j)$ (B.45) may or may not have an analytic solution. Once we have determined the parameters a_j, we need to evaluate errors on the parameters, i.e. determining the error matrix V_{ij}. We have

$$\left(V\right)_{ij}^{-1} = \frac{1}{2}\frac{\partial^2 S}{\partial a_i \partial a_j}, \quad (B.46)$$

where the second derivative is evaluated at the minimum. Note that the second derivatives forms the inverse of the error matrix. The diagonal elements V_{ii} represent the variances, while the off-diagonal elements V_{ij} with $i \neq j$ represent the covariances between the parameters a_i and a_j. Thus, we can write

$$\begin{pmatrix} \sigma_1^2 & cov(1,2) & cov(1,3) & cov(1,4) & \ldots \\ cov(2,1) & \sigma_2^2 & cov(2,3) & cov(2,4) & \ldots \\ cov(3,1) & cov(3,2) & \sigma_3^2 & cov(3,4) & \ldots \\ cov(4,1) & cov(4,2) & cov(4,3) & \sigma_4^2 & \ldots \\ \ldots & \ldots & \ldots & \ldots & \ldots \end{pmatrix}$$

B.2.3 Linear Fit

Let us consider the example of a linear or straight-line fit.

$$y = f(x) = ax + b, \quad (B.47)$$

where a and b are parameters that need to be determined. So, here

$$S = \sum_{i=1}^{n} \frac{(y_i - a \cdot x_i - b)^2}{\sigma_i^2}. \quad (B.48)$$

Appendix B: Advanced Statistical Techniques

To determine the parameters a and b, we take partial derivatives,

$$\frac{\partial S}{\partial a} = -2 \sum_{i=1}^{n} \frac{(y_i - a \cdot x_i - b) x_i}{\sigma_i^2} = 0,$$

$$\frac{\partial S}{\partial b} = -2 \sum_{i=1}^{n} \frac{(y_i - a \cdot x_i - b)}{\sigma_i^2} = 0. \quad \text{(B.49)}$$

For convenience, we define the sums

$$A = \sum_{i=1}^{n} \frac{x_i}{\sigma_i^2}, \quad B = \sum_{i=1}^{n} \frac{1}{\sigma_i^2}, \quad C = \sum_{i=1}^{n} \frac{y_i}{\sigma_i^2}, \quad D = \sum_{i=1}^{n} \frac{x_i^2}{\sigma_i^2}, \quad E = \sum_{i=1}^{n} \frac{x_i y_i}{\sigma_i^2}. \quad \text{(B.50)}$$

The two equations become

$$2(-E + aD + bA) = 0,$$
$$2(-C + aA + bB) = 0. \quad \text{(B.51)}$$

yielding the solutions.

$$a = \frac{E \cdot B - A \cdot C}{D \cdot B - A^2},$$

$$b = \frac{D \cdot C - A \cdot E}{D \cdot B - A^2}. \quad \text{(B.52)}$$

The next step is to determine the uncertainties. We form the inverse error matrix

$$V^{-1} = \begin{pmatrix} A_{11} & A_{12} \\ A_{21} & A_{22} \end{pmatrix}, \quad \text{(B.53)}$$

where

$$A_{11} = \frac{1}{2} \frac{\partial^2 S}{\partial a^2}, \quad A_{22} = \frac{1}{2} \frac{\partial^2 S}{\partial b^2}, \quad A_{12} = A_{21} = \frac{1}{2} \frac{\partial^2 S}{\partial a \partial b}. \quad \text{(B.54)}$$

Inversion yields

$$V = \frac{1}{A_{11} A_{22} - A_{12}^2} \begin{pmatrix} A_{22} & -A_{12} \\ -A_{12} & A_{11} \end{pmatrix}. \quad \text{(B.55)}$$

So, the variances are

$$\sigma_a^2 = \frac{A_{22}}{A_{11} A_{22} - A_{12}^2} = \frac{B}{B \cdot D - A^2},$$

$$\sigma_b^2 = \frac{A_{11}}{A_{11} A_{22} - A_{12}^2} = \frac{D}{B \cdot D - A^2},$$

$$cov(a, b) = \frac{-A_{12}}{A_{11} A_{22} - A_{12}^2} = \frac{-A}{B \cdot D - A^2}. \quad \text{(B.56)}$$

The next step is to evaluate the quality of the fit. This tells us that the selected fit function $f(x; a_j)$ really represents the data. We determine the χ^2 value, which is just the value of S at the minimum. If the data correspond to the function and the derivatives are Gaussian, S is expected to follow a χ^2 distribution with a mean value equal to the number of degrees of freedom (dof) ν. In the fit, we have n independent data points from which we extract m parameters, $\nu = n - m$. In the straight line fit $m = 2$, so that $\nu = n - 2$. Thus, for a good fit we expect S to be close to $n - 2$. A quick and easy test is to form the reduced χ^2,

$$\frac{\chi^2}{\nu} = \frac{S}{\nu}, \tag{B.57}$$

which should be close to one for a good fit. A more rigorous test consists of looking at the probability of obtaining a χ^2 value greater than S, i.e. $P(\chi^2) \geq S$ that requires an integration of the χ^2 distribution or use of cumulative distribution tables. A general rule of thumb is that a fit is acceptable if $P(\chi^2) \geq S > 5\%$. Another important point to consider is the regime where S is small. This implies that the points are not sufficiently fluctuating, which is most likely caused by an overestimation of the errors on data points. Since error bars represent 1σ, we expect about one third of the data point to fall outside the fit.

We may extend the method of linear least squares to certain non-linear functions, such as exponential functions. For example, we want to determine the lifetime of a radioactive decay,

$$N = N_0 \exp\left(-\frac{t}{\tau}\right). \tag{B.58}$$

By taking the logarithm, we can linearize the function.

$$\ln N = \ln N_0 - \frac{t}{\tau} \tag{B.59}$$

We set $y = \ln N$, $a = 1/\tau$ and $b = \ln N_0$. However, we need to be careful with the errors. Statistical errors are typically Poisson distributed. Thus, we first need to transform the errors using the error propagation method. We get

$$\sigma^2(\ln N) = \left(\frac{\partial \ln N}{\partial N}\right)^2 \sigma^2(N) = \frac{1}{N^2} N = \frac{1}{N}. \tag{B.60}$$

So far we have assumed that the independent variables x_i are completely free of errors. In reality, of course, this is never true, though typically errors on x_i are small with respect to those on y_i and are neglected. In cases where errors on x_i and y_i are comparable, both need to be considered, where the method of effective variance may be used in which the variance in (B.48) is replaced by

$$\sigma_i^2 \rightarrow \sigma_{y_i}^2 + \left(\frac{\partial f}{\partial x}\right)^2 \sigma_{x_i}^2, \tag{B.61}$$

where $\sigma_{x_i}^2$ and $\sigma_{y_i}^2$ are the variances of x_i and y_i, respectively. Since the derivative is usually a function of the parameters a_j, S is non-linear and numerical methods must be used to minimize S.

B.2.4 Nonlinear Fit

Non-linear fits generally require numerical procedures for minimizing S. For this purpose, a number of methods have been developed, which are just mentioned. Some discussion is given in [2]. The problem is to find the right method for the function to be fitted. Numerical minimization methods are generally iterative in nature since repeated calculations are made, while varying the parameters until the desired minimum is reached. The criteria for selecting a method are speed and stability against divergences. The methods can be classified into two broad categories, grid searches and gradient methods. Grid methods are the simplest. Here, we form a grid of equally spaced points in the variables of interest and evaluate the function at these points. The desired accuracy depends on the grid step size. For example, the simplex method is one of them.

A simplex is the simplest geometrical figure in n dimensions having $n + 1$ vertices, yielding a triangle in two dimensions, and a tetrahedron in three dimensions. Let us consider the two-dimensional case as an example. First, $n + 1 = 3$ points are chosen in some way and a simplex is formed as shown in Fig. B.3. The point with the highest value is denoted by P_H, that with the lowest value by P_L. In the next step, P_H is replaced with a better point, by reflecting it through the center-of-gravity of all points except P_H,

$$\bar{P} = \sum_i \frac{P_i - P_H}{n}, \tag{B.62}$$

yielding

$$P^* = \bar{P} + (\bar{P} - P_H). \tag{B.63}$$

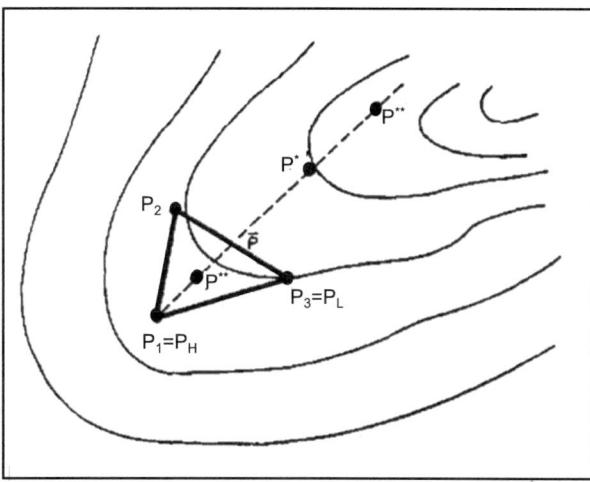

Fig. B.3 Illustration of the simplex grid method. Reprinted with kind permission from [2], © 1987, Springer. All rights reserved

If $F(P*) < F(P_L)$, a new minimum has been found and an attempt is made to do even better by trying the point

$$P^{**} = \bar{P} + 2(\bar{P} - P_H). \tag{B.64}$$

The best point is kept. If $F(P*) > F(P_H)$, the reflection is brought backwards to

$$P^{**} = \bar{P} + \frac{1}{2}(\bar{P} - P_H). \tag{B.65}$$

If this is not better than P_H, a new simplex is formed with points at $P_i = (P_i + P_L)/2$ and the procedure is restarted. In this manner, the triangle is falling towards the minimum. It is a robust technique insensitive to the type of function but can be rather slow.

Gradient methods are techniques that make use of the derivatives of the functions to be minimized. They can be obtained numerically or by functions supplied by user. The derivatives are used as guides, pointing into the direction of decreasing the function $F(x)$. Newton's method is a widely used method that is based on a second degree Taylor expansion of the function $F(x)$ at x_0.

$$F(x) = F(x_0) + \left.\frac{\partial F}{\partial x}\right|_{x_0}(x - x_0) + \frac{1}{2}\left.\frac{\partial^2 F}{\partial x^2}\right|_{x_0}(x - x_0)^2. \tag{B.66}$$

In n dimensions, this is generalized to

$$F(x) = F(x_0) + \mathbf{g}^T(x - x_0) + (x - x_0)^T \mathbf{G}(x - x_0), \tag{B.67}$$

where $\mathbf{g}$ is the vector of the first derivatives and $\mathbf{G}$ is the matrix of second derivatives, also called the hessian.

The method approximates the function $F(x)$ around x_0 by a quadratic surface. Under this assumption, it is straight forward to calculate the minimum of the n-dimensional parabola analytically.

$$x_{\min} = x_0 - \mathbf{G}^{-1}\mathbf{g}. \tag{B.68}$$

Of course, this is not the true minimum of the function, but by forming a new parabolic surface about $x_{\min}$ and calculating its minimum, a convergence to the true minimum can be obtained rather rapidly. The basic problem of this technique is that $\mathbf{G}$ has to be positive definite everywhere. Otherwise, at some point in iteration a maximum rather than a minimum may be calculated and the entire process diverges. This can be easily seen in the one-dimensional case (see (B.66)). If the second derivative is negative, we clearly have inverted parabola. Despite this defect, Newton's method is quite powerful and algorithms have been developed in which matrix $\mathbf{G}$ is artificially altered whenever it becomes negative. In this manner the iteration continues in the right direction until a region of positive-definiteness is reached. These techniques are also called quasi Newton's methods.

The disadvantage of Newton's method is that it requires an evaluation of matrix G and its inverse for each iteration, which is time consuming. In specific cases of the least squares minimization, a common procedure used with Newton's method is to linearize the fitting function, which is equivalent to approximate the hessian in following way. Rewriting

$$S = \sum_k s_k^2, \tag{B.69}$$

where $s_k = [y_k - f(x_k)]/\sigma_k$, the hessian becomes

$$\frac{\partial^2 S}{\partial x_i \partial x_j} = \frac{\partial}{\partial x_i} \frac{\partial}{\partial x_j} \sum_k s_k^2 = 2 \sum_k \frac{\partial s_k}{\partial x_i} \frac{\partial s_k}{\partial x_j} + s_k \frac{\partial^2 s_k}{\partial x_i \partial x_j}. \tag{B.70}$$

The second term in the sum can be considered as a second-order correction and is set to zero, yielding a hessian of form,

$$G_{ij} \simeq 2 \sum \frac{\partial s_k}{\partial x_i} \frac{\partial s_k}{\partial x_j}. \tag{B.71}$$

This approximation has the advantage of ascertaining positive definiteness and the result converges to the correct minimum. In general, the covariance matrix, however, will not converge to the correct covariance values, implying that errors determined from this matrix may not be correct. So, it is not a trivial task to implement a non linear least square fit. A powerful fit program called MINUIT [3], using various minimization methods including Simplex and Newton's method, was developed by James for CERN and is in the CERN library.

Another issue is to determine the global minimum among local minima. Up to now, we have assumed that $F(x)$ contains only one minimum. More generally, arbitrary functions can have many local minima in addition to a global, absolute minimum. The methods we have described are all designed to find a local minimum without regard for any possible other minima. It is up to the user to decide if the minimum found is the global minimum. Thus, we need to have an idea what the true values are, in order to start the search in the right region. Even in this case, there is no guarantee that minimization will converge onto the closest minimum. A good technique is to fix approximately known parameters and vary the remaining ones. Then, we use the results to start a second search in which all parameters are varied.

Other problems involve underflow or overflow, occurring often in exponential functions. Thus, good starting points are important. The methods discussed provide parameter values from the minimum S, but we have not given any prescription how to evaluate errors. We may obtain a clue from the linear one-dimensional case, where

$$\sigma^2 = \left| \frac{1}{2} \frac{\partial^2 S}{\partial \theta^2} \right|^{-1}. \tag{B.72}$$

If we expand S in a Taylor series about minimum, we get

$$S(\theta) = S(\theta^*) + \frac{1}{2}\frac{\partial^2 S}{\partial \theta^2}(\theta - \theta^*)^2 = S(\theta^*) + \frac{1}{\sigma^2}(\theta - \theta^*)^2. \quad (B.73)$$

At the point $\theta = \theta^* + \sigma$, we find

$$S(\theta^* + \sigma) = S(\theta^*) + 1. \quad (B.74)$$

Thus, the error on θ corresponds to the distance between the minimum and where the S distribution increases by 1. This can be generalized to the non-linear case, where the S distribution is not generally parabolic around the minimum. Finding errors for each parameter implies to get points for which the S value changes by 1 from the minimum. If S has a complicated form, this may not be easy requiring numerical methods. If the form of S can be approximated by a quadratic surface, the error matrix can be calculated and inverted as in the linear case.

B.3 Multivariate Analysis Techniques

In data analysis, the goal is to extract the signal with the highest efficiency with the smallest amount of residual background. The simplest, but least efficient method is to apply a few successive selections (cuts) on the data to enhance the signal over background. Let us consider a simple example of two arbitrary variables x, y. Figure B.4 shows the scatter plot of signal A with respect to that of background B in the x-y plane. The projections in x and y are shown on the bottom and left-hand side. With simple selection criteria for variables x and y denoted by the vertical and horizontal lines we can extract the signal. Note that we loose some signal and still have a lot of background left. We can tighten the selection to get a cleaner signal at the cost of lower efficiency. These types of analyses are called "cut and count" analyses. These analyses become easier if only uncorrelated observables are used.

We would like to combine the information in observables x and y into a single output variable q that has a larger separation between signal and background than the selection on each observable x and y. We want the maximum separation, i.e. highest efficiency and highest purity. This is schematically depicted in Fig. B.5. There are different multivariate techniques that combine different observables into one output observable. For all of them correlated observables do not cause additional problems.

B.3.1 The Fisher Discriminant

Assume we have a sample of N events and a set of k observables x_j. We can define a linear discriminating variable, the Fisher discriminant, for each event i.

$$\mathcal{F}_i(\vec{x}) = \sum_{j=1}^{k} \alpha_j x_j|_i + \beta, \quad (B.75)$$

Appendix B: Advanced Statistical Techniques

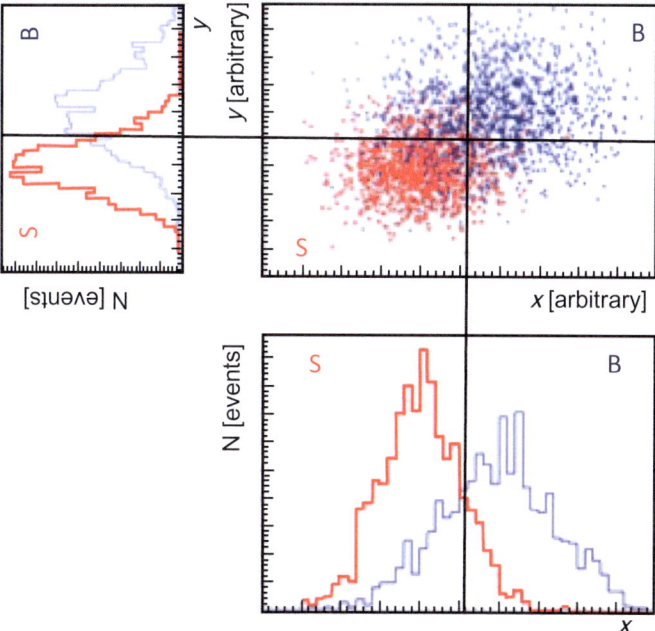

Fig. B.4 Correlation plot and projections of two variables x and y, showing signal (S) and background (B). The horizontal and vertical lines show selections to separate signal from background (left of the vertical bar and below the horizontal bar). Reprinted with kind permission from [4], © 1994, *BABAR* Collaboration. All rights reserved

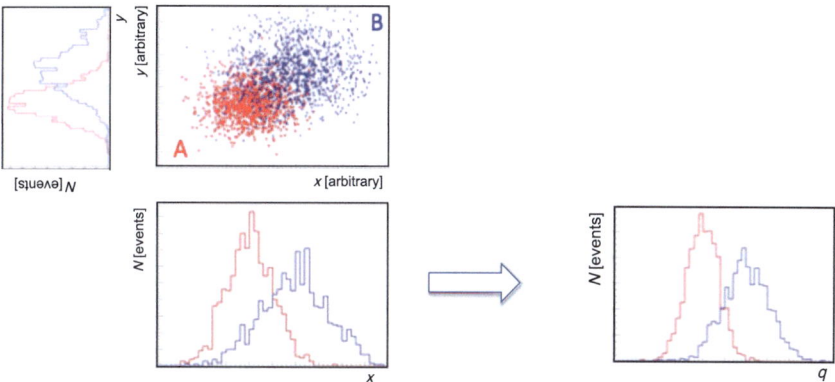

Fig. B.5 Left: The correlation plot and projections of variables x and y from Fig. B.4. Right: The output of a multivariate figure-of-merit q for the two variables x and y. Reprinted with kind permission from [4], © 1994, *BABAR* Collaboration. All rights reserved

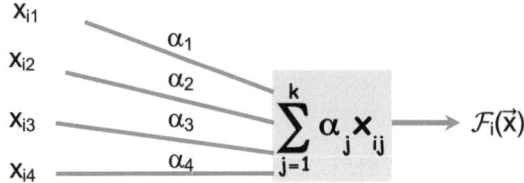

Fig. B.6 Combining four variables into a Fisher discriminant $\mathcal{F}_i(\vec{x}) = \vec{\alpha}^T \vec{x}|_i$ (see text)

where, the α_i are linear coefficients and β is an offset that may be set to zero. Let us consider an example with four observables as illustrated in Fig. B.6.

In fact, what we know are the mean $\mu(x_j)$ and variance $\sigma^2(x_j)$ for both signal and background. We can write the Fisher mean and variance of the corresponding signal and background distributions as,

$$\begin{aligned} M_S &= \alpha^T \mu_S & \sigma_S^2 &= \alpha^T \sigma_S \alpha \\ M_B &= \alpha^T \mu_B & \sigma_B^2 &= \alpha^T \sigma_B \alpha \end{aligned} \tag{B.76}$$

The goal is to maximize $|M_S - M_B|$ and to minimize Σ_S^2 and Σ_B^2. We can balance these requirements with

$$J(\alpha) = \frac{[M_S - M_B]^2}{\Sigma_S^2 + \Sigma_B^2}, \tag{B.77}$$

where

$$\begin{aligned} {[M_S - M_B]^2} &= \sum_{j,m=1}^{k} \alpha_j \alpha_m \left(\mu_S - \mu_B\right)_j \left(\mu_S - \mu_B\right)_m = \alpha^T B \alpha \\ \Sigma_S^2 + \Sigma_B^2 &= \sum_{j,m=1}^{k} \alpha_j \alpha_m \left(V_S + V_B\right)_{jm} = \alpha^T W \alpha. \end{aligned} \tag{B.78}$$

So

$$J(\alpha) = \frac{\alpha^T B \alpha}{\alpha^T W \alpha}. \tag{B.79}$$

Optimal separation is achieved for

$$\frac{\partial J(\alpha)}{\partial \alpha} = 0, \tag{B.80}$$

where

$$\alpha \propto W^{-1} \left(\mu_S - \mu_B\right). \tag{B.81}$$

This requires the determination of the inverse matrix W^{-1}. Thus, we can compute the output of the Fisher discriminant from target samples of signal and background events. Typically, we use half of Monte Carlo samples for determining α_j and the other half for validation. The solution can be obtained up to an arbitrary scale.

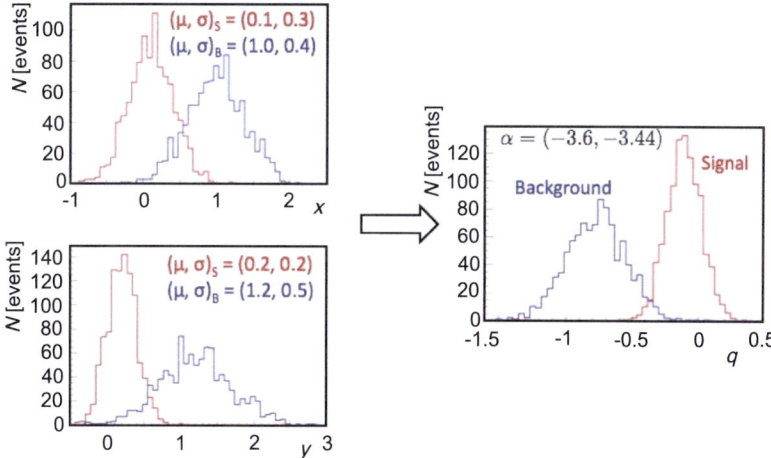

Fig. B.7 Left: Signal (red histogram) and background (blue histogram) distributions for variables x and y. The mean values and standard deviations for signal and background are shown in the figure for both variables. Right: The output distributions of the Fisher discriminant for signal and background. The linear coefficient is $\vec{\alpha} = (-3.6, -3.44)$. Reprinted with kind permission from [4], © 1994, *BABAR* Collaboration. All rights reserved

Let us consider an example. Figures B.7 (left top and left bottom) show the signal and background distributions for two observables. Figure B.7 (right) shows the signal and background distribution after using the Fisher discriminant. The signal mean minus background mean for the two distributions are

$$\left(\mu_S - \mu_B\right)_j = \left(-0.9, -1.0\right). \tag{B.82}$$

The **W** matrix is

$$W = \begin{pmatrix} 0.25 & 0 \\ 0 & 0.29 \end{pmatrix}. \tag{B.83}$$

B.3.2 Neural Networks

While the Fisher discriminant is a linear algorithm, neural networks (NN) are non-linear algorithms. The fundamental building block of a neural network is the perceptron, which is the algorithmic analog of a neuron depicted in Fig. B.8 (left). We compute a variable y, which is an n-dimensional hyperspace,

$$y = \vec{w} \cdot \vec{x} + b, \tag{B.84}$$

where y is the output, $\vec{x}$ is the input vector, $\vec{w}$ is vector of weights and b is some offset. For a binary threshold we have $Q = 1$ if $y > 0$ otherwise $Q = 0$. The perceptron

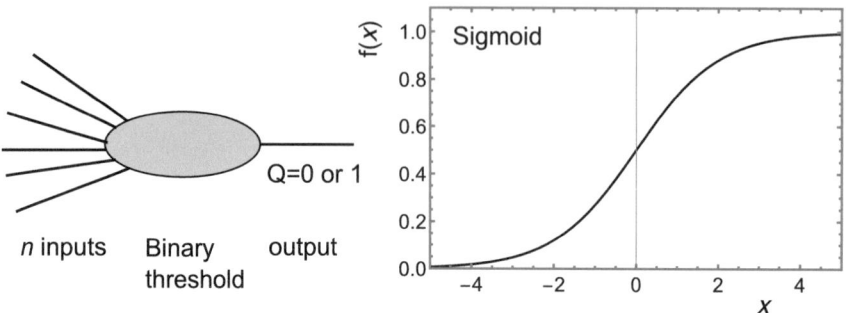

Fig. B.8 Left: Illustration of the working principle of a binary perceptron. Right: A Sigmoid function, $f(x) = 1/(1 + \exp(a \cdot x + b))$

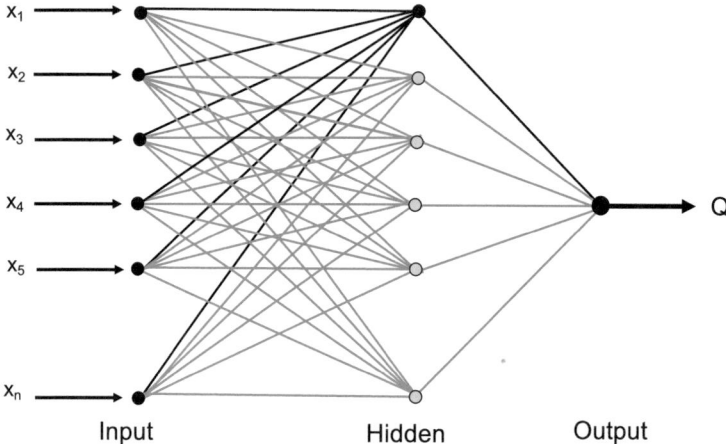

Fig. B.9 A multi-layer perceptron, having n inputs, one hidden layer and one output

basically makes a selection in the n-dimensional hyperspace. It is useful to replace the binary decision with an activation function, for example a Sigmoid or logistic function shown in Fig. B.8 (right),

$$y = f(\vec{w} \cdot \vec{x} + b). \tag{B.85}$$

Let us consider an example of a multi-layer perceptron, which has n inputs, one hidden layer and one output as depicted in Fig. B.9. The procedure for the neural-net training consists of the following steps.

- Choose an activation function for each layer and node.
- Determine weights that are used to evaluate y_i for each node.
- Check that we have not overtrained our network.

We start with a guess for weights and iterate. We set the target type t_i to $t = 1$ for signal and $t = 0$ for background. We want to train our algorithm, so we know t_i of

Appendix B: Advanced Statistical Techniques

each event e_i from the total data sample that is input to the perceptron. Sometimes we get the classification wrong and we characterize this by an error,

$$\sigma_i^e = \frac{1}{2}(t_i - y_i)^2, \tag{B.86}$$

where y_i is the output of the single perceptron. The total error Σ^e for entire data sample N is,

$$\Sigma^e = \sum_{i=1}^{N} \sigma_i^e = \sum_{i=1}^{N} \frac{1}{2}(t_i - y_i)^2. \tag{B.87}$$

First, we guess a weight w_0 and then we iterate this by $w_1 = w_0 + \Delta w$, where

$$\Delta w = -\alpha_r \frac{\partial \Sigma^e}{\partial w}, \tag{B.88}$$

where α_r is a small parameter, called the learning rate, yielding

$$\Delta \Sigma^e = -\alpha_r \left(\frac{\partial \Sigma^e}{\partial w}\right)^2. \tag{B.89}$$

This is called the gradient descent on the error and is illustrated in Fig. B.10 (left). It is a supervised learning method to determine the weights. In general, there are many nodes and thus many weight parameters. So, it is a complicated problem, like a multi-dimensional fit. We need to consider the error contribution from each node in each layer.

Let us look at this a little closer. Consider a node j that typically has a number of connections to the next layer of the multi-layer perceptron, k as depicted in Fig. B.10 (right). We need to evaluate the error from each assignment of target value, using

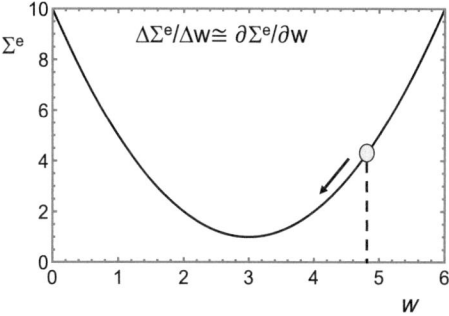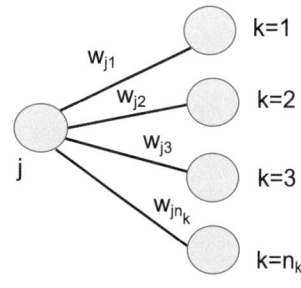

Fig. B.10 Left: The error as a function of the weight. Reprinted with kind permission from [4], © 1994, *BABAR* Collaboration. All rights reserved. Right: Connection of node j to the next layer in the multi-layer perceptron. Here, x_{input} is the input value to node j. It is either the input variable to the multi-layer perceptron or the output of the previous layer. The w_{jk} are the weights of the connection to the different k_i

the activation functions of each node. So we need to take into account the error contributions from hidden layer nodes and output layer nodes.

$$\Delta w_{jk} = -\alpha_r \Delta t_j x_k \frac{\partial \Sigma^e}{\partial w}, \tag{B.90}$$

where j are output nodes, k are hidden nodes, w_{jk} are the weights between them, α_r is the learning rate and x_k is input to node j. The Δt_j correction depends on the type of nodes. It is

$$\Delta t_j = \sum_{k\ nodes} (t_k - y_k) w_{jk} \tag{B.91}$$

for hidden nodes and

$$\Delta t_j = (t_j - y_{kj}) \tag{B.92}$$

for output nodes. The optimization algorithm works on error minimization and gradient descent. Note that this is a multi-parameter problem. There are many minima and we want to converge on the global minimum not on a local minimum. Figure B.11 shows the dependence of the error Σ^e as a function of the weight w. Determining the global minimum may not be easy.

We first need to train the multi-layer perceptrons and then validate the result on different samples. For the training, we need both signal and background samples. Typically the two samples are equal in number of events. When do we stop with training to avoid overtraining? We do a few checks by comparing the error and the error gradient to some anticipated thresholds. In addition, we compare the error to that of an independent validation sample. The amount of data needed depends on the neural network. For a network with one hidden layer and a more complicated network, the rule of thumb respectively is,

$$N_{\text{sample}} > \mathcal{O}\left(\frac{N_w}{\sigma_{\text{th}}}\right),$$
$$N_{\text{sample}} > \mathcal{O}\left(\frac{N_w}{\sigma_{\text{th}}} \log(N_{\text{node}}/\sigma_{\text{th}})\right), \tag{B.93}$$

where N_{sample} is the sample size, N_w is the number of weight parameters, N_{node} is the number of nodes and σ_{th} is error threshold. For example, for ten inputs, one

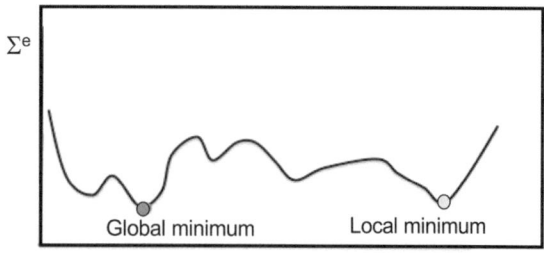

Fig. B.11 The dependence of the error Σ^e as a function of the weight w showing local minima and the global minimum. Reprinted with kind permission from [4], © 1994, *BABAR* Collaboration

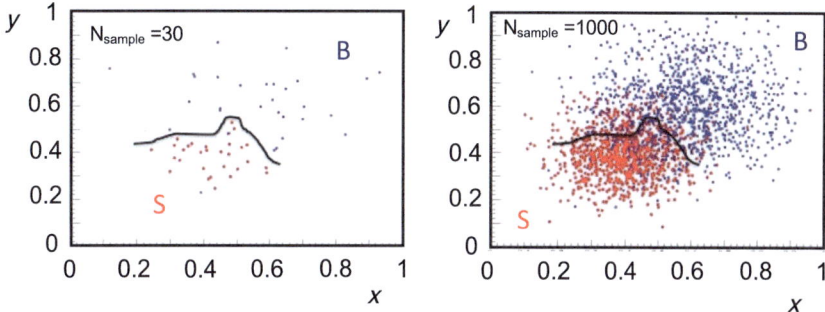

Fig. B.12 Left: Neural network separation of signal (red points) from background (blue points) trained with a low-statistics sample. The black curve is the boundary between signal and background. Right: Application of the low-statistics sample selection to a higher-statistics sample. The boundary is not the best selection here. All reprinted with kind permission from [4], © 1994, BABAR Collaboration. All rights reserved

hidden layer with ten nodes and one output, we need more than 1,100 training events. Note that overtraining should be avoided. This occurs, when the weights are tailored to specific signal and background events and are not true in general. Let us look at an example. Figure B.12 (left) shows the result of a low-statistics training. Here, the boundary is tailored to the low-statistics sample and does represent the best separation in the larger-statistics sample depicted in Fig. B.12 (right). The training is effected by statistical fluctuations. Therefore, it is important to check the training on an independent sample to ascertain that the training is good.

Let us look at an example from the BABAR experiment. We had to separate the $B \to K^{*0}\mu^+\mu^-$ signal from generic $B\bar{B}$ background and from $e^+e^- \to q\bar{q}$ continuum background. We defined two neural networks, one for separating $B \to K^{*0}(K^+\pi^-)\mu^+\mu^-$ signal from $B\bar{B}$ and one for signal from $q\bar{q}$ continuum background, using 13 discriminating variables that are depicted in Fig. B.13. The output of the neural networks is displayed by the figure of merit (FOM) that runs from zero to one, where background peaks near zero and signal peaks at one. The results for the $B \to K^{*0}(K^+\pi^-)\mu^+\mu^-$ signal selection from generic $B\bar{B}$ background and $q\bar{q}$ continuum background are displayed in Fig. B.14 (left, right), respectively.

B.3.3 Decision Trees

Another modern multivariate analysis is a decision tree depicted in Fig. B.15. We divide the data into two classes, signal and background with a binary output by applying an initial rule,

$$R(\vec{x}_1) = \vec{x} > \vec{x}_i \text{ TRUE},$$
$$= \vec{x} < \vec{x}_i \text{ FALSE}. \tag{B.94}$$

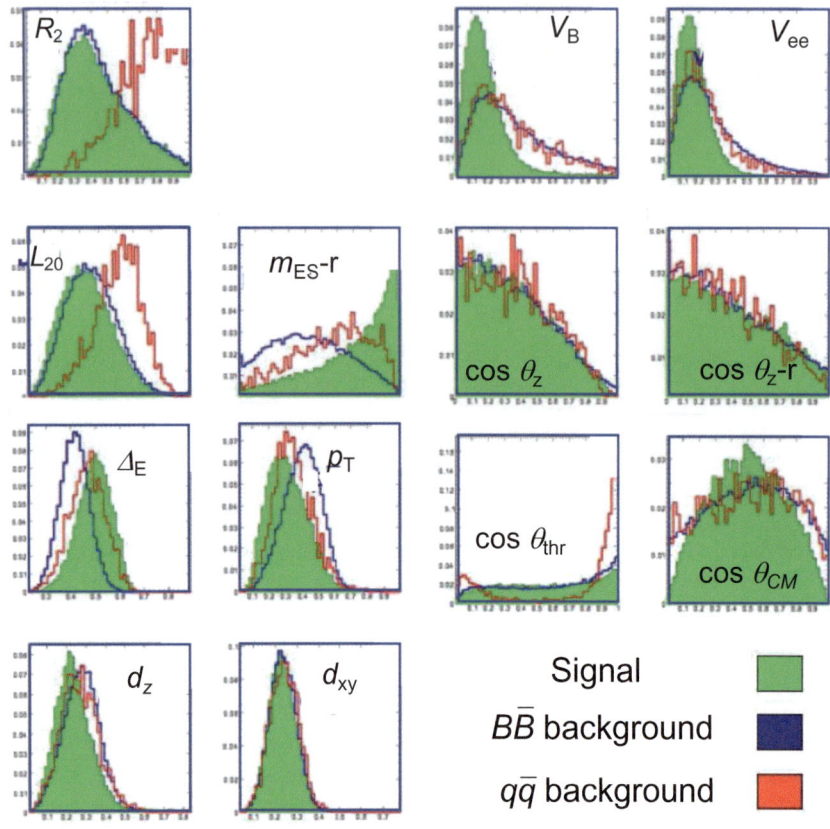

Fig. B.13 Distributions of 13 discriminating variables for the selection of $B \to K^{*0}(K^+\pi^-)\mu^+\mu^-$ signal from generic $B\bar{B}$ background and from $e^+e^- \to q\bar{q}$ continuum background. The variables are the second Fox-Wolfram moment R_2, the B vertex, the e^+e^- vertex, the second Legendre moment L_{20}, the beam-energy-constrained mass of the particles that are not associated with the signal (called other $\bar{B}$), the polar angle of the B candidate, the polar angle of the other $\bar{B}$, the energy difference of the other $\bar{B}$, the transverse momentum, $\cos\theta_{\text{thrust}}$ of the other $\bar{B}$, $\cos\theta$ of the center-of-mass momentum, distance of closest approach in z and distance of closest approach in xy [5]. In all histograms signal is shown in green, $B\bar{B}$ background in blue and $q\bar{q}$ background in red. Reprinted with kind permission from [4], © 1994, *BABAR* Collaboration. All rights reserved

Each successive layer divides the data further into signal and background. The classification for a set of selections will have a classification error. We vary the selection values to train the tree. Note that each node uses the sub-set of discriminating variables that give the best separation between signal and background. Some variables may be used by more than one node, while others are never used. The bottom of a tree looks like a sub-sample of events subjected to a cut-based analysis. There are many bottom levels of a tree, some are signal, others are background, which are defined by this algorithm. The advantage of a binary decision tree is,

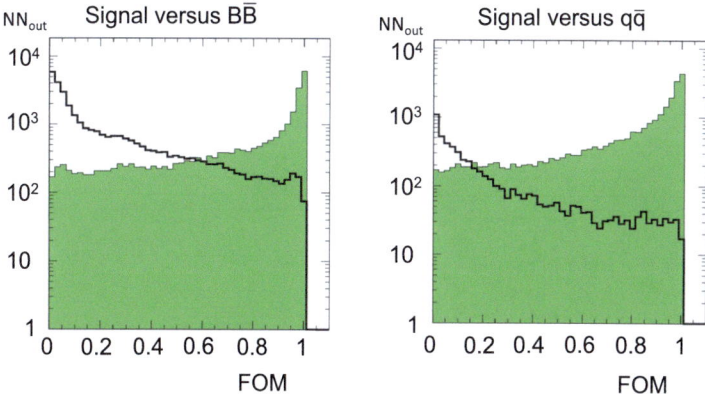

Fig. B.14 Left: The figure of merit for the neural network separating $B \to K^{*0}(K^+\pi^-)\mu^+\mu^-$ signal (green histogram) from generic $B\bar{B}$ background (black histogram). Right: The figure of merit for the neural network separating $B \to K^{*0}(K^+\pi^-)\mu^+\mu^-$ signal (green background) from $q\bar{q}$ continuum background (black histogram). Reprinted with kind permission from [4], © 1994, BABAR Collaboration. All rights reserved

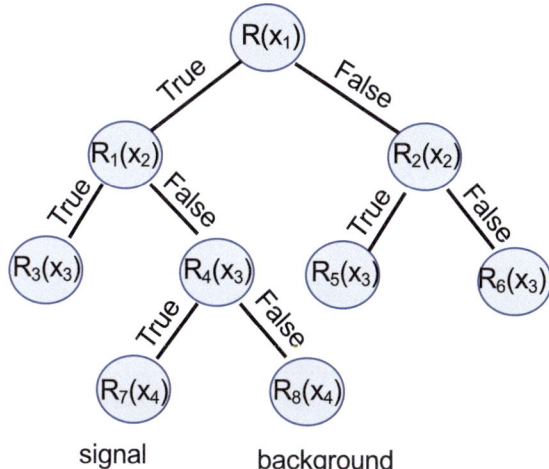

Fig. B.15 Selection procedure of a decision tree for four variables x_1, x_2, x_3, x_4. At each point we have a condition that may be true or false

- The procedure is easy to understand and results are easy to interpret.
- There is more flexibility in the algorithm when trying to separate signal from background. We obtain a better signal-to-background separation than with a simple cut-based approach.

The disadvantage is that we may have an instability with respect to statistical fluctuations in the training sample. It is possible to improve upon the binary decision tree algorithm to try and overcome the instability or susceptibility of overtraining. At each stage in training, there may be some misclassification of events. This introduces an error rate ν_{err}. To deal with it, we assign a greater event weight to mis-classified events in the next training iteration,

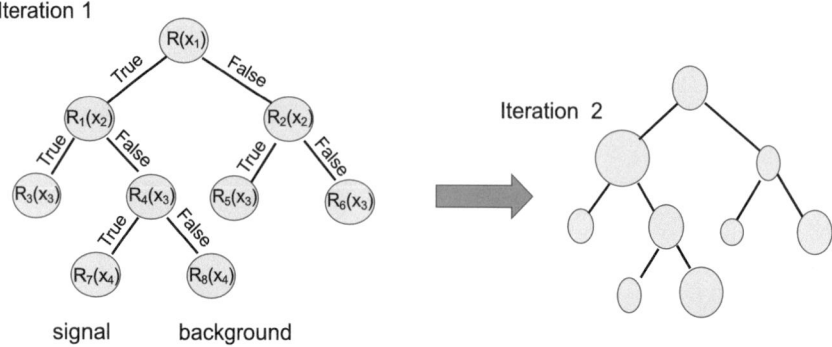

Fig. B.16 Operation principle of a boosted decision tree (see text). Reprinted with kind permission from [4], © 1994, *BABAR* Collaboration. All rights reserved

$$w_r = \frac{1 - \nu_{\text{err}}}{\nu_{\text{err}}}. \tag{B.95}$$

We reweight the whole sample so that the sum of weights remains the same and then iterate as sketched in Fig. B.16. This is called boosted decision tree (BDT) and it tends to be more stable than the normal decision tree. By re-weighting mis-classified events by w_r, the aim is to reduce the error rate of the trained tree with respect to the unboosted algorithm.

Let us look at an example from the *BABAR* experiment. In a subsequent analysis of $B \to K^{*0}(K^+\pi^-)\mu^+\mu^-$, we used a boosted decision tree to separate signal from generic $B\bar{B}$ and $q\bar{q}$ continuum background. We combined the 13 variables into two boosted decision trees. Figure B.17 (left) shows the figure-of-merit for the separation of signal from $B\bar{B}$ background. Again, the signal peaks at one, while the background peaks at zero. Figure B.17 (right) shows the $B\bar{B}$ background rejection versus signal efficiency for different multivariate procedures, including Fisher discriminant, maximum likelihood method and BDT analysis. The BDT analysis yields the best background rejection and for highest signal efficiency.

Another method is the bagged decision tree. The aim is to improve the stability of the decision tree algorithm, which is achieved by sampling the training data used to determine the solution. Then, we take the average solution of a number of re-sampled solutions. Figure B.18 depicts the outcome of this procedure. This re-sampling removes the problem of fine-tuning on statistical fluctuations. This procedure is like choosing the mean value of a cut at each level. It is more stable than using the normal decision tree.

A given decision tree may not be stable, so instead we may grow a forest. The classification of an event e_i in the signal sample or background sample for a forest is determined as the dominant result of all of the tree classifications for that event. For example, if in a forest of 100 trees, there are 80 classifications for signal and 20 classifications for background, the event is considered to be signal. Trees in a forest

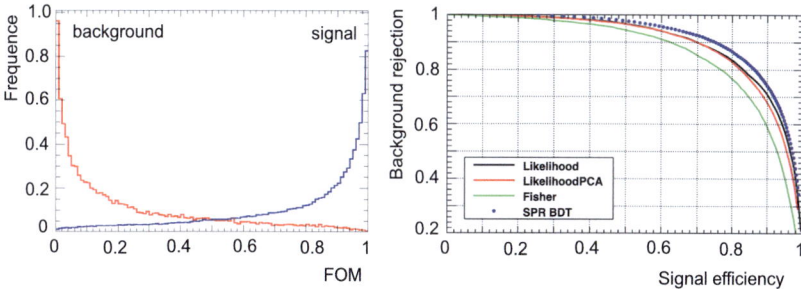

Fig. B.17 Left: Figure-of-merit of a boosted decision tree for separating $B \to K^{*0}(K^+\pi^-)\mu^+\mu^-$ signal from generic $B\bar{B}$ background. While signal peaks at one, background peaks at zero. Right: Background rejection versus signal efficiency for separating $B \to K^{*0}(K^+\pi^-)\mu^+\mu^-$ signal from generic $B\bar{B}$ background. The result is shown in comparison to those of a Fisher discriminant and maximum likelihood analyses. Reprinted with kind permission from [4], © 1994, *BABAR* Collaboration. All rights reserved

Fig. B.18 The distribution of re-sampled solutions in a bagged decision tree. Reprinted with kind permission from [4], © 1994, *BABAR* Collaboration. All rights reserved

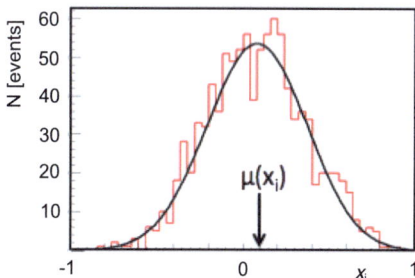

use a common training sample and are typically boosted. A finesse of the previous algorithms is to grow trees using a random subset of variables at each node. This procedure is called randomized trees. In the training, some nodes may prove to be insignificant in obtaining the final tree. It is best to obtain the tree, then remove the insignificant nodes after the fact. This procedure is called pruning. It starts at the bottom and works upward.

B.3.4 Comments

In a cut-based analysis, one has to be careful when observables are correlated. For example, an angular selection corresponds to a hidden mass selection. So, selecting an angular region and a mass region simultaneously is a wrong thing to do. In an MVA on the other hand, observables can be highly correlated, which introduces no bias in the analysis. All MVAs are superior to a simple cut-based analysis. Usually, neural networks and boosted (bagged) decision trees achieve a higher efficiency at the same background level than the Fisher discriminant. Whether the neural network or the boosted (bagged) decision tree yields a higher efficiency depends on the individual process. There is a TMVA package at CERN that lets you use various MVA procedures [6]. Other advanced textbooks on statistical analysis methods are given by Barlow [7] and Narsky and Porter [8].

References

1. The Particle Data Group (S. Navas et al.), Phys. Rev D. **110**, 030001 (2024)
2. W. R. Leo, *Techniques for Nuclear and Particle Physics Experiments* (Springer, 1987)
3. http://cdssls.cern.ch/record/2296388/files/minuit.pdf
4. Personal communication from $B\!A\!B\!A\!R$ Collaboration (K. Flood)
5. The $B\!A\!B\!A\!R$ Collaboration (J.P. Lees et al.), Phys. Rev. D **93**, 052015 (2016)
6. https://root.cern/manual/tmva/
7. R. Barlow, *Statistics* (Wiley Publishing Company, 1989)
8. I. Narsky, F.C. Porter, *Statistical Analysis Techniques in Particle Physics* (Wiley Publishing Company, 2014)